Estimator of the Variance	Approximations for Distribution of Sample Means	Variance of the Distribution of Sample Means	Notes
$x - \dfrac{x^2}{n}$	Approximately normal for $n\Pi \geq 5$ and $n(1 - \Pi) \geq 5$, use continuity correction of $+0.5$ to an upper bound or -0.5 to a lower bound; approximately Poisson when $n \geq 20$ and $\Pi \leq 0.05$ or $\Pi \geq 0.95$	$\Pi(1 - \Pi)$	...o ...he binomial ...s to the count of the num-...r of successes. The sampling distribution of the proportion is essentially the same as the binomial except that it refers to the proportion of the successes in n trials.
$\dfrac{\sum\limits_{i=1}^{n} x_i}{n}$	Approximately normal for $\mu \geq 5$, use continuity correction of $+0.5$ to an upper bound or -0.5 to a lower bound.		The interval in a Poisson process is usually measured over time (t), but it can refer to area, volume, or distance. The expected number of occurrences in the unit interval is λ and $\mu = \lambda t$.
	Approximately binomial if $n \leq 0.05\nu$. Approximately normal for $n\Pi \geq 5$ and $n(1 - \Pi) \geq 5$, use continuity correction of $+0.5$ to an upper bound or -0.5 to a lower bound.		
$s^2 = \dfrac{\sum\limits_{i=1}^{n} (x_i - \bar{x})^2}{n - 1}$	$(x - \bar{x})/\sigma$ is normally distributed for all n. $(x - \bar{x})/s$ is t distributed for all n. $(x - \bar{x})/s$ is approximately normal for $n \geq 30$.	$\sigma_{\bar{x}}^2 = \dfrac{\sigma^2}{n}\left(\dfrac{\nu - n}{\nu - 1}\right)$ $s_{\bar{x}}^2 = \dfrac{s^2}{n}\left(\dfrac{\nu - n}{\nu - 1}\right)$	The term $(\nu - n)/(\nu - 1)$ is the finite population correction factor. It is not used for an infinite or a prospective population (all future trials) and it is optional when $n \leq 0.05\nu$.
$\bar{x}^2$	Approximately normal for $n \geq 30$.		
$s^2 = \dfrac{\sum\limits_{i=1}^{n} (x_i - \bar{x})^2}{n - 1}$	Approximately normal for $n \geq 30$.	$\sigma_{\bar{x}}^2 = \dfrac{\sigma^2}{n}\left(\dfrac{\nu - n}{\nu - 1}\right)$ $s_{\bar{x}}^2 = \dfrac{s^2}{n}\left(\dfrac{\nu - n}{\nu - 1}\right)$	If $n \leq 30$, either a sign test for the population median or a Wilcoxon signed rank test can be performed. The sign test requires no parametric assumptions. The Wilcoxon signed rank test requires the assumption of a symmetrically distributed population.
$\bar{p}(1 - \bar{p})$	Approximately normal for $n\Pi \geq 5$ and $n(1 - \Pi) \geq 5$; usually a continuity correction is not applied to the proportion. Approximately Poisson when $n \geq 20$ and $\Pi \leq 0.05$ or $\Pi \geq 0.95$.	$\sigma_{\bar{p}}^2 = \dfrac{\Pi(1 - \Pi)}{n}\left(\dfrac{\nu - n}{\nu - 1}\right)$ $s_{\bar{p}}^2 = \dfrac{\bar{p}(1 - \bar{p})}{n}\left(\dfrac{\nu - n}{\nu - 1}\right)$	The term $(\nu - n)/(\nu - 1)$ is the finite population correction factor. It is not used for an infinite or a prospective population (all future trials) and it is optional when $n \leq 0.05\nu$.

(Continued on back endpaper)

STATISTICS
FOR
BUSINESS
AND
ECONOMICS

STATISTICS FOR BUSINESS AND ECONOMICS

ROBERT SANDY

INDIANA UNIVERSITY–PURDUE UNIVERSITY AT INDIANAPOLIS

McGRAW-HILL PUBLISHING COMPANY

New York St. Louis San Francisco Auckland
Bogotá Caracas Hamburg Lisbon London Madrid
Mexico Milan Montreal New Delhi Oklahoma City
Paris San Juan São Paulo Singapore Sydney
Tokyo Toronto

STATISTICS FOR BUSINESS AND ECONOMICS

2 3 4 5 6 7 8 9 0 **DOH   DOH**   9 4 3 2 1 0

ISBN 0-07-557119-6

This book was set in Bembo by York Graphic Services, Inc.
The editors were Bonnie Binkert and Judith Kromm;
the production supervisor was Stacey Alexander.
Design was done by Binns & Lubin.
R. R. Donnelley & Sons Company was printer and binder.

Cover painting: Josef Albers, *Homage to the Square: Young.* 1951.
Oil on masonite. 23⅝ inches × 23½ inches. The Metropolitan Museum of Art, George A. Hearn Fund, 1953.

Library of Congress Cataloging in Publication Data

Sandy, Robert.
    [1st ed]
    Statistics for business and economics / Robert Sandy.
      p. cm.
    Includes index.
    ISBN 0-07-557119-6
    1. Social sciences—Statistical methods.   2. Commercial
statistics.   3. Economics—Statistical methods.   4. Statistics.
I. Title.
HA29.S254 1990
519.5′024658—dc19   88-15672

To Elaine, Steven, and Rachel who have borne most of the cost

Preface

Historically, we must admit that the classical version of our key introductory college course has been an abysmal failure, semester after semester, year after year. It deserves a grade of F—yet we persist in teaching it more or less the same way we have been for forty years.

This disheartening quotation is from an article by Brian Joiner, one of the authors of MINITAB ("Let's Change How We Teach Statistics," *Chance,* 1988, no. 1). Joiner advocates two changes in the way introductory Statistics is taught: First, give students real data sets to manipulate throughout the course, and second, eliminate most of the standard material on parametric tests. This text reflects the first recommendation, but I chose not to perform major surgery on the course topics, as the second suggestion would require. I present the traditional topics, but I've done everything I can to make them intelligible and interesting.

My main objectives in writing this text were to:

☐ Explain the material in a highly intuitive and conversational manner with well-written and lively examples,

☐ Systematize the problem-solving process,

☐ Emphasize graphical methods to describe models and analyze data, and

☐ Provide real computer applications for more topics than any other text of this kind.

Wherever possible, explanations take the place of perfunctory statements such as "it can be shown that . . ." or "in advanced statistics texts it is shown that . . . ." Proofs are relegated to an appendix. The discussion of why the formula for the sample standard deviation has $n - 1$ in the denominator (Chapter 3) illustrates this intuitive approach. Other examples are the explanations of why the distribution of sample means approaches a normal distribution as the sample size increases (Chapter 7) and why the chi-square test statistic is distributed as the sum of a series of independent, squared standard normal distributions (Chapter 11).

This text emphasizes the development of the skills needed to recognize which distribution applies in a given situation and to translate a problem from words into equations. The charts inside the front and back covers present clear-cut rules that lead students to the correct formula for almost every hypothesis test or confidence interval discussed in the text.

Coverage is extensive, yet some familiar topics have been omitted. The choice was guided by two criteria. First, each topic covered should have a realistic business application. The discrete uniform distribution and kurtosis are two of the topics that were excluded because they lacked obvious applications. On the other hand, the exponential distribution, ARIMA, and probit regression, which are generally not covered in other texts, are discussed because they have good business applications. The second criterion for deciding which topics to cover was my conviction that the text should reflect the widespread use of computers in the practice of statistics in business. I believe that virtually all business-related statistical computations will soon be done with computers and that there is no benefit to covering techniques aimed solely at reducing computation time. Assuming that students using this text have access to either a computer with a statistical package or a hand-held calculator that has functions such as summation, mean and standard deviation, and a memory register, there is no need to discuss short-cut formulas and coding.

The growing use of computers also led to the inclusion of some topics that usually are *not* covered in an introductory text. For example, most statistical packages present the pooled and the unpooled variance versions of a difference-in-means test, and many computer packages have graphing routines that simplify the creation of charts such as box-and-whisker plots. These and other topics are addressed so that students can handle the options presented by a computer package or take advantage of some easy-to-understand but computationally tedious techniques.

Although the text is designed to be used in courses with or without computers, it offers many advantages if students have access to a computer. The computer applications start with Chapter 2 and continue through Chapter 17. The sections of the chapters that have computer applications are marked with either one or two computer disks in the table of contents and in the text. It is recommended that the double-disk sections be covered only in classes that have access to a computer. The rest of the text does not require use of a computer.

A computer manual is bound into the text as Appendix A. This manual contains printouts and instructions for four widely used computer packages: DATA DESK, MINITAB, SAS, and SPSSx. MYSTAT, a student version of SYSTAT, is available from McGraw-Hill either separately or in a package with the text. The documentation and examples are packaged with the MYSTAT disk rather than bound into the text. This extensive coverage is intended to help instructors integrate computers in the introductory statistics course by accommodating a wide variety of computer resources. DATA DESK works on Macintosh computers. MINITAB, SAS, and SPSSx are usually available on university mainframe computers. MYSTAT works on IBM PCs and compatibles, as well as Macintosh computers. The five packages were chosen to reflect current mainframe usage in statistics courses as well as the emerging transition to microcomputers.

Other significant features of the text include the following:

□ Over 800 exercises. Within each chapter the exercises increase in difficulty up to the highly challenging starred exercises at the end. Each section includes several drill exercises that review the mechanics of applying a formula. Exercises at the end of each section and at the end of the chapter provide a realistic context for the business use of a technique. Adding interest are exercises relating to student life and exercises having a humorous twist.

□ Solutions to the odd-numbered exercises show the intermediate steps. Alternative solutions to a problem, given the computational resources available and the validity of parametric assumptions, are detailed.

□ Optional sections within most chapters which allow instructors to tailor the course to special interests and to students' varied abilities.

□ Non-calculus proofs of most theorems in the text (Appendix C).

□ Carefully refined statistical notation. For example, all parameters are symbolized by Greek letters and all statistics are symbolized by italic Latin letters.

A *Study Guide* to accompany this text is available from the publisher. Other supplements, which are available to instructors, include a sophisticated test bank with class-tested questions. The computerized test generator prints graphs, Greek letters, and statistical notation and randomizes the order of multiple choice questions. The test bank is also available in printed form. In addition, there are over 30 data sets on disk for use with the computer applications, transparency masters, and an instructor's manual. The transparency masters include the figures in the text as well as additional worked-out examples. The instructor's manual consists of solutions to all exercises in the text plus suggested outlines for one- and two-semester courses with and without access to a computer.

ACKNOWLEDGMENTS

To a textbook author, reviewers retained by the publisher are invisible antagonists who have to be mentally wrestled into compliance with the author's goals and limitations of the publication process. I struggled with most of the reviewers' suggestions, repeatedly asking whether a given change fit my concept of the text. In the end, I accepted most of their ideas. Nevertheless, the reviewers bear no responsibility for any errors that remain. My appreciation goes to the following reviewers: William Cooke, University of Wyoming; Larry Cornwell, Bradley University; Robert Kowalczyk, Southeastern Massachusetts University; Ralph Miller, California State University—Pomona; Amitava Mitra, Auburn University; Stephen K. Pollard, California State University—Los Angeles; Ruby Ramirez, Loyola University; Michael Sklar, University of Georgia; and Carol Stamm, Western Michigan University. Robert Kowalczyk and Lee Ing-Tong, both at Southeastern Massachusetts University, reviewed the computer appendixes. Susan Reiland did an accuracy check of the manuscript. Her reputation as a meticulous and expert reviewer is entirely deserved. The McGraw-Hill Business Series Advisor, David Dannenbring of Bernard Baruch College, reviewed the manuscript at all stages, reviewed the computer appendixes and checked the solutions to the exercises. I thank him for all of his help.

The first thirteen chapters of this text have been used at IUPUI for three years. I thank my colleagues who class-tested the manuscript: Robert Calhoun, Martin Spechler, Chris Starkey, Shah Towfighi, and Gang Yi. I am also grateful to the IUPUI students who offered suggestions. I thank David Bivin, Gang Yi, and Monte Juillerat for advice on specific topics. Martin Spechler's insightful advice and unflagging encouragement were well beyond any reasonable expectation.

In conclusion, I think that my treatment as an author by my publisher has been exemplary. Courtesy, patience, and consistency were the hallmarks of all of their actions. My contacts were with Seib Adams, editorial director of the college division; June Smith, director of editing and development; Johanna Schmid, acquisitions editor; Judith Kromm, project manager; Dan Alpert, senior development editor; Anne Mahony, development editor; and Stacey Alexander, production manager. Betty Binns of Binns & Lubin and Associates and Karl Illg of IUPUI provided invaluable input on design and graphics problems. Finally, my most earnest appreciation goes to Susan Badger who worked closely with me throughout the project.

ROBERT SANDY

Indianapolis, Indiana

Contents

■ means that a computer application is provided for that topic in Appendix A, but it is not necessary to have the use of a computer in order to cover the topic.

■ ■ means that a computer application is provided at the end of the chapter for that topic and that it is recommended that the topic be covered only if a computer is available for students' use.

1

Introduction

**W**hen the first American Indians visited London, England, in the early seventeenth century, they came with instructions to count the men in the city and report their findings back to their tribe. They took sticks along, intending to keep count by their usual method—making a scratch on a stick for every man they saw. Given the number of men in London at that time, the method was hopeless. Besides quickly running out of sticks, they had no idea if they were counting the same men twice. Clearly the methods of recording and analyzing data that are adequate at one time or in one environment can be totally inadequate at another time or in another place.

1.1 DEFINING STATISTICS

Today the scope of the information required to make sound decisions to resolve problems in finance, marketing, production, and other aspects of business demands the use of modern Statistics. **Statistics**—with a capital "S"—is *the branch of mathematics devoted to the collection, display, and analysis of data to aid in decision making.* We could estimate the population of a large city, for example, by mapping the city's blocks and sampling houses in blocks selected at random to find the number of residents. We could then project these sample results upward to estimate the population of the entire city. The word "data" (plural of "datum," something given), which was used in the definition of Statistics, also requires a definition. **Data** consist of *a set of numbers that represent the measurements of a numerical characteristic or characteristics of some group.* The number of men, women, and children in each sampled block are data. Their heights, weights, and incomes are also data. A data set *can also contain terms that describe nonnumerical characteristics associated with some group.* The race of each person interviewed is a nonnumerical characteristic that could be designated by the word "white," "black," "Oriental," or "other."

The following examples will give you a better idea of why the scope of information needed for sound business decisions requires the use of modern Statistics.

Consider the problem of determining relative prices in competing supermarkets. Fifty years ago a grocery store owner could stroll through a competitor's store and get a good impression of relative prices in a few minutes. Supermarkets today carry between 15,000 and 25,000 items and in a typical week the prices of

1,000 of them will be changed. Supermarkets do occasionally check the price of every item in their competitors' stores, but this procedure is too expensive and time-consuming to do as frequently as it is needed. A monthly or even weekly pulse of relative prices is essential. Clearly, a sample of items must be taken to represent the average level of prices in a supermarket. The science of Statistics deals with such questions as how the particular items that make up the sample should be selected, how many items need to be in the sample, and how accurately the average price of the sample data will reflect the storewide average level of prices.

Ten years ago the process of brazing or joining the tubes that make up a bicycle frame was done by hand. The alignment of the tubes is critical because a bicycle with a misaligned frame will pull to the right or left when it is ridden. A skilled craftsman would check the alignment after each tube was brazed. The whole process took several hours. Today most bicycle frames are brazed by robots and the process takes a few minutes. When the frames are measured and deviations from a perfect alignment are observed, one of two factors may be responsible: the settings on the brazing robots may have drifted away from their specified values or there was an unavoidable small random variation. The tolerances are so close, in thousandths of an inch, that when a deviation is found, it is uncertain whether the robots have to be reset or if settings are correct and the deviation is unavoidable. Statistical process control is used to track the alignments of each frame as it is completed, and when the trend of measurements from frame to frame is analyzed, it is possible to identify with a high degree of certainty when the settings have drifted even slightly, so that the robots can be stopped and recalibrated.

The marketing of a new toy is much more expensive today than it was a few years ago. Instead of just relying on television commercials to make children aware of a new toy, many toy manufacturers subsidize the creation of a Saturday-morning cartoon show based on the toy. Sometimes even these efforts fail: the kids simply don't like the toy. To reduce the uncertainty inherent in marketing a new toy, manufacturers sample children to gauge their reactions. There are many millions of children in the potential market for a new toy. The science of Statistics is used to estimate the proportion of this market that would purchase the toy on the basis of the reactions of a small sample of children.

In all of these examples, samples were taken from a population and the sample was used to obtain an estimate of a parameter of the population. A **parameter** is *an overall summary measure applied to a population,* such as the average price of all items in a supermarket, the average alignment of all frames made under the current calibration of the brazing robots, or the proportion of all children who will buy the toy. A **population** is *every element in a group that is the subject of analysis.* The three populations in the above examples are the prices of all items in a supermarket, the dimensions of all future bicycle frames brazed by the robot under the current settings, and the decision to buy or not to buy a particular toy by every child in a certain age range. The word "population" is used both for the people or things being measured and for the collection of measurements. Thus the items in the supermarket and the measured or recorded prices of those items are both called populations. A **sample** is *any subset of a population*. The term **statistic** designates *a summary measure of a numerical characteristic of a sample*. The average price of all items in a sample taken from a supermarket, the average alignment of a sample of frames made under the current calibration of the brazing robots, and the proportion of children in a particular sample who will buy the toy are statistics. To repeat, a statistic is a summary measure applied to a sample while a parameter is a summary measure applied to a population.

The process of estimating a population parameter from a sample taken from the population is called **statistical inference.** This is the main purpose of the science of Statistics. A businessperson who tried to make decisions without the aid of statistical inference would be in bad shape, like the Indians who were sent to count men in London with some sticks.

Students beginning a class in Statistics are often skeptical of such claims about the importance of statistical inference in business. They feel that, unlike the Indians who visited London, we live in the information age. Events around the globe are broadcast almost instantly, and any relevant information about business issues can easily be found in each morning's *Wall Street Journal*. Although it is true that the supply of information has increased over time, the demand for information has increased even more rapidly. This entire text can be considered a defense of the claim of the increased importance of statistical inference in business. A few more examples will give you an idea of why the demand for information has increased more quickly than its supply.

The demand for information has increased as the scale of production has increased. Consider the fairly mundane matter of raising chickens. When Herbert Hoover ran for president with the slogan "A chicken in every pot," the phrase implied a certain prosperity. Chicken was once expensive in comparison with other meats. Now chickens are raised in factories where 100,000 birds are fed and watered automatically. It is a very competitive business and chicken has become the cheapest meat available. A chicken farmer makes less than a penny per bird. If a farmer could identify a cheaper feed or a cheaper way of preventing disease or some other method of saving a few cents per bird, it could represent the difference between riches and ruin. At the same time, it is too risky for a chicken farmer to try a new feed or drug on all 100,000 birds. (The technical term for such a practice is "putting all your eggs in one basket.") The only solution is to rely on a sample to gain information about the effect of a new method on the cost of raising chickens.

The demand for information has increased as the time lag between the decision to produce a product and its sale has increased. A nuclear generating plant costs billions of dollars and can take ten or more years to build. One crucial type of information needed by a utility planning to build a nuclear generating plant is the future demand for electricity. When Public Service Indiana planned the now-abandoned

Marble Hill nuclear plant, it assumed that the quantity of electricity bought by its consumers would increase by the same percentage every year. (The technical term for this method is "simple trend line.") The company's executives were painfully surprised to discover that the quantity of electricity sold did not follow a simple trend line and that they could not sell the electricity from the Marble Hill nuclear plant at a price that would cover their costs. The company's current forecasts of the quantity of electricity to be sold in the future take into consideration the price of electricity and its substitutes as well as the trends in population and economic growth.

The demand for information has increased as the life cycle of products has decreased. The term "life cycle" refers to the time that elapses from the introduction of a product until its replacement by a new product because of technological obsolescence or changing fashions. If the costs of designing and marketing a product can be recovered only over a short life cycle, it is more important that the product be immediately successful; a company doesn't get a second chance, as a new generation of competing products may appear before mistakes in its product can be corrected. Consider the Air Jordan basketball shoes sold by Nike, Inc. This shoe accounted for a large share of the company's profits while it was popular, but the company knew that the popularity of a style of athletic shoes lasted only a year. The only way to reduce the risk of marketing the new shoe was to interview a sample of consumers to gauge their reactions to the garish red-and-black Air Jordan shoe.

These examples should give you an idea of the reasons for the increased demand for information in business and how statistical inference is being used to satisfy that demand.

1.2 WHY MANY STUDENTS HAVE DIFFICULTY WITH STATISTICS

There are some things you know without having gathered any data or made any calculations. One is that many students find the study of Statistics difficult. There is no use denying that Statistics involves many formulas, symbols, terms, and even some distinctions so fine that at first glance they seem invisible. Math anxiety contributes to the difficulty. This text reduces the difficulty inherent in a Statistics course in several ways. First, the explanations are very intuitive. Words and graphs are the main vehicles for describing the why and how of statistical techniques. Mathematical proofs are relegated to an appendix. Second, every technique or term is immediately tied to a business application. Seeing a practical application of a technique helps clarify the underlying concepts and justifies its study. Third, the endpaper charts and the flow diagrams graphically show relationships between concepts and clarify the steps needed to solve a problem. Fourth, almost every exercise involves either the use of Statistics for a business decision or, for variety, some problems that are part of a student's everyday life. The theme that Statistics can be used to make or save money is heavily emphasized. These steps are not a panacea, but they will go a long way toward helping you overcome the problems that many students have had with this subject.

1.3 OUTLINE OF THE TOPICS COVERED IN THE TEXT

The text is organized as follows. Chapter 2 discusses the way data are measured and displayed in charts; Chapter 3 describes how data are summarized with numerical descriptive measures. **Descriptive statistics** is *the branch of Statistics devoted to accurate representation of a mass of data with graphs and summary measures*. Unlike inferential statistics, descriptive statistics does not attempt to use samples to predict the parameters of a population. Descriptive statistics does not look beyond the data at hand, but rather concentrates on how best to understand and present those data. Chapters 2 and 3 are the only chapters on descriptive statistics.

The next four chapters are building blocks that give you the tools needed for statistical inference. Chapter 4 is on probability theory, which is important both as the foundation of statistical inference and sample design and as a direct aid in business problems. Chapters 5 and 6 cover some mathematical distributions, such as the normal curve that can be used to describe many real populations. These distributions can be used to represent patterns that occur in many data sets related to business problems. For example, they can be used to predict the proportion of defective parts in a shipment or the probability that an appliance will fail during the warranty period or the proportion of customers requiring a certain size of clothing. Chapter 7 has two main subjects. The first is the design and collection of samples. Under this heading are discussed such issues as how to avoid bias in a survey and how to gain the most information for a given expenditure. Chapter 7 also describes the distribution of the possible values of statistics calculated from samples taken from a known population. It will be shown that there is a precise relationship between the possible values of a statistic, such as the sample mean, and the corresponding population parameter, such as the population mean. The characteristics of a population can be used to describe the characteristics of all possible samples.

Chapters 8 and 9 begin the discussion of statistical inference in regard to problems that involve one population. The idea that the characteristics of a population can be used to describe the characteristics of all possible samples will be turned around so that the characteristics of one sample will be used to describe the possible values of the population parameter. Chapter 10 extends statistical inference to problems with two populations, such as estimating the difference between the populations of men's spending and women's spending in bookstores from samples of men's and women's spending. Chapters 11 and 12 extend statistical inference to problems that involve more than two populations. Chapter 11 is on statistical inference with categorical data, such as the proportions of shoppers in three or more bookstores who are adult males, adult females, and children. Chapter 12 is on statistical inference for data consisting of an amount recorded for observation, such as the amount spent per customer in three or more bookstores.

Almost all of the material up to Chapter 13 is on univariate statistics. This means that only one dimension of a population is being considered at a time. Multivariate statistics looks at relationships among the characteristics associated with each element of the population. This technique enables us to estimate, for example, how spending in bookstores varies with customers' income. This example is bivariate because only two dimensions, income and spending in bookstores, are being considered. Chapter 13 is on statistical inference in bivariate problems. The tool used to analyze these problems is regression analysis. Chapter 14 extends regression analysis to more than two dimensions. Regression analysis could be used to determine the relationship between an individual's spending in bookstores and such variables as income, age, race, and sex.

The remaining four chapters cover important business applications of Statistics. Chapter 15 is on forecasting, or predicting the future levels of variables on the basis of their past behavior. This is a vital business application. Chapter 16, on index numbers, describes methods of grouping data to measure changes across many variables. The most prominent such index is the consumer price index, or CPI. Chapter 17, on nonparametric statistics, explains a series of techniques that extend the application of statistical inference to small samples taken from populations with unknown shapes or distributions. Many business problems require the use of such samples, and these techniques have received increased emphasis as more business users of statistics have realized that the parametric methods discussed in the preceding chapters can give unreliable results in these situations. Chapter 18 covers statistical decision analysis—formal ways to apply probability theory and information from various sources and samples to business decisions. The decision to market or not to market a particular toy in view of a variety of test results exemplifies the application of decision analysis.

Computers have made a profound impact on the practice of Statistics, and the outline of this text reflects that impact in several ways. Most chapters have an appendix that illustrates the use of computer programs, often called "canned packages," to solve some of the problems discussed in the chapter. Many real data sets are given and computer exercises are keyed to these data sets. A more subtle reflection of the impact of computers on the outline of the text is seen in the material selected to be included or excluded. Some subjects, such as pooled variance hypothesis tests, are included precisely because they appear in most canned packages and you need to know them in order to run the computer packages correctly. Other useful and powerful techniques require such long calculations that they were not worth covering until the advent of easy access to computer-based statistical packages. Two examples are autoregressive integrated moving averages, or ARIMA, for forecasting, and probit regression. Finally, some techniques, such as coding data and short-cut formulas, have been omitted because they were designed to save computational time in a paper-and-pencil age. While a computer is not needed to use this text and the few techniques that do require a computer have been flagged in the table of contents, the basic assumption that guided the selection of topics to be included was that any future business use of statistics would involve a computer.

SUGGESTED READINGS

Moore, David S. *Statistics: Concepts and Controversies*. San Francisco: W. H. Freeman, 1979.

Stigler, Stephen M. *The History of Statistics: The Measurement of Uncertainty before 1900*. Cambridge: Harvard University Press, 1986.

Tanur, J., F. Mosteller, W. H. Kruskal, R. F. Link, R. S. Pieters, and G. R. Rising, eds. *Statistics: A Guide to the Unknown*. San Francisco: Holden-Day, 1972.

NEW TERMS

data A set of numbers that represent the measurements of a numerical characteristic or characteristics of some group; a set of terms that describe nonnumerical characteristics associated with some group.

descriptive statistics The branch of Statistics devoted to accurate representation of a mass of data with graphs and summary measures.

parameter An overall summary measure applied to a population.

population Every element in a group that is the subject of analysis.

sample Any subset of a population.

statistic A summary measure of a numerical characteristic of a sample.

statistical inference The process of estimating a population parameter from a sample taken from a population.

Statistics The branch of mathematics devoted to the collection, display, and analysis of data to aid in decision making.

2

Types of Data and Describing Data Sets with Graphs

n annual budget hearing is the occasion when most college deans formally argue their case for more money. At these hearings deans describe to the university's central administration the accomplishments of their college, their priorities for the next year, and how they would use any additional funds. During his 1986 budget hearing, the dean of the college where I work had a slide of a simple bar chart projected on the wall. The chart showed the levels of spending per undergraduate student credit hour in the dean's college, the rest of the university, and another campus in the university system. This chart is shown in Figure 2.1. Why would the dean bother with a chart to relate just three numbers? The purpose of the chart was to dramatize the differences in spending across academic units. The dean did not want simply to make the central administration aware of the differences in spending per student, he wanted to beat them over the head with those numbers. One of the legitimate functions of charts is to persuade. Charts can help you convey to others important points about a data set. Another function of charts is analysis. Graphing a data set can help you discover its important features.

In this chapter you will find that certain types of data work well with certain types of charts and that different types of charts can be used to highlight different aspects of the data. If the dean had wanted to emphasize the proportions of the university system's budget going to each academic unit, for example, he could have used a pie chart instead of a bar chart. The chapter begins by

FIGURE 2.1

Dollars spent per undergraduate student credit hour at Indiana University–Purdue University at Indianapolis (IUPUI) School of Liberal Arts (SLA), at IUPUI schools other than liberal arts, and at Indiana University–Bloomington (IU-B) in 1986–1987 academic year

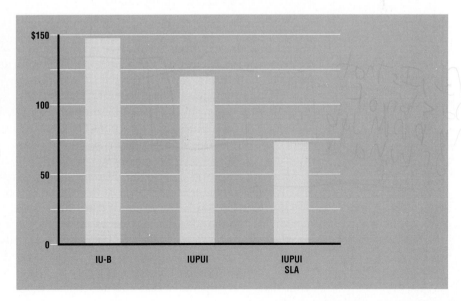

discussing four basic types of data sets—nominal, ordinal, interval, and ratio—and the distinction between continuous and discrete data. It is important that you understand these categories so that you can select the graphs and use the formulas that will be introduced in later chapters. This chapter discusses bar charts, pie charts, pictographs, frequency polygons, relative frequency polygons, cumulative frequency polygons, vertical line charts, and histograms.

2.1 DATA SOURCES AND TYPES OF DATA

2.1A DATA SOURCES

There are many sources of data. One source is a trade association, such as a state hospital association that conducts an annual survey of hospital costs; another is a private company that collects data and sells the results, such as F. W. Dodge, which issues a series of reports on the value of new construction in the United States; another consists of payroll records, invoices, receipts, and other records generated in the course of doing business; another is a local chamber of commerce that acts as a clearinghouse for wage data for local firms. The federal government has agencies that publish data, such as the Bureau of Labor Statistics of the Labor Department and the Bureau of Economic Analysis of the Commerce Department. Government data are usually inexpensive to purchase; data tapes are sold for nominal fees. At the opposite end of the cost spectrum are specially commissioned surveys, which can cost a firm millions of dollars.

After defining the problem that is to be solved, you have to find the best source or sources of data. If you have enough money and your problem justifies the time and expense, you can usually get information that is precisely focused on the issues you are studying. Most often you'll find that you have to make do with whatever data are readily available. The census data on incomes may be five years old, for example, but you cannot justify the expense of collecting newer figures; or you would like to compare the cost of each component of fringe benefits across firms in your city, but the Chamber of Commerce separates out only the cost of the medical component of fringe benefits.

Most of the sources mentioned above are known as **primary data sources,** that is, *the government agencies or firms or trade associations that collect data.* There are also many secondary sources of data, such as the *Statistical Abstract of the United States.* **Secondary data sources** are *organizations that publish data collected by other agencies or firms.* Reference librarians, especially the ones in places as lonely as the public documents sections of university libraries, are almost always helpful in finding secondary sources of data. Once you have obtained your data, it is imperative that you understand the procedures used to collect them, the definitions of the terms used in the collection process, and the potential causes of incorrect observations or data points. One horror story should suffice to impress you with the importance of these issues. A professor (he will appreciate my tactful omission of his name) compiles an annual ranking of U.S. cities according to their desirability as places to live. The rankings are based on an index computed on a wide variety of cultural, education, and economic factors. Since the rankings address an important issue and they appear to be put together in a careful and objective manner, they have received wide attention in the press over the years. One year the professor rated Houston the worst city in the United States. Houston city officials were dumfounded because their city had always done well in comparisons of this kind. They looked into the construction of the rankings and discovered that their low position was due to two errors: a simple data entry error (the wrong number was typed into the computer) and a misinterpretation of a variable (a high number that meant "good" was coded or recorded as "bad"). The professor blamed the errors on a student helper. Reasoning that if you accept the credit you should accept the blame, the city of Houston sued the professor and his university, instead of the student helper, to recover the economic losses to the city caused by the incorrect ranking. At the time of this writing, the case is still pending.

■ 2.1B TYPES OF DATA

Before you can begin to describe your data, you have to understand the differences in types of data. The four types of data, sometimes called the four measurement scales, are nominal, ordinal, interval, and ratio. **Nominal data** (also called categorical data) consist of *the number of observations that fall into specific categories.* Nominal data for two potential sites for shopping malls could be the number of males and females living within 30 minutes' driving time of each site. The categories have to be defined so that every observation—in this case an observation is a person—has to belong to only one category. **Ordinal data** consist of *observations that are ranked or ordered to indicate first place, second place, third place, and so on.* For example, cities could be ranked according to their cultural amenities, such as symphony orchestras and museums. Ranking implies nothing about the numerical distance between the observations of the characteristic being measured. The cities ranked first and second in terms of cultural amenities may have very similar resources while there may be a big gap between the second and third cities. Another example of ordinal data is seen in an attitude survey that asks people to rank a list of leisure-time activities, with spending an afternoon shopping as one of the items on the list. If someone ranked shopping as his fifth favorite leisure-time activity among the items on the list, you would know only that he valued four other activities more highly than shopping, not how much more highly he valued them.

Interval data consist of *observations measured on a scale so constructed that equal* *distances between numbers imply equal distances between the values of the characteristic being measured, and 0 is arbitrarily defined.* One example of interval data is seen in the readiness scores the army assigns to combat units. The scores run from 0 to 100. The scoring is so devised that a battalion that scores 50 in combat readiness is as far below another battalion that scores 60 as the latter is from a battalion that scores 70. A unit that scores 0 in combat readiness is not one in which every soldier is dead. Even with some men, guns, and ammunition a unit can be scored 0. If the army wanted to redefine its readiness measure, it could move the 0 point. A more familiar example of an interval scale consists of degrees Fahrenheit. The 0 on this scale was defined by Gabriel Fahrenheit as the temperature that results when equal parts of salt and snow are mixed. As with all interval data, equal numerical distances imply the same change in the characteristic being measured. For example, 50 degrees on this interval scale is as far from 60 degrees as 60 is from 70. Degrees centigrade are also data defined on an interval scale. Here 0 is the temperature at which water freezes rather than the absence of heat.

Ratio data are *observations measured on a scale so constructed that equal distances between numbers imply equal distances between the values of the characteristic being measured, and 0 represents the absence of the characteristic being measured.* People's ages are ratio data: 0 is the moment of birth, and a 70-year-old person is as much older than a 60-year-old person as a 60-year-old person is older than a 50-year-old. The distance between two cities, the number of apples in a bushel, and the time needed to drive from home to a shopping mall are all examples of ratio data. The term "ratio" is used because meaningful ratios can be computed with such data. Someone who is 80 years old is twice as old as someone who is 40. Meaningful ratios cannot be computed with interval data. The amount of heat in an object that is 80°F is *not* twice as great as the amount in an object at 40°F because the 0 on the Fahrenheit scale is set at an arbitrary level of heat.

An additional distinction in the types of data can be made for interval and ratio data. Both interval and ratio data can be either continuous or discrete. **Continuous data** are *observations that can include any value over a range of values,* while **discrete data** are *observations whose values are limited to specific points in a range of values.* The length of someone's foot measured in centimeters is one kind of continuous data. The length of the foot can be recorded in any number of digits to the right of the decimal place; the only limitation is the accuracy of the measuring device. The size of the shoe that fits that foot is a piece of discrete data. Shoes come in whole and half sizes, such as 7 and 7½. No shoe sizes exist between these two values. Although the distinction between continuous data and discrete data sounds reasonable, on reflection you may realize that no real data set can really be continuous. Every measuring device has some limit to its accuracy, and thus data on, for example, the lengths of feet must be limited to specific points in a range. Depending on the accuracy of the measuring device, the limitation may be that the data cannot take values of less than a tenth or a hundredth or a thousandth or even a ten-thousandth of a centimeter. While the underlying physical phenomena that are being measured are continuous, the recorded data are discrete. In practice the decision to treat a data set as discrete or continuous is a matter of judgment. Data that are recorded to many decimal places are treated as continuous and data recorded to a few decimal places are treated as discrete. Some Statistics texts advise students that price data should be treated as continuous while others recommend that they be treated as discrete. It depends. If you were working with data on the delivered prices of new cars and a typical observation was $14,558.12, it would make sense to treat the data as continuous even though observations would never include values between penny increments. If your data set was for the prices of "penny" candy and typical observations were 2 cents or 5 cents or 10 cents, it would make sense to treat the data as discrete.

Each type of data has its own business applications. The reason that the distinctions between the types of data need to be understood from the start is that different types of data can require different formulas, tests, or graphs. Some of the phenomena that need to be described to resolve business problems cannot be measured on a ratio scale. You'll find that you will have to use some of the other types of data when you work practical problems. Trying to work statistical problems without being aware of the type of data involved is like going to a dance with soundproof earplugs. You may know all the steps to the dances (or all the formulas for the problems) but you are bound to be out of synch with the music (the data) and look ridiculous. The exercises below will help to clarify the distinctions among the types of data.

EXERCISES FOR SECTION 2.1

2.1 Identify the type of data used in each example. If the observations are classified as ratio or interval data, identify the data as continuous or discrete.

(a) The birth dates of students in your class.

(b) The dress sizes of the women in your class.

(c) The grade point averages of students in your class.

(d) The IQ scores of students in your class.

2.2 Identify the type of data used in each example. If observations are classified as ratio or interval data, identify the data as continuous or discrete.

(a) The class standing of each student in your class—that is, freshman, sophomore, junior, or senior.

(b) The number of students in your class whose surnames begin with each letter of the alphabet.

(c) The number of letters in each student's name.

(d) Each student's response to this multiple-choice question: "I am terrified of this course," coded as 1 for "strongly agree," 2 for "agree," 3 for "uncertain," 4 for "disagree," and 5 for "strongly disagree."

2.3 Ten students are running for seats on the student council. The five who receive the most votes will win seats. Identify the type of data used in each example:

(a) The number of votes for each candidate.

(b) The order of finish of the ten candidates.

(c) The list of losers and winners.

2.4 If you knew item a in exercise 2.3, would you also know item b? If you knew item b, would you also know item c? Does this relationship hold in reverse, from c to b to a? What general principle about the relationships between types of data does this example illustrate?

2.5 The Federal Trade Commission requires companies to have proof to support claims that they make in advertisements. A claim that is vague, however, is considered puffery, and no proof is required. For example, the claim "We have low, low, low prices" does not specify what the "low" is compared to and thus it requires no proof. For each claim below, identify the type of data, if any, that would be required for proof.

(a) For the fastest relief of any antacid, take Rolaids.

(b) More doctors recommend Anacin than any other pain reliever.

(c) Coke is it.

(d) The most preferred luxury car among owners of fine automobiles is Mercedes Benz.

2.2 GRAPHIC PRESENTATION OF DATA

 2.2A INTRODUCTION

While the use of graphs to present data has always been an important part of Statistics, two recent developments have enhanced the importance of graphs. One development is the wider use of graphs in newspapers and magazines. A typical issue of *USA Today* or *Newsweek* contains some excellent graphs. This widespread use of graphs implies that most businesspeople as well as the general public will be able to read graphs and come to expect them in discussions that involve numbers. The second recent development is computer graphing programs that have greatly simplified the creation of graphs. For example, all of the graphs in this section except the pictographs were originally prepared with Cricket Software's "Cricket Graph" program, and once the data were keyed into a microcomputer, none of the graphs took more than a few seconds to make. This section will discuss graphs for nominal data, continuous interval and ratio data, and discrete interval and ratio data.

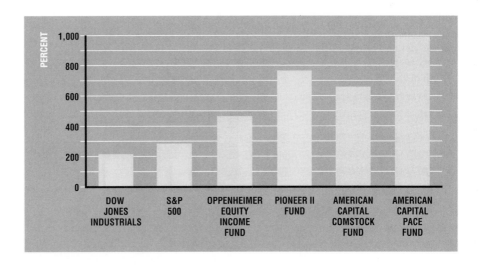

2.2B GRAPHS FOR NOMINAL DATA

The graphs most commonly used to display nominal data are bar graphs and pie charts. In Figure 2.2 the percentage changes in the net asset values of four mutual funds and in two market indexes from 1976 through 1985 are presented in a **bar chart,** *a graph in which the frequency or count of observations in each category is represented by the height or length of each bar.* The chart came from an advertisement suggesting that these mutual funds had performed better than the stock market as a whole, as measured by the two indexes. The vertical or *y* axis in the bar chart is scaled in percentage points in this example. With nominal data the order of the classes on the horizontal or *x* axis is usually arbitrary.

If the class with the highest count contains a high multiple of the number of observations in the class with the lowest count, the graph will be easier to read if the bars run horizontally. The reason is that the bars would have to be more compressed to fit vertically into a rectangle whose width is greater than its height. Graphs that are wider than they are tall seem to be preferred over graphs that are tall and narrow. Perhaps people are more used to what is termed a landscape orientation for graphs than they are to the vertical or portrait orientation. Figure 2.3 uses horizontal bars to show the frequencies of each race-sex classification of the students in a statistics class. In this example the axis showing the level for each category (the *x* axis) is scaled in the frequency or count of the number of occurrences per category.

A **pie chart**—*a graph in which the total number of observations is represented by a circle and the proportion of that total represented by each category is shown as a wedge of the circle*—is another device commonly used to represent nominal data. Figure

FIGURE 2.3
Horizontal bar chart showing students classified by race and sex

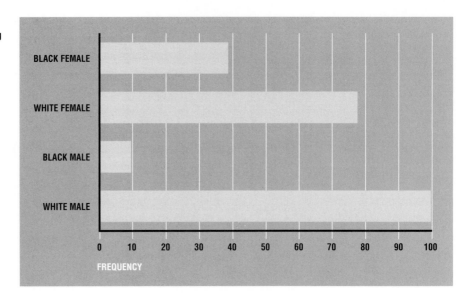

2.4 shows the same data as Figure 2.3 in a pie chart. To draw an accurate pie chart you will need a protractor to mark the angles of the pie slices. The procedure is to multiply the proportion of the data in a given category, such as 0.34 white female, by 360 degrees. The 0.34 is 34% expressed as a proportion. The product, 122.4 degrees, is the angle made by the edges of the white-female pie slice. The same procedure is followed to find the angles of the other slices.

The choice of which kind of chart to use depends on what aspect of the data you want to emphasize. While the pie chart makes the proportions clear, for example, it does not offer a clear comparison of two sets of data. Two sets of bar charts can easily be juxtaposed to emphasize the comparisons between two sets of data. Figure 2.5 shows the race-sex classifications for two sets of stu-

FIGURE 2.4
Pie chart showing students classified by race and sex

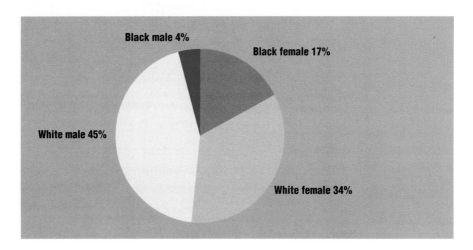

FIGURE 2.5

Two bar charts juxtaposed to show two sets of students classified by race and sex

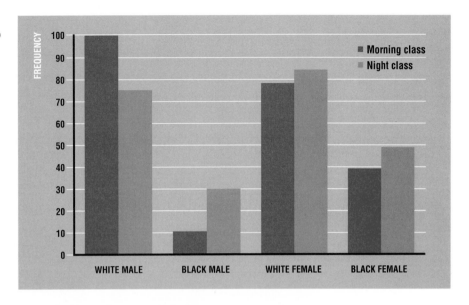

FIGURE 2.6

Pictograph showing troop strengths of Us and Them

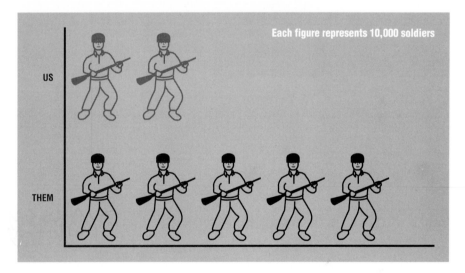

dents. Since these examples have dealt with only four classifications, the charts do not provide a great advantage over simple tables that list the count for each group. When the data are divided into many categories, these charts do offer an advantage because they provide a quick way to see the relationships in nominal data.

Another commonly used chart is the **pictograph,** *a graph in which pictures of an object represent the frequencies of that object.* A pictograph has two variants. In the first a picture is assigned a numerical value and a row of pictures sums up to the total for a category. Figure 2.6 is a pictograph of the troop strength of

FIGURE 2.7
Pictograph of troop strengths with Them 2.5 times the height of Us

two countries called Us and Them. The second variant of the pictograph scales two pictures to the heights that represent the relative troop strengths. The Them picture is 2.5 times the height of the Us picture, as in Figure 2.7. The second pictograph is highly misleading. The area of the Them figure is much larger than 2.5 times the area of the Us figure. Most casual readers would rely on the areas of the two figures to judge relative troop strengths. As the first variant of the pictograph is more likely to convey an accurate impression, the second variant should be avoided.— *unless you are in Marketing or Politics (sic)*

2.2C GRAPHS FOR CONTINUOUS INTERVAL AND RATIO DATA

Let's now switch from nominal data to continuous interval and ratio data. Grade point averages exemplify continuous ratio data. Table 2.1 lists the 75 grade point averages of the students taking a statistics class. The numbers are arranged in no particular order. No pattern emerges from an inspection of the data in Table 2.1. The first step in presenting the data in Table 2.1 is to arrange the observations in an **array**—*a list of observations in a data set arranged in ascending or descending order.* Arrays are easier to inspect than the raw data, and the creation of a graph is easier with ordered data. Table 2.2 contains the array for the data in Table 2.1.

TABLE 2.1
Grade point averages of 75 students in a statistics class

2.56	2.33	2.67	3.34	2.09	2.17	3.92	3.02	1.98	3.65	2.25	2.40	2.90	2.13	2.34	3.13
2.14	3.04	1.88	2.71	3.76	2.97	3.19	3.82	2.52	3.13	3.22	2.17	2.90	2.52	2.45	2.09
2.85	2.43	2.57	2.60	3.50	1.99	2.35	2.57	2.91	2.95	3.48	3.62	2.44	2.48	3.31	2.40
2.22	2.29	3.21	2.59	2.82	2.71	2.39	2.23	2.25	3.49	3.22	3.15	3.10	2.47	2.25	2.54
2.34	2.01	2.07	3.08	3.02	3.02	2.39	2.43	2.50	2.56	2.91					

TABLE 2.2
Data on grade point averages in ascending order

1.88	1.98	1.99	2.01	2.07	2.09	2.09	2.13	2.14	2.17	2.17	2.22	2.23	2.25	2.25	2.25
2.29	2.33	2.34	2.34	2.35	2.39	2.39	2.40	2.40	2.43	2.43	2.44	2.45	2.47	2.48	2.50
2.52	2.52	2.54	2.56	2.56	2.57	2.57	2.59	2.60	2.67	2.71	2.71	2.82	2.85	2.90	2.90
2.91	2.91	2.95	2.97	3.02	3.02	3.02	3.04	3.08	3.10	3.13	3.13	3.15	3.19	3.21	3.22
3.22	3.31	3.34	3.48	3.49	3.50	3.62	3.65	3.76	3.82	3.92					

The second step is to make a frequency distribution of the data in the array. A **frequency distribution** is *a listing of a data set which divides the data in classes and gives a count of the number of observations in each class*. If the number of classes is too small, too much detail about the data is lost. If the number of classes is too large, the details prevent you from seeing an overall pattern. There is no best number of classes. It is more a matter of trying different values and deciding on the number of classes that convey the most information. Somewhere between 5 and 15 classes is a good rule of thumb. The classes do not have to be of equal width, but the frequency distribution is much easier to interpret if they are. For this example we'll use 11 classes. The classes have to be defined so that an observation belongs in one and only one class. The frequency distribution for the above data is presented in Table 2.3.

TABLE 2.3
Frequency distribution of data on grade point averages

Class range	Frequency
1.8 and under 2.0	3
2.0 and under 2.2	8
2.2 and under 2.4	12
2.4 and under 2.6	17
2.6 and under 2.8	4
2.8 and under 3.0	8
3.0 and under 3.2	10
3.2 and under 3.4	5
3.4 and under 3.6	3
3.6 and under 3.8	3
3.8 and under 4.0	2

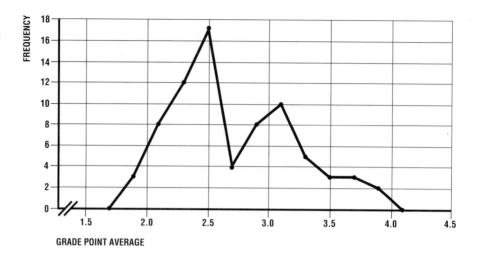

GRADE POINT AVERAGE

The chart that is generally used for continuous ratio and interval data is the **frequency polygon,** *a line connecting a series of dots, each of which is located at the midpoint of a class along the* x *axis and at the number of observations in that class along the* y *axis.* The black dots in Figure 2.8, representing grade point averages, are located at the midpoints of the various classes along the horizontal axis and at the numbers of observations in those classes along the vertical axis. The midpoint is the upper limit of a class plus the lower limit of that class divided by 2. Here the midpoints are 1.9, 2.1, 2.3, and so on. A series of straight lines connects the dots. The frequency polygon is closed or brought down to the x axis by a dot placed at the midpoint of the class below the first class that has any observations. A similar procedure closes the chart on the high side. The lowest class that has any observations is "1.8 and under 2.0." The next lowest class is "1.6 and under 1.8." The midpoint of this class is 1.7, so the leftmost dot is placed at $(x = 1.7, y = 0)$. The rightmost dot is placed at $(x = 4.1, y = 0)$. Most frequency polygons are drawn with classes of equal width because, as was mentioned before, this arrangement is the easiest to interpret. Occasionally the need to show greater detail in one section of the chart can be accommodated by use of narrower classes in that area.

Let's examine the frequency polygon. Grade point averages near 2.5 are the most common. The distribution of grade point averages tapers off smoothly to the left of 2.5 and not so smoothly to the right of 2.5. There is a second peak in the distribution around the grade point average of 3.1. When we find such a **bimodal distribution**—*a frequency distribution that has two peaks*—we may suspect that the observations have been drawn from two distinct groups. We may be dealing here with a population of not very serious students who have grade point averages near 2.5 and a second population of more serious students who have grade point averages near 3.1. Luck, ability, and differences in the

FIGURE 2.9
Relative frequency polygon
showing distribution of grade
point averages

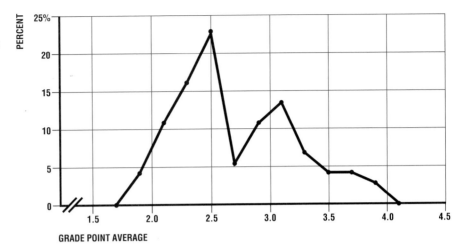

difficulty of the courses taken may account for the dispersion of grade point averages around these two values. We are only guessing now; we would need more information before our conjecture had any real support. But one of the purposes of the graph is to bring out relationships, or even to suggest possibilities, that would remain unseen in the original data.

Figure 2.9, a relative frequency distribution of the same data, looks exactly the same as Figure 2.8 except that the vertical axis is scaled in percentages rather than frequencies. A **relative frequency distribution** is *a listing of a data set that divides the data in classes and gives the proportion of the population or sample in each class.* The reason that the shape of the polygon does not change when you go from counts to percentages is that each frequency is divided by the same number. In this example the frequencies were divided by 75, the total number of students. The class "2.4 and under 2.6," for example, has 17 observations. When 17 is divided by 75, the result is 22.67%, which is the value along the *y* axis which corresponds to the dot for the class "2.4 and under 2.6."

The relative frequency polygon is a slight variation on the frequency polygon. Sometimes you will want to consider the count for each class and sometimes the percentage. For example, the regular frequency polygon, which shows the count in each range, would be best for keeping track of an inventory of hardware parts in size ranges, and the relative frequency polygon, showing the relative frequency distribution of sales by size range, would be best for making up orders.

Another slight variation of the frequency polygon is the **cumulative frequency polygon,** *a line connecting a series of dots, each of which marks the upper bound of a class; the height of each dot represents the sum of the observations in all of the classes up to and including the class marked by the dot* (see Figure 2.10). A cumulative frequency is a running total of the observations in a frequency

FIGURE 2.10
Cumulative frequency polygon showing distribution of grade point averages

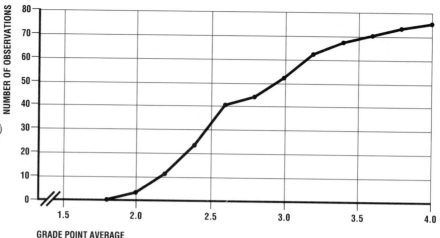

distribution. To use Figure 2.10, you find the number of students who have a grade point average of 3.0 or less, for example, by drawing a horizontal line from the dot that marks the upper bound of the class 2.8 and under 3.0 over to the y axis to find the value 52. Table 2.4 shows the frequency distribution and the cumulative frequency distribution of the data in Figure 2.10.

If you needed to focus on the number of students who are at or below a particular grade point average, the cumulative frequency polygon would be the best chart. For example, you might be considering the effect of different cut-off grade point averages for admission into an honors section. Figures 2.8, 2.9, and 2.10 should make it apparent that you have a lot of flexibility in choosing the way you will present continuous interval and ratio data.

TABLE 2.4
Frequency distribution and cumulative frequency distribution of data on grade point averages

Class range	Frequency	Cumulative frequency
1.8 and under 2.0	3	3
2.0 and under 2.2	8	11
2.2 and under 2.4	12	23
2.4 and under 2.6	17	40
2.6 and under 2.8	4	44
2.8 and under 3.0	8	52
3.0 and under 3.2	10	62
3.2 and under 3.4	5	67
3.4 and under 3.6	3	70
3.6 and under 3.8	3	73
3.8 and under 4.0	2	75

DEMOGRAPHIC STATISTICS ON THE COMPUSERVE INFORMATION SERVICE

CompuServe, a division of H&R Block Inc., is a computerized information service. Subscribers to the service can send electronic mail to other subscribers, shop by computer, look up airline schedules, and obtain information on literally thousands of other subjects. CompuServe also provides on-line access to the demographic reports of CACI, a corporation based in Arlington, Virginia. The reports issued by CACI would make a very long list, but one example will give you an idea of the potential of this system. Suppose you plan to start a regional chain of five liquor stores that will specialize in fine wines. You have identified 20 potential locations in strip malls near major shopping malls. You expect that almost all of the customers for a particular store will come from within two miles of the store. How can you identify the five best locations? You could dial CACI from your microcomputer and generate a report on each of your potential sites. CACI's Instant Demographics™ reports would describe the income, age, race, and population in the two-mile-radius circle centered on each potential location. You could also vary the radius to see how it would affect your ranking. CACI's ACORN™ (an acronym for A Classification Of Residential Neighborhoods) reports would describe the homogeneous neighborhoods within each of your 20 circles according to age and type of housing and the inhabitants' ethnic backgrounds, education, leisure activities, and spending patterns. CACI has divided and classified in "lifestyle clusters" all residential neighborhoods in the United States. The average size of a neighborhood is 350 households.

CACI uses a variety of sources for the information in these reports: census data at the city block level, proprietary household surveys (20,000 each year), other government surveys, and data collected on a confidential basis from firms that are CACI clients. Each report gives the values for the variables described for 1980, the current year, and projected five years into the future. An Instant Demographics™ report for a user-defined area currently costs $75 and an ACORN™ report costs $185.

Prices of $75 to $185 for information of this scope and timeliness suggest that the use of statistical methods and computer data bases to aid in site location or marketing and advertising plans will soon be standard operation procedure for any business larger than a lemonade stand.

2.2D GRAPHS FOR DISCRETE INTERVAL AND RATIO DATA

Histograms are generally used to depict discrete interval and ratio data. A **histogram** is *a bar chart in which the classes are placed in numerical order on the* x *axis and the area of each bar is proportional to the number of observations in its respec-*

TABLE 2.5
Frequency distribution of shoe sizes

Class range	Frequency
3½ through 5	2
5½ through 7	20
7½ through 9	38
9½ through 11	11
11½ through 13	4

tive class. As an example of discrete interval data, let's use the shoe sizes of the students in the statistics class. Table 2.5 gives the frequency distribution of shoe sizes. Note that the raw data used to make the frequency distribution are not provided. Notice that there are gaps between the class ranges. If someone had a shoe size between 5 and 5½, you would not know which class to put her in. These gaps are not a problem because shoe sizes are defined only in whole and half sizes.

Figure 2.11 shows the histogram for the frequency distribution in Table 2.5. One difference between the histogram and the simple bar chart is that in the histogram the *x* axis is scaled in the units that make up the interval or ratio scale. The *x* axis in a bar chart is merely labeled with the category of each bar. Another difference is that the *y* axis is scaled *so that the area in each bar equals the frequency for that range.* For example, there are 38 observations in the range 7½ to 9, so the bar has a width of 1½ and a height of 25.33 to make the area 38. Note that 25.33 times 1½ is 38. When the *y* axis is scaled in this manner, it is labeled "density." **Density** is *the number of observations in an interval divided by the width of the interval.* If each class has the same width, the histogram will look

FIGURE 2.11
Histogram showing distribution of shoe sizes

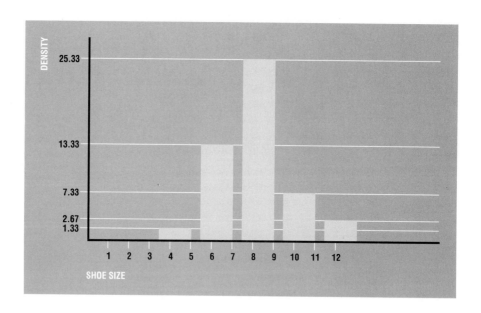

the same whether the vertical axis is scaled according to density or frequency. However, the vertical axis should always represent density when the classes have unequal widths. The great advantage of scaling the vertical axis according to density when classes are unequal is the visual information conveyed by the size of the bars. The area of each bar in the histogram tells you the total number of observations in each category.

The gaps between the bars in Figure 2.11 correspond to the gaps between the class ranges. If the data had been continuous, there would be no gaps between the bars. While histograms have often been used for continuous data, the frequency polygon is preferable because the line in the frequency polygon emphasizes the continuous nature of the data. On the other hand, the distinct bars in a histogram are more easily associated with discrete data.

If the raw data are available for a discrete interval or ratio data set, a vertical line chart can convey the exact number of observations that occur at each point in the distribution. A **vertical line chart** *shows the values of the observations on the x axis and the frequencies on the y axis. The height of each vertical line shows the number of observations at each value.* Another advantage of the vertical line chart over a histogram is that it gives even greater emphasis to the discrete nature of the data. The histogram in Figure 2.11 gives the impression that observations of shoe sizes can occur anywhere within a classification, such as between size 3½ and size 4. The vertical lines show that observations can occur only at distinct points. Figure 2.12 shows the same data organized in a vertical line chart. Note that while the raw data used to construct this chart are not given, you could use the chart to recover the raw data.

FIGURE 2.12
Vertical line chart showing distribution of shoe sizes

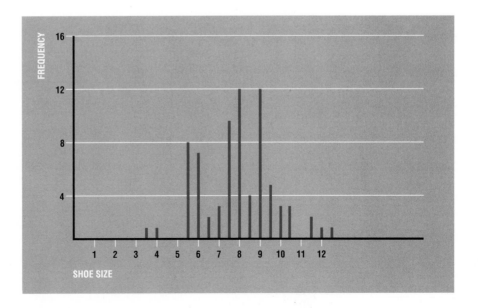

THE EFFECT OF THE MINIMUM WAGE LAW

Over the period 1973 to 1978 the minimum wage was raised from $1.60 an hour to $2.65 an hour. This differential overstates the real change in the minimum wage in those years because $1.60 was worth more in 1973 than in 1978. Adjusted for inflation, the $1.60 1973 minimum wage would have been about $2.40 in 1978. To account for inflation, both of the histograms in Figure 2.13 are in 1978 dollars. The first histogram shows the wage distribution of 16–24-year-olds in 1978 and the second histogram shows the wage distribution of the same group in 1973. The classes are in 25-cent intervals except for the end classes and those surrounding the minimum wage. The end classes include all of the observations beyond the last wage marked on the axis. Thus the last range for 1978 is not $5.90 to $6.15 but $5.90 or more. This arrangement causes an artificial peak in this last class because the observations that would be spread out over $5.90 or more are packed into the interval $5.90 to $6.15. This arrangement has been chosen to reduce the number of classes that have to be plotted on the *x* axis. An exception has been made in the 25-cent intervals around the value of the minimum wage because the impact of the minimum wage on the distribution is easier to see when the classes are broken at the minimum wage.

The minimum wage had hardly any impact on the distribution of wages in 1973. The concentration of observations at the minimum wage in 1978 shows that many more teens and young adults were affected by the minimum wage in that year. Another point that can be drawn from these histograms is that many youths worked at wages below the legal minimum. The minimum wage apparently had no spillover effect on youths employed at wages above the minimum. Only those youths who would have earned less

FIGURE 2.13
Wage distributions for 16–24-year-olds in 1978 dollars
Source: *Current Population Surveys,* May 1973 and May 1978.

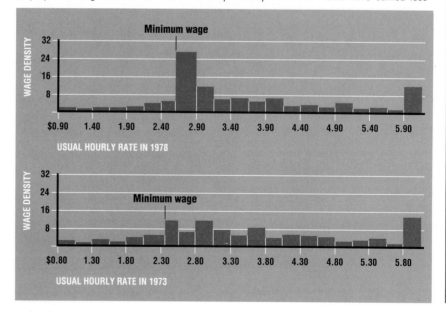

than the minimum wage rate in the absence of the legal minimum received higher wages as a result of the minimum wage. The high concentration at $2.65 an hour in 1978 combined with the low concentration in the class immediately above and the even lower concentration in the class immediately below supports this conclusion. Further, out-of-school youth employment was estimated to be about 4% below what it would have been in the absence of a legal minimum.

Adapted from Robert H. Meyer and David A. Wise, "The Effects of the Minimum Wage on the Employment and Earnings of Youth," *Journal of Labor Economics* 1 (January 1983): 66–100.

Figure 2.14 summarizes the main points concerning the appropriateness of the various types of graphs to display various types of data and the aspect of data that is emphasized by each type of graph.

FIGURE 2.14
Summary of types of graphs

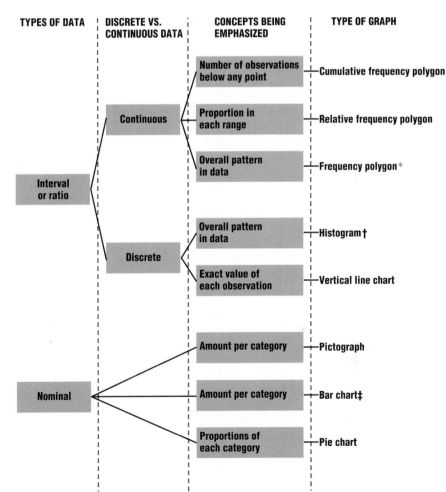

* Histograms are often used with continuous data.

† The area in any range is equal to the number of observations in the range.

‡ A bar chart can be used to show levels in two or more nominal data sets and to show negative levels.

EXERCISES FOR SECTION 2.2

2.6 Draw a vertical bar chart, a horizontal bar chart, and a pie chart to show the following data on household expenditures: food, $2,000; shelter, $4,000; clothing, $3,000; entertainment, $10,000; all other expenses, $1,000.

2.7 Draw a frequency polygon, a relative frequency polygon, and a cumulative frequency polygon to show the following continuous data. Divide the data into at least five classes.

55 90 81 70 64 39 52 75 83 99 60 47 81 67 58 75 90 89 54 58 58 69 52 70

81 52 83 62 87 62 69 91 45 66 51 73 60 86 52 71 55 41 89 56 71 45 78 34

55 51 62 52 84 50 53 79 51 60 93 58 53 64 55 71 83 69 54 59 50

2.8 Draw two histograms to show the data in exercise 2.7.

(a) Continue to assume that the data are continuous.

(b) Assume that the data can take only integer values.

2.9 Draw a histogram showing the following discrete frequency distribution. (Hint: The area in each bar has to be proportional to the number of observations.)

Class range	Frequency
0 through 9	20
10 through 19	50
20 through 29	80
30 through 35	50

2.10 Decide which type of chart would be most appropriate for each task below and explain why.

(a) To compare a company's expenditures in five categories of energy consumption before and after an energy conservation program was instituted.

(b) To summarize the time to failure of each of 1,000 electric motors used in new golf carts.

(c) To display the shares of a market controlled by five firms.

(d) To show the proportions of a clothing store's customers that wear each suit size.

2.11 Decide which type of chart would be most appropriate for each task below and explain why.

(a) To show the annual level of sales of a company's sales representatives and in the same chart show the level needed to be one of the top ten sales representatives.

(b) To show the number of hours, rounded up to a whole hour, that each of the company's sales representatives worked in the last month without losing any details about the number of representatives who worked any given number of hours.

(c) To show the percentage change in monthly sales from the previous month for each sales representative.

(d) To show the proportions of sales representatives whose annual sales are in various classes defined over the range from the lowest to the highest level.

2.12 What types of charts would be appropriate to display the following data? Explain.

Family income	Frequency
$ 0 and under 10,000	40
10,000 and under 20,000	60
20,000 and under 30,000	20
30,000 and under 100,000	30

2.13 What types of charts would be appropriate to display the following data? Explain.

Age	Frequency	Age	Frequency
17	1	24	0
18	19	25	1
19	29	26	0
20	24	27	1
21	16	28	0
22	14	29	1
23	5	30	1

SUMMARY

This chapter has described a variety of techniques for presenting data with graphs. The choice of technique depends on the type of data to be shown and on the features of the data that need to be highlighted. The types of data discussed were nominal, or data in categories; ordinal, or data in ranks; and interval and ratio, or data in levels (interval and ratio data are further classified as discrete and continuous). The types of charts presented were bar chart, pie chart, pictograph, frequency polygon, relative frequency polygon, cumulative frequency polygon, histogram, and vertical line chart. Bar and pie charts are used for nominal data. Pie charts can be used to emphasize the proportion in each category, while bar charts are best for comparisons between sets of nominal data. Frequency polygons are used for continuous interval and ratio data. The type of frequency polygon that is appropriate depends on whether the count, the proportion, or the cumulative frequency needs to be emphasized.

The main theme of the chapter is that a graph can be a powerful aid in the effort to understand the fundamental features of a data set and to convey those features to others. A secondary theme is that the use of graphs requires flexibility, as no one graph fits all data sets or serves all purposes in presenting a given data set.

SUGGESTED READINGS

Brightman, Harvey J. *Statistics in Plain English.* South-Western, Cincinnati, 1986.

Huff, Darrell. *How to Lie with Statistics.* New York: Norton, 1954.

 NEW TERMS

array A list of observations in a data set arranged in ascending or descending order.

bar chart A graph in which the frequency or count of observations in each category is represented by the height or length of each bar.

bimodal distribution A frequency distribution that has two peaks.

continuous data Observations that can include any value over a range of values.

cumulative frequency polygon A line connecting a series of dots, each of which marks the upper bound of a class; the height of each dot represents the sum of the observations in all of the classes up to and including the class marked by the dot.

density The number of observations in an interval divided by the width of the interval.

discrete data Observations whose values are limited to specific points in a range of values.

frequency distribution A listing of a data set which divides the data in classes and gives a count of the number of observations in each class.

frequency polygon A line connecting a series of dots, each of which is located at the midpoint of a class along the x axis and at the number of observations in that class along the y axis.

histogram A bar chart in which the classes are placed in numerical order on the x axis and the area of each bar is proportional to the number of observations in its respective class.

interval data Observations measured on a scale so constructed that equal distances between numbers imply equal distances between the values of the characteristic being measured, and 0 is arbitrarily defined.

nominal data The number of observations that fall into specific categories.

ordinal data Observations that are ranked or ordered to indicate first place, second place, third place, and so on.

pictograph A graph in which pictures of an object represent the frequencies of that object.

pie chart A graph in which the total number of observations is represented by a circle and the proportion of that total represented by each category is shown as a wedge of the circle.

primary data sources The government agencies or firms or trade associations that collect data.

ratio data Observations measured on a scale so constructed that equal distances between numbers imply equal distances between the values of the characteristic being measured, and 0 represents the absence of the characteristic being measured.

relative frequency distribution A listing of a data set that divides the data in classes and gives the proportion of the population or sample in each class.

secondary data sources Organizations that publish data collected by other agencies or firms.

vertical line chart Graph showing the values of the observations on the x axis and the frequencies on the y axis. The height of each vertical line shows the number of observations at each value.

 ADDITIONAL EXERCISES

An asterisk designates an exercise of greater than ordinary difficulty.

2.14 Humanoid Express, a new discount airline, would like to fit more passengers in each of its planes. If they use seats that are 28 inches wide, they can seat six across; 24-inch seats will accommodate seven across; and 20-inch seats will allow eight across. Use the following data on the widths of Humanoid's passengers to construct a cumulative relative frequency polygon and estimate the proportion of passengers that will not fit in 28-, 24-, and 20-inch seats. Graph paper will help.

Class range	Frequency
5 and under 10	50
10 and under 15	200
15 and under 20	600
20 and under 25	300
25 and under 30	100
30 and under 35	50

2.15 Identify the type of data presented in each example below, and if they are interval or ratio data, indicate whether they are continuous or discrete.

   (a) The number of cattle sold by a feedlot, graded by the U.S. Department of Agriculture as either commercial, good, choice, or prime.

   (b) The number of cattle sold by a feedlot by month for a one-year period.

   (c) The revenue from the sale of each head of cattle by the feedlot for one year.

   (d) The sex of each head of cattle sold by the feedlot.

2.16 Find a data source for the information required in each case below and identify if it is a primary source or a secondary source.

   (a) The number of people employed in your state by month.

   (b) The most recently available figure on the number of people living in your city.

(c) The road mileage between each pair of cities among the 50 largest cities in the United States.

(d) The sales of the 100 largest industrial firms in the United States.

2.17 The LMNOP Corporation would like a chart to display the sales growth of its five operating divisions during 1985 and 1986. Construct the chart that will facilitate division-by-division comparisons. The data are: Aerospace, 1985, 4.5%; Aerospace, 1986, 34%; Chemicals, 1985, 28%; Chemicals, 1986, 29%; Consumer Electronics, 1985, −20%; Consumer Electronics, 1986, −14%; Medical Supplies, 1985, 52%; Medical Supplies, 1986, −14%; Paper Products, 1985, 3%; Paper Products, 1986, 16%.

2.18 A West Coast refinery used the following stocks of oil in a production run of gasoline: 5,000 barrels of light Arabian crude oil; 20,000 barrels of high-sulfur Alaskan oil; 1,000 barrels of methyl alcohol from corn; and 6,000 barrels of Texas medium-weight crude. The respective costs of these stocks were $29, $24, $38, and $26 per barrel. Construct two pie charts, one for the amounts of each type of stock and the other for the total spent on each type of stock.

2.19 Identify a source of each of the following data sets and identify the type of data involved.

(a) The prices of corporate bonds issued by U.S. companies during the most recent month.

(b) Daily closing prices of the stocks traded on the New York Stock Exchange.

(c) The U.S. foreign trade deficit by quarter.

(d) The fraction of U.S. industrial capacity being used by month.

2.20 A survey of gas stations in a city found the prices listed below in dollars for regular gasoline. Use these data to construct the most appropriate chart for each objective:

(a) To display the exact number of gas stations charging each price.

(b) To display the percentage of gas stations in each one-cent range.

(c) To display the number of gas stations charging less than any given price.

1.09	1.092	1.099	1.089	1.076	1.089	1.085	1.079	1.096	1.084	1.094	1.072	1.085
1.095	1.096	1.099	1.099	1.099	1.10	1.093	1.078	1.086	1.087	1.086	1.098	1.099
1.079	1.084	1.087	1.096	1.078	1.094	1.099	1.089	1.083	1.082	1.093	1.096	1.099
1.095	1.096	1.092	1.091	1.086	1.092	1.12	1.099	1.094	1.093	1.093	1.099	1.11
1.10	1.099	1.098	1.099	1.087	1.085	1.085	1.089	1.097	1.092	1.093		

2.21 The Better Chip snack food company has five product lines: potato chips, pretzels, cheese curls, corn chips, and pork rinds. Its 1987 sales for these lines are, respectively, $125,000, $42,000, $14,000, $32,000, and $5,000. Construct an accurate pie chart that shows the proportion of total sales represented by each product line.

2.22 A car rental agency rents cars by the day. Over one month of operation it had the following frequencies for the number of days cars were rented: 1 day, 82; 2 days, 29; 3 days, 35; 4 days, 12; 5 days, 3; 6 days, 14; 7 days, 47; 8 days, 4; 9 days, 0; 10 days, 2; 11 days, 0; 12 days, 0; 13 days, 0; 14 days, 7; 15 or more days, 0. Construct a chart that shows the frequency for each number of days.

2.23 A real estate agency has sold 100 homes in the past six months. Use the following summary data to construct a cumulative frequency, polygon and to estimate the number of homes that sold for more than $100,000.

Range of prices	Frequency
$ 0 and under 50,000	45
50,000 and under 75,000	15
75,000 and under 90,000	11
90,000 and under 105,000	16
105,000 and under 120,000	8
120,000 and under 135,000	3
135,000 and under 150,000	1
150,000 or more	1

2.24 A food service company sells coffee, tea, hot chocolate, and carbonated soft drinks by the cup in vending machines. Measured in cups, the annual sales for these beverages are, respectively, 120,000, 10,000, 25,000, and 345,000. Construct a picto-graph that shows the level of sales by product. Note that the largest common factor of the above data is 5,000 cups, and if one picture of a cup represented 5,000 cups, you would need 69 cups (345,000/5,000) to show the level of soft drink sales. Think of a way to show the data with fewer cups.

2.25 Construct a bar chart that describes the rates of return on investment in 1986 and 1987 for five shoe companies:

	SHOE COMPANY				
	A	**B**	**C**	**D**	**E**
1986 rates	11%	14%	3%	21%	1%
1987 rates	8	10	−5	13	−6

2.26 Determine the type of data in each case below and the best chart for describing the data.

(a) The distribution of ages of all students at your college, recorded in integer years.

(b) The proportions of all students at your school who are freshmen, sophomores, juniors, seniors, and graduate students.

(c) The weights of all students at your college in pounds, recorded to the nearest one-hundredth of a pound.

2.27 Construct a frequency polygon showing the following data. Use eight equal ranges.

1.2	1.3	1.3	1.3	1.4	1.4	1.5	1.5	1.7	1.8
1.9	1.9	2.0	2.0	2.1	2.2	2.2	2.2	2.5	2.6
2.6	2.8	2.9	2.9	3.0	3.0	3.0	3.0	3.1	3.4
3.5	3.5	3.6	3.7	3.8	3.8	3.8	4.0	4.0	4.2
4.3	4.3	4.4	4.5	4.6	4.7	4.9	5.0	5.0	5.0

2.28 Use the data set in exercise 2.27 to construct a cumulative relative frequency polygon. With this polygon estimate the point in the data set that marks the top 10%

of all of the observations. What is the exact value of the observation that marks the top 10%?

2.29 Construct a histogram showing the following data on the annual salaries of professors. Label the density for each range. The highest-paid professor in the data set earns $54,000.

Salary range	Frequency
$ 0 and under 15,000	2
15,000 and under 20,000	5
20,000 and under 25,000	24
25,000 and under 30,000	30
30,000 and under 35,000	12
35,000 or more	8

2.30 (a) Plot the histogram showing the following data on the waiting time in minutes needed to reach an operator at an airline reservation number:

Range of minutes	Frequency
0 and under 10	10
10 and under 20	25
20 and under 30	10
30 and under 50	10

(b) Approximately what proportion of the callers had to wait more than 40 minutes?

2.31 (a) A consumer group has made a study of the amount of music on classical compact discs. The discs have a capacity of 70 minutes of music but many discs are released with much less. Construct a histogram of the minutes of music on the compact discs that the group measured.

Range of minutes	Frequency
0 and under 30	8
30 and under 35	59
35 and under 40	98
40 and under 45	67
45 and under 50	43
50 and under 55	30
55 and under 60	2
60 and under 65	8
65 and under 70	3

(b) The consumer group recommends that consumers not buy any compact disc that has less than 40 minutes of music as a protest against the manufacturers. According to this criterion, what proportion of the discs in the above study should not be bought?

2.32 Construct a pie chart showing the following data on sources of tax money for a local government: hotel and restaurant tax, $1,298,000; parking fines, $4,653,000; local sales tax, $18,054,000; local income tax, $35,938,000; licenses and user fees, $750,000.

2.33 Decide which type of chart would best display each set of data below and explain why.

(a) The unit sales of snowmobiles per year in the United States for the last ten years.

(b) The proportions of the funds collected by a United Way agency allocated to overhead and to each of eight local charities.

(c) The number of cars sold in the United States in 1987 by price range in $1,000 increments.

(d) The number of drivers convicted in a particular court of driving at or below the following speeds in excess of the posted limit: 5, 10, 15, 20, 25, 30, 35, 40.

2.34 Divide the following data on salesmen's commissions (in dollars) for one month at a large auto dealership into ten ranges of equal width and plot the frequency polygon. Do the data appear to be bimodal?

1,207 1,098 1,187 1,398 1,346 1,987 1,654 1,876 1,765 1,765 1,342 1,987 1,876 1,654 1,234 1,108

1,872 1,901 1,542 1,651 1,761 1,766 1,876 1,298 1,091 1,087 1,630 1,323 1,983 1,872 1,622 1,021

1,245 1,447 1,287 1,279 1,234 1,528 1,276 1,820 1,965 1,947 1,306 1,276 1,013 1,128 1,061 1,423

2.35 Draw the histogram for the following data on annual salaries of a group of bookkeepers:

Salary range	Frequency
Under $10,000	58
$10,000 and under $12,500	44
$12,500 and under $15,000	34
$15,000 and under $17,500	46
$17,500 and under $20,000	68
$20,000 and under $22,500	78
$22,500 and under $25,000	109
$25,000 and under $27,500	125
$27,500 and under $30,000	56
$30,000 and over	78

The highest-paid bookkeeper in the group had a salary of $44,000.

2.36* Construct a cumulative relative frequency polygon to show the data in exercise 2.31, on the playing times of classical compact discs, and estimate the 75th percentile of the distribution, that is, the length of time at or below which 75% of the compact discs now play. Graph paper would help.

3

Numerical Descriptive Statistics

elvin Simon and Associates is the largest developer of major shopping centers in the United States. The firm owns and operates about 100 malls across the country. Since an enormous investment is required for a new mall, a careful evaluation of the profit potential of alternative sites precedes the choice of a new location. A typical problem is to decide whether a new mall on the north side of Phoenix, Arizona, will generate a higher rate of return on investment than a new mall on the south side of Macon, Georgia. Some information about the people living near these sites is available from public sources and some has to be gathered in special surveys. The U.S. Census Bureau provides such information as the number of people who live in each area, their ages, and their incomes. To preserve the privacy of the individuals who gave them this information, the Census Bureau aggregates or adds up the individual records for areas at least as large as a city block. Even with block-by-block data, there are too many numbers to make sense of without further reduction and analysis. Other information, such as the amount of money spent on clothing, driving times to the proposed sites, and attitudes toward shopping in malls, is available only through surveys of area residents paid for by the mall developer. Before making a decision, Melvin Simon and Associates has to condense the data it has assembled into a form that clarifies the important points of comparison. No one can make a sound decision after staring at a stack of numbers the size of a large phone book. Melvin Simon and Associates needs to describe the data with summary measures.

The application of the techniques presented in this chapter is very simple and straightforward, but the payoffs can be very high. Appropriate summary measures can help you see what's really going on. The chapter begins with a discussion of summary measures for the center of a data set. Section 3.2 presents summary measures for the spread of the data. Section 3.3 is on a powerful new technique called exploratory data analysis, which can help you find either patterns or anomalies in the data.

3.1 DESCRIBING THE CENTER OF THE DATA

 3.1A THE ARITHMETIC MEAN OF INDIVIDUAL DATA AND OF DATA IN CLASSES

Sometimes the feature of a data set which is the most important to you is the location of its center. Suppose you own a dairy and are considering the purchase of either cow *A* or cow *B*. Dairies keep careful records of the daily milk production of their cows, and you could use these data to construct two frequency polygons that would show you the distributions of past daily milk production. However, the shapes of the distributions may not be an important consideration. You need to know which cow has the greatest potential production over its working life. The average daily yield is the key feature of the data you want to know. The average or mean is the most commonly used summary measure of the center of a data set. The **mean** or **arithmetic mean** is *the sum of the values of all of the observations in a data set divided by the number of observations*. Generally, means should be calculated only for interval or ratio data. The mean makes no sense if it is applied to nominal data. Occasionally you may want to calculate the average rank given to some choice, but averages applied to ordinal data can be misleading. Suppose your family members ranked their five top locations for a summer vacation and Hawaii had the highest average rank, followed by Alaska. If everyone got to make a new ranking with just Hawaii and Alaska on the list of choices, it is possible that no one would change his or her mind about the desirability of the two locations but that Alaska would come out with a higher average ranking than Hawaii. Remember that the mean of ordinal data can be ambiguous.

To reinforce the distinction between a population and a sample, there are two symbols for the mean, μ (read mu) for the population mean and $\bar{x}$ (read x-bar) for the sample mean. Formula 3.1 is for the population mean and formula 3.2 is for the sample mean.

$$\mu = \frac{\sum_{i=1}^{\nu} x_i}{\nu}$$

(3.1)

$$\bar{x} = \frac{\sum_{i=1}^{n} x_i}{n}$$

(3.2)

The letter n is the symbol for the number of observations in a sample and the Greek letter ν (read nu) is the symbol for the number of elements in a population. Throughout the text Greek letters will be used for population parameters and the roman letters for sample statistics. The capital sigma, Σ, is a math operator that means summation. The first and last term in the summation are given below and above the Σ. For example, the notation $\sum_{i=1}^{n} x_i$ means the sum of the values of $x_1, x_2, x_3, \ldots, x_n$. The letter i is a counter or index; it takes

FIGURE 3.1
The mean as the center of
balance of a data set

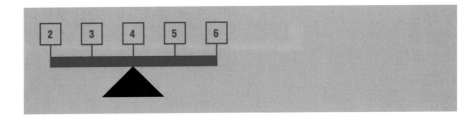

on every integer value in the range 1 to n. Thus x_1 is the first element, x_2 is the second element, and so on. A simple example will illustrate the use of the formulas. Find the mean of the data set $\{2, 3, 4, 5, 6\}$. Assume that the data set is a sample.

$$\bar{x} = \frac{\displaystyle\sum_{i=1}^{n} x_i}{n} \tag{3.2}$$

$$\bar{x} = \frac{2 + 3 + 4 + 5 + 6}{5}$$

$$\bar{x} = 4$$

If the data were a population instead of a sample, the answer would still be 4 but the symbol μ would replace the $\bar{x}$ and the symbol ν would replace the n.

The mean is the center of balance of a data set. Consider a beam scaled in the same units as the data with the lowest point in the data set marked at one end and the highest point at the other. If a weight were placed at each point on the beam which corresponded to an observation and the beam were placed over the fulcrum at the point marking the mean, the beam would stay level. Figure 3.1 illustrates this concept. The boxes labeled 2 through 6 represent equal weights located on the balance beam at the points that correspond to their labels. Clearly, if the beam were placed over a fulcrum at point 4, it would remain level. If the data set were changed to $\{2, 3, 4, 5, 6, 6\}$—that is, if another 6 were added—the mean would be 26 divided by 6, or $4\frac{1}{3}$. The balance beam for this data set would stay level only if it were placed over the fulcrum at $4\frac{1}{3}$. Figure 3.2 illustrates this concept. The mathematical property of the mean

FIGURE 3.2
Balance beam for data set
$\{2, 3, 4, 5, 6, 6\}$

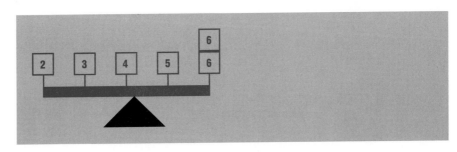

FIGURE 3.3
Balance beam for the data
set {2, 3, 4, 5, 100}

which allows the use of a balance beam as an analogue is that the sum of the
deviations from the mean of all observations must be zero. In notation, this
property is expressed as

$$\sum_{i=1}^{n} (x_i - \bar{x}) = 0$$

The proof of this property may be found in Appendix C.

The mean has some advantages as a measure of the center of the data and
one disadvantage. The disadvantage is that the mean is sensitive to extreme
observations. Suppose our data set were {2, 3, 4, 5, 100}. This data set is the
same as the first one except that 100 has replaced the 6. The new mean would
be 114 divided by 5, or 22.8. Figure 3.3 illustrates the balance beam analogy for
this data set. The black vertical bars represent identical weights placed at 2, 3,
4, 5, and 100. The beam would balance only if the fulcrum were placed at 22.8.
The reason that number 22.8 has a disadvantage when it is used to represent
the center of the data is that most of the observations are not near 22.8. The 100
is an extreme or outlying value that pulls the mean away from the rest of the
data. We'll come back to this problem of representing the center of the data
when there are extreme observations in the last part of this section, where the
median is discussed.

One of the advantages of the mean as the statistic that describes the center of
the data is that the mean is widely used and well understood. Another advantage is that if you know the means of two data sets and the number of observations in each set, you can calculate the mean of the combined data set. Label the
mean of the first data set $\bar{x}_1$ and the mean of the second data set $\bar{x}_2$ and the
respective sample sizes n_1 and n_2. An $\bar{\bar{x}}$ (read x-bar-bar) is defined as the mean
of the combined data sets. The mean of the combined data sets equals:

$$\bar{\bar{x}} = \frac{n_1\bar{x}_1 + n_2\bar{x}_2}{n_1 + n_2}$$

(3.3)

As an illustration, take the data sets {2, 3, 4, 5, 6} and {2, 3, 4, 5, 100} and use
formula 3.3 to find the mean of the combined data sets.

$$\bar{\bar{x}} = \frac{n_1\bar{x}_1 + n_2\bar{x}_2}{n_1 + n_2} \tag{3.3}$$

$$\bar{\bar{x}} = \frac{5(4) + 5(22.8)}{5 + 5}$$

$$\bar{\bar{x}} = 13.4$$

If the two data sets have the same number of observations, the mean of the combined data sets is simply the mean of the means, in this case, (4 plus 22.8) divided by 2. If the data sets do not have the same number of observations, you have to use formula 3.3, which weights the mean of each data set by the number of observations in that data set. Formula 3.3 can be extended beyond two data sets. Formula 3.3a is for k data sets, where k is any integer.

$$\bar{\bar{x}} = \frac{n_1\bar{x}_1 + n_2\bar{x}_2 + n_3\bar{x}_3 + \cdots + n_k\bar{x}_k}{n_1 + n_2 + n_3 + \cdots + n_k} \tag{3.3a}$$

Instead of weighting the $\bar{x}$'s by the sample sizes for each data set, you could use as weights the amounts of all data sets consumed in a particular application. For example, suppose the mean cost of a feedlot's corn is \$3 per 100-pound bag, the mean cost of its soy meal is \$5 per 100-pound bag, and the mean cost of its fish meal is \$12 per 100-pound bag. If a cow consumes 5 bags of corn, 10 bags of soy meal, and 1 bag of fish meal before it reaches market weight, what is the average cost per bag of feed consumed by the cow?

$$\bar{\bar{x}} = \frac{n_1\bar{x}_1 + n_2\bar{x}_2 + n_3\bar{x}_3 + \cdots + n_k\bar{x}_k}{n_1 + n_2 + n_3 + \cdots + n_k} \tag{3.3a}$$

$$\bar{\bar{x}} = \frac{5(3) + 10(5) + 1(12)}{5 + 10 + 1}$$

$$\bar{\bar{x}} = 4.81$$

The average cost per 100-pound bag of feed consumed by the cow is \$4.81.

As we saw earlier, data are often published in classes rather than as raw data or individual numbers. Unless you know the values of the individual data points, you cannot calculate the exact mean of data in classes. Formula 3.4 gives an approximation of the mean; it will be more or less accurate depending on how closely the data follow the assumption that underlies the formula. This formula is based on the assumption that the observations in any class are evenly spread out over the range of the class. This assumption allows you to treat the midpoint of each class as its center of balance. If this assumption is accurate or even if the observations are bunched up at various points but still balanced around the midpoint of each class, the approximation will be good.

$$\overline{x} \cong \frac{\sum\limits_{j=1}^{k} f_j x_j'}{n} \tag{3.4}$$

In formula 3.4, the x_j' (read x prime sub-j) stands for the midpoint of the jth class. The j is a counter that runs from 1 to k. The symbol f_j stands for the frequency of the jth class. The counter j is used instead of the counter i to remind you that classes are being counted instead of individual observations. As before, n stands for the number of observations in the sample. Unless you are told otherwise, assume that classified data are for a sample. Note the similarity of formulas 3.3a and 3.4. The midpoint of each class corresponds to the center of gravity of that class or to the mean of a sample (in the feedlot example it is the average price paid for corn or soy meal or fish meal) in formula 3.3a. The frequencies in each class correspond to the number of observations in each sample in formula 3.3a.

For an illustration of this formula we can use the 75 grade point averages that were introduced in Table 2.3, in Chapter 2. Here the grade point averages are divided into 11 classes. Table 3.1 gives the midpoints and the frequencies of each class.

To find the mean, substitute these values in formula 3.4:

$$\overline{x} \cong \frac{\sum\limits_{j=1}^{k} f_j x_j'}{n} \tag{3.4}$$

$$\overline{x} \cong \frac{3(1.9) + 8(2.1) + 12(2.3) + \cdots + 2(3.9)}{75}$$

$$\overline{x} \cong 2.713$$

The mean calculated from the raw data is 2.708, so the approximation is quite close in this example.

TABLE 3.1

Midpoints of each class and frequencies for data on grade point averages in Table 2.3

Class range	Midpoint	Frequency	Class range	Midpoint	Frequency
1.8 and under 2.0	1.9	3	3.0 and under 3.2	3.1	10
2.0 and under 2.2	2.1	8	3.2 and under 3.4	3.3	5
2.2 and under 2.4	2.3	12	3.4 and under 3.6	3.5	3
2.4 and under 2.6	2.5	17	3.6 and under 3.8	3.7	3
2.6 and under 2.8	2.7	4	3.8 and under 4.0	3.9	2
2.8 and under 3.0	2.9	8			

You may run into two problems when you try to apply formula 3.4. One is caused by an **open-ended class,** *an interval defined as all observations that are either above or below a specified value.* For example, the lowest class of the data in Table 3.1 could have been written as "2.0 or less." Without knowing the high and low points of a class, you cannot find the midpoint. In this situation you are stuck; without more information you cannot calculate the sample mean. The other problem is how to define the midpoint of a class with discrete data. Take the example of shoe sizes in Table 2.5. Should the midpoint be halfway between the class limits of 3½ and 5 or should it be halfway between the start of one class and the start of the next, such as halfway between 3½ and 5½? The correct answer is halfway between 3½ and 5. The reason is that the midpoint is used to represent the center of gravity of all of the observations in a class, and the inclusion of the empty interval between classes pulls the midpoint away from these observations. In the case of continuous data, in contrast, the midpoint of each class is halfway between the start of one class and the start of the next. In the grade point average data, for example, the first class is "1.8 and under 2.0" and the next class is "2.0 and under 2.2." The midpoint of the first class is 1.9 because, although the class never reaches 2.0, it gets infinitesimally close to 2.0.

■ 3.1B THE MODE

When statistics students are asked what the word "mode" means, the answer most frequently given is "To have vanilla ice cream on top of something." Actually, the **mode** is one of the measures of the center of a data set. It is *the observation that occurs most frequently in a data set.* Thus the response "To have vanilla ice cream on top of something" is the modal answer to the question "What is the mode?" The mode can be applied to nominal, interval, or ratio data. The applications of the mode for interval and ratio data usually involve discrete data. The data set {2, 3, 4, 5, 6} has no mode, or alternatively, since each of the five values is as frequent as any other, you could say that it has five modes. The mode of the data set {2, 3, 4, 4, 5, 6} is 4 because that number occurs more frequently than any of the others. If the data are grouped or classified, the **modal class** is *the class with the most observations.* In the data set on grade point averages in Table 2.3, the class "2.4 and under 2.6" has the most observations, so it is the modal class. You cannot determine the exact value of the mode with classified data because you do not know how the observations are distributed within each class. It even happens that the true mode falls outside the modal class. If you need one number for the mode and you have classified data, a simple rule is to *use the midpoint of the modal class.* This rule yields a good approximation of the true mode of a large data set in which the observations are smoothly distributed across their classes.

The mode can also be thought of as the single value of a data set that is most likely to be chosen if one observation is selected at random. If a business could stock or carry only one size or one type of an item, it would want it to be the modal value so that it could serve the most people. An automobile dealer purchasing a car for inventory might need to know the modal car color her customers order. A builder who is putting up a house and hoping to find a buyer after it is finished might need to know the modal number of rooms that house buyers want. A word of caution has to be given about using the sample mode as an estimate of the population mode. If a sample data set does not have a pronounced peak or concentration at one value, the sample mode can be a very unreliable indicator of the location of the population mode. The sample mode can bounce all over the population distribution from one sample to another because a few observations falling at a given point can cause the sample mode to change. The reliability of sample statistics as estimators of population parameters will be fully discussed in Chapter 7.

3.1C THE MEDIAN

The **median** is *the point in an array of data which has an equal number of observations above and below it.* The median of the data set {2, 3, 4, 5, 6} is 4. The median of the data set {2, 3, 4, 5, 100} is also 4. In each case there are two observations below 4 and two observations above 4. Unlike the mean, the median is not sensitive to extreme observations. When a data set has an odd number of observations, the median is equal to one of the observations. When a data set has an even number of observations, the median can fall between two observations. The rule for finding the median with an even number of observations is to split the distance between the two centermost observations. For example, in the data set {1, 2, 3, 4, 5, 6} the median is halfway between 3 and 4, or 3.5. The median of the data set {1, 2, 3, 3, 4, 5} is 3; in this case there is no distance between the two centermost observations to split.

The exact median of classified data cannot be determined, but it can be estimated on the assumption that the observations are evenly spread out across their classes. Let's use the data set on grade point averages to show how the median of classified data can be found. Table 2.4 is repeated in Table 3.2. Since there are 75 observations in the data set, the median is the 38th observation.

TABLE 3.2
Frequency distribution and cumulative frequency distribution of data on grade point averages

Class range	Frequency	Cumulative frequency	Class range	Frequency	Cumulative frequency
1.8 and under 2.0	3	3	3.0 and under 3.2	10	62
2.0 and under 2.2	8	11	3.2 and under 3.4	5	67
2.2 and under 2.4	12	23	3.4 and under 3.6	3	70
2.4 and under 2.6	17	40	3.6 and under 3.8	3	73
2.6 and under 2.8	4	44	3.8 and under 4.0	2	75
2.8 and under 3.0	8	52			

FIGURE 3.4
Location of the median
within the median class

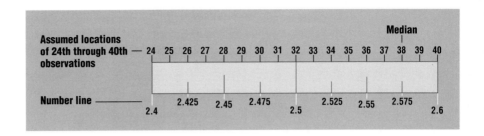

There are 37 observations below the 38th and 37 observations above the 38th. The 24th through the 40th observations are in the class "2.4 and under 2.6." This class must contain the median. With only the classified data, we cannot know the location of the 38th observation, but we will base our estimate on the assumption that the 17 observations in this class are evenly spaced over the range 2.4 to 2.6. Figure 3.4 shows the range 2.4 to 2.6 marked with 17 evenly spaced observations. If the 24th observation is placed at the beginning of the median class and the 40th observation at the end of the median class, the 38th observation—that is, the median observation—will be located at 2.575. Formula 3.5 will give you the same answer:

$$m \cong \left[\frac{\left(\frac{n+1}{2} \right) - (f_b + 1)}{(f_m - 1)} \right] w_m + L_m \qquad (3.5)$$

The symbol m stands for the sample median. The letter n stands for the number of observations. The letter f stands for "frequency," with f_m the frequency of the median class and f_b the cumulated frequency of the classes below the median class. The letter w stands for the width of a class and the letter L for the lower limit of a class. In both cases the subscript m refers to the median class. This formula applied to the data set on grade point averages yields:

$$m \cong \left[\frac{\left(\frac{n+1}{2} \right) - (f_b + 1)}{(f_m - 1)} \right] w_m + L_m \qquad (3.5)$$

$$m \cong \left[\frac{\left(\frac{75+1}{2} \right) - (23 + 1)}{(17 - 1)} \right] 0.2 + 2.4$$

$$m \cong 2.575$$

The accuracy of the estimate of the sample median given by the formula for classified data depends on the accuracy of the assumption made about the spacing of the observations within the median class. The actual median for the individual data is 2.57.

Insensitivity to extreme observations is one of the advantages of the median. The importance of this advantage depends on the data set. Many data sets do not have such extreme values that the highest observation is, for example, a thousand times the lowest. Unless your data set includes Paul Bunyon, the mean of a group of people's heights will be a reasonable measure of the center of the data. If your data set is on the assets of companies, the observations may run from a few thousand dollars to billions. The median provides a better measure of the center of such data. Another advantage offered by the median is that it allows calculation with open-ended classes (provided the median class is not open-ended). For example, if the lowest class in the grade point average data were "under 2.0" instead of "1.8 and under 2.0," it would make no difference in the above calculations. This change in definition would, however, prevent the calculation of the mean.

Here's an example of a data set in which the mean provides a misleading measure of the center and the median does not. Jury Verdict Research of Solon, Ohio, is a private research firm that publishes sample data on amounts of jury awards in major cases. According to statistics tabulated by this firm, the mean jury award in 1978 in liability cases was $1.7 million, an increase of 285% over the mean award of $596,000 in 1977. This reported jump in the mean award gave the impression that juries were becoming much more liberal. However, 1978 included one award of $127 million to a man who was severely burned in a Pinto automobile. This single high award skewed the data enough to raise the mean of the 123 recorded awards very sharply. Over the same years the increase in the median award was much more stable, about enough to keep up with inflation. True, one jury was much more liberal than usual; but juries as a whole were no more liberal than they had been the year before. The mean jury award is not a good measure of the average cost to insurance firms of all major liability cases because large jury awards tend to be reduced by judges more often than small awards. The judge in the Pinto case, for example, reduced the award to $6.7 million.

One disadvantage of the median is that it has no arithmetic properties in the sense that you can add or subtract medians from different data sets. If you know the medians of two data sets and they are not the same number, you have no way of calculating the median of the combined data sets. The median of the combined data sets is not the median of the medians or the mean of the medians or any other convenient expression. Another disadvantage of the median is that its sampling properties are much more complicated than the sampling properties of the mean. The term "sampling properties" refers to the closeness of a sample statistic to a population parameter in repeated sampling. The problem is not that the sample median is an erratic estimator of the population median (as was the case for the sample mode), but rather that the sample mean is much easier to work with. In problems involving inference from a sample to a population, the best measure of the center of the data is the mean. This topic will be discussed in more detail in Chapter 7.

3.1D COMPARING THE MEAN, MODE, AND MEDIAN IN ONE DATA SET

If a data set is perfectly symmetrical and unimodal (has one mode), the mean, mode, and median will be the same number. A symmetrical data set is one that produces a graph whose left and right halves are mirror images of each other. Figure 3.5 shows the frequency polygon of a data set that is unimodal and symmetrical.

If the data set is unimodal but not symmetrical, the mean, mode, and median will be located at different points in the distribution. If one side of the distribution is stretched out, the mean will be pulled farthest in that direction, followed by the median. Figure 3.6 shows two frequency polygons, one with a long tail to the right and the other with a long tail to the left. The locations of the mean, mode, and median are indicated in each frequency polygon. The mode is always located at the highest point in the distribution and the mean and median are pulled in the direction of the longer tail of the distribution. The mean and the median are also pulled toward the longer tail in distributions that are not unimodal.

Any distribution that is *not symmetrical around the mean* is **skewed.** A skewed distribution that has a long tail to the right is said to be skewed to the right, and a distribution that has a long tail to the left is said to be skewed to the left. The

FIGURE 3.5
Frequency polygon of a data set that is unimodal and symmetrical

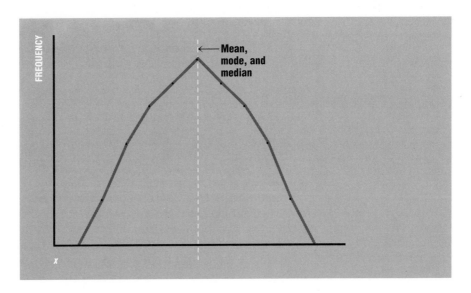

FIGURE 3.6
Frequency polygons of data sets with long right and left tails

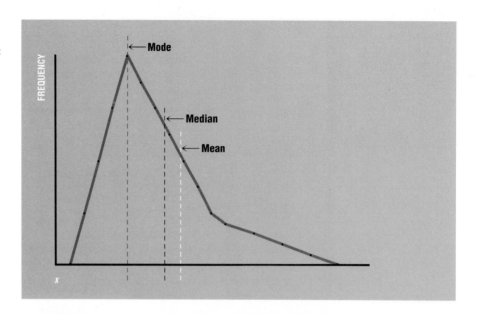

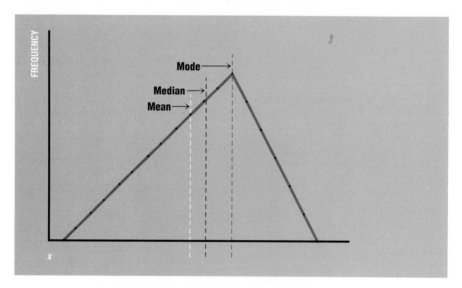

first frequency polygon in Figure 3.6 is skewed to the right and the second is skewed to the left. There are numerical measures of skewness, but they have few direct business applications. We will skip those measures and just rely on visual inspections of the frequency polygons of data sets to decide which distributions are skewed and which are the most skewed. Other things being equal, the distribution that appears to have the longest tail will be considered the most skewed.

CASE STUDY

DESCRIPTIVE STATISTICS AS ENTERTAINMENT

Descriptive statistics have always been part of the folklore of sports. A player's claim to greatness must be supported by statistics, such as yards gained per carry or batting average or proportion of free throws made. What is new is the use of computers and statisticians to create and display these statistics and the prominence such displays have gained in televised games. Steve Hirdt may be the preeminent practitioner of this new art. He is the statistician used by ABC Television for its *Monday Night Football* program. His job is to pass descriptive statistics to the announcers and prepare graphics for the screen. He contributes to the popularity of the show by using statistics to heighten the drama or to interject humor during a game. With the aid of a computer he can instantly determine an obscure statistic such as the mean number of yards gained by a team over the last five seasons in fourth-and-goal situations during the fourth quarter while running against the wind on a Monday night. Sometimes such information is simply trivia meant to amuse the fans, but often he can highlight key features of a team's strategy through the display of descriptive statistics. If you know that a team has been successful running behind its right guard in its last ten fourth-and-goal situations and never successful when passing or running another play, you have a good idea of where to look as the play develops. The fact that NFL teams now retain their own statisticians and computers is proof of the value of descriptive statistics for analyzing and predicting a team's behavior. Many famous tennis stars, too, hire statisticians to track and analyze their games. For both analysis and entertainment, the statistician has gained a new role in sports.

To summarize, each of the three measures of the center of a data set has some advantages and some disadvantages. Each of them does the best job in some situations. The mean works best for problems of statistical inference, the median for data sets with extreme observations, and the mode for problems in which the most frequently occurring observation is of interest, as when only one type or size of an item can be ordered or stocked.

EXERCISES FOR SECTION 3.1

3.1 Determine the mean, the mode, and the median of the following data set.

| 21 | 90 | 11 | 3 | 50 | 56 | 48 | 32 | 40 | 81 | 28 | 39 | 45 | 27 | 39 | 24 | 53 | 11 | 79 | 23 |
| 30 | 24 | 34 | 44 | 28 | 37 | 69 | 42 | 51 | 71 | 28 | 14 | 19 | 52 | 41 | 28 | 7 | 22 | 28 | 55 |

3.2 Estimate the mean, the median, and the mode of the following continuous data:

Class range	Frequency
0 and under 10	50
10 and under 20	150
20 and under 30	100
30 and under 40	50

3.3 Estimate the mean, the median, and the mode of the following discrete data:

Class range	Frequency
0 through 9	50
10 through 19	150
20 through 29	100
30 through 39	50

3.4 Either estimate the mean, the median, and the mode of the combined data sets of exercises 3.2 and 3.3 or, if you cannot estimate any of these terms, explain why.

3.5 Decide which measure of the center of the data is the most appropriate in each of the following cases and explain why.

(a) Data on the number of years employees had been with a company that was founded in 1920.

(b) Data on preferred locations for a company's annual sales meeting.

(c) Data on the amount of the monthly pension check of each retired employee.

(d) Data on a sample of 100 utility bills from a population of 10,000 utility bills.

3.6 (a) If a histogram of a data set was drawn so that the mean fell on the boundary between two contiguous intervals, what would you have to know about the data set for 50% of the area in the histogram to be below the mean?

(b) If the histogram was drawn so that the median fell on the boundary between two contiguous intervals, what would you have to know about the data set for 50% of the area in the histogram to be below the median?

3.7 Assume that the mean and the median of a data set are not the same. If a cardboard replica of the histogram of the data set were made, would the replica balance on a fulcrum placed under the median or under the mean? Explain.

3.8 Your checking account is with a bank in Fairbanks, Alaska, even though you live in another state, because the Fairbanks bank offers the highest interest rate on funds in a checking account. You keep a record of the number of days it takes for letters with deposits to reach your bank. The following data set contains this record: {2, 9, 2, 3, 2, 2, 3, 1, 2, 3, 3, 2, 2}. Calculate the mean, the median, and the mode of these data. Which measure of the center of the data is the most appropriate in this case?

3.2 DESCRIBING THE SPREAD OF THE DATA, THE RELATIVE SPREAD, AND THE PROPORTIONS IN THE TAILS

3.2A THE STANDARD DEVIATION WITH INDIVIDUAL DATA

You are considering the purchase of one of two competing soft drink bottling machines. Both machines can be adjusted so that the mean number of ounces placed in a bottle is 12. Both machines cost the same and have the same warranty. Is there any basis for choosing between them? What if one machine averaged 12 ounces but varied widely from bottle to bottle so that some were filled with 2 ounces of soft drink and others with 22? Most of the soft drink poured into the overfilled bottles would wind up on the floor. Underfilled bottles would require additional processing. The other machine is highly consistent, never varying by more than a tenth of an ounce. The choice is obvious. This example shows why the variability of a distribution can be an important issue. The simplest measure of variability is the **range,** which is *the distance from the lowest to the highest observation in a data set*. In the bottling machine example, the range would provide enough information to permit you to decide which machine you wanted to purchase. Sometimes the range can be a deceptive measure of the variability in a distribution. The two data sets shown in Figure 3.7 show how misleading the range can be. The two data sets in Figure 3.7 are

FIGURE 3.7
Two data sets with identical means and different ranges

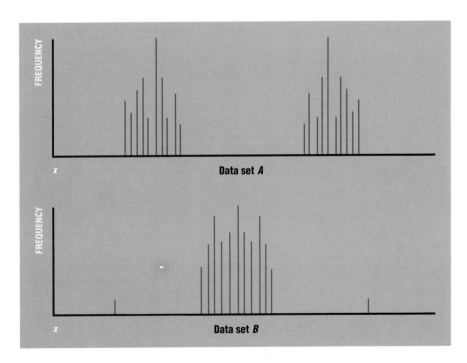

constructed so that they have the same mean but data set B has a slightly greater range than data set A. Clearly the two extreme points in data set B are outliers from the rest of the data. Which of the two sets would you describe as having the greatest average variability? The answer is as clear as the choice between the two bottling machines: data set B has, on average, less variability.

The **standard deviation** is *a numerical measure of the average variability of a data set around its mean.* Like the mean, it is usually calculated for interval and ratio data. Formulas 3.6 and 3.7 are the two formulas for the standard deviation of individual data.

$$\sigma = \sqrt{\frac{\sum_{i=1}^{\nu} (x_i - \mu)^2}{\nu}} \tag{3.6}$$

$$s = \sqrt{\frac{\sum_{i=1}^{n} (x_i - \bar{x})^2}{n - 1}} \tag{3.7}$$

The standard deviation for a population is denoted σ (the Greek lowercase letter sigma) and the standard deviation for a sample is denoted s. The difference between the two formulas involves more than a change of symbols. The denominator for the standard deviation of the population is the number of elements in the population, ν (nu), while the denominator for the standard deviation of the sample is the number of elements in the sample minus one, $(n - 1)$. Before getting into this difference, let's begin with an explanation of the formula for the standard deviation of the population.

What you are looking for is a measure of the average distance between the observations and the mean of a data set. The term $\sum_{i=1}^{\nu} (x_i - \mu)$, which is the sum of the deviations of all observations from the mean, must equal zero. Thus the sum of all of the deviations from the mean divided by the number of observations is a poor measure of the average variability in a data set because it always comes out to zero. Positive deviations cancel out the negative deviations. Squaring the term $(x_i - \mu)$ results in only positive numbers. If you divided the sum of the squared terms, $(x_i - \mu)^2$, by the number of observations, you would have *the average squared deviation between the observations and the mean,* which is the definition of the population **variance.**

(3.8)

$$\sigma^2 = \frac{\sum\limits_{i=1}^{\nu} (x_i - \mu)^2}{\nu}$$

(3.9)

$$s^2 = \frac{\sum\limits_{i=1}^{n} (x_i - \bar{x})^2}{n - 1}$$

The symbol for the variance of a population is σ^2. The symbol for the variance of a sample is s^2. It is easier to interpret the variability of a data set if the measure of variability is expressed in the same units as the original data. If your data are dollars, it is easier to understand a statement such as "The average distance between an observation and the mean is $5" than "The average squared distance between an observation and the mean is $25 squared." To get back to the same units, all you have to do is take the square root of the variance. The result is formula 3.6 for the standard deviation of the population and formula 3.7 for the standard deviation of the sample.

The reason for using different formulas for the sample variance and sample standard deviation than for the population variance and population standard deviation is complicated. It has to do with the problem of estimation. You would like a sample standard deviation to provide as accurate an estimate of the population standard deviation as possible. When you are working with a sample, you do not generally know the population mean, μ, so you have to use the sample mean, $\bar{x}$, in place of the population mean in formula 3.6. The correct term for the numerator—or, more accurately, the term that in repeated sampling would get you closest to the population standard deviation—is $(x_i - \mu)^2$, but you have to work with $(x_i - \bar{x})^2$. Here comes the tricky part. The mean of a given sample will be its center of balance and the sum of the squared deviations from $\bar{x}$ will be smaller than the sum of the squared deviations taken from any other number. For example, for most samples the sum of $(x_i - \bar{x})^2$ will be smaller than the sum of $(x_i - \mu)^2$. The exceptions are samples in which $\bar{x}$ happens to equal μ, but it very seldom does. Hence using $(x_i - \bar{x})^2$ in the numerator results in a number that is on average too small. Appendix B contains a proof that using a denominator of $(n - 1)$ instead of n just compensates for the too-small numerator, so that in repeated sampling the average value of the sample variance equals the population variance. This result is also shown by example in Chapter 7.

If the data set {2, 3, 4, 5, 6} is treated, for the sake of illustrating the formula, as a population, the variance and the standard deviation are 2 and 1.414, respectively. If the same data set is considered a sample, the variance and standard deviation are 2.5 and 1.581, respectively. The calculations for these results are shown below.

$$\sigma = \sqrt{\frac{\sum_{i=1}^{\nu}(x_i - \mu)^2}{\nu}}$$ (3.6)

$$\sigma = \sqrt{\frac{(2-4)^2 + (3-4)^2 + (4-4)^2 + (5-4)^2 + (6-4)^2}{5}}$$

$\sigma = 1.414, \ \sigma^2 = 2$

$$s = \sqrt{\frac{\sum_{i=1}^{n}(x_i - \bar{x})^2}{n-1}}$$ (3.7)

$$s = \sqrt{\frac{(2-4)^2 + (3-4)^2 + (4-4)^2 + (5-4)^2 + (6-4)^2}{4}}$$

$s = 1.581, \ s^2 = 2.5$

If you have one of the many hand-held calculators that have a key labeled "σ" or "S.D." or something else to signify the standard deviation, read your manual to find out whether your calculator is programmed for the population or the sample formula, or if it has different keys for the two formulas. Otherwise you may get the population standard deviation when you want the sample standard deviation and vice versa.

3.2B THE STANDARD DEVIATION WITH CLASSIFIED DATA

The approximate value of the standard deviation and the variance of classified data can be calculated with formula 3.10:

$$s \cong \sqrt{\frac{\sum_{j=1}^{k} f_j(x_j' - \bar{x})^2}{n-1}}$$ (3.10)

The symbol f_j is the frequency of the jth class and x_j' is the midpoint of the jth class. The grade point average data can serve as an example one more time.

$$s \cong \sqrt{\frac{\sum\limits_{j=1}^{k} f_j(x_j' - \bar{x})^2}{n - 1}} \tag{3.10}$$

$$s \cong \sqrt{\frac{3(1.9 - 2.713)^2 + 8(2.1 - 2.713)^2 + 12(2.3 - 2.713)^2 + \cdots + 2(3.9 - 2.713)^2}{74}}$$

$$s \cong 0.51$$

The standard deviation calculated from the raw data is 0.50, so the estimated standard deviation of the classified data is quite close. In the above calculations the estimated sample mean of 2.713 calculated from the classified data was used instead of the mean of the raw data. If the exact sample mean is available, it should be used. As long as the data set is a sample, the denominator should be $n - 1$. Also, for the sake of brevity, only the first three and the last terms are shown. The other midpoints and frequencies are in Table 3.1.

■ 3.2C THE COEFFICIENT OF VARIATION

The problem of the two soft drink bottling machines that began Section 3.2 was easy to solve when both machines were designed to fill the bottles to the same mean level of 12 ounces. The machine with the smallest standard deviation was preferred. The problem is more difficult if the machines are designed to emit different mean amounts. Suppose the bottling plant could sell either 12-ounce bottles of soft drink or 2-liter bottles. The machine designed to fill 12-ounce bottles has a distribution with a standard deviation of 1 ounce; the machine designed to fill the 2-liter bottles has a standard deviation of 0.2 liter. Which machine has the least variability? *This is not a problem in converting metric to English measure.* The 0.2 liter converts to 6.76 ounces, but having a standard deviation of 6.76 ounces for the metric filling machine versus 1 ounce for the other filling machine does not, in and of itself, indicate that there is greater variability in the metric machine. A 2-liter bottle is much bigger than a 12-ounce bottle and the 6.76-ounce standard deviation for the 2-liter bottle may be smaller, in relative terms, than the 1-ounce standard deviation for the 12-ounce bottle.

The **coefficient of variation** is used to compare relative variabilities across populations. It is *the standard deviation of a distribution divided by the distribution's mean.* The coefficient of variation of a population is found by formula 3.11:

$$\omega = \frac{\sigma}{\mu} \tag{3.11}$$

The coefficient of variation of a sample is found by formula 3.12:

$$w = \frac{s}{\overline{x}}$$

The Greek letter ω (read omega) stands for the population coefficient of variation. The letter w stands for the sample coefficient of variation. The coefficient of variation for the metric machine is 0.2 liter divided by 2 liters, or 0.1. The coefficient of variation for the other machine is 1 ounce divided by 12 ounces, or 0.083. According to the coefficient of variation, the metric machine has the greater relative variability. Note that it was not necessary to convert to common units to compare the variabilities of the two machines. The coefficient of variation is a dimensionless ratio; the liters in the numerator are canceled out by the liters in the denominator of the first coefficient of variation and the ounces in the numerator are canceled out by the ounces in the denominator of the second coefficient of variation. One limitation applies to use of the coefficient of variation. When the distributions being compared have negative observations, the coefficient of variation provides an unreliable way to compare variabilities across data sets. With negative values in the data sets, the means of any of the data sets could be 0 or negative and the interpretation of the coefficient as the relative variability would be lost.

3.2D TCHEBYCHEV'S INEQUALITY

Tchebychev's inequality lends itself to some business applications that make use of the mean and the standard deviation. **Tchebychev's inequality** is *a theorem that specifies either the maximum proportion of the observations in a distribution which are outside any two boundary points surrounding the population mean or the minimum proportion of the observations that are within such boundary points.* To use Tchebychev's inequality you must know the population mean and the population standard deviation. For an illustration, let's use the soft drink bottling machine that had a distribution with a mean of 12 ounces and a standard deviation of 1 ounce. Suppose any bottle that contains between 10 and 14 ounces is considered acceptable. The values are unrealistically high, but it is simpler to illustrate Tchebychev's inequality with even numbers. Tchebychev's inequality can be used to find the minimum proportion of bottles that would be acceptable or the maximum proportion of bottles that would be rejected.

The formula for Tchebychev's inequality is:

$$P_{center} \geq 1 - \frac{1}{k^2}$$

(3.13)

$$P_{tails} \leq \frac{1}{k^2}$$

(3.14)

The term P_{center} refers to the proportion of the observations between the boundaries while P_{tails} refers to the proportion of the observations outside the boundaries. The letter k refers to the number of multiples of a standard deviation between the mean and a boundary line. The formula for k is:

$$k = \frac{B - \mu}{\sigma}$$

(3.15)

B stands for one of the boundary lines and μ and σ are respectively the mean and the standard deviation of a distribution. In the bottling machine example the number of multiples of a standard deviation between the mean and either boundary is 2.

$$k = \frac{B - \mu}{\sigma}$$

(3.15)

$$k = \frac{14 - 12}{1}, \ k = \frac{10 - 12}{1}$$

$$k = 2, \ k = -2$$

The minus sign does not matter because the k term is squared in the next formula. To find the minimum proportion that would be acceptable, substitute k in formula 3.13:

$$P_{center} \geq 1 - \frac{1}{k^2}$$

(3.13)

$$P_{center} \geq 1 - \frac{1}{2^2}$$

$$P_{center} \geq 0.75$$

The minimum proportion acceptable would be 0.75. To find the maximum proportion that would be rejected, substitute k in formula 3.14:

$$P_{tails} \leq \frac{1}{k^2}$$

(3.14)

$$P_{tails} \leq \frac{1}{2^2}$$

$$P_{tails} \leq 0.25$$

The maximum proportion that would be rejected equals 0.25.

The reason that the proportion is stated as an inequality rather than an equality is that the formula does not make use of information about the full data set. Many data sets can have the same mean and the same standard deviation. In other words, these two parameters do not fully identify a data set or a distribution. Tchebychev's inequality works on the assumption that for a given mean and standard deviation, the data set being analyzed is as tail-heavy as possible. That is, as much as possible of the data set is in the tails while still yielding the specified mean and standard deviation. A proof of Tchebychev's inequality is in Appendix C.

If the data set's frequency polygon is more mound-shaped or center-heavy, the actual proportion between the boundary lines will be higher than the minimum specified by Tchebychev's inequality, and the actual proportion in the tails will be under the maximum specified by Tchebychev's inequality. If you know the exact shape of a data set's frequency polygon or the values of all of the observations, you can find the exact proportions of the distribution inside or outside any two boundary points surrounding the mean. Tchebychev's inequality is most useful when you know only the population mean and the population standard deviation of a data set. There is another version of Tchebychev's inequality for the sample mean and sample standard deviation, but it is usually not taught in an introductory course. If you are interested in this version, the list of suggested readings at the end of the chapter will direct you to a work that explains it.

Tchebychev's inequality is also used to estimate probabilities. Probability concepts are fully discussed in Chapter 4, but this application of Tchebychev's inequality is important enough to justify jumping the gun and discussing the use of probability before it is formally introduced. There is a simple correspondence between the proportion of a data set that is within a given range and the probability that an observation selected at random from the data set will be within the given range. If a minimum of 75% of the bottles filled by a machine will be in the acceptable range, the minimum probability that a randomly selected bottle will be acceptable is 0.75. The proportion of the distribution in a particular range is also the probability that a randomly selected observation will be in that range. Ordering perishable items is a good illustration of this use of Tchebychev's inequality. For instance, a flower shop sells many roses on Valentine Day and very few during the week after Valentine Day. The mean number of flowers sold on past Valentine Days was 1,000 and the standard deviation was 100. Assuming these values are the population parameters, what is the minimum probability that the flower shop will sell between 700 and 1,300 roses?

To apply Tchebychev's inequality, first determine how many multiples of a standard deviation there are between either boundary line and the mean. There are 3, because the standard deviation of 100 goes into the distance between 1,000 and 1,300 or the distance between 1,000 and 700 three times. If k equals 3, P_{center} is greater than or equal to 0.89. This value comes from substituting a k of 3 in formula 3.13. P_{center} can be interpreted either as the proportion of all Valentine Days in which the sales of roses will be between the boundaries of 700 and 1,300 or as the probability that on a randomly selected Valentine Day the sale of roses will be between 700 and 1,300. What is the maximum probability of either selling less than 700 roses or more than 1,300? P_{tails} would be less than or equal to 0.11.

Now let's rephrase the question: If the flower shop ordered 1,300 roses, what is the maximum probability of running out? The intuitive answer is one-half of 0.11, but this time intuition would be wrong. Even though you are considering only one side of the distribution, you cannot be sure that the maximum probability of having an observation on the right side of an upper boundary is one-half of the maximum probability that an observation will fall in either tail. You could be sure if you knew that the distribution was symmetric, but you don't know that. All you know is the mean and the standard deviation. It turns out that working with only one side of the distribution does not change the maximum probability. The maximum probability of selling more than 1,300 roses is 0.11. The maximum probability of selling fewer than 700 roses is 0.11, and the maximum probability of selling either fewer than 700 or more than 1,300 is also 0.11. The reason for this counterintuitive result is that you don't know if the distribution is tail-heavy at both or one of the ends, and if it is tail-heavy at one end, you don't know which one it is.

The problem can be rephrased so that the maximum probability becomes the known and the upper bound becomes the unknown; for example, how many roses should the flower shop order so that the maximum probability of running out is 0.05? Substituting 0.05 for P_{tails} in formula 3.14 and changing the inequality to an equality yields:

$$P_{tails} \leq \frac{1}{k^2} \qquad \text{(3.14)}$$

$$0.05 = \frac{1}{k^2}$$

$$k = 4.47$$

The upper boundary is 4.47 multiples of a standard deviation above the mean. The flower shop would have to order 1,447 roses so that the maximum probability of running out of roses on Valentine Day is 0.05. We obtain the 1,447 by rearranging formula 3.15:

$$k = \frac{B - \mu}{\sigma}$$

(3.15)

$$B = k\sigma + \mu$$

$$B = 4.47(100) + 1,000$$

$$B = 1,447$$

Two warnings apply to the use of Tchebychev's inequality. One is that the formula does not work if k is less than or equal to 1. The second warning is that the result is extremely conservative. For example, what Tchebychev's inequality tells you in the above problem is that the maximum probability of selling more than 1,447 roses is 0.05 no matter what the shape of the distribution of the sales of roses. If the distribution of rose sales is not highly skewed to the right, the flower shop owner can order many fewer than 1,447 roses and still have only a small probability of running out. As a rule of thumb, most mound-shaped distributions have 5% of their total area in each tail marked with a boundary line of about 1.6 or 1.7 multiples of a standard deviation from the mean. If the flower shop owner were not too worried about running out of roses, 1,160 or 1,170 would be a fairly safe number. If the problem were about some vital item, such as pints of blood for a blood bank, then it would be reasonable to rely on the conservative figure.

Finally, you may be wondering what to do with a Tchebychev problem if the boundary lines are not equidistant from the mean. If you want to find the minimum probability of an observation occurring in the center, the conservative or safer choice is to use the farthest boundary. The conservative choice for finding the maximum probability of an observation occurring in the tails is the closer boundary.

EXERCISES FOR SECTION 3.2

3.9 Calculate the standard deviation of the following data twice. First assume that the data are a population and then assume that they are a sample: {39, 62, 99, 104, 147}.

3.10 Calculate the standard deviation of the following data:

Class range	Frequency
0 and under 20	300
20 and under 40	500
40 and under 60	200

3.11 Which data set has the least relative variability, *A* or *B*? *A* = {1, 3, 6, 8, 13}; *B* = {1,030, 2,976, 5,931, 7,463, 12,652}.

3.12 If a data set has a mean of 250 and a standard deviation of 5, what is the minimum proportion of the observations in the data set which are between 225 and 275? What is the maximum proportion of the observations in the data set outside of this range?

3.13 A hot dog stand sells an average of 300 hot dogs a day. If the standard deviation of the distribution of hot dog sales is 10, how many hot dogs should the stand start the day with so that the probability of running out of hot dogs is less than 0.01?

3.3 EXPLORATORY DATA ANALYSIS

 3.3A STEM-AND-LEAF DISPLAYS

Exploratory data analysis consists of *a variety of techniques used to find patterns or anomalies in a data set or to present the key features of a data set in a manner that resists the influence of extreme values.* For example, the median is a **resistant measure** of the center of a data set—*a descriptive summary measure of a data set which is not sensitive to extreme or outlying observations.* Some of the observations in a data set can be subject to an extraneous influence, such as simple coding or keypunching errors, the inclusion of stray observations that do not belong to the population of interest, arbitrary rounding or cutoffs, and outright lies. Descriptive measures of a data set such as the mean and the standard deviation are sensitive to the extreme or wild values that such extraneous influences often introduce.

A **stem-and-leaf display** is *a method of presenting a data set so that gaps or concentrations in the data become visible.* An example will help to clarify the process of constructing a stem-and-leaf display. Table 3.3 lists the lowest prices for a stay by one person for one night quoted by all hotels and motels in Detroit. There are 52 observations in the table.

The choice of the level of detail in the data to focus on is a matter of judgment. The digits in the pennies and dimes columns are not an important feature of the data, because most observations are in round dollar values and because consideration of this level of detail is unnecessary. We will ignore these

TABLE 3.3

Prices for one person for one night at hotels and motels in the Detroit area, in dollars

65.00	38.00	54.50	27.95	25.00	32.00	84.00	47.00	45.00	32.95	70.00	37.00	64.00
26.00	40.00	45.00	34.00	47.00	61.00	66.00	43.00	46.00	62.00	56.00	47.00	52.00
28.00	28.00	26.00	32.00	94.00	40.00	57.00	36.00	30.00	54.00	60.00	23.95	23.95
23.95	23.95	25.25	50.00	65.00	34.95	70.00	32.00	32.00	26.00	25.00	33.00	100.00

FIGURE 3.8

Stem-and-leaf display of data in table 3.3 without sorting within a stem and without depth indicators

2	7568863333565
3	822742604223
4	750573670
5	462740
6	5416205
7	00
8	4
9	4
10	0

two columns and focus on the digits in the $1, $10, and $100 columns. The $10 column will be the **leading digit,** as *a digit on the stem of a stem-and-leaf display* is called, and the $1 column will be the **trailing digit,** or *a digit on a leaf of a stem-and-leaf display.* A vertical line separates the leading and trailing digits. The first step is to draw a vertical line and write every possible leading digit from the lowest to the highest in the data set along the left side of the line. The lowest observation is 23.95 and the highest is 100.00, so the leading digits run from 2 to 10. The next step is to place every trailing digit alongside its leading digit to the right of the vertical line. Figure 3.8 shows the results of these steps.

The stem-and-leaf display preserves most of the details of the data. Only the dime and penny values are lost. Each value is truncated rather than rounded. For example, the data point 27.95 is recorded as 2|7, with the 2 representing $20 and the 7 representing $7. Since there are 13 observations in the $20 stem, the row with the leading digit 2 has 13 trailing digits, one for each observation. You can immediately see from Figure 3.8 that the lowest-priced motels are the most common and that the data set appears to taper off very evenly as the prices increase. Usually stem-and-leaf displays are presented with the leaves of each stem sorted in ascending order. Figure 3.9 presents the same numbers

FIGURE 3.9

Stem-and-leaf display of data in table 3.3 without depth indicators

2	3333555666788
3	022222344678
4	003556777
5	024467
6	0124556
7	00
8	4
9	4
10	0

after sorting. It is easy to see that certain trailing digits are popular and others unpopular. Detroit motel and hotel managers seem to be avoiding 1s and 9s. There are 52 trailing digits. If they were chosen at random, one-tenth of the observations, or 5.2 observations, would take the value of each possible trailing digit, 0–9. Note that there are no 9s and only one 1. There are also only three 8s. Possibly the motel owners have avoided these numbers because they perceive that consumers have a negative reaction to them, but this is speculation.

More details about the data set can be made clear if there are two leaves for each of the previous stems. Let the stem for the trailing digits with leaves 0 through 4 be * and let the stem for the trailing digits with leaves 5 through 9 be a dot: ·. For example, the stem 2* can be followed by leaves 0, 1, 2, 3, 4, and the stem 2· by the leaves 5, 6, 7, 8, 9. Figure 3.10 also shows the depth of each line. The **depth** of a line is *the number of observations from the last observation on a line of a stem-and-leaf display to the nearest end of the display.* Which observation is the last one on a line depends on whether you are counting from the end with the low values or the high values. If you are counting from the low end, then the last observation is the farthest to the right. If you are counting from the high end, the last observation is the farthest to the left. In Figure 3.10, for example, $60 is the last observation on the line with the stem 6* and it has a depth of 12. The observation $38 is the last one on the stem 3· and it has a

FIGURE 3.10
Stem-and-leaf display of data in table 3.3 with five leaves for each stem and with depth indicators

Depth		
4	2*	3333
13	2·	555666788
22	3*	022222344
25	3·	678
(3)	4*	003
24	4·	556777
18	5*	0244
14	5·	67
12	6*	0124
8	6·	556
5	7*	00
	7·	
3	8*	4
	8·	
2	9*	4
	9·	
1	10*	0

depth of 25. The line that contains the median observation is labeled with parentheses surrounding the number of observations on that line. As a check, the number of observations on the median line plus the depths of the lines above and below must add up to the total number of observations.

Additional details become clear in Figure 3.10. First, the $84, $94, and $100 a night rooms are far away from the other observations in the Detroit market. Second, there seem to be three concentration points in the data. The first and largest is in the high 20s and low 30s. The second is centered on the high 40s and the third is centered on the low 60s. If you were building a hotel or motel in this market, you would have to be conscious of these three pricing points, which undoubtedly represent motels or hotels of different qualities. If you selected one of these pricing points as the rate for your motel and if your motel offered a quality much below that of the other firms that charged prices near that point, you would not have much repeat business.

All of these analyses may appear to be very free-wheeling; some judgments were involved in deciding on how to present the data and on how to interpret them. Exploratory data analysis requires an open mind toward trying different formats to present the data and a willingness to consider a variety of interpretations.

Here's another example of a stem-and-leaf display. A retail fish store sells live lobsters. The proportion of a lobster's weight that is meat increases as the total weight of a lobster increases. Thus, except for huge lobsters that have tough meat, a larger lobster is worth more per pound. The fish store recognizes this relationship and charges $7 a pound for lobsters in the range 1 pound and under 1.25 pounds while charging $9 a pound for lobsters that weigh more than 1.25 pounds. Each lobster is weighed before being sold and the weight is recorded by the sales clerk. Table 3.4 shows the weights of the last 120 lobsters sold at the shop.

A stem-and-leaf display for the data in Table 3.4 is shown in Figure 3.11. The stem-and-leaf display follows the same format as Figure 3.10. Since the pounds digit in all of the observations is 1, this column of data is not needed in the stem-and-leaf display. The leading digit is the one-tenth pound and the trailing digit is the one-hundredth pound. The owner of the fish store thought it was impossible for so many lobsters to weigh 1.23 and 1.24 pounds while

TABLE 3.4

Weights of 120 lobsters sold at a fish shop, in pounds

1.46	1.03	1.27	1.50	1.24	1.14	1.20	1.30	1.23	1.59	1.02	1.38	1.29	1.24	1.23	1.09
1.38	1.16	1.24	1.22	1.11	1.65	1.23	1.31	1.46	1.35	1.23	1.14	1.05	1.24	1.51	1.69
1.31	1.17	1.24	1.44	1.13	1.29	1.09	1.37	1.19	1.24	1.23	1.35	1.10	1.34	1.00	1.44
1.22	1.19	1.34	1.41	1.24	1.57	1.31	1.24	1.30	1.41	1.07	1.23	1.05	1.40	1.23	1.36
1.19	1.16	1.22	1.24	1.39	1.24	1.15	1.46	1.50	1.37	1.24	1.31	1.18	1.10	1.23	1.37
1.09	1.24	1.41	1.32	1.17	1.21	1.42	1.30	1.33	1.15	1.23	1.45	1.20	1.49	1.31	1.24
1.06	1.18	1.23	1.47	1.39	1.01	1.06	1.50	1.34	1.24	1.20	1.07	1.44	1.24	1.35	1.00
1.60	1.32	1.49	1.43	1.23	1.20	1.21	1.32								

FIGURE 3.11
Stem-and-leaf display of weights of 120 lobsters sold at a fish store

Depth

Depth	Stem	Leaves
5	0*	00123
14	0˙	556677999
20	1*	001344
31	1˙	55667788999
(35)	2*	00001122233333333333344444444444444
54	2˙	799
51	3*	000111112223444
36	3˙	55567778899
25	4*	011123444
16	4˙	5666799
9	5*	0001
5	5˙	79
3	6*	0
2	6˙	59

there were so few at or just above 1.25 pounds. He investigated what the sales clerks were doing. Customers would come in and ask for a lobster that was as close to 1.25 pounds as possible while still being $7 a pound. The sales clerks would put a lobster on the scale and if it was 1.25 pounds or a little more they would squint and record it as 1.24 or 1.23. The vision of the sales clerks improved greatly after they were told that they would be fired for short-weighting the lobsters.

3.3B BOX PLOTS

A **box plot** is *an exploratory data analysis tool that describes the central 50% of a data set, its skewness, its range, and any outlying observations.* Box plots display these characteristics of data sets in ways that are resistant to extreme values. The exception to this statement is the range, which is necessarily sensitive to extreme values. Box plots are particularly useful for comparing data sets. The data set on motel room rates can be used to illustrate a box plot. The first step is to draw a horizontal axis scaled so that the full range of the data set can be accommodated. Next, the median of the data set is marked with a vertical line. The ends of the "box" in a box diagram are two horizontal lines marking the "lower hinge" and "upper hinge" of the data set. The **lower hinge** is *the median of all of the observations below the median of the entire data set* and the **upper hinge** is *the median of all of the observations above the median of the entire data set.* The "box" enclosed by the hinges includes approximately half of the observations in the data set.

FIGURE 3.12
Box plot of motel data in table 3.3

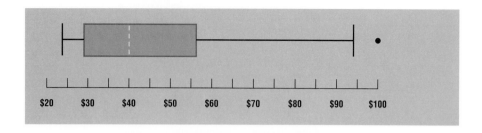

In the motel data set there are 52 observations, so the median of the entire data set is halfway between the 26th and the 27th observations. Both of these observations are $40, so the median is $40. There are 26 observations in the first half of the data set, so the median of the first half of the data set, or the lower hinge, is halfway between the 13th and the 14th observations. The 13th observation is $28 and the 14th is $30, so the lower hinge is $29. There are also 26 observations in the top half of the data set, so the upper hinge is halfway between the 13th and the 14th observations counting from the high end. The fact that the depth on the stem-and-leaf display is counted from the nearest end helps when you are finding the hinges. The 13th observation from the high end is $57 and the 14th is $56, so the point that marks the upper hinge of the data set is $56.50. Figure 3.12 shows the box plot of the motel data.

The box runs from $29.00 to $56.50. The fact that the median of $40 is closer to the lower hinge than to the upper hinge shows that the distribution is skewed to the right. The remaining features of the box plot are the outliers and the fences. An **outlier** is *an observation that is so far from the bulk of the data that it needs to be identified so that it can be checked.* It may be an error, or if the number is accurate, you may need to investigate why it is so far away from the bulk of the data. A **fence,** or outlier cutoff, is *a boundary that marks the lowest or highest observation that is not an outlier.*

Specifically, an **outlier** is *any observation that is more than three-halves of the distance between the hinge points beyond either hinge.* For example, the distance between the hinge points is $56.50 minus $29, or $27.50. Three-halves of this distance is $(3 \cdot 27.50)/2 = 41.25$. Any point more than 41.25 beyond either hinge is an outlier. Since there are no negative motel rates, there are no values to the left of $29.00 minus $41.25. On the high side an outlier would be any point beyond $56.50 plus $41.25, or $97.75. There is one point that is an outlier, the one hotel charging $100. It is labeled with a dot in Figure 3.12. It is customary to identify these outlying observations. The "whiskers" are the horizontal lines that run from the ends of the box to the fence, so on the left side the whisker goes to $23.95 and on the right side the whisker goes to $94.00.

Box plots are very resistant to wild values. The observations near the high end of the data set could be made arbitrarily large or those near the low end

FIGURE 3.13
Stem-and-leaf display of prices for one person for one night at Milwaukee motels listed in the AAA directory, in dollars

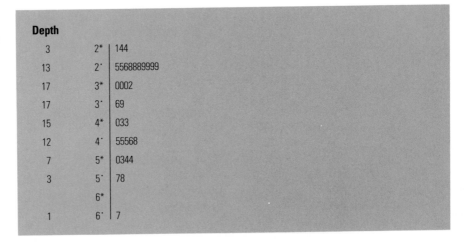

Depth		
3	2*	144
13	2˙	5568889999
17	3*	0002
17	3˙	69
15	4*	033
12	4˙	55568
7	5*	0344
3	5˙	78
	6*	
1	6˙	7

could be made arbitrarily small or negative without changing the median or the hinges. The box depends on the locations of the central 50% of the data. Since the limits of the whiskers are defined as three-halves of the distance between the hinges beyond either hinge, extreme values can pull the whiskers only a limited distance from the box.

To illustrate the use of box charts for comparing data sets, we need to construct a second box chart. Here we will use data for motels in another city. A stem-and-leaf display for the motels in Milwaukee is shown in Figure 3.13. One point has to be made about stem-and-leaf displays before we start to compare box plots. In Figure 3.13 the median falls between two lines. There are 34 observations, so the median is halfway between the 17th and the 18th observations. The 17th observation is on stem 3* and the 18th is on stem 3˙. In this situation there is no need to mark the stem containing the median with parentheses because there is no stem that contains the median. Figure 3.14 contains the box plots for the data from both Detroit and Milwaukee. The

FIGURE 3.14
Box plots of data in figures 3.10 and 3.13

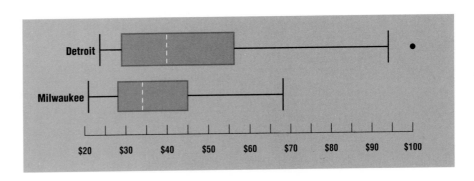

lower hinge in the Milwaukee data is at $28 and the upper hinge is at $46. The distance between the hinges is 18, so the upper outlier limit is $46 plus (3 divided by 2) times 18, or $73. The lower outlier limit is $28 minus (3 divided by 2) times 18 or $1. There are no outliers in the Milwaukee data. Since there are no outliers, the fences are located at the lowest and highest observations.

The narrower boxes and the narrower whiskers indicate that motel prices in Milwaukee have less dispersion or variability than those in Detroit. The box plots also make clear that prices are considerably lower in Milwaukee. Like Detroit's, Milwaukee's motel prices are skewed to the right. These diagrams may seem like a lot of work to reach these conclusions. If box plots are done by hand, they can be time-consuming. The time of the preparer can be justified if the data are to be shown to others who will save time when they consider them. Of course, computers can be great time-savers, and many statistical packages, such as SAS, MINITAB, and DATA DESK, draw box plots.

EXERCISES FOR SECTION 3.3

3.14 Construct a stem-and-leaf display of the following data:

254	231	209	273	280	265	230	251	201	244	257	239	281	273	262	224	253	223
228	241	236	206	214	251	237	277	235	222	209	218	253	268	227	235	216	225
254	223	240	271	221	238	206	252	225	215	246	259	241	223	222			

3.15 Construct a box plot of the data in exercise 3.14.

3.16 Your company's salesmen drive their own cars on company business and are reimbursed at a fixed rate per mile. Their instructions are to record their daily mileage to the nearest mile. Construct a stem-and-leaf display of the following data and describe what you think is happening.

20	30	12	10	20	36	10	23	20	16	11	10	12	14	20	21	20	32	30	10	20	31	25	21
20	15	10	10	13	12	32	30	21	11	20	34	21	16	10	26	11	32	33	23	20	10	10	24
34	31	22	25	26	16	20	14	20	24	31	33	31	20	10	14	16	20	24	20	32			

3.17 Your 1982 Camaro was severely damaged in an accident and the claims adjuster for the insurance company wants to pay you $3,000 in cash rather than repair the car. You agree that the car is "totaled" but you argue that $3,000 is too low. The latest issue of *Wheels and Deals* has the asking prices of 27 1982 Camaros. Construct a box plot of the following data taken from *Wheels and Deals* and decide if it supports your position.

5,500	3,400	3,500	3,666	3,800	4,500	6,020	2,880	5,000	3,780	6,200	4,750	5,200	5,600	6,333
4,999	5,375	6,600	4,675	5,000	5,800	6,375	5,250	3,000	5,200	5,600	5,800			

3.18 Mutual funds are pools of stocks and bonds selected by professional fund managers. The advantages of buying shares of a mutual fund are, first, that your risks are reduced because the fund holds stocks and bonds issued by many companies, and second, that after the initial decision of which mutual fund to buy, you are saved the trouble of studying the stock and bond markets to decide when to sell and buy and which stocks and bonds are desirable. Mutual funds are either "load" or "no-load" funds. A load is a commission paid to the salesman who sold you the shares of the mutual fund. The no-load funds are purchased directly from the company that manages the mutual fund and no commission is paid. Salesmen of load funds often argue that their commissions cost nothing because the load funds' rates of returns are higher than those of no-loads. Below are two tables with rates of return on 289 prominent mutual funds. The first table is for no-load funds and the second table is for load funds. The returns, measured as of December 1984, are based on investments of $2,000 every January from 1980 through 1984. The returns are net of any loads. The source of the data is *Consumer Reports,* July 1985. Use the box plots of the two data sets to decide if the load funds' rate of return equals that of no-load funds after the loads are deducted. Also determine if either group has a skewed distribution, and if so, in what direction. Determine if there are any outliers in each group. Which type of fund is a better investment?

No-Load Funds

28.2	27.8	23.8	23.3	22.5	20.8	20.5	20.3	20.0	19.6	19.1	19.0	17.9	17.8	17.6	16.9
16.8	16.7	16.6	16.5	16.5	16.3	16.2	15.8	15.0	15.0	14.6	14.6	14.6	14.6	14.6	14.5
14.4	14.2	13.7	13.6	13.6	13.5	13.4	13.2	13.1	12.9	12.9	12.7	12.7	12.6	12.4	12.3
12.3	12.2	12.1	12.1	11.9	11.8	10.6	10.4	10.3	10.2	9.9	9.9	9.9	9.8	9.6	9.4	9.3
9.1	9.0	9.0	8.9	8.9	8.9	8.8	8.8	8.8	8.7	8.7	8.5	8.3	8.3	8.1	8.1	8.0	8.0	7.9	7.7
7.7	7.6	7.2	7.2	7.1	6.9	6.8	6.8	6.1	6.0	6.0	5.4	5.3	5.1	5.0	4.7	4.3	4.3	4.3	4.2
4.1	3.8	3.8	3.6	2.9	2.4	2.2	2.2	1.2	0.1	0.0	−0.1	−0.7	−2.6	−8.2	−9.4	−26.1

Load Funds

27.9	22.0	18.3	17.5	16.9	16.7	15.8	15.5	15.3	15.2	15.1	15.0	14.9	14.8	14.8	14.3
14.0	13.9	13.9	13.8	13.7	13.7	13.7	13.6	13.5	13.4	13.4	13.4	13.3	13.2	13.0	12.9
12.8	12.8	12.7	12.7	12.6	12.5	12.5	12.1	12.0	12.0	11.9	11.8	11.6	11.6	11.5	11.5
11.5	11.4	11.4	11.4	11.3	11.3	11.2	11.2	11.2	11.2	11.0	11.0	10.7	10.7	10.7	10.7
10.6	10.5	10.4	10.4	10.3	10.3	10.3	10.2	10.1	10.1	10.0	10.0	10.0	9.8	9.8	9.7
9.7	9.6	9.5	9.4	9.4	9.4	9.3	9.2	9.0	9.0	9.0	8.9	8.8	8.8	8.7	8.7	8.5	8.4	8.2	8.0
8.0	7.8	7.8	7.8	7.8	7.7	7.5	7.4	7.2	7.2	7.2	7.1	7.1	7.1	7.1	7.0	6.9	6.8	6.7	6.7
6.7	6.6	6.6	6.5	6.5	6.3	6.0	5.7	5.6	5.6	5.5	5.2	5.1	5.0	4.9	4.9	4.7	4.5	4.5	4.5
4.4	4.3	4.3	4.3	4.2	4.0	4.0	3.9	3.8	3.7	3.5	3.5	3.5	3.4	3.4	2.8	2.4	2.2	1.2	0.8
0.6	0.1	−0.1	−0.2	−0.5	−0.9	−6.3

SUMMARY

Numerical descriptors of a data set have important business applications. The mean, the mode, and the median are measures of the center of a data set. Each measure of the center of the data has some advantages and some disadvantages. The mean is the most familiar; it has some convenient arithmetic properties; and in later chapters you will see that it has some desirable sampling properties that make it useful for statistical inference. The disadvantage of the mean is that it is sensitive to outlying observations. The median is not sensitive to outlying observations but it has no convenient arithmetic properties and it is less useful for statistical inference. The mode is used primarily in situations that require only one size or one style to be stocked. The two measures of the dispersion of a data set are the range and the standard deviation. The range is clearly more sensitive to outlying observations than the standard deviation. In most situations the standard deviation is the preferred measure of dispersion. However, if you had a set of boards that had to be carried through a narrow passageway, the range of sizes of the boards would be more useful information than the standard deviation. The formula for squared standard deviation or variance was also given in the chapter. While the variance is not commonly used as a descriptive statistic, in later chapters you will find that the variance has some convenient arithmetic properties and that some specific statistical tests use the variance. The coefficient of variation is used to compare dispersion across data sets. Tchebychev's inequality can be used to find the maximum proportion of a data set that can be in the tails of the distribution or the minimum proportion that can be in the center of a distribution. Finally, the chapter discussed some of the new techniques that are part of exploratory data analysis: stem-and-leaf displays and box plots. When you experiment with ways to present or summarize a data set, you should be alert to spot patterns or odd values and consider a variety of explanations for the behavior of the data. Exploratory data analysis is particularly suited to such inquiries.

 ## SUGGESTED READINGS

Hoaglin, David C., Frederick Mosteller, and John Tukey. *Understanding Robust and Exploratory Data Analysis.* New York: Wiley, 1983.

Saw, J., M. Yang, and T. Chin Mo. "Chebychev Inequality with Estimated Mean and Variance." *American Statistician* 37 (February 1983).

Velleman, Paul F., and David C. Hoaglin. *Applications, Basics, and Computing of Exploratory Data Analysis.* Belmont, Calif.: Duxbury, 1981.

 ## NEW TERMS

box plot An exploratory data analysis tool that describes the central 50% of a data set, its skewness, its range, and any outlying observations.

coefficient of variation The standard deviation of a data set divided by the distribution's mean.

depth The number of observations from the last observation on a line of a stem-and-leaf display to the nearest end of the display.

exploratory data analysis A variety of techniques used to find patterns or anomalies in a data set or to present the key features of a data set in a manner that resists the influence of extreme values.

fence A boundary that marks the lowest or highest observation that is not an outlier.

leading digit A digit on the stem of a stem-and-leaf display.

lower hinge The median of all of the observations below the median of the entire data set.

mean or **arithmetic mean** The sum of the values of all of the observations in a data set divided by the number of observations.

median The point in an array of data which has an equal number of observations above and below it.

modal class The class with the most observations.

mode The observation that occurs most frequently in a data set.

open-ended class An interval defined as all observations that are either above or below a specified value.

outlier An observation that is so far from the bulk of the data that it needs to be identified so that it can be checked; any observation that is more than three-halves of the distance between the hinge points beyond either hinge.

range The distance from the lowest to the highest observation in a data set.

resistant measure A descriptive summary measure of a data set which is not sensitive to extreme or outlying observations.

skewed Of a distribution, not symmetrical around the mean.

standard deviation A numerical measure of the average variability of a data set around its mean.

stem-and-leaf display A method of presenting a data set so that gaps or concentrations in the data become visible.

Tchebychev's inequality A theorem that specifies either the maximum proportion of the observations in a distribution which are outside any two boundary points surrounding the population mean or the minimum proportion of the observations that are within such boundary points.

trailing digit A digit on a leaf of a stem-and-leaf display.

upper hinge The median of all of the observations above the median of the entire data set.

variance The average squared deviation between the observations and the mean.

NEW FORMULAS

Population mean

$$\mu = \frac{\sum\limits_{i=1}^{\nu} x_i}{\nu} \tag{3.1}$$

Sample mean

$$\bar{x} = \frac{\sum\limits_{i=1}^{n} x_i}{n} \tag{3.2}$$

Mean of two data sets

$$\bar{\bar{x}} = \frac{n_1\bar{x}_1 + n_2\bar{x}_2}{n_1 + n_2} \tag{3.3}$$

Mean of k data set

$$\bar{\bar{x}} = \frac{n_1\bar{x}_1 + n_2\bar{x}_2 + n_3\bar{x}_3 + \cdots + n_k\bar{x}_k}{n_1 + n_2 + n_3 + \cdots + n_k} \tag{3.3a}$$

Mean of classified data

$$\bar{x} \cong \frac{\sum\limits_{j=1}^{k} f_j x_j'}{n} \tag{3.4}$$

Median of classified data

$$m \cong \left[\frac{\left(\dfrac{n+1}{2}\right) - (f_b + 1)}{(f_m - 1)}\right] w_m + L_m \tag{3.5}$$

Population standard deviation

$$\sigma = \sqrt{\frac{\sum\limits_{i=1}^{\nu} (x_i - \mu)^2}{\nu}} \tag{3.6}$$

Sample standard deviation

$$s = \sqrt{\frac{\sum\limits_{i=1}^{n} (x_i - \overline{x})^2}{n - 1}} \tag{3.7}$$

Population variance

$$\sigma^2 = \frac{\sum\limits_{i=1}^{\nu} (x_i - \mu)^2}{\nu} \tag{3.8}$$

Sample variance

$$s^2 = \frac{\sum\limits_{i=1}^{n} (x_i - \overline{x})^2}{n - 1} \tag{3.9}$$

Standard deviation with classified data

$$s \cong \sqrt{\frac{\sum\limits_{j=1}^{k} f_j(x_i' - \overline{x})^2}{n - 1}} \tag{3.10}$$

Population coefficient of variation

$$\omega = \frac{\sigma}{\mu} \tag{3.11}$$

Sample coefficient of variation

$$w = \frac{s}{\overline{x}} \tag{3.12}$$

Tchebychev's inequality for the center of a data set

$$P_{\text{center}} \geq 1 - \frac{1}{k^2} \tag{3.13}$$

Tchebychev's inequality for the tails of a data set

$$P_{\text{tails}} \leq \frac{1}{k^2} \tag{3.14}$$

Number of multiples of a standard deviation between a boundary and the mean

$$k = \frac{B - \mu}{\sigma} \tag{3.15}$$

ADDITIONAL EXERCISES

An asterisk designates an exercise of greater than ordinary difficulty.

3.19 A publisher of trade books has 4,000 titles in its catalogue. Trade books are novels, biographies, and other kinds of books that are sold in a general bookstore. From the following data on the number of copies of each title sold, estimate the median unit sales per title.

Class range	Frequency
0– 999	500
1,000– 4,999	800
5,000–24,999	700
25,000–49,999	1,500
50,000 or more	500

3.20 Refer to the information provided in exercise 3.19. If the title with the highest sales sold 1 million copies in the year to which the above data apply, what is the estimated standard deviation of the sales per title?

3.21 You own a sheep ranch with 10,000 sheep. There are 5,000 blackface sheep, 4,000 Shropshire sheep, and 1,000 Derbyshire sheep. You plan to shear all 10,000, and to estimate the average number of pounds of wool for the entire population you shear a sample of 100 sheep. The sample consists of 60 blackfaces, 35 Shropshires, and 5 Derbyshires. The respective sample means are 10, 11, and 12 pounds. What is your best estimate of the average yield for the entire population?

3.22 The helicopters used by Off Shore Services have a payload of 3,000 pounds. When carrying workers to and from drilling platforms, the helicopter pilot asks each worker to declare his weight as he boards. As soon as 3,000 pounds have accumulated, the pilot tells the workers remaining in line that they will have to wait for the next helicopter. Ignore the issue of the weight of clothing and construct a stem-and-leaf display to study the pattern of declared weights. What conclusion or conclusions can you draw?

169	177	180	200	189	178	167	180	165	173	149	158	180	186	185	190	200	210
209	208	188	147	166	186	198	148	186	200	209	186	189	168	150	145	187	172
167	160	186	170	147	178	190	146	165	190	188	170	145	160	188	180	140	144
168	190	200	179	190	149	188	190	170	166	168	179	158	167				

3.23 Suppose two countries' currencies had the following annual inflation rates: $A = \{10\%, 15\%, 12\%, 5\%, 17\%, 3\%, 11\%, 14\%, 12\%, 9\%\}$ and $B = \{200\%, 220\%, 235\%, 210\%, 195\%, 178\%, 205\%, 199\%, 230\%, 205\%\}$. The inflation rate of which country has greater relative variability? In which country would the absolute change in future rates of inflation be easier to predict? In which country would the percentage change in future rates of inflation be easier to predict?

3.24 Your company is considering marketing its refrigerators in some other countries. A refrigerator is an expensive item and you would like to identify the countries that have a substantial proportion of households with incomes high enough to afford refrigerators. In the table below are the lower fence, lower hinge, median, upper hinge, and upper fence of household incomes in five countries under consideration. Construct five box plots of this information and estimate (1) which countries have skewed income distributions, (2) which country's distribution is the most skewed, (3) which country has the highest proportion of households with incomes over $10,000, (4) which country has the highest proportion of households with incomes over $20,000.

Country	Lower fence	Lower hinge	Median	Upper hinge	Upper fence
A	$ 0	$3,000	$4,000	$ 6,000	$10,500
B	1,000	4,000	5,000	6,000	9,000
C	3,000	6,000	8,000	11,000	18,500
D	0	2,000	8,000	16,000	37,000
E	5,000	8,000	10,000	11,000	15,500

3.25 The U.S. Census generally gives annual earnings figures for an occupation in classified form along with the mean earnings for the occupation. Use the following data for the earnings of bookkeepers to estimate the midpoint of the last class and the maximum earnings of bookkeepers. The mean annual earnings of bookkeepers are $13,000.

Class range	Frequency
$ 0 and under 5,000	3,000
5,000 and under 10,000	8,000
10,000 and under 15,000	11,000
15,000 and under 20,000	9,000
20,000 and under 25,000	1,500
25,000 and over	500

3.26 A West Coast refinery used the following stocks of oil in a production run of gasoline: 5,000 barrels of light Arabian crude oil; 20,000 barrels of high-sulfur Alaskan oil; 1,000 barrels of methyl alcohol from corn; and 6,000 barrels of Texas medium-weight crude. If the respective costs of these stocks were $29, $24, $38, and $26, what is the average cost per barrel of stock used for the gasoline?

3.27 The section on Tchebychev's inequality stated that the theorem was based on the limit of a distribution that was as tail-heavy as possible given a specific mean and standard deviation. To see what this limiting case looks like, draw the most tail-heavy distribution possible for a mean of 0 and a standard deviation of 1.

3.28 You manage a trust fund for the estate of a minor. The rules set down when the fund was created limit your choice of investments to Treasury bonds, state bonds, and municipal bonds. Over the past 12 months you have tracked the monthly yields on ten-year bonds of each type. Given the following data, which bonds are the most desirable? Ignore the tax treatment of the bonds. Would your answer depend on whether the data are listed in order from month to month or listed from the highest to the lowest observation? (Note that these two ways of listing the data are not mutually exclusive.)

Treasury bonds = {12%, 12%, 11%, 11%, 11%, 10%, 10%, 10%, 10%, 10%, 9%, 9%}

State bonds = {13%, 10.5%, 10.4%, 10.4%, 10.2%, 9.9%, 9.8%, 9.6%, 9.5%, 7%, 7%, 7%}

Municipal bonds = {15%, 14%, 13%, 12%, 11%, 9%, 8%, 7%, 6%, 5%, 5%, 5%}

3.29 Construct box plots of the three data sets in exercise 3.28 and determine if there are any outliers.

3.30 An investment newsletter advises subscribers that profits of firms in the oil industry are much more variable than the profits of firms in the automobile industry. It cites the following profit rates: oil companies, {−10%, −5%, 2%, 15%, 17%, 18%}; automobile companies, {9%, 12%, 13%, 26%, 34%}. Recall that the coefficient of variation cannot be used for data with negative observations. How can you assess the claim that there is greater variability in the oil industry's profits?

3.31 Construct a stem-and-leaf display of the data in exercise 2.20. Describe any patterns that you observe.

3.32 If you knew the medians of two populations but not the actual data, under what circumstances would you be able to know the median of the combined populations?

3.33 A test of a new battery-powered car consisted of fully charging the batteries and driving the car around a test track until the batteries were exhausted. This test was done 40 times. The distances in kilometers are recorded below. Construct a stem-and-leaf display for these data and speculate on the reasons for any pattern that you observe. (Hint: Look for granularity or gaps between observations.)

640.0	657.6	644.8	656.0	644.8	641.6	648.0	659.2	643.2	651.2	640.0	654.4	644.8	649.6
644.8	648.0	651.2	654.4	659.2	644.8	656.0	644.8	641.6	649.6	659.2	644.8	643.2	
651.2	648.0	648.0	649.6	659.2	641.6	656.0	652.8	644.8	661.6	659.2	644.8	641.6	

3.34* You have a contract that gives you an option to buy 1,000 shares of stock from the company for which you work. The option expires in 30 days. The price at which you can buy the shares is $80 per share provided the current market price is under $100. If the current market price is at or above $100, the price at which you can buy the shares is $80 plus the current market price minus $100. This arrangement limits the profit that you can make per share to $20. If the current market price is below $80 a share, the option is worthless. If the distribution of possible market prices is centered on $85 and has a standard deviation of $3:

(a) What is the maximum probability that the option will be worthless?

(b) What is the maximum probability that you will reach the limit of a profit of $20 per share?

(c) What is the minimum probability that you will be within these bounds?

3.35* Modify formula 3.5 for the median of classified data so that you can find the point in a distribution below which 60% of the observations lie. Find this point in the data set in exercise 3.25.

3.36* If you used a vertical line chart to display a large data set with continuous data, what would the chart approximate? What proportion of the total area bounded by the values of the chart would be to the left of the median? How could you use the chart to identify the mode of the data set?

3.37* Two instructors, Alphonse and Bonard, are arguing about their teacher evaluations. Both have handed out a form to their students which asks the following question:

Which of the following choices most closely approximates your opinion in regard to the following statement: "Your professor inspires confidence"?
(a) Strongly agree. (b) Agree. (c) Uncertain. (d) Disagree. (e) Strongly disagree.

Alphonse got 10 a's, 5 b's, 20 c's, 10 d's, and 3 e's. Bonard got 2 a's, 23 b's, 10 c's, 10 d's, and 3 e's. They decide to use an average measure of student rating to resolve the question of who is the most inspiring. Bonard proposed the assignment of numerical values to the letters, as follows: $a = 5$, $b = 4$, $c = 3$, and so on. Alphonse insisted on treating a as 10, b as 5, c as 2.5, d as 1, and e as 0. Calculate their average ratings with both assignment schemes. Who is the most inspiring? Is there any way to resolve the issue with these data? Discuss.

3.38* If you have the box plots of two data sets, can you construct the box plot of the combined data set? Discuss.

3.39 The coach of a basketball team has complained to his players that their performance is inconsistent from game to game. Over the past 20 games they have averaged 80 points with a standard deviation of 20 points. Their opponents have averaged 65 points with a standard deviation of 18 points. Has the team been more inconsistent on defense or offense? Explain.

3.40 What is the maximum proportion of the past 20 games in which the basketball team in exercise 3.39 could have scored 120 or more points?

3.41 Construct stem-and-leaf displays of the following data sets on the prices advertised by three mail-order houses for the same 30 computer software items.

Computer Stuff: 18.99, 19.99, 21.99, 24.99, 24.99, 24.99, 24.99, 24.99, 27.99, 28.99, 30.00, 32.00, 34.60, 35.99, 40.99, 58.99, 69.99, 69.99, 69.99, 69.99, 74.99, 74.99, 74.99, 74.99, 108.99, 130.00, 132.00, 137.60, 139.99, 140.00

Peripherals Galore: 28.45, 29.23, 29.99, 29.99, 29.99, 30.87, 30.99, 31.99, 32.00, 33.10, 34.60, 35.00, 35.60, 36.99, 40.99, 51.29, 70.00, 78.99, 79.99, 79.99, 84.12, 84.31, 84.99, 94.99, 118.66, 130.00, 137.00, 137.00, 149.99, 160.00

Byte Wholesalers: 19.99, 19.99, 19.99, 20.63, 25.00, 25.00, 25.25, 25.99, 25.99, 28.26, 28.79, 28.00, 28.50, 29.99, 30.69, 38.21, 39.49, 40.14, 42.28, 59.99, 64.00, 64.24, 64.25, 64.19, 78.54, 79.48, 92.00, 97.60, 109.99, 121.00

3.42 Construct box plots of the three data sets in exercise 3.41. Describe the information that the stem-and-leaf displays and the box plots provide.

3.43 (a) Create a box plot of the data on airline fares in the following stem-and-leaf display. Label the median, the hinges, the ends of the whiskers, and any outliers.

8	7˙	00000244
22	7*	66666668888888
(19)	8˙	0000000022222222244
13	8*	888888
7	9˙	002
4	9*	6
3	10˙	0
	10*	
2	11˙	0
	11*	
	12˙	
1	12*	8

(b) If you were selecting fares at random from the above list, what is the probability that the fare selected would be at or above the median fare? That is, what proportion of all fares are above the median fare?

(c) Suppose the above data consisted of the prices quoted by various airlines for a round-trip ticket between Indianapolis and Miami. The fares vary from supersaver to first class. The stems refer to tens of dollars and the leaves to single dollars. What is the most unusual feature of these data and what is the explanation for this feature?

3.44 (a) If the median of the data set in exercise 3.2 was 12, how many observations would be in the range $10 \leq x < 12$?

(b) What is the coefficient of variation of this data set?

3.45 (a) Construct a stem-and-leaf display for the following data set on ages of job applicants at a bank:

18 18 19 19 19 19 20 20 20 20 20 20 21 21 21 21 21 21 21 21 22 22 22 22 23 23 23 23 24
24 24 24 25 25 25 26 26 27 27 27 27 28 28 28 28 28 29 29 29 29 29 29 29 29 29 29 29 29
29 29 29 29 29 29 29 29 29 30 30 31 33 34 34 35 36 38 39 39 39 39 39 40 41 42 43 47 48
49 49 50 51 53 57 58 59 60 60 61 64 65 70 71 77 80 81 85 89 90

(b) Construct a box plot of the above data set.

(c) What are the main features of the data set?

4

Probability

For the operator of a local movie theater, the choice of which feature films to show involves a great deal of uncertainty. This uncertainty is due in part to the fact that movies must be chosen before the theater operator can view them. The theater operator's decisions are usually based on written descriptions of movies. The terms of the contract between the theater operator and the movie distributor add to the uncertainty. These contract terms vary according to the expected popularity of each movie. Movies that are expected to do well have high rental fees and require a guarantee that the movie will be shown a minimum of four to six weeks; movies with lower expectations have lower rental fees, and the number of weeks it will be shown is not specified. If a movie that was expected to do well bombs out at the box office, the theater operator can lose a lot of money. If a movie that was not expected to do well turns out to be a hit, the theater operator can make a lot of money. The theater operator's decision necessarily has to do with probabilities. The theater operator has to rely on the track records of different types of movies and different actors in the theater's local market in choosing which feature films to book. Imagine having to decide if a space movie starring Clint Eastwood and Meryl Streep would have a higher probability of making a profit than a musical starring Richard Gere and Bo Derek!

While no such movies have been offered, the theater operator's problem gives a realistic idea of the need to use probabilities in a business context. Probability concepts are important not only because of their direct applications to business problems but also because probability is the foundation of statistical theory. It is the basis for the design of samples and the inferences about a population that can be made from a sample. This chapter presents three definitions of probability and some rules for working with probabilities. If, after having read the material, you can work some of the starred problems at the end of this chapter, then the chapter deserves rave reviews. If the easier problems make no sense to you after you have read the chapter, then it deserves to be panned. At this point, what probability would you assign to these outcomes?

4.1 INTRODUCTION TO PROBABILITY AND DEFINITIONS OF PROBABILITY

A **sample space** is *a list of the possible outcomes of an experiment*. A **random experiment** can be defined as *any process with an uncertain outcome*. If a theater operator decided to show a particular movie, the experiment would be showing the movie and the sample space could consist of two possible outcomes: (1) the movie will lose money and (2) the movie will at least break even. **Probability** is *a number that represents the chance that a particular outcome will occur if the experiment is conducted.* The range of possible values for probabilities is from 0 to 1. If a particular outcome is assigned a probability of 1, that assignment means that the particular outcome is certain to occur. If a particular outcome is assigned a probability of 0, that assignment means that the particular outcome is certain *not* to occur. To assign a number between 0 and 1 to the probability of a particular outcome, you need a more detailed definition of "probability."

The **a priori definition of probability** is the simplest way to assign a probability. Under this definition *every possible outcome of an experiment is assumed to be equally likely and thus the probability of any outcome is 1 divided by the total number of possible outcomes.* Tossing a die, for example, is an experiment that has six possible outcomes, the integers {1, 2, 3, 4, 5, 6}. According to the a priori definition of probability, the probability of tossing any one of these six sides is ⅙. Real dice are not perfectly balanced; one side of a real die will be at least a little bit more likely to land face up than the other sides. Strictly speaking, the a priori definition applies to an ideal die. It would also apply to an ideal coin that was equally likely to come up heads or tails or to a perfectly shuffled deck of cards from which every card was equally likely to be drawn. All of these gambling examples are appropriate to this definition of probability because this definition was first developed to solve gambling problems. Since improving your gambling skills is not one of the main objectives of this text, you may be wondering why the discussion of probability begins with a definition that is aimed at gambling problems. The reason is that the a priori definition of probability, which is based on an idealized analogue to real events, lends itself to easy explanations of the rules of probability.

Suppose you had a real die, as opposed to an ideal die, and you tossed it 100 times. Suppose the 4 was the side facing up in 50 of these 100 tosses. What is the probability that a 4 will come up on the next toss? If your intuition tells you that the answer is ½, then your intuition is using the **relative frequency definition of probability.** According to this definition, *the probability of a particular outcome of an experiment equals the proportion of the past experiments in which this outcome occurred.* If 50% of all of the previous tosses resulted in 4s, then the probability of a 4 on the next toss is 0.5. Consider the movie example. If 30% of all previous space movies shown at a particular theater lost money, then according to the relative frequency definition of probability, the probability that the next space movie will lose money is 0.3. The relative frequency definition of probability is the one that is generally used in business contexts. A business is better off relying on past experience than on some imaginary die. Clearly, the probabilities based on the relative frequency definition would be accurate only if the circumstances or conditions that had led to the past outcomes were to continue. If a theater operator kept showing space movies, the probability of losing money would not stay at 0.3 because the customers would eventually (pardon the pun) be spaced out. A second difficulty with the application of the relative frequency definition of probability is that there may be no past experiments or too few past experiments to establish a relative frequency. If, for example, a theater operator had never shown a filmed opera, he could not rely on the relative frequency definition of probability to determine the probability that a filmed opera would lose money.

Your intuition may give you an idea of the probability of a filmed opera losing money at your local theater even if it has never shown such a movie. If you do have an idea (perhaps your idea is that the probability is 1), then your intuition is using the **subjective definition of probability.** A subjective probability is *the best estimate of someone who is informed about the conditions that influence the experiment.* It refers to the degree of belief that an individual has about the chance that a particular outcome will occur. This method of assigning probabilities may appear to be very casual but there are occasions when a subjective probability has to be used. A theater operator who has never shown a filmed opera and is offered one by a distributor has to use a subjective probability.

In the rest of the chapter we'll use the a priori definition to explain probability rules, illustrate some practical uses of probability with the relative frequency definition of probability, and particularly in the last section, on Bayesian probability, give some examples in which the subjective definition of probability is useful.

EXERCISES FOR SECTION 4.1

4.1 Fred, Betty, Buck, and Harold are running for two seats on the student council. Assume that no write-in candidates are allowed and that tie votes are impossible.

(a) What is the sample space for the winners of the election?

(b) Under the a priori definition of probability, what is the probability that Buck and Harold will win?

(c) If a woman has never been elected to the student council and you are using the relative frequency definition of probability, what is the probability that Betty will get a seat on the student council?

4.2 Companies that operate communications satellites in space usually buy an insurance policy to cover the possible failure of a satellite to reach its correct orbit after being launched.

(a) If a new launch vehicle is to be used for a particular satellite, what definition of probability should be used to establish the probability of a failure?

(b) If a satellite is valued at $100 million, what are the minimum and maximum amounts that an insurance company could charge to cover this loss given the highest and lowest mathematically possible probabilities for the failure?

(c) If the insurance company relied on the a priori definition of probability to assign a probability to the chance of failure, what is the minimum amount that the insurance company would charge for insuring this risk?

4.3 You are offered your choice of two bets. The first bet is that you will win $100 if the Detroit Tigers win the American League pennant. The second bet is that you will win $100, collectible at the end of the regular baseball season, if you can reach into a barrel that contains 90 black and 10 white balls and draw a white ball.

(a) If you choose the second bet, your subjective probability of the Tigers' winning the pennant must be equal to or less than what number?

(b) If you are indifferent about the two bets, what must your subjective probability of the Tigers' winning the pennant equal?

4.4 For each situation decide which definition of probability is being used and explain why.

(a) A gambler spends three days at a casino observing a particular roulette wheel and recording the number of times the ball falls into each slot of the wheel. On the fourth day the gambler starts to bet, always on slot number 20.

(b) A state lottery allows ticket buyers to pick any five-digit number. If the number chosen by a bettor is drawn as the winning number by the state, the bettor will win $100,000. A gambler buys a lottery ticket and picks the number that represents the month, day, and year of her birth.

(c) You are at the end of a long line at a bank and you notice that one of the five tellers is gorgeous. You figure that the probability of being served by that teller is 0.2.

4.5 Suppose someone was offered his choice between a pair of bets. The first bet is that he can collect $100 if the Republicans win the next presidential election. The second bet is that he can collect $100 the day after the next presidential election if he can draw a white ball out of an urn that has 70 white balls and 30 black balls. Let's say the person chooses the first bet. The same person is offered his choice of another pair of bets. The first bet is that he can collect $100 if the Republicans win the next presidential election. The second bet is that he can collect $100 the day after the next presidential election if he can draw a white ball out of an urn that has 700 white balls and 300 black balls. This time around the person chooses the second bet. Under what circumstances, other than sudden bad news about the Republicans' prospects, would this behavior be rational?

4.2 ADDITION AND MULTIPLICATION RULES

We need to explain some terminology before describing the rules for working with probabilities. The terms **outcome** and **elementary event** mean the same thing. They refer to *any result of an experiment that cannot be divided into* *smaller components.* If our experiment involved tossing a die, the outcomes or elementary events would be {1, 2, 3, 4, 5, 6}. None of these outcomes, such as a 1, can be divided into smaller units. The term **event** refers to *a set of one or more outcomes.* The event A could be defined as any odd number from the experiment of tossing a die. The outcomes that make up event A are {1, 3, 5}. If the probabilities of all possible outcomes are added, they must sum to 1. The probabilities of all possible events do not sum to 1 because the same outcomes can be part of different events. For example, define event B as a positive integer less than 4, that is, {1, 2, 3}. The outcomes {1, 3} are part of event A and part of event B. *Two events that cannot occur as the result of the same experiment* are called **mutually exclusive events.** Outcomes are necessarily mutually exclusive. For example, you cannot roll a 1 and a 5 in a single toss of one die. *Two mutually exclusive events that have probabilities that sum to 1.0* are called **complements.** The complement of an event A is denoted $\overline{A}$ (read "not A"). In math notation, $P(\overline{A}) = 1 - P(A)$.

Events may or may not be mutually exclusive. The events A and B defined above are not mutually exclusive because the outcomes {1, 3} are in event A and in event B. Let's define an event C as getting an even number when you toss a die, that is, {2, 4, 6}. Are events A and C mutually exclusive? The answer is yes because you cannot get both an even and an odd number in a single toss of a die.

4.2A ADDITION RULE FOR TWO EVENTS

The symbol ∪ is called "union," and (A ∪ B) is called the union of events A and B. The **union** of two events A and B is *all outcomes that are either part of event* A, *part of event* B, *or part of both events* A *and* B. For example, event A includes all odd numbers that could result from tossing a die—{1, 3, 5}—and event B includes all numbers less than 4—{1, 2, 3}. The union of A and B includes the outcomes {1, 2, 3, 5}. The symbol ∩ is called "intersection," and (A ∩ B) is the intersection of events A and B. The **intersection** is *all outcomes that are both in* A *and in* B. The intersection of A and B is {1, 3}. Formula 4.1 gives the probability for the union of any two events labeled A and B. Note that this formula is not limited to the particular events A and B that are being used to illustrate these concepts. In this formula, A and B stand for any two events. This formula is called the **addition rule.**

$$P(A \cup B) = P(A) + P(B) - P(A \cap B) \hspace{2cm} \text{(4.1)}$$

Figure 4.1, which shows the six possible outcomes when a die is tossed, {1, 2, 3, 4, 5, 6}, indicates the reasoning behind the addition rule. The outcomes that are in event A, {1, 3, 5}, have circles around them and the out-

FIGURE 4.1
Possible outcomes of tossing
a die with the odd numbers
circled and the numbers less
than 4 placed in squares

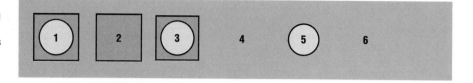

comes that are part of event B, {1, 2, 3}, have squares around them. Let's assume that we are using a fair die. There are three outcomes in event A, so under the a priori definition the probability of event A is ³⁄₆, or 0.5. Recall that under the a priori definition each outcome is equally likely, so an event that consists of half of all of the possible outcomes must have a probability of 0.5. Similarly, there are three outcomes in event B, so the probability of event B is 0.5. If the probabilities of event A and event B were added, they would equal 0.5 plus 0.5 or 1. However, the probability of obtaining event A and/or event B—in other words, the union of A and B—is not equal to 1. If the toss results in either a 4 or a 6, then neither event A nor event B will have occurred. The reason that you cannot always obtain the probability of the union of two events by simply adding the probabilities of the two events is that in arriving at the sum of the probabilities of events A and B you may count the same outcomes more than once. In Figure 4.1 the outcomes {1, 3} have both a circle and a square because they are both odd numbers and less than 4. If you added the probabilities of event A and event B, you would be counting the 1 and the 3 twice. To avoid double counting, formula 4.1 subtracts the probability of the intersection of A and B from the sum of the probabilities of A and B. The probability of the union of A and B equals 0.5 plus 0.5 minus ⅓, or ⅔. Also note that $A \cup B$ contains the outcomes {1, 2, 3, 5}, and that by the a priori definition, $P(A \cup B) = $ ⅘ or ⅔.

The die example used to explain the addition formula was based on the a priori definition of probability. However, the addition formula applies to probabilities based on the relative frequency definition and the subjective definition. All of the probability rules will apply to the three definitions of probability. Here is an example that uses the addition formula for a subjective probability. Suppose you are driving down a highway and your engine dies. You pull over to the side of the road and call a tow truck. The tow truck operator arrives, opens the hood of your car, points to a puddle of oil under the car, and says, "It looks bad. I'd say you have a fifty-fifty chance of having thrown a rod, a one-in-ten chance of having blown a head gasket, and a one-in-twenty chance of having both problems." On the basis of the tow truck operator's estimates, what is the probability that you have one problem or the other or both? To get the answer you must substitute into formula 4.1. You have to convert "fifty-fifty" to 0.5 and "one-in-ten" to 0.1 and "one-in-twenty" to 0.05. The answer is 0.55. Let A be a thrown rod and B be a blown head gasket.

$$P(A \cup B) = P(A) + P(B) - P(A \cap B) \qquad (4.1)$$

$$P(A \cup B) = 0.5 + 0.1 - 0.05$$

$$P(A \cup B) = 0.55$$

If two events are mutually exclusive, the probability of the union is just the sum of the probabilities of the two events. The intersection term in formula 4.1 drops out because the intersection of mutually exclusive events is the empty set, $\varnothing$, which has a probability of zero. For mutually exclusive events, formula 4.1 becomes formula 4.1a.

$$P(A \cup B) = P(A) + P(B) \qquad (4.1a)$$

Let's illustrate formula 4.1a with the events A (an odd number) and C (an even number). Figure 4.2 has circles around the odd numbers and squares around the even numbers. The probability of an odd number is 0.5 and the probability of an even number is 0.5, and as you can see in Figure 4.2, all of the possible outcomes are either in a circle or in a square, so the probability of having either an odd or an even number is 1.

4.2B ADDITION RULE FOR MORE THAN TWO EVENTS

You can extend formula 4.1a to any number of mutually exclusive events. Formula 4.1b shows the union of n mutually exclusive events, where n can be any integer.

$$P(X_1 \cup X_2 \cup X_3 \cup \cdots \cup X_n) = P(X_1) + P(X_2) + P(X_3) + \cdots + P(X_n) \qquad (4.1b)$$

The Xs in formula 4.1b stand for mutually exclusive events.

Here is an example that uses formula 4.1b. You are at a racetrack and you are considering how to bet on an upcoming race. Over the betting window is a display that lists the payoff per dollar bet on each of the horses in the race. You

FIGURE 4.2

Possible outcomes of tossing a die with the odd numbers circled and the even numbers placed in squares

feel lucky and you think that this is a good day to bet on a long shot—a horse that is considered so unlikely to win that if it does win, it will pay back a high multiple of the amount of money bet on it. What you would like to figure out is how to convert the posted payoffs into probabilities and then figure the probability that any horse in a group of long shots will win the race. There are eight horses in the race and three of them are clearly long shots. Old Gray Mare is paying 100 to 1, Bobtailed Nag is paying 50 to 1, and Stewball is paying 30 to 1. A horse that pays 100 to 1 will return $100 for each $1 that is bet on it. This payoff includes the dollar that was originally bet. These payoffs reflect the collective judgment of all of the bettors at the track on the chances of each horse. The track operators merely take all of the money bet on all horses in a race and, after taking off a percentage for their expenses, pay the remaining money back to the people who picked the winning horse. To simplify the problem, we'll assume that the percentage skimmed off by the track is zero. With this assumption, the reciprocals of the payoffs are the subjective probabilities of the community of bettors of each horse's winning. If a horse paid $2 back for winning, for example, the subjective probability of that horse's winning would be the reciprocal of 2, or 0.5.

As it is not obvious why the reciprocals of the payoffs are the subjective probabilities, some more examples may be helpful. Consider a race in which the total amount bet is $100, and of that total $50 was bet on a particular horse. If that horse wins, the bettors of the $50 will collect the total of $100 and the payoff will be $2 for every $1 bet. If $50 out of the $100 bet in the race is bet on a particular horse, then 50% of the betting public, as measured by their dollars, think that the horse will win, and the subjective probability of the horse's winning is 0.5. If $10 was bet on another horse in the same race, the payoff would be $10 for each $1 bet and the subjective probability of the other horse's winning would be 0.1. In general, if the track does not take a percentage, the subjective probability is the reciprocal of the payoff.

Let's get back to Stewball, Old Gray Mare, and Bobtailed Nag. If the posted payoffs were converted to the probabilities of each horse's winning and you decided to bet on each of the three long shots to win, what is the probability that one of your bets would pay off? Since ties are not allowed, no two horses can win the same race and each outcome must be mutually exclusive. To get the answer you substitute the probabilities of winning for each horse in formula 4.1b.

$$P(X_1 \cup X_2 \cup X_3) = P(X_1) + P(X_2) + P(X_3) \qquad \text{(4.1b)}$$

$$P(X_1 \cup X_2 \cup X_3) = 0.01 + 0.02 + 0.0333$$

$$P(X_1 \cup X_2 \cup X_3) = 0.0633$$

The probabilities of 0.01, 0.02, and 0.0333 are the reciprocals of 100, 50, and 30. The probability that one of the three long shots will win the race is 0.0633.

At real racetracks the percentage of the total amount bet taken by the track operators is not zero. A more typical percentage is 25%. This example was used to illustrate subjective probabilities and mutually exclusive events, not to show how actually to figure probabilities at a racetrack. If you would like to try figuring racetrack probabilities, a more realistic problem, number 4.57, is provided in the exercises at the end of the chapter.

The formula for the union of more than two events that are *not* mutually exclusive is more complicated. Let's use a Venn diagram to illustrate the union of three events that are not mutually exclusive. A Venn diagram consists of a rectangle that represents the sample space, and any object inside of the rectangle represents an event. The probability of an event equals its proportion of the area in the rectangle. A Venn diagram for three events that are not mutually exclusive is shown in Figure 4.3. Each event is represented by a circle: one with vertical lines, another with horizontal lines, and the third with diagonal lines. If the circles did not overlap, the events would be mutually exclusive. The union of the three events is the combined area of the three circles. If the total areas of the three circles were added, the following areas would be counted twice: (1) the combination of vertical and horizontal crosshatching, (2) the combination of vertical and diagonal crosshatching, and (3) the combination of horizontal and diagonal crosshatching. The black area where all three circles overlap would be counted three times. The areas that would be counted more than once are shown in Figure 4.4. If the intersection terms for each pair of circles were subtracted from the sum of the probabilities of the three circles, the black section in Figure 4.4 would be deducted three times and hence it would not be

FIGURE 4.3
Venn diagram of three events that are not mutually exclusive

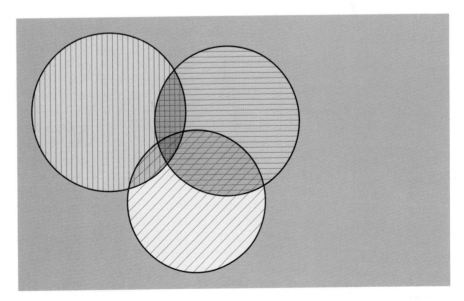

FIGURE 4.4
**Areas counted more than
once when the areas of the
three overlapping circles are
added**

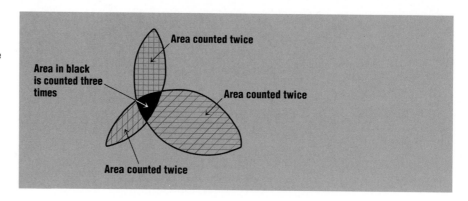

counted at all. The formula that counts every part of the three circles once and only once—that is, the union of three events that are *not* mutually exclusive—is formula 4.1c.

$$P(A \cup B \cup C) = \underbrace{P(A) + P(B) + P(C)}_{\text{area counted once}}$$

$$\underbrace{- P(A \cap B) - P(A \cap C) - P(B \cap C)}_{\text{area counted twice}} + \underbrace{P(A \cap B \cap C)}_{\substack{\text{area counted} \\ \text{three times}}}$$

(4.1c)

To illustrate the union of three events that are not mutually exclusive, let's go back to the racetrack and trot out Old Gray Mare, Bobtailed Nag, and Stewball for a second eight-horse race. Another kind of bet on a horse is called "win-place-show" or "across the board," which means that if a horse comes in first or second or third in a race, the bettor collects the payoff. Winning two win-place-show bets on different horses would not be mutually exclusive events. It is possible that two of your long shots could be among the first three horses to finish the race and it is even possible that your three long shots would be the first three horses to finish the race. Suppose a bet on Old Gray Mare to win-place-show paid 5 to 1, the same bet on Bobtailed Nag paid 2 to 1, and the same bet on Stewball paid 1.5 to 1. If we continue the assumption that the track's percentage is zero, the reciprocals of these payoffs are the probabilities that each bet will pay off, that is, 0.2, 0.5, and 0.667, respectively. Another kind of bet is to pick two horses to finish among the first three. This bet for Old Gray Mare and Bobtailed Nag paid 10 to 1, so the probability of Old Gray Mare and Bobtailed Nag both finishing in the first three was 0.1. The same bet for Old Gray Mare and Stewball paid 6.67 to 1, so the probability of winning this bet was 0.15. The same bet for Bobtailed Nag and Stewball paid 2.5 to 1, so the probability of winning this bet was 0.4. Finally, the bet for all three horses finishing in the first three places (this bet is sometimes called a trifecta) paid 20 to 1, so the probability was 0.05. With these figures you can use formula 4.1c to find the probability of winning one or more of three win-place-show bets on these horses. Let A be Old Gray Mare coming in among the first three horses, let B be Bobtailed Nag coming in among the first three horses, and let C be Stewball coming in among the first three horses.

$$P(A \cup B \cup C) = P(A) + P(B) + P(C) - P(A \cap B) - P(A \cap C) - P(B \cap C) \\ + P(A \cap B \cap C) \tag{4.1c}$$

$$P(A \cup B \cup C) = 0.2 + 0.5 + 0.667 - 0.1 - 0.15 - 0.4 + 0.05$$

$$P(A \cup B \cup C) = 0.767$$

Your chance of getting some money back is 0.767. The probability of winning one or more of a series of win-place-show bets is much higher than the probability of winning one of a series of bets that horses will come in first, but the payoffs for win-place-show bets are much lower. In this example, if either Stewball or Bobtailed Nag were the only horse among the three long shots to place in the first three, you would still lose money overall on the three win-place-show bets. Three win-place-show bets would cost $3, if we assume that the track accepts $1 bets, and if only Stewball finished in the first three, your payback would be $1.50, so you would be poorer by $1.50. If Bobtailed Nag finished in the first three, the payout would be just $2, so you would be poorer by $1.

The formulas for the unions of more than three events that are not mutually exclusive are too complicated to be discussed here. In many problems that involve finding the union of more than two events, it is easier to find the probability that none of the events will occur and then subtract that result from 1 to get the union. This procedure is the equivalent to finding the white area in Figure 4.3 and subtracting that area from 1 to get the area of the circles. We'll examine some problems that use this device after we've discussed the probability rules a little further.

4.2C MULTIPLICATION RULE

A **conditional probability** is *the probability of an event given the information that some other event has occurred or the assumption that some other event will occur.* Let's pretend that you are conducting a die-tossing experiment with some friends. Before you toss a die, the probability of getting a number less than 4 is ½. Suppose you go ahead and toss the die and you observe which face landed up. You quickly cover the die with a cloth before anyone else can see it and you announce to your friends that the side facing up is an odd number. You ask your friends if this information changes the probability that the result is a number less than 4. Your friends now know that the die must have rolled either a 1 or a 3 or a 5 and that two of these three numbers are under 4. Given the information that the die rolled an odd number or given the assumption that it will roll an odd number changes the probability of getting a number less than 4 from ½ to ⅔.

The reason that information on getting an odd number was stated as a fact about something that had already occurred and as an assumption about what will occur is that statisticians disagree on how to define probability. Some statisticians argue that it is meaningless to assign a probability to an event that has already occurred: either the top side of the die that has already been tossed is under 4 or it is not. Other statisticians argue that it makes sense to assign probabilities to events that have occurred but are not yet observed, such as a die hidden under a cloth. This disagreement does not affect the rules of probability that we are investigating in this section. I have stated the die problem both ways because I don't know which view your instructor accepts.

The notation for conditional probability is a vertical line separating the event to which the probability applies, on the left, from the event that is given, on the right. In our example the event "an odd number" is symbolized as A and the event "a number less than 4" is symbolized as B. The probability of getting a number less than 4 given that an odd number has been (or will be) tossed is written $P(B \mid A)$. Often it is better to write out a description of the events instead of using the letters A and B. For example, P(number less than 4 | odd number). With this notation you are less likely to make a mistake in interpreting the formulas. If a full expression is too long, use letters that you can associate with the names of the events, such as $P(<4 \mid \text{odd})$. The reason that the letters A and B are being used here is that this example introduces the general formula, which is expressed in terms of A and B.

The probability of getting an odd number given that a number less than 4 has been (or will be) tossed is written $P(A \mid B)$. In this example the probabilities of A given B and B given A are equal, but in general they do not have to be equal. Before you go any further, you can check your understanding by verifying that $P(A \mid B)$ equals ⅔.

According to the **multiplication rule,** the probability of the intersection of events A and B equals either *the conditional probability of* A *given* B *multiplied by the probability of* B *or the conditional probability of* B *given* A *multiplied by the probability of* A.

$$P(A \cap B) = P(A \mid B)P(B)$$

$$P(A \cap B) = P(B \mid A)P(A)$$

(4.2)

The way formula 4.2 is presented here helps to clarify why it is called the multiplication formula. The product of $P(A \mid B)$ and $P(B)$ equals the probability of the intersection term. The same formula can be rearranged to emphasize the conditional probability.

$$P(A \mid B) = \frac{P(A \cap B)}{P(B)}$$

$$P(B \mid A) = \frac{P(A \cap B)}{P(A)}$$

To clarify the multiplication formula further, let's continue with the die example. The probability of getting an odd number given that the die rolls a number under 4 is $\frac{2}{3}$. The reason is that there are three ways to get a number less than 4, and of those three ways, two are odd numbers, so the probability of getting an odd number given that the number is less than 4 must be $\frac{2}{3}$. We could get the same result by using the multiplication formula, that is, by dividing the probability of getting both an odd number and a number less than 4, $P(A \cap B) = \frac{1}{3}$, by the probability of getting a number less than 4, $P(B) = 0.5$. An intuitive explanation for this operation is that the probability of getting an odd number given that a number less than 4 has been rolled equals the proportion of all outcomes that are less than 4 and odd numbers (the 1 and the 3) out of all outcomes less than 4 (the 1, 2, and 3).

Another useful term in probability theory is **statistical dependence,** which describes *the relationship between* A *and* B *if the probability of* A *given* B *does not equal the probability of* A *alone;* or, in notation, if $P(A \mid B) \neq P(A)$, there is statistical dependence between A and B. What this means is that the knowledge that B is true affects or changes the probability of A. Knowing that a number less than 4 was rolled affects the probability of getting an odd number. **Statistical independence** characterizes *the relationship between* A *and* B *if the probability of* A *given* B *is equal to the probability of* A *alone.* In other words, statistical independence between events A and B implies that knowing that B has occurred does not change the probability of A. Let's define an event D as rolling the die so that the top face is less than 5, that is, $\{1, 2, 3, 4\}$. Before the die is tossed, the probability of getting an odd number is 0.5, and if you were given the information that the toss resulted in a number less than 5, the probability of getting an odd number would continue to be 0.5. The probability of A given D equals the probability of A alone. With statistically independent events, the probability of A alone can replace the conditional probability of A in formula 4.2. Formula 4.2a is the multiplication rule for events that are statistically independent.

$$P(A \cap B) = P(A)P(B)$$ (4.2a)

$$P(A \cap B) = P(B)P(A)$$

Formula 4.2a can easily be derived from formula 4.2. By substituting $P(A)$ for $P(A \mid B)$ or $P(B)$ for $P(B \mid A)$ in formula 4.2, you get formula 4.2a.

A Venn diagram helps us to visualize statistical independence and dependence. Let's use a different example. Suppose a particular make of car had a probability of 0.1 of requiring major warranty work during the first 90 days and that the probability that a car was built on a Friday was 0.2. If there is statistical independence between these two characteristics, the probability of the intersection term will have to be 0.02. If the intersection term is higher or lower than 0.02, there will be statistical dependence. The three Venn diagrams in Figure 4.5 show first statistical independence with an intersection term equal

FIGURE 4.5

Three Venn diagrams illustrating statistical independence and dependence between the characteristics of a car being built on a Friday and requiring major warranty work during the first 90 days

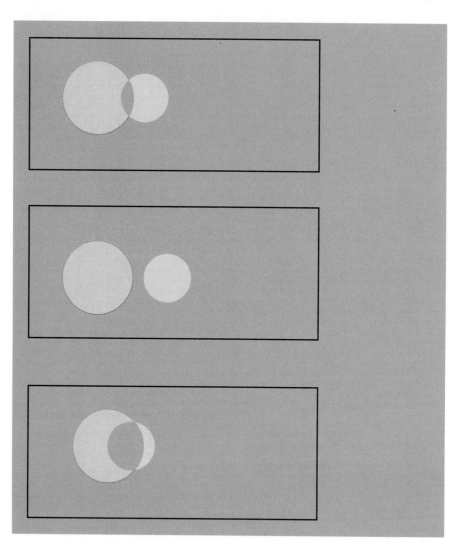

to 0.02 and then dependence with intersection terms that are less than and more than 0.02. The white circle in each of the diagrams represents the probability that a car will be built on a Friday and the shaded circle is the probability that a car will require major warranty repairs during the first 90 days. The intersection term in the top diagram equals 0.02. It illustrates the amount of overlap that is found when there is statistical independence between the two events. The second diagram shows the two new circles having no overlap. In this case the two events are mutually exclusive. If they are mutually exclusive, they must be statistically dependent. Knowing that a car was built on a Friday would affect the probability that it would require major warranty repair work. Knowing that it was built on a Friday would tell you that there is a zero probability of requiring major repair work. The third diagram shows most of the shaded circle within the white circle. Again, there is statistical dependence because knowing that a car was built on a Friday affects the probability that it will require major repair work. In this case the probability of requiring major repair work is increased if you know that the car was built on a Friday. The key point is that statistical dependence occurs if knowledge of event B makes A more likely and if knowledge of event B makes event A less likely.

Formula 4.2a can be extended to more than two statistically independent events.

$$P(X_1 \cap X_2 \cap X_3 \cap \cdots \cap X_n) = P(X_1)P(X_2)P(X_3) \cdots P(X_n)$$

(4.2b)

A simple example that uses this formula is the probability of tossing an ideal die three times and getting a 6 on each toss. The probability of getting a 6 on any one toss is ⅙. The tosses are independent because a die has no memory; knowing the result of one toss does not affect the probability of getting a 6 on the next toss. Substituting into formula 4.2b yields

$$P(X_1 \cap X_2 \cap X_3) = P(X_1)P(X_2)P(X_3)$$

(4.2b)

In this case, X_1 is the event "a 6 on the first toss," X_2 is "a 6 on the second toss," and X_3 is "a 6 on the third toss."

$$P(X_1 \cap X_2 \cap X_3) = \left(\frac{1}{6}\right)\left(\frac{1}{6}\right)\left(\frac{1}{6}\right)$$

$$P(X_1 \cap X_2 \cap X_3) = 0.00463$$

The probability of getting all 6s on three tosses is 0.00463.

The addition rule and the multiplication rule can be used in the same problem. For example, what is the probability of getting three tosses of the same

RISK ASSESSMENT FOR THE SHUTTLE'S BOOSTER ROCKETS

During the design phase of the U.S. shuttle program, a number of risk assessments were commissioned by the White House and the Department of Defense. The solid propellant booster rockets were considered a critical component because a failure of one of the two boosters during a shuttle launch could destroy the shuttle and because this type of rocket had a record of a significant proportion of failures. The historical data on the rocket boosters indicated 1 failure in every 57 firings. Since the shuttle's booster rockets were a new design that presumably reflected the lessons learned from the earlier failures, all of the commissioned studies estimated a failure rate lower than the historical record indicated. A 1981 study by J. H. Wiggins of Redondo Beach, California, estimated a 1 in 1,000 failure rate because of the improved design. A 1983 study by Sierra Energy and Risk Assessment of Sacramento, California, estimated a much higher failure rate: 1 in 100. Finally, in 1984 the Sandia National Laboratories estimated a failure rate of 1 in every 210 firings.

While it is true that these estimates were subjective probabilities and that they varied widely, all of the estimates concur in suggesting that the probability that one or more shuttles would crash due to booster rocket failure over the course of the program was quite high. The above probabilities of failure are for a single booster; if either one or both boosters failed during a launch, the shuttle would crash. If we take the most recent estimate from the Sandia Laboratories as the basis of our calculations, the probability that a shuttle will crash due to booster rocket failure in one launch will be the union of the probabilities that either the first rocket will fail or the second rocket will fail or both will fail. Assuming independence between the rockets, this union would equal

$$P(F_1 \cup F_2) = P(F_1) + P(F_2) - P(F_1 \cap F_2)$$
$$P(F_1 \cup F_2) = P(F_1) + P(F_2) - P(F_1)P(F_2)$$
$$P(F_1 \cup F_2) = 0.0047619 + 0.0047619 - 0.0000227$$
$$P(F_1 \cup F_2) = 0.0095$$

number in three trials? We already have the probability of getting three 6s in a row, but the choice of 6 is arbitrary; the same result would apply to any face of the die. Since getting three 6s and getting three of any other number are mutually exclusive events, the addition formula for mutually exclusive events can be used to get the union three 1s or three 2s or three 3s or three 4s or three 5s or three 6s. Substituting 0.00463 in formula 4.1b yields

$$P(X_1 \cup X_2 \cup X_3 \cup \cdots \cup X_n) = P(X_1) + P(X_2) + P(X_3) + \cdots + P(X_n) \qquad \text{(4.1b)}$$

The notations F_1 and F_2 stand for the failure of the first and second rockets. The probability of failure for either rocket, 0.0047619, is 1 over 210. Under the assumption of statistical independence, the probability of the intersection of two failures is a product of the probabilities of both failures.

The shuttle program was projected to launch shuttles 500 times before they were replaced by a new design. Continuing with the Sandia Laboratories estimate, what is the probability that *none* of these 500 launches would end in a crash because of a failure of the booster rockets? Assuming independence between launches, this probability is just the probability of *not* having a crash in one launch taken to the 500th power. This is just the multiplication formula for a series of independent events. The probability of *not* having a crash caused by a booster rocket in a single launch is 1 minus 0.0095 or 0.9905. Not having such a crash is the complement of having such a crash. The number 0.9905 taken to the 500th power is 0.0085. This is the probability of *not* having a failure in 500 launches; the probability of one or more failures is 1 minus 0.0085, or 0.9915. Thus if the Sandia assessment is accurate, the probability of one or more failures over the life of the program becomes a virtual certainty. Similar calculations based on the Sierra estimates of a 1-in-100 probability of a booster rocket failure yield estimates of 0.9999. Even the most optimistic estimate based on Wiggins' 1-in-1,000 probability results in a probability of 0.632 for one or more rocket-caused crashes over the life of the program.

These probabilities were not widely known before the crash of the shuttle *Challenger*, and the public seemed to perceive less risk in shuttle flights than was indicated by these probabilities based on expert opinion. NASA's internal risk assessment, issued in 1985, was that the probability of a rocket booster failure was 1 in 100,000. NASA's policies concerning who could take part in shuttle flights and the design of crew escape systems were based on their internal estimate of risk, which was much more optimistic than the independent estimates.

In this case X_1 is the event "three 1s in three tosses," X_2 is "three 2s in three tosses," X_3 is "three 3s in three tosses," and so on.

$$P(X_1 \cup X_2 \cup X_3 \cup X_4 \cup X_5 \cup X_6) = 0.00463 + 0.00463 + 0.00463$$
$$+ 0.00463 + 0.00463 + 0.00463$$

$$P(X_1 \cup X_2 \cup X_3 \cup X_4 \cup X_5 \cup X_6) = 0.02778$$

The probability of getting the same number three times in a row with an ideal die is 0.02778.

EXERCISES FOR SECTION 4.2

4.6 Let $P(A)$ equal 0.5 and $P(B)$ equal 0.3 and $P(A \cap B)$ equal 0.2. Find $P(\overline{A} \cap \overline{B})$. Draw a Venn diagram and label the area $(\overline{A} \cap \overline{B})$. [Hint: $P(\overline{A} \cap \overline{B}) = 1 - P(A \cup B)$.]

4.7 If A and B both have probabilities greater than zero and they are mutually exclusive, can A and B be statistically independent? Explain. If C and D were statistically independent, could C and D be mutually exclusive? Explain.

4.8 Find the probability of the union of A, B, and C if $P(A)$ equals 0.3, $P(B)$ equals 0.2, and $P(C)$ equals 0.1.

(a) Assume that A, B, and C are mutually exclusive.

(b) Assume now that A, B, and C are statistically independent.

4.9 Suppose you have signed up for a series of skydiving lessons that include three jumps. During your first lesson the instructor informs the class that for each jump there is a 1 in 1,000 chance that the parachute will fail to open. When you signed up for the lessons, you did not know that the sport was so dangerous. After class you ask the instructor for a refund but she refuses and points out that the probability of dying because of a parachute's failure to open in one of the three planned jumps is very low. What is this probability?

4.10 The education division of the XYZ Company trains salesmen. To determine the factors that increase the probability of successful sales presentations, the XYZ Company has observed 100,000 actual sales presentations. These presentations involved a wide variety of products and successful selling behaviors that XYZ identified as applicable to all types of sales. A successful presentation is defined as one that results in an order by the customer; an unsuccessful presentation is one that does not result in an order. One of the behaviors that was associated with a high probability of a successful outcome was the salesman's ability to induce the customer to express a specific need for the product. There were 30,000 successful presentations among the 100,000 observed. The customer identified a specific need for the product in 24,000 of the 30,000 successful presentations. The customer identified such a need in 10,000 of the 70,000 unsuccessful presentations.

(a) If a salesman can make a customer identify a specific need for the product, what is the conditional probability that the presentation will be successful?

(b) If a salesman cannot make a customer identify a specific need for the product, what is the conditional probability that the presentation will be successful?

4.3 JOINT PROBABILITY DISTRIBUTIONS

A 2-by-2 **joint probability distribution** is *a list of the probabilities of all possible outcomes of an experiment in which each outcome has two characteristics*. Consider the

TABLE 4.1

Joint probability distribution of grocery shoppers by sex and senior citizen status

	Male	Female	
Senior citizen	0.02	0.08	0.1
Not senior citizen	0.28	0.62	0.9
	0.3	0.7	1.0

experiment of observing grocery shoppers entering a store and recording two characteristics associated with each shopper: sex and senior citizen status. Each shopper is either male or female and at the same time each shopper is either a senior citizen or not a senior citizen. The possible outcomes or sample space of the experiment are: female senior citizen, male senior citizen, female not senior citizen, and male not senior citizen. Table 4.1 is a convenient way of presenting a 2-by-2 joint probability distribution. The number 0.02 refers to the proportion of all shoppers who are male and senior citizens. This number is the probability of the intersection of male and senior citizen. The proportion of all shoppers who are male is 0.3. This number is termed the marginal probability of being male because it appears in the margin of the table. Since every shopper is either male or female, the marginal probabilities of being male and being female must sum to 1.0. Since every shopper is either a senior citizen or not a senior citizen, those probabilities must also sum to 1.0.

This table is a handy way to summarize the relationships of the shoppers' characteristics. You can also use the table to calculate conditional probabilities. For example, given that a shopper is male, what is the probability that he is a senior citizen? If 30% of all shoppers are male and 2% are both male and senior citizens, the conditional probability is 2% divided by 30%, or 0.0667. This result comes from the "divided-through" version of formula 4.2:

$$P(A \mid B) = \frac{P(A \cap B)}{P(B)} \tag{4.2}$$

$$P(A \mid B) = \frac{0.02}{0.3}$$

$$P(A \mid B) = 0.0667$$

Is there statistical dependence between being male and being a senior citizen? In other words, if you know that a shopper is a male, does that knowledge affect the probability that the shopper will be a senior citizen? The answer is yes, there is statistical dependence. The marginal probability of being a senior citizen is 0.1. The conditional probability of being a senior citizen given that the shopper is male is 0.0667. For statistical *independence* to hold between being male and being a senior citizen, these two numbers would have to be the same. Here's another question to help you understand these relationships. Does statistical dependence between being male and being a senior citizen imply that there is statistical dependence between being female and being a senior citizen? The answer is yes. Since senior citizens are underrepresented among male

TABLE 4.2

Joint probability distribution for sex and senior citizen status of grocery shoppers under the assumption of statistical independence

	Male	Female	
Senior citizen	0.03	0.07	0.1
Not senior citizen	0.27	0.63	0.9
	0.3	0.7	1.0

shoppers, they must be overrepresented among female shoppers. The conditional probability of being a senior citizen given that a shopper is female is 8% divided by 70%, or 0.1143. If there were statistical independence, the proportion of female shoppers who are senior citizens would be 0.1.

If there were statistical independence between sex and senior citizen status, the four terms inside the table would have to equal the products of their respective marginal probabilities. Table 4.2 illustrates the same grocery shopper problem with statistical independence. Note that in Table 4.2 each intersection probability is the product of the respective row and column marginal probabilities and that the conditional probability for any combination of sex and senior citizen status equals the corresponding marginal probability. For example, the product of the marginal probabilities of the first row and column, 0.1 times 0.3, equals the intersection term of 0.03. The conditional probability of being a senior citizen given that someone is male equals the intersection of male and senior citizen divided by the probability of being male, or 0.03 divided by 0.3. Thus the conditional probability of being a senior citizen given that the person is male, 0.1, is the same as the marginal probability of being a senior citizen.

If your data are in frequencies rather than proportions, you can construct a joint probability table by dividing each term by the total number of observations, which in this example is the total number of shoppers. Suppose you had observed 455 shoppers and had arranged your observations in a 2-by-2 table of frequencies such as the one in Table 4.3. Table 4.4 shows the joint probability distribution that results when each term in Table 4.3 is divided by the number of observations.

Here's a more elaborate example that illustrates the use of a joint probability table. You are the credit manager for a distributor of consumer electronic

TABLE 4.3

Sex and senior citizen status of a sample of 455 grocery shoppers

	Male	Female	
Senior citizen	7	40	47
Not senior citizen	112	296	408
	119	336	455

TABLE 4.4

Joint probability distribution of sex and senior citizen status for a sample of 455 shoppers

	Male	Female	
Senior citizen	0.015	0.088	0.103
Not senior citizen	0.246	0.651	0.897
	0.261	0.739	1.000

TABLE 4.5
Marginal probabilities for the joint distribution of credit rating and payment record

	E rating	G rating	
Late			0.1
On time			0.9
	0.6	0.4	

products. Retailers typically ask you to OK the shipment of orders on a 30-day credit basis. Your experience has been that 10% of all shipments on such credit do not get paid within the required 30 days. To reduce your potential losses, you subscribe to the Dun & Bradstreet credit rating service. This service's rating of E designates a firm with an excellent credit risk; G designates a good credit risk. There are lower designations, but you have never extended credit to any firm with a lower rating. Forty percent of the firms to which you have extended credit had G ratings and the remaining 60% had E ratings. You are considering a new policy of extending credit only to firms with E ratings. If the proportion of shipments to G-rated firms that had not paid on time was 0.125, what impact would the new policy have on the proportion of late payments?

The first step in solving this problem is to set up a joint probability table. As you already know the marginal probabilities, they appear in Table 4.5. You can determine the intersection term for late and G rating by using the marginal probability of a G rating and the conditional probability of late given a G rating. Denote a late payment with the letter L. Let the E credit rating be denoted by E and the G credit rating by G. Substituting into formula 4.2 and changing the letter B to G for the G-rated firms and changing the letter A to L for the late payers, you get

$$P(L \mid G) = \frac{P(L \cap G)}{P(G)}$$

$$0.125 = \frac{P(L \cap G)}{0.4}$$

$$P(L \cap G) = 0.05$$

Once 0.05 is put into the table for the intersection of late and G rating, as in Table 4.6, it is easy to see that the intersection of late and E rating must be 0.05, because the row must sum to the marginal probability of 0.1.

TABLE 4.6
Marginal probabilities for the joint distribution of credit rating and payment record and the intersection term for late and G rating

	E rating	G rating	
Late	0.05	0.05	0.1
On time			0.9
	0.6	0.4	

The last step is to use the multiplication formula to find the probability of late given E rating.

$$P(L \mid E) = \frac{P(L \cap E)}{P(E)}$$

$$P(L \mid E) = \frac{0.05}{0.6}$$

$$P(L \mid E) = 0.083$$

If you extend credit only to E-rated firms, the proportion of late payments should drop from 0.1 to 0.0833. Before you adopt this policy, you will have to decide if the potential loss of sales to firms that are denied credit may outweigh the gain from the lower proportion of late payments. But that's another problem. The main advantage of the joint probability distribution table is that it helps you visualize the relationships between the two characteristics.

Tables for joint probability distributions can be extended beyond the 2-by-2 format to include more than two categories for each of the two characteristics. For example, the ages of shoppers can be recorded as (a) senior citizen, (b) child, and (c) anything in between. The table will now be 3-by-2. As long as only two characteristics are involved (in the grocery shopper example the two characteristics are sex and age), the joint probability distribution can easily be listed in a table. The table can be 2-by-5 or 3-by-5 or 10-by-10, depending on the number of subcategories you have for your two characteristics. But if you had a problem in which each outcome had three characteristics (such as sex, age, and marital status), you would need a cube rather than a sheet of paper for your table. You might then have a 2-by-2-by-2 table or a 2-by-3-by-5 table. If you go past two characteristics, these tables are no longer a convenient way to present joint probability distributions. For such problems you will have to rely on formulas or tree diagrams. (A tree diagram appears in Chapter 5.)

EXERCISES FOR SECTION 4.3

4.11 Let $P(A)$ equal 0.5, $P(B)$ equal 0.3, and $P(A \cap B)$ equal 0.2.

(a) Fill in a table for the joint probability distribution of A and B.

(b) Find the values of $P(A \mid B)$, $P(A \mid \overline{B})$, $P(B \mid A)$, $P(B \mid \overline{A})$, $P(\overline{A} \mid B)$, $P(\overline{A} \mid \overline{B})$, $P(\overline{B} \mid A)$, and $P(\overline{B} \mid \overline{A})$.

(c) Identify the pairs of conditional probabilities that must sum to 1.0 from among those listed in part b.

4.12 Let $P(A)$, $P(B)$, $P(C)$, and $P(D)$ be the marginal probabilities of the rows of a joint probability table and $P(I)$, $P(II)$, and $P(III)$ be the marginal probabilities of the columns of the same table. Given the following probabilities, fill in the values of the table.

$P(\text{I}) = 0.3$ $\quad\quad\quad$ $P(\text{II}) = 0.1$ $\quad\quad\quad$ $P(\text{III}) = 0.6$ $\quad\quad\quad$ $P(A) = 0.25$
$P(B) = 0.25$ $\quad\quad$ $P(C) = 0.15$ $\quad\quad$ $P(D) = 0.35$ $\quad\quad$ $P(A \mid \text{I}) = 0.333$
$P(A \mid \text{II}) = 0.5$ $\quad\quad$ $P(\text{I} \mid B) = 0.4$ $\quad\quad$ $P(\text{II} \mid B) = 0.2$ $\quad\quad$ $P(\text{I} \mid C) = 0.333$
$P(C \cap \text{II}) = 0.0$

4.13 A camera manufacturer commissioned a survey of 500 adult consumers. Each consumer was asked which type of camera he or she used most often. The responses were: instant camera, 100; disc camera, 200; single lens reflex camera, 150; and other, 50. The same consumers were also asked to identify the income range in which their annual income fell. The responses were: under $10,000, 100; $10,000–19,999, 100; $20,000–29,999, 100; and $30,000 or more, 200. Assume that income and type of camera used are independent and fill out the table for the joint probability distribution of income and type of camera.

4.14 A study of fatal automobile accidents in a particular state found that both speeding and drinking were factors in 75% of all fatal accidents. The study concluded that if drinking was a factor in a fatal accident, the probability that speeding was also a factor was 0.9375. The study also concluded that if speeding was a factor, the probability that drinking was a factor was 0.8333. What is the proportion of fatal automobile accidents in which neither speeding nor drinking was a factor?

4.15 A survey asked purchasers of new color television sets if they were replacing an old television set. If they answered yes, they were asked if the old set was at least ten years old and whether it was color or black and white. Sixty percent of the replaced machines were color sets, and of these color sets two-thirds were more than ten years old. Eighty percent of all the old sets were more than ten years old.

(a) Fill in the table for the joint probability distribution of type and age for the old television sets.

(b) Are type of television set (color versus black and white) and being more than ten years old statistically independent?

4.4 PERMUTATIONS AND COMBINATIONS

4.4A PERMUTATIONS

Many probability problems require you to count the possible number of ways the events can be ordered or selected. If the events in the sample space are equally likely, you can arrive at an answer much more quickly by using an appropriate formula than by using the addition and multiplication rules. A

permutation is *a possible order of items from a fixed list.* Suppose you were interested in the order of finish of football teams in the Big Ten. One possible permutation would be: first place, Michigan; second place, Northwestern; third place, Iowa; fourth place, Indiana; fifth place, Wisconsin; sixth place, Purdue; seventh place, Michigan State; eighth place, Illinois; ninth place, Minnesota; and last place, Ohio State.

If every possible permutation is equally likely, what is the probability that you can randomly select the correct order for a particular season? The probability is one over the number of possible permutations. In this case there are 10! permutations, or 3,628,800. "10!" is read as "ten factorial" and it means $10 \cdot 9 \cdot 8 \cdot 7 \cdot 6 \cdot 5 \cdot 4 \cdot 3 \cdot 2 \cdot 1$. The reason that there are 10! possible permutations is that you have 10 ways to choose the first team, and once that team is chosen you have 9 ways to choose the second. Thus there are 10 times 9 or 90 ways of picking the first and second teams. In general, if there are m ways of making the first pick and n ways to make the second pick, there will be m times n ways of making the first two picks. Once the first and second teams are chosen, you have 8 ways to choose the third, and so on. The probability of correctly choosing the order of finish for the 10 teams (ignoring ties) is 1/3,628,800, or 0.0000003. What about the number of permutations in a list of more or fewer than 10 items? The number of permutations of n items selected from a list of n items is $n!$, and $n! = n(n - 1)(n - 2)(n - 3) \ldots (2)(1)$.

Suppose you only wanted to find the number of permutations of the first two teams among the Big Ten football teams. The number of permutations of x items selected from a list of n items may be found by formula 4.3:

$$_nP_x = \frac{n!}{(n - x)!}$$

(4.3)

The symbol $_nP_x$ stands for the number of possible permutations of x items taken from a list of n items. The answer for the football example is $10 \cdot 9$, or 90. Substituting into formula 4.3, we get

$$_nP_x = \frac{n!}{(n - x)!}$$

(4.3)

$$_{10}P_2 = \frac{10 \cdot 9 \cdot 8 \cdot 7 \cdot 6 \cdot 5 \cdot 4 \cdot 3 \cdot 2 \cdot 1}{8 \cdot 7 \cdot 6 \cdot 5 \cdot 4 \cdot 3 \cdot 2 \cdot 1}$$

$$_{10}P_2 = 10 \cdot 9$$

$$_{10}P_2 = 90$$

The logic here is that you have ten ways to choose the first team, and once you have chosen the first team, you have nine ways to choose the second. You could also apply formula 4.3 to the problem of the number of permutations of n items selected from a list of n items. The denominator would be $(n - n)!$ or $0!$. Zero factorial is defined as 1, so that you end up with only a numerator of $n!$. Defining $0!$ as 1 may seem odd, but think of it in these terms: If you were to select 0 items from a list of n items, how many ways could you make the selection? In a mathematical sense there is only one way to select zero items; there are no choices to be made about the order of items in a list with no elements. In this case the expression $_nP_0 = n!/n!$ equals $0!$, which equals 1.

■ 4.4B COMBINATIONS

The difference between a permutation and a combination is that for a permutation the order or ranking matters while a **combination** is *a distinct set of* x *items taken from* n *items in which the order does not matter*. For example, the ranking of Indiana first and Iowa second would be a different permutation than Iowa first and Indiana second. But the pair Iowa and Indiana for the first two teams would represent one combination. How many ways can you pick the top two teams in the Big Ten without specifying their order? In other words, how many combinations of two items can be taken from a list of ten? The answer is 45. Your intuition should tell you that if there are 90 permutations and two permutations can be collapsed into one combination, there have to be half as many combinations as permutations. This result holds when x is set to 2.

Let's go over the reasoning for the number of combinations for the first three teams chosen from a list of ten. Indiana-Iowa-Purdue would be one permutation, as would Iowa-Indiana-Purdue, Indiana-Purdue-Iowa, Iowa-Purdue-Indiana, Purdue-Indiana-Iowa, and Purdue-Iowa-Indiana. There are six possible permutations for any set of three specific teams. But a set of three teams represents one combination. Thus there are one-sixth as many combinations as permutations. Since there are $10 \cdot 9 \cdot 8$ or 720 permutations for the first three teams, there are 720 divided by 6 or 120 combinations of three teams selected from a list of ten items.

In the general case the formula for combinations is:

$$_nC_x = \frac{n!}{(n - x)!x!}$$

(4.4)

The term $_nC_x$ stands for the number of combinations of x items that can be taken from a list of n items. Applying formula 4.4 to the football example for selecting two teams yields

$$_nC_x = \frac{n!}{(n-x)!x!}$$ (4.4)

$$_{10}C_2 = \frac{10 \cdot 9 \cdot 8 \cdot 7 \cdot 6 \cdot 5 \cdot 4 \cdot 3 \cdot 2 \cdot 1}{(8 \cdot 7 \cdot 6 \cdot 5 \cdot 4 \cdot 3 \cdot 2 \cdot 1)(2 \cdot 1)}$$

$$_{10}C_2 = \frac{10 \cdot 9}{2 \cdot 1}$$

$$_{10}C_2 = 45$$

Here are some examples that use permutations and combinations. Suppose you are a stock boy in a grocery store and the manager wants to test how the shelf positions of brands of breakfast cereals affect sales. There are eight brands of cereal on the shelves and the manager tells you to rearrange the brands every morning and record the number of packages of each brand sold during the day until each brand has occupied every possible position. You politely tell the manager that he may have underestimated the number of days it will take to try every possible order. There are $n!$ or 8! permutations of eight items selected from a list of eight items. This works out to 40,320 permutations, and at the pace of one a day, you would be rearranging the boxes for a little over 110 years.

Here's another permutation example. Your company manufactures revolvers and each one must have a unique identification code. To prevent tampering or erasure of the code, you make a die for each gun. Each die will be used to punch a deep impression of the identification code into one gun. You plan to make 10,000 dies for a production run of 10,000 revolvers. One code consists of letters of the alphabet and the digits 0 through 9, used without repetition. You have to determine the minimum length of the code because the dies for a shorter code will cost less than the dies for a longer code. What is the minimum length of the code that will allow 10,000 unique codes? Let the length of the code be denoted x. You want the number of permutations to be at least 10,000, and n, the number of items in your list, is 36. There are 26 letters in the alphabet and 10 digits from 0 to 9, which sum to a list of 36. Substituting into formula 4.3, you get

$$_nP_x = \frac{n!}{(n-x)!}$$ (4.3)

$$_{36}P_x \geq 10,000$$

$$\frac{36!}{(36-x)!} \geq 10,000$$

You can find the minimum x that results in at least 10,000 permutations by trying different values for x in the above equation. If x is 1, there will be 36 permutations; if x is 2, there will be 1,260 permutations; and if x is 3, there will be 42,840 permutations.

If x equals 1, $_{36}P_1 = \dfrac{36!}{(36-1)!}$

If x equals 2, $_{36}P_2 = \dfrac{36!}{(36-2)!}$

If x equals 3, $_{36}P_3 = \dfrac{36!}{(36-3)!}$

If the code contained two letters or digits, there would be fewer than 10,000 unique codes, so you must have a three-item code in order to have enough permutations.

Here's another example. A city police department used a written examination to choose 12 officers from a list of 24 for promotion to captain. Of the 24 eligible officers, 4 were black and 20 were white. When the final list was posted, the 12 selected officers were white. The black policemen's association charged that the police department had discriminated on the basis of race. What is the probability of an all-white list being selected if every possible list of 12 officers was equally likely? Since the order in which the officers were selected does not matter, this problem is about combinations instead of permutations. The probability of an all-white list being selected equals the number of all-white lists divided by the number of possible lists. The number of all-white lists is $_{20}C_{12}$ and the number of possible lists is $_{24}C_{12}$. Substituting into formula 4.4, we get

$$_nC_x = \frac{n!}{(n-x)!x!} \qquad (4.4)$$

$$\frac{_{20}C_{12}}{_{24}C_{12}} = \frac{20!/(20-12)!12!}{24!/(24-12)!12!}$$

$$\frac{_{20}C_{12}}{_{24}C_{12}} = 0.047$$

There is slightly less than a 5% chance that a selection process that was color-blind would result in an all-white promotion list. This probability does not prove that the police department discriminated, but it does raise enough concern to suggest that the testing procedure should be carefully examined.

EXERCISES FOR SECTION 4.4

4.16 A newspaper is running a contest in which readers are asked to mail in a form printed in the paper in which they can guess the exact order of finish of the first three cars in the Indianapolis 500 race. Any readers who guess correctly will share a $10,000 prize. There are 33 cars in the race. If every order of finish for the first three cars was equally likely and you randomly selected one of them, what is the probability that you would have a share of the prize?

4.17 If the contest in exercise 4.16 were changed to require you to guess the first three finishers without specifying the order, what would be the probability that you would win a share of the prize?

4.18 A pit stop at the Indianapolis 500 race usually involves five steps: changing the front tires of the car, changing the rear tires of the car, filling the tank with gas, wiping the windshield, and giving the driver a drink. Suppose these steps have to be taken sequentially rather than at the same time because the rules do not allow enough pit crew members for more than one of the steps to be taken at the same time. Seconds are crucial in the race and a pit crew wants to try every possible order of the five steps to see which one can be done fastest. How many orders of the five steps must the crew test?

4.19 (This is a true story with the numbers changed.) A superfecta bet in jai alai requires bettors to pick the exact order of finish of the first six teams among the eight teams in a contest. If no one picks correctly, the money bet in the super-fecta pool is carried to the next game and becomes part of the payoff in the next eight-team contest. In Miami the jai alai superfecta payoff grew to $100,000 because no one was able to pick the exact order of the first six teams for several weeks. A shrewd bettor figured out that if he placed bets on every possible order of six teams, he would necessarily win a share of the $100,000 payoff. This is a rare case of a "sure bet."

    (a) What would it cost the shrewd bettor to place bets on every possible order of the top six out of eight teams if each bet cost $2?

    (b) If any other bettors pick the correct order for the first six teams, the $100,000 will be shared among the winners. If all orderings of the first six teams are equally likely and 5,000 distinct orderings are chosen by other bettors, what is the probability that the shrewd bettor will have to share the payoff? (In the real bet, a group of shrewd bettors took in $2 million after betting $1.5 million and no one else picked the winning order.)

4.20 Wendy's hamburger chain ran some advertisements that were intended to emphasize that its hamburgers are cooked to order and that the exact condiments that each customer wants are put on the hamburger. To impress consumers with the individuality of each customer's hamburger, the advertisements trumpeted the number of ways in which a hamburger can be purchased: with mustard, pickles, onions, lettuce, tomatoes, and ketchup or any subset of 5 or 4 or 3 or 2 or 1 or 0 of these condiments. How many ways can you "have it your way"?

4.5 BAYES' THEOREM *skip Go to Page 116.*

4.5A BAYES' THEOREM WITH ONE EXPERIMENT

Bayes' theorem is a three-step process. You begin with a **prior probability**—*a probability about a hypothesis or state of the world before an experiment is conducted to gain additional information.* After you have conducted such an experiment—the second step—you use the results of the experiment to revise the original or prior probability. *A probability about a hypothesis or state of the world after an experiment has been conducted and the prior probability has been revised* is called a **posterior probability.** The process is easier to explain with an example, so here's an example. Many employers use chemical tests to identify employees who use illegal drugs. These employers are worried that the use of illegal drugs will cause injuries to their customers or their employees. Suppose the manager of a city bus company is concerned about marijuana use by the company's bus drivers. Before the bus company initiates any drug tests, it surveys the drivers to see what proportion use marijuana. Naturally, users of marijuana are reluctant to admit they use it because they may lose their jobs. Chapter 7 discusses a technique for conducting surveys on sensitive questions which encourages honest answers because anonymity is guaranteed. This technique is called a randomized response survey. After conducting a randomized response survey, the bus company estimates that 7% of its drivers use marijuana. This 7% is the prior probability that a randomly selected bus driver uses marijuana. The company decides that this percentage is high enough to justify the cost and the potential bad feelings caused by random urine tests for marijuana. The company announces that the tests will begin in three months and that bus drivers who test positive for marijuana use will be fired. The three-month grace period is sufficient to remove any traces of marijuana from bus drivers who give up the drug immediately. The results of the urine test will be used to revise the probability that a particular driver uses marijuana.

At this point we are back to the argument about the definition of probability for an event that has already occurred. This time around the argument cannot be avoided. Classical or objectivist statisticians argue that the probability that a particular bus driver uses marijuana is either 1 or 0. Bayesian statisticians argue that it makes sense to say that the probability that a particular bus driver uses marijuana is 0.07 until you have more information. In Bayesian terminology, the 0.07 is the prior probability and the hypothesis that you are interested in is that a randomly selected driver uses marijuana. Since we are going ahead with this section of the chapter on Bayes' theorem, it should be obvious that we'll take the Bayesian approach to defining probability.

After a prior probability is established, the next step in a Bayes problem is to conduct an experiment to gain more information. The experiment is the urine test. This experiment has two possible outcomes: the laboratory finds that the urine sample tests either positive or negative for marijuana. If the experiment were perfectly accurate, you would know whether a particular employee used marijuana. Perfect accuracy would mean that the laboratory always obtained a positive reading for individuals who used marijuana and never obtained a positive reading for an individual who did not. However, few experiments are perfect. The degree of accuracy of such chemical tests is well known because they have been repeated many times with urine samples that are known either to have or not to have traces of marijuana. If no marijuana is present, the probability that the test result will be positive is 0.01. This result is called a false positive. From this fact you can infer that the probability of a negative reading in the absence of marijuana is 0.99. A negative reading in the absence of marijuana would be called a correct negative. A false positive is not the only mistake that the laboratory can make. If someone who did use marijuana got a negative test result, it would be a false negative. The laboratory knows that the probability of a false negative is 0.05. Again you can infer the probability of the opposite result; if someone uses marijuana, the probability that the laboratory will get a positive test result is 0.95. The 0.95 is the probability of a correct positive. Let's say that the urine of a particular driver tests positive. It now appears very likely that this driver uses marijuana. The probability that the driver uses marijuana should be revised upward from the prior value of 0.07.

To do this revision you would use Bayes' theorem:

$$P(H_1 \mid E) = \frac{P(H_1)P(E \mid H_1)}{\displaystyle\sum_{i=1}^{n} P(H_i)P(E \mid H_i)} \qquad (4.5)$$

H stands for hypothesis. There are two hypotheses in this example: H_1 is that the driver uses marijuana and H_2 is that the driver does not use marijuana. E stands for experimental result, in this case that the laboratory got a positive test result. $P(H_1 \mid E)$ is the revised probability that the driver uses marijuana given the experimental result that the laboratory found the urine sample to test positive. Substituting into formula 4.5, we get

$$P(H_1 \mid E) = \frac{P(H_1)P(E \mid H_1)}{\displaystyle\sum_{i=1}^{n} P(H_i)P(E \mid H_i)} \qquad (4.5)$$

$$P(H_1 \mid E) = \frac{0.07(0.95)}{0.07(0.95) + 0.93(0.01)}$$

$$P(H_1 \mid E) = 0.877$$

TABLE 4.7

The formulas used to derive Bayes' theorem displayed in a joint probability table

	E: Lab test positive	$\bar{E}$: Lab test not positive	
H_1 true: driver uses marijuana	$P(H_1 \cap E) =$ $P(H_1)P(E \mid H_1)$		0.07
H_2 true: driver does not use marijuana	$P(H_2 \cap E) =$ $P(H_2)P(E \mid H_2)$		0.93

$P(E) = P(H_1 \cap E)$
$+ P(H_2 \cap E)$

The revised probability that the urine of a particular driver contains marijuana is approximately 0.877. Note that the information that the laboratory had a positive test result sharply increases the probability that the bus driver uses marijuana.

A table for the joint probability distribution for the presence of marijuana and the experimental results can be used to show the reasoning behind Bayes' formula, as in Table 4.7. From the multiplication rule, formula 4.2, the intersection term for H_1 and E equals $P(H_1)P(E \mid H_1)$. The equivalent relationship holds for the intersection of H_2 and E. Since a person cannot be simultaneously both a marijuana user and not a marijuana user, H_1 and H_2 are mutually exclusive. The probability of the experimental result, a positive reading from the lab, is the sum of the probabilities of all of the mutually exclusive ways of getting a positive reading: the lab can find a positive reading with a urine sample that contains marijuana or with a urine sample that does not contain marijuana. Thus $P(E)$ equals $P(H_1 \cap E)$ plus $P(H_2 \cap E)$. Using the multiplication formula again for the conditional probability of H_1 given E yields

$$P(H_1 \mid E) = \frac{P(H_1 \cap E)}{P(E)}$$

Substituting $P(H_1)P(E \mid H_1)$ for $P(H_1 \cap E)$ yields

$$P(H_1 \mid E) = \frac{P(H_1)P(E \mid H_1)}{P(E)}$$

Substituting $P(H_1 \cap E) + P(H_2 \cap E)$ for $P(E)$ yields

$$P(H_1 \mid E) = \frac{P(H_1)P(E \mid H_1)}{P(H_1 \cap E) + P(H_2 \cap E)}$$

The last step is to substitute $P(H_1)P(E \mid H_1)$ for $P(H_1 \cap E)$ and $P(H_2)P(E \mid H_2)$

for $P(H_2 \cap E)$ in the denominator. This substitution yields formula 4.5, except that the summation over H_i is replaced by the sum of H_1 and H_2:

$$P(H_1 \mid E) = \frac{P(H_1)P(E \mid H_1)}{P(H_1)P(E \mid H_1) + P(H_2)P(E \mid H_2)}$$

If there are more than two hypotheses, the denominator is summed over H_i:

$$P(H_1 \mid E) = \frac{P(H_1)P(E \mid H_1)}{\sum_{i=1}^{n} P(H_i)P(E \mid H_i)} \tag{4.5}$$

As was shown above, Bayes' theorem can be derived from the multiplication and addition rules. In words, the logic of Bayes' theorem is that the probability that H_1 is true given an experimental result E equals the probability of E and H_1 occurring divided by the sum of the probabilities of all the ways E can occur. It is not the mathematical derivation of formula 4.5 that troubles some statisticians about Bayes' theorem; rather it is the application of probabilities to events that have already occurred and the frequent use by Bayesian statisticians of subjective probabilities. You can make up your own mind on the soundness of the Bayesian approach after you work the problems at the end of this section.

■ 4.5B BAYES' THEOREM WITH MORE THAN ONE EXPERIMENT

After having done the experiment and calculated the posterior probability, you might still feel uncomfortable about firing an employee who had tested positive; after all, there is still a chance of about 12 in 100 that the driver is not a marijuana user. What can the employer do? The same experiment can be conducted a second time; a urine specimen can be sent to another lab, for example, or a different experiment can be conducted. A more accurate test for the presence of marijuana in urine is the gas chromatograph test. This test is much more expensive than the simple urine test, so it is usually not used unless the simple urine test has a positive reading. The probability of a false positive for the gas chromatograph test is 0.0001 and the probability of a false negative is 0.002. If, after the simple urine test produced a positive reading, the gas chromatograph also produced a positive reading, what would be the new posterior probability that the bus driver used marijuana? The setup for a second Bayes test is similar to the first; the main difference is that the prior probability for the second equation is the posterior probability calculated after the first experiment. In other words, the prior probability that the driver is a marijuana user given the information that the simple urine test was positive is 0.877. After a positive reading from the gas chromatograph test, the posterior probability that the bus driver uses marijuana is

$$P(H_1 \mid E) = \frac{P(H_1)P(E \mid H_1)}{\sum_{i=1}^{n} P(H_i)P(E \mid H_i)}$$

(4.5)

$$P(H_1 \mid E) = \frac{0.877(0.998)}{0.877(0.998) + 0.123(0.0001)}$$

$$P(H_1 \mid E) = 0.99999$$

After the gas chromatograph test has yielded a positive reading, there is only a remote possibility that the driver does not use marijuana. However, the bus company could continue the process with a third and a fourth test. Even if every test resulted in a positive reading, the probability that the driver did not use marijuana would never be zero. This gas chromatograph example clearly conveys the Bayesian framework: new information can be used to revise the probability that a particular hypothesis—or what Bayesians call a "state of the world"—is true.

Here's a last example that uses Bayes' theorem. Suppose you are a textbook publisher and you are considering making a bid on a project for a statistics textbook. Your in-house editorial staff has reviewed a few chapters of the proposed textbook which the author submitted and they have recommended that you bid on the project. Your experience has been that 40% of the textbooks that are favorably reviewed by the in-house editorial staff and subsequently published make a profit. Thus your prior probability of making a profit if you go ahead and bid on the project and the bid is accepted is 0.4. To gain more information, you show the chapters to some professors and ask them for their reaction. Your experience has been that 95% of the proposed books that go on to make a profit get favorable reviews by the professors. However, 10% of the proposed books that do not make a profit also get favorable reviews by the professors. If the reaction of the professors to the project is unfavorable, what is the revised or posterior probability that the book will make a profit?

Let A be defined as making a profit and F as a favorable reaction by the professors. Then $\overline{F}$ is an unfavorable response by the professors. What you need to find is $P(A \mid \overline{F})$. Substituting into formula 4.5, you get

$$P(H_1 \mid E) = \frac{P(H_1)P(E \mid H_1)}{\sum_{i=1}^{n} P(H_i)P(E \mid H_i)}$$

(4.5)

$$P(A \mid \overline{F}) = \frac{P(A)P(\overline{F} \mid A)}{P(A)P(\overline{F} \mid A) + P(\overline{A})P(\overline{F} \mid \overline{A})}$$

The term $P(A)$, the prior probability of making a profit, equals 0.4, so the term $P(\overline{A})$, the prior probability of not making a profit, must equal 0.6. Given that a profit will be made, the probability of a favorable response by the professors is 0.95. This term is $P(F \mid A)$. You need $P(\overline{F} \mid A)$, the probability of an unfavorable response given that the book will make money, for the above equation. This probability is 0.05. Note that the conditional probabilities sum to 1.0 while the condition is held constant, that is, $P(\overline{F} \mid A)$ and $P(F \mid A)$ sum to 1.0. The conditional probability of a favorable response given that the book will not make money, $P(F \mid \overline{A})$, equals 0.1, so $P(\overline{F} \mid \overline{A})$ must equal 0.9. Another point to note when you set up the Bayes formula is that the experiment yields only one result, either the professors give the book a favorable response or they give it an unfavorable response. You have to be careful to put the correct values in the above equation and not mix different experimental results. The correct values substituted in the equation are

$$P(A \mid \overline{F}) = \frac{0.4 \cdot 0.05}{0.4 \cdot 0.05 + 0.6 \cdot 0.9}$$

$$P(A \mid \overline{F}) = 0.036$$

The unfavorable response by the professors revises the prior probability of 0.4 sharply downward to a posterior probability of 0.036. With this posterior probability you might want to put the current project in your circular file and start looking for another project to bid on.

The three examples covered so far allow for only two possible states of the world. In the case of both the simple urine test and the gas chromatograph test, the bus driver could either be a marijuana user or not be a marijuana user. In the textbook case, either the new book will make a profit or it will not make a profit. Many problems allow for more than two states of the world. You may have a defective part and be unsure which of three suppliers provided the part. A famous painting may be a modern forgery, an original, or a work done by the assistants of an old master. Any one of five construction companies may have the winning bid on a construction project. You can analyze these examples in a Bayesian framework by obtaining more information to revise the probability of each state of the world. Formula 4.5 can be used with any number of hypotheses.

EXERCISES FOR SECTION 4.5

4.21 Suppose the probability of E given H_1 equaled 0.5 and that the probability of E given H_2 equaled 0.5. If the prior probability, H_1, equaled 0.9, what would the posterior probability of $H_1 \mid E$ equal? What would the posterior probability of $H_1 \mid \overline{E}$ equal? Do you gain any information by conducting such an experiment?

4.22 Suppose the probability of E given H_1 equaled 1.0 and the probability of E given H_2 equaled 0. If the prior probability, H_1, equaled 0.9, what would the posterior probability of $H_1 \mid E$ equal? What would the posterior probability of $H_1 \mid \overline{E}$ equal? Do you gain any information by conducting the experiment with the conditional probabilities of 1.0 and 0?

4.23 You are a stock market analyst and you specialize in advising clients about takeover bids. A takeover bid is an attempt by someone outside of a company whose stock is publicly traded to purchase a controlling interest in that company by offering to buy shares of its stock at a price above the current market price. You review the balance sheets of companies whose stock is publicly traded and you assign to each firm a probability for the event "a takeover bid in the next 90 days." The companies that have most of the following characteristics have a high probability of being takeover targets: cash rich, little debt, no antitrust or public regulation barriers to takeovers, widely held stock, and assets undervalued in the stock market. On the basis of these factors you assign a prior probability of 0.3 that a takeover bid will be launched against Standard Oil of Indiana in the next 90 days. You keep your eye on this stock, and shortly after you assign your prior probability you hear a rumor that large blocks of Standard Oil of Indiana stock are being purchased by anonymous buyers. Often such rumors are false; even if they are true, they do not guarantee that a takeover bid is about to be made. Your experience has been that when a takeover bid is made, such rumors have surfaced in 70% of the cases. When no takeover bid is made, such rumors surface in 20% of the cases. What is the posterior probability that a takeover bid against Standard Oil of Indiana will be made in the next 90 days?

4.24 Suppose you are a woman who is worried about being pregnant. You consult a book titled *The Home Medical Companion*. According to this book, if you neglected to take any precautions against getting pregnant, and if you wake up every morning feeling nauseous, and if your period is four weeks late, the probability of being pregnant is 0.995. To gain more information, you go to the drugstore and purchase an EZ Home Pregnancy Testing Kit. According to the instructions that come with the kit, given that someone is pregnant, the probability of a positive indication of pregnancy from the kit is 0.999. Given that someone is not pregnant, the probability of a negative indication of pregnancy from the kit is 0.98.

(a) Choose the experimental result of either a positive or a negative indication of pregnancy from the kit and calculate the posterior probability of being pregnant.

(b) The chemistry of these kits can be adjusted to reduce the probability of a false positive reading at the expense of increasing the probability of a false negative reading. Which of the two types of mistake is the EZ Home Pregnancy Testing Kit Company more eager to avoid, and why would the company be more eager to avoid one error than to avoid the other?

4.25 Suppose you were considering the purchase of a particular painting attributed to the Hoosier artist T. C. Steele. Forgeries of Steele's paintings are so common that there are reputed to be four fake Steeles for every real one. To gain more information, you take the painting under consideration to an art museum and have its curator of American paintings examine it. The curator claims that in 90% of the cases in which she is asked to examine a painting that is a genuine Steele, she correctly identifies it as genuine. But in 15% of the cases in which she has been asked to examine a forgery, the curator has mistakenly identified it as a genuine Steele. After examining your painting, the curator says that she thinks it is a forgery. What is the posterior probability that it is a forgery?

SUMMARY

This chapter introduced the three definitions of probability: the a priori, the relative frequency, and the subjective. The rules for working with probabilities, such as the addition rule, the multiplication rule, and Bayes' theorem, apply to probabilities assigned under any of these definitions. Joint probabilities were defined and joint probability tables were introduced as an aid in problem solving. Other aids in problem solving in regard to events that are equally likely include such counting techniques as the formulas for permutations and combinations.

The main themes of the chapter are that probability theory has important business applications and that it is the foundation of statistical inference. Another theme is the disagreements between Bayesian and classical statisticians. While classical probability theorists argue against the use of subjective probabilities and the assignment of probabilities to events that have already occurred, Bayesian probability theorists argue that these interpretations of probability have many practical business applications. The Bayesian process of acquiring additional information and adjusting one's expectations seems to match the way people think. This Bayesian framework will receive more attention in Chapter 18, on decision theory.

SUGGESTED READINGS

Hymans, Saul H. *Probability Theory with Applications to Econometrics and Decision Making*. Englewood Cliffs, N.J.: Prentice-Hall, 1967.

NEW TERMS

a priori definition of probability Every possible outcome of an experiment is assumed to be equally likely and thus the probability of any outcome is 1 divided by the total number of possible outcomes.

addition rule The probability of the union of A and B equals the sum of the probability of A and the probability of B minus the probability of the intersection of A and B.

combination A distinct set of x items taken from n items in which the order does not matter.

complements Two mutually exclusive events that have probabilities that sum to 1.0.

conditional probability The probability of an event given the information that some other event has occurred or the assumption that some other event will occur.

event A set of one or more outcomes.

intersection All outcomes that are both in A and in B.

joint probability distribution A list of the probabilities of all possible outcomes of an experiment in which each outcome has two characteristics.

multiplication rule The probability of the intersection of A and B equals the conditional probability of A given B multiplied by the probability of B or the conditional probability of B given A multiplied by the probability of A.

mutually exclusive events Two events that cannot occur as the result of the same experiment.

outcome or elementary event Any result of an experiment that cannot be divided into smaller components.

permutation A possible order of items from a fixed list.

posterior probability A probability about a hypothesis or state of the world after an experiment has been conducted and the prior probability has been revised.

prior probability A probability about a hypothesis or state of the world before an experiment is conducted to gain additional information.

probability A number that represents the chance that a particular outcome will occur if the experiment is conducted.

random experiment Any process with an uncertain outcome.

relative frequency definition of probability The probability of a particular outcome of an experiment equals the proportion of the past experiments in which this outcome occurred.

sample space A list of the possible outcomes of an experiment.

statistical dependence The relationship between A and B if the probability of A given B does not equal the probability of A alone.

statistical independence The relationship between A and B if the probability of A given B is equal to the probability of A alone.

subjective definition of probability The best estimate of someone who is informed about the conditions that influence the experiment.

union For events labeled A and B, all outcomes that are either part of event A, part of event B, or part of both events A and B.

NEW FORMULAS

Addition rule

$$P(A \cup B) = P(A) + P(B) - P(A \cap B) \tag{4.1}$$

Addition rule with mutually exclusive events

$$P(A \cup B) = P(A) + P(B) \tag{4.1a}$$

Addition rule with more than two mutually exclusive events

$$P(X_1 \cup X_2 \cup X_3 \cup \cdots \cup X_n) = P(X_1) + P(X_2) + P(X_3) + \cdots + P(X_n) \tag{4.1b}$$

Addition rule with three events that are not mutually exclusive

$$P(A \cup B \cup C) = P(A) + P(B) + P(C) - P(A \cap B) - P(A \cap C) - P(B \cap C) + P(A \cap B \cap C) \tag{4.1c}$$

Multiplication rule

$$P(A \mid B) = \frac{P(A \cap B)}{P(B)} \tag{4.2}$$

Multiplication rule with two statistically independent events

$$P(A) = \frac{P(A \cap B)}{P(B)} \tag{4.2a}$$

Multiplication rule with more than two statistically independent events

$$P(X_1 \cap X_2 \cap X_3 \cap \cdots \cap X_n) = P(X_1)P(X_2)P(X_3) \cdots P(X_n) \tag{4.2b}$$

Permutations of x items taken from n items

$$_nP_x = \frac{n!}{(n-x)!} \tag{4.3}$$

Combinations of x items taken from n items

$$_nC_x = \frac{n!}{(n-x)!x!} \tag{4.4}$$

Bayes' theorem

$$P(H_1 \mid E) = \frac{P(H_1)P(E \mid H_1)}{\sum_{i=1}^{n} P(H_i)P(E \mid H_i)} \tag{4.5}$$

ADDITIONAL EXERCISES

An asterisk designates an exercise of greater than ordinary difficulty.

4.26 An attorney once gave the following argument to a jury. "Mr. X could have died from a heart attack or from poison. We don't know for sure. The coroner could not establish the exact cause of death. Without certainty in this matter, we must assume that the probability that poison caused Mr. X's death is 0.5. Now my client had access to the poison and so did the butler. We can't be certain about which of them gave Mr. X the poison, so we must assume that the probability that it was my client is 0.5. Ladies and gentlemen of the jury, I submit that if these two probabilities are taken together, the probability that my client murdered Mr. X is only 0.25."

   (a) What definition of probability is the attorney using? Explain.

   (b) What definition would be more appropriate in this case? Explain.

4.27 You've forgotten the number needed to open your combination lock. The lock has a dial marked with the numbers 1 through 36, and you know that a sequence of three numbers is used to open the lock. You hate to throw the lock away, so you want to try every sequence of three numbers until you find the right one.

   (a) If a number is used only once, how many sequences are possible?

   (b) If a number can be used more than once, how many sequences are possible?

4.28 The Sunshine Promotions Company is planning an outdoor rock concert. The site it plans to use is available on May 5 or June 12. Sunshine would like to pick the date with the least chance of rain. Historical records show that it has rained once in every five May 5s and once in every three June 12s. To get more information, Sunshine consults a long-range weather forecast service. This service uses a computer to analyze wind and temperature data and its prediction record is quite good. For days when it rained, the model predicted rain in 90% of the trials. For days when it did not rain, the model predicted no rain in 82% of the trials. The model predicts no rain for June 12 and rain for May 5. Which date should Sunshine Promotions use?

4.29 A local Red Cross office has enough resources for one major emergency per year in its community. A major emergency might be a flood, a tornado, or an earthquake. In the past, floods have occurred with an average frequency of once every 20 years, tornados have occurred with an average frequency of once every 10 years, and earthquakes with an average frequency of once every 8 years. If the occurrences of floods, tornados, and earthquakes are independent, what is the probability that the local Red Cross office will run out of resources in a year?

4.30 In a paternity case the alleged father has type A+ blood, which occurs in 30% of the male population; Rh+ blood, which occurs in 60% of the male population; and an inheritable antibody that occurs in 5% of the male population. The child who is the subject of the suit has each of these characteristics and the mother has none of them. If these characteristics are independent, what proportion of the male population has the same three characteristics observed in the alleged father and in the child?

4.31 You own an architectural firm and you are considering drawing up a set of plans for a large office building that a developer plans to erect. The architectural firm that will ultimately design the building will be selected on the basis of a contest in which the plans of competing architects are judged. You don't like to work under these conditions because all of your effort could be wasted if you did not win the contest. You have found out from the supervisor of the contest that three other firms are interested in entering. Without knowing anything about the other firms, you would assign a prior probability of 0.25 to the chance that your firm will win if it enters the contest. You feel that you don't have enough information to decide whether to enter. You approach a friend who works for a rival architectural firm that is your greatest competitor and ask him if his firm is one of three that are planning to submit designs for the contest. Your friend says yes. On the occasions when you have won such contests in the past, this rival has competed against you 10% of the time; and on the occasions when you have lost such contests, this rival has competed against you 40% of the time. What is the posterior probability that you will win the contest?

4.32 An advertising firm plans to set up a team of five employees chosen from ten present employees to handle the account of a new client. The compatibility of the personalities of the team members is an important factor in determining the success of the team. If the advertising firm planned to consider every possible team, how many would it have to review?

4.33 If you had a joint probability distribution table with three rows and four columns and you knew the marginal probabilities for each row and column, what is the minimum number of intersection terms you would have to know in order to fill out the entire table?

4.34 Your company is planning to screen all employees for the use of illegal drugs with surprise urine tests. The samples of urine will be tested for cocaine, heroin, and marijuana. The cocaine test has a 5% false positive rate, the heroin test has a 1% false positive rate, and the marijuana test has a 2% false positive rate. A false positive occurs when the test indicates that a drug was used when in reality it has not been used. Assume that there is statistical independence between the outcomes of tests for different drugs on one sample and between tests for the same drug on successive samples.

   (a) What is the probability that someone will be falsely identified as a user of one or more of the drugs in one set of tests?

   (b) What is the probability that someone who has falsely been identified as a drug user will be falsely identified again in a second set of tests?

   (c) What is the probability that someone will be falsely identified as a drug user in two tests in a row? How does this question differ from the question in part b?

4.35 For each example explain which definition of probability is being used.

   (a) Since 15% of all students in introductory statistics get a letter grade of A for the course, the probability that you will get an A is 0.15.

   (b) Even if 15% of all students get an A for the course, you assign a probability of 0.75 to the outcome of your getting an A.

   (c) Professor Softheart has never given a student who regularly attended class an F for the course, no matter how little the student might know about statistics. Therefore, if you regularly attend class, the probability of getting an F from Professor Softheart is 0.0.

4.36 You are a used truck dealer and you have appointments with three customers at different times during the same day. Each customer is interested in the same truck.

Your experience has been that the probability that a customer will buy a truck after looking it over during an appointment is 0.3. What is the probability that the truck will be sold by the end of the day? Assume that only the three customers with appointments will look at the truck.

4.37 The Naval Avionics Center designs the guidance system for cruise missiles. These designs have a top-secret classification and at the end of a workday each desk at the center is cleared of all copies of the plans. There is a separate vault for the plans for each component of the guidance system, and each checkin or checkout of every numbered copy of the plans is recorded in a logbook. One morning a set of plans for the radar component of the guidance system is discovered to be missing. Past experience has shown that 99 times out of 100 the missing plans have merely been misfiled. The one time in 100 represents plans that are really lost. On this morning the person who last checked out the plans did not come in to work and did not answer his home phone. The head of security thinks that if the missing worker really stole the plans, the probability of his not coming to work this morning is 1.0. If the missing worker has not stolen the plans, the probability of his not coming in to work equals his past absentee rate of 0.02. What is the revised probability that the plans were stolen?

4.38 Charles Lindbergh chose a single-engine plane for his historic flight across the Atlantic Ocean. He thought that a single-engine plane would have a better chance than a two-engine plane. The reason was that if one of the two engines in a two-engine plane broke down, the plane would not be able to fly across the Atlantic. If each engine failure is independent and engine failure is the only cause of a crash landing, is the probability of a two-engine plane crash-landing in the ocean twice the probability of a one-engine plane crash-landing in the ocean?

4.39 Because consumers tend to buy what is available on their retailers' shelves, manufacturers of consumer products have to pay as much attention to persuading retailers to carry their products as they do to convincing consumers that their products are desirable. Manufacturers of consumer electronics generally reach store owners in two ways. One way is to rent space at the annual consumer electronics show in Chicago and the other is to have a representative visit stores to demonstrate the products. The Pure Sound Speaker Company uses both methods. A survey of the retailers who carry Pure Sound speakers shows that 70% of them saw the speakers at the annual show and that 60% of them were visited by a manufacturer's representative. Also, 50% of all of the retailers both saw the speakers at the show and were visited by a representative.

(a) What percentage of Pure Sound's retailers neither were visited by a representative nor saw the speakers at the show?

(b) The Pure Sound company is considering dropping its program of visits by manufacturer's representatives because it is an expensive way to meet retailers. If the company goes ahead and drops the program, what is the probability that a retailer who has been on the list of those visited by a manufacturer's representative will be among the retailers who see the speakers at the next annual show? Assume that the same retailers who went to the show this year will go next year. A table for the joint probability distribution may help you solve this problem.

4.40 A domestic airline is considering a new promotional fare, ten stops in ten U.S. cities over a 21-day period for $500. The promotion is aimed at foreign tourists and price-sensitive American tourists who want to see a lot of the United States in a short time. To discourage business travelers from using the fare, tickets will have to be purchased for at least two travelers taking the same route, and the entire itinerary will have to be specified at the time of purchase. The airline serves 46 U.S. cities. To cost out the plan, every possible itinerary is fed into a computer along with cost data for flying passengers between each pair of cities. For how many itineraries will the computer have to calculate the cost?

4.41 Your family has gone out to dinner at a Chinese restaurant. Everyone has agreed to order from the family-plan menu, but the discussion of which dishes to order has dragged on for over half an hour and no consensus seems to be in sight. You point out that the family will starve if the discussion goes on much longer, and to get their attention you tell them how many different meals could be ordered on the family plan, which involves picking two dishes from list *A* and three dishes from list *B*. If list *A* has 20 dishes and list *B* has 25 dishes, what number did you tell them?

4.42 For each example decide whether there is statistical independence or dependence between the two events and explain why.

(a) The first and the second flip of a fair coin result in heads.

(b) The first and the second flip of an ordinary coin result in heads.

(c) A grocery shopper is a senior citizen and a woman.

(d) A student is ill on a particular day and a midterm exam is to be given on that day.

4.43 Your shipping clerk has just told you that a rush order for steaks for one of your restaurants can't be delivered. You ask why and the clerk tells you that a forklift driver rolled over the box of steaks and that they now look more like hamburgers than steaks. You ask the clerk which restaurant the steaks were meant for and he answers that no one knows because the papers were on top of the box that was crushed. You are too embarrassed to call every restaurant and ask them all if they were the one who called in a rush order, so you want to call the most likely one first. Your company has six restaurants, which we'll call *A*, *B*, *C*, *D*, *E*, and *F*. The percentages of orders for steaks that come from your restaurants are respectively 30%, 20%, 20%, 10%, 15%, and 5%. You know that the order came in on Monday, and the probabilities of observing the experimental result "order on Monday" for your restaurants are 0.05, 0.15, 0.40, 0.60, 0.20, and 0.70, respectively. In what order should you call the restaurants?

4.44 A hospital receives a shipment of ten canisters of oxygen for its operating rooms and then gets a phone call from an employee of the oxygen company to say that one defective canister was shipped by mistake. The hospital needs to identify four good canisters in a short time for four pending operations, and a technician is told to test each canister in turn until the defective one is found. What is the probability that the empty canister will be among the first four tested?

4.45 An extended service contract for a car is a form of insurance that a consumer can buy to limit the costs of repairs. For example, the standard warranty on a new car is for 12 months. Dealerships typically offer a service contract that will cover the costs of repairs needed between the end of the warranty period and 48 months. If the estimated cost of the service contract were based on the amount of repairs required by the average car owner, the estimate might be too low or too high. If consumers who planned to make heavy use of their cars were more likely to purchase the service contract, the amount of repairs would be higher than the average for all new-car purchasers. If cautious consumers who took extra care of their cars were more likely to purchase the service contracts, the amount of repairs would be lower than the average for all new-car purchasers. The table below summarizes the experience of 1,000 new-car purchasers in regard to whether they had purchased a service contract and whether they needed a major repair during the period from the 13th to the 48th month of new-car ownership.

	No major repairs	Major repairs
No service contract	350	250
Service contract	150	250

(a) Do these sample data suggest that purchasers of service contracts have a higher or a lower probability of requiring major repairs than all car purchasers?

(b) What are the conditional probabilities of requiring a major repair given that a service contract was purchased and given that a service contract was not purchased?

4.46 An insurance salesman with State Farm Insurance has five customers whose homeowners' policies are up for renewal in August. The salesman's experience has been that 20% of the homeowners will not renew their policies. This time all five of the customers do not renew.

(a) If each customer acts independently and past experience accurately predicts the probability of each customer's not renewing, what is the probability that all five will not renew their policies?

(b) Give several reasons why it is unlikely that the customers are acting independently.

4.47 Technical analysts of the stock market use trends or patterns in the past prices of stocks to predict the direction of future price changes. Suppose a technical analyst noted that for 12 of the last 15 month-long periods in which Standard & Poor's index of 500 leading stocks rose continuously, the stocks of steel companies rose in price continuously over the following month. This observation suggests that buying steel stocks during a run up in the Standard & Poor index will be profitable. However, the prices of steel stocks in a given month also seem to be related to the previous month's prices of auto stocks. Of the last 30 occasions in which there was a month-long decline in the price of steel stocks, the price of auto stocks had fallen in 25 of the preceding months. Of the last 45 occasions when there was a month-long increase in the price of steel stocks, the price of auto stocks rose in 40 of the preceding months. If the Standard & Poors index has risen continuously over the last month and the price of auto stocks has fallen, what is the probability that steel stocks will rise continuously over the next month?

4.48 An environmental group claims that 10% of the salmon swimming to a spawning ground fail to swim over a flood-control dam. The group also claims that because there are five identical flood-control dams between the mouth of the river and the salmon spawning grounds, the probability that a salmon that starts at the mouth of the river will reach the spawning ground is 0.5. Is there something fishy about this probability?

4.49 In a class of 50 students, what is the probability that at least one of them has the same birthday as you do? Assume that no one was born on February 29.

4.50 The U.S. Customs Service occasionally does a complete search of all of the luggage on an arriving international flight. These searches have shown that 20% of all passengers returning to the United States understate the value of purchases they have made abroad. The Customs Service has developed a profile of passengers who are likely to understate the value of their purchases. For example, among those who understated the value of their purchases, 70% either wore sunglasses or avoided direct eye contact with the customs agent when they submitted their declaration. Among the passengers who did not understate the value of their purchases, 15% either wore sunglasses or avoided direct eye contact with the customs agent when they submitted their declaration. Another clue was the amount of the declaration. Among the passengers who understated the value of their purchases, the percentage who declared $1,000 or more was 85%. Among the passengers who did not understate the value of their purchases, the percentage who declared $1,000 or more was 5%. If a passenger is wearing sunglasses and declares purchases of more than $1,000, what is the probability that he is understating the value of his purchases? Assume that among the passengers who understate the value of their purchases, the two experiments of eye contact and the amount declared are independent.

4.51 Your older brother calls you to tell you that he and his best friend plan to visit your college this weekend and that he would like you to fix both of them up with dates with your sorority sisters. You put notes in everyone's mailbox to explain the situation and to ask if anyone has Saturday night free and is interested. You get 25 responses from sorority sisters who say they are available. How many different arrangements for dates do you have to choose from?

4.52 General Dynamics Corporation is a defense contractor that has many weapons contracts with the Department of Defense. Twelve of these contracts are due for Defense Department audits. Of these 12 contracts, two involve charges that could be considered questionable. The company hopes that these two contracts will not receive close scrutiny. The Defense Department auditors begin by announcing that two contracts will be chosen for detailed reviews and the other ten will get standard reviews. The two that are chosen are the ones that General Dynamics is most worried about. The company suspects that someone inside General Dynamics tipped off the auditors. What is the probability that the two contracts involving questionable charges were selected by chance?

4.53 A mail-order firm estimates that 20% of the households that receive its catalogue will make an order. Five catalogues were printed and mailed with incorrect prices.

   (a) What is the probability that exactly two of the households that received these catalogues will make orders?

   (b) What assumption do you have to make in order to answer part a?

4.54 Which is more likely: (a) one or more 6s in five tosses of a fair die or (b) two or more 6s in ten tosses of a fair die?

4.55* Your construction firm is one of many that has placed a bid for a large city sewer project. When the bids are opened, the city manager finds that three firms, Acme and Excelsior and yours, made identical bids on the project, and that these bids are also the lowest. By city law the choice of which firm will get the contract when there are identical low bids must be decided by lot. You figure that your firm has a probability of $\frac{1}{3}$ of getting the contract. With a troop of Boy Scouts in his office as witnesses, the city manager puts the names of the three firms in a hat and pulls out the winner. After the selection you call him and ask if your firm won. He answers that he will reveal the winner at a public meeting that afternoon. You argue with him, but he won't budge. Then you say, "Look, I know that at least one of the other firms had to lose out. If you can't tell me the name of the winner, at least tell me the name of one firm that lost." The city manager says OK, and tells you that Acme is one of the losers. Now you are excited, for you figure that with Acme eliminated, the winner has to be either you or Excelsior, and since each of the remaining firms is equally likely to win, the probability of your firm's winning has to be $\frac{1}{2}$. Have you made a mistake? Since at least one of the other two firms had to lose, does knowing the name of one that lost the bid change the probability of your firm's winning? Explain.

4.56* What is the probability that two or more students in a class of 20 will have the same birthday? Assume that no one was born on February 29.

4.57* Here's the racetrack problem that was promised on page 89. A total of $1 million was bet in the win pool in a particular eight-horse race. The track takes 25% of $1 million, so the amount of money that will be paid to the bettors who picked the winning horse is $750,000. The tote board posts the following payoffs at the start of the race:

Horse	Payoff per Dollar Bet to Win
Alhambra	1.875
Black Beauty	8.333
Flicka	75.000
Secretariat	15.000
Silver	7.500
Trigger	3.750
Victory	15.000
Whimsy	7.500

Find the subjective probability of each horse winning the race.

4.58* A robot welding machine is used to make welds on a motorcycle frame. The frame requires 15 distinct welds, labeled A through O. The robot can do any three of the 15 at the same time. For example, ABC, BCD, or AGN. If the robot welder is to be tested to see which order of welds will complete the frame in the shortest time, how many sets of three welds have to be tried?

4.59* Tay-Sachs is a fatal genetically transmitted disease found only among Jews of Eastern European ancestry. Children who have the disease die by the age of five. For a child to inherit the disease, both parents have to carry the Tay-Sachs gene as a recessive trait. If both parents are carriers of the recessive gene, the probability of a child's having the disease is 0.25. Suppose a married Jewish couple of Eastern European ancestry are worried about having a child with this disease. They both are tested for carrying the recessive gene and the test results are positive for one of them and negative for the other. If 20% of all Eastern European Jews are carriers of the recessive gene and the test that the couple took has a false positive rate of 0.01 and a false negative rate of 0.02, what is the probability that a child of theirs will have Tay-Sachs disease?

4.60* Your company is considering marketing a new perfume called Lust and your experience has been that nine out of ten new fragrances fail in the marketplace. One way to gain more information about the probability of a new perfume's success is a focus group discussion. In this discussion six women are asked to discuss perfume and to sample and give their opinions of some well-known perfumes and some test fragrances. Among the fragrances that have succeeded in the marketplace and were given a focus discussion test, 70% were singled out as the most desirable fragrance among those the women sampled. Among the fragrances that have failed in the marketplace and were given a focus discussion test, 20% were singled out as the most desirable fragrance among those the women sampled. You decide to market Lust only if the probability of its success in the marketplace is 0.95 or higher. If Lust is singled out as the most desirable fragrance in every focus group in which it is tested, what is the minimum number of focus groups that will have to be used to get a posterior probability of 0.95?

5

Discrete Random Variables

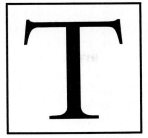

he planners of a 1980 U.S. military mission to rescue 52 Americans held hostage in Teheran, Iran, knew that at least six working helicopters were needed to complete the mission. The helicopters were complicated machines that often broke down. By sending more than six helicopters, they could increase the chances that at least six would be available to complete the mission. However, the more helicopters sent, the greater the risk that they would be detected. From the service records of the helicopters the planners knew that, given normal weather conditions, the probability of a helicopter's breaking down on a flight of the length required by the mission was 0.05. They decided to send enough helicopters to have at least a 99% chance that six or more could complete the mission. The planners did not anticipate a sandstorm.

The analysis required to determine the number of helicopters was based on the binomial distribution, which will be introduced in this chapter. Even though they are less dramatic than a hostage rescue mission, many business applications involve similar uncertainties and trade-offs. The chapter begins with the concepts of random variable, probability distribution, and expected value, which are needed to formulate certain distributions or patterns that can be applied to events with uncertain outcomes. These distributions include the binomial, the Poisson, and the hypergeometric. The names of the distributions may sound esoteric, especially the one with "hyper" in front of it. Stay calm; even though they are powerful tools for analyzing problems, their applications are straightforward. With a little study you'll avoid sandstorms.

5.1 INTRODUCTION TO RANDOM VARIABLES

5.1A DEFINITIONS OF DISCRETE AND CONTINUOUS RANDOM VARIABLES

A **random variable** is *a function that assigns numerical values to the possible outcomes of an experiment.* Many experiments have such categorical outcomes as defective versus not defective and male versus female. These outcomes can be made into a random variable by the assignment of a numerical value to each category, such as 0 to represent male and 1 to represent female. The symbol for a random variable is a capital letter. A lowercase letter is the symbol for an individual value of the random variable. Tossing a die is an experiment that creates a random variable with the integer values 1 through 6. This random variable could be symbolized by an X and each individual value by x_i. Thus x_1 would be 1, x_2 would be 2, and so on.

The experiment that creates the random variable does not have to be a process that you initiate or control. The number of traffic fatalities in a state over an upcoming holiday weekend or the interest rate on home mortgages six months from now are random variables. These are discrete random variables. This distinction among random variables parallels the distinction made between discrete and continuous data in Chapter 2. A **discrete random variable** is *a random variable whose numerical values are limited to specific values within its range.* Discrete random variables generally take integer values. The die and traffic fatality examples are integer random variables because you can't have a fractional fatality or a fractional number of dots on a die. Mortgage rates do take fractional values in tenths of a percentage point, such as 12.5% or 12.6%, but the fact that banks usually do not quote annual home mortgage rates in terms smaller than tenths of a decimal point makes the mortgage rate a discrete random variable.

A **continuous random variable** is *a random variable that can take any value over a continuous range of values.* The example of shoe sizes and foot lengths in Chapter 2 can illustrate the distinction between types of random variables. The length of the right foot of a randomly selected adult male in the United States is a continuous random variable because a person's right foot can be of any length over the range from the smallest to the largest right foot in the United States. A foot can be measured with any degree of accuracy to the nearest tenth, hundredth, or thousandth of an inch. The only limitation on the number of decimal places is the accuracy of the measuring equipment. Shoe size, in contrast, is a discrete random variable because shoes come only in whole and half sizes. Chapter 6 is devoted to continuous random variables; they are mentioned here only to establish the distinction between the two types of random variables. This chapter is devoted to discrete random variables.

■ 5.1B PROBABILITY DISTRIBUTIONS

Every possible value of a discrete random variable can be assigned a probability. The **probability distribution** or probability function of a discrete random variable is *a list of all the possible values of the random variable and the associated probabilities.* This listing could take the form of an equation, a table, or a chart. Assuming a fair die, the equation for the die-toss experiment would be

$$P(x_i) = \frac{1}{6} \qquad x_i = \{1, 2, 3, 4, 5, 6\}$$

TABLE 5.1
Values of the random variable generated by tossing a die and their associated probabilities

Table 5.1 lists the possible values of the die and their corresponding probabilities, and Figure 5.1 presents the same information. Since the list of possible values of a random variable includes all of the possible outcomes of the experiment that generates the variable, the probabilities must sum to 1.0.

x_i	1	2	3	4	5	6
$P(x_i)$	$\frac{1}{6}$	$\frac{1}{6}$	$\frac{1}{6}$	$\frac{1}{6}$	$\frac{1}{6}$	$\frac{1}{6}$

FIGURE 5.1
Values of the random variable generated by tossing a die and their associated probabilities

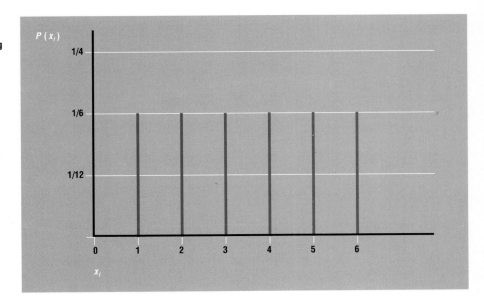

EXERCISES FOR SECTION 5.1

5.1 Indicate whether the following examples are discrete or continuous random variables. If the variable is discrete, state whether it takes fractional values or only integer values.

(a) The shoe size of a randomly selected European man. European shoe sizes take only integer values.

(b) The time taken by a flight from New York to London in hours.

(c) The number of flights between New York and London in a 24-hour period.

(d) The pump price of a gallon of gasoline in dollars.

5.2 Indicate whether the following examples are discrete or continuous random variables. If the variable is discrete, state whether it takes fractional values or only integer values.

(a) The age of a grocery store shopper in years.

(b) The daily quoted price of a stock traded on the New York Stock Exchange.

(c) The number of days of sick leave taken in the previous year by an employee of a particular company.

(d) The gross national product of the United States.

5.3 Let X be the random variable created by the experiment of tossing a pair of fair dice, where X equals the sum of the faces that land on top.

(a) Use a table to show the probability distribution.

(b) Use a graph to show the probability distribution.

5.4 Let X be a random variable that represents the number of candy bars sold by a vending machine in a day. X has the range 0 to 500. If each value of X is equally likely:

    (a) What is the probability that exactly 250 candy bars will be sold?

    (b) What is the probability that more than 250 candy bars will be sold?

    (c) What is the probability that an odd number of candy bars will be sold?

5.5 You own a local moving company that has special equipment for moving pianos. Your company has enough equipment to handle three pianos in a day. The number of requests to move pianos is a random variable with the following distribution:

x_i	0	1	2	3	4	5	6	7 or more
$P(x_i)$	0.4	0.4	0.1	0.05	0.02	0.02	0.01	0.00

What is the probability that you will run out of equipment? What is the probability that some of your equipment will be idle?

5.2 THE EXPECTED VALUE, STANDARD DEVIATION, AND VARIANCE OF A RANDOM VARIABLE

 5.2A THE EXPECTED VALUE OF A RANDOM VARIABLE

The **expected value** is *the average value of a random variable.* If you ran an experiment many times, summed the outcomes of each trial, and divided that sum by the number of trials, the end result would be the expected value. Let's use as an illustration a bet based on tossing a coin: if the coin comes up tails, you lose $1, and if it comes up heads, you win $1. The values that the random variable can take are plus or minus $1. To get the expected value of this random variable you could toss the coin, say, 1,000 times, total up your net winnings and losses, and divide the total by 1,000. Suppose the coin came up heads 505 times and tails 495 times. The expected value per toss is the net winnings of $10 divided by 1,000 trials, or $0.01. The meaning of expected value is that *with the coin used in this experiment* you would expect to average a gain of $0.01 per toss over many trials. This result is based on the relative frequency definition of probability. If you knew the probability of each outcome of an experiment, you could calculate the expected value without having to run the experiment. For example, if you could assume that you had a fair coin, the expected value of the above bet would be zero.

In mathematical terms, the expected value of a random variable is the weighted average of the values of the random variable where each weight is the probability that each value will occur. The following two equations show that average net winnings over many trials are the same as the weighted average of the possible outcomes of one experiment. The first equation has the total net winnings of $10 in the numerator and the trials in the denominator. The second equation, derived from the first, shows the weighted average of the two possible outcomes of winning and losing $1, where the weights are the probability of winning and the probability of losing. In this case the probabilities of winning and losing are based on the relative frequency definition of probability. Both expressions result in an expected value of $0.01.

$$\text{Expected value} = \frac{505(\$1) + 495(\$-1)}{1,000} \qquad \text{(average net winnings)}$$

$$\text{Expected value} = \underbrace{\frac{505}{1,000}}_{P(H)}(\$1) + \underbrace{\frac{495}{1,000}}_{P(T)}(\$-1) \qquad \begin{array}{l}\text{(weighted average of} \\ \text{each outcome)}\end{array}$$

Expected value = $0.01

The symbol $E(X)$ is used for the expected value of the random variable X. A capital X represents the random variable and the lowercase x_i represents the individual values of X. The general formula for the expected value of X is formula 5.1:

(5.1)

$$E(X) = \sum_{i=1}^{n} x_i P(x_i)$$

To apply this formula, let's use the random variable with the possible values {2, 3, 4, 5, 6}, each with a probability of 0.2. Substituting into formula 5.1 yields

$$E(X) = \sum_{i=1}^{n} x_i P(x_i) \qquad \text{(5.1)}$$

$$E(X) = 2(0.2) + 3(0.2) + 4(0.2) + 5(0.2) + 6(0.2)$$

$$E(X) = 4$$

The expected value is 4. From this example you can see that the arithmetic mean is the same as the expected value when the x_i's are equally likely. Suppose you took the same x_i values of {2, 3, 4, 5, 6} but made the 2 have a probability of 0.1 and the 6 have a probability of 0.3. The other numbers stay at a probability of 0.2. Again substituting into formula 5.1, we get

$$E(X) = \sum_{i=1}^{n} x_i P(x_i) \qquad\qquad \text{(5.1)}$$

$$E(X) = 2(0.1) + 3(0.2) + 4(0.2) + 5(0.2) + 6(0.3)$$

$$E(X) = 4.4$$

The expected value is 4.4. When the value 6 is more likely and the value 2 less likely, the expected value is increased. This example also shows that the expected value is not the same as the arithmetic mean when the probabilities of all outcomes are not equal.

Expected values can be used to guide your decision making when you face a problem with several choices and each choice has a set of payoffs that have known probabilities. If your decision problem is one in which the same experiment will be repeated many times and you know the expected payoff for any choice, the choice with the highest expected payoff would be your best decision. Here are some examples of business problems for which you can use expected values:

1 A company that can sell many life insurance policies needs to set a premium for the insurance. The company knows the probability of death for any purchaser of the insurance.

2 A company must decide on the value of the prizes that will be offered in a contest designed to encourage consumers to purchase the company's product. For a chance to win a prize, consumers must mail a proof of purchase of the company's product.

3 A gambling casino estimates the revenue per dollar bet from many bets placed at a roulette wheel.

Since the operation of a roulette wheel is easy to understand, let's use it as an illustration. American casinos use roulette wheels with 38 slots with a 0, a 00, and the numbers 1 through 36. If a bettor picks a numbered slot between 1 and 36 and the ball falls into the chosen slot, the bettor gets back the money he or she bet plus $35 in addition for each $1 bet. If the ball falls into one of the zero slots, the casino takes all the money that has been bet. Figure 5.2 shows such a roulette wheel. How much will the casino take in per $1 bet? Table 5.2 lists the possible outcomes and the associated probabilities. The probabilities are based on the assumption that each slot on the roulette wheel is equally likely. The $35 is negative because it represents a loss to the casino. To get the expected value to the casino per $1 bet, substitute the figures in Table 5.2 in formula 5.1:

$$E(X) = \sum_{i=1}^{n} x_i P(x_i) \qquad\qquad \text{(5.1)}$$

$$E(X) = \$1\frac{37}{38} + (\$-35)\frac{1}{38}$$

$$E(X) = \$0.053$$

FIGURE 5.2
Roulette wheel with 36
numbered slots and 0 and 00
slots

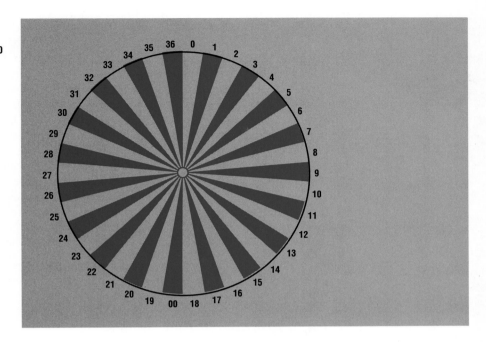

TABLE 5.2
Casino's revenue per
dollar bet and
associated
probabilities

x_i	$P(x_i)$
$1	37/38
−35	1/38

The casino will keep an average of $0.053 for every $1 bet.

The expected values are important to the bettors as well as to the casino. A rational gambler (if that isn't a contradiction in terms) would prefer a casino that, on average, took a smaller bite out of each dollar bet. In Las Vegas and Reno the casinos compete by advertising and/or varying their expected values. It sounds stupid to say "You'll lose your money more slowly in our casino," so the advertisements say something like "Our slot machines are the winningest slots in Reno, they pay back 99 cents per dollar bet." In a further effort to compete, the casinos change the rules of such games as blackjack to lower the expected values for the casino. The early morning is a slow time for casinos, and some of them attract weary but bargain-conscious bettors by letting the bettor instead of the casino win all ties.

Expected values are a reliable guide to decision making when an experiment can be repeated many times, but they may not be a good guide when an experiment cannot be repeated. Your best choice depends on how well you can afford to bear risk and your attitude toward risk. Here's an example of an experiment that cannot be repeated. Suppose you had bought $100,000 of a new issue of Long Island Lighting Company bonds that had a 20-year maturity and a coupon rate of 17⅛%. You read an article in *The Wall Street Journal* which predicts that there is a 10% chance that the utility will file for bankruptcy. You immediately call your broker and find out that you can sell the bonds for $82,000. Should you sell or should you keep the bonds? Because of the assets the utility owns, your broker estimates that if the utility goes bankrupt, the bonds can be sold for $30,000; but if the utility can finance the completion of the Shoreham nuclear plant and if the plant is allowed to operate, the bonds can probably be sold for $115,000. The reason that the bonds will sell above the par value of $100,000 is that the coupon rate of 17⅛% is above the current market interest rate. The possible selling prices of the bonds and the probabilities of bankruptcy and survival can be used to compute an expected value for the decision to hold on to the bonds. Using formula 5.1, you can expect the value of the bonds to be $106,500 if you hold on to them.

$$E(X) = \sum_{i=1}^{n} x_i P(x_i) \tag{5.1}$$

$$E(X) = (\$115,000 \cdot 0.9) + (\$30,000 \cdot 0.1)$$

$$E(X) = \$106,500$$

If you could repeat the experiment of holding the bonds many times, you would average a value of $106,500 per experiment. An expected value of $106,500 looks like a better deal than the $82,000 you can get by selling the bonds now. But you can't repeat the experiment. Either the utility will file for bankruptcy and you'll get $30,000 or it will survive the crisis and you'll get $115,000. Clearly the decision about holding the bonds depends on more than the expected value. The same choice would not be right for everyone. If you were retired and the interest on the bonds were your only source of income you'd be risk-averse and sell the bonds to limit your losses. Someone who is **risk-averse** is *unwilling to pay as much as the expected value of the possible winnings from a bet for the right to make the bet.* Someone who is **risk-neutral** is *willing to pay exactly the expected value of the possible winnings from a bet for the right to make the bet.* If you were younger and wealthier, you might be risk-seeking, in which case you would probably hold on to the bonds. Someone who is **risk-seeking** is *willing to pay more than the expected value of the possible winnings from a bet for the right to make the bet.*

Here is another example of the use of expected values. An insurance company sells one-year insurance policies to many homeowners, each of whom faces a probability of 0.001 that the house will burn down. Each home is insured for $100,000. Ignore the salesman's commission and the possibility of a partial loss. If a house does not burn down, the insurance company's cash position is the amount of the premium. If a house does burn down, the company's cash position is the premium minus $100,000. What amount should the insurance company charge as a premium to just cover the payments on houses that burn down? In other words, what premium will result in an expected value of 0 for the insurance company's cost per policy? The substitution of 0 for the expected value in formula 5.1 and $100,000 and 0.001 for the probability of a fire yields:

$$E(X) = \sum_{i=1}^{n} x_i P(x_i) \tag{5.1}$$

$$E(X) = 0$$

$$0 = \$(\text{Premium} - 100,000)0.001 + (\text{Premium})0.999$$

$$\text{Premium} = \$100.00$$

Note that the $100,000 has a negative sign because it is a loss to the insurance company. The premium level at which the insurance company would collect as much in total premiums as it had to pay out in benefits is $100. Premiums *set so that the insurance company collects as much in total premiums as it pays out in benefits* are called **actuarially fair** insurance rates. Despite such overhead costs as paperwork and sales staffs, some insurance companies charge less than actuarially fair rates on commercial fire and casualty policies. The reason for charging these low rates is that they can earn a profit by investing the premium money while they wait for fires to break out. The real risk in setting such low rates is not in underestimating the number of fires but rather in overestimating the interest that the invested premium money will earn. When many businesses are insured, the proportion that will have fires can be estimated with considerable precision, but future interest rates are notoriously difficult to forecast.

5.2B THE VARIANCE AND STANDARD DEVIATION OF A RANDOM VARIABLE

The **variance of a random variable**—*a measure of the average variation of the outcomes of a random experiment around the mean value*—can also be used in business problems. The formula for the variance of a random variable is similar to the formula for the population variance (3.8):

$$\sigma^2 = \frac{\sum_{i=1}^{v} (x_i - \mu)^2}{v} \tag{3.8}$$

PROCTER & GAMBLE'S GEMSTONE PROMOTION

In 1986 Procter & Gamble devised an extremely successful and attention-getting promotion. The company included a gemstone in each specially marked package of Spic and Span® powder, Camay® soap, Bounce® fabric softener, and Safeguard® deodorant soap. Each gemstone weighed one-third of a carat and was either a diamond, an emerald, a sapphire, or a garnet. The following statement appeared on a package of Safeguard soap:

(1) The approximate retail value of each gem is as follows: diamond—$500; emerald—$250; sapphire—$50; and garnet—$5. The total number of gems in specially marked packages is approximately 13,675,000: diamonds—1,368; emeralds—2,736; sapphires—50,596; and garnets—13,620,300. The approximate odds of winning are: diamond—1 in 10,000; emerald—1 in 5,000; sapphire—1 in 270. All packages not containing a diamond, emerald, or sapphire contain a garnet.

(2) NO PURCHASE NECESSARY FOR AN OPPORTUNITY TO WIN A DIAMOND, EMERALD, OR SAPPHIRE. For a free chance to win a diamond, emerald, or sapphire, send a self-addressed, stamped envelope to Gem Write-In Request, P.O. Box 4614, Blair NE 68009. Residents of the State of Washington ONLY need not affix postage to the return envelope. Limit one request per envelope. Requests must be received by September 14, 1986. Odds of winning are the same as in (1) above.

   The reason that Procter & Gamble had to offer a "free" chance to win a diamond, an emerald, or a sapphire at the same odds it offered the purchasers of its products is that many states consider that a lottery in which a payment has to be made for the right to obtain a chance to win is a form of illegal gambling. Of course the "free" chance offered by Procter & Gamble is not free because you have to provide your own envelopes and, unless you happen to live in the state of Washington, two 22-cent postage stamps. The time needed to mail the request also has some opportunity cost. While the above description of the promotion relies on the word "approximate" or "approximately" three times, let's assume that the stated values and odds are accurate. With this assumption, the expected value of the gems from a mail-in request net of the postage cost is about minus 15 cents: $E(X) = \$500(0.0001) + 250(0.0002) + 50(0.003704) - \$0.44$, $E(X) = \$-0.1548$. Note that the garnet is not included in the mail-in request. Even if you lived in the state of Washington, the mail-in request would not be worthwhile unless you were very risk-seeking because there would still be some cost to the envelope and your time. If you wanted to avoid the mail-in request for the gems, you could purchase one of the 13,675,000 specially marked packages. The purchase of one of these packages seems like a no-lose proposition, even if you have no use for the soap. You are assured of at least the $5 garnet when you buy one of the packages. Unfortunately, mounting the garnet in a modest ring could cost $100. To most consumers the garnet was worthless. It seems likely that if consumers had more information about the cost of mounting gemstones, the Procter & Gamble promotion would not have been so successful.

The formula for the variance of a random variable X is:

(5.2)

$$\sigma^2(X) = \sum_{i=1}^{\nu} [x_i - E(X)]^2 P(x_i)$$

For a given set of x_i's, the two formulas will result in the same number only if the x_i's have the same probability. The standard deviation of a random variable is the positive square root of the variance. The standard deviation is easier to interpret because it is measured in the same units as the random variable. In the bond example the variance would be in dollars squared and the standard deviation would be in dollars.

Here's a business application of the standard deviation of a random variable. Suppose you are considering two locations for a ski resort. You hire an expert on forecasting snowfall who gives you the following table of the probability distribution of daily snowfall during the skiing season. Note that while number of inches of snowfall is really a continuous random variable, it does little harm to treat it as discrete in this example. Both locations have an expected value of 4.5 inches of snowfall per day. Since the expected values are the same, they cannot be used to help you choose between the locations. However, the location with the more consistent level of snowfall is more desirable because it would cost less to groom the slopes for skiing. Table 5.3 tells you that location B has less dispersion or more clustering than location A. With this tighter clustering around the mean, location B has a smaller standard deviation than location A. Table 5.4 is a worksheet for calculating the standard deviations for locations A and B. The standard deviation of random variable A is 2.872 and the standard deviation of random variable B is 1.118.

	Inches of snow	Location A	Location B
TABLE 5.3 Estimated probabilities of daily snowfall at two potential ski resort locations	0	0.1	0.0
	1	0.1	0.0
	2	0.1	0.0
	3	0.1	0.25
	4	0.1	0.25
	5	0.1	0.25
	6	0.1	0.25
	7	0.1	0.0
	8	0.1	0.0
	9	0.1	0.0

TABLE 5.4

Worksheet for the standard deviation of a random variable

a_i*	$[a_i - E(A)]^2$	$P(a_i)$	$[a_i - E(A)]^2 P(a_i)$	b_i†	$[b_i - E(B)]^2$	$P(b_i)$	$[b_i - E(B)]^2 P(b_i)$
0	20.25	0.1	2.025	0	20.25	0.0	0.0
1	12.25	0.1	1.225	1	12.25	0.0	0.0
2	6.25	0.1	0.625	2	6.25	0.0	0.0
3	2.25	0.1	0.225	3	2.25	0.25	0.5625
4	0.25	0.1	0.025	4	0.25	0.25	0.0625
5	0.25	0.1	0.025	5	0.25	0.25	0.0625
6	2.25	0.1	0.225	6	2.25	0.25	0.5625
7	6.25	0.1	0.625	7	6.25	0.0	0.0
8	12.25	0.1	1.225	8	12.25	0.0	0.0
9	20.25	0.1	2.025	9	20.25	0.0	0.0
Sum			8.250				1.250
Square root of sum			2.872				1.118

*Individual value at location A.
†Individual value at location B.

EXERCISES FOR SECTION 5.2

5.6 Sertoma is a charity that raises money by running "casino nights" in which participants can bet money in return for prizes that have been donated by local merchants. The Sertoma roulette wheel has 36 numbered slots and 0, 00, 000, and 0000 slots. The payoffs are the same as on a regular roulette wheel. What is the expected value to the bettor of a $1 bet on one of the 36 numbers?

5.7 After the outbreak of the Iran–Iraq war, the insurance premiums on a policy that would pay $25 million to the owners of a supertanker lost in the Persian Gulf rose from $250,000 to $1 million for one trip through the Gulf. Assuming that the insurance rates were actuarially fair, by how much did the probability of loss of a supertanker increase?

5.8 Suppose you opened a box of breakfast cereal and found a card that could be mailed to the company which would give you a chance to win one of three cash prizes. In very fine print on the side of the box was a statement that the chance of winning the first prize of $10,000 was 1 in 1 million, the chance of winning the second prize of $1,000 was 1 in 10,000, and the chance of winning the third prize of $10 was 1 in 1,000. Suppose you valued the time it would take you to mail the card at 50 cents and the postage would cost 22 cents. If you were risk-neutral, would it be worthwhile to mail in the card?

5.9 You are a stockbroker and you are advising a client to purchase one of two stocks, labeled A and B. The following table gives the possible changes in the prices of the two stocks over the next year and the probability associated with each price change. Which stock, if either, is more desirable? What missing information could guide your choice?

Price change	Probability for stock A	Probability for stock B
−2.00	0.2	0.1
−1.00	0.2	0.1
0.00	0.2	0.6
1.00	0.2	0.1
2.00	0.2	0.1

5.10 At Indianapolis Pacers basketball games a local supermarket chain sponsors a contest with a prize of $10,000. A fan, selected at random from the crowd, is given one chance to make a basket from center court during the halftime break. To cover the possible prize money, the supermarket chain pays a fixed fee of $150 to an insurance company for each attempt. What is the maximum value that the insurance company could have assigned to the probability that a randomly selected fan would make a basket from center court?

5.11 Find the expected value of a specially marked package of Safeguard soap as described in the case study in section 5.2. Assume the garnet is worthless.

5.12 Suppose you receive a letter from the Happy Time Resort that begins "YOU MAY ALREADY HAVE BEEN SELECTED AS A WINNER OF A $10,000 DREAM VACATION." This sentence gets your attention. The rest of the letter states that there are three other prizes, and that you are guaranteed to have won one of the four prizes. The three other prizes are a $300 color television set, a $100 luggage set, and a coupon worth $5 off the price of your next dinner at a participating Howard Johnson's restaurant. To collect your prize, you and your spouse must spend a day at the Happy Time Resort, during which you will be given a presentation about Happy Time condominiums. However, you will be under no obligation to purchase a condominium.

(a) If the expected value of your prize equaled $10, what is the maximum probability of winning the $10,000 dream vacation? the color television set? the luggage set?

(b) If the expected value of your prize equaled $10 and 99% of all participating couples received the restaurant coupon as a prize, what is the maximum probability of winning the $10,000 dream vacation? the color television set? the luggage set?

5.13 You are filling out your income tax return and you are thinking about claiming the $10,000 annual cost associated with your yacht as a business expense. Your reasoning is that you have occasionally entertained clients on your yacht. Your accountant warns you that there is a 0.25 probability of being audited, and that if you are audited there is a 0.9 probability that the deduction will be disallowed. If the yacht deduction is disallowed, you will owe 33% of the $10,000 as back taxes and a fine of $10,000.

(a) Assuming that you are risk-neutral and the morality of claiming a doubtful business expense does not trouble you, should you claim the yacht as a business expense?

(b) What fine would the Internal Revenue Service have to set to make you indifferent to the choices of claiming and not claiming the deduction?

5.3 ARITHMETIC PROPERTIES OF THE EXPECTED VALUE AND VARIANCE OF A RANDOM VARIABLE

5.3A ARITHMETIC PROPERTIES OF EXPECTED VALUES

Many business problems involve more than one random variable. Suppose, for example, 45 people had dinner reservations at your restaurant. The amount spent per person is a discrete random variable. The spending of all 45 customers is the sum of 45 random variables. If you knew the expected value and the standard deviation per customer, could you assign a probability to the event "the total money spent by all of the customers is less than $400"? You could if you knew the arithmetic properties of the expected values and variances of combinations of random variables.

Expected values can be added or subtracted. Suppose the expected value of the money spent on a lunch in your restaurant was $5 and the expected value of a dinner was $10. The expected value for a lunch and dinner would be $15. This idea is expressed in mathematical notation in formulas 5.3 and 5.4:

$$E(X + Y) = E(X) + E(Y)$$

(5.3)

$$E(X - Y) = E(X) - E(Y)$$

(5.4)

Suppose there was a 1% tax on meals at your restaurant. The expected value per dinner of the revenue generated by the tax would be 1% of $10 or 10 cents. The expected value of a random variable multiplied by a constant equals the constant times the expected value.

$$E(bX) = bE(X)$$

(5.5)

In formula 5.5 b stands for any constant. If the restaurant had a cover charge, then a constant term would be added to each bill. The expected value of a constant plus a random variable equals the constant plus the expected value of the random variable.

$$E(b + X) = b + E(X)$$

(5.6)

Formulas 5.3 through 5.6 can be expanded to cover more than two random variables or more than two constant terms.

5.3B ARITHMETIC PROPERTIES OF THE VARIANCES OF RANDOM VARIABLES

The variances of random variables are additive if the random variables are statistically independent. You *cannot* use formulas 5.7 and 5.7a if the random variables are not statistically independent:

$$\sigma^2(X + Y) = \sigma^2(X) + \sigma^2(Y)$$

(5.7)

$$\sigma^2(X - Y) = \sigma^2(X) + \sigma^2(Y)$$

(5.7a)

Note that the variance of the difference between two statistically independent random variables equals the sum of the variance of the first and the variance of the second. (For a proof of formulas 5.7 and 5.7a, see Appendix C.) These formulas can be expanded to more than two random variables as long as all of the variables are statistically independent. *The standard deviations of random variables are not additive.* If you need the standard deviation of the sum of two or more random variables, first add their variances and then take the positive square root of the sum of the variances. Finally, adding a constant does not change the variance of a random variable, but multiplying by a constant does.

$$\sigma^2(b + X) = \sigma^2(X)$$

(5.8)

$$\sigma^2(bX) = b^2\sigma^2(X)$$

(5.8a)

Let's come back to the question about the probability of the 45 customers spending less than \$400 for dinners. The expected value for 45 customers would be 45 times the expected value of the amount spent by one customer. Substituting into equation 5.3, we get

$$E(X + Y) = E(X) + E(Y) \qquad (5.3)$$

$$E(X_1 + X_2 + X_3 + \cdots + X_{45}) = E(X_1) + E(X_2) + E(X_3) + \cdots + E(X_{45})$$

$$E(X_1 + X_2 + X_3 + \cdots + X_{45}) = 10 + 10 + 10 + \cdots + 10$$

$$E(X_1 + X_2 + X_3 + \cdots + X_{45}) = 450$$

If the money spent by each customer were independent of the spending of other customers, the variance of combined spending would be 45 times the variance of the spending per customer. Suppose the variance per customer was $2.25 squared. The corresponding standard deviation per customer would be $1.50. Substituting into formula 5.7, we get

$$\sigma^2(X + Y) = \sigma^2(X) + \sigma^2(Y) \qquad (5.7)$$

$$\sigma^2(X_1 + X_2 + X_3 + \cdots + X_{45}) = \sigma^2(X_1) + \sigma^2(X_2) + \sigma^2(X_3) + \cdots + \sigma^2(X_{45})$$

$$\sigma^2(X_1 + X_2 + X_3 + \cdots + X_{45}) = 2.25 + 2.25 + 2.25 + \cdots + 2.25$$

$$\sigma^2(X_1 + X_2 + X_3 + \cdots + X_{45}) = 101.25$$

The variance for 45 customers is 101.25. The standard deviation for the 45 customers is the square root of 101.25, which is 10.06. Without knowing the shape of the distribution of sales for the 45 customers, you cannot find the exact probability of having sales of less than $400. Since you know the expected value and the standard deviation, however, you can use Tchebychev's inequality to find the maximum probability for the proportion of the distribution in one or both tails or equivalently the percentage of the total area under the distribution in one or both tails. Recall Tchebychev's formula from Chapter 3:

$$P_{\text{tails}} \leq \frac{1}{k^2} \qquad (3.14)$$

The letter k stands for the number of multiples of a standard deviation between the mean or expected value and your boundary value. In this problem k equals -4.97:

$$k = \frac{x_i - \mu}{\sigma} \qquad (3.15)$$

$$k = \frac{400 - 450}{10.06}$$

$$k = -4.97$$

Substituting this value for k in equation 3.14 yields a probability of 0.04:

$$P_{\text{tails}} \leq \frac{1}{k^2} \tag{3.14}$$

$$P_{\text{tails}} \leq \frac{1}{(-4.97)^2}$$

$$P_{\text{tails}} \leq 0.04$$

The interpretation of this probability is that, at most, the probability that the total spending of the 45 customers will come to less than $400 is 0.04. With this result you should be fairly confident that the total spending will be $400 or more. Remember the assumption of statistical independence among the customers. If you looked through your restaurant window and saw a bus pull up with 45 high school band members, who generally travel on a tight budget, you might want to throw these calculations out the window.

EXERCISES FOR SECTION 5.3

5.14 X and Y are random variables with the following distributions:

$P(x_i) = 0.1$ for $x_i = 15, 16, 17, \ldots, 24$
$P(y_i) = 0.05$ for $y_i = 21, 22, 23, \ldots, 40$

What is the expected value of $X + Y$?

5.15 Using the distributions in exercise 5.14, find the expected value of $5X + (Y - 100)$.

5.16 Assume that X and Y are independent random variables. For the distributions in exercise 5.14 find:

   (a) The standard deviation of $X - Y$.
   (b) The standard deviation of $5X + (Y - 100)$.

5.17 The company you work for plans to produce 100 new business jets. Your company is considering contracting out the tail-wing assembly to another firm if it will cost less to do so than to make the tail wings itself. You have been given the job of negotiating a contract. You need to have a good idea of the in-house cost of assembling a tail wing. From company records you know that the cost of making the tail wing is five times the direct labor cost of the assembly plus an overhead cost of $1,000 per wing. Your hourly labor costs have been fixed by a union contract at $25 per hour. The engineering department gives you the following probability distribution for the hours required for assembly of one wing:

x_i	20	25	30	35	40
$P(x_i)$	0.1	0.2	0.3	0.2	0.2

Find the expected value of the costs for 100 tail wings and the standard deviation of the costs for 100 tail wings.

5.18 Refer to the information provided in exercise 5.17. Another company has offered to assemble the 100 tail wings for $100,000. Use Tchebychev's inequality to find the maximum probability that the sum of the in-house costs for 100 tail wings is greater than $100,000.

5.4 THE BINOMIAL DISTRIBUTION

5.4A CIRCUMSTANCES REQUIRED FOR THE USE OF THE BINOMIAL DISTRIBUTION

The chapter introduction mentioned that the binomial distribution was used to determine the optimal number of helicopters in the mission to rescue the hostages in Iran. If you've been wondering what a binomial distribution is, your curiosity is about to be satisfied. The **binomial distribution** is *a distribution that describes the probabilities of all of the possible outcomes of a series of experiments, when each experiment in the series is identical to every other and has two possible outcomes.* In the hostage rescue example, the experiment could be defined as sending a helicopter on a journey of a given length. The two possible outcomes are that the helicopter breaks down or that it does not break down. If eight helicopters were sent on the mission, you would have a series of eight identical experiments or trials. The binomial distribution could be used to determine the probability of any number of breakdowns from zero to a maximum of eight. For example, the binomial distribution could be used to calculate the probability that two or fewer helicopters would break down. Since situations involving experiments that have two possible outcomes are so common, you'll find that you will use the binomial distribution more often than any other discrete distribution. An experiment such as tossing a coin or asking someone out on a date has two possible outcomes. Situations with two possible outcomes abound in the business world: a part is either defective or not; an applicant accepts a job offer or not; a shopper who looks at a window display decides to come into the store or not.

In addition to the requirement that each experiment have two possible outcomes, two other requirements must be met by the *series* of experiments before the binomial distribution can be used to describe the probability of each outcome. One of these requirements is that *each experiment or trial in the series must be statistically independent.* If each helicopter was statistically independent, a breakdown by any one helicopter would not change the probability of a breakdown of any other helicopter. In the coin-toss example, each trial is statistically independent: if a coin comes up heads in one toss, that result will not affect the probability of a head on the next toss. The other requirement is that *the probability of a given outcome of an experiment must be the same in all repetitions of the experiment.* This requirement is met by the coin toss because the probability of a head stays the same as you keep tossing the coin. Suppose you kept sending the *same* helicopter on 800-mile trips without doing any maintenance work on it. As the helicopter's parts wore out, the probability of a breakdown on a trip would increase. In this case the third requirement of the binomial distribution would not be met.

5.4B FORMULA FOR THE BINOMIAL DISTRIBUTION AND THE LOGIC BEHIND IT

The formula that gives the probability of x successes in n trials for a random variable that fits the assumptions of the binomial distribution is:

$$P(x) = \frac{n!}{(n-x)!x!}\Pi^x(1-\Pi)^{n-x} \tag{5.9}$$

The letter n stands for the number of trials or repetitions of an experiment. The letter x stands for the number of occurrences of a particular outcome, such as a head in the coin toss or a breakdown of a helicopter. The letter x is also called the number of successes. The term "success" can be applied arbitrarily to either of the two outcomes, so that a helicopter breakdown, for example, can be called a success. The expression $P(x)$ means the probability of x occurrences or successes. The exclamation point means factorial; for example, 5! is $5(4)(3)(2)(1) = 120$. The capital Greek letter Π (pi) stands for the probability of a success in one trial.

To understand the reasoning behind this formula, let's solve a simple coin problem, first using a tree diagram and then the formula. The problem is to determine the probability of getting exactly two heads in three tosses of a fair coin. The tree diagram for this problem is seen in Figure 5.3. There are eight possible sequences of heads and tails in three tosses. The sample space for this series of experiments is *HHH, HHT, HTH, HTT, THH, THT, TTH, TTT*. Since the coin is assumed to be fair, every sequence is equally likely. Three of the eight sequences have exactly two heads, so the probability of getting exactly two heads in three tosses must be three-eighths or 0.375.

Now let's do it over with formula 5.9:

$$P(x) = \frac{n!}{(n-x)!x!}\Pi^x(1-\Pi)^{n-x} \tag{5.9}$$

$$P(2) = \frac{3!}{1!2!}(0.5)^2(1-0.5)^1$$

$$P(2) = 0.375$$

You can see that the formula works, but why does it work? The formula can be separated into two parts. One part gives you the probability of one sequence that contains two heads and a tail. The other part gives you the number of sequences that have exactly two heads. Using the multiplication formula for

FIGURE 5.3
Tree diagram for tossing a fair coin three times

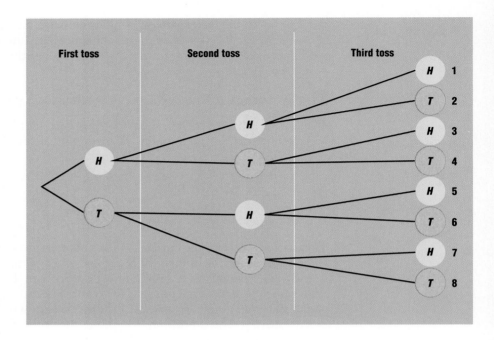

statistically independent events, we find the probability of the sequence a head on the first toss, a head on the second toss, and a tail on the third toss:

$$P(H_1 \cap H_2 \cap T_3) = P(H_1) \cdot P(H_2) \cdot P(T_3)$$

where H_1 means a head on the first toss, H_2 a head on the second toss, and T_3 a tail on the third toss. This probability is $0.5 \cdot 0.5 \cdot 0.5$, or 0.125. The part of the binomial formula that corresponds to the probability of the sequence of two heads followed by a tail is $(\Pi)^x(1 - \Pi)^{n-x}$. This sequence is just one of the three sequences that contain exactly two heads. Since all sequences with exactly two heads are equally likely and all are mutually exclusive, the probability of two heads in three trials equals the probability of one sequence with two heads times the number of sequences that contain two heads. The basis for this result is the addition formula with mutually exclusive events. The sequences are mutually exclusive because you cannot get more than one sequence out of a set of three tosses. The part of the binomial formula that corresponds to the number of sequences with two heads is $n!/[(n - x)!x!]$, which you should recognize as the formula for combinations, $_nC_x$. Let's look at equation 5.9 again and summarize what the formula does:

$$P(x) = \frac{n!}{(n - x)!x!} \quad (\Pi)^x(1 - \Pi)^{n-x} \tag{5.9}$$

NUMBER OF SEQUENCES
WITH x SUCCESSES IN n TRIALS

PROBABILITY OF ONE
SEQUENCE WITH x SUCCESSES

In our description of the binomial formula we have been assuming a fair coin, but such an assumption is not required for binomial problems. All that is necessary is that the coin be consistent, that the probability of a head be the same in all trials. Suppose the coin was not fair, that the probability of a head on any toss was 0.6 and the probability of a tail on any toss was 0.4. Under this assumption it is no longer true that all of the eight possible sequences are equally likely. The probability of three heads in a row, for example, is (0.6)(0.6)(0.6) or 0.216. The probability of three tails in a row is (0.4)(0.4)(0.4) or 0.064. Even though it is no longer true that all possible sequences are equally likely, it is still true that all sequences that contain exactly two heads are equally likely: *HHT* or (0.6)(0.6)(0.4) equals *HTH* or (0.6)(0.4)(0.6) equals *THH* or (0.4)(0.6)(0.6), or 0.144. Since each sequence with exactly two heads is as likely as any other sequence with exactly two heads and all are mutually exclusive, the probability of getting exactly two heads with this coin is three times 0.144, or 0.432.

Now let's apply the formula to the helicopter problem. Recall that the military planners sent eight helicopters on the mission. If the probability is 0.95 that a helicopter will not break down, what is the probability that six or more of the eight helicopters sent will complete the trip? Substitute 6, 7, and 8 for x in equation 5.9 to get the probability of six or seven or eight successes.

$$P(x) = \frac{n!}{(n - x)!x!}\Pi^x(1 - \Pi)^{n-x} \tag{5.9}$$

$$P(6) = \frac{8!}{2!6!}0.95^6(0.05^2) \qquad P(6) = 0.0515$$

$$P(7) = \frac{8!}{1!7!}0.95^7(0.05^1) \qquad P(7) = 0.2793$$

$$P(8) = \frac{8!}{0!8!}0.95^8(0.05^0) \qquad P(8) = 0.6634$$

The probability of six or more successes is the sum of the above three probabilities, 0.9942. These probabilities can be added because they represent mutually exclusive events. This probability would be accurate only under the assumption of normal weather conditions.

Now let's try a business example. You are the human resources manager for a large corporation. The most common type of claim filed by the company's employees under the workers' compensation system is for a back injury. Your company has had a policy of contesting back injury claims unless they can be tied to a specific accident at the workplace. Your position is that without such a tie it is impossible to prove that a back injury is work-related. You need to set aside enough money to pay for the injury claims that are upheld by the Workers' Compensation Board. The board has upheld 30% of the past claims for back injuries that could not be tied to an accident. You have 12 such cases pending before the board and you decide to reserve enough money to cover three claims. What is the probability that four claims or more will be upheld? You have to substitute the numbers 4, 5, 6, 7, 8, 9, 10, 11, and 12 for x in formula 5.9 and 0.30 for Π. Since the probabilities of all of the possible successes in 12 trials must sum to 1, you can save time by subtracting the probabilities of 0, 1, 2, and 3 from 1. The event $x \leq 3$ is the complement of $x \geq 4$.

$$P(0) = \frac{12!}{12!0!}0.3^0(0.7^{12}) \qquad P(0) = 0.0138$$

$$P(1) = \frac{12!}{11!1!}0.3^1(0.7^{11}) \qquad P(1) = 0.0712$$

$$P(2) = \frac{12!}{10!2!}0.3^2(0.7^{10}) \qquad P(2) = 0.1678$$

$$P(3) = \frac{12!}{9!3!}0.3^3(0.7^9) \qquad P(3) = 0.2397$$

The probability of 0 through 3 successes is 0.4925. The probability of 4 or more successes is 1.0 minus 0.4925, or 0.5075. A chart of the probability distribution for this problem may help your decision making because you can see how the probability of a given number of successes falls rapidly as the number increases. Figure 5.4 is a histogram that plots the number of successes

FIGURE 5.4
Histogram of binomial probability distribution when $n = 12$ and $\Pi = 0.3$

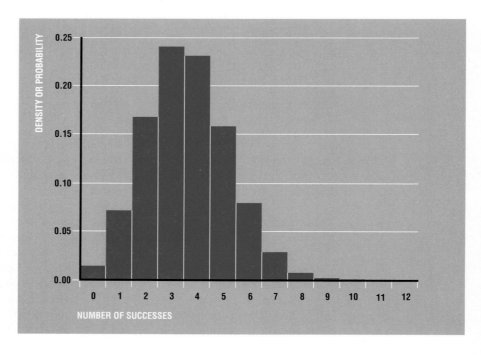

CASE STUDY

THE MAN WITH THE GOLDEN ARM

Nicholas Caputo has served as the clerk of Essex County, New Jersey, for over two decades. Part of Mr. Caputo's duties has been to hold a drawing to determine whether the Democrats or the Republicans would be listed first on the county's ballots. In the last 40 of the 41 drawings conducted by Mr. Caputo, the Democrats won the drawing. Being listed first on the ballot is considered an advantage, particularly in a close contest, because some voters tend to vote for anyone whose name heads the list. Not surprisingly, Mr. Caputo is a Democrat. Some Republicans, skeptical about his uncanny luck, have begun calling Mr. Caputo "the man with the golden arm." Essex County Republican officials sued Mr. Caputo, and in August 1985 the New Jersey Supreme Court ruled in their favor, advising Mr. Caputo to change the way he conducts the drawings to stem "further loss of public confidence in the integrity of the electoral process." The New Jersey Supreme Court relied on the binomial distribution to estimate the probability that a fair drawing would result in 40 or more successes for the Democrats in 41 elections. The court found that the probability was less than 1 in 50 billion. The court noted that, "confronted with these odds, few persons of reason will accept the explanation of blind chance."

Here are the calculations showing that the probability is less than 1 in 50 billion:

$$\Pi = 0.5, \, n = 41, \, x \geq 40$$

$$P(x) = \frac{n!}{(n-x)!x!} \Pi^x (1 - \Pi)^{n-x}$$

$$P(x \geq 40) = \frac{41!}{(41-40)!40!} 0.5^{40}(1-0.5)^1 + \frac{41!}{(41-41)!41!} 0.5^{41}(1-0.5)^0$$

$$P(x \geq 40) = 1.91 \times 10^{-11}$$

The Π of 0.5 was based on the assumption that there were two parties in each drawing, Democrats and Republicans, and that in a fair drawing each party was equally likely to be drawn first. The number 2×10^{-11} is 1 divided by 50 billion in scientific notation. The probability of 40 or more successes in 41 trials of 1.91×10^{-11} is slightly below 1 in 50 billion.

on the *x* axis and the corresponding probabilities on the *y* axis. Since the area in each rectangle in the histogram equals the proportion of the total area in the distribution represented by each number of successes, the density of the histogram is the same as the probability. The area in each rectangle corresponds to the probability of the number of successes listed on the *x* axis. Figure 5.4 applies only to the binomial distribution with 12 trials and a probability of success of 0.3 for each trial. As the value of Π gets closer to 0.5, the binomial distribution becomes more symmetrical. Also, as the number of trials increases, the distribution begins to look like a smooth curve instead of a series of rectangles.

5.4C USING A TABLE TO FIND PROBABILITIES FOR THE BINOMIAL DISTRIBUTION

You can use the binomial formula for any binomial problem, but for some problems the formula would take too much time. If you wanted to find the probability of 300,000 or more successes in 1 million trials, you would have to compute the probabilities of getting 300,000, 300,001, 300,002, 300,003, and so on out to 1 million successes and then add up these probabilities to get the probability of 300,000 or more successes. A problem like this would take a long time, even if you used the complementary event $x \leq 299,999$ or used a very fast computer. Since the binomial distribution is so important, we will explore three shortcuts that avoid computations with the binomial formula. With the formula and these shortcuts you will be able to do any binomial problem. Two of the shortcuts are approximation rules that allow you to apply other distributions to the binomial problems. In some circumstances the normal distribution provides a good approximation of the binomial and in some other circumstances the Poisson distribution provides a good approximation. We will get to these approximation rules after we discuss the normal and Poisson distributions. The third shortcut is the table for the binomial distribution in Appendix D. A section of Table D1 is reproduced in Table 5.5. The Π's on the top of the table refer to the probability of success in any one trial. The portion of the table that is reproduced here covers the case of 12 trials, that is, $n = 12$. The figures in the column headed x, at the left of the table, refer to the number of successes. For example, the probability of zero successes in 12 trials with a Π of 0.3 is 0.0138. The limitations of Table D1 are that it does not go past an n of 20 and does not cover all Π values. If you have a problem involving 20 trials or fewer and the probability of success is one of the listed Π values, the table can save you some time.

5.4D THE EXPECTED VALUE AND THE STANDARD DEVIATION OF A BINOMIAL RANDOM VARIABLE

Formulas 5.10 and 5.11, for the expected value and the standard deviation of a binomial distribution, can be derived from formulas 5.1 and 5.2, for the expected value and variance of a random variable. If you used formula 5.1 for the binomial distribution with an n of 12 and a Π of 0.3, you would find that the expected value equaled 3.6:

TABLE 5.5

Section of Table D1

						Π				
n	x	0.1	0.15	0.2	0.25	0.3	0.35	0.4	0.45	0.5
12	0	0.2824	0.1422	0.0687	0.0317	0.0138	0.0057	0.0022	0.0008	0.0002
	1	0.3766	0.3012	0.2062	0.1267	0.0712	0.0368	0.0174	0.0075	0.0029
	2	0.2301	0.2924	0.2835	0.2323	0.1678	0.1088	0.0639	0.0339	0.0161
	3	0.0852	0.1720	0.2362	0.2581	0.2397	0.1954	0.1419	0.0923	0.0537
	4	0.0213	0.0683	0.1329	0.1936	0.2311	0.2367	0.2128	0.1700	0.1208
	5	0.0038	0.0193	0.0532	0.1032	0.1585	0.2039	0.2270	0.2225	0.1934
	6	0.0005	0.0040	0.0155	0.0401	0.0792	0.1281	0.1766	0.2124	0.2256
	7	0.0000	0.0006	0.0033	0.0115	0.0291	0.0591	0.1009	0.1489	0.1934
	8	0.0000	0.0001	0.0005	0.0024	0.0078	0.0199	0.0420	0.0762	0.1208
	9	0.0000	0.0000	0.0001	0.0004	0.0015	0.0048	0.0125	0.0277	0.0537
	10	0.0000	0.0000	0.0000	0.0000	0.0002	0.0008	0.0025	0.0068	0.0161
	11	0.0000	0.0000	0.0000	0.0000	0.0000	0.0001	0.0003	0.0010	0.0029
	12	0.0000	0.0000	0.0000	0.0000	0.0000	0.0000	0.0000	0.0001	0.0002

$$E(X) = \sum_{i=1}^{n} x_i P(x_i) \tag{5.1}$$

$$E(X) = 0(0.0138) + 1(0.0712) + 2(0.1678) + 3(0.2397) + 4(0.2311)$$
$$+ 5(0.1585) + 6(0.0792) + 7(0.0291) + 8(0.0078) + 9(0.0015)$$
$$+ 10(0.0002) + 11(0.0000) + 12(0.0000)$$

$$E(X) = 3.6$$

The interpretation of this expected value for back injury claims is that if you could repeat the process of defending your company against sets of 12 claims many times and for each hearing the probability that the claim would be upheld was 0.3, then you would average 3.6 claims upheld per set of 12. Note that 3.6 is 30% of 12. The intuition behind this result is that if you have a probability of 0.3 of losing any one claim, you can expect to lose 30% of all of the claims. Thus the expected value of the binomial distribution, or equivalently the expected number of successes, is:

$$E(X) = n\Pi \tag{5.10}$$

If we use formula 5.10, $E(X) = n\Pi$, or 12(0.3), or 3.6. The intuition behind the formula for the standard deviation is not so clear. The formula for the standard deviation can be derived from formula 5.2 (see Appendix C).

$$\sigma(X) = \sqrt{n\Pi(1 - \Pi)} \qquad \text{(5.11)}$$

When n equals 12 and Π equals 0.3, the standard deviation is 1.587. Substituting into formula 5.11, we get

$$\sigma(X) = \sqrt{n\Pi(1 - \Pi)} \qquad \text{(5.11)}$$
$$\sigma(X) = \sqrt{12(0.3)(1 - 0.3)}$$
$$\sigma(X) = 1.587$$

EXERCISES FOR SECTION 5.4

5.19 If for a binomial random variable the probability of success in one trial is 0.25, what is the probability of:

(a) Eight successes in 20 trials?

(b) Eight or more successes in 20 trials?

(c) Twelve or fewer successes in 20 trials?

(d) Twelve or fewer failures in 20 trials?

5.20 X is a binomial random variable with Π equal to 0.7 and n equal to 10. Find:

(a) $P(7)$.

(b) $E(X)$ and $\sigma(X)$.

(c) $P(x > 5)$.

(d) $P(x \leq 4)$.

5.21 You run a company that conducts bicycle tours. Part of your tour service is a "sag" wagon that follows the bicyclists and picks up riders who have a mechanical breakdown or are just exhausted. Your sag wagon has room for four bicyclists and their bikes. Your experience has been that the probability of a rider using the sag wagon is 0.1. If you limit your tours to 20 riders, what is the probability that you will run short of space in your sag wagon?

5.22 Suppose you need 18 256-bit memory chips to assemble the mother board of a microcomputer. You plan to order more than 18 memory chips because you know that some of the chips may be defective.

(a) If the probability of any one memory chip's being defective is 0.05, what is the probability of getting 18 or more good chips in a shipment of 20 chips?

(b) How many chips do you need to order to have a probability of at least 0.95 of having 18 good chips? (*Hint:* This exercise cannot be done with the binomial table.)

5.23 You see a television commercial that states that "seven out of ten physicians surveyed prefer the ingredients in pain reliever X to plain aspirin." You write the company because you are curious about the survey, and to your amazement, you find that the survey was of exactly ten physicians. Suppose the population of physicians was indifferent as to the advertised product and plain aspirin, that is, 50% preferred aspirin and 50% preferred the advertised product. What is the probability that a survey of ten randomly selected physicians would find seven or more who preferred the advertised product?

5.5 THE POISSON DISTRIBUTION

 5.5A CIRCUMSTANCES REQUIRED FOR THE USE
OF THE POISSON DISTRIBUTION

The **Poisson distribution** is *a distribution that describes the probabilities of discrete, independent events spaced over time, area, volume, or length,* such as the probability of x phone calls coming into a switchboard over some time period. For the Poisson distribution to describe accurately the probability of the number of occurrences x over some time interval:

1 *Each occurrence must be independent of any other occurrence.*

2 *The expected number of occurrences must be proportional to the time period.*

If 10 phone calls are expected in a 10-minute interval, for example, then 5 must be expected in a 5-minute interval and 20 calls must be expected in a 20-minute interval. Another assumption which will make sense if you go through the derivation of the Poisson formula in Appendix C is that:

3 *The probability of more than one occurrence during an infinitesimally small interval is zero.*

The derivation involves dividing the time interval into smaller and smaller pieces until each interval follows the binomial distribution, because in an infinitesimally small interval there is either one occurrence or none. For the Poisson distribution to work, two phone calls cannot come in at the same instant. Suppose your variable was the number of customers arriving per minute at a drive-in movie. The Poisson distribution would not apply because the customers arrive in groups as each car pulls up to the entrance.

4 *Finally, all intervals of the same size must have the same expected number of calls.*

If the average number of calls in a 10-minute interval in the morning is 10 and in the afternoon is 15, the Poisson distribution will not work. Note how the Poisson differs from the binomial: For the Poisson it is possible to have any number of occurrences between zero and infinity for a time interval of a given length provided that the interval is not infinitesimally small. The binomial refers to trials, and any trial or experiment must have only two possible outcomes.

An example of the application of the Poisson distribution to occurrences spaced over an area is the number of trees damaged by acid rain per square mile of a forest. Again the assumptions of independence and proportionality must hold. Thus damage to any one tree caused by acid rain must be independent of damage to any other tree, and if you average 100 damaged trees in a square mile, you should average 200 damaged trees in two square miles. Also, all square miles of the forest should be equally likely to be damaged by acid rain. An example of the application of the Poisson distribution to events spaced over volume is the number of naval mines per kilometer cubed of ocean in the vicinity of a blockaded port. The key to all of these applications is that there must be a constant rate of occurrences expected per unit of time or area or volume.

5.5B USING THE POISSON DISTRIBUTION

Here's a business example. You own a deep-sea charter fishing boat and you advertise that if a customer does not catch a fish during a four-hour charter trip you will refund the charter fee (some of the charter boats based in Miami make such an offer). You average one fish per hour on such trips. The rate for a unit interval of time (one hour in this example) is called lambda (λ). The number of fish expected to be caught in a four-hour charter is 4 times λ, or 4. The symbol for the expected number of occurrences over the interval that the problem refers to is μ. Now you can solve for the probability that you will have to refund the charter fee to a customer, that is, the probability that no fish will be caught in a four-hour charter. Note that to use the Poisson distribution, the probability of catching one fish must be independent of the occurrence of catching another fish. If you fished for a type of fish that swam in schools, such as mackerel, this assumption would be violated. If you went after sailfish or another fish that swam alone, then the assumption would be reasonable. Given the large number of fish in the sea, the number you catch during a charter will not change the expected number of fish that you can catch per four-hour charter. Finally, the probability of catching two fish at the same time is zero. The formula for the Poisson distribution is:

$$P(x) = \frac{\mu^x e^{-\mu}}{x!}$$

(5.12)

$P(x)$ refers to the probability of x occurrences. The letter μ is the expected number of occurrences in the interval specified in the problem. The letter e stands for the base of the natural logarithm, approximately 2.718. (For a proof of this formula, see Appendix C.) Substituting a μ of 4 and an x of 0 in formula 5.12 gives the probability of catching 0 fish during a 4-hour charter in which the average number of fish caught is 4.

$$P(x) = \frac{\mu^x e^{-\mu}}{x!} \qquad (5.12)$$

$$P(0) = \frac{4^0 e^{-4}}{0!}$$

$$P(0) = \frac{(1)e^{-4}}{(1)}$$

$$P(0) = 0.0183$$

Recall that 0! equals 1 and that any number taken to the power 0 equals 1. The probability of 0 occurrences is 0.0183, or approximately 2%. If the offer of a money-back guarantee attracts enough customers, giving approximately 2% of the customers their money back may be worthwhile.

Here's a second example that uses the Poisson distribution. A machine that winds steel cables for bridges averages three snags per mile of cable. If a bridge requires 0.26 mile of cable, what is the probability that the machine will wind this length without a snag? In this example a unit interval is a mile and λ equals 3. The mean number of occurrences in the interval considered in the problem, μ, is 0.26λ, or 0.78. Substituting this value into formula 5.12, we get

$$P(x) = \frac{\mu^x e^{-\mu}}{x!} \qquad (5.12)$$

$$P(0) = \frac{0.78^0 e^{-0.78}}{0!}$$

$$P(0) = \frac{(1)e^{-0.78}}{1}$$

$$P(0) = 0.46$$

The probability that a length of cable of 0.26 mile will have 0 snags is 0.46.

5.5C CHOOSING BETWEEN THE POISSON AND BINOMIAL DISTRIBUTIONS

Sometimes you may have difficulty deciding whether a random variable follows the binomial or the Poisson distribution. The same problem can be described as a set of trials or as occurrences per period of time. Suppose 100 trucks per hour went over a certain route and an average of 5 trucks per hour had a breakdown. Should each truck be considered a separate trial and the binomial distribution applied, or should the Poisson distribution with a mean rate of 5 occurrences per hour be used? Use the binomial distribution. In a binomial process the number of nonoccurrences can be computed; in this example, 95 trucks on average do *not* break down. In a Poisson process, the number of nonoccurrences is infinite. In the switchboard example, you could either say that the number of calls that were *not* made is infinite or that the question is meaningless. In the fishing example, the number of fish in the ocean that were *not* caught during a four-hour charter could be considered infinite.

■ 5.5D THE EXPECTED VALUE, STANDARD DEVIATION, AND GRAPH OF THE POISSON DISTRIBUTION

The expected value of a Poisson random variable is μ and the standard deviation is the square root of μ. The distribution becomes more symmetrical as μ increases and it also looks more like a smooth curve. The probability distributions for the Poisson where $\mu = 4$, $\mu = 10$, and $\mu = 20$ are plotted in Figures 5.5, 5.6, and 5.7, respectively.

Table D2 in Appendix D lets you save the time otherwise needed to compute Poisson probabilities provided that the value of μ you need is given in the table. The table gives selected values of μ in the range 0.1 to 20. As the table is

FIGURE 5.5
Histogram of Poisson distribution with mean of 4

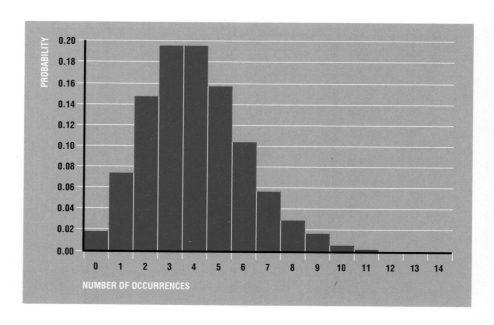

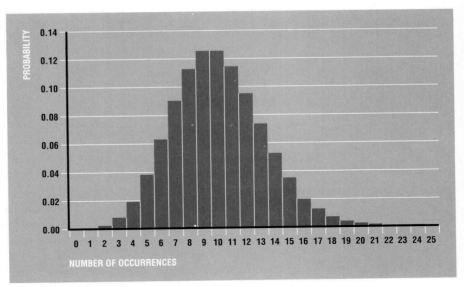

FIGURE 5.6
Histogram of Poisson distribution with mean of 10

FIGURE 5.7
Histogram of Poisson distribution with mean of 20

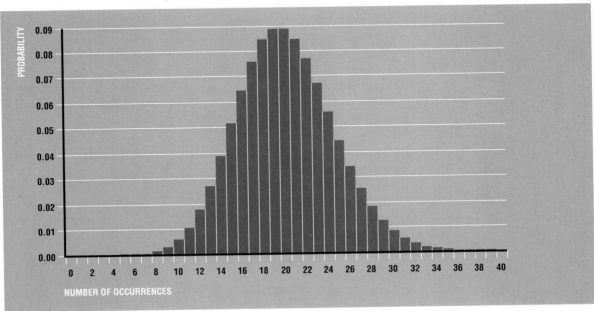

easy to read, it is not necessary to reproduce a portion of it here and explain its use. In some circumstances (a sufficiently large μ) the Poisson distribution can be approximated by the normal distribution. This topic will be described in Chapter 6, after we have discussed the normal distribution.

EXERCISES FOR SECTION 5.5

5.24 If X is a random variable that follows a Poisson distribution with a mean of 10, what is the probability that the number of occurrences is less than 5?

5.25 If X is a random variable that follows a Poisson distribution with a mean of 15, what is the probability that the number of occurrences is greater than or equal to 7?

5.26 You own a construction company and you are leveling a large area of land that you plan to use for single-family homes on quarter-acre plots. A survey of the area indicates that there are large rocks scattered under the ground, with an average of three rocks per acre. These rocks must be removed before homes can be built. You would like to set up a model home before hiring someone to remove the rocks. What is the probability that a randomly selected quarter-acre plot will have no rocks?

5.27 Suppose you own a shirtmaking company. The weaving machines that make the oxford cloth used in your shirts cannot make perfect cloth because some knots or blemishes occur whenever the yarn snags or tears. The cloth averages one knot or blemish per square yard. Your standard for rejecting a two-square-yard piece of cloth needed for making one shirt is four or more blemishes. What proportion of all two-square-yard pieces will be accepted?

5.28 Suppose you own a factory that uses lead in its manufacturing process. The Occupational Safety and Health Administration (OSHA) limits the amount of airborne lead in a factory to five particles per cubic meter of air. You are about to be inspected by OSHA and you know that you have an average of four particles of lead per cubic meter of air. What is the probability that OSHA will find that your air quality violates the standard, assuming that the inspector tests two cubic meters of air?

5.6 THE POISSON APPROXIMATION TO THE BINOMIAL DISTRIBUTION

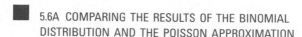

5.6A COMPARING THE RESULTS OF THE BINOMIAL DISTRIBUTION AND THE POISSON APPROXIMATION

The binomial distribution can be approximated by the Poisson distribution in some circumstances. The approximation gets better as n, the number of trials, increases and the further Π is from 0.5. Let's take a large n and a small Π as an example. Suppose you have 100 trials and the probability of success in each trial is 0.01. To find the probability of zero successes, you can substitute into the binomial formula, number 5.9:

$$P(x) = \frac{n!}{(n-x)!x!}\Pi^x(1-\Pi)^{n-x} \tag{5.9}$$

$$P(0) = \frac{100!}{100!0!}(0.01^0)(0.99^{100})$$

$$P(0) = (1)(1)(0.99^{100})$$

$$P(0) = 0.366$$

The answer is 0.366. To use the Poisson approximation, set μ equal to the expected value of the binomial, $n\Pi$. In this example μ equals 1. Consulting Table D2, you find that the probability of 0 occurrences with a μ of 1 is 0.3679. You can see that the approximation is close but not perfect.

5.6B THE APPROXIMATION RULE FOR APPLYING THE POISSON TO BINOMIAL PROBLEMS

A good rule of thumb is that the Poisson approximation to the binomial distribution is used when n is greater than or equal to 20 *and* Π is less than 0.05 or greater than 0.95. Figures 5.8, 5.9, and 5.10 show the Poisson approximation to the binomial when Π equals 0.5, 0.25, and 0.05, respectively. In each case the number of trials for the binomial is 20. The black columns represent the

FIGURE 5.8

Poisson approximation to binomial when $\Pi = 0.5$ and $n = 20$

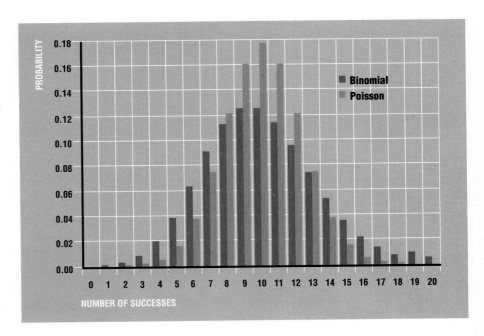

FIGURE 5.9
Poisson approximation to binomial when $\Pi = 0.25$ and $n = 20$

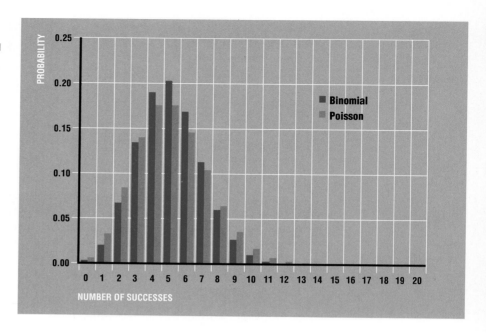

FIGURE 5.10
Poisson approximation to binomial when $\Pi = 0.05$ and $n = 20$

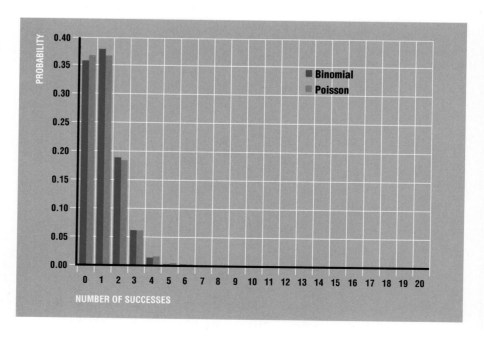

probabilities for the binomial distribution and the white columns represent the probabilities for the Poisson approximation. A bar chart is used instead of a vertical line chart so that it will be easier to keep track of the two distributions. You can see that the fit of the Poisson to the binomial improves as the Π value declines.

Let's try a business example. You are a candy wholesaler and you buy candy bars in boxes of 1,000 bars. Your experience is that 2 of every 1,000 bars have to be thrown out because the outer wrappers are missing. One of your workers tells you that 10 candy bars have been discarded from a box of 1,000 because of missing wrappers. You suspect that the worker has been eating or stealing some candy bars. What is the probability of having 10 or more unwrapped bars in a box of 1,000? In this example μ equals $n\Pi$, or 1,000(0.002). This works out to a μ of 2. If you check Table D2, you will find that the probability of 10 or more occurrences when μ is 2 is 0. This result suggests that something is going on, because if the true population proportion is 0.002 and each candy bar is an independent trial, you should not get 10 bad bars in 1,000.

You may be wondering why in the computer age it is necessary to discuss the Poisson approximation to the binomial. After all, a computer supposedly can do a binomial problem very quickly. If the problem is large enough, however, the computer can be tied up for days. Say that you had 10 million trials and the probability of success in any trial was 1 in 1 million. Determining the probability of some range of successes, such as 10 to 1,000, would be a huge problem if you worked it out directly with the binomial formula. Instead, you could look the answer up in the Poisson table under a μ of 10.

EXERCISES FOR SECTION 5.6

5.29 Let X be a binomial random variable with a Π of 0.01. Find the probability that x is less than 3 when n equals 100.

5.30 Let X be a binomial random variable with a Π of 0.99. Find the probability that x is greater than 97 when n equals 100.

5.31 You are the vice-president of a gambling casino charged with the responsibility of avoiding counterfeit money. You know that one-tenth of 1% of all $1 bills in circulation are counterfeit. In a day's operation you have accumulated 10,000 $1 bills. There are 15 counterfeit bills among the 10,000. Should you warn your tellers that someone is trying to pass counterfeit $1 bills in the casino? In other words, what is the chance that 15 or more counterfeit $1 bills would randomly turn up in a set of 10,000 $1 bills?

5.32 You manage a tire factory and you know that the probability of any one tire's being out of round is 0.02. You are worried that your inspectors are not checking carefully for out-of-round tires. In a batch of 500 tires they have found 4 bad tires. What is the probability of having 4 or fewer bad tires out of 500?

5.33 A factory that makes toaster ovens checks each oven as it comes off the assembly line. A proportion of defective ovens of 0.003 or less is considered acceptable. If the proportion of defective ovens is any higher than 0.003, the assembly line is supposed to be stopped so that the cause or causes of the defects can be corrected. Suppose the assembly line was so adjusted that the probability of any oven's being defective was 0.001. If 1,000 ovens were tested, what is the probability that the assembly line would be ordered to stop to correct a mistake, that is, that 3 or more ovens would be defective?

5.7 THE HYPERGEOMETRIC DISTRIBUTION   Skip Go to page 168

5.7A CIRCUMSTANCES REQUIRED FOR THE USE OF THE HYPERGEOMETRIC DISTRIBUTION

The hypergeometric distribution is closely related to the binomial distribution. Both are *discrete distributions of events that have two possible outcomes for each trial*. With the binomial, the probability of a success is the same for every trial; the **hypergeometric distribution,** however, is *a distribution that describes the probability of any number of successes in n trials when the trials are drawn without replacement from a finite population.* Here's an example that will clarify the difference. A gumball machine has 100 gumballs, of which 10 are licorice-flavored. You have 5 pennies and you are going to spend them on gumballs. What is the probability that you will get exactly 5 licorice-flavored gumballs? The probability of drawing a licorice-flavored gumball on the first try is 0.1, because 10 out of the 100 are licorice. If you return your gumball to the machine after each try and the gumballs are mixed, the probability will always be 0.1 and the binomial distribution fits the problem; but if you are sampling without replacement from a finite population, the hypergeometric distribution applies. If you are successful on the first try, only 9 licorice gumballs are left out of the remaining 99, so the probability on the second try is no longer 0.1. Given a success in the first try, the probability of drawing a licorice gumball on the second try is 9 divided by 99. Given successes on the first two tries, the probability of a licorice gumball on the third try is 8 divided by 98, and so on. The probability of 5 licorice gumballs in a row is:

$$\left(\frac{10}{100}\right)\left(\frac{9}{99}\right)\left(\frac{8}{98}\right)\left(\frac{7}{97}\right)\left(\frac{6}{96}\right)$$

or 0.0000033. This result is based on use of the multiplication formula for a series of dependent events. The formula for the hypergeometric distribution gives the same answer.

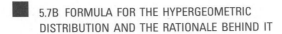

5.7B FORMULA FOR THE HYPERGEOMETRIC DISTRIBUTION AND THE RATIONALE BEHIND IT

The formula for the hypergeometric distribution is:

$$P(x) = \frac{{}_\gamma C_x \; {}_{(\nu-\gamma)}C_{(n-x)}}{{}_\nu C_n}$$

(5.13)

The letter n, as in the binomial distribution, refers to the number of trials. In the gumball example there are 5 trials, one for each penny. The x is also the same; it refers to the number of successes. The problem is to determine the probability of exactly 5 successes where a success is drawing a licorice gumball. The ν or nu refers to the number of elements in the population. There are 100 gumballs in the machine. The γ or gamma refers to the number of successes in the population. There are 10 licorice gumballs in the population.

Let's substitute the values from the gumball problem in formula 5.13 and then discuss the logic behind the formula:

$$P(x) = \frac{{}_\gamma C_x \; {}_{(\nu-\gamma)}C_{(n-x)}}{{}_\nu C_n}$$

(5.13)

$$P(5) = \frac{{}_{10}C_5 \; {}_{90}C_0}{{}_{100}C_5}$$

$$P(5) = 0.0000033$$

After some arithmetic you'll get the answer: 0.0000033. The denominator in the fraction represents the total number of possible samples of size 5 drawn from a population of size 100. If you have exactly 5 successes in five trials, you must have 0 failures. The first term in the numerator represents the number of ways in which 5 licorice gumballs can be drawn from the 10 in the population, and the second term in the numerator represents the number of ways that 0 failures, or nonlicorice gumballs, can be drawn from the 90 failures in the population. To see the intuition behind the hypergeometric formula, let's take another look at the formula labeled so that the parts are clear. Figure 5.11 shows formula 5.13 with labels to indicate these three parts. Another way to describe the hypergeometric formula is to say that the numerator is the number of samples of size n containing x successes and the denominator is the number of samples of size n that can be drawn from a population of size ν.

It is certainly easier to work with the multiplication formula than to use the hypergeometric formula for the gumball problem. But if the observations drawn are not all successes or all failures, the hypergeometric formula provides a much easier way of solving the problem than using the probability rules

FIGURE 5.11
Explanation of formula 5.13

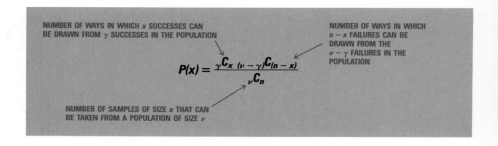

directly. Let's try to find the probability of getting 2 licorice balls in a sample of size 5. Again substituting in formula 5.13, we get

$$P(x) = \frac{{}_\gamma C_x \; {}_{(\nu-\gamma)}C_{(n-x)}}{{}_\nu C_n}$$

(5.13)

$$P(2) = \frac{{}_{10}C_2 \; {}_{90}C_3}{{}_{100}C_5}$$

$$P(2) = \frac{10!/8!(2!) \; 90!/87!(3!)}{100!/95!(5!)}$$

$$P(2) = \frac{45(117,480)}{75,287,520}$$

$$P(2) = 0.07$$

The answer is 0.07. There are $_{10}C_2$ or 45 ways to pick the 2 licorice gumballs out of the 10. There are $_{90}C_3$ or 117,480 ways to pick the 3 failures out of the 90 in the population. Thus there are 45 times 117,480 ways to pick a sample of size 5 that has exactly 2 licorice gumballs and 3 nonlicorice gumballs. The total number of samples of size 5 that are possible is $_{100}C_5$, or 75,287,520. The number of samples with 2 licorice gumballs divided by the number of samples of size 5 gives you the probability of 0.07. Note that this result uses the a priori definition of probability, that is, all samples of size 5 are equally likely.

Let's try another example. You work for Levi Strauss and Company and you are in charge of the program to prevent imitation Levi® blue jeans from being sold as genuine Levis. You have found some Hong Kong–made copies of your product which look very much like the originals. You want to see if your retailers can distinguish the fake jeans from the real ones. You plan to ask some retailers to try to identify 4 fakes in a set of 10 jeans. Find the probability of randomly selecting 2 fakes in a sample of 4. The denominator is the number of samples of size 4 that can be drawn from a population of 10. The first combination is the number of ways of selecting 2 successes from the 4 successes in the population. The second combination is the number of ways 2 failures can be selected from the 6 failures in the population. Substituting in formula 5.13, you get

$$P(x) = \frac{_{\gamma}C_x \; _{(\nu-\gamma)}C_{(n-x)}}{_{\nu}C_n} \tag{5.13}$$

$$P(2) = \frac{_4C_2 \; _6C_2}{_{10}C_4}$$

$$P(2) = \frac{[4!/(4-2)!2!] \; [6!/(6-2)!2!]}{10!/(10-4)!4!}$$

$$P(2) = 0.429$$

The answer is 0.429.

5.7C EXPECTED VALUE, VARIANCE, AND TABLE OF THE HYPERGEOMETRIC DISTRIBUTION

Table D10 provides a shortcut for solving hypergeometric problems. The table gives the individual probabilities for the hypergeometric distribution of populations of 20 or fewer. The columns of the table are headed n, γ, x, $P(x)$, and Cumulative $P(x)$. Suppose the population had 20 elements and there were 10 successes in the population. What is the probability of getting 3 or fewer successes in a sample of size 5? Looking up the respective columns for these figures, you find the answer under the cumulative probability column: 0.8483.

The expected value of the hypergeometric is the same as the binomial, $n\Pi$, where Π is the proportion of successes in the population.

$$E(X) = n\Pi \tag{5.10}$$

The variance for the hypergeometric distribution is:

$$\sigma^2(X) = n\Pi(1-\Pi)\frac{\nu-n}{\nu-1} \tag{5.14}$$

If the sample is a small proportion of the population, the term $(\nu-n)/(\nu-1)$ will be very close to 1. In this case the variance of the hypergeometric is the same as the variance of the binomial. If the sample size is under 5% of the population, the binomial provides a good approximation to the hypergeometric. The term $(\nu - n)/(\nu - 1)$ is the finite population correction factor. The discussion of the finite population correction factor will be deferred until Chapter 7 because it is easier to explain in the context of sampling distributions. For the time being it is sufficient to mention that the binomial distribution applies to either infinite populations or finite populations sampled with replacement, and the hypergeometric applies to finite populations sampled without replacement. This is why this term is called the finite population correction factor.

EXERCISES FOR SECTION 5.7

5.34 If a population has 10 type A's and 10 type B's, what is the probability of drawing 4 or more type A's in a sample of size 5?

5.35 If a population has 10 type A's and 70 type B's, what is the probability of drawing 4 or more type A's in a sample of size 6?

5.36 You are an attorney defending three men accused of robbing a bank. Your clients have been placed in a police lineup with three decoys. A witness to the robbery identified two of your clients and one of the decoys as bank robbers. At the trial you want to show the jury that this identification could easily be the result of a chance selection. What is the probability that a random selection of three of the six men in the lineup would result in the picking of two of your clients and a decoy?

5.37 You are the scoutmaster for a Boy Scout troop. You are planning a camping trip and you tell one of your scouts to check each of the batteries in a box of 20 with a battery tester. You need 6 good batteries for a lantern. Your scout checks the batteries and reports that 5 of the 20 are bad. Unfortunately, he does not separate the good and bad batteries. You do not realize that the 5 bad batteries are mixed in with the 15 good ones. What is the probability that when you randomly select 6 batteries for the lantern, 1 or more of the 6 will be bad?

5.38 You own a company that rebuilds septic tanks. You plan to run a door-to-door selling campaign in an area that is known to have problems with its septic tanks. The county sanitation authorities have taken an aerial infrared survey and identified 200 homes out of the 1,000 in the area as having leaking septic tanks. You don't know which homes are the ones with the bad tanks. If your salesmen randomly select 25 homes for a sales presentation, what is the probability that 6 or more of them will have leaking septic tanks?

 SUMMARY

The chapter began by describing a random variable as one whose numerical values represent all of the possible outcomes of an experiment. Discrete random variables take only certain values and continuous random variables take any value over a range. The concepts of expected value and standard deviation of a random variable have many business applications, such as insurance pricing and decisions about the purchase of risky assets. Many business problems involve combinations of random variables, so the next section discussed the rules for the expected values of sums and differences of random variables and for the variances of sums and differences of random variables. The binomial distribution is a convenient way to find the probabilities associated with experiments that have two possible outcomes when the probability of "success," which is arbitrarily defined as one of the two outcomes, is the same in every repetition of the experiment. As these assumptions are often found to fit business problems, the binomial is the most frequently used discrete probability distribution. The Poisson distribution applies to experiments that can result in a discrete number of occurrences measured over volume or time or length or area. The Poisson distribution fits experiments in which no two occurrences can come at the same time or location, each occurrence is statistically independent, and the expected number of occurrences is proportional to the volume or time or length or area. The Poisson provides a good approximation to the binomial when Π is close to 0 or 1 and n is large. The suggested rule of thumb is to use the Poisson to approximate the binomial when n is greater than or equal to 20 and Π is either less than 0.05 or greater than 0.95. The hypergeometric distribution applies to experiments that have two possible outcomes, but in contrast to the binomial, the successes are being drawn without replacement from a finite population with a fixed number of successes and failures. Consequently, with the hypergeometric distribution the probability of drawing a success is not constant across trials.

SUGGESTED READING

Larson, H. J. *Introduction to Probability Theory and Statistical Inference*. New York: Wiley, 1969.

NEW TERMS

actuarially fair Of insurance premiums, set so that the insurance company collects as much in total premiums as it pays out in benefits.

binomial distribution A distribution that describes the probabilities of all of the possible outcomes of a series of experiments, when each experiment in the series is identical to every other and has two possible outcomes.

continuous random variable A random variable that can take any value over a continuous range of values.

discrete random variable A random variable whose numerical values are limited to specific values within its range.

expected value The average value of a random variable.

hypergeometric distribution A distribution that describes the probability of any number of successes in n trials when the trials are drawn without replacement from a finite population.

Poisson distribution A distribution that describes the probabilities of discrete, independent events spaced over time, area, volume, or length.

probability distribution A list of all of the possible values of the random variable and the associated probabilities.

random variable A function that assigns numerical values to the possible outcomes of an experiment.

risk-averse Unwilling to pay as much as the expected value of the possible winnings from a bet for the right to make the bet.

risk-neutral Willing to pay exactly the expected value of the possible winnings from a bet for the right to make the bet.

risk-seeking Willing to pay more than the expected value of the possible winnings from a bet for the right to make the bet.

variance of a random variable A measure of the average variation of the outcomes of a random experiment around the mean value.

 NEW FORMULAS

Expected value

$$E(X) = \sum_{i=1}^{n} x_i P(x_i) \tag{5.1}$$

Variance of a random variable

$$\sigma^2(X) = \sum_{i=1}^{\nu} [x_i - E(X)]^2 P(x_i) \tag{5.2}$$

Expected value of the sum of two random variables

$$E(X + Y) = E(X) + E(Y) \tag{5.3}$$

Expected value of the difference between two random variables

$$E(X - Y) = E(X) - E(Y) \tag{5.4}$$

Expected value of a constant times a random variable

$$E(bX) = bE(X) \tag{5.5}$$

Expected value of a constant plus a random variable

$$E(b + X) = b + E(X) \tag{5.6}$$

Variance of the sum of two independent random variables

$$\sigma^2(X + Y) = \sigma^2(X) + \sigma^2(Y) \tag{5.7}$$

Variance of the difference between two independent random variables

$$\sigma^2(X - Y) = \sigma^2(X) + \sigma^2(Y) \tag{5.7a}$$

Variance of a constant plus a random variable

$$\sigma^2(b + X) = \sigma^2(X) \tag{5.8}$$

Variance of a constant times a random variable

$$\sigma^2(bX) = b^2\sigma^2(X) \tag{5.8a}$$

Binomial distribution

$$P(x) = \frac{n!}{(n - x)!x!}\Pi^x(1 - \Pi)^{n-x} \tag{5.9}$$

Expected value of a binomial random variable and a hypergeometric random variable

$$E(X) = n\Pi \tag{5.10}$$

Standard deviation of a binomial random variable

$$\sigma(X) = \sqrt{n\Pi(1 - \Pi)} \tag{5.11}$$

Poisson distribution

$$P(x) = \frac{\mu^x e^{-\mu}}{x!} \tag{5.12}$$

Hypergeometric distribution

$$P(x) = \frac{{}_{\gamma}C_x \; {}_{(\nu-\gamma)}C_{(n-x)}}{{}_{\nu}C_n} \tag{5.13}$$

Variance of a hypergeometric random variable

$$\sigma^2(X) = n\Pi(1 - \Pi)\frac{\nu - n}{\nu - 1} \tag{5.14}$$

ADDITIONAL EXERCISES

An asterisk designates an exercise of greater than ordinary difficulty.

5.39 Your employer offered you two medical insurance plans, A and B. The A plan has a deductible of $400 and a copayment of 20% for all expenses between $400 and $250,000. The B plan has no deductible and no copayment. The maximum annual payment is $250,000. Plan A is entirely paid by your employer and plan B costs you $600 a year. If your medical bills averaged $1,000 over the past ten years, which plan should you choose? Would your willingness to bear risk affect this choice?

5.40 Futures markets exist for many commodities, such as wheat and corn. A "future" is a contract to deliver a specified amount of a commodity at a specified price on a specified date. A 30-day corn future, for example, might require 100,000 bushels of corn to be delivered to a grain elevator at the end of 30 days for $3.50 a bushel. Suppose you had just sold a contract with the above terms and you did not own any corn. Your plan is to wait 30 days and buy corn on the spot market. Commodities traded on the "spot market" are bought and sold for immediate delivery. You decide to enter into such a contract because you expect that the spot market price at the end of 30 days will be below $3.50. If at the end of 30 days you can buy corn at any price below $3.50, you can pocket the difference between the spot price and $3.50. If the spot price is above $3.50, however, you will have to make up the difference out of your pocket. Given the following subjective probabilities for the price of corn on the due date of the contract, what is the contract worth to you?

Price	2.00	2.25	2.50	2.75	3.00	3.25	3.50	3.75	4.00
Probability	0.05	0.05	0.10	0.10	0.40	0.10	0.10	0.05	0.05

5.41 You are a bookie who takes bets on football games. For each game you establish a point spread; for example, Michigan 10 points over Ohio State. If a bettor picks Michigan over Ohio State for $25, you will have to pay the bettor $50 if Michigan beats Ohio State by more than 10 points. If a bettor takes the Ohio State side of the bet and Ohio State either wins or loses by fewer than 10 points, you will have to pay that bettor back the money he or she bet plus an additional dollar for each dollar bet. If the spread in the scores is exactly 10 points, you get to keep all the money bet. The art of bookmaking consists of setting point spreads that cause as much money to be placed on one side of a bet as on the other. If this pattern of betting develops, then in effect the losers wind up paying the winners and the bookmaker faces no risk (except perhaps from the police) and has a chance of keeping all the money bet. Suppose you had offered a 10-point spread in favor of Michigan and you had collected $1,000 for the Michigan side of the bet and $2,000 for the Ohio State side of the bet. Your subjective probabilities of the outcome of the game are 0.4 that Michigan will win by more than 10 points, 0.1 that Michigan will win by exactly 10 points, and 0.5 that the spread will be less than 10 points in Michigan's favor. What is the expected value to you of these bets?

5.42 You are a dentist and you schedule appointments during an eight-hour day at half-hour intervals, so you see 16 patients a day. Your experience has been that 15% of your patients miss their appointments. To make sure that you use all of your office time, you plan to schedule three extra patients at the beginning of the day and make them wait until you have an open time slot. What is the probability that you will have 16 or fewer patients in one day? What is the probability that you will run short of time slots? Can you think of reasons why the assumption of independence in regard to patients may not be reasonable?

5.43 You work for a company that makes cold remedies and you want to test a new cough medicine. You would like to get the opinions of at least 10 people who have sore throats. To establish the impact of your new medicine, it is necessary that it be taken from the onset of the sore throat. Thus you cannot survey people and ask them if they now have a sore throat; you have to give the medicine to healthy people and hope that at least 10 of them will develop sore throats within a reasonable time. Suppose you knew that the probability that a randomly selected healthy person will develop a sore throat in a 12-week period during the winter was 0.7. How many people should you include in your sample to have a probability of 0.72 that 10 or more of them will get sore throats over a 12-week period?

5.44 The same drug company as in exercise 5.43 has a new medicine for rashes caused by poison ivy. Again the company wants to give the medicine to a group of healthy individuals and observe the reactions of anyone who uses the drug from the onset of the rash. The probability that a randomly selected healthy person will contract poison ivy over a 12-week period in the summer is 0.001. How many people should be included in the sample so that the probability of getting at least one case of poison ivy is equal to 0.99?

5.45 You are a certified public accountant and you feel that the Internal Revenue Service is deliberately harassing you by frequently auditing tax returns that you have prepared. Ten of the last 20 returns you prepared have been audited. If the Internal Revenue Service audits 5% of all returns, what is the probability that a random selection would have resulted in audits of 10 or more tax returns of 20 you prepared?

5.46 Your medical insurance company offered you a "catastrophic" insurance plan. Your present coverage is limited to $250,000 a year. The catastrophic plan covers expenses between $250,000 and $1 million. The price of this insurance is only $400. Before buying the insurance, you check on the probability of having expenses above $250,000. Suppose only one of every 10,000 insured individuals incurs medical expenses in excess of $250,000 in a year. Should you buy the catastrophic insurance? Explain the reason or reasons behind your answer.

5.47 You are the admissions officer at a university. Your university has room for a freshman class of 5,000. When you admit students, you know that some of them will not attend your university. Your experience has been that the probability of an admitted student attending your university is 0.75. To compensate for this slippage, you plan to admit more than 5,000 students. If you admit 6,500 students, what is the probability that more than 5,000 will accept? (Use Tchebychev's inequality for this problem; in Chapter 6 we'll discuss the same problem with the normal approximation to the binomial.)

5.48 You are making a chestnut stuffing for a turkey. Your recipe calls for 150 chestnuts. Your experience has been that one of every ten chestnuts is spoiled. If you buy

200 chestnuts, what is the probability that you will have enough for your stuffing? (Use Tchebychev's inequality for this problem; in Chapter 6 we'll cover the same problem with the normal approximation to the binomial.)

5.49 You own a barbershop that contains four barber chairs. If a haircut takes 30 minutes and your barbershop averages four customers per half-hour interval, what is the probability that one or more chairs will be empty for a half-hour? Assume that the customers arrive independently.

5.50 You run a city ambulance service. You have ten ambulances on duty. You average five calls per hour. If an ambulance is called, it is usually occupied for an hour. What is the probability that in a one-hour period you will get more than ten calls? Why might the Poisson distribution *not* be appropriate for this problem?

5.51 The number of deliveries of babies at a hospital averaged eight during a 24–hour stretch. Consider twins, triplets, or more siblings as one delivery. Assume that each delivery is independent. For what time period would the probability of exactly one delivery equal 0.1?

5.52 You operate a drawbridge on a river. Ships have the right of way over traffic crossing the bridge. Whenever a ship that is too tall to pass beneath the bridge approaches, you must stop all traffic over the bridge and lift the drawbridge until the ship has passed. You average one ship per hour. If you left your post and a ship crashed into the bridge, you would be in a great deal of trouble. A ship has just passed by and you really want to take a five-minute coffee break. What is the probability that no ships will come in the next five minutes?

5.53 You manufacture video tapes. Dust particles on the tape can distort the images recorded on the tape and damage the recording and playback heads of a video tape player. Your factory maintains a "clean room" environment to limit the amount of dust in the air, but you cannot reduce the dust to zero. You average 0.001 particle of dust per foot of video tape. Each 200-foot-long cassette tape is checked for dust particles, and if there are four or more particles in the cassette, it is destroyed. Assume that the occurrence of dust particles is independent. What proportion of all tapes are destroyed?

5.54 You own a pizza parlor. You get orders for 500 pizzas a day. The probability that any one pizza will be ordered with anchovies is 0.01. You have enough anchovies for only ten pizzas. What is the probability of running out of anchovies?

5.55 Telephone lines have electronic "noise" that can interrupt or distort transmissions. During conversations the noise is usually not bothersome, but when data are being transmitted between computers, the noise can change characters and ruin programs or information sent by phone. You want to transmit a data file that has 10,000 characters (letters or numbers) and the probability that any character will be changed by noise is 0.00003. What is the probability that the file will be transmitted with no errors?

5.56 The probability that a gem-quality diamond is flawless is 0.02. The term "flawless" means that no particles are visible within the diamond when it is observed with 10-power magnification. The term "gem quality" means that no defects are visible to the naked eye, the color of the diamond is sufficiently close to clear or "white," and the weight is above 0.25 carat. You are a diamond wholesaler and you have bought a bag of 5,000 unsorted gem-quality diamonds. After inspecting each diamond, you discover that there are only four flawless stones. You suspect that the bag was not unsorted; perhaps the seller removed some of the flawless stones and substituted stones with flaws. What is the probability of observing four or fewer flawless stones if the bag was unsorted?

5.57 You are a county sheriff and part of your job is to repossess property that has been purchased on credit when the court declares the purchaser delinquent in making payments. You repossess cars, stereos, televisions, and other household goods. The one thing you hate to repossess is a gun, because sometimes the owner objects—violently. Your office has 20 repossession orders for a certain day and you will be randomly assigned ten of them. You find out that two of the orders are for guns. What is the probability that you will have to repossess one or more guns?

5.58 You belong to a motorcycle club that rides only Harley–Davidson motorcycles. Each year the local police department auctions off Harley–Davidson police bikes that have 100,000 miles on their odometers. Some of the bikes have bad engines and transmissions and others are in good repair. Your club would like to buy five bikes. You have to buy the bikes "as is" and you cannot test them to identify the good ones. You know the motorcycle mechanic for the police department, and he tells you that four of the 15 bikes that will be sold require major repairs. What is the probability that your club will purchase five good bikes?

5.59 You work for the Federal Office of Contract Compliance and your job is to make sure that federal contractors do not discriminate on the basis of sex. You suspect a defense contractor of discriminating in favor of men in hiring. The contractor has advertised three positions for lathe operators. You examine the records of all of the applicants for the positions and you observe that eight men and three women have applied. You consider all of the applicants to be qualified for the positions. The contractor hired three of the men. What is the probability that only men would be hired if the contractor had not been discriminating on the basis of sex?

5.60 As part of the examination to qualify as diplomate in radiology, interns are asked to examine ten X rays of lungs. Five of the X rays are of patients with lung cancer and the other five have different diseases. The interns are asked to identify the five X rays that exhibit lung cancer. Correct identification of four or more is considered a passing score. What is the probability that an intern will pass if he or she is sure about only two of the X rays and has to guess about the others?

5.61* You have taken a job as bellboy in a hotel. The main source of your income is supposed to be tips for carrying guests' suitcases to and from their rooms. When you were hired, the bell captain told you that you could expect to carry a set of bags on average once every 20 minutes. You've spent one hour on the job and you have not carried any bags.

    (a) If the average rate was accurate, what is the probability of carrying no bags in one hour?

    (b) You've thought about quitting your bellboy job. You decide to wait and see if you continue to have no bags to carry. How long should you wait while observing no customers so that the probability of that outcome would be equal to 0.001, assuming that the rate quoted by the bell captain was accurate?

5.62* You are the promoter for an automobile race. You want to buy insurance to cover your losses in the event that rain forces you to cancel the race and refund everyone's money. According to the weather service, the probability of rain is 0.2. You know that insurance companies charge a 10% load on this type of insurance; that is, the expected value of the insurance policy for the company will be 10% of the premium. What will the premium be if the payout in the event of rain is $10,000?

5.63* You own a seafood restaurant that is famous for its fresh sole. Each day you sell from 10 to 15 sole dinners. The probability is the same for each number of sole dinners. To maintain your reputation for freshness, you purchase sole daily and throw away any that are left over. If each sole costs you $2, all other costs of preparing sole dinners do not vary with the number of sole dinners served, and the price of your dinners is $8, how many sole should you purchase each day to maximize the expected value of your profits?

5.64* You own a 200-unit apartment building. Your tenants must pay their rent by the fifth day of the month. A payment on the sixth day of the month is considered one day late. On the basis of your tenants' past payment behavior, you have the following probability distribution for the number of days a payment will be late:

x_i	0	1	2	3	4	5	6	7
$P(x_i)$	0.3	0.1	0.1	0.1	0.1	0.1	0.1	0.1

If you charge your tenants a late fee of $5 a day, what is the maximum probability that you will collect more than $3,000 in late fees in one month from all of your tenants? Assume that each tenant's payment behavior is independent.

5.65* In 1987 the Ralston company tried to outdo the gemstone promotion of Procter and Gamble. This time the promotion was to give away money. Boxes of Almond Delight cereal announced "FREE MONEY IN EVERY BOX!" Here are the rules listed on the side of the box:

ALMOND DELIGHT
GREAT CASH GIVEAWAY
OFFICIAL RULES

1. INSTANT WIN: Packed inside of every specially-marked package of Almond Delight™ brand cereal is currency from any one of seven countries (eighteen foreign and four United States denominations). The exchange rate of the currency packed inside any one package ranges from less than 1¢ up to $500. For a free chance to win United States currency, send a hand-printed, self-addressed, stamped envelope (WA and VT State residents need not include return postage) before June 30, 1987, to: Great Cash Giveaway, P.O. Box 8538, Westport, CT 06888. Limit one request per household, group or organization per day. NO PURCHASE NECESSARY TO PLAY OR WIN PRIZES.

2. IN-PACKAGE CURRENCY: Various types of foreign currency with an average U.S. exchange rate of less than 1¢ are packed inside 4.2 million specially-marked packages of Almond Delight. Some of these currencies are no longer in circulation and are, therefore, delivered along with currently circulated currency. An additional 100,550 specially-marked packages carry United States currency valued as follows:

Value	# Distributed	Odds of winning
$1.00	90,000	1 in 47
$5.00	10,000	1 in 427
$50.00	500	1 in 8,534
$500.00	50	1 in 85,340

(a) What is the expected value of the mail-in lottery net of the cost of two 22¢ stamps?

(b) In light of the present value calculated in part a, do the restrictions on the mail-in lottery (hand-lettered, limit one per day, etc.) make any sense? Explain.

(c) What is the expected value of the currency in the cereal box?

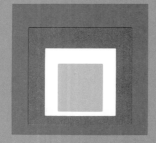

6

Continuous Random Variables and the Normal and Exponential Distributions

The importance of quality control is widely acknowledged in the U.S. automobile industry. The Ford Motor Company steering gear plant in Indianapolis provides an example of the use of statistics for quality control. The plant manufactures the two main parts of a steering gear, worms and rollers, to very precise tolerances. The diameter of a worm is supposed to be one-ten-thousandth of an inch smaller than that of a roller. The reason for the precision is that if the difference between the diameters is much less than that amount, a steering wheel will feel stiff in a driver's hands and it will be difficult to steer a car equipped with such a steering gear. Alternatively, if the difference between the diameters is much more than that amount, the steering wheel will feel loose and a driver will be able to turn the steering wheel without turning the tires. The manufacturing process entails unavoidable variation in diameters from part to part. The machinery can be adjusted to set the average size of the parts and the variability from part to part, but not to make every part come out the same. Each completed worm and roller is carefully measured and color-coded with a dab of paint to indicate how much it is over or under the desired size. Along one wall of the factory is a series of bins for rollers and along another wall is a series of bins for worms, each series of bins arranged in ascending order of diameters. If you looked at the bins for rollers you would see the pattern shown in Figure 6.1. The bins for the worms display an identical pattern. The pattern is a rough approximation of the normal distribution. It is the most common and the most important distribution in statistics. The fact that the distributions of worm and roller diameters follow a normal curve helps the Ford Motor Company make sure that no worms or rollers are wasted while all of the assembled steering gears have a precise fit. When the machinery is set so that the average diameter of a worm is one-ten-thousandth of an inch less than the average diameter of a

FIGURE 6.1
Parts bins for rollers at Ford steering gear plant

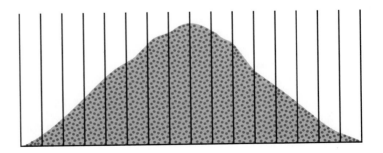

roller and so that the standard deviation of the distribution of worm diameters equals the standard deviation of the distribution of roller diameters, every worm can be lined up with an appropriate roller. The assemblers merely match them by color code.

This chapter begins with a discussion of continuous random variables. An explanation of the normal distribution follows. Next we shall see why the normal distribution is so common and so important, and we shall explore some of its business applications. The normal approximations to the binomial and the Poisson are discussed next. Finally, another continuous distribution that has important business applications, the exponential distribution, will be discussed.

6.1 CONTINUOUS RANDOM VARIABLES

6.1A INTRODUCTION

A continuous random variable was defined in Chapter 5 as the outcome of an experiment that could take any value over some range of values. In this section we will discuss how continuous random variables are represented by equations, tables, and graphs, and we will see that the probability that a continuous random variable will take values between two points on the x axis corresponds to the area under the graph of the density function between those two points.

As in the case of discrete random variables, the probabilities for all of the possible values of a continuous random variable must sum to 1. However, the probability that a continuous random variable equals any specific value is 0. We can clarify this point by considering the size of a glove and the size of the hand that wears it. Many people wear size 8 gloves. Thus the probability that someone randomly selected from the U.S. population wears a size 8 glove must be greater than zero. However, no one in the United States has a hand that measures *exactly* 8 inches around the palm. This assertion may sound bizarre but on reflection it is plausible. Having *exactly* an 8-inch hand means that no discrepancy is tolerated, not a millionth or even a billionth of an inch. By increasing the accuracy of the measurement, you could guarantee that no hand measured exactly 8 inches. Continuous random variables have zero probabilities at any specific point, such as exactly 8 inches. A continuous random variable can have only a nonzero probability over a range of values, such as from 7 to 8 inches.

6.1B THE CONTINUOUS UNIFORM DISTRIBUTION

The most simple continuous random variable is the **continuous uniform distribution,** *a continuous distribution in which all equal intervals in the range of the* *distribution have the same probability.* The continuous uniform distribution has few, if any, business applications because almost no physical phenomena fit this description. Because of its simplicity, however, it provides a good introduction to continuous distributions.

Suppose the time needed to play a video game varied from 5 to 10 minutes and that every range of completion times that have the same width have the same probability. For example, every range of completion times 1-minute wide has a probability of 0.2. The equation for this continuous uniform random variable is:

$$f(x) = 0.2 \qquad 5 \leq x \leq 10$$

The notation in the continuous case is different from the notation in the discrete case. Since X can take any value over the range 5 to 10 minutes, it no longer makes sense to count or iterate the x_i's as distinct values. The number of possible values in this range is infinite, so it is not even possible to count them. The expression $f(x)$, or the **probability density function,** is *a function that assigns a number called "density" to every value in the range of a continuous distribution in such a way that the total area under the density function equals 1 and the area under the density function over any section of the range equals the probability that the continuous random variable will be observed in that section.* One way to interpret the function in this example is that the completion time for any one-minute interval has a probability of 0.2. The same information is displayed in Table 6.1 and Figure 6.2. It would not matter if the ranges in Table 6.1 were rewritten so that they included the integer values—for example, $5 \leq x \leq 6$—because the probability that a continuous random variable will equal any exact value is zero. In our example the probability that the random variable will equal exactly 5 or exactly 6 is zero. As is the case with any random variable, the sum of the probabilities of the entire range of X must equal 1. In Table 6.1 the probabilities sum to 1 and in Figure 6.2 the enclosed area has a probability equal to 1. You can check that the area within the rectangle equals 1 by multiplying the base of 5 by the height of 0.2. The probability of any range within the limits of $x = 5$ and $x = 10$—say, from 5 to 7—would correspond to the area under the density function that is between $x = 5$ and $x = 7$. In this case the probability

TABLE 6.1

Continuous uniform random variable with range of 5 to 10

Range of x	Probability that x is within that range
$5 < x < 6$	0.2
$6 < x < 7$	0.2
$7 < x < 8$	0.2
$8 < x < 9$	0.2
$9 < x < 10$	0.2

FIGURE 6.2
**Probability distribution of
continuous uniform
distribution with range of 5
to 10**

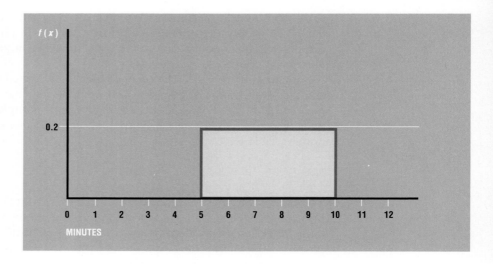

would be 0.4 because 40% of the area under the curve is between 5 and 7. Note that, as in the histograms in Chapter 2, the density of a continuous random variable must be defined so that the total area under the density function equals 1. The area under the density function between any two points, such as 5 and 7, represents the proportion of the distribution that is between those points. This proportion is the same as the probability that the random variable will fall in that range.

The general formula for the probability density function of the continuous uniform distribution is:

$$f(x) = \frac{1}{b - a} \qquad \text{for } a \leq x \leq b \tag{6.1}$$

where b is the upper limit of the distribution and a is the lower limit

The expected value, $E(X)$, is the midpoint of the distribution because it is symmetrical. In notation the expected value is $(b + a)$ divided by 2.

$$E(X) = \frac{b + a}{2} \tag{6.2}$$

The variance of a continuous uniform distribution equals:

$$\sigma^2(X) = \frac{(b-a)^2}{12}$$

(6.3)

This formula is derived in Appendix C.

6.1C ADDITIONAL EXAMPLES OF CONTINUOUS RANDOM VARIABLES

The correspondence between the area under the curve and probabilities is not limited to the continuous uniform distribution. For all continuous random variables the area under the curve between any two points is the same as the probability that the random variable will take values in that range. Figure 6.3 shows the distribution of another continuous random variable. This time the shape does not matter. Two points are marked off as x_1 and x_2 and the area between those two points is labeled A. If A equaled, say, 0.1, then the probability that the random value would fall between x_1 and x_2 would equal 0.1.

Finding the expected value or the variance of a continuous random variable usually requires some calculus. To avoid using calculus, we will not derive the expected value or variance from the probability density function. When these terms are needed for particular probability density functions, they will be presented without derivation. For some additional practice with continuous random variables, consider a continuous random variable X that has a range of 0

FIGURE 6.3
Area under the curve of a portion of the distribution of a random variable

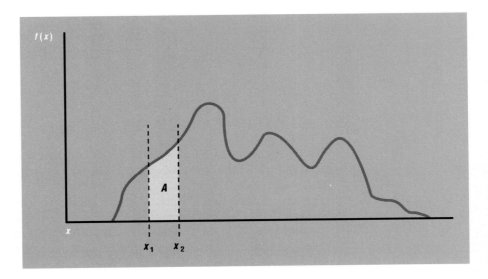

FIGURE 6.4
Probability density function
of a random variable with a
range of 0 to 20, the highest
point at $x = 0$, the lowest
point at $x = 20$, and a
straight line between

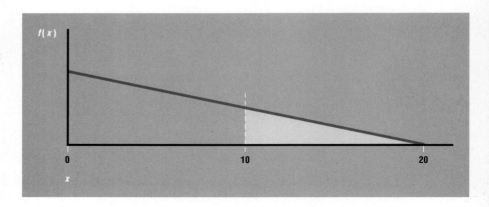

to 20. The graph of X is a straight line with its highest point on the y axis and its lowest point when it reaches the x axis at $x = 20$. Draw X, find its probability density function, and find the probability that the random variable will take values greater than 10. Figure 6.4 contains the graph of the probability density function. Note that the x and y axes and the straight line of the graph of the probability density function form a triangle. This fact can be used to find the equation of the probability density function. Recall that the area under the probability density function must equal 1 and that the area of the triangle equals one-half the base times the height. The base of the triangle is 20, so the height of the triangle or the y intercept of the line equals:

$$\text{area} = \frac{1}{2}\ \text{base} \times \text{height}$$

$$\text{height} = \frac{2 \times \text{area}}{\text{base}}$$

$$\text{height} = \frac{2 \times 1}{20}$$

$$\text{height} = 0.1$$

The next step is to find the probability density function. It is the function of the straight line that runs from $(x = 0, y = 0.1)$ to $(x = 20, y = 0)$. This function is $f(x) = 0.1 - 0.005x$, for $0 \le x \le 20$, because the y intercept is 0.1 and the slope is -0.005. You may recall from a previous math class that the slope of a straight line is the change measured along the y axis for every unit increase in X. In this case the y coordinate falls 0.005 for every one-unit increase in X. The last part of the question was to find the probability that X is greater than or equal to 10. Since the probability for a section of a continuous random variable equals the area under that section of the curve, what you need to find is the area to the right of 10, or the shaded area in Figure 6.4. Note that this area is also a triangle. This result allows you to use the formula for the area of a triangle again. The shaded area is one-half the base times the height of the shaded triangle, or ½(10)(0.05). It may not be obvious why the height of the shaded triangle is 0.05, but if you look at the diagram you can see that the entire triangle that runs from $x = 0$ to $x = 20$ has a height of 0.1 and that the triangle for the shaded area running from $x = 10$ to $x = 20$ is exactly half as high as the larger triangle. The height of the shaded triangle could also be found by substituting 10 for x in the function $f(x) = 0.1 - 0.005x$. The probability that the random variable will be greater than 10 is 0.25.

EXERCISES FOR SECTION 6.1

6.1 Let X be a continuous uniform random variable over the range 30 to 40. Find the probability density function of X and the expected value and variance.

6.2 Let X be a continuous random variable such that $f(x) = 0.1 - 0.005x$ (the triangle in Figure 6.4). Find the median of the distribution. Note that the median of a continuous distribution has 50% of the area on either side. The mean or expected value would have 50% of the area on either side only if the distribution were symmetrical.

6.3 You take your laundry to a laundromat and pay a quarter each time you run the dryer. After much study you have determined that the length of time that the dryer runs on one quarter is a continuous uniform random variable bounded by 6 and 12 minutes. If a pile of your wet laundry needs 7 minutes to dry, what is the probability that you will need to use two quarters?

6.4 The time a family spent at a theme park that is open for 10 hours a day is a continuous random variable with the following distribution:

$f(x) = 0.05 \qquad 0 < x < 2$
$f(x) = 0.1 \qquad 2 < x < 4$
$f(x) = 0.2 \qquad 4 < x < 6$
$f(x) = 0.1 \qquad 6 < x < 8$
$f(x) = 0.05 \qquad 8 < x < 10$

    (a) Graph the probability density function.

    (b) What is the probability that a family will stay 7 or more hours?

6.5 The time needed to assemble a plastic model of an airplane is a continuous uniform random variable with a mean of 20 minutes and a variance of 12. What is the probability that it will take more than 22 minutes to assemble the plane?

6.2 THE NORMAL DISTRIBUTION

6.2A INTRODUCTION

The **normal distribution** is *a family of continuous distributions that have similar shapes. All normal curves are symmetrical and have one mode. Normal curves are often called bell-shaped curves because they look like the silhouette of a bell.* Two parameters define a particular normal curve: the mean and the standard deviation. Figure 6.5 contains three overlapping normal curves with identical standard deviations and different means, while Figure 6.6 has three overlapping normal curves with identical means and different standard deviations. Note that as μ increases, the normal curve moves farther to the right. Thus μ_1 is less than μ_2, which is less than μ_3. Note also that as the standard deviation increases, the normal curve becomes wider. Thus σ_1 is less than σ_2, which is less than σ_3.

6.2B WHY THE NORMAL DISTRIBUTION IS SO COMMON

After reading the chapter introduction, you may be wondering why normal curves are so common. The worm-and-roller example can be used to provide an intuitive explanation. Many factors (such as temperature changes, variations in the chemical composition of the metal used to make the parts, changes in the hardness of the cutting tools caused by wear, and differences in the positions of the parts in the cutting machines) can account for minute variations in the diameters of the worms and rollers. Most rollers or worms come out near the average because these factors tend to balance or "wash out," that is, some are high while others are low. Further, the factors themselves tend to have normal or at least mound-shaped distributions, so that most of the time

FIGURE 6.5

Three normal curves with identical standard deviations and different means

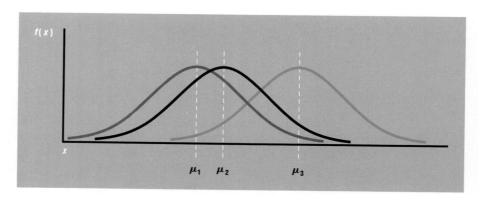

FIGURE 6.6
Three normal curves with identical means and different standard deviations

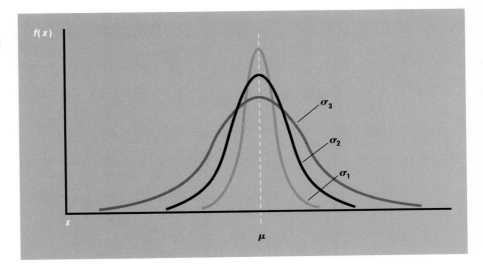

each factor will be near its average value. The occurrence of an oversized or undersized part is less and less likely the greater the deviation of a diameter from the mean value is, because it is unlikely that all of the causes of these minute variations will work strongly in one direction. Moreover, deviations below the mean are as likely as deviations above the mean, so the resulting distribution of diameters is symmetrical. Put together, all of these considerations result in a normal curve.

Curves that can be closely approximated by the normal distribution show up in many physical and natural phenomena. The heights of adult males, the number of ounces of a soft drink dispensed by a filling machine, and the lives of a particular model of car battery are all approximately normally distributed. Although the normal distribution shows up in many situations, don't make the mistake of assuming that every continuous distribution is normal. Legal or institutional considerations may truncate what would otherwise be a normal distribution. For example, the distribution of wage rates of men aged 16–24 is truncated by the federal minimum wage of $3.35 per hour, as we saw in Chapter 2. Other distributions are known generally to be skewed. For example, income distributions are highly skewed to the right in almost all societies, from the United States to the Soviet Union to Sweden. There is no way to be sure that a particular distribution closely follows a normal curve without collecting some data and testing to see if the normal distribution provides a good fit. We'll discuss some accurate tests for normality in Chapter 10 and Chapter 17, but a crude test for normality is to check the percentage of observations within plus or minus one and plus or minus two standard deviations of the mean. If a distribution is normal, you will expect approximately 68% of the observations to be within one standard deviation and approximately 95% to be within two standard deviations. If your actual percentages depart very far from these values, your distribution probably does not follow a normal curve.

■ 6.2C AREAS UNDER A NORMAL CURVE

So far a normal curve has been described in pictures and words. The probability density function of the normal distribution gives the exact description:

$$f(x) = \frac{1}{\sqrt{2\pi}\,\sigma}\, e^{-1/2[(x-\mu)/\sigma]^2} \tag{6.4}$$

In this formula x can take any value from $-\infty$ to $+\infty$, so the normal distribution is not bounded. The function approaches the x axis asymptotically. As a practical matter, very few interesting variables range from $-\infty$ to $+\infty$. However, observations more than three standard deviations away from the mean have such a low probability in the normal distribution that it provides a good fit for many variables that are bounded. The symbol π (read "pi") stands for the ratio of the circumference to the diameter of a circle, approximately 3.1416. The letter e stands for the base of the natural logarithms, approximately 2.7182818. As with other continuous random variables, the total area under the probability density function equals 1. The probability that a randomly selected observation from a normal population will be between any two points, a and b, corresponds to the shaded area in Figure 6.7.

It is not necessary to use the probability density function to find the area

FIGURE 6.7
Area under a normal curve between two points

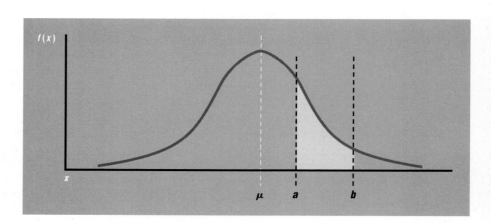

between any two points. All normal curves have the property that the same percentage of the total area lies between the same multiples of standard deviations from the mean. For example, suppose point *a* was one-half of a standard deviation above the mean and point *b* was two standard deviations above the mean. The shaded area would be 28.57% of the area under the curve. You could vary the mean and the standard deviation, but as long as *a* was one-half of a standard deviation above the mean and *b* was two standard deviations above the mean, the shaded area would remain 28.57% of the total. For example, suppose your mean was 100 and your standard deviation was 10, then 28.57% of the area would be between 105 and 120. If your mean was 50 and your standard deviation was 5, then 28.57% of the total area would be between 52.5 and 60. Finding the probability that observations will be between any two points requires that you express the two points as multiples of a standard deviation. Table D4, in Appendix D, gives the proportion of the total area between the mean and any multiple of a standard deviation.

Let's see how the table works and where the 28.57% came from. The left-hand column of Table D4 is headed *z*. This letter stands for the number of multiples of a standard deviation between the mean and any point on a normal curve. The formula for the *z* score is:

$$z = \frac{x - \mu}{\sigma}$$

(6.5)

The letter *x* stands for any point on the normal curve. Suppose you had a random variable that was normally distributed with a mean of 100 and a standard deviation of 10. The point 105 would be one-half of a standard deviation to the right of the mean and have a *z* score of 0.50.

$$z = \frac{105 - 100}{10}$$

$$z = 0.50$$

In Table D4 find the *z* score of 0.50 in the left-hand column. This row contains all of the proportions for the *z* scores from 0.50 to 0.59 at intervals of 0.01. The first row of the table is headed 0.00, 0.01, 0.02, 0.03, . . . , 0.09. These headings refer to the value of the second digit to the right of the decimal place. The proportion of the total area between the mean and a *z* score of 0.50 is found at the intersection of the row labeled 0.5 and the column labeled 0.00. This row of Table D4 is reproduced in Table 6.2. The proportion of the total

TABLE 6.2
Portion of Table D4, standard normal distribution

z	0.00	0.01	0.02	0.03	0.04	0.05	0.06	0.07	0.08	0.09
0.5	0.1915	0.1950	0.1985	0.2019	0.2054	0.2088	0.2123	0.2157	0.2190	0.2224

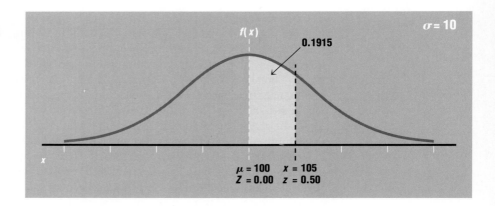

area between the mean of 100 and the point 105 is 0.1915. This area is shown in Figure 6.8. You can use the same procedure to find the proportion of the total area that is between 100 and 120, or equivalently, the area between the mean and a z score of 2.00. The proportion is 0.4772. Figure 6.9 shows this area. The area between 100 and 120 is marked off by a brace and labeled 0.4772. Now you can guess where the 28.57% came from. The proportion of the total area between 105 and 120 is $0.4772 - 0.1915$, which is 0.2857. This area is left unshaded in Figure 6.9. The number 28.57% is the same area expressed as a percentage instead of a proportion.

6.2D THE STANDARD NORMAL DISTRIBUTION

There are two equivalent ways to interpret the z score formula. What we have done so far is to consider the z score as the number of multiples of a standard deviation between the mean and any boundary point on a normal distribution.

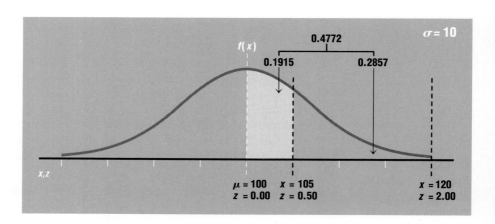

The other interpretation is that the z score transforms any normal distribution into the standard normal curve. The **standard normal curve** is *the particular normal curve that has a mean of 0 and a standard deviation of 1*. Subtracting the mean from every x value of a normal distribution centers the distribution on 0, and dividing the difference between the mean and each x value by the standard deviation widens or narrows the distribution so that it has a standard deviation of 1. On a standard normal distribution the x values and the z scores are the same. For example, take an x equal to 1 on a standard normal curve. If you compute the z score by subtracting the mean of 0 from 1 and divide the difference by the standard deviation of 1, you'll get a z score of 1. For another example, take an x equal to -1 on a standard normal curve. If you compute the z score by subtracting the mean of 0 from -1 and divide the difference by the standard deviation of 1, you'll get a z score of -1. Thus the z scores can be interpreted as points along the standard normal curve as well as the number of multiples of the standard deviation between the mean and any x value. Table D4 actually gives areas under the standard normal curve. Under the alternative interpretation of the z score, the process of using the z score formula converts a normal distribution to the standard normal distribution. After the conversion you can conveniently read in Table D4 the areas or probabilities that correspond to the section of the original normal curve. These two interpretations of the z score formula (as a conversion to the standard normal distribution and as the number of multiples of a standard deviation between the mean and an x value) are two ways of thinking about the same process; there is no difference in the way problems are handled under the two interpretations.

Table D4 provides areas only for the right side of a standard normal curve. The symmetry property of normal distributions can be used to find areas on the left side of a curve. For example, for a normal distribution with a mean of 100 and a standard deviation of 10, the probability of observing values between 80 and 95 equals the probability of observing values between 105 and 120. Areas on the left side of a normal curve have negative z scores, but you can ignore the negative sign when you look up an area in Table D4. Let's try another example using the normal distribution with a mean of 100 and a standard deviation of 10. For this distribution, what is the probability that an observation will fall between 98 and 115? This area is shown in Figure 6.10. The point 115 in the distribution has a z score of 1.50 and the point 98 has a z score of -0.20. These z scores can be obtained by formula 6.5:

$$z = \frac{x - \mu}{\sigma} \tag{6.5}$$

$$z = \frac{98 - 100}{10} \qquad z = \frac{115 - 100}{10}$$

$$z = -0.20 \qquad z = 1.50$$

FIGURE 6.10
Area under a normal curve
between z scores of −0.20
and +1.50

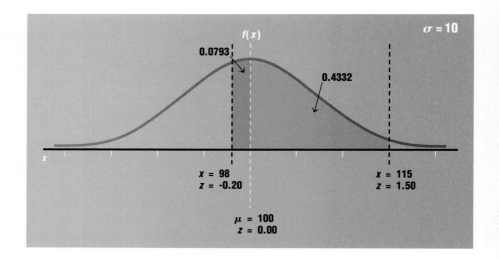

By checking Table D4, you'll find that the proportion of the total area between the mean and a z score of −0.20 is 0.0793 and the proportion between the mean and a z score of 1.50 is 0.4332. The proportion of the total area between z scores of −0.20 and 1.50 is the sum of these proportions, 0.5125. Equivalently, the proportion of the total area for the above distribution between 98 and 115 is 0.5125.

The fact that 50% of the area under the curve lies on either side of the mean can be used to find areas in the tail of a distribution. For example, Figure 6.11 shows the area to the left of 90. The point 90 has a z score of −1.00, and according to Table D4, the proportion of the area between the mean and a z score of 1.00 is 0.3413. To find the area to the left of 90, just subtract 0.3413 from 0.5. The result, as shown in Figure 6.11, is 0.1587.

Now for some business examples that use the normal curve. Suppose you were a manufacturer of clothing for teenage males and you knew that the distribution of waist sizes for teenage males in the United States was normal with a mean of 26 inches and a standard deviation of 5 inches. Pants for teenage males come in even sizes, such as 22, 24, 26, 28, 30. Each size covers a 2-inch range centered on the number of inches that corresponds to the size number. For example, size 28 should fit waists between 27 and 29 inches. Your firm has an order for corduroy pants from a retail chain which calls for 10% of the pants to be size 38. The chain does not specialize in large sizes, and you think the order is a mistake. Assuming that the distribution of the chain's customers is the same as the distribution of the population of teenage males, what proportion of the chain's teenage male customers would wear a size 38? The z score for the point 37 inches is 2.20 and the z score for the point 39 inches is 2.60. You obtain these values by substituting in formula 6.5:

FIGURE 6.11
Area under a normal curve to the left of a z score of −1.00

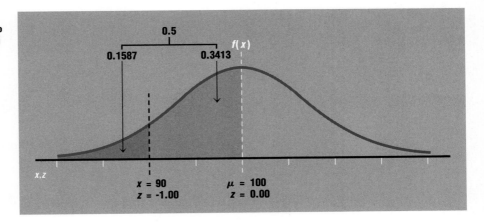

$$z = \frac{x - \mu}{\sigma} \tag{6.5}$$

$$z = \frac{39 - 26}{5} \qquad z = \frac{37 - 26}{5}$$

$$z = 2.60 \qquad z = 2.20$$

The proportion of the total area between the mean and a z score of 2.20 is 0.4861 and the corresponding area for the z score of 2.60 is 0.4953. The area between these z scores is $0.4953 − 0.4861$, or 0.0092. Thus slightly less than 1% of all teenage males would wear size 38 pants, and it appears that the order was a mistake.

Here's another example. Suppose you were a journalist following the story of the air traffic controllers who were hired to replace those fired by President Reagan. Some of the fired controllers claimed that the Federal Aviation Administration lowered its standards for newly hired controllers by making the test they had to take easier so that more of them would be available to operate the air traffic system. Suppose in past tests the distribution of scores was approximately normal with a mean of 60 points and a standard deviation of 5. Seventy points are needed to pass the exam. Suppose 40% of the newly hired air traffic controllers passed the exam. If the difficulty of the test had not been lowered, what score would be the expected lower bound or limit for the top 40% of all scores?

This question reverses what is the known and what is the unknown variable. In earlier examples, the points on the distribution were known and you had to find the proportions. Here you know that the proportion is the top 40% of the distribution and you have to find the corresponding point on the x axis.

FIGURE 6.12

Area under a normal curve to the right of a z score of 0.253

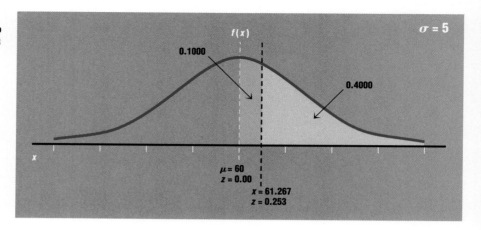

FIGURE 6.12

Area under a normal curve to the right of a z score of 0.253

Figure 6.12 describes this problem. Since 50% of the area is to the right of the mean, the point in the distribution that has 40% of the area to its right must have 10% of the area between that point and the mean. If you look in Table D4 for the probability 0.1000, you won't find exactly that value. The two closest probabilities are 0.0987 and 0.1026. The z scores for these probabilities are 0.25 and 0.26, respectively. The simplest thing to do in this situation is to use the z score for the probability that is closest to 0.1000, or 0.25. If you need greater precision, you'll have to interpolate the value between the z scores of 0.25 and 0.26. Here's how the interpolation works. The distance between 0.0987 and 0.1026 is 0.0039. The distance between 0.0987 and 0.1000 is 0.0013. The correct z score is 0.0013/0.0039, or one-third of the distance from 0.25 to 0.26. We'll round the z score to 0.253. Now that you have the z score that corresponds to 40% of the area in the right-hand tail of the distribution, you can substitute into the z score formula to find the point on the distribution.

$$z = \frac{x - \mu}{\sigma}$$ (6.5)

$$x = \mu + z\sigma$$

$$x = 60 + 0.253(5)$$

$$x = 61.265$$

The point on the distribution that has 40% to its right is 61.265. Since 40% of the new recruits got scores of 70 or more, there may be some substance to the charges that the test was made easier. This example takes a simple approach to this problem. Clearly, you would have more confidence in your conclusion if the result had been derived from a large number of recruits, say several thousand, instead of a small number, say a dozen. In Chapter 9, on hypothesis testing, we'll take a more sophisticated approach to these types of problems which adjusts for the sample size.

EXERCISES FOR SECTION 6.2

6.6 Let X be a normally distributed random variable with a mean of 30 and a standard deviation of 15.

(a) What proportion of the distribution is below 0?

(b) What proportion of the distribution is between 8 and 17?

(c) What proportion of the distribution is above 36?

(d) What value of the x axis marks the top 20% of the distribution?

6.7 Let X be a normally distributed random variable with a mean of 120 and a standard deviation of 3.

(a) What is the probability that X will be above 125?

(b) What is the probability that X will be between 121 and 124?

(c) What is the probability that X will be between 118 and 124?

(d) What is the probability that X will either be between 115 and 116 or 122 and 124?

6.8 Suppose the distribution of sardine lengths was approximately normal with a mean of 4 centimeters and a standard deviation of 0.5 centimeter. Sardines that are too long or too short for a sardine can are ground up for cat food. The range of sizes that can be used in a sardine can is from 3 to 5 centimeters. What proportion of all sardines are ground up for cat food?

6.9 You are planning a weekend camping trip to an isolated area. You need to pack paper diapers for your baby. You don't want to carry too many diapers because your car is short of space; on the other hand, you would hate to run out. You have checked your baby's consumption pattern for paper diapers on weekends, and it is approximately normal with a mean of 30 and a standard deviation of 10. How many diapers should you pack so that you have a 1% chance of running out?

6.10 The distribution of low daily temperatures in winter in central Florida is approximately normal with a mean of 50 degrees and a standard deviation of 8 degrees. If the daily temperature falls below 30 degrees, the orange crop will suffer frost damage. What is the probability that on a winter day the orange crop will suffer frost damage?

6.11 Refer to the information provided in exercise 6.10. If the distributions of low daily temperatures over the winter in Florida are independent, what is the probability that the orange crop will suffer frost damage over a 90-day winter period? (Hint: Calculate the probability of the intersection of 90 independent events: not getting frost damage on each day.)

6.12 You speculate in foreign currencies. You hold a 30-day option to buy pounds sterling at $1.50 a pound and you have sold a 30-day contract to sell

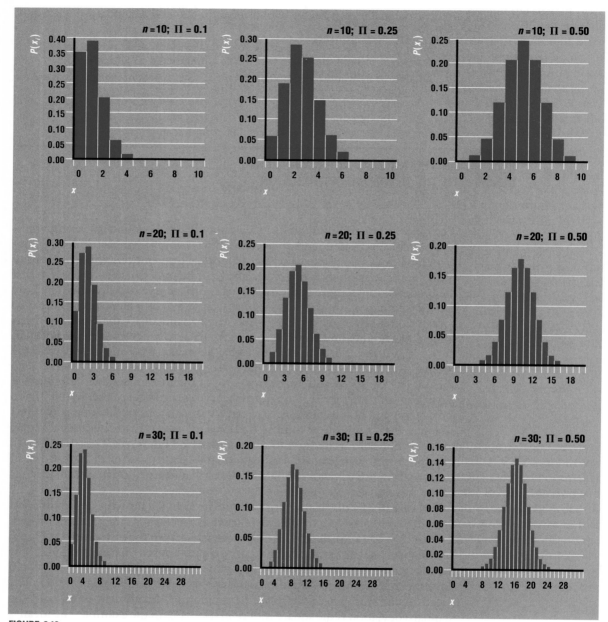

FIGURE 6.13
Histograms for nine binomial distributions with *n* equal to 10, 20,
and 30 and with Π equal to 0.1, 0.25, and 0.5

FIGURE 6.14
Normal approximation to
binomial when $\Pi = 0.5$ and
$n = 20$

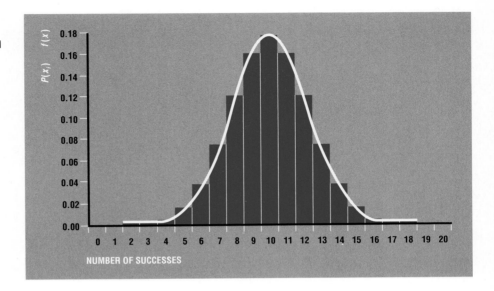

Note that the y axis is labeled both $P(x_i)$ and $f(x)$ because a discrete and a continuous distribution are being plotted on the same graph. Suppose you wanted to use the normal approximation to find the probability of 12 or more successes in 20 trials. The exact answer taken from the binomial table is 0.2517. What you need to approximate is the shaded area in Figure 6.15. The area of each rectangle in this histogram of the binomial distribution represents the

FIGURE 6.15

Area that represents the
probability of 12 or more
successes for the binomial
distribution when $\Pi = 0.5$
and $n = 20$

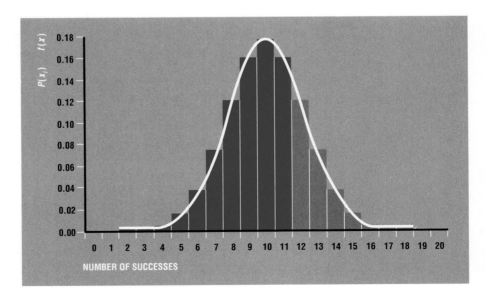

FIGURE 6.16
Area to the right of 12 under the normal curve

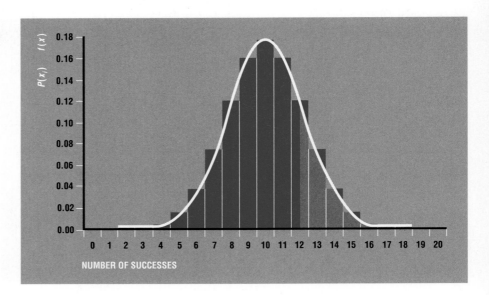

probability of the corresponding number of successes. Since each rectangle has a width of one unit, the height or density of each rectangle is the same as the probability. The area in the rectangles labeled 12 through 20 is a graphical representation of the probability of 12 or more successes. Suppose you used the area under the normal curve to the right of 12 as the area that would approximate the shaded columns in Figure 6.15. Similarly, the area under the normal curve is a graphical representation of probability.

The area under the normal curve to the right of 12 is shaded light blue in Figure 6.16. Notice the discrepancies between the areas in Figures 6.15 and 6.16. The left half of the column that is centered on 12 is not part of the shaded area in Figure 6.16. A small triangle in the upper right corner of that column is also missing from the area under the normal curve. So that you can see these discrepancies more clearly, a section of Figure 6.16 has been circled and enlarged in Figure 6.17. The little triangle that is missing, labeled *a* in the enlargement, has approximately the same area as the triangle labeled *c*. The absence of the *a* area is compensated for by the inclusion of the *c* area under the normal curve, because the *c* area is not part of the column for the 13th success. The area labeled *b*, however, is missing, and it represents half of the area of the column for the 12th success. The adjustment for the missing *b* area is called the continuity correction. Instead of taking the area to the right of 12 under the normal curve, you take the area to the right of 11.5. The **continuity correction** is *an adjustment to the z score formula used for normal approximations to the binomial and the Poisson. The adjustment allows a smooth distribution (the normal) to approximate one that looks like a series of steps (either the binomial or the Poisson). The correction involves subtracting one-half from the number of successes for a lower bound or adding*

FIGURE 6.17
Enlargement of a section of Figure 6.16

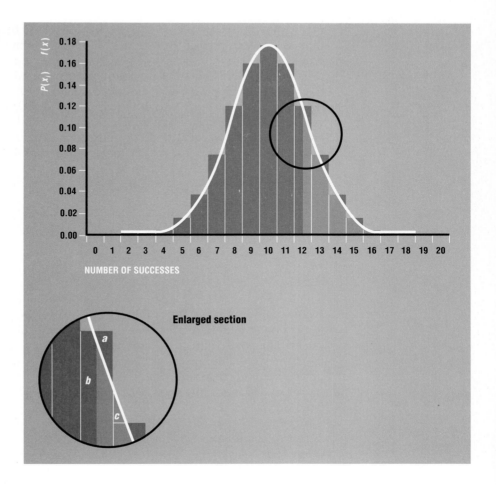

NUMBER OF SUCCESSES

Enlarged section

one-half to the number of successes for an upper bound. Recall that the mean of the binomial is $n\Pi$ and the standard deviation is $\sqrt{n\Pi(1 - \Pi)}$. Thus the mean would be 10 and the standard deviation would be 2.236. In this example the z score associated with the point 11.5 is 0.67 and the probability that corresponds to the right tail for this z score is 0.2514. You can see that the approximation is quite close because the exact answer is 0.2517. If the continuity correction had not been used, the answer would have been 0.1867.

Let's try some business examples. Suppose a brokerage house has an analyst who watches oil company stocks. The analyst forecasts the direction of changes in stock prices on a weekly basis. You want to review the analyst's performance and decide whether her advice is worth anything. Over the past 10 years, the analyst has made 500 predictions about stock prices and 300 of the predictions were correct. What is the probability that someone who just flipped a coin to predict the direction of the change in the price of each stock would get 300 or more successes in 500 trials? On the assumption that oil stock prices are equally likely to go up or down in a randomly selected week, predictions based on flipping a coin should be right 50% of the time. The expected number of successes in 500 trials would be $n\Pi$, or 250, and the standard deviation would be the square root of $n\Pi(1 - \Pi)$, or 11.18. Since $n\Pi$ and $n(1 - \Pi)$ are greater than or equal to 5, the normal can be used to approximate the binomial. The z score for 300 or more successes is 4.43:

$$z = \frac{x - \mu}{\sigma} \tag{6.5}$$

$$z = \frac{299.5 - 250}{11.18}$$

$$z = 4.43$$

The reason that 299.5 is used in place of 300 is that the continuity correction requires the subtraction of one-half from a lower bound. The value 300 is a lower bound because we are looking for the probability of 300 or more successes. The probability of the area to the right of a z score of 4.43 is approximately zero, so the analyst has been able to predict stock price changes with greater accuracy than someone who was flipping a coin. Clearly the advice is worth something. How much you might be willing to pay for it would depend on the accuracy and the charges of competing investment services.

Here's another example. You run a photo-processing laboratory and you do not charge your customers for bad prints. The decision as to what is a bad print is left to the customer. Your experience has been that 10% of all prints are returned for refunds. In one day of processing you make 10,000 prints. What is the probability that 9% or less will be returned? In this example $n\Pi$ is 1,000 and the standard deviation is 30. Again $n\Pi$ and $n(1 - \Pi)$ are greater than or equal to 5, so the normal approximation can be used. The z score for 900 or fewer bad prints is approximately -3.32:

$$z = \frac{x - \mu}{\sigma} \tag{6.5}$$

$$z = \frac{900.5 - 1,000}{30}$$

$$z = -3.32$$

The reason that 900.5 was used instead of 900 is that one-half has to be added to an upper bound. The value 900 is an upper bound because we want the probability of 900 or fewer bad prints. The probability of getting z scores below -3.32, or 900 or fewer bad prints, is 0.0005.

EXERCISES FOR SECTION 6.3

6.16 Let X be a binomial random variable with a Π of 0.5 and an n of 20.

(a) What is the probability of 7 or more successes? Use both the normal approximation and the binomial table.

(b) What is the probability of more than 9 successes and fewer than 15? Use both the normal approximation and the binomial table.

(c) What is the probability of fewer than 8 successes? Use both the normal approximation and the binomial table.

(d) What is the probability of 8 or fewer successes? Use both the normal approximation and the binomial table.

6.17 Let X be a binomial random variable with a Π of 0.25 and an n of 100.

(a) What is the probability of between 22 and 30 successes?

(b) What is the probability of fewer than 26 successes?

(c) What is the probability of more than 28 successes?

(d) What is the probability of 28 or fewer successes?

6.18 You manage a supermarket that uses optical scanning devices to record the prices of items at the cash register. The scanners misread a tiny percentage of all items. The misreadings are random and it is as likely that the incorrectly assigned prices are too high as that they are too low. You've accumulated 5,000 tapes from customers who pointed out that they were incorrectly billed because the scanners misread one of their purchases. Of the 5,000, 2,800 were identified as being misread over the actual price. If the machine is as likely to read too high as too low and the 5,000 errors complained about by customers were a random sample of all misread items, what is the probability that 2,800 or more out of the 5,000 would be misread over the actual price? Why is it extremely unlikely that the 5,000 items that your customers complained about are a random sample of all misread items?

6.19 A machine that molded automobile tires had a defect rate of 1%. Assuming that each tire molding is independent, what is the probability that the number of defects will be greater than or equal to 90 and less than or equal to 120 in a production run of 10,000 tires?

6.20 You run a blood laboratory that extracts components from human blood. A particular antibody is found in 10% of the population of blood donors. You need 20 pints of blood that contain the antibody in order to extract enough of the antibody to treat a particular patient. If you have 150 pints of blood on hand, what is the probability that you will have enough pints that contain the antibody?

6.21 Insurance investigators become suspicious of fraud when a cluster of claims come from one location. Recently, marine insurance companies found that many claims were filed for wood-hulled cabin cruisers. Unscrupulous people would buy an old boat for $10,000 and then hunt for a marine appraiser who was willing to value the boat at $40,000. After getting this appraisal, they would take the boat out where no one was looking and poke a hole in the hull. In such situations, statistical tests can be used to signal an unusual level of claims activity. Suppose a marine insurance company's historical experience with wood-hulled cabin cruisers is that 5% of the boats they have insured have claims in any one year. In a particular state the insurance company has 3,482 such boats insured and over the past year there were 246 claims. If each claim was independent and the probability of a claim on any boat is 0.05, what is the probability that the company would have 246 or more claims?

6.22 **(a)** A market researcher wants to conduct a survey among households that purchase 9 Lives brand cat food. She knows that 15% of all households purchase this brand of cat food. If her initial sample consisted of 1,600 randomly selected households and her target was at least 250 households that purchased 9 Lives, what is the probability that she would reach her target?

(b) How many households need to be in the initial sample so that the probability of getting at least 250 9 Lives purchasers is 0.9?

6.23 **(a)** In Seoul, South Korea, high school students are randomly assigned to high schools and to teachers within the schools. Every student takes the same comprehensive exam to determine if he or she will be allowed to enter college. The top 30% on the exam are allowed to enter college. Teachers' salaries depend on the proportion of their students who pass the exam, with teachers who have a large percentage of their students passing getting high bonuses. If each teacher was responsible for 60 students and the government wanted to reward the top 10% of teachers, in terms of percentage of students who passed the exam, how many of the 60 students would need to pass for the teacher to get a bonus?

(b) What is wrong with using the binomial distribution for this problem?

6.4 THE NORMAL APPROXIMATION TO THE POISSON DISTRIBUTION (OPTIONAL)

Since the material in this section closely parallels the last section, we can be brief. As the expected number of occurrences for a Poisson distribution increases, the Poisson becomes more symmetrical, more like a smooth curve, and thus looks increasingly like the normal distribution. A good rule of thumb is to use the normal to approximate the Poisson when μ is greater than or equal to 5. Figure 6.18 plots the Poisson with μ equal to 5. The area that represents 8 or more occurrences is shaded. For the normal approximation to the Poisson,

FIGURE 6.18
Poisson distribution when
$\mu = 5$

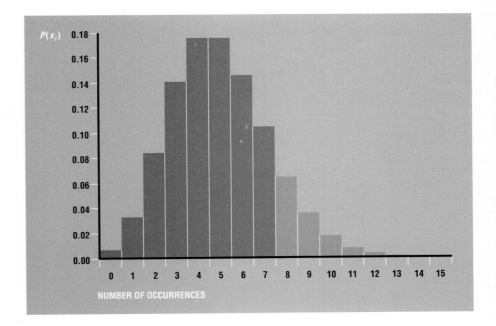

the normal curve with the same mean and standard deviation as the Poisson distribution in Figure 6.18 gives the closest fit. In this case the mean is 5. Recall from Chapter 5 that the standard deviation of a Poisson distribution is the square root of the mean, in this case the square root of 5. Figure 6.19 superimposes this normal curve on the Poisson distribution in Figure 6.18 and shades the area to the right of 8 under the normal curve. The fit is fairly rough, but as μ gets larger than 5, the fit becomes much better. You can see that the area to the right of 8 on the normal curve misses one-half of the column that represents 8 occurrences; so the continuity correction applies to the normal approximation to the Poisson as well as to the normal approximation to the binomial. The z score is $(7.5 - 5)/\sqrt{5}$, or 1.12. The proportion of the total area to the right of a z score of 1.12 is 0.1314. The exact probability of 8 or more occurrences taken from the Poisson table is 0.133.

Here are some business applications. You own a car wash that can wash 100 cars in an hour. The distribution of cars arriving at the car wash follows a Poisson distribution with a mean of 80 per hour. In what proportion of all one-hour periods will more cars arrive than can be serviced, that is, 101 or more cars? The normal distribution can be used to approximate the Poisson because μ is greater than or equal to 5. The normal curve for the approximation is centered on 80 and has a standard deviation equal to the square root of 80. The z score is $(100.5 - 80)/8.94$, or 2.29. One-half is subtracted from 101 because 101 is a lower bound. The proportion of the total area to the right of a z score of 2.29 is 0.011.

FIGURE 6.19
Normal approximation to the
Poisson when $\mu = 5$

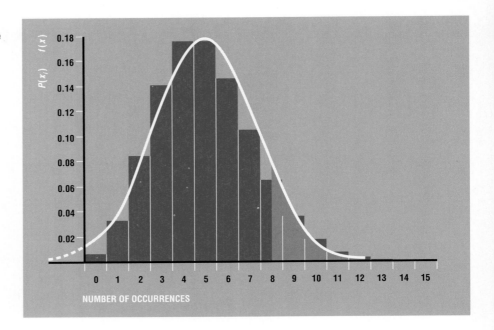

Suppose an average of 20 ships enter the port of San Francisco per 12-hour daylight period and the arrival of each ship is independent. Each pilot can handle one ship in a 12-hour period. How many pilots should the port authority have so that the probability that one or more ships will have to wait overnight is less than or equal to 1%? The z score for 1% in the right-hand tail is 2.326. Let x be the number of pilots. Substituting into the z score formula, we get

$$z = \frac{x - \mu}{\sigma} \qquad \text{(6.5)}$$

$$x = \mu + z\sigma$$

$$x = 20 + 2.326\sqrt{20}$$

$$x = 30.402$$

The solution for x is 30.402. The port authority cannot hire a fractional pilot, so it must hire enough of them to keep the x value above 30.402. If 31 are hired, the x value in the z score formula, after adjustment for the continuity correction, is 30.5, because the boundary is a lower limit. If 30 pilots are hired, the x value in the z score formula will be 29.5. Thus the port authority will have to hire 31 pilots to keep the probability that a ship will have to wait overnight equal to or below 1%.

EXERCISES FOR SECTION 6.4

6.24 Let X be a Poisson random variable with a mean of 30.

(a) What is the probability that there will be more than 35 occurrences?

(b) What is the probability that there will be between 20 and 40 occurrences?

(c) What is the probability that there will be 20 or more or 40 or fewer occurrences?

(d) What is the probability that there will be 36 or fewer occurrences?

6.25 Let X be a Poisson random variable with a mean of 10.

(a) Find the probability of 5 or fewer occurrences using both the Poisson table and the normal approximation.

(b) Find the probability of 12 or more occurrences using both the Poisson table and the normal approximation.

(c) Find the probability of more than 8 and fewer than 11 occurrences using both the Poisson table and the normal approximation.

(d) Find the probability of 20 or fewer occurrences using both the Poisson table and the normal approximation.

6.26 Some consumers are surprised to learn that the U.S. Department of Agriculture has a standard of allowed contaminants per pound of hot dogs. The contaminants generally are insect parts, and the reason that the standard is not set to zero is that it is very expensive to make a microscopic search of all of the ingredients of hot dogs. The reason for the standard is to force manufacturers to maintain reasonable standards of cleanliness. Suppose the USDA standard was 20 contaminants per pound and a hot dog manufacturer knew that the average level of contaminants per pound of his hot dogs was 15. Assume that the number of contaminants per pound follows a Poisson distribution. What is the probability that a randomly selected pound of hot dogs will have more than the allowed number of contaminants?

6.27 You have been asked to be a statistical consultant to a diamond mine. The mine's owners are worried that, despite their elaborate precautions, some employees are stealing diamonds. The diamond-bearing ore averages 25 diamonds per ton. As the ore is crushed, workers inspect the crushed rocks for diamonds, and they are supposed to turn in any that they find. According to the mine's records, the number of diamonds turned in from the last 100 tons of crushed ore was 2,300. Assuming that the number of diamonds per ton follows a Poisson distribution, what is the probability that 2,300 or fewer diamonds would be found in 100 tons of ore?

6.28 The police department of a large city decided to combat violent crime by putting extra patrols in high crime areas. The average number of violent crimes over a weekend is 50. During the first weekend in which the new policy is in effect, the number of violent crimes falls to 48. The mayor claims that the program is working. If the number of violent crimes on a weekend follows a Poisson distribution and the policy has no effect on the average rate of violent crimes, what is the probability of observing 48 or fewer in one weekend?

6.5 THE EXPONENTIAL DISTRIBUTION (OPTIONAL)

Have you ever owned an appliance that broke just after the warranty ran out? It was not just a case of bad luck. To make sure that almost none of their units will fail during the warranty period, many firms use the exponential distribution to calculate the length of the warranty period. The **exponential distribution** is *a distribution used to calculate the probability that the first or next occurrence of an event will be in a given interval.* The warranty example illustrates the use of the exponential distribution to find the probability of the first occurrence of an event when the rate of occurrences for similar events is known. Another application of the exponential distribution is waiting times between events, such as the time between emergency calls for an ambulance service or the time between the arrivals of taxis at a taxi stand.

The exponential distribution is closely related to the Poisson distribution. *If the distribution of occurrences is Poisson with a mean of μ, then the time between occurrences follows an exponential distribution with a mean of $1/\mu$.* Note that the past does not matter for the Poisson or for the exponential distribution because all occurrences are independent. If one occurrence just happened, it would have no effect on the time to the next occurrence. This property may go against your intuition. Suppose you were at a taxi stand waiting for a cab. If the expected time between arrivals of cabs was five minutes, you could expect to wait five minutes if a cab had just pulled away or if no cabs had come to the stand in the last half hour. The reason is that the cabs have nothing to do with each other; they arrive independently. The concept is the same as flipping a coin: if a fair coin has come up heads three times in a row, the probability of a head on the next toss is still one-half. The coin has no memory and the taxi stand has no memory. This property of independence of events allows you to apply the exponential distribution to the time between occurrences or from any point in time to the next occurrence.

To see the relationship between the Poisson and the exponential more clearly, let's use an example. Suppose the number of breakdowns per year for washing machines followed a Poisson distribution with a mean of 0.25 per year. Because it is Poisson, the expected number of breakdowns must be proportional to the time period covered. For example, you would expect 0.5 breakdown over two years, one breakdown over four years, and two breakdowns over eight years. Recall that the rate of occurrences in a unit interval is λ, in this case, 0.25 per year. The term μ refers to the expected number of breakdowns over the interval considered in the problem.

If you expect one breakdown per four-year interval, then the average time between breakdowns is four years. No matter which units you use to express the rate of breakdowns, the expected time interval between breakdowns will stay the same. For example, if μ is 0.25 per year, then $1/\mu$ equals four one-year intervals. If μ is 0.5 per two-year interval, then $1/\mu$ equals two two-year intervals. If μ is 2 per eight-year interval, then $1/\mu$ is 0.5 eight-year intervals. Any way you do it, the average time between breakdowns comes out to four years.

The probability density function of the exponential is:

$$f(x) = \mu e^{-\mu x}$$

(6.6)

To keep the notation consistent between the exponential and the Poisson, μ in formula 6.6 is the expected *rate* of occurrences, which is the same definition of μ as in the Poisson distribution. *The mean as well as the standard deviation of the exponential is $1/\mu$.*

$$E(X) = \frac{1}{\mu}$$

(6.7)

$$\sigma(X) = \frac{1}{\mu}$$

(6.8)

where μ is the mean rate of the corresponding Poisson process

The graph of the probability density function for μ set to 0.25 is plotted in Figure 6.20.

The exponential distribution approaches the x axis asymptotically. The area under the curve sums to 1 and the area under any section of the curve between points a and b is the probability that the first breakdown will occur between time a and time b. The area to the right of point a is the probability that the first breakdown will occur after time a. The exponential distribution always slopes downward, so the probability that the first breakdown will occur after point a decreases as a increases. What this means is that it is less likely that a new washing machine will survive without a breakdown for five years than that it will survive for four years without a breakdown. In turn, it is less likely that a new washing machine will survive without a breakdown for four years than that it will survive for three years without a breakdown, and so on for any pairs of time periods in which the first number is greater than the second.

Table D5 in Appendix D gives the area to the left of a point a. If the warranty on a new washing machine was one year, what is the probability that it would

FIGURE 6.20
Exponential distribution when
$\mu = 0.25$

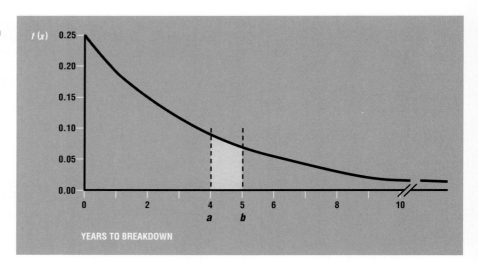

not break down during the warranty period? Since the time interval is one year, μ and λ are the same, or 0.25. The table is similar to the normal table because the first column under the heading μ gives the digits one place to the right and to the left of the decimal point. The first row gives the digits two places to the right of the decimal point. If you look up 0.2 in the first column and 0.05 in the first row, you will find that the intersection of this row and column is 0.2212. Thus 22.12% of all new washers would break down in their first year of use provided that all of the assumptions given in the example were accurate. The percentage that do not break down in the first year is $100 - 22.12$, or 77.88%. Let's use the table to find the probability that a washer will last eight years without a breakdown. The expected rate of breakdowns per eight-year interval is 8 times a λ of 0.25, or 2. With a μ of 2, the probability of a breakdown within eight years is 0.8647. The probability that a breakdown will not occur within the first eight years of use is 0.1353.

All of the above examples have been limited to applications involving time, either the time to the next occurrence or the time between occurrences. Problems that refer to distance, area, or volume can also be analyzed with the exponential distribution because the exponential can be applied to any Poisson process. For example, suppose engine failures on diesel trucks occurred on average once every 200,000 miles. What is the probability that a new engine will last 250,000 miles without a breakdown? The expected number of occurrences per mile, λ, is 1 divided by 200,000 or 0.000005. The expected number of occurrences in 250,000 miles is 0.000005 times 250,000, or 1.25. Using the value 1.25 for μ in the exponential table results in a probability of 0.7135. Note that for the exponential distribution to be accurate in this context, the expected number of failures must be proportional to the distance. If the expected failure rate per mile *increased* as an engine accumulated miles, the exponential distribution would not give accurate results.

EXERCISES FOR SECTION 6.5

6.29 Let X be a Poisson random variable with a mean of 0.5 per year.

(a) What is the probability that the time period between the next two occurrences will be greater than five years?

(b) What is the probability that the time period until the next occurrence will be greater than five years?

(c) What is the probability that the next occurrence will take place in less than ten years?

(d) What is the probability that the next occurrence will take place in four to six years?

6.30 Let X be a Poisson random variable with a mean of 0.2 per week.

(a) What is the probability that the time period between the next two occurrences will be greater than five weeks?

(b) What is the probability that the time period until the next occurrence will be less than five weeks?

(c) What is the probability that the time period between the next two occurrences will be less than 12 weeks?

(d) What is the probability that the time period until the next occurrence will be between 7 and 14 weeks?

6.31 The main computer used by a company for payrolls and inventory control "crashed," on average, once a week. A crash is defined as an unintended interruption of service. Assume that the crashes are independent. What is the probability that the time between the next two crashes will be less than two weeks?

6.32 Suppose the average time between major U.S. airline crashes caused by mechanical failure or pilot error is 65 days. A crash in this context is an accident involving one or more planes. What is the probability that the next crash of a major airline will occur between 30 and 40 days from now? Suppose you were planning to take a trip by air and you decided to wait until after the next crash before leaving to reduce the chance that your plane would crash. Would waiting do you any good?

6.33 Chrysler Corporation cars carried a power-train warranty of 50,000 miles. For the first 50,000 miles the rate of power-train failures per mile was 0.00001. What was the probability that a Chrysler car would have a first failure during the warranty period?

6.34 In 1987 a series of television commercials featured Chrysler's chairman, Lee Iacocca. Mr. Iacocca claimed that Chrysler had the most reliable cars made in the United States because it was willing to back its cars with the longest warranty in the industry. General Motors and Ford responded by offering a 60,000-mile warranty. Chrysler in turn responded with a 70,000-mile warranty. If the rate of power-train failures remained at the 0.00001 per mile figure of exercise 6.33, what is the increase in the proportion of all Chrysler cars that would need a warranty repair because of the longer Chrysler warranty?

6.35 You are planning to build a factory in a location that has averaged one flood every 20 years. You expect the factory to have a 12-year useful life and you have a choice of building a floodproof factory or an ordinary factory. While the decision depends on the potential damage that a flood can cause and on the cost of building a floodproof building, it also depends on the probability that there will be a flood during the factory's useful life. What is this probability?

6.36 A truck rental company averages three rental contracts a week. The company is out of trucks and it will not get one back for two days. What is the probability that no customers will ask to rent a truck until one becomes available?

SUMMARY

The chapter began with a discussion of continuous random variables and the concept of a probability density function. A continuous random variable was defined as the numerical outcomes of an experiment that can take any value over a range of numbers. The probability density function plots the density of the continuous random variable. The density under the continuous curve is so defined that the total area under the curve equals 1. The continuous uniform distribution was used as a simple vehicle to introduce these concepts.

Most of the chapter was devoted to the normal distribution, including the z score, how to find areas under a normal curve, and how to use the normal to approximate both the binomial and the Poisson. The normal distribution is important because it approximates many physical phenomena and it shows up often in statistical theory. The z score was interpreted as the number of multiples of a standard deviation between the mean and any point on a normal distribution. The z score formula transforms any normal distribution into the standard normal distribution. The standard normal distribution has a mean of 0 and a standard deviation of 1. The subject of finding areas under the normal curve was discussed in great detail. You should be able to find areas in the center or the tail of a distribution and areas that represent a slice of the distribution. Given an area on a normal distribution, you should also be able to find the boundary line or lines that mark that area. The normal approximations to the binomial and the Poisson let you easily do large problems that would bog down even a large computer if the calculations were done directly with the binomial or Poisson formula. The rule of thumb for approximating the binomial with the normal is that

$n\Pi$ and $n(1 - \Pi)$ have to be greater than or equal to 5, and the rule of thumb for the normal to Poisson approximation is that μ has to be greater than 5. Since for both approximations you are using a smooth curve to approximate a stair-step function, the continuity correction improves the accuracy of the approximation.

The last topic in the chapter was the exponential distribution, which can be applied to any Poisson process. The exponential distribution is used to find the probability that the first or next occurrence of an event will be within a given interval. This distribution is often applied to warranty problems.

NEW TERMS

continuity correction An adjustment to the z score formula used for normal approximations to the binomial and the Poisson. The adjustment allows a smooth distribution (the normal) to approximate one that looks like a series of steps (either the binomial or the Poisson). The correction involves subtracting one-half from the number of successes for a lower bound or adding one-half to the number of successes for an upper bound.

continuous uniform distribution A continuous distribution in which all equal intervals in the range of the distribution have the same probability.

exponential distribution A distribution used to calculate the probability that the first or next occurrence of an event will be in a given interval. If the distribution of occurrences is Poisson with a mean of μ, then the time between occurrences follows an exponential distribution with a mean of $1/\mu$.

normal distribution A family of continuous distributions that have similar shapes. All normal curves are symmetrical and have one mode. Normal curves are often called bell-shaped curves because they look like the silhouette of a bell.

probability density function A function that assigns a number called "density" to every value in the range of a continuous distribution in such a way that the total area under the density function equals 1 and the area under the density function over any section of the range equals the probability that the continuous random variable will be observed in that section.

standard normal curve The particular normal curve that has a mean of 0 and a standard deviation of 1.

NEW FORMULAS

Probability density function of the continuous uniform distribution

$$f(x) = \frac{1}{b - a}$$

(6.1)

Expected value of the continuous uniform distribution

$$E(X) = \frac{b + a}{2}$$

(6.2)

Variance of the continuous uniform distribution

$$\sigma^2(X) = \frac{(b - a)^2}{12}$$

(6.3)

Probability density function of the normal distribution

$$f(x) = \frac{1}{\sqrt{2\pi}\,\sigma}\, e^{-1/2[(x - \mu)/\sigma]^2}$$

(6.4)

z score formula

$$z = \frac{x - \mu}{\sigma}$$

(6.5)

Probability density function of the exponential distribution

$$f(x) = \mu e^{-\mu x}$$

(6.6)

Expected value of the exponential distribution (where μ is the mean of the corresponding Poisson process)

$$E(X) = \frac{1}{\mu}$$

(6.7)

Standard deviation of the exponential distribution

$$\sigma(X) = \frac{1}{\mu}$$

(6.8)

 ADDITIONAL EXERCISES

An asterisk designates an exercise of greater than ordinary difficulty.

6.37 A filling machine for beer bottles can be set for any mean value but it cannot be set to have every bottle contain the same amount of beer. Assume that the distribution of ounces per bottle is normal with a standard deviation of 0.5. At what mean level should the filling machine be set so that the probability that a bottle contains less than 16 ounces is 0.01?

6.38 The distribution of the weights of chicken eggs is normal with a mean of 1 ounce and a standard deviation of 0.2 ounce. Eggs classified as "jumbo" sell at a higher price per ounce than other sizes. The heaviest 5% of all eggs are classified as jumbo. What is the minimum weight for a jumbo egg?

6.39 A college restricts its admissions to students whose SAT scores are in the top 10% of all scores. You are wondering if you can be admitted. You have a combined verbal and math SAT score of 1,300. If the distribution of SAT scores is normal with a mean of 1,080 and a standard deviation of 48, is your score high enough?

6.40 The worldwide demand for wheat is almost perfectly inelastic because wheat consumption is not sensitive to changes in the price of wheat. The supply of wheat varies from year to year, primarily because of changes in weather. The current equilibrium price of wheat is $4.40 a bushel based on an annual production of 480 million metric tons. The expected size of next year's crop is 500 million metric tons, but it could vary because of the weather. Suppose the distribution for next year's crop is normal, with a mean of 500 million metric tons and a standard deviation of 15 million metric tons. Assuming that the demand for wheat does not shift, what is the probability that the price of wheat will fall?

6.41 You run a phone solicitation service. Your company is hired to call people for small donations. You are planning a campaign for a candidate for the U.S. Senate. Your phone operators will ask for $10 donations, and for each $10 given, your company keeps $4. The key to making a solicitation campaign a success is a list of likely donors. The candidate has a list of 10,000 households that have been contacted in the past by candidates of the same party and 20% of them have given $10. Assuming that the probability that any household on the list will donate $10 is 0.2, what is the probability that your company will earn more than $8,200?

6.42 A cigarette vending machine sells packs of cigarettes for $1. The machine inadvertently swallows 1% of all of the dollars put into it. Swallowing $1 means giving a customer no cigarettes and no money back. If 100 customers put in $1 each and every attempted purchase is an independent trial, what is the probability that the machine will generate $4 or more in unearned income in a day?

6.43 The Yenta matchmaking service claims that 50% of all of the couples it introduces eventually marry. You've canvassed 50 friends who have used the service, and 30 of them claim that there is no chance that they will marry the person to whom Yenta introduced them. Assume that friends you canvassed are a random sample of all of the customers of the service and that none of your friends was introduced to another of your friends. What is the probability that 30 or more of your friends will not eventually marry the person to whom they were introduced if the success rate claimed by Yenta is accurate?

6.44 Your brother-in-law spent his family's grocery money on lottery tickets. He bought 200 tickets for $1 each. Your sister wants to sell you the tickets for what they cost so that they'll have something to eat. Suppose a winning ticket pays $10 and the probability that any ticket is a winner is 0.05. If you agree to buy the tickets, what is the probability that you will break even or come out ahead when the winning tickets are announced?

6.45 You run the box office at a concert hall and you have booked a concert that is sold out. The hall seats 2,500 and the selection of seats is on a first-come, first-served basis. Assuming that each ticketholder's behavior is independent and that the probability that a ticketholder will not show up at the concert is 0.04, how many seats can you oversell so that the probability of running out of seats in the hall is 0.05?

6.46 You own an international construction company and you are planning to bid on clearing a section of war-torn Beirut, Lebanon. Part of the clearing process is the removal and demolition of unexploded artillery shells. A survey of five randomly selected hectares of land in the green-line district of Beirut found an average of six unexploded shells per hectare. Assuming that this rate is accurate and that the locations of the shells in the construction area are independent, you plan your bid on the basis of 650 shells for a 100-hectare area. What is the probability that you have underestimated the number of unexploded shells?

6.47 A city of a million people has an average of 14 deaths per day. Assuming that the deaths are independent, what is the probability that there will be more than 5,000 deaths in a year? If you ran a funeral home that had one-tenth of the funeral business, what is the probability that in a year you would have more than 450 funerals?

6.48 The first baby born in a new year in a city usually gets his or her picture in the local paper and donations of free baby products from local firms. Suppose you or your spouse gave birth to a baby at 12:10 A.M. on January 1. You won't know if your baby was the first until all of the other local hospitals are checked, but you know that a baby is born in your city, on average, once every 20 minutes. What is the probability that someone beat you?

6.49 You have entered a car in a classic motor car race. "Classic" for the purposes of this race means a car that was manufactured before World War I. These cars are not very reliable; they break down, on average, once every 10 miles. If the race is 60 miles long, what is the probability that your car will go the distance without a breakdown?

6.50 You are a track inspector for a railroad. Cracks in tracks that are ten years old average one per mile. Given that you've gone two miles without observing a crack, what is the probability that a crack will be observed in the next mile of track? Assume that the cracks are independent.

6.51 (This is the same exercise as 5.47 in Chapter 5.) You are the admissions officer at a university. Your university has room for a freshman class of 5,000. When you admit students, you know that some of them will not attend your university. Your experience has been that the probability that an admitted student will attend your university is 0.75. To compensate for this slippage, you plan to admit more than 5,000 students. If you admit 6,500 students, what is the probability that more than 5,000 will accept? (Use Tchebychev's inequality for this problem and the normal approximation to the binomial.)

6.52 (This is the same exercise as 5.48 in Chapter 5.) You are making a chestnut stuffing for a turkey. Your recipe calls for 150 chestnuts. Your experience has been that one of every ten chestnuts is spoiled. If you buy 200 chestnuts, what is the probability that you will have enough for your stuffing? (Use Tchebychev's inequality for this problem and the normal approximation to the binomial.)

6.53* You manufacture 1,000 high-security steel door frames and 1,000 steel doors. The door frames are supposed to be exactly 3 feet wide, but they came out normally distributed with a mean of 3 feet and a standard deviation of 0.01 foot. The doors are supposed to be exactly 2.95 feet wide, but they came out with a normal distribution centered on 2.95 feet with a standard deviation of 0.009 foot. When you match the door frames with the doors, how many frames and how many doors will be left over without a matching part?

6.54* You are a professor and you have to make copies of a one-page exam for 40 students in your class. Your copying machine makes bad copies 20% of the time. Each copy is an independent trial. You have to get the copies immediately because it is just before exam time and you don't have time to check for bad copies. How many copies should you make so that the probability of getting at least 40 good copies is 0.99?

6.55* The Jarvik III mechanical heart is so reliable that it averages only one breakdown per 100 years of service. Suppose 10,000 patients use this mechanical heart. Find the range of the number of failures in a year which represents the central 50% of the distribution of failures.

6.56* Suppose your company owned a subsidiary in a less developed country which manufactured a toxic gas used to make insecticide. You are concerned that poor maintenance at the plant could cause a gas leak that could kill the people and animals who live near the plant. For a fatal leak to occur, the valve on the tank that holds the gas has to stick open, the automatic water sprayer that detoxifies the escaping gas has to fail, and the emergency alarm system that will summon a crew of technicians also has to fail. If any one of these three safeguards is working, fatalities will be prevented. Through extensive tests on similar valves you have established that they fail, on average, once in a year of use. The water sprayers fail to operate, on average, once in each five years of use. The alarm fails, on average, once in every two years of use. Assuming that the three systems are independent and that the maintenance workers in the plant do not repair a system once it has failed, what is the probability that a fatal gas leak will occur in the next five years?

6.57* You take your laundry to a laundromat and pay a quarter each time you run the dryer. After much study you have determined that the length of time that the dryer runs on one quarter is a continuous uniform random variable bounded by 6 and 12 minutes. If a pile of your wet laundry needs 19 minutes of dryer time and you will use only one dryer, what is the maximum probability that you will need more than three quarters? Assume that the amount of time the dryer gives you for each quarter is independent.

7

Sampling Methods, Sampling Distributions, and the Properties of Estimators

state recently passed a law requiring the use of automobile safety seats for children under four. Since enforcing laws costs money, the state wanted to know if the law would do any good. A survey was commissioned to estimate the proportion of young children riding in safety seats before and after the law went into effect. Unfortunately, neither the employees of the state agency that commissioned the survey nor the not-for-profit child-safety lobbying group that carried out the survey knew what they were doing. The survey used teams of interviewers stationed in the parking lots of suburban shopping centers throughout the state on Saturday afternoons. Their rationale was that this location at this time was the best for finding small children in cars. Large signs that read "Child Safety Seat Survey—Free Gift" were placed near the exits of the parking lots. The gifts were intended to entice motorists to stop and be questioned. Would this survey design provide a sample that was representative of all drivers in the state who had young children? The answer would be yes if all drivers with young children were equally likely to shop at a suburban mall on a Saturday afternoon, equally likely to be enticed by free gifts, and equally likely to want to stop to voice their opinions about child safety seats. In other words, the answer to the question is no. The survey was worthless for its stated purpose of estimating the proportion of young children who rode in safety seats.

This chapter presents the basics of sampling theory, particularly how to get a sample that is representative of a population. It covers strategies for collecting samples which can save time or money in a variety of situations. We will be discussing simple random samples, stratified samples, cluster samples, and systematic samples. The concept of a sampling distribution, the sampling distribution of the mean, and the sampling distribution of the proportion come next. The last section discusses the desirable properties of estimators, which are used to predict population parameters from sample data. These desirable properties are unbiasedness, efficiency, and consistency. It will be a quick ride through a lot of material, so fasten your seat belts.

7.1 TYPES OF SAMPLES

 ### 7.1A REASONS FOR SAMPLING

Samples are collected to obtain information about a population. There are several reasons for using a sample instead of a **census,** which is *a survey of an entire population.* One obvious reason is destructive testing. A manufacturer of flash cubes who tested every one of his flash cubes to see how many would fail

CASE STUDY

RESEARCH ON RESEARCH

The survey research industry is big business. According to the industry trade group, CASRO, the top 100 firms had research-generated revenues of $1.79 billion in 1985. Some of these firms are well known to the public because their polls are widely quoted in newspapers and on television. The A. C. Nielsen Company, Louis Harris and Associates, and the Gallup organization are in this category. Most of the 100 largest firms, however, are not household names; they do business primarily with corporations. One of these less well-known firms, Walker Research, conducts a biennial survey on the number of times that consumers have participated in and their attitudes toward research studies. Walker Research is the 14th largest consumer research firm, with $20.7 million in revenues in 1985.

For its Industry Image Study, Walker Research interviews by telephone 500 randomly selected male and female heads of households. Most of the results described here are from the June 1986 survey. Seventy percent of the respondents said that they had previously been polled and 33% said that they had been polled during the past year. As you can see in Figure 7.1, both of these percentages have shown a rising trend.

The consumer research industry depends on the goodwill of people it asks to be surveyed. Unhappy experiences with surveys could eventually turn off so many consumers that it could become impossible to create samples that were representative of the general public. One experience that is usually an unhappy one is a sales pitch disguised as a survey. In the 1986 poll, 40% of the respondents said that they had been exposed to such sales pitches. A related area of concern for the industry is distrust among consumers. Some 68% of the respondents agreed with the statement that "the true purpose of the survey is not disclosed" and only 62% agreed that "survey research firms maintain the confidentiality of answers." One of the purposes of Walker Research's Industry Image

to flash would know the population proportion of failures but would have only burned or defective flash cubes to sell. Another reason is that the information provided by a census may cost more than it is worth. A shampoo manufacturer might want to know the proportion of the population of users of its brand of shampoo who would prefer a new scent for its brand. Identifying and surveying each user among the U.S. population would cost much more than this information would be worth. A closely related reason is the timeliness of the information. Sometimes it is better to have a good idea of what is going on right away than to wait a long time for precise information. Wheat prices are sensitive to changes in the size of the wheat crop, which in turn is sensitive to changes in weather and the number of acres planted. If you wait until after the

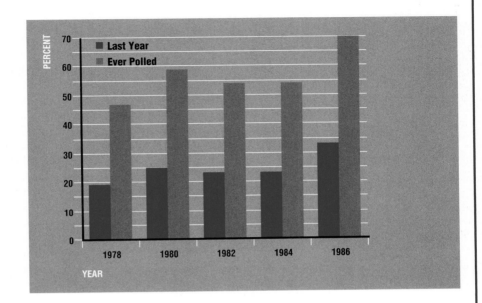

FIGURE 7.1
Participation levels in polls among a representative sample of heads of households

Study is to track these attitudes and experiences over time so that the industry can monitor its efforts to maintain the public's goodwill. The industry promotes cooperation with survey research through its American Opinion Week campaign. The campaign has used advertisements with the slogan "Your opinion counts." Given the frequency of interviews, the income generated by the industry, and the importance of information collected for marketing decisions, it is quite true that your opinion counts.

harvest, you'll know the exact size of the crop, but timely information may save you money. Most commodity brokerage firms hire experts who travel around the country sampling wheat fields to update on a daily basis their estimate of the size of the crop. Finally, a true census is often impossible. The U.S. Decennial Census is supposed to count everyone in the United States, but despite its best efforts, it misses some people, such as illegal aliens who fear to be found, street people who have no permanent address, and illiterates who cannot fill out the survey form. A well-designed sample that made a particular effort to find and interview such people would give a more accurate estimate of the U.S. population than the U.S. Census now does, but the U.S. Constitution requires that a census be taken every ten years.

7.1B PROBABILITY VERSUS NONPROBABILITY SAMPLES

Even though sampling is cheaper than a census, sampling costs time and money and it is important that, for a given level of spending, the sample be designed to obtain estimates of the population parameters which are as accurate as possible. The description of the types of samples in this section will focus on this issue. The most basic distinction in the types of samples is between probability samples and nonprobability samples. A **nonprobability sample** is *a sample in which the probability that an element of the population will be drawn is not known.* A **probability sample** is *a sample in which each element of the population has a known and nonzero chance of being selected.* The advantage of probability samples is that the results of taking a sample can be projected to the population. "Projected" means that you can make precise statements about the probability that the sample statistic is within a given distance of the population parameter—the probability, for example, that the sample mean is within a given distance of the population mean. Nonprobability samples are not projectable. Most of our discussion will be on probability samples, but to give you an idea of the difference between probability and nonprobability samples, we'll discuss a few examples of nonprobability samples. A **convenience sample** is what the name implies: *a sample selected on a nonrandom basis to reduce the time and cost of drawing the sample.* You've seen newscasts in which a reporter stops people in the street and asks their opinions about some current event. The people who are walking around downtown at noon and are willing to stop and talk to a TV reporter are not representative of the entire city's population. Convenience samples are prone to **self-selection bias:** *systematic differences between the sample statistics and the population parameters which occur when the choice of being in or out of a sample is left to each person who is selected as a potential member of the sample and the selection process is not random.* Newspapers and magazines often ask their readers to write or mail in their opinions on a current issue. The readers who respond are the ones who feel most strongly about the issue, so that it is possible, for example, to find a majority of the respondents opposing abortion while a majority of the population favors abortion.

Another nonprobability sample is a **judgment sample,** or *a sample in which the elements of the population to be selected are decided by expert opinion.* This procedure can be fine if you have a reliable and knowledgeable expert. A subtle problem is that biases can inadvertently creep into the expert's selection process. Suppose you want to test a computer program designed for fourth-graders. You want to see how children with average reading ability do with the program. To save time and effort, you get several teachers to identify students of average reading ability in their classes. The teachers will generally have the expertise to identify the average students, but they may subconsciously select good students so that they themselves will look good, or they may select some real troublemakers just to get them out of the room for awhile, or they may think that you really want students with computer experience. The list of subconscious and even conscious reasons could be continued. The problem with judgment samples is that you have no idea which motivations may be at work. Don't treat these criticisms of judgment samples as a warning never to use them. There are occasions when constraints of time or money make them the only practical alternative and when a disinterested and knowledgeable expert can produce a sample that leads to accurate estimates.

■ 7.1C SIMPLE RANDOM SAMPLES AND A RANDOM-NUMBER TABLE

The most common probability sample is a simple random sample. Suppose you were selecting one student from a fourth-grade class. A simple random sample of size 1 requires that every student in the class is equally likely to be chosen. If you wanted to select five students from the class, every possible group of five students should be equally likely to be chosen. In general, a **simple random sample** *is a sample of size* n *in which all possible groups of size* n *are equally likely to be chosen.* The beauty of simple random samples is that high and low values of the population tend to balance each other so that a sample statistic is usually close to the population parameter. The sample mean, for example, tends to be very close to the value of the population mean. We'll be more precise about this point in the next two sections.

Ensuring that every group of size *n* is equally likely to be selected is often difficult. You have to begin with a **sampling frame,** which is *a list of all of the elements in the population.* So you need a list of the names of all of the students in the fourth-grade class. Next you need a mechanism for selecting your sample which guarantees that every group of size *n* is equally likely to be chosen. Suppose you put the names on pieces of paper, put the papers into a hat, and reached in blindfolded. All sorts of things could go wrong: kids with names near the end of the alphabet might be near the top of the pile; long names might be on bigger pieces of paper; the slips of paper bearing names adjacent on the list might stay close to each other in the hat, so that if one name were drawn, the other would be likely to be drawn. It can take considerable effort to design a physical selection process that ensures equal probabilities.

The draft lottery conducted on national television in 1970 is a famous (or perhaps infamous) example of a selection process that was supposed to be a simple random sample but did not turn out that way. The draft lottery was used to determine which eligible men would be drafted. The selection process consisted of pouring balls with January birth dates into a wooden box, followed by the February balls, followed by the March balls, and so on, up to the December balls. After each month's worth of balls was poured into the box, all of the balls were mixed with a cardboard divider. Thus the December balls were only mixed in once and the November balls only mixed in twice. When the balls were poured into a deep bowl for the final selection, the December and November dates were near the top and the January dates were near the bottom. The early lottery numbers were "rich" or overrepresented with December and November dates. Few January or February dates were drawn early, so most people with such birth dates were not drafted. My own birth date is in January and I was eligible for the draft during that lottery, so I greatly appreciated the bungling.

You may be asking yourself, "Isn't there an easier way of doing this?" There is, but as it's not so dramatic as drawing balls out of a bowl, it isn't good television. The easier way is to use a random-number table. You'll find one in Table D6, in Appendix D. The single-digit numbers in a random-number table are generated by a computer program designed to ensure that each digit is independent and that the values 0 through 9 are equally likely. Suppose you had an alphabetical list of 40 students in a fourth-grade class and you assigned each student a number from 1 to 40. To draw a sample of five students using a random-number table, you need to find five two-digit numbers in the range 1 through 40. You can begin anywhere on the table and go in any direction. The only wrong way to use a random-number table is to search for some favored numbers. You should use a pattern: start in the first row, for instance, and take the first two digits in the row as the first number, the next two digits in the row as the second number, and so on. Skip any two-digit numbers above 40 and skip any number that has already been picked. Keep searching until you have five two-digit numbers in the range 1 through 40. The students with those numbers will be your sample. This procedure gives you a sample without replacement. If you did not skip any numbers that had already been picked, the sample would be with replacement. With a one-page random-number table it is easy to run out of numbers if you are trying to draw a large sample. At one time statisticians used published tables as large as the New York City phone book to draw large random samples, but as computers have become more readily available, they have replaced random-number tables in practical work.

■ 7.1D PROBABILITY SAMPLES THAT ARE ALTERNATIVES TO SIMPLE RANDOM SAMPLES

There are three problems with simple random samples. First, there are occasions when it is either impossible or prohibitively expensive to construct a **sampling frame,** that is, *a list of every member of the population.* Second, a simple random sample does not take advantage of any prior information about the population. For a given sample size, prior information about the population can generally be used to gain a more accurate estimate. Third, a simple random sample draws observations scattered across all of the data. It may be possible to save money by structuring the sampling procedure so that the actual sample is not spread across all of the population while retaining the accuracy of a simple random sample.

Let's cover the three problems in reverse order. Suppose you are an accountant for a large national chain of women's clothing stores. You are given the assignment of sampling 1% of all sales slips from the past year to check for accuracy. A simple random sample requires that you collect all of the sales slips, assign each one a number, and use a computer or a random–number table to draw a 1-in-100 sample. If the sales slips are kept at the local stores and you have many stores in the chain, you are talking about a lot of work. Suppose that instead you visit each store and begin by randomly selecting a number from 1 to 100. If the number 61 comes out, you pull the 61st sales slip from the beginning of the file, the 161st, the 261st, the 361st, and so on until the end of the file. This procedure is called a systematic sample. A **systematic sample** is *a sample drawn by dividing the population into equal groups, randomly selecting a number equal to or smaller than the size of the groups, and drawing the observation in each group which corresponds to that number.*

If the population cannot be divided into equal groups of the size required, then the last group contains the number of observations that are left over. If there were 5,060 sales slips and you wanted a 1-in-100 sample, you would have 50 groups of 100 and one group of 60. If your randomly selected number for drawing your sample was 61 and you wanted at least a 1-in-100 sample, you would have a problem. The last group would not have enough observations to let you draw the 61st. The procedure in this case is to "wrap around." The 61st observation in the last group is the first observation in the first group. Systematic samples are equivalent to simple random samples if there is no pattern or periodicity in the data. Suppose most of your stores sold to an average of 100 customers in a day and that the sales slips written near closing time had more errors than other sales slips. If the random number used for systematic sampling was in the range 95 to 100, your sample would tend to overstate the extent of errors; and if the random number was lower than 95, the sample would tend to understate the extent of errors. Patterns in the data are unusual, but they are something you should watch out for if you want to use systematic sampling.

Another method of restricting the sample so that it does not spread over the entire data set is cluster sampling. A **cluster sample** is *a sample drawn by identifying subgroups of the population as clusters on the basis of geographical proximity and then drawing a random selection of clusters.* Suppose there were 400 stores in your chain and you did not have time to visit all of them. You could randomly select 40 from which to draw your sample of sales slips. Each of the 40 selected stores is a cluster in which the sales slips are kept together. Once you visited a store, you could check all of the sales slips in that store or use a simple random sample to select the individual sales slips to be checked or use systematic sampling. In other situations you could select city blocks in which to interview households rather than individual homes scattered throughout a city, or in a national election you could select precincts in which to interview voters as they left the polling booths instead of selecting individual voters scattered across the country. The only danger with cluster sampling is that if too few clusters are drawn, some atypical precincts, blocks, or stores can lead to an inaccurate estimate.

The issue of using prior information must also be considered. Suppose the local stores differed in some characteristic that was known to affect the accuracy of their sales slips. For example, some of the stores might use electronic cash registers that read a code number off the tag on each item and automatically create a sales slip, while other stores relied on the sales clerks to key the prices and item codes on the cash register. The former method is known to cause fewer errors than the latter. Suppose half of your 400 stores use one of the methods and half use the other. If you take a random sample of 40 stores for your clusters, you may by chance over- or underrepresent one type of store. It is even possible, albeit highly unlikely, that all of the stores chosen in the sample will be of the same type. To use the information about the types of stores, you could draw a sample of 20 stores from the 200 that use the tag-reading cash registers and 20 stores from the 200 that use the ordinary cash registers. In this case you have done stratified sampling. A **stratified sample** is *a sample derived by dividing the population into strata or layers and drawing a random sample from each stratum.* Stratified sampling can be applied to any number of strata. The variant of stratified sampling used in this example is proportional allocation. Each stratum of the sample was allocated the proportion of the total sample that corresponds to the proportion of that stratum in the population. If 25% of the stores had the tag-reading cash registers, then 25% of the sample would be taken from those stores. There are other allocation schemes for stratified sampling which depend on knowing even more about the population. If you knew the population standard deviation in each stratum, for example, or if you knew two or more characteristics that affect accuracy (such as whether the store has tag-reading cash registers or whether the sales clerks are primarily part-time or full-time employees), you could take advantage of the information with a stratified sample. One of the golden rules of statistics is to use all of the information that you have. We won't go into the mechanics of the more elaborate allocation schemes; they can be found in any text on sampling theory.

Inability to create a sampling frame is the most intractable problem. Suppose you wanted to interview a random sample of white, male employed blue-collar workers in a city on the subject of job safety. No one has a list of all such workers. You might visit firms and try to get their permission to interview their employees, but you would tend to miss blue-collar workers whose jobs kept them outdoors, workers in very small firms, self-employed workers, and workers in firms with such poor safety records that the bosses would refuse to let you interview their employees. One of the data sets used in the author's research is a citywide survey of white, male employed blue-collar workers on the subject of job safety. The designers of this survey randomly selected homes and apartments in the city and asked the residents if any of them were white, male employed blue-collar workers and willing to be interviewed. This design ensured that every member of the population was equally likely to be asked to be interviewed, since all members of the population lived in an apartment or a house. Residents of mental institutions or jails were not part of the population of interest. The surveyors had to visit an average of 33 residences to get one usable interview. Creating a random sample can sometimes be expensive. In general, if you cannot make a sampling frame for the population of interest, you have to find a larger population that is identifiable (in this example it is the list of all residences) and includes the population of interest.

7.1E NONRESPONSE BIAS

Another source of bias in surveys is nonresponse bias, which differs from self-selection bias *in that the respondents were randomly selected*. **Nonresponse bias** consists of *systematic differences between the sample statistics and the population parameters which occur when some potential respondents in a randomly selected sample refuse to be interviewed*. With both types of bias there is a strong possibility that the decision to be interviewed is not independent of characteristics of the population under study. Most commercial contracts for professional interview firms, such as Market Opinion Research in Detroit and National Opinion Research Center in Chicago, specify both the number of completed interviews and the response rate. If the contract calls for a 90% response rate and the actual response rate is running below this figure, the polling company has to use follow-up questionnaires or personal interviews or even literally beg some of the people drawn in the sample to be interviewed.

The purpose of these contract clauses is to limit the effect of nonresponse bias. Suppose a survey of 1,000 families with children of elementary school age was about preferences between two home computers, the Atari and the Commodore. If the survey's response rate was 90% and the results indicated that 70% of the respondents favored the Atari (that is, 630 families out of 900 interviewed), how far off could the sample results be from those of a survey that had a 100% response rate? If all of the 100 missing respondents had favored the Atari, the sample proportion would have been 73% (730 out of 1,000 families), and if all of the missing respondents had favored the Commodore, the sample proportion would have been 63% (630 out of 1,000 families).

There isn't time to get too deeply into sampling theory in an introductory statistics course. The above discussion of types of samples and sources of bias in sampling was meant to give you an overview of the main issues. Some important practical topics, such as how to word questions so as to elicit more honest answers, or how to repeat questions in an interview to check for consistent responses, have been skipped altogether. Almost all of the subsequent examples are based on simple random samples because the simple random samples are the easiest to analyze and also the most common.

CASE STUDY

A RANDOMIZED RESPONSE SURVEY

Suppose your company wanted to estimate the extent of recreational drug usage among its employees. Since this activity is illegal, employees would be very sensitive about answering any questions on this subject. Even if they were assured that their responses could not be traced back to any individual, they might be suspicious and not answer the questions truthfully. Such suspicions are not entirely groundless. There have been a few instances when unethical survey firms secretly marked mailed survey forms so that the actual respondents could be identified while the cover letters accompanying the forms promised the respondents anonymity. A randomized response survey is a technique that can be used to elicit truthful answers to sensitive questions. The technique consists of having each respondent randomly select the question he or she will answer. For example, the company might survey 100 employees and ask each of them to fill out the survey form in private. There could be two questions: one innocuous question, such as "Is your favorite flavor of ice cream vanilla?" and one sensitive question, such as "Have you used marijuana, cocaine, or heroin in the past 30 days?" Each employee in the sample would be asked to flip a coin and then answer the ice cream question if the coin came up heads or the drug-use question if the coin came up tails. The employees would be safe: even if the survey forms carried their names in invisible ink, it would be impossible for anyone to know which of the two questions they had answered.

You may be wondering what benefit a company can gain from having honest answers to its questions if it does not know which question has been answered. The key is that the company already knows the proportion of the population that prefers vanilla ice

EXERCISES FOR SECTION 7.1

7.1 In 1936 *Literary Digest* magazine conducted a poll in order to predict who would win the presidential election. Ten million survey forms were mailed to the magazine's subscribers, to people who had registered a car in any state, and to people whose names were listed in any U.S. telephone directory. Some 2.5 million survey forms were filled out and returned to the magazine. The poll predicted that Alfred Landon would win by a large majority. Franklin Roosevelt won instead, and the magazine went bankrupt. Comment on the problems in the design of the survey.

7.2 In the 1950s a refrigerator manufacturer conducted a door-to-door survey to ask housewives if they planned to buy a refrigerator in the next year. The interviewers assured the respondents that they were not salesmen and that the information was being sought for product planning purposes. The company cream, and with this information it can estimate the proportion that answered yes to the question about drug use. Suppose 60% of the population is known to prefer vanilla ice cream and 50 of the 100 employees in the survey responded yes to the question they answered. What proportion of the company's employees answered yes to the drug question?

Assuming that the coins used to decide which question was answered were reasonably fair, about 50% or 50 of the 100 employees must have answered the ice cream question and about 50 of the employees must have answered the drug question. The number of yes answers among the employees who answered the ice cream question would be approximately 60% of 50, or 30. Recall that 60% of the population is known to prefer vanilla ice cream. The remaining 20 yes answers out of the total of 50 yes answers is an estimate of the number of yes answers to the drug question. If 20 of the 50 employees who would have been expected to answer the drug question answered it with a yes, the estimated proportion of recreational drug users is 0.40.

There are other methods of letting the respondents determine which question they will answer besides flipping a coin. The decision could be based on whether they were born on an odd or an even day, or whether the last digit of their social security number is odd or even, or whether they pulled a white or a black ball out of a bag. Almost anything can serve for the innocuous question as long as its proportion in the population is known. Randomized response surveys are expensive in the sense that half of the answers are to the wrong question and those answers are essentially thrown away. The expense can be justified, however, by the greater accuracy. People are much more likely to answer sensitive questions truthfully if their privacy can be assured.

was overjoyed when it tabulated the results of the survey: it seemed that refrigerator purchases were to be at an all-time high. Production was geared up to meet the anticipated demand, but it never materialized. Comment on any problems in the design of the survey.

7.3 Suppose you wanted to test-market a new shampoo for dogs in five pet-grooming shops. Use the Yellow Pages to identify the sampling frame and the random-number table in Appendix E to draw your sample. Describe the pattern you used in the random-number table and list the five selected pet-grooming shops.

7.4 You run a company that pipes background music into stores. You want to estimate the effect of playing rock music on the average spending of supermarket shoppers by interviewing 100 randomly selected shoppers who visit a particular supermarket when rock music is being played and asking them to examine their checkbooks to see if they spent more on their current shopping trip than on their last shopping trip. You know that 70% of all grocery shoppers are women and that the average spending of female grocery shoppers is $70 while the average spending of male grocery shoppers is $35. How should you allocate your sample?

7.5 You have commissioned a survey of new homeowners in the United States to estimate their preferences regarding Venetian blinds versus drapes. You plan a national probability sample of 5,000 respondents and you have specified a response rate of 95%. The completed survey achieves exactly the minimum response rate of 95%. What is the maximum difference between the proportion of respondents who favor Venetian blinds in the sample and the proportion who favor Venetian blinds among all 5,000 potential respondents if 82% of the respondents said that they favored Venetian blinds?

7.6 As an experiment, the Eat Here chain of restaurants had its waiters hand out 10,000 survey forms to customers after they had paid their bills and ask each customer to fill out and mail back the stamped form. The key question on the form was "Would you come back to our restaurant for another meal?" Of the 10,000 forms, 8,500 were mailed back, and of those 8,500, 8,000 customers answered no while 500 answered yes. What are possible limits of the proportion of customers who would come back among the 10,000 given the survey form?

7.7 When Isaac Schlimazel was a high school senior, his main goal in selecting a college was to get away from home and to have a good time. He bought a guidebook on colleges which described the results of the following survey: A questionnaire was placed in every college newspaper in the country and one of the questions was "Is your college a great party school?" The survey asked students to fill it out and mail the completed survey to the editors of the guidebook. The number of positive responses to the party-school question received from each college divided by the number of students enrolled at the college was used to rank colleges. The top party school turned out to be Cal Tech. Isaac immediately applied to Cal Tech and was accepted. When he got there he decided that there must have been some terrible mistake, for the classes were hard and the students spent most of their time studying. Can you explain what might have happened?

7.8 During the 1980 presidential election, when Ronald Reagan, on the Republican ticket, ran against Jimmy Carter, on the Democratic ticket, the usually reliable exit polls sharply underestimated Reagan's margin of victory in predominantly blue-collar-union precincts. A similar phenomenon occurred in these precincts during the primary elections for the 1964 presidential nominations, when the exit polls underestimated the proportion of votes going to George Wallace, a candidate who at the time favored racial segregation. What happened to throw the exit polls off and how could the survey firms have come up with more accurate estimates?

7.9 Survey firms have long recognized that answers to sensitive questions can vary depending on who is asking the question. For example, white interviewers who ask blacks if they believe racial prejudice is a serious problem at their workplace will get quite different responses than black interviewers who ask the same question. Black interviewers who ask whites the same question will also get different responses than a white interviewer with a white respondent. Sometimes the survey firm cannot set up the interviews so that only whites question whites and blacks question blacks, or women question women, and so on. To get around this difficulty, an interview firm can ask respondents to take the survey form to a private room and answer the question about racial prejudice with a yes or a no if the last digit of their social security number is less than 5. If the last digit of their social security number is 5 or more and their birth date is an even number, they are to answer yes; otherwise the respondents are to answer no. Suppose 200 blacks were interviewed in this manner and there were 120 yes answers, and that 200 whites were interviewed and there were 101 yes answers. Estimate the population proportions of blacks and whites who believe that racial prejudice is a serious problem at their workplace.

7.10 A state civil rights commission wants to estimate the proportion of female state employees who have been sexually harassed on the job. The commission draws a sample of female state employees and asks each employee in the sample to take a bag with three balls to a private room. While in the room the employees are to draw one of the balls from the bag. If the ball drawn is white, they are to write the answer "yes" on the survey. If the ball drawn is black, they are to write the answer "no" on the survey. If the ball drawn is green, they are to answer yes or no to the question "Have you ever been sexually harassed on the job?" If 45 of 100 employees sampled have answered yes, what is the estimated proportion of employees who have been sexually harassed on the job?

7.2 SAMPLING DISTRIBUTIONS AND THE CENTRAL LIMIT THEOREM

7.2A SAMPLING DISTRIBUTIONS

A **sampling distribution** is *a function that gives the probability of every possible value of a sample statistic for a specified population and sample size*. For example, your statistics class could be the population and the sample statistic could be the mean age of a sample of five students. Suppose your class had 100 students and the sampling was done without replacement. The sampling distribution would be the function that would specify all of the possible sample means and their probabilities. There would be $_{100}C_5$ possible samples, and each of those samples would have a sample mean age. This is a large number. Later in the chapter, in section 7.4, you'll find the sampling distributions of some small populations listed. In addition to the sample mean, some possible sample statistics are the sample median, the sample variance, and the sample proportion. For sampling distributions that are finite and discrete, every possible value of the sample statistic could be listed along with its associated probability. For continuous sampling distributions, the possible values of the statistic would have to be listed in ranges along with the probability associated with each range. Unless otherwise indicated, sampling is done *without replacement,* that is, after an item is drawn from the population it is not put back into the population for the next draw. Sampling *with replacement* is useful in statistical theory but of no practical value. The reason that sampling with replacement is of no practical value is that once an observation has been drawn and its characteristics observed and recorded, it would be a waste of money to draw the same observation and go through the same recording procedure. An exception occurs when an interviewing firm is trying to prevent "curbstoning." This expression refers to interviewers who are paid according to the number of interviews they complete, and succumb to the temptation to write up fictitious interviews. The expression comes from the image of the interviewer sitting on a curb and writing down what he thinks are plausible answers on a stack of interview forms. Instead of sampling with replacement to reinterview some of the respondents, interview firms generally draw a small second sample from the set of completed interviews and then send a supervisor to reinterview the people in that sample-within-a-sample. Any curbstoning by one of the original interviewers can easily be recognized when some of the presumed respondents deny having been interviewed before.

To see the difference that sampling with and without replacement makes in the numbers of possible samples, let's take a simple numerical example. Consider the population {1, 2, 3, 4, 5}. There are 25 possible samples of size 2 when sampling is done with replacement because there are 5 ways to choose the first number and 5 ways to choose the second. There are 20 possible samples when the sampling is done without replacement because there are 5 ways to choose the first number and then, after the first is chosen, 4 ways to choose the second.

7.2B THE CENTRAL LIMIT THEOREM

When you take a sample of size n, the values of each observation can be considered a random variable. Consider the example of samples of size 2 taken with replacement from the population {1, 2, 3, 4, 5}. The first number drawn, X_1, is a random variable that can take the values 1 through 5 and the second number drawn, X_2, is a random variable that can take the same values. Another way to describe X_1 and X_2 is to call them identical independent random variables. The most important theorem in statistics is the central limit theorem, which describes the effect of an increase in sample size on the shape of a sampling distribution. The simplest version of the **central limit theorem** states that *the random variable that equals the mean of* n *independent and identical random variables approaches a normal distribution as* n *increases.* If the X_1, X_2, X_3, . . . , X_n random variables were the possible values of each observation in a sample of size *n*, they would necessarily be independent if the sample was taken with replacement. What all this means is that if you are sampling with replacement from almost any population, *the distribution of the sample mean will be approximately normal for large enough values of* n. As a rule of thumb, *an* n *of 30 is considered large enough*.

The great importance of this theorem is that even if you know nothing about the shape of a population distribution, you can be sure that the sampling distribution of the mean of that population will be approximately normal for samples of size 30 or more. The benefit of knowing that the sampling distribution of the $\overline{x}$ is normal is that we can use the *z* score formula and the normal table to find the probability that the sample mean will fall in any given range. The normality of the sampling distribution of $\overline{x}$ is the key to making statistical inferences about the possible values of the population mean. Without the central limit theorem, inferences from the sample to the population would be much more difficult.

The simplest version of the central limit theorem assumes that the sampling is done with replacement (a proof is beyond the scope of this text), but more complicated versions of the central limit theorem apply to sampling without replacement. You can ignore the requirement of sampling with replacement because these central limit theorems will generally hold even if there is some dependence between the X_i's. Thus, *for all practical purposes, the sampling distribution of the mean for samples from any population will be approximately normal if the sample size is 30 or more.* Figure 7.2 shows three examples of parent populations and the sampling distribution of the mean for samples of sizes 2, 20, and 100 from each population. The figure demonstrates that the sampling distribution of the mean converges to a normal curve for a variety of population distributions. The three population distributions in Figure 7.2 have the same means and standard deviations but they have different shapes. The first population distribution is normal with a mean of 0 and a standard deviation of 1. The second population distribution is the continuous uniform distribution with a mean of 0 and a standard deviation of 1. Its range is from −1.732 to +1.732. The third population distribution consists of two spikes equidistant from 0 so that the mean of the distribution is 0 and the standard deviation is 1. The spikes would be located at −1 and +1.

The three population distributions are in the first row of Figure 7.2. The second row consists of the sampling distribution of the mean when samples of size 2 are taken with replacement. The respective sampling distributions are directly below the population distributions from which the samples are taken. The horizontal axis for the first row is labeled *x* because the values refer to the

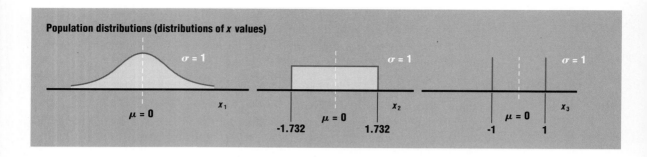

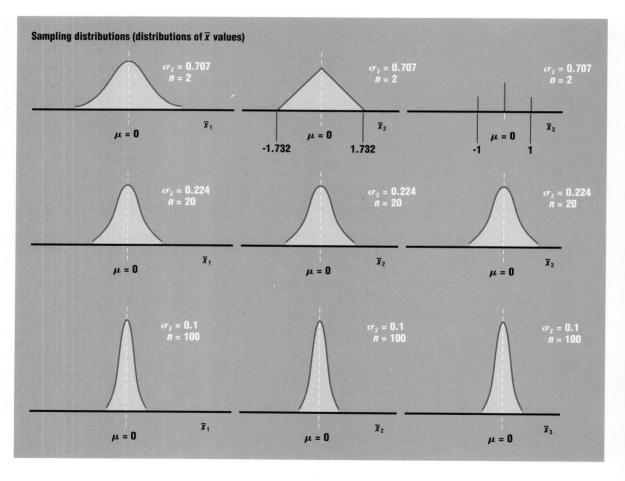

FIGURE 7.2
Three population distributions and the sampling distributions of
the mean for samples of sizes 2, 20, and 100

population distribution. The horizontal axes for the second, third, and fourth rows are labeled $\bar{x}$ because the values refer to the sampling distribution of the mean. Note that in each case the sampling distribution of the mean is centered on the population mean. The center of each distribution is 0. This result does not depend on the shape of the population or the sample size. *The expected value of the sampling distribution of the mean always equals the population mean.* This result is proved in Appendix C. In mathematical notation this result can be written as:

$$E(\bar{x}) = \mu$$

(7.1)

In formula 7.1, μ refers to the population mean and $E(\bar{x})$ is the expected value of the sampling distribution of $\bar{x}$. Note also that in the second row of Figure 7.2, the three standard deviations of the sampling distributions of the mean are all equal. The symbol for the standard deviation of the sampling distribution of the mean is $\sigma_{\bar{x}}$. In each case when the sample size is 2, the value of $\sigma_{\bar{x}}$ is 0.707. The reason that the standard deviations of the sampling distributions of the mean are the same is that *for a given population standard deviation and a given sample size, the standard deviation of the sampling distribution of the mean does not depend on the shape of the population.*

The most important point to notice about the second row is that even for a sample of size 2, the distributions of the sample means are more tightly centered on the population mean than are the population distributions. The population distribution that consists of two equal spikes had no values at the mean of 0, but in half of all samples of size 2, the -1 and $+1$ will be drawn, so the resulting sample mean will be 0. For the continuous uniform distribution it is extremely unlikely that two observations that are both near the same boundary will be drawn. A sample mean near $+1.732$ or -1.732 is much less likely than a sample mean near the middle of the population distribution. The sampling distribution of the mean for the continuous uniform distribution with a sample of size 2 is a triangle that is symmetrical around the population mean. Finally, note that the sampling distribution of the mean for a normal population is a normal curve. This is a general result; *the sampling distribution of the mean will always be normal when the population is normal, regardless of the sample size.*

The third row of Figure 7.2 shows the respective sampling distributions of the mean when the sample size is 20. The three sampling distributions appear to be identical. Actually, the one on the left is exactly normally distributed and the other two are approximately normally distributed. They are all close approximations to normal curves centered on 0 and having standard deviations of 0.224. This is a remarkable result. How could samples taken from the two spikes result in an almost smooth curve that is normally distributed? The sampling distribution of the mean from the two spikes is not exactly normal, but it is close enough for all practical purposes. The reason for the almost smooth curve is that there are 21 distinct values of $\bar{x}$ for the $(2)^{20}$ possible samples of size 20 taken with replacement from a population that has two distinct values. The ends of the 21 closely spaced vertical lines look like the outline of a smooth curve. Finally, the probability declines rapidly for values of the sample mean as these values move farther away from the population mean. Extreme observations of the sample mean are rare because high and low values of the population tend to "wash out" when large samples are drawn. The last row shows what happens as the sample size increases. The three curves for the sampling distribution of the mean seem to be identical normal curves with less dispersion than the curves in the third row. The standard deviation for all three curves in the fourth row is 0.1.

The three populations used in Figure 7.2 were symmetrical. If a population is not symmetrical, the convergence to a normal curve still occurs, but it requires a larger sample size. The rule of thumb of 30 or more observations is quite conservative. No matter how skewed the population distribution, the distribution of the sample mean will be very close to normal. When distributions are symmetrical, the distribution of sample means will be close to normal with a sample size of only 10 or 20.

■ 7.2C THE STANDARD DEVIATION OF THE DISTRIBUTION OF SAMPLE MEANS

There is a simple relationship between the population standard deviation and the standard deviation of the distribution of sample means. Appendix C shows that:

$$\sigma_{\bar{x}} = \frac{\sigma}{\sqrt{n}}$$

(7.2)

In other words, the standard deviation of the distribution of sample means equals the population standard deviation divided by the square root of the sample size. Formula 7.2 holds if the sampling is done either with replacement from a finite population or with an infinite population. *The standard deviation of the sampling distribution of the mean* is also called the **standard error of the mean.** Roughly 68% of all sample means will be within one standard error of the population mean if the sample size is large enough to make the distribution of sample means approximately normal. The standard error of the mean can be thought of as the average absolute distance between the sample mean and the population mean. This average distance declines as the sample size increases; that is, the sample mean tends to be closer to the population mean with larger samples. Because of the square root in the denominator of formula 7.2, however, the decline is not in proportion to the increase in the sample size.

Doubling the sample size from 2 to 4 for any of the three distributions in Figure 7.2 would reduce the standard error from 0.707 to 0.5.

$$\sigma_{\bar{x}} = \frac{\sigma}{\sqrt{n}} \qquad (7.2)$$

For $n = 2$,   For $n = 4$,

$$\sigma_{\bar{x}} = \frac{1}{\sqrt{2}} \qquad \sigma_{\bar{x}} = \frac{1}{\sqrt{4}}$$

$$\sigma_{\bar{x}} = 0.707 \qquad \sigma_{\bar{x}} = 0.5$$

One consequence of this less than proportionate reduction in the standard error for increases in the sample size is that if the cost of a survey is proportional to the sample size, doubling the cost does not double the precision of the survey. Suppose the distribution of family income in the United States had a known standard deviation of $1,000. A sample of 1 million families with replacement would yield a standard error of $1, that is, $1,000 divided by the square root of 1 million. In contrast, a sample of 1,000 families would provide a standard error of only $31.62.

It was stated in the last section that sampling with replacement has no practical purpose other than to illustrate statistical theory. No one who planned to interview 1,000 families in the United States would go to the trouble of drawing the name of one family, putting it back in the barrel, and then drawing the name of the next family. Formula 7.2 has to be modified when you are dealing with a finite population and samples are taken without replacement. The only change is a term called the finite population correction factor, which was introduced in the section on the hypergeometric distribution in Chapter 5. The modified formula that is to be used when samples are taken without replacement from a finite population is:

$$\sigma_{\bar{x}} = \frac{\sigma}{\sqrt{n}} \sqrt{\frac{\nu - n}{\nu - 1}} \qquad (7.3)$$

Recall that ν is the number of elements in the population, or the population size. This formula is always correct for finite populations sampled without replacement, but as a rule of thumb, formula 7.3 is optional if the sample size is less than 5% of the population. The reason for this rule of thumb is that when n is a small percentage of ν, the correction factor is very close to 1, so it can be ignored. To see this result, try an n of 5 and a ν of 100 (this is a 5% sample of the population). Substituting into just the finite population correction portion of formula 7.3 yields

$$\sqrt{\frac{\nu - n}{\nu - 1}} = \sqrt{\frac{100 - 5}{100 - 1}}$$

$$\sqrt{\frac{\nu - n}{\nu - 1}} = 0.9796$$

The numbers used in this example justify the rule of thumb that if n is 5% of ν or less, it is not necessary to use the finite population correction factor be-cause the correction to $\sigma_{\bar{x}}$ is so small as to be trivial. If n is a smaller percentage of ν than 5%, the correction factor will be closer to 1 than 0.9796. The tern n/ν is called the sampling fraction. Whenever you are sampling without replace-ment from a finite population, you should check the sampling fraction in order to determine if you need to use the finite population correction factor. If the sampling fraction is less than 0.05 (5%), use of the finite population correction factor is not necessary. If the sampling fraction is greater than or equal to 0.05, then the use of the finite population correction factor is required.

Let's try a few problems that make use of the central limit theorem and the standard error. Suppose the population of air travelers has an average weight of 170 pounds with a standard deviation of 10 pounds. This distribution is known to be normal. You have a small plane that seats 10 passengers and can safely carry a load of 2,000 pounds. Assuming that your passengers' weights are independent, what is the probability that the total weight of 10 passengers will be over 2,000 pounds? The assumption of independence is questionable because heavy and lightweight passengers tend to come in groups or families, but we'll stick to it for this problem. Questions about the total weight can be changed into questions about the mean weight. A total weight of 2,000 pounds implies a mean weight of 200 pounds, since $\bar{x}$ equals $\sum_{i=1}^{n} x_i/n$, or 2,000/10. Since the population is normal, the sampling distribution of the mean will be normal, no matter how small the sample size. Figure 7.3 superimposes the sampling distribution of the mean on the population distribution. The stan-dard error equals 3.16, because it is the population standard deviation of 10 divided by the square root of the sample size of 10. The z score of a sample mean of 200 is 9.49. Substituting in formula 6.5, we get

$$z = \frac{\bar{x} - \mu}{\sigma_{\bar{x}}} \tag{6.5}$$

$$z = \frac{200 - 170}{3.16}$$

$$z = 9.49$$

Since this problem involves the sampling distribution of the mean, $\bar{x}$ is used in place of x and $\sigma_{\bar{x}}$ in place of σ. This z score of 9.49 is off the table for the normal curve, so the probability that the mean weight of a sample of 10 inde-

FIGURE 7.3

Sampling distribution of the mean of passengers' weights superimposed on the population distribution

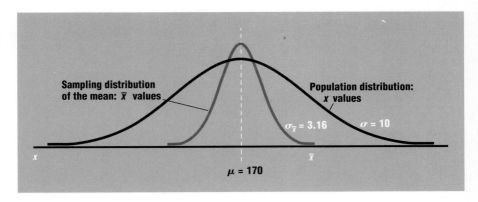

pendent passengers will be 200 or more is essentially zero. Be careful if a team of sumo wrestlers tries to buy tickets.

It is important to keep clear in your mind the distinction between the population distribution and the distribution of the sample mean. The population distribution can tell you either the proportion of all airline passengers who weigh more than 200 pounds or the probability that one randomly selected passenger will weigh more than 200 pounds. The sampling distribution of the mean can tell you the probability that the mean of a sample of 10 passengers will exceed 200 pounds. It is also important to keep in mind that there is a different sampling distribution for the mean for every sample size. If n were changed, the probability that the mean weight of the passengers would exceed 200 pounds would also change.

Here's another example. You are a cab driver and the population distribution of fares in your cab has a mean of $10 with a standard deviation of $3. If you know that you will have 50 fares in a week, what is the probability that your total earnings will be over $510? In this example you do not need the assumption of normality in the population because the sample size is over 30. The standard error of the mean is $0.42. You obtain this figure by substituting into formula 7.2. The reason that the finite population correction factor is not needed is that the population of all possible future fares can be considered infinite.

$$\sigma_{\bar{x}} = \frac{\sigma}{\sqrt{n}}$$ (7.2)

$$\sigma_{\bar{x}} = \frac{3}{\sqrt{50}}$$

$$\sigma_{\bar{x}} = 0.42$$

If the total earnings from 50 fares is \$510, the mean earnings must be \$10.20. The z score for a sample mean of \$10.20 or more is 0.48. Substituting into formula 6.5, you get

$$z = \frac{\bar{x} - \mu}{\sigma_{\bar{x}}} \qquad (6.5)$$

$$z = \frac{10.2 - 10.0}{0.42}$$

$$z = 0.48$$

The area to the right of this z score is approximately 0.32. Thus the probability that the total earnings from 50 fares will be \$510 or more is 0.32. Note that the reason that the z score can be used in this problem is that the central limit theorem states that with a sample of 50, the sampling distribution of the mean will be approximately normal.

Here's the last example for this section. You run a mail-order business for gardening supplies. Your catalogue lists 300 items. The mean of the distribution of prices for the items in the catalogue is \$8.40 and the standard deviation is \$7.00. In mathematical notation, μ equals \$8.40, σ equals \$7.00, and v equals 300. According to your records, 60 of the items in your catalogue have lost unit sales compared to the previous year. These 60 items have an average price of \$10.00. In notation, your $\bar{x}$ is \$10.00 and your n is 60. You suspect that consumers are giving up some of the higher-priced items in your catalogue. If the consumers' decisions regarding the items that lost unit sales had nothing to do with price, what is the probability that the mean of a sample of 60 items would be \$10.00 or more? This question is asking what is the probability that a chance or random selection would result in a sample of 60 items with a mean of \$10.00 or more.

The distribution of sample means will be approximately normal because the sample size is greater than 30. The standard deviation of the distribution of sample means can be obtained by formula 7.3:

$$\sigma_{\bar{x}} = \frac{\sigma}{\sqrt{n}} \sqrt{\frac{v - n}{v - 1}} \qquad (7.3)$$

$$\sigma_{\bar{x}} = \frac{7.00}{\sqrt{60}} \sqrt{\frac{300 - 60}{300 - 1}}$$

$$\sigma_{\bar{x}} = 0.81$$

The reason for using the finite population correction factor is that the population is finite and the sample size is more than 5% of the population ($n/v = 60/300 = 0.20 > 0.05$). The z score for the boundary of \$10.00 is:

$$z = \frac{\bar{x} - \mu}{\sigma_{\bar{x}}}$$ (6.5)

$$z = \frac{10.00 - 8.40}{0.81}$$

$$z = 1.98$$

The area to the right of a z score of 1.98 is 0.024. Thus there is a small probability that a random selection would have a mean price of $10.00 or more. The odds that this outcome will occur by chance are remote. It appears that your customers are shying away from items with higher than average prices.

EXERCISES FOR SECTION 7.2

7.11 Let X be a normally distributed random variable with a mean of 89 and a standard deviation of 12.

(a) What is the probability that the mean of a sample of size 19 will be between 85 and 93?

(b) What is the probability that the mean of a sample of size 19 will be between 80 and 98?

(c) What is the probability that the mean of a sample of size 40 will exceed 91?

(d) What is the probability that the mean of a sample of size 5 will be less than 91?

7.12 Let X be a random variable with a mean of 1,200 and a standard deviation of 20.

(a) What is the probability that the mean of a sample of size 80 will exceed 1,202?

(b) What is the probability that the mean of a sample of size 30 will be between 1,202 and 1,204?

(c) What is the probability that the mean of a sample of size 50 will be less than 1,202?

(d) If the probability that the mean of a sample of size n will exceed 1,201 is 0.25, what must n equal?

7.13 The population of orders for printing jobs at a printshop is approximately normal with a mean of 200 pages and a standard deviation of 40 pages. The shop is almost out of paper and it has five orders that must be finished before a shipment of paper can be expected. If the shop has 1,200 sheets of paper left, what is the probability that the five orders will not exhaust the stock of paper?

7.14 A city highway department has enough salt for 100 inches of accumulated snowfall. The distribution of daily snowfall in the city during the winter is normal with a mean of 2 inches and a standard deviation of 0.4 inch. For what number of days does the probability of running out of salt equal 0.1?

7.15 Suppose your company audited its long-distance phone calls and found that for one month there were 800 calls with a mean charge of $4.55 and a standard deviation of $2.00. Your department has made 100 of the calls and the mean for your department was $5.11. You are wondering if the difference in mean charge for a call for the entire company and for your department could be due to chance. What is the probability that 100 randomly selected calls out of the 800 would have a mean of $5.11 or more?

7.16 An economic historian who is studying the U.S. transportation system in the 19th century found an article written in 1850 which claimed that the average price of shipping by horse-drawn cart was 10 cents per ton per mile. The historian has a random sample of 80 shipping bills from that year and she calculates the average ton–mile rates used in these bills. The average is 15 cents with a sample standard deviation of 3 cents. If the 1850 article was correct about the population mean rate and the standard deviation found in the historian's sample was used as an estimate of the population standard deviation, what is the probability that a sample mean as large as the one found by the historian or larger would be drawn?

7.17 A cattle feedlot operator has a contract with a meatpacker which calls for the delivery of 50 cattle with an average age of 12 months or more with no cattle aged less than 10 months or more than 24 months. The feedlot operator has 500 head of cattle with an average age of 13 months and a standard deviation of one month. None of the 500 cattle in the feedlot are younger than 10 months or over 24 months. If the 50 cattle to be sent to the meatpacker were selected at random, what is the probability that the contract provision in regard to the average age would not be fulfilled?

7.18 The distribution of cash withdrawals from the automatic teller machine at the First National Bank has a mean of $50 with a standard deviation of $7. To reduce the incentives for robbery, the bank puts money into the machine every 12 hours and it keeps the amount deposited fairly close to the expected total withdrawals for a 12-hour period. If 100 withdrawals were expected in each 12-hour period and each withdrawal was independent, how much should the bank put into the machine so that the probability of running out of money was 0.05?

7.19 The number of salmon that a commercial fisherman off the coast of Alaska can catch in one day is strictly regulated. Suppose a salmon license entitled the fisherman to 25 salmon per day. The population of salmon weights is normally distributed with a mean weight of 5 pounds and a standard deviation of 1 pound. The wholesale price of the fish is $1.50 per pound. If a fisherman was certain to catch the quota of 25 fish and the cost of fuel and material for fishing for one day was $125, what is the probability that on a given day a salmon fisherman would not cover these out-of-pocket costs?

7.20 Suppose the average yield of the stocks listed in the Standard & Poor's index of 500 leading companies was 15% over the past year with a standard deviation of 3%. The yield is made up of dividends, changes in the price of the stock, and stock splits. Your investment adviser suggested a portfolio of 10 stocks from the Standard & Poor's list at the beginning of the year and the average yield of these stocks was 9%. If you had chosen 10 stocks at random from the list of 500, what is the probability that you would have done as badly as or worse than your investment adviser? Assume that the distribution of yields is approximately normal.

7.3 THE SAMPLING DISTRIBUTION OF THE PROPORTION

7.3A THE RELATIONSHIP BETWEEN THE SAMPLING DISTRIBUTION OF THE PROPORTION AND THE BINOMIAL DISTRIBUTION

The sampling distribution of the proportion is something you've already seen before in a slightly different form. It is just the binomial distribution except that the number of successes is expressed as a proportion instead of as a count. If a binomial process has x successes in n trials, the sample proportion of successes, $\bar{p}$, is x *divided by* n. A count of 225 successes in 1,000 trials, for example, is the same thing as a proportion of 0.225. Similarly, the expected value of the sample proportion, $E(\bar{p})$, equals the expected value of the number of successes in a binomial distribution, $(n\Pi)$, *divided by* n. In notation, $E(\bar{p}) = \Pi$.

Recall formula 5.11, for the standard deviation of the number of successes in a binomial distribution:

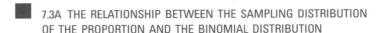

$$\sigma(X) = \sqrt{n\Pi(1 - \Pi)} \tag{5.11}$$

Let's treat the 0.225 as the population proportion, Π, and assume that there is an infinite population. If the probability of success in any trial was 0.225, the standard deviation of the number of successes in a binomial distribution with 1,000 trials would be

$$\sigma(X) = \sqrt{1,000(0.225)(0.775)}$$

$$\sigma(X) = 13.21$$

The number 13.21 is in units that measure the count or the number of successes. The standard deviation for the sampling distribution of the proportion is the same except that the standard deviation measures changes in the proportion of successes, so that the standard deviation of the distribution of sample proportions for 1,000 trials is one one-thousandth of $\sigma(X)$, or 0.01321. If there were 100 trials, the standard deviation of the distribution of sample proportions would be one one-hundredth of $\sigma(X)$, or 0.1321. In general, the standard deviation of the distribution of sample proportions is $\sigma(X)$ *divided by n,* where $\sigma(X)$ is the standard deviation of a binomial random variable with n trials. The formula for the standard deviation of the sampling distribution of the proportion of successes for samples taken from infinite populations or for samples taken with replacement is:

$$\sigma_{\bar{p}} = \sqrt{\frac{\Pi(1 - \Pi)}{n}}$$

(7.4)

In formula 7.4, Π can be called either the population proportion of successes or the probability of success in any one trial. These interpretations are equivalent. If this formula is applied to a Π of 0.225 and a sample of size 1,000, you can see that the standard deviation of the sample proportion, $\sigma_{\bar{p}}$, is one one-thousandth of $\sigma(X)$.

$$\sigma_{\bar{p}} = \sqrt{\frac{\Pi(1 - \Pi)}{n}}$$

(7.4)

$$\sigma_{\bar{p}} = \sqrt{\frac{0.225(0.775)}{1,000}}$$

$$\sigma_{\bar{p}} = 0.01321$$

In terms of the binomial distribution, going up one standard deviation from the mean would be going from 225 successes to 238.21 successes in 1,000 trials. In terms of the sampling distribution of the proportion of successes, going up one standard deviation from the mean would be going from a proportion of 0.225 to a proportion of 0.23821. These are just two different ways to express the same change.

If the sample was drawn from a finite population without replacement, the formula for the standard deviation of the sampling distribution would be

$$\sigma_{\bar{p}} = \sqrt{\frac{\Pi(1 - \Pi)}{n}} \sqrt{\frac{\nu - n}{\nu - 1}}$$

(7.5)

Note the inclusion of the finite population correction factor in formula 7.5. Another point to note is that formula 7.5 is $1/n$ times the formula for the standard deviation of the hypergeometric, formula 5.14. To summarize, the standard deviation of the proportion of successes is $1/n$ times the standard deviation of the corresponding number of successes in a binomial distribution for infinite populations or samples taken with replacement and it is $1/n$ times the standard deviation of the corresponding number of successes in a hyper-geometric distribution for samples taken without replacement from a finite population. The same rule of thumb about when it is not necessary to use the finite population correction factor applies to the standard deviation of the pro-portion as the standard deviation of the mean. In either case, if the sample size is less than 5% of the population, it is not necessary to use the finite population correction factor. Note that it is always correct to use the finite population correction factor for samples taken without replacement from a finite popula-tion; the rule of thumb is just a time-saver because the correction factor makes very little difference when the sample is a tiny fraction of the population.

7.3B NORMAL APPROXIMATION TO THE SAMPLING DISTRIBUTION OF THE PROPORTION

The same rule of thumb for applying the normal approximation to the bino-mial applies to the sampling distribution of the proportion. Specifically, $n\Pi$ and $n(1 - \Pi)$ must be greater than or equal to 5. Let's run through an example with an infinite population and then an example with a finite population sam-pled without replacement. A commercial factor buys accounts receivables from companies at a discount. Suppose a small company, Acme, had sent a $1,000 bill to another firm, Scoflaw, with a 30-day payment requirement. If Acme needed the cash immediately and it did not want to bear the risk of having Scoflaw delay the payment, Acme could sell the bill to a commercial factor. The commercial factor would pay $950 for the bill and then wait 30 days to collect the money from Scoflaw. The commercial factor serves as a risk pool. Since the commercial factor buys bills from many companies such as Acme, the standard deviation of the sample proportion of bills that will be paid late is much smaller than the standard deviation of the proportion for a small company such as Acme. Suppose the population proportion of bills that are late is 0.2 and Acme has 25 bills outstanding, including the one to Scoflaw. If Acme decided to hold on to its 25 bills, what values would represent the central 50% of the sampling distribution of the proportion of late payments? Note that the population can be considered infinite because the population consists entirely of future possible trials of sending a bill with a 30-day pay-ment requirement. Note also that the normal approximation to the sampling distribution of the proportion can be used because $n\Pi$ or $25(0.2)$ equals 5 and $n(1 - \Pi)$ equals 20. The rule of thumb is that $n\Pi$ and $n(1 - \Pi)$ both have to be greater than 5. The last point to note is that each of the bills must be an independent trial if this procedure is to be accurate. If there was a wave of bankruptcies among Acme's debtors, this assumption would not hold.

Figure 7.4 shows the normal approximation sampling distribution of the proportion. According to Figure 7.4, the range of sample proportions from

FIGURE 7.4
Sampling distribution of the
proportion of late payments
to Acme company for 25
bills, each of which has a
probability of 0.2 of being
paid late

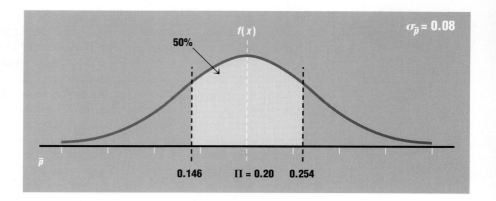

0.146 to 0.254 bounds the central 50% of the sampling distribution of the proportion. Thus the probability is 0.5 that the sample proportion of late payments among the 25 bills will be within the range 0.146 to 0.254. To find these values, we first calculate the standard deviation for the sampling distribution of the proportion and then use the z score formula to find the central 50%. We can find the standard deviation of the sampling distribution of the proportion by substituting into formula 7.4:

$$\sigma_{\bar{p}} = \sqrt{\frac{\Pi(1 - \Pi)}{n}} \tag{7.4}$$

$$\sigma_{\bar{p}} = \sqrt{\frac{0.2(0.8)}{25}}$$

$$\sigma_{\bar{p}} = 0.08$$

The z score that corresponds to 25% of the area between the mean and a boundary line to the right of the mean is 0.67. We can obtain this z score from the normal table by looking for the area closest to 0.2500. If you wanted to have a more accurate z score, you could interpolate between the z scores of 0.67 and 0.68. Substituting a z score of 0.67 into the z score formula will give the right boundary and substituting a z score of -0.67 will give the left boundary.

$$z = \frac{\bar{p} - \Pi}{\sigma_{\bar{p}}} \tag{6.5}$$

$\bar{p} = \Pi + z\sigma_{\bar{p}}$ \qquad\qquad $\bar{p} = \Pi + z\sigma_{\bar{p}}$

$\bar{p} = 0.2 + 0.67(0.08)$ \qquad $\bar{p} = 0.2 + (-0.67)(0.08)$

$\bar{p} = 0.254$ \qquad\qquad\quad $\bar{p} = 0.146$

Note that we have modified the z score formula in this problem by replacing μ with Π, x with $\bar{p}$, and σ with $\sigma_{\bar{p}}$.

Now let's see what happens to the sampling distribution of the proportion as the sample size increases. If the commercial factor had 1,000 bills, what values would represent the central 50% of the sampling distribution of the proportion of late payments? The math is the same except that the n of 25 is replaced by an n of 1,000. Figure 7.5 shows the results. The central 50% of the sampling distribution of the proportion of late payments is bounded by the sample proportions 0.191 and 0.209. Because of the large number of bills carried by the commercial factor, the proportions that represent the central 50% of the distribution of sample proportions fall into a much smaller range than the values calculated for Acme. As long as the assumptions of independence and a Π of 0.2 hold, the commercial factor has a much better idea of the proportion of late payments that it may face than does a small company such as Acme.

Here is another example that requires the use of the finite population correction factor. A refrigerator manufacturer has discovered that one of its inspectors was improperly trained, and that substandard compressors had passed his inspection and had subsequently been put into finished refrigerators. Apparently 100 substandard compressors got into a production run of 2,000 refrigerators. Identifying the refrigerators with substandard compressors is impossible without taking all 2,000 machines apart. The company feels that it might be cheaper to bear the ill will created by sending out the defective machines to retailers and to have the consumers eventually discover which ones are substandard than to go to the expense of tearing down the 2,000 machines. However, the company is worried that if too high a proportion of defective machines reaches any one of its retailers, it may permanently lose that retailer. The company decides that it cannot afford to let any retailer get a proportion of defective refrigerators higher than 0.075. If the selection of refrigerators sent to each retailer is random and a retailer is shipped 100 refrigerators, what is the

FIGURE 7.5

Sampling distribution of proportion of late payments to commercial factor for 1,000 bills, each of which has a probability of 0.2 of being paid late

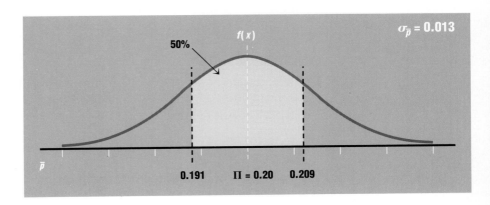

probability that the proportion of refrigerators that are defective will exceed this limit?

Note that the proportion of successes (defective refrigerators) in the population is 100 divided by 2,000, or 0.05. For a retailer with a shipment of 100 refrigerators, the rule of thumb for applying the normal approximation to the sampling distribution of the proportion would be met because $n\Pi$ is 100(0.05), or 5, and $n(1 - \Pi)$ is 100(0.95), or 95. We find the standard deviation of the distribution of the sample proportion by substituting into formula 7.5:

$$\sigma_{\bar{p}} = \sqrt{\frac{\Pi(1 - \Pi)}{n}} \sqrt{\frac{\nu - n}{\nu - 1}} \qquad (7.5)$$

$$\sigma_{\bar{p}} = \sqrt{\frac{0.05(0.95)}{100}} \sqrt{\frac{2,000 - 100}{2,000 - 1}}$$

$$\sigma_{\bar{p}} = 0.0212$$

Note that the finite population correction factor was used because the sample size was at least 5% of the population. The probability of obtaining a sample proportion above the limit of 0.075 can be obtained by the z score formula.

$$z = \frac{\bar{p} - \Pi}{\sigma_{\bar{p}}} \qquad (6.5)$$

$$z = \frac{0.075 - 0.05}{0.0212}$$

$$z = 1.18$$

The area to the right of a z score of 1.18 is approximately 0.12. Thus the refrigerator company faces a fairly high risk that a retailer who orders 100 refrigerators will get a shipment with a proportion of defective refrigerators above the limit of 0.075. It may pay the refrigerator company to tear down the assembled machines and find the substandard compressors.

EXERCISES FOR SECTION 7.3

7.21 If a sample of size 100 is taken from a population of size 1,000 with replacement and the population contains 300 successes:

    (a) What is the probability that the sample proportion of successes will be 0.35 or more?

    (b) What is the probability that the sample proportion of successes will be between 0.4 and 0.6?

    (c) What is the probability that the sample proportion of successes will be less than or equal to 0.2?

(d) What proportion of successes marks the upper 10% of the sampling distribution of the proportion?

7.22 If a sample of size 90 is taken from a population of size 1,000 without replacement and the population contains 400 successes:

(a) What is the probability that the sample proportion of successes will be 0.35 or more?

(b) What is the probability that the sample proportion of successes will be between 0.4 and 0.6?

(c) What is the probability that the sample proportion of successes will be less than or equal to 0.2?

(d) What proportion of successes marks the upper 10% of the sampling distribution of the proportion?

7.23 A pharmaceutical company that makes a drug given to women who suffer from nausea during pregnancy is facing a class–action suit brought by women who claim that the drug causes birth defects. There are records on 1,000 women who were given the drug and delivered full-term infants. Among these infants, 48 had serious birth defects. If the population proportion of serious birth defects resulting from all full-term pregnancies is 0.025, what is the probability of observing a sample proportion of 0.048 or more in a sample of 1,000?

7.24 An advertising campaign for a new perfume has the goal of reaching 50% of all women between the ages of 40 and 60 who have incomes above $50,000 per year. The word "reaching" means that these women can recall having seen either a print ad or television commercial for the perfume. Suppose a national probability sample of 300 women with the relevant age and income characteristics is drawn to see how the campaign is succeeding. Among the 300 women, only 120 can recall having seen an ad or commercial for the new perfume. If the population proportion was 0.50, what is the probability of observing a sample proportion of 0.40 or less in the sample of 300 women?

7.25 Retailers keep track of the proportion of items that are returned after purchase because the expense of the paperwork and labor required to refund the purchase price and sales tax and to handle the returned item is considerable. Items can be returned because customers have bought the wrong size or simply changed their minds. Such returns are unavoidable in retailing. However, some items are returned because customers have found the identical item at a lower price in another store. Retailers can control returns caused by uncompetitive prices by adjusting prices. On the other hand, if a retailer's prices were so low that no customer ever found a better price on any item, the retailer might go broke. Let's say a large department store considers that a reasonable target for returns due to uncompetitive prices is 10% of all returns. The store surveys a sample of all shoppers who return items and asks them whether they are returning the items because they have found a lower price at another store. Of

500 shoppers surveyed, 70 answer yes. If the population proportion was the desired 10%, what is the probability that the sample proportion would be 0.14 or more?

7.26 **(a)** Random blood tests conducted by the National Center for Disease Control for the presence of antibodies to HTLV-III, the AIDS virus, have established that 1% of the adult population has been exposed to the virus. Blood donation centers now routinely test all donated blood for the presence of these antibodies. One regional blood center found only 45 positives among 10,000 donors. What is the probability of observing a sample proportion at or below 0.45% when the expected proportion is 1%?

(b) What might account for the discrepancy between the observed and the expected proportions?

7.27 A seed company would like to develop a blue tomato. The company's plant geneticist estimates that 1 of every million seeds exposed to alpha particles will develop blue tomatoes. The geneticist also estimates that a pool of 30 genetically independent seeds that will bear blue tomatoes are needed to breed other desirable qualities, such as taste, resistance to disease, and long shelf life. If 50 million seeds are exposed to alpha particles, what is the probability that there will be 30 or more successful seeds?

7.28 A company that markets classical music on compact discs has found that about 25% of its new titles are failures. These titles do not sell enough copies to cover production and marketing costs. If the company adds 40 titles to its list in one year, what is the probability that the proportion of failures will exceed 30%?

7.4 THE PROPERTIES OF POINT ESTIMATORS (OPTIONAL)

 7.4A INTRODUCTION

Most of the material in Chapters 4 through 7 can be considered building blocks needed for statistical inference. We have arrived at the penultimate step. The goal is to use samples to infer some of the characteristics of a population. In this process we use estimates and estimators. An **estimate** is *a number that is calculated from a specific sample that is used to predict the value of a population parameter.* If you took a sample and the mean of that sample was $12.50, then $12.50 could be used as a point estimate or predictor of the population mean. An **estimator** is *a random variable that can take different values depending on the particular sample that is drawn.* For example, the sample mean is a random variable that can be used as an estimator of the population mean. Similarly, the sample proportion can be used as an estimator of the population proportion, and the sample variance can be used as an estimator of the population variance. Again, estimates are specific numbers derived by applying a formula to a specific sample, and estimators are the formulas that can be applied to any sample.

7.4B UNBIASED, EFFICIENT, AND CONSISTENT ESTIMATORS

More than one estimator can be used to predict a particular population parameter. For example, the sample mean or the sample median could be used to predict the population mean. In introductory statistics classes, the best choice among possible estimators of a particular parameter is generally easy to identify because in each case the obvious estimator happens to be the best possible estimator. For example, the sample mean is the best estimator of the population mean, the sample variance is the best estimator of the population variance, and the sample proportion is the best estimator of the population proportion. In other words, the estimator whose name corresponds to the same term in the population is the best available estimator. It is "best" in terms of the criteria of unbiasedness, efficiency, and consistency. These three criteria for choosing among estimators represent desirable properties for an estimator. Estimators that are unbiased, efficient, and consistent are desirable because they tend to give accurate predictions.

An **unbiased estimator** is *an estimator that has an expected value equal to the parameter being estimated.* Each of the "best" estimators listed above is unbiased. For an illustration, let's take the population $\{1, 2, 3, 4, 5\}$, which has a mean of 3. Suppose you took samples, with replacement, of size 2 from this population. The decision to take the sample with replacement was made for convenience, as it avoids the use of the finite population correction factor. There are 25 possible samples. These samples are listed in Table 7.1.

TABLE 7.1
Sampling with replacement from the population $\{1, 2, 3, 4, 5\}$ when $n = 2$

x_1, x_2	$\bar{x}$	$(s_i)^2$	s_i	x_1, x_2	$\bar{x}$	$(s_i)^2$	s_i
1, 1	1.0	0.0	0.0000	3, 1	2.0	2.0	1.4142
1, 2	1.5	0.5	0.7071	3, 2	2.5	0.5	0.7071
1, 3	2.0	2.0	1.4142	4, 4	4.0	0.0	0.0000
1, 4	2.5	4.5	2.1213	4, 5	4.5	0.5	0.7071
1, 5	3.0	8.0	2.8284	4, 1	2.5	4.5	2.1213
2, 2	2.0	0.0	0.0000	4, 2	3.0	2.0	1.4142
2, 3	2.5	0.5	0.7071	4, 3	3.5	0.5	0.7071
2, 4	3.0	2.0	1.4142	5, 5	5.0	0.0	0.0000
2, 5	3.5	4.5	2.1213	5, 1	3.0	8.0	2.8284
2, 1	1.5	0.5	0.7071	5, 2	3.5	4.5	2.1213
3, 3	3.0	0.0	0.0000	5, 3	4.0	2.0	1.4142
3, 4	3.5	0.5	0.7071	5, 4	4.5	0.5	0.7071
3, 5	4.0	2.0	1.4142				

$E(\bar{x}) = 3.0 \quad E(s^2) = 2.0 \quad E(s) = 1.1314$

Table 7.1 also lists the means, variances, and standard deviations of these 25 samples. If you add up the 25 sample means and divide by 25, you'll get the expected value of the sampling distribution of the mean. The mean of the sample means is 3, which is also the population mean. In this example the sample mean is an unbiased estimator of the population mean. This is a general result; the sample mean is always an unbiased estimator of the population mean. For comparison, a biased estimator of the population mean would be the lowest value of each pair of numbers in a sample. For this estimator the expected value would be 2.2. In a graph, the distribution of an unbiased estimator would be centered on the parameter and the distribution of a biased estimator would be centered on another value (see Figures 7.6 and 7.7).

By following the same procedure used to get the expected value of the mean, you can see that the sample variance is an unbiased estimator. In Table 7.1 the expected value of the sample variance is 2, the same as the population variance. If you run through this same procedure for the standard deviation, you'll get an expected value of 1.1314. Yet the population standard deviation is the square root of 2, approximately 1.414. This example shows that the *sample standard deviation is a biased estimator of the population standard deviation*. This bias is less of a problem than it appears to be, because the bias is generally quite small. The exact amount of bias depends on the population distribution and the sample size. The bias falls quickly as the sample size increases. For a normal population and a sample of size 2, the expected value of the sample standard

FIGURE 7.6
Distribution of $\bar{x}$ with samples of size 2 taken with replacement from the population {1, 2, 3, 4, 5}

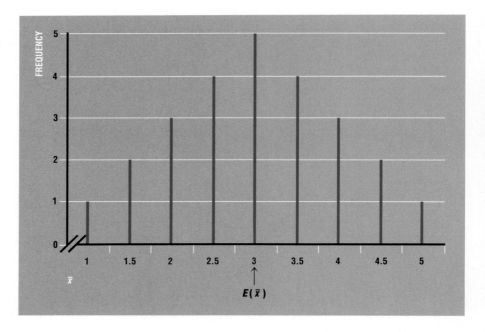

FIGURE 7.7
Distribution of the lowest value in samples of size 2 taken from the population {1, 2, 3, 4, 5}

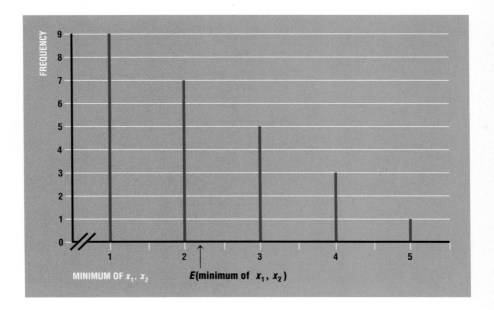

deviation will be about 20% below the population standard deviation. If the sample size goes to 30, the bias is under 1%. Appendix B includes a discussion of an adjustment to the sample standard deviation which will eliminate the bias when the population is normal.

Unbiasedness is a handy property for an estimator; your intuition should tell you that it is a good idea to work with an estimator that comes out right on average. However, being unbiased isn't everything. There can be more than one unbiased estimator of a parameter. Efficiency is a criterion that helps you choose between unbiased estimators. A more efficient unbiased estimator is one that has a tighter fit around the parameter than another estimator. In mathematical terms, a **relatively efficient estimator** is *an estimator that has a sampling distribution with a smaller variance than another estimator.* An unbiased estimator of the population mean for symmetrical populations is the sample median. However, the variance of the sample median is about 57% higher than the variance of the sample mean (see Appendix C for a proof), so if you were using the sample median as an estimator of the population mean, you would need 157 observations to give you as much information as 100 observations would yield if you used the sample mean. You can save money on interviewing, data entry, and computer time by using a more efficient estimator.

Figures 7.8 and 7.9 illustrate a relatively efficient estimator. The two figures have the same range of 1 to 5, but the vertical lines in the graph of the sampling distribution of the mean (Figure 7.8) are much more tightly packed around the

FIGURE 7.8
Sampling distribution of $\bar{x}$
with samples of size 3 taken
with replacement from the
population {1, 2, 3, 4, 5}

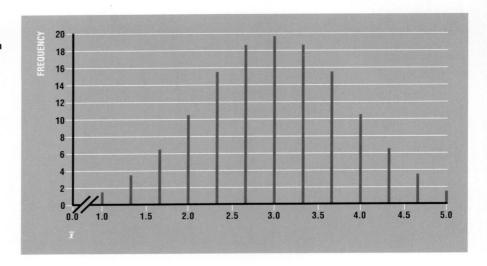

FIGURE 7.9
Sampling distribution of the
median with samples of size
3 taken with replacement
from the population
{1, 2, 3, 4, 5}

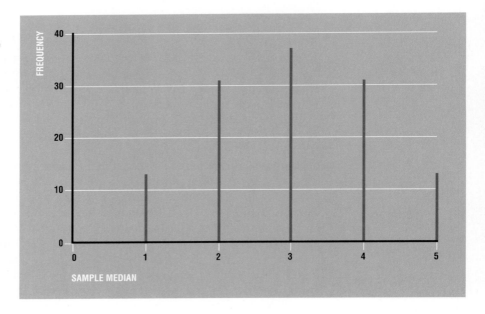

population mean of 3 than the vertical lines in the chart for the sampling distribution of the median (Figure 7.9). The comparison is even easier to see when the charts are superimposed. Figure 7.10 shows the sampling distribution of the mean and the median for a continuous distribution on the same scale.

FIGURE 7.10
Sampling distributions of the
mean and the median for a
continuous population

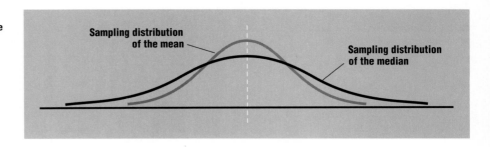

FIGURE 7.11
Sampling distribution of s
when $n = 2$

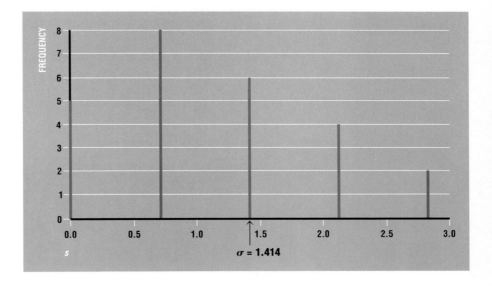

For some problems, no one has been able to find an unbiased estimator or
even to figure out which estimator is most efficient. Another criterion that you
can fall back on in such circumstances is consistency. A **consistent estimator**
is *an estimator such that its expected value gets closer to the parameter being estimated as
the sample size increases and the variance of its sampling distribution decreases as the
sample size increases.* The sample standard deviation is one consistent estimator.
Its bias and its variance get smaller as the sample size increases. Figures 7.11,
7.12, and 7.13 illustrate the consistency of the sample standard deviation. The
three charts plot the sampling distribution of the standard deviation with samples of sizes 2, 3, and 4, respectively, from the population {1, 2, 3, 4, 5}. The
samples are taken with replacement. You can see that the distribution has a
tighter fit around the population standard deviation of 1.4142 as the sample size
increases.

FIGURE 7.12
Sampling distribution of s when $n = 3$

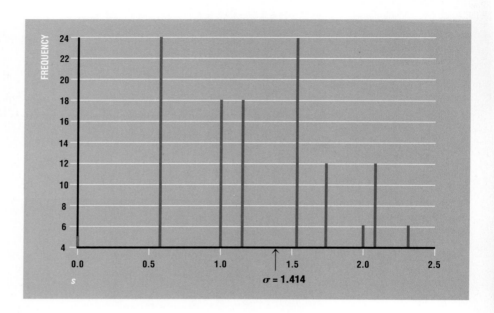

FIGURE 7.13
Sampling distribution of s when $n = 4$

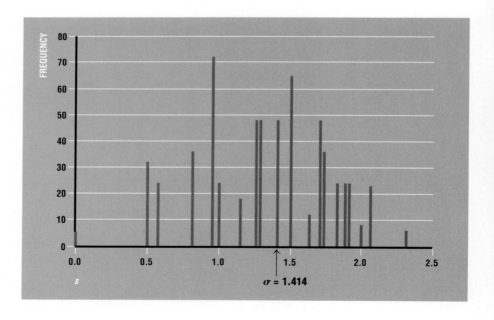

If you have a biased estimator and a small sample, you don't have much information. If your estimator is consistent, you at least know that the estimator converges on the parameter as the sample size increases. The lesson from all this is to make sure you have a large sample size if the only criterion your estimator can satisfy is consistency. The dot on the y axis of each figure corresponds to the frequency of samples with a standard deviation of 0. This frequency cannot be shown as a vertical line because it is on the axis. Note that the vertical lines have a tighter fit and are centered closer to the population standard deviation of 1.4142 as the sample size increases. Vertical line diagrams are used because the data are discrete, and the discrete data are chosen to afford a simple example that you can easily check. The diagram for the sampling distribution of the standard deviation of a continuous distribution shows the concept of consistency more clearly. In Figure 7.14, three sampling distributions of the standard deviation for a continuous distribution are superimposed.

If you need to choose among possible estimators, a commonly used criterion is the **mean square error,** which is *the expected value of the squared difference between an estimator and the population parameter.* In notation the mean square error is

---

Mean square error $= E(\theta - \hat{\theta})^2$

(7.6)

The symbol θ stands for any population parameter, such as μ, Π, or σ. The symbol $\hat{\theta}$ stands for any estimator. The idea behind this criterion for choosing

FIGURE 7.14
Sampling distributions of the standard deviation for a large, medium, and small sample

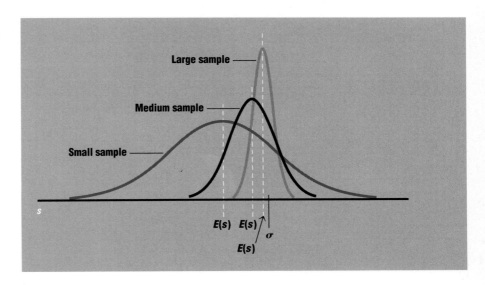

among estimators is that the cost associated with an error in estimating a population parameter increases in proportion to the square of the error. In other words, small errors cost little and large errors are very costly. The mean square error depends on both the amount of bias and the efficiency of an estimator. The amount of bias is the difference between the parameter and the expected value of an estimator. In Appendix C it is shown that the mean square error equals the squared bias plus the variance of an estimator. Since the math needed to determine the mean square error of many estimators is complicated, the exercises at the end of this section include a two-starred exercise on this subject (exercise 7.33) for students who like a challenge.

EXERCISES FOR SECTION 7.4

Asterisks designate an exercise of greater than ordinary difficulty.

7.29 The average of the lowest and highest values in a sample can be used as an estimator of the population mean. In samples of size 2, this estimator is the same as the sample mean; for larger samples it is not the same. Plot the sampling distribution of this estimator for samples of sizes 3 and 4 from the population {1, 2, 3, 4, 5} and determine if this estimator is as good as the sample mean according to the criteria of unbiasedness, efficiency, and consistency.

7.30 According to the criteria of unbiasedness, efficiency, and consistency, is the average absolute deviation from the sample mean as good an estimator of the population standard deviation as the sample standard deviation? The average absolute deviation is defined as

$$\frac{\sum_{i=1}^{n} |x_i - \bar{x}|}{n}$$

where the two vertical lines mean absolute value. To answer this question, use the population {1, 2, 3, 4, 5} and plot the sampling distribution of the average absolute deviation from the sample mean for samples of sizes 3 and 4.

7.31 Indicate whether each of the following statements refers to an estimate or an estimator.

(a) The mean time for a consumer to assemble a Huffy youth bicycle is estimated to be 75 minutes.

(b) The mean of a sample of 100 customers can be used to estimate the average time needed to assemble a bicycle.

(c) The sample median is inefficient.

(d) The variance of a sample of delivery times based on 12 observations was 44 minutes squared.

7.32 Comment on the following statements.

(a) Since the sample mean is an unbiased estimator of the population mean, the sample mean of 75 minutes for the time needed to assemble a bicycle must be the correct value of the population mean.

(b) Since the sample standard deviation is a biased estimator of the population standard deviation, it should not be used to estimate the population standard deviation.

7.33** You own a travel agency and you are planning a new service: custom computerized itineraries. A client fills out a questionnaire on his or her interests and budget, and your custom itinerary will list the best restaurants, museums, and night spots in a city along with the times your client can expect to avoid crowds and get discounted prices. Before making the heavy investment required to inaugurate this service, you want to estimate the proportion of your customers who would use the service. You could either mail 250 surveys to customers or at the same cost telephone 50 of them. Assume that the phone survey has no bias. The mail survey may be biased, because people who like the idea of a computerized itinerary will be more likely to mail the survey form back. Your subjective judgment is that there is a 0.8 probability that someone who likes the idea will mail the survey back, and that there is a 0.4 probability that someone who does not like computerized itineraries will mail the survey back. According to the mean square error criterion, are you better off with the mail survey or the phone survey?

 SUMMARY

This chapter has discussed some crucial topics. Sample design and sampling techniques are practical issues, and their study can pay off in terms of more accurate surveys at lower cost. The chapter described the distinction between probability and nonprobability samples. Among the latter are judgment and convenience samples. Among the former are simple random samples, stratified samples, and cluster samples. The rest of the chapter was more theoretical. The central limit theorem is the theoretical foundation for most of statistical inference. The precise relationship between the mean and standard deviation of a known population and the distribution of sample means from that population will be turned around in Chapter 8 and used to assign probabilities that a sample mean will be within any given distance of an unknown population mean. Another theoretical discussion was on the properties of estimators. Most of the estimators used in an introductory statistics course are unbiased, efficient, and consistent; but in the case of the sample standard deviation, only the property of consistency holds.

 SUGGESTED READINGS

Cochran, William G., *Sampling Techniques,* 3rd ed. New York: Wiley, 1977.

Roj, Des, *Sampling Theory.* New York: McGraw-Hill, 1968.

NEW TERMS

census A survey of an entire population.

central limit theorem The random variable that equals the mean of n independent and identical random variables approaches a normal distribution as n increases.

cluster sample A sample drawn by identifying subgroups of the population as clusters on the basis of geographical proximity and then drawing a random selection of clusters.

consistent estimator An estimator such that its expected value gets closer to the parameter being estimated as the sample size increases and the variance of its sampling distribution decreases as the sample size increases.

convenience sample A sample selected on a nonrandom basis to reduce the time and cost of drawing the sample.

estimate A number that is calculated from a specific sample that is used to predict the value of a population parameter.

estimator A random variable that can take different values depending on the particular sample that is drawn.

judgment sample A sample in which the elements of the population to be selected are decided by expert opinion.

mean square error The expected value of the squared difference between an estimator and the population parameter.

nonprobability sample A sample in which the probability that an element of the population will be drawn is not known.

nonresponse bias Systematic differences between the sample statistics and the population parameters which occur when some potential respondents in a randomly selected sample refuse to be interviewed.

probability sample A sample in which each element of the population has a known and nonzero chance of being selected.

relatively efficient estimator An estimator that has a sampling distribution with a smaller variance than another estimator.

sampling distribution A function that gives the probability of every possible value of a sample statistic for a specified population and sample size.

sampling frame A list of all of the elements in the population.

self-selection bias Systematic differences between the sample statistics and the population parameters which occur when the choice of being in or out of a sample is left to each person who is selected as a potential member of the sample and the selection process is not random.

simple random sample A sample of size n in which all possible groups of size n are equally likely to be chosen.

standard error of the mean The standard deviation of the sampling distribution of the mean.

stratified sample A sample derived by dividing the population into strata or layers and drawing a random sample from each stratum.

systematic sample A sample drawn by dividing the population into equal groups, randomly selecting a number equal to or smaller than the size of the groups, and drawing the observation in each group which corresponds to that number.

unbiased estimator An estimator that has an expected value equal to the parameter being estimated.

 NEW FORMULAS

Expected value of the sampling distribution of the mean

$$E(\bar{x}) = \mu \tag{7.1}$$

Standard deviation of the distribution of the sample mean for an infinite population or for a finite population that has been sampled with replacement

$$\sigma_{\bar{x}} = \frac{\sigma}{\sqrt{n}} \tag{7.2}$$

Standard deviation of the distribution of the sample mean for finite populations sampled without replacement

$$\sigma_{\bar{x}} = \frac{\sigma}{\sqrt{n}} \sqrt{\frac{\nu - n}{\nu - 1}} \tag{7.3}$$

Standard deviation of the distribution of the sample proportion for an infinite population or for a finite population that has been sampled with replacement

$$\sigma_{\bar{p}} = \sqrt{\frac{\Pi(1 - \Pi)}{n}} \tag{7.4}$$

Standard deviation of the distribution of the sample proportion for finite populations sampled without replacement

$$\sigma_{\bar{p}} = \sqrt{\frac{\Pi(1 - \Pi)}{n}} \sqrt{\frac{\nu - n}{\nu - 1}} \tag{7.5}$$

Mean square error

$$\text{Mean square error} = E(\theta - \hat{\theta})^2 \tag{7.6}$$

ADDITIONAL EXERCISES

An asterisk designates an exercise of greater than ordinary difficulty.

7.34 Your company plans to bring ten job candidates to its headquarters for interviews. You are the vice-president for human resources, and the charges for the hotel and plane fares for the candidates come out of your budget. Your experience has been that the distribution of costs for job candidates has a mean of $1,200 and a standard deviation of $300. Assume that the distribution is normal. How much should you budget for the ten candidates so that there is a 5% chance of running out of money?

7.35 You are a waiter in a restaurant that has a policy of pooling all tips; that is, you are required to put your tips into a pot and the total in the pot is divided equally among all of the waiters. Yesterday your share of the pot was $28. You suspect that some of your fellow waiters have pocketed some of their tips. You don't know what the true distribution of tips for all the waiters is, but you have kept a record of your own tips, and your daily tips average $30 with a standard deviation of $4. If the other waiters average the same tip income as you and there were 40 waiters altogether, what is the probability that on one day you would receive a share of $28 or less?

7.36 An article in the sports section of your local paper made the point that it was no longer surprising to find an average time of under 4 minutes for all of the contestants in a mile-long college foot race. According to the article, 10% of all college varsity runners who compete in mile-long races can run the mile in under 4 minutes. The article also stated that the distribution of times for completing the mile among this group has a standard deviation of 0.25 minutes. If a mile race had ten randomly selected contestants from among all college milers, what is the probability that the mean time for the ten runners would be 4 minutes or less? Assume that the distribution of running times is normal.

7.37 Your company sells frozen breaded fish filets. You continually monitor your fish purchases for mercury contamination. Suppose the Food and Drug Administration (FDA) had an action limit of 50 parts per billion (PPB) for mercury. An action limit is the maximum amount of an ingredient that the FDA will allow before taking a product off the market. Your own monitoring of mercury content has established a mean of 12 PPB with a standard deviation of 3 PPB. The FDA wants to recall all of your product because its laboratory found that a sample of 10 boxes of fish had a mean of 55 PPB. You are certain that the FDA lab made a mistake. If your monitoring of the population distribution was accurate, what is the probability that a sample of size 10 would have a mean of 55 PPB or more?

7.38 Refer to the information provided in exercise 7.37. The FDA has agreed that there may be something fishy about either your lab results or its own. To check the FDA lab, 100 packages of your frozen fish filets are sent to an independent laboratory, which finds that the mean is 13 PPB with a standard deviation of 5 PPB. The same 100 packages are then sent to the FDA lab for a second check. The FDA lab wants to save money on the recheck, so it tests only 10 of the 100 packages. The FDA has agreed that it will rescind its recall order if its lab misses the PPB level established by the independent lab by more than plus or minus 3 standard errors. What points on the sampling distribution of the mean represent 13 PPB plus and 13 PPB minus 3 standard errors? If the FDA lab finds a PPB level of 18, should it rescind the recall order?

7.39 Hospitals record most items given to patients, such as medicine and disposable thermometers, at the time they are used, so that they will know how much the patient costs the hospital. It is not worth keeping an exact record of inexpensive items such as adhesive bandages. Nevertheless, hospitals have to allocate the costs of such minor items across its patients by some method. Suppose the use of adhesive bandages was known to have a standard deviation of 3 per patient-day for all categories of patients. To estimate the mean bandage use for various categories of patients, the accounting department wants the nursing staff to record actual bandage use by five randomly selected patients in each category. What is the probability that the sample mean will be within one bandage of the population mean for any category? Assume that the distribution of bandage use per patient is normal.

7.40 Suppose the reading scores of third-graders on a national exam had a mean of 100 and a standard deviation of 20. You are considering purchasing a home in a new city, and an important factor in your decision is the quality of the neighborhood school. You visit various schools and ask the principal if you may see the mean score of all of the school's third-graders on the national test. Not all principals are willing to show you these scores; generally the ones that ranked high are the most willing. One school that is near an attractive home and convenient to your workplace has a mean score of 97 for all 62 third-graders in the school. The principal of this school tells you that there is a great deal of variability in the test at that age level, and that the difference between the mean of 97 for her third-graders and the population mean of 100 for all third-graders may be due to chance variation; some children may have felt ill on the test day, for example, or been distracted by events going on in the school. If the true mean of the school is 100, what is the probability that a sample mean of 97 or less will be observed? What is the flaw in using the standard deviation of 20 points to find this probability?

7.41 You want to survey the U.S. adult population about preferences in regard to changes in the tax law. The subject is too complicated for a phone survey because you want to explore detailed answers to changes in the rules governing various types of deductions. Because of budget limitations, you've settled on a mail survey with a sample size of 10,000, but you know that the response rate to long questionnaires sent in the mail is very low, and you are worried about nonresponse bias. What fairly low-cost method can be used to check for nonresponse bias?

7.42 You plan a survey of shoppers at a suburban mall. You pay a fee to the mall operators and instruct your interviewers to stop every 20th person walking through the mall and ask some questions about plans for major household purchases. The interviewers are to record the sex and estimated age of each respondent. When you get the survey results, you notice that 95% of the respondents were women. Your experience has been that about 75% of the shoppers in this mall are women. The sample size was too large for the difference in proportions to have been caused by chance. Something went wrong. What is the most likely cause?

7.43 You want to estimate the mean number of acres per farm planted in corn in a particular agricultural county. The county has 300 farms and you decide on a 10% sample. You randomly select a number from 1 to 30 and it turns out to be 13. You sample the 13th, 43rd, 73rd, and so on farms on a list that is in alphabetical order. The mean acreage of your sample is 350 acres. When you check the previous year's actual acreage, you discover that the mean for all 300 farms was 1,080 acres. You know that corn acreage has not dropped that much. There has been no wholesale conversion to other crops or abandonment of large farms. What could have gone wrong? What kind of sampling other than systematic sampling might give you a more accurate estimate?

7.44 Is the formula $\Sigma(x_i - \bar{x})^2/n$ an unbiased estimator of the population variance? If it is a biased estimator, does its expected value over- or underestimate the population variance?

7.45 If you took a sample from a population with a known mean μ, you could calculate the sum of the squared deviations of the sample values from μ or from the sample mean, that is, $\Sigma(x_i - \mu)^2$ or $\Sigma(x_i - \bar{x})^2$. Explain why one of these terms must be greater than the other if $\bar{x}$ does not equal μ. What is the implication of this result for the formula for the sample variance?

7.46 Joe is one of 30 salesmen at the Levitz Furniture Factory Warehouse and Showroom. All of the salesmen are in a contest to win a new Cadillac. The salesmen meet the customers as they walk into the showroom in rotation, so every salesman sees about the same number of customers. The winner of the contest will be the salesman who has the highest percentage of customers making a purchase. The storewide proportion of customers that make purchases is 0.25. Joe has made sales to 40 of the 100 customers he has met during the contest. If a customer's decision about making a furniture purchase was independent of the salesman, what would be the probability that any other salesman who met 100 customers would do better than Joe?

7.47 The Gotham city libraries charge 10 cents a day for overdue books. The overall proportion of borrowed books that become overdue is 0.10. As an experiment, the library raises the charge to 20 cents a day at one of its branches and observes that among the 300 books borrowed at this branch, the proportion overdue is 0.08. If each borrower acts independently and the probability of not returning the books by the due date is not affected by the higher charge, what is the probability of observing a sample proportion of 0.08 or less at that branch?

7.48* You are a dietitian for a cruise line and you need to estimate the mean daily food consumption in calories for a population that consists of all the people who have booked passage on a three-month cruise to Antarctica. The population consists of three groups: senior citizens, college students, and teenagers. You have enough resources to survey 30 randomly selected people from this population. You know that the population sizes of the three groups are 1,000, 200, and 100, respectively. You also know that the population standard deviations for food consumption for the three groups are 100, 300, and 400, respectively. How should you allocate your sample across the three groups so that the sample mean will be as accurate an estimator of the population mean as possible?

7.49* The U.S. Department of Transportation monitors highway speeds by state. If more than 50% of the drivers in a state go faster than 65 miles an hour on roads posted at 65, the department will cut off a substantial portion of federal highway dollars intended for that state. The monitoring procedure is based on a sample because it is impossible to check the speed of every driver at every moment. You are a highway official. Your state has been found not to be in compliance with the 65–mile-an-hour limit and is threatened with the loss of some of its highway money. According to the U.S. Department of Transportation, the average speed of 1,000 sampled drivers was 67.1 miles per hour. You argue that there is a high probability that the population mean speed could be 65 miles per hour and that the observed 67.1 is due to several errors. You point out first that the radar units used in the monitoring have an error distribution that is normal, centered on 0, and has a standard deviation of 4 miles per hour. Second, you point out that motorists' speedometers have errors as well, that the distribution of speedometer errors is normal with a mean of 0 and a standard deviation of 8 miles per hour. Finally, you point out that there is variation in the speeds of the motorists, and that the 67.1 sample average could be due to the chance selection of a sample that travels faster than the average of all state drivers. Assume that the standard deviation of drivers' speeds is 12 miles per hour. Since millions of dollars are at stake, you want to think of as many reasons as possible. If the true mean speed of your state's motorists as recorded on their speedometers was 65 miles per hour, what is the probability that a sample mean of 67.1 or greater would be observed in a sample of 1,000?

7.50* Is the following statement true? "The median of all possible sample medians for a given sample size and a given population equals the population median."

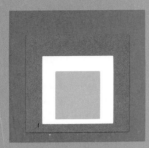

8

Confidence Intervals

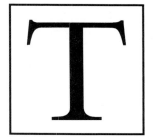he Westinghouse Corporation factory in Muncie, Indiana, makes transformers for generating plants. In 1976 the plant was struck for eight weeks by the 1,200-member International Brotherhood of Electrical Workers. The union contended that the prestrike wage rates were much lower than the wage rates offered by other factories in the surrounding area. The company contended that Westinghouse's wage rates were higher than those of competing firms. To settle the strike, the company proposed to hire a consulting firm to do an area wage survey for each of the job classifications in the plant. Both sides agreed to use the results of the survey to set new wage rates. Most Westinghouse classifications were found to be slightly underpaid in comparison with similar classifications in other plants, but one group, consisting mainly of the 60 safety inspectors, was found to be overpaid. Because everyone except these 60 workers was satisfied with the agreement to use an area wage survey to set wage rates, this method was used again in 1978, 1980, and 1982. In each of these subsequent surveys the same group of workers was found to be overpaid. They received no wage increase from 1976 to 1982 except for a cost-of-living adjustment, or COLA. Frustrated and puzzled by these results, the 60 workers sued the company, their union, and the consulting firm.

The mean wage rate for a job classification found by the survey was used to set the wage rates in the factory. However, a number such as $12.50 per hour, provided by a single sample and used as the estimate of the true population mean wage rate, gives you no sense of how close it may be to the true rate. If the consultants told both sides that they were about 95% sure that the true mean was between $12.40 and $12.60, then the $12.50 could be considered a fairly reliable estimate. If the consultants told both sides that they were about 95% sure that the true mean was between $10.00 and $15.00, then the estimate of $12.50 per hour could be considered much less reliable.

Before you can rely on any estimate, you need to have a sense of how accurate it is. A range of possible values and an associated level of confidence serve this purpose. Such statements are called confidence intervals. The consultants hired by Westinghouse never gave anyone confidence intervals for their estimates. For the safety inspector classification, the confidence intervals were enormous. According to the formula for confidence intervals, at a 95% level of confidence you could state that the mean area wage rate was between negative $7.50 and positive $32.50. If the consultants had told the union and the company about this level of accuracy, the estimates would have been recognized as worthless and they would not have been used to set wage rates. Unfortunately, the union negotiators, the company negotiators, the safety inspectors, and the consultant who conducted the survey had no background in statistics. Ignorance can be very expensive.

This chapter begins with an introduction to statistical inference and the two main techniques of confidence intervals and hypothesis tests. The bulk of the chapter is devoted to explaining how to construct confidence intervals for the population mean and the population proportion in a variety of circumstances: when populations are small or large; when population standard deviations are known or unknown; and when populations are normal or not normal. The last section discusses the minimum sample size needed to construct a confidence interval of a given width.

8.1 CONFIDENCE INTERVALS FOR THE POPULATION MEAN WHEN THE POPULATION IS NORMAL AND THE POPULATION STANDARD DEVIATION IS KNOWN

8.1A INTRODUCTION TO CONFIDENCE INTERVALS

In Chapter 7 you studied the relationship between a known population and the possible sample means taken from that population. If you knew the mean and the standard deviation of a population, you could find the probability that the sample mean was between any two values. In fact, however, you are rarely in a position to know the expected value or the standard deviation of a population. Most of the time you will be doing the exact opposite of what you did in Chapter 7. Instead of going from the population to the sample, you will use the sample to make some inferences about the population. This change in objectives does not mean that Chapter 7 was a waste of time. You'll be using the theory presented in Chapter 7, but essentially turning it on its head. If the population can tell you about the sample, then the sample can tell you about the population.

There are two closely related procedures for making inferences about the population from a sample: confidence intervals and hypothesis tests. Which you should use depends on whether or not you have a prior value in mind for the population parameter. If, for example, you think the mean number of soft drinks consumed by college students in a week is less than or equal to 20 cans, you can use a sample to test that hypothesis. On the other hand, when you construct a confidence interval, you are letting the data "speak to you." You don't have a preconceived value that you want to test, and all you want is a sense of how close the sample mean or proportion may be to the corresponding population parameter. This chapter is devoted to confidence intervals for one population and Chapter 9 is on hypothesis tests for one population. Chapter 10 discusses both confidence intervals and hypothesis tests for problems that involve two populations.

Confidence intervals can provide more useful information than a single-number estimate. You could use an estimate as the one best number for predicting the population parameter—for example, the mean of a particular sample to estimate the population mean. A **point estimate** is *the one number that is the best estimate of a parameter*. The problem with a point estimate is that you have no idea how far off it may be, that is, how far it deviates from the true parameter. The \$12.50 estimate for the mean wage rate of safety inspectors which was given in the chapter introduction does not provide any sense of the accuracy of the estimate.

8.1B BASIC FORMULA FOR CONFIDENCE INTERVALS AND THE LOGIC BEHIND IT

A **confidence interval** for the population mean is *a lower and an upper bound within which the population parameter is expected to be located along with an associated level of confidence*. An example might be "The mean number of soft drinks consumed per week by college students is between 20.5 ± 4 at the 95% level of confidence." The formula for a confidence interval always has three parts: (1) the center or sample statistic that is the point estimate of the population parameter, (2) the standardized factor based on the confidence level, such as the z score, and (3) the standard deviation of the sampling distribution of the sample statistic, such as the standard error of the mean. For example, the center of the interval given by equations 8.1a and 8.1b is $\bar{x}$, the factor that reflects the level of confidence is $z_{\alpha/2}$, and the standard deviation of the sampling distribution is $\sigma_{\bar{x}}$. The meaning of the term $z_{\alpha/2}$ will be explained in a few pages.

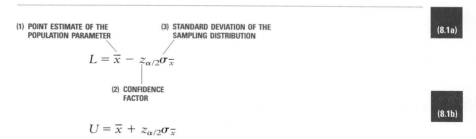

(1) POINT ESTIMATE OF THE POPULATION PARAMETER   (3) STANDARD DEVIATION OF THE SAMPLING DISTRIBUTION   (8.1a)

$$L = \bar{x} - z_{\alpha/2}\sigma_{\bar{x}}$$

(2) CONFIDENCE FACTOR

(8.1b)

$$U = \bar{x} + z_{\alpha/2}\sigma_{\bar{x}}$$

where L is the lower bound and U is the upper bound.

Note: $\sigma_{\bar{x}} = \dfrac{\sigma}{\sqrt{n}}$, or $\dfrac{\sigma}{\sqrt{n}}\sqrt{\dfrac{\nu - n}{\nu - 1}}$

depending on whether the finite population correction factor is required.

For brevity, the formula for confidence intervals will be written as

$$L, \ U = \bar{x} \pm z_{\alpha/2}\sigma_{\bar{x}}$$

(8.1c)

Let's continue with the example of soft drink consumption and see where this formula comes from and what it means to have a 95% level of confidence. Suppose you knew the true population distribution shape, its mean, and its standard deviation. Say that the population was approximately normally distributed with a mean level of soft drink consumption among college students of 20 cans per week with a standard deviation of 10. On the basis of the theory you learned in Chapter 7, you know that the distribution of sample means for samples of size n will be normally distributed, centered on the population mean, and have a standard deviation of $10/\sqrt{n}$. Let's make the math simple and set n equal to 400. Then $\sigma_{\bar{x}}$ would equal 0.5. This sampling distribution of the mean is shown in Figure 8.1.

Ninety-five percent of the area under a standard normal curve centered on the mean lies within the z scores of ± 1.96. Here's how the z score of 1.96 is obtained. Note that the 95% is in the middle of the distribution, and therefore half of the 95% is on each side of μ. That leaves 0.95/2, or 0.4750, as the area between the mean and the upper bound. By looking up the probability of 0.4750 in the normal table, you'll find that the z score that corresponds to this probability is 1.96. By the symmetry property of the normal distribution, the z scores for the central 95% of a normal distribution are ± 1.96. Since the distribution in Figure 8.1 has a standard deviation of 0.5, 95% of the area must lie between 19.02 and 20.98. These two points are plus and minus 1.96 multiples of the standard deviation of 0.5 away from the population mean. Two vertical lines have been drawn in Figure 8.1 at these values and the areas outside these vertical lines are shaded. Ninety-five percent of all possible samples of size 400 have means that fall between 19.02 and 20.98. Five percent of all possible samples have means that fall outside of this range. Again, since the

FIGURE 8.1

Sampling distribution of the mean for soft drink consumption by college students

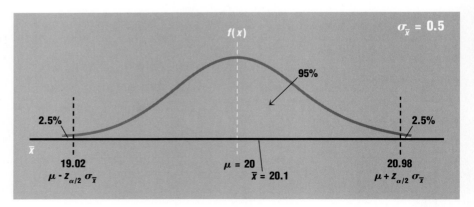

FIGURE 8.2
Confidence interval centered
on a sample mean of 20.1

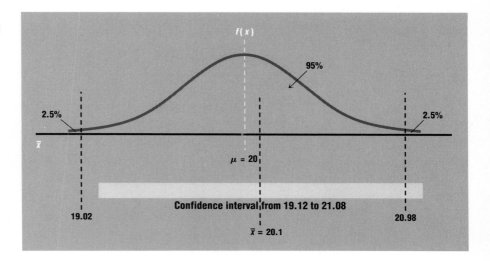

distribution is symmetrical, 2.5% of the total area must be in each of the tails marked by these vertical lines. Suppose you took a sample of size 400 and the sample mean turned out to be 20.1. To construct a confidence interval around this sample mean, you would have to use formula 8.1c. The three parts to constructing a confidence interval would be:

$$L, U = \bar{x} \pm z_{\alpha/2}\sigma_{\bar{x}} \qquad \text{(8.1c)}$$

$$L, U = 20.1 \pm 1.96(0.5)$$

$$L = 19.12 \qquad U = 21.08$$

This confidence interval has been drawn under the sampling distribution of the mean in Figure 8.2. It is the shaded bar beneath the curve. Since 20 is between 19.12 and 21.08, this particular confidence interval includes the true mean of 20. If the sample mean had been a value outside the area marked off by the two vertical lines, either below 19.02 or above 20.98, and if a confidence interval were centered on one of these outlying values, then the confidence interval would not have included the population mean. The reason is that the width of the confidence interval is equal to the distance between the vertical lines. Try a sample mean of 21.5.

$$L, U = \bar{x} \pm z_{\alpha/2}\sigma_{\bar{x}} \qquad \text{(8.1c)}$$

$$L, U = 21.5 \pm 1.96(0.5)$$

$$L = 20.52 \qquad U = 22.48$$

FIGURE 8.3
Confidence interval centered
on a sample mean of 21.5

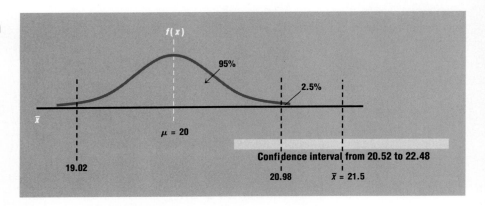

This confidence interval is shown below the sampling distribution of the mean in Figure 8.3, and you can see that it has the *same width* as the other confidence interval and that it does *not* include the population mean.

What is the probability that a confidence interval will miss the true mean and what is the probability that it will include the true mean? Every time you take a sample that has a mean between the vertical lines, the resulting confidence interval will include the true mean, and every time you take a sample that has a mean outside the vertical lines, the resulting confidence interval will not contain the true mean. Since the vertical lines were drawn so that 95% of all possible samples will have means that will fall between them, the probability that a confidence interval will include the true mean is 0.95 and the probability that a confidence interval will not include the true mean is 0.05. A 95% level of confidence does not imply that for any *particular* confidence interval you have constructed you are 95% sure that *it* contains the true mean. The 95% is really a statement about the procedure you are following. It implies that you are following a procedure that will lead to confidence intervals that contain the true mean for 95% of all possible samples. Figure 8.4 shows many possible confidence intervals constructed from a series of samples taken from the same population. Most of the confidence intervals catch the true mean ($\mu = 20$) and a few miss it.

■ 8.1C CHOICE OF A CONFIDENCE LEVEL

The α in formula 8.1 is your chance of being wrong. It is written $z_{\alpha/2}$ because the chance of being wrong is divided between the two tails; in the above example you have a 2.5% chance of getting a confidence interval that misses the true mean because it is too high and a 2.5% chance of getting one that misses the true mean because it is too low. Your level of confidence is the chance that you are right, $(1 - \alpha)$, or 95%. The level of confidence does not

FIGURE 8.4
Many confidence intervals

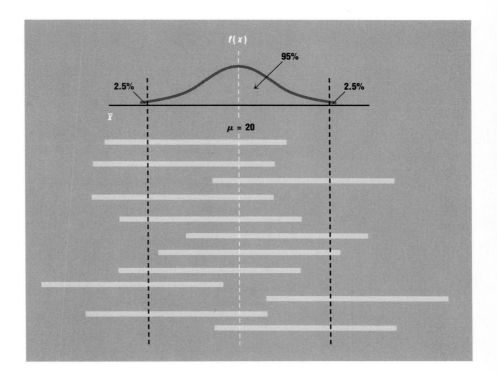

have to be set at 95%, although this number is commonly used. If you want to be more certain that your confidence interval contains the true mean, just increase the confidence level. For example, the $z_{\alpha/2}$ for a 99.9% level of confidence would equal 3.29. This z score corresponds to having 0.05% of the area in each tail. Applying this z score to the last example results in the following confidence interval:

$$L, U = \bar{x} \pm z_{\alpha/2}\sigma_{\bar{x}} \qquad \text{(8.1c)}$$

$$L, U = 21.5 \pm 3.29(0.5)$$

$$L = 19.86 \qquad U = 23.15$$

With the higher confidence level, the true mean of 20 is now inside the confidence interval. You may ask, "Why not always construct confidence intervals at high levels of confidence and be very sure that the interval contains the true mean?" Why settle for 99.9%? Why not 99.99999%? The answer is that your confidence interval is getting wider as you increase the confidence level. As you raise the level of confidence, you have to use higher numerical values for the z score in formula 8.1. This is why the confidence interval widens. This process is an example of being more and more certain of less and less until you are absolutely sure of nothing. For example, it does you no good to be absolutely sure that the true mean is between $-\infty$ and $+\infty$. For a given sample size and a given variance in the population that you are sampling, the confidence interval can give you only so much precision. In other words, the informational content of a sample of a given size is fixed. Your choice of how to present that information is fairly arbitrary. Habit seems to be the only reason that 95% is commonly used. Since the variability of the population that you are sampling is fixed, the only way to get more precision from a confidence interval is to increase the sample size on which the interval is based. Increasing the sample size costs money, so you have to choose between the value of the added information and the cost of a larger sample. Section 8.4 describes how you can find the minimum sample size for a confidence interval of a specified width and a specified confidence level.

Formula 8.1 will always work for constructing confidence intervals if the population is normal and the standard deviation is known. You can always use the z score with a normal population and known σ because (as we saw in Chapter 7) the sampling distribution of the mean will be exactly normal no matter how small the sample size. It is an unusual situation to know the population standard deviation and not to know the population mean. This situation is mostly a textbook case that helps get across the explanation of how some of the more realistic examples work. You also have to know that your population is normally distributed. Some rough tests for normality were described when the normal distribution was introduced in Chapter 6. More powerful tests will be described in Chapters 11 and 17. If it is unlikely that your sample came from a normal population or something close to a normal population and your sample is less than 30, *do not* use equation 8.1. If the population is not normal, your results can be way off. How far off depends on how much the population distribution departs from normality.

EXERCISES FOR SECTION 8.1

8.1 Let X be a normally distributed random variable with a standard deviation of 5.

(a) Given that the mean of a sample of size 10 taken from this distribution is 26, construct a confidence interval for the population mean at the 95% level of confidence.

(b) Given that the mean of a sample of size 30 taken from this distribution is 26, construct a confidence interval for the population mean at the 95% level of confidence.

(c) Given that the mean of a sample of size 10 taken from this distribution is 26, construct a confidence interval for the population mean at the 90% level of confidence.

(d) Given that the mean of a sample of size 30 taken from this distribution is 26, construct a confidence interval for the population mean at the 99% level of confidence.

8.2 Let X be a random variable that represents observations drawn from a population that is normally distributed, has a standard deviation of 20, and has a population size of 1,000.

(a) Given that the mean of a sample of size 10 taken from this population is 26, construct a confidence interval for the population mean at the 95% level of confidence.

(b) Given that the mean of a sample of size 100 taken without replacement from this population is 26, construct a confidence interval for the population mean at the 95% level of confidence.

(c) Given that the mean of a sample of size 100 taken with replacement from this population is 26, construct a confidence interval for the population mean at the 95% level of confidence.

(d) Given that the mean of a sample of size 100 taken without replacement from this population is 26, construct a confidence interval for the population mean at the 99% level of confidence.

8.3 You are the credit manager for a large corporation and you are interested in determining the mean number of business days that elapse between the mailing out of a bill and receipt of payment. This information can help you manage your corporation's cash, because if you knew how much money was coming into the corporation and how much had to be spent, you could determine how much money the corporation could tie up in accounts that paid high interest rates. Another reason that this information is valuable is that an increase in the mean time between billing and payment might indicate that the corporation's customers were losing sales. Your corporation sends out about 10,000 bills each month, so it would be a waste of money for you to use the entire population to calculate the mean time; an estimate based on a sample would be accurate enough. Suppose you took a sample of 300 bills and found that the mean time was 23 business days. From past records on payments you have a value for the population standard deviation of 8 days. Assume that the distribution of days between billing and payment is normal.

(a) Construct a confidence interval at the 95% level of confidence for the population mean.

(b) Construct a confidence interval at the 99% level of confidence for the population mean.

(c) Suppose the sample size was changed to 2,000; now construct a confidence interval at the 90% level of confidence.

8.4 A tire manufacturer is testing a new compound for tread wear. Tires made with the compound are placed on a machine that simulates road wear, and the amount of tread left after the equivalent of 40,000 road miles is recorded. The mean amount of tread left after a test of 10 tires was 3.6 millimeters. Construct a confidence interval for the population mean amount of tread left after 40,000 miles at the 99% level of confidence. Assume that the population distribution of tread remaining is normally distributed with a known standard deviation of 0.2 millimeters.

8.5 You own a lawn care company that sprays fertilizer and weed killer on consumers' lawns. You are considering opening a branch of your company in a particular city and one factor in your decision is the mean annual expenditure on lawn care per residence with a privately owned lawn. You hire a market survey firm to interview a random sample of 100 family heads who own their own lawns and estimate, at the 95% level of confidence, the mean expenditure in the population. Your contract with the market survey firm provides that you will pay $25 per completed interview. Assume that the population distribution of expenditures is normal and that the population standard deviation is known to be $68. Suppose the report from the market survey firm states that the 95% confidence interval has a lower limit of $56.15 and an upper limit of $93.85. The report is accompanied by a bill for 100 interviews. Explain why the above confidence limits might lead you to suspect that the market survey firm did not complete 100 interviews.

8.6 Is the following statement correct? "The probability equals 0.95 that the mean expenditure on lawn care is between $56.15 and $93.85 for the population sampled in exercise 8.5." Explain.

8.7 A telephone company is planning to introduce a measured service billing system. Under this system the charge for a local phone call depends, in part, on the duration of the call. To estimate the revenue that will be generated by any charge based on duration of calls, the telephone company needs to estimate the mean duration of phone calls. Assume that the time taken for phone calls is normally distributed with a known standard deviation of 0.25 minute.

    (a) Construct a confidence interval at the 99% level of confidence for the population mean duration if a sample of 1,000 calls has a sample mean of 4.3 minutes.

    (b) If the above sample was taken when the measured billing system was not in effect, what is wrong with using the above confidence interval to estimate the potential revenue from a measured billing system?

8.8 An automobile dealer plans to order enough cars to have a 90-business-day inventory. Over the past 30 business days the dealer has sold an average of 20 cars a day.

    (a) Construct a confidence interval for the population mean number of cars sold per day at the 95% level of confidence. Assume that the distribution of cars sold per day is approximately normal with a standard deviation of 5.

    (b) If the dealer ordered 90 times the upper limit of the confidence interval for the daily mean that you calculated in part a, would the probability of running out of cars in 90 days be 0.025? Discuss.

8.2 CONFIDENCE INTERVALS FOR THE POPULATION MEAN WHEN THE POPULATION STANDARD DEVIATION IS NOT KNOWN

 ### 8.2A CONFIDENCE INTERVALS FOR THE POPULATION MEAN WHEN THE POPULATION STANDARD DEVIATION IS NOT KNOWN AND THE POPULATION IS NORMAL

If you don't know the population standard deviation, you need to modify formula 8.1. A simple way to change the formula would be to substitute the sample standard deviation for the population standard deviation. As I mentioned in Chapter 7, however, the sample standard deviation is a biased estimator of the population standard deviation. If the population is normally distributed, the expected amount of bias can be calculated and your confidence interval can be adjusted so that it has the correct width. The sample standard deviation will, on average, underestimate the population standard deviation, so you need to widen your confidence interval to adjust for the bias. William Steeley Gosset published an article in 1908 titled "The Probable Error of a Mean," which provided a solution for this problem. As his employer, Guinness Stout, wouldn't let him publish under his own name, he signed the article "A Student." Gosset's solution involves a distribution called the Student t. The mathematics of this distribution are beyond this text, but simply stated, the t score replaces the z score in formula 8.1 and it automatically corrects for the bias in the sample standard deviation.

Table D7 in Appendix D contains the t scores that correspond to having 10%, 5%, 2.5%, 1%, and 0.5% of the area in the right tail. The Student t distribution is a family of distributions whose shapes vary slightly with the degrees of freedom. The degrees of freedom are $n - 1$ because treating the sample mean as an estimate of the population mean in the formula for the sample standard deviation uses up one degree of freedom. What this means is that if you were given all but one of the values of a sample of size n and told the value of the sample mean, you could determine the value of the missing observation. In this sense, once the sample mean is specified, only $n - 1$ of the observations are free to vary. The Student t is a bit wider than the standard normal distribution, but it is still symmetrical and centered on zero. Figure 8.5 illustrates this difference by showing Student t distributions with 10 and 15 degrees of freedom superimposed on a standard normal curve.

In Table D7 you can find the t score that has 2.5% of the area in the right tail. The t score for 10 degrees of freedom is 2.228, so it is easy to see that the Student distribution has to be wider than the standard normal because the z score that corresponds to having 2.5% of the area in the right tail is 1.96. If you run your finger down the column headed "2.5%" in Table D7, you see that the Student distribution converges on the normal. The t scores get closer and closer to 1.96 as the degrees of freedom increase, and with an infinite number of degrees of freedom, the t score is 1.96.

FIGURE 8.5
Student *t* distributions with
10 and 15 degrees of freedom
superimposed on a standard
normal distribution

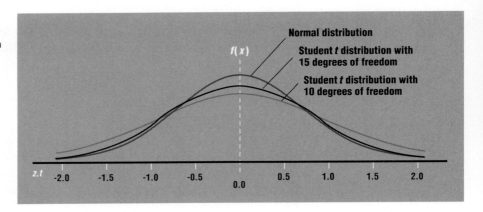

It is much more common to know that the population is normal but not to know the population standard deviation than to know both facts. The latter situation was discussed in section 8.1. Now formula 8.1 has to be modified slightly; whenever σ is not known and the population is known to be normal, the Student *t* has to be used in place of the *z* score. When the standard deviation of the distribution of the sample mean, $\sigma_{\bar{x}}$, cannot be calculated because the population standard deviation is not known, the estimated standard deviation of the distribution of the sample mean based on the sample standard deviation, $s_{\bar{x}} = s/\sqrt{n}$, must be used in place of $\sigma_{\bar{x}}$. Formula 8.1 is transformed into

(8.2)

where

$$s_{\bar{x}} = \frac{s}{\sqrt{n}} \quad \text{or} \quad \frac{s}{\sqrt{n}}\sqrt{\frac{\nu - n}{\nu - 1}}$$

depending on whether the finite population correction factor is required.

Note: Degrees of freedom are $n - 1$.

If you can obtain the *t* score from either a table or a computer program, you should always use it when you work with normal populations and unknown σ's. However, if the sample size is 30 or more and you cannot obtain the correct *t* score, you can use the *z* score in place of the *t* score because the normal distribution provides a good approximation to the Student t when the sample is this large. Confidence intervals based on formula 8.2 are considered "robust" by statisticians. A **robust estimator** is *an estimator that will still give accurate estimates when the assumptions of normality are violated*. Thus if your population distribution is fairly close to normal, you can still use formula 8.2. Symmetrical distributions that have more observations in the tails than would be found in a normal distribution are OK, but skewed distributions may lead to unreliable estimates. Here are some examples that use formula 8.2.

You own the Thermo-Sash Window Company and you employ 200 salesmen throughout the United States to sell windows to be installed in new homes or as replacements in older homes. Your salesmen visit building materials wholesalers and take their orders for shipment within three to six months. During the first few weeks of every quarter your salesmen report their sales for the previous quarter. These sales figures are totaled and the sum is factored into your production plans for the quarter. Suppose your production staff needed an estimate of a quarter's sales before all of the salesmen had reported. Let's say 20 salesmen had reported an average $30,000 per salesman with a sample standard deviation of $16,000. How could you get a 95% confidence interval for the sales total for all 200 salesmen? You have to assume that the first 20 salesmen to report are a representative sample of the population. If the early reporters were the salesmen with the best or worst records, your estimate could be way off. This situation would be an example of self-selection bias. You could check to find out if the first 20 salesmen to report in previous quarters had sales within one standard error of the remaining 180 salesmen. If they did, their sales could be said to be within one standard error. If this check works out, you are pretty safe. If the distribution of sales was normal in past quarters, you are fairly safe in assuming that the population distribution will be normal for the current quarter.

To get a confidence interval on the total sales, first get confidence limits on the mean level of sales and then multiply those limits by the total number of salesmen. The formula for a confidence interval on the mean would be:

$$L, U = \bar{x} \pm t_{\alpha/2}s_{\bar{x}} \qquad (8.2)$$

$$L, U = \bar{x} \pm t_{\alpha/2}\left(\frac{s}{\sqrt{n}}\right)\sqrt{\frac{\nu - n}{\nu - 1}}$$

$$L, U = 30,000 \pm 2.093\left(\frac{16,000}{\sqrt{20}}\right)\sqrt{\frac{200 - 20}{200 - 1}}$$

$$L = 22,878 \qquad U = 37,122$$

The lower limit for the mean would be $22,878 and the upper limit would be $37,122. The t score of 2.093 was based on an $\alpha/2$ of 2.5% with 19 degrees of freedom. Because the sample was more than 5% of the population, the finite population correction factor was used in the above formula. The original problem was to find a 95% confidence interval for the total sales rather than a confidence interval for the mean sales per salesman. To obtain a confidence interval for the total sales, just multiply $22,878 and $37,122 by 200 salesmen.

Now suppose you are a mining engineer and you are negotiating a contract to deliver coal from a new mine to a nearby generating plant. The new mine has been worked for 25 days and the average daily output is 23,000 tons with a sample standard deviation of 4,000 tons. The contract you are working on is for the next 100 days' output of the mine. The owners of the generating plant want you to guarantee the amount of coal that will be delivered. If you guarantee more than the mine actually produces, your coal company will have to make up the difference by shipping coal from a distant mine. If you underestimate what the new mine will actually produce, your coal company may have to sell the extra coal at a low price on the spot market. You would like to have an accurate estimate, say at the 95% confidence level. You use formula 8.2 to find the confidence limits for the daily output:

$$L, \ U = \bar{x} \pm t_{\alpha/2} s_{\bar{x}} \tag{8.2}$$

$$L, \ U = \bar{x} \pm t_{\alpha/2}\left(\frac{s}{\sqrt{n}}\right)$$

$$L, \ U = 23,000 \pm 2.064\left(\frac{4,000}{\sqrt{25}}\right)$$

$$L = 21,349 \qquad U = 24,651$$

The confidence limits for 100 days' output are 100 times these figures.

The above illustration used the t score for an α of 2.5% in each tail with 24 degrees of freedom. Several assumptions have to hold if these confidence limits are to be accurate. One is that the distribution of daily coal output is normal. Another assumption is that the population mean does not change over time. If the coal seam became narrower as mining continued, average daily output could decline. It is also possible that the productivity of a new mine could increase as the miners gained experience. Confidence intervals are not a technique that can be applied mechanically. You have to understand the environment that you are working in and know if the assumptions you are making are reasonable.

8.2B CONFIDENCE INTERVALS FOR THE POPULATION MEAN WHEN THE POPULATION STANDARD DEVIATION IS NOT KNOWN AND THE POPULATION IS NOT NORMAL

The most common situation that you will face when you need to construct a confidence interval is having a nonnormal population and an unknown population standard deviation. In this situation you must depend on the central limit theorem. You need a sample of size 30 or more so that the distribution of sample means will be approximately normal. The formula for the confidence interval with a nonnormal population is formula 8.3.

(1) POINT ESTIMATE OF   (3) ESTIMATED STANDARD   (8.3)
     THE POPULATION   DEVIATION OF THE SAMPLING
     PARAMETER   DISTRIBUTION

$$L, U = \overline{x} \pm z_{\alpha/2} s_{\overline{x}}$$

(2) CONFIDENCE
   FACTOR

where

$$s_{\overline{x}} = \frac{s}{\sqrt{n}} \quad \text{or} \quad \frac{s}{\sqrt{n}}\sqrt{\frac{\nu - n}{\nu - 1}}$$

depending on whether the finite population correction factor is required.

Note: This formula should be used only when $n \geq 30$.

Let's say you own a fruit-packing firm that ships gift packages of Washington State pears. You are considering a new container for shipping fruit and you want to estimate the mean number of pears that will fit into the container. You pack 45 containers and find that the mean number of pears per container is 12 with a standard deviation of 3. What is the confidence interval at the 95% level of confidence for the mean number of pears?

$$L, U = \overline{x} \pm z_{\alpha/2} s_{\overline{x}} \qquad\qquad \text{(8.3)}$$

$$L, U = 12 \pm 1.96\left(\frac{3}{\sqrt{45}}\right)$$

$$L = 11.12 \qquad U = 12.88$$

This confidence interval can be used to estimate the mean cost of the pears needed to fill each container.

8.2C CHART THAT SUMMARIZES CONFIDENCE INTERVALS FOR THE POPULATION MEAN

Figure 8.6 may clarify the situations that arise when you construct confidence intervals for the population mean. While Figure 8.6 shows eight logically distinct cases, there are only four procedures to follow: use the z score with σ, use the z score with s, use the t score, or give up.

	σ KNOWN		σ NOT KNOWN	
	$n < 30$	$n \geq 30$	$n < 30$	$n \geq 30$
POPULATION NORMAL	Use z score because the distribution of $\bar{x}$ is normal $$\sigma_{\bar{x}} = \frac{\sigma}{\sqrt{n}}$$	Use z score because the distribution of $\bar{x}$ is normal $$\sigma_{\bar{x}} = \frac{\sigma}{\sqrt{n}}$$	Use t score $$s_{\bar{x}} = \frac{s}{\sqrt{n}}$$	Use t score if it is available, but z score provides a good approximation $$s_{\bar{x}} = \frac{s}{\sqrt{n}}$$
POPULATION DISTRIBUTION NOT KNOWN	Can't be done	Use z score; $\bar{x}$ is approximately normal from central limit theorem $$\sigma_{\bar{x}} = \frac{\sigma}{\sqrt{n}}$$	Can't be done	Use z score; $\bar{x}$ is approximately normal from central limit theorem $$s_{\bar{x}} = \frac{s}{\sqrt{n}}$$

EXERCISES FOR SECTION 8.2

8.9 Let X be a normally distributed random variable with an unknown mean and an unknown standard deviation.

(a) Construct a confidence interval at the 95% level of confidence for the population mean when a sample of size 12 has a mean of 24 and a standard deviation of 5.

(b) Construct a confidence interval at the 90% level of confidence for the population mean when a sample of size 6 has a mean of 24 and a standard deviation of 5.

(c) Construct a confidence interval at the 95% level of confidence for the population mean when a sample of size 36 has a mean of 24 and a standard deviation of 5.

(d) Construct a confidence interval at the 80% level of confidence for the population mean when a sample of size 36 has a mean of 24 and a standard deviation of 5.

8.10 Let X be a random variable with an unknown mean, an unknown standard deviation, and an unknown distribution.

(a) Construct a confidence interval at the 95% level of confidence for the population mean when a sample of size 60 has a mean of 68 and a standard deviation of 2.

(b) Construct a confidence interval at the 90% level of confidence for the population mean when a sample of size 60 has a mean of 68 and a standard deviation of 2.

(c) Construct a confidence interval at the 80% level of confidence for the population mean when a sample of size 60 has a mean of 68 and a standard deviation of 2.

(d) Construct a confidence interval at the 95% level of confidence for the population mean when a sample of size 6 has a mean of 68 and a standard deviation of 2.

8.11 You own a store that rents video cassettes of movies. Your display space is very limited and you want to stock only titles that will be rented frequently. You can display only 200 titles. To get a base figure to judge the performance of individual titles, you check your records for 20 randomly selected movies for a one-month period to see how many times each movie was rented. The sample mean was 8 with a sample standard deviation of 3. You've plotted the data and the distribution appears to be normal. Construct a confidence interval at the 95% level of confidence for the population mean number of rentals per movie per month.

8.12 Your company self-insures its fleet of cars against collisions. You need to estimate the mean repair costs per accident so that you can set aside enough money to cover these costs. Suppose in the past month there have been seven accidents and the mean repair cost was $2,400 with a sample standard deviation of $400. Construct a confidence interval at the 90% level of confidence for the population mean of the repair costs per accident. What assumptions have to hold if this estimate is to be accurate?

8.13 You own a company that recovers sunken treasure. You know the location of a sunken Spanish galleon, and preliminary dives and some magnetometer tests indicate that the ship's treasure chests burst open when the ship sank and that the coins are scattered over the ocean floor. To get a better idea of the number of gold coins that can be recovered in the area, you mark off the ocean floor with plastic pipes in grids of 10 square yards. The area that you think contains some treasure is 10,000 square yards. You need to convince some investors that financing the salvage operation will be a good investment. You randomly select 100 of the grids and sift through them for coins. The sample mean is 2 with a sample standard deviation of 3. Construct a confidence interval at the 95% level of confidence for the total number of coins that can be recovered.

8.14 A chair manufacturer has designed a new line of upholstered chairs that are fire-resistant. Fire-resistant chairs can be sold in some institutional markets, such as hotels and hospitals. To estimate the mean temperature at which these chairs start to burn, the manufacturer tests six chairs, one at a time, in a burn chamber. The temperature of the chamber is raised 1 degree at a time until each chair bursts into flame. The temperatures for the six chairs were 420°, 441°, 478°, 482°, 487°, and 495°. Assume that the distribution of temperatures at which chairs of this design burst into flame is normally distributed. Construct a confidence interval for the population mean temperature at the 90% confidence level.

8.15 (a) A camera manufacturer has introduced a new 35mm SLR camera that focuses automatically. The manufacturer would like to estimate the mean number of rolls of film used over the first year of ownership. These data would help the company predict the proportion of cameras that will require warranty repair. Because the new camera is easier to use than previous models, it is possible that less serious photographers will buy it. The less serious photographers would tend to shoot fewer rolls of film than the serious ones. On the other hand, it is possible that the new model is so easy to use that photographers find that they want to use it more often. The camera manufacturer needs a survey to see if the mean number of rolls has gone up or down with the new model. The manufacturer contacts 50 people who have sent in warranty registration cards and asks them to keep a diary of their camera use. Only 24 of those 50 people maintain their diaries for a year. Among the 24, the mean number of rolls is 14 with a standard deviation of 4. Assume that the population distribution of the number of rolls shot per camera is normal. Construct a confidence interval for the population mean at the 99% level of confidence.

(b) Why might the confidence level you constructed in part a be inaccurate? Why might the assumption of normality be unreasonable?

8.16 A private company that markets commemorative gold and silver coins has commissioned an artist to design a gold coin to commemorate the bicentennial of the U.S. Constitution. After a die has been made from the design, the company wants to estimate the mean weight of the coins that will be made with the die. Twenty sample coins are made. These coins have a mean weight of 1.0003 ounces with a standard deviation of 0.005 ounce. Construct a confidence interval at the 99% level of confidence for the population mean weight.

8.3 CONFIDENCE INTERVALS FOR THE POPULATION PROPORTION WHEN THE NORMAL APPROXIMATION IS USED

You own an advertising agency that represents a fast-food chain and you have the results of a random survey of the chain's customers. Let's say that 225 out of the 1,000 people interviewed said that they could recall seeing a specific television commercial that your agency produced. The sample proportion of 0.225 is a point estimate of the population proportion. Like all point estimates, it gives you no idea how far off the point estimate may be from the population value. For that you need an interval estimate. However, to construct confidence intervals on the proportion, you have to use the normal approximation to the sampling distribution of the proportion.

Since the sampling distribution of the proportion is closely related to the binomial distribution, the rule for using the normal distribution to approximate the sampling distribution of the proportion is the same as the rule for the normal approximating the binomial: $n\Pi$ and $n(1 - \Pi)$ have to be greater than 5. Since you don't know Π, substitute the sample proportion, $\bar{p}$. In this example the normal is a good approximation because both $n\bar{p}$ and $n(1 - \bar{p})$ are much higher than 5. Specifically, $n\bar{p}$ is $1,000(0.225)$ or 225 and $n(1 - \bar{p})$ is $1,000(0.775)$, or 775.

The formula for a confidence interval on the population proportion closely corresponds to the formulas used for confidence intervals on the population mean. The three elements are (1) the sample proportion as the center or sample statistic, (2) the z score as the standardized score, and (3) the estimated standard deviation of the distribution of sample proportions as the standard deviation of the sample statistic. The formula is formula 8.4.

(1) POINT ESTIMATE OF THE POPULATION PARAMETER

(3) ESTIMATED STANDARD DEVIATION OF THE SAMPLING DISTRIBUTION

(8.4)

$$L, U = \bar{p} \pm z_{\alpha/2} s_{\bar{p}}$$

(2) CONFIDENCE FACTOR

where

$$s_{\bar{p}} = \sqrt{\frac{\bar{p}(1 - \bar{p})}{n}} \quad \text{or} \quad \sqrt{\frac{\bar{p}(1 - \bar{p})}{n}} \sqrt{\frac{\nu - n}{\nu - 1}}$$

depending on whether the finite population correction factor is required.

Note: Formula 8.4 should be used only when $n\bar{p}$ and $n(1 - \bar{p}) \geq 5$.

The term $\bar{p}$ is the sample proportion. The term $s_{\bar{p}}$ is the estimated standard deviation of the distribution of sample proportions. The estimate is based on the sample proportion. You cannot use the term $\sigma_{\bar{p}}$ in the formula for a confidence interval because it requires that you know Π, the population proportion. If you knew Π, you would have no reason to construct a confidence interval for Π. To get $s_{\bar{p}}$ instead of $\sigma_{\bar{p}}$, replace Π with the sample proportion in formula 7.4:

$$s_{\bar{p}} = \sqrt{\frac{\bar{p}(1 - \bar{p})}{n}}$$

(8.5)

or, when the finite population correction factor is required,

$$s_{\bar{p}} = \sqrt{\frac{\bar{p}(1 - \bar{p})}{n}} \sqrt{\frac{\nu - n}{\nu - 1}}$$

Let's apply formula 8.5 to the survey of the customers of the fast-food chain. Now let's treat the 0.225 as a sample value and construct a 95% confidence interval around that estimate.

$$L, U = \bar{p} \pm z_{\alpha/2} s_{\bar{p}} \tag{8.4}$$

$$L, U = 0.225 \pm 1.96 \sqrt{\frac{0.225(0.775)}{1,000}}$$

$$L = 0.199 \qquad U = 0.251$$

The upper and lower limits for a confidence interval for the population proportion at the 95% level of confidence are 0.251 and 0.199.

Now let's say you manufacture a device that warns parents whose infants are at risk for sudden infant death syndrome that their sleeping child has stopped breathing. The device consists of a sensor that is strapped around the child's chest, some electronic components, and an alarm. Considerations of humanity, perhaps aided by the possibility of product liability lawsuits, have made you very concerned that the device must not fail in home use. Even the most reliable suppliers of electronic components have not been able to provide you with shipments in which every component tested triggered the alarm when it was supposed to and did not trigger it when it was not supposed to. You test every electronic component before it is built into your machine and you return all defective components to their manufacturer. For production planning purposes you need to estimate the proportion of defective components in any shipment so that you can anticipate how many will have to be returned and how much you have to overorder in future shipments.

You have a shipment of 1,000 electronic components and you have so far tested 100 of them. Of the 100 tested, 6 were defective. In this example $n\bar{p}$ and $n(1 - \bar{p})$ are over 5, so the normal approximation can be used. An important assumption is that each trial is independent, that is, a defect in one component is not related to a defect in another component. If there was a production problem or a design problem in the manufacture of the component, then you could get a run of bad parts and your estimate could be way off. If the assumption of independence is satisfied, then a confidence interval at the 95% level would be

$$L, U = \bar{p} \pm z_{\alpha/2} s_{\bar{p}} \tag{8.4}$$

$$L, U = 0.06 \pm 1.96 \sqrt{\frac{0.06(0.94)}{100}} \sqrt{\frac{1,000 - 100}{999}}$$

$$L = 0.016 \qquad U = 0.104$$

The last term in the equation is the finite population correction factor, which was needed because the sample was more than 5% of the population. With these upper and lower limits for the population proportion, you may decide to order about 10% more electronic components than the number of devices you plan to assemble.

CASE STUDY

PORNOGRAPHY AS DEFINED BY A COMMUNITY'S STANDARDS

The owner of a local video store in Indianapolis was recently convicted of violating a state ordinance against distributing pornography after he rented the X-rated movie *Barbara Broadcast* to an undercover police officer. The judge at the trial allowed the defense to introduce the results of a survey of community residents regarding their attitudes toward renting X-rated movies at video stores. The survey was introduced to show that the general public tolerated this activity and that under the community standard definition of pornography, the movie was not pornographic. The survey was sent to about 1,200 local residents and only about 700 returned the survey. While this response rate is high for a mailed survey that offers no financial reward for its return, the response rate does raise a serious question about potential nonresponse bias. Also, an expert witness for the prosecution argued that the wording and order of the questions pushed the responses toward favoring X-rated videos. However, these substantial issues were not the ones that caught the jury's attention. The prosecutor pointed out during the trial that the number of completed interviews sent out was a small fraction of the population, about 0.7%. The foreman of the jury was quoted in the local paper to the effect that the sampling fraction was so small that the jury felt that the survey was completely unreliable. He said that the jury placed no credence in the survey because of its small sampling fraction. Apparently the finding of the survey, that over 60% of the respondents favored the right to rent X-rated videos, had no impact on the jury.

An estimate of the population proportion of 0.6 from a large population based on a sample of size 700 would have a fairly tight confidence interval. At a 95% level of confidence, the upper and lower bounds of a confidence interval would be 0.6 ± 0.036. We can find these limits by substituting into formula 8.4:

$$L, U = \bar{p} \pm z_{\alpha/2} s_{\bar{p}}$$ (8.4)

$$L, U = 0.6 \pm 1.96 \sqrt{\frac{0.6(0.4)}{700}}$$

If we put aside the issues of nonresponse bias and the wording of the questions, it seems clear that a substantial proportion of Indianapolis residents want to be able to rent X-rated videos. The issue here is not a defense of X-rated movies. The point is that with large populations the sampling fraction will tell you nothing about the accuracy of a survey. The accuracy of the survey depends on the sample size.

Here's another example of constructing a confidence interval on the population proportion. You head a syndicate that owns a stallion that has been a successful Thoroughbred race horse. From your perspective, the stallion is valuable because of the stud fees that horse breeders are willing to pay (according to the *New York Times* of April 17, 1984, the stud fee for Secretariat is $80,000). The custom in the industry is that if a breeder pays a stud fee and the breeder's mare does not become pregnant, the stud fee is returned. If you knew the proportion of matings that would result in a pregnancy and the number and price of matings projected for your stallion, you could forecast the syndicate's income. Suppose your stallion had mated with 40 mares and 32 of them had become pregnant. A 95% confidence interval for the population proportion would be

$$L, \ U = \bar{p} \pm z_{\alpha/2} s_{\bar{p}} \tag{8.4}$$

$$L, \ U = 0.8 \pm 1.96 \sqrt{\frac{0.8(0.2)}{40}}$$

$$L = 0.676 \qquad U = 0.924$$

For this confidence interval to be accurate you have to assume that age or disease will not affect your stallion's performance.

You may ask, "What can I do if $n\bar{p}$ or $n(1 - \bar{p})$ is less than 5? If I can use the *t* score for small samples for the mean, why can't I use it for the proportion?" The analogy doesn't hold because a requirement for using the Student distribution is that the population being sampled is normal or at least close to normal. This requirement cannot be met by any population distribution that represents successes and failures. If the successes were given a numerical value of 1 and the failures a numerical value of 0, then the population would look like two spikes or vertical lines at 1 and 0 (see Figure 8.7). If the population consisted of women and men and the women were coded as 0's and the men as 1's and 70% of the population was female, the distribution of the population would look like Figure 8.7. The relative heights of the vertical lines in the figure would not change as the population size varied as long as the proportions of men and women remained constant.

If you can't use the *z* score or the *t* score, what's left? There are some methods for constructing confidence intervals for the population proportion based directly on the sampling distribution of the proportion. There are two problems associated with confidence intervals constructed directly from the sampling distribution of the proportion. One is that the resulting intervals tend to be so wide that they often provide little useful information. Suppose you were a marketing executive and you read a report that stated that the estimated proportion of the population which had heard of your product was between 0.02 and 0.98 at the 95% level of confidence. Your response would be "I don't need a survey to tell me that." The other problem is that the procedure is

FIGURE 8.7
Population distribution of the proportion

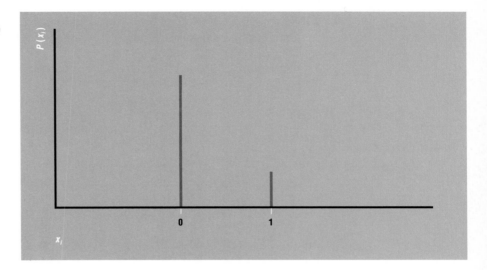

cumbersome. A starred exercise, number 8.47, has been included in the exercises at the end of the chapter for any students who want to try to figure out on their own how to find a confidence interval for the population proportion without using the normal approximation. One obvious solution to the difficulties of using the sampling distribution of the proportion is, if feasible, to increase the sample size so that $n\bar{p}$ and $n(1 - \bar{p})$ are at least 5. Another solution is to use a specially constructed graph in which the limits of the confidence interval can be read off for any sample size and sample proportion. Such tables can be found in most reference books of statistical tables, such as *Biometrika Tables for Statisticians* (Cambridge: Cambridge University Press, 1954), or Mary G. Natrella's *Experimental Statistics,* National Bureau of Standards Handbook 91 (Washington, D.C.: U.S. Government Printing Office, 1963).

EXERCISES FOR SECTION 8.3

8.17 If a sample of 100 observations has 40 successes,

(a) Construct a confidence interval for the population proportion of successes at the 95% level of confidence.

(b) Construct a confidence interval for the population proportion of successes at the 90% level of confidence.

(c) Construct a confidence interval for the population proportion of failures at the 90% level of confidence.

(d) Construct a confidence interval for the population proportion of successes at the 95% level of confidence. Assume that the population has 1,000 elements.

8.18 If a sample of size 200 has 50 successes,

(a) Construct a confidence interval for the population proportion of successes at the 95% level of confidence.

(b) Construct a confidence interval for the population proportion of successes at the 80% level of confidence.

(c) Construct a confidence interval for the population proportion of failures at the 99% level of confidence.

(d) Construct a confidence interval for the population proportion of successes at the 95% level of confidence. Assume that the population has 500 elements.

8.19 You own a soft drink company and you are considering replacing the sweetening agent in your diet sodas, saccharin, with aspartame. Aspartame is more expensive than saccharin, and you want to estimate the proportion of diet soft drink consumers who prefer the aspartame-sweetened soda to the saccharin before you switch to the new product. Suppose you sampled 300 randomly selected consumers who regularly drink diet soft drinks and gave each of them a sample of each type of soft drink in a blind taste test. The result was that 245 preferred the aspartame-sweetened soda. Construct a confidence interval at the 95% level of confidence for the population proportion of diet soft drink consumers who prefer aspartame.

8.20 You are a wine importer and you have found a French wine merchant who wants to sell you 10,000 bottles of 1947 Château Laffite Rothschild. Because of the wine's age, some of the bottles may have turned to vinegar, but there is no way to tell if a particular bottle is spoiled until it is opened. The French wine merchant is willing to let you open some of the bottles to estimate the proportion that have spoiled. You decide to have a party and open 100 bottles. Fifteen of the 100 turn out to be spoiled. Construct a confidence interval for the population proportion at the 95% level of confidence.

8.21 You work in an advertising agency that represents the California Pistachio Growers' Association. Your job is to make consumers aware of the existence of California pistachio nuts as an alternative to Iranian pistachio nuts and to promote the consumption of pistachios. You know that very few consumers are aware of California pistachios. To get a base point against which the success of your advertising campaign can be judged, you commission a national survey with 1,000 respondents. Thirty of the 1,000 say that they have heard of California pistachio nuts. Construct a confidence interval at the 90% level of confidence for the population proportion.

8.22 An army ordnance depot has 50,000 nerve gas canisters in storage. These canisters fit into artillery shells. To estimate the proportion of canisters that are usable, the depot takes a sample of 500 canisters and tests the gas in each one for chemical decomposition. Of the tested canisters, 28 were found to be unusable. The test on the canisters is a destructive test, that is, after the test the canisters cannot be used.

(a) At the 95% level of confidence, what is the range of the number of canisters that would be expected to be usable?

(b) Do part a over but change the number of canisters tested to 5,000, of which 280 were found to be unusable.

8.23 The men's clothing buyer for a chain of department stores has ordered 10,000 shirts from a Hong Kong manufacturer. The shirts carry the label of the department store. The department store does not sell "seconds" with its label, so each defective shirt has to have the label torn out and be sold to an odd-lot dealer. When 1,000 of the original 10,000 shirts were sent to the individual department stores, the clerks or purchasers identified 100 as defective.

(a) At the 95% level of confidence, what is the range of the number of shirts among the 9,000 that have not been released to the stores that can be expected to be defective?

(b) What assumption do you have to make to do this exercise?

8.24 A canoe rental agency takes phone reservations for weekend rentals of its canoes. Of the past 400 phone reservations, 40 were no-shows.

(a) Construct a confidence interval for the population proportion of no-shows at the 90% level of confidence.

(b) The canoe rental company is considering changing its policy to accept only mailed-in reservations with a $10 deposit. They try this new policy out on an experimental basis for one weekend and they get one no-show out of eight reservations. Can you construct a confidence interval for the population proportion of no-shows under the new policy? Explain.

8.4 MINIMUM SAMPLE SIZES FOR A GIVEN CONFIDENCE INTERVAL

Sampling is usually expensive. If you are not careful, you can waste money by getting too large or too small a sample. "Too large" means that you can construct a confidence interval with much more precision than you need. Suppose you had to price a part and you could specify the price to the penny. You need an estimate of the cost of the part before you can price it. With a large enough sample, you could construct a confidence interval, at the 95% level of confidence, with a range of plus or minus one-millionth of a penny. What good would it do you to have that level of precision? Similarly, you can construct a confidence interval with much less precision than you need. What good would it do you to have a confidence interval with a range of plus or minus $10? You may be saying to yourself, "Who would do anything that stupid?" Unfortunately, it happens, particularly when the person deciding on the sample size is spending someone else's money. People often use irrelevant rules of thumb to decide on a sample size, such as "Sample 10% of the population" or "Sample an even number like 1,000."

If you know the precision that you'll need, it's fairly easy to find the minimum sample size required to give you that precision. Let's go back to formula 8.1c:

$$L, \ U = \bar{x} \pm z_{\alpha/2}\sigma_{\bar{x}} \tag{8.1c}$$

The half-width of the confidence interval is $z_{\alpha/2}\sigma_{\bar{x}}$. You want to take a sample sufficiently large so that you can expect that 95 times out of every 100 times that you draw such a sample, the population mean will be within a penny of the sample mean. Let's call the penny your "tolerance," or tolerated error, and symbolize it by the Greek letter tau, τ. What you have to do is set $z_{\alpha/2}\sigma_{\bar{x}}$ equal to τ and solve for n.

$$\tau = z_{\alpha/2}\sigma_{\bar{x}}$$

$$\tau = z_{\alpha/2}\frac{\sigma}{\sqrt{n}}$$

$$n = \left(\frac{z_{\alpha/2}\sigma}{\tau}\right)^2 \tag{8.6}$$

Suppose you know that the population standard deviation of the distribution of costs of making the part that you are trying to price is 50 cents. Then, substituting into formula 8.6, you get

$$n = \left(\frac{1.96(50)}{1}\right)^2$$

$$n = 9,604$$

Your sample size would have to be 9,604 to have a probability of 0.95 of estimating the true average cost within one cent. Note that if the answer had come out to be 9604.2 (or, for that matter, any value between 9,604 and 9,605), you would have had to round up to 9,605 in order to ensure the desired level of accuracy.

What can you do if you don't know σ, the population standard deviation? You could rely on past surveys; sometimes the standard deviation of a distribution is fairly stable over time. You could try a range of hypothetical values for σ and see what impacts different plausible values have on the required sample size. This procedure makes much more sense than picking a round number for the sample size. Another approach that is occasionally taken is to use an estimate of the range of the data, based on the highest and lowest known values, to get a planning value for the standard deviation. A rough rule is that if the population is approximately normal, the standard deviation will be between one-fourth and one-sixth of the range. Finally, you could take a small pilot sample to estimate the standard deviation. If a great deal of money is at stake, it may be worth the effort to take a pilot survey so that you'll have a better idea of the sample size needed for the full survey. To distinguish the known population standard deviation, σ, from a standard deviation that is simply a planning value, we'll give the latter the symbol $\hat{\sigma}$. If a planning value is used for the population standard deviation, then formula 8.6 can be rewritten as

$$n = \left(\frac{z_{\alpha/2}\hat{\sigma}}{\tau} \right)^2 \tag{8.6}$$

If you had a pilot survey, the sample standard deviation from that pilot survey, s, could be set equal to $\hat{\sigma}$ and substituted in formula 8.6.

8.4A MINIMUM SAMPLE SIZE NEEDED TO ESTIMATE THE POPULATION PROPORTION

Now let's try to find the minimum sample size for the proportion. The procedure is similar to that of finding the minimum sample size for the mean. In formula 8.7 you'll need the standard deviation of the distribution of sample proportions instead of the standard deviation of the distribution of sample means. In the formula for the mean (formula 8.6), you could use either a planning value of the standard deviation or, if it was available, the population standard deviation. The reason for this choice is that for the sampling distribution of the mean it is possible to know the population standard deviation and not to know the population mean. This choice does not come up when you are finding the minimum sample size for the proportion. Since the population proportion is the only information needed to calculate the standard deviation of the proportion, once n is specified, *if you knew the population standard deviation, you would also know the population proportion.* If you knew the population proportion, you would have no reason to collect a sample to estimate it. We will work only with planning values of the population proportion and the standard deviation for these problems. Formula 8.6 is modified to read

$$n = \frac{\hat{\Pi}(1 - \hat{\Pi})z_{\alpha/2}^2}{\tau^2} \tag{8.7}$$

The term $\hat{\Pi}$ is the planning value of the population proportion.

A MEDICAID MUDDLE

Medicaid is a state-run program that pays doctors, hospitals, pharmacies, nursing homes, and other health-care providers for serving indigent patients. Since a hospital can accumulate many Medicaid bills over the course of a year, it would probably cost a state more to check every bill than it could hope to recover from any overcharges it might find. Most states rely on sample audits to keep health-care providers honest. Indiana has contracted this auditing function out to the Indiana Blue Cross. The contract between Blue Cross and the state specifies that a particular table in a text on auditing be used to decide on the sample size needed for each hospital. This table gives the minimum sample size needed to find the population proportion within 7 percentage points at the 90% level of confidence. The Blue Cross auditors would collect the specified sample, identify all overcharges, and then divide the overcharges by the total amount in the sample bills. For example, the auditors might find $100,000 of overcharges in a sample that had $1 million worth of bills. If the total Medicaid billing of the hospital for a year was $25 million, the auditors would ask that 10%, or $2.5 million, be returned to the state.

When I was consulted about the sample sizes needed for these audits, it became clear that both the state agency and the Blue Cross auditors were confused about what the 7 percentage points in the table referred to. They thought that the 7 percentage points referred to the amount by which the estimated overcharges could be off. That is, if a hospital's total estimated overcharges came to $5 million, the true overcharges would be somewhere between $4.65 million and $5.35 million, at the 90% level of confidence ($5 million plus or minus 7%). It should be clear that the table provided the minimum

Here's an example that uses formula 8.7. You own an accounting firm that specializes in preparing tax returns for individuals. You have offices around the United States staffed by bookkeepers who have had at least two weeks' training in preparing tax returns. You guarantee that if any of your clients is assessed a penalty because of an error made by one of your employees, your firm will pay it. You need to set aside some money to cover this guarantee. To start, you would like a good estimate of the proportion of returns that will have errors in the taxpayer's favor which will result in a penalty. How many returns do you need to sample and check for such errors to estimate the true proportion within 1 percentage point at the 95% level of confidence? For a planning value of the population proportion, you could take the rate from the previous year. Let's say that 1.5% or 0.015 of last year's returns had such errors. Now you have all the facts you need to substitute into equation 8.7:

sample size needed to estimate the proportion of all bills that contained overcharges within 7 percentage points of the true population proportion, but that this estimated proportion is not the same as the fraction of all *dollars* that are overcharges. One bill could be for $1 and have no overcharges and another bill could be for $25,000 and have $24,000 of overcharges.

After the two agencies agreed that their procedure estimated the proportion of all bills with any overcharges, a second problem came up: what to do after an audit revealed that a small proportion of all bills had overcharges? Suppose an audit found overcharges on only 2% of a hospital's bills. The state's position was that with a tolerance of plus or minus 7 percentage points, it was possible that the true proportion of overcharges among all bills was zero and that a hospital with such a low sample proportion of bills with overcharges should not be asked to reimburse the state. Blue Cross took a more aggressive position about reimbursement, perhaps because its contract gave it a share of any overcharges that were collected. Who was right? The Blue Cross position was correct. The proportion of all bills in the sample that contain overcharges is the best point estimate of the population proportion. Furthermore, if any bills with overcharges are found in the sample, you know that the population proportion cannot be zero. The state's position would have given Medicaid providers a perverse incentive. An unscrupulous Medicaid provider would try to have very few bills with overcharges but make sure that those few overcharges were whoppers. It would be unlikely that those few bills would be caught in the audit, and even if they were, the state would argue that the true proportion of all bills with overcharges might be zero! This analysis should convince you that the people who wrote the contract on behalf of the state and Blue Cross would have benefited from a course in Statistics.

$$n = \frac{\hat{\Pi}(1 - \hat{\Pi})z_{\alpha/2}^2}{\tau^2} \tag{8.7}$$

$$n = \frac{0.015(0.985)(1.96)^2}{(0.01)^2}$$

$$n = 567.60$$

Since you can't check a fraction of a tax return, you have to round up to the next whole number to achieve the accuracy specified in the problem. In this case you will have to check 568 returns.

What can you do if you don't have a planning value for the population proportion? The most conservative procedure is to use 0.5 for $\hat{\Pi}$. The number 0.5 is the most conservative because you can be certain that your sample size is

TABLE 8.1
Values of $\hat{\Pi}(1 - \hat{\Pi})$ for different values of $\hat{\Pi}$

$\hat{\Pi}$	$(1 - \hat{\Pi})$	$\hat{\Pi}(1 - \hat{\Pi})$
0.1	0.9	0.09
0.2	0.8	0.16
0.3	0.7	0.21
0.4	0.6	0.24
0.5	0.5	0.25
0.6	0.4	0.24
0.7	0.3	0.21
0.8	0.2	0.16
0.9	0.1	0.09

large enough to achieve the desired accuracy. Setting $\hat{\Pi}$ at 0.5 results in the largest sample for a given τ and a given confidence level. The reason that this procedure works is that, unlike the standard deviation of the mean, the standard deviation of the proportion cannot go above a certain value for each n. It reaches a maximum value when $\hat{\Pi}$ is at 0.5. This result can be seen if you try different values for $\hat{\Pi}$ and $(1 - \hat{\Pi})$. Table 8.1 illustrates this point.

Using a $\hat{\Pi}$ of 0.5 instead of 0.005 in the tax return problem described above changes the required sample size from 192 to 9,604. Sometimes being very conservative can be very expensive.

Another alternative is a two-step procedure. The first step is a small pilot survey to give you a planning value for the proportion and the second step is the full sample needed to give you the required accuracy.

EXERCISES FOR SECTION 8.4

8.25 Let X be a random variable with an unknown mean and a known variance of 40.

(a) What is the minimum sample size needed to estimate μ with a tolerance of 10 at the 95% level of confidence?

(b) What is the minimum sample size needed to estimate μ with a tolerance of 10 at the 90% level of confidence?

(c) What is the minimum sample size needed to estimate μ with a tolerance of 1 at the 95% level of confidence?

(d) What is the minimum sample size needed to estimate μ with a tolerance of 1 at the 99% level of confidence?

8.26 Let X be a binomial random variable with an unknown Π.

(a) What is the minimum sample size needed to estimate Π with a tolerance of 0.02 at the 95% level of confidence if the planning value for Π is 0.34?

(b) What is the minimum sample size needed to estimate Π with a tolerance of 0.01 at the 90% level of confidence if the planning value for Π is 0.34?

satisfied, they were not discussed because the resulting confidence intervals tend to be so wide that they and the method are fairly cumbersome. If you have to construct a confidence interval for the population proportion without the normal approximation, you can use one of the special tables cited in the chapter.

One idea that received careful development is that the level of confidence associated with a confidence interval is a statement about the accuracy of the procedure being used rather than the probability that a particular interval includes the population parameter. The chapter also discussed minimum sample sizes for a given tolerated error at a specified confidence level.

NEW TERMS

confidence interval A lower and an upper bound within which the population parameter is expected to be located along with an associated level of confidence.

point estimate The one number that is the best estimate of a parameter.

robust estimator An estimator that will still give accurate estimates when the assumptions of normality are violated.

NEW FORMULAS

Confidence interval for the population mean with a normal population and a known population standard deviation

$$L,\ U = \bar{x} \pm z_{\alpha/2}\sigma_{\bar{x}} \tag{8.1c}$$

Confidence interval for the population mean with a normal population and an unknown population standard deviation

$$L,\ U = \bar{x} \pm t_{\alpha/2}s_{\bar{x}} \tag{8.2}$$

Confidence interval for the population mean with a nonnormal population and an unknown population standard deviation when $n \geq 30$

$$L,\ U = \bar{x} \pm z_{\alpha/2}s_{\bar{x}} \tag{8.3}$$

Confidence interval for the population proportion

$$L,\ U = \bar{p} \pm z_{\alpha/2}s_{\bar{p}} \tag{8.4}$$

Estimated standard deviation of the sampling distribution of the proportion

$$s_{\bar{p}} = \sqrt{\frac{\bar{p}(1 - \bar{p})}{n}} \tag{8.5}$$

Minimum sample size needed to estimate the population mean within a given tolerance and at a given level of confidence

$$n = \left(\frac{z_{\alpha/2}\sigma}{\tau}\right)^2 \tag{8.6}$$

Minimum sample size needed to estimate the population proportion within a given tolerance and at a given level of confidence

$$n = \frac{\hat{\Pi}(1 - \hat{\Pi})z_{\alpha/2}^2}{\tau^2} \tag{8.7}$$

ADDITIONAL EXERCISES

Asterisks designate exercises of greater than ordinary difficulty.

8.33 You sell product liability insurance to manufacturers of mobile homes and you want to estimate the proportion of long-term mobile home owners who have or have had cancer. The cancer could be blamed on the formaldehyde used in mobile homes and the companies that you insure could be liable. Sales and housing records indicate that there are approximately 1 million people who have lived five or more years in a mobile home. You take a sample of 3,000 current residents of mobile homes and find that 40 currently have cancer. Construct a confidence interval at the 95% level for the population proportion based on this sample proportion. Explain why this sample would not provide a reliable estimate of the potential liability of mobile home manufacturers.

8.34 From the viewpoint of advertisers, the Nielsen television ratings don't provide exactly the information they need. Advertisers would like to know how many people are watching their commercials. The Nielsen ratings just estimate the proportion of all households that have their TV sets turned on and tuned to a particular channel. The problem is that many TV sets are turned on when no one is looking at them. You are planning a service to compete with Nielsen which will give advertisers exactly the information they want. Your plan is to pay households to allow you to place a video recorder and camera on top of their TVs. By reviewing the video tapes of what is happening in front of a TV set, you can tell if anyone is watching a particular commercial. You need to know how many homes have to be signed up for this program. Suppose your service wants to estimate the population proportion within one-tenth of a percentage point at the 95% level of confidence. How many homes do you need to sign up? What factors could cause your estimate to be unreliable?

8.35 You own a company that makes custom bodies for large trucks, such as dump trucks, cherry pickers, cement trucks, and garbage trucks. For tax purposes you have to certify the weight of each completed truck. If the weight you certify is off, your company can be fined. The scale you use for weighing the trucks has a sampling error that has a normal distribution with a mean of zero and a standard deviation of 50 pounds. Assume that the errors on successive weighings of the same truck are independent. If you wanted to certify the weight of each completed truck so that it was within one pound of the actual weight 95 out of every 100 times that you ran through the procedure, how many times should you weigh each truck before taking the average of the sampled weights as the certified value of a truck's weight?

8.36 You own a company that makes peanut butter. Since peanuts are an agricultural product, it is impossible to purchase batches of peanuts that are entirely free of rodent hairs and insect parts. The U.S. Department of Agriculture has established standards for the maximum allowable contamination of peanut butter. When a shipment of peanuts arrives at your plant, you have to sample the peanuts for contamination. If they are too contaminated, you can throw them away or mix them in with a cleaner batch. Suppose you had sampled 5 pounds of peanuts in a particular shipment and had observed a sample mean of 8 contaminants per pound with a sample standard deviation of 3. Assume that the distribution of contaminants per pound is normal. Construct a confidence interval at the 95% level of confidence for the mean number of contaminants per pound for the entire shipment of 1,000 pounds.

8.37 You are the manufacturer of a new turboprop commuter plane and you want to estimate its mean fuel consumption per nautical mile. You have the plane fly a 300-mile test route 10 times and record the fuel consumption. The sample mean was 900 gallons with a sample standard deviation of 45 gallons. Construct a confidence interval at the 95% level of confidence for the mean level of fuel consumption for the 300-mile trip and then express the confidence limits on a per mile basis. Assume that the distribution of fuel consumption is normal.

8.38 You are the manufacturer of stereo speakers and you want to compare consumer preferences for a new speaker design versus the design that you are currently marketing. Returned warranty cards allow you to identify owners of your present design. You invite 100 owners of your current design to a blind comparison of the two speakers, not telling them that one of the speakers that they will be listening to is the same as the one they have at home. You ask each listener to identify the speaker that he or she prefers. The result is that 90 prefer the new design. Construct a confidence interval for the proportion of all potential purchasers who would prefer the new design. Is there a flaw in your sample design?

8.39 You work at the central office of a chain of Mexican fast-food restaurants that has 2,000 units nationwide. The chain is considering the introduction of a new menu item, turkey mole. Before introducing the item across the chain, the company would like to estimate the mean sales per unit of the item. Fifty units are selected at random for a test-marketing program. If the test markets indicate that consumer interest in turkey mole is too low, your company will not introduce the product and the heavy promotional expenses associated with the introduction of a new menu item will be saved. The mean weekly sales of turkey mole in the test restaurants is $254. On the basis of past test marketings, the population standard deviation is assumed to be $14. Assume that the population distribution of weekly turkey mole sales is normally distributed. Construct a confidence interval for the population mean at the 95% level of confidence.

8.40 You are the relocation officer for a worldwide company. Many of the moves that you help employees plan involve overseas destinations for less than a year's stay. For these short-term trips your company provides employees with furnished homes with every detail covered down to the forks and spoons. However, the employees need to ship some personal goods, such as records, books, and clothing. To budget your company's moving expenses, you need to estimate the shipping costs for these personal goods. The shipping companies charge for a minimum of 500 pounds even if the shipment weighs less than 500 pounds. Because of this practice, you know that the employees will send at least 500 pounds of personal goods. What you would like is a confidence interval that is one-sided—that is, a confidence interval that states that the mean weight of a shipment is less than or equal to some upper limit, U, at the 95% level of confidence. You examine the records for 14 short-term moves and find that the sample mean for these shipments was 683 pounds with a sample standard deviation of 130 pounds. What is the upper limit for a one-sided confidence interval? Assume that the distribution of shipping weights is normal.

8.41 You are the fish buyer for a company that sells frozen seafood to supermarket chains. Part of your job is to bid on boatloads of shrimp when they arrive at the dock. The selling price of shrimp per pound depends on the number of shrimp in a pound, large shrimp being more valuable than shrimpy shrimp. You have to estimate the mean number of shrimp per pound for a boatload before you know how much to bid. You randomly select 50 pounds of shrimp and count the number in each pound. The sample mean is 24 with a sample standard deviation of 6. Construct a confidence interval at the 95% level of confidence for the mean number of shrimp per pound in the boatload.

8.42* Suppose you took a sample of size 100 from some population. The sample mean was 88 and the sample standard deviation was 10. Using these figures, you have constructed a confidence interval with a lower limit of 87.5 and an upper limit of 88.5. What level of confidence did you use?

8.43 You are in charge of health insurance for your company. Your insurance carrier has just sent you a schedule of "usual and customary physicians' fees" for common medical services. Your contract with the insurance company requires that the carrier pay 100% of these usual and customary fees and only 80% of any charges above those levels after an employee has paid a deductible of $500 per family. In examining the schedule, you notice that the fee for delivering a baby seems much lower than the rates prevailing in your city. According to the insurance carrier, the usual and customary charge is $400. You check your records for claims filed for this service in the past year and you find that you have 20 claims. The mean physician's charge was $950 with a sample standard deviation of $246. Construct a confidence interval at the 95% level of confidence for the mean charge by physicians in your city for delivering a baby. What assumptions have to hold if this estimate is to be accurate?

8.44 You manage a factory that makes riding mowers. To cut costs, you plan to reduce all of your inventories to the equivalent of two days' production and rely on a "just in time" system for deliveries of your parts. If this system is to work, your parts suppliers must ship parts so that they arrive precisely on time. Otherwise your loading docks will be clogged and you will run out of space to store the parts. Your most important supplier is the firm that makes the engines for the mowers. You negotiate a contract with the engine supplier which provides for three truckloads of engines per day to arrive at 7:00 A.M., 11:00 A.M., and 1:00 P.M. To fulfill these terms the engine supplier has to know how long it takes for a truck to travel from the engine plant to your mower assembly plant. On the basis of 50 test runs, it is established that the sample mean time is 6 hours with a sample standard deviation of 12 minutes. Construct a confidence interval for the mean time at the 99% level of confidence.

8.45 You are an attorney and your law firm is negotiating a contract with the county government to represent indigent clients in nonfelony criminal cases. The county wants to pay a flat rate per client and you would like to bill on a per hour basis. After much haggling, both sides agree to a flat rate equal to $50 times the mean number of hours per case for similar clients served by your firm in the past year. According to your firm's records, you served 350 such clients last year with a mean time of 12 hours with a sample standard deviation of 14 hours. Since it is impossible to have negative hours, what does a higher value for the standard deviation than the value of the sample mean suggest about the population? To get an idea of how far off the sample mean may be from the population mean, construct a confidence interval at the 95% level of confidence for the population mean.

8.46 You are an insurance underwriter in the new field of space insurance. You write policies to cover the launching of commercial satellites. You need to estimate the probability that a new satellite carried by the space shuttle will fail to reach geosynchronous earth orbit. The owner of the satellite, the government of Japan, wants to purchase insurance against this failure in the amount of $50 million. Suppose 12 out of the last 50 satellites launched by space shuttles failed to reach orbit. Construct a confidence interval at the 99% level of confidence for the proportion of all future satellites to be carried by the shuttle which will fail to reach orbit, assuming that the technology involved stays the same. Use these confidence limits to compute the upper and lower bounds for an actuarially fair insurance premium for the $50 million policy.

8.47* Suppose you had a sample of size 20 and had observed 4 successes and 16 failures. With these figures you could not use the normal approximation for the sampling distribution of the proportion because $n\bar{p}$ is less than 5. Use the binomial table to construct a confidence interval for the population proportion at the 95% level of confidence.

8.48* You own a firm that makes mechanical hearts. You've run extensive bench tests on the durability of hearts and you've established that the time between startup and failure for one of your machines is normally distributed with a mean of 377 days and a standard deviation of 16 days. Bench tests cannot duplicate the conditions inside the human body, and you plan to test the machine in large dogs to get a better idea of how long it will last in humans. This testing is both time-consuming and expensive. Nevertheless, accurate estimates are important because doctors will rely on these estimates when they make decisions about operations to implant the machine or to replace an implanted machine. A more precise estimate of the durability of the machine will help sell the machine. Suppose the standard deviation established in the bench tests is the correct standard deviation for the dog population and the time to failure is also normally distributed in the dog population. Suppose further that the cost of the test per dog is $4,000. Also suppose that the benefits to your company of acquiring more precise information on the estimated life of the machine can be expressed as a function of the width of the confidence interval, specifically, $MB = 1250 + 61x + 2x^2$, where MB is the marginal benefit of having a 95% confidence interval of a given width and x is the width of the confidence interval (x must be greater than zero). How many dogs should you test?

8.49* You want to estimate the mean commission earned by stockbrokers at rival brokerage houses in your city so that you can gauge the performance of your stockbrokers. You approach some executives at the rival houses and offer to exchange earnings data. You obtain the annual commission earnings of 27 stockbrokers out of a population of 345 stockbrokers, and when you plot the data you observe that they appear highly skewed. By taking the logarithm of each observation, you can transform the data so that they will be approximately normal. Use these transformed data to construct a confidence interval at the 95% level of confidence for the mean annual commission earnings. (Note that this problem would be easier on a computer.)

Stockbroker	Commission earnings
1	$16,000
2	16,500
3	17,500
4	18,500
5	20,000
6	21,400
7	22,600
8	23,000
9	23,700
10	24,200
11	24,600
12	25,000
13	25,500
14	26,800
15	27,400
16	30,000
17	33,000
18	37,000
19	39,500
20	41,400
21	45,400
22	49,500
23	53,200
24	58,700
25	64,500
26	76,900
27	87,000

8.50** You work for a company that makes mechanical heart valves. The valves are considered defective if they fail before completing 1 million cycles of opening and closing. You would like to estimate the population proportion of defective valves. Testing to see if a valve is defective destroys the valve. The cost for testing a valve is $100. If you underestimate the defect rate, your company may not have enough liability insurance. But if you overestimate the defect rate, your company may spend money on an unnecessary redesign of the valve. Assume that the cost of any error in the estimate of the proportion of defective valves is $5,000 per percentage point of error in either direction. If your planning value for the population proportion is 0.1, how many valves should you test?

8.51** An engineer who is designing a production line for motorcycles needs to estimate the mean time taken by an industrial robot to assemble, align, and weld the parts of a frame. If the robot takes more than one minute for this procedure, the assembly will be pulled off the production line and finished by hand. Thus the engineer knows that the mean time for the robot to work on the procedure must be less than or equal to one minute. What the engineer needs is a one-sided confidence interval that gives the lower limit for the population mean. The robot has been tested 10 times and the mean amount of time taken for the procedure in these tests was 0.92 minute with a sample standard deviation of 0.09 minute.

(a) At the 90% level of confidence, what is the lower limit for a confidence interval for the mean time taken by the robot on the procedure?

(b) How does this example violate the assumptions required for constructing a confidence interval?

9

Hypothesis Tests with One Population

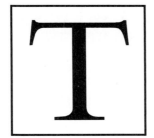

T he county prosecutor in Indianapolis received an anonymous tip that a construction company that had built a city road had defrauded the city. The contract between the city and the company called for an asphalt road with an average thickness of 10 inches. The company could save money on materials by building the road with an average thickness of less than that amount. The prosecutor needed evidence to persuade a jury that the company had defrauded the city by making the average thickness of the road less than 10 inches. Digging up the entire road to establish its average thickness was impractical. Besides the expense, it would have proved highly embarrassing if the road had turned out to have the correct average thickness. Instead, a sample of 100 cores 1 inch in diameter was taken from the road. The mean thickness of the cores was 9.5 inches. Should this sample mean be treated as sufficient evidence to prosecute the company? Alternatively, is 9.5 inches so close to 10 inches that the true average might be 10 inches? In other words, might the difference between the sample mean and 10 inches be due to random variation or sampling error? These questions exemplify the hypotheses that can be used to set up a formal hypothesis test. The reason that a hypothesis test was used, rather than a confidence interval, was that the prosecutor wanted to show that a population mean of 10 inches was very unlikely given the observed sample mean of 9.5 inches. In the actual case, the statistical evidence must have been pretty convincing, because the jury decided that the company had cheated the city and several of the company's officers were given prison terms.

This chapter discusses the mechanics of hypothesis tests on the population mean and on the population proportion for problems that involve one population. The chapter emphasizes the similarities between confidence intervals and hypothesis tests. The formula for finding the decision rule for every hypothesis test has three parts, corresponding to the three parts of the formula for a confidence interval. If you've mastered confidence intervals, this material should flow easily. Some material that does not parallel that of Chapter 8 is the discussion of the two types of possible errors in a hypothesis test and the probability of making each type of error. A confidence interval can have only one error: either it includes the population parameter or it does not. Errors in hypothesis tests are more complicated: one error is acceptance of a hypothesis that is really false and the other error is rejection of a hypothesis that is really true. This chapter will discuss how to calculate the probabilities of these errors and how to reduce the chances of making them.

9.1 FORMULATION OF A HYPOTHESIS TEST, NULL AND ALTERNATIVE HYPOTHESES, AND TYPE I AND TYPE II ERRORS

9.1A THE NULL AND ALTERNATIVE HYPOTHESES

The difference between a hypothesis test and a confidence interval is that with a hypothesis test you have a prior value for the parameter and you are using the sample data to see if they will cause you to reject the hypothesis that the prior value is correct. With a confidence interval you don't have a prior value and you just want to estimate a parameter. Every hypothesis test involves a pair of hypotheses: you must formulate a null and an alternative hypothesis. The prior value is generally used to form the null hypothesis. In our road example, the null hypothesis would be "The mean thickness of the road is 10 or more inches"; that is, the construction company is innocent. While the contract with the construction company calls for only a 10-inch road, the prosecutor would not charge the company with fraud if it had made the road thicker than the contract specified. The city would be better off with a thicker road, and it is also implausible that the road would be thicker than 10 inches because of the company's incentive to save on materials. The reason that the null hypothesis was stated in terms of 10 or more inches is that the null and the alternative hypotheses are defined to cover all possible values of the parameter. The alternative hypothesis would be "The mean thickness of the road is less than 10 inches"; that is, the construction company is guilty. Note that the alternative hypothesis can be inferred from the null. The null can also be inferred from the alternative.

Setting up null and alternative hypotheses can be a little tricky. The **null hypothesis** is *the claim that a population parameter equals the supposed or "maintained" value or values, or that it equals the "no effect" value or values.* The **alternative hypothesis** is *the claim that a population parameter has a value or values other than those specified in the null hypothesis;* it specifies *the suspected or "challenging" values of the population parameter.* If the prosecutor considered it fraudulent to make a road with an average thickness more *or* less than 10 inches, then the null hypothesis would be "The mean thickness of the road equals 10 inches," and the alternative hypothesis would be "The mean thickness of the road is not equal to 10 inches." Another example might help. Suppose you suspected that a new cold medicine did not reduce the average duration of a cold. If the average duration of a cold without any medicine was seven days, the null hypothesis would be "The mean duration of a cold with the medicine is seven or more days" (in other words, the medicine is ineffective). The alternative hypothesis that could be inferred from this null is "The mean duration of a cold with the medicine is less than seven days" (the medicine is effective). The word "null" means "nothing" or "no favorable impact," so it is easy to see why seven or more days was used for the null hypothesis. The interpretation of no effect or no favorable impact can be used to decide how to set up many hy-

BOXING AS A METAPHOR FOR THE NULL AND ALTERNATIVE HYPOTHESES

Some statisticians have argued that the terms "null hypothesis" and "alternative hypothesis" should be dropped because they do not accurately describe the hypotheses. Their preferred terms are "maintained" rather than null hypothesis and "challenging" rather than alternative hypothesis. To illustrate why the new terms are more descriptive, consider the example of a boxing match between the heavyweight champion of the world and a challenger. The maintained or null hypothesis is that the better boxer is the reigning heavyweight champion, because he has won many fights and successfully defended his title. The challenging or alternative hypothesis is that the challenger is a better fighter. If a match appears to be close, the judges always give the benefit of the doubt to the champion. *The challenger has to score a knockout to win the match, or the burden of proof is on the challenger.* Suppose the challenger wins by a knockout and becomes the new champion. The new maintained hypothesis is that he is the better of the two fighters. The benefit of the doubt is now in the new champion's favor, and to retake his title in a rematch, the former champion must score a knockout. The example could be expanded to a third or fourth match, but we shouldn't beat a good metaphor to death.

pothesis tests. While the road example does not easily lend itself to this interpretation, you could say that you were testing for fraud and that the range of 10 or more inches was the no-fraud-effect thickness of the road.

A more general interpretation of the null and alternative hypotheses is that when you set up a hypothesis test, you want to choose the hypotheses so that the burden of proof is on the alternative hypothesis. In other words, you will accept the null unless, given the sample data, it is very unlikely that the null is true. Assume that the prosecutor is more interested in avoiding the conviction of an innocent company than he is concerned about letting a guilty company go free. Given these objectives, the null hypothesis should be that the company is innocent of fraud or, in numerical terms, that the thickness of the road is 10 or more inches. This setup for the null allows you to specify an upper limit for the probability of rejecting the null when it is true. You can make the upper limit of the probability of rejecting the null as low as you want, usually 0.05 or 0.01, and thus put the burden of proof on the alternative hypothesis. This interpretation of selecting the null and alternative hypotheses so as to put the burden of proof on the alternative hypothesis will be clearer when the probabilities of falsely rejecting and falsely accepting the null hypothesis are discussed in the next section.

■ 9.1B FORMULA FOR A HYPOTHESIS TEST WITH A
NORMAL POPULATION AND A KNOWN POPULATION
STANDARD DEVIATION AND THE RATIONALE BEHIND IT

Let's continue with the road contract example. Recall that the road was supposed to have an average thickness of 10 inches, that 100 cores from randomly
selected locations were sampled, and that the sample mean was 9.5 inches.
Since the machines that lay asphalt are not perfectly accurate, the city contract
specified only the average thickness, not a perfectly level layer of asphalt exactly 10 inches thick. From industry sources the prosecutor knew that the
equipment used to lay the asphalt had normally distributed errors around a
mean thickness that could be set by the operator of the machinery. The standard deviation of the distribution of errors was 1.5 inches.

To set this problem up as a hypothesis test, you start by specifying the null
and the alternative hypotheses in the form of equations. *The null hypothesis
always includes an equality sign or a "greater than or equal" sign or a "less than or
equal" sign, and the alternative has a strict inequality*. In the example below, the
"greater than or equal" sign is in the null hypothesis.

$$H_0: \mu \geq 10 \qquad H_1: \mu < 10$$

H_0 stands for the null hypothesis and H_1 for the alternative hypothesis.

Let's suppose for a moment that the road contractor had set his machinery to
average 10 inches. In Chapter 7 you learned that the distribution of the sample
mean has an expected value equal to the population mean and that the standard
deviation of the distribution of sample means equals the population standard
deviation divided by the square root of the sample size. If the null hypothesis
was true, the distribution of sample means of cores taken from the road would
be centered on 10 inches and the distribution would have a standard deviation
of 1.5 inches divided by the square root of 100, or 0.15 inch. Figure 9.1 illustrates the sampling distribution of the mean under the assumption that the
population mean is 10 inches. Again, this diagram is the pattern of sample
means you would expect if the contractor had set his asphalt-laying machinery
at 10 inches. Just by glancing at the pattern you can see that 9.5 inches is way
off. You can be a bit more formal than this eyeball test. A hypothesis test for
the population mean is really a template that describes the distribution of the
sample mean under the assumption that the null hypothesis is true and provides a decision rule for rejecting the null hypothesis. This template is shown
by the broad arrows in Figure 9.1. Let's say that you wanted to be 95% sure
that, if the contractor had set his machinery at 10 inches, you would not falsely
accuse him of cheating. Statisticians tend to describe hypothesis tests in terms
of the chance of mistakenly rejecting the null hypothesis instead of the chance
of correctly accepting the null hypothesis. If there is a 95% chance of not
rejecting the null hypothesis when it is true, then there is a 5% chance of

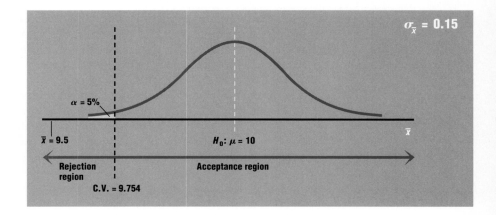

rejecting the null hypothesis when it is true. *The probability of rejecting the null hypothesis when it is true* is called the **significance level** of the test.

To find the decision rule—that is, the value of the sample mean below which you will reject the null hypothesis—at a significance level of α, you would use formula 9.1. This formula applies to normal populations with known standard deviations.

Critical value for a left-tailed hypothesis test

(9.1)

where

$$\sigma_{\bar{x}} = \frac{\sigma}{\sqrt{n}} \quad \text{or} \quad \frac{\sigma}{\sqrt{n}}\sqrt{\frac{\nu - n}{\nu - 1}}$$

depending on whether the finite population correction factor is required.

The letters C.V. stand for the **critical value** on which the decision rule is based: *a value in the range of the sampling distribution which marks the boundry between the acceptance and rejection regions of the null hypothesis.* The decision rule is: *If the sample mean is in the range bounded by the critical value or values which also includes the null hypothesis value of the parameter, then the null hypothesis should not* *be rejected.* The symbol μ_0 stands for the value of the mean specified in the null hypothesis. The formulas for the critical value or values in hypothesis tests are always *centered on the value of the parameter specified in the null hypothesis.* Confidence intervals, in contrast, are centered on the sample mean. This is an example of a one-tailed test in the left tail. It's one-tailed because you are interested only in seeing if the contractor short-changed the city; you are not worried about the possibility that the asphalt may be thicker than 10 inches. The direction of the inequality symbol for the alternative hypothesis tells you which side the tail is on. If the alternative hypothesis is that the mean is less than 10 inches, it is a left-tailed test. If the alternative hypothesis were that the mean is greater than 10 inches, you would have a right-tailed test. The only difference in the formula for a right-tailed test is that the minus sign in formula 9.1 is replaced by a plus sign.

Critical value for a right-tailed hypothesis test

$$\text{C.V.} = \mu_0 + z_\alpha \sigma_{\bar{x}} \qquad\qquad\qquad (9.1a)$$

where

$$\sigma_{\bar{x}} = \frac{\sigma}{\sqrt{n}} \quad \text{or} \quad \frac{\sigma}{\sqrt{n}}\sqrt{\frac{\nu - n}{\nu - 1}}$$

depending on whether the finite population correction factor is required.

Note that if the alternative hypothesis is that the mean is not equal to 10 inches, you have a two-tailed test. The formula and a discussion of two-tailed tests is coming up in a few pages.

   In both of the above formulas you use z_α instead of $z_{\alpha/2}$ because you are working with one tail. The α is your significance level. With a left-tailed test, if the sample mean comes out to the right of the critical value, then you should not reject the null hypothesis. In a left-tailed test, the area to the right of the critical value is called the "acceptance region." The **acceptance region** is *the range marked by the critical value or values which contains the null hypothesis value of the population parameter.* If your sample mean comes out to the left of the critical value, then you should reject the null hypothesis. In a left-tailed test the area to the left of the critical value is called the "rejection region." The **rejection region** is *the range marked by the critical value or values which does not contain the null hypothesis value of the population parameter.* With a 5% significance level C.V. = 9.754. You can get this number by substituting into equation 9.1.

$$\text{C.V.} = \mu_0 - z_\alpha \sigma_{\bar{x}} \qquad\qquad\qquad (9.1)$$

$$\text{C.V.} = 10 - 1.64(0.15)$$

$$\text{C.V.} = 9.754$$

Since the sample mean of 9.5 inches falls in the rejection region, your conclusion is that you can reject the null hypothesis that the population mean is greater than or equal to 10 inches at the 5% significance level. The 5% significance level means that if the null hypothesis was true and you repeated the experiment of taking sample cores many times, you would expect to mistakenly reject the true null hypothesis 5% of the time.

9.1C CHOOSING THE SIGNIFICANCE LEVEL OF THE HYPOTHESIS TEST

You may be asking yourself, "Why settle for a 5% chance of being wrong? Why not 1% or even 0.01%?" A similar point came up in Chapter 8. Again, you can reduce your chance of being wrong, but only at a price. What is different here is that there are two kinds of errors. *Rejection of a null hypothesis when it is true* is a **Type I error.** The probability of a Type I error is the significance level of the test, or α. In this example, if you made a Type I error, you would be trying to convict an innocent road contractor. The other possible error is *acceptance of a null hypothesis when it is false.* This is a **Type II error.** The symbol for the probability of a Type II error is β, the Greek letter beta. In this context it means the probability of letting the guilty road contractor go free. Figure 9.2 summarizes the two types of decisions you can reach and the two types of errors you can make. If you hold your sample size constant, reductions in the probability of a Type I error must come at the expense of increases in the probability of a Type II error. Conversely, for a given sample size, reductions in the probability of a Type II error must come at the expense of increases in the probability of a Type I error.

To see what happens when the probability of a Type I error is reduced, let's redo the problem with a significance level of 0.01%. The only change is to substitute a z score of 3.71 for 0.01% of the area in one tail in place of the z score of 1.64 for 5% of the area in one tail.

$$\text{C.V.} = \mu_0 - z_\alpha \sigma_{\bar{x}} \qquad\qquad (9.1)$$

$$\text{C.V.} = 10 - 3.71(0.15)$$

$$\text{C.V.} = 9.444$$

FIGURE 9.2

Types of decisions in a hypothesis test and types of errors

STATISTICAL DECISION	STATE OF NATURE	
	Null is true	**Null is false**
Reject the null hypothesis	Type I error	Correct decision
Do not reject the null hypothesis	Correct decision	Type II error

FIGURE 9.3
Sampling distribution of the
mean when the null
hypothesis is true and a
hypothesis test with a 0.01%
significance level

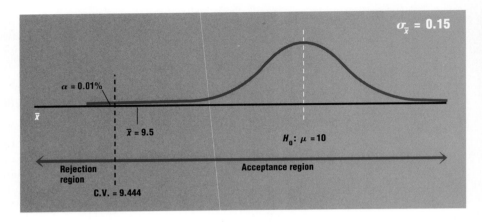

Figure 9.3 illustrates the hypothesized sampling distribution of the mean and the new critical value. Now you would accept the null hypothesis that the population mean is greater than or equal to 10. By moving the boundary line to the left, however, you have increased the chance of a Type II error. As you can see in Figure 9.3, the acceptance region has been extended considerably to the left and the rejection region has been correspondingly reduced. The above test has been set up so that there is a minuscule chance of a Type I error; but if the true mean is even a little below 10 inches, there is a high probability of a Type II error, that is, acceptance of the null hypothesis by mistake.

The probability of a Type II error depends on the true value of the population mean. The closer the true mean thickness of the asphalt is to 10 inches while remaining below that number, the more likely one is to conclude mistakenly that the true mean really was 10 inches. If the true mean was 3 or 4 inches, the probability of making a Type II error would be very low. If the true mean thickness was 9.9 inches, the probability of making a Type II error would be very high. Since you don't know the true mean, you can't determine the exact probability of a Type II error. All you can do is try a range of values for the true mean and calculate what the probability of a Type II error would be in each case. We'll do this in section 9.2.

Custom or habit is the reason that most hypothesis tests are set up with a significance level of 5%. For many problems it is impossible to determine the exact costs of making a Type I error or a Type II error, so you cannot determine the best possible significance level for the hypothesis test. If you knew these costs and the costs of drawing samples of various sizes, you could optimize the hypothesis test. In other words, you could find the α, β, and n that would minimize your expected costs or maximize your expected profits. As such costs are often unknown, some general examples will offer a few rough guidelines on significance levels. Suppose you designed a medical test that could detect a certain kind of cancer that can be cured if it's found and treated in time. The chemistry of the test could be altered to trade false positives against false negatives. A false positive would be a diagnosis of cancer when the patient does not have cancer and a false negative would be a diagnosis of no cancer when the patient does have it. The cost of a false negative is much higher than that of a false positive. While you are not exactly doing people a favor by telling them they have cancer when they don't have it, if you make the other mistake, you will kill them. This example could be considered a hypothesis test with the null hypothesis of "cancer present," in which a significance level of 5% is too high. You would probably want to push the chemistry of the test as far as it could go to reduce Type I errors at the expense of Type II errors.

Here's another example. Your company uses a transistor in making portable radios. Your contract with your supplier specifies that the proportion of defective transistors is to be no more than 0.03. Before accepting a shipment of transistors, you plan to test a random sample of transistors. This procedure is equivalent to a hypothesis test in which the null hypothesis is that the population proportion is less than or equal to 0.03. The cost of rejecting a shipment depends on how many transistors you have in stock and the time needed to get another shipment. The cost of accepting a shipment with too high a failure rate depends on the cost of servicing potential warranty claims by purchasers of your radios. If your factory would be shut down for a week waiting for another shipment of transistors, you might be very tolerant about accepting a shipment in which the sample proportion of defectives was higher than 0.03. In this situation a fairly high α level could be used. If the potential warranty claims were the greatest costs and the factory had a supply of transistors, a low α would make sure that you would be unlikely to accept a shipment that did not meet the contract terms. Most of the hypothesis tests done in this chapter will use the traditional significance level of 5%, but remember that there is nothing special about that number. The economic circumstances of a particular problem should be considered before the significance level is set.

So far we have discussed the formula for a one-tailed hypothesis test on the mean for a normal population with a known standard deviation and a choice of significance level. Now we will discuss the formula for a two-tailed test in the same circumstances.

■ 9.1D TWO-TAILED HYPOTHESIS TEST FOR THE POPULATION MEAN WITH A NORMAL POPULATION AND A KNOWN POPULATION STANDARD DEVIATION

Although it is unrealistic we can use the road problem as an example of a two-tailed test. The null hypothesis is that the population mean equals 10 inches. The test could also be stated in terms of the alternative hypothesis, that the population mean does not equal 10 inches. Recall that the null always includes the equality sign. The equations for these null and alternative hypotheses are:

$$H_0: \mu = 10 \qquad H_1: \mu \neq 10$$

The formula for finding the critical values for the test is

Critical values for a two-tailed hypothesis test

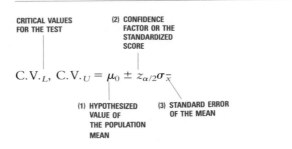

(9.2)

where

$$\sigma_{\bar{x}} = \frac{\sigma}{\sqrt{n}} \quad \text{or} \quad \frac{\sigma}{\sqrt{n}}\sqrt{\frac{\nu - n}{\nu - 1}}$$

depending on whether the finite population correction factor is required.

There are two differences between formulas 9.1 and 9.2. The first is that, since *a two-tailed hypothesis test has two critical values,* a $\pm$ symbol replaces the minus sign. The second difference is that $z_{\alpha/2}$ replaces z_α because the opportunities to reject the null falsely are now spread over two tails. Let's use the significance level of 5% again and substitute into formula 9.2 to find the critical values:

$$C.V._L, C.V._U = \mu_0 \pm z_{\alpha/2}\sigma_{\bar{x}} \tag{9.2}$$

$$C.V._L, C.V._U = 10 \pm 1.96(0.15)$$

$$C.V._L = 9.706 \qquad C.V._U = 10.294$$

The critical values for the test are 9.706 and 10.294. The acceptance region is the range bounded by these critical values and the rejection regions are the two ranges outside these two critical values. Figure 9.4 shows the sampling distribution of the mean under the assumption that the null hypothesis is true and the acceptance and rejection regions. As you can see, the sample mean of 9.5 inches would lead you to reject the null hypothesis that the population mean equals 10 inches.

9.1E CHECKLIST FOR THE STEPS IN A HYPOTHESIS TEST

Since there are many steps in a hypothesis test, it is easy to forget some of them. Students taking exams occasionally do all of the hard steps of a hypothe-

FIGURE 9.4

Sampling distribution of the mean when the null hypothesis is true and a two-tailed hypothesis test with a 5% significance level

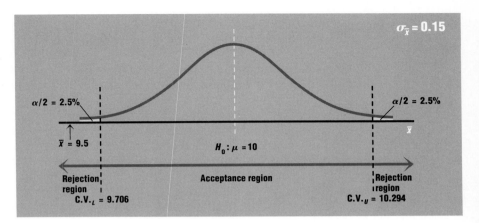

sis test correctly, such as establishing the correct null and alternative hypotheses and finding the critical value for the test, but then forget the easy step of deciding if the null is rejected or not. The basic purpose of the hypothesis test is to come to some decision about the hypothesis. Instructors tend to deduct points on exams when students omit this step, even though it is an easy step to do. The checklist of steps shown in Figure 9.5 can help you avoid these calamities.

Let's review these steps with our road construction example. Step 1 is to define the null hypothesis as a mean population thickness of 10 or more inches and to define the alternative hypothesis as a mean population thickness of less than 10 inches. The null is selected to place the burden of proof on the "suspected" hypothesis of under 10 inches. The second step is to pick the estimator and determine its distribution. The sample mean is used to estimate the population mean because it is an efficient and unbiased estimator of the population mean. Given the assumptions of a known population standard deviation and normality in the population distribution of road thickness, the distribution of the sample mean is determined to be normal with a standard deviation of $\sigma_{\bar{x}}$. Step 3 is to choose a significance level for the test. The significance level of 5%

FIGURE 9.5

Checklist of the steps in a hypothesis test

1. Define the null and alternative hypotheses.

2. Select the estimator and determine its distribution.

3. Choose a significance level for the test.

4. Define the decision rule for rejecting or not rejecting the null hypothesis.

5. Perform the computation.

6. Make the statistical decision.

7. Make the administrative decision.

is used because it is the most common, and in this example it is impossible to assign costs to Type I and Type II errors and determine an optimum significance level. Step 4 is to define the decision rule for the test. In this case the decision rule is to reject the null hypothesis if the sample mean of the 100 cores taken from the road is below a point on the hypothesized distribution of $\bar{x}$ which marks the lowest 5% of the distribution. Step 5 is to calculate this point, the critical value for the test. Step 6 is to check to see if the sample mean falls in the rejection region, which it does, and then to conclude that the null hypothesis is rejected. Step 7 is the decision of what to do with the results of the test. In the actual case, the prosecutor decided to indict the officers of the construction company and put them on trial.

Except for the plotting of Type II errors in section 9.2, most of the rest of this chapter corresponds closely to Chapter 8. The correspondence is so close that the explanations can be much briefer and will take much less time than those in Chapter 8. Most of the formulas in Chapter 8 will carry over to Chapter 9 after we substitute the null hypothesis value of the mean or proportion for the sample mean or sample proportion.

EXERCISES FOR SECTION 9.1

An asterisk designates an exercise of greater than ordinary difficulty.

9.1 Test the hypothesis that the mean of the population equals 25 if the population is normally distributed and has a standard deviation of 3, and:

(a) The mean of a sample of size 15 is 26. Use a 5% significance level.

(b) The mean of a sample of size 40 is 23. Use a 10% significance level.

(c) The mean of a sample of size 60 is 24.6. Use a 1% significance level.

(d) The mean of a sample of size 10 is 24. Use a 20% significance level.

9.2 Test the alternative hypothesis that the mean of a population exceeds 25 if the population is normally distributed and has a standard deviation of 3, and:

(a) The mean of a sample of size 15 is 26. Use a 5% significance level.

(b) The mean of a sample of size 40 is 27. Use a 10% significance level.

(c) The mean of a sample of size 60 is 25.6. Use a 1% significance level.

(d) The mean of a sample of size 10 is 24. Use a 20% significance level. (Hint: For this problem you do not need to calculate a critical value.)

9.3 For each example write the null and the alternative hypotheses and state whether it is a right-, left-, or two-tailed test.

(a) Test if the mean thickness of the asphalt equals 10 inches.

(b) Test if the mean thickness of the asphalt does not equal 10 inches.

(c) Test if the mean thickness of the asphalt is greater than 10 inches.

(d) Test if the mean thickness of the asphalt is greater than or equal to 10 inches.

(e) Test if the mean thickness of the asphalt is less than 10 inches.

(f) Test if the mean thickness of the asphalt is less than or equal to 10 inches.

9.4 William Blackstone, an 18th-century English jurist and legal scholar, once said that he would rather let 100 guilty men go free than to convict one innocent man. If "The accused is innocent" is considered a null hypothesis, was Judge Blackstone more worried about Type I or Type II errors?

9.5 Suppose you did a one-tailed hypothesis test with a left tail and you accepted the null hypothesis. If you kept the significance level the same and changed the test to a two-tailed test, would you necessarily accept or necessarily reject the null hypothesis, or is the result uncertain? Use a graph to explain your conclusion.

9.6 For each of the following situations determine the appropriate null and alternative hypotheses and explain your reasoning.

(a) You own a company that upholsters leather furniture. You buy entire cowhides that are supposed to be classified as first quality. This means that the average number of blemishes is less than or equal to 2 per square yard of hide. From a shipment of 100 hides you have cut up 20 hides and found an average of 2.2 blemishes per square yard. You suspect that the shipment does not meet the quality rating.

(b) In the Communist Chinese legal system the accused is presumed to be guilty. An accused person who can offer no convincing evidence of innocence is convicted. Jo-en Koo has been accused by the prefecture of police of black-market profiteering.

(c) The Gillette Company's chemists have formulated a new underarm deodorant. A panel of volunteers try the deodorant and then are rated for underarm odor by sensory consultants. The company has to decide whether to replace its current formulation with the new formulation.

(d) A genetic engineering company has created a bacterium that is designed to protect strawberries from frost damage. After a hard frost, random samples of strawberries from fields treated with the bacteria and from fields that were not treated by the bacteria are brought to the company's laboratory.

9.7 For each of the following situations determine the appropriate null and alternative hypotheses and explain your reasoning.

(a) A pickle processor is considering replacing its glass jars with a more expensive but unbreakable plastic jar. It test-markets the new jar in selected stores.

(b) The city manager of a small town in Michigan has been offered a new chemical that is cheaper than road salt and supposed to be just as effective in ridding the roads of ice. The manager plans to buy some of the new chemical and put it on some roads while using salt on other roads.

(c) A public school system that is troubled by drug use in high schools starts a program of inviting former convicts and drug addicts to describe the effects of drug use on their lives to the high school students. The school system monitors the subsequent incidence of drug problems in the schools that have been visited by the teams of former convicts and addicts and compares these numbers with those of schools that were not in the program.

(d) During spring training a football coach evaluates the performance of his quarterbacks. He has one veteran quarterback who was the starter in the previous season and two rookies trying out for the position of starting quarterback.

9.8* For each of the following situations determine the appropriate null and alternative hypotheses and explain your reasoning.

(a) The manufacturer of a hard-disc drive for a microcomputer guarantees the drive against all defects for 90 days. You are the head of data processing for a large corporation that plans to buy 500 of the drives and you are considering the purchase of an extended warranty plan from the disc-drive manufacturer which would cover the first year of operation on each drive. You estimate that the expected life of a drive until the first breakdown would have to be less than 120 days before it would be worthwhile to buy the warranty plan at the price quoted by the manufacturer.

(b) You and your roommates are planning to buy a keg of beer for a party. You could buy a premium brand or an inexpensive brand. You think that no one will be able to notice the difference, so it is not worth spending extra money on the premium beer. Some of your roommates disagree. Since you have studied statistics, you propose a blind taste test with a few bottles of each brand.

(c) You are the buyer for a regional car rental company. You have decided to buy 20 station wagons for your rental fleet. After selecting the model and options that you want, you get price quotes over the phone from 100 car dealers. The mean quoted price for the 20 cars is $249,000 with a sample standard deviation of $600. The best price you've been quoted is $2,000 below the sample mean. You have to decide if you should spend more time calling other dealers or take this best offer.

9.2 PLOTTING THE PROBABILITY OF A TYPE II ERROR (OPTIONAL)

9.2A FINDING THE PROBABILITY OF A TYPE II ERROR AND THE POWER CURVE FOR A HYPOTHESIS TEST

As we noted in section 9.1, there is no unique value for the probability of a Type II error because the probability of a Type II error is different for every possible value of the population mean. Recall that the probability of a Type II error has the symbol β. Instead of finding a unique value for β, you can determine how the β varies as a function of the possible values of the population mean. We can use the road example again; this time we'll go back to a left-tailed test. To show the procedure involved, let's arbitrarily assume that the true mean equals 9.8 inches. We'll use the symbol μ_1 to denote the mean of an alternative distribution. This notation parallels the use of μ_0 for the mean of the hypothesized distribution. A Type II error occurs when the null hypothesis is accepted when it is really false. Suppose our null was that the mean was 10 or more inches. This hypothesis was the first example in the chapter. The

FIGURE 9.6

Sampling distribution of the mean centered on a value other than the null hypothesis and the probability of a Type II error

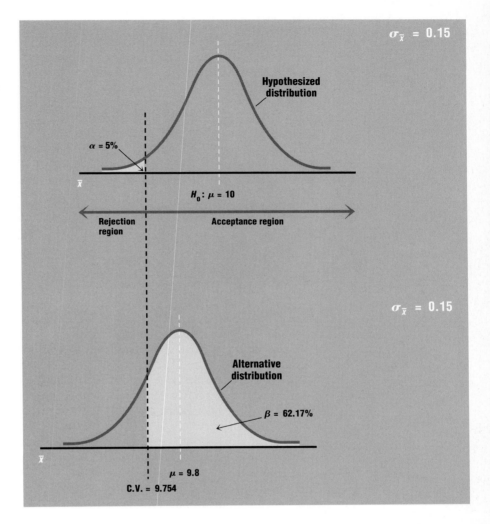

critical value for this test with the significance level set at 5% was 9.754. Thus, given that the true mean was 9.8 inches, a Type II error would occur if a sample mean fell on the high or acceptance side of 9.754. Figure 9.6 shows the sampling distribution of the mean centered on a μ_1 of 9.8 inches and the critical value of 9.754. The proportion of the total area under the distribution which is within the acceptance region is β.

To clarify the relationship between the sampling distribution of the mean when the null hypothesis is true versus when some alternative hypothesis is true, Figure 9.6 shows the sampling distribution for both situations. The sampling distribution of the mean when the null hypothesis is true is above the sampling distribution of the mean centered on the alternative value of 9.8 inches. The only thing that changes in the second distribution is the mean, from 10 to 9.8. Note that the critical value stays at 9.754. If you sum the probabilities for the shaded sections, you will obtain the β for a mean of 9.8 inches, in this case 0.6217. These two sections make up the area under the sampling distribution of the mean which is in the acceptance region. The probability to the right of 9.8 equals 0.5 because of the symmetry of the distribution. The probability between 9.8 and 9.754 equals 0.1217. We can find this probability by taking the z score for the point 9.754 (it is -0.31) and looking up the corresponding area in Table D4.

$$z = \frac{\text{C.V.} - \mu_1}{\sigma_{\bar{x}}}$$

$$z = \frac{9.754 - 9.8}{0.15}$$

$$z = -0.31$$

Thus if the mean road thickness was 9.8 inches, you would have a 0.6217 probability of mistakenly accepting the null of 10 or more inches. Note that 9.8 inches was picked arbitrarily as a number to illustrate how to calculate β levels.

There is one possible value for the population for which you won't have to do any work to find β. Consider a population mean of 9.754. Since this is also the critical value of the test, any sample mean to the right of the population mean will result in a Type II error. Since the distribution of sample means is symmetrical around the population mean, the probability of a Type II error must be 0.5. In general, β for a one-tailed test will equal 0.5 if the population mean equals the critical value of the test.

The procedure for finding the β level can be repeated for a variety of possible values of the population mean. Figure 9.7 shows a series of possible alternative values for the population mean and shades in the area representing the β under each distribution. Figure 9.8 shows the β's plotted as a function of the population mean. In Figure 9.8 the possible values of the mean are plotted on the x axis. The vertical axis is scaled as $1 - \beta$ because this arrangement makes the graph easier to interpret or more intuitive. With $1 - \beta$ on the vertical axis, higher numbers mean that you are "better off," since you would ordinarily like to avoid a Type II error. The term $1 - \beta$ is the probability of *not* making a Type II error. *For any specified value of the population parameter, the probability of not making a Type II error* is called the **power of a hypothesis test.**

Note that the highest value on the x axis of Figure 9.8 is 10. The reason for stopping at 10 is that if the population mean is 10 or more, it is no longer

FIGURE 9.7
Probability of a Type II error
for a series of alternative
values of the population
mean

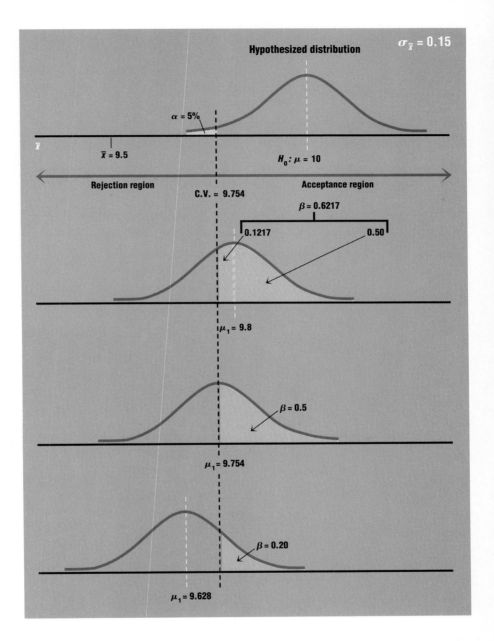

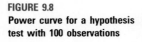

FIGURE 9.8
Power curve for a hypothesis test with 100 observations

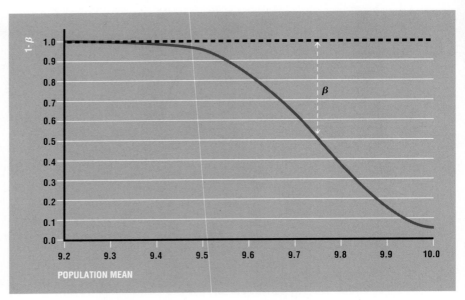

possible to make a Type II error. The null hypothesis is that the mean is 10 or more, so if the sample mean falls in the acceptance region when the population mean is 10 or more, then no error is being made. However, if the true mean was infinitesimally close to 10 but below it, the null hypothesis would be false. A picture of this situation would look the same as Figure 9.1, in which the sampling distribution of the mean is centered on 10. The probability of falling in the acceptance region is infinitesimally close to 0.95, so the β level for a population mean very close to but below 10 would be 0.95 and the power of the test would be 0.05.

Study Figure 9.8 for a moment. If the population mean is below 9.4, you are almost certain not to make a Type II error. As the possible values of the mean approach 10, it is more and more likely that you will make a Type II error. The power of the test falls as the population mean approaches the value of the null hypothesis. It is very hard for a test to distinguish a population mean of 9.99 from a hypothesized mean of 10. The limiting value for the power of the test is 0.05 as the mean approaches 10.

The idea that a hypothesis test cannot distinguish between a population with a μ equal to the null hypothesis value and a population with a μ close to the null hypothesis value helps to explain a bit of awkward terminology that many statisticians use. When the sample mean falls in the acceptance region, many statisticians prefer to say that "the null hypothesis is not rejected" rather than that "the null hypothesis is accepted." This may seem to be much ado about nothing, but here's the reasoning. If the sample mean were 9.8, your best point estimate of the population mean would be 9.8 inches. It is a little presumptuous to say that the mean must be 10 or more inches if the sample mean comes out below 10 inches. The statement "The null hypothesis is accepted" implies that you either know or believe that the true mean is 10 or more inches. The statement "The null hypothesis is not rejected" emphasizes that the sample mean is not far enough away from the null hypothesis value to lead you to reject the null. The sample does not provide sufficient evidence to reject the null. The latter statement reminds everyone that your acceptance of the null is tentative; you'll accept the null provisionally simply because the sample mean is close enough to the hypothesized value.

9.2B THE PROBABILITY OF A TYPE II ERROR AS A FUNCTION OF THE SAMPLE SIZE

An ideal test would have a zero probability of a Type II error. The power curve of an ideal test would be a straight line at a zero probability of a Type II error or equivalently at a power of 1.00. If you increase your sample size, your test becomes more powerful and it looks more like the straight line of an ideal test. As a test becomes more powerful, it is less and less likely that you will accept the null hypothesis when it is false. To give you an idea of how the power curve changes as the sample size increases, Figure 9.9 plots the power curves

FIGURE 9.9

Power curves for the hypothesis test with samples of size 100 and 200

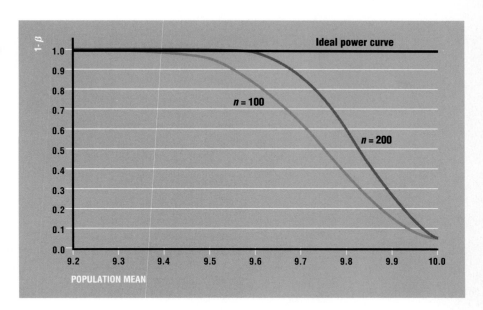

for samples of size 100 and 200 for the road-thickness hypothesis test. Note that the lower limit for the power of the test is 0.05 for both curves. If the prosecutor had used 200 cores instead of 100 cores, the chance of letting a guilty construction company go free would have been substantially reduced over the range of population means from 9.5 to 9.9 inches. For population means below 9.5 inches, the probabilities of a Type II error are essentially 0 on both curves. For population means close to 10 inches, both curves are close to a β of 0.95 or a power of 0.05. One conclusion that you can draw from Figure 9.9 is that more money in the form of a larger sample size gives you more power. If you don't have any more money, you can get a more powerful test by increasing the α level (for example, from an α of 0.05 to an α of 0.10). Recall that a more powerful test is one that is less susceptible to a Type II error. Section 9.2C shows how reductions in the β level of a test can be traded for increases in the α levels.

■ 9.2C THE PROBABILITY OF A TYPE II ERROR AS A FUNCTION OF THE SIGNIFICANCE LEVEL

Let's continue with the road example and keep the one-tailed hypothesis test with the null set at 10 or more inches. What would happen to the power curve of the test if the α or significance level were set at 20%? The new hypothesis test would have a critical value of 9.874. We obtain this number by substituting into formula 9.1:

$$\text{C.V.} = \mu_0 - z_\alpha \sigma_{\bar{x}} \qquad\qquad (9.1)$$

$$\text{C.V.} = 10 - 0.84(0.15)$$

$$\text{C.V.} = 9.874$$

To get the power curve for this test, you have to find the β levels for a variety of possible values of the population mean and then plot out the results. For the sake of additional practice, let's go over the procedure again, using an alternative population mean of 9.8. Figure 9.10 shows the sampling distribution of the mean centered on 9.8 inches and the critical value of 9.874. The shaded area under the curve is the probability of a Type II error. The z score for the point 9.874 is 0.49 and the area to the right of this z score is 0.3121. Note that the β level is much lower now that the significance level has been raised to 20%. When 9.8 was used as the population mean and the significance level was 5%, the β level was 0.6217. Figure 9.11 plots the power curves for the hypothesis test with the significance levels set at 5% and 20%. When the α level is increased to 20%, the power of the test is raised along most of the range, from about 9.3 to 10. Of course, if the null hypothesis is true, you are four times as likely to reject it by mistake with an α level at 20% than with an α level at 5%, so this increased power does not come free.

FIGURE 9.10

Sampling distribution of the mean centered on 9.8 inches and the probability of a Type II error when the critical value is 9.874

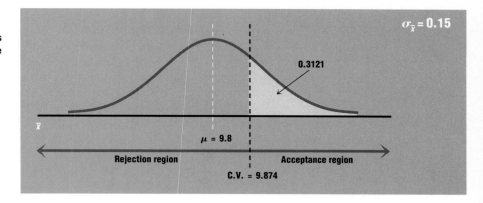

$\sigma_{\bar{x}} = 0.15$

0.3121

$\mu = 9.8$

Rejection region   Acceptance region

C.V. = 9.874

FIGURE 9.11

Power curves for the hypothesis test with significance levels of 5% and 20%

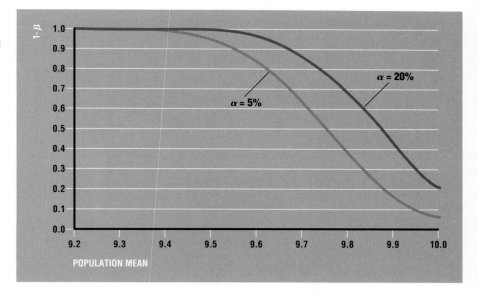

$\alpha = 20\%$

$\alpha = 5\%$

POPULATION MEAN

Here's another example that makes use of the α and β levels of a hypothesis test. The Kellogg factory in Battle Creek, Michigan, makes 20-ounce boxes of Kellogg's Corn Flakes. The filling machine runs at the rate of 10,000 boxes per hour. There is an inherent or unavoidable variability in the amount of cereal poured into each box. If the filling machine is correctly calibrated, the standard deviation of the weight of cereal per box will be 1 ounce and the mean will be 20 ounces. The calibration of the filling machine can drift away from the setting of 20 ounces. Samples are taken over each hour to check for any drift. This problem can be described as a hypothesis test in which the null hypothesis is that the mean setting is 20 ounces and the alternative hypothesis is that the mean setting is not 20 ounces. The management wants to stop the production line if the mean setting drifts up or down 0.5 ounce from 20.

The reason that the management is concerned about both high and low settings is that both errors can cost the company money. If the settings are too high, the company is giving cereal away, and if the settings are too low, it may have to pay a fine and lose some of its customers' goodwill. The plant statistician proposes that the significance level of the test be 5% and the probability of a Type II error be 10% for population means of 19.5 or 20.5 ounces. How large a sample must be taken to achieve these probabilities?

We can solve this problem by plotting the power curves for a series of sample sizes until we find the sample size for which β equals 0.10 with the α set at 5%. However, an algebraic solution is quicker and more elegant than plotting. Before we do the algebra, a glance at Figure 9.12 will help you visualize the problem. It shows one sampling distribution of the mean centered on the null hypothesis value of 20 ounces and another sampling distribution of the mean centered on the alternative value of 20.5 ounces. The required sample size would be the same if 19.5 were used for the alternative population mean instead of 20.5, so there is no reason to do the problem from both sides. A critical value has been drawn so that 2.5% of the hypothesized distribution is to the right of the critical value and 10% of the distribution centered on 20.5 ounces is to the left of the critical value.

FIGURE 9.12
Sampling distributions of the mean centered on 20 and 20.5 ounces

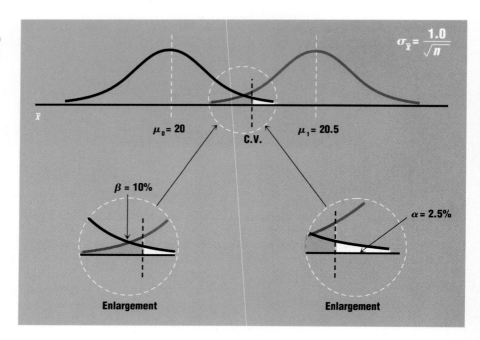

The z score for 2.5% of the area in the right tail is 1.96 and the z score for 10% of the area in the left tail is -1.28. The boundary line for both z scores is the critical value of the test. Both the critical value and the sample size for the test are unknown. You have two unknowns, but you also have two equations. Substituting into the z score formula for each sampling distribution of the mean, you get

$$z = \frac{C.V. - \mu_0}{\sigma/\sqrt{n}} \qquad\qquad z = \frac{C.V. - \mu_1}{\sigma/\sqrt{n}}$$

$$z = \frac{C.V. - 20}{1/\sqrt{n}} \qquad\qquad z = \frac{C.V. - 20.5}{1/\sqrt{n}}$$

$$C.V. = 1.96\frac{1}{\sqrt{n}} + 20 \qquad C.V. = -1.28\frac{1}{\sqrt{n}} + 20.5$$

$$1.96\frac{1}{\sqrt{n}} + 20 = -1.28\frac{1}{\sqrt{n}} + 20.5$$

$$3.24\frac{1}{\sqrt{n}} = 0.5$$

$$n = 41.99$$

$$C.V. = 20.30$$

The number of cereal boxes you would need to sample each hour is 41.99, but since you can't sample a fractional cereal box, you would have to round up to 42 boxes. With 42 boxes sampled, the actual α and β levels would be slightly below 5% and 10%, respectively.

Working through the above formulas can be clumsy, and errors are likely. Formula 9.3 can save some steps:

$$n = \frac{(z_{\alpha/2} + z_\beta)^2 \sigma^2}{(\mu_0 - \mu_1)^2} \qquad\qquad (9.3)$$

Note: $z_{\alpha/2}$ and z_β are always positive numbers, whether the hypothesis test involves the left or right tail. For a one-tailed test use z_α instead of $z_{\alpha/2}$.

The term z_α refers to the positive z score for a given α level, such as 1.64 for an α level of 5%. The term z_β refers to the positive z score for a given β level, such as 1.28 for a β level of 10%.

EXERCISES FOR SECTION 9.2

An asterisk designates an exercise of greater than ordinary difficulty.

9.9 For a hypothesis test at the 5% significance level in which the null hypothesis is that the mean is less than or equal to 30 and the population standard deviation is 5, find the probability of a Type II error if:

(a) The population mean is 32 and the sample size is 40.

(b) The population mean is 31 and the sample size is 40.

(c) The population mean is 30.5 and the sample size is 40.

(d) The population mean is 30.25 and the sample size is 40.

9.10 Use the four β levels found in exercise 9.9 plus the values of the mean at which the β levels are 0.5 and 0.05 to plot the power curve of the test.

9.11 Repeat exercise 9.9 with the sample size changed to 60. Then superimpose the power curve for this test on the power curve drawn in exercise 9.10.

9.12 For a hypothesis test at the 10% significance level in which the null hypothesis is that the mean is equal to 30 and the population standard deviation is 5, find the probability of a Type II error if:

(a) The population mean is 32 and the sample size is 40.

(b) The population mean is 31 and the sample size is 40.

(c) The population mean is 30.5 and the sample size is 40.

(d) The population mean is 30.25 and the sample size is 40.

9.13 Use the four β levels found in exercise 9.12 plus the β levels for population means of 28, 29, 29.5, and 29.75 to plot the power curve of the test.

9.14 Repeat exercise 9.13 with the sample size changed to 60. Then superimpose the power curve for this test on the power curve drawn in exercise 9.13.

9.15 What is the minimum sample size needed to achieve an α level of 5% and, assuming a population mean of 31, a β level of 20% for the hypotheses described in exercise 9.9?

9.16* You own an intracity message and small package delivery service. Your drivers generally park in metered spaces and run into a building to make their deliveries. Your policy has been not to have your drivers put coins in the meters and to pay any parking fines that they incur. Your reasoning has been that it is cheaper to pay occasional fines than to feed the meters, but you would like to test this proposition. Suppose the time between parking in a metered space and the arrival of a meter maid to write a ticket follows an exponential distribution. Given the factors of the time needed to make a delivery, the charge for a parking ticket, and the charge for the meter, if expected time until the arrival of a meter maid is 10 or more minutes, you are better off not feeding the meters.

(a) How many observations of the arrival times of a meter maid should you make to test the hypothesis that the expected arrival time is 10 or more minutes at the 5% significance level?

(b) If the true mean arrival time was 9 minutes, what would be the β level of this test?

9.3 HYPOTHESIS TESTS ON THE POPULATION MEAN: POPULATION STANDARD DEVIATION NOT KNOWN

If the population is normal, the standard deviation is not known, and the sample size is under 30, you have to use the t score in place of the z score in formula 9.1. The reasons for this substitution were discussed in Chapter 8. Formula 9.1 changes to

Critical value for a left-tailed hypothesis test with σ unknown

CRITICAL VALUE   (2) CONFIDENCE
FOR THE TEST   FACTOR OR THE
                    STANDARDIZED
                    SCORE

$$\mathrm{C.V.} = \mu_0 - t_\alpha s_{\bar{x}}$$

(1) HYPOTHESIZED   (3) ESTIMATED
    VALUE OF   STANDARD ERROR
    THE POPULATION   OF THE MEAN
    MEAN

(9.4)

where

$$s_{\bar{x}} = \frac{s}{\sqrt{n}} \quad \text{or} \quad \frac{s}{\sqrt{n}}\sqrt{\frac{\nu - n}{\nu - 1}}$$

depending on whether the finite population correction factor is required.

The corresponding formula for a right-tailed hypothesis test is

Critical value for a right-tailed hypothesis test with σ unknown

$$\mathrm{C.V.} = \mu_0 + t_\alpha s_{\bar{x}}$$

(9.4a)

where

$$s_{\bar{x}} = \frac{s}{\sqrt{n}} \quad \text{or} \quad \frac{s}{\sqrt{n}}\sqrt{\frac{\nu - n}{\nu - 1}}$$

depending on whether the finite population correction factor is required.

If your sample size is greater than or equal to 30, you can use the z score in place of the t score, although the z score is an approximation and the t score is the correct value. If your sample size is less than 30, you have to check to see if the population is normal. If the population distribution is skewed or departs sharply from normality, then you are statistically out of luck. In this situation there is no way to construct a hypothesis test on the mean.

Here's an example of a hypothesis test with a normal population and an unknown standard deviation. Suppose you own a chain of pizza parlors that delivers most of the pizzas it sells. You are planning an advertising campaign stressing the quickness of your delivery service. You will advertise that the average time for a home delivery is 20 minutes. Before making this claim, you want to check to see if the mean time really is 20 minutes. You randomly select 15 calls on different days and record the mean time between the call and the delivery. The sample mean is 21 minutes with a sample standard deviation of 5 minutes. Use a 5% significance level for your test.

This test is one-tailed because no one will complain if a pizza is delivered in less than 20 minutes. The null and alternative hypotheses are

$$H_0: \mu \leq 20 \qquad H_1: \mu > 20$$

Suppose you tested to see if the sampled delivery times came from a normal distribution and you accepted the hypothesis that the population was normal. This test is discussed in Chapter 11. To find the critical values for the test you have to substitute into formula 9.4a. With a sample size of 15 there are 14 degrees of freedom.

$$\text{C.V.} = \mu_0 + t_\alpha s_{\bar{x}} \qquad\qquad (9.4a)$$

$$\text{C.V.} = 20 + 1.761\left(\frac{5}{\sqrt{15}}\right)$$

$$\text{C.V.} = 22.27$$

Your acceptance region is 22.27 or less. Since the sample mean of 21 falls in this range, you will not reject the null hypothesis that the population mean is 20 minutes or less.

Let's see how the seven steps described in Figure 9.5 apply to this hypothesis test.

1 Define the null and alternative hypotheses.
A right-tailed test was used because you were concerned that the pizzas might take longer to deliver than the planned time of 20 minutes.

2 Select the estimator and determine its distribution.
The estimator chosen was the sample mean. It will be normally distributed if the population distribution is normally distributed.

3 Choose a significance level for the test.
The customary significance level of 5% was chosen because the costs of making a Type I and a Type II error were not specified.

4 Define the decision rule for rejecting or not rejecting the null hypothesis.
The decision rule was to reject the null hypothesis if the sample mean was above the critical value as defined by formula 9.4a. This formula was used because the population standard deviation was not known.

5 Perform the computation.
The critical value was 22.27.

6 Make the statistical decision.
There was not enough evidence to reject the null hypothesis that the mean time was less than or equal to 20 minutes.

7 Make the administrative decision.
Presumably you will go ahead with the advertising campaign.

When your population is not normal, you need a sample of at least 30 to do hypothesis tests on the mean. With a large sample, the distribution of sample means will be approximately normal no matter what the shape of the population distribution. The only change required in formula 9.4 or 9.4a is to substitute the z score for the t score.

Here's an example using the z score in place of the t score. You are a fruit wholesaler and you have purchased 400 bushels of apples from an orchard. You paid a premium price for the apples because they were supposed to be extra-large; 52 or fewer apples should make one bushel. Your workers clean and sort the apples and place them in trays before selling them to retailers. Fifty of the 400 bushels have been sorted, and in these 50 bushels the mean number of apples was 54 with a sample standard deviation of 9 apples. You want to test the hypothesis at the 5% significance level that the mean number of apples per bushel in the 400 bushels is 52 or fewer. Your null hypothesis is H_0: $\mu \leq 52$. Your alternative hypothesis is H_1: $\mu > 52$. In this problem you have to use the finite population correction factor because the sample size is more than 5% of the population. Substituting into formula 9.4a with the z score in place of the t score, you get

$$\text{C.V.} = \mu_0 + z_\alpha s_{\bar{x}} \tag{9.4a}$$

$$\text{C.V.} = 52 + 1.64\left(\frac{9}{\sqrt{50}}\right)\sqrt{\frac{400 - 50}{400 - 1}}$$

$$\text{C.V.} = 53.96$$

Figure 9.13 shows the hypothesized distribution of the sample mean and the critical value for the hypothesis test. Since the sample mean of 54 is in the rejection region, you reject the null hypothesis. Note that if you had not used the finite population correction factor, you would have accepted the null hypothesis. The critical value calculated without the finite population correction factor is 54.09. If you made this mistake, it wouldn't be a Type I or a Type II error, it would just be a matter of using the wrong formula. Note also that it is important that the 50 bushels in the sample were a random selection of all of the apples. If the apples had been first sorted by size and the smallest packed first, the test would have had no validity.

FIGURE 9.13
**Hypothesized distribution of
the sample mean and the
hypothesis test**

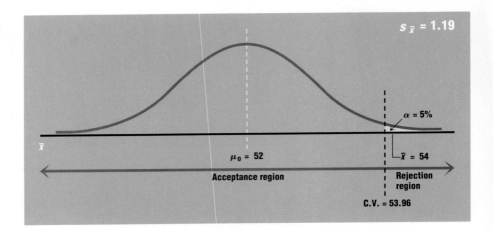

Let's see how the seven steps described in Figure 9.5 apply to this hypothesis test.

1 Define the null and alternative hypotheses.
A right-tailed test was used because you were concerned about getting smaller apples than you had paid for. Smaller apples imply more per bushel than the 52 specified in the contract.

2 Select the estimator and determine its distribution.
The estimator chosen was the sample mean. It will be normally distributed because the central limit theorem guarantees that the sample mean will be approximately normal for a sample of size 50.

3 Choose a significance level for the test.
The customary significance level of 5% was chosen because the costs of making Type I and Type II errors were not specified.

4 Define the decision rule for rejecting or not rejecting the null hypothesis.
The decision rule was to reject the null hypothesis if the sample mean was above the critical value as defined by formula 9.4a. This formula was used because the population standard deviation was not known, but because of the large sample size, the z score was used in place of the t score. Also, the finite population correction factor was used because the sample size was larger than 5% of the population.

5 Perform the computation.
The critical value was 53.96.

6 Make the statistical decision.
The null hypothesis that the mean number of apples per bushel was less than or equal to 52 was rejected.

7 Make the administrative decision.
Presumably you will demand a price cut from the orchard.

EXERCISES FOR SECTION 9.3

9.17 Test at the 5% significance level the hypothesis that a population mean equals 60, assuming that the population is normally distributed and that:

(a) A sample of size 10 taken from this population has a mean of 58 and a sample standard deviation of 5.

(b) A sample of size 15 taken from this population has a mean of 58 and a sample standard deviation of 5.

(c) A sample of size 10 taken from this population has a mean of 58 and a sample standard deviation of 10.

(d) A sample of size 10 taken from this population has a mean of 58 and a sample standard deviation of 20.

9.18 Test at the 5% significance level the hypothesis that a population mean is less than 40, assuming that the population is normally distributed and that:

(a) A sample of size 20 taken from this population has a mean of 38 and a sample standard deviation of 6.

(b) A sample of size 25 taken from this population has a mean of 38 and a sample standard deviation of 6.

(c) A sample of size 12 taken from this population has a mean of 38 and a sample standard deviation of 15.

(d) A sample of size 12 taken from this population has a mean of 38 and a sample standard deviation of 1.

9.19 You are a professional photographer and you use a battery pack to power your strobe lights when you take flash pictures. You have purchased a new type of battery pack that is supposed to average 100 flashes before it needs recharging. You need to know if this number is accurate so that you can plan your photo sessions. You keep a record of the number of flash pictures taken every time the battery pack is recharged. Suppose that you've recharged the pack 10 times and that the mean number of flashes was 115 with a sample standard deviation of 8. Assume that the distribution of flashes per charge is normal. Test at the 5% significance level the hypothesis that the mean number of flashes per charge is equal to or greater than 100. (Hint: You won't have to look up the *t* score.)

9.20 You own a firm that drills oil wells. You've purchased 100 drill bits that are supposed to last an average of 2 or more hours each while cutting through hard rock. So far you have used 8 of them and they lasted an average of 1 hour and 45 minutes with a sample standard deviation of 16 minutes. At the 1% significance level, test the hypothesis that the population of 100 drill bits will last an average of 2 hours or more. Assume that the distribution of the drill bits' lives is normally distributed.

9.21 You are an auditor for a firm that manufactures industrial fasteners. You suspect that an employee in the billing department has been undercharging customers in return for under-the-table payments. You review all of the bills prepared by the suspected employee and you find 25 bills that have errors, such

HONESTY TESTS

Honesty tests are paper-and-pencil tests that employers give to job applicants to determine if they would be reliable and honest employees. The tests typically ask some direct questions about past behavior, such as "Have you ever stolen from a previous employer?" and some attitudinal questions, such as "Did you ever think that you had a good reason for cheating a company out of some money?" The tests are designed by psychologists to uncover an individual's attitude toward honesty. Potential employees can appear to be dishonest if they never admit to having thoughts about doing something illegal or if they claim that they always tell the truth. Conversely, answering the question "Have you ever stolen from a previous employer?" with a yes is enough to guarantee that the applicant will not be hired. Surprisingly, some applicants do answer yes to this and similar questions. If nothing else, the honesty tests can save the employer from hiring someone who is awfully stupid.

About 2.5 million honesty tests were given in 1986 in the United States, and the use of these tests has been growing. In 1976 only about 1.2 million such tests were given. Many retailers view the tests as an effective mechanism for reducing employee theft. They are cheap to administer and they avoid some of the legal difficulties of polygraph tests. However, the effectiveness of these tests in identifying potential thieves is widely disputed. The companies that market these tests claim that they are nearly as effective as polygraph tests, but there are few independent studies of their effectiveness. Many firms regard reductions in their "shrinkage," or missing inventories, as evidence of the effects of the tests. Consider a grocery chain that averaged $4,000 of shrinkage per month per store before it adopted the honesty tests. In the year after it introduced the tests, the average shrinkage per month per store was $3,200 with a sample standard deviation of $600. If the 12 months after the adoption of honesty tests are treated as a

as the wrong price for an item or a smaller number of items than were actually shipped. If these errors were inadvertent, there would be about as much money in errors favoring your company as there was favoring your customers. If the errors were planned, there would be more money in the errors favoring the customers. The sample standard deviation of the errors is $175 and the mean error is $289 in favor of the customers. Test at the 5% significance level the hypothesis that the mean error is less than or equal to zero when positive errors favor the customer. Assume that the distribution of errors is normal. Explain why this hypothesis test would not be valid if you discovered that before you had checked the records, some of the errors were caught by your customers and the bills corrected. Also explain why an average error of zero dollars would not guarantee that there were no kickbacks.

random sample of all possible months in which the honesty tests could be used, is this change enough to permit us to reject the null hypothesis that honesty tests do not reduce shrinkage? Let's use an α of 5%. The null and alternative hypotheses are

$H_0: \mu \geq 4{,}000$ $H_1: \mu < 4{,}000$

Because the sample is small and the population standard deviation is not known, the only way to do the test is to use the t score formula. The use of this formula requires the assumption of normality in the distribution of shrinkage per month. The formula and the substitutions are:

$$\text{C.V.} = \mu_0 - t_\alpha s_{\bar{x}} \tag{9.4}$$

$$\text{C.V.} = 4{,}000 - \frac{1.796\,(600)}{\sqrt{12}}$$

$$\text{C.V.} = 3{,}689$$

Since the sample mean of $3,200 is below the critical value of $3,689, the null hypothesis that the tests do not reduce shrinkage is rejected.

Is this type of before-and-after study good evidence of the effectiveness of honesty tests? The answer depends on whether or not other factors changed over the same period. If the company increased its monitoring of employees, or if the penalties for employee theft increased, or if there were increased publicity about employee theft, these changes might have had more to do with the reduction in shrinkage than the honesty tests. A formal experiment would be a better way to assess the effectiveness of the tests. For example, the grocery chain might use the tests in some stores selected at random and not use them in its other stores. We'll discuss hypothesis tests for two population problems in Chapter 10.

9.4 HYPOTHESIS TESTS ON THE POPULATION PROPORTION WITH THE NORMAL APPROXIMATION

There is an important difference between hypothesis tests on the population proportion and hypothesis tests on the population mean. With tests on the mean, if you did not have the population standard deviation, you substituted the sample standard deviation. For confidence intervals on the proportion, you did the same thing: you calculated a sample standard deviation on the basis of the sample proportion. For a hypothesis test on the proportion, however, it would be illogical to use the sample proportion to estimate the standard deviation of the distribution of sample proportions. Remember that a hypothesis test is a template that gives you a picture of the sampling distribution under the assumption that the null hypothesis is true. The standard deviation of the sampling distribution of the proportion can be determined if you know the center of the distribution. Since you always center the formula for the critical value or values on the value of the null hypothesis, when you work with proportion, the value of the null hypothesis determines the value of the standard deviation. The message you should be picking up is this: *For hypothesis tests on the proportion, do not use the sample proportion to calculate the standard deviation; use the standard deviation based on the value of the null hypothesis.*

Suppose you wanted to test the hypothesis that a population proportion equaled 0.8 and you had a sample proportion from a sample of size 50 which equaled 0.7. H_0 would be $\Pi = 0.8$ and H_1 would be $\Pi \neq 0.8$. If the null hypothesis were true, the standard deviation of the distribution of sample proportions would equal (from formula 7.4)

$$\sigma_{\bar{p}} = \sqrt{\frac{\Pi(1 - \Pi)}{n}} \qquad (7.4)$$

$$\sigma_{\bar{p}} = \sqrt{\frac{0.8(1 - 0.8)}{50}}$$

If the null hypothesis was correct, the standard deviation must be based on 0.8 rather than the sample value of 0.7. To keep this distinction clear, let's define symbols for the hypothesized value for the population proportion, Π_0, and for the standard deviation of the distribution of the sample proportion based on that hypothesized value, $\sigma_{\bar{p}_0}$. The sampling distribution of the proportion will be approximately normal if $n\Pi_0$ and $n(1 - \Pi_0)$ are both greater than 5. The standard deviation of the distribution of sample proportions based on the null hypothesis is

$$\sigma_{\bar{p}_0} = \sqrt{\frac{\Pi_0(1 - \Pi_0)}{n}} \qquad (9.5)$$

or, if the finite population correction factor is needed:

$$\sigma_{\bar{p}_0} = \sqrt{\frac{\Pi_0(1 - \Pi_0)}{n}} \sqrt{\frac{\nu - n}{\nu - 1}}$$

Now you have the information you need to do hypothesis tests on the proportion. The formula for the critical values is

Critical values for a two-tailed test on the population proportion

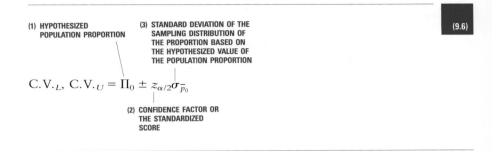

(9.6)

(1) HYPOTHESIZED POPULATION PROPORTION

(3) STANDARD DEVIATION OF THE SAMPLING DISTRIBUTION OF THE PROPORTION BASED ON THE HYPOTHESIZED VALUE OF THE POPULATION PROPORTION

$$\text{C.V.}_L, \text{C.V.}_U = \Pi_0 \pm z_{\alpha/2}\sigma_{\bar{p}_0}$$

(2) CONFIDENCE FACTOR OR THE STANDARDIZED SCORE

The left- and the right-tailed versions of the formula for the critical value for a hypothesis test on the population proportion are

Critical value for a left-tailed hypothesis test on the population proportion

$$\text{C.V.} = \Pi_0 - z_{\alpha}\sigma_{\bar{p}_0}$$

(9.7)

Critical value for a right-tailed hypothesis test on the population proportion

$$\text{C.V.} = \Pi_0 + z_{\alpha}\sigma_{\bar{p}_0}$$

(9.7a)

Let's go ahead and construct and test the hypothesis that the population proportion equals 0.8. Use an α level of 5%. Substituting into formula 9.6 yields

$$\text{C.V.}_L, \text{C.V.}_U = 0.8 \pm 1.96\sqrt{\frac{0.8(1-0.8)}{50}}$$

$$\text{C.V.}_L = 0.689 \qquad \text{C.V.}_U = 0.911$$

The acceptance region would be in the range 0.689 to 0.911. With a sample proportion of 0.70, the null hypothesis would not be rejected.

Here's a business example of hypothesis tests on the population proportion. You manufacture mink coats and you have purchased a truckload of 10,000 mink pelts. Your contract with the mink breeder specified that at least 10% of the pelts would be silver minks. Your workers have sorted and cleaned 1,000 of the pelts and they have found 85 silver pelts. You want to test at the 5% significance level the hypothesis that 10% or more of the entire shipment are silver pelts. You can use the normal approximation because $n\Pi_0$ and $n(1-\Pi_0)$ are both greater than 5. Your null hypothesis is that Π is greater than or equal to 0.1 and your alternative hypothesis is that Π is less than 0.1.

$$H_0: \Pi \geq 0.1 \qquad H_1: \Pi < 0.1$$

You would use formula 9.7. Since your sample is more than 5% of your population, you have to use the finite population correction factor. Your critical value is 0.0852:

$$\text{C.V.} = \Pi_0 - z_\alpha \sigma_{\bar{p}_0} \tag{9.7}$$

$$\text{C.V.} = 0.1 - 1.64 \left[\sqrt{\frac{0.1(1-0.1)}{1,000}} \right] \left[\sqrt{\frac{10,000 - 1,000}{10,000 - 1}} \right]$$

$$\text{C.V.} = 0.0852$$

Since your sample proportion is 85 divided by 1,000, or 0.085, you will reject the null hypothesis. Figure 9.14 shows a diagram of the hypothesized sampling distribution of the proportion for this test.

Let's see how the seven steps described in Figure 9.5 apply to this hypothesis test.

1 Define the null and alternative hypotheses.
A left-tailed test was used because you were concerned that the proportion of silver mink pelts would be less than the 10% specified in the contract.

2 Select the estimator and determine its distribution.
The estimator chosen was the sample proportion. It will be approximately normally distributed if the null hypothesis of a population proportion of 0.10 is true and the sample size is 1,000 because $n\Pi_0$ and $n(1 - \Pi_0)$ are both above 5.

3 Choose a significance level for the test.
The customary significance level of 5% was chosen because the costs of making Type I and Type II errors were not specified.

4 Define the decision rule for rejecting or not rejecting the null hypothesis.
The decision rule was to reject the null hypothesis if the sample mean was below the critical value as defined by formula 9.7. The finite population correction factor was used because the sample size was more than 5% of the population.

FIGURE 9.14
Hypothesized sampling distribution of the proportion

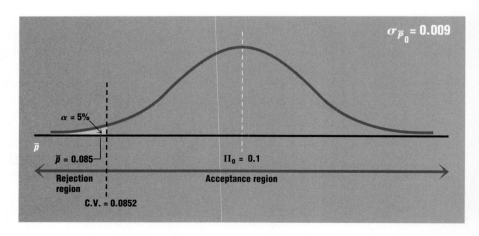

CASE STUDY

UNITED PARCEL SERVICE SORTERS

United Parcel Service, Inc. (UPS) relies on part-time help to sort its packages. Most of the part-timers are college students. UPS has strict performance standards for all of its employees: the sorters are expected to handle 1,124 packages per hour with no more than one mistake per 2,500 packages. The company has to establish standards for disciplining sorters who may have too large a proportion of mistakes. Consider, for example, the rule of handing out a written warning if one or more mistakes are found when a particular sorter's work is sampled for one hour. An important issue in establishing such a rule is the accuracy of the sampling. What is the probability that an employee who averages the tolerated proportion of mistakes would chance to exceed this limit? In other words, what is the probability that UPS would be making a Type I error of rejecting the null hypothesis of a mistake rate of no more than 1 in 2,500 when it is true?

Assuming that each package being sorted is an independent trial with the same probability of a mistake, the number of mistakes would follow a binomial distribution. With a low probability of a "success" in a trial and a large number of trials, the binomial distribution could be closely approximated by the Poisson. There are 1,124 packages or trials in an hour, and the probability of incorrectly sorting any package is 0.0004 (the 0.0004 is just 1 over 2,500 in decimal form). The expected number of errors in one hour would be $n\Pi$, or 1,124(0.0004). This works out to 0.4496 expected errors. We can find the probability of exactly zero errors by substituting into the Poisson formula (formula 5.12). One minus the probability of zero errors is the probability of one or more errors. The probability of finding one or more errors in one hour for a sorter who averages 0.4496 mistakes per hour is α, the chance of making a Type I error under the proposed rule of warning a sorter who has one or more mistakes during the survey hour.

$$P(x \geq 1) = 1 - P(x = 0)$$

$$P(x \geq 1) = 1 - \frac{\mu^x e^{-\mu}}{x!}$$

$$P(x \geq 1) = 1 - \frac{0.4496^0 (e^{-0.4496})}{0!}$$

$$P(x \geq 1) = 0.362$$

The rule of warning employees if one or more mistakes are found during a survey hour results in an α of 0.362. If an employee who is meeting the company's standards has roughly a 0.36 probability of being reprimanded, the company's personnel relations may suffer. The sorters will have little respect for a supervisory system that demands high accuracy on their part but metes out discipline in a highly inaccurate fashion.

5 Perform the computation.
The critical value was 0.0852.

6 Make the statistical decision.
The null hypothesis that the population proportion was greater than or equal to 0.10 was rejected.

7 Make the administrative decision.
You will ask for a price cut on the minks or return them for a refund.

EXERCISES FOR SECTION 9.4

9.22 Test the hypothesis that the population proportion equals 0.6 at the 5% significance level if:
- **(a)** The sample proportion is 0.62 from a sample of size 200.
- **(b)** The sample proportion is 0.55 from a sample of size 200.
- **(c)** The sample proportion is 0.59 from a sample of size 1,000.
- **(d)** The sample proportion is 0.595 from a sample of size 2,000.

9.23 Test the hypothesis that the population proportion exceeds 0.09 at the 1% significance level if:
- **(a)** The sample proportion is 0.1 from a sample of size 200.
- **(b)** The sample proportion is 0.13 from a sample of size 200.
- **(c)** The sample proportion is 0.095 from a sample of size 1,000.
- **(d)** The sample proportion is 0.092 from a sample of size 2,000.

9.24 You own a clothing mill that is the target of a unionization drive. The union asked your employees not to show up for work on a particular day to show their solidarity with the union. In the past the probability that any employee would not show up for work was 0.1. On the day in question 40 of 200 employees did not show up for work. The union claims that its call for support was successful because twice the expected number of employees did not show up. You claim that the higher-than-average absenteeism was due to chance. Test the hypothesis that the solidarity day was a random sample from a population of days that has a true proportion of no-shows of 0.1 or less. Use a significance level of 5%.

9.25 You raise chickens and find that 15% of all of the eggs laid by your hens never hatch because the eggs are infertile. You are testing a new feed supplement called "Chicken Power" which is supposed to reduce the proportion of infertile eggs. Out of 300 eggs hatched by hens given the feed supplement, the number of infertile eggs is 35. Test at the 5% significance level the hypothesis that the true proportion of infertile eggs laid by hens given "Chicken Power" is greater than or equal to 0.15.

9.26 A medical insurance company, Golden Rule Insurance, has an advertising campaign based on the claim that it processes forms for reimbursement faster than its competitors. Specifically, it claims that 90% of all forms filed with its office are processed in one day. You have been hired by a competitor to test this claim. You monitor the processing of 1,000 randomly selected claims sent in over a six-month period and determine that 823 were processed in one day. Use an α level of 5% and test the hypothesis that the true proportion is 0.9 or more.

9.27 A rating service for local radio stations claims that 40% of the prime drive-time audience listens to WXIX. You are the advertising manager of a competing radio station and you doubt that WXIX's market share is that large. Formulate and test the appropriate hypothesis at the 5% significance level if 380 people out of 1,000 surveyed say that they listen to WXIX.

9.28 Your company manufactures outboard motors for boats. You have been losing market share to some Japanese competitors and you think that the consumer's perception of higher quality in the Japanese products is the main cause. To test this hypothesis, you send surveyors to a series of regional boat and camping shows that attract consumers who may be interested in purchasing outboard motors. Each person polled is asked whether he or she thinks that the Japanese outboard motors offer higher quality than your brand. Of 1,000 people polled, 550 answer yes. Formulate and test the appropriate hypothesis at the 5% significance level.

9.29 The Breck Tulip Company sells tulip bulbs grown in Holland by mail order to U.S. customers. The company has a standard that at least 98% of all bulbs sold to consumers should, if properly planted in the fall, grow to maturity the following spring. The company has developed a new variety of tulip called Iridescent Black. It has 10,000 bulbs of this variety ready for sale. In a test planting of 1,000 of the bulbs, 35 failed to grow. Formulate and test the appropriate hypothesis at the 5% significance level.

9.5 HYPOTHESIS TESTS WITH *P* VALUES (OPTIONAL)

The *P* or probability value is another method of presenting hypothesis tests that is generally used by statistical programs on computers. The *P*-value method is a simple variation on the method of presenting hypothesis tests used so far in this chapter. This section is designated "optional" only because it can be skipped without a loss of continuity. In all of the hypothesis tests you've done so far, you have specified an α level in advance of the test. The **P value** is *the highest α level at which a null hypothesis may be accepted.* Let's go back to the example of the asphalt road and use the *P*-value method. Recall that the null hypothesis was that the road was 10 or more inches thick and that a sample of 100 cores had a mean of 9.5 inches. The population standard deviation was

FIGURE 9.15
Sampling distribution of $\bar{x}$
when H_0 is true and P value
for a sample mean of 9.5

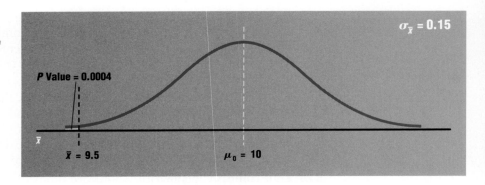

known to be 1.5 inches. The proportion of the total area of the hypothesized distribution of the sample mean to the left of 9.5 inches is the P value. This area is the largest α level for which you can accept the null hypothesis. Figure 9.15 shows this area.

We obtain the P value by calculating the z score for the value 9.5 and looking up the area in the tail in Table E4. The z score is $(9.5 - 10)/0.15$, or -3.33. The advantage of the P-value method is that it provides more information than a conventional hypothesis test. With the conventional method, if you were told that the null was rejected when the α level was set to 5%, you would not know if it had just barely been rejected or been rejected by a wide margin. With the P value you know exactly the highest α level at which you can accept the null. The reason that most computer programs are set up to use this method is that programs can automatically provide a P value even if the z score is outside the range of most published tables. If the user has an α level in mind, it is easy to compare it with the P value to decide whether to reject or accept the null hypothesis. Suppose in the above example you had planned on an α level of 5%. The P value of 0.0004 would cause you to reject the null hypothesis. If you had planned an α level of 0.01%, or 0.0001 (the value used in Figure 9.3), a P value of 0.0004 would lead you to accept the null hypothesis.

Here's an example of a P value for the population proportion with a one-tailed test. A discotheque was accused of applying discriminatory standards of admission to blacks. Some black patrons complained to the State Human Rights Commission that they were more likely to be asked for ID cards or refused admission for violation of the dress code. When confronted with this complaint, the owner of the discotheque replied that his employees used their own discretion in asking for identification and enforcing the dress code on the basis of the appearance of a patron and that race was not a factor in these decisions. An anonymous employee of the Human Rights Commission visited the discotheque and observed that 13 out of 100 patrons inside the discotheque were black. If the population proportion of blacks in the local area was 0.2 and if blacks were as likely to try to patronize the discotheque as whites, what P value would result from a test of the hypothesis that the discotheque discriminated? The null and alternative hypotheses are:

H_0: $\Pi \geq 0.2$ H_1: $\Pi < 0.2$

Figure 9.16 shows the hypothesized distribution of the sample proportion and the P value. We obtain the P value of 0.04 by finding the z score for the point 0.13 and looking up the area to the left of that z score in Table E4. The z score is $(0.13 - 0.2)/0.04$, or -1.75. This area is 0.0401. We obtain the standard deviation of the hypothesized distribution of sample proportions, 0.04, by substituting 0.2 for Π_0 in formula 9.5:

$$\sigma_{\bar{p}_0} = \sqrt{\frac{\Pi_0(1 - \Pi_0)}{n}} \tag{9.5}$$

$$\sigma_{\bar{p}_0} = \sqrt{\frac{0.2(1 - 0.2)}{100}}$$

$$\sigma_{\bar{p}_0} = 0.04$$

If the Human Rights Commission used a significance level of 5%, it would have enough evidence to reject the null hypothesis that the discotheque did not discriminate.

The procedure for obtaining the P value with a two-tailed test is more complicated. The P value is the area outside the sample mean or proportion on one side of the hypothesized distribution plus the corresponding area on the opposite side of the distribution. Here's an example of the P value for a two-tailed test on the population mean.

You are the benefits and compensation director for a *Fortune* 1,000 corporation. You have a national medical insurance plan for your company's employ-

FIGURE 9.16
Hypothesized distribution of sample proportion and *P* value for $\bar{p}$ of 0.13

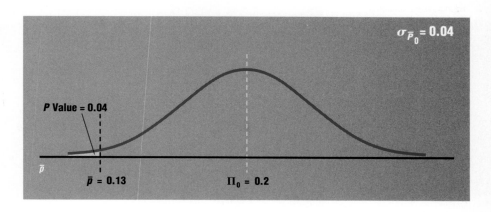

ees which provides for the payment of either 100% of the usual and customary rates for a variety of surgical procedures or the actual charge, whichever is lower. You retain the records of the actual charges by procedure, and for most surgical procedures the average actual charge is close to the usual and customary rate defined by the insurer. You have noticed a discrepancy, however, between the average actual charge and the usual and customary rate for childbirths by Caesarean section. The usual and customary rate is set at $1,200 and the average of the actual charges is only $1,150. You are concerned about discrepancies in either direction. Since your insurance fees are experience-rated, if the usual and customary rate were set too high, your company would be paying too much. In effect you would be covering part of the upper half of the distribution of actual charges, when your intention was to cover only the lower half of the distribution. If the insurer's usual and customary rates were set too low, your employees might view your compensation package as inadequate and be more willing to jump to other employers.

Suppose the $1,150 average actual charge for Caesarean sections was based on 32 cases and that the sample standard deviation of actual charges was $300. If the 32 cases were considered a random sample of national physicians' fees for Caesarean sections and your null hypothesis was that the usual and customary rate allowed by the insurer was correct, what would be the P value for this hypothesis test? The null and alternative hypotheses are

$$H_0: \mu = 1,200 \qquad H_1: \mu \neq 1,200$$

Figure 9.17 shows the hypothesized sampling distribution of the mean and the P value for the test. The P value in Figure 9.17 is the shaded area in the two tails. The z score for the upper bound of the left tail is -0.94.

FIGURE 9.17
Hypothesized sampling distribution of mean actual charge for Caesarean sections

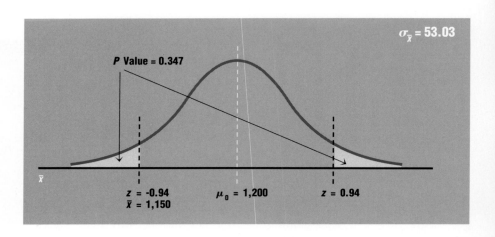

$$z = \frac{\bar{x} - \mu_0}{s_{\bar{x}}}$$

$$z = \frac{1{,}150 - 1{,}200}{300/\sqrt{32}}$$

$$z = -0.94$$

Note that the z score can be used in place of the t score even though the population's distribution is not specified and the population's standard deviation is not known because the sample size is over 30. If it were available, the area under the Student distribution would be more accurate than the area under the standard normal provided that the population was normally distributed. If you had a small sample with an unknown population distribution, you could not determine the P value.

The area to the left of a z score of -0.94 is 0.1736. The corresponding area in the right tail is the area bounded on the left by $1{,}250$. It would have a z score of $+0.94$ and the area would also be 0.1736. The combined shaded areas equal approximately 0.347. Recall that the P value is the largest α level that you can specify and still accept the null hypothesis. If you specified an α that was smaller than 34.7%, say 5%, you would have to accept the null hypothesis that the mean actual charge for Caesarean sections was $1,200.

EXERCISES FOR SECTION 9.5

9.30 Find the P value for the hypothesis test H_0: $\Pi = 0.5$, H_1: $\Pi \neq 0.5$:
(a) If the sample proportion is 0.45 for a sample of size 100.
(b) If the sample proportion is 0.55 for a sample of size 100.
(c) If the sample proportion is 0.6 for a sample of size 50.
(d) If the sample proportion is 0.4 for a sample of size 50.

9.31 Find the P value for the hypothesis test H_0: $\mu = 100$, H_1: $\mu \neq 100$:
(a) If the sample mean is 95 and the sample standard deviation is 20 for a sample of size 100.
(b) If the sample mean is 105 and the sample standard deviation is 20 for a sample of size 100.
(c) If the sample mean is 99 and the sample standard deviation is 10 for a sample of size 50.
(d) If the sample mean is 101 and the sample standard deviation is 10 for a sample of size 50.

9.32 According to a national survey, the price of a new family home averages $115,000. You hypothesize that the average price in your city exceeds the national average. To test this hypothesis, you record the prices of 100 new homes in your city and find the sample mean to equal $120,000 with a sample standard deviation of $30,000. What is the P value for your hypothesis test?

9.33 Suppose you saw an ad for a cruise ship that said that its customers were so satisfied with their vacations that, on average, 80% of all of the cruise passengers were repeat customers. The ad sounds convincing, so you book passage for two weeks of island hopping. When you get on board you survey the passengers in your dining area and find that 40 out of 70 are repeat passengers. If you were to test the hypothesis that the true proportion of all passengers who are repeat customers was 0.8, what would be the P value for the test?

9.34 You have been waiting at a taxi stand for a cab for half an hour. The stand has a sign that states that cabs arrive, on average, at 10-minute intervals. If you were testing the hypothesis that the sign is a correct description of cab arrivals, what would your P value be? Assume that the cabs are independent. (Hint: Don't use the normal distribution.)

SUMMARY

This chapter has discussed hypothesis tests for one population. Hypothesis tests should be used when you have a prior value for a parameter. Both the similarities and the differences in the formulas for hypothesis tests and confidence intervals were emphasized. The similarities are that each formula has three parts and that the second and third parts are usually the same across confidence intervals and hypothesis tests. The differences are that confidence intervals are always centered on the value of the sample statistic, but the formula for the critical value of a hypothesis test always replaces the sample statistic with the null hypothesis value of the parameter. Another important difference is that the standard deviation of the distribution of sample proportions is based on the sample proportion for a confidence interval and on the hypothesized proportion for a hypothesis test. The chapter also discussed how changing the sample size or the significance level affects the power of a test. While almost all of the examples used 5% as a significance level, it was pointed out that this significance level is arbitrary. If the exact costs of making Type I and Type II errors are known, they should be used as guides to set the significance level and the sample size. Unfortunately, this information is usually not available, and rough guesses must be made. Finally, the P-value method of hypothesis tests was discussed. This method, a slight variant of the usual method, has the advantage of conveying more information.

Chapter 10 discusses hypothesis tests and confidence intervals for problems that involve two populations. Again, there is great similarity in the formulas for hypothesis tests and confidence intervals in the two-population case.

NEW TERMS

acceptance region The range marked by the critical value or values which contains the null hypothesis value of the population parameter.

alternative hypothesis The claim that a population parameter has a value or values other than those specified in the null hypothesis; the suspected or "challenging" values of the population parameter.

critical value A value in the range of the sampling distribution which marks the boundary between the acceptance and rejection regions of the null hypothesis. A two–tailed hypothesis test has two critical values.

null hypothesis The claim that a population parameter equals the supposed or "maintained" value or values, or that it equals the "no effect" value or values.

power of a hypothesis test For any specified value of the population parameter, the probability of not making a Type II error.

P value The highest α level at which a null hypothesis may be accepted.

rejection region The range marked by the critical value or values which does not contain the null hypothesis value of the population parameter.

significance level The probability of rejecting the null hypothesis when it is true.

Type I error Rejection of a null hypothesis when it is true.

Type II error Acceptance of a null hypothesis when it is false.

 NEW FORMULAS

Critical value for a left-tailed hypothesis test for the population mean when the population standard deviation is known and the population distribution is normal

$$\text{C.V.} = \mu_0 - z_\alpha \sigma_{\bar{x}} \tag{9.1}$$

Critical value for a right-tailed hypothesis test for the population mean when the population standard deviation is known and the population distribution is normal

$$\text{C.V.} = \mu_0 + z_\alpha \sigma_{\bar{x}} \tag{9.1a}$$

Critical values for a two-tailed hypothesis test for the population mean when the population standard deviation is known and the population distribution is normal

$$\text{C.V.}_L, \text{C.V.}_U = \mu_0 \pm z_{\alpha/2} \sigma_{\bar{x}} \tag{9.2}$$

Minimum sample size for a hypothesis test on the population mean with a given significance level and power

$$n = \frac{(z_{\alpha/2} + z_\beta)^2 \sigma^2}{(\mu_0 - \mu_1)^2} \tag{9.3}$$

Critical values for a left-tailed hypothesis test for the population mean when the population standard deviation is not known and the population distribution is normal

$$\text{C.V.} = \mu_0 - t_\alpha s_{\bar{x}} \tag{9.4}$$

Critical value for a right-tailed hypothesis test for the population mean when the population standard deviation is not known and the population distribution is normal

$$\text{C.V.} = \mu_0 + t_\alpha s_{\bar{x}} \tag{9.4a}$$

Standard deviation of the sampling distribution of the proportion based on the hypothesized value of the population proportion

$$\sigma_{\bar{p}_0} = \sqrt{\frac{\Pi_0(1 - \Pi_0)}{n}} \tag{9.5}$$

or, if the finite population correction factor is needed,

$$\sigma_{\bar{p}_0} = \sqrt{\frac{\Pi_0(1 - \Pi_0)}{n}} \sqrt{\frac{\nu - n}{\nu - 1}}$$

Critical values for a two-tailed hypothesis test on the population proportion using the normal approximation to the sampling distribution of the proportion

$$\text{C.V.}_L, \text{ C.V.}_U = \Pi_0 \pm z_{\alpha/2}\sigma_{\bar{p}_0} \tag{9.6}$$

Critical value for a left-tailed hypothesis test on the population proportion

$$\text{C.V.} = \Pi_0 - z_\alpha \sigma_{\bar{p}_0} \tag{9.7}$$

Critical value for a right-tailed hypothesis test on the population proportion

$$\text{C.V.} = \Pi_0 + z_\alpha \sigma_{\bar{p}_0} \tag{9.7a}$$

 ADDITIONAL EXERCISES

An asterisk designates an exercise of greater than ordinary difficulty.

9.35 Suppose you did a two-tailed test and you rejected the null hypothesis. If the significance level was kept constant but you changed the test to a one-tailed test, would you necessarily reject the null hypothesis? necessarily accept the null hypothesis? or are the results uncertain? Use a graph to explain your answer.

9.36 You work for a New England utility company that operates a nuclear generating plant. The plant discharges into the ocean the water used to cool the reactor. For public relations purposes you would like to demonstrate that the warm water dumped into the ocean has increased the average daily lobster catch in the area near the plant. The average catch per day before the advent of the nuclear plant was 346 lobsters. You observe that on 16 days when lobsters were harvested during the year after the plant started to discharge its heated water there was an average catch of 351 lobsters with a sample standard deviation of 36. Test at the 5% significance level the hypothesis that the mean catch is less than or equal to 346. Assume that the distribution of lobsters caught per day is normal.

9.37 You are a quality control engineer at a television factory. At your factory the production line for picture tubes makes 5,000 tubes per hour. If the production line is working properly, the proportion of defective tubes is less than or equal to 0.005. You test a sample of 100 picture tubes each hour and rate each tested picture tube as good or defective. If you had a decision rule to stop the production line when more than one tube out of the 100 was defective and you considered this decision rule a test of the hypothesis that the true defect rate equals 0.005 or less, what would be the α level of this test?

9.38 You work for a company that sells a nondairy dessert topping called Dessert Whip. You have planned a coupon campaign to attract new consumers to your product. Your problem is that the coupons may be used by loyal customers instead of new ones. If 50% or more of the customers who used the coupons were new customers, you would consider the coupon campaign to be cost-effective. To see if this 50% ratio may be achieved, you test the coupons in one city. You randomly survey 200 supermarket customers who turn in a newspaper coupon for Dessert Whip. The number who claim to be new customers is 110. If your null hypothesis is that the true population proportion is 0.5 or less, what is the P value of your test?

9.39 You are considering a new package for toothpaste. Instead of a tube, the package is a pump that squirts toothpaste onto a brush. You are worried that the pump package is more likely to leak than the traditional tube. Your experience is that, on average, one in every 100,000 tubes leaks. You leave 1 million pump packages in a warehouse for several months and then have every package examined for leaks. The number with leaks is 13. If your null hypothesis is that the true proportion of leaks is less than or equal to 0.00001, what is the P value for your test?

9.40 In exercise 9.39 the sample proportion of leaks was 0.000013. If this sample value was the true population proportion and you had set up a test for the null hypothesis at the 5% significance level, what would be the β level of your test?

9.41 Suppose it has been hypothesized that an ore-crushing machine jams, on average, once in every hour of operation. Assume that the jams are cleared immediately and that each jam is independent. You observe that the machine has jammed ten times in the last two hours of operation. If you constructed a one-tailed test of the hypothesis that the true rate is one per hour or less, what would be the P value of your test? (Hint: Don't use the normal distribution.)

9.42 You have waited at a taxi stand for 30 minutes without seeing a cab. The stand has a sign that states that the average waiting time for a cab is five minutes. If you were using your waiting time as a test of the hypothesis that the true mean waiting time is five minutes or less, what would be the P value of your test? (Hint: Don't use the normal distribution.)

9.43 Miller Beer, a U.S. brewery, has purchased the rights to market a famous German beer, Löwenbräu, in the United States. Instead of importing the Löwenbräu, Miller makes a beer in Texas which it calls Löwenbräu. Consumer's Union once ran a test to see if consumers could distinguish between the domestic version of Löwenbräu and Miller beer. The test consisted of giving 25 people three unmarked glasses, two that contained Miller and one that contained Löwenbräu. Each person was asked to taste all three glasses and to identify the one that tasted different from the other two.

Formulate the null and alternative hypotheses. At a 5% significance level, what would be the critical value of the test? (In the actual test the null hypothesis that consumers could not distinguish the Löwenbräu could not be rejected.)

9.44 You have developed a new type of lawn mower that resembles a hovercraft. It can skim over lawns and mow a quarter-acre lot in five minutes. You need to sell your machine in the consumer market for at least $5,000 to make a profit. To see if consumers might pay this much for a lawn mower, you demonstrate the prototype of the mower to consumers who come to a lawn mower store to buy a riding mower. The customers are told that the machine is not yet for sale and they are asked what they would pay for such a machine if it were available. The average price offered by 40 customers was $3,000 with a standard deviation of $2,500. Test at the 5% significance level the hypothesis that the mean price consumers would be willing to pay for your mower is $5,000 or more.

9.45 You have purchased 1,000 smoke alarms for a large hotel. You have installed 60 of them and 3 failed to sound an alarm when tested with the smoke from a match. If you were testing the null hypothesis that the true proportion of defective fire alarms was 0.01 or less in the population of 1,000 alarms, what would be the P value for your test?

9.46 A filling machine for shampoo has a known population standard deviation of 0.1 ounce. The machine is supposed to put 10 ounces of shampoo in a bottle. If you can reject the hypothesis that the mean amount of shampoo is 10 ounces at the 15% significance level, you will have the machine recalibrated. If a sample of 20 bottles has a mean of 10.01 ounces, should the machine be recalibrated? Assume that the population distribution is normal.

9.47 Until 1985 the National Basketball Association used a $20 gold coin to select which of two teams would have the first pick in the draft of college players. The same coin has been used in every toss. If the coin had been used 25 times and 17 of the tosses were heads, test at the 5% significance level the hypothesis that the coin is fair.

9.48 You are the manager of some toll stations on a highway. To find out if your booth operators pocket any of the money they handle, you compare the revenue each operator generates with the average of all operators. You fire any attendant who consistently generates less revenue than the average. Suppose the distribution of tolls collected per day per booth is normal with a mean of $300 per day and a standard deviation of $35. You've hired a new person and recorded the following daily revenues for his toll station: $225, $235, $275, $360, $359, $175, and $150. If you are willing to accept a 5% chance of mistakenly firing someone who is honest, should you fire him?

9.49 You own a bakery and you have an automatic scaling machine that cuts dough into loaf-sized pieces. The distribution of the pieces is normal with a standard deviation of 0.3 pound. The mean is set at 2 pounds, but the calibration of the machine can drift over time and the pieces come out consistently high or low. You hand-scale some of the pieces from time to time to see if the calibration is accurate. Suppose you hand-scaled 10 pieces and the sample mean was 3.12 pounds. Test at the 1% significance level the hypothesis that the mean population weight is 2 pounds.

9.50 You are the head of a public utility that sells electricity. Your daily sales have had a normal distribution with a mean of 500,000 kilowatts and a standard deviation of 120,000 kilowatts. Your research department has issued a report stating that there has been a significant decline in the mean level of electricity consumption while the standard deviation has stayed constant. This conclusion is based on a mean consumption of 478,000 kilowatts during the last 30 days. If you set α at 1%, would the decline in consumption from the historic average be statistically significant?

9.51 Suppose the population mean for electricity consumption described in exercise 9.50 really had declined to 478,000 kilowatts. If you took another 30-day sample, what would be the chance of making a Type II error, that is, accepting the null hypothesis that the population mean is 500,000 kilowatts or more?

9.52* Suppose you hypothesized that the mean retail price of a competitor's line of dresses was $100 or more, and if this hypothesis proved to be correct you would raise your price from its present level of $90 to $100 to match your competitor. From past surveys you know that the distribution of your competitor's retail prices is normally distributed with a standard deviation of $10. Suppose you wanted to set the α level for your hypothesis test at 1% and you wanted the β level to be 10% if the true mean was $90. What boundary line should you use for your hypothesis test and how large should your sample be?

9.53* Your company operates a toll-free telephone complaint service and it normally receives 60 calls per hour. You think that your company's product quality control has improved and that the flow of calls has been slowing down. In the last 36 hours that the service was open, there were 2,060 calls. At the 5% level of significance test the hypothesis that the rate of calls is equal to or greater than the previous rate. (Hint: The number of calls follows a Poisson distribution.)

9.54* Calculate the probability of a Type II error, β, for the asphalt example at the beginning of the chapter, using the 5% significance level and the values of 9.5, 9.6, 9.7, 9.8, and 9.9 as population means. Plot the β levels as a function of the population mean. If the sample size was increased to 500, how would this plot change?

10

Confidence Intervals and Hypothesis Tests with Two Populations

rganizational fund-raisers have a keen interest in how the wording of a request for a donation affects the probability and the amount of a donation. One compliance-increasing ploy is to "legitimize" small donations by telling a potential donor that any amount, even $1, will help the cause. While such a statement may encourage someone who is embarrassed about donating a small amount of money to go ahead and donate, it can also make potential donors lower their donations by giving them the impression that a small amount is expected. A test of the hypothesis that the proportion of donors is increased by a statement legitimizing small donations and of the hypothesis that the average amount donated is decreased by such a statement was conducted by Joel Brockner, Beth Guzzi, Julie Kane, Ellen Levine, and Kate Shaplen, who published their findings in "Organizational Fundraising: Further Evidence on the Effect of Legitimizing Small Donations," *Journal of Consumer Research* 11 (June 1984): 611–614. All but Brockner were undergraduates when the study was done.

The article describes how college students called randomly selected residents in Lexington, Massachusetts, on behalf of the National Reye's Syndrome Foundation (Reye's syndrome is a fatal childhood disease). Some of the calls used the statement "Even a dollar will help"; others did not. The latter calls were referred to as the "control condition." The authors concluded that in future fund-raising campaigns (1) the percentage of individuals who would donate would be raised by a statement legitimizing small donations, and (2) among people who would donate, the mean amount would not differ between populations that were solicited with and without a statement legitimizing small donations. "From the organization's perspective," they wrote, "the bottom line was that more than 20 times as much money was raised in the $1 condition [as] in the control condition."

The two sets of hypotheses in these tests are examples of two-population problems. The two populations are the distribution of donations from all potential donors with and without the statement legitimizing small donations. Many problems can be distilled into two treatments, two brands, two suppliers, and so forth. Two-population problems may constitute the most common business use of statistics. Most of the material in Chapters 8 and 9 will easily carry over to this chapter, but be forewarned that some of the optional material on pooled variances has been known to cause cardiac arrest and hair loss. A small donation of your time to study the material will make a great difference. Even a few minutes will help.

10.1 CONFIDENCE INTERVALS AND HYPOTHESIS TESTS ON THE DIFFERENCE BETWEEN TWO POPULATION MEANS: STANDARD DEVIATIONS KNOWN AND INDEPENDENT SAMPLES

This section describes the procedure for constructing confidence intervals or hypothesis tests for the difference between the means of two populations when the populations are normal, the population standard deviations are known, and the samples used for making the confidence intervals are independent. Ordinarily, you won't know the standard deviations of both populations. This case is discussed primarily to help you understand what happens in more complicated cases.

Imagine that you own a company that specializes in raising campaign funds for political candidates. Your company organizes phoneathons, direct mailings, banquets, door-to-door solicitations, and the like. A candidate for the U.S. Senate has hired your company and she has a list of 50,000 potential donors. You know from past experience that a phone solicitation will raise more money per contact than a mail solicitation. You are interested in estimating the difference between the average contributions generated by these two methods. You send letters to 100 randomly selected potential donors and you telephone 300 randomly selected donors. The mean contribution from the 100 people who get the letters is $5.25 and the sample standard deviation is $1.40. The mean contribution of the people who receive phone calls is $6.40 and the sample standard deviation is $2.11. A confidence interval would give you a better idea of the possible difference in mean contributions from the two methods than just the point estimate of the difference of $1.15.

Let's suppose that you know the populations are normal and that you know the population standard deviations from phone and mail solicitations from previous surveys. Let's say the population standard deviation for the mail is $1.50 and for the phone is $2.00. Naturally, if you know the population values, you should use them instead of the sample values when you construct the confidence interval.

10.1A THE SAMPLING DISTRIBUTION OF THE DIFFERENCE IN MEANS

What you need to know before you can construct the confidence interval for the difference between two population means is the **sampling distribution of the difference in means,** or *the probability density function of the difference between the means of two independent samples.* If both populations are normal, the distribution of the sample mean for each population will also be normal. Further, the distribution of the differences between the sample means taken from each population, $\bar{x}_1 - \bar{x}_2$, will also be normal. In general, the sums or differences of two normal random variables create another normal random variable. (It may be noted that any linear transformation of a normal random variable creates another normal random variable. In other words, if you take a nonzero constant and perform any operation with it—multiplication, division, addition, or subtraction—on every element of a normal random variable, the result is another normal random variable.)

The sampling distribution of the difference in means is centered on the difference between the population means, $\mu_1 - \mu_2$. This idea is expressed as

$$E(\bar{x}_1 - \bar{x}_2) = \mu_1 - \mu_2$$

(10.1)

The variance of the sampling distribution of the difference in means is

$$\sigma^2_{\bar{x}_1 - \bar{x}_2} = \frac{\sigma_1^2}{n_1} + \frac{\sigma_2^2}{n_2}$$

(10.2)

In Chapter 5, section 3, you learned that the variance of the sum or the difference of two independent random variables equals the sum of the variances of each variable. *Formula 10.2 holds only when the two distributions are independent.* Since the people who received phone and mail solicitations were randomly selected from the list of potential donors, the two groups have to be independent. The standard deviation of the distribution of the difference in sample means is just the square root of the variance. Remember, variances for independent random variables can be added but standard deviations cannot. Formula 10.3 is for the standard deviation of the distribution of the difference in sample means from independent samples:

$$\sigma_{\bar{x}_1 - \bar{x}_2} = \sqrt{\frac{\sigma_1^2}{n_1} + \frac{\sigma_2^2}{n_2}}$$

(10.3)

Figure 10.1 illustrates the sampling distribution of the means taken from two normal populations with known standard deviations. The two populations have different means and different standard deviations. The purpose of the confidence interval is to estimate the true difference between μ_1 and μ_2. The key element needed to construct the confidence interval is the sampling distribution of the difference in means derived from the sampling distribution of the mean for each population.

Conceptually, the sampling distribution of the difference in means is derived as follows: a sample of size n_1 is drawn from the first population and the sample mean calculated; then a sample of size n_2 is drawn from the second population and the sample mean calculated; then the difference between the two sample

FIGURE 10.1
The sampling distributions of
the mean and the population
distributions of two normal
populations with known
standard deviations

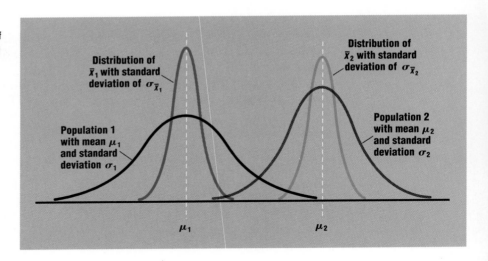

means is recorded. This procedure is repeated until every possible set of two sample means has been drawn. The plot of all of the differences between the two sample means with the frequencies plotted on the vertical axis is the sampling distribution of the difference in means. It is shown in Figure 10.2. It is centered on $\mu_1 - \mu_2$ and its standard deviation is $\sigma_{\bar{x}_1 - \bar{x}_2}$. Since an infinite number of sets of two sample means may be drawn from two normal populations, you cannot actually draw every possible set. The plot of the differences in sample means would approach the normal curve in Figure 10.2 as the number of sets of sample means increased.

In Figure 10.2, two tails have been marked off by vertical lines and labeled $\alpha/2$. If α were 5%, then 95% of all possible differences in sample means would have values between the two vertical lines.

FIGURE 10.2
Sampling distribution of the
difference in means

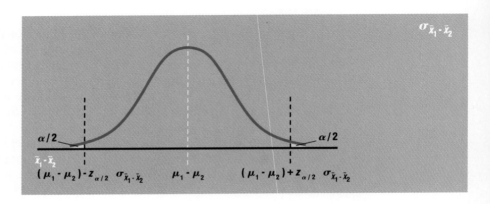

10.1B FORMULA FOR THE LIMITS OF A CONFIDENCE INTERVAL FOR THE DIFFERENCE IN TWO POPULATION MEANS WITH NORMAL POPULATIONS AND KNOWN STANDARD DEVIATIONS

Now you have the pieces to put together a confidence interval. Formula 10.4 gives the upper and lower limits of a confidence interval for the difference between two means.

Confidence interval for the difference in population means

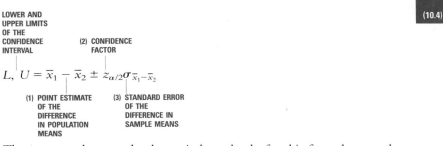

LOWER AND UPPER LIMITS OF THE CONFIDENCE INTERVAL   (2) CONFIDENCE FACTOR

$$L, \ U = \bar{x}_1 - \bar{x}_2 \pm z_{\alpha/2}\sigma_{\bar{x}_1 - \bar{x}_2}$$

(1) POINT ESTIMATE OF THE DIFFERENCE IN POPULATION MEANS   (3) STANDARD ERROR OF THE DIFFERENCE IN SAMPLE MEANS

(10.4)

The two samples must be drawn independently for this formula to apply.

The format and the logic of this confidence interval are exactly the same as in Chapter 8. Let's substitute the numbers from the mail versus phone example in formula 10.4.

$$L, \ U = \bar{x}_1 - \bar{x}_2 \pm z_{\alpha/2}\sigma_{\bar{x}_1 - \bar{x}_2}$$ (10.4)

$$L, \ U = 5.25 - 6.40 \pm 1.96\sqrt{\frac{(1.50)^2}{100} + \frac{(2.00)^2}{300}}$$

$$L = -1.52 \qquad U = -0.78$$

The interpretation of these limits is that, at the 95% level of confidence, the average payment from a mailed solicitation is estimated to be between $1.52 and $0.78 less than the average payment from a phone solicitation. You may be puzzled as to why the confidence limits have negative values. If you had defined the phone results as $\bar{x}_1$ instead of $\bar{x}_2$, the numbers would have been positive but the confidence interval would have had the same width. The choice of the variable to be called $\bar{x}_1$ or $\bar{x}_2$ is arbitrary.

10.1C HYPOTHESIS TESTS ON THE DIFFERENCE IN POPULATION MEANS

The main difference between confidence intervals and hypothesis tests for the difference in population means is that with a hypothesis test you have a prior value in mind for the difference in population means while with a confidence interval you are only trying to estimate the difference in population means.

This section discusses hypothesis tests for normal populations with known standard deviations. Again, this setup is mainly a teaching device to help with the introduction of more complicated cases.

If you specify a null hypothesis for that difference *and* the null hypothesis is true, then the sampling distribution of the difference in means will be centered on the null hypothesis value of the difference in means. Again, $E(\bar{x}_1 - \bar{x}_2)$ will equal $\mu_1 - \mu_2$. If the two population standard deviations are known, then the standard deviation of the difference in sample means will be $\sigma_{\bar{x}_1 - \bar{x}_2}$, as in formula 10.3.

Let's try an example that illustrates a difference-in-means hypothesis test. You are an engineer who works for a company that manufactures brake linings for cars. You have a new compound for the linings which costs the same to manufacture as your old brake lining. Your company will switch to the new compound if it will enable cars to stop in a shorter distance. Suppose you knew that the standard deviation of stopping distances was the same for the old type of brake lining and brake linings made with the new material and that it equaled 100 feet for a stop begun at 60 miles per hour. You also know that the distribution of stopping distances will be normal. You test 25 randomly selected cars at 60 miles per hour with the old linings and 25 other cars with the new linings. The mean stopping distance with the old linings is 365 feet, while the mean distance with the new linings is 350 feet. (Note that if you tested each car with both types of linings, the samples would not be independent. Hypothesis tests with dependent samples are discussed in section 10.4.) Your null hypothesis is H_0: $\mu_1 - \mu_2 \geq 0$. Your alternative hypothesis is H_1: $\mu_1 - \mu_2 < 0$. Let's use a 5% significance level. The symbol μ_1 is the population mean for the new compound and μ_2 is the population mean for the old compound. The expression $(\mu_1 - \mu_2)_0$ is the hypothesized difference in population means. The formula you would use is

Critical value for a left-tailed hypothesis test on the difference in population means with normal populations and known population standard deviations

HYPOTHESIZED
DIFFERENCE IN MEANS

(10.5)

$$\text{C.V.} = (\mu_1 - \mu_2)_0 - z_\alpha \sigma_{\bar{x}_1 - \bar{x}_2}$$

To do a right-tailed test, substitute $+$ for $-$ before the z. A two-tailed test requires the use of $\pm$ in place of $-$ and $\alpha/2$ in place of α.

Not all of the parts of formula 10.5 are labeled. Since formula 10.5 is so similar to formula 10.4, for confidence intervals, only the key difference is labeled so that it will be highlighted. The key difference is that the first term is the hypothesized difference in population means rather than the observed difference in sample means. Also, since it is a one-tailed test, the z score is noted as z_α rather than $z_{\alpha/2}$ and the plus-or-minus symbol changes to a minus. As you

FIGURE 10.3
Hypothesized sampling
distribution of the difference
in population means and the
critical value for the
hypothesis test

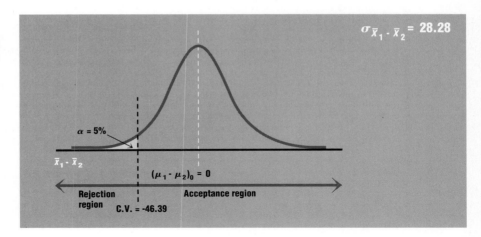

might expect, the formula for a right-tailed test simply substitutes a plus for the minus in formula 10.5. A two-tailed test uses a $\pm$ and a $z_{\alpha/2}$ in place of z_α.

Substituting into formula 10.5 and assuming a 5% significance level yields

$$\text{C.V.} = 0 - 1.64\sqrt{\frac{100^2}{25} + \frac{100^2}{25}}$$

$$\text{C.V.} = -46.39$$

Figure 10.3 shows the template for this hypothesis test.

The critical value for the test is -46.39. Since the difference in sample means was -15, you cannot reject the null hypothesis that the stopping distances are the same or longer with the new compound. You should also consider the possibility of a Type II error. A Type II error would occur if the true mean stopping distance of the new compound was less than that of the old compound but you had accepted the null hypothesis that the stopping distances were equal or greater with the new compound. Since there is so much variability in stopping distances relative to the mean stopping distance, your sample size of 25 cars for each type of lining may be too small. To really design the best hypothesis test, you would have to know what it would cost your company if you went ahead with the new compound and then found that it really did not stop cars in a shorter distance. You would also have to know what it would cost your company to stay with the old compound. Finally, you would have to know the cost of increasing your sample size. If you had this information, you could figure the optimal α and β levels and the optimal sample size. Clearly, these figures are difficult to obtain; sometimes you have to make what you hope is an educated guess.

Let's see how the seven steps described in Figure 9.5 apply to the above hypothesis test.

1 Define the null and alternative hypotheses.

A left-tailed test was used because the burden of proof was placed on the new brake linings. You wanted to switch to the new lining only if you were strongly convinced that it could stop cars in a shorter distance. With the variables set up as ($\mu_{new} - \mu_{old}$), a shorter distance for the new brake linings would be a negative number.

2 Select the estimator and determine its distribution.

The estimator chosen was the difference in sample means. It will be normally distributed if both populations are normally distributed.

3 Choose a significance level for the test.

The customary significance level of 5% was chosen because the costs of making Type I and Type II errors were not specified.

4 Define the decision rule for rejecting or not rejecting the null hypothesis.

The decision rule was to reject the null hypothesis if the difference in sample means was below the critical value as defined by formula 10.5.

5 Perform the computation.

The critical value was -46.39.

6 Make the statistical decision.

There is not enough evidence to reject the null hypothesis that the new brake linings perform no better than the old brake linings.

7 Make the administrative decision.

Continue with the old brake linings.

It should be noted that it is possible to do a hypothesis test for a difference in means with nonnormal populations and small samples, but it requires a different technique than that presented above. This test, the Wilcoxon-Mann-Whitney test, is discussed in Chapter 17, section 2. Also note that it is not possible to do a confidence interval to estimate the difference between population means with nonnormal populations and small samples.

EXERCISES FOR SECTION 10.1

10.1 Let X and Y be two normally distributed random variables with the means μ_x and μ_y and variances of 20 and 30, respectively. Construct a confidence interval for $\mu_x - \mu_y$ at the 95% level of confidence if:

(a) A sample of size 10 taken from X has a mean of 34 and a sample of size 20 taken from Y has a mean of 25.

(b) A sample of size 20 taken from X has a mean of 44 and a sample of size 20 taken from Y has a mean of 65.

(c) A sample of size 50 taken from X has a mean of 44 and a sample of size 70 taken from Y has a mean of 65.

(d) A sample of size 200 taken from X has a mean of 44 and a sample of size 200 taken from Y has a mean of 65.

10.2 Let X and Y be two normally distributed random variables with the means 60 and 54 and variances of 20 and 30, respectively.

(a) If a sample of size 5 is taken from X and a sample of size 10 is taken from Y, what is the probability that $\bar{x} - \bar{y}$ is greater than 7?

(b) If a sample of size 10 is taken from X and a sample of size 12 is taken from Y, what is the probability that $\bar{x} - \bar{y}$ is greater than 5?

(c) If a sample of size 15 is taken from X and a sample of size 18 is taken from Y, what is the probability that $\bar{x} - \bar{y}$ is greater than 0.5?

(d) If a sample of size 55 is taken from X and a sample of size 98 is taken from Y, what is the probability that $\bar{x} - \bar{y}$ is greater than 6.5?

10.3 You work for a company that sells automobile insurance. Your company plans to market a "no smokers" automobile insurance policy because the management thinks that nonsmokers will generate a lower average level of annual claims per policy than smokers. The company's reasoning is that smokers take more risks than nonsmokers in all facets of their lives. Your job is to survey a sample of the company's present policyholders and estimate the difference in the average annual claims per policy for nonsmokers and smokers. Assume that the population standard deviations for annual claims for both groups are known to equal $300. You survey 500 policyholders and find that 200 are smokers and that the average annual claim for this group is $2,000. The remaining 300 respondents are nonsmokers and their average annual claim equals $1,500. Construct a confidence interval for the population difference in average annual claims at the 95% level of confidence.

10.4 You work in the marketing department of a national chain of department stores. You know that advertisements of sales that specify a discount on regular prices will attract customers to your stores. You are interested in how much more store traffic is generated when the discount is increased. Suppose you ran an experiment in which your stores in 40 randomly selected cities advertised a clearance on fall dresses at 30% off the regular price. Another 40 randomly selected stores advertised a clearance on fall dresses at 40% off the regular price. The mean number of customers making purchases of the marked-down dresses in the 30%–off stores was 800, while the corresponding figure for the other stores was 975. Assume that the standard deviations for both populations were known to equal 200 customers. Construct a confidence interval at the 90% level of confidence for the difference in the number of customers per store.

10.5 Your company is considering moving its headquarters from one city to another. The company has a policy of compensating any employee who has to pay more for housing equivalent to his or her old home in the new city. The compensation is a cash payment equal to the difference in the prices of the old and new homes. What you need is an estimate of the difference in mean prices of equivalent homes in the two cities. You hire a real estate agency that has offices in both cities and you ask them to appraise 50 homes in the neighborhoods in which your company executives currently reside and 50 homes in similar neighborhoods in the new city. The mean price of the 50 homes in the old city is $90,000 and the mean price of the 50 homes in the new city is $125,000. Assume that both population standard deviations are known to equal $20,000. If 2,000 executives may be moved, estimate a confidence interval at the 95% level of confidence for the total compensation required.

10.6 Let X be a normally distributed random variable with a standard deviation of 5 and let Y be a normally distributed random variable with a standard deviation of 10. Test the hypothesis that the mean of X equals the mean of Y if:

(a) A sample of size 20 taken from X has a mean of 13 and a sample of size 15 taken from Y has a mean of 14. Use a 5% significance level.

(b) A sample of size 20 taken from X has a mean of 13 and a sample of size 15 taken from Y has a mean of 14. Use a 10% significance level.

(c) A sample of size 30 taken from X has a mean of 13 and a sample of size 50 taken from Y has a mean of 14. Use a 5% significance level.

(d) A sample of size 30 taken from X has a mean of 16 and a sample of size 50 taken from Y has a mean of 18. Use a 10% significance level.

10.7 Let X be a normally distributed random variable with a standard deviation of 20 and let Y be a normally distributed random variable with a standard deviation of 30. Test the hypothesis that the mean of X exceeds the mean of Y if:

(a) A sample of size 20 taken from X has a mean of 13 and a sample of size 15 taken from Y has a mean of 12. Use a 5% significance level.

(b) A sample of size 20 taken from X has a mean of 13 and a sample of size 15 taken from Y has a mean of 10. Use a 10% significance level.

(c) A sample of size 30 taken from X has a mean of 13 and a sample of size 50 taken from Y has a mean of 14. Use a 5% significance level.

(d) A sample of size 30 taken from X has a mean of 16 and a sample of size 50 taken from Y has a mean of 14. Use a 10% significance level.

10.8 There are two kinds of avocados, black and green. You are offered either kind by a fruit vendor at two different prices. You plan to buy several truckloads. You will use the avocados as an ingredient in a cosmetic, so you don't care about the taste; you want to buy the type that has the lowest cost per ounce. However, the vendor quotes you prices on a per avocado basis. Suppose green avocados had a known standard deviation of 2 ounces and black avocados had a known standard deviation of 3 ounces. If the mean weight of the green avocados is more than 5 ounces above the mean weight of the black avocados, it will be cheaper for you to buy the green ones. You take a sample of both types. The sample mean of 30 green avocados is 22 ounces. The sample mean of 20 black avocados is 16 ounces. At the 5% significance level, can you reject the null hypothesis that the mean difference is less than or equal to 5 ounces? Assume that both population distributions are normal.

10.9 You are developing a new hybrid corn for popcorn. One factor that you are trying to breed into the new corn is more kernels per ear. Suppose the number of kernels is normally distributed and known to have a standard deviation of 15. You assume that this standard deviation will be the same in your new corn. You have a sample of 20 ears of your new corn and 30 ears of the last variety that you bred. The new corn has an average of 144 kernels per ear and the older variety has a mean of 138 kernels per ear. At the 5% significance level, test the hypothesis that the population mean number of kernels in the new corn exceeds the population mean in the older variety.

10.10 Suppose you manufacture a bolt and a washer. The washer is supposed to be, on average, 0.01 inch larger in diameter than the bolt. Assume that both population distributions are normal and that the known standard deviations for both diameters equals 0.001. Test the hypothesis that the true difference equals 0.01 inch at the 1% significance level if a sample of 40 bolts has a mean diameter of 0.191 and a sample of 50 washers has a mean diameter of 0.2.

10.11 The Stroh brewery was considering a new 12-ounce bottle for its beer. To test the effect of changing the shape of its bottle on unit sales, the brewery conducted a test in which the 100 outlets in a chain of liquor stores were divided into two groups of equal size. The first group retained the old bottle and the second group used the new bottle. The shelf space for each design was the same at every outlet. The average sale per store during the test period was 350 cases for the old design and 360 for the new design, with sample standard deviations of 35 and 40, respectively. If the 100 outlets were considered a representative sample of all stores that carry Stroh's beer, should the hypothesis that the new design will outsell the old design be rejected at a 5% significance level? Why might the observed gain of 10 cases per store be an unreliable predictor of the potential benefit of the new design?

10.12 A national temporary help firm, Manpower Inc., has offices in most U.S. cities. The firm's offices are located in the central business districts, but the firm is considering a new policy of locating offices in suburban areas. Along with the important issue of whether suburban locations would improve or hinder the operation of their offices, the company wants to know if suburban offices are likely to cost less in rent than offices in central business districts. The company has an office space broker find a rate for suburban locations in 50 cities, and these rates are compared with the rates that Manpower is paying in downtown locations in the same cities. The mean suburban rate per square foot per month is $4.25 and the mean downtown rate is $4.40. The respective sample standard deviations are 50 cents and 40 cents. Test at the 5% significance level the hypothesis that suburban locations are cheaper than downtown locations. How does this problem violate the assumptions required for hypothesis tests in this section?

10.13 A local telephone company is considering the adoption of a new, more expensive type of switching equipment that is supposed to be more reliable than its old equipment. The company's engineers calculate that if the difference in mean time between breakdowns is more than 100 hours in favor of the new switches, then the higher cost of the new switches can be justified. The company runs a test on 50 old and 50 new switches and observes a mean time between breakdowns of 1,200 hours for the old design and 1,250 for the new design. The respective sample standard deviations are 100 hours and 50 hours. Test at the 5% significance level the hypothesis that the difference in mean time between breakdowns between the new and old designs is 100 hours or less. Why is this test very easy?

10.2 CONFIDENCE INTERVALS AND HYPOTHESIS TESTS ON THE DIFFERENCE BETWEEN TWO POPULATION MEANS: STANDARD DEVIATIONS NOT KNOWN AND INDEPENDENT SAMPLES

10.2A CONFIDENCE INTERVALS WITH LARGE SAMPLES

When you are confronted with two population confidence intervals and hypothesis tests, it is much more common not to know whether the populations are normally distributed and what the population standard deviations are than the examples in section 10.1 may suggest. We'll start section 10.2 with the case of large samples without the assumption of normality and without knowing the population standard deviations. Next we'll consider how you can work problems with small samples without knowing the population standard deviations, as long as you can assume that both populations are normally distributed.

If you don't know the population standard deviations, you can substitute the sample standard deviations into formula 10.3. Because of the central limit theorem, sample sizes of 30 or more from both populations will guarantee that the distributions of both sample means will be approximately normal. If $\bar{x}_1$ and $\bar{x}_2$ are normally distributed, the difference in sample means, $(\bar{x}_1 - \bar{x}_2)$, will also be normally distributed. Thus, when you have large samples you don't have to have normality for the two population distributions in order to construct a difference-in-means confidence interval. The formula for a difference-in-means confidence interval with large samples is

Confidence interval for the difference in population means with unknown population standard deviations and large samples

(10.6)

LOWER AND
UPPER LIMITS
OF THE
CONFIDENCE   (2) CONFIDENCE
INTERVAL   FACTOR

$$L,\ U = \bar{x}_1 - \bar{x}_2 \pm z_{\alpha/2} s_{\bar{x}_1 - \bar{x}_2}$$

(1) POINT ESTIMATE   (3) ESTIMATED
    OF THE   STANDARD ERROR
    DIFFERENCE   OF THE
    IN POPULATION   DIFFERENCE IN
    MEANS   SAMPLE MEANS

where

$$s_{\bar{x}_1 - \bar{x}_2} = \sqrt{\frac{s_1^2}{n_1} + \frac{s_2^2}{n_2}}$$

(10.7)

This formula holds only if the samples are independent.

Here's an example that uses formulas 10.6 and 10.7. Hotels that book large conventions do not quote the same room rate to every potential client. The room rate varies with the number of rooms that the convention will book and with the hotel's expected sales of food and drinks to people attending the convention. The expected sales can vary widely among conventions. Suppose the Hyatt Regency in Chicago was considering the room rates it should quote in bids for two conventions of roughly equal size which planned to meet July 10–12, 1991: the economists' convention and the funeral home directors' convention. Because these conventions are the same size, the difference in the room rates would depend on the difference in expected spending per economist and funeral home director. Generally, a hotel chain has records of past spending by the people attending particular conventions, but the Hyatt chain has never hosted these groups. To estimate the difference in mean spending, the hotel hires a market research company to poll attendees at the current convention of each group. The market research firm persuades 36 economists and 36 funeral home directors to keep diaries of their in-hotel spending. The respective sample means and sample standard deviations in dollars are $\bar{x}_1 = 48$, $\bar{x}_2 = 120$, $s_1 = 12$, $s_2 = 12$. What are the upper and lower limits of a confidence interval at the 95% level of confidence for the difference in population means?

$$s_{\bar{x}_1 - \bar{x}_2} = \sqrt{\frac{s_1^2}{n_1} + \frac{s_2^2}{n_2}}$$

(10.7)

$$s_{\bar{x}_1 - \bar{x}_2} = \sqrt{\frac{12^2}{36} + \frac{12^2}{36}}$$

$$s_{\bar{x}_1 - \bar{x}_2} = 2.83$$

$$L, U = \bar{x}_1 - \bar{x}_2 \pm z_{\alpha/2} s_{\bar{x}_1 - \bar{x}_2}$$

(10.6)

$$L, U = (48 - 120) \pm (1.96)(2.83)$$

$$L = -77.55 \qquad U = -66.45$$

The interpretation of the confidence interval is that at the 95% level of confidence, the mean per person in-hotel spending is about $66 to $78 less for the economists than for the funeral home directors over a three-day convention. The Hyatt can be fairly sure that if it bids a much lower room rate for the funeral home directors than for the economists, it will make up the difference on drinks and food.

The formula for a hypothesis test with large samples and unknown population standard deviations is very similar to formula 10.6, for confidence intervals with large samples and unknown population standard deviations. The main differences are that instead of the upper and lower limits of the confidence interval, the formula finds the limits of the acceptance region, and that the first term is the null hypothesis value for the difference in means. The formula for a right-tailed test would be

Critical value of a right-tailed hypothesis test on the difference in population means with large samples and unknown population standard deviations

(10.8)

HYPOTHESIZED DIFFERENCE IN MEANS

$$\text{C.V.} = (\mu_1 - \mu_2)_0 + z_\alpha s_{\bar{x}_1 - \bar{x}_2}$$

To do a left-tailed test, use $-$ in place of $+$ before the z. A two-tailed test requires the use of $\pm$ in place of $+$ and $\alpha/2$ in place of α.

Here's an example of a hypothesis test with large samples. You run a feedlot for cattle and you can purchase ordinary corn or high-protein corn for cattle feed. You know that your cattle will gain weight faster on the high-protein corn, but the high-protein corn costs more. You have determined that if your cattle can gain more than 100 pounds more on high-protein corn than on regular corn during their first year, you can justify the additional cost of the high-protein corn. Your null and alternative hypotheses are

$$H_0: (\mu_1 - \mu_2) \le 100 \qquad H_1: (\mu_1 - \mu_2) > 100$$

where the subscript 1 refers to the high-protein corn and the subscript 2 refers to the regular corn.

You feed 100 cattle the ordinary corn and 100 cattle the high-protein corn. Each animal is kept on the experimental diet from birth to its first birthday. The cattle on the high-protein diet weigh an average of 1,400 pounds with a sample standard deviation of 44 pounds; the cattle that have been fed the ordinary corn weigh an average of 1,275 pounds with a sample standard deviation of 45 pounds. Let's use the customary 5% significance level. Substituting these values into equations 10.7 and 10.8 yields

$$s_{\bar{x}_1 - \bar{x}_2} = \sqrt{\frac{s_1^2}{n_1} + \frac{s_2^2}{n_2}}$$

(10.7)

$$s_{\bar{x}_1 - \bar{x}_2} = \sqrt{\frac{(44)^2}{100} + \frac{(45)^2}{100}}$$

$$s_{\bar{x}_1 - \bar{x}_2} = 6.29$$

$$\text{C.V.} = (\mu_1 - \mu_2)_0 + z_\alpha s_{\bar{x}_1 - \bar{x}_2} \qquad\qquad (10.8)$$

$$\text{C.V.} = (100) + 1.64(6.29)$$

$$\text{C.V.} = 110.32$$

The observed difference in sample means in the experiment was 125 pounds. This value is to the right of the critical value of this test, so the null hypothesis that the difference in average weight gains will be 100 pounds or less can be rejected. Your decision will be to buy the high-protein corn for cattle feed.

Let's see how the seven steps described in Figure 9.5 apply to this hypothesis test.

1 Define the null and alternative hypotheses.

A right-tailed test was used because you had to be convinced that the higher-priced high-protein feed was worth the extra money. The setup for your hypothesis test was the mean weight gain on high-priced feed minus the mean weight gain on low-priced feed.

2 Select the estimator and determine its distribution.

The estimator chosen was the difference in sample means. It will be approximately normally distributed if both sample sizes are above 30.

3 Choose a significance level for the test.

The customary significance level of 5% was chosen because the costs of making Type I and Type II errors were not specified.

4 Define the decision rule for rejecting or not rejecting the null hypothesis.

The decision rule was to reject the null hypothesis if the sample mean was above the critical value as defined by formula 10.8.

5 Perform the computation.

The critical value was 110.32.

6 Make the statistical decision.

The null hypothesis that the difference in population means was 100 or less was rejected.

7 Make the administrative decision.

You would buy the high-protein feed.

10.2B CONFIDENCE INTERVALS AND HYPOTHESIS TESTS WITH SMALL SAMPLES WITHOUT POOLING THE VARIANCES (OPTIONAL)

With small samples the procedure for constructing a confidence interval gets a little more complicated. You have to use the t score in place of the z score in formula 10.6 because with small samples the sample standard deviations can substantially underestimate the population standard deviations. Again, the reasoning for using the t score is the same as in Chapter 8. You also have to have normality for both population distributions to construct a confidence interval

CASE STUDY

CABLE TELEVISION AS A MARKET RESEARCH TOOL

In the primitive days before cable TV, a firm that had to choose between two television commercials for a product might test-air one version of the commercial in an isolated community and test-air the other version in a second community. Changes in sales in the two communities could be used to estimate the relative effectiveness of the two commercials. All sorts of problems could creep into such an analysis and distort the results. The selected communities might not have the same demographic or income characteristics. Even if they matched, it might be impossible to find isolated communities that were representative of all consumers. The television commercial might have aired at a poor time in one of the cities and at a good time in the other. Sales might have gone up or down in one of the cities for reasons unrelated to the television commercial. From a marketing research standpoint, too many factors that could affect the results were unknown.

The use of cable television can help market researchers avoid most of these difficulties. One firm that is a pioneer in this field is Chicago-based BehaviorScan. The company seeks the cooperation of local cable television companies, local consumers, and local retailers. Consumers sign up with BehaviorScan for such prizes as free vacations and dinners. After signing up with BehaviorScan, they fill out a questionnaire on their demographics and family income. The questionnaire even includes such details as whether the family owns a dog so that only the families that own dogs will be shown test dog food commercials. The next step in the process is to insert the test commercials into the cable television signals. BehaviorScan can aim specific commercials at a random selection of households. Typically the test commercials are substituted for other commercials of the same company on network shows. The consumers don't know which commercials are test commercials and which ones are being shown nationally. Each participating household has an electronic device attached to its television set to record the shows being watched. Finally, when the consumer visits a participating retailer, he or she shows a card at the cash register which tells the checkout clerk to record and transmit to BehaviorScan a record of all the customer's purchases.

With this system BehaviorScan can try two new dog food commercials only in households that own dogs and that have identical demographics and income. Among these households it can limit its test to the ones that saw one of the commercials. BehaviorScan can measure the impact of the commercials on these households' spending on any brand of dog food. It seems that everything is wired into this system except the dog.

for the difference in population means. A further complication with small samples with unknown population variances is that there are two different methods for confidence intervals and hypothesis tests. The choice between these methods depends on whether the two populations have the same variance. The formula based on the assumption of common variances is slightly easier, but this assumption is not necessary and often it cannot be justified. With this common variance assumption you have to pool the two sample variances into one estimate of the population variance. Let's first investigate the procedure without common variances. It is based on Satterthwaite's approximation of the degrees of freedom for the t score when the two populations are normal and the population variances are not known. The degrees of freedom are found by formula 10.9.

Satterthwaite's approximation of the degrees of freedom for Student distribution for confidence intervals or hypothesis tests on two population means with normal populations, small samples, and two unknown population variances

$$df = \frac{[(s_1^2/n_1) + (s_2^2/n_2)]^2}{\dfrac{(s_1^2/n_1)^2}{n_1 - 1} + \dfrac{(s_2^2/n_2)^2}{n_2 - 1}}$$

(10.9)

Note that you can get fractional degrees of freedom. Many statistical computer programs will give the correct t score for fractional degrees of freedom. You can interpolate between the integer degrees of freedom when you have a fractional t score. The interpolation usually makes a trivial difference in the answer. Instead, we'll adopt the rule of rounding down to the lower integer when there are fractional degrees of freedom.

Here's an example. Suppose you own an accounting firm and you are planning to buy a computer program that can compute individual tax returns. You have a choice of two programs, EasyTax and MicroTax. Since both programs are copy-protected, the vendor is willing to let you borrow them and try them out. You want to know if either of the programs will be easier for your employees to learn and use. You give eight employees the EasyTax program and six the MicroTax. All the employees work on the same number of tax returns, each set of returns containing the same data on earnings, deductions, and so forth. The mean time for the employees who use the EasyTax program is 3.6 hours with a sample standard deviation of 0.8 hour. The mean time for the employees who use the MicroTax program is 2.8 hours with a sample standard deviation of 0.5 hour. Before you can use formula 10.10 to construct a confidence interval for the difference in population means, you have to establish that the two samples came from normal populations. Let's assume that this has been done (Chapter 11 will show you how). Before you can get the t score needed in formula 10.10, you have to find the degrees of freedom by using formula 10.9:

$$df = \frac{[(s_1^2/n_1) + (s_2^2/n_2)]^2}{\frac{(s_1^2/n_1)^2}{n_1 - 1} + \frac{(s_2^2/n_2)^2}{n_2 - 1}} \qquad (10.9)$$

$$df = \frac{[((0.8)^2/8) + ((0.5)^2/6)]^2}{\frac{[(0.8)^2/8]^2}{8 - 1} + \frac{[(0.5)^2/6]^2}{6 - 1}}$$

$$df = 11.73$$

The degrees of freedom are 11.73. We'll use the t score for 11 degrees of freedom with 2.5% of the area in the tail, 2.201. The formula for a confidence interval for the difference in population means is

Confidence interval for the difference in population means with small samples, unknown population variances, and normal populations

$$L, U = \bar{x}_1 - \bar{x}_2 \pm t_{\alpha/2} s_{\bar{x}_1 - \bar{x}_2} \qquad (10.10)$$

If the variances are not pooled, the degrees of freedom for the Student distribution are given by formula 10.9, for Satterthwaite's approximation. If the variances are pooled, the degrees of freedom are $n_1 + n_2 - 2$.

Substituting into formula 10.10, you get

$$L, U = (3.6 - 2.8) \pm 2.201 \sqrt{\frac{(0.8)^2}{8} + \frac{(0.5)^2}{6}}$$

$$L = 0.03 \qquad U = 1.57$$

The interpretation of these confidence limits is that at the 95% level of confidence, the difference between the population means for the EasyTax and the MicroTax programs is between 0.03 and 1.57 hours. Since this confidence interval is always positive, you can be fairly sure that the EasyTax will take longer for your employees to learn and run than the MicroTax. The decision on which program to buy could also involve such factors as the price, the availability of support services, and the possibility of upgrades.

Now let's investigate a hypothesis test for the difference in population means without pooled variances. Suppose you were interested in testing at the 5% significance level the hypothesis that acid rain affects the growth of trees. You plant 20 genetically identical trees in two greenhouses, 10 in greenhouse A and 10 in greenhouse B. The trees are raised for a year in a controlled environment in which each tree receives the same amount of sunlight and sprayed-on "rain." The A trees get acidic rainwater and the B trees get neutral rainwater. Your null and alternative hypotheses are

$$H_0: \mu_1 - \mu_2 = 0 \qquad H_1: \mu_1 - \mu_2 \neq 0$$

At the end of a year the A trees have had a mean increase in diameter of 0.5 inch with a sample standard deviation of 0.02 inch. The B trees have had a mean increase in diameter of 0.6 inch with a sample standard deviation of 0.03 inch. Assume that the distribution of tree growth is normal. It is possible that the acid rain will affect the population variance as well as the population mean, so you can't presume that the variances are the same. Unless you have a reason for treating the variances as equal, you should not test for their equality and you should not pool them.

The standard deviation of the distribution of the difference in sample means is given by formula 10.7:

$$s_{\bar{x}_1-\bar{x}_2} = \sqrt{\frac{s_1^2}{n_1} + \frac{s_2^2}{n_2}} \qquad (10.7)$$

$$s_{\bar{x}_1-\bar{x}_2} = \sqrt{\frac{0.02^2}{10} + \frac{0.03^2}{10}}$$

$$s_{\bar{x}_1-\bar{x}_2} = 0.011$$

Now what you need is the t score and the degrees of freedom. For the degrees of freedom use formula 10.9:

$$df = \frac{[(s_1^2/n_1) + (s_2^2/n_2)]^2}{\dfrac{(s_1^2/n_1)^2}{n_1 - 1} + \dfrac{(s_2^2/n_2)^2}{n_2 - 1}} \qquad (10.9)$$

$$df = \frac{[(0.02^2/10) + (0.03^2/10)]^2}{\dfrac{(0.02^2/10)^2}{10 - 1} + \dfrac{(0.03^2/10)^2}{10 - 1}}$$

$$df = 15.68$$

The degrees of freedom are approximately 15.68. We'll use the rule of rounding down to 15 degrees of freedom rather than interpolating the fractional t score. The t score for 15 degrees of freedom and 2.5% in the right tail is 2.131. Now you have almost everything you need for the hypothesis test. What's missing is the formula for the critical values. You should have enough experience with these formulas to know the components of the two–tailed hypothesis test formula for small samples, normal populations, and unknown population standard deviations. The formula is number 10.11:

$$\text{C.V.}_L, \ \text{C.V.}_U = (\mu_1 - \mu_2)_0 \pm t_{\alpha/2} s_{\bar{x}_1-\bar{x}_2} \qquad (10.11)$$

$$\text{C.V.}_L, \ \text{C.V.}_U = 0 \pm 2.131(0.011)$$

$$\text{C.V.}_L = -0.0234 \qquad \text{C.V.}_U = 0.0234$$

Since the observed difference in sample means was -0.1, you should reject the null hypothesis and conclude that acid rain has an effect on tree growth. The lower growth observed for the trees given acidic water suggests that acid rain reduces tree growth. However, the test was a two-tailed test, so your only conclusion is that there is a *difference* in growth rates. If your null had been that trees with acid rain grew the same as or better than trees without acid and you had rejected this null, your conclusion would have been that acid rain reduces tree growth. Note that at the same significance level the critical value for a one-tailed test would be to the right of the lower critical value for a two-tailed test. Thus if the null is rejected on the left side of the critical value for a two-tailed test, it must also be rejected in a one-tailed test at the same significance level. In the acid rain example, after you had rejected the null in the two-tailed test, you could automatically, holding the significance level constant, reject the null that acid rain has no effect or a positive effect on tree growth.

Let's see how the seven steps described in Figure 9.5 apply to this hypothesis test.

1 Define the null and alternative hypotheses.
A two-tailed test was used because you were concerned that acid rain could either improve or reduce tree growth. The null hypothesis was that the difference in growth between trees with and without acid rain would be zero.

2 Select the estimator and determine its distribution.
The estimator chosen was the difference in sample means. It will be normally distributed if both population distributions are normal. Since there is no reason for the variances of the two distributions to be the same, the standard deviation of the distribution of the difference in means was based on formula 10.7.

3 Choose a significance level for the test.
The customary significance level of 5% was chosen because the costs of making Type I and Type II errors were not specified.

4 Define the decision rule for rejecting or not rejecting the null hypothesis.
The decision rule was to reject the null hypothesis if the difference in sample means was above the upper critical value or below the lower critical value as defined by formula 10.11. The degrees of freedom for the t score (used because of the small sample sizes) was based on Satterthwaite's approximation.

5 Perform the computation.
The critical values were -0.0234 and $+0.0234$.

6 Make the statistical decision.
The null hypothesis that the population means were equal was rejected.

7 Make the administrative decision.

With this information a forest products company might avoid planting trees in areas subject to acid rain or avoid leasing such areas. Public policy makers could also use the information in regulating the emission of industrial pollutants that cause acid rain.

■ 10.2C CONFIDENCE INTERVALS AND HYPOTHESIS TESTS WITH SMALL SAMPLES WITH POOLED VARIANCES (OPTIONAL)

Pooling the sample variances—that is, combining the two sample variances into one estimate of the population variance—is both a mathematical and a logical issue. You should first decide if there is a logical reason to expect that the two populations will have the same variances. If you decide that there is, then you should test to see if you can accept the hypothesis that the two variances are the same. Some computer programs automatically do a test for the equality of population variances when they are instructed to do a difference-in-means hypothesis test. When would it be reasonable to assume that the two populations have the same variances? If you thought that both samples came from the same population, that would imply equal variances.

Now let's investigate an example of a confidence interval when the two population variances are assumed to be the same. Suppose you have two consumer products, let's say Post Raisin Bran and generic raisin bran, and you want to construct a confidence interval for the difference in mean number of raisins per box. You know that the two products are packed by the same type of equipment and that the average number of raisins per box can be set by the operator. Because of the identical machines, you know that the standard deviations of the distributions of raisins for the two brands are the same.

If you did not know that the two standard deviations had to be equal, you could test to see if they were equal. The test is actually for the equality of population variances and it is discussed in Appendix B. The advice to use a test in Appendix B may seem a little awkward. The reason that the test for equality of two variances is not included in this chapter is that it requires the use of a distribution that will be introduced in Chapter 12. While instructors tend to discuss most of the procedures for working with means and proportions, the tests for variances are less commonly discussed. Three types of tests on population standard deviation and variances have been placed in Appendix B. This format gives you a convenient reference. Think of this appendix as "Everything You Always Wanted to Know about Variances but Were Afraid to Ask." This format also allows the instructor to introduce these tests at the point at which he or she feels that they best fit into the course.

Before constructing your confidence interval for the difference in the mean number of raisins, you have to pool the sample variances. Now you have two numbers, s_1^2 and s_2^2, which are both estimates of the one population variance, σ^2. The **pooled estimate of the population variance** is *a weighted average of two sample variances, in which the weights are the degrees of freedom for each sample.*

The pooled variance equals

$$s_{pool}^2 = \frac{(n_1 - 1)s_1^2 + (n_2 - 1)s_2^2}{n_1 + n_2 - 2}$$

(10.12)

The pooled variance is substituted in formula 10.7 in place of s_1^2 and s_2^2.

$$s_{\bar{x}_1 - \bar{x}_2} = \sqrt{\frac{s_{pool}^2}{n_1} + \frac{s_{pool}^2}{n_2}}$$

(10.13)

Now you have all the formulas needed to construct a confidence interval. Let's say that you have randomly selected 12 Post Raisin Bran boxes and 7 generic raisin bran boxes and counted the number of raisins in each box. The Post boxes have a mean of 149 raisins with a sample standard deviation of 26, and the generic boxes have a mean of 109 with a sample standard deviation of 48. To get the pooled estimate of the variance, use formula 10.12:

$$s_{pool}^2 = \frac{(n_1 - 1)s_1^2 + (n_2 - 1)s_2^2}{n_1 + n_2 - 2}$$

(10.12)

$$s_{pool}^2 = \frac{(12 - 1)(26)^2 + (7 - 1)(48)^2}{12 + 7 - 2}$$

$$s_{pool}^2 = 1,250.59$$

To find the standard deviation of the distribution of the difference in sample means, you have to use formula 10.13:

$$s_{\bar{x}_1 - \bar{x}_2} = \sqrt{\frac{s_{pool}^2}{n_1} + \frac{s_{pool}^2}{n_2}}$$

(10.13)

$$s_{\bar{x}_1 - \bar{x}_2} = \sqrt{\frac{1,250.59}{12} + \frac{1,250.59}{7}}$$

$$s_{\bar{x}_1 - \bar{x}_2} = 16.82$$

$s_{\bar{x}_1 - \bar{x}_2}$ would equal 16.82. You are almost ready to construct the confidence interval. To get the t score, you have to know the degrees of freedom. When you pool the variances, the formula for the degrees of freedom is much simpler than the formula for degrees of freedom with small samples and no pooling. The degrees of freedom are $n_1 + n_2 - 2$. The reason for subtracting 2 from the pooled sample size is that you lose one degree of freedom for each population mean that you replace with a sample mean in the formula for the standard deviation. The degrees of freedom equal 17. At a 95% level of confidence, the t score would be 2.110. To get the confidence interval, substitute the t score and $s_{\bar{x}_1 - \bar{x}_2}$ in formula 10.10:

$$L, \ U = \bar{x}_1 - \bar{x}_2 \pm t_{\alpha/2} s_{\bar{x}_1 - \bar{x}_2} \tag{10.10}$$

$$L, \ U = (149 - 109) \pm 2.11(16.82)$$

$$L = 4.51 \qquad U = 75.49$$

Here's how to interpret this confidence interval. If the distributions of raisins per box of Post Raisin Bran and generic raisin bran are normally distributed with equal variances, then at the 95% level of confidence, the estimated mean difference is between 4.51 and 75.49 raisins per box. Since all the values in the confidence interval are positive, you can be fairly sure that the Post Raisin Bran has, on average, more raisins per box.

Let's try an example of a hypothesis test with a pooled variance. Suppose you work for a consumer magazine and you suspect that two brands of suntan lotion, one expensive and the other lower-priced, are made in the same factory out of the same chemicals. If your suspicions are correct, then the mean tanning protection of both brands *and* the variances will be the same. Since analyzing the chemical properties of samples of both brands is too expensive, you plan a test on human skin. Thirty volunteers expose a portion of their anatomy that ordinarily receives no sunlight to an hour under a sun lamp. Fifteen apply brand A lotion on their skin and 15 apply brand B. You use a light meter to measure the amount of light reflected from the skin before and after the exposure. No change in the light meter readings would mean that the suntan lotion provided a perfect screen; the greater the change in the readings, the less protection afforded by the lotion. The mean change in the readings for brand A is -3 with a sample standard deviation of 6, and the mean change in the readings for brand B is -4 with a sample standard deviation of 8.

Your null hypothesis is H_0: $\mu_1 - \mu_2 = 0$. Your alternative hypothesis is H_1: $\mu_1 - \mu_2 \neq 0$. Since brand A could be better than brand B or vice versa, you need a two-tailed test. Let's assume you've tested to see if the population variances are equal and you cannot reject the null hypothesis of equality. To get the pooled variance, you use formula 10.12:

$$s_{pool}^2 = \frac{(n_1 - 1)s_1^2 + (n_2 - 1)s_2^2}{n_1 + n_2 - 2} \tag{10.12}$$

Substituting the values for s_1^2 and s_2^2 from the suntan lotion problem in equation 10.12 yields

$$s_{pool}^2 = \frac{(15 - 1)6^2 + (15 - 1)8^2}{15 + 15 - 2}$$

The pooled variance equals 50. To get the standard deviation of the distribution of the difference in sample means, substitute into formula 10.13:

$$s_{\bar{x}_1 - \bar{x}_2} = \sqrt{\frac{s_{pool}^2}{n_1} + \frac{s_{pool}^2}{n_2}}$$

(10.13)

$$s_{\bar{x}_1 - \bar{x}_2} = \sqrt{\frac{50}{15} + \frac{50}{15}}$$

$$s_{\bar{x}_1 - \bar{x}_2} = 2.582$$

You use formula 10.14 to set up the hypothesis test.

Critical values for a two-tailed hypothesis test on the difference in population means with a pooled estimate of the population standard deviation

	HYPOTHESIZED DIFFERENCE IN MEANS	ESTIMATED STANDARD ERROR OF THE DIFFERENCE IN SAMPLE MEANS

$$\text{C.V.}_L, \text{C.V.}_U = (\mu_1 - \mu_2)_0 \pm t_{\alpha/2} s_{\bar{x}_1 - \bar{x}_2}$$

(10.14)

where $df = n_1 + n_2 - 2$. To do a left- or right-tailed test, use t_α in place of $t_{\alpha/2}$ and $-$ or $+$ in place of $\pm$.

The formula for the critical values of the hypothesis test with pooled variance is about the same as formula 10.11, for unpooled variance. The difference is the degrees of freedom. Let's use a 5% significance level. The Student distribution would have 28 degrees of freedom. Substituting into formula 10.14 yields:

$$\text{C.V.}_L, \text{C.V.}_U = 0 \pm 2.048(2.582)$$

$$\text{C.V.}_L = -5.288 \qquad \text{C.V.}_U = 5.288$$

The observed difference in the sample means is 1, which is in the acceptance region. Thus you cannot reject the null hypothesis that the two lotions have the same effect on the skin.

Let's see how the seven steps described in Figure 9.5 apply to this hypothesis test.

1 Define the null and alternative hypotheses.

A two-tailed test was used because you wanted to pick up any difference between the two lotions.

2 Select the estimator and determine its distribution.

The estimator chosen was the difference in sample means. It will be normally distributed if both population distributions are normal.

3 Choose a significance level for the test.

The customary significance level of 5% was chosen because the costs of making Type I and Type II errors were not specified.

4 Define the decision rule for rejecting or not rejecting the null hypothesis.
The decision rule was to reject the null hypothesis if the difference in sample means was either below the lower critical value or above the upper critical value as defined by formula 10.14. Since the two population distributions were hypothesized to be identical, the pooled estimate of the population variance was used in the formula for the critical values.

5 Perform the computation.
The critical values were plus and minus 5.288.

6 Make the statistical decision.
The null hypothesis that population means were equal could not be rejected.

7 Make the administrative decision.
The consumer magazine would publish its findings and claim that the quality of the lower-priced brand of suntan lotion was equal to that of the more expensive lotion.

Admittedly, the math is complicated for either the pooled or the unpooled versions of the small-sample difference-in-means confidence intervals or hypothesis tests. But remember, there are occasions when each observation drawn for a sample is extremely expensive. If that is the case, it's nice to know that there are some formulas around that will work with small samples. Another point to remember is that these small-sample confidence intervals depend on the normality of the populations. If the sample data look as though they did not come from two normal populations, then you are statistically out of luck. There is no way to construct a confidence interval with small samples. You can, however, use the Wilcoxon-Mann-Whitney test (Chapter 17) for a difference-in-means hypothesis test with small samples and nonnormal populations.

EXERCISES FOR SECTION 10.2

10.14 Let X and Y be two normally distributed random variables with the means μ_x and μ_y and unknown variances. Construct a confidence interval for $\mu_x - \mu_y$ at the 95% level of confidence if:

(a) A sample of size 10 taken from X has a mean of 34 and a standard deviation of 3 and a sample of size 12 taken from Y has a mean of 25 and a standard deviation of 5.

(b) A sample of size 10 taken from X has a mean of 34 and a standard deviation of 3 and a sample of size 12 taken from Y has a mean of 25 and a standard deviation of 5. Assume that the two population variances are equal.

(c) A sample of size 100 taken from X has a mean of 34 and a standard deviation of 3 and a sample of size 120 taken from Y has a mean of 25 and a standard deviation of 5.

(d) A sample of size 5 taken from X has a mean of 34 and a standard deviation of 8 and a sample of size 7 taken from Y has a mean of 25 and a standard deviation of 5.

10.15 Let X and Y be two normally distributed random variables with the means μ_x and μ_y and unknown variances. Construct a confidence interval for $\mu_x - \mu_y$ at the 99% level of confidence if:

(a) A sample of size 12 taken from X has a mean of 24 and a standard deviation of 9 and a sample of size 14 taken from Y has a mean of 25 and a standard deviation of 3.

(b) A sample of size 14 taken from X has a mean of 44 and a standard deviation of 6 and a sample of size 11 taken from Y has a mean of 20 and a standard deviation of 1. Assume that the two population variances are equal.

(c) A sample of size 50 taken from X has a mean of 20 and a standard deviation of 15 and a sample of size 250 taken from Y has a mean of 14 and a standard deviation of 18.

(d) A sample of size 3 taken from X has a mean of 15 and a standard deviation of 4 and a sample of size 8 taken from Y has a mean of 32 and a standard deviation of 7.

10.16 You own a national chain of photographic portrait studios. After the photographs from a particular sitting are developed, you would like your customers to order as many prints as possible. Your problem is to devise methods to encourage the customers to order many copies. You run an experiment based on two methods, the soft sell and the hard sell. The soft sell consists of telling customers how lovely the photographs turned out, what a bargain multiple copies provide, and how much the grandparents would appreciate copies of the pictures. The hard sell consists of telling the customers that the proofs will be burned unless they order copies immediately, that their children won't love them when they grow older if they don't have family portraits to look at, and that one's friends and associates often find out about people who are too cheap to buy copies. You try the soft sell on 12 customers and the hard sell on 10 customers. The respective sample means of orders are $125 and $155; the sample standard deviations are $8 and $10. Assume that the two methods result in distributions of orders with equal standard deviations. Construct a confidence interval at the 95% level of confidence for the difference in population means. Next, construct another confidence interval without assuming that the standard deviations are equal. What assumptions do you have to make to construct either of these confidence intervals?

10.17 Do exercise 10.16 again, using the same sample means and standard deviations but changing the sample sizes to 120 and 100. Construct the confidence interval and describe the changes required as the sample size increases.

10.18 You are an attorney and you represent a firm that is being sued for manufacturing a car with rear brakes that tend to lock suddenly and cause accidents. To support your client's claim that the automobile it manufactures has rear brakes at least as safe as the brakes in cars made by competing firms, you survey 10,000 owners of the car in question and 10,000 owners of cars in the same weight and price range made by competing firms. You get 9,950 responses from the first group and 9,975 from the second group. You ask all respondents how many miles they have on their car and how many times their rear brakes have suddenly locked. You then compute a rate of brake locks per mile by dividing the latter figure by the former. Your client's car had a sample rate of 0.00002 brake lock per mile with a sample standard deviation of 0.000005, while the competing cars had a sample rate of 0.000021 with a sample standard deviation of 0.000002. Construct a confidence interval at the 95% level of confidence for the difference in population rates of brake locks per mile. Before going to trial, you hire a statistician to advise you, and she tells you that something is fatally wrong with the evidence you've put together and that you had better not present it. Why does the statistician tell you this?

10.19 Let X and Y be normally distributed random variables. At the 5% significance level, test the hypothesis that the mean of X equals the mean of Y:

(a) If a sample of size 10 taken from X has a mean of 20 and a standard deviation of 4 and a sample of size 8 taken from Y has a mean of 22 and a standard deviation of 8.

(b) Do part a again, this time assuming that the two population standard deviations are equal.

(c) If a sample of size 6 taken from X has a mean of 20 and a standard deviation of 7 and a sample of size 8 taken from Y has a mean of 25 and a standard deviation of 8.

(d) Do part c again, this time assuming that the two population standard deviations are equal.

10.20 Let X and Y be normally distributed random variables. At the 1% significance level, test the hypothesis that the mean of X is less than or equal to the mean of Y:

(a) If a sample of size 10 taken from X has a mean of 20 and a standard deviation of 4 and a sample of size 8 taken from Y has a mean of 22 and a standard deviation of 8.

(b) Do part a again, this time assuming that the two population standard deviations are equal.

(c) If a sample of size 6 taken from X has a mean of 20 and a standard deviation of 7 and a sample of size 8 taken from Y has a mean of 25 and a standard deviation of 8.

(d) Do part c again, this time assuming that the two population standard deviations are equal.

10.21 Test the hypothesis that the mean number of minutes it takes to be served a hamburger, a soft drink, and fries at McDonald's is less than the mean number of minutes it takes to be served the same items at Burger King. The data for the test are: on 10 visits to McDonald's you have been served in an average time of 1.3 minutes with a standard deviation of 0.2 minute, and on 15 visits to Burger King you have been served in an average time of 1.2 minutes with a standard deviation of 0.4 minute. Assume that both population distributions are normal and use a 5% significance level.

10.22 Do exercise 10.21 again, this time assuming that the two population standard deviations are equal.

10.23 On 100 round trips between your home and your workplace, the sample mean time for the trip to work was 33 minutes with a standard deviation of 3 minutes and that of the trip home was 35 minutes with a standard deviation of 8 minutes. You hypothesize that the difference between the time it takes to drive to work and the time it takes to drive home is zero. Test this hypothesis at the 5% significance level.

10.24 The Cummins Engine Company has designed a new diesel truck engine in which ceramic technology is used for the pistons and the cylinder walls. The ceramic engine can burn much hotter than a conventional metal engine, and this higher temperature leads to more efficient use of fuel. One of the design goals of the project was to ensure that the ceramic engine would last at least as long as the conventional metal truck engine made by Cummins. Cummins has bench-tested 10 of the ceramic engines and 10 of the conventional engines until each engine failed. The average time to failure was 6,500 hours for the ceramic engines and 6,300 hours for the conventional engines. The respective sample standard deviations are 150 and 200 hours. Assume that the population distribution of time to engine failure is approximately normal. Test at the 5% significance level the hypothesis that the mean time to failure for the ceramic engine exceeds the mean time to failure for the conventional engine. Is the assumption of normality for the distribution of the time to failure reasonable?

10.25 A hotel chain gives its maids small chocolate mints to place on the pillows after a room has been cleaned. An executive of the hotel is curious as to whether these chocolates affect the tips that guests leave for the maids. Since maids may be reluctant to reveal what they pick up in tips, the executive cleans 20 rooms, leaving the customary chocolates in only 10 of them. The average tip in the chocolateless rooms was $3.00 with a sample standard deviation of 50 cents. The average tip in the rooms with the chocolate mints was $3.25 with a sample standard deviation of 75 cents. The mints cost the hotel 10 cents each. Test at the 5% significance level the hypothesis that the mints add more than 10 cents to the maid's average tip. Assume that the distributions of tips with and without the chocolates are normal and have equal variances. If the chocolates did not cover their cost in adding to the maid's tips, would this be a sufficient reason to drop the chocolates?

10.26 Your company manufactures tractors and sells them worldwide. You would like to sell a special low-cost tractor in India, but imported tractors face a stiff tariff set by the Indian government. The Indian tariff on tractor parts, however, is zero. Because of this differential tax treatment, you are considering first assembling the tractors in the United States, then breaking them down and shipping the tractors in parts, and finally reassembling them in India. To see which arrangement is more favorable, you assemble ten tractors in the United States and ship them to India as completed tractors and also have ten tractors assembled in India by an Indian contractor. The extra tariff per U.S.-finished tractor is 10 million rupees. If your costs for the U.S.-finished tractors were more than 10 million rupees less than the costs for the tractors assembled in India, you would be better off paying the high tariff. The mean cost for the ten U.S.-finished tractors is 14 million rupees with a sample standard deviation of 3 million rupees. These figures are net of the tariff. The mean cost for the ten Indian-assembled tractors is 25.5 million rupees with a sample standard deviation of 7 million rupees. Test at the 5% significance level the hypothesis that there is more than a 10-million–rupee spread between the mean costs of Indian-assembled tractors and U.S.-assembled tractors.

10.3 CONFIDENCE INTERVALS AND HYPOTHESIS TESTS ON THE DIFFERENCE BETWEEN TWO POPULATION PROPORTIONS

10.3A CONFIDENCE INTERVALS ON THE DIFFERENCE BETWEEN TWO POPULATION PROPORTIONS

You are the subscription manager for a magazine and you are considering two plans designed to encourage your subscribers to renew their subscriptions. Plan A is a low-cost plan that involves sending renewal notices to subscribers shortly before their subscriptions expire. Plan B is a high-cost plan; you will offer a premium, such as a record album, a digital watch, or a silk-screened T-shirt with your magazine's logo, to subscribers who renew by a certain date. You are sure that the high-cost plan will induce a greater proportion of the subscribers to renew, but you would like to estimate the difference between the proportions of subscribers who renew under plans A and B. After all the whistles and bells in section 10.2, you'll be pleased to know that constructing confidence intervals on the difference between two population proportions is straightforward. If the distribution of sample proportions for the first population is normal and the distribution of the sample proportions for the second population is also normal, then the sampling distribution of the difference in sample proportions will also be normal. The rule for using the normal approximation for the sampling distribution of the proportion is that $n\bar{p}$ and $n(1 - \bar{p})$ are greater than or equal to 5. In the two-sample case, $n_1\bar{p}_1$, $n_2\bar{p}_2$, $n_1(1 - \bar{p}_1)$, and $n_2(1 - \bar{p}_2)$ all must be greater than or equal to 5. The expected value of the sampling distribution of the difference in proportions, $E(\bar{p}_1 - \bar{p}_2)$, is $(\Pi_1 - \Pi_2)$, the difference in population proportions. The **sampling distribution of the difference in proportions** is defined as *the probability density function of the difference between the proportions of successes in two independent samples.*

    Let's say that you send out 100 renewal notices to randomly selected subscribers whose subscriptions are about to expire and at the same time send out 100 premium offers to a second randomly selected group whose subscriptions are about to expire. Suppose 40 of the 100 people who receive letters renew and 77 of the 100 people who are offered premiums renew. To construct a confidence interval for the difference in population proportions, use formula 10.15:

Confidence interval for the difference in population proportions

$$L,\ U = \bar{p}_1 - \bar{p}_2 \pm z_{\alpha/2} s_{\bar{p}_1 - \bar{p}_2}$$

(10.15)

$s_{\bar{p}_1 - \bar{p}_2}$ is similar to $\sigma_{\bar{x}_1 - \bar{x}_2}$. The terms $\bar{p}_1(1 - \bar{p}_1)$ and $\bar{p}_2(1 - \bar{p}_2)$ take the place of σ_1^2 and σ_2^2 in formula 10.3:

$$s_{\bar{p}_1 - \bar{p}_2} = \sqrt{\frac{\bar{p}_1(1 - \bar{p}_1)}{n_1} + \frac{\bar{p}_2(1 - \bar{p}_2)}{n_2}}$$

(10.16)

$$s_{\bar{p}_1 - \bar{p}_2} = \sqrt{\frac{0.4(1 - 0.4)}{100} + \frac{0.77(1 - 0.77)}{100}}$$

$$s_{\bar{p}_1 - \bar{p}_2} = 0.065$$

$$L,\ U = \bar{p}_1 - \bar{p}_2 \pm z_{\alpha/2} s_{\bar{p}_1 - \bar{p}_2}$$

(10.15)

$$L,\ U = (0.4 - 0.77) \pm 1.96(0.065)$$

$$L = -0.497 \qquad U = -0.243$$

The interpretation of the confidence interval is as follows: at the 95% level of confidence, plan A is estimated to result in renewals between 49.7 to 24.3 percentage points lower than plan B. Whether plan A or B is a better deal for the magazine depends on the difference between the costs of the plans on a per renewal basis and on the value to the magazine of each renewal.

 10.3B COMPARISON OF DIFFERENCE-IN-PROPORTIONS AND DIFFERENCE-IN-MEANS CONFIDENCE INTERVALS

There are several reasons that difference-in-proportions confidence intervals are easier than difference-in-means confidence intervals. One reason is that there is no way to compute difference-in-proportions confidence intervals with small samples. "Small" in this context is not an n of less than 30 but an $n\bar{p}$ or $n(1 - \bar{p})$ of less than 5; that is, the normal approximation can't be used. Another reason is that the question of whether or not to pool the variances does not apply. With binary data (proportions), two sampling distributions that have the same variances have the same means or the mean of one must be 1 minus the mean of the other. There is a one-to-two correspondence from variances to means. The variance for the sample proportion is $\sigma_{\bar{p}}^2 = \Pi(1 - \Pi)/n$. If you know $\sigma_{\bar{p}}^2$, you know that Π can take only two values. For example, if σ^2 was 0.0021 and n was 100, Π could be either 0.7 or 0.3; that is, $\sigma_{\bar{p}}^2$ equals either $0.7(1 - 0.7)/100$ or $0.3(1 - 0.3)/100$. In the two-sample case, pooling the sample variances into one estimate of the population variance would be justified if you knew that the two population proportions were identical or if you knew that the two population proportions summed to 1. Of course, if you knew either of these things, you would have little reason to construct a confidence interval to estimate the difference between the population proportions. In other words, *don't pool the sample variances for difference-in-proportions confidence intervals*.

10.3C HYPOTHESIS TESTS ON THE DIFFERENCE BETWEEN TWO POPULATION PROPORTIONS

We'll start this section with an example. You own a retail furniture store and in the course of your business some of your customers have given you checks that the bank refused to honor because there were insufficient funds in your customers' accounts. You are considering two methods of collecting these debts: you can either hire an attorney to send a letter threatening a lawsuit or send a dunning letter on your company's stationery. The attorney charges a percentage of the money collected but you know that a letter from an attorney is more effective than a letter from your store. You have decided that if the proportion of debts that are paid in response to the attorney's letter is greater than 20 percentage points above the proportion paid in response to the store letters, you will be better off using the attorney exclusively. You want to test at the 5% significance level the hypothesis that the difference between the success rates is 20 or more percentage points in favor of the attorney. Define the population proportion of successes for the attorney as Π_1 and the population proportion of successes for the dunning letters as Π_2. The null and alternative hypotheses are:

$$H_0: \Pi_1 - \Pi_2 \leq 0.2 \qquad H_1: \Pi_1 - \Pi_2 > 0.2$$

You send out 100 dunning letters on your letterhead and have the attorney send out 100 letters to other customers on his letterhead. Of the 100 customers who are sent your letter, 35 pay up, and of the 100 who are sent the attorney's letter, 58 pay up. Your formula for the hypothesis test is

Critical value for a right-tailed test on the difference in population proportions

(10.17)

HYPOTHESIZED DIFFERENCE IN POPULATION PROPORTIONS

ESTIMATED STANDARD DEVIATION OF THE DIFFERENCE IN SAMPLE PROPORTIONS

$$\text{C.V.} = (\Pi_1 - \Pi_2)_0 + z_\alpha s_{\bar{p}_1 - \bar{p}_2}$$

To do a left-tailed test, substitute $-$ for $+$ in this formula. A two-tailed test requires the use of $\pm$ in place of $+$ and $z_{\alpha/2}$ in place of z_α.

Recall that the expected value of the sampling distribution of the difference in proportions, $E(\bar{p}_1 - \bar{p}_2)$, is $(\Pi_1 - \Pi_2)$, the difference in population proportions. The expression $(\Pi_1 - \Pi_2)_0$ in formula 10.17 is the hypothesized difference in population proportions and the last term in the formula for estimated standard deviation of the difference in sample proportions is from formula 10.16:

$$s_{\bar{p}_1 - \bar{p}_2} = \sqrt{\frac{\bar{p}_1(1 - \bar{p}_1)}{n_1} + \frac{\bar{p}_2(1 - \bar{p}_2)}{n_2}} \qquad (10.16)$$

Substituting into equation 10.17 yields

$$\text{C.V.} = (\Pi_1 - \Pi_2)_0 + z_\alpha s_{\bar{p}_1 - \bar{p}_2} \qquad (10.17)$$

$$\text{C.V.} = 0.2 + 1.64\sqrt{\frac{0.58(0.42)}{100} + \frac{0.35(0.65)}{100}}$$

$$\text{C.V.} = 0.313$$

Figure 10.4 shows the template for this hypothesis test.

Since the observed difference in sample proportions is 0.23, you cannot reject the null hypothesis that the true difference is less than or equal to 20 percentage points. In this case the administrative decision would be to send your own dunning letters.

Let's see how the seven steps described in Figure 9.5 apply to this hypothesis test.

1 Define the null and alternative hypotheses.

A right-tailed test was used because you had to be convinced that the proportion of successes from the attorney was at least 20 percentage points above the proportion of successes from dunning letters on your own stationery.

2 Select the estimator and determine its distribution.

The estimator chosen was the difference in sample proportions. It will be approximately normally distributed if the sampling distributions for both propor-

FIGURE 10.4
Acceptance and rejection regions for hypothesis test on two types of dunning letters

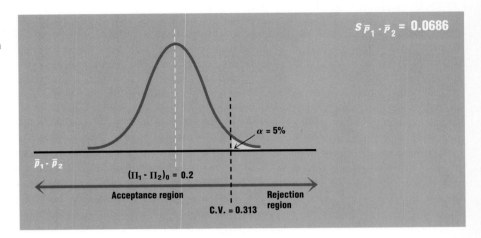

tions are approximately normally distributed. That is, if $n_1\bar{p}_1$, $n_1(1 - \bar{p}_1)$, $n_2\bar{p}_2$, and $n_2(1 - \bar{p}_2)$ all are greater than or equal to 5.

3 Choose a significance level for the test.
The customary significance level of 5% was chosen because the costs of making Type I and Type II errors were not specified.

4 Define the decision rule for rejecting or not rejecting the null hypothesis.
The decision rule was to reject the null hypothesis if the difference in sample proportions was above the critical value as defined by formula 10.17.

5 Perform the computation.
The critical value was 0.313.

6 Make the statistical decision.
The null hypothesis that the difference in population proportions was less than or equal to 0.2 could not be rejected.

7 Make the administrative decision.
You would use your own dunning letters.

■ **10.3D POOLING THE VARIANCES IN A DIFFERENCE-IN-PROPORTIONS HYPOTHESIS TEST**

The issue of pooling variances also comes up in difference-in-proportions tests. The decision rule on whether to pool or not is simpler for difference-in-proportions tests than it is for difference-in-means tests. If your null hypothesis is that the difference in population proportions is (1) zero, (2) less than or equal to zero, or (3) greater than or equal to zero, then you must pool. If the null is centered on any value other than zero, don't pool. What's going on here is that the assumption that the two populations have the same proportions implies that the distributions of sample proportions have the same variances given

GETTING THE ENVELOPE OPENED

Direct-mail advertisers are in an ideal position to use statistical tests. The mailing lists are all computerized and it is a simple matter to draw random samples from these lists and to test different approaches. A direct-mail advertiser's first problem is to get consumers to open the envelopes and read the sales literature. Many consumers throw away unopened any envelope that looks like a sales promotion. Advertisers constantly experiment with the outside of their envelopes to make them irresistible to consumers. The reason for the constant experimentation is that after awhile, consumers learn each new gimmick. One of these gimmicks is to make the envelopes look as though they were sent by a government agency. The envelope is made of brown paper and the return address says something like *United States Committee for Internal Revenue*. Beneath the return address is some tiny print to the effect that this committee has no connection with any government agency. Another gimmick is to stamp the words URGENT—DATED MATERIAL in red across the front of the envelope. The third gimmick is to use an envelope with a tab that the consumer has to pull or tear. Even something as simple as tilting the printed bulk-mail permit seems to increase the response rate.

Some direct-mail advertisers have found that many consumers will not open anything sent as bulk mail. The question of whether first-class postage is cost-effective can easily be answered by a statistical test. Suppose a direct-mail marketer planned to send out a million letters to a list of consumers who had previously purchased items through the mail. The letters would contain an advertisement for a porcelain Christmas-tree ornament.

equal sample sizes. The formula for a pooled estimate of the two population proportions is

$$\bar{p}_{\text{pool}} = \frac{n_1\bar{p}_1 + n_2\bar{p}_2}{n_1 + n_2}$$

(10.18)

Here's an example that illustrates pooling for a difference-in-proportions hypothesis test. A scrap metal firm has found that it has very high workplace injury rates and that its spending on workers' compensation insurance far exceeds the industry average. A consultant from the firm's insurance carrier inspects the workplace and suggests that alcohol consumption may be the cause of the high injury rate. The consultant masquerades as a new worker and he finds that many employees go to bars and drink during the lunch hour. The scrap metal firm has no on-site cafeteria. As an experiment, some of the work-

If the response rate for the letters sent with first-class stamps was more than 5 percentage points above the response rate for letters sent by bulk mail, the higher cost of the first-class stamps would be justified. The null and alternative hypotheses would be $H_0: \Pi_1 - \Pi_2 \le 0.05$ and $H_1: \Pi_1 - \Pi_2 > 0.05$. The Π_1 refers to the population proportion of successes of envelopes with first class stamps and Π_2 refers to the population proportion of successes of envelopes with bulk mail permits. A small pilot mailing would provide an accurate test. Let's use a 5% significance level and a sample size of 200 for each type of mail. Suppose the proportion of bulk-mail letters that were successful was 6% and the proportion of first-class letters that were successful was 13%. Could the null hypothesis be rejected? Substituting into formula 10.17:

$$\text{C.V.} = (\Pi_1 - \Pi_2)_0 + z_\alpha s_{\bar{p}_1 - \bar{p}_2} \tag{10.17}$$

$$\text{C.V.} = 0.05 + 1.64\sqrt{\frac{0.06(0.94)}{200} + \frac{0.13(0.87)}{200}}$$

$$\text{C.V.} = 0.098$$

In this case the observed difference of 7 percentage points (13 less 6) is not enough evidence to lead you to reject the null hypothesis that first-class stamps are not cost-effective. This result would apply only to these particular circumstances. Another advertisement might get a better response with first-class stamps, or another mailing list might get a better response. Direct mail advertising provides a never-ending field for trying statistical tests.

ers, selected at random, are given coupons good for free lunches at nearby fast-food restaurants that do not serve alcohol. The proportion of worker–days that result in an injury for workers with the coupons is 0.02. There are 1,000 coupon worker–days observed in this experiment. During the same period there were 2,000 worker–days without coupons and the proportion of these days that resulted in injuries was 0.03. The scrap metal firm wants to test at the 5% significance level the hypothesis that the coupons had no effect in reducing the proportion of worker–days with injuries. If Π_1 is the population proportion of injuries with lunch coupons and Π_2 is the population proportion of injuries without coupons, the null and alternative hypotheses are

$$H_0: \Pi_1 - \Pi_2 \ge 0 \qquad H_1: \Pi_1 - \Pi_2 < 0$$

Since the test is centered on a hypothesized difference of 0, you have to pool the sample proportions. The pooled estimate of the population proportion of worker–days with injuries is

$$\bar{p}_{\text{pool}} = \frac{n_1\bar{p}_1 + n_2\bar{p}_2}{n_1 + n_2} \tag{10.18}$$

$$\bar{p}_{\text{pool}} = \frac{1,000(0.02) + 2,000(0.03)}{1,000 + 2,000}$$

$$\bar{p}_{\text{pool}} = 0.0267$$

Use $\bar{p}_{\text{pool}}$ in place of $\bar{p}_1$ and $\bar{p}_2$ in formula 10.16. The rest of the test is the same as before.

$$\text{C.V.} = (\Pi_1 - \Pi_2)_0 - z_\alpha s_{\bar{p}_1 - \bar{p}_2} \tag{10.17}$$

$$\text{C.V.} = 0 - 1.64\sqrt{\frac{0.0267(1 - 0.0267)}{1,000} + \frac{0.0267(1 - 0.0267)}{2,000}}$$

$$\text{C.V.} = -0.01$$

The critical value is now -0.01. The null hypothesis of a true difference in proportions of less than zero is just barely rejected at the 5% significance level because the observed difference in proportions was -0.01. While the coupons seem to have a slight, albeit statistically significant, effect in reducing injuries, it is not clear that the firm would maximize profits by using them. From a bottom-line point of view, the cost of the coupons would have to be made up by a reduction in the costs of injuries. Since the setup for this example is so close to the last hypothesis (the only difference being the pooling), we'll skip the review of the steps for a hypothesis test.

EXERCISES FOR SECTION 10.3

10.27 Let X be a random variable that can have the value 0 for a failure and 1 for a success and let the probability of a success be 0.5. Let Y be a random variable that can have the value 0 for a failure and 1 for a success and let the probability of a success be 0.49.

(a) If a sample of size 50 is taken from the population that makes up X and a sample of 100 is taken from the population that makes up Y, what is the probability that the proportion of successes in the X group will exceed the proportion of successes in the Y group?

(b) If a sample of size 20 is taken from the population that makes up X and a sample of 30 is taken from the population that makes up Y, what is the probability that the proportion of successes in the X group will be less than the proportion of successes in the Y group?

(c) If a sample of size 50 is taken from the population that makes up X and a sample of 100 is taken from the population that makes up Y, what is the probability that the absolute value of the difference in proportions of successes will be less than 0.01?

(d) If a sample of size 25 is taken from the population that makes up X and a sample of 35 is taken from the population that makes up Y, what is the probability that the absolute value of the difference in the proportion of successes will be greater than 0.05?

10.28 Construct a confidence interval at the 90% level of confidence for the difference in the proportion of successes in two populations:

(a) If a sample of size 100 taken from the first population had 30 successes and a sample of size 80 from the second population had 20 successes.

(b) If a sample of size 60 taken from the first population had 30 successes and a sample of size 80 from the second population had 40 successes.

(c) If a sample of size 60 taken from the first population had 20 successes and a sample of size 60 from the second population had 30 successes.

(d) If a sample of size 20 taken from the first population had 10 successes and a sample of size 60 from the second population had 40 successes.

10.29 Your company produces television sets in several plants in the United States. You have had problems with quality control. Your television sets are getting a bad word–of–mouth reputation and too many sets are being returned under the terms of the warranty. You have devised a plan to improve quality control. The plan involves keeping a record that identifies the employee who assembled every component in every television set shipped. If a set is returned under the warranty, the defect is traced back to the employee who was responsible and that employee is required to repair the set while everyone else in the plant watches. You test the plan by trying it out on the day shift at one plant and comparing the results with the night shift. Before you instituted the plan, the day and night shifts had the same proportions of defects. You know that your plan will reduce the proportion of defects but it will also increase record-keeping costs and you may have to raise wages to hold on to employees who regard the plan as a deterioration in working conditions. To see if the plan will be cost-effective, you need to estimate the difference in the proportions of defective sets under the plan and under the old quality control system. Suppose that out of 5,580 TV sets shipped out by the day shift you observed 136 defective sets that were returned under the warranty, and that out of 5,630 TV sets shipped by the night shift you observed 224 defective sets. Estimate the difference in population proportions using a 95% level of confidence.

10.30 You run the payroll and accounting office for a large firm. For each hourly employee, your clerks key into the company computer the straight-time and overtime hours. These figures are used to compute the weekly payroll. If there is a keypunching error in favor of the company, the affected employee is likely to notice it and bring the error to the attention of your department. If the error is in favor of the employee, however, he or she is much less likely to complain. To hold down the number of keypunching errors, you have a high-cost program and a low-cost program. In the high-cost program, every figure is keypunched twice by different clerks and then the two sets of figures are compared. Any disparities are resolved by a review of the employee's time cards. In the low-cost program, a computer program flags any unusual number of hours, "unusual" being either much higher or lower than an employee's historic average. To see if the high-cost program is worthwhile, you have 500 paychecks reviewed under the high-cost program and another 500 paychecks reviewed under the low-cost program. The proportion of checks found to be in error by the high-cost program is 0.05 while the proportion of checks found to be in error by the low-cost program is 0.03. Construct a confidence interval at the 99% level of confidence for the difference in the proportions of errors detected under the two programs.

10.31* You work for an insurance company that sells homeowners' insurance. Your company plans to offer a discount to owners of new homes (five years old or newer) because new homes are less likely to have fires. To get an idea of how large a discount the company can afford to offer, you need an estimate of the difference in the proportions of newer and older homes that have fires. At the 95% level of confidence, find the lower limit of a one-sided confidence interval for the difference in population proportions if 25 out of 1,000 newer homes insured by your company have fires and 55 out of 1,100 older homes insured by your company have fires. Explain what this confidence interval means.

10.32 Suppose that you hypothesized that the difference in the proportions of men and women who use newspaper coupons when they shop for groceries was −0.2. You have a sample of 100 male grocery shoppers and 30 of them use coupons. You also have a sample of 200 female grocery shoppers and 120 of them use coupons. Test the hypothesis at a 5% significance level.

10.33 Do exercise 10.32 again, this time testing the null hypothesis that the true difference in proportions is zero.

10.34 Test the hypothesis that the difference in population proportions of senior citizens and non–senior citizens who travel by air at least once a year is equal to zero at the 1% significance level if 300 out of a sample of 1,000 senior citizens have traveled at least once in the past year and 450 out of a sample of 1,100 non–senior citizens have also traveled at least once in the past year.

10.4 HYPOTHESIS TESTS ON THE DIFFERENCE BETWEEN TWO POPULATION MEANS WITH PAIRED SAMPLES

Let's go back to the suntan lotion problem in section 10.2c. It's possible that the experiment of giving 15 people lotion *A* and 15 people lotion *B* could give inaccurate results. By chance a few of the people in the *A* group could be more sensitive to sunlight than the average person and a few people in the *B* group could be less sensitive to sunlight than the average person. What you think is a difference in the sun-screening ability of the two lotions may be a difference in the mean sensitivity to sunlight in your two groups. If the two groups were large and they were randomly selected, you would not have to worry about any difference between the two groups, but with only 15 people in each group, even one or two individuals who have atypical sensitivity can make the two groups unequal. A solution for this problem is to give each person in the experiment both lotions, one on the left side of the body and one on the right. If you did this, you would eliminate the possibility that the results could be biased by having two groups with different skin sensitivities.

Giving both lotions to each person makes the results for the A and B lotions dependent. Sensitive skins will show a large change for both lotions and non-sensitive skins will show a small change for both lotions. You won't have the 28 degrees of freedom that you had in the previous suntan lotion test. Recall that the degrees of freedom were $n_1 + n_2 - 2$ because one degree of freedom was lost when the sample mean of the first group was used to estimate the population mean of the first population and another degree of freedom was lost when the sample mean of the second group was used to estimate the population mean of the second population. With dependent samples your test is on *the mean difference in the pairs of observations.* **Dependent samples** (also called **paired, correlated,** or **matched samples**) are *samples in which there is statistical dependence between the pairs of observations drawn from two populations.* With 30 pairs of observations you will have 29 degrees of freedom. One degree of freedom will be lost because you have to use the sample mean of the paired differences to estimate the population mean of the paired differences. Let's define the symbol **delta** (δ) as *the mean difference between all pairs of observations taken from two populations.* In this example your null hypothesis is that δ equals zero and your alternative hypothesis is that δ does not equal zero. Table 10.1 lists the values for each pair of observations and the differences for each pair. In math notation the hypotheses are:

$$H_0: \delta = 0 \qquad H_1: \delta \neq 0$$

Let's use d_i to represent the sample differences. The mean of the sample differences would be

$$\bar{d} = \frac{\sum\limits_{i=1}^{n} d_i}{n}$$

(10.19)

$$\bar{d} = \frac{33}{30}$$

$$\bar{d} = 1.1$$

TABLE 10.1	Individual	Lotion A	Lotion B	$A - B$	$d_i - \bar{d}$	$(d_i - \bar{d})^2$
Changes in light	1	−6	−6	0	−1.1	1.21
meter readings for	2	−2	−3	1	−0.1	0.01
lotions A and B	3	−5	−7	2	0.9	0.81
	4	−1	−1	0	−1.1	1.21
	5	−3	−4	1	−0.1	0.01
	6	−3	−3	0	−1.1	1.21
	7	−8	−9	1	−0.1	0.01
	8	−5	−7	2	0.9	0.81
	9	−4	−7	3	1.9	3.61
	10	−2	−3	1	−0.1	0.01
	11	−4	−8	4	2.9	8.41
	12	−4	−5	1	−0.1	0.01
	13	−1	−1	0	−1.1	1.21
	14	−4	−5	1	−0.1	0.01
	15	−5	−6	1	−0.1	0.01
	16	−2	−2	0	−1.1	1.21
	17	−3	−5	2	0.9	0.81
	18	−5	−7	2	0.9	0.81
	19	−4	−6	2	0.9	0.81
	20	−6	−6	0	−1.1	1.21
	21	−3	−3	0	−1.1	1.21
	22	−6	−8	2	0.9	0.81
	23	−1	−4	3	1.9	3.61
	24	−2	−2	0	−1.1	1.21
	25	−3	−4	1	−0.1	0.01
	26	−4	−4	0	−1.1	1.21
	27	−6	−7	1	−0.1	0.01
	28	−8	−9	1	−0.1	0.01
	29	−5	−6	1	−0.1	0.01
	30	−4	−4	0	−1.1	1.21
				$\overline{33}$		$\overline{32.70}$

Note that n is the number of pairs. The mean difference is 1.1.

The estimated population standard deviation for the distribution of the paired differences, s_d, is

$$s_d = \sqrt{\frac{\sum_{i=1}^{n}(d_i - \bar{d})^2}{n-1}}$$

(10.20)

The standard error of the paired difference equals

$$s_{\bar{d}} = \frac{s_d}{\sqrt{n}}$$

(10.21)

The distinction between s_d and $s_{\bar{d}}$ parallels the distinction between s, the estimated population standard deviation, and $s_{\bar{x}}$, the standard error of the mean. The term s_d is the estimated standard deviation of d_i distribution, where the d_i's are the paired differences in observations. The term $s_{\bar{d}}$ is the estimated standard deviation of $\bar{d}$ distribution, the mean of the paired differences in a sample. In this example $s_{\bar{d}}$ equals 0.1939. This result is shown in the calculations below:

$$s_d = \sqrt{\frac{\sum_{i=1}^{n}(d_i - \bar{d})^2}{n-1}}$$

(10.20)

$$s_d = \sqrt{\frac{32.70}{30-1}}$$

$$s_d = 1.062$$

$$s_{\bar{d}} = \frac{s_d}{\sqrt{n}}$$

(10.21)

$$s_{\bar{d}} = \frac{1.062}{\sqrt{30}}$$

$$s_{\bar{d}} = 0.1939$$

The value for the sum of the squared deviations around $\bar{d}$ was taken from Table 10.1. Now you can do the hypothesis test. The formula you need is

Critical values of a two-tailed hypothesis test on the difference in population means with paired samples

HYPOTHESIZED
DIFFERENCE
BETWEEN MEANS

(10.22)

$$\text{C.V.}_L, \text{C.V.}_U = \delta_0 \pm t_{\alpha/2} s_{\bar{d}}$$

STANDARD ERROR
OF THE DIFFERENCE BETWEEN
PAIRED SAMPLE MEANS

where $df = n - 1$. To do a left- or right-tailed test, use t_α in place of $t_{\alpha/2}$ and $-$ or $+$ in place of $\pm$.

The symbol δ_0 is the hypothesized difference between the population means. Substituting 0 for δ_0 and using a 5% significance level, we get

$$\text{C.V.}_L, \text{C.V.}_U = 0 \pm 2.045(0.1939)$$

$$\text{C.V.}_L = -0.397 \qquad \text{C.V.}_U = 0.397$$

Since the observed mean difference was 1.1, you should reject the null hypothesis and conclude that the lotions do not have the same sun-screening effect. Apparently lotion B does a better job.

Let's see how the seven steps described in Figure 9.5 apply to this hypothesis test.

1 Define the null and alternative hypotheses.
A two-tailed test was used because you wanted to see if there was any difference between the two lotions.

2 Select the estimator and determine its distribution.
The estimator chosen was the mean difference between pairs of observations. It will be normally distributed if both population distributions are normally distributed.

3 Choose a significance level for the test.
The customary significance level of 5% was chosen because the costs of making Type I and Type II errors were not specified.

4 Define the decision rule for rejecting or not rejecting the null hypothesis.
The decision rule was to reject the null hypothesis if the sample mean was below the lower critical value or above the upper critical value as defined by formula 10.22.

5 Perform the computation.
The critical values were plus and minus 0.397.

6 Make the statistical decision.
The null hypothesis that the two population means were equal was rejected.

7 Make the administrative decision.
The magazine would advise its readers that lotion A provided less protection from the sun.

EXERCISES FOR SECTION 10.4

10.35 Let X and Y be normally distributed random variables with the means μ_X and μ_Y. The following pairs of dependent observations were drawn from the two distributions:

x_i: 14, 44, 68, 90, 120
y_i: 44, 64, 92, 99, 150

(a) At the 5% significance level, test the hypothesis that μ_X minus μ_Y is less than -20.

(b) At the 5% significance level, test the hypothesis that μ_X minus μ_Y equals -20.

(c) At the 5% significance level, test the hypothesis that μ_X minus μ_Y exceeds -20.

(d) At the 5% significance level, test the hypothesis that μ_X minus μ_Y is not equal to -20.

10.36 Let X and Y be normally distributed random variables with the means μ_X and μ_Y. The following pairs of dependent observations were drawn from the two distributions:

x_i: 14, 16, 20, 21, 25
y_i: 10, 8, 12, 14, 15

(a) At the 10% significance level, test the hypothesis that μ_X minus μ_Y is less than 5.

(b) At the 10% significance level, test the hypothesis that μ_X minus μ_Y equals 5.

(c) At the 10% significance level, test the hypothesis that μ_X minus μ_Y exceeds 5.

(d) At the 10% significance level, test the hypothesis that μ_X minus μ_Y is not equal to 5.

10.37 A tire company wants to test at the 5% significance level the hypothesis that the tread wear on a new type of tire will be less than the tread wear on an old model in 30,000 miles. Ten cars are used for the test, with the left front wheel of each car carrying the new design and the right front wheel of each car carrying the old design. Use the following data to conduct the test. Assume that the distributions of tread wear for both types are normal.

Millimeters of tread wear

Old design	New design	Old design	New design
1.4	1.2	3.4	2.5
2.1	2.0	2.5	2.3
1.7	1.7	1.7	1.5
2.9	2.6	2.4	2.1
1.0	1.1	2.0	1.6

10.38 A company is looking for ways to reduce its medical insurance and it considers offering free hypnosis to employees who are interested in reducing the consumption of cigarettes. Before offering the hypnosis to all employees, the company tries a small pilot study in which six employees record their cigarette consumption over a two-month period, with a hypnosis session occurring at the end of the first month. Use the following data to test at the 5% significance level the hypothesis that cigarette smoking is reduced by hypnosis.

Number of cigarettes consumed

First month	Second month
900	650
545	524
1,200	0
1,400	1,500
350	340
890	210

10.39 A new type of crank for bicycles called the Powercam™ relies on a pushing motion rather than a circular motion. The company that markets this crank plans a test of five riders who will travel a 100-mile route once with the Powercam™ and once with a conventional crank. The company would like to show that the new crank reduces the time needed to cover the 100 miles. Use a 5% significance level and assume that both population distributions are normal. The times in hours for each rider are as follows:

	Rider				
	1	**2**	**3**	**4**	**5**
Powercam™	4.4	4.6	4.7	4.8	4.9
Conventional crank	5.1	5.3	5.1	4.9	5.2

10.40 You run a grocery store and you are considering offering a 10% discount to senior citizens. You plan an experiment to see if this discount increases sales per senior citizen. The experiment consists of observing the spending of six randomly selected senior citizens and then inviting them to come back to your store at the end of a month and shop with a 10% discount. Assume that the distributions of spending before and during the discount are normal. Use the following data to test your hypothesis at the 5% significance level. Also, can you find any flaws in this experimental design?

Shopper	Prediscount sale	Discount sale
1	$35	$40
2	25	35
3	40	40
4	45	40
5	35	40
6	20	20

10.41 You run a secretarial pool and you are converting from regular typewriters to microcomputers with word processing programs. You are considering two word processing programs: Wordbuster and Ineptwriter. To see which program is faster, you randomly select five secretaries from your pool and have them type the same test letter with each program. The results are recorded below. Test at the 5% significance level the hypothesis that Wordbuster is faster than Ineptwriter. Assume that the distributions of times for typing the letters with both programs are normal.

Secretary	Minutes with Wordbuster	Minutes with Ineptwriter
1	6	7
2	4	6
3	8	8
4	5	6
5	4	7

Also, can you find any flaws in this experimental design?

10.42 The U.S. Army is considering replacing the laser aiming system used for the 105-millimeter canon in the Abrahms tank. To test the new system against the old system, a tank is run through a course twice at high speed while firing shells at a series of targets. The first time through the course the tank has the old aiming system and the second time it has the new system. The distance between each target and the striking point of the shell is recorded during the test. The mean difference is 8 feet in favor of the new system and the standard error of the sample differences is 3 feet. There are 24 targets in the course.

(a) Test at the 5% significance level the hypothesis that the new system is more accurate than the old system.

(b) Why is this test a paired difference-in-means test?

(c) Are there any flaws in this research design?

 SUMMARY

Computers can be very helpful when you have to do a hypothesis test for two populations. Many statistical programs automatically give the results for both a one-tailed and a two-tailed test, give the P value, and give the results with and without pooled variance and with and without paired samples. All of these results save you computation time but they don't save you thinking time. You have to know whether a one- or two-tailed test is appropriate, what a reasonable significance level is, whether the variances ought to be pooled, and whether the samples are dependent. Otherwise, the computer is of little use. Most introductory statistics textbooks skip the issue of pooling for a difference-in-means test. Given the complexity of the formula for Satterthwaite's approximation and the subtle judgment required to decide whether or not to pool for a difference-in-means test, you may be sympathetic to the approach taken by those other texts. The decision to discuss these issues here was motivated by the fact that many of you will use computers for these tests, and if these issues were not introduced, you would wind up staring at your printout and have no idea which version of the test you ought to use.

For review, here are the four ways to set up a difference-in-means hypothesis test when the population standard deviations are not known. The first way is when both samples are large; you can use the z score and the test is the same as in section 10.1 with the sample standard deviations replacing the population standard deviations in formula 10.3. The second way is when you have small samples and the populations are not normal. You would use the Wilcoxon-Mann-Whitney test from Chapter 17. The third way is to use a t score with pooled variances. The fourth way is to use a t score without pooled variances.

Chapters 11 and 12 basically extend hypothesis testing to situations in which there are more than two populations. Chapter 11 has multipopulation tests for categorical data and Chapter 12 has multipopulation tests for data in levels. Some new distributions will be required to handle these problems, but the framework of significance levels, β levels, acceptance regions, rejection regions, and null and alternative hypotheses will carry over.

 NEW TERMS

delta (δ) The mean difference between all pairs of observations taken from two populations.

dependent samples (also called **paired, correlated,** or **matched samples**) Samples in which there is statistical dependence between the pairs of observations drawn from two populations.

pooled estimate of the population variance A weighted average of two sample variances, in which the weights are the degrees of freedom for each sample.

sampling distribution of the difference in means The probability density function of the difference between the means of two independent samples.

sampling distribution of the difference in proportions The probability density function of the difference between the proportions of successes in two independent samples.

 NEW FORMULAS

Expected value of the difference in sample means

$$E(\bar{x}_1 - \bar{x}_2) = \mu_1 - \mu_2 \tag{10.1}$$

Variance of the sampling distribution of the difference in means

$$\sigma^2_{\bar{x}_1-\bar{x}_2} = \frac{\sigma_1^2}{n_1} + \frac{\sigma_2^2}{n_2} \tag{10.2}$$

Standard deviation of the distribution of the difference in sample means

$$\sigma_{\bar{x}_1 - \bar{x}_2} = \sqrt{\frac{\sigma_1^2}{n_1} + \frac{\sigma_2^2}{n_2}}$$

(10.3)

Lower and upper limits of a confidence interval for the difference in population means with normal populations and known population standard deviations

$$L, \ U = \bar{x}_1 - \bar{x}_2 \pm z_{\alpha/2}\sigma_{\bar{x}_1 - \bar{x}_2}$$

(10.4)

Critical value for a left-tailed hypothesis test on the difference in population means with normal populations and known population standard deviations

$$\text{C.V.} = (\mu_1 - \mu_2)_0 - z_\alpha \sigma_{\bar{x}_1 - \bar{x}_2}$$

(10.5)

Lower and upper limits of a confidence interval for the difference in population means with large samples and unknown population standard deviations

$$L, \ U = \bar{x}_1 - \bar{x}_2 \pm z_{\alpha/2}s_{\bar{x}_1 - \bar{x}_2}$$

(10.6)

Estimated standard deviation of the sampling distribution of the difference in means

$$s_{\bar{x}_1 - \bar{x}_2} = \sqrt{\frac{s_1^2}{n_1} + \frac{s_2^2}{n_2}}$$

(10.7)

Critical value of a right-tailed hypothesis test on the difference in population means with large samples and unknown population standard deviations

$$\text{C.V.} = (\mu_1 - \mu_2)_0 + z_\alpha s_{\bar{x}_1 - \bar{x}_2}$$

(10.8)

Satterthwaite's approximation of the degrees of freedom for Student distribution for the sampling distribution of the difference in means with independent samples, normal populations, and small sample sizes

$$df = \frac{[(s_1^2/n_1) + (s_2^2/n_2)]^2}{\dfrac{(s_1^2/n_1)^2}{n_1 - 1} + \dfrac{(s_2^2/n_2)^2}{n_2 - 1}}$$

(10.9)

Lower and upper limits of a confidence interval for the difference in population means with normal populations, small samples, and unpooled variances. The degrees of freedom are given by formula 10.9.

$$L, \ U = \bar{x}_1 - \bar{x}_2 \pm t_{\alpha/2}s_{\bar{x}_1 - \bar{x}_2}$$

(10.10)

Critical values for a two-tailed hypothesis test on the difference in population means with

normal populations, small samples, and unpooled variances. The degrees of freedom are given by formula 10.9.

$$\text{C.V.}_L, \text{C.V.}_U = (\mu_1 - \mu_2)_0 \pm t_{\alpha/2} s_{\bar{x}_1 - \bar{x}_2} \tag{10.11}$$

Pooled estimate of the population variance

$$s_{\text{pool}}^2 = \frac{(n_1 - 1)s_1^2 + (n_2 - 1)s_2^2}{n_1 + n_2 - 2} \tag{10.12}$$

Estimated standard deviation of the difference in sample means based on the pooled estimate of the population variance

$$s_{\bar{x}_1 - \bar{x}_2} = \sqrt{\frac{s_{\text{pool}}^2}{n_1} + \frac{s_{\text{pool}}^2}{n_2}} \tag{10.13}$$

Lower and upper critical values for a two-tailed hypothesis test on the difference in population means with a pooled estimate of the population variance. The degrees of freedom are $n_1 + n_2 - 2$.

$$\text{C.V.}_L, \text{C.V.}_U = (\mu_1 - \mu_2)_0 \pm t_{\alpha/2} s_{\bar{x}_1 - \bar{x}_2} \tag{10.14}$$

Lower and upper limits of a confidence interval for the difference in population proportions using the normal approximation to the sampling distribution of the difference in sample proportions

$$L, U = \bar{p}_1 - \bar{p}_2 \pm z_{\alpha/2} s_{\bar{p}_1 - \bar{p}_2} \tag{10.15}$$

Standard deviation of the difference in sample proportions

$$s_{\bar{p}_1 - \bar{p}_2} = \sqrt{\frac{\bar{p}_1(1 - \bar{p}_1)}{n_1} + \frac{\bar{p}_2(1 - \bar{p}_2)}{n_2}} \tag{10.16}$$

Critical value of a right-tailed hypothesis test on the difference in population proportions

$$\text{C.V.} = (\Pi_1 - \Pi_2)_0 + z_{\alpha} s_{\bar{p}_1 - \bar{p}_2} \tag{10.17}$$

Pooled estimate of the population proportion for a hypothesis test in which the null hypothesis is that the difference in the population proportions of successes is either (1) zero, (2) greater than or equal to zero, or (3) less than or equal to zero

$$\bar{p}_{\text{pool}} = \frac{n_1 \bar{p}_1 + n_2 \bar{p}_2}{n_1 + n_2} \tag{10.18}$$

Mean paired difference with dependent samples

$$\bar{d} = \frac{\sum\limits_{i=1}^{n} d_i}{n} \tag{10.19}$$

Standard deviation of the distribution of the differences between paired observations

$$s_d = \sqrt{\dfrac{\sum\limits_{i=1}^{n} (d_i - \bar{d})^2}{n - 1}}$$ (10.20)

Standard error of the mean paired difference

$$s_{\bar{d}} = \dfrac{s_d}{\sqrt{n}}$$ (10.21)

Critical values of a two-tailed hypothesis test on the difference in population means with paired samples

$$\text{C.V.}_L, \; \text{C.V.}_U = \delta_0 \pm t_{\alpha/2} s_{\bar{d}}$$ (10.22)

 ## ADDITIONAL EXERCISES

An asterisk designates an exercise of greater than ordinary difficulty.

10.43 You own a national company that leases office space and provides tenants with building maintenance services. You charge a premium for laying carpeting on the floors of your offices in place of vinyl flooring. You are not sure, however, what the difference in annual costs is between your two types of floor covering. According to their manufacturers' guarantees, the carpets have a three-year life and the vinyl flooring has a ten-year life. With these guarantees you can estimate the replacement costs of the two types of floor covering. What you need more information on is the maintenance costs. You tell your building managers to keep careful records on the costs of maintaining each type of floor covering. A sample of 40 of your 1,000 building managers report an annual average of $4 a square foot for maintaining the carpeting with a sample standard deviation of 55 cents and an annual average of $1 a square foot for maintaining the vinyl flooring with a sample standard deviation of 24 cents. Construct a confidence interval at the 95% level of confidence for the difference between the annual average costs of maintaining the two types of floor covering.

10.44 During the 1950s the federal government conducted some above-ground tests of atomic bombs in Nevada. You are an attorney who represents the citizens of some Nevada towns that were located downwind of the test site. To get an idea of the number of citizens who may be entitled to compensation, you want to estimate the difference between the proportions of the downwind and upwind populations who have had or still have leukemia. You obtain the medical records of people who lived downwind and upwind of the test site when the tests were conducted. Suppose that 25 out of a sample of 2,000 individuals in the downwind population had leukemia and that 15 out of a sample of 3,000 individuals in the upwind population had leukemia. Construct a confidence interval at the 95% level of confidence for the difference between the population proportions.

10.45* You own a company that manufactures track shoes. Your chemists have developed a new compound for the sole of a track shoe which will make sprinters run faster. You want to advertise the new shoe with the claim "Go x% faster with the Cheeta Hot Sole running shoe." To get an idea of how much faster the new sole will be, you recruit 25 world-class sprinters to run the 100-meter dash, 10 of them wearing their own shoes and 15 of them wearing your new shoe. The 10 have a mean time of 9.8 seconds with a sample standard deviation of 0.4 second. The 15 have a mean time of 9.6 seconds with a sample standard deviation of 0.2 second. Using a one-sided confidence interval at the 90% level of confidence, what is the largest percentage that you could claim as a reduction in running time for the 100-meter dash for your new shoe? Assume that the distributions of running times for both groups are normal. Also assume that the population standard deviations of the two groups are equal.

10.46 You work in the marketing department of a major motion picture studio. You have noticed a trend toward advertising new theatrical movies on television instead of the traditional advertisements in newspapers. The television advertisements are more expensive but they also increase a movie's box office revenue. You are interested in estimating the difference between the mean box office revenues per city for the two types of advertising. To estimate this difference, you conduct an experiment in which the same movie is shown in 90 cities, 45 of which will have heavy TV advertising and 45 of which will have the usual newspaper advertisements. The cities are chosen so that the two groups match each other in mean number of inhabitants and standard deviations. Suppose the movie in question generated a mean of $400,000 in the TV cities and a mean of $324,000 in the newspaper cities with respective sample standard deviations of $45,000 and $65,000. Use these figures to construct a confidence interval at the 95% level of confidence for the difference between the mean revenues for the two types of advertising. Does your experiment have any weaknesses as a guide for deciding whether to advertise a movie on television or in newspapers?

10.47 Very little information on the quality of some consumer products is available to the consumer. When consumers buy a new mattress, for example, they have no way of telling how durable it will be. In such circumstances, consumers often rely on price as a rough guide to quality. However, mattresses of a given quality can sell at different prices. You own a company that makes mattresses and you have a design for a new, very durable mattress that uses Kelvar springs. You are considering setting the suggested retail price at one of two levels: either at a pricing point competitive with other firms' top-of-the-line mattresses or at a lower level competitive with most popular mattresses. You know that you will sell more mattresses at the lower price, but are the additional sales sufficient to offset the lower price? You want to estimate the mean difference between the per outlet monthly sales of mattresses at the two price levels. You select 40 outlets to offer the mattress at the high price and 40 different outlets to offer the mattresses at the lower price. Suppose the sample mean for the high-priced mattresses was 26 and the sample mean for the low-priced mattresses was 21 with sample standard deviations of 5 and 7, respectively. Either construct the confidence interval for the difference in mean monthly sales or explain why you have no reason to construct the confidence interval.

10.48 You design the uniforms worn by the employees of a national chain of fast-food restaurants. You are considering two fabrics: a low-cost and less durable fabric that is a blend of cotton and polyester and a high-cost and highly durable fabric that is a blend of nylon and wool. You order ten of each type and have employees at a particular restaurant wear them. After each use the uniforms are laundered and inspected for wear. When the uniforms are considered too deteriorated for further use, they are thrown out and the number of times they have been worn is recorded. You want to estimate the difference between the mean numbers of launderings of the uniforms made of the two fabrics before they were discarded. If the low-cost fabrics lasted an average of 70 launderings with a sample standard deviation of 15 and the high-cost fabrics lasted an average of 90 launderings with a sample standard deviation of 28, what would be the upper and lower limits of a confidence interval for the difference in population means at the 99% level of confidence, assuming that both populations are normally distributed?

10.49* Your company is marketing a new ketchup that is thicker and pours more slowly than Heinz ketchup. You are planning an advertising campaign around the slogan "Don't buy a ketchup with the runs! Ours pours x% slower than Heinz." You need to estimate the mean difference in pouring time for your advertisements. You plan a test that consists of pointing a full open bottle at a 45° angle into a container on a scale that has a timer to indicate when 12 ounces have poured out. If you want to estimate the mean difference with a tolerance of one second at the 95% level of confidence and you use a planning value of 30 seconds for the standard deviations of both types of ketchup, how many bottles of each type will you have to sample? (Note: This is a teach-yourself problem. No problems involving a minimum sample size have been discussed in Chapter 10.)

10.50 You are the general manager of a team in the National Basketball Association. You know that when opponents who have losing records visit your city, your attendance drops below the attendance you can expect when your opponents have winning records. You are planning some promotions for next season, such as two-for-one tickets, children admitted free with an adult, free T-shirts, and after-game concerts. You would like to bring the average attendance in the poor-opponent games up to the average attendance in the good-opponent games. You would like an estimate of the difference between the average attendances at the two types of games before you plan your promotions. Suppose in the last four seasons you had 45 home games against opponents with losing records. These games had an average attendance of 6,000 with a standard deviation of 2,000. The remaining 35 games, against teams with winning records, had an average attendance of 14,000 with a standard deviation of 4,000. Construct a confidence interval for the difference in mean attendance at the 99% level of confidence. Suppose you go ahead with the planned promotions for the games against teams with poor records and achieve equal average attendances at the two types of games. What could happen that might make you very unhappy even though you had reached the stated objective of having the average attendance at games against poor opponents the same as the average attendance at games against good opponents?

10.51 You work for a company that markets over-the-counter medicines. Your company has a new formulation for a cream that reduces acne. You want to estimate the mean difference in the reduction of acne between the new and old formulations. You design a test in which 20 teenagers who have acne are given applications of the new formula on the left sides of their faces and the old formula on the right sides of their faces. The reduction in the number of pimples is recorded for each side of each subject's face. You estimate the mean difference using the method in section 10.2, write up the results, and report them to the director of marketing. She reads the report, gets angry, and calls you a pimple-brain. What went wrong?

10.52 The Burger King chain of fast-food restaurants has a problem with high labor costs. Although the chain's franchisees pay relatively low wages to the mostly teenage work force, the labor costs have been pushed up by the training and recruiting costs (about $500 per worker) and very high turnover rates (the average duration of employment is four months). The chain has begun an experimental program of educational assistance which allows employees to accumulate up to $2,000 of college tuition credits over two years of employment. In the pilot test of the program with 2,000 participants, the proportion of workers who quit before working a full year was 0.22 versus a chainwide proportion of 0.97. However, the success of the pilot program may not be reproduced in a companywide program because the 2,000 participants may have been self-selected to overrepresent workers interested in attending college. Suppose the Burger King chain needed at least a reduction of 20% in the proportion of workers who quit in the first year to justify the cost of the tuition credit program. In a second pilot study of the program, the tuition credits were given to 2,000 employees selected at random while a control group of 1,000 other employees, also selected at random, were not given the credits. The proportion of employees with tuition credit who quit within a year was 0.44 and the proportion without the credit who quit was 0.68. Test at the 5% significance level the hypothesis that the tuition credit program reduces the proportion of quits by 20% or more.

10.53 **(a)** You want to test the hypothesis that the PCB levels in two rivers are the same. You take 10 samples of water from the first river and 12 samples from the second. The sample mean from the first river is 25 parts per billion and the sample mean for the second river is 30 parts per billion. Use a 5% significance level and assume that the population distributions are normal. Assume that the standard deviation for the first river is known to equal 15 and that the standard deviation for the second river is known to equal 8.

(b) Work the same problem again, this time assuming that the two standard deviations are sample values.

(c) Work the same problem again, this time assuming that the two population standard deviations are equal.

10.54 You are a Broadway producer. Your experience has been that musicals that are not canceled immediately after the opening run longer than plays that are not immediately canceled. However, musicals cost much more to stage than plays. You want to estimate the difference in the mean number of performances for musicals and plays that were not immediately canceled. You take a random sample of five musicals and six plays. The musicals ran for an average of 458 performances with a sample standard deviation of 254, and the plays ran for an average of 309 performances with a sample standard deviation of 123. Construct a confidence interval at the 95% level of confidence for the difference in population means. What assumptions do you have to make to work this problem?

10.55* You are an executive with a U.S.-based steel company. If you can prove that Japanese steel companies "dump" their steel in the U.S. market, you may be able to get tariffs imposed on Japanese steel. To "dump" a product is to sell it in a foreign market at a price below the cost to produce it. You want to estimate the difference between American and Japanese costs per ton of steel. Since the Japanese steel companies pay world market prices for capital and such materials as iron ore, limestone, and coal, the key factors you have to estimate are the differences in labor costs per hour and the differences in output per worker-hour. You take a sample of 500 steelworkers in the United States and find that the sample mean for their wages plus fringe benefits is $28 per hour with a sample standard deviation of $3. You also take a sample of 400 Japanese steelworkers and find that the sample mean of their compensation per hour is $21 with a sample standard deviation of $9. Next you observe that the output per hour of the 500 American workers has a sample mean of 1.4 tons of steel with a sample standard deviation of 0.2, while the output per hour of the 400 Japanese workers has a sample mean of 1.7 tons with a sample standard deviation of 0.5. If you ignore all other costs, such as shipping, what is the lowest relative price at which the Japanese could sell steel without dumping? Assume that your firm sells steel at a price equal to the cost to produce it.

10.56* You want to reestimate the improvement in running times attributable to the new track shoes described in exercise 10.45. You are worried that the 15 runners who used your shoe might by chance variation be faster runners than the 10 who used their own shoes. You could eliminate that possibility by having every runner use both his own shoe and your new shoe. The results for all of the runners are listed below. Now construct a one-sided confidence interval for the difference between the mean running times of the old shoes and your product. Assume that the distributions of running times are normal.

Runner	Seconds with own shoe	Seconds with your shoe	Runner	Seconds with own shoe	Seconds with your shoe
1	9.9	9.7	14	9.7	9.5
2	9.7	9.7	15	9.6	9.5
3	9.8	9.7	16	9.9	9.7
4	9.6	9.5	17	10.2	10.3
5	9.9	9.6	18	9.6	9.7
6	10.1	10.0	19	9.9	9.6
7	9.9	10.0	20	9.9	9.7
8	9.9	9.6	21	9.8	9.8
9	9.8	9.8	22	9.9	9.6
10	9.9	9.7	23	9.9	9.5
11	9.9	9.8	24	9.7	9.5
12	9.8	9.5	25	9.8	9.5
13	9.7	9.7			

10.57* You are a loan officer for a finance company and you are interested in identifying characteristics of a loan applicant which could be used to predict whether· the applicant would later default on his or her loan. In the past you have noticed that applicants who do not have a phone are likely to default, and that applicants who cannot remember their addresses when they fill out the loan form are also likely to default. You've never had an applicant who did not have a phone and at the same time could not remember his or her address. Apparently any potential applicants with such characteristics have been unable to locate your bank. You want to estimate the difference in the proportion of applicants with either characteristic who default on loans with a tolerance of 1 percentage point at a 95% confidence level. You have no planning values for the population proportions, but you think that the two proportions are fairly close. What sample sizes should you use? (Note: This is a teach-yourself problem. No problems involving a minimum sample size have been discussed in Chapter 10.)

10.58* You work for a tobacco company and you are told to estimate the difference between the proportions of men and women who smoke Old Coffin Nails cigarettes. You decide to do your research in some bars, because you know most people in bars smoke. You also know that you need at least five observations of men and of women who smoke Old Coffin Nails so that you can apply the normal approximation to the sampling distribution of the proportion; that is, so that $n\overline{p}$ is greater than or equal to 5. Your plan is to ask people you see smoking in some randomly selected bars which brand they are using until you've found five men and five women who smoke Old Coffin Nails. Although this research design has some personal social advantages, it has some serious statistical flaws. Describe those flaws.

10.59* You are a speculator in commodity futures who specializes in wheat. A typical transaction might involve a contract to sell a million bushels of wheat in 30 days at a price of $4.10 a bushel. If at the end of 30 days wheat sells on the spot market at $4.00 a bushel, you can buy your million bushels and immediately resell them for 10 cents more per bushel than you paid. Your profit from the transaction will be $100,000. Of course if the spot price of wheat rises to $4.20, your loss will be $100,000. Suppose you've written 100 contracts each of which involves selling a million bushels of wheat. Each contract is written so that it falls due on a different day. What you would like to do is estimate the minimum cash reserve you will need to cover your potential losses on those 100 contracts. In the past you've written 40 similar contracts with a mean selling price of $4.10 and a standard deviation of 15 cents. The spot market prices when these contracts came due averaged $4.00 with a standard deviation of 55 cents. Assume that the spot and contract prices are independent. Also assume that past prices are unbiased estimators of future prices. Estimate a lower bound for the difference between the total sales for the 100 contracts and the spot market purchases using a 95% significance level. (Hint: This exercise requires a one-sided confidence interval.)

10.60 You own a franchised national chain of exercise spas for women which feature exercising to music, weightlifting, and diet counseling. In your advertisements and mailings you have targeted women in the 40–50 age group. When you discuss the patronage habits of your clients with your franchise holders, however, they often point out that younger women are more profitable because once they've signed a membership contract, they come to the spas less frequently and thus require less space and service. Before retargeting your ad campaigns, you want to estimate the mean number of monthly visits by age group. You survey 3,000 clients who are in their 20s and 30s and 3,000 clients who are in their 40s and 50s. The mean number of visits for the younger women is 3.6 and the mean number of visits for the older women is 3.9.

Assume that the population standard deviation for younger women is known to be 1.5 and the population standard deviation for older women is known to be 1.1. Construct a confidence interval at the 90% level of confidence for the difference between the population means. What other information would you need before making the decision about retargeting your advertising?

10.61* In exercise 10.60 the sample sizes of both groups of women were 3,000. If you had budgeted a survey of 6,000 interviews and you knew the population standard deviations were as specified in exercise 10.60, how could you have altered the sample sizes to gain more information from the survey?

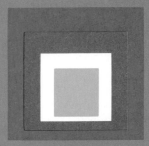

11

The Chi-Square Distribution and Tests for Independence and Goodness of Fit

hen a company decides to reformulate one of its products, there are great risks as well as potentially great benefits. If the product has been losing ground to competitors, the reformulation may reverse this trend. On the other hand, a reformulation can offend loyal customers and lose sales. One way to reduce this risk and gain information is to test-market the reformulated product in several cities. A survey of consumer brand preferences in some cities that have the new formulation and in others that have the old can provide quick feedback on consumer reactions. For example, Scotties toilet paper was recently reformulated to have more and thinner sheets in a roll. Suppose the manufacturer, Kimberly-Clark, test-marketed the new formulation for six months in Indianapolis and Denver. These cities are often used to test new or reformulated products because they are considered "typical." A phone survey is conducted in these cities and in Atlanta and Seattle as well, for comparison purposes: 1,000 married women in each city are asked which brand of toilet paper they would most prefer if all brands sold at the same price. If the new formulation had no effect on consumer preferences and preferences were known to be the same across cities before the reformulation, the number of women who said that they preferred Scotties, Charmin, Cottonelle, Northern, and so on should be the same in each city. For example, if 20% of all women across the four cities, or 800 of the 4,000 women surveyed, said that they preferred Scotties and the reformulation had no effect on brand preferences, you would expect 20% or 200 women to prefer Scotties in each of the four cities. However, even if the reformulation had no effect on preferences, chance or sampling variation could cause you to find slightly more or slightly fewer than 200 women saying that they preferred Scotties in a city. The problem is to decide if any departure from the expected frequencies is due to chance or due to a real impact caused by the reformulation of the product.

This problem deals with cross-classified categorical data. Each woman surveyed has two characteristics, city and brand preference. If there were 5 brands and 4 cities, the response of each woman surveyed would fall into one of the 20 possible brand-city classifications. The solution to the statistical problem of distinguishing between chance variation across the cities and a real impact due to the reformulation requires the use of a probability distribution called the chi-square distribution. With this distribution you can assign a probability to whether the deviations between the observed and the expected frequencies of preferences for each brand in each city are due to chance variation or to sampling error.

This chapter discusses two important uses of the chi-square distribution. The problem described above is a test of homogeneity across populations. Another test, for independence of two characteristics within a population, is discussed in the same section. It follows the same procedure as the homogeneity test. The other important use is in tests for goodness of fit. In a goodness-of-fit test, observed and expected frequencies are compared to test, for example, if the population follows the distribution specified in a business contract. Another goodness-of-fit test can be used to determine if a population distribution follows one of the common probability distributions, such as a normal distribution.

11.1 TESTS FOR GOODNESS OF FIT

11.1A INTRODUCTION

We'll begin the discussion of tests for goodness of fit with an example that illustrates how a set of observed frequencies is compared to a set of expected frequencies that are based on a business contract. Let's say you run a chain of women's clothing stores and you have ordered 10,000 red polka-dot dresses from a manufacturer. The dresses come in the following sizes: 5, 7, 9, 11, 13, 15, 17. You ordered 500 size 5s, 1,500 size 7s, 2,000 size 9s, 2,000 size 11s, 2,000 size 13s, 1,500 size 15s, and 500 size 17s. A semitrailer pulls up to your warehouse with the 10,000 dresses and the driver asks you to sign a form indicating that you have accepted them. You are worried that the actual frequencies may not match your order because the manufacturer might have tried to save material by selling you more small dresses and fewer large ones. If he did, you may be stuck with dresses that you cannot sell and have unhappy customers who cannot find dresses in their sizes. You tell the truck driver that you want to check to see if the shipment has the correct frequencies before you sign the form. The driver says, "Sure, buddy, go ahead, but my rig costs $400 an hour while it's sitting in your lot." You quickly throw open the doors of the trailer and discover that the dresses are not sorted by size. To establish that the contract has been fulfilled, you would have to check the size of each of the 10,000 dresses.

The solution to this dilemma is to take a sample of, say, 100 dresses. If the shipment met the contract terms, you would average, in repeated samples of 100 dresses, 5 size 5s, 15 size 7s, 20 size 9s, 20 size 11s, 20 size 13s, 15 size 15s, and 5 size 17s. That is, the proportion of the sample in each dress size would, on average, equal the corresponding population proportion. Expressed as a hypothesis test, the null and alternative hypotheses would be

H_0: $\Pi_5 = 0.05$, $\Pi_7 = 0.15$, $\Pi_9 = 0.20$, $\Pi_{11} = 0.20$, $\Pi_{13} = 0.20$, $\Pi_{15} = 0.15$,
    $\Pi_{17} = 0.05$
H_1: At least one Π is incorrect

The subscripts in the null hypothesis refer to the dress sizes. If the observed frequencies in your sample equaled the expected frequencies, then you would certainly accept the hypothesis that the shipment met the contract terms. In this case you could say that there was a "good fit" or even a perfect fit between the sample frequencies and the expected frequencies. On the other hand, if your observed frequencies deviated sharply from your expected frequencies, say 100 size 5 dresses in your sample of 100, you would certainly reject the hypothesis that the shipment met the contract terms. In most situations the decision to reject or accept will not be as obvious as these two extremes. In a nonobvious case you would calculate a test statistic based on formula 11.1.

$$\chi^2 = \sum_{i=1}^{k} \frac{(o_i - e_i)^2}{e_i}$$

(11.1)

The symbol χ is the Greek letter chi, the o_i is the observed frequency of the ith class (in this example a particular dress size), and e_i is the expected frequency of the ith class. The k refers to the number of classes or comparison terms. If the observed frequency of each dress size matched the expected frequency, the test statistic would equal 0. As the differences between the o_i's and the e_i's increased, the χ^2 term would grow larger. If χ^2 had a large value, you would tend to reject the null hypothesis, that the population proportions matched the contract terms.

Let's go over the reasoning behind this formula. If the null hypothesis were true, in repeated samples the o_i for each dress size would, on average, equal the e_i. Thus the term $(o_i - e_i)$ would have an expected value of 0. Second, if the null hypothesis were true, the observation of extra dresses in a particular size would be as likely as the observation of too few dresses. Thus the expression $(o_i - e_i)$ should be symmetrical around the value of 0. Next, a large deviation between an o_i and an e_i should be much less likely than a small deviation. The probability distribution of $(o_i - e_i)$ should taper sharply in either direction from its center of 0. In this example the e_i's are integers. Thus the difference between o_i and e_i is an integer. What distribution looks like the above description? If you guessed a binomial distribution with a Π of 0.5, you would be right. The only difference is that the binomial will not have negative values while $(o_i - e_i)$ is centered on 0 and does have negative values. Figure 11.1 shows the sampling distribution of $(o_i - e_i)$ for the size 9 dresses.

As in the binomial distribution, the distribution of $(o_i - e_i)$ *will be approximately normal for any e_i of 5 or more.* The e_i is equivalent to the term $n\Pi$ in the binomial distribution. Take size 9 dresses for an illustration of this point. You are conducting 100 trials when you sample 100 dresses, and Π_9, the probability that any dress is a size 9, is approximately 0.20. The terms $n\Pi_9$ and e_9 are both the expected number of size 9 dresses, which equals 20. The reason that the Π_9

FIGURE 11.1
Sampling distribution of
$(o_i - e_i)$ **when the expected
frequency is 20**

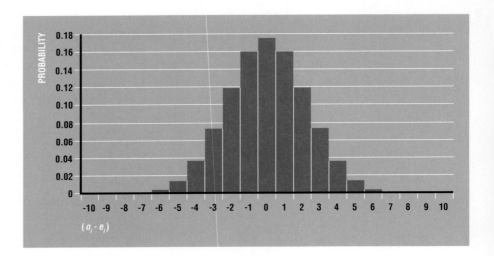

is only approximately equal to 0.20 is that you are sampling from a finite population without replacement. However, the large population makes the approximation very close.

The $(o_i - e_i)$ terms are squared before they are added in formula 11.1 because the sum of all of the differences between the observed and the expected frequencies must be 0. If the $(o_i - e_i)$ terms were not squared, you would have an uninformative test statistic. Under the assumption that the null hypothesis is true, the numerator of formula 11.1 would, in repeated samples, be the sum of a series of squared normal distributions. The last part of the formula to explain is the division by e_i. This division scales each of these normal distributions so that they have a standard deviation of 1. Thus, under the assumption that the null hypothesis is true, the test statistic χ^2 is *distributed as the sum of a series of independent, squared standard normal random variables*. Recall that the standard normal distribution has a mean of 0 and a standard deviation of 1.

The distribution that consists of *the sum of a series of independent squared standard normal random variables* is called the **chi-square distribution.** The sole parameter of this distribution is the number of squared standard normal random variables being summed, or, equivalently, the degrees of freedom. In the dress example there are seven dress sizes but not all seven comparisons are independent. If you knew how many dresses were observed in the first six sizes and you knew that the sample consisted of 100 dresses, you would also know how many dresses there were in the remaining size. The number of independent comparisons in the dress example is six. If the null hypothesis is true, the test statistic χ^2 will follow a chi-square distribution with 6 degrees of freedom. Figure 11.2 shows the chi-square distribution with 6 degrees of freedom and the critical value of the hypothesis test with a significance level of 5%.

The chi-square distribution has a lower bound of 0 because the squared

FIGURE 11.2
Chi-square distribution with 6 degrees of freedom

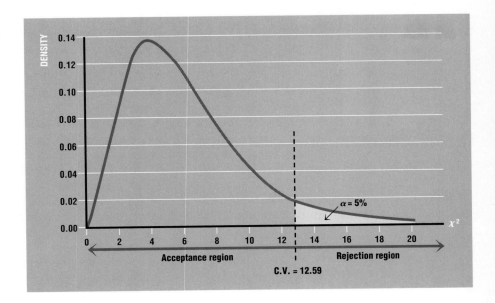

terms must be nonnegative. There is no upper bound. The distribution is skewed to the right and its mean is the degrees of freedom; in this case the mean is 6. The standard deviation is the square root of the product of 2 times the degrees of freedom, in this case the square root of 12. The template for accepting or rejecting the null hypothesis is based on this chi-square distribution. The boundary line on a chi-square distribution with 6 degrees of freedom which puts 5% of the area in the right tail is 12.59. The number 12.59 is the critical value for a hypothesis test with a significance level of 5%. The number 12.59 can be obtained from Table D8, in Appendix D; look under the column headed "0.05" and the row for 6 degrees of freedom.

With this template in hand, your next step is to collect a sample of 100 dresses and calculate the value of the χ^2 test statistic and see if it falls in the acceptance or the rejection region. Suppose you had found 7 size 5 dresses, 18 size 7 dresses, 22 size 9 dresses, 20 size 11 dresses, 20 size 13 dresses, 13 size 15 dresses, and no size 17 dresses. These observed frequencies and the expected frequencies of 5, 15, 20, 20, 20, 15, and 5 are substituted into formula 11.1:

$$\chi^2 = \sum_{i=1}^{k} \frac{(o_i - e_i)^2}{e_i} \qquad (11.1)$$

$$\chi^2 = \frac{(7-5)^2}{5} + \frac{(18-15)^2}{15} + \frac{(22-20)^2}{20} + \frac{(20-20)^2}{20} + \frac{(20-20)^2}{20}$$
$$+ \frac{(13-15)^2}{15} + \frac{(0-5)^2}{5}$$

$$\chi^2 = 6.87$$

CASE STUDY

BEATING A STATE LOTTERY

State lotteries are generally poor bets in the sense that the expected value per dollar bet is very low. Typically, only 50 to 60 cents out of each $1 bet is returned to the bettors. Many lotteries allow ticket buyers to select their own numbers, say a four-digit number. The four numbers selected by a ticket buyer are recorded on the lottery commission's computer and printed on the ticket. After a random drawing of four digits is held by the lottery commission, the prize money is divided equally among the ticketholders who have selected the winning number. While all four-digit numbers are equally likely to be drawn by the lottery commission's computer or drawing machine, the ticket-buying public is not equally likely to pick all of the four digits. If you picked a winning number that many other people had picked, your prize would be small. If you picked a winning number that few other people had picked, your prize would be large. Herman Chernoff, a professor at Harvard, has analyzed the Massachusetts daily lottery, and on the basis of winning numbers and payoffs between 1976 and 1980, he found that by choosing numbers that few people pick, you could turn the lottery into a favorable bet. The unpopular numbers should give you, on average, $1.086 back for each dollar bet.

The chi-square distribution can be used to test the hypothesis that not all four-digit numbers are equally likely to be picked by ticket buyers. Since there are 10(10)(10)(10) or 10,000 possible four-digit numbers, this problem is too large to use as an example in the text. A scaled-down example would be to test if all leading digits are equally likely to be chosen by ticket buyers. The null and alternative hypotheses would be:

With a test statistic of 6.87 you will not reject the null hypothesis (at a 5% significance level) that the contract terms were fulfilled. Finding no size 17 dresses is suspicious, but since you were expecting only 5 dresses in this size and otherwise the overall pattern is close to what you expected, the evidence is not strong enough to reject the null.

The expected frequencies in the above example came from a business contract, but there are many other sources of expected frequencies. Sometimes it is important to test if a set of sample data came from a binomial, a Poisson, a normal, or an exponential distribution. We will discuss goodness–of–fit problems based on theoretical distributions in sections 11.1B and 11.1C. Determining the degrees of freedom and the expected frequencies is more complicated in these tests than in the contract-based cases.

$H_0: \Pi_1 = \Pi_2 = \Pi_3 = \Pi_4 = \Pi_5 = \Pi_6 = \Pi_7 = \Pi_8 = \Pi_9 = \Pi_{10} = 0.1$

H_1: The proportion of at least one of the leading digits is not 0.1

You have data on the leading digits from a random sample of 1,000 lottery tickets. The ten possible leading digits are 0, 1, 2, 3, 4, 5, 6, 7, 8, and 9. The respective numbers of tickets with each leading digit are 50, 110, 115, 120, 130, 108, 97, 80, 120, 70. If all were equally likely to be chosen, the expected frequency for each digit would be 0.1(1,000), or 100. Substituting these values into formula 11.1, we get

$$\chi^2 = \sum_{i=1}^{k} \frac{(o_i - e_i)^2}{e_i} \tag{11.1}$$

$$\chi^2 = \frac{(50 - 100)^2}{100} + \frac{(110 - 100)^2}{100} + \frac{(115 - 100)^2}{100} + \frac{(120 - 100)^2}{100} + \frac{(130 - 100)^2}{100}$$
$$+ \frac{(108 - 100)^2}{100} + \frac{(97 - 100)^2}{100} + \frac{(80 - 100)^2}{100} + \frac{(120 - 100)^2}{100} + \frac{(70 - 100)^2}{100}$$

$$\chi^2 = 58.98$$

The chi-square test statistic is 58.98. With ten comparison terms there would be nine independent comparisons, or 9 degrees of freedom. The critical value at a 5% significance level is 16.92, so the null hypothesis is rejected. Apparently the leading digit 0 and the leading digit 9 are unpopular among ticket buyers and the leading digit 4 is popular. Perhaps the bettors feel that a number in the middle of the range is more likely to be drawn.

11.1B TESTS AGAINST DISTRIBUTIONS WITH KNOWN PARAMETERS

There are two ways to set up a hypothesis test when the null hypothesis is that data were generated by a probability distribution such as the binomial, the normal, the Poisson, or the exponential. One way is to assume that the population parameter or parameters are known and the other is to use the sample data to estimate the parameters. Here we will assume that the parameters are known or specified in the null hypothesis.

A "flip flash" is a device that holds ten flash cubes in two ranks of five each. It's called a flip flash because after the first five cubes are used, the device is flipped over to shoot five more flash pictures. The manufacturer of the flip flash plans to guarantee double your money back if one cube in a flip flash fails,

TABLE 11.1

Probabilities for a binomial distribution with an *n* of 10 and a II of 0.01

x	P(x)	x	P(x)
0	0.9044	6	0.0000
1	0.0914	7	0.0000
2	0.0042	8	0.0000
3	0.0001	9	0.0000
4	0.0000	10	0.0000
5	0.0000		

triple your money back if two cubes fail, quadruple your money back if three cubes fail, and so on. The projected costs of this guarantee are based on the estimate that the proportion of cubes that will fail is 0.01 and that each cube is an independent trial. Under these assumptions, the number of failures in a flip flash would follow a binomial distribution with a II of 0.01 and an *n* of 10. Before going ahead with the guarantee program, the company wants to test if its estimate of the proportion of failures and its assumption of independence of cubes in a flip flash are correct. To conduct this test, the manufacturer tries a large number of flip flashes and records how many of them have no failures in ten trials, how many have one failure in ten trials, and so on up to ten failures in ten trials. These observed frequencies are compared with the number of failures that would be expected from a binomial process with a II of 0.01 and an *n* of 10. The probabilities of 0 through 10 failures can be obtained from the binomial table. These values are reproduced in Table 11.1.

The company tests 10,000 flip flashes. How many of them should have 0 failures? The answer is 0.9044 times 10,000, or 9,044. The expected number of flip flashes with exactly one failure is 914 and the expected number of flip flashes with exactly two failures is 42. The expected number of flip flashes with exactly three failures is one and the expected number of flip flashes with more than three failures is zero. These figures are not quite correct because the table stops at four places to the right of the decimal point, but they are accurate enough for our purposes. The probabilities in the binomial table do not add to 1.0 because of rounding. If the observed frequencies were 9,010 for 0 failures, 934 for 1 failure, 47 for 2 failures, 8 for 3 failures, and 1 for 4 failures, should the null hypothesis that the failure pattern follows a binomial distribution with a II of 0.01 be rejected? Before substituting into formula 11.1, recall that the e_i's must be 5 or more. Not all of the e_i's in this example are greater than or equal to 5; in particular, the expected frequency of 3 failures is 1, and that of any number above 3 is 0. You have to collapse the comparisons so that no e_i is under 5. The 2, the 3, and the 4 or more comparisons could be collapsed into 2 or more. The expected frequency of 2 or more failures is 43. We obtain this number by adding the 1 for exactly 3 failures and the 42 for exactly 2 failures. Table 11.2 shows the observed and expected frequencies of the collapsed comparisons. There are three sets of comparisons in Table 11.2, but only two of

x	Observed frequency	Expected frequency
0	9,010	9,044
1	934	914
2 or more	56	43

TABLE 11.2
Observed and expected frequencies of the number of failures in 10,000 flip flashes

them are independent. If you knew any two of the observed frequencies, you could determine the third. Thus there are 2 degrees of freedom.

The template for the hypothesis test is given as Figure 11.3. Note that the χ^2 distribution is not mound-shaped when there are only 2 degrees of freedom; it falls continuously over its range. Figure 11.3 shows the critical value for a test at a 5% significance level. The last step is to substitute into formula 11.1:

$$\chi^2 = \sum_{i=1}^{k} \frac{(o_i - e_i)^2}{e_i} \tag{11.1}$$

$$\chi^2 = \frac{(9,010 - 9,044)^2}{9,044} + \frac{(934 - 914)^2}{914} + \frac{(56 - 43)^2}{43}$$

$$\chi^2 = 4.50$$

Since the test statistic of 4.50 is to the left of the critical value of 5.99, the null hypothesis cannot be rejected. In view of the large sample size used to test the

FIGURE 11.3
Chi-square distribution with 2 degrees of freedom

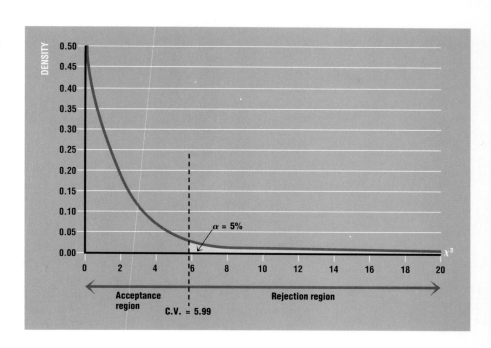

hypothesis, there is little chance of a Type II error. The company can be fairly confident of the cost estimates it based on a Π of 0.01 and the assumption of independence of flash cubes.

For the sake of review, here are the seven steps described in Figure 9.5 as applied to this hypothesis test.

1 Define the null and alternative hypotheses.

The null hypothesis was that the distribution of failures per flip flash was binomial with a Π of 0.01. A goodness-of-fit test is always a right-tailed test because discrepancies between the observed and expected frequencies add to the test statistic.

2 Select the estimator and determine its distribution.

The estimator chosen was the χ^2 test statistic. It will be approximately chi-square-distributed if the null hypothesis of a population proportion of 0.01 is true and if the expected frequencies are at least 5.

3 Choose a significance level for the test.

The customary significance level of 5% was chosen because the costs of making Type I and Type II errors were not specified.

4 Define the decision rule for rejecting or not rejecting the null hypothesis.

The decision rule was to reject the null hypothesis if the χ^2 test statistic was above the critical value as defined by the upper 5% of the chi-square distribution.

5 Perform the computation.

The critical value taken from the chi-square distribution with 2 degrees of freedom was 5.99. The test statistic was 4.50.

6 Make the statistical decision.

The null hypothesis that the distribution of failures per flip flash was binomial with a Π of 0.01 was not rejected.

7 Make the administrative decision.

You will go ahead with the money-back guarantee.

The next example is a goodness-of-fit test for the normal distribution. A company designs and sells intelligence tests used by personnel offices to screen potential employees. The company has written a new test with the intention that, for the population of individuals with a 12th-grade education, the distribution of test scores will be normal with a mean of 100 and a standard deviation of 10. To make sure that the intelligence test meets these criteria, the company has administered the test to 10,000 individuals with 12th-grade educations. Table 11.3 shows the frequencies of the resulting test scores in various ranges.

To find the expected frequency for each of the ranges in Table 11.3, you need to find the proportion of the area under the normal curve with a mean of 100 and a standard deviation of 10 which corresponds to each range. Each

TABLE 11.3

Test scores for 10,000 individuals with a 12th-grade education

Range	Frequency	Range	Frequency
Less than 75	20	100 and under 105	1,950
75 and under 80	160	105 and under 110	1,470
80 and under 85	455	110 and under 115	930
85 and under 90	902	115 and under 120	430
90 and under 95	1,520	120 and under 125	151
95 and under 100	2,000	125 or more	12

proportion will then be multiplied by the overall frequency of 10,000. Expressing the limits of the ranges as z scores will help you to find these proportions. Table 11.4 shows the z scores, the proportion of the area under the normal curve, and the expected and observed frequencies for each range.

To get the z score for each range, simply use the z score formula. For example, the z score for the point 105 is $(x - \mu)/\sigma$, or $(105 - 100)/10$, which equals 0.5. Once you have the z scores, you can use the normal table to find the proportion of the area in each range. For example, you find the proportion of the area under a normal curve between a z score of 0.5 and a z score of 1.0 by subtracting the area between the mean and 0.5 from the area between the mean and 1.0. The proportion of the total area between the mean and the z score of 0.5 is 0.1915, and the proportion of the total area between the mean and a z score of 1.0 is 0.3413. Thus the proportion of the total area between the z scores of 0.5 and 1.0 is 0.1498, which is the value in Table 11.4. This procedure is repeated for each range of test scores. Figure 11.4 shows the area under a standard normal curve between a z score of 0.5 and 1.0.

The next step in the test is to set up the template for rejecting or not rejecting the null hypothesis. There are 12 comparisons in Table 11.4, but if 11

TABLE 11.4

z scores, areas under standard normal curve, expected and observed frequencies of the 10,000 test scores, and the χ^2 comparison terms, by range

Range	z scores	Proportion	e_i	o_i	$o_i - e_i$	$(o_i - e_i)^2$	$(o_i - e_i)^2/e_i$
Less than 75	Under −2.5	0.0062	62	20	−42	1,764	28.45
75 and under 80	−2.5 to −2.0	0.0166	166	160	−6	36	0.22
80 and under 85	−2.0 to −1.5	0.0440	440	455	15	225	0.51
85 and under 90	−1.5 to −1.0	0.0919	919	902	−17	289	0.31
90 and under 95	−1.0 to −0.5	0.1498	1,498	1,520	22	484	0.32
95 and under 100	−0.5 to 0.0	0.1915	1,915	2,000	85	7,225	3.77
100 and under 105	0.0 to 0.5	0.1915	1,915	1,950	35	1,225	0.64
105 and under 110	0.5 to 1.0	0.1498	1,498	1,470	−28	784	0.52
110 and under 115	1.0 to 1.5	0.0919	919	930	11	121	0.13
115 and under 120	1.5 to 2.0	0.0440	440	430	−10	100	0.23
120 and under 125	2.0 to 2.5	0.0166	166	151	−15	225	1.36
125 or more	Over 2.5	0.0062	62	12	−50	2,500	40.32

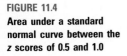
FIGURE 11.4
Area under a standard
normal curve between the
z scores of 0.5 and 1.0

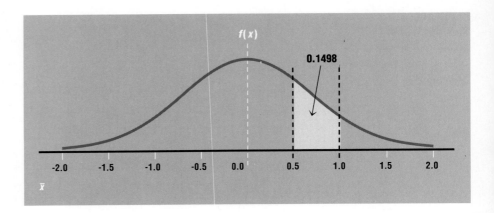

of the observed frequencies are known, the 12th can be determined. Thus there are 11 degrees of freedom. Figure 11.5 shows the acceptance and rejection regions for a hypothesis test with 11 degrees of freedom and a 5% significance level.

The last step in the hypothesis test is to substitute the observed and expected frequencies into formula 11.1:

$$\chi^2 = \sum_{i=1}^{k} \frac{(o_i - e_i)^2}{e_i} \tag{11.1}$$

$$\chi^2 = \frac{(20 - 62)^2}{62} + \frac{(160 - 166)^2}{166} + \frac{(455 - 440)^2}{440} + \cdots + \frac{(12 - 62)^2}{62}$$

$$\chi^2 = 76.78$$

Since 76.78 is in the rejection region, you will reject the null hypothesis that the distribution of test scores is normal with a mean of 100 and a standard deviation of 10. By inspecting the 12 comparisons in Table 11.4, you can see that the observed and expected frequencies that contribute the most to the χ^2 test statistic are the extreme ranges of under 75 and over 125. Since both of these extremes have observations below what would be expected from the null hypothesis, the company might concentrate on redesigning the test so that more individuals would fall in these ranges. As the test is currently designed, it does not seem to be able to assign high and low scores to extremely intelligent and extremely unintelligent individuals, respectively.

In section 11.1C we will discuss tests in which the population parameter or parameters have to be estimated from the sample data. These tests differ from the tests we have been discussing in two respects. One difference is that you have to calculate a sample mean or a sample proportion or a sample standard deviation; the other is that the rule for the degrees of freedom is more complicated.

FIGURE 11.5

Acceptance and rejection regions for a hypothesis test using a chi-square distribution with 11 degrees of freedom and a 5% significance level

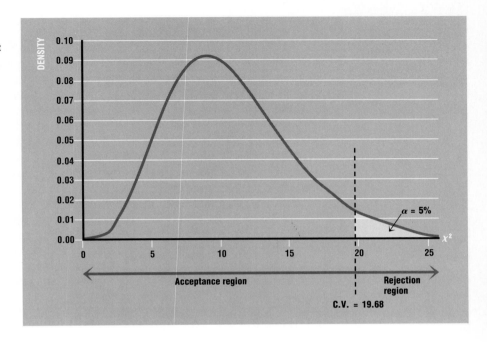

■ 11.1C TESTS AGAINST DISTRIBUTIONS WITH ESTIMATED PARAMETERS

A fast-food chain wants to test the hypothesis that the number of cars per minute which arrive at a drive-up window follows a Poisson distribution. It can use this information to plan the staffing of the drive-up window and to design restaurants with more than one drive-up window. Note that no mean rate of occurrences is specified. The null hypothesis is simply that the distribution is Poisson. The chain records the number of arrivals over 750 one-minute intervals. Table 11.5 shows the resulting data.

Since there is no hypothesized value of the population mean to test against,

TABLE 11.5

Frequency of arrivals per minute at a drive-up window of a fast-food restaurant over 750 one-minute intervals

Number of arrivals	Frequency of arrivals
0	46
1	250
2	260
3	100
4	50
5	34
6	10
7 or more	0

MTBF FOR ATMs

The letters MTBF stand for "mean time between failures" and the letters ATM stand for "automatic teller machine." The Diebold Corporation, the largest manufacturer of ATMs, has recently introduced a new design that is supposed to increase the MTBF. ATMs are complicated machines that have been failure-prone. Because of the high cost of maintaining these machines, they have only limited appeal to many small banks. Realizing that most transactions are simply withdrawals of cash, Diebold has reduced the possible types of transactions that the machine can conduct and it has reduced the number of moving parts in the machine. The goal is to increase the MTBF from 100 hours to 200 hours.

Suppose the company placed one of the new models in a test site and recorded its performance until 25 failures occurred. If each failure was independent, the recorded times between failures could be compared with the exponential distribution with a mean of 200 hours between occurrences to see if the new model met the company's goals. The expected rate of occurrences in a unit interval, one hour, would be $\lambda = 0.005$, or 1/200. The exponential distribution has to be broken into ranges of time so that comparisons can be made between the expected and observed frequencies. Since the e_i has to be at least 5 and there are 25 observed failures, the most powerful test would be to break the exponential distribution into 5 time ranges in which the expected frequency is 5 for each range. Note that the ranges are not of equal length; rather, the expected frequency for each range has to equal 5. If the e_i for any range was less than 5, the test could not be done. If e_i was greater than 5 for any range, the comparisons would not follow the shape of the exponential distribution as closely as the available data allow.

Here is how the five ranges are determined. First you have to find the values of λt or μ that correspond to the 20th, 40th, 60th, and 80th percentiles of the exponential distribution. The letter t refers to time in hours. These values are, respectively, 0.22, 0.51, 0.92, and 1.61. To see how these numbers were determined, consider the 20th percentile, or 0.2000. If you look at the probabilities in Table D5, you find that the two closest values are 0.1975 for a μ of 0.22 and 0.2055 for a μ of 0.23. Rather than interpolate, you take 0.22 as the closest value. Next consider the 40th percentile, or 0.4000. The two closest values are 0.3995 for a μ of 0.51 and 0.4055 for a μ of 0.52. Again, you take the value closer to 0.4000 as an approximation of the 40th percentile, that is, the probability of 0.3995 with a μ of 0.51. Next consider the 60th percentile, or 0.6000. The two closest values in the table are 0.5975 for a μ of 0.91 and 0.6015 for a μ of 0.92. Again you take the value closer to 0.6000 as an approximation of the 60th percentile, that is, the probability of 0.6015 with a μ of 0.92. Finally, consider the 80th percentile. The two closest values are 0.7981 for a μ of 1.60 and 0.8001 for a μ of 1.61. Again you take the value closer to 0.8000 as an approximation of the 80th percentile, that is, the probability of 0.8001 with a μ of 1.61.

The four μ's determined in this manner break the exponential distribution into five ranges, each of which has approximately 20% of the area under the exponential distribution. Figure 11.6 shows the exponential ranges that mark off five equal areas. To convert these ranges into times, recall that μ equals λt. You have the μ and the λ, so you can solve for t. For example, the μ of 0.22 marks the upper limit of the first 20% of the exponential distribution. Substituting the λ of 0.005 in the equation $\mu = \lambda t$ yields

$0.22 = t(0.005)$. The t equals 44. Thus, if the null hypothesis of an MTBF of 200 is true, 5 of the 25 occurrences should be in the range of 0 to under 44 hours. Similarly, 5 occurrences would be expected in the ranges 44 and under 102 hours, 102 and under 184 hours, 184 and under 322 hours, and 322 or more hours. We find each of the ranges by substituting the μ's that correspond to 40th, 60th, and 80th percentiles of the exponential distribution and the expected rate of occurrences in a unit interval, $\lambda = 0.005$, in the formula $\mu = \lambda t$.

If the observed frequencies of occurrences in the experiment were 2 for the range 0 and under 44, 2 for the range 44 and under 102 hours, 0 for the range 102 and under 184 hours, 5 for the range 184 and under 322 hours, and 16 for the range 322 or more hours, should the null hypothesis that the distribution of time to failure is exponential with a mean of 200 hours be accepted at a 1% significance level? Substituting into formula 11.1, we get

$$\chi^2 = \sum_{i=1}^{k} \frac{(o_i - e_i)^2}{e_i} \qquad (11.1)$$

$$\chi^2 = \frac{(2-5)^2}{5} + \frac{(2-5)^2}{5} + \frac{(0-5)^2}{5} + \frac{(5-5)^2}{5} + \frac{(16-5)^2}{5}$$

$$\chi^2 = 32.8$$

The chi-square test statistic is 32.8. There are 4 degrees of freedom and the critical value of the test with a significance level of 1% is 13.28. The null of an MTBF of 200 is rejected. However, if you examine the data, the greatest discrepancy is that too many failures occurred in the range 322 hours or more. Apparently the null hypothesis that the distribution of time to failure is exponential with a mean time of 200 hours is rejected because the new machine exceeds the company goal. Note that the goodness-of-fit test rejects the null if the mean is over or under the hypothesized 200 hours.

FIGURE 11.6
Exponential distribution broken into five ranges that have equal areas

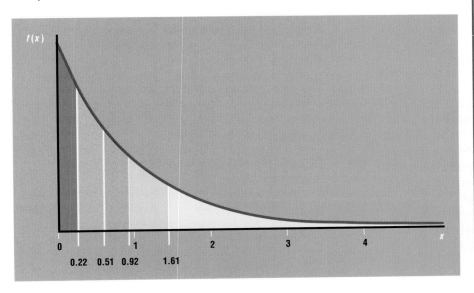

sample data must be used to estimate a mean rate. To find the sample mean, or the average number of arrivals per minute, you must divide the total number of arrivals by 750 minutes. You calculate the total number of arrivals by multiplying the frequencies in Table 11.5 by the corresponding rate per minute. For example, 0 arrival in a minute was recorded 46 times, or a total of 0 arrivals; 1 arrival in a minute was recorded 250 times, or a total of 250 arrivals; 2 arrivals in a minute were recorded 260 times, or a total of 520 arrivals, and so on. This procedure is analogous to taking a weighted mean in which the observations are 0 through 6 arrivals and the weights are the number of times 0 through 6 arrivals were observed. The sample mean equals

$$\bar{\bar{x}} = \frac{n_1\bar{x}_1 + n_2\bar{x}_2 + n_3\bar{x}_3 + \cdots + n_k\bar{x}_k}{n_1 + n_2 + n_3 + \cdots + n_k} \tag{3.3a}$$

$$\bar{\bar{x}} = \frac{0(46) + 1(250) + 2(260) + 3(100) + 4(50) + 5(34) + 6(10)}{750}$$

$$\bar{\bar{x}} = 2$$

With the sample mean of 2 arrivals per minute, you can now use the Poisson table to find the proportion of occurrences in which there are expected to be 0, 1, 2, 3, 4, 5, 6, or 7 or more occurrences. These proportions are reproduced in Table 11.6, along with the expected and the observed frequencies of arrivals. We find the expected frequencies by multiplying the proportion for each class by the total frequency. For example, the e_i for the class of 0 arrivals is (0.1353)(750), or 101.48. The other e_i values are found in the same way. Note that a Poisson distribution with a mean of 2 has a probability of 0.0045 for 7 or more arrivals. Since you are testing against this Poisson distribution, you cannot ignore this part of the distribution even though there were 0 observations of 7 or more arrivals per minute interval. The next step is to collapse comparisons so that every e_i will be at least 5. We do this by making the last comparison 6 or more arrivals. The expected frequency for this comparison is 12.38 and the observed frequency is 10.

TABLE 11.6

Proportions for a Poisson distribution with a mean of 2 and the expected and observed frequencies of arrivals for 750 one-minute intervals

Number of arrivals	Proportion	Expected frequency	Observed frequency
0	0.1353	101.48	46
1	0.2707	203.03	250
2	0.2707	203.03	260
3	0.1804	135.30	100
4	0.0902	67.650	50
5	0.0361	27.080	34
6 ⎱ 6 or more 7 or more ⎰	0.0120 ⎱ 0.0165 0.0045 ⎰	9.000 ⎱ 12.38 3.38 ⎰	10 ⎱ 10 0 ⎰

After the comparisons are collapsed, seven are left. However, there are not 6 degrees of freedom! Using the sample data to estimate the population mean cost one additional degree of freedom. If you knew the frequencies of any five of the seven comparisons, plus the fact that there are 750 observations and the fact that the sample mean number of occurrences is 2, you could find the frequencies of the other two comparisons. Finding the other two frequencies is a simple matter because you have two linear equations with two unknowns. Because you can determine the remaining two frequencies once you know any five of them, there are 5 degrees of freedom. In general, every time you estimate a population parameter from the sample data, it costs you one degree of freedom. The formula for the degrees of freedom is

$$df = k - h - 1 \tag{11.2}$$

where df is the degrees of freedom, k is the number of comparisons after collapsing, and h is the number of population parameters estimated from the sample data. In problems in which the population parameter or parameters are specified in the null hypothesis, the h would equal 0 and the degrees of freedom would simply be the number of comparisons minus 1. The graph of the acceptance and rejection regions for a chi-square distribution with 5 degrees of freedom at a significance level of 5% would look about the same as Figure 11.2, in which the degrees of freedom were 6. We'll skip the graph and just use the critical value from the chi-square table. This critical value is 11.07. If the test statistic is greater than 11.07, the null hypothesis, that the distribution of arrival times follows a Poisson distribution, will be rejected. The last step is to substitute into formula 11.1:

$$\chi^2 = \sum_{i=1}^{k} \frac{(o_i - e_i)^2}{e_i} \tag{11.1}$$

$$\chi^2 = \frac{(46 - 101.48)^2}{101.48} + \frac{(250 - 203.03)^2}{203.03} + \frac{(260 - 203.03)^2}{203.03} + \cdots + \frac{(10 - 12.38)^2}{12.38}$$

$$\chi^2 = 73.22$$

The null hypothesis will be rejected.

Here's another example of a hypothesis test against the normal distribution. This example is more complicated than the previous example on the normal distribution for an intelligence test because no population parameters are given and because it uses raw data instead of classified data. A tour service in a large city runs a bus tour of the major sights in the city from 9:00 A.M. to noon. To plan the number and sizes of the buses that it needs for its tour service, the

TABLE 11.7

Numbers of tickets sold for city bus tour per day over 100 days

131	133	135	142	143	147	152	152	154	154	156	160	160	163	163	166	167	167	168
168	171	171	172	172	174	175	175	176	177	177	180	181	182	182	183	183	184	186
187	188	188	190	191	191	192	193	194	196	198	198	199	199	200	201	202	203	204
204	205	206	206	207	207	208	208	209	209	209	209	210	210	211	211	213	213	214
215	216	216	217	217	218	219	219	221	223	223	224	224	226	227	231	237	244	248
249	250	252	254	261														

company wants to estimate the shape of the distribution of the daily number of passengers who purchase tickets for the tour. The company thinks that even though this distribution is discrete, it can be approximated by a normal distribution. Table 11.7 shows the number of tickets sold per day over a 100-day period. The numbers are listed in ascending order.

Since no prior mean or standard deviation is specified, these parameters must be estimated from the sample data. The sample mean of the data in Table 11.7 is 195.0 and the sample standard deviation is 28.7. The next step is to group the data into classes so that the expected frequencies of a normal distribution with a mean of 195.0 and a standard deviation of 28.7 can be compared with the observed frequencies in each class. To construct the most powerful test for a given significance level, you should have as many classes as possible, provided that each class has an expected frequency of at least 5. With many small classes the pattern of expected frequencies will closely follow the shape of a normal curve. With a few large classes the comparison with the normal curve will be much rougher. Figure 11.7 illustrates this concept by

FIGURE 11.7

Expected frequencies for a hypothesis test against the normal distribution with 8 and 20 classes

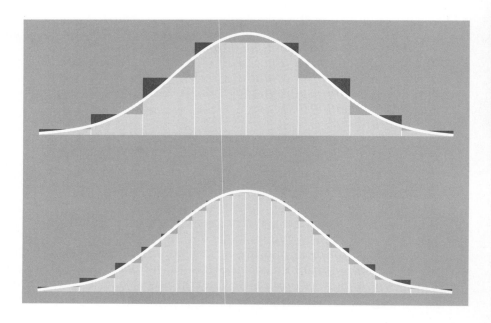

showing the expected frequencies for a pair of tests in which a normal distribution has been divided into 8 and 20 classes. A normal curve is superimposed on both sets of expected frequencies so that the comparison between the expected frequencies and the normal curve will be easier to see. The first graph shows the 8 classes and the second shows the 20. The 8 expected frequencies are based on classes with equal widths, which are common in grouped data. The set of 20 expected frequencies have classes of unequal widths. Note that the classes widen as they get farther from the mean. The reason for the unequal class widths is that each class contains 5% of the total area. With the 100 observations in our example, 20 classes that each contain 5% of the area, or an expected frequency of 5, will give the closest possible fit to a normal curve and therefore the most powerful test of the null hypothesis. The z score for the lower limit of the rightmost class is 1.64, which results in an area of 5% in the right tail. The z score of the lower limit of the next class is 1.28, because 5% of the area is between z scores of 1.28 and 1.64. Table 11.8 gives the z scores for the limits of each class, the upper and lower limit of each class, and the expected and observed frequencies in each class.

TABLE 11.8
z scores and upper and lower limits for each class and the expected and observed frequencies

z scores	Range	Expected frequency	Observed frequency
$-\infty$ to -1.64	Less than 147.9	5	6
-1.64 to -1.28	147.9 and under 158.3	5	5
-1.28 to -1.04	158.3 and under 165.2	5	3
-1.04 to -0.84	165.2 and under 170.9	5	7
-0.84 to -0.67	170.9 and under 175.8	5	7
-0.67 to -0.52	175.8 and under 180.1	5	3
-0.52 to -0.39	180.1 and under 183.8	5	5
-0.39 to -0.25	183.8 and under 187.8	5	4
-0.25 to -0.13	187.8 and under 191.3	5	3
-0.13 to 0.00	191.3 and under 195.0	5	4
0.00 to 0.13	195.0 and under 198.7	5	3
0.13 to 0.25	198.7 and under 202.2	5	5
0.25 to 0.39	202.2 and under 206.2	5	6
0.39 to 0.52	206.2 and under 209.9	5	10
0.52 to 0.67	209.9 and under 214.2	5	5
0.67 to 0.84	214.2 and under 219.1	5	8
0.84 to 1.04	219.1 and under 224.8	5	5
1.04 to 1.28	224.8 and under 231.7	5	3
1.28 to 1.64	231.7 and under 242.1	5	1
1.64 to $+\infty$	More than 242.1	5	7

The next step is to set up the template for the hypothesis test. There are 20 comparisons in Table 11.8 but we lost 2 degrees of freedom by estimating the population mean and the population standard deviation. To get the degrees of freedom, substitute into formula 11.2:

$$df = k - h - 1 \qquad\qquad (11.2)$$

$$df = 20 - 2 - 1$$

$$df = 17$$

With 17 degrees of freedom and a significance level of 5%, the critical value of the test is 27.59.

The last step is to substitute the expected and observed frequencies into formula 11.1:

$$\chi^2 = \sum_{i=1}^{k} \frac{(o_i - e_i)^2}{e_i} \qquad\qquad (11.1)$$

$$\chi^2 = \frac{(6-5)^2}{5} + \frac{(5-5)^2}{5} + \frac{(3-5)^2}{5} + \cdots + \frac{(7-5)^2}{5}$$

$$\chi^2 = 17.2$$

Since the χ^2 test statistic is 17.2 and the critical value of the test is 27.59, the null hypothesis, that the number of ticket buyers per morning follows a normal distribution, cannot be rejected.

After this last example you ought to be able to handle any goodness-of-fit problem. To summarize, the steps are:

1 Estimate the population parameters if they are not given in the null hypothesis.

2 Classify the data, if they are continuous and not already classified, so that the expected frequency of each class equals 5.

3 Calculate the degrees of freedom.

4 Find the critical value for the test.

5 Calculate the χ^2 test statistic.

6 Decide whether to reject or not to reject the null hypothesis.

Before we leave this section, a word of warning. Using the sample values in a goodness-of-fit test "loads" the test in favor of accepting the null hypothesis. It is difficult to calculate the probability of a Type II error, but it can be fairly high for a goodness-of-fit test in which the sample values are used to estimate the population parameters. The conclusion that a particular data set fits a particular population distribution should be treated with caution.

EXERCISES FOR SECTION 11.1

11.1 What is the critical value for a goodness–of–fit test if there are:
  (a) 10 degrees of freedom and the significance level is 5%?
  (b) 10 degrees of freedom and the significance level is 10%?
  (c) 20 degrees of freedom and the significance level is 5%?
  (d) 30 degrees of freedom and the significance level is 1%?

11.2 Section 11.1 contains an example of a goodness–of–fit test for the binomial about flip flashes and another test for the normal distribution about an intelligence test. Do these hypothesis tests again, this time assuming that the proportion of failures in flip flashes of 0.01 and the mean and standard deviation of test scores of 100 and 10 are all sample values. Use a 5% significance level.

11.3 Test at a 5% significance level the hypothesis that the following data came from a normal distribution with a mean of 10 and a standard deviation of 2.

Range	Frequency	Range	Frequency
Under 5	10	11 and under 14	35
5 and under 7	20	14 and under 15	15
7 and under 11	70	15 and over	8

11.4 You own a do-it-yourself frame shop. You are a franchisee of a regional chain and you use its training manual to set prices and order materials. According to your manual, 70% of all customers make metal frames, 20% of all customers make wood frames, and 10% mat their pictures on cardboard without any frames. After three months in business you have recorded the following numbers for types of frames sold: metal, 7,700; wood, 1,800; mat on cardboard, 500. Test at the 5% significance level the hypothesis that your customers' distribution of frame purchases matches the distribution suggested in the manual. In other words, test if the pattern 7,700, 1,800, and 500 could be a random sample from the population with the proportions 0.70, 0.20, and 0.10.

11.5 An appliance repair service would like to test at the 5% significance level the hypothesis that the distribution of calls for repairs per day is Poisson with a mean of 5. Use the following data, which were collected over a 48-day period, to conduct the test.

Number of calls per day	Frequency	Number of calls per day	Frequency
0	0	6	6
1	1	7	4
2	5	8	3
3	7	9	2
4	9	10	1
5	10	11 or more	0

11.6 A firm processes human milk for babies who are allergic to cow's milk and whose mothers cannot nurse them. The firm serves a three-county area and it needs to plan its inventories according to the expected number of births per month in the area. The firm wants to test the hypothesis that the expected number of births is the same over the months of a year. Test this hypothesis at the 10% significance level if the observed number of births over a particular 12-month period are 440, 380, 389, 431, 375, 383, 507, 460, 457, 436, 482, and 499.

11.7 A real estate agency's experience has been that the average time between the listing of a home and its sale is 60 days. The agency thinks that the current market is more favorable than its past experience, and it is considering advising new clients that the expected time will be shorter. The last 50 homes sold by the agency were listed for the following number of days:

Range of days	Number of homes sold in each range	Range of days	Number of homes sold in each range
0 and under 20	20	60 and under 80	4
20 and under 40	12	80 and under 100	4
40 and under 60	6	100 and over	4

Assuming that the sale of each home is independent, would the above data lead you to reject the hypothesis that the mean time for the sale of a home is 60 days at the 5% significance level? Why might the assumption of independence of sales be unwarranted in this market?

11.8 Your company puts out fires in oil wells around the world. Each oil well fire can be considered independent. For planning purposes you would like to know if the distribution of oil well fires over time is Poisson. Use the following data for the past 40 months to test this hypothesis at the 5% significance level.

Fires per month	Frequency	Fires per month	Frequency
0	4	4	3
1	9	5	4
2	10	6	3
3	7		

11.2 TESTS FOR EQUALITY AMONG PROPORTIONS, INDEPENDENCE, AND HOMOGENEITY

 11.2A TESTS FOR EQUALITY AMONG PROPORTIONS

Hypothesis tests for the difference between two proportions were discussed in Chapter 10. These tests would be on the equality of two population proportions in which the null hypothesis was $H_0: \Pi_1 - \Pi_2 = 0$. Recall that such tests required a pooled estimate of the population variance. If you had more than two populations and you wanted to test the hypothesis that all of them had the same proportion, you could do a series of pairwise tests. For example, if you had four populations labeled A, B, C, and D, you could set up hypothesis tests for the pairs AB, AC, AD, BC, BD, CD. In general, there are $_nC_2$ distinct pairs for n populations. There is a serious problem with this pairwise approach to testing the hypothesis that all of the population proportions are equal. If you set the α level for each test at 5% and you conducted the test six times for the pairs AB, AC, AD, BC, BD, CD, you would have a very high probability of concluding that the population proportions were not all equal, even if they were equal. The reason for this high probability of rejecting the null when it is true, or making a Type I error, is that if you repeat a test with a 5% chance of making a Type I error many times, you are bound eventually to make a Type I error. Each time you conduct one of the pairwise tests you could consider the test itself a binomial process with a 5% chance of "success." Your chance of making a Type I error is the probability of one or more successes in six trials, or 0.2649. If you had ten populations and you wanted to test for the equality of each population proportion with a pairwise procedure, you would have $_{10}C_2$ or 45 possible pairs. The probability of a Type I error would equal the probability of one or more successes in a binomial process with a Π of 0.05 and an n of 45. This probability is 0.9006. Thus, as the number of populations that will be compared gets larger, it becomes increasingly likely that you will reject the null hypothesis that all population proportions are equal when it is true. The moral of this discussion is *do not use a series of pairwise tests to test the equality of a set of population proportions*.

The correct procedure for these problems is to use the chi-square distribution. Here's an example. Nine teams of 20 workers each are used to assemble heat-seeking air-to-air missiles. Each missile is inspected by the Air Force before it is accepted. The management of the missile company wants to test at the 5% significance level the hypothesis that the population proportions of acceptable and unacceptable missiles are the same across the nine teams. The null and alternative hypotheses are

$H_0: \Pi_1 = \Pi_2 = \Pi_3 = \cdots = \Pi_9$
H_1: at least one of the proportions differs from the others

Note that the null would not be true if any one or more of the proportions were not equal. Thus the alternative hypothesis is not that there are nine distinct values for the population proportion of either acceptable or unacceptable missiles, but rather that there is more than one distinct value. Table 11.9 gives the recent experience of the nine teams. This type of table, often called a **contingency table,** is *a cross-listing of the frequencies of two categorical variables.* The concept is the same as that of the joint probability tables that you used in Chapter 4.

TABLE 11.9
Number of
acceptable and
unacceptable
missiles for each of
nine production
teams

RATING	TEAM NUMBER									Total
	1	2	3	4	5	6	7	8	9	
Acceptable	360	400	370	330	390	320	380	360	330	3,240
Unacceptable	40	0	30	70	10	80	20	40	70	360
Total	400	400	400	400	400	400	400	400	400	3,600

The overall sample proportion of acceptable missiles is the total number acceptable divided by the total number of missiles assembled, or 3,240 divided by 3,600. Thus the overall proportion acceptable is 0.9. The overall sample proportion of unacceptable missiles is 0.1. Without a prior hypothesis about the true population proportion acceptable, the sample proportion must be used to estimate the population proportion. To get the expected frequencies of acceptable and unacceptable missiles for each team, each column total is multiplied by the overall proportions. In this case, (400)(0.9) yields an expected frequency of 360 for the number of acceptable units by team 1. Since in this example the column totals are all 400, all of the expected frequencies of acceptable missiles are equal to 360. The overall proportion of 0.1 unacceptable missiles yields an expected frequency of 40 for each team. Thus you expect that each team does as well as the average of all of the teams. Table 11.10 shows these expected frequencies along with the original information.

The next step is to find the critical value for the test. In this example, there are 8 degrees of freedom. The reason that there are 8 degrees of freedom is that if 8 of the 18 observed frequencies are known, the other 10 can be determined. The loss of the 10 degrees of freedom is attributable to the fact that the column and row totals are used to estimate the population proportions. Table 11.11 shows the same information as Table 11.10, but with 10 of the observed frequencies missing.

We can find the number of unacceptable missiles for each of the first eight teams by subtracting the number acceptable from the row totals; 400 less 360, for example, yields 40 unacceptable for the first team. We can find the number of acceptable missiles for the ninth team by adding up the number of acceptable missiles for teams 1 through 8 and subtracting that number from the row total of 3,240. Once the number of acceptable missiles for the ninth team is

TABLE 11.10
Expected and
observed
frequencies of
acceptable and
unacceptable
missiles

RATING	TEAM NUMBER									Total
	1	2	3	4	5	6	7	8	9	
Acceptable o_i	360	400	370	330	390	320	380	360	330	3,240
e_i	360	360	360	360	360	360	360	360	360	
Unacceptable o_i	40	0	30	70	10	80	20	40	70	360
e_i	40	40	40	40	40	40	40	40	40	
Total	400	400	400	400	400	400	400	400	400	3,600

TABLE 11.11

Example showing that eight observed frequencies and the row and column totals can be used to find the missing observed frequencies

RATING	TEAM NUMBER									Total
	1	2	3	4	5	6	7	8	9	
Acceptable o_i	360	400	370	330	390	320	380	360	?	3,240
Unacceptable o_i	?	?	?	?	?	?	?	?	?	360
Total	400	400	400	400	400	400	400	400	400	3,600

determined, we can find the number of unacceptable missiles for the ninth team by subtracting the acceptables from the column total. A general formula for the degrees of freedom for tabular data in which the row and column totals are used to estimate the population proportions is

$$df = (r - 1)(c - 1)$$

(11.3)

where r is the number of rows in the table and c is the number of columns. A row is a line of observed frequencies, so there are 2 rows and 9 columns. If you substitute those values in formula 11.3, you will get 8 degrees of freedom. With 8 degrees of freedom and a 5% significance level, the critical value of the test is 15.51.

The next step is to substitute the observed and expected frequencies in formula 11.1.

$$\chi^2 = \sum_{i=1}^{k} \frac{(o_i - e_i)^2}{e_i}$$

(11.1)

$$\chi^2 = \frac{(360 - 360)^2}{360} + \frac{(400 - 360)^2}{360} + \frac{(70 - 40)^2}{40} + \cdots + \frac{(330 - 360)^2}{360}$$

$$\chi^2 = 17.78$$

The null hypothesis that the proportions of acceptable and unacceptable missiles are the same for the nine teams must be rejected. Having reached this conclusion, the management of the missile assembly plant should investigate why some teams are doing better than others. Apparently teams 4, 6, and 9 are doing poorly and need help first. Teams 2 and 5 are doing well and could be used as models. Perhaps a comparison of the two extremes would reveal what is being done right and what is being done wrong.

The rule of thumb that e_i has to equal at least 5 still applies to these chi-square tests. This rule is fairly conservative; some statisticians suggest that approximation to the χ^2 distribution will be sufficiently close if e_i is 1 or more. We'll stick to 5. Again, if some of the e_i's are less than 5, you can do the test by collapsing categories.

TABLE 11.12
Observed frequencies of late and on-time deliveries by three overnight services

DELIVERY STATUS	DELIVERY SERVICE			Total
	U.S. Postal Service	Federal Express	Purolator	
Late deliveries o_i	10	10	20	40
On-time deliveries o_i	70	90	200	360
Total	80	100	220	400

Here is another example of a test of equality of population proportions. The difference between this example and the last is that here the column totals are not equal. Your company manufactures a poison antidote and you frequently have to send packages of your product by one of the overnight delivery services. You have used the U.S. Postal Service's Express Mail, Federal Express, and Purolator. You would like to test at a 1% significance level the hypothesis that each of these services is equally reliable in delivering your packages on time. You have sent 80 packages via the U.S. Postal Service, 100 via Federal Express, and 220 via Purolator. The respective numbers of packages that were not on time were 10, 10, and 20. Table 11.12 presents the observed frequencies.

The total number of late deliveries is 40, or 10% of the 400 deliveries made. If the null hypothesis of equal population proportions were true, the expected frequency of late deliveries by any of the services would be 10% of the number of packages that they handled. For example, 10% of the 80 packages handled by the Postal Service is 8. Similarly, across the three services, 90% of all packages were delivered on time. The expected frequency of the number of packages delivered on time by each service is 90% of the number that each service handled. For example, 90% of 80 yields 72 on-time packages for the Postal Service. Table 11.13 shows the expected and observed frequencies. The

TABLE 11.13
Observed and expected frequencies of late and on-time deliveries by three overnight services

DELIVERY STATUS	DELIVERY SERVICE			Total
	U.S. Postal Service	Federal Express	Purolator	
Late deliveries o_i e_i	10 8	10 10	20 22	40
On-time deliveries o_i e_i	70 72	90 90	200 198	360
Total	80	100	220	400

formula for the degrees of freedom is $(r - 1)(c - 1)$, or $(2 - 1)(3 - 1)$. The degrees of freedom for this problem are 2. With the specified significance level of 1%, the critical value for the test is 9.21.

The next step is to substitute into formula 11.1 to determine the value of the χ^2 test statistic:

$$\chi^2 = \sum_{i=1}^{k} \frac{(o_i - e_i)^2}{e_i} \tag{11.1}$$

$$\chi^2 = \frac{(10 - 8)^2}{8} + \frac{(10 - 10)^2}{10} + \frac{(20 - 22)^2}{22} + \frac{(70 - 72)^2}{72} + \frac{(90 - 90)^2}{90}$$
$$+ \frac{(200 - 198)^2}{198}$$

$$\chi^2 = 0.76$$

The χ^2 test statistic is well below the critical value of 9.21, so there is not enough evidence to reject the null hypothesis of equal population proportions.

■ 11.2B TESTS FOR INDEPENDENCE AND HOMOGENEITY

When the procedure used in section 11.2A is extended beyond a two-way classification of the variable that is the subject of the hypothesis test, the test is no longer called a test of equality among proportions. It is either a test of independence or a test of homogeneity. For a test of independence you have one population, say all shoppers at a particular supermarket, and you are testing whether two characteristics in the population, say method of payment and time of shopping, are independent. With a test of homogeneity you have more than one population, say shoppers at supermarkets in different cities, and you are testing to see if a characteristic, say the proportion who use each of three methods of payment, is the same, or "homogeneous," across populations. With a test of independence, neither the column totals nor the row totals of the contingency table are determined in advance. A sample of shoppers would be drawn from the one store, and column and row totals would depend on the number of shoppers in that sample who use each method of payment and the number who shop during each time segment. In a test of homogeneity, the columns usually refer to the samples drawn from the two populations, and these amounts are determined in advance. The row totals still depend on the number of people in the samples who use each method of payment summed across the different stores.

Here is an example of a test of independence. A supermarket wants to test at the 5% significance level the hypothesis that the method of payment for groceries is independent of the time of purchase. There are three methods of payment: check, cash, and credit card. With a test of independence, there is no basis for choosing which variable will go across the top of the contingency

TABLE 11.14
Contingency table for time of purchase and method of payment in a supermarket

	TIME PERIOD			
PAYMENT METHOD	Morning	Afternoon	Evening	Total
Check o_i	750	1,500	750	**3,000**
Cash o_i	125	300	75	**500**
Credit card o_i	125	200	175	**500**
Total	1,000	2,000	1,000	4,000

TABLE 11.15
Expected and observed frequencies of method of payment by time of purchase

	TIME PERIOD			
PAYMENT METHOD	Morning	Afternoon	Evening	Total
Check o_i	750	1,500	750	**3,000**
e_i	750	1,500	750	
Cash o_i	125	300	75	**500**
e_i	125	250	125	
Credit card o_i	125	200	175	**500**
e_i	125	250	125	
Total	1,000	2,000	1,000	4,000

table and which along the side. With a test of homogeneity, the populations go across the top of the table. Table 11.14 is a contingency table that shows some sample data for method of payment and time of purchase. The overall sample proportion of payments by check is the row total of 3,000 divided by the grand total of 4,000 customers, or 0.75. The overall sample proportion of payments by cash is the row total of 500 divided by the grand total of 4,000, or 0.125. The proportion of payments by credit card is also 0.125. To get the expected frequency for each cell in the contingency table, multiply the column total by the proportion for the type of payment. For example, the expected frequency for payment by check in the morning is 0.75 times 1,000, or 750. Table 11.15 shows the expected and observed frequencies.

The next step is to find the critical value for the test. Substitution into formula 11.3 yields

$$df = (r - 1)(c - 1) \qquad (11.3)$$
$$df = (3 - 1)(3 - 1)$$
$$df = 4$$

With 4 degrees of freedom and a significance level of 5%, the critical value of the test is 9.49.

The last step is to substitute the observed and expected frequencies in formula 11.1:

$$\chi^2 = \sum_{i=1}^{k} \frac{(o_i - e_i)^2}{e_i}$$

(11.1)

$$\chi^2 = \frac{(750 - 750)^2}{750} + \frac{(1,500 - 1,500)^2}{1,500} + \frac{(750 - 750)^2}{750} + \cdots + \frac{(175 - 125)^2}{125}$$

$$\chi^2 = 60$$

Since the test statistic of 60 exceeds the critical value of 9.49, the null hypothesis, that the method of payment by grocery shoppers is independent of the time of purchase, must be rejected. Note that all of the deviations from expectations are in the categories of cash or credit card in the afternoon or the evening. Specifically, there is less use of cash and more of credit cards than expected in the evening, while this pattern is reversed in the afternoon. The supermarket could use this information to plan when cash should be removed from the store safe or to see if the different methods of payment imply differences in shopping patterns across times of the day which can be exploited for greater sales.

Here is an example of a homogeneity test. The French Peasant is a chain of gourmet restaurants that has locations in four cities. The restaurant offers a fixed–price dinner consisting of a choice of hors d'oeuvre, salad, soup, entrée, and dessert. The five entrées currently offered by the restaurant are rack of lamb, Châteaubriand, fillet of sole Véronique, duck à l'orange, and lobster thermidor. For ordering purposes, the restaurant chain would like to test the hypothesis that the population proportions of each type of entrée are the same across its four locations. Use the data in Table 11.16, taken from three weeks of operation at each location, to test this hypothesis at the 5% significance level.

TABLE 11.16
Observed frequencies of entrée orders by restaurant location

ENTREE	LOCATION 1	2	3	4	Total
Lamb	25	30	40	55	150
Beef	15	20	25	40	100
Sole	25	35	35	55	150
Duck	10	15	15	10	50
Lobster	25	25	60	40	150
Total	100	125	175	200	600

TABLE 11.17
Expected and observed frequencies of entrée orders by restaurant location

ENTREE	LOCATION 1	2	3	4	Total
Lamb					
o_i	25	30	40	55	150
e_i	25	31.25	43.75	50	
Beef					
o_i	15	20	25	40	100
e_i	16.67	20.83	29.17	33.33	
Sole					
o_i	25	35	35	55	150
e_i	25	31.25	43.75	50	
Duck					
o_i	10	15	15	10	50
e_i	8.33	10.42	14.58	16.67	
Lobster					
o_i	25	25	60	40	150
e_i	25	31.25	43.75	50	
Total	100	125	175	200	600

The procedure for finding the expected frequencies in a homogeneity test is exactly the same as the procedure in a test for independence. If 150 out of 600 diners, or 25%, ordered lamb, then if the four restaurants have homogeneous populations, 25% of the orders in each restaurant will be expected to be lamb. Table 11.17 shows the expected and observed frequencies.

The degrees of freedom for a homogeneity test are also determined with the $(r-1)(c-1)$ formula. In this case, the degrees of freedom are 12. The critical value at a 5% significance level is 21.0. The next step is to substitute into the formula for the χ^2 test statistic:

$$\chi^2 = \sum_{i=1}^{k} \frac{(o_i - e_i)^2}{e_i} \qquad (11.1)$$

$$\chi^2 = \frac{(25 - 25)^2}{25} + \frac{(30 - 31.25)^2}{31.25} + \frac{(40 - 43.75)^2}{43.75} + \cdots + \frac{(40 - 50)^2}{50}$$

$$\chi^2 = 20.02$$

The χ^2 test statistic is just below the critical value, so there is not sufficient evidence to reject the null of population homogeneity at the 5% significance level.

EXERCISES FOR SECTION 11.2

11.9 Test the null hypothesis that $\Pi_1 = \Pi_2 = \Pi_3 = \Pi_4 = \Pi_5$ at the 5% significance level if samples of 100, 200, 200, 250, and 250 had sample proportions of 0.5, 0.55, 0.43, 0.5, 0.5, respectively.

11.10 Test at the 1% significance level the hypothesis that the population proportions of furniture buyers who use and do not use in-store credit are the same across race-sex classifications if in a sample of 1,000 furniture buyers 400 are white men, 400 are white women, 100 are black men, and 100 are black women, and the respective numbers of purchasers who used in-store credit are 50, 40, 20, and 18.

11.11 Test the hypothesis that the population proportions of operations that do and do not result in malpractice lawsuits are the same across types of surgery if 300 heart surgeries sampled resulted in 10 lawsuits, 600 brain surgeries resulted in 24 lawsuits, 200 appendectomies resulted in 5 lawsuits, and 200 tonsillectomies resulted in 10 lawsuits. Use a 5% significance level.

11.12 Use the sample data in the following table to test at the 5% significance level the hypothesis that the brand of beer preferred by a man who drinks beer at least once a week is independent of his personality profile.

PERSONALITY PROFILE	BEER BRAND						
	Budweiser	Miller	Strohs	Coors	Schlitz	Old Milwaukee	Other
Family man	50	40	30	20	10	15	60
Party man	60	50	25	18	8	12	70
Sportsman	80	60	40	30	12	22	100
Other	45	35	32	16	5	11	60

11.13 Use the sample data in the following table to test the hypothesis that preferences for goat cheese are the same across populations. Use a 5% significance level.

GOAT CHEESE PREFERENCE	POPULATION			
	U.S.	Greece	France	Great Britain
Hate goat cheese	800	200	300	600
Like goat cheese	100	500	300	500
Other	100	200	300	400

11.14 An underwriter for an automobile insurance company is evaluating a new type of discount offered by the company. Teenagers who have good grades in high school are offered lower rates on auto insurance. On the basis of 500 policy-years, the underwriter would like to test the hypothesis that high school grades are independent of accident experience. Use a 5% significance level and the following data.

GRADE CATEGORY	NUMBER OF CLAIMS			Total
	None	One	More than one	
Good grades	120	140	40	300
Lower grades	50	80	70	200
Total	170	220	110	500

11.15 A professor is curious about the relationship between the type of calculator a student uses in a statistics class and the course grade the student receives. The professor asks each student to write down the name of his or her calculator on the final exam. The calculators are then classified as "sophisticated," "general," and "simple" on the basis of the number of statistical functions that they support. At the 5% significance level, can the hypothesis that the type of calculator is independent of the student's grade be rejected? If the null hypothesis is rejected and more A students than expected have sophisticated calculators, should the professor tell students in subsequent classes that a sophisticated calculator will improve their grades?

TYPE OF CALCULATOR	COURSE GRADE					Total
	A	B	C	D	F	
Sophisticated	110	130	205	135	20	600
General	20	35	80	44	21	200
Simple	20	35	65	21	59	200
Total	150	200	350	200	100	1,000

11.16 An automobile dealer who currently carries Chevrolet and Honda cars is considering adding the Yugo line to his dealership. The dealer would carry the Yugo if it complemented the other two lines of cars. If the same buyers were interested in more than one of these makes, the addition of the Yugo line would cause the dealer to bear a franchise fee and higher inventory and training costs without adding to his sales. To test whether the car lines appeal to different types of buyers, the dealer mails local residents a coupon good for half off on tickets to a baseball game if they come in for a test ride. One thousand coupons each were mailed for test rides in a Chevy, Honda, and Yugo. Of these coupons, 500 were used for the Chevy, 400 for the Honda, and 300 for the Yugo. Every person who showed up for the test ride was classified as either single man, single woman, married man, or married woman.

    (a) Use the data on the next page to test the hypothesis that the same types of people are interested in all three cars. Pick your own significance level.

    (b) Is this a test of independence or homogeneity?

    (c) What, if any, are the flaws in this research design?

TYPE OF CUSTOMER	CAR MAKE			Total
	Chevy	Honda	Yugo	
Single man	200	150	150	**500**
Single woman	50	80	70	**200**
Married man	200	120	80	**400**
Married woman	50	50	0	**100**
Total	500	400	300	**1,200**

 SUMMARY

The chapter began with a discussion of why the test statistic $(o_i - e_i)^2/e_i$ summed over the k comparison terms would be approximately distributed as the sum of a series of independent squared standard normal random variables if the expected value of o_i equaled e_i for every comparison. A second requirement for this approximation to hold was that the e_i's equal at least 5. The sum of a series of independent squared standard normal random variables is the chi-square distribution.

The chi-square distribution has several important uses. Many of the hypothesis tests throughout this text rely on the assumption that the populations from which the samples have been drawn are normally distributed. While a goodness-of-fit test for the normal distribution can take a great deal of time with a hand-held calculator, the time can be justified if an accurate test is needed. Statistical computer packages can be used to avoid the tedious calculations. Sophisticated statistical packages automatically check for normality when a test that requires normally distributed populations is used. Goodness-of-fit tests can be used to test whether the terms of a contract or a guarantee are being fulfilled. They can also be used to test whether a process follows another common distribution, such as the binomial or the Poisson. These tests have important business applications; if a firm can determine that the arrival of customers follows a distribution such as the Poisson, it can make use of that knowledge in planning. The degrees of freedom for a goodness-of-fit test are the number of comparisons after cells have been collapsed minus the number of population parameters that have to be estimated from the sample data and minus 1, because only $k - 1$ comparisons among k can be independent.

Another important use of the chi-square distribution is to test for homogeneity or independence in cross-classified categorical data. These tests have many applications in marketing, quality control, banking, and so on. In these tests the degrees of freedom are computed as the product of the number of columns minus 1 times the number of rows minus 1. The exercises below will introduce you to some of these applications.

 NEW TERMS

chi-square distribution The sum of a series of independent squared standard normal random variables.

contingency table A cross-listing of the frequencies of two categorical variables.

NEW FORMULAS

χ^2 test statistic

$$\chi^2 = \sum_{i=1}^{k} \frac{(o_i - e_i)^2}{e_i} \qquad (11.1)$$

Degrees of freedom for a goodness-of-fit test

$$df = k - h - 1 \qquad (11.2)$$

Degrees of freedom for a contingency table

$$df = (r - 1)(c - 1) \qquad (11.3)$$

ADDITIONAL EXERCISES

An asterisk designates an exercise of greater than ordinary difficulty.

11.17 Test at the 5% significance level the hypothesis that the population proportions of consumers who prefer Classic Coke to new Coke is the same across cities if 300 out of 1,000 people surveyed in Detroit preferred Classic Coke, 325 out of 1,000 people surveyed in Dallas preferred Classic Coke, and 350 out of 1,000 people in San Francisco preferred Classic Coke.

11.18 Public blood banks typically have a difficult time persuading people to donate blood. They make about ten calls for every donation. To save costs on calling people, a blood bank would like to identify characteristics that make it more likely that someone will agree to donate. One variable that is hypothesized to affect the willingness to donate is whether a person has ever received blood from the blood bank. Use the following 3-by-2 classification of sample data to test at the 5% significance level the hypothesis that having received blood from the bank is independent of donation behavior: received blood and donated, 100; received blood, promised to donate, but did not show up, 50; received blood and refused to donate, 400; did not receive blood and donated, 800; did not receive blood, promised to donate, but did not show up, 350; did not receive blood and refused to donate, 1,400.

11.19 Airline flights can be classified as either departing on time, departing late, or canceled. If your company booked many flights and you knew the proportion of flights in the airline industry which fell into these three classifications, you could test whether your experience with a particular airline could be considered a random sample drawn from a population that had the industry proportions. For example, if the industry proportions were 0.8, 0.17, and 0.03, you had the records of 1,000 flights by your employees with that airline, and there were 812, 165, and 23 flights in the respective categories, should you reject or not reject the null hypothesis?

11.20 To test consumer satisfaction with its super-premium ice cream, Haagen-Däz prints a tollfree complaint number on production runs of 10,000 pints of ice cream at each of five plants. The tollfree numbers are different for the five plants and a log is kept of the number of complaints coming in to each number. Test at the 5% significance level the hypothesis that the proportion of consumers who complain is the same across the five plants if the number of complaints logged are 100, 120, 110, 130, and 100.

11.21 The manufacturer of Fruit of the Loom™ men's briefs uses inspectors to judge the quality of its finished briefs. When an inspector finds that a garment is acceptable, he or she puts a tag on the garment that reads "Inspected by No. 5" or whatever number the inspector is assigned. The manufacturer would like to know if the various inspectors are applying the same standard when they decide which garments are acceptable. Suppose there are 10 inspectors and that in a particular week the respective proportions of garments found acceptable by the inspectors were 0.88, 0.89, 0.90, 0.90, 0.90, 0.91, 0.91, 0.92, 0.92, and 0.93. The respective number of garments inspected were 4,500, 5,500, 6,000, 7,000, 8,000, 8,000, 8,000, 8,000, 8,000, 8,000.

(a) Formulate the appropriate null and alternative hypotheses, select a significance level, and conduct the hypothesis test.

(b) Can you suggest a different research design that would give the manufacturer a better idea of whether the inspectors are applying a consistent standard?

11.22 A college football coach claims that the criticism of the football program for failure to emphasize academics is unwarranted. Among all nonathletes at the university in the 1980 freshman class, 40% have not graduated, 40% graduated with their class, 10% graduated a year after their class, and 10% graduated more than a year after their class. Among the 50 football players who were freshmen in 1980, the corresponding percentages are 50%, 30%, 10%, and 10%. The coach claims that a random sample from the general student body could easily have yielded the percentages found on the football team. Test the coach's claim at a 5% significance level.

11.23 An office supply firm gives small presents, such as desk calendars bearing the company's name and phone number, to its regular customers. The presents are supposed to remind customers of the office supply company and lead to more orders. At Christmas the company gives out its low-cost calendars to one-third of its customers, a Cross™ pen and pencil desk set to one-third of its customers, and nothing to the remaining third. At the end of a year the office supply firm classifies each customer's orders as declining, constant, or increasing.

(a) Use the following data to test the hypothesis that the type of present given by the office supply company does not affect a customer's orders.

(b) Is this a test of homogeneity or of independence?

(c) Interpret the results of the test. What administrative decision should the office supply company make?

SALES RESPONSE	GIFT		
	Calendar	Pen set	Nothing
Increasing	45	50	55
Constant	15	12	20
Declining	40	38	25

11.24 An investigator for the Securities and Exchange Commission suspects that insider trading may have driven up the price of a particular stock. Insider trading occurs when, for example, a corporation's officers learn that their corporation plans to make a take-over bid and then purchase shares of the company that is the object of the take-over bid. Before going to the expense and trouble of tracing every purchase order, the investigator wants to check whether there was an unusual number of purchases of the stock in question. She observes that the average rate of purchases of blocks of 1,000 or more shares was one per hour. Such large blocks of shares tend to be bought by institutional investors, such as pension funds. On the day before the take-over bid was announced, the following times between purchases were observed:

Time between large purchases of stock	Frequency
Under 1 hour	10
1 hour and under 2 hours	8
2 or more hours	3

If there was no insider trading and each purchase was an independent event, could the above data be a sample from a Poisson process with a mean of 1? Pick your own significance level.

11.25 Many of the manufacturers of floppy discs offer a "lifetime" guarantee on their discs. The guarantee is only to replace a defective disc and does not cover the value of any data lost on the disc. These guarantees cost the disc manufacturers very little because the discs are so inexpensive that most consumers don't bother to mail in the defective disc and seek a replacement. One company, Maxell, has offered a two-for-one guarantee, which it hopes will convince consumers that its discs are more reliable. If the two-for-one guarantee encourages consumers to mail in their defective discs, the company will have much greater costs. Suppose the company wanted to estimate the potential costs. It gave 1,000 boxes of 10 discs each to consumers and asked them to use the discs for six months. Use the following data to test at the 5% significance level the hypothesis that the distribution of the number of failures per box is binomial.

Failures per box	Frequency	Failures per box	Frequency
0	950	2	10
1	40	3 or more	0

11.26 Your company manufactures chocolate candy bars and you are interested in how the color of the wrapper affects consumers' decisions. You set up a test in a shopping mall in which consumers are offered a choice of either a Hershey™ bar in its familiar brown wrapper or one of your candy bars. You run the test 400 times with your candy bar wrapped in red, blue, brown, and yellow in 100 of the tests. The proportions who choose your bar are: blue, 0.58; red, 0.50; brown, 0.50; and yellow, 0.44. Use a 5% significance level.

11.27 This exercise continues exercise 11.26. Each consumer who chooses one of your candy bars is asked to rate it as "excellent," "good," "fair," or "poor." The candy bars under the wrappers are identical, so the purpose of this question is to see if the color of the wrapper affects perceptions of the taste of the bar.

(a) Use the following data to test this hypothesis at the 5% significance level.

(b) Is there anything wrong with this research design?

RATING	COLOR				Total
	Red	Blue	Brown	Yellow	
Excellent	12	12	12	14	50
Good	10	10	6	14	40
Fair	14	14	16	16	60
Poor	22	14	16	0	52
Total	58	50	50	44	202

11.28 Movie distribution companies often show a movie in selected theaters for one night in what is called a "sneak preview." The reactions of the audiences to the movie during a sneak preview help the distribution company decide how much to spend to promote the movie and into how many theaters to try to book it. One distribution company is concerned that sneak preview audiences may vary across the country and that a favorable response in one region will not guarantee that the movie will be well received in other regions. To check whether the reactions of sneak preview audiences are the same across regions, the distribution company shows the same movie in Los Angeles, Detroit, Salt Lake City, and Minneapolis. Everyone in the audience is asked to rate the movie as *A*, "must see again"; *B*, "will recommend to friends but won't see again"; or *C*, "will not recommend."

(a) Use the following data to test the hypothesis at the 5% significance level.

(b) Is this a test of homogeneity or independence?

RATING	CITY			
	Los Angeles	Detroit	Salt Lake City	Minneapolis
A	600	620	500	605
B	400	388	345	414
C	200	210	178	183

11.29 The movie distribution company in exercise 11.28 is concerned about the reliability of sneak previews not only across regions but also as a predictor of the reactions of regular audiences. The distribution company thinks that the people who show up for sneak previews often are movie buffs whose reactions will differ from those of the general public. Use the data on reactions of sneak preview audiences and regular audiences to the movie *Star Trek V* to see if these samples could have come from populations with the same proportions. The *A*, *B*, and *C* responses are the same as in exercise 11.28. Pick your own significance level.

RATING	AUDIENCE	
	Sneak preview	Regular
A	3,346	3,027
B	1,148	1,361
C	456	830

11.30 The Personal Computers Peripherals Corporation (PCPC) is designing a new hard disc for microcomputers. To assess the potential warranty claims, PCPC bench-tests 50 units until they fail.

(a) Use the following data on days until breakdown to determine the mean time until failure and test the hypothesis that the distribution of time until failure is exponential.

2	2	3	4	4	6	6	8	8	10
12	15	18	23	25	25	27	28	29	30
32	37	41	49	50	52	53	67	69	70
74	75	79	89	100	110	121	122	133	143
161	168	173	177	209	213	215	258	273	302

(b) If PCPC offered a 90-day warranty on its hard discs, what proportion would be expected to require warranty service?

11.31 An economist claims that trends in the past price of a foreign currency cannot be used to predict future changes in the currency's price. For example, the economist claims that the change in the closing price of the Canadian dollar from one day to the next in New York is approximately normally distributed with a mean of 0. The economist records the following 100 changes in closing prices of the Canadian dollar recorded in U.S. pennies. Can the claim be rejected at a 5% significance level?

−3	−2	0	1	0	−1	1	−2	1	1
2	4	0	2	−5	−5	3	−1	4	−3
0	−1	−3	−5	4	−1	−8	1	−8	−1
−4	2	0	0	1	−3	0	−3	−1	3
0	−3	−4	2	−3	5	−3	1	2	5
3	0	2	−4	0	3	−1	−1	1	3
1	1	1	−1	1	0	−1	2	−4	−5
4	2	−2	1	2	4	−3	−4	0	4
−2	0	−4	5	2	2	−2	−2	−2	4
−2	−4	1	0	0	−8	−4	0	2	−1

11.32 The media manager of a large consumer products firm is concerned about the accuracy of low-cost polls that ask people which television shows they watch. The Nielsen survey provides a high-cost alternative based on either diaries kept by house-holds or electronic records. The latter are created by a monitoring device that records the channels to which a television set is tuned. Suppose that according to Nielsen's electronic records, the ratings share (defined as the percentage of homes with televi-sion sets turned on) was as follows for a particular time slot on a particular Sunday evening: *Masterpiece Theatre,* 5%; *The New Gilligan's Island,* 15%; and the ABC movie, *Rocky VII,* 30%. The remaining 50% of the sets were tuned to videos, cable channels, or the test pattern. In the same time slot on the same Sunday evening, the low-cost poll asked 1,000 people if they were watching television, and among the 500 who answered yes, the proportions who said they watched *Masterpiece Theatre, The New Gilligan's Island,* and the ABC movie were, respectively, 8%, 12%, and 33%.

Could the discrepancies between the Nielsen electronic records and the low-cost poll be due to sampling variation? Use a 5% significance level.

11.33* The trade association for carbonated beverage bottlers wants to test the hypothesis that the proportion of families who purchase soft drinks in cans is not affected by a state law requiring deposits on all soft drink cans. The association has conducted a telephone survey of 100 families in every state and it plans to do a series of pairwise tests for every possible combination of two states. The null of overall equality will be rejected if any of the pairwise tests results in its rejection.

    (a) If the pairwise tests are at a 1% significance level, what is the significance level for the overall test?

    (b) Suggest a better design for testing the hypothesis which uses the same data.

11.34* You are the marketing director for the AAA Travel Service and want to test a hypothesis about the percentages of consumers who prefer the following types of vacations over the course of a year: (1) one long vacation, (2) two medium-length vacations, (3) three short vacations, (4) more than three very short vacations. Your hypothesis is that 40% of all consumers prefer option 1, 10% prefer option 2, 15% prefer option 3, and 35% prefer option 4. When you survey 100 randomly selected individuals, you find that 43 prefer option 1, 4 prefer option 2, 16 prefer option 3, and 37 prefer option 4.

    (a) Do a chi-square test of your hypothesis at the 5% significance level.

    (b) Test at the 5% significance level the hypothesis that the proportion of consumers who prefer option 2 is 10%, using the method for a test on the population proportion discussed in Chapter 9.

    (c) One of the above tests will tell you to reject the null hypothesis and the other will not. Which test is appropriate for this problem? Why?

12

Analysis of Variance

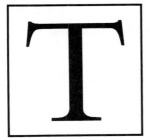

he Federal Home Loan Bank Board (FHLBB) regulates savings and loan associations and serves as a "banker's bank" for its member savings and loans. The member savings and loan associations deposit money in fixed-interest-rate accounts with the FHLBB and in turn the FHLBB reinvests these funds in liquid, low-risk assets. The typical level of deposits held by the FHLBB is about $100 million. This money is split up into smaller funds that are managed by different individuals. Every business day the managers of the separate funds buy and sell assets, such as 30- or 90-day Treasury notes or 24-hour interbank notes. The goal of the managers of the separate funds is to obtain the highest possible yield while limiting their investments to the types of assets approved by the FHLBB. The fund managers can earn a high yield if they can anticipate the direction of daily changes in interest rates or if they can arbitrage differences in the yields of assets with different maturities. Since the funds they manage are so large, a tiny difference in a manager's average daily yield can make a big difference in the profits of the FHLBB.

Suppose the FHLBB relies on four managers who act independently with separate funds. The FHLBB wants to test the hypothesis that the average yields of its four separately managed funds are equal. If this hypothesis can be rejected, the FHLBB will look for new managers for the funds that have done relatively poorly and reward the managers of the more successful funds. There is more to this problem than simply calculating the average yield of each of the four funds over some time period. The average yield over some time period is really just a point estimate of what each manager could earn in the future. What the FHLBB is interested in testing is the hypothesis that the future earnings potentials or abilities of the four managers are the same. Since there are random variations in the movement of interest rates, over a short period a high-ability manager could do poorly and a low-ability manager could do well. The FHLBB needs to know if the observed differences in the four managers' yields could be due to such chance variations or to real differences in their abilities.

In Chapter 10 you learned how to test the hypothesis that the means of *two* populations were equal. What you have in this problem is *four* populations, namely, potential yields of the four managers employed by the FHLBB. What you could do in this problem is conduct a series of tests comparing the yields of all possible combinations of two managers taken from the four employed by the FHLBB. For example, you could test whether the mean yield of manager 1 equaled the mean yield of manager 3, or if the mean yield of manager 4 equaled the mean yield of manager 1, and so on. There are six possible sets of tests for comparing the four managers, two at a time. There is something fundamentally wrong with this procedure: if the null hypothesis were true, you would be very likely to make a Type I error. Suppose that the significance level of each of the six tests was set at 5% and that the null hypothesis of equal population means was true. The probability of making a Type I error—that is, concluding that the mean yields were not all equal when in fact they were equal—would be close to 0.26. Why is the probability of a Type I error so high? Think of each of the six hypothesis tests for sets of two managers as binomial trials with a Π of 0.05 and "success" defined as rejecting the null hypothesis that two managers have the same mean yields when that null is true. You would reject the null that all four managers have equal mean yields if you found *one or more* of the six tests showed a significant difference. This procedure is the equivalent to obtaining one or more successes in six trials of a binomial process with a Π of 0.05, which according to the binomial table has a probability of 0.2649. Now you can see why you would be very likely to reject the null hypothesis of equal yields if you did a series of two population tests. If there were more than four managers of the funds, you would be even more likely to make a Type I error. The idea here is that if you keep repeating a test with a 5% chance of making an error, it is very likely that you will eventually make an error.

The correct procedure for this problem is **one-way analysis of variance,** which is *a joint test for the equality of more than two population means.* With this procedure you can test the hypothesis that all population means are equal and hold the significance level to any desired amount. In Chapter 10 you learned how to test the hypothesis that $\mu_1 - \mu_2$ equals 0 or, equivalently, that $\mu_1 = \mu_2$. In this chapter you will learn how to test the hypothesis that $\mu_1 = \mu_2 = \mu_3 = \cdots = \mu_k$, where k is any integer. The alternative hypothesis is that *one or more of the population means is not equal to the other population means.*

Section 12.1 introduces one-way analysis of variance and discusses the F distribution. Section 12.2 discusses confidence intervals and hypothesis tests in one-way analysis of variance. Section 12.3, on two-way analysis of variance with a factorial design and two-way analysis of variance with a randomized block design, is optional. Two-way analysis of variance with a factorial design allows you to test two sets of hypotheses over the same data set. For example, you could test the hypothesis that the four managers have the same average yield and the hypothesis that yields are the same on days when interest rates rise or fall. Two-way analysis of variance with a factorial design also allows you to test for interactions among your factors; for example, do some managers have significantly higher mean yields during periods with rising interest rates while others do well when interest rates are falling? This last hypothesis could be true even if the mean yields across managers and across rising and falling interest rates were the same. Two-way analysis of variance with a randomized block design is the multipopulation counterpart to the paired difference-in-means test you studied at the end of Chapter 10. With this design the interaction effects between the factors are eliminated and you would test only for a difference in yields across managers or across days with rising or falling interest rates. Two-way analysis of variance with either the factorial design or the randomized block design is a good deal more complicated than one-way analysis of variance, so don't be alarmed if some of these ideas are not yet clear to you. This preview is just meant to give you an idea of where you are going. Some computer programs go up to thirty-way analysis of variance, but we'll stop at two, because problems in which more than two factors are used to explain the variation in the variable of interest can be more easily handled in the multiple regression framework discussed in Chapter 14.

12.1 ONE-WAY ANALYSIS OF VARIANCE AND THE *F* DISTRIBUTION

12.1A THE NULL AND ALTERNATIVE HYPOTHESES AND ASSUMPTIONS NEEDED FOR ONE-WAY ANALYSIS OF VARIANCE

Suppose there were four distributions, X_1, X_2, X_3, and X_4 representing the daily yields of the funds controlled by the four managers. The distributions would have means of μ_1, μ_2, μ_3, and μ_4, respectively. The null and alternative hypotheses that the FHLBB would like to test are:

H_0: $\mu_1 = \mu_2 = \mu_3 = \mu_4$ H_1: Not all of the means are equal

One-way analysis requires three assumptions about the distributions of X_1, . . . , X_n. The first assumption is that each population is normally distributed and the second assumption is that the variances of all of the populations are equal. The third assumption is that the observations taken from each population are independent. One-way analysis of variance is fairly robust to moderate departures from the first two assumptions. If the populations were not quite normally distributed or if the variances were not exactly equal, the test would still be very likely to lead you to reject the null hypothesis of equal means if the null was false. However, the test is very sensitive to violations of the assumption of independence. For example, if the managers' yields moved together over time because they were all influenced by overall yields in the market, one-way analysis of variance would not provide a good test of equality of the mean yields across managers. You would have to control for marketwide interest rates before you would have a reliable test. If there are sharp departures from normality or from equal variances, you must use the nonparametric counterpart to one-way analysis of variance, the Kruskal-Wallis test for equality of ranks. Nonparametric tests require less stringent assumptions about the population distributions. This test is discussed in Chapter 17, on nonparametric tests, as is the Lilliefors test for normality. The Bartlett test for equality of population variances may be found in Neter (Suggested Readings).

Figure 12.1 shows two situations. In the upper panel there are four normal distributions with equal variances and *different* means and in the lower panel

FIGURE 12.1

Population distributions when
the null hypothesis of equal
population means is false
and when it is true

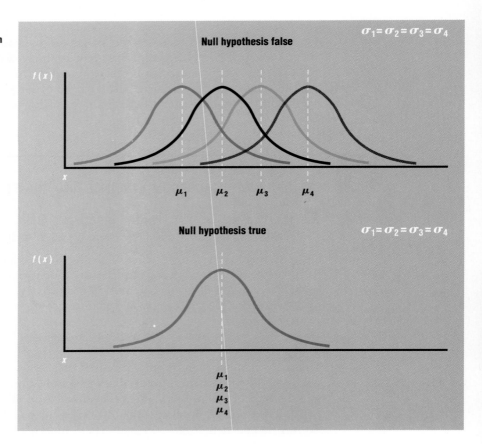

FIGURE 12.1

Population distributions when
the null hypothesis of equal
population means is false
and when it is true

there are four normal curves with equal variances and *equal* means. The four
distributions in the lower panel are superimposed. The purpose of this figure is
to illustrate two possible states of the world. If the null hypothesis and the
parametric assumptions were true, the lower panel would represent the distributions of daily yields of the four managers. While there are many possible
arrangements showing one or more of the four managers' mean yields differ-
ing from the others, the upper panel shows all four as distinct distributions
because this arrangement shows the greatest contrast between states of the
world when the null or the alternative hypothesis is true.

12.1B THE *F* TEST STATISTIC AND THE LOGIC BEHIND IT

You might be tempted to use four sample means from the populations X_1, X_2,
X_3, and X_4 to decide if the null hypothesis of equal population means should be
rejected, implying that the state of the world is probably something like the

upper panel, or not to reject the null hypothesis, implying that the state of the world is probably described by the lower panel. As an intuitive first approach to the problem, if the four sample means are very close to each other, you would lean toward believing the null hypothesis; and if they were far apart, you would lean toward rejecting the null and think that the upper panel is the accurate picture. This description of whether to reject or accept the null does not really help you because if you don't know the common variance of the four populations, you have no idea of what differences between the sample means might be considered close or far apart. Suppose that the four funds managers' average daily yields expressed on an annualized basis were 10.01%, 10.02%, 10.02%, and 10.04%. These numbers look very close, but suppose the population standard deviations of the distributions of daily yields were 0.001%. Then the distance between the lowest sample mean of 10.01% and the highest sample mean of 10.04% would be 30 multiples of a standard deviation—an enormous distance on a normal curve.

Analysis of variance tests are done without knowledge of the population standard deviations, so you would not be able to use the population standard deviation to decide if the four sample means appear to be close together or far apart. The method that you do use is to compute two estimates of the population variance, one estimate based on the variance within each of the four samples that you collect and the other based solely on the four sample means. The test statistic in analysis of variance, called the *F* test statistic, is the ratio of these two estimates of the population variance. The estimate of the population variance based solely on the sample means is the numerator of the ratio and the estimate based on the within-sample variances is the denominator. The two estimates of the population variance will tend to have the same value if the null hypothesis is true, that is, when the lower panel in Figure 12.1 accurately describes the world. The ratio of the two estimated variances—that is, the test statistic—will be close to 1 if the lower panel in Figure 12.1 is the accurate description. The ratio will tend to be much higher than 1 if the upper panel describes the state of the world. To understand why the ratio of the two estimates of the population variance respond this way, we'll have to look at the formula for the *F* test statistic.

The formula for the *F* test statistic is*

$$F = \frac{\sum_{j=1}^{k} n_j (\bar{x}_j - \bar{\bar{x}})^2/(k-1)}{\sum_{j=1}^{k} s_j^2/k}$$

(12.1)

*If the sample sizes are not equal computer programs use $\sum_{j=1}^{k}\sum_{i=1}^{nj}(x_{ij}-\bar{x}_j)^2 / \left(\sum_{j=1}^{k} n_j - k\right)$ for the denominator of *F*. As this change usually makes little difference, we will use the unweighted average in hand calculations.

The letter k in formula 12.1 stands for the number of samples being tested. In the Federal Home Loan Bank Board example there are four samples, one for each of the managers. The symbol $\bar{\bar{x}}$, which is called the grand mean, is the weighted mean of the four sample means with the respective sample sizes as the weights. The formula for the grand mean was first presented in Chapter 3.

$$\bar{\bar{x}} = \frac{n_1\bar{x}_1 + n_2\bar{x}_2 + n_3\bar{x}_3 + \cdots + n_k\bar{x}_k}{n_1 + n_2 + n_3 + \cdots + n_k} \tag{3.3a}$$

The symbol n_j stands for the sample size within each of the four samples. The subscript j is also used with the symbols $\bar{x}$ and s^2. In both cases the subscript is used to keep track of the samples. For example, $\bar{x}_1$ would be the mean of the sample taken from the first manager, $\bar{x}_2$ would be the mean of the sample taken from the second manager, and so on. The within-sample variance is the term s_j^2. The formula for within-sample variance is a modification of the familiar one for the sample variance from Chapter 3.

(12.2)

$$s_j^2 = \sum_{i=1}^{n_j} \frac{(x_{ij} - \bar{x}_j)^2}{n_j - 1}$$

Note that x_{ij} refers to the observations within one sample, that is, for a particular fund manager. With the four samples in this example you will have four within-sample variances. Each of these within-sample variances is an estimate of one population variance, σ^2. Since each of the within-sample variances is an estimate of the same term, they can be averaged to get one estimate of the population variance. The denominator of the F test statistic is simply the average within-sample variance. In the FHLBB example, you would add the four within-sample variances and divide the total by 4 to get an estimate of the population variance. In effect, you are pooling the four sample variances to get one best estimate. This procedure is similar to the pooling of two sample variances, discussed in Chapter 10.

The logic behind the numerator of the F test statistic is more complicated. Here you are working with the sampling distribution of the mean instead of the population distribution. Let's look at the picture of the sampling distribution of the mean, starting with the assumption that the null hypothesis of equal population means is true. The lower panel of Figure 12.1 is reproduced in Figure 12.2, but superimposed on the population distributions are four sampling distributions of the mean. Recall that in the lower panel of Figure 12.1 there were four identical population distributions with means μ_1, μ_2, μ_3, and μ_4; now if the sample sizes are the same, there are also four identical sampling distributions for $\bar{x}_1$, $\bar{x}_2$, $\bar{x}_3$, and $\bar{x}_4$. The four equal sample sizes are not neces-

FIGURE 12.2
Sampling distributions of the mean when the null hypothesis is true

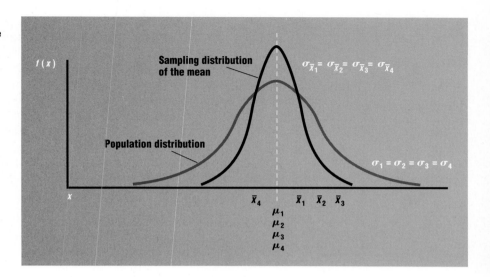

sary for the problem, but the use of equal sample sizes makes the figure much simpler.

The positions of $\bar{x}_1$, $\bar{x}_2$, $\bar{x}_3$, and $\bar{x}_4$ mark four possible values of the sample means taken from the four populations. Note that the sample mean for the fourth population is to the left of the other three. It was put there to show that when the null hypothesis is true, four sample means could be in any order. These sample means could be used as observations in the formula for the sample variance to estimate the variance of the distribution of the sample mean, $(\sigma_{\bar{x}})^2$. This procedure treats the four sample means as individual observations drawn from the distribution of possible sample means. The formula for the estimated variance of the sampling distribution of the mean is part of the numerator of formula 12.1:

$$F = \frac{\sum\limits_{j=1}^{k} n_j\,(\bar{x}_j - \bar{\bar{x}})^2/(k-1)}{\sum\limits_{j=1}^{k} s_j^2/k}$$

THE COLOR LETTERS SHOW THE ESTIMATED VARIANCE OF THE SAMPLING DISTRIBUTION OF $\bar{x}$ WHEN THE NULL HYPOTHESIS OF EQUAL POPULATION MEANS IS TRUE.

The estimated variance of the sampling distribution of the mean is shown to exclude the term n_j. The reason that the denominator is $k-1$ is that the estimated variance is based on k sample means.

What you need in the numerator is an estimate of the *population* variance, rather than the variance of the sample mean. Recall that the denominator is an estimate of the population variance and that the F test statistic is the ratio of two estimates of the population variance. The function of the tern n_j in this formula is to scale up the estimated variance of the distribution of the sample mean to the estimated variance of the population mean. The scaling uses the relationship between the variance of the sample mean and the variance of the population which you first saw in Chapter 7, namely, that $(\sigma_{\bar{x}})^2$ equals σ^2/n or, equivalently, that σ^2 equals $n(\sigma_{\bar{x}})^2$. The values of n may differ across the four samples, so the letter n bears the subscript j to remind you to use the sample size that corresponds to each sample.

Now we have the full description of the F test statistic. The formula is labeled below:

$$F = \frac{\sum_{j=1}^{k} n_j (\bar{x}_j - \bar{\bar{x}})^2/(k-1)}{\sum_{j=1}^{k} s_j^2/k} \quad\quad \begin{array}{l} \text{\footnotesize ESTIMATED POPULATION} \\ \text{\footnotesize VARIANCE BASED ON THE} \\ \text{\footnotesize VARIATION AMONG SAMPLE} \\ \text{\footnotesize MEANS} \\[1em] \text{\footnotesize ESTIMATED POPULATION} \\ \text{\footnotesize VARIANCE BASED ON THE} \\ \text{\footnotesize AVERAGE WITHIN-SAMPLE} \\ \text{\footnotesize VARIANCE} \end{array} \quad\quad (12.1)$$

If the null hypothesis of equal population means was correct, this test statistic should have a value close to 1. Both the numerator and the denominator would be unbiased estimates of the common population variance.

So far we have described the formula when the null hypothesis is correct. How would the picture change if the null hypothesis was false? The denominator would still give an unbiased estimate of the population variance but the numerator would, on average, be too high if the four populations were not centered on the same value. Figure 12.3 shows why.

FIGURE 12.3
Sampling distributions of the mean when the null hypothesis is false

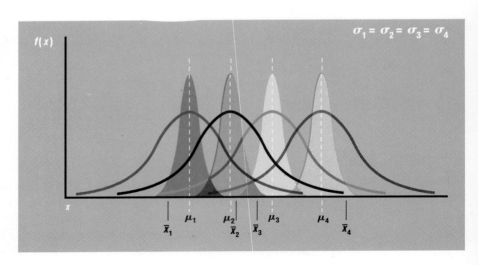

In Figure 12.3 we see four population distributions centered on different population means but with identical population standard deviations. The four sampling distributions of the mean are shaded. These four distributions would also have identical standard deviations if the four sample sizes were equal. Again, it was not necessary for the problem to work with identical sample sizes, but they make the figure easier to follow. Beneath the four sampling distributions of the mean are possible values of $\bar{x}_1$, $\bar{x}_2$, $\bar{x}_3$, and $\bar{x}_4$. These four sample means appear much farther apart than the four sample means in Figure 12.2. If the values of these four sample means were plugged into the formula for the *F* test statistic, the numerator would *overestimate* the population variance and the *F* test statistic would be much higher than 1. Thus a high value for the *F* test statistic would lead you to reject the null hypothesis that the four population means are equal.

■ 12.1C THE *F* DISTRIBUTION

The last sentence stated that you would reject the null hypothesis when the test statistic had a high value. "High value" is pretty vague. You may be wondering where you get the critical value that you need to compare to the test statistic to decide if you will reject or not reject the null hypothesis. What you have to use is the *F* distribution. The **F distribution** is defined as *the ratio of two chi-square distributions, each divided by its degrees of freedom*. The probability density function of the *F* distribution is

$$f(x) = \frac{\chi^2_{\text{numerator}}/df_{\text{numerator}}}{\chi^2_{\text{denominator}}/df_{\text{denominator}}}$$

(12.3)

The terms $df_{\text{numerator}}$ and $df_{\text{denominator}}$ stand for the degrees of freedom of the chi-square random variables in the numerator and the denominator. The two chi-square random variables are labeled "numerator" and "denominator" to help you keep track of them.

The squared term in the numerator of the formula for the *F* test statistic, $(\bar{x}_j - \bar{\bar{x}})^2$, follows a chi-square distribution. The reasoning is the same as in Chapter 11. Assuming that the null hypothesis of equal population means is true, on average, $\bar{x}_j$ will equal $\bar{\bar{x}}$ and the deviations between them will tend to be symmetrical and be less likely as the size of the deviation increases. Similar reasoning applies to the s_j^2 terms in the denominator. The numerator has $k - 1$ degrees of freedom and the denominator has $(\Sigma_{j=1}^{k} n_j - k)$ degrees of freedom. The reason that the numerator has $k - 1$ degrees of freedom is that you used the k sample means to find the grand mean, $\bar{\bar{x}}$, which cost one degree of freedom. The expression $\Sigma_{j=1}^{k} n_j$ is the total number of observations across the k samples. One degree of freedom was lost in each of those samples when the sample mean was used in the formula for the within-sample variance, so there are $(\Sigma_{j=1}^{k} n_j - k)$ degrees of freedom in the denominator.

Table D9, in Appendix D, shows the critical values for both 5% and 1% of the area in the right tail. The F distribution is a family of distributions whose shapes vary with the degrees of freedom in the numerator and denominator. The degrees of freedom for the numerator are listed across the tops of the tables and the degrees of freedom for the denominator are listed down the left side. For example, the critical value for 5% of the area in the right tail with 6 degrees of freedom in the numerator and 5 in the denominator is 4.95. Table 12.1 reproduces a section of the F table and shows the critical value of 4.95. Note that the critical value for a given area in the right tail is not the same if you mix up the numerator and the denominator degrees of freedom. For example, with 5 degrees of freedom in the numerator and 6 in the denominator, the critical value for 5% of the area in the right tail is 4.39 rather than the 4.95 obtained when the numerator and denominator degrees of freedom were reversed.

Figure 12.4 shows the probability density function of an F distribution with 3 degrees of freedom in the numerator and 30 degrees of freedom in the denominator. These values were picked because, as you'll see in a moment, they fit the test on the FHLBB's managers. If you wanted to use a significance level of 5% for your test, your critical value would be the point on the F distribution with the appropriate degrees of freedom which has 5% of the area to its right. The rejection region would be the area to the right of this critical value. The acceptance and rejection regions are marked in Figure 12.4. The F distribution is skewed to the right and has a minimum value of 0. The reason that its minimum value is 0 is that both the numerator and the denominator must be positive because they are squared. Its mean is usually close to 1 and its standard deviation is a complicated expression. The formulas for the mean and the standard deviation are

TABLE 12.1

Portion of table for critical values of F distribution which mark 5% of area in right tail

DENOMINATOR DEGREES OF FREEDOM	NUMERATOR DEGREES OF FREEDOM									
	1	2	3	4	5	6	7	8	9	10
1										
2										
3										
4										
5	6.61	5.79	5.41	5.19	5.05	4.95	4.88	4.82	4.77	4.74
6										
7										
8										
9										

FIGURE 12.4

***F* distribution with 3 degrees of freedom in the numerator and 30 degrees of freedom in the denominator**

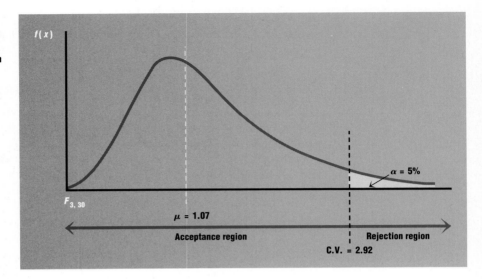

$$\mu = \frac{df_{\text{denominator}}}{df_{\text{denominator}} - 2} \qquad df_{\text{denominator}} > 2$$

(12.4)

$$\sigma = \sqrt{\frac{2 df_d^2 (df_n + df_d - 2)}{df_n (df_d - 2)^2 (df_d - 4)}} \qquad df_d > 4$$

(12.5)

The labels "numerator" and "denominator" were changed to n and d in formula 12.5 to make it more compact. These formulas are given for the sake of completeness (you've had the mean and the standard deviation of all of the other distributions discussed) but you won't need them to do analysis of variance. Figure 12.4 is a template for the hypothesis test of equal population means. Note that if you knew that you would have 4 samples (one for each fund manager) and 34 observations across those samples (there are 34 observations in Table 12.2), the degrees of freedom for the F distribution would be 3 for the numerator and 30 for the denominator. The template is what you expect the sampling distribution of your F test statistic to look like if the null hypothesis is true. The template can be made before you collect your sample and calculate your F test statistic. What the template tells you is that if the null hypothesis is true, the probability of getting an F test statistic above 2.92 is 0.05. The number 2.92 came from Table D9 in Appendix D. If 0.05 is the probability of a Type I error that you will allow, then you should reject the null of equal population means for any value of the F test statistic above the critical value of 2.92.

■ 12.1D CALCULATIONS FOR ONE-WAY ANALYSIS OF VARIANCE

Now you are finally ready to collect a sample and do the test. Table 12.2 lists daily yields expressed as annual rates of return for the four managers. The sample means and sample standard deviations of the four managers' yields are also shown. Again, since there are 4 populations and 34 observations, the degrees of freedom are 3 for the numerator and 30 for the denominator. The grand mean is the weighted average of the four sample means:

$$\bar{\bar{x}} = \frac{n_1\bar{x}_1 + n_2\bar{x}_2 + n_3\bar{x}_3 + \cdots + n_k\bar{x}_k}{n_1 + n_2 + n_3 + \cdots + n_k} \tag{3.3a}$$

$$\bar{\bar{x}} = \frac{10(15.3) + 10(15.2) + 7(19.29) + 7(19.43)}{10 + 10 + 7 + 7}$$

$$\bar{\bar{x}} = 16.94$$

The next step is to substitute into formula 12.1 to get the F test statistic:

$$F = \frac{\sum\limits_{j=1}^{k} n_j \, (\bar{x}_j - \bar{\bar{x}})^2/(k-1)}{\sum\limits_{j=1}^{k} s_j^2/k} \tag{12.1}$$

$$F = \frac{[10(15.3 - 16.94)^2 + 10(15.2 - 16.94)^2 + \cdots + 7(19.43 - 16.94)^2]/3}{(9.79 + 19.29 + 20.24 + 14.29)/4}$$

$$F = 2.92$$

With an F test statistic of 2.92 and a critical value of 2.92, you would just reject the null hypothesis that the four population means are equal. This test statistic is an approximation. If you use a statistics package on a computer you will get a different value, 2.97. Recall that the formula in computer programs adjusts the denominator for different sample sizes. At this point you have rejected, by a slight margin, the null that the four managers have the same rates of return. To identify the managers who are significantly above or below the grand mean, you would have to do an additional test, called Tukey's honest significant difference test. We'll discuss this test shortly, but at this point you might benefit from a second example showing these calculations.

Analysis of variance is often used to design and analyze controlled experiments in which a researcher can determine how each observation in the data set

TABLE 12.2

Daily yields for funds managers of FHLBB expressed on annualized basis (percent)

	Manager 1	Manager 2	Manager 3	Manager 4
	20%	18%	29%	26%
	16	24	19	21
	15	18	17	19
	14	12	19	17
	11	13	16	14
	15	13	19	18
	20	14	16	21
	17	11		
	14	19		
	11	10		
Sample mean	15.30	15.20	19.29	19.43
Sample variance	9.79	19.29	20.24	14.29
Sample size	10	10	7	7

will be treated or assigned. The example about the yields of the funds managers of the FHLBB was not a controlled experiment; the data were the yields of the four managers over various periods as recorded in the course of business. To see how a controlled experiment works, consider the example of a restaurant chain's waiter-training program. The company has three methods of training new waiters: (1) no training, (2) having the trainee observe an experienced waiter for one day, and (3) sending the trainee to a week-long program that includes video tapes and lectures on the techniques of being a waiter. Each of these methods can be called a "treatment," and the null hypothesis would be that all three treatments have the same effect on the quality of the work. The word "treatment" comes from agricultural research, which was the context in which analysis of variance was originally developed. A "treatment" consisted of various amounts of fertilizer or other chemicals or types of seeds that were used on a plot of land. *In an experimental context,* a **treatment** is *an arrangement of experimental units or a process to which they are subjected; in a nonexperimental context,* it is *a classification or category in which the observations are placed.* In a controlled experiment, members of a pool of applicants for waiter jobs could be randomly assigned to one of the three treatments and their quality scored after they had been on the job for a few weeks. *A research design in which each unit in a controlled experiment is randomly assigned to a treatment* is called a **completely randomized design.** A completely randomized design provides the simplest way to set up a controlled experiment.

Suppose 30 applicants were randomly assigned to the three treatments, 10 to each treatment. If the mean quality scores were 70, 72, and 81 for the three treatments and the sample variances were 10, 14, and 16, respectively, could you reject at the 5% significance level the null hypothesis that the three treatments all had the same population mean quality scores? The F test statistic would be

$$F = \frac{\sum_{j=1}^{k} n_j (\bar{x}_j - \bar{\bar{x}})^2/(k-1)}{\sum_{j=1}^{k} s_j^2/k} \qquad (12.1)$$

$$F = \frac{[10(70 - 74.33)^2 + 10(72 - 74.33)^2 + 10(81 - 74.33)^2]/2}{(10 + 14 + 16)/3}$$

$$F = 25.75$$

Note that the grand mean is the same as the simple average of the three sample means because the three sample sizes are equal. There are $k - 1$ or 2 degrees of freedom in the numerator and $\sum_{j=1}^{k} n_j - k$ or 27 degrees of freedom in the denominator. The critical value for an F distribution with 2 and 27 degrees of freedom and 5% of the area in the tail is 3.35. Since the F test statistic of 25.75 is to the right of the critical value, you know that the null hypothesis of equal population means is rejected at the 5% significance level.

Two points can be made about this problem, one practical and the other statistical. The practical point is that a statistically significant difference in the mean quality scores across treatments does not prove that the most expensive treatment, in this case the week–long training program, is cost–effective. Note that the waiters who went through the most expensive training program did garner the highest scores. As a manager you have to cost out the various training programs and put a dollar value on the benefit of the higher quality service that the expensive treatment provides before you can decide if it is worthwhile. The statistical point is the problem of selection bias. Even though the assignment of trainees to the various treatments was random, there may be a problem of selection bias because some of the trainees may drop out of the program. Suppose many trainees given the zero training treatment find it rough going and drop out before their quality scores are recorded. You might conclude that the zero training treatment is the best one simply because only trainees who have a natural flair for the work stay with that treatment long enough to be scored. If you adopted the zero training treatment as a chainwide policy, your restaurant chain might not be able to find enough innately gifted waiters and the average quality of service would be very low as your restaurants were continually breaking in new waiters. While there are some techniques to control for selection bias, they are beyond the scope of an introductory text. The only practical advice at this level is to be aware of this problem and to design your experiment so that the number of missing observations is minimized. In this restaurant example you might score the waiters' service quality shortly after the training ended to reduce the number of dropouts.

CASE STUDY

NO PAIN, NO GAIN

A multitude of published articles describe experiments that test the effects of various treatments on job performance or attitudes. This line of research is in the bailiwick of psychologists and management professors. Professor Mel E. Schnake of Eastern Illinois University ("Vicarious Punishment in a Work Setting," *Journal of Applied Psychology* 71, no. 2 [1986]: 343–345) tested the effect of punishment on job performance. Sixty students (mostly women) were recruited for a temporary clerical job that paid $5 an hour. The job consisted of checking the New York Stock Exchange's closing quotations and recording on a special coding sheet the name, volume, and closing price of any stock of which 200,000 or more shares had changed hands on a particular day. The work was done over two successive weekends. The 60 students were randomly assigned to eight groups called (1) control, time 1; (2) threat, time 1; (3) punishment, time 1; (4) control, time 2; (5) threat, time 2; (6) punishment, time 2; (7) threat, time 2 only; and (8) punishment, time 2 only. "Time 1" refers to the first weekend and "time 2" to the second.

A "ringer" was added to each group, that is, a student who had been coached to behave in a certain way. The ringers worked slowly, an average of 2.5 coding sheets over the experiment compared to an average of 4.8 sheets for the experimental subjects. At the halfway point of both the first and second periods, the supervisor in each room collected everyone's coding sheets and checked the amount that had been done. The "threat" consisted of telling the ringer in front of everyone in the room that she was working much more slowly than the others and that if she did not shape up, her pay would be cut to $3.50 an hour. The "punishment" consisted of telling the ringer in front of everyone that she was working much more slowly than the others and that her pay thereafter would be $3.50 an hour. The ringers were instructed to say nothing and keep on working.

The null hypothesis that the mean output per student was the same across the eight groups could be rejected at a 1% significance level. Witnessing another student being either threatened or punished *raised* output, and the effect persisted into the second weekend among the groups that witnessed a threat or a punishment during the first weekend.

Generalizing from such artificial experiments can be hazardous. The students may have felt that the ringers were shirking so notoriously that they deserved what they got. Also, the ringers were not known to the subjects. Real employees might have reacted differently if a well-liked co-worker had been disciplined. At the end of the article, Schnake suggests that the effect of punishment on workers' performance is a neglected area and that it should be carefully examined in real work settings.

EXERCISES FOR SECTION 12.1

12.1 What value marks the 95th percentile of an F distribution with the following numerator and denominator degrees of freedom?
  (a) 3–5.
  (b) 5–3.
  (c) 1–4.
  (d) 1–∞.
  (e) ∞–∞.

12.2 What value marks the 99th percentile of an F distribution with the following numerator and denominator degrees of freedom?
  (a) 8–16.
  (b) 4–12.
  (c) 1–4.
  (d) 1–∞.
  (e) 8–8.

12.3 (a) Use the following data to test at the 5% significance level the null hypothesis that $\mu_1 = \mu_2 = \mu_3$ versus the alternative hypothesis that at least one of the population means does not equal the other population means.

Sample 1	Sample 2	Sample 3
3	5	4
6	4	5
4	6	6
5	4	6
4	6	5
3	5	6
4	5	5
5	4	
4		

(b) What assumptions about the population distributions are required if this test on population means is to be accurate?

12.4 (a) Use the following data to test at the 5% significance level the null hypothesis that $\mu_1 = \mu_2 = \mu_3$ versus the alternative hypothesis that at least one of the population means does not equal the other population means.

Sample 1	Sample 2	Sample 3	Sample 1	Sample 2	Sample 3
8	14	15	13	14	9
3	1	2˙	10	7	13
11	2	1	3	3	
9	12	8	12		
2	3	5			

(b) What assumptions about the successive observations are required if this test on population means is to be accurate?

12.5 Because of consumers' health concerns and a declining per capita consumption of beef, the U.S. Department of Agriculture is promoting the breeding of low–cholesterol cattle. Four universities have submitted for testing carcasses of cattle that they have bred under USDA contract. The USDA wants to see if the four breeds have the same mean amount of cholesterol.

(a) State the null and alternative hypotheses in words and algebraically.

(b) Use the following data in grams of cholesterol per kilogram of trimmed muscle tissue for each carcass measured to test these hypotheses at the 5% level.

Breed 1	Breed 2	Breed 3	Breed 4
35	41	40	50
44	58	42	56
45	60	39	51
48	53	38	50
45	47	46	47

12.2 CONFIDENCE INTERVALS AND HYPOTHESIS TESTS IN ONE-WAY ANALYSIS OF VARIANCE

12.2A CONFIDENCE INTERVALS FOR ONE POPULATION MEAN

After you have rejected the null hypothesis that all population means are equal, it may be useful to estimate confidence intervals for each population mean. The procedure for constructing confidence intervals for a single population mean in an analysis of variance context is a straightforward extension of the procedure in Chapter 8. However, a series of one-population confidence intervals brings up a problem similar to the one described in the chapter introduction. There it was pointed out that if a series of hypothesis tests were done for all possible pairs of population means, the probability of making a Type I error could be very high. Here we have an analogous problem: if a series of confidence intervals is constructed, each at the 95% level of confidence, the probability that all of the confidence intervals will simultaneously contain the true population parameter is less than 95%. First we'll discuss the procedure for constructing a single confidence interval and then we'll discuss the confidence level that applies to the series of intervals.

Suppose you wanted to construct a confidence interval at the 95% level of confidence for the mean population yield of the first manager of the FHLBB's fund using the data from section 12.1. Since you have assumed that there is one population variance, you should use the pooled estimate of the population

∨

variance from all four samples when you construct your confidence interval. The best estimate of the population variance is the denominator of the F test statistic. The formula for the confidence interval is

$$L,\ U = \bar{x}_j \pm t_{\alpha/2}\sqrt{\dfrac{\sum\limits_{j=1}^{k} s_j^2/k}{n_j}}$$

(12.6)

The expression $\sum_{j=1}^{k} s_j^2/k$ is the pooled estimate of the within-sample population variance that appears in the denominator of the F test statistic. The degrees of freedom for the t score in formula 12.6 is the same as the degrees of freedom in the denominator of the F test statistic, $(\sum_{j=1}^{k} n_j - k)$, or 30 in this example. The advantage of using the pooled within-sample estimate of the population variance to construct the confidence interval over using the variance within only one sample is that the degrees of freedom are higher, so that you have a lower t score and a narrower confidence interval. If only the ten observations for the first manager were used to construct a confidence interval for the mean yield of that manager, the t score for a 95% level of confidence would be 2.262. With 30 degrees of freedom from the pooled estimate of the population variance, the t score is 2.042. A confidence interval based on the mean of one population in this example would be about 11% wider than a confidence interval based on the pooled estimate of the variance. This procedure is another application of the golden rule of statistics: use all of the data that you have.

Let's go through the calculations for a confidence interval at the 95% level of confidence for the first manager's mean yield, using formula 12.6:

$$L,\ U = \bar{x}_j \pm t_{\alpha/2}\sqrt{\dfrac{\sum\limits_{j=1}^{k} s_j^2/k}{n_j}}$$

(12.6)

$$L,\ U = 15.3 \pm 2.042\sqrt{\dfrac{15.9}{10}}$$

$$L = 12.73 \qquad U = 17.87$$

At the 95% level of confidence, the lower and upper limits of the confidence interval for the mean population yield of the first manager are 12.73 and 17.87.

If four 95% confidence intervals were constructed, one each for the population means of the four fund managers, the probability that all four intervals would contain the respective population means simultaneously would not be 95%. Since the intervals are independent, the confidence level for the four intervals would be 0.95 to the fourth power. This expression is the intersection of four independent events, each with a probability of 0.95. The overall level of confidence would be only 81.5%. In general, if k confidence intervals are constructed, each with a confidence level of $1 - \alpha$, the overall level of confidence for the k intervals would be $(1 - \alpha)^k$.

12.2B CONFIDENCE INTERVALS FOR THE DIFFERENCE BETWEEN TWO POPULATION MEANS

When you construct a confidence interval for the difference between two population means, you can again take advantage of the information from all of the samples in estimating the common population variance. Recall that the term from the denominator of the F test statistic, $\sum_{j=1}^{k} s_j^2/k$, is an unbiased estimator of the common population variance. This term replaces the pooled estimate of the sample variances, s_{pool}^2, from formula 10.13, used for the standard error of the difference between two sample means when you could assume that the two populations had the same variance. Formula 10.13 is repeated below so that you can compare it with the formula that uses the additional information from all of the samples, formula 12.7:

$$s_{\bar{x}_1-\bar{x}_2} = \sqrt{\frac{s_{pool}^2}{n_1} + \frac{s_{pool}^2}{n_2}}$$

(10.13)

$$s_{\bar{x}_1-\bar{x}_2} = \sqrt{\frac{\sum_{j=1}^{k} s_j^2/k}{n_1} + \frac{\sum_{j=1}^{k} s_j^2/k}{n_2}}$$

(12.7)

In effect, you would pool k sample variances instead of the two sample variances pooled in Chapter 10.

The next step in constructing a confidence interval is to use the standard formula for a difference-in-means confidence interval. Recall that one of the initial assumptions required for one-way analysis of variance was that the population distributions be normal. When you have normal populations and unknown population standard deviations, the formula for a confidence interval for the difference in population means is

$$L, U = \bar{x}_1 - \bar{x}_2 \pm t_{\alpha/2}s_{\bar{x}_1-\bar{x}_2}$$

(10.10)

The remaining issue is the degrees of freedom. Recall that in Chapter 10 the degrees of freedom when you pooled the variances were $n_1 + n_2 - 2$. You lost one degree of freedom for each sample mean that you used to estimate a population mean. The equivalent rule here is that the degrees of freedom are $(\sum_{j=1}^{k} n_j - k)$; you lose one degree of freedom for each of the k sample means used in the estimate of the pooled variance, $\sum_{j=1}^{k} s_j^2/k$.

Now that we have all of the pieces, we can proceed with an example. Suppose the FHLBB wanted to construct a confidence interval at the 95% level for the difference in the population mean rates of return for the first and the fourth fund managers. The standard error of the difference in sample means would be

$$s_{\bar{x}_1 - \bar{x}_2} = \sqrt{\frac{\sum\limits_{j=1}^{k} s_j^2/k}{n_1} + \frac{\sum\limits_{j=1}^{k} s_j^2/k}{n_2}} \tag{12.7}$$

$$s_{\bar{x}_1 - \bar{x}_2} = \sqrt{\frac{15.9}{10} + \frac{15.9}{7}}$$

$$s_{\bar{x}_1 - \bar{x}_2} = 1.97$$

The last step is to substitute into formula 10.10 for the limits of a confidence interval:

$$L, \ U = \bar{x}_1 - \bar{x}_2 \pm t_{\alpha/2} s_{\bar{x}_1 - \bar{x}_2} \tag{10.10}$$

$$L, \ U = (15.3 - 19.43) \pm 2.042(1.97)$$

$$L = -8.15 \qquad U = -0.11$$

Again, the construction of a series of confidence intervals for the difference between the population means for every pair of managers would raise the issue of the overall confidence level, the probability that all of the confidence intervals will contain the population differences in means simultaneously. There are $_4C_2$ or six possible differences in means for the four fund managers. The overall confidence level would be $(1 - \alpha)^6$, where $1 - \alpha$ is the confidence level for one interval. In this example, the overall confidence level would be 73.5%.

Since the above confidence interval does not contain 0, you may suspect that you could reject the null hypothesis that the true difference in the mean rates of return for the first and fourth managers was 0 at a 5% significance level. If the interval was based on the information from only two samples, this surmise would be correct. However, hypothesis testing is more complicated when you are pooling information from many samples. In the next section we'll discuss Tukey's honest significant difference test, which is one of several tests for differences in means in the context of a one-way analysis of variance.

■ 12.2C TUKEY'S HONEST SIGNIFICANT DIFFERENCE TEST

There are many procedures for conducting hypothesis tests on individual population means or on the difference between pairs of population means in an analysis–of–variance context. A discussion of the advantages and disadvantages of each of these tests is beyond the scope of this text. The suggested readings at the end of the chapter refer to some of these tests. We will discuss here one commonly used test, Tukey's honest significant difference test. This test (named for its originator, John W. Tukey) lets you identify the pairs of means that are significantly different from the others. The words "significantly different" are used in a statistical sense, not as they are used in everyday speech. In ordinary English, a "significant difference" is an important or substantial difference. A significant difference in a statistical sense can be very small or even trivial, but it is a difference that, however small, is unlikely to have occurred by chance in the sample data. Tukey's test should be used only after the null of equal treatment means has been rejected. Tukey's original test was for samples of equal size, but it was extended to cover samples of unequal sizes by Spøtvoll and Stoline. Since it is more general, this is the version we'll use. To apply the test you need at least three treatments and a significance level of 5% or less. Every pair of observed differences between treatment means is compared with the term HSD.

$$\text{HSD} = q\sqrt{\frac{\sum\limits_{j=1}^{k} s_j^2/k}{\min(n_j)}}$$

(12.8)

The term in the numerator under the radical is the average within–sample variance. The expression in the denominator of the radical, $\min(n_j)$, is the smaller of the two samples being compared. The term q is the point that marks the upper 5% of a distribution called the **Studentized range,** which is *the distribution of the difference between the highest and lowest observations of a series of* k *observations drawn from* k *independent normal distributions with identical means and standard deviations, divided by the estimated population standard deviation.* Dividing the range—that is, the difference between the highest and the lowest observations—by s, the estimated population standard deviation, "Studentizes" the range. For Tukey's honest significant difference test, the degrees of freedom of the Studentized range are k and $\sum_{j=1}^{k} n_j - k$. In effect, the q gives you a limit on how far apart the highest and lowest observations from a series of identical normal curves are likely to be; the range divided by the standard deviation will exceed q in only 5% of all samples.

Let's use the rates of returns of the FHLBB's fund managers to illustrate Tukey's test. There are four sample means, so there are $_4C_2$ or six possible comparisons between pairs of sample means. The six differences in sample means are:

1–2	$|15.3-15.2| = 0.1$
1–3	$|15.3-19.29| = 3.99$
1–4	$|15.3-19.43| = 4.13$
2–3	$|15.2-19.29| = 4.09$
2–4	$|15.2-19.43| = 4.23$
3–4	$|19.29-19.43| = 0.14$

The vertical lines around the differences mean absolute value. The sample sizes for the four managers were 10, 10, 7, and 7, respectively. Thus the minimum sample size is 7 for five of the six comparisons and it is 10 for the comparison between the first and second managers. The q for the Studentized range with k and $\Sigma_{j=1}^{k} n_j - k$, or 4 and 30 degrees of freedom, is 3.85. This value can be found in Table D17, in Appendix D. To find the honest significant difference, you have to use formula 12.8. We'll calculate two HSDs, one for the minimum sample size of 10 and another for the minimum sample size of 7.

$$HSD = q\sqrt{\frac{\sum\limits_{j=1}^{k} s_j^2/k}{\min(n_j)}}$$

$$HSD = 3.85\sqrt{\frac{(9.79 + 19.29 + 20.24 + 14.29)/4}{10}}$$

$$HSD = 4.86$$

$$HSD = 3.85\sqrt{\frac{(9.79 + 19.29 + 20.24 + 14.29)/4}{7}}$$

$$HSD = 5.80$$

Since the HSD of 4.86 is greater than the observed difference of 0.1 between the sample means of the first and the second fund managers, the null that μ_1 equals μ_2 cannot be rejected. The other five comparisons are less than the HSD of 5.8 calculated with the minimum sample size of 7. Thus there is not enough evidence to reject the null that any of the pairs of population means are equal at a 5% significance level. The one-way analysis of variance test tells you that the four population means are probably not all equal, but Tukey's test tells you that you do not have enough data to tell which pairs of means are significantly different.

EXERCISES FOR SECTION 12.2

12.6 **(a)** Construct confidence intervals at the 95% level of confidence for the three population means in exercise 12.3.

(b) What is the overall level of confidence that these three confidence intervals will hold simultaneously?

12.7 **(a)** Construct confidence intervals at the 95% level of confidence for the three population means in exercise 12.4.

(b) What is the overall level of confidence that these three confidence intervals will hold simultaneously?

12.8 **(a)** Construct confidence intervals at the 99% level of confidence for the differences between all of the possible pairs of three population means in exercise 12.3.

(b) What is the overall level of confidence that these confidence intervals will hold simultaneously?

(c) Construct confidence intervals at the 95% level of confidence for the differences between all of the possible pairs of three population means in exercise 12.4.

(d) What is the overall level of confidence that these confidence intervals will hold simultaneously?

12.9 If you rejected the null hypothesis of equal population means for the data in exercise 12.4, use Tukey's honest significant difference test to identify the pairs of means that differ significantly at the 5% level.

12.10 If you rejected the null hypothesis of equal population means for the data in exercise 12.5, use Tukey's honest significant difference test to identify the pairs of means that differ significantly at the 5% level.

12.3 TWO-WAY ANALYSIS OF VARIANCE (OPTIONAL)

 12.3A INTRODUCTION TO TWO-WAY ANALYSIS OF VARIANCE

To introduce two-way analysis of variance, we'll start with a new example. The Internal Revenue Service uses an annual sample of 50,000 personal tax returns to estimate compliance with the tax laws. Each tax return in this sample undergoes an intensive audit. The IRS checks the documentation on every deduction and verifies that all taxable income that can be traced has been declared. At the end of one of these intensive audits, the IRS has a very good idea of how far the original return deviated from the truth. With the data from this survey the IRS can test the hypothesis that the average amounts of unpaid taxes per return are the same across regions of the country. If the IRS finds that compliance varies by region, it can target its enforcement resources where they will generate the most tax revenue.

The simplest test of the hypothesis of equal means across regions would be the procedure discussed in section 12.1, one-way analysis of variance. However, a person's incentive to underpay taxes depends on his or her tax bracket—the higher the tax bracket, the more a person saves per dollar of hidden income or dollar of unwarranted deductions. It is possible that, by chance, the IRS could draw relatively more high-bracket taxpayers in one region. When it subsequently conducted the one-way analysis of variance test, it might mistakenly conclude that the region was a hotbed of tax evaders. What the IRS really would be observing was the effect of tax brackets on compliance.

Since both region and tax bracket may affect the amount of underpayment, the appropriate technique is **two-way analysis of variance,** which is *a joint test for the equality of more than two population means used when two factors are suspected of affecting the variable of interest.* Note that even if the IRS were not interested in the impact of tax brackets on compliance, it would still be better off using two-way analysis of variance. First, if the null of equal compliance across regions were true, the two-way test would be less likely to encourage the Type I error of rejecting the null. This error corresponds to attributing the effect of the tax brackets to the region because people in high brackets are overrepresented in the sample of one region. Second, if the null hypothesis of equal compliance across regions were false, the two-way test would be less likely to encourage the Type II error of accepting the null hypothesis. The extraneous variation created by the tax-bracket effect could increase the average within-sample variability, the term that goes into the denominator of the F test statistic. An increase in the denominator of the F test statistic would reduce the test statistic. Anything that lowers the F test statistic makes it more likely that it will fall on the acceptance side of the critical value of the test. To summarize, even if your only concern is to test a hypothesis about the effect of *one* variable, if you think that *two* variables have an influence, use two-way analysis of variance, because it will reduce the probability of either a Type I or a Type II error.

12.3B FACTORIAL DESIGNS VERSUS RANDOMIZED BLOCK DESIGNS

There are a number of ways of doing two-way analysis of variance. We will discuss two of the most common methods, factorial designs and randomized block designs. After exploring the distinctions between these two methods, we will investigate the assumptions needed for two-way analysis of variance and then the calculations for the factorial design and the randomized block design.

The **factorial design** is *a technique used in two-way analysis of variance in which there are* h *treatments,* k *blocks, and* m *observations or experimental units per treatment-block combination, or* hkm *independent observations.* The factorial design requires that you have more than one observation for every two-way classification. For example, you must have more than one tax return for every combination of region and tax bracket that you will test.

With the randomized block design you have only one observation for every two-way classification. The **randomized block design** is *a technique used in two-way analysis of variance in which the experimental units within a block are randomly assigned to the treatments.* One obvious advantage of the randomized block design is that it requires fewer data. But this design has two disadvantages. The first disadvantage is that a randomized block design can be used only in controlled experiments. The researcher has to be able to assign each experimental unit randomly to one of the two-way classifications. Since the IRS cannot assign taxpayers at random to a region of the country or a tax bracket, the randomized block design cannot be used for our tax example. To see the context in which it could be used, consider an agricultural example. The effect on plant growth of various types of fertilizer and of various types of soils could be tested with a randomized block design. One variable would be called the "treatment" and the other would be called the "block." A **block** is *a factor or variable other than the treatment variable which is thought to affect the values of the variable of interest.* The experimental units that the researcher would randomly assign to the various fertilizer–soil type pairs are the individual plants. If the researcher were interested in testing hypotheses about the effects of both the fertilizers and the soil types, the variable that would be called the block and the one that would be called the treatment would be chosen arbitrarily. Even if the researcher were interested only in the effects of the fertilizers, two-way analysis of variance would provide a better test than one-way analysis of variance. Two-way analysis of variance allows the researcher to recognize and account for the effects of soil types while conducting a test on fertilizers. In this case, the fertilizers would be the treatment and the soil types, the block.

The second disadvantage of a randomized block design is that you cannot test for an **interaction effect**—*any difference between the population mean of a treatment-block combination and the sum of the treatment and combination effects.* If you have multiple observations for each two-way classification, you can test whether some blocks and treatments interact. For example, is tax compliance relatively low in the South except for high-bracket taxpayers? Or is tax compliance relatively high in the Midwest except for low-bracket taxpayers? If the possible interaction effects are of no interest and if you can randomly assign the experimental units to the various treatments and blocks, you are better off with a randomized block design than with a factorial design. If you cannot use random assignment or if the possible interactions are important, you have to use the factorial design.

Because the math is easier, we will stick to what is called a **balanced factorial design:** *a research design in which an equal number of observations or experimental units is found in every treatment-block combination.* With our tax example a balanced design requires that the same number of observations be taken from each region–bracket combination.

FIGURE 12.5
Population distribution when
the null hypotheses of no
treatment effects, block
effects, and interaction
effects are true

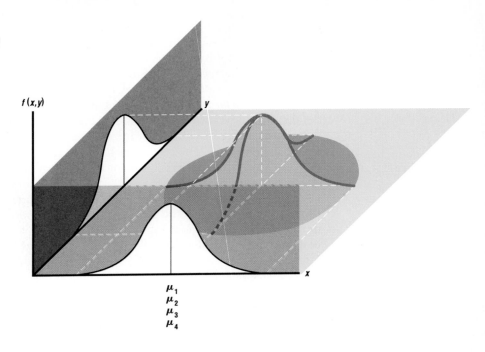

$f(x,y)$

μ_1
μ_2
μ_3
μ_4

12.3C ASSUMPTIONS REQUIRED FOR TWO-WAY ANALYSIS OF VARIANCE

Most of the assumptions required for two-way analysis of variance are extensions of the assumptions required for one-way analysis of variance. For one-
way analysis of variance you must have normal population distributions, and
each population distribution has to have the same variance. If these assumptions hold or even if they are closely approximated, one-way analysis of variance provides an accurate test for the null hypothesis of the equality of the
population means. For two-way analysis of variance the population distributions have to be normal when measured on either the treatment axis or the
block axis. Figure 12.5 shows what this assumption of normality looks like
when there are no treatment effects, no block effects, and no interaction effects. The figure looks like a hill. In our tax example the population distribution of underpayments on personal tax returns would be plotted on the x axis,
and the means labeled μ_1, μ_2, μ_3, and μ_4 represent the mean underpayment in
different regions of the country, our treatment variable. The silhouette of the
hill is projected onto the x–$f(x)$ plane and it is a normal curve. This silhouette
corresponds to the four superimposed normal curves with identical variances
at the bottom of Figure 12.1. The population distributions of underpayment
are also plotted along the y axis. Again a silhouette of the hill projected onto the

FIGURE 12.6
Population distribution when the null hypotheses of no block effects and interaction effects are true and the null hypothesis of no treatment effect is false

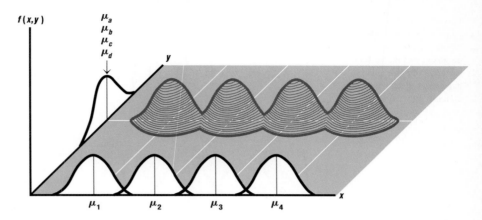

$y-f(y)$ plane shows one normal curve. Assume that there are four tax brackets. What is different here is that the second projection represents four superimposed normal distributions of tax underpayment for each tax bracket rather than each region. The population variances of these four distributions are identical and equal to the population variances of the four distributions in the four regions. A symmetrical hill, as in Figure 12.5, is what would be found if there were no interaction effects. If there were lumps or depressions on the surface of the hill, certain combinations of region and tax bracket would interact and be associated with more or less tax compliance.

Figure 12.6 shows the assumptions of normality and equal variances holding. The difference is that while there are no block or interaction effects—the projection on the $y-f(x)$ plane still shows one normal curve—there are treatment effects. In other words, in our tax example, the mean underpayment of taxes varies by region. Figure 12.6 looks like four identical hills in a row. Running along the x axis are four distinct normal curves, each with the same variance. The projection of the silhouettes of these four normal curves on the $x-f(x)$ plane corresponds to the top half of Figure 12.1, where the null hypothesis of equal means was false. A figure showing what happens when there are block effects with no treatment effects would be essentially the same as Figure 12.6, but the four normal curves would be projected onto the $y-f(y)$ plane instead of the $x-f(x)$ plane.

Figure 12.7 shows block and treatment effects but no interaction effects. The figure looks like 16 identical hills on a 4-by-4 grid. Here four normal curves are projected onto both the $x-f(x)$ and the $y-f(y)$ planes.

Figures 12.5, 12.6, and 12.7 by no means exhaust the possibilities of showing treatment, block, and interaction effects while retaining the assumptions of normality and equal variances. When treatment or block effects are shown in Figures 12.6 and 12.7, for example, each region and tax bracket has a distinct

FIGURE 12.7
Population distribution when the null hypotheses of no treatment and block effects are false and the null hypothesis of no interaction effect is true

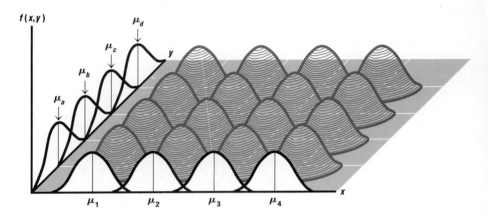

mean. The null hypotheses of no treatment and no block effects would still be false if one of the four means on the x–$f(x)$ or the y–$f(y)$ plane had a value different from the others. Four distinct means are shown to make it easier to visualize the four population distributions of either treatments or blocks. None of these figures has shown any interaction effect. There are many ways to show interaction effects. Figure 12.8 shows treatment, block, and interaction effects.

Figures 12.5, 12.6, 12.7, and 12.8 give you an idea of what the parametric assumptions of normality and equal variances imply when the null hypotheses of no treatment, block, or interaction effects are either true or false. Like one-way analysis of variance, the two-way procedure is fairly robust to moderate departures from normality and equality of the variances. A nonparametric procedure called Friedman analysis of variance can be used if the departures

FIGURE 12.8
Population distribution when the null hypotheses of no treatment, block, and interaction effects are false

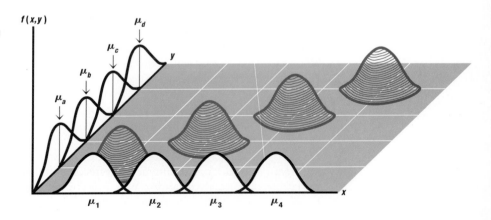

volunteers' backs. Within a block (on a given person's back) the variation in suntans is due entirely to differences in the lotions. The advantage of the randomized block design is that random variation across blocks cannot influence the results of the test on the treatments.

Here's an example of a randomized block design. Your company is considering the purchase of one of four competing brands of a scanning device. The scanner reads text consisting of numbers and letters on paper and then creates an electronic code that can be used to store the text in a computer. Before making a decision, you test the four brands on the same document printed in four different typefaces: Courier, Times Roman, Garamond, and Palatino. The treatments are the four brands of scanners and the blocking factor is the typeface. Presumably all of the scanners will have an easier time with a simple typeface, such as Courier, than with a more decorative typeface, such as Palatino.

Your null hypothesis is that the mean numbers of characters misread by the four scanners are the same. You could also test the hypothesis that all of the typefaces have the same mean number of misread characters. Sometimes a test on the blocking factor is not carried out because the effect of the blocks is of no interest. The blocking factor is used merely to improve the accuracy of the test on the treatments. Use a significance level of 5%. Table 12.5 has the data for the experiment.

The formulas for two-way analysis of variance with a randomized block design are similar to the formulas for a factorial design. There is no formula to test for interactions because you have to have repetitions of each treatment-block combination to test for interaction. Basically, to use a randomized block design you have to assume that there are no interaction effects or that they are very small. For treatments, the formula for the F test statistic is

$$F = \frac{k \sum_{i=1}^{h} (\bar{x}_{i.} - \bar{\bar{x}})^2/(h-1)}{\sum_{j=1}^{k} \sum_{i=1}^{h} (x_{ij} - \bar{x}_{i.} - \bar{x}_{.j} + \bar{\bar{x}})^2/(h-1)(k-1)}$$

(12.12)

TABLE 12.5
Numbers of text characters misread by four brands of optical scanners in documents in four typefaces

BLOCK	TREATMENT				BLOCK MEAN
	Brand 1	Brand 2	Brand 3	Brand 4	
Courier	45	31	49	46	42.75
Times Roman	52	58	54	50	53.50
Garamond	56	62	61	60	59.75
Palatino	83	81	84	88	84.00
TREATMENT MEAN	59	58	62	61	

For blocks the F test statistic is

$$
F = \frac{h \sum_{j=1}^{k} (\bar{x}_{.j} - \bar{\bar{x}})^2/(k-1)}{\sum_{j=1}^{k} \sum_{i=1}^{h} (x_{ij} - \bar{x}_{i.} - \bar{x}_{.j} + \bar{\bar{x}})^2/(h-1)(k-1)}
$$

(12.13)

The numerators of formulas 12.12 and 12.13 should be clear. They are the same as the numerators for the F test statistics in the factorial design except that the term m, the number of repetitions per treatment-block combination, has been dropped. The denominators of formulas 12.12 and 12.13 require more explanation. Since there is only one observation per sample—that is, per treatment-block combination—it is impossible to calculate the average within-sample variance. If there are no interaction effects, the denominator provides an unbiased estimate of the common population variances. The expression $(x_{ij} - \bar{x}_{i.} - \bar{x}_{.j} + \bar{\bar{x}})$ nets out the block and treatment effects and only the residual or error variation is left. The numerator and denominator degrees of freedom for formula 12.12 are $(h-1)$ and $(h-1)(k-1)$. The corresponding terms for formula 12.13 are $(k-1)$ and $(h-1)(k-1)$.

The calculations for the optical scanner example are, for treatments:

$$
F = \frac{k \sum_{i=1}^{h} (\bar{x}_{i.} - \bar{\bar{x}})^2/(h-1)}{\sum_{j=1}^{k} \sum_{i=1}^{h} (x_{ij} - \bar{x}_{i.} - \bar{x}_{.j} + \bar{\bar{x}})^2/(h-1)(k-1)}
$$

(12.12)

$$
F = \frac{4[(59-60)^2 + (58-60)^2 + (62-60)^2 + (61-60)^2]/3}{[(45-59-42.75+60)^2 + (31-58-42.75+60)^2 + \cdots + (88-61-84-60)^2]/3(3)}
$$

$$
F = 0.51
$$

and for blocks:

$$
F = \frac{h \sum_{j=1}^{k} (\bar{x}_{.j} - \bar{\bar{x}})^2/(k-1)}{\sum_{j=1}^{k} \sum_{i=1}^{h} (x_{ij} - \bar{x}_{i.} - \bar{x}_{.j} + \bar{\bar{x}})^2/(h-1)(k-1)}
$$

(12.13)

$$F = \frac{4[(42.75 - 60)^2 + (53.5 - 60)^2 + (59.75 - 60)^2 + (84 - 60)^2]/3}{[(45 - 59 - 42.75 + 60)^2 + (31 - 58 - 42.75 + 60)^2 + \cdots + (88 - 61 - 84 + 60)^2]/3(3)}$$

$F = 46.87$

The numerator and denominator degrees of freedom for the F test statistic on treatments are 3 and 9. In this problem the numerator and denominator degrees of freedom for the F test statistic on blocks are also 3 and 9. The critical value of an F distribution with 3 and 9 degrees of freedom and a 5% significance level is 3.86. The null of no treatment effects cannot be rejected; the null of no block effects is rejected at the 5% significance level. Since the error rates of the four scanners appear to be the same, you could choose on the basis of price or service or some other criterion.

EXERCISES FOR SECTION 12.3

12.11 Use the following data to test the null hypotheses of no treatment, block, or interaction effects at the 5% significance level.

	Treatment 1	Treatment 2	Treatment 3	Treatment 4
Block 1	34, 54, 82	42, 54, 59	52, 57, 63	48, 59, 69
Block 2	20, 40, 60	33, 40, 67	25, 26, 35	23, 29, 36
Block 3	80, 92, 98	90, 95, 100	76, 83, 96	87, 90, 93

12.12 Use the following data to test the null hypotheses of no treatment, block, or interaction effects at the 5% significance level.

	Treatment 1	Treatment 2	Treatment 3	Treatment 4
Block 1	12, 13, 14, 16, 20	13, 16, 16, 18, 22	22, 23, 24, 24, 25	20, 21, 23, 25, 26
Block 2	56, 57, 60, 61, 64	58, 59, 61, 61, 65	46, 47, 49, 50, 58	54, 54, 55, 56, 56
Block 3	37, 37, 39, 44, 46	27, 28, 29, 34, 37	25, 30, 32, 34, 35	38, 38, 39, 40, 41

12.13 Use the following data to test the null hypothesis of no treatment effects at the 1% significance level. The design is randomized blocks.

	Treatment 1	Treatment 2	Treatment 3	Treatment 4
Block 1	5	6	5	7
Block 2	22	24	24	27
Block 3	8	7	9	10
Block 4	48	47	44	50

12.14 The management of a symphony orchestra wants to test the hypothesis that the expected numbers of tickets sold will be the same across types of soloists they plan to engage. The most common types of soloists the symphony hires are pianists, violinists, flutists, and cellists. It also gives orchestral concerts without any soloists. Ticket sales are sensitive to the program: modern works are bad for the box office and well-known romantic pieces seem to sell well. The following table shows ticket sales for all of the concerts during the past season, listed by type of soloist and categorized by the popularity of the music programmed. Test at the 5% significance level the hypotheses that the type of soloist or the absence of a soloist does not affect ticket sales and that there is no interaction between the type of soloist and the popularity of the music being played.

	Pianists	Violinists	Flutists	Cellists	No soloist
Highly popular	2,200, 2,200	2,100, 2,050	1,800, 1,900	2,000, 2,000	2,200, 2,200
Moderately popular	1,700, 1,850	1,650, 1,850	1,500, 1,400	1,700, 1,550	1,500, 1,650
Modern	1,200, 1,150	1,050, 1,100	1,000, 1,100	1,300, 1,450	350, 250

12.15 Elizabeth Arden is a manufacturer of women's health and beauty products. The company is considering four formulations of a new shade of hair dye. An important factor in a hair dye is its lasting power, defined as the number of days before treated hair becomes indistinguishable from untreated hair. Elizabeth Arden's labs hire four long-haired women to test the new dyes. The test consists of dividing each woman's hair into five sections and treating four of the sections with one of the different dyes. The women shampoo their hair daily and the number of days until each treated section cannot be distinguished from the untreated section is recorded. Use the following data to test at the 5% significance level the hypothesis that the four hair dyes have the same mean lasting power.

	Dye 1	Dye 2	Dye 3	Dye 4
Woman 1	18	24	30	25
Woman 2	34	36	30	29
Woman 3	21	18	17	19
Woman 4	41	48	45	50

SUMMARY

One- and two-way analyses of variance are used to test for equality of more than two population means. These techniques can be used with data from controlled experiments or with ordinary data. When ordinary data are used, the assignment of the experimental units is not under the control of the researcher; the data merely record past experience. When experimental data are used, the assignment of the experimental units is under the researcher's control. In one-way analysis of variance the observations or experimental units are classified by treatment. The term "treatment" originally referred to an agricultural experiment; now it means any breakdown of data into groups that are suspected to have different population means. In two-way analysis the data are cross-classified by treatment and block. The term "block" originally referred to a controlling factor applied to the experimental units, such as the type of soil used to grow plants. In a nonexperimental context a "block" is another way of classifying data: the observations within a block are considered "homogeneous" or having evolved under similar conditions. The advantage of having a blocking variable is that, even if you are not interested in hypotheses about the impact of the blocking variable, the accuracy of the test on the treatments is improved.

Several assumptions about the population and the data are required for analysis of variance. It is important to have independence among the observations taken from any treatment. The distributions of the observations around their respective treatment and block means should be normal and have the same variance. Analysis of variance is robust with respect to moderate departures from the assumptions of normality and equal variance.

The basic idea behind all analysis-of-variance techniques is to compare two different estimates of the population variance. One estimate, always placed in the denominator of the F test statistic, is an unbiased estimator of the population variance calculated from the within-sample variation. The other estimate, always placed in the numerator, is an unbiased estimate of the population variance if and only if the null hypothesis being tested is true. Otherwise the expected value of the numerator is larger than the denominator and the F test statistic tends to be greater than 1. The null hypotheses that can be tested are that there are no treatment effects, no block effects, and no interaction effects. An absence of treatment effects means that the population means are equal across treatment categories. Similarly, an absence of block effects means that the population means are equal across block categories. An absence of interaction effects means that the mean of each population defined as a treatment-block combination equals the sum of its treatment effect and its block effect.

The F test statistic calculated for an analysis-of-variance test is compared to the $1 - \alpha$ percentile of the F distribution, that is, the point in the distribution that has α in the right tail. The α is the significance level of the test. Since the F distribution is a family of distributions that is described by degrees of freedom in both the numerator and the denominator, a table that provided many α values would be long. Only the 95th and 99th percentiles are provided in the F table (Table D9) in Appendix D.

The estimated population variance can be used to construct confidence intervals for each population mean and the difference between each possible pair of population means. These intervals have a simultaneous confidence level of $(1 - \alpha)^k$, where k is the number of confidence intervals that have been constructed. These intervals are much narrower than the confidence intervals constructed with the information from only one sample (one treatment) or from two samples for the difference in means. The reason that they are narrower, or more precise, is that the use of all of the data to estimate the population variance provides a more accurate estimator of the population variance. Tukey's honest significant difference test can be used to test for equality of all possible pairs of population means. Examples of confidence intervals and hypothesis tests were discussed for only one-way analysis of variance, but these methods extend in a straightforward manner to two-way analysis of variance. Some of the starred exercises give you a chance to figure out how the extensions work.

The factorial design allows tests for interaction effects; the randomized block design saves on the number of data needed. A randomized block design applies only in an experimental context because the experimental units have to be randomly assigned. All of the examples of two-way analysis of variance in this chapter used balanced designs. The number of observations in each treatment-block combination has to be the same if a design is to have balance. This restriction was imposed to simplify the math. Many statistical packages on a computer easily handle unbalanced designs.

 ## SUGGESTED READINGS

Mendenhall, William. *Introduction to Linear Models and the Design and Analysis of Experiments.* Belmont, Calif.: Wadsworth, 1968.

Neter, John, and William Wasserman. *Applied Linear Statistical Models.* Homewood, Ill.: Richard D. Irwin, 1974.

 ## NEW TERMS

balanced factorial design　A research design in which an equal number of observations or experimental units is found in every treatment-block combination.

block　A factor or variable other than the treatment variable which is thought to affect the values of the variable of interest.

completely randomized design　A research design in which each unit in a controlled experiment is randomly assigned to a treatment.

factorial design　A technique used in two-way analysis of variance in which there are h treatments, k blocks, and m observations or experimental units per treatment-block combination, or hkm independent observations.

F distribution　The ratio of two chi-square distributions, each divided by its degrees of freedom.

interaction effect　Any difference between the population mean of a treatment-block combination and the sum of the treatment and block effects.

one-way analysis of variance　A joint test for the equality of more than two population means.

randomized block design　A technique used in two-way analysis of variance in which the experimental units within a block are randomly assigned to the treatments.

Studentized range　The distribution of the difference between the highest and lowest observations of a series of k observations drawn from k independent normal distributions with identical means and standard deviations, divided by the estimated population standard deviation.

treatment In an experimental context, an arrangement of experimental units or a process to which they are subjected; in a nonexperimental context, a classification or category in which the observations are placed.

two-way analysis of variance A joint test for the equality of two population means used when two factors are suspected of affecting the variable of interest.

 NEW FORMULAS

F test statistic for one-way analysis of variance

$$F = \frac{\sum_{j=1}^{k} n_j \,(\bar{x}_j - \bar{\bar{x}})^2/(k-1)}{\sum_{j=1}^{k} s_j^2/k} \tag{12.1}$$

Estimated within-sample variance

$$s_j^2 = \sum_{i=1}^{n_j} \frac{(x_{ij} - \bar{x}_j)^2}{n_j - 1} \tag{12.2}$$

Probability density function of the F distribution

$$f(x) = \frac{\chi^2_{\text{numerator}}/df_{\text{numerator}}}{\chi^2_{\text{denominator}}/df_{\text{denominator}}} \tag{12.3}$$

Mean of the F distribution

$$\mu = \frac{df_{\text{denominator}}}{df_{\text{denominator}} - 2} \qquad df_{\text{denominator}} > 2 \tag{12.4}$$

Standard deviation of the F distribution

$$\sigma = \sqrt{\frac{2df_d^2(df_n + df_d - 2)}{df_n(df_d - 2)^2(df_d - 4)}} \qquad df_d > 4 \tag{12.5}$$

Confidence interval for one population mean

$$L,\ U = \bar{x}_j \pm t_{\alpha/2}\sqrt{\frac{\sum_{j=1}^{k} s_j^2/k}{n_j}} \tag{12.6}$$

Standard error of the difference between two population means

$$s_{\bar{x}_1 - \bar{x}_2} = \sqrt{\frac{\sum\limits_{j=1}^{k} s_j^2/k}{n_1} + \frac{\sum\limits_{j=1}^{k} s_j^2/k}{n_2}}$$

(12.7)

Critical value for Tukey's honest significant difference test

$$\text{HSD} = q\sqrt{\frac{\sum\limits_{j=1}^{k} s_j^2/k}{\min(n_j)}}$$

(12.8)

F test statistic for the null hypothesis of equal treatment means

$$F = \frac{km \sum\limits_{i=1}^{h} (\bar{x}_{i..} - \bar{\bar{x}})^2/(h-1)}{\sum\limits_{i=1}^{h} \sum\limits_{j=1}^{k} \sum\limits_{l=1}^{m} (x_{ijl} - \bar{x}_{ij.})^2/kh(m-1)}$$

(12.9)

F test statistic for the null hypothesis of equal block means

$$F = \frac{hm \sum\limits_{j=1}^{k} (\bar{x}_{.j.} - \bar{\bar{x}})^2/(k-1)}{\sum\limits_{i=1}^{h} \sum\limits_{j=1}^{k} \sum\limits_{l=1}^{m} (x_{ijl} - \bar{x}_{ij.})^2/kh(m-1)}$$

(12.10)

F test statistic for the null hypothesis of no interaction effects

$$F = \frac{m \sum\limits_{i=1}^{h} \sum\limits_{j=1}^{k} (\bar{x}_{ij.} - \bar{x}_{i..} - \bar{x}_{.j.} + \bar{\bar{x}})^2/(k-1)(h-1)}{\sum\limits_{i=1}^{h} \sum\limits_{j=1}^{k} \sum\limits_{l=1}^{m} (x_{ijl} - \bar{x}_{ij.})^2/kh(m-1)}$$

(12.11)

F test statistic for the null hypothesis of no treatment effects in a randomized block design

$$F = \frac{k \sum\limits_{i=1}^{h} (\bar{x}_{i.} - \bar{\bar{x}})^2/(h-1)}{\sum\limits_{j=1}^{k} \sum\limits_{i=1}^{h} (x_{ij} - \bar{x}_{i.} - \bar{x}_{.j} + \bar{\bar{x}})^2/(h-1)(k-1)}$$

(12.12)

F test statistic for the null hypothesis of no block effects in a randomized block design

$$F = \frac{h \sum\limits_{j=1}^{k} (\bar{x}_{\cdot j} - \bar{\bar{x}})^2/(k-1)}{\sum\limits_{j=1}^{k} \sum\limits_{i=1}^{h} (x_{ij} - \bar{x}_{i\cdot} - \bar{x}_{\cdot j} + \bar{\bar{x}})^2/(h-1)(k-1)}$$

(12.13)

ADDITIONAL EXERCISES

Asterisks designate exercises of greater than ordinary difficulty.

12.16 The U.S. Army conducted a competitive test of three models of "smart" artillery shells. The test consisted of firing a shell at a target 30 miles away and electronically guiding the shell. The distance between the target and the landing point of each shell was recorded. The test was supposed to consist of ten firings of each model, but two shells of one model failed to fire and one shell of a second model failed to fire. Thus the actual sample sizes were 8, 9, and 10. Given the mean distances of 248, 295, and 287 feet for the three models and the respective sample standard deviations of 85, 109, and 118 feet, test at the 5% significance level the hypothesis that the three models of smart artillery shells have the same mean error. Assume that the population standard deviations of the distances between the target and the landing points are the same for all three models of artillery shell.

12.17 The federal Manpower Development and Training Administration (MDTA) wanted to test the hypothesis that three treatments for disadvantaged workers—(1) job-training programs provided by public agencies, (2) a voucher system for training, and (3) a job-referral service with no training—would have the same impact on the workers' earnings. With publicly provided job training, a city or state agency chooses which skills the trainees will learn and then hires a local agency to provide the training. A voucher system lets the disadvantaged worker choose a skill and a school, and a voucher is used to pay the tuition fee. A referral service tries to find jobs for disadvantaged persons. The MDTA also wanted to test the hypothesis that the impact on earnings is the same across race and sex groups. Suppose the MDTA randomly assigned five trainees to each of the three treatments and from each of four race-sex categories. Use the following data to test the hypotheses of no treatment and block effects and no interaction effects, all at the 5% significance level. The observations are the difference in earnings in dollars for each trainee between the year after ending the training or referral and the year before the worker applied to the program.

> White men, public training: 3,208, 4,526, −1,299, 7,532, 2,884
>
> White women, public training: 6,549, 2,312, 6,759, 9,878, 8,326
>
> Black men, public training: −3,488, −7,658, 8,965, 778, 5,439
>
> Black women, public training: 2,207, 7,650, 6,761, 1,216, 643
>
> White men, voucher system: 5,543, 7,654, 342, −120, 8,321
>
> White women, voucher system: 10,932, 12,080, 9,342, 9,076, 9,876

Black men, voucher system: 201, −98, −76, 9,872, 3,422

Black women, voucher system: 3,421, 8,762, 4,321, 3,321, 7,632

White men, referral service: 2,341, 3,423, 9,873, 3,428, 6,532

White women, referral service: 2,231, 4,342, 6,754, 7,863, 4,657

Black men, referral service: 3,421, 3,243, −432, 987, 3,232

Black women, referral service: 6,543, 432, −654, 6,521, 8,765

12.18 A consumer research firm suspected that the use of a computer for an interview would yield different answers than a mailed survey or a personal interview. The firm used shopping mall intercepts to select at random 300 white married women between the ages of 30 and 50. Each woman was offered $10 to participate in the survey. One hundred women were given a questionnaire to fill out and return by mail; 100 were personally interviewed; and 100 were left alone in a booth with a computer that prompted them with questions. The key question in the survey was each woman's weight. The sample mean for the mailed questionnaires was 122.4 pounds and the sample standard deviation was 12.8; the sample mean for the personal interviews was 117.6 pounds and the sample standard deviation was 14.1; and the sample mean for the computer interview was 135.0 pounds and the sample standard deviation was 15.9. Test at the 5% significance level the hypothesis that the mean weights identified by the three methods would be the same for the population of all white married women between the ages of 30 and 50.

12.19 A utility company wants to test the hypothesis that electricity consumption among retail stores in its service area will be the same under three treatments: (1) an experimental peak-load pricing system, (2) the current flat-rate pricing system, and (3) a system with an increasing marginal price as a function of annual usage. To reduce random variation in electricity usage, the company picks nine stores in three sets of three. The stores within each set consumed the same amount of electricity during the previous year. One store in each set is assigned to one of the three treatments. The measured consumption of each store in millions of kilowatts over a one-year period were:

Prior use low, peak-load pricing system: 30,987

Prior use medium, peak-load pricing system: 89,765

Prior use high, peak-load pricing system: 132,376

Prior use low, flat-rate pricing system: 24,321

Prior use medium, flat-rate pricing system: 82,121

Prior use high, flat-rate pricing system: 123,876

Prior use low, increasing marginal price system: 26,546

Prior use medium, increasing marginal price system: 76,541

Prior use high, increasing marginal price system: 126,540

Test at the 5% significance level the hypothesis that the three treatments would result in the same mean level of electricity consumption by all stores in the utility's service area.

12.20 Rand McNally & Company publishes road maps of the United States, Canada, and Mexico. To estimate the highway mileage between cities, the company has drivers equipped with accurate odometers drive between the cities. The drivers are instructed to drive down the middle of the right lane of the highway. If a driver drifts away from the middle of the lane, the recorded mileage may vary up or down, depending on whether the drift is toward the outside or the inside of any curves. Rand McNally tests the interdriver consistency by having five drivers travel each of the same five routes.

(a) Use the following data to test at the 5% significance level the hypothesis that there are no treatment (each driver is a treatment) effects on the recorded mileage.

(b) What type of research design has Rand McNally constructed?

	Driver 1	Driver 2	Driver 3	Driver 4	Driver 5
Route 1	103.2	103.7	104.1	103.3	103.5
Route 2	88.6	88.3	88.9	88.2	88.4
Route 3	51.7	51.7	51.9	51.4	51.6
Route 4	45.9	45.8	46.2	46.1	45.7
Route 5	32.1	32.4	32.6	32.1	32.2

12.21 Exercise 12.16 described an army test on three models of "smart" artillery shell. Use the data in that exercise to construct a confidence interval for the mean distance between the target and the landing point at a 95% level of confidence for the model that had the best accuracy in the test.

12.22 Exercise 12.18 described an experiment on the effects of three survey methods on the mean weight reported by white married women between the ages of 30 and 50 who were intercepted in shopping malls. Construct confidence intervals for the differences in population mean weights reported under each possible set of two methods. Use an overall level of confidence for these intervals of 0.729.

12.23 Use Tukey's honest significant difference test to identify all pairs of methods that would report different weights for the experiment in exercise 12.18 at the 5% significance level.

12.24 A firm that sells commodity brokerage services is considering three plans for paying its salesmen. Plan 1 is to allow salesmen either a minimum of $500 per week or a 10% commission on their sales. Plan 2 is to have no minimum and a 12% commission. Plan 3 is to offer a sliding commission scale that ranges from 8% to 20%, depending on the amount of sales. To test whether the mean earnings per salesman is the same across the three plans, the firm randomly assigns ten salesmen to each plan. Over a one-month test period the respective sample means and standard deviations for sales per salesman were: plan 1, $11,400 and $2,659; plan 2, $16,744 and $877; and plan 3, $11,432 and $503. Test at the 5% significance level the hypothesis that the population mean monthly sales per salesman is the same for all three plans.

12.25 Construct a confidence interval at the 95% level of confidence for the population mean monthly sales per salesman under plan 2 in exercise 12.24.

12.26 Construct a series of confidence intervals for the differences in population mean monthly sales per salesman between all possible pairs of plans described in exercise 12.24. Use an overall level of confidence of 0.857.

12.27 Use Tukey's HSD test to identify the pairs of plans described in exercise 12.24 which have statistically significant differences. Use a 5% significance level.

12.28 Automobile manufacturers have a labor rate book that specifies the amount of time a mechanic is expected to need to complete every possible repair on every car model that the manufacturer produces. Dealerships use this rate book to determine their labor charges. For example, the book might list six hours as the time needed to replace a clutch on a Toyota Celica, and if a particular Toyota dealer charges $30 an hour for labor, the total labor charge to replace a clutch will be $180. The dealers typically charge for the book time rather than the actual time it takes to do the job. Experienced mechanics often beat the book time and inexperienced mechanics usually take longer. The rates in these books are based on experiments in which mechanics are observed making the listed repairs. Suppose Toyota was conducting these experiments among three types of mechanics: (1) Japanese mechanics, (2) American mechanics trained at a Toyota mechanics' school, and (3) American mechanics who were not trained at a Toyota mechanics' school. A blocking factor used in these experiments is the number of years of experience each mechanic has had. If the mean time needed to complete the task is the same for all mechanics, Toyota could save money in setting up future rate books by testing only its Japanese mechanics. In the experiment, each cell of type of mechanic and number of years of experience has eight repetitions. Use the following data on the time needed to replace a clutch on a Celica to test at the 5% significance level the hypothesis that the mean repair time is the same for all three types of mechanics. Also test for any interaction effects among type of mechanic and years of experience.

Japanese mechanic, one year of experience: 7.1, 7.8, 6.2, 8.0, 9.2, 6.6, 8.1, 7.9

American mechanic trained at Toyota school, one year of experience: 8.3, 8.3, 7.9, 4.7, 5.3, 6.1, 5.9, 9.0

American mechanic not trained at Toyota school, one year of experience: 5.2, 5.7, 9.2, 8.3, 9.4, 7.9, 8.6, 8.3

Japanese mechanic, five years of experience: 5.3, 4.9, 4.8, 5.1, 5.0, 4.6, 5.5, 5.0

American mechanic trained at Toyota school, five years of experience: 5.2, 5.8, 6.6, 5.0, 5.6, 5.7, 6.1, 6.0

American mechanic not trained at Toyota school, five years of experience: 4.9, 6.2, 6.6, 6.7, 6.9, 7.0, 5.9, 6.5

Japanese mechanic, ten years of experience: 4.9, 4.9, 5.0, 4.9, 5.1, 5.0, 5.1, 5.2

American mechanic trained at Toyota school, ten years of experience: 5.2, 5.0, 4.9, 4.9, 4.8, 5.2, 4.7, 4.9

American mechanic not trained at Toyota school, ten years of experience: 5.3, 5.2, 5.4, 5.4, 5.5, 5.6, 5.9, 5.1

12.29* In exercise 12.28 it is unlikely that one of the assumptions required for an F test will hold. Identify the suspect assumption and explain why it is suspect.

12.30** Use Tukey's HSD test to identify in exercise 12.28 the pairs of cells—type of mechanic and years of experience—that have different population means. Set the significance level at 5%.

12.31* The school board in a community that was running out of room in its elementary schools and did not have enough money to build a new school wanted to try year-round schooling. With year-round schooling, children go to school for the same number of days as children who have a traditional school year, but they have a series of short vacations rather than one long summer vacation. These vacations are staggered so that the same rooms can be used by different classes. The staggering of the vacations is what cuts down on the number of schools needed. Before adopting the year-round plan in all of its schools, the board decided to offer it on a pilot basis in a few schools. Any parents who wanted their children to participate in the year-round school experiment could sign up. The advocates of year-round schools claimed that, besides saving money, the plan enhanced students' performance because students would forget less material over a series of short vacations than over a long summer vacation. In year-round schools the teachers could spend more time teaching new material and less time reviewing what had been forgotten. To test for this effect, the school board recorded the reading scores of all third-grade students in the school system the year before the experiment began and at the end of the second year of the experiment. Family income was used as a blocking factor. The gain in reading scores was higher for the year-round students than for the traditional students, and an F test for the null of equal means could be rejected at the 1% significance level. The school board interpreted these results as showing that community-wide adoption of year-round schooling would improve overall reading scores. What might be wrong with this conclusion?

13

Simple Regression and Correlation

any local telephone companies in the United States are changing their method of billing for residential phones from a flat fee to "local measured service" (LMS). Under the flat-fee system, subscribers are charged for the phone lines to their residences but not for each individual local call made. With LMS, the monthly bill consists of a fee for the phone line plus charges for "billing units," which depend on the duration and distance of local calls. A phone company that adopts LMS has to decide on the rate it will charge for each billing unit. The advantage of LMS to the phone company is that it makes customers sensitive to their phone use and it can reduce the amount of capital (lines and switching equipment) that the phone company has to maintain. The advantage of LMS to some customers is that it can reduce their phone bills. Of course, a relatively heavy phone user will wind up paying more under LMS.

The phone company knows that the rate it charges for a billing unit and the number of billing units generated per household are correlated. That is, if one of the variables changes, the other also changes. The closeness of the association or co-movement between two variables in a population data set is measured by the **population correlation coefficient,** which is *a measure of the linear association between two random variables calculated from population data.* The degree of association between two variables in a sample data set is measured by the **sample correlation coefficient,** which is *a measure of the linear association between two random variables calculated from sample data.* If phone use goes down as the charge per billing unit is raised, the correlation is negative or inverse. Correlation coefficients are used to measure the closeness of the association between two random variables. Part of this chapter is on correlation analysis.

However, the rate charged by the phone company for a billing unit is *not* a random variable; it is not determined by chance. The phone company does not toss a die to determine what rate it will charge. The charge per billing unit is set by the phone company and the state regulatory commission. The amount of phone use per household is a random variable. While phone use is sensitive to the billing rate, within a given household it varies from month to month for myriad reasons. A local phone company does not need to know why an individual household's phone use goes up or down from month to month. *For the purpose of setting local phone rates,* the company doesn't even need to know an individual household's phone use (though it does need to know the individual household's phone use to figure its bill). To estimate the revenue it would receive at different potential rates per billing unit, the phone company needs to know the *average* use per household in its service area for each rate.

Simple regression is *a technique that is used to describe the effect of changes in an explanatory variable on a dependent or explained variable.* In regression analysis one variable is set or controlled (the charge per billing unit) and the other variable is a random variable (the use per household); that is, it is determined at least partly by chance. Regression analysis can be done for population data or for a sample. Most of the examples in this chapter and in Chapter 14 will be linear or straight-line regression, in which a unit change in one variable (e.g., number of billing units in a household) is assumed to be associated with a constant change in the average level of the other variable (e.g., charge per billing unit). For example, if there were a linear relationship between the charge per billing unit and the average number of billing units, an increase in the charge per billing unit from 5 to 6 cents would be associated with the same *change* in the average number of billing units as an increase from 12 to 13 cents. *The change in the mean value of the dependent variable associated with a unit change in the independent variable calculated with population data*—in this case, the change in the average number of billing units per additional penny charged for a billing unit—is called the **population slope coefficient,** β. *The change in the mean value of the dependent variable associated with a unit change in the independent variable calculated with sample data* is called the **sample slope coefficient,** b. The phone company would like to know the value of β.

Regression analysis is one of the most widely used techniques in business and economic analysis. It can be applied to such problems as forecasting the trend of sales over time, estimating the demand for a commodity as a function of its price, and predicting how changes in wage rates affect employee turnover. In this chapter we will explore the mechanics of regression and correlation analysis, such as finding the slope coefficient and the correlation coefficient. We will also discuss statistical inference in regression and correlation, such as confidence intervals and hypothesis tests for the population slope coefficient. The techniques of statistical inference in regression analysis can be applied only when certain restrictive assumptions about the population are true. We'll discuss these assumptions and in the last (optional) section find out what can be done when, as often is the case, some of them are violated. In Chapter 14 we'll discuss multiple regression, a technique that is closely related to simple regression. Multiple regression is used to show the amount of change in one characteristic in a population (for example, number of billing units used per household) which is associated with changes in several other characteristics (charge per billing unit, size of household, income of household, etc.).

13.1 INTRODUCTION TO REGRESSION AND CORRELATION

13.1A DESCRIBING THE RELATIONSHIP BETWEEN TWO VARIABLES WITH A LEAST-SQUARES LINE

If you are working with the data for a population, regression and correlation are purely descriptive techniques. Regression and correlation analysis can be used descriptively to summarize how two variables in a population are related. We'll begin with this descriptive use of regression and correlation analysis and then discuss the inferential use of regression and correlation. The inferential use of regression or correlation analysis is the use of sample data to estimate the population parameters, such as the population slope coefficient and the correlation coefficient. For our first example, let's update a data set we have already used. In Chapter 2, the GPAs for 75 students in a statistics class were listed in Table 2.1. Let's treat this group of students as a population and consider how GPA is related to their weekly study hours. Table 13.1 shows each student's study hours and GPA.

Figure 13.1 shows a scatter diagram of the data in Table 13.1. A **scatter diagram** is *a graph in which each observation is represented by a dot. The explanatory or independent variable is scaled on the* x *axis and the dependent or explained variable is scaled on the* y *axis.* In this example the number of hours of study per week is treated as the explanatory variable and plotted on the x axis, while the "explained" or dependent variable, the GPA, is plotted on the y axis. The choice variable—that is, the variable that can be set or controlled—is always the explanatory variable in regression analysis. The random variable is always the explained or dependent variable. The chapter introduction mentioned that correlation analysis applies when both variables are random and regression analy-

TABLE 13.1

Grade point averages and weekly study hours of 75 students in a Statistics class

Hours	GPA	Hours	GPA	Hours	GPA	Hours	GPA
2	1.88	1	1.98	0	1.99	5	2.01
9	2.07	4	2.09	3	2.09	14	2.13
6	2.14	9	2.17	9	2.17	7	2.22
12	2.23	9	2.25	11	2.25	5	2.25
10	2.29	12	2.33	24	2.34	11	2.34
10	2.35	8	2.39	13	2.39	13	2.40
12	2.40	10	2.43	7	2.43	9	2.44
10	2.45	14	2.47	11	2.48	18	2.50
15	2.52	13	2.52	13	2.54	16	2.56
6	2.56	19	2.57	12	2.57	20	2.59
14	2.60	15	2.67	14	2.71	10	2.71
12	2.82	16	2.85	17	2.90	15	2.90
13	2.91	15	2.91	7	2.95	20	2.97
16	3.02	19	3.02	19	3.02	20	3.04
18	3.08	21	3.10	12	3.13	16	3.13
22	3.15	21	3.19	20	3.21	26	3.22
17	3.22	20	3.31	14	3.34	23	3.48
19	3.49	16	3.50	18	3.62	22	3.65
28	3.76	25	3.82	21	3.92		

FIGURE 13.1

Scatter diagram of hours of study and GPA scores

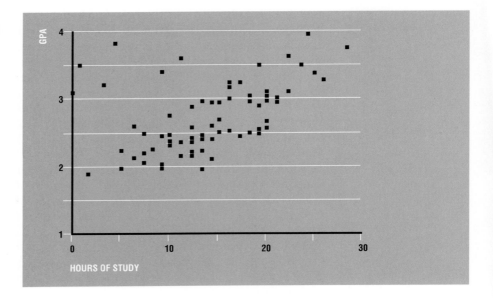

sis applies when only one of the variables is random. Since students choose the number of hours they study each week, the appropriate technique for this example is regression analysis.

When one variable is chosen to be the explanatory variable and the other the dependent, it is easy to fall into the trap of thinking that changes in the explanatory variable "cause" changes in the dependent variable. While it seems plausible that an increase in hours of study would cause a student's GPA to rise, this direction of causation is basically an assumption. It is possible that the direction of causality is reversed. Bright students who would have high GPAs with or without studying might think that study time makes a difference and study long hours by mistake. (Admittedly this explanation is implausible. You may be asking yourself, "If they're so bright, why can't they figure out that studying doesn't matter?") Finally, there may be no causal relationship between the two characteristics. By pure chance, in this class the students who have high GPAs may also be the ones who tend to study a lot. While there are some causality tests that can be applied to certain types of data, these tests are usually discussed in econometrics courses rather than in introductory statistics. We'll just have to assume that we know the direction of causation when we pick the variable that belongs on the x axis and the one that belongs on the y axis in regression analysis.

The scatter diagram in Figure 13.1 shows a direct relationship between hours of study and GPA because as the hours of study increase, the GPA tends to rise. If the scatter diagram had tended to drift down as the number of hours of study increased, the relationship would be inverse. If the scatter diagram tended to stay at the same level as the hours of study increased, there would be no apparent effect of hours of study on GPA.

The **regression line** is the straight line that best fits the scatter diagram. It is used to summarize the overall trend or movement in the data. The most commonly used criterion is that it is *the straight line that minimizes the sum of the squared deviations between the predicted and observed values of the dependent variable.* The value of observation *i* measured on the *y* axis in Figure 13.2 is labeled y_i, and the value of the corresponding point on the regression line—the predicted value—is labeled $\hat{y}_i$. The vertical distance between these two points is the deviation. Each dot represents an observation and one of them is labeled y_i for illustration. A vertical line has been drawn from the observation labeled y_i to the regression line. The point at which this vertical line intersects the regression line is labeled $\hat{y}_i$. When the data set is a sample, the circumflex over the *y* stands for *estimated mean of* Y *given* x—in our example, the estimated average GPA given the hours of study. When the set is a population, the same symbol refers to a point on the regression or best-fitting line. If the population were distributed symmetrically above and below the regression line, then $\hat{y}_i$ would be the population mean of Y given *x*, or $\mu_{Y|x}$.

There are several reasons for using the line that minimizes the sum of the squared vertical distances between all of the y_i's and the corresponding $\hat{y}_i$'s as the best-fitting line. One reason is that the sampling properties of lines defined in this way provide estimators of the population slope and intercept which are efficient and unbiased, given some assumptions about the population which will be discussed in section 13.1C. A second reason for this definition of the best-fitting line is that squaring the deviations puts greater weight on observations that are far from the line. Distant observations tend to "pull" the line toward themselves. Consider an alternate criterion for the best-fitting line, minimizing the sum of the *absolute values* of the deviations instead of minimiz-

FIGURE 13.2
Deviation between an observation and the regression line

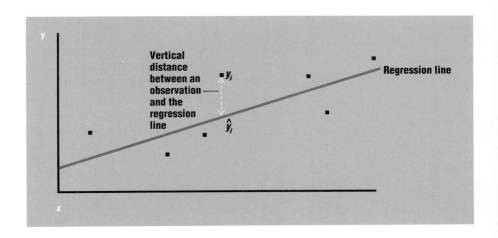

ing the sum of the squared deviations. Consider two deviations of 10 and 12 from either a line defined by the absolute deviation criterion or from a regression line defined by the squared deviation criterion. If the absolute deviation criterion were used to define a best-fitting line, the observation with a deviation of 12 would have 20% more weight in determining the best-fitting line than the observation with a deviation of 10. If the squared deviation criterion were used, an observation with a deviation of 12 would have 44% more weight than an observation with a deviation of 10 (144 versus 100). The greater weight on more distant observations under the squared deviation criterion can sometimes be an advantage and sometimes a disadvantage. If a distant or outlying observation is accurate, the regression line will fully reflect the possibility of having such extreme values. On the other hand, if a distant observation is due to an error, the greater weight placed on it will give a false picture of the true relationship between x and the average level of Y. We'll stick to the sum of squared deviations criterion because of its desirable sampling properties, although there are some alternatives, such as the method of resistant lines, which place less weight on distant observations than on close observations. There are some references to resistant lines in the suggested readings at the end of this chapter.

The sum of the squared vertical distances between all of the y_i's and the corresponding $\hat{y}_i$'s is

$$\text{SSE} = \sum_{i=1}^{\nu} (y_i - \hat{y}_i)^2$$

(13.1)

The symbol SSE stands for sum of squared errors, which is another term for the squared vertical distances between all of the y_i's and the corresponding $\hat{y}_i$'s. The symbol ν stands for the number of observations in the population. A data point or observation is an x–y pair. The reason that these deviations are called errors is that they indicate how far off the regression line is from each observation. If there were a perfect fit between the regression line and the observations, the SSE term would be 0. The best-fitting line or *the* regression line is the one that minimizes the above term, or has the least error.

A pair of formulas, derived in Appendix C, provides a simple method for finding the best-fitting line. Since any straight line can be described by its slope and its y intercept, the two formulas for the regression line determine the population slope coefficient, β, and the population intercept, α. The Greek letters α, alpha, and β, beta, have to do double duty. The probability of making a Type II error was given the symbol β in Chapter 9, and you have used α as the significance level of a test in the last four chapters. This double duty is common usage in statistics. The same two formulas can be used to find the sample slope coefficient, b, and the sample intercept, a.

$$\beta = \frac{\sum_{i=1}^{\nu} x_i y_i - \nu \mu_x \mu_y}{\sum_{i=1}^{\nu} x_i^2 - \nu(\mu_x)^2}$$

(13.2)

$$\alpha = \mu_y - \beta \mu_x$$

(13.3)

The symbols μ_x and μ_y stand for the population mean of x and Y, respectively. The differences between the formulas for the population slope and intercept and the formulas for the sample slope and intercept are that the symbol for the number of observations in a population, ν, is replaced by the symbol for the number of observations in a sample, n, and the symbols for the population means of x and Y are replaced by the symbols for the sample means of x and Y, namely, $\bar{x}$ and $\bar{y}$.

$$b = \frac{\sum_{i=1}^{n} x_i y_i - n \bar{x} \bar{y}}{\sum_{i=1}^{n} x_i^2 - n(\bar{x})^2}$$

(13.4)

$$a = \bar{y} - b \bar{x}$$

(13.5)

The population regression line takes the form $\hat{y}_i = \alpha + \beta x_i$ and the sample regression line takes the form $\hat{y}_i = a + b x_i$. Applying the above formulas to the population data on GPAs and hours of study yields the regression line $\hat{y} = 1.833 + 0.064 x_i$. The interpretation of this population regression line is that the *average* GPA tends to increase by 0.064 for every additional hour of study per week.

Figure 13.3 shows the regression line superimposed on the scatter diagram. Since there are 75 data points in Figure 13.3, it would take too much space to use these data to illustrate the computation of the slope and intercept terms.

FIGURE 13.3
Regression line for hours of study and GPA

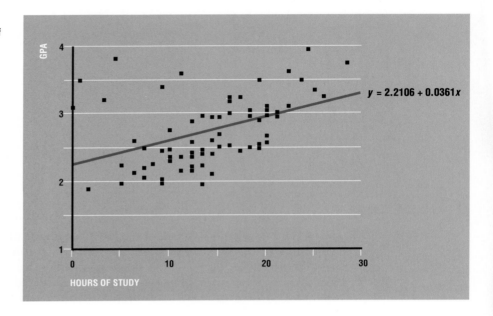

Instead, let's use a small data set. Let x be $\{1, 2, 3, 4, 5\}$ and let Y be $\{20, 14, 10, 8, 8\}$. Table 13.2 contains a worksheet for finding the terms needed in formulas 13.2 and 13.3. The mean of x is 15 divided by 5, or 3. The mean of Y is 60 divided by 5, or 12. Substituting into formulas 13.2 and 13.3, we get

$$\beta = \frac{\sum_{i=1}^{\nu} x_i y_i - \nu \mu_x \mu_y}{\sum_{i=1}^{\nu} x_i^2 - \nu(\mu_x)^2} \tag{13.2}$$

$$\beta = \frac{150 - 5(3)(12)}{55 - 5(9)}$$

$$\beta = -3$$

$$\alpha = \mu_y - \beta \mu_x \tag{13.3}$$

$$\alpha = 12 - (-3)(3)$$

$$\alpha = 21$$

The regression line is $\hat{y} = 21 - 3x$. In this example there is an inverse relationship between x and Y. This inverse relationship can easily be seen in

TABLE 13.2
Worksheet for terms needed in regression formulas

x_i	y_i	$x_i y_i$	$(x_i)^2$
1	20	20	1
2	14	28	4
3	10	30	9
4	8	32	16
5	8	40	25
15	60	150	55

Figure 13.4, which plots the scatter diagram for these data and the regression line.

Getting a regression line for a given data set is a matter of mechanically substituting into a formula; there is no thought involved. This is why such routine work is best done by a computer or a hand-held calculator. After you have the regression line, you may ask: How good is the fit? Figure 13.4 gives you an idea of the quality of the fit between the regression line and the observations. Since the observations are close to the line, the fit appears to be good. In the scatter diagram for the data on GPAs versus hours of study in Figure 13.3, the fit was not so close. In section 13.1B we will discuss two measures of the linear association between variables and the regression line, namely, the coefficients of determination and correlation. These two measures are related (one measure is the square of the other) but they have different interpretations.

FIGURE 13.4
Scatter diagram and regression line of data in Table 13.2

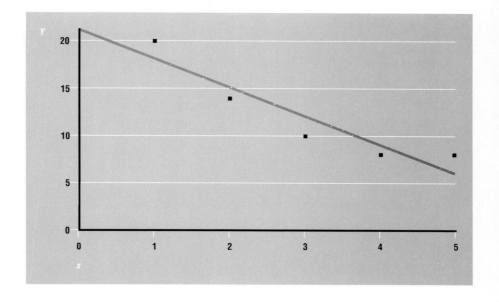

13.1B THE COEFFICIENTS OF DETERMINATION AND CORRELATION

The **coefficient of determination** measures *the percentage of the total sum of squared deviations of the dependent variable around its mean which is accounted for by the independent variable.* The symbol for the population coefficient of determination is ρ^2 and the symbol for the sample coefficient of determination is r^2. Both measures can have values from 0 to 1. If the coefficient of determination for a particular data set were 0.75, you could say that 75% of the variation of the Y variable around its mean was explained by changes in x. If the coefficient of determination for a particular data set were 0.50, you could say that 50% of the variation of the Y variable around its mean was explained by changes in x.

Before we can give the formula for the coefficient of determination, the terms "total squared deviation around the mean of Y" and "explained variation" have to be defined. The total variation in Y around its mean is

$$\text{SST} = \sum_{i=1}^{v} (y_i - \mu_y)^2$$

(13.6)

If the data set is a sample, the sample mean, $\bar{y}$, replaces the population mean in formula 13.6. Figure 13.5 illustrates the deviations between the mean of Y and each observation in the small data set of Table 13.2. We can find the sum of the squared deviations between the mean of Y and the observations by substituting into formula 13.6:

FIGURE 13.5

Deviations between y_i and μ_y—the total deviations

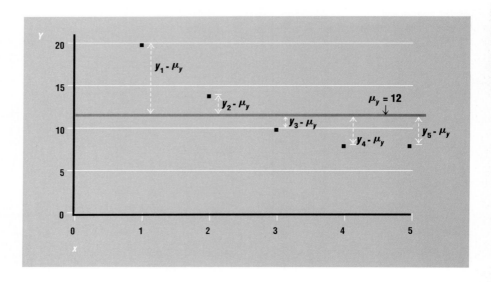

$$SST = \sum_{i=1}^{\nu} (y_i - \mu_y)^2 \tag{13.6}$$

$$SST = (20 - 12)^2 + (14 - 12)^2 + (10 - 12)^2 + (8 - 12)^2 + (8 - 12)^2$$

$$SST = 104$$

The SST of 104 can be partitioned into the sum of squares accounted for by the regression and the sum of squares not accounted for by the regression. The unaccounted-for or unexplained variation is the SSE, or sum of squared errors, given by formula 13.1. Statisticians reserve the term "variation" for squared deviations. The accounted-for or explained variation is the sum of the squared deviations between the mean of Y and the regression line. This term is called the sum of squares for the regression, or SSR:

$$SSR = \sum_{i=1}^{\nu} (\hat{y}_i - \mu_y)^2 \tag{13.7}$$

If the data set were a sample, the population mean of Y would be replaced by the $\bar{y}$ in formula 13.7.

Figure 13.6 shows the deviations between the regression line and the mean

FIGURE 13.6

Deviations between $\hat{y}_i$ and μ_y—the explained deviations

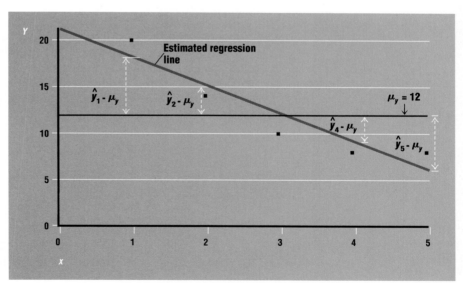

of Y. Note that the deviation between the regression line and the mean of Y is zero for the third observation. The population regression line always goes through the center of the data: μ_x, μ_y, or for sample data the point $\bar{x}, \bar{y}$. In this case the regression line passes through the point $(3,12)$. The unexplained deviations consist of the distances between each y_i and the corresponding $\hat{y}_i$. For example, the y value of the first data point, y_i, is 20 and the predicted value, $\hat{y}_i$, is 18, so the unexplained deviation for the first data point is 2.

To find SSR, substitute into formula 13.7:

$$SSR = \sum_{i=1}^{\nu} (\hat{y}_i - \mu_y)^2 \qquad (13.7)$$

$$SSR = (18 - 12)^2 + (15 - 12)^2 + (12 - 12)^2 + (9 - 12)^2 + (6 - 12)^2$$

$$SSR = 90$$

We obtain the values of $\hat{y}$ required for formula 13.7 by substituting the different x values into the equation for the regression line. For example, we obtain the $\hat{y}$ of 18 when x equals 1 by substituting the x value of 1 into the equation: $\hat{y} = 21 - 3x$. It was mentioned above that the sum of squares total consists of the sum of squared errors plus the sum of squares for the regression. In notation:

$$SST = SSE + SSR \qquad (13.8)$$

The idea behind formula 13.8 is that the total variation of Y relative to the mean of Y can be broken into two parts: (1) the variation explained by the regression line or, equivalently, accounted for by the x variable; and (2) the variation that the regression line does not account for. We can check if formula 13.8 is true in this example by calculating the SSE. Then we will have all three terms.

$$SSE = \sum_{i=1}^{\nu} (y_i - \hat{y}_i)^2 \qquad (13.1)$$

$$SSE = (20 - 18)^2 + (14 - 15)^2 + (10 - 12)^2 + (8 - 9)^2 + (8 - 6)^2$$

$$SSE = 14$$

The sum of 90 and 14 is 104, so the SSE plus the SSR equals the SST. Now you have the terms needed for the formula for the coefficient of determination. It is simply the SSR divided by the SST, that is, the explained variation divided by the total variation.

$$(13.9)$$

$$\rho^2 = \frac{\sum\limits_{i=1}^{\nu} (\hat{y}_i - \mu_y)^2}{\sum\limits_{i=1}^{\nu} (y_i - \mu_y)^2}$$

The formula for the sample coefficient of determination is

$$(13.10)$$

$$r^2 = \frac{\sum\limits_{i=1}^{n} (\hat{y}_i - \bar{y})^2}{\sum\limits_{i=1}^{n} (y_i - \bar{y})^2}$$

The coefficient of determination for the data in Table 13.2 is 90 divided by 104, or 0.865. Thus 86.5% of the variation in Y around its mean is accounted for by changes in x or explained by the regression line. Conversely, 13.5% of the variation in Y around its mean is not accounted for by the regression line.

The range of possible values for the coefficient of determination is from 0 to 1. Negative numbers are impossible because both the numerator and the denominator are squared, and the numerator cannot be larger than the denominator because the explained variation cannot be larger than the total variation. You can't explain more variation than there is. If the regression line accounts for all of the variation in Y, then the numerator will equal the denominator and the coefficient of determination will be 1. If the regression line accounts for none of the variation in Y, the SSR will be 0 and the coefficient of determination will be 0. Figure 13.7 shows two scatter diagrams and their respective regression lines, one with a coefficient of determination of 1 and the other with a coefficient of determination of 0. Note that if all of the observations were on a horizontal line, the total sum of squares around $\bar{y}$ would be 0 and the formulas for the population and sample coefficients of determination would not work because you cannot divide by 0. In this case the coefficient of determination is defined as 0.

One advantage of the coefficient of determination as a measure of the fit of a regression line is that it is easy to interpret. A coefficient of determination of 0.8 means that x accounts for or explains 80% of the variation in Y. Be careful not to confuse "account for" or "explain" with "cause." A high coefficient of determination does not prove that changes in x cause changes in Y. Causality may flow in the opposite direction, or another unobserved variable may be the

FIGURE 13.7

Regression lines with coefficients of determination of 1 and 0

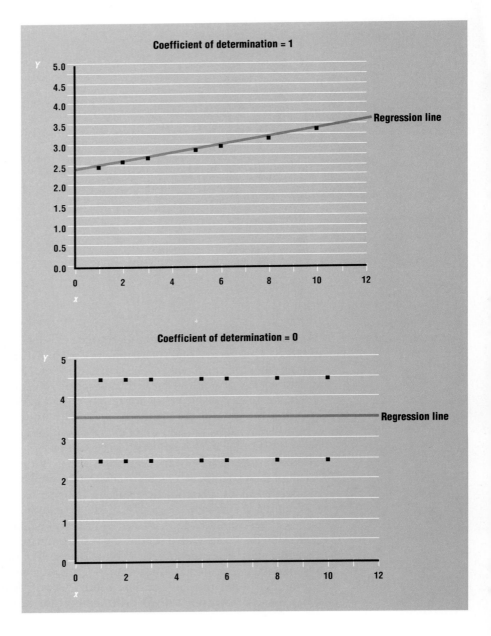

cause of the changes in both x and Y, or the observed pattern in x and Y may be due entirely to chance. Some statistics texts have used the regression in which sunspot activity is the explanatory variable and gross national product is the dependent variable as a classic example of spurious or chance association. Around the beginning of this century, a prominent English economist, William Stanley Jevons, claimed that sunspots caused variations in economic activity. His claims were not well received and subsequently he went insane. However, Jevons' work on sunspots is no longer a good example of a spurious regression because some astrophysicists now argue that variation in sunspot activity affects the earth's weather. When Jevons did his sunspot study, the economy was tied much more closely to weather patterns than it is today. Perhaps Jevons was right after all. Please don't think that this example shows that whenever you have a high coefficient of determination, there must be some hidden mechanism of causation. It is easy to get a high coefficient of determination without any causality. Take the example of the grade point averages of the 75 students in Table 13.1. There are millions of measurable characteristics of these students, such as the distance from the tips of their thumbs to the ends of their elbows or the number of hamburgers they consumed in the past year or the range of musical notes they can sing. If you were to try out every one of those characteristics as an explanatory variable, you would be bound to find some that result in high coefficients of determination even though they have nothing to do with GPAs. Occasionally someone tries to pass off the results of a grab-bag effort to find a high coefficient of determination as evidence that some unsuspected variable is really responsible for changes in the dependent variable. Don't be fooled by such a masquerade.

Now we are ready to move on to the coefficient of correlation. While the coefficient of correlation is mathematically related to the coefficient of determination (it is either the positive or negative square root of the coefficient of determination), the two are used in different contexts. These contexts of the coefficients of determination and correlation require some discussion, because the difference is a subtle one. With the coefficient of determination and with regression in general, the x variable is *not* a random variable. The x variable is considered a constant that can be set at various values or controlled exactly. The Y variable, in contrast, is a random variable that has a probability distribution for every level of x. This is the reason that x has been a lowercase letter and Y has been a capital. Thus, for a given x the Y variable can take many values. In the example of study hours versus GPA, the hours of study are controlled or determined by each student. For a given level of hours of study there is a range of possible GPA values. There are several reasons for this range in the dependent variable. One reason is that some other omitted variable, such as intensity of study time or the student's IQ, may affect GPA. Another reason is measurement error in the Y variable. A third reason is that the "world" or "reality" is in part stochastic, or is subject to random variation. Even if there were no omitted variables and no measurement error, two students with identical hours of study might have different GPAs. Consider two identical twins who study the same number of hours. They must have the same intelligence because they have the same genes and the same environment or upbringing. Suppose they even took the same classes. Will they necessarily have the same GPA? The answer is no. Neither of them may know the answer to a question on a multiple-choice test and have to guess; one may guess right and the other

CASE STUDY

THE CONSUMPTION FUNCTION

Every student who takes a course on the principles of macroeconomics learns about the consumption function. This concept was introduced by John Maynard Keynes, an English economist, in the 1930s. Keynes argued that the amount people spent on goods and services was a stable fraction of their disposable income, that is, the income left after taxes. Keynes based his argument on his view of human nature—that as incomes rose, people would discover that they needed to buy more things, and they would wind up saving the same proportion of income as before. Keynes called this allocation of any increase in income the "fundamental psychological rule of any modern community."

Keynes argued further that investment spending was much less stable than consumption spending, and that given this difference, most of the year-to-year variation in employment was caused by changes in investment. Is Keynes's theory about the stability of the consumption function supported by economic data? A regression analysis with consumption as the dependent variable and disposable income as the explanatory variable would show whether Keynes was correct about the stability of the consumption function. The scatter diagram and estimated regression line in Figure 13.8 are for the years 1929 through 1986. As you can see, the fit is almost perfect. The coefficient of determination is close to 1.00; that is, close to 100% of the variation in the economy-wide personal consumption is accounted for by variations in disposable income. The interpretation of the slope coefficient for the regression line, $b = 0.914$, is that, on average, 91 cents out of every additional dollar of personal income goes to additional spending on personal consumption. At least in regard to the consumption function, Keynes was correct.

FIGURE 13.8

Scatter diagram of disposable personal income and personal consumption expenditures, United States, 1929–1986 (billions of dollars)

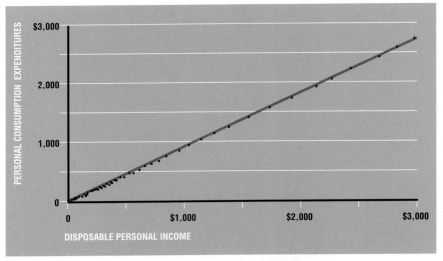

Source: *Economic Report of the President*, 1986.

wrong. They may be equally susceptible to colds, but one twin may be sick when there is an important quiz and the other twin may be sick when nothing important is due. To restate the main point, in regression analysis the x variable is controlled or determined, and for any given value of the x variable, the Y has a probability distribution.

In correlation analysis both the X and Y variables are considered random variables. Note the switch to uppercase for X. Here we are not dealing with one independent and one dependent variable. Each variable is treated the same. Correlation analysis applies to situations in which there is no suggestion of causality or in which the direction of causality is uncertain. The issue in correlation analysis is simply association. Is the movement in the X variable associated with movement in the Y variable? Are relatively high values of the X variable associated with relatively high or relatively low values of the Y variable? The choice of which of the two variables will be called X and which will be called Y is arbitrary. While an additional formula for the coefficient of correlation is not necessary because you could use formula 13.10 for the coefficient of determination and then take either the positive or the negative square root to find the coefficient of correlation, we'll introduce an additional formula that clarifies the interpretation of the coefficient of correlation by defining it in terms of z scores. One way to describe an X or a Y value as relatively high or low is to calculate its z score. The notation z_{x_i} means the z score of observation x_i. The formula for the population coefficient of correlation expressed in terms of z scores is

(13.11)

$$\rho = \frac{\sum_{i=1}^{\nu} z_{X_i} z_{Y_i}}{\nu}$$

If relatively high values of X are associated with relatively high values of Y in formula 13.11, the numerator will be positive and will have a value close to ν. The resulting coefficient of correlation will be close to $+1$. If relatively high values of X are associated with relatively low values of Y in formula 13.11, the numerator will be negative and will have a value close to $-\nu$. The resulting coefficient of correlation will be close to -1. The upper limit on the coefficient of correlation is 1 and the lower limit is -1. In Appendix C it is shown that the numerator of formula 13.11 equals either ν or $-\nu$ when x and y are perfectly correlated. In other words, the proof shows that if x and y are perfectly correlated, the z scores of the y variable are the same as the z scores of the x variable, and every term in the product of the two z scores cancels out except the ν that appears in the denominator of the standard deviation formula. (Can you explain why some of the low values of X and Y must have negative z scores? The

**TABLE 13.3
Worksheet for
coefficient of
correlation**

x_i	$z(x_i)$	y_i	$z(y_i)$	$z(x_i)z(y_i)$
1	−1.414	20	1.754	−2.480
2	−0.707	14	0.439	−0.310
3	0.0	10	−0.439	0.0
4	0.707	8	−0.877	−0.620
5	1.414	8	−0.877	−1.240
15		60		−4.650

reason is that some of the low values of X and Y must be below the means of X and Y.)

Table 13.3 shows the z scores of the data set in Table 13.2. The standard deviation of x is 1.414 and the standard deviation of Y is 4.561. We'll continue to treat the data set as a population. If we take the results of this worksheet and substitute into formula 13.11, we get

$$\rho = \frac{\sum_{i=1}^{\nu} z_{X_i} z_{Y_i}}{\nu} \qquad (13.11)$$

$$\rho = \frac{-4.650}{5}$$

$$\rho = -0.93$$

The coefficient of correlation is −0.93. As a check you can take the negative square root of the coefficient of determination of 0.865 and see if it equals −0.93. The reason for the negative square root is that the relationship between X and Y is inverse. If the scatter diagram sloped upward—in other words, if the relationship were direct—the positive square root would be used. Note that the sign of the slope of the regression line, either b or β, is the same as the sign of the coefficient of correlation. A positive slope coefficient tells you that the sign of the correlation coefficient is positive and a negative slope coefficient tells you that the sign of the correlation coefficient is negative.

If the data set is a sample, the formula for the coefficient of correlation is

(13.12)

$$r = \frac{\sum_{i=1}^{n} z_{X_i} z_{Y_i}}{n - 1}$$

The coefficient of correlation does not have as clear-cut an interpretation as the coefficient of determination. A coefficient of correlation of −0.93 does not

CASE STUDY

ARE EARTHQUAKE DEATHS INCREASING OVER TIME?

Simple regression analysis can be used to examine the question of whether something has been growing or declining over time. Describing the course or path of events over time is called time-series analysis. Time-series analysis will receive a full discussion in Chapter 15, but the technique examined in this chapter, simple regression, is the simplest and most direct method of time-series analysis. For example, is there a trend in the number of deaths caused by major earthquakes in the 20th century? After the 1985 Mexico City earthquake, the UPI published a table listing the dates of major earthquakes and the numbers of deaths associated with them. The scatter diagram and regression line for those data appear in Figure 13.9. There are 42 data points with the number of deaths in an earthquake plotted on the vertical axis and the year on the horizontal axis. The regression line shows a downward trend but the coefficient of correlation is only -0.181, which suggests that there is not a strong association between time and the number of deaths in earthquakes. If you were interested only in the time period 1900 to 1986—in other words, if this time period were treated as the population—you could conclude that there was a slight downward trend in the number of deaths in earthquakes over the period. If you wanted to forecast future trends in the number of deaths in earthquakes, the period 1900 to 1986 would be considered a sample. In section 13.2 we'll learn how to test the null hypothesis that the population correlation coefficient between time and the number of deaths in earthquakes is 0, based on the sample coefficient of correlation of -0.181. The extreme outlier in 1976, which occurred in Tangshan, China, suggests that forecasting the number of deaths in future earthquakes may be very difficult.

FIGURE 13.9
Scatter diagram and regression line for deaths caused by major 20th-century earthquakes in which the explanatory variable is the year

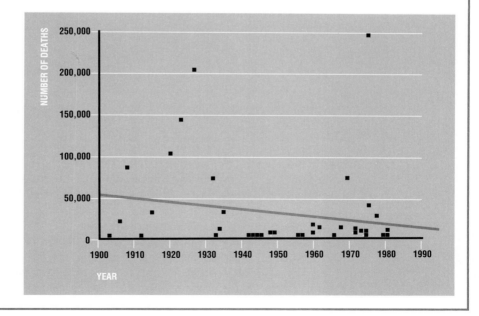

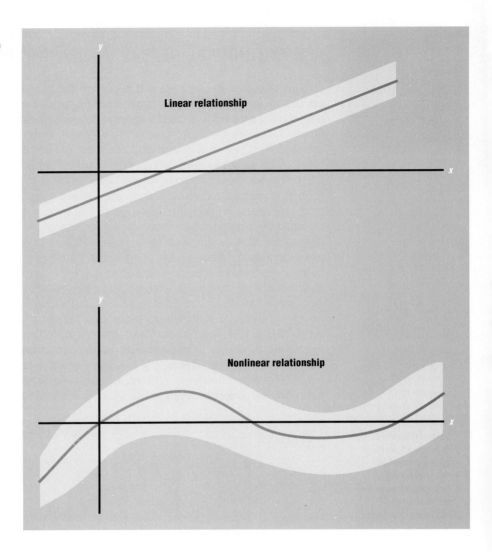

imply that 93% of the variation in one variable is associated with variation in the other variable. All you can say is that as the coefficient of correlation approaches the value of 1 or −1 from a beginning value of 0, the strength of the direct or inverse association increases.

One last important point about the coefficient of correlation is that it is a measure of *linear* association. It is possible for two random variables to be closely associated—that is, movements in one variable are associated with movements in the other—without the relationship being linear. For example,

y could equal the square of x. The coefficient of correlation won't pick up nonlinear associations; that is, the calculated coefficient of correlation will be low or even zero.

13.1C ASSUMPTIONS ABOUT THE POPULATION REQUIRED FOR STATISTICAL INFERENCE

Although we have discussed the formulas for the *sample* slope, intercept, coefficient of determination, and coefficient of correlation, we have not discussed how these sample statistics can be used as estimators of the corresponding population parameters. Up to now, we have concentrated on the descriptive uses of regression and correlation. While the inferential use of these techniques is much more important, it was easier to begin with the concept of a regression line as a descriptive measure and wait until this point to discuss the assumptions about the characteristics of the population that allow the sample statistics to be used as estimates of the population parameters.

The most basic of these assumptions is that there is a linear relationship between the mean values of the dependent and independent variables. Since the Y variable is assumed to be stochastic, there must be a range of y values for a given value of the x variable. The assumption of linearity implies that *the mean of* Y *is a linear function of* x. Figure 13.10 shows two population distributions: one in which the assumption of linearity is true and another in which the assumption of linearity is false. The range of y values is shown as a band around the mean of Y given x to emphasize that there is a distribution of y values for any given value of x. For the moment we won't say anything about the assumptions regarding the distribution of the y values around the mean values of Y. If you took a series of samples from the population in which the relationship between X and the mean value of Y was linear and used these samples to estimate the population slope and intercept, the expected values of the estimated slope and intercept would be the population slope and intercept. If you took a series of samples from a population in which the relationship between X and the mean value of Y given x was nonlinear, the expected regression line would be close to the mean value of Y for some values of x and far away for other values of x.

Figure 13.11 shows the result of estimating a linear regression on a sample taken from the nonlinear population. You cannot track a nonlinear relationship with a straight-line regression. Of course, when you are doing statistical inference, you don't know if the population relationship is linear or nonlinear. One check that you can do before running a regression with sample data is to look at the scatter diagram and see if it seems to follow a linear trend. If the scatter diagram appears to be nonlinear, some "fixes" discussed in section 13.4 will allow you to track the mean values of Y as x varies.

The second assumption is that the standard deviation of the distribution of Y given x is the same for all values of x. For any given x, you can determine the

FIGURE 13.11
Regression line for a
population in which the
mean value of *Y* is not a
linear function of *x*

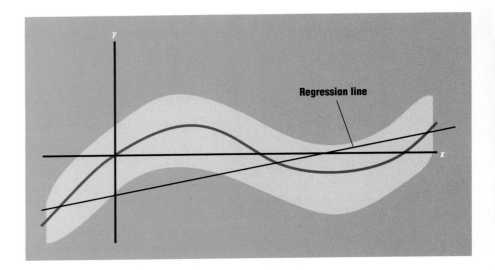

mean of Y and then calculate the standard deviation of the Y values around that
mean, holding the x constant. If this assumption were true and the relationship
between the mean of Y and x were linear, the population distribution would
look like the first population in Figure 13.10. Suppose the variability of the Y
values increased as x increased. Instead of the band that represents the population paralleling the population regression line, it would spread out as x increased. Figure 13.12 shows such a population. Populations for which *the
standard deviations of the distributions of* Y *given* x *are all equal* are termed **homo-**

FIGURE 13.12
Population in which the
mean of *Y* is a linear
function of *x* and the
variability of the *Y* values
increases as *x* increases

scedastic. Populations for which *the standard deviations of the distributions of* Y *given* x *are not all equal* are termed **heteroscedastic.**

Recall that outlying observations are given greater weight than close observations when the true regression line is estimated. The problem posed by a population such as the one in Figure 13.12 is that a few outlying observations drawn from the area in which Y has a high variance can pull the estimated regression line far away from the true regression line. Note that this is not a problem of bias. The symmetry of the population y values around the mean of Y given x in Figure 13.12 guarantees that in repeated sampling the estimated slope will average out to the population slope. The issue with a heteroscedastic population is one of efficiency. Observations taken from the narrow part of the cone—that is, when the variability in Y is low—contain more information about the mean value of Y than do observations taken from the wide part of the cone. The "fix" for this problem requires weighting the observations that contain more information more heavily. We'll discuss weighted least squares in the optional section 13.4. The problem caused by the lack of constancy in the variability in Y for all values of x is most serious when you have a small sample. If you have thousands or even hundreds of observations, your estimated slope will be close to the true regression line even if the assumption of equal standard deviations is violated. If you have a few dozen observations, it is very likely that a few high observations drawn from the wide part of the cone will not be balanced by low observations from the wide part of the cone or vice versa. Again, when you are working with sample data you won't know if the population has the same or different standard deviations for the distributions of Y given x. You will have to use a scatter diagram of the sample data to decide if the second assumption is violated.

A third assumption about the population is needed to make confidence intervals or to do hypothesis tests. (We'll do confidence intervals and hypothesis tests for the population slope coefficient in section 13.2.) The third assumption is that the distribution of the y values around their respective means is normal. Since we already have two dimensions in our figures for the x and y axes, the probability density function of Y will have to be plotted in a third dimension. Figure 13.13 illustrates the assumption of normality in the distribution of Y given x. The shaded area is the x–y plane. The true or population regression line is the dashed line labeled "$\hat{y}_i = \alpha + \beta x_i$." The cross section in the plane y–f(y) is a normal curve. Note that the cross section visible on the right side of Figure 13.13 is the same as the cross section visible on the left side. The reason that these normal curves have the same shape is the assumption that all of the distributions of Y given x have the same standard deviation. If a slice parallel to the y–f(y) plane were cut through Figure 13.13 at any x value, the cross section of a normal curve would appear and it would have the same standard deviation as the curves on the left and right sides. If you are working with a small sample and the assumption of normality is violated, the confidence intervals or hypothesis tests to be discussed in section 13.2 will not be reliable. If your sample

FIGURE 13.13
**Three-dimensional figure
showing Y as normally
distributed around the mean
of Y**

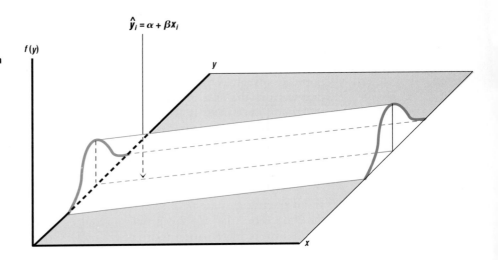

$\hat{y}_i = \alpha + \beta x_i$

$f(y)$

is large enough, the sampling distribution of the sample slope will be approximately normal even if the distribution of the Y values is not normal. There are some possible "fixes" for the problem of nonnormality in the distribution of the Y values, but they are beyond the scope of this book.

A fourth assumption about the population, called the absence of serial correlation, is usually violated only in time-series data, so we needn't discuss it just yet.

Figures 13.10 through 13.13, illustrating the violations of the assumptions needed for statistical inference, all show the population distributions of Y given x. In practice you won't know the population distributions and you will have to work with the sample data. One simple technique to check visually for violations of these assumptions is to plot the residuals of the regression against the x values. The residuals are the differences between the actual and the predicted values of Y, or $y_i - \hat{y}_i$. Plotting the residuals can highlight departures from the assumption of linearity or the assumption of homoscedasticity better than a scatter diagram of the x and Y values. Figure 13.14 shows four graphs. The first graph is a scatter diagram of the x and Y values in which the assumption of linearity seems to hold. The second graph is a plot of the residuals against the x values for the same data. The pronounced curve in this plot shows that the relationship between x and Y is not linear. The third graph is a scatter diagram of a second data set in which the assumption of homoscedasticity seems to hold. The fourth graph is a plot of the residuals against the x values of this second data set. The fourth graph of the residuals shows the heteroscedasticity in the data much more clearly than the scatter diagram.

Now that we have the assumptions required for statistical inference, we can proceed with confidence intervals and hypothesis tests. One term that is needed for most of these tests is an estimate of the standard deviation of the

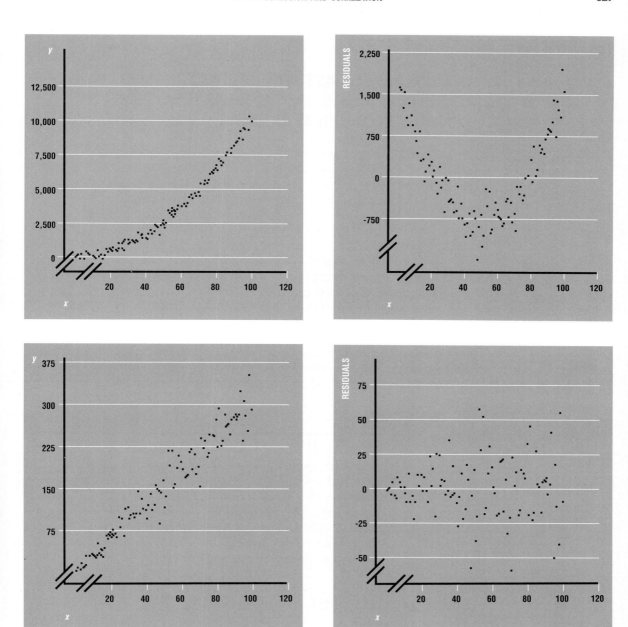

FIGURE 13.14
Scatter diagrams of two data sets that violate the assumptions of
linearity and homoscedasticity and plots of the residuals from the
estimated regression line

DESCRIBING THE ASSUMPTIONS NEEDED FOR STATISTICAL INFERENCE IN TERMS OF A NOISE-GENERATING RANDOM VARIABLE

A more compact way to describe the assumptions needed for statistical inference in regression analysis is to consider the observed values of Y to be made up of two components: (1) a nonrandom component that depends on the relationship between the mean value of Y and x and (2) a random component that depends on a random variable, ε (read "epsilon"), which generates some "noise" around the mean of Y given x. The observed values of Y equal the mean of Y given x plus the value of the random ε. In notation, $y_i = \mu_{Y|x} + \varepsilon_i$, where $\mu_{Y|x}$ is the mean of Y given x. Except for the assumption of linearity, the assumptions required for statistical inference can be stated in terms of the random variable ε. These assumptions are that (1) ε is normally distributed with a mean of 0 and a standard deviation of σ_ε; (2) ε is independent of the value of x; and (3) each ε_i is independent of the values of the other ε_i's. The third assumption will be discussed in detail in Chapter 15. Mathematically, the two ways of stating the assumptions required for statistical inference are equivalent. The advantage of describing the assumptions in terms of a noise-generating function is that the description is more compact. Since the concept of a noise-generating function is a little harder to comprehend, the first description in the text talked only about the distribution of the y values around the mean of Y given x.

distribution of the y values around the mean of Y given x (or, equivalently, an estimate of the standard deviation of ε, σ_ε, described in the accompanying box). You'll find that there is a close correspondence between the methods and concepts of statistical inference in regression analysis and the methods discussed in Chapters 8, 9, and 10. Again, every confidence interval or hypothesis test will have three parts: a center, a standardized score, and a measure of the standard deviation of the variable being tested or estimated.

EXERCISES FOR SECTION 13.1

13.1 For the data set $x = \{1, 3, 6, 9, 10\}$ and $Y = \{30, 45, 47, 51, 55\}$, find the regression slope, intercept, coefficient of determination, and coefficient of correlation.

13.2 For the data set $x = \{15, 14, 10, 8, 5\}$ and $Y = \{14, 35, 60, 71, 85\}$, find the regression slope, intercept, coefficient of determination, and coefficient of correlation.

13.3 Decide whether correlation or regression analysis is more appropriate for describing each of the following relationships:

(a) The relationship between an adult's age and the time it takes her to run a mile.

(b) The relationship between the rate of growth in gross national product and the rate of growth in a company's sales.

(c) The relationship between the number of rabbits and the number of deer in a wildlife preserve.

(d) The relationship between the number of deer hunters in a forest and the number of deer.

13.4 Decide whether correlation or regression analysis is more appropriate for describing each of the following relationships:

(a) The relationship between a person's height and weight.

(b) The relationship between a country's annual imports and exports in goods and services.

(c) The relationship between a country's population and its gross national product.

(d) The relationship between a mother's height and a baby's height at birth.

13.5 The Happy Time Resort Company owns a chain of resort hotels. The management of the company is concerned that its earnings are sensitive to changes in the economy and they are considering the purchase of another company to reduce the variability of their earnings. Three companies have been proposed: Acme Bread Bakers, Grim Auction and Salvage, and Six Banners Theme Park Corporation. The table below lists each company's annual earnings in millions of dollars over the years 1977 to 1987. Use the coefficient of correlation between Happy Time's earnings and each of the other companies' earnings to decide which one Happy Time should purchase.

Year	Happy Time	Acme Bread Bakers	Grim Auction and Salvage	Six Banners Theme Park
1977	50	10	15	20
1978	55	10	14	22
1979	40	11	22	12
1980	38	10	29	9
1981	48	9	20	15
1982	58	10	13	24
1983	60	10	10	26
1984	40	9	18	15
1985	56	11	13	17
1986	65	10	9	28
1987	75	10	8	30

13.2 CONFIDENCE INTERVALS AND HYPOTHESIS TESTS ON THE POPULATION SLOPE AND INTERCEPT

▇ 13.2A THE STANDARD ERROR OF THE ESTIMATE

The population standard deviation of the distribution of Y for a given x has the symbol σ_e. Recall that this standard deviation is assumed to be the same for every value of x. The symbol s_e stands for *the estimated standard deviation of the distribution of the dependent variable for values of the independent variable.* This term is usually called the **standard error of the estimate.** The term is used in the formulas for a variety of confidence intervals and hypothesis tests in regression analysis. Since all of the standard deviations of Y given x are supposed to be the same for the possible values of x, you could take one value of x, find the corresponding point on the estimated regression line, and calculate a standard deviation around that point using whatever Y values there were for the given x value. For example, if you had some sample data on hours of study and GPA, you could pick some x such as 15 hours a week, find the $\hat{y}$ given that x is set at 15 hours, and then use all of the observations of GPAs for students who studied 15 hours a week to calculate s_e. The problem with this procedure is that you would be ignoring the rest of the data for students who have studied more or less than 15 hours a week. One of the cardinal rules of statistics is to use all of the available data. To use all of the data, you must, in effect, collapse all of the normal curves for the various values of x into one curve. Since these normal curves have the same standard deviations, you can collapse them into one curve. Figure 13.15 helps to explain what is meant by collapsing all of the normal curves into one curve. All of the normal curves are squeezed together, metaphorically speaking, along the axis of the estimated regression line.

FIGURE 13.15

Normal curves representing the distribution of Y given x collapsed into one curve

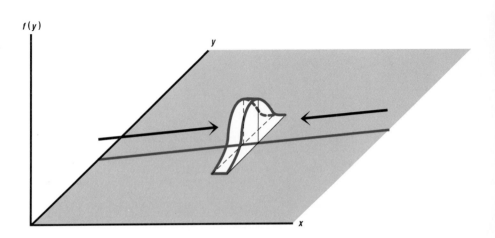

TABLE 13.4

Sales in millions of dollars and population in millions of people in eight cities with offices selling steam cleaning equipment and a worksheet for calculating the regression coefficients

Population	Sales	$x_i y_i$	$(x_i)^2$
2.0	$ 3.0	6.0	4.00
1.0	1.0	1.0	1.00
1.5	2.0	3.0	2.25
0.5	0.8	0.4	0.25
1.5	3.0	4.5	2.25
1.0	1.5	1.5	1.00
3.0	5.0	15.0	9.00
2.0	4.0	8.0	4.00
12.5	$20.3	39.4	23.75

The formula for the standard error of the estimate is

$$s_e = \sqrt{\frac{\sum_{i=1}^{n}(y_i - \hat{y}_i)^2}{n-2}}$$ (13.13)

The reason for the $n-2$ in the denominator is that one degree of freedom is lost when the sample mean of x is used in place of the population mean, μ_x, and another degree of freedom is lost when the sample mean of Y is used in place of the population mean, μ_Y. Recall that the formulas for the estimated regression line used the terms $\bar{x}$ and $\bar{y}$.

To illustrate the formula for the standard error of the estimate, let's use a new example. Your company sells machines that use steam to clean industrial equipment. You have sales offices in eight cities and you would like to estimate the potential sales of an office in another city. With population as the explanatory variable, you could run a regression on the sales per city which could then be used to predict the sales in a new city. If the assumptions behind regression analysis hold and the population of the new city is within or at least close to the range of populations of the cities in the data set, the estimated sales for the new city should be accurate. Table 13.4 provides the data on sales and population for the eight cities and a worksheet for calculating the regression coefficients. The x variable is population and the Y variable is sales.

Recall that $\bar{x}$ equals the summation of x_i divided by n, or 12.5 divided by 8. The term $\bar{y}$ equals 20.3 divided by 8. Substituting these results into formulas 13.4 and 13.5 for the sample slope and intercept yield

$$b = \frac{\sum_{i=1}^{n} x_i y_i - n\bar{x}\bar{y}}{\sum_{i=1}^{n} x_i^2 - n(\bar{x})^2}$$ (13.4)

$$b = \frac{39.4 - 8(1.563)(2.538)}{23.75 - 8(1.563)^2}$$

$$b = 1.82$$

$$a = \bar{y} - b\bar{x} \qquad\qquad (13.5)$$

$$a = 2.538 - 1.82(1.563)$$

$$a = -0.31$$

The estimated regression line is $\hat{y} = -0.31 + 1.82x$. Suppose the new city that you were considering had a population of 2.25 million people. Substituting into the above estimated regression equation yields an estimated sales of $\hat{y} = -0.31 + 1.82(2.25)$, or \$3.78 million. This number is a point estimate. Confidence intervals on the predicted value of y given x, to be discussed in section 13.3, can give you an idea of the accuracy of the point estimate for a single city. Note that the negative intercept term in the estimated regression line does not mean that in a city with a population of 0 the sales would be negative \$0.31 million. Negative sales are impossible. Since there are no observations of cities near the value of 0 population, the intercept term should be thought of as the point where the regression line cuts the y axis rather than a prediction of the sales at 0 population. Figure 13.16 shows the scatter diagram for the population and sales data and the estimated regression line.

There appears to be a close fit between the points on the scatter diagram and

FIGURE 13.16
Scatter diagram and estimated regression line of sales of steam cleaning equipment and populations of cities

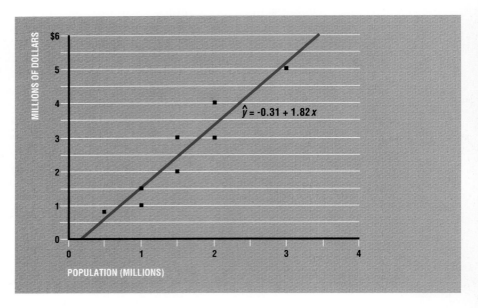

TABLE 13.5
Worksheet for calculating the coefficient of determination

x_i	$\hat{y}_i = a + bx_i$	$\bar{y}$	$(\hat{y}_i - \bar{y})$	$(\hat{y}_i - \bar{y})^2$	y_i	$\bar{y}$	$(y_i - \bar{y})$	$(y_i - \bar{y})^2$
2.0	3.33	2.54	0.79	0.624	3.0	2.54	0.46	0.212
1.0	1.51	2.54	-1.03	1.061	1.0	2.54	-1.54	2.372
1.5	2.42	2.54	-0.12	0.014	2.0	2.54	-0.54	0.292
0.5	0.60	2.54	-1.94	3.764	0.8	2.54	-1.74	3.028
1.5	2.42	2.54	-0.12	0.014	3.0	2.54	0.46	0.212
1.0	1.51	2.54	-1.03	1.061	1.5	2.54	-1.04	1.082
3.0	5.15	2.54	2.61	6.812	5.0	2.54	2.46	6.052
2.0	3.33	2.54	0.79	0.624	4.0	2.54	1.46	2.132
				13.974				15.382

the estimated regression line. The coefficient of determination for the regression line in Figure 13.16 is 0.91, so by that measure, the regression line does fit the scatter diagram well. Population accounts for 91% of the variation in sales across cities. Table 13.5 shows the calculations for the coefficient of determination. The coefficient of determination equals the explained variation, 13.974, divided by the total variation, 15.382.

$$r^2 = \frac{\sum_{i=1}^{n} (\hat{y}_i - \bar{y})^2}{\sum_{i=1}^{n} (y_i - \bar{y})^2} \qquad (13.10)$$

$$r^2 = 0.91$$

To determine the standard error of the estimate, you have to calculate $\hat{y}_i$ for every city in the sample and then substitute into formula 13.13. Table 13.6 provides the necessary worksheet.

TABLE 13.6
Predicted values of sales in each city and a worksheet for the standard error of the estimate

x_i	$\hat{y}_i = a + bx_i$	y_i	$(y_i - \hat{y}_i)$	$(y_i - \hat{y}_i)^2$
2.0	3.33	3.0	-0.33	0.1089
1.0	1.51	1.0	-0.51	0.2601
1.5	2.42	2.0	-0.42	0.1764
0.5	0.60	0.8	0.20	0.0400
1.5	2.42	3.0	0.58	0.3364
1.0	1.51	1.5	-0.01	0.0001
3.0	5.15	5.0	-0.15	0.0225
2.0	3.33	4.0	0.67	0.4489
				1.3933

Substituting into formula 13.13, we get

$$s_e = \sqrt{\frac{\sum\limits_{i=1}^{n}(y_i - \hat{y}_i)^2}{n-2}} \qquad (13.13)$$

$$s_e = \sqrt{\frac{1.393}{6}}$$

$$s_e = 0.48$$

The standard error of the estimate is 0.48. This term is used in all of the formulas for confidence intervals and hypothesis tests in regression analysis. In section 13.2B we'll see how it is used for confidence intervals and hypothesis tests on the population slope.

13.2B CONFIDENCE INTERVALS AND HYPOTHESIS TESTS ON THE POPULATION SLOPE

In section 13.2A the estimated change in sales for a unit increase in city population was 1.82. This value is a point estimate and, provided that the assumptions of linearity and homoscedasticity are satisfied, it is the best possible estimate of β, the population slope. However, a point estimate provides no sense of how far it may be from the population parameter. To get a feel for the accuracy of your estimate, you have to construct a confidence interval. The word "population" has been used in two different senses in this paragraph. The population of a city—that is, the number of people—constitutes the values of the explanatory variable. The statistical population is the set of all cities that could be locations of the steam cleaning company's sales offices.

The formula for a confidence interval on the population slope is

$$L, U = b \pm t_{\alpha/2}s_b \qquad (13.14)$$

where $df = n - 2$

The format of formula 13.14 should be familiar. The confidence interval is centered on the value of the estimator b, the standardized score or reliability factor is $t_{\alpha/2}$, and the standard deviation of the distribution of the estimator b is s_b. The formula for s_b is

$$s_b = \frac{s_e}{\sqrt{\sum\limits_{i=1}^{n}(x_i)^2 - n(\bar{x})^2}} \qquad (13.15)$$

Formula 13.14 depends on the assumption of normality in the distribution of Y given x. Note that the degrees of freedom for the t score used in formula 13.14 is $n - 2$. We have all of the information needed to construct a confidence interval, so the next step is to pick a confidence level and substitute into formulas 13.15 and 13.14. Let's use a confidence level of 0.95.

$$s_b = \frac{s_e}{\sqrt{\sum\limits_{i=1}^{n} (x_i)^2 - n(\bar{x})^2}} \tag{13.15}$$

$$s_b = \frac{0.48}{\sqrt{23.75 - 8(1.563)^2}}$$

$$s_b = 0.23$$

$$L, U = b \pm t_{\alpha/2}s_b \tag{13.14}$$

$$L, U = 1.82 \pm 2.447(0.23)$$

$$L = 1.26 \qquad U = 2.38$$

At the 95% level of confidence, the population slope is estimated to be between 1.26 and 2.38. In terms of this example, choosing a city with a million more people than some other city would, at the 95% level of confidence, add somewhere between \$1.26 million and \$2.38 million to the local office's average annual sales. If the costs of opening a new sales office were the same for all cities and population were the only important determinant of sales, the company would obviously want to start new offices in the largest available cities. Knowing the slope of the regression line would be useful in helping the company plan the amount of steam cleaning equipment and the employees needed to support the new sales offices.

Now let's use the same data for a hypothesis test on the population slope. The most common null hypothesis is that the population slope equals 0—in other words, that the explanatory variable has no effect on the dependent variable. In this example the null hypothesis is that population has no effect on sales. If for some reason a particular value of the population slope were important, you could test against the specified value. For example, a franchise contract could specify that sales would be expected to go up by \$2 million a year for every increase of a million persons in the city's population. Formula 13.16 gives the critical values for a two-tailed hypothesis test.

$$C.V._L, C.V._U = b_0 \pm t_{\alpha/2}s_b \tag{13.16}$$

where $df = n - 2$.

To do a one-tailed test, use $+$ or $-$ in place of $\pm$ and α in place of $\alpha/2$.

FIGURE 13.17
**Acceptance and rejection
regions for the test on the
null hypothesis that the
population slope equals 0**

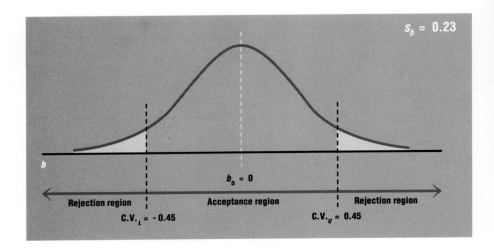

$s_b = 0.23$

b

$b_0 = 0$

Rejection region

Acceptance region

Rejection region

$C.V._L = -0.45$

$C.V._U = 0.45$

The term b_0 stands for the null hypothesis value of the population slope. Substituting into the above formula and using a significance level of 10% yield

$$C.V._L, C.V._U = b_0 \pm t_{\alpha/2}s_b \tag{13.16}$$

$$C.V._L, C.V._U = 0 \pm 1.943(0.23)$$

$$C.V._L = -0.45 \qquad C.V._U = 0.45$$

The critical values for the test are -0.45 and 0.45 and the sample slope is 1.82, so the null hypothesis of a population slope of 0 is rejected at the 10% significance level. Figure 13.17 shows the template for the hypothesis test based on the distribution of b, the sample slope.

In some problems the value of the population slope is an important issue in and of itself and in other problems the estimated slope is used only to estimate the likely values of Y for a given x. For example, if the populations of all the cities in which your company had sales offices were growing, you might use the slope coefficient to estimate future sales on the basis of projected population growth. The key issue would be how much sales grew as population increased. In this case the slope coefficient is the important term. If you were planning to open a new sales office in a city with a population of 2.25 million, you would be interested in estimating sales in a city of that size. Here the slope coefficient is being used to estimate the value of Y given a particular x. We'll discuss this problem in section 13.3.

EXERCISES FOR SECTION 13.2

13.6 Calculate the standard error of the estimate for the data in Table 13.2. Treat the data as a sample.

13.7 Calculate the standard error of the estimate for the following sample data: $x_i = \{3, 7, 2, 9, 8, 4\}$, $y_i = \{0.4, 0.9, 0.5, 1.1, 1.0, 0.0\}$.

13.8 Continue to treat the data in Table 13.2 as a sample and test the hypothesis that the population slope equals 0 at the 5% significance level.

13.9 Test the hypothesis that the population slope for the data from which the x and y values were drawn for exercise 13.7 exceeds 0 at the 1% significance level.

13.10 A large city holds a monthly auction of all abandoned cars that have been towed by the police to a city-owned lot. All of the cars available go to the person who offers the highest bid for the entire set. The following table shows the number of cars offered per auction and the price per car.

Number of cars	Price per car
100	$125
125	118
150	80
80	140
130	100
100	110

(a) In economic terms, what would the estimated regression of the price per car on the quantity purchased be called?

(b) Test the hypothesis that the population slope coefficient equals 0 at the 5% significance level.

13.11 A homeowner whose property fronts Lake Michigan is worried that the lake will flood over his retaining wall. The lake has been rising steadily for the last eight years. The homeowner has measured the distance between the level of the lake and the top of his retaining wall each July.

(a) Use the following data to estimate the year that the property will be flooded if a higher wall is not built.

(b) Construct a confidence interval at the 95% confidence level for the slope of the regression line.

Inches between lake and top of wall	Year
22	1980
21	1981
20	1982
18	1983
17	1984
14	1985
12	1986
9	1987

13.12 Mary Avon Amway is a manufacturer of women's cosmetics. The company markets its products door to door, primarily through part-time representatives. The company has experimented with its commission structure to see how changes in the commission affect the level of sales per representative.

(a) Plot the data in the table below.

(b) Use the data to estimate the effect of a 1-percentage-point change in the commission rate on the average level of sales per representative.

(c) Construct a confidence interval for the population slope at the 99% level.

(d) Explain why the estimated regression line might not provide a reliable estimate of the average sales per representative if the commission were set at 20%.

Annual sales per representative	Commission rate
$3,000	10%
3,400	10
3,200	10
3,100	10
2,900	10
3,600	12
3,300	12
3,500	12
3,600	12
3,700	12
3,800	15
3,900	15
3,900	15
4,000	15
4,200	17
4,100	17

13.13 A manufacturer of brown paper bags used by supermarkets wants to estimate the effect of changes in the number of employees working during a shift on the average output per shift. The employees set the machinery before each run, feed rolls of paper into the machinery, load and stack the finished bags, and clear any jams in the machinery.

(a) Use the following data to estimate the effect of an increase in one employee per shift on the average output per shift.

(b) Construct a confidence interval for the population slope at the 95% level.

(c) What is the name of the economic law that tells you that the additional output per added employee must decline if enough employees are added to a factory of a given size.

Output of paper bags (reams)	Number of employees per shift	Output of paper bags (reams)	Number of employees per shift
5,435	5	7,490	8
6,227	5	6,902	8
5,548	5	7,839	8
4,986	5	7,947	8
6,310	6	8,611	9
6,497	6	8,205	9
5,832	7	8,487	9
5,571	7	8,878	9
7,555	7	8,971	9
6,836	7		

13.3 CONFIDENCE INTERVALS AND HYPOTHESIS TESTS ON THE MEAN OF Y, THE PREDICTED VALUES OF Y, AND THE COEFFICIENT OF CORRELATION

13.3A CONFIDENCE INTERVALS AND HYPOTHESIS TESTS ON THE PREDICTED VALUE OF Y GIVEN x

The problem of estimating the annual sales for an office of the steam cleaning equipment company in a new city was described at the end of section 13.2. Recall that the new city has a population of 2.25 million. Applying this population to the estimated regression equation of $\hat{y}_i = -0.31 + 1.82x_i$ yields a point estimate of $3.79 million. Note that it is important to be consistent about the units. Since the original data used to estimate the regression line (Table 13.4) are given with the decimal point to the right of the millions column for the population variable, the number to be substituted into the estimated regression equation is 2.25, not 2,250,000. If you use 2,250,000 in the above equation, you will get the wrong answer.

A point estimate provides no sense of how far off it may be from the population parameter. The formula for a confidence interval for the predicted value of y given x is

$$L, U = \hat{y}_i \pm t_{\alpha/2}s_e \sqrt{1 + \frac{1}{n} + \frac{(x^* - \bar{x})^2}{\sum\limits_{i=1}^{n}(x_i)^2 - n\bar{x}^2}} \qquad \text{(13.17)}$$

where $df = n - 2$

The only symbol that is new in formula 13.17 is x^*. It refers to the given value of x. Let's ignore the expression under the radical for a moment. Formula 13.17 is complicated and it will be easier to explain a part at a time. The confidence interval is centered on $\hat{y}_i$, the point on the regression line that corresponds to the given x (which is a population of 2.25 million in this example). As you would expect, the standardized score is the t score with $n - 2$ degrees of freedom. The s_e term is not divided by the familiar square root of n. The reason is that we are not constructing a confidence interval for $\mu_{Y|x}$, that is, we are not constructing a confidence interval for the mean sales of all cities with a population of 2.25 million. The steam cleaning equipment company is planning to open only one office in one city of this size. The company would like to have a 95% confidence interval on the sales of this one office, not on the average sales of all cities of this size. The confidence interval on the *predicted* value of Y is an interval that predicts the range of likely values for *one observation of* Y given the value of x. When you construct a confidence interval for $\mu_{Y|x}$, you have to work with the sampling distribution of the mean of Y. When you construct a confidence interval for the *predicted value of* Y *given* x, you have to work with the sampling distribution of Y given x for samples of size 1. Of course the distribution of all possible samples of size 1 is the same as the population distribution. This distinction is illustrated in Figure 13.18. The

FIGURE 13.18
Sampling distribution of *Y* given *x* and of *ŷ* given *x*

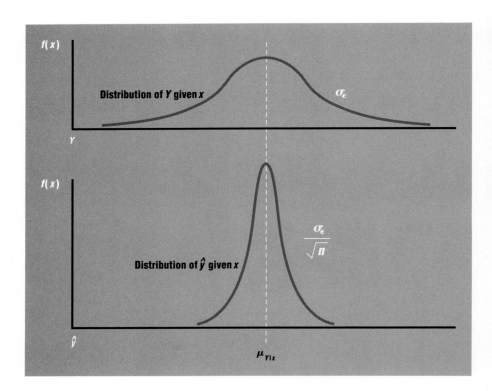

distinction between the two distributions is the same as the distinction between the population distribution of X and the sampling distribution of $\bar{x}$ found in Chapter 7. To construct a confidence interval on the predicted value of Y given x, you have to work with the distribution of Y given x, the wider of the two distributions in Figure 13.18.

Now we can come back to the expression under the radical that you were asked to ignore temporarily:

$$\sqrt{1 + \frac{1}{n} + \frac{(x^* - \bar{x})^2}{\displaystyle\sum_{i=1}^{n}(x_i)^2 - n\bar{x}^2}}$$

This term is an adjustment factor for values of x^* that are far away from $\bar{x}$. If x^* equaled $\bar{x}$ and n was large, the adjustment factor would be approximately 1. This is because the third term would equal zero and would drop out of the sum while the second term $(1/n)$ would approach zero as n got larger and it would also drop out. As x^* gets farther from $\bar{x}$, the adjustment factor gets larger. The reason is that as the distance between x^* and $\bar{x}$ increases, small errors in estimating the population slope can be magnified when the incorrect slope is used to estimate the point on the population regression line.

Figure 13.19 illustrates how the spread between the population regression line and an estimated regression line increases as the distance between x^* and $\bar{x}$ increases. The spread between $\mu_{Y|x}$ and $\hat{y}$ given x will increase as the distance between x^* and $\bar{x}$ increases whenever the estimated slope and the estimated intercept do not equal the population slope and intercept. If the estimated slope

FIGURE 13.19

A population and a sample regression line showing that the spread between $\mu_{Y|x}$ and $\hat{y}$ given x increases as the distance between x^* and $\bar{x}$ increases

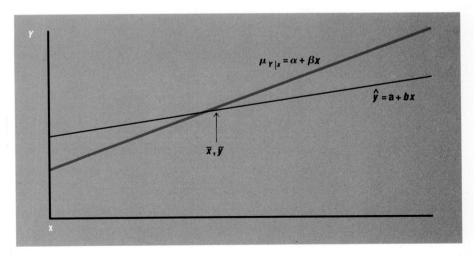

FIGURE 13.20
Upper and lower bounds of
confidence intervals at the
95% level of confidence for
the predicted value of Y

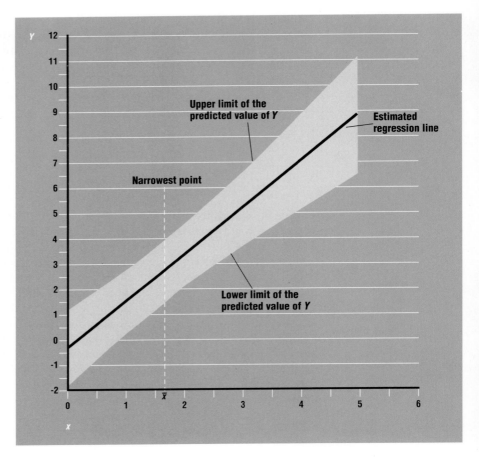

and intercept equaled the population slope and intercept, the two lines would
be superimposed. If the estimated slope equaled the population slope but the
estimated intercept did not equal the population intercept, the two lines would
be parallel. Since the estimated slopes and intercepts tend to deviate at least
slightly from the population values for almost all samples, you can generally
expect that the estimated regression line will be closer to the population line as
you approach the center of the sample data, $\bar{x}$, $\bar{y}$.

To adjust for this greater uncertainty about the position of the population
regression line as the distance between x^* and $\bar{x}$ increases, the last part of the
formula widens the confidence interval for the predicted value of Y given x as
the distance between x^* and $\bar{x}$ increases. In Figure 13.20 an estimated regression line is surrounded by two curves that represent the upper and lower
bounds of a 95% confidence interval for the predicted value of Y. Note that the
vertical distance between the curved lines is minimized when x^* equals $\bar{x}$.

Now that you have an idea about the reasoning behind the formula for the confidence interval for the predicted value of Y given x, let's go ahead with the problem of constructing such an interval at the 95% level of confidence for the sales of steam cleaning equipment for a city with 2.25 million people. Recall that the point estimate was \$3.79 million. Substituting into formula 13.17 yields

$$L, \ U = \hat{y}_i \pm t_{\alpha/2} s_e \sqrt{1 + \frac{1}{n} + \frac{(x^* - \overline{x})^2}{\sum\limits_{i=1}^{n} (x_i)^2 - n\overline{x}^2}} \tag{13.17}$$

$$L, \ U = 3.79 \pm 2.447(0.482)\sqrt{1 + \frac{1}{8} + \frac{(2.25 - 1.563)^2}{23.75 - 8(1.563)^2}}$$

$$L = 2.54 \qquad U = 5.04$$

The limits of the confidence interval are \$2.54 million and \$5.04 million. The company could use this information in several ways. If some minimum amount of sales was necessary to justify the new office, say \$2 million a year, the above confidence interval would assure the company that this minimum was likely to be obtained. Another use of the confidence interval is for planning staff or production; the company can use the range of likely sales levels for the new office in hiring decisions and when ordering equipment.

In section 13.3B we'll discuss the formula for the confidence interval on the mean of Y given x. The formulas for confidence intervals on the mean of Y given x and the predicted values of Y given x are almost the same; the key issue is knowing which formula is appropriate for a particular problem.

■ 13.3B CONFIDENCE INTERVALS AND HYPOTHESIS TESTS ON THE MEAN VALUE OF Y GIVEN x

The formula for a confidence interval on the mean value of Y given x, $\mu_{Y|x}$, is

$$L, \ U = \hat{y}_i \pm t_{\alpha/2} s_e \sqrt{\frac{1}{n} + \frac{(x^* - \overline{x})^2}{\sum\limits_{i=1}^{n} (x_i)^2 - n\overline{x}^2}} \tag{13.18}$$

where $df = n - 2$

If you compare formula 13.18, for confidence intervals on the mean of Y given x, with formula 13.17, for confidence intervals on the predicted value of Y given x, you will see that they are almost the same. The only difference is that the first part of the expression under the square root symbol is $1/n$ instead of

$(1 + 1/n)$. This change makes the confidence interval for the mean of Y given x narrower than the confidence interval for the predicted value of Y given x, since $1/n$ will always be smaller than $(1 + 1/n)$. Intuitively, the reason for the narrower confidence interval on the mean of Y given x is that you are more certain about the possible locations of the mean of Y given x than you are about a single observation of Y. Other than being narrower (how much narrower depends on the n), the confidence intervals on the mean of Y given x have the same features as the confidence intervals on the predicted value of Y given x. If Figure 13.20 were redrawn for confidence intervals on the mean of Y given x, the narrowest point of the confidence intervals would still be at x^* equal to $\bar{x}$. The curves of the upper and lower bounds of the confidence intervals for the mean of Y would be similar in shape to the curves for the confidence intervals on the predicted value of Y, but those for the mean would be closer to the estimated regression line. Figure 13.21 shows these changes.

FIGURE 13.21

Upper and lower confidence intervals of the mean of *Y* given *x* and of the predicted value of *Y* given *x*

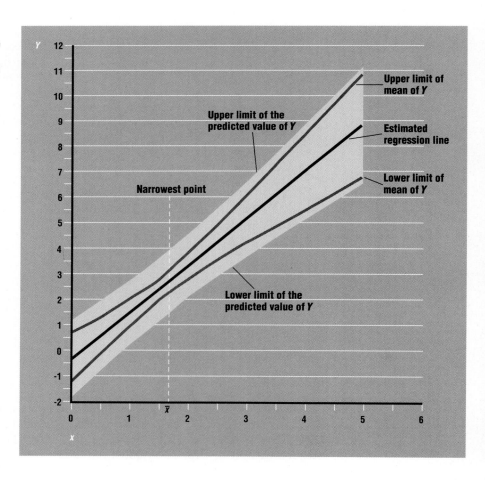

The key issue with these two formulas is understanding when to use the one for the confidence interval on the mean of Y given x and when to use the one for the confidence interval on the predicted value of Y given x. In the last example the formula for the predicted value of Y given x was used because the steam cleaning equipment company wanted to open one sales office in a city of a particular size, so it was interested in the likely value of one observation drawn from the population of Y given x. If a company planned to repeat a process and it needed to know the average result for many repetitions of the process, the confidence interval on the mean of Y given x would be appropriate. Here's an example that illustrates this second objective.

Airline pilots' salaries vary with the type of plane that they fly. The larger planes are more complicated and require more training and experience. An airline plans to purchase a new type of plane that carries 100 passengers and it wants to hire 50 pilots with the proper training and experience. The company needs to set a salary for these pilots near the average of pilots' salaries for planes of this size. If the salary is set too low, qualified pilots will not apply, and if the salary is set too high, the company will lose profits. If there were other planes of the same size, the company could simply find the average salary of pilots who fly 100-seat planes. As there are no 100-seat planes currently in service, a regression of pilots' salaries on the number of seats will estimate the relationship between the size of plane and the mean salary of pilots. The x variable would be the number of seats and the y variable the salary. This estimated regression can be used to make a point estimate of the average salary of pilots of 100-seat planes. Since there are no such planes in service, this point estimate is an interpolated value based on the salary of pilots of smaller and larger planes. Finally, a confidence interval on the mean value of Y given x can be constructed to get a sense of how close the point estimate of average salaries might be to the population mean of Y given x. Suppose the airline collected data from a random sample of 1,000 pilots (the raw data are not shown here) and calculated the following terms:

$a = 11,733.218$ $\qquad\qquad$ $b = 277.126$

$\bar{x} = 237$ $\qquad\qquad$ $\bar{y} = 77,412$

$$\sum_{i=1}^{n} (x_i)^2 = 78,236,000 \qquad \sum_{i=1}^{n} x_i y_i = 2.4461976(10^{10})$$

$$\sum_{i=1}^{n} (y_i)^2 = 7.7296907(10^{12}) \qquad s_e = 7024.24$$

$r^2 = 0.972$

The sum of the squared x_i values and the sum of the products of every pair of $x_i y_i$ values are big numbers—too big to fit into most hand-held calculators unless they accept scientific notation. Many calculators have a button labeled "EE." These calculators let you enter a number such as $7.7296907(10^{12})$. This number is 7.7296907 times 10^{12}, or 7.7296907 times $1,000,000,000,000$.

The point estimate of the mean salary of pilots of 100-seat airplanes is

$$\hat{y} = 11{,}733.218 + 277.126(100)$$

$$\hat{y} = 39{,}445.784$$

To construct a confidence interval on the mean of Y given x at the 95% level of confidence, you have to substitute into formula 13.18:

$$L, \ U = \hat{y}_i \pm t_{\alpha/2} s_e \sqrt{\frac{1}{n} + \frac{(x^* - \bar{x})^2}{\displaystyle\sum_{i=1}^{n} (x_i)^2 - n\bar{x}^2}} \tag{13.18}$$

$$L, \ U = 39{,}445.784 \pm 1.96(7024.24) \sqrt{\frac{1}{1{,}000} + \frac{(100 - 237)^2}{78{,}236{,}000 - 1{,}000(237)^2}}$$

$$L = 38{,}853.53 \qquad U = 40{,}038.03$$

Note that with 998 degrees of freedom, the Student distribution is about the same as the normal distribution. This is the reason for using the t score of 1.96. The lower limit for a 95% confidence interval on the mean of Y given x is \$38,853.53, and the upper limit is \$40,038.03. This spread of roughly \$1,200 gives the airline a good idea of what the average industry salary will be for its 100-seat planes once this type of plane is in wider service. Knowing the average salary does not solve the airline's problem about deciding what salary to set. Some companies knowingly choose a high-salary profile in the hope of holding down turnover. Lower turnover saves money on training and recruiting. Other companies choose a low-salary profile because their managements think that they can save more on the low salaries than they will lose on training and recruiting costs. A salary near \$40,000 will probably attract enough qualified pilots, and this figure can be modified up or down if the airline thinks that a high- or low-salary profile will be advantageous.

To clarify the different purposes of confidence intervals on the mean of Y given x versus the predicted value of Y given x, consider the following variation on the above problem. Suppose you were an economist and you had to estimate the lost earnings of a U.S. Air Force pilot who had died in an automobile accident. The estimate was needed by the pilot's estate for an insurance settlement. The pilot was qualified to fly jets that could carry 350 passengers and he was about to retire from the Air Force and take a job with a commercial airline. The above regression could be used to make a point estimate of the mean salary of pilots qualified to fly 350-seat planes. Again, a point estimate provides no sense of how far off it may be from the actual value. To construct a confidence interval on the dead pilot's lost earnings in commercial aviation, should you use the formula for the confidence interval on the mean of Y given x or the predicted value of Y given x? The answer is to use the formula for the predicted value of Y given x, because you are dealing with one observation drawn from the distribution of pilots' salaries, holding the size of the plane constant.

So far you've done only confidence intervals for the mean and predicted values of Y. A hypothesis test for the mean of Y given x requires two changes in formula 13.18. The hypothesized value of the population mean replaces the point estimate of the mean of Y given x, and the symbols L and U, for the lower and upper limits of the confidence interval, are replaced by the symbol C.V., for the critical value or values of the hypothesis test. Let μ_0 be the hypothesized value of the population mean of Y given x. Then the formula for a two-tailed hypothesis test is

$$\text{C.V.}_L, \text{C.V.}_U = \mu_0 \pm t_{\alpha/2} s_e \sqrt{\frac{1}{n} + \frac{(x^* - \overline{x})^2}{\sum\limits_{i=1}^{n} (x_i)^2 - n\overline{x}^2}} \qquad (13.19)$$

where $df = n - 2$.

To do a one-tailed test, use $+$ or $-$ in place of $\pm$ and α in place of $\alpha/2$.

To illustrate hypothesis tests on the mean of Y given x, we'll need a new example. A manufacturer of kerosene-burning space heaters has a research team working on the design of a low-pollution space heater. The research team has come up with a design that it thinks will produce much less pollution than existing designs. The team runs a series of tests on this design. Each test consists of keeping the heater going continuously in a closed room and then measuring the amount of air pollutants in the room. After conducting the tests, the research team is told that the company will market the new design if the mean amount of air pollutants accumulated over ten hours of continuous operation is less than half of the government standard of 10 grams. In other words, the design will be marketed if the mean amount is under 5 grams. Can the test data in Table 13.7 be used to reject at the 5% significance level the null hypothesis that the mean amount of pollutants is 5 or more grams after ten hours of operation?

All of the sums needed to calculate the slope, the intercept, the standard error of the estimate, the coefficient of determination, and the critical value for the hypothesis test are displayed in Table 13.8. The numbers are carried to so many places because they were calculated by a computer. It would take too long to check all of the numbers, but you should verify the values on one row to make sure that you understand the formulas.

TABLE 13.7

Hours of operation on high setting in a closed room and grams of air pollutants from an experimental kerosene space heater

Hours of operation	Grams of pollutants	Hours of operation	Grams of pollutants
2	2.7	9	3.7
2	3.1	10	4.0
3	2.4	11	4.4
4	2.5	11	4.3
4	2.8	12	5.1
5	2.8	13	5.0
5	3.1	13	5.2
5	2.9		
5	3.4		
6	3.1		
6	3.3		
7	3.4		
8	3.5		

TABLE 13.8

Sums needed for hypothesis test on grams of pollutants from experimental space heater

x_i	y_i	$(x_i)^2$	$x_i y_i$	$\hat{y}_i$	$(y_i - \hat{y}_i)^2$	$(y_i - \bar{y})^2$	$(\hat{y}_i - \bar{y})^2$
2	2.7	4	5.4	2.3894464	0.09644354	0.697225	1.31229305
2	3.1	4	6.2	2.3894464	0.50488642	0.189225	1.31229305
3	2.4	9	7.2	2.6162886	0.04678076	1.288225	0.84403064
4	2.5	16	10.0	2.8431308	0.11773875	1.071225	0.47868299
4	2.8	16	11.2	2.8431308	0.00186027	0.540225	0.47868299
5	2.8	25	14.0	3.0699730	0.07288542	0.540225	0.21625011
5	3.1	25	15.5	3.0699730	0.00090162	0.189225	0.21625011
5	2.9	25	14.5	3.0699730	0.02889082	0.403225	0.21625011
5	3.4	25	17.0	3.0699730	0.10891782	0.018225	0.21625011
6	3.1	36	18.6	3.2968152	0.03873622	0.189225	0.05673200
6	3.3	36	19.8	3.2968152	1.0143E-05	0.055225	0.05673200
7	3.4	49	23.8	3.5236574	0.01529115	0.018225	0.00012865
8	3.5	64	28.0	3.7504996	0.06275005	0.001225	0.04644008
9	3.7	81	33.3	3.9773418	0.07691847	0.027225	0.19566627
10	4.0	100	40.0	4.2041840	0.04169111	0.216225	0.44780723
11	4.4	121	48.4	4.4310262	0.00096263	0.748225	0.80286295
11	4.3	121	47.3	4.4310262	0.01716787	0.585225	0.80286295
12	5.1	144	61.2	4.6578684	0.19548035	2.449225	1.26083344
13	5.0	169	65.0	4.8847106	0.01329165	2.146225	1.82171870
13	5.2	169	67.6	4.8847106	0.09940741	2.772225	1.82171870
141	**70.7**	**1,239**	**554.0**		**1.54101247**	**14.1455**	**12.6044861**

The slope of the estimated regression line is

$$b = \frac{\sum\limits_{i=1}^{n} x_i y_i - n \bar{x} \bar{y}}{\sum\limits_{i=1}^{n} x_i^2 - n(\bar{x})^2}$$

(13.4)

$$b = \frac{554 - 20(7.05)3.535}{1{,}239 - 20(7.05)^2}$$

$$b = 0.227$$

The intercept of the estimated regression line is

$$a = \bar{y} - b\bar{x}$$

(13.5)

$$a = 3.535 - 0.227(7.05)$$

$$a = 1.94$$

The predicted value of Y given x equal to 10 is

$$\hat{y} = a + bx$$

$$\hat{y} = 1.94 + 0.227(10)$$

$$\hat{y} = 4.21$$

The standard error of the estimate is

$$s_e = \sqrt{\frac{\sum\limits_{i=1}^{n} (y_i - \hat{y}_i)^2}{n - 2}}$$

(13.13)

$$s_e = \sqrt{\frac{1.541}{20 - 2}}$$

$$s_e = 0.293$$

Now you have all of the pieces for the hypothesis test. The hypothesis test should be left-tailed; you want to put the burden of the proof on the hypothesis that the design meets the company's goals. In notation the null and alternative hypotheses are

$$H_0: \mu_{Y|10} \geq 5 \qquad H_1: \mu_{Y|10} < 5$$

Substituting into a left-tailed version of formula 13.19 yields

FIGURE 13.22
Sampling distribution of the
predicted value of Y given x
and the acceptance and
rejection regions for the null
hypothesis that the mean of
Y is greater than or equal to
5 when x is 10

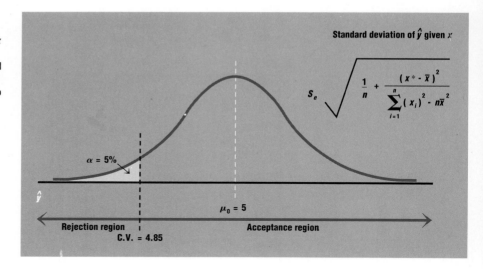

$$\text{C.V.} = \mu_0 - t_\alpha s_e \sqrt{\frac{1}{n} + \frac{(x^* - \bar{x})^2}{\displaystyle\sum_{i=1}^{n}(x_i)^2 - n\bar{x}^2}} \qquad (13.19)$$

$$\text{C.V.} = 5 - 1.734(0.293)\sqrt{\frac{1}{20} + \frac{(10 - 7.05)^2}{1{,}239 - 20(7.05)^2}}$$

$$\text{C.V.} = 4.85$$

The predicted value of Y given an x of 10 is 4.21, which is below the critical value of the test. At the 5% significance level, the null hypothesis, that the mean number of grams of pollution after ten hours of continuous operation is greater than or equal to 5, is rejected.

Figure 13.22 shows the template for this test.

13.3C CONFIDENCE INTERVALS AND HYPOTHESIS TESTS ON THE COEFFICIENT OF CORRELATION

Recall that the coefficient of correlation is used to describe the degree of association between two variables. If the populations of the two variables are normally distributed, there is a simple test for the null hypothesis that the population correlation coefficient equals 0. The null hypothesis of zero correlation is the most common. It means that the two random variables are not linearly related. It is mathematically equivalent to a hypothesis test on the population slope with a null of 0. Note that the test works only for the null of 0. In notation, the null and alternative hypotheses would be:

$$H_0: \rho = 0 \qquad H_1: \rho \neq 0$$

The formula for the test against the null hypothesis that ρ equals 0 is

$$\text{C.V.}_L, \text{C.V.}_U = 0 \pm t_{\alpha/2} s_r \qquad \text{(13.20)}$$

where $df = n - 2$

where the standard deviation of the distribution of sample correlation coefficients, s_r, equals

$$s_r = \sqrt{\frac{1 - r^2}{n - 2}} \qquad \text{(13.21)}$$

The degrees of freedom for the t distribution are $n - 2$. One degree of freedom is lost when $\bar{x}$ is used to estimate μ_x and another degree of freedom is lost when $\bar{y}$ is used to estimate μ_y. Note that the center of formula 13.20 is written as 0 instead of ρ_0, the hypothesized population correlation coefficient. The reason is that the test should be applied only to the null hypothesis of a population correlation coefficient equaling 0.

Here's an example that illustrates the use of these formulas. A record company produces rock videos to promote the sales of its albums. Note that neither the air time nor the number of albums sold is directly controlled by the record company; both are random variables. Whether air time for the video promotes record sales or sales of records promote video air time is not clear; the causation may run in both directions because the programmers of TV video shows may favor the videos of big-selling albums and record buyers may favor the albums that get the most air time. This is a case in which causation can run in both directions. Without some more information it would be impossible to disentangle the separate effects. However, the sample coefficient of correlation can be used to describe the degree of association between air time and album sales. Suppose the record company released 20 albums in a particular year and the sample coefficient of correlation between minutes of air time and numbers of records sold was 0.51. At the 5% significance level, can you reject the null hypothesis that the population coefficient of correlation between air time and number of albums sold is 0? Substituting into formula 13.20 yields

$$\text{C.V.}_L, \text{C.V.}_U = 0 \pm t_{\alpha/2}s_r \qquad\qquad (13.20)$$

$$\text{C.V.}_L, \text{C.V.}_U = 0 \pm 2.101\sqrt{\frac{1 - (0.51)^2}{20 - 2}}$$

$$\text{C.V.}_L = -0.43 \qquad \text{C.V.}_U = 0.43$$

The acceptance region is bounded by the critical values -0.43 and 0.43. The sample correlation coefficient of 0.51 is in the rejection region. The null hypothesis that the population correlation between air time and album sales is 0 must be rejected. It seems that record sales and the amount of air time move together. Since the company does not know the direction of causality, it cannot be sure if promoting record sales will increase air time or if purchasing air time will increase record sales. The record company could promote record sales with advertising and buy air time or produce flashier videos that will encourage more air time. If the causality runs in both directions, the two efforts will reinforce each other. Figure 13.23 contains the template for this hypothesis test.

There are two alternatives to the above procedure for tests on the population correlation coefficient. If the two populations are not normally distributed, the rank correlation test can be applied. This test will be described in Chapter 17, on nonparametric statistics. The other alternative test is Fisher's z transformation, which is used when the null hypothesis is a value other than 0. This test can be applied only if the X and Y variables are normally distributed. We won't go into Fisher's z transformation, but the suggested readings at the end of the chapter discuss it.

FIGURE 13.23

Sampling distribution of the coefficient of correlation and the acceptance and rejection regions for the null hypothesis that ρ equals 0

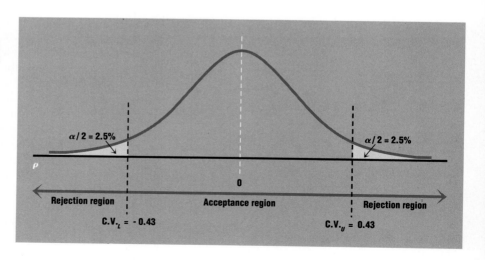

EXERCISES FOR SECTION 13.3

13.14 Let x be {12, 17, 33, 36, 45, 46} and Y be {30, 18, 36, 12, 9, 5}, where the x values are constants and the Y values are randomly drawn observations from the population of Y given x. Assume that Y given x is normally distributed and that Y given x has the same standard deviation for all values of x. Finally, assume that the mean of Y given x is a linear function of x.

(a) Construct a confidence interval for the predicted value of Y given x at the 99% level of confidence, with x set at 30.

(b) Construct a confidence interval for the mean value of Y given x at the 99% level of confidence, with x set at 30.

(c) At the 5% significance level, test the hypothesis that the mean of Y when x is 30 is less than or equal to 15.

13.15 Assume everything you had to assume in exercise 13.14. Let x be {15, 17, 19, 22, 29, 32} and Y be {24, 30, 35, 38, 39, 40}.

(a) Construct a confidence interval for the predicted value of Y given x at the 95% level of confidence, with x set at 20.

(b) Construct a confidence interval for the mean value of Y given x at the 99% level of confidence, with x set at 20.

(c) At the 5% significance level, test the hypothesis that the mean of Y when x is 20 is greater than or equal to 30.

13.16 Let X and Y be normally distributed random variables in which each observation of X is paired with an observation of Y. Let a random sample of pairs of observations be: $X = $ {20, 46, 70, 100, 122, 134}; $Y = $ {14, 10, 16, 9, 8, 7}. At the 5% significance level, test the hypothesis that the population correlation coefficient equals 0.

13.17 The B & B Gear Company has experimented with different speeds for its assembly line for transmissions for small tractors. The object of these experiments is to observe how the number of defective transmissions per hour is related to the speed of the line. Use the following data to regress the number of defects on the assembly-line speed. The company plans to set the assembly-line speed at 60 transmissions per hour. Given this speed, estimate the mean number of defects per hour and construct the appropriate confidence interval for this estimate at the 95% level of confidence.

Assembly-line speed	Number of defects observed	Assembly-line speed	Number of defects observed
55	8	65	18
55	11	70	16
60	14	70	17
60	12	70	20
65	14	75	19

13.18 (a) For the data in exercise 13.17, test at the 5% significance level the hypothesis that the population correlation coefficient is 0.

(b) Test at the 5% significance level the hypothesis that the mean number of defects per hour exceeds 16 when the assembly-line speed is set at 70.

13.19 The following data show the weekly sales for a museum gift shop by hours of operation. Use these data to construct confidence intervals at the 90% level of confidence for the mean weekly sales for 40 hours of operation and for one week's sales for 40 hours of operation.

Hours of operation per week	Weekly sales	Hours of operation per week	Weekly sales
20	$1,404	36	$1,793
20	1,320	36	1,917
24	1,435	40	2,095
24	1,569	40	3,974
25	1,599	40	2,121
25	1,538	40	2,500
30	1,636	56	3,116
30	1,643	56	2,980
30	1,600	56	3,421
36	1,694		

13.20 (a) Calculate the coefficients of determination and correlation for the data in exercise 13.19 and test at the 5% significance level the hypothesis that the population correlation coefficient is 0.

(b) Test at the 5% significance level the hypothesis that the population slope is 0.

13.21 The "hemline theory" of the stock market may be the silliest theory to date. The theory states that the Dow-Jones average of 30 industrial stocks will decline if women start to wear longer skirts and that it will rise when skirts get shorter. Proponents of the theory keep their eyes on Paris fashions and subscribe to *Women's Wear Daily* in order to forecast a bear (falling) or bull (rising) stock market.

(a) Use the following data on hemlines relative to the middle of the knee and the Dow-Jones average to test at the 5% significance level the hypothesis that the population correlation coefficient is 0.

(b) If the null of no correlation cannot be rejected, would a promotion of micromini skirts raise the Dow-Jones industrial average? Explain.

Hemline length (inches below or above knee)	Dow-Jones industrial average	Hemline length (inches below or above knee)	Dow-Jones industrial average
-6	940	2	1,650
-3	1,220	3	1,500
-2	1,350	4	1,650
1	1,400	6	1,800

13.4 LOGARITHMIC TRANSFORMATIONS FOR ESTABLISHING LINEARITY AND WEIGHTED LEAST SQUARES (OPTIONAL)

 ### 13.4A LOGARITHMIC TRANSFORMATIONS

So far we have used examples in which the relationship between the mean of Y and x was assumed to be linear. Many problems do not fit this assumption, but it is possible to transform the data so that the technique of linear regression can be applied to the transformed data. Consider the problem of the price of wine as a function of the age of the wine. As the wine gets older, it tastes better and is more scarce, and the price reflects this scarcity as well as the better taste. The price must also be high enough to compensate the owner of the wine for the opportunity cost of the money tied up in the wine during each year that the wine ages. Instead of a linear function of age, the price of wine tends to be an exponential function of the wine's age, as is evident in the retail prices of red Bordeaux listed in Table 13.9.

Figure 13.24 is a scatter diagram of the data listed in Table 13.9. The scatter diagram shows a pronounced curvature in the relationship between a wine's age and its price. But if the same data were plotted on semilogarithmic paper, the relationship would be close to a straight line. With semilogarithmic (or semilog) paper, the x axis is in a linear scale and the y axis is in logs. The reason for the straight line is that equal changes in a logarithmic scale represent equal percentage changes in a linear scale. Thus if there is a constant rate of growth of wine prices as a function of age, the relationship will appear to be linear when it is plotted on the semilog paper. Figure 13.25 plots the same data on semilog paper.

What you need to do if you suspect that there is a constant percentage change in the mean level of your dependent variable associated with a unit change in the independent variable is to regress the dependent variable on the

TABLE 13.9	Age	Price	Age	Price	Age	Price
Age in years and	25	$2,100	10	$125	7	$34
retail prices of	22	1,325	10	89	6	39
cases of red	20	800	9	79	6	25
Bordeaux wine	19	700	9	70	6	18
	17	550	8	59	5	24
	17	460	8	55	5	8
	16	400	8	51	4	16
	13	220	7	60	4	11
	12	170	7	39	4	9
	10	100				

FIGURE 13.24
Scatter diagram of ages and prices of red Bordeaux wine

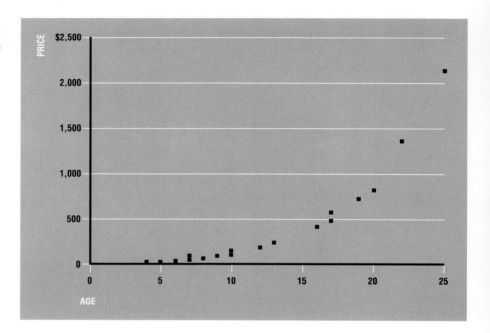

FIGURE 13.25
Scatter diagram of wine prices and age on semilog paper

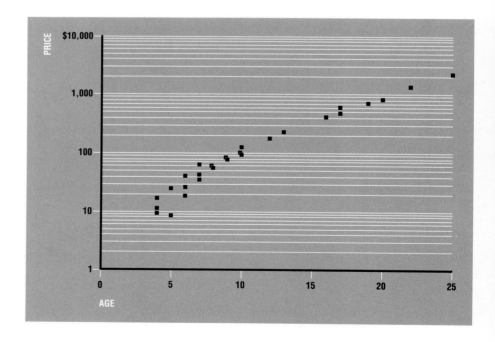

log value of the independent variable. In notation the exponential relationship is

$$y_i = \alpha e^{\beta x_i} \varepsilon_i \qquad \text{(13.22)}$$

Formula 13.22 shows the observed value of y given x is equal to a constant term, α, times e taken to the βx power, times the value of a noise-generating random variable, ε_i. This noise-generating function was described in a boxed section earlier in the chapter. If you take the natural logs of both sides of formula 13.22, the resulting formula looks like the familiar additive intercept, slope, and error-generating function of linear regression.

$$\ln(y_i) = \ln(\alpha) + \beta x_i + \ln(\varepsilon_i)$$

If the term $\ln(\varepsilon_i)$ is normally distributed with the same standard deviation for all values of x, the usual confidence intervals and hypothesis tests can be applied to the estimated regression.

The various sums needed to calculate the estimated slope and intercept, the standard error of the estimate, the coefficient of determination, and the critical values for any hypothesis tests are given in Table 13.10. Again, these values are provided to show you where the sums came from and to let you test your understanding of the formulas. As you can see from the table, calculating all of these sums would take a long time. The best method is to use a computer for these calculations or even a hand-held calculator that can do linear regression. There are some short-cut formulas that can be used if you don't have either of these resources, but if you are going to tackle a large problem, you would be much better off with a computer or a powerful calculator. The Texas Instruments model TI-35 STAT does linear regression and costs about $10.

The estimated slope is 0.254:

$$b = \frac{\displaystyle\sum_{i=1}^{n} x_i y_i - n \overline{x}\,\overline{y}}{\displaystyle\sum_{i=1}^{n} x_i^2 - n(\overline{x})^2} \qquad \text{(13.4)}$$

$$b = \frac{1,552.98 - 28(10.5)(4.474)}{4,024 - 28(10.5)^2}$$

$$b = 0.254$$

Note that the y values substituted into formula 13.4 are in logs. For example,

TABLE 13.10
Sums needed for wine price versus age regression analysis

x_i	y_i	$\ln(y_i)$	x_i^2	$x_i\ln(y_i)$	est. $\ln(y_i)$	$[\ln(y_i) - \text{est. } \ln(y_i)]^2$	$[\text{est. } \ln(y_i) - \overline{\ln(y)}]^2$	$[\ln(y_i) - \overline{\ln(y)}]^2$
25	2,100	7.6497	625	191.2423	8.1507	0.25103	13.5173	10.0842
22	1,325	7.1892	484	158.1617	7.39	0.04035	8.50255	7.37143
20	800	6.6846	400	133.6922	6.8829	0.03933	5.80231	4.88623
19	700	6.5511	361	124.4705	6.6294	0.00613	4.64506	4.31372
17	550	6.3099	289	107.2686	6.1223	0.03522	2.71632	3.37012
17	460	6.1312	289	104.2309	6.1223	8E-05	2.71632	2.74597
16	400	5.9915	256	95.86343	5.8687	0.01507	1.94482	2.30231
13	220	5.3936	169	70.11716	5.108	0.08157	0.40182	0.84548
12	170	5.1358	144	61.62958	4.8545	0.07915	0.14466	0.43781
10	100	4.6052	100	46.0517	4.3474	0.06647	0.01607	0.01717
10	125	4.8283	100	48.28314	4.3474	0.23133	0.01607	0.12545
10	89	4.4886	100	44.88636	4.3474	0.01996	0.01607	0.00021
9	79	4.3694	81	39.32503	4.0938	0.07599	0.14466	0.01096
9	70	4.2485	81	38.23646	4.0938	0.02393	0.14466	0.05091
8	59	4.0775	64	32.6203	3.8402	0.05631	0.40182	0.15729
8	55	4.0073	64	32.05867	3.8402	0.02792	0.40182	0.2179
8	51	3.9318	64	31.45461	3.8402	0.00839	0.40182	0.29409
7	60	4.0943	49	28.66041	3.5867	0.25773	0.78757	0.14424
7	39	3.6636	49	25.64493	3.5867	0.00591	0.78757	0.65702
7	34	3.5264	49	24.68452	3.5867	0.00364	0.78757	0.89827
6	39	3.6636	36	21.98137	3.3331	0.10919	1.3019	0.65702
6	25	3.2189	36	19.31325	3.3331	0.01305	1.3019	1.57566
6	18	2.8904	36	17.34223	3.3331	0.19603	1.3019	2.50829
5	24	3.1781	25	15.89027	3.0796	0.0097	1.94482	1.67981
5	8	2.0794	25	10.39721	3.0796	1.00024	1.94482	5.73453
4	16	2.7726	16	11.09035	2.826	0.00285	2.71632	2.89524
4	11	2.3979	16	9.591581	2.826	0.18328	2.71632	4.31075
4	9	2.1972	16	8.788898	2.826	0.39536	2.71632	5.1843
294	**7,636**	**125.3**	**4,024**	**1,553**	**125.3**	**3.235**	**60.241**	**63.476**

the 4.474 substituted for $\bar{y}$ is the sum of the third column, labeled $\ln(y_i)$, divided by the number of pairs of observations, 28.

The estimated intercept is 1.81:

$$a = \bar{y} - b\bar{x} \tag{13.5}$$

$$a = 4.474 - 0.254(10.5)$$

$$a = 1.81$$

The standard error of the estimate is 0.353:

$$s_e = \sqrt{\frac{\sum\limits_{i=1}^{n} (y_i - \hat{y}_i)^2}{n-2}} \tag{13.13}$$

$$s_e = \sqrt{\frac{3.236}{28-2}}$$

$$s_e = 0.353$$

The coefficient of determination is 0.95:

$$r^2 = \frac{\sum\limits_{i=1}^{n} (\hat{y}_i - \bar{y})^2}{\sum\limits_{i=1}^{n} (y_i - \bar{y})^2} \tag{13.10}$$

$$r^2 = \frac{60.261}{63.476}$$

$$r^2 = 0.949$$

This regression accounts for 95% of the variation in logs of prices. The log regression provides a good fit to the data.

To go back to prices in dollars from the log of prices you have to take the antilog. For example, at the end of 13 years the estimated log of the price of a case of wine is 5.108. This number came from the sixth column of Table 13.10. The antilog of 5.108 is e to the 5.108 power, or $165.34.

After reviewing the results of the above regression, you may be tempted to buy some wine as an investment. For example, you could buy a case of ten-year old wine for $77.28 ($e$ to the 4.3474 power) and sell it at the end of three years for $165.34. There are two potential problems, one statistical and the other not. The nonstatistical problem is the margin between wholesale and retail prices. The prices quoted are at retail. If you buy at retail and have to sell at wholesale, you may lose money. The statistical problem is selection bias. Red Bordeaux wines are a broad category, and only the best wines tend to be kept for long periods. The data may reflect both the appreciation of these wines and the fact that the cheaper wines are used up first. If you don't know which of these wines will appreciate, your investment may turn sour.

While these considerations make the regression results a dubious guide for investment, you can still use the example for practice with confidence intervals or hypothesis tests. Let's construct a confidence interval at the 99% level on the predicted price of a case of wine that is ten years old. Substituting into formula 13.17 yields

$$L, \ U = \hat{y}_i \pm t_{\alpha/2} s_e \sqrt{1 + \frac{1}{n} + \frac{(x^* - \bar{x})^2}{\sum\limits_{i=1}^{n}(x_i)^2 - n\bar{x}^2}} \tag{13.17}$$

$$L, \ U = 4.347 \pm 2.779(0.353)\sqrt{1 + \frac{1}{28} + \frac{(10 - 10.5)^2}{4{,}024 - 28(10.5)^2}}$$

$$L = 3.349 \qquad U = 5.345$$

The lower and upper limits of the confidence interval are in the logs of prices. The antilogs of 3.349 and 5.345 are \$28.47 and \$209.56. The confidence interval suggests that when you go to buy one case of wine, you will find a wide variety of prices, even if the age is held constant.

To summarize the main point of using the logarithmic transformation: it allows you to estimate a regression line for data that exhibit constant rates of growth or equal percentage changes in the dependent variable for every unit change in the independent variable.

■ 13.4B WEIGHTED LEAST SQUARES

If the standard deviations of the distribution of Y given x are not equal for all values of x, you need to weight the observations drawn where Y has a low variance more heavily than the observations where Y has a high variance. Figure 13.12 illustrates a simple case in which the variance of the Y distributions increases in proportion to the value of x squared. Let the variance of ε_i or σ_ε^2 equal Cx_i^2, where C is a constant term. The observed values of y would equal

$$y_i = \alpha + \beta x_i + \varepsilon_i$$

If the data are transformed by dividing every term by x_i, for example,

$$\frac{y_i}{x_i} = \frac{\alpha}{x_i} + \beta + \frac{\varepsilon_i}{x_i}$$

FIGURE 13.26

Scatter diagram of taxi odometer miles at beginning of each month and monthly maintenance costs

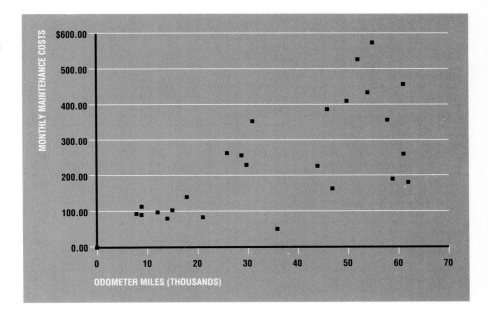

the new error term will have a constant variance or be homoscedastic. The new error term is ε_i/x_i. The variance of ε_i/x_i equals

$$\text{Variance}\left(\frac{\varepsilon_i}{x_i}\right) = \frac{1}{x_i^2} \text{ variance } (\varepsilon_i) = \frac{Cx_i^2}{x_i^2} = C$$

Recall that the variance of a constant times a random variable is the constant squared times the variance of the random variable. Recall also that in regression analysis the x is a constant. Since the new error term has a variance equal to C, the standard deviation of the transformed Y variable must be the same for all values of x. This transformation is equivalent to weighting every Y value by the value of the corresponding x.

Let's run through an example to see how the transformation works. The owner of a fleet of taxicabs would like to estimate the monthly repair costs per taxi as a function of the number of miles on a taxi's odometer. He knows that the variability in repair costs increases as the miles accumulate. Figure 13.26 shows the scatter diagram for the data on repair costs. There is a definite upward trend in the sample data and the presence of heteroscedasticity is evident.

The assumption that this increase in variability (variance) is proportional to the number of miles squared allows the use of weighted least squares for the transformation. Table 13.11 shows the raw and the transformed sample data and the sums needed to calculate the various terms for the regression. Scientific

TABLE 13.11

Beginning-of-month odometer miles and monthly repair costs of a fleet of taxis and the transformed variables and sums needed for regression analysis

x_i	y_i	NEW Y y_i/x_i	NEW x $1/x_i$	$(1/x_i)^2$	$(y_i/x_i)(1/x_i)$	EST. NEW Y $\hat{y}_i/x_i$	SSR $(\hat{y}_i/x_i - \bar{y}/\bar{x})^2$	SSE $(y_i/x_i - \hat{y}_i/x_i)^2$	SST $(y_i/x_i - \bar{y}/\bar{x})^2$
8	93.54	11.69	1.25E − 01	1.56E − 02	1.46E + 00	1.09E + 01	1.27E + 01	5.72E − 01	1.87E + 01
9	113.91	12.66	1.11E − 01	1.23E − 02	1.41E + 00	1.03E + 01	8.79E + 00	5.38E + 00	2.79E + 01
9	91.59	10.18	1.11E − 01	1.23E − 02	1.13E + 00	1.03E + 01	8.79E + 00	2.55E − 02	7.87E + 00
12	98.10	8.18	8.33E − 02	6.94E − 03	6.81E − 01	9.14E + 00	3.11E + 00	9.23E − 01	6.45E − 01
14	79.20	5.66	7.14E − 02	5.10E − 03	4.04E − 01	8.62E + 00	1.56E + 00	8.79E + 00	2.94E + 00
15	104.54	6.97	6.67E − 02	4.44E − 03	4.65E − 01	8.42E + 00	1.09E + 00	2.09E + 00	1.62E − 01
18	138.72	7.71	5.56E − 02	3.09E − 03	4.28E − 01	7.94E + 00	3.18E − 01	5.24E − 02	1.12E − 01
21	83.68	3.98	4.76E − 02	2.27E − 03	1.90E − 01	7.59E + 00	4.86E − 02	1.30E + 01	1.15E + 01
26	262.65	10.10	3.85E − 02	1.48E − 03	3.89E − 01	7.20E + 00	3.07E − 02	8.44E + 00	7.45E + 00
29	257.23	8.87	3.45E − 02	1.19E − 03	3.06E − 01	7.02E + 00	1.21E − 01	3.41E + 00	2.24E + 00
30	231.18	7.71	3.33E − 02	1.11E − 03	2.57E − 01	6.98E + 00	1.58E − 01	5.34E − 01	1.12E − 01
31	354.18	11.43	3.23E − 02	1.04E − 03	3.69E − 01	6.93E + 00	1.97E − 01	2.02E + 01	1.64E + 01
36	51.81	1.44	2.78E − 02	7.72E − 04	4.00E − 02	6.73E + 00	4.06E − 01	2.80E + 01	3.52E + 01
44	225.98	5.14	2.27E − 02	5.17E − 04	1.17E − 01	6.52E + 00	7.31E − 01	1.91E + 00	5.00E + 00
46	387.42	8.42	2.17E − 02	4.73E − 04	1.83E − 01	6.47E + 00	8.06E − 01	3.80E + 00	1.10E + 00
47	162.07	3.45	2.13E − 02	4.53E − 04	7.34E − 02	6.45E + 00	8.43E − 01	9.03E + 00	1.54E + 01
50	410.89	8.22	2.00E − 02	4.00E − 04	1.64E − 01	6.40E + 00	9.47E − 01	3.31E + 00	7.15E − 01
52	525.29	10.10	1.92E − 02	3.70E − 04	1.94E − 01	6.37E + 00	1.01E + 00	1.40E + 01	7.45E + 00
54	432.22	8.00	1.85E − 02	3.43E − 04	1.48E − 01	6.33E + 00	1.08E + 00	2.79E + 00	4.00E − 01
55	572.74	10.41	1.82E − 02	3.31E − 04	1.89E − 01	6.32E + 00	1.11E + 00	1.68E + 01	9.25E + 00
58	355.24	6.12	1.72E − 02	2.97E − 04	1.06E − 01	6.28E + 00	1.19E + 00	2.39E − 02	1.56E + 00
59	189.29	3.21	1.69E − 02	2.87E − 04	5.44E − 02	6.27E + 00	1.22E + 00	9.36E + 00	1.73E + 01
61	260.09	4.26	1.64E − 02	2.69E − 04	6.99E − 02	6.24E + 00	1.27E + 00	3.92E + 00	9.66E + 00
61	457.26	7.50	1.64E − 02	2.69E − 04	1.23E − 01	6.24E + 00	1.27E + 00	1.57E + 00	1.54E − 02
62	179.79	2.90	1.61E − 02	2.60E − 04	4.68E − 02	6.23E + 00	1.30E + 00	1.11E + 01	2.00E + 01
907	**6,118.62**	**184.30**	**1.06E + 00**	**7.20E − 02**	**9.00E + 00**	**1.84E + 02**	**5.01E + 01**	**1.69E + 02**	**2.19E + 02**

notation is used so that all of the terms will fit in one table. To see how to interpret this notation, take the sum of squares total, or SST. This value is listed in the lower right corner of Table 13.11 as 2.19E + 02. This value translates into $2.19(10)^2$, or 219.

The estimated regression line takes the form

$$\frac{\hat{y}}{x} = b + a\left(\frac{1}{x}\right)$$

You have to be careful with these terms. The former slope is now the intercept and the former intercept divided by x is now the slope. The terms "New Y" and "New x" have been placed at the top of the table to help you keep track of the transformed variables. The procedure for getting the estimated slope, intercept, and other terms for regression analysis is the same as before, except that the new Y and x have to be used in place of the original Y and x. To help you keep track, we'll put a prime on the slope and intercept terms whenever they refer to the transformed variables. For example, the estimated slope and intercept are

$$b' = \frac{\sum\limits_{i=1}^{n} x_i y_i - n\bar{x}\,\bar{y}}{\sum\limits_{i=1}^{n} x_i^2 - n(\bar{x})^2} \tag{13.4}$$

$$b' = \frac{9.00 - 25(0.0425)(7.3719)}{0.072 - 25(0.0425)^2}$$

$$b' = 43.22$$

$$a' = \bar{y} - b'\bar{x} \tag{13.5}$$

$$a' = 7.372 - 43.22(0.0425)$$

$$a' = 5.53$$

Please note that while the original formulas for the slope and intercept were used in the above substitutions, the a' and b' refer to the slope and intercept for the new Y and x, that is, Y/x and a/x.

To get a point estimate of the average monthly maintenance costs for a taxi with a given number of miles, you have to work back from the estimated regression equation to the original terms. Take a taxi with 30,000 miles as an example. The new x variable has the value of $1/x$ or $1/30$ when x is 30. The estimated value of the new Y variable is

$$\frac{\hat{y}}{x} = b + a\left(\frac{1}{x}\right)$$

$$\frac{\hat{y}}{x_i} = 5.53 + 43.22\left(\frac{1}{30}\right)$$

$$\frac{\hat{y}}{x_i} = 6.97$$

Here we've gone back to the a and b of the original notation instead of the Y/x and a/x. To work back from $\hat{y}/x_i$ to $\hat{y}$, you have to multiply $\hat{y}/x_i$ by x_i. The point estimate for the mean maintenance costs for a taxi with 30,000 miles would be 30 times 6.97, or $209.

The confidence intervals and hypothesis tests discussed earlier in the chapter apply to weighted least-squares regression. You'll find some examples in the exercises. Other than inspecting a scatter diagram, we have not discussed how you can determine if a data set is heteroscedastic. There is a statistical test for heteroscedasticity, but it is usually introduced in econometrics courses. The method of visual inspection is reliable enough and reasonable for the level of this text. We also have not discussed weighting schemes other than the use of the x_i's as weights. Other weighting schemes are discussed in an econometrics text cited in the suggested readings at the end of the chapter.

To review, the main feature of weighted least squares is that it provides a more efficient or more accurate estimate of the population parameters when the data are heteroscedastic. Heteroscedasticity is a common problem. In our taxi example, new taxis would rarely need repairs, so the maintenance costs would vary little from taxi to taxi. Maintenance expenses would be confined mainly to gas and oil. Older taxis are subject to big repair bills, but they don't come in a steady pattern over time; an old taxi can go for months needing only minor maintenance and then need a new transmission. The increasing variability in the dependent variable as the independent variable increases fits many other situations.

EXERCISES FOR SECTION 13.4

13.22 A state board of health would like to estimate the number of AIDS cases in the state by the end of 1988. The estimate is needed to determine the resources that will be needed for AIDS patients. AIDS is a disease that attacks the body's immune system. Plot the scatter diagram of the following monthly data on the number of AIDS cases and use the appropriate regression technique to make a point estimate of the number of cases for December 1988.

Year	Month	Number of cases	Year	Month	Number of cases
1985	January	2	1985	November	14
	February	3		December	14
	March	4	1986	January	19
	April	4		February	23
	May	5		March	28
	June	6		April	34
	July	6		May	40
	August	7		June	48
	September	9		July	55
	October	11		August	63

Year	Month	Number of cases	Year	Month	Number of cases
1986	September	72	1987	May	200
	October	82		June	220
	November	90		July	251
	December	114		August	278
1987	January	126		September	316
	February	139		October	358
	March	160		November	412
	April	175		December	440

13.23 Use weighted least squares to estimate the population regression line for the following data set. Assume that the variance in the distribution of Y given x is proportional to the value of x squared.

x_i	y_i	x_i	y_i
20	44	59	65
37	51	62	107
40	68	65	57
49	53	70	132
58	79	72	86

13.24 Construct a confidence interval at the 95% level of confidence for the point estimate of the number of AIDS cases in December 1988 which you made for exercise 13.22.

13.25 Construct confidence intervals at the 95% level of confidence for the mean and predicted values of Y given x for the data set in exercise 13.23, with x set at 60.

13.26 The state board of health referred to in exercise 13.22 would like a new estimate that takes into account the greater variability in the number of AIDS cases over time as new groups of people are affected and different methods of spreading the disease occur. Assume that this greater variability is proportional to the square of the time that has elapsed from the beginning of the data. Estimate the regression and a point estimate for the number of cases in December 1988 and construct a confidence interval for that point estimate at the 95% level of confidence.

13.27 A travel agents' association is interested in estimating the effect of per capita income on family spending for vacations. The trade association would like to know if low-income groups might purchase travel agents' services if a marketing program were aimed at them.

(a) Use the data below to plot the annual spending and family income.

(b) If the plot does not appear to indicate a linear relationship, use the appropriate technique to estimate the effect of family income on vacation spending.

Per capita family income	Annual travel spending	Per capita family income	Annual travel spending
$3,000	$10	$ 5,400	$ 99
3,000	10	6,000	160
3,400	20	7,000	1,250
3,600	30	7,500	2,900
3,800	45	7,500	3,000
4,500	40	7,500	3,500
4,600	45	9,000	4,400
4,700	57	9,500	4,500
5,100	65	10,000	4,900
5,300	80	10,000	4,500

13.28 (a) Run a linear regression for the data in Table 13.11, on maintenance costs for taxis, and then use linear regression to construct a confidence interval for the mean maintenance costs for a taxi with 30,000 miles on the odometer. Construct the same confidence interval for the weighted regression.

(b) Which confidence interval is narrower? Explain why one is narrower than the other.

13.29 Use the data in exercise 13.27 to test at the 5% significance level the hypothesis that the mean vacation spending for families with an annual per capita income of $5,000 is $900.

 SUMMARY

Simple regression is a powerful technique that can be used to describe the relationship between two variables. For purely descriptive purposes, no assumptions about the population distributions of Y given x have to be made. The only requirement for the descriptive use of regression is that a linear relationship must be a reasonable way to describe the data. However, regression analysis is almost always used to make inferences beyond the sample data. The least-squares estimates of the population slope and intercept are efficient and unbiased if the distributions of Y given x are normal and have the same standard deviations and if the mean of Y is a linear function of x. This standard deviation is estimated by the term s_e, the standard error of the estimate. All of the formulas for confidence intervals and hypothesis tests in regression analysis require this term. The formulas for confidence intervals and hypothesis tests in regression analysis have the same format as the formulas in earlier chapters on confidence intervals and hypothesis testing. The chief difference is an adjustment term for the increased uncertainty about the location of the mean of Y given x as the value of x gets further from $\bar{x}$.

In contrast to regression analysis, correlation analysis requires that X be a random variable. Correlation analysis is used when neither variable is thought to be the cause of changes in the other or when the direction of causation is uncertain. The optional section of the chapter discussed two techniques that extend the use of regression analysis to two situations in which some of the assumptions required for statistical inference do not hold. These assumptions are the linearity between the mean of Y given x and the constancy of the standard deviations for the distributions of Y given x. Logarithmic regression can be used when it appears that the underlying relationship between x and Y is such that a percentage change in Y is associated with a unit change in x. Weighted least squares can be used for nonconstant standard deviations in the distributions of Y given x. The only illustration of weighted least squares given covered the situation when the variances of distributions of Y given x are proportional to the value of x squared.

In the next two chapters regression analysis will be extended to cover more than one explanatory variable (Chapter 14) and the important business problem of forecasting (Chapter 15). This material is built up very systematically from the material in Chapter 13, so make sure you understand this chapter before you go on.

 NEW TERMS

coefficient of determination The percentage of the total sum of squared deviations of the dependent variable around its mean which is accounted for by the independent variable.

heteroscedastic A data set that is not homoscedastic.

homoscedastic A data set in which the distribution of the dependent variable has the same variance for all values of the independent variable.

population correlation coefficient A measure of the linear association between two random variables calculated from population data.

population slope coefficient The change in the mean value of the dependent variable associated with a unit change in the independent variable calculated with population data.

regression line The straight line that minimizes the sum of the squared deviations between the predicted and observed values of the dependent variable.

sample correlation coefficient A measure of the linear association between two random variables calculated from sample data.

sample slope coefficient The change in the mean value of the dependent variable associated with a unit change in the independent variable calculated with sample data.

scatter diagram A graph in which each observation is represented by a dot. The explanatory or independent variable is scaled on the x axis and the dependent or explained variable is scaled on the y axis.

simple regression A technique that is used to describe the effect of changes in an explanatory variable on a dependent or explained variable.

standard error of the estimate The estimated standard deviation of the distribution of the dependent variable for values of the independent variable.

SUGGESTED READINGS

Daniel, Wayne W., and James C. Terrell. *Business Statistics*. Boston: Houghton Mifflin, 1986. Fisher's z transformation is discussed on p. 398.

Park, J. Ewart, James S. Ford, and Chi-Yuan Lin. *Applied Managerial Statistics*. Englewood Cliffs, N.J.: Prentice-Hall, 1982. Fisher's z transformation is discussed on p. 377.

Velleman, Paul F., and David C. Hoaglin. *Applications, Basics, and Computing of Exploratory Data Analysis*. Boston: Duxbury Press, 1981. Resistant lines are discussed in Chapter 5.

 NEW FORMULAS

Sum of squared errors

$$\text{SSE} = \sum_{i=1}^{\nu} (y_i - \hat{y}_i)^2$$

(13.1)

Population slope coefficient

$$\beta = \frac{\sum_{i=1}^{\nu} x_i y_i - \nu \mu_x \mu_y}{\sum_{i=1}^{\nu} x_i^2 - \nu (\mu_x)^2}$$

(13.2)

Population intercept

$$\alpha = \mu_y - \beta \mu_x$$

(13.3)

Sample slope coefficient

$$b = \frac{\sum_{i=1}^{n} x_i y_i - n \bar{x} \, \bar{y}}{\sum_{i=1}^{n} x_i^2 - n (\bar{x})^2}$$

(13.4)

Sample slope

$$a = \bar{y} - b \bar{x}$$

(13.5)

Sum of squares total

$$\text{SST} = \sum_{i=1}^{\nu} (y_i - \mu_y)^2$$

(13.6)

Sum of squares for the regression

$$\text{SSR} = \sum_{i=1}^{\nu} (\hat{y}_i - \mu_y)^2 \tag{13.7}$$

Partition of the sum of squares total

$$\text{SST} = \text{SSE} + \text{SSR} \tag{13.8}$$

Population coefficient of determination

$$\rho^2 = \frac{\displaystyle\sum_{i=1}^{\nu} (\hat{y}_i - \mu_y)^2}{\displaystyle\sum_{i=1}^{\nu} (y_i - \mu_y)^2} \tag{13.9}$$

Sample coefficient of determination

$$r^2 = \frac{\displaystyle\sum_{i=1}^{n} (\hat{y}_i - \overline{y})^2}{\displaystyle\sum_{i=1}^{n} (y_i - \overline{y})^2} \tag{13.10}$$

Population coefficient of correlation

$$\rho = \frac{\displaystyle\sum_{i=1}^{\nu} z_{X_i} z_{Y_i}}{\nu} \tag{13.11}$$

Sample coefficient of correlation

$$r = \frac{\displaystyle\sum_{i=1}^{n} z_{X_i} z_{Y_i}}{n - 1} \tag{13.12}$$

Standard error of the estimate

$$s_e = \sqrt{\frac{\displaystyle\sum_{i=1}^{n} (y_i - \hat{y}_i)^2}{n - 2}} \tag{13.13}$$

Confidence interval for the population slope coefficient

$$L, \ U = b \pm t_{\alpha/2} s_b \tag{13.14}$$

Standard deviation of the distribution of sample slopes

$$s_b = \frac{s_e}{\sqrt{\sum_{i=1}^{n} (x_i)^2 - n(\overline{x})^2}} \tag{13.15}$$

Critical values of a hypothesis test on the population slope

$$\text{C.V.}_L, \ \text{C.V.}_U = b_0 \pm t_{\alpha/2} s_b \tag{13.16}$$

Confidence interval for the predicted value of Y given x

$$L, \ U = \hat{y}_i \pm t_{\alpha/2} s_e \sqrt{1 + \frac{1}{n} + \frac{(x^* - \overline{x})^2}{\sum_{i=1}^{n} (x_i)^2 - n\overline{x}^2}} \tag{13.17}$$

Confidence interval for the mean of Y given x

$$L, \ U = \hat{y}_i \pm t_{\alpha/2} s_e \sqrt{\frac{1}{n} + \frac{(x^* - \overline{x})^2}{\sum_{i=1}^{n} (x_i)^2 - n\overline{x}^2}} \tag{13.18}$$

Critical values for a hypothesis test on the mean of Y given x

$$\text{C.V.}_L, \ \text{C.V.}_U = \mu_0 \pm t_{\alpha/2} s_e \sqrt{\frac{1}{n} + \frac{(x^* - \overline{x})^2}{\sum_{i=1}^{n} (x_i)^2 - n\overline{x}^2}} \tag{13.19}$$

Critical values for the hypothesis test that the population correlation coefficient equals 0

$$\text{C.V.}_L, \ \text{C.V.}_U = 0 \pm t_{\alpha/2} s_r \tag{13.20}$$

Standard deviation of the distribution of sample correlation coefficients

$$s_r = \sqrt{\frac{1 - r^2}{n - 2}} \tag{13.21}$$

Exponential relationship between Y and x

$$y_i = \alpha e^{\beta x_i} \varepsilon_i \tag{13.22}$$

ADDITIONAL EXERCISES

An asterisk designates an exercise of greater than ordinary difficulty.

13.30 A management consultant whose specialty was executive compensation was asked to make recommendations for the salary of the president of a medium-sized lumber products company. The consultant surveyed competing lumber products firms, asking for executives' salaries and the number of employees. Use the following data to estimate the effect of increasing a firm's work force on its president's salary.

Number of employees	President's salary	Number of employees	President's salary
589	$88,000	1,892	$145,000
690	76,000	2,143	150,000
892	102,340	2,560	120,000
1,134	56,000	2,987	169,000
1,238	112,000	3,004	175,000
1,596	125,000	3,547	180,000

13.31 What percentage of the total variation in presidents' salaries in exercise 13.30 is accounted for by variation in employment?

13.32 Construct a confidence interval at the 95% level of confidence for the population slope coefficient for the effect of employment on presidents' salaries, using the data in exercise 13.30.

13.33 The lumber products firm that hired the consultant in exercise 13.30 has 1,642 employees. Estimate the salary of the president of this firm at a 95% level of confidence.

13.34 Construct a confidence interval at the 95% level of confidence for the population mean of presidents' salaries in exercise 13.30, given an employment level of 1,642.

13.35 Plot the residuals for a regression run on the following data set and decide which, if any, of the assumptions needed for statistical inference are violated.

x: 1, 2, 3, 4, 5, 6, 7, 8, 9, 10, 11, 12, 13, 14, 15, 16, 17, 18, 19, 20, 21, 22, 23, 24, 25

y: -11.28, -13.23, -12.62, -12.93, -14.20, -15.04, -14.54, -14.94, -16.15, -16.41, -15.26, -16.63, -16.16, -16.05, -17.10, -17.28, -17.30, -16.20, -17.06, -16.72, -16.49, -18.12, -17.64, -17.10, -17.54

13.36 An employer that has large manufacturing plants located in 15 cities found that its costs for medical insurance per employee varied substantially from plant to plant. The company observed that after these costs were adjusted for differences in the cost of living, most of the variation in medical insurance costs was due to variation in the frequency of surgery among its employees. The frequency of surgery measured by the number of operations per 1,000 employees per city was strongly correlated with the availability of hospital beds. The simple correlation between hospital beds per

1,000 population in the 15 cities and the number of surgeries per year per 1,000 employees was 0.84.

(a) Test at the 5% significance level the hypothesis that the true correlation between the availability of hospital beds and the frequency of surgery is zero.

(b) Which type of analysis is appropriate for this problem, correlation analysis or regression analysis?

13.37 A used-truck dealer has kept records on the 584 semitractors he has sold in the past year. Using these records, he ran a regression in which the dependent variable was the price of the semitractor and the independent variable was the number of miles on each semitractor. The estimated regression equation was $\hat{y}_i = 45,000 - .015x_i$. The standard error of the estimate was 1,450 and the mean mileage on the semitractors in the sample was 1 million. The dealer has a chance to pick up 100 trucks at once from a bankrupt trucking company. To be able to make a profit on the transaction, the dealer has to be able to sell the 100 trucks for at least $30,500 per truck. Assuming that the 100 trucks each have 1 million miles, if the dealer is willing to accept a 5% chance of not making a profit on the transaction, should he go ahead with the deal? This problem can be restated as a hypothesis test in which the null is that the mean selling price for semitractors with 1 million miles is $30,500 or more.

13.38 The alumni association of a college would like to estimate the effect of increases in family income on the amount that members of the alumni association donate to the school. Make a scatter diagram of the following data. Use the appropriate regression technique to estimate this relationship, given the pattern observed in the scatter diagram.

Annual donation	Family income (thousands of dollars)	Annual donation	Family income (thousands of dollars)
$ 25	$25	$ 250	$ 60
25	26	300	72
25	23	300	67
30	28	280	69
30	27	345	70
35	30	425	71
40	29	500	75
45	32	750	82
50	32	1,200	90
50	35	1,000	80
100	37	1,500	95
125	40	1,750	98
150	44	2,000	95
160	50	2,500	100
190	65		

13.39 Plot the residuals for a regression run on the following data set and decide which, if any, of the assumptions needed for statistical inference are violated.

x: 1, 2, 3, 4, 5, 6, 7, 8, 9, 10, 11, 12, 13, 14, 15, 16, 17, 18, 19, 20, 21, 22, 23, 24, 25

y: 12.22, 10.44, 37.01, 47.22, 40.03, 51.53, 35.01, 58.83, 110.97, 24.67, 162.04, 99.08, 114.93, 9.63, 76.87, −12.13, 2.34, 73.87, 97.51, 28.45, −211.26, −60.50, −46.55, 142.71, −55.69

13.40 A high school guidance counselor was concerned that after-school jobs would hurt students' grades. The counselor polled all of the school's seniors to find out how many hours each of them worked and ran a regression in which the dependent variable was the previous semester's grade point average and the explanatory variable was hours of work per week. The estimated regression equation was $\hat{y}_i = 3.62 - 0.02x_i$. There were 346 students in the sample and the standard error of the slope coefficient was 0.004. Test at the 5% significance level the hypothesis that the true effect of hours of work on grade point average is zero. Might causality run from grades to hours of work rather than from hours of work to grades? Explain.

13.41 Most, if not all, economics textbooks show supply and demand curves with the quantity on the x axis and the price on the y axis. If you could identify the quantities demanded at various prices in a market, say the market for umbrellas, and you wanted to estimate the relationship between quantity demanded and price, should you use regression analysis or correlation analysis? If your decision is to use regression analysis, which variable, price or quantity, should be the independent variable?

13.42 As an experiment the manager of a supermarket has varied the price of the house brand of frozen orange juice concentrate each Saturday over 14 successive weeks. During this period the prices of the national brands were constant. The store never ran out of the house brand of orange juice, so the observed sales should be an estimate of average demand at each price. Use the following data to estimate the demand function for orange juice, with price as the explanatory variable.

Price per 12-ounce can (cents)	Quantity purchased	Price per 12-ounce can (cents)	Quantity purchased
50	138	51	135
55	120	52	134
60	111	65	115
62	116	49	140
55	124	45	137
48	136	56	122
45	140	59	116

13.43 Using the data in exercise 13.42, test at the 5% significance level the hypothesis that changes in price have no effect on the sales of frozen orange juice concentrate.

13.44 Using the data in exercise 13.42, construct a confidence interval at the 95% level of confidence for the population slope coefficient.

13.45 Using the data in exercise 13.42, construct a confidence interval at the 95% level of confidence for the sales of frozen orange juice concentrate over one day if the price was set at 55 cents.

13.46 Using the data in exercise 13.42, construct a confidence interval at the 95% level of confidence for the mean sales of frozen orange juice concentrate if the price was set at 50 cents.

13.47 Using the data in exercise 13.42, test at the 10% significance level the hypothesis that the population correlation coefficient equals 0.

13.48 A fruit wholesale company buys bananas in boxcar lots at the Port of New York. The manifest on each cargo lists the date the bananas left their home port. The shipper has kept records of the proportion of all bunches of bananas that have started to change from green to yellow in each shipment. Use the following data to estimate the effect of an additional day between the shipping and the delivery dates on the proportion of bunches of bananas in a boxcar which have started to change color.

Days between shipment and delivery	Proportion of bunches that have started to change from green to yellow
7	0.08
4	0.12
5	0.11
9	0.16
12	0.18
11	0.13
11	0.14
8	0.10
9	0.10
6	0.07
8	0.06
13	0.15
5	0.05
5	0.04

13.49 Using the data in exercise 13.48, test at the 5% significance level the hypothesis that changes in the number of days between the date of shipment and the date of delivery have no effect on the proportion of banana bunches that are starting to turn yellow.

13.50 Using the data in exercise 13.48, construct a confidence interval at the 95% level of confidence for the population slope coefficient.

13.51 Using the data in exercise 13.48, construct a confidence interval at the 95% level of confidence for the proportion of banana bunches that are starting to turn yellow in one boxcar of bananas which have been at sea ten days between shipment and delivery.

13.52 Using the data in exercise 13.48, construct a confidence interval at the 95% level of confidence for the average proportion of banana bunches that have started to turn yellow for all shipments that have been at sea eight days between departure and delivery.

13.53 Using the data in exercise 13.48, test at the 10% significance level the hypothesis that the population correlation coefficient equals 0.

13.54 An electric utility would like to estimate the relationship between the average temperature per day in the summer and the amount of electricity its customers consume. Use the following data to find the least–squares coefficients.

Average daily temperature (degrees Fahrenheit)	Millions of kilowatts of electricity consumed
77	10.0
84	12.1
85	13.1
90	14.2
92	15.6
91	14.1
81	9.7
88	10.7
79	8.1
86	11.5
78	8.4
93	9.9
105	16.3
95	12.7

13.55 Using the data in exercise 13.54, test at the 5% significance level the hypothesis that changes in the average daily temperature have no effect on the quantity of electricity consumed.

13.56 Using the data in exercise 13.54, construct a confidence interval at the 95% level of confidence for the population slope coefficient.

13.57 The electrical utility described in exercise 13.54 has a generating capacity of 17 million kilowatts. If more than that amount is needed, the utility has to buy additional electricity from other utilities at a high price. If the utility had a weather report that predicted an average temperature of 95 degrees for one day, what is the probability that the utility would have to purchase electricity from other utilities?

13.58 Using the data in exercise 13.54, construct a confidence interval at the 95% level of confidence for the electricity consumption for days with 90° average temperatures.

13.59 Using the data in exercise 13.54, test at the 10% significance level the hypothesis that the population correlation coefficient equals 0.

13.60* An auction company that sells farm property uses appraisers to estimate the selling prices of the properties that it auctions. While the company has observed a high correlation between the appraised values and the final auction prices, the variability around the appraised prices is greater when the appraised value is high than when it is low. Use the following data to estimate the effect of higher appraised values on the final auction price.

Appraised value	Final auction price	Appraised value	Final auction price
$ 20,000	$ 22,000	$120,000	$138,000
25,000	25,000	125,000	99,000
30,000	31,000	130,000	169,000
35,000	33,000	135,000	88,000
40,000	44,000	140,000	178,000
45,000	40,000	145,000	82,000
50,000	56,000	150,000	199,000
55,000	52,000	155,000	205,000
60,000	69,000	160,000	135,000
65,000	72,000	165,000	225,000
70,000	60,000	170,000	176,000
75,000	79,000	175,000	194,000
80,000	92,000	180,000	279,000
85,000	89,000	185,000	345,000
90,000	109,000	190,000	105,000
95,000	129,000	195,000	197,000
100,000	75,000	200,000	245,000
105,000	133,000	205,000	286,000
110,000	154,000	210,000	356,000
115,000	126,000	215,000	358,000

13.61* The auction house described in exercise 13.60 has a new property that has been appraised at $215,000. What is the probability that it will sell for more than $400,000?

14

Multiple Regression

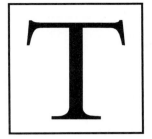

The meatpacking process begins with the slaughter, evisceration, and skinning of the animals. Fifteen years ago, the meatpacker would then ship the dressed carcasses to retailers. Retailers would age sides of beef (half carcasses) in their own coolers before selling cuts to consumers. Now meatpackers prepare "boxed beef"; they age the beef and then cut the carcasses up into convenient portions that are shipped in cardboard boxes. The retailer trims the individual cuts and wraps them for sale to consumers. The boxed beef revolution, led by Iowa Meatpackers Inc., has sharply reduced labor costs in beef processing. The work is now done on an assembly-line basis. Another advantage of boxed beef is that retailers can buy boxes according to their customers' preferences. When a traditional butcher in a poor neighborhood bought sides of beef, he was hard-pressed to find customers for the expensive cuts and he was often short of inexpensive portions suitable for ground beef. A traditional butcher in an affluent neighborhood would have the opposite problem. Now the tenderloins find their way to the affluent and the chucks migrate to the poor.

Aging beef used to be a simple matter for a traditional butcher. A new side of beef would go to the back of the locker and be rotated forward as the older sides were cut up. Hanging the beef improved its flavor and tenderness, but the water loss reduced the weight of the carcass. Packers of boxed beef make a very small profit on each of the thousands of cattle they slaughter daily. A method that would reduce the weight loss during the aging process could have a considerable impact on a meatpacker's profits, because the boxed beef is sold by weight. The weight loss during aging depends on two factors—the length of time that the carcass hangs and the temperature inside the cooler. By experimenting with different temperatures and hanging times, the meatpacker can observe how each variable affects weight loss.

Multiple regression is *an estimation technique that predicts the effect on the average level of the dependent variable of a unit change in any independent variable while the other independent variables are held constant.* With this technique the meatpacker can find the combination of time and temperature that will sufficiently age the beef and minimize the weight loss. You'll find that multiple regression is a logical extension of simple regression. Most of the topics—multiple regression as a descriptive technique; the overall fit of the multiple regression line; the assumptions needed for statistical inference in multiple regression; and hypothesis tests and confidence intervals in multiple regression—all have close counterparts in simple regression. These connections will be clear after the material on multiple regression has been cut up and boxed into convenient portions. At the end of the chapter we'll introduce some topics that were not discussed in Chapter 13: dummy variables and probit regression.

14.1 INTRODUCTION TO MULTIPLE REGRESSION

14.1A MULTIPLE REGRESSION AS A DESCRIPTIVE TECHNIQUE

A comparison of analysis of variance and multiple regression can help explain some of the uses and some of the advantages of multiple regression. If you have skipped Chapter 12, don't worry—it's not a prerequisite for the material in this chapter, and the comparison will still be helpful. The meatpacker's decision described at the beginning of the chapter could have been evaluated with two-way analysis of variance. The time and temperature would be the treatment and block variables for the two-way design. For this and similar problems, however, multiple regression has some important advantages. Time and temperature can be varied continuously. To find out the exact time and temperature that will sufficiently age the beef while minimizing weight loss, you need to know the functional relationship between changes in time and average weight loss and changes in temperature and average weight loss. Two-way analysis of variance cannot be used to estimate this functional relationship. All you can do with two-way analysis of variance is test to see if changes in time or temperature have an effect on average weight loss, and if they do, you can test to see if there is a statistically significant difference in weight loss between various tested temperature-and-time combinations. Analysis of variance is best suited to problems in which the treatments or blocks (explanatory variables) are categorical variables. In Chapter 12 we applied this technique to various tax brackets, different types of fertilizers, different training programs for waiters, and so on. While it is possible to break a continuous variable, such as time or temperature, into ranges, so that you could work the meatpacker's problem with analysis of variance, you would be throwing information away. Remember the golden rule of statistics—use all of the information available. Finally, another advantage of multiple regression over analysis of variance is that problems with more than two explanatory variables are much easier to handle with multiple regression.

Two-way analysis of variance can be used for formal experiments, in which each experimental unit is assigned to a treatment-block combination, or applied to nonexperimental data that are not randomly assigned to treatment-block combinations. The same flexibility holds in multiple regression. The meatpacker's problem is a formal experiment. The meatpacker can assign each experimental unit—a carcass—to some combination of time and temperature. Nonexperimental data are more common in business and economic applications of multiple regression. For example, an economist might want to estimate the effect of years of job experience and years of education on wage rates. The economist cannot assign years of education or experience to the workers in her sample. However, provided that some assumptions are satisfied, multiple regression will provide unbiased and efficient estimates of the effect of both education and experience on wage rates. We'll discuss these assumptions in section 14.1D.

Multiple regression gives you the ability to isolate or "pull out" the effect of changes in one variable in a nonexperimental context in which many explanatory variables can change their values from observation to observation. You can estimate the effect of an additional year of education under the assumption that years of job experience are held constant without ever actually holding years of job experience constant. You can also estimate the effect of an additional year of job experience under the assumption that years of education are held constant without ever actually holding years of education constant. The great power of multiple regression analysis is that it lets business people and social scientists obtain estimates as if they had run a controlled experiment in which each explanatory variable was allowed to vary in turn while the others remained fixed.

We will discuss the use of multiple regression as a descriptive technique before we discuss the inferential use of multiple regression. You would use multiple regression as a descriptive technique if you were interested only in measuring the relationship between the explanatory variables and the dependent variable in a particular data set. For the sake of reducing the computations, we'll work with only two explanatory variables and a small data set. When simple regression was used as a descriptive technique, you found the straight line that minimized the sum of the squared deviations between the observed values of the dependent variable and the line. In other words, you found the "best-fitting" line. In multiple regression with two independent variables, you have to find the plane that minimizes the sum of the squared deviations between the observed values of the dependent variable and the plane. In other words, you have to find the "best-fitting" plane.

Let Y be a random variable that is the dependent variable and x and z be explanatory variables that have no random variation. The explanatory variables can be set or controlled by the researcher in a formal experiment or by someone else when data are nonexperimental. Table 14.1 has a small population data set that will be used to show the steps in the calculations. Since one purpose of these data is to illustrate the calculations in multiple regression, it would be helpful to keep the data set as small as possible and still be able to complete the calculations. However, a second purpose is to give you a good idea of the data "cloud" in three dimensions and the plane that minimizes the sum of squared deviations between the data points and the plane. The 20 observations in Table 14.1 are a compromise between ease of calculation and a good picture.

Figure 14.1 shows the x, z, and y axes and the data cloud or scatter diagram for the data in Table 14.1. It is difficult to see a three-dimensional perspective drawn in two dimensions. What we are trying to find is the plane that cuts through the data cloud and minimizes the sum of the squared deviations between points in the cloud and the plane. The estimated regression plane is used

TABLE 14.1

Data to illustrate the calculations of the least-squares plane

x_i	z_i	y_i	x_i	z_i	y_i
−19	−30	−153.540	110	10	581.384
2	−70	−115.086	12	144	328.586
−37	−20	−224.901	13	18	89.985
14	−20	12.190	29	49	269.471
6	−38	−36.564	15	26	120.827
6	−10	15.169	16	30	151.344
60	−6	321.968	16	16	123.894
8	−2	22.050	16	38	152.603
9	2	58.806	19	42	196.902
10	69	208.011	20	46	188.234

to describe the average change in the dependent variable for a unit change in x and for a unit change in z. Figure 14.2 shows the original data cloud and the best-fitting plane. Here we see four of the y_i's and the corresponding $\hat{y}_i$'s. The blue boxes are the data points and the distances from the data points to the regression plane are shown by vertical lines. For the two points above the plane the vertical lines are solid and for the two points below the regression plane the vertical lines are broken. The points at which the vertical lines intersect the regression plane, the $\hat{y}_i$'s, are shown in color. The term that has to be

FIGURE 14.1

Scatter diagram plotted in x-y-z space

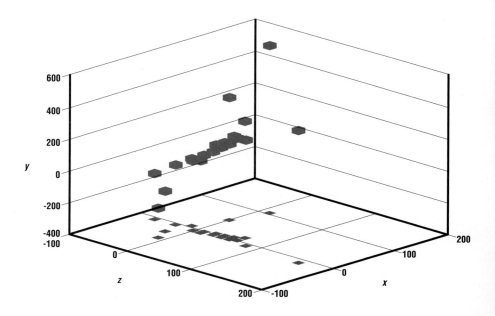

FIGURE 14.2
**Estimated regression plane
for the data set in Table 14.1**

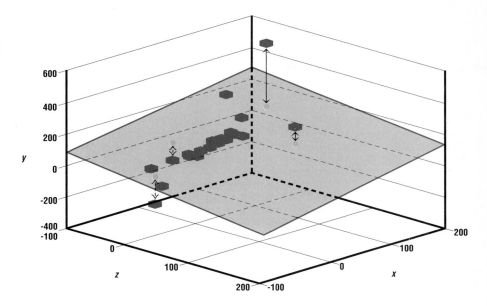

minimized to find the best-fitting plane is exactly the same as in simple regression. The best-fitting plane minimizes the squared deviations between the plane and each data point; that is, it minimizes the sum of squared errors:

$$SSE = \sum_{i=1}^{\nu} (y_i - \hat{y}_i)^2 \tag{13.1}$$

In section 14.1 we'll see the formulas and the calculations for minimizing this term.

■ 14.1B CALCULATIONS FOR LEAST-SQUARES REGRESSION PLANE

The formulas for the intercept, the slope coefficient for x, and the slope coefficient for z are derived in Appendix C. The calculations shown here are too cumbersome and too prone to error to be done by hand or even with most hand-held calculators. Multiple regression belongs to computers. The purpose of running through one set of calculations is to show you what the computer does and to give you a better idea of the logic behind least-squares estimation. With the aid of calculus, Appendix C shows how the following "normal" equations will minimize the sum of squared errors. These formulas are for population data.

$$\sum_{i=1}^{\nu} y_i = \nu\alpha + \beta_1 \sum_{i=1}^{\nu} x_i + \beta_2 \sum_{i=1}^{\nu} z_i$$

(14.1)

$$\sum_{i=1}^{\nu} x_i y_i = \alpha \sum_{i=1}^{\nu} x_i + \beta_1 \sum_{i=1}^{\nu} x_i^2 + \beta_2 \sum_{i=1}^{\nu} x_i z_i$$

(14.2)

$$\sum_{i=1}^{\nu} z_i y_i = \alpha \sum_{i=1}^{\nu} z_i + \beta_1 \sum_{i=1}^{\nu} x_i z_i + \beta_2 \sum_{i=1}^{\nu} z_i^2$$

(14.3)

The normal equations for sample data are

$$\sum_{i=1}^{n} y_i = an + b_1 \sum_{i=1}^{n} x_i + b_2 \sum_{i=1}^{n} z_i$$

(14.4)

$$\sum_{i=1}^{n} x_i y_i = a \sum_{i=1}^{n} x_i + b_1 \sum_{i=1}^{n} x_i^2 + b_2 \sum_{i=1}^{n} x_i z_i$$

(14.5)

$$\sum_{i=1}^{n} z_i y_i = a \sum_{i=1}^{n} z_i + b_1 \sum_{i=1}^{n} x_i z_i + b_2 \sum_{i=1}^{n} z_i^2$$

(14.6)

The α and a are, respectively, the population intercept and the sample intercept. The intercept is where the regression plane cuts the y axis. The β_1 and b_1 are, respectively, the population slope and sample slope for the x variable. The interpretation of this slope coefficient is the change in the average value of Y for each unit change in x while z is held constant. The β_2 and b_2 are, respectively, the population slope and sample slope for the z variable. Which explanatory variable is called x and which is called z is arbitrary. The interpretation of the second slope coefficient parallels the interpretation of the first slope coefficient; it is the change in the average value of Y for each unit change in z while x is held constant. The ν and the n are the number of elements in the population and the sample. The normal equations are derived directly from the condition of minimizing the sum of squared errors, but if you substituted the various summation terms into them, you would still have a lot of work to get the intercept and the two slopes. There are three equations and three unknowns, so you would have to solve for the three unknowns. The normal equations are rearranged below so that the unknown terms are isolated. The formulas shown are for a population, but changing the formulas to accommodate sample data is easy. Just substitute n for ν and the sample means of x, y, and z for the respective population means.

$$\alpha = \mu_Y - \beta_1\mu_x - \beta_2\mu_z \tag{14.1a}$$

$$\beta_1 = \frac{\sum_{i=1}^{\nu}(z_i - \mu_z)^2 \sum_{i=1}^{\nu}(x_i - \mu_x)(y_i - \mu_Y) - \sum_{i=1}^{\nu}(x_i - \mu_x)(z_i - \mu_z)\sum_{i=1}^{\nu}(z_i - \mu_z)(y_i - \mu_Y)}{\sum_{i=1}^{\nu}(x_i - \mu_x)^2 \sum_{i=1}^{\nu}(z_i - \mu_z)^2 - \left[\sum_{i=1}^{\nu}(x_i - \mu_x)(z_i - \mu_z)\right]^2} \tag{14.2a}$$

$$\beta_2 = \frac{\sum_{i=1}^{\nu}(x_i - \mu_x)^2 \sum_{i=1}^{\nu}(z_i - \mu_z)(y_i - \mu_Y) - \sum_{i=1}^{\nu}(x_i - \mu_x)(z_i - \mu_z)\sum_{i=1}^{\nu}(x_i - \mu_x)(y_i - \mu_Y)}{\sum_{i=1}^{\nu}(x_i - \mu_x)^2 \sum_{i=1}^{\nu}(z_i - \mu_z)^2 - \left[\sum_{i=1}^{\nu}(x_i - \mu_x)(z_i - \mu_z)\right]^2} \tag{14.3a}$$

Table 14.2 contains the various sums needed to find the intercept and the two slope coefficients. Test your understanding by trying to duplicate a few rows in the table.

Substituting the sums from Table 14.2 into formulas 14.1a, 14.2a, and 14.3a yields the intercept and two slope coefficients:

$$\beta_1 = \frac{\sum_{i=1}^{\nu}(z_i - \mu_z)^2 \sum_{i=1}^{\nu}(x_i - \mu_x)(y_i - \mu_Y) - \sum_{i=1}^{\nu}(x_i - \mu_x)(z_i - \mu_z)\sum_{i=1}^{\nu}(z_i - \mu_z)(y_i - \mu_Y)}{\sum_{i=1}^{\nu}(x_i - \mu_x)^2 \sum_{i=1}^{\nu}(z_i - \mu_z)^2 - \left[\sum_{i=1}^{\nu}(x_i - \mu_x)(z_i - \mu_z)\right]^2} \tag{14.2a}$$

$$\beta_1 = \frac{39,344.2(88,638.837) - 4,091.5(97,017.769)}{15,573.75(39,344.2) - [4,091.5]^2}$$

$$\beta_1 = 5.185$$

TABLE 14.2
Sums needed to calculate the multiple regression coefficients for the population data in Table 14.1

x_i	z_i	y_i	$(x_i - \mu_x)^2$	$(z_i - \mu_z)^2$	$(x_i - \mu_x)(y_i - \mu_Y)$	$(z_i - \mu_z)(y_i - \mu_Y)$	$(x_i - \mu_x)(z_i - \mu_z)$
−19	−30	−153.540	1,242.563	1,998.09	9,486.016	12,029.075	1,575.675
2	−70	−115.086	203.063	7,174.09	3,286.794	19,536.244	1,206.975
−37	−20	−224.901	2,835.563	1,204.09	18,129.915	11,814.236	1,847.775
14	−20	12.190	5.063	1,204.09	232.598	3,587.183	78.075
6	−38	−36.564	105.063	2,777.29	1,559.340	8.017.291	540.175
6	−10	15.169	105.063	610.09	1,029.074	2,479.817	253.175
60	−6	321.968	1,914.063	428.49	9,030.039	−4,272.499	−905.625
8	−2	22.050	68.063	278.89	771.511	1,561.726	137.775
9	2	58.806	52.563	161.29	411.512	720.856	92.075
10	69	208.011	39.063	2,948.49	−577.778	5,019.739	−339.375
110	10	581.384	8,789.063	22.09	43,670.416	−2,189.344	−440.625
12	144	328.586	18.063	16,718.49	−905.332	27,543.403	−549.525
13	18	89.985	10.563	10.89	83.139	−84.418	−10.725
29	49	269.471	162.563	1,176.49	1,962.277	5,278.910	437.325
15	26	120.827	1.563	127.69	−6.576	59.444	−14.125
16	30	151.344	0.063	234.09	−8.944	547.394	−3.825
16	16	123.894	0.063	1.69	−2.082	10.825	−0.325
16	38	152.603	0.063	542.89	−9.259	862.946	−5.825
19	42	196.902	7.563	745.29	223.671	2,220.443	75.075
20	46	188.234	14.063	979.69	272.504	2,274.498	117.375
325	**294**	**2,311.333**	**15,573.750**	**39,344.20**	**88,638.837**	**97,017.769**	**4,091.500**

$$\beta_2 = \frac{\sum_{i=1}^{\nu}(x_i - \mu_x)^2 \sum_{i=1}^{\nu}(z_i - \mu_z)(y_i - \mu_Y) - \sum_{i=1}^{\nu}(x_i - \mu_x)(z_i - \mu_z)\sum_{i=1}^{\nu}(x_i - \mu_x)(y_i - \mu_Y)}{\sum_{i=1}^{\nu}(x_i - \mu_x)^2 \sum_{i=1}^{\nu}(z_i - \mu_z)^2 - \left[\sum_{i=1}^{\nu}(x_i - \mu_x)(z_i - \mu_z)\right]^2}$$ (14.3a)

$$\beta_2 = \frac{15,573.75(97,017.769) - 4,091.5(88,638.837)}{15,573.75(39,344.2) - [4,091.5]^2}$$

$$\beta_2 = 1.927$$

$$\alpha = \mu_Y - \beta_1\mu_x - \beta_2\mu_z$$ (14.1a)

$$\alpha = 115.567 - 5.185(16.250) - 1.927(14.700)$$

$$\alpha = 2.983$$

The equation that describes the population regression plane is

$$\hat{y} = \alpha + \beta_1 x + \beta_2 z$$

(14.7)

For the data set in Tables 14.1 and 14.2, the regression plane is

$$\hat{y} = 2.983 + 5.185x + 1.927z$$

The value 2.983 is where the plane cuts the y axis. With z held constant, a unit increase in x is associated with an average increase of 5.185 in Y. With x held constant, a unit increase in z is associated with an average increase of 1.927 in Y.

This example of the calculations for the coefficients in multiple regression has two independent variables. The procedure for problems with more than two independent variables is similar. The least-squares solution for the coefficients of the regression equation always minimizes the sum of the squared deviations between either all Y values and the population mean of Y with population data or the observed Y values and the estimated mean of Y with sample data. However, the algebraic notation for the formulas with more than two independent variables is unmanageable. The formulas for the regression coefficients when there are more than two independent variables are usually expressed in matrix notation. The calculations for more than two independent variables are extremely difficult to do by hand. At this point you probably don't need any further convincing that a computer should be used for multiple regression, so we won't discuss the formulas for doing calculations with more than two independent variables.

14.1C THE COEFFICIENT OF MULTIPLE DETERMINATION AND THE ADJUSTED COEFFICIENT OF MULTIPLE DETERMINATION

The coefficients that we have calculated give you no sense of how accurately the regression plane fits the y values. In simple regression, the statistic that gave a measure of the accuracy of the regression line was the coefficient of determination. The **coefficient of multiple determination** is the multiple-regression counterpart to the coefficient of determination used in simple regression: it expresses *the percentage of the total variation in the dependent variable accounted for by the independent variables*. The symbols for the coefficient of multiple determination are ρ^2 for population data and R^2 for sample data (in simple regression the symbol for the sample coefficient of determination was a lowercase r^2). In multiple regression with two independent variables, if all of the data points were on the regression plane, there would be a perfect fit. The regression plane would exactly describe the data. If there were a perfect fit between the data and the regression plane, the ρ^2 or R^2 would equal 1. The range of possible values for the coefficient of multiple determination is from 0 to 1. With a coefficient of multiple determination of 0, the independent variables would account for none of the variation in the dependent variable.

The formulas for the population and sample coefficients of multiple determination are the same as the corresponding formulas in Chapter 13. The concept behind the formulas is also the same: it is still the explained variation divided by the total variation. The two formulas from Chapter 13 are repeated below but with an uppercase R for the sample coefficient of multiple determination.

$$\rho^2 = \frac{\sum\limits_{i=1}^{\nu} (\hat{y}_i - \mu_y)^2}{\sum\limits_{i=1}^{\nu} (y_i - \mu_y)^2} \tag{13.9}$$

$$R^2 = \frac{\sum\limits_{i=1}^{n} (\hat{y}_i - \bar{y})^2}{\sum\limits_{i=1}^{n} (y_i - \bar{y})^2} \tag{14.8}$$

Since the symbol changed from r^2 to R^2, the formula for the sample coefficient of *multiple* determination has a different number than the formula for the sample coefficient of determination in Chapter 13. The formula for the population coefficient of *multiple* determination is not renumbered because it has the same symbol as the population coefficient of determination. The numerators in both formula 13.9 and formula 14.8 are the sum of squares for the regression and the denominators are the sum of squares total.

Let's continue with the example data set to see how closely the regression plane fits the y values. Table 14.3 gives the y and $\hat{y}$ values and the sums of squared deviations needed to calculate the coefficient of multiple determination. We obtain the $\hat{y}$'s by substituting the x and z values for the given y_i into the formula for the regression plane. The first $\hat{y}$ in the table, for example, equals

$$\hat{y} = \alpha + \beta_1 x + \beta_2 z \tag{14.7}$$

$$\hat{y} = 2.983 + 5.185(-19) + 1.927(-30)$$

$$\hat{y} = -153.342$$

TABLE 14.3

Sums needed to calculate the coefficient of multiple determination for the data set in Table 14.1

y_i	$\hat{y}_i$	$(y_i - \hat{y}_i)^2$	$(y_i - \mu_Y)^2$	$(\hat{y}_i - \mu_Y)^2$
−153.540	−153.339	0.041	72,418.488	72,310.170
−115.086	−121.511	41.283	53,200.451	56,205.708
−224.901	−227.410	6.292	115,918.386	117,632.740
12.190	37.045	617.813	10,686.812	6,165.579
−36.564	−39.117	6.517	23,143.769	23,927.045
15.169	14.829	0.116	10,079.647	10,148.156
321.968	302.546	377.181	42,601.330	34,961.424
22.050	40.612	344.559	8,745.339	5,618.136
58.806	53.504	28.111	3,221.736	3,851.731
208.011	187.774	409.546	8,545.993	5,213.893
581.384	592.642	126.737	216,986.196	227,601.054
328.586	342.642	197.570	45,377.246	51,563.205
89.985	105.072	227.608	654.404	110.138
269.471	247.764	471.190	23,686.468	17,476.094
120.827	130.856	100.574	27.673	233.759
151.344	143.748	57.704	1,280.022	794.174
123.894	116.775	50.676	69.338	1.460
152.603	159.161	43.005	1,371.690	1,900.448
196.902	182.423	209.614	6,615.365	4,469.835
188.234	195.315	50.142	5,280.588	6,359.862
2,311.333	2,311.333	3,366.279	649,910.941	646,544.609

The slight discrepancy between the tabled value of $\hat{y}$ and the above answer is due to rounding.

The coefficient of multiple determination is

$$\rho^2 = \frac{\sum\limits_{i=1}^{\nu} (\hat{y}_i - \mu_y)^2}{\sum\limits_{i=1}^{\nu} (y_i - \mu_y)^2} \qquad (13.9)$$

$$\rho^2 = \frac{646,544.609}{649,910.941}$$

$$\rho^2 = 0.995$$

In this data set the x and z variables account for 99.5% of the variation in the Y variable around its mean.

The coefficient of multiple determination can overstate the fit or the explanatory power of the estimated regression equation when the number of independent variables is a large fraction of the number of observations. To see why such overstatement can occur, consider the example of estimating a student's GPA in Chapter 13. The independent variable used there was hours of study. In a multiple regression context, more than one explanatory variable could be used. For a given set of students, the denominator of the coefficient of multiple determination, SST, is fixed. If you added independent variables to the regression, the numerator of the coefficient of determination would either stay the same (if the correlation between the added independent variable and GPA were 0) or it would increase (if the correlation between the added independent variable and GPA were not 0). The addition of independent variables never reduces the numerator of the coefficient of multiple determination. Recall that there were thousands of measurable characteristics for each student (for example, the distance between each student's elbow and the tip of the thumb, the range of notes the student could sing, the number of hamburgers consumed in the last week). If you keep adding even implausible independent variables to the regression, some of them will have a chance correlation with GPA and their addition will increase the R^2, because the addition of these variables increases the numerator of the coefficient of multiple determination without changing the denominator. If you have enough variables, you can get a high R^2.

An adjustment to the coefficient of multiple determination eliminates the possible false impression of a high explanatory power when the number of independent variables is a large fraction of the number of observations. The **adjusted coefficient of multiple determination** is *the fraction of the total variation in the dependent variable which is accounted for by the independent variables adjusted for the number of independent variables*. Suppose the data set with 75 students in Chapter 13 was used for a multiple regression with 20 independent variables and the R^2 was 0.90. The SST was 18.1796. If the R^2 was 0.9, the SSR must be 90% of the SST, or 16.36164. We'll use these figures to illustrate the difference between the coefficient of multiple determination and the adjusted coefficient of multiple determination. The formula for the adjusted coefficient of multiple determination is

$$\bar{R}^2 = \frac{\sum\limits_{i=1}^{n} (\hat{y}_i - \bar{y})^2 / (n - k - 1)}{\sum\limits_{i=1}^{n} (y_i - \bar{y})^2 / (n - 1)} - \frac{k}{n - k - 1}$$

(14.9)

The symbol for the adjusted coefficient of multiple determination is $\overline{R}^2$. The adjusted coefficient of multiple determination is generally used for sample data, so we won't bother with an adjusted population coefficient of multiple determination. The letter k refers to the number of independent variables in the multiple regression—20 in the GPA example. Gaining intuitive understanding of why this formula works is not easy. Crudely put, the adjustment accounts for the degrees of freedom in estimating the overall Y variance $(n - 1)$ and the degrees of freedom for estimating the variance of the Y distributions for any given values of the independent variables $(n - k - 1)$. When we substitute into formula 14.9, the adjusted coefficient of multiple determination equals

$$\overline{R}^2 = \frac{16.36164/(75 - 20 - 1)}{18.1796/(75 - 1)} - \frac{20}{75 - 20 - 1}$$

$$\overline{R}^2 = 0.863$$

The adjusted coefficient of multiple determination is a little below the original R^2 of 0.90. To see the effect of increasing the number of independent variables on the adjusted coefficient of multiple determination, examine Table 14.4, which shows a series of $\overline{R}^2$'s. The SST and SSR and the sample size are held at the same values as in the above calculation, but the number of independent variables runs from 5 to 70 in increments of 5. Here you see what would happen to the adjusted coefficient of determination if independent variables that were uncorrelated with the dependent variable were added to the regression equation. The $\overline{R}^2$ clearly falls. In contrast, the unadjusted coefficient of determination would stay at 0.90. In practice, most potential independent variables will have a nonzero correlation with the dependent variable. The adjusted coefficient of multiple determination will tell you how much explanatory power (or, put more modestly, how great an association) there is between a potential independent variable and the dependent variable. If a potential explanatory variable is added to the regression equation and the adjusted coefficient of determination falls, the added variable provides no help in describing the variation in the dependent variable.

TABLE 14.4	k	$\overline{R}^2$	k	$\overline{R}^2$
Effect of increasing the number of independent variables on the coefficient of multiple determination	5	0.893	40	0.782
	10	0.884	45	0.745
	15	0.875	50	0.692
	20	0.863	55	0.611
	25	0.849	60	0.471
	30	0.832	65	0.178
	35	0.810		

You may infer from this discussion that if you can find a set of explanatory variables that has a high *adjusted* coefficient of multiple determination, you have found *the* causal variables in the data set—the set of independent variables that really determine the behavior of the dependent variable. The point about not proving causality cannot be overemphasized. Even with a high adjusted coefficient of multiple determination, you do not know whether independent variables cause the dependent variable. Unless you are working with a controlled experiment, causality could flow in the opposite direction; it could be due to unobserved variables that are correlated with both the dependent and the independent variables; or, the observed association could be due to chance. If you had enough potential explanatory variables or even "explanatory" variables that were created by a random number program in a computer, you could hunt for a set of explanatory variables that would give you a high adjusted coefficient of multiple determination.

If the number of independent variables is a small fraction of the number of observations, the adjusted coefficient of multiple determination will be about the same as the unadjusted coefficient of multiple determination. Most computer programs give you both measures of the fit of a regression equation.

Here are the main points that we have discussed so far: (1) With either population or sample data, the coefficients of the regression plane—or, with more than two independent variables, the regression equation—describes the average change in the dependent variable for a unit change in an independent variable with the other independent variables held constant. (2) For either sample or population data, the coefficient of multiple determination measures the percentage of the variation in the dependent variable accounted for by the independent variables. This measure tells you how good the fit is between the data points and the regression equation. (3) The adjusted coefficient of multiple determination is a more accurate measure of fit when there are more than a few independent variables.

Multiple regression is used primarily for statistical inference. Statistical inference with multiple regression is the use of sample data to estimate the slope and intercept coefficients for a population and to construct confidence intervals and hypothesis tests for those coefficients and for the regression equation as a whole. Before we can begin to discuss statistical inference, we have to know the assumptions behind the multiple linear regression model.

■ 14.1D ASSUMPTIONS REQUIRED FOR STATISTICAL INFERENCE

Most of the assumptions required for statistical inference in multiple regression carry over from simple regression. Recall that in simple regression the population mean of Y had to be a linear function of x. This was the assumption of linearity. In a two-variable multiple regression the mean of Y has to be a linear function of z while x is held constant and a linear function of x while z is held constant.

FIGURE 14.3
Plane showing the mean of
Y intersecting a plane in
which *x* is held constant at
x*: the intersection of the
two planes is the mean of *Y*
as a function of *z*

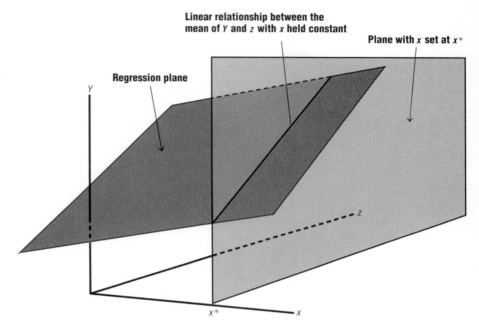

Figure 14.3 illustrates the assumption of linearity. The thick line where the two planes intersect shows the change in the average value of *Y* for a change in *z* while *x* is held constant. One requirement for the population data in multiple regression is that the relationship between the average value of *Y* and the value of *z* while *x* is held constant be linear. The relationship between the average value of *Y* and the value of *x* while *z* is held constant must also be linear. If these assumptions hold, the coefficients of the least-squares estimates of the regression plane will be unbiased estimators of the population slope coefficients. With two independent variables, if the mean values of *Y* were not described by a plane, the linearity assumption would not be satisfied. Figure 14.4 shows what happens.

The two issues with nonlinearity are how to recognize when it is present and what to do after you decide that it is present. With simple regression, the recommended method for detecting nonlinearity was to look at the scatter diagram. The equivalent approach in multiple regression is to look at the data cloud from all angles and see if the data points appear to follow a plane or a curved surface. Until 1986 this method of checking for nonlinearity could not be recommended because it required the construction of a physical model of the data cloud. Occasionally researchers would go to the trouble and expense of making a three-dimensional model of a data cloud with small styrofoam balls and wood or wire rods. In 1986 a program called MacSpin was released

FIGURE 14.4

Surface for the mean of *Y* showing a nonlinear relationship between the mean of *Y* and *z* while *x* is held constant

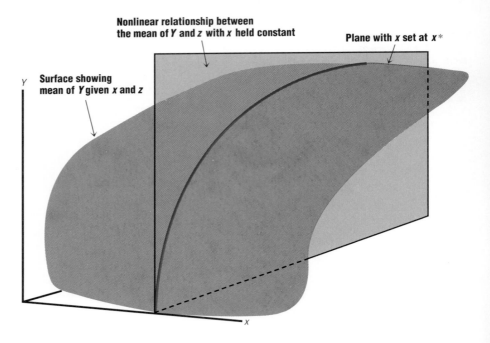

which lets you view a data cloud easily and rotate it so that it can be seen from any angle. The program can plot more than two independent variables, but we'll stick to two. At this time MacSpin works only on a Macintosh computer, but it or some competitors will undoubtedly be adapted to other computers. Figure 14.5 shows a data cloud viewed from several angles. The nonlinearity becomes apparent when the cloud is viewed from the correct angle. To help visualize what the program does, think of a data cloud as a slice cut out of an apple. In the first view the slice is held so that it appears as wide as possible. Going from views 1 through 4, you are slowly rotating the apple slice in your hand until you can see the slice in profile at its narrowest view. In the first view the nonlinearity in the cloud cannot be seen; here it appears that the data cloud fits the assumption required for statistical inference. In the fourth view, however, the nonlinearity of the data cloud is clear.

   If you do not have access to a program that can draw and rotate the data cloud, some other techniques can detect nonlinearity. Unfortunately, these techniques are neither as visual nor as intuitive as rotating the data cloud. One technique is to plot the **residuals** $(y_i - \hat{y}_i)$ of the multiple regression equation— *the differences between the observed and predicted values of the dependent variable*— first as a function of x and second as a function of z. If the residuals or prediction errors show any pattern—if the residuals rise as x rises, for example, or if they rise and then fall as z rises—nonlinearity may be the cause. Another

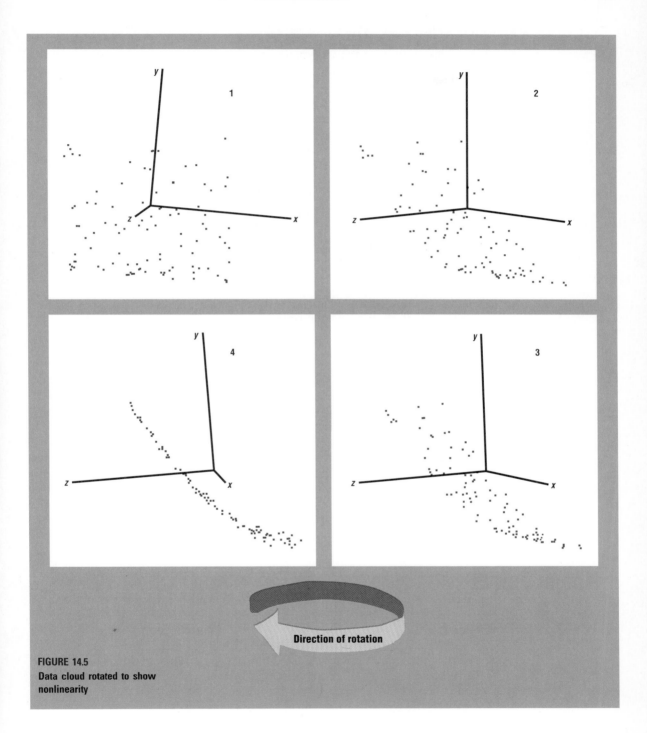

FIGURE 14.5
Data cloud rotated to show
nonlinearity

technique for detecting nonlinearity is a **partial scatter diagram**—*a scatter diagram of the dependent variable plotted against one independent variable while the effects of all other independent variables are held constant.* To make a partial scatter diagram you plot the values of Y as a function of x net of the effect of z and then plot the values of Y as a function of z net of the effect of x. This technique is described in an econometrics text by Henri Theil, cited among the suggested readings at the end of the chapter.

After you have determined that there is a nonlinear relationship between the mean of Y and the values of x and z, you have to decide how to proceed to estimate the effect of changes in x and z on the mean of Y. An ordinary multiple regression is not suitable, as it cannot track a nonlinear relationship. The potential solution is either to transform the data or to change the estimation technique. It is often possible to create a linear relationship in the data with a simple transformation, such as taking the natural log of each variable. Logarithmic transformations were discussed in Chapter 13. We'll discuss a few more examples of simple transformation later in this chapter. If the surface of the mean of Y appears to be so convoluted or irregular that a simple transformation will not make it linear, such estimation techniques as nonlinear least squares will allow you to estimate the mean of Y for any x and z. These techniques are usually introduced in econometrics or advanced statistics courses.

The next assumption required for statistical inference is homoscedasticity. What this means in a multiple regression context is that the distribution of Y values has to have the same variance for all combinations of x and z. Homoscedasticity implies that if x is held constant to some x^*, the distribution of Y values will have the same variance for all values of z. Also, when z is held constant to some z^*, the distribution of Y values will have the same variance for all values of x. The graphs to illustrate homoscedasticity and heteroscedasticity in multiple regression are about the same as those used to show the same concepts in simple regression. The dark shaded area represents the distribution of the Y values about their mean. Figure 14.6 shows homoscedasticity and Figure 14.7 shows heteroscedasticity. The figures show only x held constant. Figures showing z held constant would look the same except that the labels for x and z would be switched.

To detect heteroscedasticity you could use a program such as MacSpin, which would allow you to view cross sections of the data cloud while holding x constant, essentially the view of the shaded areas in Figures 14.6 and 14.7. Another graphical technique is to plot the residuals against each of the independent variables separately. If the residuals appear to narrow or spread out as the value of an independent variable increases, the cause is usually heteroscedasticity. The Goldfeld–Quant test for heteroscedasticity is described in econometrics texts (see Suggested Readings). There are some "fixes" for heteroscedasticity, but they are beyond the scope of this book.

There are several reasons for the assumption of homoscedasticity for statisti-

FIGURE 14.6

Homoscedasticity in multiple regression: the shaded area in the $x = x*$ plane is the distribution of the Y values about the mean of Y

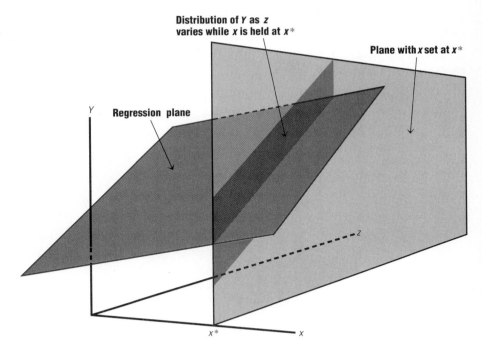

cal inference in multiple regression. As in simple regression, the standard error of the estimate is used for a variety of confidence intervals and hypothesis tests. If there is heteroscedasticity, the standard error of the estimate will not be a consistent estimator of the population standard deviation of Y for a given x and z. The tests that rely on the standard error of the estimate will not be accurate. Second, if the population is heteroscedastic and the sample size is small, the estimated slope coefficients can be thrown off by a few extreme observations taken from x–z combinations at which Y has a high variance.

The next assumption required for statistical inference is normality for the Y distributions at any combination of x–z values. Short of a hologram, a figure to illustrate the normality assumption would be impossible to draw because Y, x, and z take up three axes, and a fourth axis is needed to plot the density of Y. This normality assumption is not important if you have a large sample.

So far the assumptions required for statistical inference in multiple regression have been extensions of the assumptions in simple regression. The next assumption does not have a counterpart in simple regression because it refers to the relationship between the independent variables. The assumption is that the independent variables are uncorrelated. With two independent variables, the assumption is that x and z are uncorrelated. With more than two independent variables, the assumption is that all pairs of independent variables are uncorrelated and that all possible combinations of independent variables are

FIGURE 14.7

Heteroscedasticity in multiple regression: the shaded area in the $x = x^*$ plane is the distribution of the Y values about the mean of Y

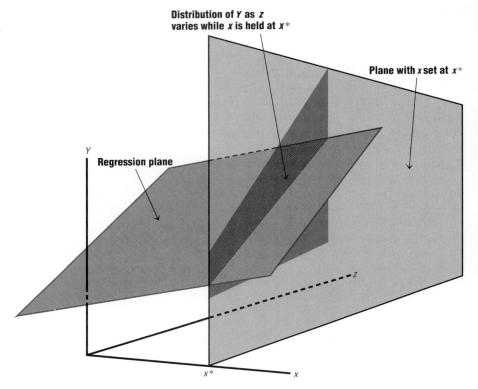

Distribution of Y as z
varies while x is held at $x*$

Plane with x set at $x*$

Y

Regression plane

uncorrelated with other independent variables. For example, if the independent variables were x and z and w, the sum of x and z could not be correlated with w. *Any correlation among independent variables or among sets of independent variables* is called **multicollinearity.** If two of the independent variables were perfectly correlated, there would be no statistical way to distinguish their separate effects. If two variables always moved together, it would not matter which one you put in the regression. If you tried to put both variables in the regression, you would discover that the normal equations for the slope coefficients could not be solved because perfect multicollinearity causes a division by zero in the formula.

Before looking at a figure that illustrates the mathematical problem caused by perfect multicollinearity, let's consider a fairly trivial example that illustrates the concept. Suppose you were a professional basketball coach and you had a theory that the players who wore the biggest shoes were the most effective players (if you can recall the shoe-size examples in Chapters 2 and 6, you may be thinking that statistical applications of shoe sizes are endless). When it came to drafting players, you paid close attention to their shoe sizes. Most other coaches thought that a player's height was the key explanatory variable. To test your theory, you collect data on some measure of player effectiveness and

FIGURE 14.8
Data cloud with perfect
multicollinearity

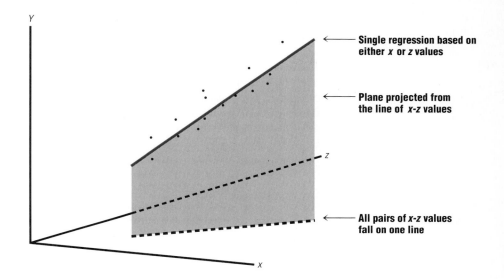

Single regression based on
either *x* or *z* values

Plane projected from
the line of *x-z* values

All pairs of *x-z* values
fall on one line

shoe sizes and heights. Of course, the tallest players tend to have the biggest feet. The correlation may not be perfect, but it is very close to perfect. If the correlation were perfect, it would not matter if you drafted players on the basis of height or shoe size.

The mathematical difficulty caused by perfect multicollinearity is that there is no one plane that minimizes the sum of squared errors. Figure 14.8 illustrates this problem. It shows a plane projected upward from a line of *x-z* values. Since all of the observations of *x* and *z* fall along this line, all of the observations of *Y* must be in this plane. Inside this plane it is possible to use simple regression to see how much the mean of *Y* changes for every unit change in either *x* or *z*, but it is impossible to say what would happen to *Y* if either *x* or *z* moved off the one line. Recall that multiple regression tries to find the one plane that minimizes the sum of squared errors. Any plane that contained the simple regression line would minimize SSE. The normal equations for slope coefficients for *x* and *z* cannot be solved because the number of planes that contain the regression line is infinite.

Multicollinearity is rarely perfect, but the problem of having potential independent variables that are highly correlated is common. In this case you can find estimates of the slope coefficients, but these estimates will be unreliable. The problem with these estimates is not bias but rather wide standard errors. One solution is to have a large sample; how large depends on the extent of the multicollinearity. Techniques for detecting multicollinearity and for obtaining more efficient estimates are usually introduced in econometrics courses. The detection is fairly straightforward if there are only two independent variables; just check for a high coefficient of correlation between *x* and *z*. We'll skip the details of testing for multicollinearity with more than two independent variables and the estimation techniques, such as ridge regression, which can compensate for multicollinearity.

The last of the assumptions needed for statistical inference is that the observations be independent. For example, when observations are independent, if one observation of the dependent variable has a high value, that information will tell you nothing about the probability that the next observation will have a high or low value. This assumption is commonly violated in time-series data. **Time-series data** are *observations that are made over time*. Time-series data often have some pattern or momentum, such as one high value followed by more high values. If this assumption is violated, the data are said to have **serial correlation;** that is, they show *a nonzero correlation among residuals*. Since this problem is almost exclusively an issue in time-series data, we'll defer the discussion of how to detect serial correlation and what to do about it until Chapter 15, on the analysis of time-series data.

After reviewing the assumptions required for statistical inference, you may be wondering if any real data set ever met all of these requirements. It seems like a long and exacting list. Data based on formal experiments usually satisfy all of the assumptions. Consider the example of estimating the effects of the amount of chrome and the amount of manganese on the hardness of steel alloy. The researcher has alloys of steel made with various amounts of chrome and manganese and then measures their hardness on the Rockwell scale. By selecting the levels of the independent variables, the researcher can make sure that the independent variables are not correlated. Each observation is independent, so there is no serial correlation. The distribution of hardness for any combination of chrome and manganese would be expected to be approximately normal because many small or unobserved factors can affect hardness. Finally, the data set should be homogeneous because these unobserved factors or the stochastic variation should be the same across all combinations of chrome and manganese. The multiple regression model was originally applied to experimental data that fit these assumptions.

Nonexperimental data are much less likely to satisfy the assumptions. Since isolating the effect of each independent variable in nonexperimental data is so valuable, however, a great deal of effort has been devoted to finding estimation techniques that can compensate for violations of these assumptions. These techniques are the main topics covered in econometrics courses. **Econometrics** can be defined as the study of estimation techniques for data that violate the assumptions of the multiple regression model. The usual definition is that it is *the application of statistical techniques to economic and business problems*. The two definitions differ very little, because most of econometrics is centered on regression analysis, and it seems that all data sets for economic and business problems violate some of the assumptions of the multiple regression model.

Don't feel that you have to take an econometrics course to work real problems with nonexperimental data. Statistical inference in multiple regression is fairly robust to departures from these assumptions. The main purpose of reviewing these assumptions is to help you understand the requirements for statistical inference and to be aware of its limitations. The blind use of any statistical technique can be very dangerous.

EXERCISES FOR SECTION 14.1

14.1 For the data in the following table, calculate the intercept, the x and z slope coefficients, and the coefficient of multiple determination.

x_i	z_i	y_i	$(x_i - \bar{x})^2$	$(z_i - \bar{z})^2$	$(x_i - \bar{x})(y_i - \bar{y})$	$(z_i - \bar{z})(y_i - \bar{y})$	$(x_i - \bar{x})(z_i - \bar{z})$
1	10.00	15.178	30.25	10.656	145.815	87.248	17.954
2	10.50	21.480	20.25	7.641	90.944	56.464	12.439
3	11.03	23.109	12.25	5.014	65.034	42.093	7.838
4	11.58	28.499	6.25	2.849	32.977	22.632	4.220
5	12.16	32.350	2.25	1.230	14.001	10.600	1.664
6	12.76	39.553	0.25	0.252	1.069	1.180	0.251
7	13.40	45.674	0.25	0.019	1.992	0.515	0.068
8	14.07	49.655	2.25	0.651	11.947	6.251	1.210
9	14.77	51.012	6.25	2.281	23.305	13.752	3.776
10	15.51	58.889	12.25	5.058	60.197	38.194	7.872
11	16.29	61.440	20.25	9.149	88.875	59.083	13.611
12	17.10	66.386	30.25	14.739	135.825	93.978	21.115
78	159.17	493.225	143.00	59.445	671.995	431.636	91.945

14.2 For the data in the following table, calculate the intercept, the x and z slope coefficients, and the coefficient of multiple determination.

x_i	z_i	y_i	$(x_i - \bar{x})^2$	$(z_i - \bar{z})^2$	$(x_i - \bar{x})(y_i - \bar{y})$	$(z_i - \bar{z})(y_i - \bar{y})$	$(x_i - \bar{x})(z_i - \bar{z})$
1.00	10.00	7.47	0.61	144,058.20	38.06	18,522.04	296.05
1.10	16.00	18.20	0.46	139,539.60	25.89	14,221.05	254.01
1.21	25.60	26.45	0.32	132,459.60	17.00	10,852.99	207.45
1.33	40.96	10.77	0.20	121,514.99	20.48	15,860.85	156.87
1.46	65.54	23.49	0.10	104,982.48	10.49	10,621.05	103.68
1.61	104.86	23.52	0.03	81,048.40	5.57	9,323.60	48.40
1.77	167.77	38.55	0.00	49,186.37	0.18	3,929.94	2.22
1.95	268.44	39.35	0.03	14,667.63	−2.88	2,049.18	−20.59
2.14	429.50	57.53	0.13	1,596.00	0.45	50.34	14.38
2.36	687.19	90.25	0.34	88,589.57	19.71	10,113.81	172.63
2.59	1,099.51	137.90	0.66	504,043.20	66.12	57,954.03	575.07
2.85	1,759.22	201.77	1.14	1,875,995.91	155.69	199,286.99	1,465.55
21.37	4,674.59	675.25	4.03	3,257,681.95	356.75	352,785.85	3,275.72

14.3 Determine the intercept, the slope coefficients, and the coefficient of multiple determination for the following data:

x_i	z_i	y_i
4	12	9
6	8	13
6	7	14
7	5	16
8	6	19
11	4	25

14.4 Determine the intercept, the slope coefficients, and the coefficient of multiple determination for the following data:

x_i	z_i	y_i
1	20	4
3	18	14
4	17	14
9	16	20
10	16	24
16	9	44

14.5 **(a)** Plot the residuals of the regression equation for the data in exercise 14.1 as a function of x and as a function of z, and decide if there is evidence of either nonlinearity or heteroscedasticity.

(b) Calculate the correlation coefficient of the independent variables in exercise 14.1 and decide if there is multicollinearity in the data set.

(c) Plot the residuals of the regression equation for the data in exercise 14.2 as a function of x and as a function of z, and decide if there is evidence of either nonlinearity or heteroscedasticity.

(d) Calculate the correlation coefficient for the independent variables in exercise 14.2 and decide if there is multicollinearity in the data set.

14.6 Calculate two simple regressions for the data in exercise 14.1, one with Y as the dependent variable and x as the independent variable and the other with Y as the dependent variable and z as the independent variable. Are the slope coefficients on x and z the same in the simple regression as in the multiple regression? If you were interested only in estimating the effect of x on Y, which slope coefficient for x should you use? Explain.

14.7 Calculate two simple regressions for the data in exercise 14.2, one with Y as the dependent variable and x as the independent variable and the other with Y as the dependent variable and z as the independent variable. Are the slope coefficients on x and z the same in the simple regression as in the multiple regression? If you were interested only in estimating the effect of z on Y, which slope coefficient for z should you use? Explain.

14.2 CONFIDENCE INTERVALS AND HYPOTHESIS TESTS IN MULTIPLE REGRESSION

 14.2A TESTING THE HYPOTHESIS THAT THE INDEPENDENT VARIABLES HAVE NO EFFECT ON THE DEPENDENT VARIABLE

In simple regression the test of the hypothesis that the one independent variable has no effect on the dependent variable was the hypothesis test that the slope coefficient equaled zero. In multiple regression the equivalent test is that all of the population slope coefficients are zero. You need a joint test rather than a series of tests against each of the slope coefficients in turn. In notation the null and alternative hypotheses are

H_0: $\beta_1 = \beta_2 = \beta_3 = \cdots = \beta_k = 0$

H_1: Not all slope coefficients equal zero

So far we have used the letters x and z for the independent variables. A notation that uses a letter for each independent variable is satisfactory for a small number of variables, but it is impractical if there are many independent variables. A more general notation for the multiple regression model with any number of independent variables is

$$\hat{y} = \alpha + \beta_1 x_1 + \beta_2 x_2 + \beta_3 x_3 + \cdots + \beta_k x_k$$

The following test statistic is used to test this hypothesis:

(14.10)

$$F = \frac{\sum\limits_{i=1}^{n} (\hat{y}_i - \overline{y})^2 / k}{\sum\limits_{i=1}^{n} (y_i - \hat{y}_i)^2 / (n - k - 1)}$$

If the null hypothesis that the population slopes all equal zero is true, the test statistic will follow an F distribution with k degrees of freedom for the numerator and $n - k - 1$ degrees of freedom for the denominator. The F distribution was introduced in Chapter 12. If you did not cover that chapter, you should at least read the section on the F distribution. The numerator of the F test statistic is the sum of squares for the regression divided by the number of independent variables in the regression, k. The denominator of the test statistic is the sum of squared errors divided by the number of observations minus k minus 1. If the null hypothesis is true, the numerator is an unbiased estimator of the common variance of the Y distributions, σ_ε^2, for all possible combinations of independent variables. If the null hypothesis is not true, its expected value will exceed σ_ε^2. The denominator of the F test statistic is an unbiased estimator of the σ_ε^2 whether or not the null hypothesis is true.

We'll use a new example to illustrate formula 14.10. You work for an advertising agency that has set up a national advertising campaign to market a new brand of dog food, Wonder Woofs. The advertising campaign consists of radio spots in the 100 top media markets (cities) over a one-month period. The number of times the ads are presented, the duration of each ad in seconds, and the rating points of the shows that carry the ads are under the control of your advertising agency. To estimate the effect of each of these variables on the future per capita sales of Wonder Woof, you pick different combinations of these variables for each of the 100 cities. The coefficient of multiple determination for the regression of per capita sales on the three variables is 0.125. The SSR is 25 and the SSE is 175. At the 5% significance level, test the hypothesis that none of the independent variables has an effect on per capita sales. Substituting into formula 14.10 yields

$$F = \frac{\sum\limits_{i=1}^{n} (\hat{y}_i - \bar{y})^2 / k}{\sum\limits_{i=1}^{n} (y_i - \hat{y}_i)^2 / (n - k - 1)} \qquad (14.10)$$

$$F = \frac{25/3}{175/(100 - 3 - 1)}$$

$$F = 4.57$$

The critical value for an F distribution with 3 and 96 degrees of freedom with a 5% significance level is somewhere between 2.76 for 60 degrees of freedom in the denominator and 2.68 for 120 degrees of freedom in the denominator. There is no reason to interpolate the correct critical value because the F test statistic of 4.57 is greater than both of these values. The null that all of the slope coefficients are equal to 0 is rejected at the 5% significance level. A coefficient of multiple determination of only 0.125 suggests that the model does not include all of the important variables that affect per capita dog-food sales. If the number of dogs per person varied across cities, this variable would be a good candidate for inclusion in the regression equation. Perhaps per capita dog-food sales is also sensitive to income. Even though the advertising agency is interested only in the variables it can control, the inclusion of the other likely variables should improve the estimate of the effect of the variables of interest in the study. A final point is that a low R^2 does not mean that the independent variables do not have a statistically significant effect on the dependent variable.

Because of the hypothesis test, you are confident that one or more of these variables—the number of times the ads are presented, the duration of each ad, and the rating points of the shows that carry the ads—has a statistically significant effect. What the low R^2 tells you is that these variables can account for only a small fraction of the variation in per capita sales, and that it is likely that some important variables are missing from the regression equation.

14.2B CONFIDENCE INTERVALS AND HYPOTHESIS TESTS ON THE POPULATION SLOPE COEFFICIENTS

After rejecting the null hypothesis that all of the population slope coefficients are equal to zero, you have some reassurance that the independent variables you have chosen have some explanatory power. In most problems you want to know which of the independent variables have nonzero population slopes; that is, which of the independent variables have an effect on the dependent variable. The formula for a hypothesis test on the jth independent variable's population slope is

$$\text{C.V.}_L, \text{C.V.}_U = (\beta_j)_0 \pm t_{\alpha/2} s_{b_j}$$

(14.11)

where the degrees of freedom equal $n - k - 1$. n is the number of observations and k is the number of independent variables. To do a one-tailed test, use $+$ or $-$ in place of $\pm$ and α in place of $\alpha/2$.

The term $(\beta_j)_0$ is the null hypothesis value for the population slope of the jth independent variable. Generally the null would be that the population slope is zero. Occasionally a test against some other value is needed. The last term in the formula is the standard error of the sample slope of the jth independent variable. The formula for this term is very complicated if there are more than two independent variables. For the sake of illustrating the concepts, we'll discuss the formulas for the standard errors of the slope with two independent variables and assume that you'll have access to a computer to get these terms when there are more than two independent variables. Let x and z be the two independent variables. The formulas for the standard errors of the sample slopes of x and z both use *the estimated standard deviation of the distributions of* Y *values about the mean of* Y *for any combinations of values of the independent variables*—the **standard error of the estimate.** In the simple regression, the standard error of the estimate was the square root of the fraction SSE over $n - 2$. Recall that the reason for the division by $n - 2$ was that one degree of freedom was lost in estimating the intercept and one degree of freedom was lost in estimating the population slope coefficient for x. In a multiple regression involving two independent variables, the standard error of the estimate is the square root of the fraction SSE over $n - 3$. Here three degrees of freedom are lost—one for the intercept, one for the sample slope of x, and one for the sample slope of z. The formula for the standard error of the estimate with two independent variables is

$$s_e = \sqrt{\frac{\sum_{i=1}^{n} (y_i - \hat{y}_i)^2}{n - 3}}$$

(14.12)

Here are the formulas for the standard errors of the slope coefficients:

$$s_{b_x} = \sqrt{\frac{s_e^2 \sum_{i=1}^{n} (z_i - \overline{z})^2}{\sum_{i=1}^{n} (x_i - \overline{x})^2 \sum_{i=1}^{n} (z_i - \overline{z})^2 - \left[\sum_{i=1}^{n} (x_i - \overline{x})(z_i - \overline{z})\right]^2}}$$

(14.13)

$$s_{b_z} = \sqrt{\frac{s_e^2 \sum_{i=1}^{n} (x_i - \overline{x})^2}{\sum_{i=1}^{n} (x_i - \overline{x})^2 \sum_{i=1}^{n} (z_i - \overline{z})^2 - \left[\sum_{i=1}^{n} (x_i - \overline{x})(z_i - \overline{z})\right]^2}}$$

(14.14)

Formulas 14.13 and 14.14 can help you understand why the sample slopes are unreliable estimators of the population slopes when the degree of multicollinearity is high. As the correlation of x and z increases, the denominators of formulas 14.13 and 14.14 approach zero. When the value of the denominator is close to zero, the standard error of the sample slope gets very large, and it approaches infinity as the denominator approaches zero. When the standard error of the slope is a mile wide, you have no idea if the sample slope is anywhere near the population slope.

To illustrate the application of these formulas, we'll use a new example and work through every detail. You are a researcher for a Wall Street financial services firm and you have been assigned the task of deciding how the price of a stock and its quarterly dividend affect the future return on the stockholders' investment. You randomly select 100 stocks traded on various exchanges

TABLE 14.5
Sums needed for regression analysis where Y is return on investment in percent, x is beginning stock price in dollars, and z is beginning quarterly dividend in dollars for 100 randomly selected stocks

x_i	z_i	y_i	$(x_i - \bar{x})^2$	$(z_i - \bar{z})^2$	$(x_i - \bar{x})(y_i - \bar{y})$	$(z_i - \bar{z})(y_i - \bar{y})$	$(x_i - \bar{x})(z_i - \bar{z})$	$\hat{y}_i$	$(y_i - \hat{y}_i)^2$	$(y_i - \bar{y})^2$	$(\hat{y}_i - \bar{y})^2$
0.50	0.10	0.121	2,211.351	13.724	239.816	18.892	174.209	2.814	7.250	26.005	5.793
1.45	0.10	3.919	2,122.906	13.717	59.956	4.819	170.643	2.860	1.121	1.693	5.568
2.40	0.10	2.236	2,036.266	13.709	134.710	11.053	167.079	2.908	0.452	8.911	5.348
⋮	⋮	⋮	⋮	⋮	⋮	⋮	⋮	⋮	⋮	⋮	⋮
92.65	3.24	8.875	2,036.266	0.319	164.902	−2.063	−25.478	7.448	2.037	13.356	4.961
93.60	3.16	9.533	2,122.906	0.416	198.697	−2.780	−29.700	7.493	4.161	18.600	5.166
94.55	3.08	7.668	2,211.351	0.525	115.097	−1.774	−34.074	7.539	0.017	5.991	5.375
4,752.50	380.46	522.052	75,200.81	1,582.048	3,856.372	343.869	6,294.612	522.052	207.555	405.735	198.180

which vary in price from under \$1 to over \$100 and vary in quarterly dividends from 0 to \$5. You track these stocks for a year and record the end-of-year prices, the dividends, and all stock splits. With these data you determine the return on investment (ROI) for each stock. Since 100 observations make the table for the sums needed for regression analysis too long, we'll list only the first three and last three entries in each column of Table 14.5. The dots in the table stand for the missing entries. We'll assume that the data conform to the assumptions required for statistical inference in multiple regression.

The first step in the regression analysis is to get the estimated slope and intercepts. Substituting into formulas 14.4a, 14.5a, and 14.6a yields

$$b_1 = \frac{\sum_{i=1}^{n}(z_i - \bar{z})^2 \sum_{i=1}^{n}(x_i - \bar{x})(y_i - \bar{y}) - \sum_{i=1}^{n}(x_i - \bar{x})(z_i - \bar{z}) \sum_{i=1}^{n}(z_i - \bar{z})(y_i - \bar{y})}{\sum_{i=1}^{n}(x_i - \bar{x})^2 \sum_{i=1}^{n}(z_i - \bar{z})^2 - \left[\sum_{i=1}^{n}(x_i - \bar{x})(z_i - \bar{z})\right]^2} \qquad \text{(14.4a)}$$

$$b_1 = \frac{1,582.048(3,856.372) - 6,294.612(343.869)}{75,200.81(1,582.048) - (6,294.612)^2}$$

$$b_1 = 0.0496$$

$$b_2 = \frac{\sum_{i=1}^{n}(x_i - \bar{x})^2 \sum_{i=1}^{n}(z_i - \bar{z})(y_i - \bar{y}) - \sum_{i=1}^{n}(x_i - \bar{x})(z_i - \bar{z}) \sum_{i=1}^{n}(x_i - \bar{x})(y_i - \bar{y})}{\sum_{i=1}^{n}(x_i - \bar{x})^2 \sum_{i=1}^{n}(z_i - \bar{z})^2 - \left[\sum_{i=1}^{n}(x_i - \bar{x})(z_i - \bar{z})\right]^2} \qquad \text{(14.5a)}$$

$$b_2 = \frac{75,200.81(343.869) - 6,294.612(3,856.372)}{75,200.81(1,582.048) - (6,294.612)^2}$$

$$b_2 = 0.020$$

$$a = \bar{y} - b_1\bar{x} - b_2\bar{z} \tag{14.6a}$$

$$a = 5.22 - 0.0496(47.525) - 0.02(3.805)$$

$$a = 2.787$$

We obtained the means of x, z, and Y by dividing the column totals for x, z, and Y by 100. Recall that there were 100 stocks in the sample. The interpretation of these coefficients is: (1) for every \$1 increase in the price of a stock at the beginning of the year, *with the quarterly dividend held constant,* the average ROI over the course of a year is estimated to increase 0.0496 percentage points; (2) for every \$1 increase in the last quarter's dividend before the beginning of the year, *with the beginning price held constant,* the average ROI is estimated to increase 0.02 percentage point; (3) the regression plane intercepts the y axis at 2.787 percentage points. The intercept should not be interpreted as the expected rate of return of a stock with a zero price and a zero dividend. There are no stocks selling at zero, so this interpretation is not valid.

In this admittedly small sample, the point estimates of population slope coefficients suggest that higher-priced stocks and stocks that have higher dividends will have a higher ROI. The next step is to calculate the coefficient of multiple determination to see how well the two independent variables describe or fit the pattern of ROI in the data. Substituting into formula 14.8 yields

$$R^2 = \frac{\sum_{i=1}^{n} (\hat{y}_i - \bar{y})^2}{\sum_{i=1}^{n} (y_i - \bar{y})^2} \tag{14.8}$$

$$R^2 = \frac{198.180}{405.735}$$

$$R^2 = 0.488$$

The interpretation of the coefficient of multiple determination is that about 49% of the variation in ROI is accounted for by the variable's initial price and quarterly dividend. This coefficient of multiple determination suggests that some other variables that affect ROI may be missing from the regression equation. Actually, an R^2 of 0.488 is fairly respectable for nonexperimental data. Higher coefficients of multiple determination are usual for experimental data because the experimental design weeds out extraneous sources of variation. Higher coefficients of multiple determination are also common with time-series data, because many economic time series move together with the general ebb and flow of the economy. One last point on coefficients of multiple determination—the adjusted coefficient of multiple determination will be about the same as the unadjusted coefficient because the number of independent variables (2) is a small fraction of the number of observations (100). The adjusted coefficient of multiple determination is 0.478.

The third step is to test the hypothesis that the independent variables jointly have no effect on the dependent variable. We'll use a 5% significance level for this test. Substituting into formula 14.10 yields

$$F = \frac{\sum\limits_{i=1}^{n} (\hat{y}_i - \bar{y})^2/k}{\sum\limits_{i=1}^{n} (y_i - \hat{y}_i)^2/(n - k - 1)} \tag{14.10}$$

$$F = \frac{198.180/2}{207.555/(100 - 2 - 1)}$$

$$F = 46.309$$

The degrees of freedom for the F test statistic are 2 and 97. The critical value for an F distribution with 2 and 97 degrees of freedom and a 5% significance level is somewhere between 3.07 and 3.15. The null hypothesis that the independent variables jointly have no effect on the dependent variable is rejected. With this high F test statistic, the null would be rejected at any plausible significance level.

The last step is to test the null hypothesis that either of the population slope coefficients is zero. Again we'll use a 5% significance level. To substitute into the formula for the critical values for the hypothesis test, we'll need the standard error of the estimate and then the standard error of each slope coefficient. The standard error of the estimate is

$$s_e = \sqrt{\frac{\sum\limits_{i=1}^{n} (y_i - \hat{y}_i)^2}{n - 3}} \tag{14.12}$$

$$s_e = \sqrt{\frac{207.555}{100 - 3}}$$

$$s_e = 1.463$$

The standard errors of the slope coefficients are

$$s_{b_x} = \sqrt{\dfrac{s_e^2 \sum\limits_{i=1}^{n} (z_i - \bar{z})^2}{\sum\limits_{i=1}^{n} (x_i - \bar{x})^2 \sum\limits_{i=1}^{n} (z_i - \bar{z})^2 - \left[\sum\limits_{i=1}^{n} (x_i - \bar{x})(z_i - \bar{z})\right]^2}} \tag{14.13}$$

$$s_{b_x} = \sqrt{\dfrac{(1.463)^2(1{,}582.048)}{75{,}200.81(1{,}582.048) - (6{,}294.612)^2}}$$

$$s_{b_x} = 0.00653$$

$$s_{b_z} = \sqrt{\dfrac{s_e^2 \sum\limits_{i=1}^{n} (x_i - \bar{x})^2}{\sum\limits_{i=1}^{n} (x_i - \bar{x})^2 \sum\limits_{i=1}^{n} (z_i - \bar{z})^2 - \left[\sum\limits_{i=1}^{n} (x_i - \bar{x})(z_i - \bar{z})\right]^2}} \tag{14.14}$$

$$s_{b_z} = \sqrt{\dfrac{(1.463)^2(75{,}200.81)}{75{,}200.81(1{,}582.048) - (6{,}294.612)^2}}$$

$$s_{b_z} = 0.0450$$

We're just about done. To find the critical values for the two hypothesis tests, substitute into formula 14.11. Recall that the degrees of freedom are $n - k - 1$.

$$\text{C.V.}_L, \text{C.V.}_U = (\beta_j)_0 \pm t_{\alpha/2}s_{b_j} \tag{14.11}$$

test for β_1

$$\text{C.V.}_L, \text{C.V.}_U = 0 \pm 1.96(0.00653)$$

$$\text{C.V.}_L = -0.0128 \qquad \text{C.V.}_U = 0.0128$$

test for β_2

$$\text{C.V.}_L, \text{C.V.}_U = 0 \pm 1.96(0.045)$$

$$\text{C.V.}_L = -0.088 \qquad \text{C.V.}_U = 0.088$$

The critical values for the tests on the population slopes are compared with the sample slopes. The sample slope for the price of the stock, x, of 0.0496 is outside the acceptance region of -0.0128 to 0.0128. The null that the initial price of the stock does not affect its ROI is rejected at the 5% significance level. The sample slope for the quarterly dividend, z, of 0.020 is inside the acceptance region of -0.088 to 0.088. The null that the quarterly dividend has no effect on

DO TEACHER EVALUATIONS MAKE A DIFFERENCE?

Most students are familiar with end-of-semester evaluation forms. You may have wondered whether your instructors read them and whether they helped improve teaching. John A. Centra did an experiment with teacher evaluations at five colleges that had not used teacher evaluations before the experiment ("Effectiveness of Student Feedback in Modifying College Instruction," *Journal of Education Psychology* 65, no. 3 [1973]: 395–401). Centra's hypothesis was that instructors who thought they were good teachers and discovered from the teacher evaluations that their students disagreed would be motivated to improve their teaching. To test this hypothesis, he had all instructors fill out an evaluation form on themselves and had one of each instructor's classes also fill out an evaluation form. The results of the student evaluations were given to the instructors. At the end of the next semester, a second evaluation form was distributed to students in one of each instructor's classes.

Regression analysis was used to estimate the following equation:

$$R_2 = a + bR_1 + c(I - R_2)$$

where R_2 is the estimated rating on an item at the end of the second semester of the experiment, R_1 is the student rating on an item at the middle of the first semester, and I is the instructor's self-rating at the beginning of the first semester. The letters a, b, and c are the estimated intercept and slope coefficients. This equation was estimated for each of 17 questions. It was also run separately for instructors who rated themselves less favorably than their students and for instructors who rated themselves more favorably than their students. Thus the equation was estimated 34 times. If the hypothesis (that instructors who are unpleasantly surprised by their students' ratings will improve their teaching) is correct, the slope coefficient, c, should be positive and statistically significant for all instructors for whom I was greater than R_2. Centra did not hypothesize that instructors who were pleasantly surprised by their students' ratings would slack off and get lower ratings in the second semester.

The slope coefficient, c, had the expected positive sign in 13 of the 17 equations for instructors who had rated themselves more highly than their students. In 5 of these 13 equations the null hypothesis that the population slope coefficient was less than or equal to zero could be rejected at the 5% significance level. "Thus the major hypothesis of this study—that student feedback would effect changes in teachers who had rated themselves more favorably than their students had rated them—was generally supported by the regression analysis."

Apparently teacher evaluations can make a difference, provided the instructor is unduly optimistic about how good a teacher he or she is. Students usually do not know if their instructors rate themselves more or less favorably than their students do. You never know, the evaluations just may do some good.

the ROI cannot be rejected. If you were using the information gleaned from this sample to guide your stock-buying strategy for the financial services firm, you would recommend buying stocks with high per share prices and ignore the most recent quarterly dividends.

You have only done a hypothesis test on the population slopes. Constructing a confidence interval for the slope of either independent variable takes only slight modification of the procedure for a hypothesis test. Just substitute that sample slope for the null hypothesis value of the population slope in formula 14.11 and change the critical values to lower and upper limits for the confidence interval.

EXERCISES FOR SECTION 14.2

14.8 Use the data in exercise 14.1 to test at the 5% significance level the hypothesis that the independent variables have no effect on the dependent variable.

14.9 Use the data in exercise 14.2 to test at the 5% significance level the hypothesis that the independent variables have no effect on the dependent variable.

14.10 Use the data in exercise 14.3 to test at the 5% significance level the hypothesis that the population slope of the x variable is 0.

14.11 Use the data in exercise 14.3 to test at the 5% significance level the hypothesis that the population slope of the z variable is 0.

14.12 You run a construction company that digs the holes and hauls away the dirt for large buildings. The number of days needed for each job depends on the number of truckloads of dirt to be removed and the miles between the construction site and the place where the dug-up dirt is dumped. You pay your crew by the day and you need to estimate the number of days required for each hole so that you can cost out these jobs. With the data for 16 construction jobs listed in the table on the next page, you have the following results for the estimated regression equation: intercept, 0.852; slope on the truckloads variable, 0.0085; slope on the miles variable, 0.408; coefficient of multiple determination, 0.846; SSE, 31.376; SST, 204; standard error of the slope of the truckload variable, 0.0012; and standard error of the slope of the miles variable, 0.115.

(a) At the 5% significance level, test the hypothesis that the variables miles and truckloads have no effect on days per job.

(b) Construct confidence intervals at the 95% confidence level for the slope coefficients on the miles and truckloads variables.

(c) Estimate the number of days for a job with 300 truckloads and 14 miles. If the population slopes were at the high end of the confidence intervals, how many days would be needed for a job with 300 truckloads and 14 miles? If the population slopes were at the low end of the confidence intervals, how many days would be needed for a job with 300 truckloads and 14 miles?

Truckloads	Miles	Days	Truckloads	Miles	Days
150	2	3	120	12	6
350	5	8	200	3	6
100	10	3	400	8	9
500	1	4	1,300	9	16
250	2	5	70	2	1
300	6	7	120	4	4
1,000	1	8	190	3	3
60	1	1	190	3	4

14.3 DUMMY VARIABLES, INTERACTION TERMS, AND TRANSFORMATIONS TO ESTABLISH LINEARITY (OPTIONAL)

14.3A DUMMY VARIABLES

Earlier in the chapter I said that multiple regression analysis was a powerful technique for analyzing data because it could extract the effect that one independent variable has on a dependent variable *as if* all other independent variables were held constant, without the need actually to hold the other independent variables constant. While this is the main reason for the popularity of multiple regression, a second reason is that it can easily be modified to account for various kinds of data. We'll discuss three such modifications, to account for nonlinearities in the data, categorical instead of interval or ratio data in some or all of the independent variables, and independent variables that interact.

Dummy variables are *indicator variables for categorical or nominal data which take the value of 1 or 0, depending on whether an observation does or does not belong in a specified category.* The examples we have discussed so far all used data in levels (interval or ratio data) for the independent variables. Many problems come with mixed data: some characteristics of each observation are measured with data in levels and some characteristics are described with data in categories. The use of dummy variables to accommodate these mixed data is best explained with an example.

You work for an insurance company that sells automobile insurance. You need timely and accurate estimates of the prices of used cars in order to set insurance premiums and to settle claims for cars that have been "totaled." Guidebooks are published on a quarterly basis, listing the average auction prices of used cars by mileage, region of the country, manufacturer, car model, and year. The guidebooks also provide adjustments to the selling price for such major options as automatic transmissions and air conditioning. The problem with the guidebooks is that the prices of used cars are volatile; they can move up or down before the next guidebook is printed. If interest rates fall and the economy is improving, for example, many people will buy new cars, and the used-car market will be flooded with trade-ins. Such conditions depress used-car prices. On the other hand, if the economy went into a recession and interest rates rose sharply, the prices of used cars would jump.

If you had more current data on the auction prices of used cars, say weekly data, you could use a multiple regression equation to estimate the price of any used car, even if the weekly auction data did not include all makes, models, and mileage levels of cars. The data for the regression equation would be mixed—mileage is ratio data, and whether a car has air conditioning or not is an example of categorical data. Let's start with only the independent variables mileage and air conditioning status and then expand the regression equation to account for the other factors. An observation of a car with 20,000 miles and no air conditioning would be recorded as "20,000 0." An observation of a car with 30,000 miles and air conditioning would be recorded as "30,000 1." The 0 stands for no air conditioning and the 1 for the presence of air conditioning. We'll assume that the data fit the requirements for statistical inference in multiple regression. Actually, the linearity assumption is not valid, because the decline in prices of used cars as a function of the mileage is greatest for the beginning miles and tapers off at high mileages. We'll discuss transformations for linearity later in the chapter.

All of the calculation procedures and statistical tests are the same with dummy variables and data in levels. The main issue for dummy variables is the interpretation of their coefficients. Suppose you had observations of the mileages and air conditioning status of 1,000 used cars sold at auction in the last week and you estimated the following regression equation:

$$\hat{y} = 11{,}285 - 0.14x + 856z$$

The x variable is the mileage and the z variable is the air conditioning status. The interpretation of the coefficients is: (1) the average auction price of a used car goes down 14 cents for each additional mile, *when air conditioning status is held constant;* (2) the average auction price goes up \$858 for a car that has air conditioning over the price of a car that has no air conditioning, *when the mileage is held constant;* (3) a car that has 0 miles and no air conditioning would sell at a value of intercept of \$11,285. While it is possible to have a car with 0 miles sold at an auction, if there are no such cars in the data set, this interpretation of the intercept should be treated cautiously, because it extrapolates away from the known values of the independent variables. Another limitation of the interpretation of the coefficients is that you cannot go too far in the opposite direction of high mileage. Suppose the car with the highest mileage sold in the auction had 60,000 miles on the odometer. If you plugged the values for a car with 100,000 miles and no air conditioning in the estimated regression equation, the predicted selling price would be \$-2,715. Obviously, no cars sell for negative dollars. The difficulty is that the regression equation cannot be used to

FIGURE 14.9

Two-dimensional graph of used-car data

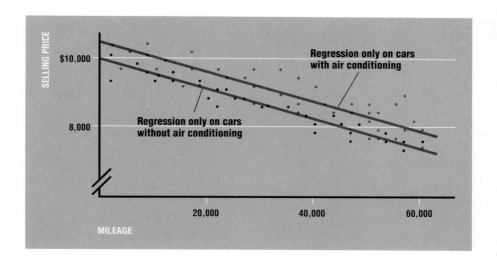

extrapolate estimates far from the values of the observations in the data set.

Figure 14.9 shows the data for the used-car prices with mileage on the x axis and selling price on the y axis. The data points indicated by blue dots represent cars with air conditioning; the ones indicated by black dots represent cars without air conditioning. If two simple regressions were run for the data on used-car sales, one for air conditioned cars and the other for cars without air conditioning, the difference between the intercepts of the two simple regressions would be the effect of air conditioning on the selling price. In this example the difference between the two intercepts is $856. If the effect of mileage on selling price is the same for both types of cars in the population, the two simple regression estimates of the population slope will be about the same; the two simple regression lines will be parallel or at least close to parallel.

A multiple regression with a dummy variable gives a more precise estimate of the population slope in these circumstances than two simple regressions. The standard error of the slope on the mileage variable will be smaller in the multiple regression than in either simple regression because there are more observations for the multiple regression. However, if the slopes for the mileage variables are not the same for both types of cars, the multiple regression will give misleading results. Figure 14.10 shows different estimated slopes for the separate regressions. If, after plotting the scatter diagrams and running two simple regressions, you decide that the data look more like Figure 14.10 than Figure 14.9, you should rely on two simple regressions to estimate the selling prices of used cars.

Suppose the coefficient of multiple determination for the multiple regression equation was 0.22. The interpretation of this R^2 is that the variables mileage and air conditioning status account for 22% of the variation in auction prices.

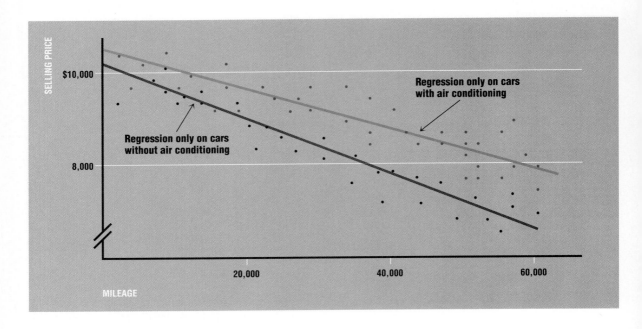

SELLING PRICE

$10,000

Regression only on cars
with air conditioning

Regression only on cars
without air conditioning

8,000

20,000   40,000   60,000

MILEAGE

FIGURE 14.10

**Two-dimensional graph of
used-car data with different
sample slopes for the two
simple regressions**

The relatively low R^2 suggests that some other independent variables that
affect auction prices are missing from the regression equation. Let's expand the
regression equation to take account of the manufacturer of the car. The manufacturers recorded in the data set are American Motors, Chrysler, Ford, General Motors, Honda, Nissan, Toyota, and Volkswagen. Each car has one and
only one manufacturer. The dummy variables in this context are slightly different from the dummy variable for the air conditioning. In the air conditioning example, one dummy variable was coded 1 if an air conditioner was present and 0 otherwise. The interpretation of the slope coefficient of this dummy
variable was how much more a car with air conditioning would sell for than a
car without air conditioning. Suppose you created a dummy variable for
American Motors cars coded as 0 for a car that was not manufactured by
American Motors and as 1 for a car that was manufactured by American
Motors. The interpretation of the slope coefficient would *not* be how much
more or less an American Motors car sold for at auction relative to a car with
no manufacturer—all of the cars have some manufacturer. In estimating the
regression equation, you treat one of the manufacturers as the base. A dummy
variable is included in the regression for every manufacturer except the base
manufacturer. Usually the group with the largest number of observations is
used as the base. In this data set the General Motors cars are the base. The
interpretation of the slope coefficient for the dummy variables for any other
manufacturer are relative to the base of the omitted manufacturer. Note that
when a variable can be broken into more than two categories, such as manu-

facturer, the dummy variable representing one of the categories must be omitted from the regression equation. If one category is not omitted, there will be perfect multicollinearity and the equation cannot be estimated.

Consider the following estimated regression equation:

$$\hat{y} = 12,941 - 0.14x + 870z - 620\text{AMC} - 220\text{CHRYSLER} + 710\text{FORD}$$
$$+ 1,100\text{HONDA} + 920\text{NISSAN} + 1,400\text{TOYOTA} - 200\text{VOLKSWAGEN}$$

AMC, CHRYSLER, FORD, and the rest stand for dummy variables for the respective manufacturers. Note that there is no dummy variable for General Motors in this regression equation. The x and z are the mileage and the air conditioning status of each car. The interpretation of the estimated slope coefficient on the dummy variable AMC is that a car manufactured by American Motors will sell on average for $620 less than a car manufactured by General Motors, *when the mileage and the air conditioning status are held constant*. The interpretations of the coefficients of the other dummy variables are similar.

Here's an interesting subsidiary issue. Suppose you ran a hypothesis test on each of the slope coefficients and the null of zero population slope coefficient could not be rejected for the AMC dummy variable. If you had to predict the price of an AMC car at auction, should you throw out the dummy variable for AMC cars because it was not statistically significant? In effect, throwing out the AMC dummy variable would be pretending that all AMC cars were made by General Motors. Even though the difference between the point estimate of $-620 and 0 is not statistically significant (perhaps because there are too few AMC cars in the data), it is your best estimate of the effect of a car's being manufactured by AMC on its selling price. As long as you think that the manufacturer is an important variable, you should retain the AMC dummy or any other manufacturer dummy when you predict the selling price of a car, whether or not the dummy is statistically significant.

Occasionally students fall into the trap of coding such categories as manufacturers as one variable that would take the value 1 for AMC, 2 for Chrysler, 3 for General Motors, and so on. You cannot treat categorical data as interval or ratio data. A regression that used such a variable would make no sense.

14.3B INTERACTION TERMS

Multiple regression is based on the assumption that the effect of each of the independent variables on the average level of the dependent variable is additive. Consider a regression with two independent variables. This additivity assumption implies that if both x and z increase by one unit, the impact on the mean of Y is the sum of the effect of the increase in x and the effect of the increase in z. In notation the population regression equation is

$$\hat{y} = \alpha + \beta_1 x + \beta_2 z \tag{14.7}$$

There are occasions when the independent variables interact in their effect on the mean of Y. When there is interaction between the independent variables, the extent of the change in the mean of Y caused by a unit change in x will depend on the level of z, and the extent of the change in the mean of Y caused by a unit change in z will depend on the level of x. Note that interaction does not imply that the independent variables are correlated. The independent variables may not be correlated at all, but the effects of changes in x on the mean of Y could still depend on the level of z and vice versa. A simple functional form for the population regression equation that allows for interaction is

$$\hat{y} = \alpha + \beta_1 x + \beta_2 z + \beta_3 xz \qquad \text{(14.15)}$$

If there are more than two independent variables, say three, symbolized by x, z, and w, the interaction terms will be zx, zw, and xw. The independent variables could be dummy variables or ordinary independent variables. For our example, we'll stick to two independent variables for data in levels.

You are a researcher for the Congressional Budget Office and you have been asked to model the factors that determine the number of undocumented aliens crossing over the border from Mexico to the United States who are caught per month. You hypothesize that the number of aliens found per month by the Border Patrol depends on the actual flow of aliens across the border (the more aliens crossing, the more will be caught) and on the number of agents watching the border (the more agents, the more aliens caught). These variables clearly interact, because when there are many agents and many aliens, it is easier for the agents to catch the aliens. You do not have direct evidence of the number of aliens attempting to cross the border each month; the actual flow is not recorded because the undocumented aliens do not want to be identified and deported. A common procedure in this situation is to use another variable that you think is highly correlated with the variable that cannot be measured. *A variable that is correlated with an independent variable that cannot be measured* is called a **proxy variable.** In this example the proxy variable is the number of unemployed people in Mexico. The idea behind the proxy variable is that the flow out of Mexico is determined primarily by conditions in the Mexican labor market.

In this situation, the mean of Y, the number of aliens caught per month, no longer falls on a plane. The interaction term creates a surface with a twist. Think of a thin but stiff piece of plastic the shape of a license plate. If you held it in both hands and gave it a twist without breaking it, you would have the type of surface described by equation 14.15. Figure 14.11 shows such a surface. The coefficient on the interaction term determines the amount of twist to the surface, representing the mean of Y. As the coefficient goes up, so does the amount of twist. To understand the picture, consider the situation of many

FIGURE 14.11

Mean of _Y_ given _x_ and _z_ when _x_ and _z_ interact

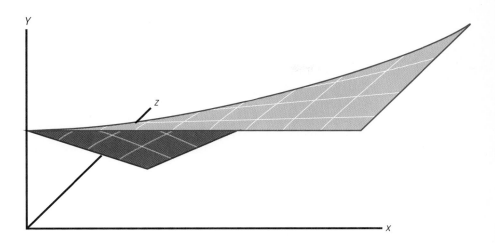

Border Patrol agents (high _x_) and a low number of unemployed in Mexico (low _z_). The rate of arrests would be low because, while a high percentage of aliens attempting to cross would be apprehended by the large number of agents, the flow of aliens across the border would be low. The lower right-hand corner of the surface would correspond to this situation. Now consider a large number of unemployed in Mexico and a large number of Border Patrol agents. In this situation the rate of alien arrests would be relatively high. This situation corresponds to the upper right-hand corner of the surface.

Suppose the estimated regression equation were

$$\hat{y} = 3{,}500 + 5.11x + 0.002z + 0.0000041xz$$

The interpretation of the intercept term is simply where the surface of the mean of _Y_ cuts the _y_ axis. The interpretation of 3,500 aliens being arrested per month when there are no border agents and no unemployed persons in Mexico is not valid for two reasons. First, it is not possible to have any arrests without agents, and second, even if it were possible, the problem of extrapolating a value far away from the actual data would invalidate this interpretation. There are no observations in the data near zero agents and zero unemployed people in Mexico. We can no longer describe the effect of a unit change in _x_ on the mean of _Y_ without specifying the level of _z_. For example, suppose the number of unemployed in Mexico were 2 million. Adding one border agent would cause the estimated mean number of arrests to go up 5.11 for _x_ alone and 8.2 for the interaction term (0.0000041 times 2,000,000). The total effect of any additional agent on the estimated mean number of arrests when _z_ is 2 million is 13.31 per month. If the number of unemployed in Mexico were 20 million instead of 2 million, the effect of an additional agent would be 5.11

for x alone and 82 for the interaction term (0.0000041 times 20,000,000). The total effect of adding one agent on the estimated mean number of arrests would be 87.11 when the number of unemployed in Mexico was 20 million.

Clearly, the interaction term is important in this situation. The addition of one agent causes a much larger number of arrests when the flow of aliens attempting to cross the border is great than when it is small. If the regression were estimated without the interaction term, the estimated effect of adding one agent could be way off. The word "important" is used here in terms of the size of the coefficient on the interaction term. Given plausible variations in z, the coefficient on the interaction term is large enough to show important differences in the effect of adding one agent. The usual hypothesis tests and confidence intervals for multiple regression apply to regression equations with interaction terms. You should test the null hypothesis that the coefficient of the interaction term is zero before using the term to make estimates of the mean of Y given x and z.

▨ 14.3C TRANSFORMATIONS TO ESTABLISH LINEARITY

There are many possible ways to transform the data in order to establish linearity. You can, for instance, take the logarithms of all of the variables, the square roots of all of the variables, or the squares of all of the variables. Transformations are also used for other purposes, such as to make heteroscedastic data homoscedastic. The subject of transformations in multiple regression analysis is too wide to cover in this text. We'll go over a logarithmic transformation—one of the most commonly used transformations—so that you will at least have an introduction to the subject. Economists call *the function that describes the relationship between the inputs and output of any production process* a **production function.** Capital and labor could be two inputs of a production process and the output would be finished goods. The law of diminishing returns (it's also called the law of variable proportions) implies that the relationship between the mean level of output and an increase of one unit of input is not linear. Adding labor while keeping capital fixed will cause a diminution in the additional output per unit of labor. Returns will similarly diminish with an increase in capital or any other factor of production. If you've had an introductory economics course, this terminology should sound at least vaguely familiar.

One functional form for production functions is the Cobb–Douglas production function. Let output be Y, labor inputs into a production process be x, and capital inputs into a production process be z. In the Cobb–Douglas production function, the mean of Y equals

$$\hat{y} = \alpha x^{\beta_1} z^{\beta_2}$$

(14.16)

FIGURE 14.12
Mean of Y given x and z for a Cobb-Douglas–type production function

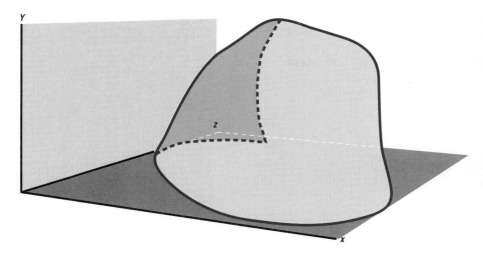

When we take the natural logs of both sides, formula 14.16 is transformed into a linear equation:

$$\ln(\hat{y}) = \ln(\alpha) + \beta_1 \ln(x) + \beta_2 \ln(z)$$

One peculiarity of this model is the assumption required for the distributions of Y given x and z. The usual assumption in multiple regression is that the distributions have equal variances (homoscedasticity). Here the assumption is that the distributions of the natural log of Y given x and z have equal variances. This assumption is required for statistical inference for the log form of formula 14.16. Figure 14.12 shows the surface of the mean of Y before the transformation. After the transformation, the surface of the natural log of the mean of Y is a plane.

Here is an example that uses a logarithmic transformation to estimate a production function. The World Bank provides loans to less developed countries (LDCs) to finance projects that will promote economic growth. Many of the loans finance the construction of what economists call "infrastructure." The term refers to public capital, such as roads, bridges, railroads, ports, and hydroelectric projects, which are needed to support general economic activity. The World Bank is interested in estimating the production function for the construction of roads. If it knew the parameters of this production function, the bank could advise LDCs on the most cost-effective methods of road construction, and more mileage would be obtained from the available funds. Suppose the bank had the data in Table 14.6 on the amount of labor, the amount of capital, and the number of miles of asphalt roads completed in projects financed by the bank in various LDCs. The labor is in worker-days and the

TABLE 14.6

Labor and capital costs for constructing a mile of asphalt road in various LDCs

Country	Labor (worker-days)	Capital (millions of dollars)	Miles
Sri Lanka	3,000	2.56	11.75
India (1)	4,400	3.00	2.98
India (2)	3,200	8.91	70.23
India (3)	200	11.10	47.05
India (4)	3,790	15.00	11.05
Nigeria	8,000	1.82	42.41
Costa Rica	400	4.09	10.12
Bangladesh	7,040	0.22	3.90
Botswanna	569	6.90	38.44
Haiti	4,668	1.50	28.24
Indonesia (1)	15,000	3.40	47.50
Indonesia (2)	10,000	2.96	47.70
Indonesia (3)	8,754	3.88	23.34
Zaire	3,267	4.91	42.01
North Yemen	254	16.00	65.37

capital is the interest and depreciation charges on the equipment used in each project measured in millions of dollars. The estimated regression equation for these data is:

$$\ln(\hat{y}) = 1.61 + 0.41 \times \ln(x) + 0.72 \times \ln(z)$$

The interpretation of the coefficients is: (1) the natural log of the average number of miles completed increases by 0.41 for a unit increase in the natural log of the number of worker-days; (2) the natural log of the average number of miles completed increases by 0.72 for a unit increase in the natural log of the number of millions of dollars of equipment used on the project; (3) the surface of the mean of the natural log of miles completed crosses the y axis at 1.61.

The coefficient of determination is 0.27, so the logs of labor and capital account for 27% of the variation in the log of miles completed. The low coefficient of multiple determination suggests that either some variables that affect the number of miles completed are missing or the Cobb–Douglas functional form for the production function does a poor job of describing the production process. Some candidates for additional variables would be a measure of how the terrain affected the difficulty of road construction and some measure of the quality of a worker-day of labor. Another possible explanation is that the records are not accurate; they may reflect ghost employees or equipment paid for but not delivered. Another explanation for the low coefficient of multiple determination is that the process of road construction may be highly stochastic, that is, subject to wide random variation.

The standard errors on the slope coefficients for labor and capital are, respectively, 0.84 and 0.36. With these data it is not possible to reject the hypothesis that the amount of labor has no effect on the mileage completed at a 5% significance level. The test is left to the reader.

To get the estimated effects expressed in units of miles, worker-days, and millions of dollars, you have to take the antilogs. A project with 1,000 worker-days and $1 million, for example, would be expected to complete 86 miles:

$$\ln(\hat{y}) = 1.61 + 0.41 \times \ln(1,000) + 0.72 \times \ln(1.00)$$

$$\ln(\hat{y}) = 4.45$$

We obtained the 86 miles by taking the antilog of 4.45.

The techniques discussed in section 14.3—dummy variables, interaction terms, and logarithmic transformations—are part of the broader topic of model building. The wide variety of possible models makes multiple regression a flexible tool for estimating parameters with multivariate data. For example, you could create many different models by combining some of these techniques. You could use dummy variables, interaction terms, and logarithmic transformations in the same regression. The great flexibility of multiple regression analysis is beneficial, but it is also dangerous. The benefit comes from being able to model many relationships and types of data. The danger occurs when a researcher keeps trying alternative techniques or models until he or she is satisfied with the R^2. Instead of resulting in the model that describes the population relationships, this hunt for a good fit may result in a model that follows the contours of every stray observation in the sample data.

The choice of the model to be estimated must be guided by an understanding of the data. Consider our examples. It is clear that the number of agents watching the border and the number of aliens attempting to cross the border have an interactive effect on the number of arrests. With production functions, it is well known that additions of only one input have a nonlinear effect on the level of output. Some statisticians argue that a model should be formed before the raw data are examined, and that the data should be used only after the model is formed, to test its validity. This level of statistical purity is rarely maintained in actual work. Most research is an amalgam of prior knowledge about the appropriate model or specification and some hunting for an improved fit. The issue here is similar to the point raised earlier in the chapter on the choice of independent variables. The earlier point was that if you have enough variables, you can find a set of independent variables that will give you a high R^2 for any dependent variable. Many computer programs offer a technique called stepwise regression, which speeds up the process of hunting for the best-fitting set of independent variables. This technique eliminates the need for any thinking in the selection of independent variables. Every potential variable is fed into a computer and the algorithm hunts for the set of indepen-

USING MULTIPLE REGRESSION FOR A SALARY ADJUSTMENT

I teach at Indiana University–Purdue University at Indianapolis (IUPUI). IUPUI was created in 1969 out of Indiana University's and Purdue University's extension programs in Indianapolis. The fiscal and academic relationships between IUPUI and the older universities are so Byzantine that I will not attempt to describe them. The new school grew rapidly from an initial enrollment of about 10,000 students to about 23,000 in 1985. In response to this growth, the trustees of Indiana University endorsed a policy of "equal excellence" for the original campus of Indiana University at Bloomington and IUPUI. "Equal excellence" implies, among other things, equal salary systems. In 1975 I was asked to compare the salary levels at the School of Liberal Arts at IUPUI and in the corresponding departments of the School of Arts and Science at IU-Bloomington. One simple way to make the comparison was to take the means of salaries at Liberal Arts at IUPUI and the corresponding departments at IU-Bloomington. The average at IUPUI was about $25,000 and the average at IU-B was about $40,000. Giving every faculty member at IUPUI Liberal Arts $15,000 would wipe out the discrepancy. The administration did not accept this proposal. Since IUPUI was a very new school, most of the faculty were younger than the faculty at IU-B and had lower academic rank and less experience. Salaries also varied across academic departments, and the proportion of faculty by department varied across the two universities.

The procedure used to make the comparison was multiple regression. Data on 576 faculty members at IU-B were used to estimate the system of compensation in effect at IU-B. The estimated regression equation was approximately

$$\hat{y} = 10,000 + 250\text{EXP} + 10,000\text{FULL-PROF} + 5,000\text{ASSOC-PROF}$$
$$+ 1,000\text{ASST-PROF} + 5,000\text{ECON} + 3,000\text{SOC} + 2,500\text{GEOG}$$
$$+ 2,000\text{ANTH} + 1,000\text{POLI} + 500\text{SPAN} + 500\text{FREN} + 500\text{GER}$$
$$+ 250\text{HIST} + 100\text{PHIL} + 0\text{RELI}$$

The dependent variable is the academic-year or ten-month salary. The variable EXP referred to the years of experience of a professor since acquiring his or her Ph.D. FULL-PROF was a dummy variable that had the value 1 if a professor had the rank of full professor and 0 otherwise. ASSOC-PROF was a dummy variable that had the value 1 if a professor had the rank of associate and 0 otherwise. ASST-PROF was a dummy variable that had the value 1 if a professor had the rank of assistant and 0 otherwise. One rank had to be omitted from the regression equation for it to run. The omitted rank was lecturer. The remaining variables were dummies that represented a professor's department: economics, sociology, geography, anthropology, political science, Spanish, French, German, history, philosophy, and religious studies. The omitted department was English.

The interpretation of the intercept is that an instructor in the English Department at the rank of lecturer with no experience was expected to have an academic-year salary of $10,000 in 1975. The interpretation of the coefficient on the experience variable was that an additional year of experience since earning a Ph.D. added $250 to the expected academic-year salary, with the effect of rank and department held constant. The interpretation of the coefficient for a particular rank dummy variable is the difference between the expected salary for that rank and the expected salary for lecturers, with the effect of experience and department held constant. For example, other things being equal, a full professor's expected academic-year salary would be $10,000 more than a lecturer's salary; an associate professor's expected academic-year salary would be $5,000 more than a lecturer's salary; and an assistant professor's expected academic-year salary would be $1,000 more than a lecturer's salary. The interpretation of the dummy variables for academic departments is the difference between the expected academic-year salary for a particular department and the expected salary of the omitted department, English. For example, the expected salary for an economist is $5,000 more than that of an English professor, when rank and experience are held constant.

To see how these dummy variables work together, consider the expected salary of a full professor of economics with 20 years of experience. The expected salary would be $10,000 for the intercept plus $5,000 for the 20 years of experience plus $10,000 for the rank of full professor plus $5,000 for the Economics Department effect. According to the regression equation estimated on the IU-B data, the expected academic-year salary for a full professor of economics with 20 years' experience in 1975 was $30,000.

The variables chosen to represent the system of compensation in effect at IU-B did a good job of predicting salaries. Most of the IU-B professors' actual salaries were within $100 of their predicted salaries, and the greatest discrepancy was $1,500. If the regression equation is accepted as an accurate representation of the salary system at IU-B in the specified departments, the comparison with the faculty at IUPUI is easy. The experience, ranks, and departments of the IUPUI faculty were substituted into the regression equation and an estimated salary for each IUPUI faculty member was calculated. This estimated salary represented what the IUPUI faculty member would have been paid according to the IU-B salary system. The presumption in these calculations is that unmeasured characteristics related to salary determination of faculty members did not vary across universities.

For the IUPUI Liberal Arts faculty, this story had a happy ending. The difference between the actual IUPUI faculty salaries and their predicted salaries was presented to the IUPUI administration and they found the money to make up the shortfalls. The regression equation was used to allocate the extra funds. (No one in the administration checked the regression equation for accuracy. If I had anticipated this trusting behavior, the estimated regression equation might have favored lecturers—my rank at the time—much more than it did.)

dent variables with the highest adjusted R^2. The point here is that if you have a long enough list of possible models for a given set of independent variables, such as mixed interaction terms and transformations, you can find some model with a high R^2.

EXERCISES FOR SECTION 14.3

14.13 Estimate the slope coefficients, the intercept, the coefficient of multiple determination, the adjusted coefficient of multiple determination, the standard error of the estimate, and the standard errors of the slope coefficients for the following data:

x_i	z_i	y_i	$(x_i - \bar{x})^2$	$(z_i - \bar{z})^2$	$(x_i - \bar{x})(y_i - \bar{y})$	$(z_i - \bar{z})(y_i - \bar{y})$	$(x_i - \bar{x})(z_i - \bar{z})$	$\hat{y}_i$
3	0	6.37	5.06	0.25	5.57	1.24	1.13	6.83
5	1	6.53	0.06	0.25	0.58	−1.15	−0.13	5.91
6	1	7.03	0.56	0.25	−1.35	−0.90	0.38	7.89
4	0	8.97	1.56	0.25	−0.16	−0.06	0.63	8.81
5	1	5.70	0.06	0.25	0.78	−1.57	−0.13	5.91
6	0	13.70	0.56	0.25	3.65	−2.43	−0.38	12.77
7	0	14.46	3.06	0.25	9.84	−2.81	−0.88	14.75
1	0	2.90	18.06	0.25	25.27	2.97	2.13	2.86
9	0	18.73	14.06	0.25	37.07	−4.94	−1.88	18.71
5	1	6.18	0.06	0.25	0.67	−1.33	−0.13	5.91
6	1	8.43	0.56	0.25	−0.31	−0.21	0.38	7.89
7	1	9.35	3.06	0.25	0.89	0.25	0.88	9.87
6	1	7.35	0.56	0.25	−1.12	−0.75	0.38	7.89
6	1	8.83	0.56	0.25	0.00	0.00	0.38	7.89
4	0	8.02	1.56	0.25	1.03	0.41	0.63	8.81
2	0	5.13	10.56	0.25	12.04	1.85	1.63	4.85
9	1	13.44	14.06	0.25	17.26	2.30	1.88	13.83
10	0	21.02	22.56	0.25	57.84	−6.09	−2.38	20.69
1	1	−1.89	18.06	0.25	45.59	−5.36	−2.13	−2.02
3	0	6.61	5.06	0.25	5.01	1.11	1.13	6.83
105	10	176.86	119.75	5.00	220.15	−17.48	3.50	176.86

14.14 Estimate the slope coefficients, the intercept, the coefficient of multiple determination, the adjusted coefficient of multiple determination, the standard error of the estimate, and the standard errors of the slope coefficients for the following data:

x_i	z_i	y_i	$(x_i - \bar{x})^2$	$(z_i - \bar{z})^2$	$(x_i - \bar{x})(y_i - \bar{y})$	$(z_i - \bar{z})(y_i - \bar{y})$	$(x_i - \bar{x})(z_i - \bar{z})$	$\hat{y}_i$
3	0	6.55	5.06	0.25	12.67	2.82	1.13	7.24
5	1	14.30	0.06	0.25	−0.53	1.06	−0.13	13.36
6	1	14.05	0.56	0.25	1.40	0.94	0.38	15.33
4	0	9.45	1.56	0.25	3.41	1.36	0.63	9.21
5	1	13.05	0.06	0.25	−0.22	0.44	−0.13	13.36
6	0	14.56	0.56	0.25	1.78	−1.19	−0.38	13.15
7	0	14.69	3.06	0.25	4.40	−1.26	−0.88	15.12
1	0	3.34	18.06	0.25	37.56	4.42	2.13	3.30
9	0	19.09	14.06	0.25	25.91	−3.45	−1.88	19.07
5	1	13.77	0.06	0.25	−0.40	0.79	−0.13	13.36
6	1	16.14	0.56	0.25	2.97	1.98	0.38	15.33
7	1	16.52	3.06	0.25	7.60	2.17	0.88	17.30
6	1	14.52	0.56	0.25	1.76	1.17	0.38	15.33
6	1	16.75	0.56	0.25	3.43	2.29	0.38	15.33
4	0	8.02	1.56	0.25	5.20	2.08	0.63	9.21
2	0	5.70	10.56	0.25	21.06	3.24	1.63	5.27
9	1	20.66	14.06	0.25	31.81	4.24	1.88	21.24
10	0	21.52	22.56	0.25	44.39	−4.67	−2.38	21.04
1	1	5.67	18.06	0.25	27.66	−3.25	−2.13	5.48
3	0	6.92	5.06	0.25	11.83	2.63	1.13	7.24
105	10	255.27	119.75	5.00	243.67	17.80	3.50	255.27

14.15 The standard error for each coefficient in the following estimated regression equation is given below it in parentheses. The estimated regression equation is based on 32 observations.

$$\hat{y} = 114.29 + 20.05x + 5.08z - 16.66xz$$
$$\quad (14.9) \quad (5.11) \quad (2.09) \quad (9.95)$$

(a) Determine the estimated value of the dependent variable when x is 10 and z is 15.

(b) Determine the estimated value of the dependent variable when x is 20 and z is 25.

(c) Determine the limits of a 90% confidence interval on the slope of x.

(d) Determine the limits of a 90% confidence interval on the slope of z.

(e) Test at the 5% significance level the hypothesis that there is no interaction between x and z.

14.16 The standard error for each coefficient in the following estimated regression equation is given below it in parentheses. The estimated regression equation is based on 22 observations.

$$\hat{y} = \underset{(102.55)}{864.30} + \underset{(8.52)}{12.86x} - \underset{(7.78)}{14.94z} - \underset{(3.87)}{11.71xz}$$

(a) Determine the estimated value of the dependent variable when x is 10 and z is 15.

(b) Determine the estimated value of the dependent variable when x is 20 and z is 25.

(c) Determine the limits of a 90% confidence interval on the slope of x.

(d) Determine the limits of a 90% confidence interval on the slope of z.

(e) Test at the 5% significance level the hypothesis that there is no interaction between x and z.

14.17 Use a logarithmic transformation to estimate the regression equation for the following data:

x_i	z_i	y_i	x_i	z_i	y_i
1	4	25	4	6	500
3	8	200	7	10	2,250
2	9	600	6	9	1,460

14.18 Use a logarithmic transformation to estimate the regression equation for the following data:

x_i	z_i	y_i	x_i	z_i	y_i
1	3	20	4	10	390
2	4	75	7	2	160
2	5	240	6	1	140

14.19 The operator of a traveling carnival that operates in small towns wanted to estimate the effect of weather, day of the week, and city population on the sales of ride tickets. The following regression was estimated for 100 days of operation.

$$\text{SALES} = 300 - 200\text{RAIN} + 20\text{TUES} + 20\text{WED} + 40\text{THURS} + 200\text{FRI} + 600\text{SAT} + 500\text{SUN} + 0.00045\text{POP}$$

The SALES variable is the daily sales of ride tickets. RAIN is a dummy variable that takes the value of 1 if it rains and 0 otherwise. The variables TUES through SUN are dummies for days of the week and POP is the population of the town in which the carnival is operating. The standard errors of the slope coefficients are: RAIN, 60; TUES, 45; WED, 80; THURS, 18; FRI, 55; SAT, 91; SUN, 126; POP, 0.00012.

(a) What are the estimated sales for a rainy Monday in a town with 1,000 people?

(b) Which slope coefficients are not statistically significantly different from 0 at the 5% significance level?

14.20 (This exercise continues exercise 14.19.) If you thought that the effect of rain on sales might vary with the day of the week, how would you specify the regression equation?

14.21 An economist studying the world petroleum market estimates the following regression equation:

ln(BARRELS) = 9.48 + 0.392PRICE + 0.00237RESERVES

The variable BARRELS refers to annual production of oil in barrels per day. The variable price is average price in a calendar year in 1967 U.S. dollars. The variable RESERVES is the proven world reserves of petroleum in millions of barrels.

(a) What would be the estimated annual worldwide production of petroleum expressed as a rate of barrels per day if the average price over a year were $12 in 1967 dollars and the proven reserves were 1,540 million barrels?

(b) Holding reserves constant, can you determine how much annual production would increase per dollar increase in price? Explain.

14.4 PROBIT REGRESSION (OPTIONAL)

The technique of dummy variables discussed in section 14.3 allowed you to estimate models with categorical *independent* variables. Multiple regression with dummy *independent* variables can be considered an extension of analysis of variance. With analysis of variance you could test if a dependent variable had the same mean in all of its categories. With multiple regression with dummy variables, you could estimate how much the mean of the dependent variable changes as you go from category to category. In this section we will work with categorical *dependent* variables. Many business problems require the use of categorical *dependent* variables. Again, multiple regression can be considered an extension of a technique that you have already learned. When you used the chi–square distribution to test for independence, you tested the hypothesis that the population proportion of a dependent variable was the same at all levels of an independent variable. For example, you could test whether the proportion of households that had purchased new cars in the previous year was the same at all income levels. The multiple regression extension of this problem, probit regression, allows you to estimate the proportion of households that purchase new cars at each level of household income. **Probit regression** is *an estimation technique in which the dependent variable is a binary categorical variable and the relationship between the values of the independent variables and the proportion of successes in the dependent variable is assumed to follow a cumulative normal distribution.* The ability to estimate the change in the average level of the dependent variable when either some or all of the independent variables are categorical (dummies on the right side of the equal sign) and/or when the dependent variable is categorical (a dummy on the left side of the equal sign) demonstrates the great power and flexibility of multiple regression analysis. Consider the following two statements: (1) "The estimated proportion of households that purchased a new car in the previous year is 0.18 for households with an annual income of $30,000 and 0.22 for households with an annual income of $40,000"; (2) "The null hypothesis that the proportion of households that purchase new cars is the same across all income levels is rejected at the 5% significance level." The information needed to make the first statement requires multiple regression analysis with a categorical dependent variable, while the information for the second statement could be acquired through a chi-square test for independence. Clearly, the first statement is more useful.

Categorical variables can have any number of categories, but to simplify the calculations, we'll stick to examples in which the categorical dependent variable is limited to two categories, such as buy or not buy a new car. Also, to simplify the calculations, our example will have only one right-hand or independent variable. The computer appendix to this chapter has some examples with more than one right-hand variable.

You are the vice-president for development of a public television channel. This is a fancy title that means that you are in charge of begging for money. Most of the funds for your station come from memberships. You would like to estimate how a person's educational level affects the probability that he or she will pay for a membership in your television station. You can use this information to target your appeals for new memberships. You have data on years of education for a sample of your viewing audience and whether or not they have paid for a membership. Table 14.7 shows the data. The sample was kept small to reduce the calculations.

TABLE 14.7

Membership status of viewers of a public television station, by educational level

Years of education	Membership status	Years of education	Membership status
6	No	14	Yes
8	No	16	No
9	No	16	Yes
10	No	16	Yes
11	No	16	Yes
12	No	18	Yes
12	No	18	Yes
12	Yes	18	Yes
12	Yes	20	Yes
12	Yes	22	Yes
14	No		

The dependent variable is membership status and the independent or explanatory variable is years of education. A viewer who had paid for a membership would be coded as 1; a viewer who had not paid for a membership would be coded as 0. If you ran a least-squares regression for these coded data, the results would violate the rules of probability. You would get predicted proba-

FIGURE 14.13
Simple regression of membership status on education

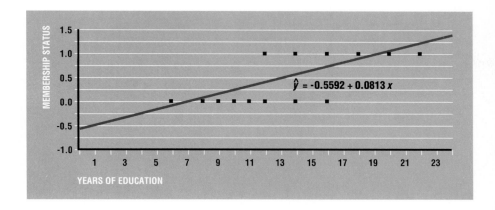

$$\hat{y} = -0.5592 + 0.0813\,x$$

bilities below 0 and above 1. Figure 14.13 shows why. The dots are coded as either 0 or 1 on the vertical axis. The vertical axis can be interpreted as either the estimated proportion of the viewers in an educational level who have paid for memberships or the estimated probability that a randomly selected individual at each educational level will have paid for a membership. You get negative probabilities and probabilities greater than 1 from the least-squares regression line. Clearly, linear regression is a poor estimation technique when the dependent variable is categorical.

Probit regression is a preferable estimation technique (there is a good alternative called logit regression, but we won't go into it). The first assumption required for probit regression is that there is an underlying continuous variable that the researcher cannot observe. In this example the underlying continuous random variable could be called "positive attitude toward membership in the local public television station." Denote the underlying continuous random variable with the symbol y^*. This continuous random variable is assumed to be normally distributed. There is some threshold value of the continuous random variable that will cause a person to pay for a membership. This attitude varies across individuals for two reasons: random variation from person to person and differences in the independent variable (years of education, in this example). The second assumption required for probit regression is that the attitude variable (or whatever the underlying continuous variable is) is a linear function of the independent variable. All you can observe in the data is whether someone has reached the threshold value (she bought a membership) or has not reached the threshold value (she did not buy a membership). Figure 14.14 shows a normal distribution of the underlying variable "positive attitude toward membership in the local public television station" and indicates the threshold point on the distribution for some educational level, labeled x. You would like to estimate the probability that this threshold value will be reached for every educational level or, equivalently, the proportion of all individuals at

FIGURE 14.14
Distribution of the
unobserved variable "positive
attitude toward membership
in the local public television
station" and threshold value
at which membership will be
purchased

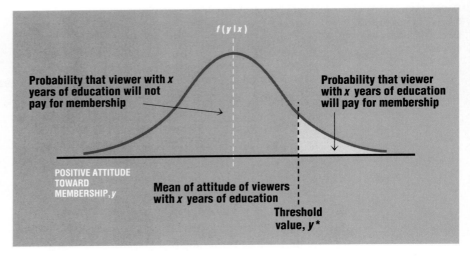

an educational level who pay for memberships. In this case you are estimating how the dark area in Figure 14.14 varies as a function of education. Another name for the dark *area* as a function of the threshold value—*the area to the left of any point on a normal distribution*—is the **cumulative normal distribution.**

The graph of the cumulative normal distribution provides a good fit for the type of data in Figure 14.13. The lowest value of the cumulative normal distribution is 0 at $-\infty$ (probabilities cannot go below 0) and the highest is 1 at $+\infty$ (the total area cannot exceed 1). The cumulative normal distribution starts out flat, rises as the threshold value approaches the mean of the attitude variable, and then flattens out again when the threshold value passes the mean. Figure 14.15 shows the graph of the cumulative normal distribution.

If you compare the plot of the scatter diagram in Figure 14.13 and the cumulative normal distribution in Figure 14.15, the appropriateness of the cumulative normal for regressions with binary dependent variables should be clear. Every normal distribution and the counterpart cumulative normal distribution are uniquely defined by their means and their standard deviations. Determining the "best-fitting" cumulative normal distribution requires that you find the "best" mean and standard deviation. The concept of "best fitting" is the same as in simple regression: it is the curve that minimizes the sum of squared prediction errors, or SSE.

The problem with finding the best-fitting cumulative normal distribution is that you cannot use ordinary least squares. The method to use is **maximum likelihood estimation,** *a technique for finding the best estimates of the population parameters, so called because the estimates sought define the population from which a sample has the greatest likelihood of matching the actual observations.*

FIGURE 14.15

Plot of the cumulative normal distribution

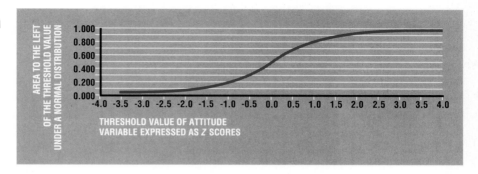

Here are the steps for hunting for the mean and standard deviation that are most likely to yield the actual sample. First you pick some mean and standard deviation as a guess for the parameters of the best-fitting cumulative normal curve. Let's use 12 years of education for the mean and 2 for the population standard deviation. Note that these numbers were pulled out of thin air; you have no idea if they are the parameters of the best-fitting cumulative normal distribution. Next you calculate the z score for every observed number of years of education in Table 14.7, given these numbers for μ and σ. For example, the first observation in Table 14.7 is 6 and the z score is -3, $(6 - 12)/2$. You will have 21 z scores, one for each observation in the data set. For every person who did not buy a membership you use the z score and a normal table to calculate the white area in Figure 14.14. With the assumed parameters of 12 and 2, the dark area is an estimate of the probability that someone with the observed years of education would not buy a membership. For every person in the sample who did buy a membership, you would use the z score to calculate the respective shaded area, the estimated probability that someone with a given education would buy a membership. Now you have 21 probabilities, one for each person. Note that each person is assumed to act independently. The probability that 21 people with the given years of education will have the exact pattern of buying or not buying memberships observed in Table 14.7 is the product of the 21 individual probabilities. The reason for using the product is just the multiplication rule for independent events. This probability (the product of the 21 events) is called the likelihood of observing the actual sample given the assumed mean and standard deviation. Thus, the likelihood for a given μ and σ is the product of the n probabilities of buying (for everyone in the sample who was observed to buy) or not buying (for everyone in the sample who was observed not to buy).

The likelihood for the assumed mean of 12 and the assumed standard deviation of 2 is 4.6735E-05. This likelihood is in the upper left corner of Table 14.8,

TABLE 14.8
Values of the likelihood function for various means and standard deviations of the cumulative normal distribution

	STANDARD DEVIATION								
Mean	2.0	2.5	3.0	3.5	4.0	4.5	5.0	5.5	6.0
12.0	4.6735E-05	0.00010088	0.00013216	0.00013649	0.00012498	0.00010766	9.0048E-05	7.444E-05	6.143E-05
12.5	9.3448E-05	0.00016092	0.00018382	0.00017327	0.00014876	0.00012225	9.8712E-05	7.9442E-05	6.4213E-05
13.0	0.00012167	0.00018931	**0.0002030**	0.00018342	0.00015285	0.00012294	8.727E-05	7.7728E-05	6.2288E-05
13.5	0.00010294	0.00016379	0.00017764	0.00016161	0.00013538	0.00010936	8.727E-05	6.9679E-05	5.6039E-05
14.0	5.6326E-05	0.00010385	0.00012281	0.0001183	0.00010322	8.5971E-05	7.0266E-05	5.7197E-05	4.6739E-05
14.5	1.9794E-05	4.8041E-05	6.6908E-05	7.1816E-05	6.7659E-05	5.967E-05	5.097E-05	4.2969E-05	3.6125E-05
15.0	4.424E-06	3.9115E-06	2.8642E-05	3.6091E-05	3.8086E-05	3.6537E-05	3.3292E-05	2.9531E-05	2.5866E-05
15.5	6.2144E-07	3.9115E-06	9.6059E-06	1.499E-05	1.8393E-05	1.9724E-05	1.957E-05	1.8561E-05	1.7153E-05
16.0	5.4101E-08	6.8026E-07	2.5161E-06	5.1369E-06	7.6128E-06	9.3818E-06	1.035E-05	1.0666E-05	1.0533E-05

under the mean of 12 and the standard deviation of 2. Recall that the 12 and 2 were picked arbitrarily. To find the *maximum* likelihood estimates of the mean and standard deviation, you would have to repeat the procedure I have described for a series of values for the mean and standard deviation. Table 14.8 shows the results of such a search—the likelihood for all combinations of means and standard deviations taken from the set {12, 12.5, 13.0, 13.5, 14.0, 14.5, 15.0, 15.5, 16.0} for means and {2.0, 2.5, 3.0, 3.5, 4.0, 4.5, 5.0, 5.5, 6.0} for standard deviations. The table has 81 likelihoods. The highest value in the table is indicated by bold numbers. It is for the mean 13 and the standard deviation 3. Since each entry represents the product of 21 areas, each requiring the calculation of a z score, the number of calculations is enormous. Recall that this problem was kept small to reduce the calculations. You can check your understanding of how the likelihood term is calculated by trying to reproduce one of the values in the table. In practice, the hunt for the maximum likelihood values for the mean and standard deviation is done with a statistical computer program. The actual values obtained by such a program which maximize the likelihood of observing the sample data are 12.975 for the mean and 2.90 for the standard deviation.

We'll defer the discussion of measures of the goodness of fit of the cumulative normal distribution derived with maximum likelihood estimation until the computer appendix. The focus here is the concept of maximum likelihood estimation and the use of probit regression. Using the maximum likelihood estimates of the population mean and standard deviation, you can find the probability that a person of a given educational level will purchase a membership in the public television channel. For example, what is the probability that someone with a tenth-grade education will purchase a membership? This probability is the shaded area in Figure 14.14 with x set at 10 and, on the basis of the best value in Table 14.8, the mean and standard deviation set at 13 and 3, respectively (the values of 12.975 and 2.90 derived from the computer program would be more accurate). The z score is -1 and the dark area to the left of a z score of -1 is 0.1587. The estimated proportion of viewers with 10 years of education who will buy memberships is 0.1587. The corresponding estimated probabilities for viewers with 11, 12, and 13 years of education are 0.2514, 0.3707, and 0.5. Figure 14.16 also shows the distribution of y^* given an x of 13 and the area to the right of the threshold value. Note that the change in probability is not a linear function of the change in years of education. This nonlinearity is the reason that the least-squares technique of multiple regression did not provide accurate estimates.

The straight line in Figure 14.16 labeled $\hat{y}^* = -4.4752 + 0.3449x$ is the estimated linear relationship between the underlying continuous attitude variable, y^*, and the years of education. All computer packages present the results of probit regressions in terms of the coefficients of this line rather than the mean and standard deviation of the best-fitting cumulative normal distribution. In this case, the method of presentation used in computer programs aids in an intuitive understanding of the model. There is a simple correspondence between the slope and intercept of the line and the mean and standard deviation for the best-fitting normal distribution. The value of x when the line crosses the threshold value of y^* (approximately 13 in this example) is the maximum likelihood estimate of the mean of the normal curve in which the cumulative

FIGURE 14.16
Calculating the probabilities of membership in a public television station with the probit model

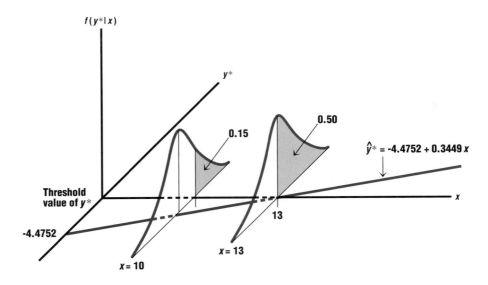

Source: The above graph is modeled after one in *Econometric Modeling and Economic Research,* William E. Becker, ed., Kluwer Academic Publishers, Boston, 1986

density function provides the best fit to the data. The inverse of the slope coefficient of the straight line (1 over 0.3449, or approximately 3 in this example) is the maximum likelihood estimate of the standard deviation of the normal curve in which the cumulative density function provides the best fit to the data. The explanation of why the mean is the x intercept of the line and the standard deviation is the inverse of the slope coefficient of the line requires us to go back to the formula for the standard normal curve:

$$f(x) = \frac{1}{\sqrt{2\pi}\sigma} e^{-1/2[(x-\mu)/\sigma]^2} \tag{6.1}$$

The probit model replaces the expression in parentheses by the intercept and slope for a straight line:

$$(a + bx)$$

The two expressions are equivalent if b equals the inverse of σ and a equals $-\mu/\sigma$:

$$\left(\frac{x-\mu}{\sigma}\right) = a + bx$$

$$\left(\frac{x-\mu}{\sigma}\right) = \frac{-\mu}{\sigma} + \frac{1}{\sigma}x$$

$$\left(\frac{x-\mu}{\sigma}\right) = \frac{-\mu + x}{\sigma}$$

$$\left(\frac{x-\mu}{\sigma}\right) = \frac{x-\mu}{\sigma}$$

EXERCISES FOR SECTION 14.4

14.22 For the following data set find the likelihood if the mean is set at 5 and the standard deviation at 1.

x	Yes or no
3	No
4	No
5	No
6	Yes
9	Yes

14.23 For the following data set find the likelihood if the mean is set at 6 and the standard deviation at 1.5.

x	Yes or no
3	No
4	No
5	No
6	Yes
9	Yes

14.24 Given the maximum likelihood estimates of the population mean and standard deviation of 15 and 6, respectively:

(a) What is the probability of a yes for an x of 16?

(b) What is the probability of a no for an x of 17?

14.25 Given the maximum likelihood estimates of the population mean and standard deviation of 80 and 20, respectively:

(a) What is the probability of a yes for an x of 75?

(b) What is the probability of a no for an x of 45?

14.26 One method for estimating the probabilities of binary categorical variables as functions of a continuous independent variable avoids the use of maximum likelihood estimation. The method is very simple: (1) run a linear regression on the data with the yes responses coded as 1's and the no responses as 0's; (2) define the probabilities of yes responses as 0 for estimated values of the dependent variable at or below 0; (3) define the probabilities of yes responses as 1 for estimated values of the regression line at or above 1; (4) if the predicted value of the binary dependent variable is between 0 and 1, define the probability of a yes response as the predicted value of the dependent variable.

(a) Using this method and the data for education versus public TV station membership in Table 14.7, draw the probability of a yes response as a function of x.

(b) What is wrong with this method of assigning probabilities?

 SUMMARY

Multiple regression analysis is both powerful and flexible. It can be used to analyze controlled experiments or it can be applied to nonexperimental data. In a nonexperimental context you can estimate the effect on the dependent variable of changing one of the independent variables while the other independent variables are held constant. You can make this estimate even though the other independent variables were never controlled or held constant in the nonexperimental data.

The assumptions required for statistical inference are linearity, homoscedasticity, independence among the right-hand variables, and no serial correlation. An understanding of these assumptions is important for two reasons. First, the assumptions clarify how statistical inference works in multiple regression. Second, one or more of these assumptions are often violated in sets of nonexperimental data. The flexibility of multiple regression analysis can be seen in its ability to be modified to account for such violations. The chapter discussed two such modifications: the use of interaction terms when some of the right-hand variables are not independent and the use of logarithmic transformations when the linearity assumption is violated.

The flexibility of multiple regression analysis can also be seen in its ability to mix different types of data. With dummy variables you can use categorical data as explanatory variables. Probit analysis is a technique that lets the dependent variable be a dummy variable.

Many of the assumptions required for statistical inference are the same in multiple and simple regression. Many of the techniques of statistical inference also carry over from simple regression. For example, hypothesis tests and confidence intervals on the slope coefficients are similar in simple and multiple regression. The coefficient of multiple determination is closely related to the coefficient of determination used in simple regression. One exception to this correspondence between tests in simple and multiple regression is more apparent than real. In multiple regression analysis, a joint test of the hypothesis that all of the slope coefficients are zero—that is, that the independent variables have no effect on the dependent variable—uses the F distribution. In simple regression the equivalent hypothesis is that either ρ or β equals zero. Actually, you could do an F test in simple regression for the null hypothesis that the independent variable has no effect on the dependent variable. Since two other ways to test this hypothesis were discussed in Chapter 13 and since some classes skip the F distribution, the F test was not discussed in Chapter 13.

 SUGGESTED READINGS

Kennedy, Peter. *A Guide to Econometrics.* Cambridge, Mass.: M.I.T. Press, 1985.

Kmenta, Jan. *Elements of Econometrics.* New York: Macmillan, 1971.

Theil, Henri. *Introduction to Econometrics.* Englewood Cliffs, N.J.: Prentice-Hall, 1978.

Vigderhous, Gideon. "Probit Analysis of Radio Ad Awareness." *Journal of Advertising Research* 17 (April 1977):21–26.

 NEW TERMS

adjusted coefficient of multiple determination The fraction of the total variation in the dependent variable which is accounted for by the independent variables adjusted for the number of independent variables.

coefficient of multiple determination The percentage of the total variation in the dependent variable accounted for by the independent variables.

cumulative normal distribution The area to the left of any point on a normal distribution.

dummy variables Indicator variables for categorical or nominal data which take the value of 1 or 0, depending on whether an observation does or does not belong in a specified category.

econometrics The application of statistical techniques to business and economic problems.

maximum likelihood estimation A technique for finding the best estimates of the population parameters, so called because the estimates sought define the population from which a sample has the greatest likelihood of matching the actual observations.

multicollinearity Any correlation among individual independent variables or among sets of independent variables.

multiple regression An estimation technique that predicts the effect on the average level of the dependent variable of a unit change in any independent variable while the other independent variables are held constant.

partial scatter diagram A scatter diagram of the dependent variable plotted against one independent variable while the effects of all other independent variables are held constant.

probit regression An estimation technique in which the dependent variable is a binary categorical variable and the relationship between the values of the independent variables and the proportion of successes in the dependent variable is assumed to follow a cumulative normal distribution.

production function The function that describes the relationship between the inputs and output of any production process.

proxy variable A variable that is correlated with an independent variable that is difficult to measure.

residuals The differences between the observed and predicted values of the dependent variable.

serial correlation A nonzero correlation among residuals.

standard error of the estimate The estimated standard deviation of the distributions of Y values about the mean of Y for any combinations of values of the independent variables.

time-series data Observations that are made over time.

 NEW FORMULAS

Normal equations for intercept and slope of x and z with population data

$$\sum_{i=1}^{\nu} y_i = \nu\alpha + \beta_1 \sum_{i=1}^{\nu} x_i + \beta_2 \sum_{i=1}^{\nu} z_i \tag{14.1}$$

$$\sum_{i=1}^{\nu} x_i y_i = \alpha \sum_{i=1}^{\nu} x_i + \beta_1 \sum_{i=1}^{\nu} x_i^2 + \beta_2 \sum_{i=1}^{\nu} x_i z_i \tag{14.2}$$

$$\sum_{i=1}^{\nu} z_i y_i = \alpha \sum_{i=1}^{\nu} z_i + \beta_1 \sum_{i=1}^{\nu} x_i z_i + \beta_2 \sum_{i=1}^{\nu} z_i^2 \tag{14.3}$$

Normal equations for population data solved for the intercept and the slope coefficients

$$\alpha = \mu_Y - \beta_1 \mu_x - \beta_2 \mu_z \tag{14.1a}$$

$$\beta_1 = \frac{\sum_{i=1}^{\nu} (z_i - \mu_z)^2 \sum_{i=1}^{\nu} (x_i - \mu_x)(y_i - \mu_Y) - \sum_{i=1}^{\nu} (x_i - \mu_x)(z_i - \mu_z) \sum_{i=1}^{\nu} (z_i - \mu_z)(y_i - \mu_Y)}{\sum_{i=1}^{\nu} (x_i - \mu_x)^2 \sum_{i=1}^{\nu} (z_i - \mu_z)^2 - \left[\sum_{i=1}^{\nu} (x_i - \mu_x)(z_i - \mu_z) \right]^2} \tag{14.2a}$$

$$\beta_2 = \frac{\sum_{i=1}^{\nu} (x_i - \mu_x)^2 \sum_{i=1}^{\nu} (z_i - \mu_z)(y_i - \mu_Y) - \sum_{i=1}^{\nu} (x_i - \mu_x)(z_i - \mu_z) \sum_{i=1}^{\nu} (x_i - \mu_x)(y_i - \mu_Y)}{\sum_{i=1}^{\nu} (x_i - \mu_x)^2 \sum_{i=1}^{\nu} (z_i - \mu_z)^2 - \left[\sum_{i=1}^{\nu} (x_i - \mu_x)(z_i - \mu_z) \right]^2} \tag{14.3a}$$

Normal equations for intercept and slope of x and z with sample data

$$\sum_{i=1}^{n} y_i = an + b_1 \sum_{i=1}^{n} x_i + b_2 \sum_{i=1}^{n} z_i \tag{14.4}$$

$$\sum_{i=1}^{n} x_i y_i = a \sum_{i=1}^{n} x_i + b_1 \sum_{i=1}^{n} x_i^2 + b_2 \sum_{i=1}^{n} x_i z_i \tag{14.5}$$

$$\sum_{i=1}^{n} z_i y_i = a \sum_{i=1}^{n} z_i + b_1 \sum_{i=1}^{n} x_i z_i + b_2 \sum_{i=1}^{n} z_i^2 \tag{14.6}$$

Normal equations for sample data solved for the intercept and the slope coefficients

$$b_1 = \frac{\sum_{i=1}^{n} (z_i - \bar{z})^2 \sum_{i=1}^{n} (x_i - \bar{x})(y_i - \bar{y}) - \sum_{i=1}^{n} (x_i - \bar{x})(z_i - \bar{z}) \sum_{i=1}^{n} (z_i - \bar{z})(y_i - \bar{y})}{\sum_{i=1}^{n} (x_i - \bar{x})^2 \sum_{i=1}^{n} (z_i - \bar{z})^2 - \left[\sum_{i=1}^{n} (x_i - \bar{x})(z_i - \bar{z}) \right]^2} \tag{14.4a}$$

$$b_2 = \frac{\sum_{i=1}^{n} (x_i - \bar{x})^2 \sum_{i=1}^{n} (z_i - \bar{z})(y_i - \bar{y}) - \sum_{i=1}^{n} (x_i - \bar{x})(z_i - \bar{z}) \sum_{i=1}^{n} (x_i - \bar{x})(y_i - \bar{y})}{\sum_{i=1}^{n} (x_i - \bar{x})^2 \sum_{i=1}^{n} (z_i - \bar{z})^2 - \left[\sum_{i=1}^{n} (x_i - \bar{x})(z_i - \bar{z}) \right]^2} \tag{14.5a}$$

$$a = \bar{y} - b_1\bar{x} - b_2\bar{z}$$

$(14.6a)$

Population regression relating the mean of _Y_ to _x_ and _z_

$$\hat{y} = \alpha + \beta_1 x + \beta_2 z$$

(14.7)

Coefficient of multiple determination

$$R^2 = \frac{\sum_{i=1}^{n} (\hat{y}_i - \bar{y})^2}{\sum_{i=1}^{n} (y_i - \bar{y})^2}$$

(14.8)

Adjusted coefficient of multiple determination

$$\bar{R}^2 = \frac{\sum_{i=1}^{n} (\hat{y}_i - \bar{y})^2/(n - k - 1)}{\sum_{i=1}^{n} (y_i - \bar{y})^2/(n - 1)} - \frac{k}{n - k - 1}$$

(14.9)

_F_ test statistic for the null hypothesis that the independent variables have no effect on the dependent variable

$$F = \frac{\sum_{i=1}^{n} (\hat{y}_i - \bar{y})^2/k}{\sum_{i=1}^{n} (y_i - \hat{y}_i)^2/(n - k - 1)}$$

(14.10)

Critical values for a hypothesis test on one population slope

$$\text{C.V.}_L, \ \text{C.V.}_U = (\beta_j)_0 \pm t_{\alpha/2} s_{b_j}$$

(14.11)

Standard error of the estimate when there are two independent variables

$$s_e = \sqrt{\frac{\sum_{i=1}^{n} (y_i - \hat{y}_i)^2}{n - 3}}$$

(14.12)

Standard error of the slope of the x variable

$$s_{b_x} = \sqrt{\frac{s_e^2 \sum\limits_{i=1}^{n} (z_i - \bar{z})^2}{\sum\limits_{i=1}^{n} (x_i - \bar{x})^2 \sum\limits_{i=1}^{n} (z_i - \bar{z})^2 - \left[\sum\limits_{i=1}^{n} (x_i - \bar{x})(z_i - \bar{z})\right]^2}}$$ (14.13)

Standard error of the slope of the z variable

$$s_{b_z} = \sqrt{\frac{s_e^2 \sum\limits_{i=1}^{n} (x_i - \bar{x})^2}{\sum\limits_{i=1}^{n} (x_i - \bar{x})^2 \sum\limits_{i=1}^{n} (z_i - \bar{z})^2 - \left[\sum\limits_{i=1}^{n} (x_i - \bar{x})(z_i - \bar{z})\right]^2}}$$ (14.14)

Regression equation when x and z are assumed to interact

$$\hat{y} = \alpha + \beta_1 x + \beta_2 z + \beta_3 xz$$ (14.15)

Regression equation when changes in x and z are assumed to be associated with equal percentage changes in Y

$$\hat{y} = \alpha x^{\beta_1} z^{\beta_2}$$ (14.16)

 ADDITIONAL EXERCISES

An asterisk designates an exercise of greater than ordinary difficulty.

14.27 A state highway engineer would like to predict the number of chuckholes per mile on state highways in order to estimate the costs of highway repairs. The engineer has ten one-mile stretches of highway monitored for the number of 18-wheel trucks per hour and the number of cycles of above- and below-freezing temperatures over a winter. Use the resulting data, shown below and on the next page, to estimate the effect of increasing the number of 18-wheel trucks passing each hour and additional cycles of above- and below-freezing temperatures on the number of chuckholes per mile.

Chuckholes per mile	Trucks per hour	Freezing cycles
36	45	8
44	21	11
49	76	7
56	65	10
70	121	6
70	43	9

Chuckholes per mile	Trucks per hour	Freezing cycles
89	98	12
91	231	16
99	120	13
120	109	18

14.28 Determine the coefficient of multiple determination and the adjusted coefficient of multiple determination for the data in exercise 14.27.

14.29 At the 5% significance level, test the hypothesis that all of the independent variables in exercise 14.27 jointly have no effect on the dependent variable.

14.30 At the 5% significance level, test the hypotheses that each of the individual slope coefficients in exercise 14.27 equals zero.

14.31 A tax assessor needs to estimate the value of 450 houses in a particular area. He has data on the selling prices of ten of these houses, their square footage, and the sizes of their lots. Use the data below to estimate the effect of an increase in square footage and an increase in lot size on selling price.

Selling price (thousands of dollars)	Living space (square feet)	Lot size (acres)
$ 55	1,400	0.30
56	1,350	0.40
65	1,500	0.50
78	1,750	0.25
79	1,500	0.35
79	1,650	0.55
89	1,860	0.75
110	1,750	0.80
118	1,800	0.75
128	2,000	0.90

14.32 Determine the coefficient of multiple determination and the adjusted coefficient of multiple determination for the data in exercise 14.31.

14.33 At the 5% significance level, test the hypothesis that all of the independent variables in exercise 14.31 jointly have no effect on the dependent variable.

14.34 At the 5% significance level, test the hypotheses that each of the individual slope coefficients for the data in exercise 14.31 equals zero.

14.35 A church official who wanted to gain a better understanding of the determinants of contributions to his church gave an anonymous questionnaire to randomly selected church members. The questionnaire asked each family the amount that they had donated in the previous year, their annual income, and the number of times they attended Sunday service. Use the results of the survey, shown on the next page, to estimate the effect of an increase in family income and an increase in attendance on the amount donated.

Amount donated	Family income (thousands of dollars)	Sunday services attended
$115	$ 24	49
125	23	36
128	27	52
140	14	48
145	28	32
156	20	50
157	48	30
160	34	26
160	47	39
170	49	15
181	105	34
194	36	44
202	51	32
214	56	39
225	80	47

14.36 Determine the coefficient of multiple determination and the adjusted coefficient of multiple determination for the data in exercise 14.35.

14.37 At the 5% significance level, test the hypothesis that all of the independent variables in exercise 14.35 jointly have no effect on the dependent variable.

14.38 At the 5% significance level, test the hypotheses that each of the individual slope coefficients for the data in exercise 14.35 equals zero.

14.39 A regression on 20 observations with five independent variables has a coefficient of multiple determination of 0.8. What is the adjusted coefficient of multiple determination?

14.40 A cereal products company has a new formulation of a high-energy breakfast cereal aimed at athletically inclined young adults. The company has tested different packages of the cereal to determine the effect of the package on unit sales. The different types of packages were placed in 90 supermarkets for a week and the unit sales in each supermarket were recorded. The 90 supermarkets were chosen so that they matched in terms of size and demographic characteristics. The packages varied in color (red, blue, and silver) and in design (with photographs of either a diver, a runner, or a skier on the cover). Ten stores were used for each color-design combination. The following regression equation was estimated on the data generated by the experiment:

$$\text{SALES} = 121 - 26*\text{BLUE} + 83*\text{SILVER} + 14*\text{RUNNER} - 3*\text{SKIER} \qquad R^2 = 0.83$$
$$\qquad\qquad (19.2) \qquad\quad (14.5) \qquad\qquad (16.1) \qquad\qquad (5.1)$$

where SALES is the estimated unit sales per week in one supermarket, BLUE is a dummy variable that equals 1 if the package was blue and 0 otherwise, SILVER is a dummy variable that equals 1 if the package was silver and 0 otherwise, RUNNER is a dummy variable that equals 1 if the photograph is of a runner and 0 otherwise, and SKIER is a dummy variable that equals 1 if the photograph is of a skier. The numbers in parentheses are the standard errors of the respective slope coefficients.

(a) What are the estimated unit sales per supermarket for a package that has a red cover with a diver on the front?

(b) Would the estimated change in unit sales caused by switching the photograph from a diver to either a runner or a skier differ significantly from 0 at the 5% significance level?

(c) Which photograph should the company use? Explain your choice.

14.41 An agricultural chemicals company has a new herbicide effective against broad-leaf weeds. To estimate the effect of different dosages of the herbicide, the company runs an experiment in 100 test plots scattered throughout the Midwest. The locations are chosen so that the expected yield per acre will be the same in each site. The company tries different dosages of the herbicide at the sites and observes the rainfall and corn yield. The estimated regression has an interaction term because the company thinks that the herbicide will be more effective at higher levels of rainfall. The estimated regression is:

$$\text{YIELD} = 211 + 0.13*\text{DOSAGE} + 18*\text{RAIN} + 0.00295*\text{RAIN-DOSAGE} \quad R^2 = 0.92$$
$$\quad\quad\quad\quad (0.02) \quad\quad\quad (4.19) \quad\quad\quad\quad (0.00048)$$

where YIELD is the estimated number of bushels of corn per acre for a given dosage and rainfall, DOSAGE is the number of pounds of the herbicide applied per acre, RAIN is the number of inches of rainfall over the growing season, and RAIN-DOSAGE is the product of the RAIN and DOSAGE levels at each test site.

(a) If the recommended dosage is 100 pounds per acre and the average level of rainfall across the sites is 16 inches, what is the expected yield?

(b) Does the interaction term differ significantly from 0 at the 5% level?

(c) At the average level of rainfall, what is the estimated effect of increasing the dosage from 100 pounds to 110 pounds?

14.42 An education researcher is interested in estimating the effect of additional resources on the reading levels of fifth-grade students in a school system. Some school systems have longer school years than others and some have a higher teacher-to-student ratio than others. Because she thinks that both lengthening the school year and increasing the teacher-to-student ratio are subject to the law of diminishing returns, the model fitted is the log of each variable. The data are for a sample of 238 school districts. The estimated equation is

$$\ln(\text{LEVEL}) = 0.2 + 0.46*\ln(\text{DAYS}) + 0.21*\ln(\text{RATIO}) \quad\quad R^2 = 0.41$$
$$\quad\quad\quad\quad\quad (0.13) \quad\quad\quad\quad (0.05)$$

where DAYS is the length of the school year in days and RATIO is the number of teachers in the district who teach kindergarten and the first five grades divided by the number of students in those grades, and ln(LEVEL) is the estimated natural log of the reading level expressed in years. The numbers in parentheses are the respective standard errors for the two slope coefficients.

(a) At the 5% significance level, does either the length of the school year or the teacher-to-pupil ratio have a significant positive effect on reading levels?

(b) If the cost to a school system that had a 200-day school year and 0.05 teacher per student was the same for either an increase of five days in the school year or an increase of 0.01 in the teacher-to-student ratio and the objective of the system was to raise its reading scores, which change should it make?

14.43 You are the personnel director for a firm that manufactures funeral caskets. Some female employees have sued your firm for sex discrimination. The basis of the suit is that the wages of female employees are much lower than the wages of male employees with similar levels of education and experience. You have run a regression in which the dependent variable is an employee's annual wage and the explanatory variables are years of education, years of employment at your firm, and a dummy variable that equals 1 if the employee is a man. The data set consists of all 546 full-time employees of the firm. The estimated regression equation is

$$\hat{y} = 12{,}000 + 250^*ED + 326^*EXP + 1{,}069^*MAN \qquad R^2 = 0.26$$
$$\qquad\quad\ (35.2) \qquad\ (24.0) \qquad\quad (356.09)$$

where $\hat{y}$ is the estimated annual wage, ED is the number of years of education, EXP is the number of years at your firm, and MAN is the dummy variable that equals 1 if the employee is a man or 0 if the employee is a woman. The numbers in parentheses are the standard errors of the slope coefficients.

(a) If the data set were treated as a sample of all possible future wage outcomes given the system of salary compensation in effect at your firm, could you reject at the 5% significance level the null hypothesis that when education and experience are held constant, females have the same or higher annual wages than males?

(b) Do these results prove in a statistical rather than a legal sense that your firm pays higher salaries to some employees because they are men?

14.44* A researcher for the U.S. Department of Justice wants to estimate the effect an increase in the number of prior felony convictions has on the probability that someone convicted of murder will receive the death sentence. The researcher collects the following sample of murder convictions:

Prior felony convictions	Death penalty status	Prior felony convictions	Death penalty status
0	No	5	No
0	No	6	Yes
1	No	6	Yes
1	Yes	7	No
1	No	7	Yes
1	No	7	Yes
2	No	8	Yes
2	No	8	No
2	No	8	No
3	No	9	Yes
3	No	9	Yes
3	Yes	10	Yes
4	No	10	Yes
4	No	10	Yes
4	No	11	Yes
5	Yes	12	Yes
5	No		

Assume that there is a linear relationship between the number of prior felony convictions and an underlying variable called "jury's willingness to impose the death penalty after a murder conviction." Estimate the probability that someone convicted of murder will be given a death sentence for each number of prior felony convictions in the above data set.

14.45* A manufacturer of diesel engines is experimenting with ceramic pistons. These pistons can withstand higher temperatures than metal pistons, and engines with these pistons are more fuel-efficient. The manufacturer has tried different baking times for the pistons and then observed if the piston fails in a million-mile bench test. The following data show the results of this experiment.

Baking time (hours)	Status	Baking time (hours)	Status	Baking time (hours)	Status
1	Failed	15	Failed	28	Did not fail
2	Failed	16	Failed	29	Did not fail
3	Failed	17	Did not fail	30	Did not fail
4	Did not fail	18	Did not fail	31	Did not fail
5	Failed	19	Did not fail	32	Did not fail
6	Failed	20	Failed	33	Did not fail
7	Failed	21	Failed	34	Did not fail
8	Failed	22	Did not fail	35	Did not fail
9	Failed	23	Did not fail	36	Did not fail
10	Failed	24	Did not fail	37	Did not fail
11	Failed	25	Did not fail	38	Did not fail
12	Did not fail	26	Did not fail	39	Did not fail
13	Failed	27	Did not fail	40	Did not fail
14	Failed				

Assume that over the range of baking times in the experiment there is a linear relationship between baking time and an underlying variable called "failure resistance." How many hours should the pistons be baked so that the probability of not failing is 0.99?

14.46* Actuarial firms analyze pension plans and certify whether a plan has enough funds to cover its future obligations. The analysis is complicated because the investment earnings of the funds are uncertain and the timing and duration of retirements and the number of retirees covered by a plan are also uncertain. Actuarial firms typically hire college graduates with good math, statistics, or economics backgrounds and train them in actuarial work. These trainees have to take a series of exams before they are certified as actuaries by a national board. An actuary has to be certified before he or she can sign off on the analysis of a pension plan. The exams are difficult and most trainees do not get certified. The following data show the SAT scores of a sample of trainees and their status ten years after first being employed by an actuarial firm.

SAT scores	Status	SAT scores	Status	SAT scores	Status
1050	Not certified	1124	Not certified	1272	Not certified
1070	Not certified	1127	Not certified	1284	Certified
1076	Not certified	1130	Not certified	1293	Certified
1080	Certified	1145	Certified	1300	Certified
1089	Not certified	1160	Not certified	1316	Certified
1092	Not certified	1189	Certified	1328	Certified
1097	Not certified	1207	Certified	1342	Certified
1098	Certified	1211	Certified	1361	Certified
1101	Not certified	1235	Certified	1407	Certified
1104	Not certified	1251	Certified		

Assume that there is a linear relationship between SAT scores and an underlying variable called "aptitude for actuarial work." If an actuarial firm wanted to set a cut-off point on SAT scores so that newly hired trainees would have at least a 0.75 probability of being certified at the end of ten years, what should this cut-off point be?

14.47* A sales office for a manufacturer of corporate jet planes has made a survey of jet plane ownership in a sample of small to mid-sized firms in the petroleum exploration industry. The data below indicate the number of employees in the firm and whether or not it owns a jet plane.

Number of employees	Ownership status
250	No
278	No
303	No
345	Yes
350	No
369	No
375	No
390	No
401	Yes
421	Yes
489	Yes
492	No
499	Yes
505	Yes
521	Yes
536	Yes
588	Yes
612	Yes
625	Yes

Assume that there is a linear relationship between the number of employees and an underlying variable called "need for corporate jet." Estimate the effect of increasing employment on the probability that a small to mid-sized oil exploration firm will own a corporate jet. The sales office plans to visit a firm that recently grew from 300 to 400 employees and does not have a corporate jet. On the basis of the above data, what does this increase in employment do to the probability that this firm will eventually own a jet?

15

Forecasting Time Series

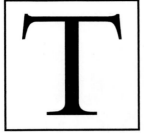

he advertisement for the U.S. Dollar Index™ shown on p. 652 ran in the *Wall Street Journal* of January 2, 1986. The index measures the value of the U.S. dollar against ten other major currencies. The squiggly line running across the ad is a graph of the time series for this index. The ad suggests that two types of customers may be interested in purchasing units of the U.S. Dollar Index™ futures: traders and companies with foreign currency exposure. To help you understand what the ad is selling, let's go over the objectives of the two types of potential buyers, starting with companies with foreign currency "exposure."

A U.S. company can have foreign currency exposure whenever it plans to make or receive a fixed payment in a foreign currency. Consider, for example, a U.S. commodity broker who sells soybeans to a French company and agrees to accept a fixed number of French francs in payment in 30 days. If the exchange value of the franc falls relative to the dollar over the next 30 days, the U.S. broker will lose money. This potential loss is called "exposure." Another example of exposure is a U.S. clothing wholesaler who buys some clothing from a Japanese firm with a fixed payment in Japanese yen due at the end of 30 days. If the yen rises unexpectedly relative to the dollar before the payment is due, the clothing wholesaler will lose money. Exposure to foreign currency fluctuations is very common among U.S. businesses. Almost all companies buy products from foreign firms or sell products to foreign firms, and some time usually elapses between the order and the payment. The exposure problem is more acute for U.S. companies that operate foreign subsidiaries. Currency fluctuations can have large impacts on the reported profits of these firms, even if their sales are stable within each country.

By buying units of the U.S. Dollar Index™ futures (essentially a contract for future delivery of fixed amounts of the ten foreign currencies that make up the index), a U.S. firm can limit its exposure to swings in the U.S. dollar relative to the currencies of the United States' main trading partners. The U.S. Dollar Index™ futures market and similar markets for individual foreign currency futures serve an important function in limiting the risks of currency transactions.

The second type of potential customer listed in the ad is a "trader." A trader is someone who has or can borrow money and who thinks that he or she can "beat the market." Here the exposure is not part of some other business with a foreign firm or subsidiary. A trader's business is anticipating fluctuations in the price of the dollar. For example, suppose you were a trader and you thought that the ten foreign currencies that make up the index will rise 5% relative to the dollar over the next six months while the "market" seemed to think that the value of those currencies in U.S. dollars would stay the same. What the "market" thinks is indicated by the current price for a contract for future delivery of the currencies that make up the U.S. Dollar Index™. Suppose you were right, and you had bought a contract for delivery of $50 million worth of

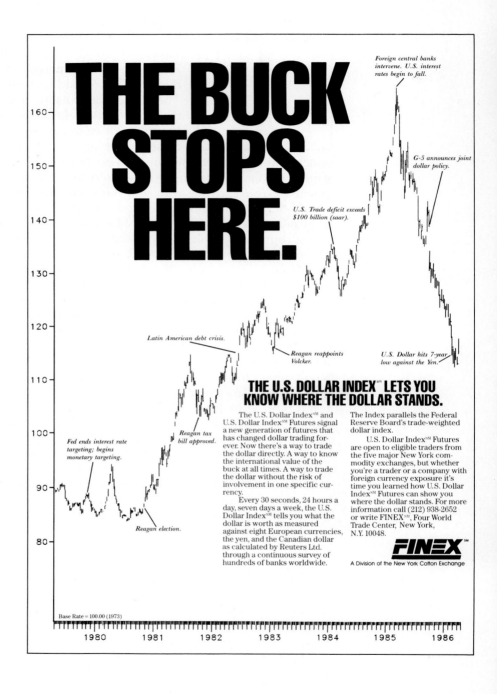

THE BUCK STOPS HERE.

Foreign central banks intervene. U.S. interest rates begin to fall.

G-5 announces joint dollar policy.

U.S. Trade deficit exceeds $100 billion (saar).

Latin American debt crisis.

Reagan reappoints Volcker.

U.S. Dollar hits 7-year low against the Yen.

Fed ends interest rate targeting; begins monetary targeting.

Reagan tax bill approved.

Reagan election.

THE U.S. DOLLAR INDEX™ LETS YOU KNOW WHERE THE DOLLAR STANDS.

The U.S. Dollar Index℠ and U.S. Dollar Index℠ Futures signal a new generation of futures that has changed dollar trading forever. Now there's a way to trade the dollar directly. A way to know the international value of the buck at all times. A way to trade the dollar without the risk of involvement in one specific currency.

Every 30 seconds, 24 hours a day, seven days a week, the U.S. Dollar Index℠ tells you what the dollar is worth as measured against eight European currencies, the yen, and the Canadian dollar as calculated by Reuters Ltd. through a continuous survey of hundreds of banks worldwide.

The Index parallels the Federal Reserve Board's trade-weighted dollar index.

U.S. Dollar Index℠ Futures are open to eligible traders from the five major New York commodity exchanges, but whether you're a trader or a company with foreign currency exposure it's time you learned how U.S. Dollar Index℠ Futures can show you where the dollar stands. For more information call (212) 938-2652 or write FINEX, Four World Trade Center, New York, N.Y. 10048.

FINEX℠
A Division of the New York Cotton Exchange

Base Rate = 100.00 (1973)

1980 1981 1982 1983 1984 1985 1986

the ten currencies for delivery in six months. At the end of the six months you could take delivery of the bundle of ten currencies and then sell them for 5% more than you had paid for them. It seems like an easy way to pocket $2.5 million (5% of $50 million) less the broker's fee for the futures contract.

Your ability to outguess the market is the key to this easy money. Somehow you have to forecast where the squiggly line will be in six months. Even if you were not a professional currency speculator, forecasting those squiggles could be important. A company that is interested only in making a profit on the sales of its product still has to decide if it is worthwhile to buy units of the U.S. Dollar Index™ futures to limit its exposure to foreign currency fluctuations. If the company's financial officers think that the chances of the dollar's fall or rise beyond the futures price quoted by the index are small, the "insurance" against currency fluctuations offered by the futures contracts may not be worth the broker's fee. A company that buys and/or sells abroad cannot avoid forecasting the future value of the dollar. If the company chose to ignore the problem and act as if the dollar would stay the same relative to foreign currencies, it would implicitly be making a forecast.

Two points are being made about this ad: (1) speculators can use accurate forecasts to make profits and (2) firms that have no intrinsic interest in speculation must make forecasts if they buy or sell abroad. This second point can be extended far beyond those firms that have significant foreign exchange transactions. Any firm that buys or sells for future delivery needs to make forecasts. A local home contractor bidding on a house may have to forecast the prices of labor and material six months in the future. A bakery that knows it will need flour six months from now can wait six months and buy the flour on the spot market or fix its future price today by buying a contract for future delivery. Forecasting is a vital business use of statistics.

This chapter describes techniques for modeling and forecasting time-series data. We'll start with a technique that can give reliable estimates if the pattern in the data is fairly simple. This technique, the classical model of forecasting, breaks the motion of the time series into a trend and seasonal and random components. Another approach to forecasting is to smooth the data so that most of the random variation is removed. We'll discuss two smoothing techniques, moving averages and exponential smoothing. Next we'll describe some newer techniques, autoregression and ARIMA, which generally provide more reliable estimates for a wide variety of patterns in the data.

15.1 INTRODUCTION TO FORECASTING

 15.1A STRUCTURAL MODELS VERSUS NAIVE MODELS

The graph of the time series in the ad for the U.S. Dollar Index™ futures is labeled with events that are considered important determinants of the recent path of the price of the dollar. A **time series,** as we saw in Chapter 14, is *a set of measurements on a variable taken over some time period*. Usually a set of time-series data is taken at regular intervals, such as every month, and presented in chronological order. One point is labeled "Reagan election," another is "Reagan reappoints Volcker," and a third is "Foreign central banks intervene." Presumably each of these events affected either the future supply of or demand for dollars. A **structural model** of a time series is *a model that attempts to describe the key behavioral relationships among the agents that affect the path of a time series*. A structural model would describe the supply of and demand for each currency over time in terms of the actions of consumers, businesses, and monetary authorities in each country. For example, a structural model would describe how consumers and businesses react to U.S. trade deficits, interest-rate differentials across countries, and changes in tax laws.

A **naive model,** in contrast, is *a model based on the assumption that there is some pattern over time in the forecast variable and that this pattern can be used to predict the future path of the forecast variable*. In this example a naive model would use the pattern of past behavior of the price of the dollar to predict the future path of the dollar. The essential naiveté of such models comes from the assumption that the future will repeat the past. The simplest naive models use only the past path of the variable that is to be forecast to estimate its future path. More complicated naive models use the paths of several variables to estimate the future path of the variable that must be forecast.

At first glance it may seem that the structural models would be better for forecasting. And they would be if you could identify the parameters of the underlying processes that determine the path of the forecast variable. In other words, if you had the correct structural model, you would have excellent forecasts. Unfortunately, it is extremely difficult to understand and describe the underlying structure of the variables that determine the path of a variable with as complicated a structure as the price of the dollar. Naive models of economic phenomena tend to outperform structural models. Here "outperform" means that they have less forecast error over a series of forecasts.

Naive and structural models really have different purposes. If you were an economist working for the Federal Reserve Board or for the monetary authority of one of the major trading partners of the United States, a structural model would be a necessary tool to help you advise policy makers. Before the Fed or some foreign central bank intervenes in currency markets, it must have an idea of the effect of its intervention. A structural model can be used to predict the reaction of a system to new policies or rules. A naive model cannot answer such "What if we did this?" questions because the basic assumption behind a naive model is that the future will repeat the past. The purpose of a naive model is strictly to forecast. A currency speculator might have an intellectual interest in why and how the world monetary system works, but for the purposes of currency speculation, an accurate forecast of the future price of the dollar is all that is needed.

Naive models are the preferred technique for making forecasts unless there is a major change in the system. If the world economy went back to the gold standard, you could throw away any forecasts of the path of the dollar made on the basis of past fluctuations. If the underlying regime or rules are stable, the naive forecasts can do a better job of catching the patterns, or describing the movement of the squiggly line in our example.

Here is a second example that illustrates the different uses of naive and structural models. A refrigerator manufacturer may subscribe to an econometric service that runs a structural model of the economy, such as Chase Econometrics or Wharton or Data Resources Inc., to estimate the effect of a new tax law on the sales of its refrigerators. The new tax law may discourage home construction and thus discourage refrigerator sales. The same firm may use a naive model to forecast sales for short-term production planning. The naive model should outperform the structural model as long as there are no fundamental changes in the factors that affect refrigerator sales.

In this chapter we will stick to naive models. Structural models are usually introduced in econometrics courses. Moreover, we will stick to the simplest naive models, which predict the future path of one variable only in terms of the past motion in that variable. Don't be misled by the word "naive"; models that use only the path of the forecast variable to predict its future motion are often complicated. Tracking all of the squiggles may require an elaborate mathematical form.

Before we discuss any specific models, another point must be made. Some phenomena do not have a useful past in the sense that the pattern of their past behavior provides no clue to their future behavior. A case in point is the case study in Chapter 13 on the time path of deaths caused by major earthquakes. The time series of this variable appeared to be completely random. A structural model of the geological causes of earthquakes would be more helpful in forecasting deaths from earthquakes (and presumably preventing them) than a naive time-series model.

A third example may help clarify the point of the benefits of a structural model when there is no pattern in the time series of the forecast variable. Many gamblers subscribe to tip sheets that predict the point spreads in professional football games. Some of the most sophisticated tip sheets use structural models to predict point spreads. For example, variables indicating whether a game will be played in an indoor stadium, on artificial turf, in rainy weather, and so on are all factored into the structural model because these conditions are thought to favor the team that has a better field goal kicker, a better running game, a more experienced offensive line, and so on. The accuracy of the forecasts of the point spreads depends on the accuracy and timeliness of the data used to estimate the coefficients of the model and on the correct understanding of the determinants of the point spreads. Again, creating such a structural model is a formidable task, and only a few tip sheets have a track record of making predictions accurate enough to permit a bettor to make even the thinnest consistent profit from gambling. To make a profit a gambler needs to beat the point spread in approximately 54% of his bets. The reason for the 54% is that bookies charge a fee for making the bets. Only a few tip services consistently make the required 54%, and then only by a small margin.

If there were a consistent pattern to the actual point spreads in professional football games, a naive model could replace all the work of collecting the data on so many variables and doing the estimating required by complex structural models. All the information you would need to make money betting on professional football would be contained in the past pattern of points scored by each team. The better-performing tip sheets all seem to rely on structural models. I say "seem" because no one really knows how some of the tip sheets come up with their predictions. The implication of the widespread use of structural models is that the pattern of points scored by each team over time does not contain enough information to allow a bettor to make a profit from naive forecasts.

■ 15.1B THE CLASSICAL TIME-SERIES MODEL: TREND, SEASONAL, AND RANDOM COMPONENTS

A **trend** is *a consistent long-term change in the average level of the forecast variable per unit of time.* In the data on the U.S. Dollar Index™ there is a clear upward trend from the middle of 1980 to January 1985, followed by a downtrend in the balance of 1985 and the beginning of 1986. The trend in a data series can be estimated by the slope coefficient for a simple regression in which the dependent variable is the forecast variable and the explanatory variable is time. The time variable is usually coded as 1 for first observation and incremented by 1 for each subsequent observation. In notation a simple time-trend model is

$$(\mu_Y)_t = \alpha + \beta t \qquad \textbf{(15.1)}$$

where t stands for time and $(\mu_Y)_t$ is the average value of the forecast variable at time t. Formula 15.1 is the population relationship between the average level of the forecast variable and time. In the actual data, the observations of Y are not always at the mean of Y. At any given time there is some random variation in the distribution of Y values about the mean of Y. If the random variation has a mean of zero and no discernible pattern, it is called "white noise" in time-series analysis. The term "white noise" was originally used by audio engineers to describe a distribution of sound waves, but in statistical usage "noise" refers to unpredictable errors rather than sound.

Given these assumptions, the observed value of Y at any time t is

$$y_t = \alpha + \beta t + \varepsilon_t \qquad \text{(15.2)}$$

where y_t is the observed value of Y at time t and ε_t is the value of the white noise random variable at time t. Thus the observed value of Y at time t is made up of a deterministic component $(\alpha + \beta t)$ and a random component (ε_t). The random component is also called the error random variable. The estimated regression equation takes the form

$$\hat{y}_t = a + bt \qquad \text{(15.3)}$$

Formula 15.3 is so simple that it rarely captures the motion of the forecast variable. Few variables seem to move with a steady trend over time plus some white noise variation. The next step up in complexity is to account for any seasonality in the data. **Seasonality** is *any consistent pattern in a forecast variable which is repeated within a year*. "Seasonality" may refer to the four seasons of the year, but it can also refer to repetition of a pattern by months, by weeks, or even by days of the week. In general, you can test for seasonality over periods as long as the steps in your data. Daily data let you test for seasonality by day, week, month, or quarter. Weekly data let you test for seasonality by week, month, or quarter. Monthly data let you test for seasonality by month or quarter.

There are many techniques for accounting for seasonality. The one that we will discuss is an extension of techniques introduced in Chapter 14, on dummy variables. Formula 15.4 is used to estimate the effect of a trend variable and the seasons summer, fall, winter, and spring on the forecast variable.

$$\hat{y}_t = a + b_1 t + b_2 \text{SUMMER} + b_3 \text{FALL} + b_4 \text{WINTER} \qquad \text{(15.4)}$$

SUMMER is a dummy variable that takes the value 1 if the observation is from June, July, or August and 0 otherwise. FALL is a dummy variable that takes the value 1 if the observation is from September, October, or November and 0 otherwise. WINTER is a dummy variable that takes the value 1 if the observation is from December, January, or February and 0 otherwise. Recall that a categorical independent variable that can fall into k categories can be represented by $k - 1$ dummy variables. The interpretation of the coefficient of the SUMMER dummy variable is the amount the average level of the forecast

TABLE 15.1
Monthly data series with quarterly seasonal variation and a trend

Month	Y_t	Summer	Fall	Winter	Month	Y_t	Summer	Fall	Winter
Jan. 1980	17.9510620	0	0	1	Jan. 1983	41.2069946	0	0	1
Feb.	12.3628784	0	0	1	Feb.	39.1055054	0	0	1
Mar.	30.5022339	0	0	0	Mar.	54.9564087	0	0	0
Apr.	23.7701904	0	0	0	Apr.	57.0106812	0	0	0
May	32.6019287	0	0	0	May	64.4423950	0	0	0
June	44.0553711	1	0	0	June	65.0931641	1	0	0
July	41.2958008	1	0	0	July	65.0320190	1	0	0
Aug.	35.8683838	1	0	0	Aug.	75.7678711	1	0	0
Sept.	31.4153931	0	1	0	Sept.	69.5277710	0	1	0
Oct.	30.2015991	0	1	0	Oct.	68.0521362	0	1	0
Nov.	37.9528320	0	1	0	Nov.	64.4771973	0	1	0
Dec.	20.4280640	0	0	1	Dec.	47.3981079	0	0	1
Jan. 1981	19.2693848	0	0	1	Jan. 1984	49.1222412	0	0	1
Feb.	29.5070068	0	0	1	Feb.	51.9472046	0	0	1
Mar.	41.9702759	0	0	0	Mar.	66.8104370	0	0	0
Apr.	35.3911255	0	0	0	Apr.	65.3736206	0	0	0
May	36.2760864	0	0	0	May	67.1116699	0	0	0
June	46.4499878	1	0	0	June	73.9680664	1	0	0
July	53.3684570	1	0	0	July	77.7380371	1	0	0
Aug.	48.4372559	1	0	0	Aug.	75.0952881	1	0	0
Sept.	40.9648560	0	1	0	Sept.	70.0191895	0	1	0
Oct.	41.6187378	0	1	0	Oct.	73.2142090	0	1	0
Nov.	47.4394897	0	1	0	Nov.	80.3727295	0	1	0
Dec.	33.7165405	0	0	1	Dec.	62.8626099	0	0	1
Jan. 1982	30.3598633	0	0	1	Jan. 1985	64.8701904	0	0	1
Feb.	33.6261108	0	0	1	Feb.	60.0744873	0	0	1
Mar.	47.6664673	0	0	0	Mar.	80.6606812	0	0	0
Apr.	49.7985596	0	0	0	Apr.	72.0569336	0	0	0
May	48.8030273	0	0	0	May	75.4645386	0	0	0
June	60.3606567	1	0	0	June	89.3360718	1	0	0
July	58.5155151	1	0	0	July	82.8785034	1	0	0
Aug.	63.0642334	1	0	0	Aug.	85.0259766	1	0	0
Sept.	49.1163086	0	1	0	Sept.	81.2671387	0	1	0
Oct.	53.0062134	0	1	0	Oct.	85.3163452	0	1	0
Nov.	61.1741943	0	1	0	Nov.	80.8408936	0	1	0
Dec.	43.7494019	0	0	1	Dec.	72.4388062	0	0	1

variable increases in the summer relative to spring, the omitted dummy variable. The coefficients of FALL and WINTER also represent changes in the average level of the forecast variable relative to its spring level. If you wanted to estimate a monthly seasonality, you would need 11 dummies labeled JANUARY, FEBRUARY, . . . , NOVEMBER. Here the omitted dummy variable is DECEMBER. Note that the choice of which dummy to omit is arbitrary.

Table 15.1 shows an artificial monthly time-series data set that meets all of the assumptions required for statistical inference. Few real data sets satisfy the requirements. Here we see how the seasonal dummy variables are coded.

A graph gives a much better idea of any pattern in the data than a table of the raw data. Figure 15.1 shows the graph of the data in Table 15.1. The upward trend is the general rise in the data from the first to the last observation. In each year there is a pattern of low values in winter and high values in summer, with intermediate values in spring and fall.

The next step is to run a multiple regression program to estimate the coefficients of the time variable and the seasonal dummies. The estimated coefficients are

$$\hat{y}_t = 25.846 + 0.793t + 8.214\text{SUMMER} + 1.647\text{FALL} - 13.053\text{WINTER} \qquad R^2 = 0.970$$
$$\quad\;\; (0.0194) \qquad (1.135) \qquad\;\; (1.140) \qquad\quad (1.134)$$

The standard error of the slope coefficient is shown in parentheses under each term. If you tested the null hypothesis that the population slope was 0 for each of the four slope coefficients at the 5% significance level, you would accept the null only for the population slope of FALL. The test for the FALL coefficient is

FIGURE 15.1
Graph of the data in Table 15.1

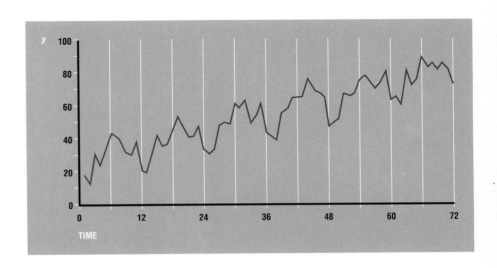

$$\text{C.V.}_L, \text{C.V.}_U = (\beta_j)_0 \pm t_{\alpha/2}s_{b_j} \tag{14.11}$$

$$\text{C.V.}_L, \text{C.V.}_U = 0 \pm 1.96(1.14)$$

$$\text{C.V.}_L = -2.23 \qquad \text{C.V.}_U = 2.23$$

The sample slope of 1.647 is within the acceptance region of -2.23 to 2.23, so there is not enough evidence to reject the null that the slope coefficient for the FALL variable is 0. Note that the interpretation of the slope coefficient for the FALL dummy variable is the difference between the average level of Y during fall and spring—SPRING being the omitted dummy. Your conclusion is that there is not enough evidence to say that the average level of Y differs between fall and spring.

Suppose you wanted to forecast a future value of Y, say for January 1988. The variable t would take the value 97 in January 1988. The t variable starts at 1 for January 1980 and goes up by 1 for each month. The values of the dummy variables SUMMER, FALL, and WINTER are 0, 0, and 1, respectively. You obtain the predicted value of y for January 1988 by substituting these values into the estimated regression equation:

$$\hat{y}_t = 25.846 + 0.793t + 8.214\text{SUMMER} + 1.647\text{FALL} - 13.053\text{WINTER}$$

$$\hat{y}_t = 25.846 + 0.793(97) + 8.214(0) + 1.647(0) - 13.053(1)$$

$$\hat{y}_t = 89.714$$

The point estimate for the value of Y in January 1988 is 89.714. You may be wondering what to do if you have to forecast the value of Y for the month of September, October, or November. Recall that the slope coefficient for the FALL variable did not differ significantly from 0 at the 5% level. Should you drop the FALL variable from the regression equation or use its coefficient of 1.647? Use the coefficient. For a forecast you want to use your best point estimate of the value of Y. While the true effect of FALL relative to SPRING may be 0, your best point estimate of the effect of FALL relative to SPRING is the estimated coefficient of 1.647.

15.1C TESTING FOR SERIAL CORRELATION

We assumed that error random variables, ε_t, for the above regression had no serial correlation. If the data set had been a real time series, the assumption would be unwarranted. Positive serial correlation is common in the error random variables of a regression equation for economic time series even if you use dummy variables to control for seasonality. If something causes one observation to be high, the same factor is likely to persist and affect subsequent observations. Similarly, if something causes one observation to be low, subsequent observations are likely to be low as well. For example, a company's sales may rise because a competitor has gone bankrupt. After awhile new competitors may enter the industry or existing firms may add to their capacity, but the boost in sales should persist longer than one month.

Before using a classical estimated regression equation (the type estimated in section 15.1B) to forecast values of Y, you have to check the validity of the assumption that there is no serial correlation in the error random variables. Recall that serial correlation was defined in Chapter 13 as a nonzero correlation between any of the error random variables. If positive serial correlation is present in the data, the standard errors of the slope coefficients will be biased downward. This means that on average they will be too low. Undetected, positive serial correlation will give you the impression that your model is much more precise than it is, and you will tend to reject the null hypothesis that individual slope coefficients are zero when the null is true. In other words, positive serial correlation causes you to make Type I errors in selecting the variables to include in your model. Note, however, that serial correlation, either positive or negative, does not bias the estimated slope coefficients.

You do not know the actual parameters of the error random variables. In practice you have to use the residuals of the estimated regression equation as estimates of the ε_t's. A residual, as defined in Chapter 14, is the difference between an observed value of Y and a predicted value of Y. If there is no serial correlation in the error terms, there should be no pattern to the residuals. The residuals should approximate the white noise assumed for the regression model. If there is positive serial correlation in the error, residuals with high values will tend to be followed by more residuals with high values, and residuals with low values will tend to be followed by more residuals with low values. In other words, each residual is followed by another of similar value. If there is negative serial correlation among the error random variables, residuals with high values will tend to be followed by residuals with low values. In other words, there tends to be an alternating pattern in the values of the residuals: high, low, high, low. Figure 15.2 shows three plots of residuals with no serial correlation, positive serial correlation, and negative serial correlation. If you connect the dots in order of time, it will help you to see the distinctions between positive and negative serial correlation and no serial correlation.

When there is positive serial correlation in the error random variables, the occurrence of a high residual is followed by other high residuals and the occurrence of a low residual is followed by other low residuals. Generally the residuals following the first high or low one in a group tend to have smaller absolute value. If the residuals grew following a shock of a high or low observation, the series would explode. Its values would go off the graph. Economic time series that explode are rare; hyperinflation, as in Germany after World War I, is one example. When there is negative serial correlation in the error random variables, the occurrence of a high residual is followed by a low residual and the occurrence of a low residual is followed by a high residual. The damping or

FIGURE 15.2
Three plots of residuals of
classical regression
equations showing no serial
correlation, positive serial
correlation, and negative
serial correlation

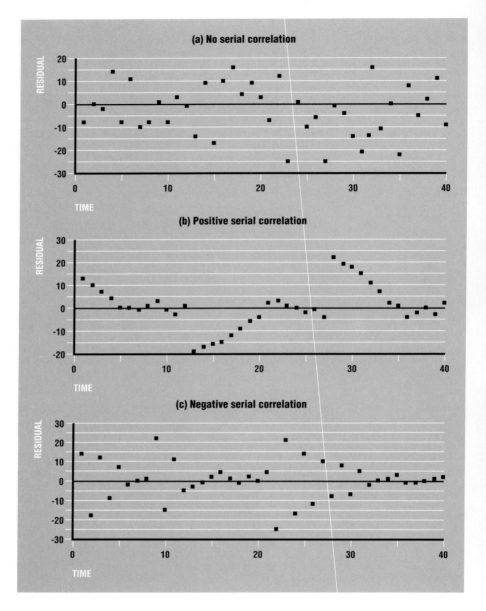

FIGURE 15.2
Three plots of residuals of classical regression equations showing no serial correlation, positive serial correlation, and negative serial correlation

decaying in effect of an unusually high or low residual is also clear in the plot exhibiting negative serial correlation.

Table 15.2 shows the residuals of the regression equation for the data in Table 15.1. It is difficult to inspect a long column of numbers and decide if positive or negative serial correlation is present in the residuals. A plot of the

TABLE 15.2
Residuals of the regression on the data in Table 15.1

Observed value of Y	Predict value of Y	Residual	Observed value of Y	Predicted value of Y	Residual
17.95106	13.58663	4.364433	41.20699	42.14175	−0.9347572
12.36288	14.37983	−2.016947	39.10551	42.93495	−3.829441
30.50223	28.22571	2.276524	54.95641	56.78083	−1.824421
23.77019	29.01891	−5.248716	57.01068	57.57403	−0.5633469
32.60193	29.81211	2.789824	64.44240	58.36723	6.075172
44.05537	38.81923	5.236137	65.09316	67.37435	−2.281189
41.29580	39.61243	1.683369	65.03202	68.16755	−3.135529
35.86838	40.40563	−4.537243	75.76787	68.96075	6.807121
31.41539	34.63182	−3.216427	69.52777	63.18694	6.340832
30.20160	35.42502	−5.223419	68.05214	63.98014	4.072002
37.95283	36.21822	1.734615	64.47720	64.77334	−0.2961426
20.42806	22.31180	−1.883739	47.39811	50.86692	−3.468815
19.26938	23.10500	−3.835617	49.12224	51.66012	−2.537880
29.50701	23.89820	5.608809	51.94720	52.45332	−0.506115
41.97028	37.74408	4.226192	66.81044	66.29920	0.5112381
35.39113	38.53728	−3.146156	65.37362	67.09240	−1.718781
36.27608	39.33048	−3.054394	67.11167	67.88560	−0.7739258
46.44999	48.33760	−1.887615	73.96807	76.89272	−2.924660
53.36846	49.13080	4.237656	77.73804	77.68592	5.211639E-02
48.43726	49.92400	−1.486744	75.09529	78.47912	−3.383827
40.96486	44.15019	−3.185337	70.01919	72.70531	−2.686127
41.61874	44.94339	−3.324654	73.21421	73.49851	−0.2843018
47.43949	45.73659	1.702904	80.37273	74.29171	6.081017
33.71654	31.83018	1.886366	62.86261	60.38530	2.477314
30.35986	32.62337	−2.263512	64.87019	61.17849	3.6917
33.62611	33.41657	0.2095375	60.07449	61.97169	−1.897205
47.66647	47.26245	0.4040108	80.66068	75.81757	4.843109
49.79856	48.05566	1.742905	72.05693	76.61077	−4.553841
48.80303	48.84885	−4.582596E-02	75.46454	77.40397	−1.93943
60.36066	57.85598	2.504677	89.33607	86.41109	2.92498
58.51552	58.64918	−0.1336632	82.87850	87.20429	−4.32579
63.06423	59.44238	3.621857	85.02598	87.99749	−2.971512
49.11631	53.66857	−4.552258	81.26714	82.22369	−0.9565582
53.00621	54.46177	−1.455551	85.31635	83.01689	2.299454
61.17419	55.25496	5.919231	80.84089	83.81009	−2.969193
43.74940	41.34855	2.400848	72.43881	69.90367	2.535133

FIGURE 15.3
Plot of the residuals in Table 15.2

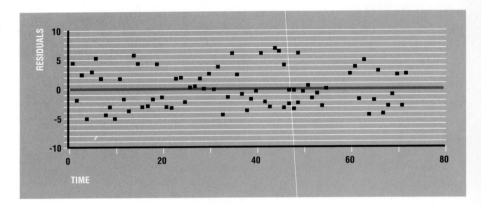

residuals, as in Figure 15.3, sometimes gives clearer evidence of the presence of serial correlation. Even with the plot, however, it is difficult to decide whether there is serial correlation in these data. There seems to be some negative serial correlation from around the 18th to the 36th observation, but it is not a sharp pattern. What we need is a statistical test.

The **Durbin–Watson test** is *a test for serial correlation which analyzes the residuals of the estimated regression line.* The Durbin–Watson test statistic is

$$\text{DW} = \frac{\sum\limits_{t=1}^{n-1} (e_{t+1} - e_t)^2}{\sum\limits_{t=1}^{n} e_t^2}$$

(15.5)

The e's stand for the residuals from the estimated regression. The residual at time t is an estimate of the mean of the noise-generating random variable, ε, at time t. The Greek letter epsilon, ε, is used to denote the random variable that creates the variation around the population regression line, while e refers to the residuals around the sample regression line. If there is positive serial correlation among the error random variables, successive residual terms will be closer together than would be expected when the error random variables are independent. If there is negative serial correlation, successive residuals will be farther apart than would be expected when the error random variables are independent. The effect of positive serial correlation is to reduce the value of the numerator relative to the value of the denominator in formula 15.5, while the effect of negative serial correlation is to increase the value of the numerator in the formula.

Appendix C contains a proof that in the absence of serial correlation the expected value of the Durbin–Watson test statistic is 2. The range of possible values for DW is from 0 to 4. Values for DW above 2 would make you suspect negative serial correlation, and values below 2 would make you suspect positive serial correlation. Table D11 provides the critical values, DW_L and DW_U, of the Durbin–Watson test statistic. There are three possible tests that use these critical values: (1) against the alternative hypothesis of positive serial correlation; (2) against the alternative hypothesis of negative serial correlation; and (3) against the two-tailed alternative hypothesis of either positive or negative serial correlation.

(1) For a test for positive serial correlation, the null hypothesis of no serial correlation or negative serial correlation will be rejected if the test statistic is less than the tabled value of DW_L. If DW is greater than DW_U, the null is accepted. If DW is either between DW_L and DW_U or equal to DW_L or DW_U, the test is inconclusive. (2) For a test for negative serial correlation, the null hypothesis of no serial correlation or positive serial correlation will be rejected if the test statistic is greater than 4 minus the tabled value of DW_L. If DW is less than 4 minus DW_U, the null is accepted. If DW is equal to or between 4 minus DW_U and 4 minus DW_L, the test is inconclusive. (3) For a test for negative or positive serial correlation, the null hypothesis of no serial correlation will be rejected if the test statistic is less than the tabled value of DW_L or more than 4 minus DW_L. If DW is between DW_U and 4 minus DW_U, the null is accepted. Finally, if the test statistic is outside these ranges, the test is inconclusive. Figure 15.4 summarizes the decision rules for these hypothesis tests.

The tabled values of DW_L and DW_U depend on the number of observations in the time series, n, and on the number of independent variables, k. The critical values are given for one-tailed tests at the 1% and 5% significance levels. For two-tailed tests the significance levels are twice as high as Table D11 indicates.

Let's use the residuals in Table 15.2 to calculate the Durbin–Watson test statistic and test against negative serial correlation among the error random variables at the 5% significance level. Usually you would test for positive serial correlation, because positive serial correlation is more common in time-series data and because it biases the standard errors downward. Since there is no hint of positive serial correlation in Figure 15.3 and there seems to be negative serial

FIGURE 15.4
Decision rules for tests of serial correlation

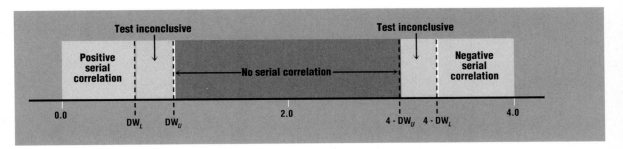

TABLE 15.3
Calculations for the
Durbin-Watson test
statistic

e_t	$e_t - e_{t-1}$	$(e_t - e_{t-1})^2$	$(e_t)^2$
4.364433	—	—	19.0482754
−2.016947	−6.38138	40.7220107	4.0680752
2.276524	4.293471	18.4338932	5.18256152
⋮	⋮	⋮	⋮
2.299454	3.2560122	10.6016154	5.2874887
−2.969193	−5.268647	27.7586412	8.81610707
2.535133	5.504326	30.2976047	6.42689933
		1,447.36514	774.754186

correlation, we'll use the Durbin-Watson statistic to test for negative serial correlation. To save space, only the sums and the first three and last three values are shown in Table 15.3.

When we substitute the calculated values into the formula for the Durbin-Watson test statistic, we get

$$\text{DW} = \frac{\sum_{t=1}^{n-1} (e_{t+1} - e_t)^2}{\sum_{t=1}^{n} e_t^2} \tag{15.5}$$

$$\text{DW} = \frac{1,447.365}{774.754}$$

$$\text{DW} = 1.868$$

The tabled critical values for a 5% significance level, 70 observations, and 4 independent variables are $\text{DW}_L = 1.49$ and $\text{DW}_U = 1.74$. The closest value to the sample size of 72 in Table D11 is 70. The decision rule for a test for negative serial correlation is

Reject H_0 if DW $> 4 - \text{DW}_L$
        (1.868)   (4 − 1.49)
        (1.868)   (2.51)

Accept H_0 if DW $< 4 - \text{DW}_U$
        (1.868)   (4 − 1.74)
        (1.868)   (2.26)

Draw no conclusion if $4 - \text{DW}_U \leq$ DW $\leq 4 - \text{DW}_L$
                (4 − 1.74)   (1.868)   (4 − 1.49)
                (2.26)   (1.868)   (2.51)

The condition for accepting the null of no negative serial correlation among the error random variables is satisfied. At the 5% significance level the null hypothesis of no negative serial correlation is accepted.

The Durbin–Watson test applies to structural models that use other variables as right-hand or independent variables. However, there are some limitations on the use of the Durbin–Watson test for serial correlation. The test will not give accurate answers if lagged observations of the dependent variable are used as independent variables. If lagged values of the forecast variable are used as independent variables, the DW test statistic is biased toward 2.0, and you are likely to conclude that no serial correlation exists even if it is present. In section 15.2 we'll investigate the use of lagged values of the forecast variable as independent variables.

Recall that the data set in this example was made up so that it would satisfy all of the standard assumptions in multiple regression analysis. This result would be unusual in a real time-series data set. The most plausible data sets for the classical time-series model are ones in which a strong seasonal pattern is well known. For example, the time-series for transatlantic air travel has a well-known pattern of high summer and Christmas-season travel. A time series for an agricultural product is also a good candidate for the classical time-series model. If you find a real data set that seems to fit all of the assumptions, the classical model should give a reliable forecast. The transatlantic air-traffic example can be expanded to point out the limitations of naive forecasts. If the economic conditions that govern transatlantic air traffic, such as the price of fuel and national income in Europe and the United States, are stable, the classical model should give a reliable forecast. A structural model would be required to predict passenger traffic if changes in these conditions are expected. Finally, no model is proof against random events. The sharp decline in transatlantic air traffic in the summer of 1986 because of the fears of Libyan attacks on American tourists would be hard for any model to forecast.

The classical model can be modified to accommodate time series that depart from some of the other assumptions required for statistical inference in regression analysis. If the linearity assumption seems to be violated while the other assumptions hold up, you can transform the data to establish linearity (the most common transformation is to take the logs of the Y values, as we saw in Chapter 13). If the assumption of homoscedasticity is violated, you can use weighted least squares (also discussed in Chapter 13). However, violations of the assumption of no serial correlation often do not have a simple "fix" because the correlations between the error or noise terms can be complicated.

The quarterly seasonality in the data set in Table 15.1 was an example of positive serial correlation. In a regression with time as the only independent variable, the error term for any one month is correlated with the error terms for all months in the same season. The "fix" is a set of seasonal dummy variables. The residuals of the regression with the seasonal dummies show no serial correlation. This seasonal pattern is common; many other patterns are also common in time-series data. For example, the error term for a given month may be correlated only with the error term for the same month in previous years, or only with the error term of the previous month. Regression models for such patterns of serial correlation that yield uncorrelated residuals are fairly straightforward: for the year-to-year correlations you can use 11 monthly dummy variables; for the preceding-month correlations you can regress the first differences in the forecast variable on the time trend. (We'll discuss differencing in more detail in section 15.4.)

Suppose the correlation of the error random variables evolved or shifted over time, so that at the beginning of the data series the correlation was with the previous month's error random variable and as the series progressed to more recent data points the correlation for one month's error random variable was with month 2 or 3 or 4 or 5 before the data point. It is much more difficult to find a regression model to describe an evolving pattern of serial correlation than to find one that describes a stable pattern. Later in the chapter you will find out how to recognize error structures and how to fit the best regression models.

This focus on the residuals may seem like a lot of bother. Who cares about the residuals, anyway? You want to know the future values of the forecast variable. The point is that knowledge of the correlations among the error random variables can be exploited to make more accurate forecasts. The residuals of the regression equation should contain only white noise. If there is a predictable component to the residuals, the model can be reformulated to include the predictable component, so that it can help you predict future values of the forecast variable. Otherwise you will be ignoring useful information. The techniques of autoregressive models and ARIMA attempt to formulate models that exploit any consistent variation in the residuals of the regression equation. These techniques, especially ARIMA, are quite complicated. Before discussing any of them, we'll investigate an entirely different approach to forecasting. Smoothing is a way of forecasting which tries to remove most of the variation without trying to model it. In effect, smoothing treats most of the variation in the forecast variable as random. Smoothing techniques have the defect of being highly arbitrary, as you never know how much of the variation to smooth away. When you do data smoothing, you can't be sure if you have stripped away the noise that obscures the underlying pattern in the data or if you have smoothed out a consistent pattern that could be useful information. The chief advantage of data smoothing is that it is simple. Perhaps because of this advantage, it is still widely used.

EXERCISES FOR SECTION 15.1

15.1 Use a simple trend-line model to estimate the value of Y at time $t = 20$ for the following data series, starting at $t = 1$: 43, 48, 49, 37, 49, 52, 56, 57, 59, 61, 70, 72, 78, 71, 85, 90, 91, 90, 94.

15.2 Use the following estimated regression equation on monthly data to predict the value of Y at time $t = 100$. Assume that the data are monthly and that the first month in the data series is January.

$$\hat{y}_t = 35 + 4t + 6\text{SPRING} - 2\text{SUMMER} + 9\text{WINTER}$$

15.3 Use the data in exercise 15.1 to test for positive serial correlation at the 5% significance level.

15.4 (a) Use the following estimated regression equation on monthly data to predict the value of Y at time period $t = 110$. Assume that $t = 1$ is May.

$$\hat{y}_t = 14.93 + \underset{(0.076)}{2.00t} + \underset{(0.001)}{0.41\text{SPRING}} - \underset{(0.08)}{0.94\text{SUMMER}} - \underset{(0.083)}{0.09\text{WINTER}}$$

(b) If the estimated regression is based on 100 observations, which estimated slope coefficients differ significantly from 0 at the 5% level?

(c) The Durbin-Watson test statistic for the above regression is 1.80. Test the hypothesis that there is negative correlation at the 5% significance level.

15.5 The following estimated regression equation is on the daily number of patients at an emergency clinic. The variables MONDAY, TUESDAY, WEDNESDAY, THURSDAY, FRIDAY, and SATURDAY are dummy variables for days of the week.

$$\hat{y}_t = 14.93 + \underset{(1.61)}{0.4t} + \underset{(7.90)}{35.2\text{MONDAY}} + \underset{(12.9)}{21.1\text{TUESDAY}} - \underset{(14.2)}{19.0\text{WEDNESDAY}}$$
$$+ \underset{(9.9)}{13.6\text{THURSDAY}} + \underset{(12.3)}{45.6\text{FRIDAY}} + \underset{(14.8)}{89.7\text{SATURDAY}}$$

(a) Is there any trend in the number of patients per day?

(b) On which days does the average number of patients differ significantly from the average number on Sunday?

(c) If the Durbin-Watson test statistic for the above regression is 2.91 and the regression is based on 140 observations, can the null of no serial correlation be rejected at the 5% significance level?

15.2 DATA SMOOTHING

 ### 15.2A MOVING AVERAGES

A **moving average** is *a transformation of a time series by which an observation is replaced by the average of the replaced observation and some adjacent observations*. The technique of smoothing a time series with a moving average has several purposes. It can be used to reduce noise in a time series and let you focus on any underlying patterns; it can be used for making forecasts; and it can be used to deseasonalize data. Since we have already discussed the use of dummy variables to capture seasonal variation, we won't go into this third use of moving averages—called the ratio to moving average method of deseasonalizing a time

	PERIOD									
	1	**2**	**3**	**4**	**5**	**6**	**7**	**8**	**9**	**10**
Time series	6.70	7.80	9.30	5.10	4.50	8.30	9.00	8.20	6.20	5.60
Three-item moving average	—	7.93	7.40	6.30	5.97	7.27	8.50	7.80	6.67	—
Five-item moving average	—	—	6.68	7.00	7.24	7.02	7.24	7.46	—	—

series. Moving averages can be either weighted or unweighted and either centered or not centered. Let's start with an unweighted and centered moving average and then go on to some more complex types.

To construct a centered, unweighted moving average, we replace an observation in a time series by the average of that observation and a specified number of preceding and following observations. For example, to construct a five-item centered and unweighted moving average of a time series, you would replace an observation by the average of the observation plus the two preceding and the two following observations. To construct a three-item centered and unweighted moving average of a time series, you would replace an observation by the average of the observation plus the preceding and the following observation.

Table 15.4 shows the five- and three-item moving averages of a short time series. The first value in the five-item moving average, 6.68, replaces the third observation, 9.30, in the original time series. It was obtained by summing the observations 6.70, 7.80, 9.30, 5.10, and 4.50 and dividing that sum by 5. Clearly, the moving average must have fewer data points than the original time series. A five-item centered moving average loses two data points from each end of the original time series. The three-item centered moving average loses one data point from each end of the series. In general, an x-item centered moving average would lose $(x - 1)/2$ data points from each end of the series. For example, a seven-item centered moving average would lose $(7 - 1)/2$ or 3 data points from each end. For the sake of convenience, the number of items in a centered moving average is usually an odd number. If an even number were used, the data points in the moving average would not line up under the data points in the original series. In a six-item moving average, for example, the first data point would be halfway between the third and the fourth observations in the original series.

To see how the moving average reduces the noise in a time series, consider Figure 15.5. It shows a real time series, the corporate Aaa bond rate by month over the period January 1981 through December 1986. A five-item centered, unweighted moving average is superimposed on the original data series. As you can see, the moving average is much smoother than the original series. This is the reason that moving averages are called a "smoothing" technique. Note also that the first two and the last two data points in the original time series are lost in the centered moving average. The smoothing effect is increased as the number of items in the moving average goes up. There is no

FIGURE 15.5
Average annual yields of new issues of corporate Aaa-rated bonds by month, 1981–1986, and the same series smoothed with a centered average of five observations

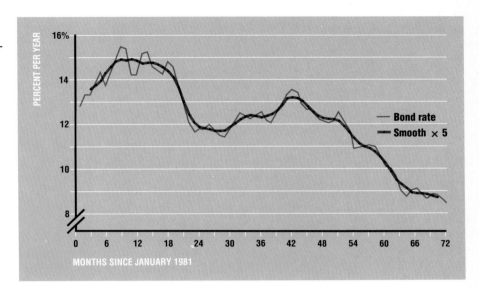

clear-cut rule on how many items should go into the moving average. If too many items are used, any pattern in the data is suppressed along with the noise. For example, if you used a 72-item moving average for the time series shown in Figure 15.5, the resulting moving average would have only one data point. The moving average would be smooth but not very informative. One determinant of the number of items in the moving average is based on whether the moving average is being used to suppress any seasonality in the data. For example, if the data are a daily time series, a seven-item moving average can be used to remove any daily seasonality. Similarly, a moving average of 12 items would remove any monthly seasonality if the data were monthly. A moving average of 52 items can smooth out any weekly seasonality in a time series of weekly data.

The primary purpose of a centered moving average is to remove noise or seasonal variation in a time series. If there is a great deal of noise or pronounced seasonality in a time series, any trends or long-term patterns or cycles can be obscured. A last-term moving average is more practical for making forecasts with moving averages. In a last-term moving average of order x, an observation is replaced by the average of that observation and the preceding $x - 1$ observations. Table 15.5 shows a three- and five-item last-term moving aver-

TABLE 15.5
A ten-period time series and its three- and five-item last-term moving averages

	PERIOD									
	1	**2**	**3**	**4**	**5**	**6**	**7**	**8**	**9**	**10**
Time series	6.70	7.80	9.30	5.10	4.50	8.30	9.00	8.20	6.20	5.60
Three-item moving average	—	—	7.93	7.40	6.30	5.97	7.27	8.50	7.80	6.67
Five-item moving average	—	—	—	—	6.68	7.00	7.24	7.02	7.24	7.46

age on the short time series shown in Table 15.4. With a last-term moving average, all of the lost data points are at the beginning of the series. Other than being shifted over, the values in the moving average are the same as in the centered moving average. The reason for making this shift is that it is easier to use the moving average for forecasting when the values are lined up in this fashion. The moving average forecast for any period is the value of the moving average for the preceding period. If you were using the three-item moving average, for example, the forecast for the fifth period would be 7.40. The forecast based on a three-item moving average for the 11th period would be 6.67. If you have to forecast more than one step ahead, the forecast stays the same; that is, the forecasts made at the end of the 10th period for the 12th, 13th, and 14th periods would all be 6.67. Until you have more information, your forecast for all future periods is the value of the most recent moving average term.

The moving average method of forecasting is about as simple as you can get. There are several problems with a moving average forecast. One is its essential arbitrariness. Why should the last three or the last five or the last x observations determine the forecast and any preceding observations have no effect? Why should each of the last three or the last five or the last x observations have equal weight in determining the forecast? The answer is convenience. You have to stop the moving average somewhere; and treating observations as having equal weight is the easiest way to assign the weights. Another problem with a moving average forecast is that it will always understate a trend. Another way to put it is that a moving average will always lag behind a trend. Figure 15.6 shows a 12-item last-term moving average for a time series with a clear upward trend. While the 12-item moving average removes the

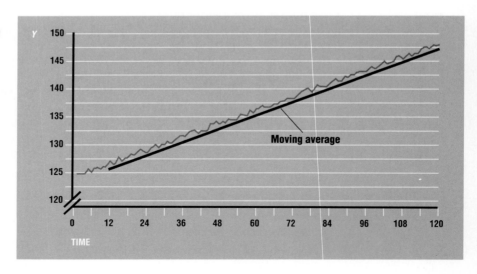

FIGURE 15.6
Twelve-item last-term moving average for a time series with an upward trend

apparent noise in the original time series, the moving average is always below the original series. Clearly, a moving average is a poor tool for forecasting a series that has a clear trend. There are some techniques for using moving averages to make forecasts with data that have a trend. For example, you can use regression analysis to estimate the trend in a time series and then use a moving average on the residuals of the regression to obtain a forecast. We will skip these techniques. The chief benefit of moving averages is their simplicity, and there is no point to learning a complex method of using moving averages to forecast when more powerful techniques are available at no greater learning cost.

In view of the difficulties, you may be wondering when it would make sense to use moving averages for forecasting. Say a bank hires temporary clerical help on the weekends to handle the crediting of checks. If too few temporary employees are hired, the bank will lose the opportunity to earn interest on money deposited in checking accounts. If too many are hired, the bank will waste money on their wages. The bank has to commit itself to hire a specified number of temporaries by Friday afternoon. Its forecast of the number of checks that will arrive over the weekend is a moving average of checks that arrived over the last five business days. The moving average forecast makes sense in this context, because the number of checks per day does not have a clear trend and the flow of checks in the immediate past is a good predictor of the number to be expected on any weekend. The actual method used by some banks to forecast the number of checks is a moving average plus an adjustment for holiday weekends.

■ 15.2B EXPONENTIAL SMOOTHING

Exponential smoothing is a forecasting method that is based on some assumptions that are more plausible than the ones implied by the moving average forecast method. Recall that in a last-term moving average of order x, each of the last x observations has equal weight in determining the forecast, and that any observations that precede the last x observations have no effect on the forecast. It seems more plausible that recent observations are more important predictors of the value of the next observation than observations far in the past. It seems more plausible that there should be no arbitrary cutoff on which observations are used to make the forecast. **Exponential smoothing** is *a method of smoothing a time series by which an observation is replaced by the weighted sum of all past observations, the weights declining exponentially as the observations recede in time*. Formula 15.6 expresses this definition. Let the subscript t refer to the time period and x_t to the value of the time series at time t. Let a be any number between zero and 1. The smoothed value of the time series at time t equals

$$\bar{x}_t = ax_t + a(1-a)x_{t-1} + a(1-a)^2 x_{t-2} + a(1-a)^3 x_{t-3} + \cdots \tag{15.6}$$

The equation will have as many terms as there are observations in the time series. If a were set at 0.8, then the formula would read

$$\bar{x}_t = 0.8x_t + 0.16x_{t-1} + 0.032x_{t-2} + 0.0064x_{t-3} + \cdots$$

As you can see, each succeeding x value has less weight in determining the smoothed value of the time series at time t, and there is no arbitrary cutoff on which prior observations go into the equation. Formula 15.6 can be rearranged to make another point about exponential smoothing. The smoothed value for time $t - 1$ equals

$$\bar{x}_{t-1} = ax_{t-1} + a(1 - a)x_{t-2} + a(1 - a)^2 x_{t-3} + a(1 - a)^3 x_{t-4} + \cdots$$

This equation is just formula 15.6 moved back one observation. Multiplying the above equation by $(1 - a)$ yields

$$a\bar{x}_{t-1} = a(1 - a)x_{t-1} + a(1 - a)^2 x_{t-2} + a(1 - a)^3 x_{t-3} + a(1 - a)^4 x_{t-4} + \cdots$$

Finally, subtracting the above equation from formula 15.6 results in formula 15.7:

$$\bar{x}_t = (1 - a)\bar{x}_{t-1} + ax_t \tag{15.7}$$

The interpretation of formula 15.7 is that the smoothed value at time t is the weighted average of the smoothed value for the previous period and the actual value at time t. The a in formula 15.7 is called the smoothing constant. Recall that a must be between 0 and 1. If the smoothing constant is set close to 1, the smoothed value at time t will depend almost entirely on the value of the time series at time t. If the smoothing constant is set close to 0, the smoothed value at time t will depend mostly on the previous values of the time series which were used to determine the smoothed value at time $t - 1$. The forecast for period $t + 1$ and all subsequent periods is simply the exponentially smoothed value at time t. There is no exact rule that provides the best smoothing constant for forecasting. While exponential smoothing seems more plausible as a forecasting tool than a moving average, exponential smoothing still has an arbitrary element. Typically smoothing constants are in the range of 0.2 to 0.8. A high value for the smoothing constant may make the smoothed value at time t too sensitive to unusually large or small observations. In other words, a high value for the smoothing constant may make the smoothed value too sensitive to noise in the data. On the other hand, a low value for the smoothing constant may make the smoothed value at time t too stable. A low value for the smoothing constant may cause the smoothed value at time t to be insensitive to turning points in the time series.

TABLE 15.6

Weekly sales at your chain of record stores of Bruce Springsteen's five-record album and the exponentially smoothed values with smoothing constants of 0.5 and 0.1

Week	Unit sales	Exponentially smoothed series with $a = 0.5$	Exponentially smoothed series with $a = 0.1$	Week	Unit sales	Exponentially smoothed series with $a = 0.5$	Exponentially smoothed series with $a = 0.1$
1	25	25.00	25	26	7	15.62	16.74
2	13	19.00	23.80	27	22	18.81	17.27
3	22	20.50	23.62	28	28	23.41	18.34
4	31	25.75	24.36	29	13	18.20	17.81
5	19	22.38	23.82	30	28	23.10	18.82
6	13	17.69	22.74	31	7	15.05	17.64
7	16	16.84	22.07	32	19	17.03	17.78
8	10	13.42	20.86	33	22	19.51	18.20
9	28	20.71	21.57	34	25	22.26	18.88
10	22	21.36	21.62	35	7	14.63	17.69
11	4	12.68	19.85	36	16	15.31	17.52
12	16	14.34	19.47	37	13	14.16	17.07
13	22	18.17	19.72	38	19	16.58	17.26
14	19	18.58	19.65	39	4	10.29	15.94
15	7	12.79	18.38	40	31	20.64	17.44
16	10	11.40	17.55	41	7	13.82	16.40
17	25	18.20	18.29	42	7	10.41	15.46
18	13	15.60	17.76	43	10	10.21	14.91
19	4	9.80	16.39	44	28	19.10	16.22
20	16	12.90	16.35	45	10	14.55	15.60
21	7	9.95	15.41	46	22	18.28	16.24
22	22	15.97	16.07	47	28	23.14	17.42
23	16	15.99	16.06	48	22	22.57	17.87
24	19	17.49	16.36	49	10	16.28	17.09
25	31	24.25	17.82	50	28	22.14	18.18

Here is an example that illustrates both the calculations used in exponential smoothing and how the choice of the smoothing constant affects the forecast. You own a chain of record stores and your electronic cash registers give you week-by-week information on the sales of every record title in your inventory. You use these sales figures to decide on the number of copies of a record you will stock. The long-term popularity of any record title is uncertain, and a decline in sales over a few weeks may be due to chance variation or to a genuine decline in the popularity of the title. Suppose the five-record set of Bruce Springsteen's greatest hits had the weekly unit sales shown in Table 15.6

at your stores. Consider the column headed "Exponentially smoothed series with $a = 0.5$." The first exponentially smoothed value in the series is just the initial weekly sales figure of 25 units. The next number in this column, 19.00, can be found by substituting into formula 15.7:

$$\overline{x}_t = (1 - a)\overline{x}_{t-1} + ax_t \tag{15.7}$$

$$\overline{x}_t = 0.5(25) + 0.5(13)$$

$$\overline{x}_t = 19.00$$

The remaining values in the column can also be found by substituting into formula 15.7. For example, the third value in the column is

$$\overline{x}_t = 0.5(19) + 0.5(22)$$

$$\overline{x}_t = 20.50$$

If the smoothing constant were changed to 0.1, the above equation for the second smoothed value would change to

$$\overline{x}_t = 0.9(25) + 0.1(13)$$

$$\overline{x}_t = 23.80$$

FIGURE 15.7
Time series of weekly sales of Bruce Springsteen's five-record album and two exponentially smoothed series

The smoothing constant of 0.1 places more weight on the prior values of the series than the smoothing constant of 0.5.

This difference can be seen in Figure 15.7. The smoothed series with a smoothing constant of 0.5 is marked by dots and the series with a smoothing

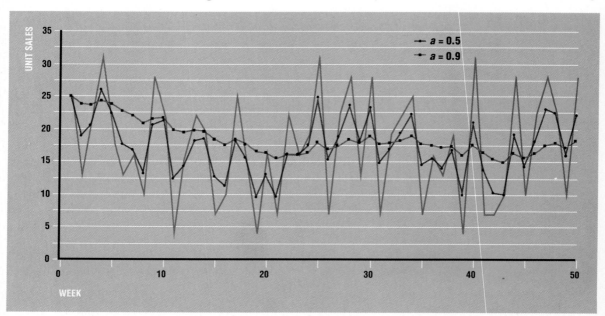

CASE STUDY

AIRLINE INVENTORY MANAGEMENT

People sitting next to each other on commercial airline flights often pay widely different fares. A factor of 8 between the lowest and the highest price in the coach section of the plane is not unusual. The passengers with the lowest fare have bought maxisaver or supersaver tickets, and the passengers with the highest fare bought regular coach tickets. Typically the high-fare passengers are business travelers who cannot plan their trips far in advance and are locked out of the fares that require advance purchase. The low fares often go to people such as students and grandmas, who are willing to cope with the elaborate restrictions in order to save money. A maxisaver fare might require a round trip with a weekend stay but no Friday departure. The goal of inventory management is to hold back enough seats for business travelers, who tend to buy tickets the day of the flight. The airline wants to make sure that no one who is willing to pay a high price is turned away while at the same time filling any seats that would otherwise go empty with the price-sensitive passengers. American Airlines is reputed to be the most successful airline at inventory management. The company employs 90 full-time inventory managers, who track on their computer screens the reservations for more than half a million flights up to 330 days in the future.

   If a particular flight has a history of being filled by business passengers, the inventory manager sets the number of maxisaver tickets that will be offered at zero. If the flight has a history of few business passengers, the inventory manager may initially allow half of the seats to be offered at the maxisaver fare. There may be a half-dozen intermediate fares on the same flight, with varying levels of restrictions. As more reservations come in for the flight, the inventory manager gets a better idea of the number of tickets of each type that will eventually be sold. Inventory managers can change the mix of tickets offered up to two hours before the flight. Some of the spaces offered at a low fare may be switched to regular coach, for example, if there is an indication that business demand for the flight will pick up. The exact method used by American Airlines to forecast the number of passengers who will buy tickets in each fare class is a secret. While the company won't reveal the details of its forecast methods, it does acknowledge the impact of accurate forecasts on its bottom line. Effective yield management can raise an airline's revenues by up to 20%.

constant of 0.1 by squares. The line for the original time series has no marks. Clearly the exponentially smoothed series with the smoothing constant of 0.1 is much less sensitive to changes in the original series. The choice of a smoothing constant is a matter of costs and benefits. If you were worried about getting stuck with records you could not sell, you would want to use a high value for the smoothing constant so that the forecast would immediately catch any decline in the popularity of a title. Similarly, you would want to use a high value for the smoothing constant if you were worried about the lost sales if you understocked a title that was gaining in popularity. If shipping costs to your store were an important factor (a better example here might be cement instead of records), or if discounts on volume purchases were substantial, you would be less concerned about having your inventories match the weekly pattern of sales. Another consideration in selecting a smoothing constant is the amount of noise in the series. If there is a great deal of random variation in the original series, you should use a low smoothing constant to suppress this noise in the smoothed series. By the way, the earliest commercial computer programs for automatic ordering in record stores were notoriously bad. Most new records start out with a rising trend in sales. The early programs relied on a simple trend line to determine the next week's order. The problem was that the programs had no way to predict a turning point, and the record store owner was guaranteed a binful of unsellable records when a title's popularity inevitably declined. A simple trend line is a poor forecasting technique for ordering records. Exponential smoothing or a more sophisticated method or even human judgment would do better.

EXERCISES FOR SECTION 15.2

15.6 Calculate the centered three-term and seven-term moving averages for the following time series and plot the moving averages on a single graph with the original series:

9 17 15 13 15 15 13 16 12 12 13 12 10 18 12 17 14 10 10
9 15 9 11 16 12

What are the forecast values of the series based on the three-term and seven-term moving averages?

15.7 Calculate the last-term four-term and eight-term moving averages for the following time series and plot the moving averages on a single graph with the original series:

6 11 3 6 2 9 7 3 8 6 5 3 5 3 9 3 8 5 6 1 1 3 3 3
8 4 6 1 7 5 5 8 6 7 6 4 5 2 4 6

What are the forecast values of the series based on the four-term and eight-term moving averages?

15.8 Exponentially smooth the following time series, using smoothing constants of 0.5 and 0.7:

18 9 11 14 18 16 15 11 12 21 12 16 14 16 14 11 19 16 12
18

Plot the original and the smoothed series. What are the forecast values of the series based on the two smoothing constants?

15.9 Exponentially smooth the following time series, using smoothing constants of 0.2 and 0.8:

−27 30 −43 24 49 42 58 −2 16 2 18 −41 −14 −9 14 39 33
−26 53 4 −8 8 −16 63 −25 20 38 −9 31 9 −19 81 17 36
48 21 29 18 9 −20 −40 −34 −38 −12 −20 28 36 28 −1 75

Plot the original and the smoothed series. What are the forecast values of the series based on the two smoothing constants?

15.10 Decide whether a moving average or exponential smoothing is the most appropriate technique for describing any patterns in each of the time series indicated below. Decide also between a high degree and a low degree of smoothing. (Hint: Moving averages with a sufficient number of terms can remove seasonal variation and may be better suited to describing long-term patterns in strongly seasonal data. Also, more noise in the data would suggest the use of more items in the moving average or a higher smoothing constant.)

   (a) The weekday sales of fresh fish in a supermarket.

   (b) The quarterly level of sales for a chain of jewelry stores.

   (c) The number of births per day in a hospital maternity ward.

15.3 AUTOREGRESSIVE AND MOVING AVERAGE MODELS

15.3A FIRST-ORDER AUTOREGRESSIVE MODELS

In the classical regression model the structure of a trend and seasonal variation is imposed on the data. You have to use your knowledge or at least your best guess about the underlying process to decide if the appropriate model should be estimated with monthly or quarterly dummies. You have to assume that there is a simple trend component to the time series. An autoregressive model imposes a much more flexible structure on the data. An autoregressive model can "capture" a much more complex pattern than a classical time-series model. Essentially the method used to formulate an autoregressive model is to hunt for a set of lags of the forecast variable which have statistically significant effects on the current value of the forecast variable. An **autoregressive model** is so called because it consists of *the regression of current values of the forecast variable on the past or lagged values of the forecast variable*. The **order** of an autoregression is *the number of lags of the forecast variable used as independent variables*. A **first-order autoregressive model** is *a regression in which the first lag of the forecast variable is used as an explanatory variable*. The mean of the forecast variable at time t in a first-order autoregressive model equals

$$(\mu_Y)_t = \alpha + \beta y_{t-1}$$

(15.8)

The observed value of Y in each time period does not always equal the mean or expected value of the forecast variable, μ_Y, for that time period. There is some random variation around the mean, which is assumed to be described by a white-noise error random variable. The term ε_t is the white-noise error random variable for the distribution of the forecast variable at time t. The observed value of Y at time t equals

$$y_t = \alpha + \beta y_{t-1} + \varepsilon_t$$

(15.9)

The regression equation used to estimate the population slope and intercept in formula 15.9 is

$$\hat{y}_t = a + b y_{t-1}$$

(15.10)

This form of the regression equation is equivalent to a simple trend line with time as the explanatory variable. The sample slope, b, is the same as the estimated change in the average level of Y per unit time interval. The intercept term, a, is the estimated average level of Y one period before the first observation in the time series.

 15.3B HIGHER-ORDER AUTOREGRESSIVE MODELS

A **pth-order autoregressive model** is *a regression in which* p *lags of the forecast variable are used as independent variables.* The formulas for the mean of Y at time t and the observed value of Y at time t are the same as formulas 15.8 and 15.9 except that there are p lags of observed value of Y to the right of the equal sign. The regression equation for estimating the intercept and slope coefficients for a pth-order autoregressive model is

$$\hat{y}_t = a + b_1 y_{t-1} + b_2 y_{t-2} + b_3 y_{t-3} + \cdots + b_p y_{t-p}$$

(15.11)

An equation of the type in formula 15.11 is a more flexible way of capturing the motion or pattern in a time series than the classical model. The key assumption required for use of this type of formula is that there is a linear relationship between each lag of Y and Y. The idea is that the population slope coefficients on the lagged forecast variable are a function of only the length of the lag. In other words, there is no evolution in the lag structure. An evolutionary time series is one in which the population slope coefficient on, say the fourth lag varies from the beginning to the end of the series. An evolving time series cannot be described by a simple autoregressive model. For an example of an evolutionary time series, consider a time series on the production of cattle. Weather conditions, such as a cold spell or a drought, could be the cause of the random or unpredictable component of the series. Since cattle in any one period are an input for production of cattle in the next period, there is an obvious correlation between the numbers of cattle produced across periods. The evolutionary aspect of the series may be caused by a change in the coefficients on the lags over time as breeders improve their technology so that a higher percentage of cows calve in a period and a lower percentage of calves and mature cattle die of disease. In section 15.4 we'll discuss methods of fitting the regression model to a time series that has an evolutionary pattern of correlations.

When you estimate a regression of the type in formula 15.11, you have to decide how many lags to use. If you have too few lags, you will not capture all of the important correlations. If you have too many lags, the estimates of the slope coefficients on the lags that are important will be inefficient. The extraneous lags included in the regression will cause the sampling distributions of the slope coefficients to be wider than necessary. Of course, you don't know how many lags are too few or too many.

One rule of thumb is to try to run regressions for all sets of lags from one lag to $\sqrt{n} + 10$ lags, where n is the number of observations in the data set. Suppose you had a data set with 100 observations. Then $\sqrt{n} + 10$ would equal 20. Using this rule of thumb, you would run 20 regressions. The first regression would have only the first lag of the forecast variable as the independent variable. The second regression would have the first and second lags as independent variables, and so on, until the 20th regression equation would have lags 1 through 20 as independent variables. Next you would examine the coefficients in the 20 regressions. With this large sample size, a slope coefficient that has an absolute value greater than twice its standard error can be considered to differ significantly from 0 at the 5% level. This is a shortcut for hypothesis tests on the slope coefficients. Next you examine the coefficients and observe if any lags tend to be significant across most of the regression equations. The regression equation that you use for forecasting should include all of the lags having slope coefficients that consistently differ significantly from 0. An example will clarify this procedure.

Suppose the data in Table 15.7 were the number of sessions sold per month by a national chain of weight-loss clinics, in units of 10,000 sessions. When new customers first purchase the services of a local clinic, they tend to come to a string of weekly sessions to initiate their weight-loss program. Many customers come back after dropping out of the program to renew their weight-loss efforts.

TABLE 15.7

Monthly clinic sessions per month sold by a national chain of weight-loss clinics

Sessions per month (units of 10,000)	Month	Sessions per month (units of 10,000)	Month	Sessions per month (units of 10,000)	Month
11.90	1	12.53	33	14.03	65
12.01	2	12.86	34	14.43	66
11.33	3	12.68	35	15.04	67
11.99	4	12.55	36	14.87	68
12.84	5	13.74	37	14.55	69
12.61	6	13.50	38	14.58	70
11.65	7	13.13	39	14.43	71
12.26	8	12.74	40	15.03	72
11.33	9	13.75	41	15.93	73
11.42	10	14.06	42	15.36	74
12.13	11	13.37	43	14.38	75
12.68	12	13.70	44	15.29	76
13.26	13	13.73	45	15.74	77
12.84	14	13.97	46	16.12	78
12.96	15	13.17	47	16.17	79
11.88	16	14.48	48	14.94	80
11.85	17	14.19	49	14.38	81
11.89	18	14.76	50	15.15	82
11.76	19	15.05	51	15.92	83
12.79	20	14.36	52	16.02	84
12.26	21	14.47	53	16.62	85
13.53	22	14.29	54	16.43	86
12.91	23	14.76	55	15.20	87
12.96	24	13.60	56	15.67	88
12.19	25	14.53	57	15.85	89
12.02	26	14.62	58	16.51	90
12.21	27	13.59	59	16.11	91
11.97	28	14.73	60	17.02	92
12.44	29	15.47	61	16.25	93
12.67	30	14.06	62	16.16	94
13.31	31	13.79	63	16.51	95
12.47	32	14.23	64	16.22	96

FIGURE 15.8
Plot of data in Table 15.7

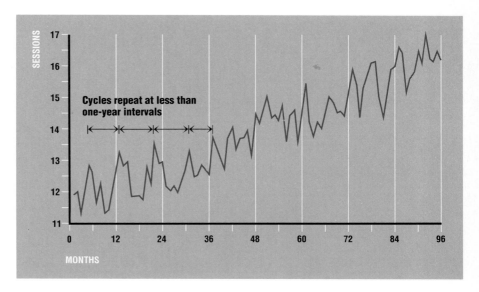

The data in Table 15.7 are plotted in Figure 15.8. While there is a pattern to the data over time, it is not an annual pattern; the peaks in these data do not come at one-year intervals, as in Figure 15.1. The vertical lines in Figure 15.8 marking every 12th month make this clear. As you doubtless suspect, this time series is a good candidate for an autoregressive model. There are 96 observations in the data set, so the rule of thumb on the number of regressions (i.e., $\sqrt{n} + 10$) suggests running regressions up to the 19th lag.

Table 15.8 lists the coefficients and standard errors for the 19 regressions. Each coefficient that has an absolute value more than twice its standard error is printed in bold. The standard errors are in parentheses below the coefficients.

The slope coefficient on the first lag is significant in every regression. The coefficients of the seventh, eighth, and ninth lags are significant in almost every regression that contains these lags. Beyond the ninth lag, only the slope coefficient on the thirteenth lag is significant in one of the equations. Given these results, the best regression for forecasting is probably the one with lags 1 through 9.

To use this regression equation to forecast one period ahead, you have to substitute the observed values of Y for the nine months starting with the most recent month into the estimated regression equation:

$$\hat{y}_{t+1} = -0.029245 + 0.48119y_t - 0.068388y_{t-1} - 0.026968y_{t-2} + 0.010267y_{t-3}$$
$$+ 0.011299y_{t-4} + 0.18958y_{t-5} + 0.34631y_{t-6} - 0.27144_{t-7} + 0.34765y_{t-8}$$

$$\hat{y}_{t+1} = -0.029245 + 0.48119(16.22) - 0.068388(16.51) - 0.026968(16.16)$$
$$+0.010267(16.25) + 0.011299(17.02) + 0.18958(16.11) + 0.34631(16.51)$$
$$-0.27144(15.85) + 0.34765(15.67)$$

$$\hat{y}_{t+1} = 16.4870$$

TABLE 15.8
Coefficients and standard errors for autoregressions with lags of 1 to 19

Constant	L1	L2	L3	L4	L5	L6	L7	L8	L9	L10	L11	L12	L13	L14	L15	L16	L17	L18	L19
1.2596	**0.91267**	—	—	—	—	—	—	—	—	—	—	—	—	—	—	—	—	—	—
(0.61947)	(0.00445)																		
0.98638	**0.71726**	**0.21589**	—	—	—	—	—	—	—	—	—	—	—	—	—	—	—	—	—
(0.63405)	(0.10331)	(0.1037)																	
0.86561	**0.6671**	0.048849	**0.22651**	—	—	—	—	—	—	—	—	—	—	—	—	—	—	—	—
(0.63556)	(0.103)	(0.12482)	(0.10356)																
0.64105	**0.63847**	0.03694	0.11904	0.16596	—	—	—	—	—	—	—	—	—	—	—	—	—	—	—
(0.6529)	(0.10631)	(0.12485)	(0.12473)	(0.10756)															
0.32486	**0.5958**	0.037217	0.099341	0.031279	**0.22041**	—	—	—	—	—	—	—	—	—	—	—	—	—	—
(0.65377)	(0.10563)	(0.12399)	(0.12243)	(0.12478)	(0.10697)														
-0.022508	**0.50723**	0.02532	0.06251	0.032077	-0.021613	**0.4069**	—	—	—	—	—	—	—	—	—	—	—	—	—
(0.61829)	(0.10109)	(0.11495)	(0.11515)	(0.11575)	(0.11637)	(0.10208)													
-0.062293	**0.39184**	0.050775	0.054948	-0.015158	-0.00412	**0.24857**	**0.29208**	—	—	—	—	—	—	—	—	—	—	—	—
(0.60228)	(0.10578)	(0.11063)	(0.11019)	(0.11241)	(0.11154)	(0.11228)	(0.10703)												
-0.030112	**0.43381**	0.078061	0.045371	-0.014399	0.01023	**0.24406**	**0.34014**	-0.12256	—	—	—	—	—	—	—	—	—	—	—
(0.61576)	(0.1125)	(0.11505)	(0.11149)	(0.11293)	(0.1136)	(0.11295)	(0.11641)	(0.11243)											
-0.029245	**0.48119**	-0.068388	-0.026968	0.010267	0.011299	0.18958	**0.34631**	**-0.27144**	**0.34765**	—	—	—	—	—	—	—	—	—	—
(0.58427)	(0.10588)	(0.11465)	(0.10785)	(0.10611)	(0.10613)	(0.10708)	(0.10883)	(0.11453)	(0.10631)										
0.023446	**0.48069**	-0.06904	-0.02677	0.018872	0.016295	0.18776	**0.34144**	**-0.2693**	**0.35764**	-0.022013	—	—	—	—	—	—	—	—	—
(0.6022)	(0.11608)	(0.12093)	(0.11673)	(0.11034)	(0.10804)	(0.10847)	(0.11079)	(0.11729)	(0.12168)	(0.11641)									
-0.021679	**0.48133**	-0.12395	0.014672	-0.030963	-0.008297	0.1937	**0.34484**	**-0.2756**	**0.37725**	-0.088197	0.13668	—	—	—	—	—	—	—	—
(0.61683)	(0.11663)	(0.12927)	(0.12186)	(0.11794)	(0.11164)	(0.10941)	(0.11121)	(0.11814)	(0.12324)	(0.12324)	(0.11721)								
-0.089928	**0.47717**	-0.11873	0.00445	-0.023609	-0.016084	0.18005	**0.34173**	**-0.2725**	**0.37929**	-0.085582	0.11451	0.046258	—	—	—	—	—	—	—
(0.63676)	(0.11928)	(0.13128)	(0.13212)	(0.12435)	(0.12062)	(0.11387)	(0.11327)	(0.11991)	(0.12558)	(0.13353)	(0.13353)	(0.11988)							
-0.10034	**0.4797**	-0.07641	-0.0144	0.10022	-0.10839	**0.29211**	**0.37312**	**-0.30066**	**0.35553**	-0.048029	0.05245	0.1694	**-0.25318**	—	—	—	—	—	—
(0.6258)	(0.1151)	(0.12735)	(0.1276)	(0.12848)	(0.12118)	(0.11729)	(0.11173)	(0.11599)	(0.12124)	(0.12959)	(0.12904)	(0.12795)	(0.11562)						
-0.031523	**0.43216**	-0.047056	**0.37438**	**0.8784**	**-0.43933**	0.24319	**0.44015**	**-0.25421**	**0.33065**	-0.041327	0.060531	0.14517	-0.17013	-0.16945	—	—	—	—	—
(0.31523)	(0.12132)	(0.12903)	(0.12806)	(0.1288)	(0.12995)	(0.12266)	(0.12252)	(0.12063)	(0.12339)	(0.12975)	(0.12984)	(0.12933)	(0.12957)	(0.11977)					
-0.15723	**0.45385**	-0.042043	-0.013558	0.077747	-0.030974	0.20879	**0.45926**	**-0.29382**	**0.30347**	-0.048507	0.059994	0.14921	-0.16418	-0.22429	0.12945	—	—	—	—
(0.65124)	(0.12372)	(0.13275)	(0.10466)	(0.12999)	(0.1307)	(0.13272)	(0.12689)	(0.13556)	(0.12584)	(0.13102)	(0.13031)	(0.13059)	(0.13092)	(0.13169)	(0.12519)				
-0.10729	**0.45865**	-0.042389	-0.003475	0.075235	-0.034449	0.20853	**0.44895**	**-0.28909**	**0.29261**	-0.048977	0.06704	0.14642	-0.17135	-0.21869	0.12777	0.00405	—	—	—
(0.6723)	(0.1272)	(0.13808)	(0.13489)	(0.13306)	(0.13346)	(0.1351)	(0.13764)	(0.14267)	(0.14349)	(0.13412)	(0.13326)	(0.13269)	(0.13392)	(0.13462)	(0.13991)	(0.12808)			
-0.003335	**0.44671**	-0.04149	0.030437	0.13678	-0.065347	0.18461	**0.4616**	**-0.36821**	**0.34698**	-0.15406	0.04617	0.17711	-0.18427	-0.2491	0.15249	-0.070182	0.16703	—	—
(0.6757)	(0.12559)	(0.13798)	(0.13624)	(0.13499)	(0.1331)	(0.13378)	(0.13603)	(0.14753)	(0.14571)	(0.14626)	(0.13246)	(0.13194)	(0.13221)	(0.13394)	(0.13855)	(0.13902)	(0.1277)		
0.042801	**0.42794**	-0.037646	0.034513	0.14366	-0.047114	0.17853	**0.45785**	**-0.3567**	**0.32047**	-0.13603	0.018301	0.17454	-0.17118	-0.25578	0.13555	-0.060112	0.1546	0.031709	—
(0.69922)	(0.13174)	(0.14014)	(0.14004)	(0.14126)	(0.13888)	(0.13676)	(0.13817)	(0.15166)	(0.15866)	(0.1552)	(0.14984)	(0.13457)	(0.13581)	(0.13629)	(0.14335)	(0.14197)	(0.14152)	(0.13347)	
0.11512	**0.41097**	-0.079673	0.043959	0.15322	-0.023361	0.22722	**0.44162**	**-0.35776**	**0.34368**	-0.20157	0.066241	0.10139	-0.17798	-0.22586	0.11084	-0.10069	0.17974	-0.0040754	0.10248
(0.71089)	(0.13204)	(0.14306)	(0.14003)	(0.14302)	(0.14298)	(0.14069)	(0.13898)	(0.15151)	(0.15979)	(0.16422)	(0.15606)	(0.14964)	(0.13633)	(0.13759)	(0.14409)	(0.14437)	(0.14231)	(0.14445)	(0.13341)

FIGURE 15.9

Actual and forecast values of the number of sessions sold per month

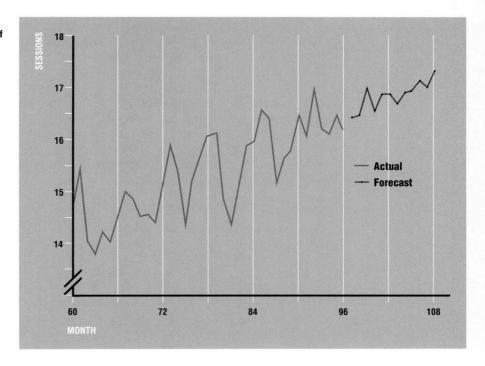

The forecast for one step ahead of the current time is 16.487 units of sessions (a unit is 10,000 sessions). To forecast two steps ahead, the forecast value of Y replaces the current value of Y in the above equation, and every observed value of Y is shoved back one step so that the last value (15.67) is dropped from the equation. With an extension of this procedure you can use the estimated regression equation to forecast any number of steps ahead of the current value. Let j be some number of steps ahead of the current period. The forecast of Y at time $t + j$ requires that you substitute the $t + j - 1$ to $t + j - 9$ forecast or observed values of Y in the estimated regression equation.

Figure 15.9 shows the actual data and the forecast values for steps $t + 1$ through $t + 12$. The accuracy of the forecast decreases as the number of steps between the current month and the forecast month increases. Judgment should be used before you rely on far-ahead forecasts based on an autoregressive model.

One characteristic of autoregressive models is that a shock (an unusually low or high value in the series) affects an infinite sequence of subsequent values of the series. The forecast values in Figure 15.9 were based on the last nine observations of the actual data. The peak observed in the actual data in month 92 is reflected by peaks in the forecast values for months 99 and 106. If the forecasts were carried forward for more steps, other peaks would appear at every seventh step. An autoregressive model has a long memory. A simple numerical example illustrates this point. Suppose the estimated regression equation for a first-order autoregressive model had the values:

$$\hat{y}_t = 2 + 0.5y_{t-1}$$

Suppose the observed value of Y at time t was much higher than the other values in the series, the past values in the series were in a range close to the intercept value of 2, and the current value of Y took a jump to 100. The future values of the series would stay far above the intercept term for many steps. The calculations below, showing a forecast one through six steps ahead, use the regression equation estimated above:

$\hat{y}_{t+1} = 2 + 0.5y_t$	$\hat{y}_{t+1} = 2 + 0.5(100)$	$\hat{y}_{t+1} = 52$
$\hat{y}_{t+2} = 2 + 0.5\hat{y}_{t+1}$	$\hat{y}_{t+2} = 2 + 0.5(52)$	$\hat{y}_{t+2} = 28$
$\hat{y}_{t+3} = 2 + 0.5\hat{y}_{t+2}$	$\hat{y}_{t+3} = 2 + 0.5(28)$	$\hat{y}_{t+3} = 16$
$\hat{y}_{t+4} = 2 + 0.5\hat{y}_{t+3}$	$\hat{y}_{t+4} = 2 + 0.5(16)$	$\hat{y}_{t+4} = 10$
$\hat{y}_{t+5} = 2 + 0.5\hat{y}_{t+4}$	$\hat{y}_{t+5} = 2 + 0.5(10)$	$\hat{y}_{t+5} = 7$
$\hat{y}_{t+6} = 2 + 0.5\hat{y}_{t+5}$	$\hat{y}_{t+6} = 2 + 0.5(7)$	$\hat{y}_{t+6} = 5.5$

The forecast values $\hat{y}_{t+1}$ to $\hat{y}_{t+6}$ decline at each step. The series approaches the value of the intercept 2 as the number of forecast steps increases. The rate of decline depends on the slope coefficient. If the slope coefficient were above 1, the values in the forecast series would grow rather than decline. However, the estimated slope coefficients for almost all economic and business time series are below 1. If the slope coefficient were above 1, the series would explode, or run off the top of the graph paper.

The lingering impact of an initial shock is also seen in autoregressive models with an order higher than 1 (the order is the number of autoregressive terms in the regression), such as the weight-loss clinic example. Suppose the 7th lag had a slope coefficient of 0.5. The calculations showing the lingering impact of an initial shock in the 7th, 14th, 21st, 28th, 35th, and 42d steps past the current period would be the same as the above calculations for the first-order autoregressive model. Shocks in an autoregressive model have staying power.

There are autoregression estimation techniques that adjust for the serially correlated errors. These techniques provide more efficient estimates, but least squares estimation of an autoregressive model is much simpler. Whether or not the autoregressive representation of the time series is accurate depends on the underlying behavior that makes up the time series. If decreasing percentages of the people who attended the weight-loss clinics came back for new sets of treatments four or five times at approximately seven-month intervals, the autoregressive model would be a good representation of the time series. In effect, the customers' bulges determine the bulges in the time series. If the typical customer repeated a set of treatments only once, then the use of an autoregressive model would make you predict too many return customers.

In practice the behavioral or physical process that creates the time series is rarely known, and you are unlikely to know whether an autoregressive model can capture all of the systematic motion in the time series. While the autoregressive model is more flexible than the classical time-series model, it still is an imposition of a structure on the data. An a priori structure may or may not pick up all of the patterns of motion in the time series. In section 15.4 we'll discuss a newer method of time-series forecasting which imposes much less structure. This method, called ARIMA, uses some diagnostic tools to find the structure or model that will eliminate any predictable component of motion from the residuals. With ARIMA the goal is to squeeze out of the data all of the possible information about systematic or patterned motion in the time series.

EXERCISES FOR SECTION 15.3

15.11 Use the following estimated regression equation to forecast the next ten steps if the current-period value of Y is 88.

$$\hat{y}_t = 14 - 0.7y_{t-1}$$

15.12 Use the following estimated regression equation to forecast the next ten steps if y_{t-1} is 60, y_{t-2} is 57, y_{t-3} is 56, y_{t-4} is 50, and y_{t-5} is 59.

$$\hat{y}_t = 14 - 0.5y_{t-1} + 0.1y_{t-2} + 0.2y_{t-3} + 0.1y_{t-4} + 0.3y_{t-5}$$

15.13 The following table shows the regression coefficients for all autoregressions from 1 to 18 lags for the data in Table 15.1. Decide which equation is the best for making a forecast and explain why. Recall that the boldface coefficients are statistically significant at a 5% level compared to a null hypothesis value of zero.

Constant	$t-1$	$t-2$	$t-3$	$t-4$	$t-5$	$t-6$	$t-7$	$t-8$	$t-9$	$t-10$	$t-11$	$t-12$	$t-13$	$t-14$	$t-15$	$t-16$	$t-17$	$t-18$
3.889	**0.915**	—	—	—	—	—	—	—	—	0.439	—	—	—	—	—	—	—	—
3.116	**0.796**	0.128	—	—	—	—	—	—	—	0.212	**0.375**	—	—	—	—	—	—	—
2.736	**0.764**	0.054	0.109	—	—	—	—	—	—	0.157	0.121	**0.611**	—	—	—	—	—	—
2.192	**0.752**	0.053	0.043	0.087	—	—	—	—	—	0.156	0.123	**0.606**	0.017	—	—	—	—	—
3.320	**0.760**	0.047	0.033	0.194	−0.120	—	—	—	—	0.067	0.156	**0.610**	−0.052	**0.316**	—	—	—	—
3.944	**0.746**	0.070	0.034	0.193	−0.036	−0.104	—	—	—	0.041	0.141	**0.633**	−0.041	0.280	0.188	—	—	—
1.008	**0.799**	0.093	−0.094	0.165	0.056	**−0.557**	**0.598**	—	—	0.051	0.147	**0.621**	−0.040	0.299	0.179	0.024	—	—
−0.798	**0.610**	0.281	−0.067	0.095	−0.020	**−0.580**	0.298	**0.355**	—	0.031	0.165	**0.625**	0.051	0.310	0.204	0.012	0.062	—
−1.322	**0.563**	0.233	0.021	0.098	−0.039	−0.057	0.261	0.260	0.150	0.032	0.164	**0.630**	−0.052	0.308	0.211	0.015	0.059	0.028
−2.452	**0.497**	0.163	−0.145	**0.368**	−0.037	**−0.629**	**−0.244**	0.177	−0.104									
−5.036	**0.408**	0.147	−0.203	**0.295**	0.171	**−0.582**	0.120	0.208	−0.146									
−8.524	0.174	0.059	−0.170	**0.237**	0.061	**−0.231**	0.016	0.047	−0.068									
−8.462	0.164	0.076	−0.192	**0.241**	0.070	**−0.248**	0.026	0.053	−0.079									
−10.368	0.161	−0.136	−0.134	0.123	0.081	−0.188	−0.067	0.138	−0.073									
−12.471	0.133	−0.123	−0.253	0.081	0.094	−0.197	−0.068	0.140	−0.038									
−12.597	0.139	−0.151	−0.247	0.063	0.110	−0.216	−0.062	0.146	−0.056									
−13.300	0.133	−0.152	−0.282	0.068	0.068	−0.204	−0.083	0.149	−0.049									
−13.737	0.131	−0.151	−0.289	0.064	0.068	−0.223	−0.087	0.150	−0.051									

15.14 Using the information in exercise 15.13, determine the values of the forecasts for the first six periods after the last observation in Table 15.1.

15.15 What problems, if any, would be posed by fitting an autoregressive model to data that generated each of the graphs below?

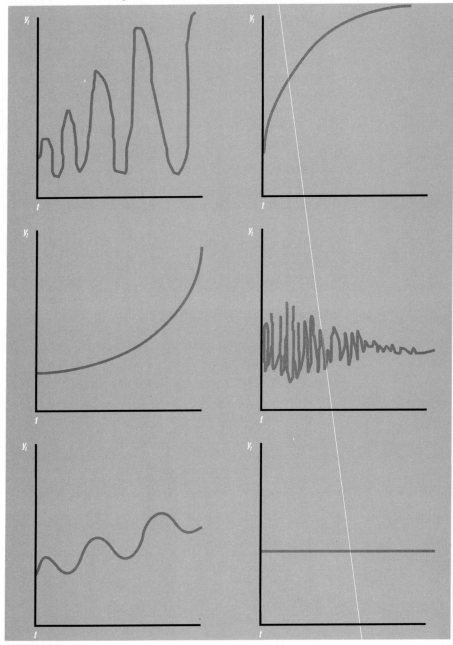

15.4 ARIMA MODELS (OPTIONAL)

15.4A STATIONARY TIME SERIES

ARIMA is *an acronym for autoregressive integrated moving average, a forecasting model that breaks the pattern of the forecast variable into trend, autoregressive, and moving average components.* ARIMA analysis is often called Box-Jenkins analysis, after its originators, George Box and Gwilym Jenkins. After *the patterns in a time series are decomposed into the three main types or sources of variation, a model is found that will capture each type of motion.* "Integrated" refers to the procedure for capturing any trends in the series. We'll review examples of linear and quadratic trends. We discussed the autoregressive pattern in section 15.3. The moving average pattern will be described in section 15.4B.

The first step in ARIMA analysis is the "integrated" part, capturing or describing any trends. This step involves transforming a time series so that it is stationary. A **stationary time series** is *a time series with a fixed mean and a constant variance of the error random variables for all time periods, in which the autocorrelations are constant for a given lag.* For example, the correlation between the current-term and the fourth lags is the same for all subsets of the time series. If a series has stable correlation coefficients on the lag terms, the transformation to a stationary time series requires that you remove any trend and any heteroscedasticity from the data. The reason for beginning with this transformation is that a stationary series allows you to focus your search for patterns on the residuals of autoregressive motion and the moving average motion without being misled by trends or heteroscedasticity. In effect, you account for any simple patterns in the data before trying to capture anything that is complex.

There are many possible transformations to make a time series stationary, depending on what is happening in the original time series, but the simplest and the most common transformation is differencing the data.

Suppose the time-series process could be described by a simple linear trend:

$$y_t = \alpha + \beta t + \varepsilon_t$$

The first difference is:

$$y_t - y_{t-1} = (\alpha + \beta t) - [\alpha + \beta(t - 1)] + \varepsilon_t - \varepsilon_{t-1}$$

$$y_t - y_{t-1} = \alpha - \alpha + \beta(t - t + 1) + \varepsilon_t - \varepsilon_{t-1}$$

$$y_t - y_{t-1} = \beta + \varepsilon_t - \varepsilon_{t-1}$$

The expected value of the first difference is just the slope coefficient because the expected values of ε_t and ε_{t-1} are both 0. The first difference series, $y_t - y_{t-1}$, can be renamed w_t. The w_t series is stationary. It has a fixed mean of β and a constant variance, and the correlations on all lags of Y are stable.

A series that is quadratic instead of linear can be made stationary by second differencing. A quadratic series takes the form

$$y_t = \alpha + \beta_1 t + \beta_2 t^2 + \varepsilon_t$$

The first difference of this series is

$$y_t - y_{t-1} = (\alpha + \beta_1 t + \beta_2 t^2 + \varepsilon_t) - [\alpha + \beta_1(t-1) + \beta_2(t-1)^2 + \varepsilon_{t-1}]$$

$$y_t - y_{t-1} = \alpha - \alpha + \beta_1(t - t + 1) + \beta_2 t^2 - \beta_2(t-1)^2 + \varepsilon_t - \varepsilon_{t-1}$$

$$y_t - y_{t-1} = \beta_1 + \beta_2 t^2 - \beta_2(t-1)^2 + \varepsilon_t - \varepsilon_{t-1}$$

$$y_t - y_{t-1} = \beta_1 + \beta_2(t^2 - t^2 + 2t - 1) + \varepsilon_t - \varepsilon_{t-1}$$

$$y_t - y_{t-1} = \beta_1 + 2\beta_2 t - \beta_2 + \varepsilon_t - \varepsilon_{t-1}$$

Recall that the first difference series, $y_t - y_{t-1}$, has the symbol w_t. The second difference of the original series is $w_t - w_{t-1}$.

$$w_t - w_{t-1} = (\beta_1 + 2\beta_2 t - \beta_2 + \varepsilon_t - \varepsilon_{t-1}) - [\beta_1 + 2\beta_2(t-1) - \beta_2 + \varepsilon_{t-1} - \varepsilon_{t-2}]$$

$$w_t - w_{t-1} = \beta_1 - \beta_1 - \beta_2 + \beta_2 + 2\beta_2(t - t + 1) + \varepsilon_t - \varepsilon_{t-1} - \varepsilon_{t-1} + \varepsilon_{t-2}$$

$$w_t - w_{t-1} = 2\beta_2 + \varepsilon_t - \varepsilon_{t-1} - \varepsilon_{t-1} + \varepsilon_{t-2}$$

The second difference of the quadratic time series is stationary. The expected mean is a constant, $2\beta_2$, and the variance is the same for all t.

The method for checking if first or second or more differencing creates a stationary series is to run a regression with t as the independent variable and the differenced series as the dependent variable. A plot of the residuals of this regression will show if there is any remaining trend or any heteroscedasticity. Here's an example that shows a stationary series after second differencing. Table 15.9 shows a time series; the residuals of a simple regression for the time series; the first difference of the series; the residuals of the simple regression with the first differences used as the dependent variable and time as an independent variable; the second differences; and the residuals of the simple regression with the second differences used as the dependent variable and time as an independent variable.

The results in Table 15.9 are summarized in Figure 15.10. In Figures 15.10a, c, and e the linear regressions are superimposed on the original time series, the first difference, and the second difference. Figures 15.10b, d, and f show the corresponding residuals and the values predicted by the linear regressions. By the first difference the trend found in the original series is almost entirely gone. The second differences are included in Table 15.9 and Figure 15.10f to illustrate the computations. Figures 15.10b, d, and f highlight some of the effects of differencing the original series. The residuals have a clear pattern in Figure 15.10b and look more like a white-noise series in Figures 15.10d and f. Another point to notice in the plots of the residuals is that they do not consis-

TABLE 15.9

Time series and first and second differences plus the residuals of simple regressions of the original series and the first and second difference series on a trend variable

t	y_t	Residuals of linear regression on y_t	First difference	Residuals of linear regression on first difference	Second difference	Residuals of linear regression on second difference
1.00	8.52	−4.95				
2.00	13.23	−2.76	4.71	0.51		
3.00	14.25	−4.26	1.02	−2.99	−3.69	−3.08
4.00	19.30	−1.73	5.05	1.23	4.03	4.58
5.00	24.00	0.45	4.70	1.08	−0.35	0.13
6.00	25.89	−0.17	1.89	−1.54	−2.81	−2.39
7.00	29.34	0.75	3.45	0.22	1.56	1.92
8.00	33.57	2.46	4.23	1.19	0.78	1.07
9.00	38.64	5.02	5.07	2.22	0.84	1.07
10.00	40.28	4.13	1.64	−1.01	−3.43	−3.26
11.00	42.47	3.81	2.19	−0.27	0.55	0.65
12.00	45.93	4.75	3.46	1.20	1.27	1.31
13.00	45.52	1.82	−0.41	−2.48	−3.87	−3.90
14.00	47.73	1.52	2.21	0.33	2.62	2.53
15.00	51.24	2.51	3.51	1.83	1.30	1.15
16.00	53.16	1.91	1.92	0.43	−1.59	−1.81
17.00	54.20	0.43	1.04	−0.26	−0.88	−1.16
18.00	53.74	−2.55	−0.46	−1.56	−1.50	−1.85
19.00	51.77	−7.05	−1.97	−2.88	−1.51	−1.92
20.00	55.23	−6.10	3.46	2.75	5.43	4.96

tently widen or narrow as t increases. Thus there is no apparent heteroscedasticity.

First or second differencing should be enough to remove any trends in most economic and business data. If too many differences of the original series are used in the ARIMA estimation, the estimates will be inefficient. The effect is the same as putting irrelevant independent variables into a multiple regression equation.

Now we'll go on to another pattern used to capture time-series motion, moving averages.

15.4B MOVING AVERAGE MODELS

The moving average component of ARIMA is different from the moving averages described in section 15.2. Instead of a moving average of the past observations, ARIMA uses what amounts to a weighted moving average of the past errors in the forecast. This approach is unique to ARIMA. The intuition behind a moving average of past error terms is that some shocks (unusu-

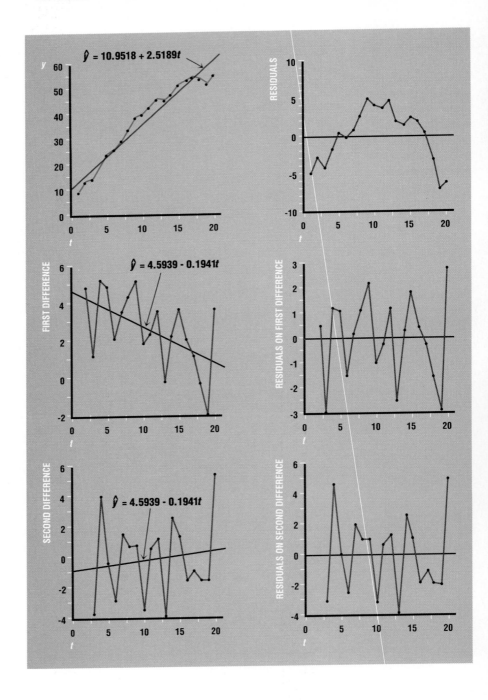

ally high or low values in the time series) can cause unusual values for a few more periods. This "short memory" contrasts with the "long memory" of an autoregressive process. Recall that the assumption behind the moving average forecast method was that a certain number of past values of the forecast variable would have an effect on the future value of the forecast variable, and that there was an abrupt cutoff to this effect. In ARIMA models a certain number of past error terms have an effect on the forecast variable, and there is an abrupt cutoff to this effect. A shock that would have a temporary effect on a time series might be the extra sales made by a firm while a competitor is shut down by a strike; another might be an extremely hot day that causes a utility to sell a large amount of electricity, and is likely to be followed by some other days that are hotter than average.

A **first-order moving average model** is *a model in which the current value of the forecast variable is a function of the forecast error for the previous period.* The observed value of Y at time t equals

$$y_t = \alpha + \varepsilon_t - \theta\varepsilon_{t-1}$$

(15.12)

In the moving average model the past values of Y do not affect the future values of Y. Only the past values of the error random variable affect the future values of Y.

A **qth-order moving average model** is *a model in which q lags of the forecast error affect the current value of the forecast variable.*

$$y_t = \alpha + \varepsilon_t - \theta_1\varepsilon_{t-1} - \theta_2\varepsilon_{t-2} - \theta_3\varepsilon_{t-3} - \cdots - \theta_q\varepsilon_{t-q}$$

(15.13)

In a qth-order moving average model, a shock at time t has an impact on the value of Y in each of the q steps ahead of time t. The current-period shock affects only q future values of Y. In an autoregressive model, in contrast, a shock affects an infinite sequence of future values of Y.

Understanding the physical or behavioral processes that generate moving averages or autoregressive patterns will help you see how to apply ARIMA analysis. Recall that the example of an autoregressive process was the weight-loss clinics at which most customers paid for a set of weekly sessions, then dropped out of the program and returned some months later for a second set of clinic sessions. If a decreasing fraction of the customers came back for a third or fourth set of sessions, an autoregressive model would describe the pattern in the time series. The series' long memory could manifest itself in another way if the current customers periodically recommended new customers. If a weight-loss clinic's customers came back only for a second set of sessions, a moving average model would describe the pattern in the time series.

FIGURE 15.10

Y_t, first and second differences of Y_t, and the residuals of linear regression on Y_t and the first and second differences of Y_t

Here are more examples of moving average and autoregressive processes.

Suppose you own a towing company that specializes in towing semitrailers in the Midwest. Each order to tow a semitrailer is an independent event. Your experience has been that the time needed to send a towing rig to a semitrailer and tow it to a repair shop can be one, two, or three days, depending on where a truck breaks down and where the company that owns the truck has a repair shop. You don't keep a backlog of orders. If you don't have an available rig to send to a broken semitrailer, the owners find another towing company. The orders for towing on a given day help you forecast the time series for the number of rigs in use for the next three days. Beyond three days, the current orders tell you nothing about the future. Is this a moving average process or an autoregressive process? The answer is moving average. In this example the order of the moving average process is three.

You own a lemonade stand and you sell an average of five glasses of lemonade per hour. You've run out of lemons to make more lemonade but you don't want to close up your stand when your bucket of lemonade is empty. As an expedient, every time you sell a glass of lemonade you replace it with a glass of water. The amount of lemon juice in the bucket decreases every hour, but it will never reach zero unless you dump out the bucket or are arrested. Is the time series on the amount of lemon juice in the bucket a moving average or an autoregressive process? It is autoregressive. A shock of a few more or a few less customers in one hour would affect the levels of lemon juice in all subsequent hours. Since your customers are unlikely to come back, the process is first-order autoregressive.

You own a motel near a national park. Motel bookings show some persistence, that is, some of the guests will spend more than one day at your motel. Also, some of the guests take week-long trips inside the national park and then come back to your motel for a night's stay before heading home. Is the time series on the number of guests per day an autoregressive process or a moving average process? The answer is both. A shock on a given day persists with an exponential decay on subsequent days as some guests linger in the motel, and the shock has a one-time echo seven days later because some customers will spend one day at the motel before taking the week-long trip inside the park and then return to the motel. The process is first-order autoregressive and first-order moving average. The order refers to the number of autoregressive parameters and the number of moving average parameters that have to be estimated. Even though the moving average effect is on the seventh step, it is first-order because only one parameter needs to be estimated. There are no moving average effects for steps 1 through 6.

The motel example shows how moving average processes and autoregressive processes can be combined. The use of both processes and the differencing to describe the patterns in a time series give you great flexibility. Almost any pattern in an economic or business time series can be described with differencing, autoregressive, and moving average components. Usually you won't have to go beyond second differencing, second-order autoregressive terms, and second-order moving average terms to remove any predictable pattern from the residuals. Seasonal variations can be modeled with seasonal ARIMA models, but we will not discuss these models.

Multiple regression cannot be used to estimate ARIMA models in which the moving average order is greater than zero. In other words, if there is any moving average effect, you cannot use multiple regression. The reason is that moving average terms are inherently nonlinear. The estimation techniques (there are several) for ARIMA are too complicated for an introductory statistics text. The suggested readings at the end of the chapter describe the estimation techniques. In the computer application for this chapter (Appendix A15) we'll go over the estimation of the parameters in ARIMA models with MINITAB, and SAS. Here our objective is just to understand the concepts behind ARIMA.

The key issue in ARIMA analysis is deciding what order of autoregressive and moving average terms best describes the time series. This issue is the subject of section 15.4C.

■ 15.4C MODEL SELECTION AND FORECASTING

The main diagnostic tool for selecting a model in ARIMA analysis is the **correlogram,** *a plot of autocorrelations as a function of the lags of the forecast variable*. The kth autocorrelation, r_k, is calculated with formula 15.14:

$$r_k = \frac{\sum\limits_{t=1}^{n-k} (y_t - \bar{y})(y_{t+k} - \bar{y})}{\sum\limits_{t=1}^{n} (y_t - \bar{y})^2}$$

(15.14)

The logic behind the formula for the kth autocorrelation is the same as for the formula for simple correlation between two random variables, discussed in Chapter 12. The reason for a different formula here is that there are fewer observations on the lagged value of Y than on the current value of Y, but in Chapter 12 the two random variables had to have the same number of observations. Table 15.10 shows a small numerical example that illustrates the calculations for the coefficient of autocorrelation.

TABLE 15.10
Numerical example illustrating the calculation of the coefficient of autocorrelation with $k = 1$

y_t	y_{t+1}	$(y_t - \bar{y})(y_{t+1} - \bar{y})$	$(y_t - \bar{y})^2$
3	−3	−9	9
−3	1	−3	9
1	−3	−3	1
⋮	⋮	⋮	⋮
0	−1	0	0
−1	—	—	1
0		**−9**	**40**

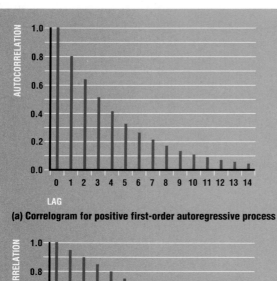

(a) Correlogram for positive first-order autoregressive process

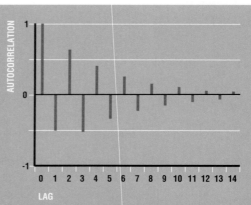

(b) Correlogram for negative first-order autoregressive process

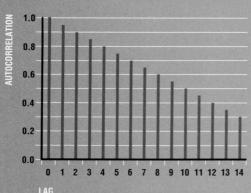
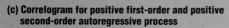

(c) Correlogram for positive first-order and positive second-order autoregressive process

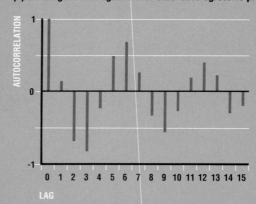

(d) Correlogram for positive first-order and negative second-order autoregressive process

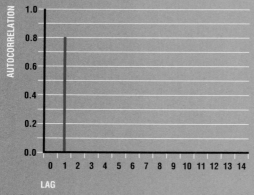
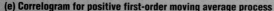

(e) Correlogram for positive first-order moving average process

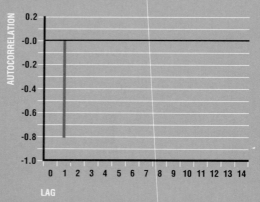

(f) Correlogram for negative first-order moving average process

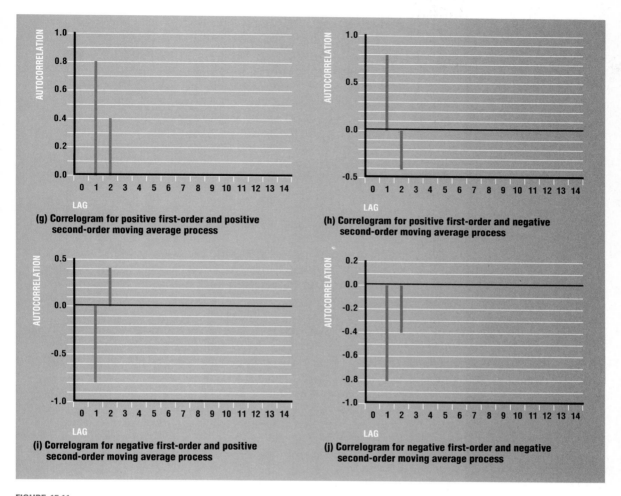

(g) Correlogram for positive first-order and positive second-order moving average process

(h) Correlogram for positive first-order and negative second-order moving average process

(i) Correlogram for negative first-order and positive second-order moving average process

(j) Correlogram for negative first-order and negative second-order moving average process

FIGURE 15.11

Correlograms for first- and second-order autoregressive and moving average processes

The mean of Y is 0 ($\bar{y} = \Sigma y_t/n$). The first-order autocorrelation coefficient, r_1, is -9 divided by 40, or -0.225. The calculations of autocorrelation coefficients and the plotting of their values are usually done by computer.

You can estimate the order of any autoregressive and moving average processes in a time series by inspecting the correlogram, because autoregressive and moving average models of different orders have distinctive correlograms. Figure 15.11 shows the correlograms for first-order autoregression with positive and negative serial correlation, second-order autoregression with positive and negative serial correlation, first-order moving average, and second-order moving average. The first point to notice in these correlograms is that the autocorrelations of all of the autoregressive processes approach 0 slowly as the order of the autocorrelation increases. The correlograms for the various moving average processes stop abruptly. With a negative first-order autoregressive process, the autocorrelations approach 0 but they oscillate from positive to negative values. The autocorrelations for an autoregressive process that is positive first-order and negative second-order follow a sine wave pattern that dampens as the order of the autocorrelations increases. The correlograms for the moving average processes are either one or two spikes, depending on whether it is a first- or second-order moving average process. The spikes go up or down depending on whether the autocorrelations are negative or positive.

If a time series combines autoregressive and moving average processes, the graphs of the autocorrelations show both the exponential decay characteristic of an autoregressive process and the spike or spikes that are characteristic of a moving average process.

When you are confronted with real data sets, it is often difficult to decide on the order of the moving average terms and the autoregressive terms after you compare the sample correlogram with the idealized correlograms in Figure 15.11. Real data can be messy and difficult to interpret. Another tool that can be used to determine the order of the moving average and autoregressive terms is a **partial autocorrelation**—*the correlation between two variables when the effects of all other variables are held constant; in a time-series context, it is the correlation between the forecast variable and a given lag of the forecast variable when the effects of all shorter lags are held constant.* A coefficient of partial correlation for the kth lag measures the correlation between y_t and y_{t-k} with the effect of all shorter lags held constant. The concept behind a partial autocorrelation can be conveyed by analogy with the coefficient of multiple determination. Recall that in a multiple regression the coefficient of multiple determination is the explained variation divided by the total variation of the dependent variable around its mean. Partial determination is the proportion of the total variation explained by *one variable* while the other variables are held constant. To obtain a coefficient of partial determination, you would run a multiple regression with and without the *one variable*. The addition to the explained variation caused by the addition of one variable divided by the unexplained variation in the regression which does not have the added variable is the partial coefficient of determination. In other words, the partial coefficient of determination is the proportion of unexplained variation which can be picked up by the additional variable. Taking the square root of the coefficient of partial determination gives you a partial correlation coefficient. In an ARIMA model the variables are the lagged values of the forecast variable.

Unlike the autocorrelations, the partial autocorrelations cannot be calculated from the original time series by a simple formula. The calculations for the required explained and unexplained variations with each added lag term are so lengthy that it is impractical to calculate even small examples by hand. You will have to rely on a computer for the partial autocorrelations. The key point about partial autocorrelations is that they can help you identify the order of the autoregressive and moving average terms in the ARIMA model. The partial autocorrelation for an autoregressive process of order 1 will have one spike, either positive or negative. The partial autocorrelation for an autoregressive process of order 2 will have two spikes, either both positive or both negative or mixed. If the partial autocorrelations come to an abrupt halt, you can be fairly certain that you have an autoregressive process. Conversely, the partial autocorrelations for a mixed autoregressive–moving average process or a pure moving average process decline exponentially.

Now that we have all of the pieces for at least a rudimentary ARIMA analysis, let's run through one complete example. We'll use the monthly time series on corporate Aaa bond rates that were plotted in Figure 15.5 (Table 15.11). These data do not have a seasonal component, so we can avoid the issue of deseasonalizing the data. The patterns in the data are complicated, however, and this example pushes at the limit of what can be analyzed given our rudimentary description of ARIMA. Recall that the original time series had two trends. The bond rates rose in the beginning, peaked in September

TABLE 15.11
Mean monthly corporate Aaa bond rates for January 1981 through June 1987

Month	Bond rate	Month	Bond rate	Month	Bond rate
Jan. 1981	12.81	Jan. 1983	11.79	Jan. 1985	12.08
Feb.	13.35	Feb.	12.01	Feb.	12.13
Mar.	13.33	Mar.	11.73	Mar.	12.56
Apr.	13.88	Apr.	11.51	Apr.	12.23
May	14.32	May	11.46	May	11.72
June	13.75	June	11.74	June	10.94
July	14.38	July	12.15	July	10.97
Aug.	14.89	Aug.	12.51	Aug.	11.05
Sept.	15.49	Sept.	12.37	Sept.	11.07
Oct.	15.40	Oct.	12.25	Oct.	11.02
Nov.	14.22	Nov.	12.41	Nov.	10.55
Dec.	14.23	Dec.	12.57	Dec.	10.16
Jan. 1982	15.18	Jan. 1984	12.20	Jan. 1986	10.05
Feb.	15.27	Feb.	12.08	Feb.	9.67
Mar.	14.58	Mar.	12.57	Mar.	9.00
Apr.	14.46	Apr.	12.81	Apr.	8.79
May	14.26	May	13.28	May	9.09
June	14.81	June	13.55	June	9.13
July	14.61	July	13.44	July	8.88
Aug.	13.71	Aug.	12.87	Aug.	8.72
Sept.	12.94	Sept.	12.66	Sept.	8.89
Oct.	12.12	Oct.	12.63	Oct.	8.86
Nov.	11.68	Nov.	12.29	Nov.	8.68
Dec.	11.83	Dec.	12.13	Dec.	8.49

FIGURE 15.12
Plot of the data in Figure 15.5

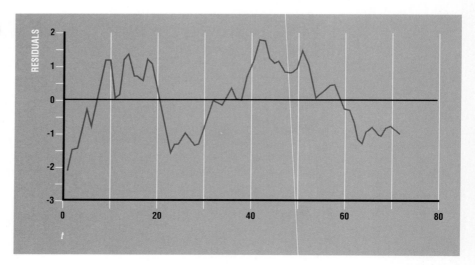

1981, and declined fairly steadily thereafter. A plot of the residuals of a regression on the original series (Figure 15.12) shows a distinct pattern.

First-differencing the data does not remove all of the trends. Figure 15.13 contains the plot of the first difference and the residuals of the first difference. The second differences appear to be a stationary series. While there is an upward trend in the second differences, the variation around that trend is quite wide. We'll use the second differences in the ARIMA model.

To decide the order of the autoregressive terms and moving average terms, you need to look at the plot of the autocorrelations for the second-difference data. Figure 15.14 plots the autocorrelations. The damped sine wave in the autocorrelations suggests that the process is second-order autoregressive with different signs on the first- and second-order terms. The long tail to the partial autocorrelations suggests that the process is also moving average. The damped sine wave pattern to the partial autocorrelations also suggests that the process is second-order moving average.

A computer program is needed to estimate these coefficients. MINITAB estimates the first-order autoregressive coefficient as 0.229 and the second-order autoregressive coefficient as -0.231. The estimated first-order moving average coefficient is 0.842 and the estimated second-order moving average coefficient is 0.166. All four coefficients are significant at the 5% level. ARIMA programs will also forecast future values of the series on the basis of the estimated coefficients. The one-step-ahead forecast (for January 1987) is 8.33, while the actual value is 8.36. Very little can be made of a comparison of one forecast value and the actual observation. Many more observations beyond the original data set would be required to gauge the accuracy of the forecasts. The computer application for this chapter (Appendix A15) discusses ARIMA procedures in different software packages.

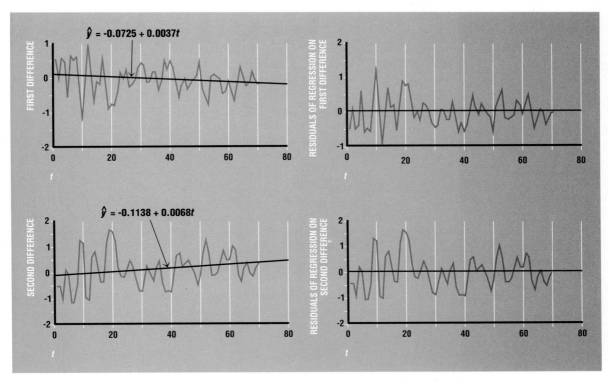

FIGURE 15.13
Plots of the first differences, the residuals of a regression on the
first differences, the second differences, and the residuals of a
regression on the second differences

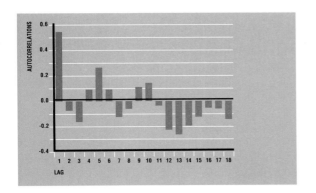

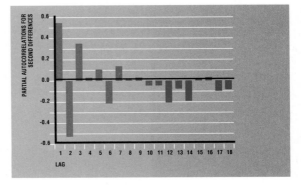

FIGURE 15.14
Autocorrelations and partial autocorrelations for the second
differences of the bond rate data

The discussion of identification of an ARIMA process has been kept rudimentary, and the discussion of partial autocorrelation was quite brief. The important seasonal multiplicative ARIMA model has not even been touched on. But you should have a good idea of what's involved in ARIMA analysis now, and the suggested readings at the end of the chapter will help you to put ARIMA to practical work.

ARIMA has some advantages and some disadvantages. Many commonly used forecasting techniques fit the data to arbitrary or ad hoc patterns. ARIMA has the advantage of extreme flexibility. It can accommodate a wide variety of patterns in the data. ARIMA also has the advantage of using all of the information in the data in an attempt to find a model that removes any predictable component in the residuals. A simple trend forecast, in contrast, ignores any possible pattern in the residuals. One disadvantage of ARIMA is its complexity. Some judgment and experience are required to select the best model. Finally, ARIMA shares the disadvantages of all single-variable naive models. It uses only the past values of the forecast variable, while other variables may contain important information about the future values of the forecast variable. (Some extensions of ARIMA, such as vector autoregression, allow the use of other variables.) A naive model cannot make an accurate forecast when the structure or rules governing the behavior of the forecast variable change.

EXERCISES FOR SECTION 15.4

15.16 Calculate simple trend-line regressions for the following data set and for the first difference of the data. Does first-differencing remove the trend? 3, 6, 9, 12, 15, 17, 21, 24, 27, 30, 33, 37, 39, 41.

15.17 Calculate simple trend-line regressions for the following data set and for the first and second differences of the data. Does first- or second-differencing remove the trend? 1, 5, 12, 20, 30, 42, 56, 72, 90, 110, 132, 156, 182.

15.18 Calculate the first and second autocorrelations for the following data: 18, 16, 13, 9, 2, 1, 0, −24, −21, −20, −17, −14, −8, −4, −2, 0, 0.

15.19 Decide if each of the processes below is autoregressive, moving average, or a combination of autoregressive and moving average. Explain why.

(a) The number of pages of ads in a monthly magazine. Note that the same ads often run for more than one issue because the magazine offers a discount for multiple-issue ads. Assume that the discount applies only to ads booked for two to six months.

(b) The population of deer in a state park measured every July. The population is subject to shocks of unusually good or bad weather, which can cause variations in the number of deer that survive each winter.

(c) The weekly sales of a hit album. Assume that each person who buys the album has a nonzero probability of recommending the album to some friends in subsequent weeks and that some of those friends will buy the album. The process tails off as the pool of friends of the album's owners who have not received a recommendation shrinks.

15.20 Decide if each of the processes below is autoregressive, moving average, or a combination of autoregressive and moving average. Explain why.

> **(a)** Claims of product tampering per week in the United States, such as an anonymous caller claiming to have put cyanide into a particular brand of pain-reliever capsules. A news story about product tampering seems to inspire some imitators for a few weeks and then such activity dwindles until there is another death or a big product recall.

> **(b)** The number of television sets each day in a repair shop. The number of sets brought into the shop in any one day is independent of the numbers brought in on other days, but after a set is brought in, it may stay in the shop for one to ten days.

> **(c)** The number of books borrowed per day at a local public library. Some of the people who return books will borrow other books on the same trip to the library. Also, whenever science projects or term papers are assigned at the local high schools, borrowing is heavy for a few days.

SUMMARY

Very few time series have a pattern of motion that is simple enough to be captured by a simple trend line. A regression with trend line plus seasonal dummies, however, is a practical technique for a time series with a strong trend and a consistent seasonal pattern. Agricultural production and retail sales are among the time series that have very regular seasonal components. The Durbin-Watson test can be used after such a regression is run to check if some pattern of serial correlation exists in the residuals. The Durbin-Watson test should not be used for regressions in which lagged values of the forecast variable are independent variables. The Durbin-Watson test for serial correlation can be used if variables other than lagged values of the forecast variable are independent variables. If there is serial correlation, defined as a nonzero correlation between error terms, then the usual hypothesis tests in multiple regression are not accurate.

Data smoothing is an approach to forecasting which attempts to suppress any noise in the data. Two smoothing methods are moving averages and exponential smoothing. The idea behind both methods is that the suppression of noise in the time series will reveal any underlying patterns. While these methods are highly arbitrary, they have the advantages of being simple and workable. These methods continue to be popular.

In contrast to data smoothing, autoregressive methods try to exploit any pattern in the residuals to help predict the forecast variable. One technique that searches for serial correlations (correlations between successive error terms) or autocorrelations between errors terms (correlations between error terms that are any fixed number of steps apart) consists of autoregressive models. In an autoregressive model the independent variables are all lags of the forecast variable. The suggested procedure for determining how many lags to employ has two parts. First, run all regressions up to the $\sqrt{n} + 10$ lags, adding one lag for each regression. Second, identify the slope coefficients that consistently differ significantly from 0 at the 5% significance level.

The estimated regression equation that should be used for forecasting is the smallest one that contains the significant slope coefficients.

The new forecasting technique called ARIMA can capture much more complex patterns than can be captured by trend lines with seasonal dummies or by autoregressive models. In fact, these patterns are just special cases of the ARIMA model. The procedure for estimating an ARIMA model is first to remove any trend in the data by differencing. A time series that has no trend and no heteroscedasticity can be considered stationary. Once the time series is transformed so that it is stationary, the correlogram of the transformed series is plotted to check if the series follows the characteristic patterns of autoregressive or moving average processes or a mixed autoregressive–moving average process. The correlogram is the main tool for identifying the appropriate ARIMA model. The estimation technique for ARIMA models is nonlinear, and multiple regression cannot be used.

All of the time-series models discussed in the chapter are naive models in the sense that they assume that past patterns will be repeated in the future. Structural models attempt to explain the behavioral or physical processes that govern a time series. Another limitation of all of the models discussed is that they rely on only the current and past values of the forecast variables. There are more advanced naive models that make use of the information in other variables. While ARIMA analysis can be difficult, it has better forecasting accuracy than models that impose a specific pattern on the data. Because of this advantage, ARIMA analysis is rapidly gaining in popularity as a forecasting technique.

 ## SUGGESTED READINGS

C. W. J. Granger. *Forecasting in Business and Economics*. New York: Academic Press, 1980.

Spyros Makridakis, Steven C. Wheelwright, and Victor McGee. *Forecasting: Methods and Applications*. 2d ed. New York: Wiley, 1983.

Charles R. Nelson. *Applied Time Series Analysis for Managerial Forecasting*. San Francisco: Holden-Day, 1973.

Walter Vandaele. *Applied Time Series and Box-Jenkins Models*. New York: Academic Press, 1983.

 ## NEW TERMS

ARIMA An acronym for autoregressive integrated moving average, a forecasting model that breaks the pattern of the forecast variable into trend, autoregressive, and moving average components.

autoregressive model The regression of current values of the forecast variable on the past or lagged values of the forecast variable.

correlogram A plot of autocorrelations as a function of the lags of the forecast variable.

Durbin-Watson test A test for serial correlation which analyzes the residuals of the estimated regression line.

exponential smoothing A method of smoothing a time series by which an observation is replaced by the weighted sum of all past observations, the weights declining exponentially as the observations recede in time.

first-order autoregressive model A regression in which the first lag of the forecast variable is used as the explanatory variable.

first-order moving average model A model in which the current value of the forecast variable is a function of the forecast error for the previous period.

moving average A transformation of a time series by which an observation is replaced by the average of the replaced observation and some adjacent observations.

naive model A model based on the assumption that there is some pattern over time in the forecast variable and that this pattern can be used to predict the future path of the forecast variable.

order The number of lags of the forecast variable used as independent variables.

partial autocorrelation The correlation between two variables when the effects of all other variables are held constant; in a time-series context, the correlation between the forecast variable and a given lag of the forecast variable when the effects of all shorter lags are held constant.

pth-order autoregressive model A regression in which p lags of the forecast variable are used as independent variables.

qth-order moving average model A model in which q lags of the forecast error affect the current value of the forecast variable.

seasonality Any consistent pattern in a forecast variable which is repeated within a year.

stationary time series A time series with a fixed mean and a constant variance of the error random variables for all time periods, in which the autocorrelations are constant for a given lag.

structural model A model that attempts to describe the key behavioral relationships among the agents that affect the path of a time series.

time series A set of measurements on a variable taken over some time period.

trend A consistent long-term change in the average level of the forecast variable per unit of time.

 ## NEW FORMULAS

Mean of Y at time t in a simple trend model

$$(\mu_Y)_t = \alpha + \beta t \tag{15.1}$$

Observed value of Y at time t in a simple trend model

$$y_t = \alpha + \beta t + \varepsilon_t \tag{15.2}$$

Predicted value of Y at time t in a simple trend model

$$\hat{y}_t = a + bt \tag{15.3}$$

Predicted value of Y at time t in a simple trend model with seasonal dummies

$$\hat{y}_t = a + b_1 t + b_2\text{SUMMER} + b_3\text{FALL} + b_4\text{WINTER} \tag{15.4}$$

Durbin-Watson test statistic

$$\text{DW} = \frac{\sum\limits_{t=1}^{n-1} (e_{t+1} - e_t)^2}{\sum\limits_{t=1}^{n} e_t^2} \tag{15.5}$$

Exponentially smoothed value of a time series at time t

$$\overline{x}_t = ax_t + a(1-a)x_{t-1} + a(1-a)^2 x_{t-2} + a(1-a)^3 x_{t-3} + \cdots \tag{15.6}$$

Exponentially smoothed value of a time series at time t expressed in terms of the current observation and the most recent forecast

$$\overline{x}_t = (1-a)\overline{x}_{t-1} + ax_t \tag{15.7}$$

Mean of Y at time t in a first-order autoregressive model

$$(\mu_Y)_t = \alpha + \beta y_{t-1} \tag{15.8}$$

Observed value of Y at time t in a first-order autoregressive model

$$y_t = \alpha + \beta y_{t-1} + \varepsilon_t \tag{15.9}$$

Predicted value of Y at time t in a first-order autoregressive model

$$\hat{y}_t = a + b y_{t-1} \tag{15.10}$$

Predicted value of Y at time t in a pth-order autoregressive model

$$\hat{y}_t = a + b_1 y_{t-1} + b_2 y_{t-2} + b_3 y_{t-3} + \cdots + b_p y_{t-p} \tag{15.11}$$

Observed value of Y at time t in a first-order moving average model

$$y_t = \alpha + \varepsilon_t - \theta \varepsilon_{t-1} \tag{15.12}$$

Observed value of Y at time t in a qth-order moving average model

$$y_t = \alpha + \varepsilon_t - \theta_1 \varepsilon_{t-1} - \theta_2 \varepsilon_{t-2} - \theta_3 \varepsilon_{t-3} - \cdots - \theta_q \varepsilon_{t-q} \tag{15.13}$$

Estimated coefficient of autocorrelation on the kth lag

$$r_k = \frac{\sum_{t=1}^{n-k} (y_t - \bar{y})(y_{t+k} - \bar{y})}{\sum_{t=1}^{n} (y_t - \bar{y})^2} \qquad (15.14)$$

ADDITIONAL EXERCISES

An asterisk designates an exercise of greater than ordinary difficulty.

15.21 The following data set is a hypothetical time series on the annual per capita consumption of coffee. 509, 500, 486, 481, 475, 468, 459, 446, 441, 436.

(a) Use a simple trend line to forecast the level of coffee consumption for the year after the time series ends.

(b) Construct a confidence interval at the 95% level for this forecast. (Hint: This is a confidence interval on the predicted value of Y given x. The topic was discussed in Chapter 13.)

15.22 A video rental store has estimated the following equation for the number of tapes rented per day:

$\hat{y}_t = 314 + 0.6t - 80\text{MONDAY} - 124\text{TUESDAY} - 164\text{WEDNESDAY} + 36\text{THURSDAY} + 460\text{FRIDAY} + 149\text{SATURDAY}$

where t is the number of days the store has been open. The store has been open a year. All of the coefficients are significant at the 5% level. Forecast the number of tapes that would be rented on day 366, assuming that day 366 is a Sunday.

15.23 If a regression on a time series with 100 observations with a trend term as the independent variable resulted in a Durbin-Watson test statistic of 0.9, could the null hypothesis of no positive serial correlation in the residuals be rejected? If it was rejected, what implication would this have for any forecasts made on the basis of this regression?

15.24 Construct a five-item centered moving average for the following monthly data on orders for mufflers for Porsche 911s at a national chain of muffler shops: 5, 16, 15, 12, 18, 9, 3, 11, 13, 12, 7, 3, 15, 12, 3, 17, 2, 5, 14, 1. Plot the original series and the moving average. What is the forecast for the first period after the time series ends?

15.25 Exponentially smooth the muffler order data in exercise 15.24, using a smoothing constant of 0.5. What is the one-step-ahead forecast? Which method of smoothing is more sensitive to the most recent observation, exponential smoothing with a smoothing constant of 0.5 or a five-item moving average?

15.26 Calculate a first-order autoregressive model for the following time series: 7, 10, 15, 21, 27, 32, 36, 41, 45, 52, 56, 60, 66, 71, 77, 81, 86, 91, 95, 101. Explain what the intercept and slope coefficients mean.

15.27 The following equation was estimated on a monthly time series of the tons of iron ore held in inventory by an integrated steel manufacturer:

$$\hat{y}_t = 12{,}502 + 0.384y_{t-1} + 0.049y_{t-2} + 0.029y_{t-3} + 0.012y_{t-4} + 0.022y_{t-5} - 0.528y_{t-6}$$

The inventories for the past six months were: $t-1 = 87{,}645$, $t-2 = 75{,}431$, $t-3 = 99{,}259$, $t-4 = 51{,}713$, $t-5 = 85{,}303$, $t-6 = 58{,}125$. Construct a forecast for the next six months.

15.28 Construct the correlogram for the data in exercise 15.26 for the first four lags of the forecast variable.

15.29 If the correlograms and partial correlograms for the first differences of a time series were as follows, what ARIMA model would you estimate? Explain your reasoning.

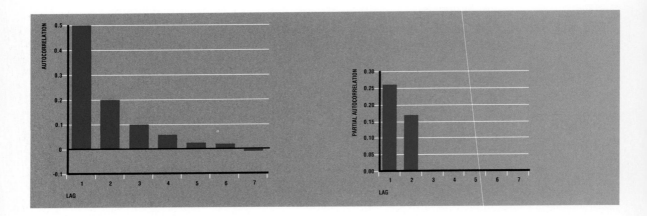

15.30 An ice cream parlor that had been opened for ten months was burned down in an accidental fire. The sales for these months are as follows: $2,476, $2,881, $3,065, $3,240, $3,389, $3,451, $3,356, $3,512, $3,608, $3,623. The business had an insurance policy that covered its lost sales while the ice cream parlor was being rebuilt. It took six months to rebuild the parlor. Use a simple trend line to estimate the lost sales. Is this a reasonable method for estimating the lost sales? Explain.

15.31 The ministry of tourism on a Caribbean island has estimated the following regression on a quarterly time series of the number of tourists visiting the island:

$$\hat{y}_t = 1{,}034 + 121t - 456\text{SUMMER} + 5{,}903\text{WINTER} + 19\text{FALL}$$

All of the coefficients except the one for the FALL dummy variable are significant at the 5% level. If there are 26 quarters in the time series and the next quarter is summer, what is the forecast of the number of tourists who will visit the island?

15.32 If the regression in exercise 15.31 had a Durbin-Watson test statistic of 1.69, could the null hypothesis of no positive serial correlation be rejected at the 5% significance level?

15.33 Plot the original time series and centered moving averages of three, five, and seven items from the following data set on the same graph: 1, 8, 9, 11, 12, 10, 8, 6, 5, 3, 6, 10, 7, 2, 11, 4, 3, 7, 8, 4. Which moving average appears to provide the best representation of any pattern in the data?

15.34 Exponentially smooth the following time series, using smoothing constants of 0.9, 0.6, and 0.3: 9, 9, 31, 45, 15, 55, 8, 33, 27, 48, 31, 6, 10, 13, 1, 28, 30, 43, 12, 4. Plot the original series and the three smoothed series and decide which smoothing constant appears to provide the best representation of any pattern.

15.35 At the 5% significance level, test the hypothesis that there is no time trend to the following series: 8, 20, 11, 22, 23, 3, 15, 9, 20, 23, 25, 25, 18, 25, 8, 20, 21, 12, 22, 29.

15.36 A mail-order company that specializes in Christmas-tree ornaments has estimated the following regression equation on a quarterly time series of its past sales in dollars over 36 quarters:

$$\hat{y}_t = 10{,}034 + 344t + 456\text{SUMMER} + 28{,}970\text{WINTER} + 39{,}799\text{FALL}$$
$$\quad\quad (445.7)\quad\quad (216.3)\quad\quad\quad (167.0)\quad\quad\quad\quad (8{,}904.8)$$

The standard errors of the slope coefficients are in parentheses. At the 5% significance level, test the hypotheses that there is no time trend to the series and that the summer, winter, and fall quarters do not differ from the spring quarter.

15.37 Test the hypothesis that there is no negative serial correlation in the sales of the company described in exercise 15.36 if the Durbin-Watson test statistic for the above regression was 2.26.

15.38 Use a four-item moving average to forecast the next value of the following quarterly time series on the number of cars using a ferry service: 146, 253, 547, 819, 109, 255, 508, 839, 112, 224, 500, 835, 128, 223, 528, 840, 114, 249, 523, 803. Explain why this method is not well suited for making such a forecast.

15.39 The following time series is an annual series on the average prices of movie tickets in a city: 0.48, 1.24, 1.40, 1.43, 1.94, 1.67, 1.55, 2.20, 2.48, 2.64, 2.55, 2.57, 3.38, 3.44, 3.17, 4.09, 3.50, 3.88, 4.50, 4.10.

  (a) Use a first-order autoregressive model to forecast the average price for the next year in the series.

  (b) Test the hypothesis that the population slope coefficient equals 0.2 at the 5% significance level.

  (c) Test the hypothesis that there is no serial correlation at a 10% significance level.

  (d) If there was serial correlation, what would this imply about the appropriateness of forecasting with a first-order autoregressive model?

15.40 First difference the following time series and run a regression on the first differences: 11, 19, 23, 17, 32, 23, 36, 16, 20, 35, 33, 24, 46, 29, 40, 51, 42, 57, 49, 49. Plot the residuals of this regression. Does first differencing result in a stationary series?

15.41 If the correlograms and partial correlograms for the first differences of a time series were as follows, what ARIMA model would you estimate? Explain your reasoning.

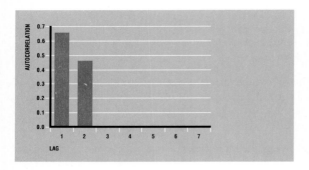

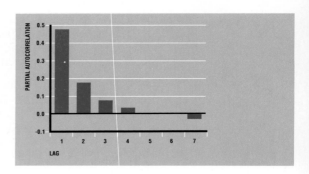

15.42 Decide whether the process in each of the following examples is autoregressive, moving average, or a combination of both. Explain your reasoning.

(a) The daily number of patients in the intensive care unit of a hospital. Stays in intensive care average four days and few patients come back for a second stay.

(b) Attendance at games of a professional baseball team with an outdoor stadium. Attendance levels are a function of the weather and a team's won-lost record.

(c) Sales of insulin at a drugstore. Assume that diabetics usually buy a week's supply at a time.

15.43 An art museum has kept a record of its daily attendance over the past ten years. Using these data, it has estimated the following regression:

$$\hat{y}_t = 345 + 2t - 245\text{MONDAY} - 145\text{TUESDAY} - 34\text{WEDNESDAY} - 308\text{THURSDAY}$$
$$- 145\text{FRIDAY} - 24\text{SATURDAY}$$

If the first day of the museum's 11th year is a Sunday, what is the forecast for attendance on that day?

15.44 Use a five-item moving average to forecast the next value of the following time series: 21, 22, 22, 22, 23, 24, 24, 25, 26, 27, 27, 27, 28, 29, 30, 30, 31, 31, 32, 32. What is wrong with using this forecast method for these data?

15.45 The operator of an insecticide service has recorded the following number of termite jobs per month the service has done: 59, 51, 52, 54, 59, 58, 55, 55, 52, 60, 54, 52, 57, 56, 60, 53, 53, 57, 52, 57. Plot the time series and forecast the number of jobs that will be required in the next month using a simple time trend, a four-item moving average, exponential smoothing with a smoothing constant of 0.3, and a first-order autoregressive model. Is there any reason to prefer any of these methods over any of the others?

15.46 Decide whether the process in each of the following examples is autoregressive, moving average, or a combination of both. Explain your reasoning.

(a) A quarterly time series on the number of inmates in prison in one state. Most sentences are longer than three months and many inmates return to prison for other crimes after serving out their current sentences.

(b) An annual time series on the Florida orange crop. The amount of oranges grown per year depends on the acreage planted, diseases, and weather. Assume that trees lost to disease or frost take five years to replace.

(c) An annual series on new-car sales. New-car sales depend on whether the economy is expanding or in a recession. Recessions and expansions tend to last several years. Also, consumers tend to keep a car a fixed number of years. For example, some consumers plan to keep a new car for three years and others plan to keep a new car for ten years.

15.47 Suppose you ran an autoregressive equation on the first 12 lags of a variable. If you wanted to test for the presence of serial correlation, you could use a Durbin-Watson test, use a test on the slope coefficient of the first lag, or plot the correlogram. Decide which method would be best and explain why.

15.48 First- and second-difference the following time series and run regressions on the first and second differences: 95, 55, 99, 79, 97, 181, 186, 235, 270, 282, 344, 336, 351, 418, 543, 532, 595, 664, 811, 905. Plot the residuals of each of these regressions. Does either first or second differencing result in a stationary series?

15.49 If the correlograms and partial correlograms for the first differences of a time series were as follows, what ARIMA model would you estimate? Explain your reasoning.

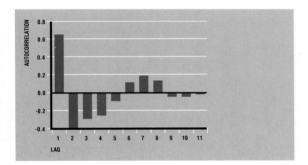 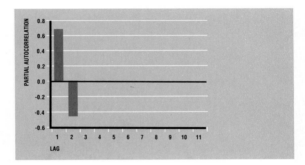

15.50* In the examples in the text on forecasting with an autoregressive model, you first determined the highest lag that was consistently significant and then used the regression with that lag and all shorter lags to make the forecast. An alternative method would be to use only the coefficients of the lags that were statistically significant to make the forecast. Does the method used in the text make more sense? Explain.

16
Index Numbers

T he federal government publishes many series of index numbers. An **index number** is *a measure of the value of a variable in relation to its value at some other location or in some other time period.* The most famous of the federal government's index numbers is the consumer price index (CPI). The CPI compares the prices of a set of goods and services over time. The approximately 400 goods and services in this index were chosen to represent all of the purchases made by urban households.

The function served by index numbers is vital—there is no other way to describe how sectors of the economy change over time or across regions. The fact that the release of such government indexes as the CPI and the index of leading economic indicators is often headline news gives you some idea of the importance of index numbers. All sorts of business decisions and government policies rely on the information provided by index numbers.

Unfortunately, some of the federal government's index numbers confuse rather than inform. Intelligent use of index numbers requires that you understand how they are defined and calculated and the procedures specific to any index you plan to use. Here are two examples of problems in index numbers published by the federal government. One index is the cost of commercial construction. Economists were puzzled in the 1970s because the rapid rise of this index relative to other cost indexes suggested that productivity in commercial construction was declining. This suggestion did not square with the fact that the industry had adopted many improvements in construction equipment and techniques. It was difficult for economists to formulate a theory of growing ineptness in the construction industry. The decline in productivity, however, was more apparent than real. To understand what was going on with the index, imagine two office buildings side by side, one erected in 1906 and the other in 1987. The old building is faced with stone and has carved ornamental stonework over the doors and windows; the new building is faced with glass and metal. They are both office buildings, but in their details they are very different. One approach to an index number on construction costs over time is to estimate how much it would cost today to reproduce the 1906-style building. Another approach is to estimate how much it would have cost in 1906 to build the 1987-style building. Obviously, no one builds 1906-style office buildings today, and it would have been impossible to build the current office building back in 1906. Cost comparisons over long time periods are very difficult because there are fundamental changes in the goods and services. One reason (perhaps the entire reason) that the federal government's index of commercial construction costs went up so fast during the 1970s is that the weight or importance of each item in the index was based on 1906 construction practices. In particular, ornamental stonework was an important part of the index. The prices of carved statues and decorations are astronomical because the process is very labor-intensive. The federal government's index shows how expensive it would be today to build an office building covered with ornamental stonework. This index is of little use for explaining how much construction costs changed from 1987 to 1988.

The federal government also publishes an index on the cost of raising children. To get an idea of what goods and services parents buy for their children, the government ran a survey on family expenditures on children in 1961. Patterns of expenditures on children have changed quite a bit since 1961, almost as much as office buildings since 1906. Consider these items: the index is based on the assumption that no restaurant meals will be purchased before a child reaches the age of 6; another assumption is that no child-care services, such as day care, will be purchased. It should be apparent that these assumptions are unrealistic. The index on the costs of raising children will give you a poor understanding of changes in recent years.

These examples were not meant to disparage all federal government indexes. The point is that the creation of an accurate index is difficult and that the user of any published index has to be aware of any potential pitfalls in its construction.

16.1 INTRODUCTION TO INDEX NUMBERS

16.1A WEIGHTED INDEX NUMBERS: LASPEYRES' INDEX, PAASCHE'S INDEX, AND FISHER'S IDEAL INDEX

Comparisons of changes in either prices or quantities are easy if only one commodity is involved. Suppose you had purchased 30 bars of a particular brand of soap in 1987 and the price averaged 50 cents. The corresponding numbers for 1986 were 33 bars and 48 cents. Three simple ratios could be used to describe the changes. **Price relatives** express *the ratio of prices of one commodity in the current period and the base period.* **Quantity relatives** express *the ratio of quantities of one commodity in the current period and the base period.* **Expenditure relatives** express *the ratio of expenditures on one commodity in the current period and the base period.* The three ratios are:

$$PR_t = \frac{P_t}{P_0} \qquad \text{(16.1)}$$

$$QR_t = \frac{Q_t}{Q_0} \qquad \text{(16.2)}$$

$$ER_t = \frac{P_t Q_t}{P_0 Q_0} \qquad \text{(16.3)}$$

PR, QR, and *ER* stand for price relatives, quantity relatives, and expenditure relatives, respectively. The subscript 0 stands for the base period and the subscript *t* for some time period *t*. If the comparison were across regions rather than over time, the subscripts would stand for some base region and some region *t*. The price, quantity, and expenditure relatives for your soap are

$$PR_t = \frac{P_t}{P_0} \qquad \text{(16.1)}$$

$$PR_t = \frac{0.50}{0.48}$$

$$PR_t = 1.042$$

$$QR_t = \frac{Q_t}{Q_0} \qquad \text{(16.2)}$$

$$QR_t = \frac{30}{33}$$

$$QR_t = 0.909$$

$$ER_t = \frac{P_t Q_t}{P_0 Q_0} \qquad \text{(16.3)}$$

$$ER_t = \frac{0.50(30)}{0.48(33)}$$

$$ER_t = 0.947$$

According to these calculations, from 1986 to 1987 the average price of the soap rose 4.2%, the quantity that you purchased declined 9.1%, and your expenditures on soap fell by 5.3%. (You may recall from introductory economics the concept of elasticity. Your demand for soap is price elastic; the percentage change in quantity is greater than the percentage change in price. If the demand is elastic, the total expenditure declines as the price rises. The observation that your demand is elastic assumes that your income and all other determinants of the demand for soap are held constant.)

The difficulties in making comparisons begin as soon as you have more than one item. You have to construct a **composite index,** *an index that consists of more than one item.* Let's narrow the focus to comparisons of prices over time.

TABLE 16.1

Prices, quantities, and expenditures for three items, 1986 and 1987

Item	1986			1987		
	P_{0i}	Q_{0i}	E_{0i}	P_{ti}	Q_{ti}	E_{ti}
Soap	0.48	33	15.84	0.50	30	15.00
Toothpaste	0.90	12	10.80	0.92	11	10.12
Haircut	5.00	14	70.00	6.00	10	60.00

The discussion of quantity indexes parallels the discussion of price indexes. To illustrate the concepts, Table 16.1 lists the prices, quantities, and expenditures for three items in 1986 and 1987. The i refers to the ith item.

There are many ways to compare all three goods over the two years. One way is to compare the changes in the average prices of the three goods over the two years. A second way is to compare expenditures on the three goods over the two years. A third way is to compute what it would cost you to make your 1986 purchases in 1987 prices and then compare the ratio of the actual expenditures in 1986 with the cost of the same bundle of purchases in 1987 prices. Finally, you could reverse the last procedure and compute what it would cost you to make your 1987 purchases at 1986 prices and then compare the ratio of the actual expenditures in 1987 with the cost of the same bundle of purchases in 1986 prices. Some more comparisons are possible, but this list is long enough to make the point that there are many ways to compare price changes over time.

Some of the ways convey more information than others. The simple comparison of the average prices may not be helpful. The average price of the three items in 1986 is (0.48 + 0.90 + 5.00)/3, or $2.13. The average price in 1987 is (0.50 + 0.92 + 6.00)/3, or $2.47. If these three items were the only things in your budget, you could say that your personal rate of inflation was $2.47 divided by $2.13, or 1.16. By this procedure you have constructed a **simple unweighted index,** *an index in which the sum of the prices in the current period is divided by the sum of the prices in the base period.* The increase in the average price was 16%. However, this comparison of average prices treats each item in the budget as equally important. If the three items were weighted in some way that reflected their importance in your budget, your personal rate of inflation could be more or less than 16%.

For most analyses the second comparison, total spending across years, would also not be helpful. *The ratio of current-period over base-period spending* is called a **weighted aggregate index.** The weights on the prices are the quantities of the various items in the base and current periods. Total spending went from $96.64 to $85.12. The ratio is 0.88. The decrease in total spending indicated by this ratio does not tell you much, however, because the medium used for the comparison, dollars, was not the same thing in 1986 and 1987. The 1987 dollar had less purchasing power than the 1986 version. Your spending went down, but you also bought less of every item.

The third and fourth methods are commonly used to compute price indexes. The third method, called **Laspeyres' index,** is *an index that uses weights* *based on some base period's quantities. A Laspeyres index can be used to determine what it would cost to buy the base period's market basket at the prices prevailing in the current period.* The formula for a Laspeyres index number is

$$L_t = \frac{\sum\limits_{i=1}^{n} Q_{0i}P_{ti}}{\sum\limits_{i=1}^{n} Q_{0i}P_{0i}} \times 100 \qquad \text{(16.4)}$$

L_t stands for the value of a Laspeyres index at time t. The n is the total number of items in the index. A Laspeyres index uses the original quantities purchased as weights to determine the change in prices over time. It answers the question "By what percentage would my spending have to go up to allow me to purchase the same bundle of goods that I purchased in the base period?" The examples described at the beginning of the chapter are Laspeyres' indexes. One index answered the question of how much it would cost to build the 1906-style building today and the other index answered the question of how much it would cost to raise a 1961-style child today. Here are the calculations for the Laspeyres index for the data in Table 16.1:

$$L_t = \frac{33(0.50) + 12(0.92) + 14(6.00)}{33(0.48) + 12(0.90) + 14(5.00)} \times 100$$

$$L_t = 115.42$$

According to a Laspeyres index, the rate of inflation is 15.42%.

The opposite procedure is to use the current quantities of the various items purchased to weight the beginning and ending prices. This type of index is called a Paasche's index. **Paasche's index** is *an index that uses weights based on the most recent period's quantities. Paasche's index can be used to determine what it would cost to buy the current period's market basket at the prices that prevailed in some prior period.* The formula for a Paasche index is

$$\text{Paasche}_t = \frac{\sum\limits_{i=1}^{n} Q_{ti}P_{ti}}{\sum\limits_{i=1}^{n} Q_{ti}P_{0i}} \times 100 \qquad \text{(16.5)}$$

DU PONT'S COST OF MATERIALS AND ENERGY INDEX

E. I. Du Pont de Nemours & Co., Inc., a large chemical company based in Wilmington, Delaware, has been using an index to track its materials and energy costs for over 50 years. This index offers the company several advantages over a broad federal government price index, such as the producer price index. One advantage of Du Pont's index is that it is tailored to its specific purchases. A second advantage is that the index reflects the actual costs of materials used in production rather than purchases of materials that are held in inventory. Only stocks drawn from inventory go into the calculations of the index. A third advantage is that the prices are actual delivered prices. Government producer price indexes are often based on list prices and do not reflect delivery costs. The differences between the Du Pont index and some wider government indexes can be seen in Figure 16.1. These differences highlight the benefit of a custom-made index.

Du Pont's index uses 200 items that represent 85% of the expenditures made by the company's Energy and Materials Department. The 200 items are divided into four groups: (1) basic materials, such as titanium ores and sulfur; (2) hydrocarbons derived from the refining of petroleum; (3) chemicals, such as complex organics and chlorine; (4) all forms of energy.

In the formula for calculating the index, the base year is always the year before the current year. In effect, the Du Pont index is a moving Laspeyres index as the base period moves from year to year. Since Du Pont can determine at low cost exactly what it purchases each year, the annual change in the base weights that its formula requires is not expensive. The technical term for the formula is a chain index.

Du Pont uses its index mainly as an analytical tool. A rising or a falling index can

Paasche$_t$ stands for the value of a Paasche index at time t. The calculations for a Paasche index for the data in Table 16.1 are

$$\text{Paasche}_t = \frac{30(0.50) + 11(0.92) + 10(6.00)}{30(0.48) + 11(0.90) + 10(5.00)}$$

$$\text{Paasche}_t = 114.56$$

According to the Paasche index, your own personal rate of inflation was 14.56%. In this example the results of the Paasche and Laspeyres indexes were fairly close, 14.56% versus 15.42%. The reason was that the amounts or weights differed little in the two formulas. While a budget made of only three items is highly artificial, it is plausible that a Paasche index and a Laspeyres index would have about the same value when only one year separates the base

FIGURE 16.1

Percent change in costs of energy and materials, Du Pont cost index, an economy-wide price index, and producer price index of crude materials, 1977–1981

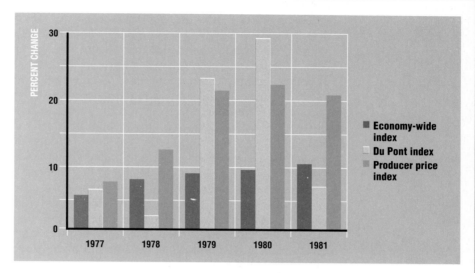

help explain variations in the company's earnings. The index can be used to catch major shifts in the strategy of Du Pont's suppliers. A third use of the index is to gauge the performance of Du Pont's buyers and planners. A version of the index based on 100 items is used for forecasting costs four quarters and four years ahead. The forecast version of the index helps Du Pont to set prices for its products, decide on which products to market, and establish negotiating targets for its buyers.

Source: "How to Build Your Own Materials Cost Index," *Purchasing*, June 10, 1982, pp. 51–55.

and the end of the series. If the weights or consumption bundles in the two years were widely different, the Paasche and Laspeyres indexes would give widely different answers. This result tends to occur when there are many years between the index's base year and ending year.

Since Paasche's index uses the current period's consumption patterns as weights, a new survey would be needed each year to determine the current pattern of consumption. Laspeyres' index requires a survey to determine the weights only in the base year. Thus a Paasche index is much more expensive to construct than a Laspeyres index. However, it is not true that a Paasche index would necessarily be preferred to a Laspeyres index if the cost of constructing it were immaterial. There is nothing so special about the most recent pattern of consumption that it should be preferred for determining the weights in a price index. The problem is that the prices you are using to compute your index influence your pattern of consumption. People shy away from items that have gone up in price and they tend to buy more of items that have fallen in price, reflecting what economists call the substitution effect. Relative prices help determine the pattern of consumption. The pattern of consumption is always shifting toward the goods that are bargains in a particular year. This shift in consumption patterns in response to price changes could be seen where energy prices rose sharply in response to the 1974 oil embargo. Consumers responded in myriad ways to the sharply higher energy prices: by buying smaller cars, putting more insulation in their homes, living closer to work, buying wood stoves. Insulation and wood stoves were not important parts of household budgets before 1974. Conversely, a large sedan equipped with a V-8 engine is not an important part of household budgets today. A price index can be used to determine how much it would cost to buy the base-period bundle of goods at current prices (Laspeyres) or how much it would cost to buy the current bundle of goods in base-period prices (Paasche). Neither type of index can tell you how much money you would need in order to be as well off today as you were in the base period because at today's prices you would not choose the base-period bundle and at the base-period prices you would not choose the current-period bundle. What you would like the index to help you determine is how much better or worse off you are in terms of your standard of living. This is a difficult issue.

Fisher's ideal index, *an index that uses weights based on quantities of both the current period and the base period,* gets around the problem of choosing either the base-period or the current-period consumption patterns. The formula for a Fisher ideal index is

$$F_t = \sqrt{L_t \times \text{Paasche}_t} \tag{16.6}$$

F_t stands for the value of a Fisher ideal index at time t. For the data in Table 16.1, the Fisher ideal index is

$$F_t = \sqrt{115.42(114.56)}$$

$$F_t = 114.99$$

When you treat your base-period consumption pattern and your ending period consumption pattern as equally important, your estimated personal rate of inflation is 14.99%.

Fisher's ideal index has some desirable properties as a measure of price changes. However, it is the most expensive to construct, because both Laspeyres' and Paasche's indexes have to be calculated before Fisher's ideal index can be computed. Few published indexes are Fisher ideal indexes.

So far we have concentrated on the definitions and the calculations of price indexes. Section 16.1B discusses the interpretation and use of index numbers.

16.1B INTERPRETING INDEXES

Suppose you have a Laspeyres index and you want to make comparisons between years other than the base and the ending year. For example, the CPI is a Laspeyres index with the base year of 1967 (the base year is the year in which the index has a value of 100). Let's focus on one component of the CPI and two years, say medical care in 1978 and 1983. The CPI for medical care is 235.3 in 1978 and 384.1 in 1983. What these numbers mean is that you would have needed $235.30 in 1978 to buy the same bundle of medical services that cost $100.00 in 1967, and by 1983 you would have needed $384.10. By what percentage did the price of medical services go up over the period 1978 to 1983? The percentage is the change in the index from 1978 to 1983 divided by the value of the index in 1978. The answer is 63.2%. The all–item percent change in prices for the same period was 48%. The higher rate of inflation in medical services than for all items in the CPI made inflation in medical services a major political issue. Since 1983 the rate of inflation in medical services has gotten much closer to the overall rate of inflation, in part because of changes in federal rules on how hospitals are paid.

One issue that clouds the interpretation of a price index is quality changes. The Bureau of Labor Statistics (the agency that issues the CPI) tries to adjust for changes in quality over time, but it is very difficult. Let's continue with the medical services component of the CPI as an example of the problems of quality changes. One item that is used in the index is the cost of a day in a hospital. However, the equipment and techniques used in hospitals have changed dramatically over time. If changes in quality can make a patient more comfortable, or more likely to regain his or her health, or recover in fewer days, then a "day" in a hospital is not the same service that it was before those changes. The Bureau of Labor Statistics can adjust for shifts in consumption toward goods or services of higher quality if the two types of goods or services are available at the same time. Black-and-white television sets, and color sets were commonly purchased in one period but now almost all television sets purchased are color sets. The differential in prices between the black-and-white and the color sets can be used to estimate the difference in quality. If one type of good is entirely supplanted by another, the difference in manufacturing costs can be used to estimate the difference in quality. For example, people used to have iceboxes to keep food cold. The last iceboxes were replaced by refrigerators in the late 1940s and early 1950s. The difference in the costs of manufacturing an icebox and a refrigerator could be used to estimate the difference in quality of the two goods. Subtle changes in quality can stump statisticians. A stay in a hospital in 1950 was quite different from a stay today, but the change in quality was the product of many small changes. You just have to be cautious when you interpret an index on anything that can have subtle or small incremental changes in quality. Another good example of this problem is automobiles. They need less service today and they have features that were not available a few years back, such as computer-generated warnings for doors ajar or low oil pressure. It is difficult to adjust an index of automobile prices for these quality differences.

Another interpretive issue for Laspeyres indexes is the choice of a base year. The choice of an unusual year, such as a war year, could sharply change the weights of the items in the index. Wars are often accompanied by price controls, rationing, and lines for desired consumer goods, and an item such as beef may have a small weight in the consumption bundle simply because it could not be purchased. Usually it is not difficult to pick a typical year for the base, but you should be alert to attempts to present a misleading index by selection of an unusual base year.

There are many other interpretive issues. An index may not be accurate for all geographical regions. An index may rely on list prices rather than on actual prices. An index may have too small a sample, so that the estimated prices can be inaccurate. An index may rely on information provided by large stores or large companies; the observed prices might be different if a representative sample of all types of stores or companies were included. It is easy to extend this list. In spite of these often insurmountable problems, index numbers are important business tools.

■ 16.1C SPLICING OVERLAPPING INDEXES AND USING AN INDEX TO DEFLATE AN EXPENDITURE SERIES

Occasionally an index is revised with new weights. It is possible to get one consistent series from the old and the new series as long as the two series have at least one overlapping period. Table 16.2 shows hypothetical old and new indexes and a spliced index created from them. From 1982 on, the spliced index is the same as the new index. In the common year for the old and new indexes, 1982, the ratio of the old to the new is 125/100, or 1.25. To get the values of the spliced index for the years 1970 to 1981, the values of the old index are divided by this ratio. For example, the spliced index in 1981 equals 120.3/1.25, or 96.2.

While the math for splicing indexes is easy, the procedure should be used with caution. Old and new indexes have different weights and they may be based on different commodities. There is often no good way to make comparisons over long time periods.

TABLE 16.2

Old and new indexes and a spliced index

Year	Old index (1972 = 100)	New index (1982 = 100)	Spliced index (1982 = 100)
1970	92.3	—	73.8
1971	95.6	—	76.5
1972	100.0	—	80.0
1973	101.7	—	81.4
1974	103.4	—	82.7
1975	111.8	—	89.4
1976	112.0	—	89.6
1977	114.9	—	91.9
1978	115.2	—	92.2
1979	118.6	—	94.9
1980	119.9	—	95.9
1981	120.3	—	96.2
1982	125.0	100.0	100.0
1983	—	105.1	105.1
1984	—	108.5	108.5
1985	—	110.5	110.5
1986	—	120.6	120.6
1987	—	134.0	134.0

Another simple manipulation of index numbers is to use them to deflate a quantity series. Table 16.3 lists the purchases of a household for a series of years and the corresponding values of the consumer price index. The last column, "Constant-dollar expenditures," measures the family's spending in base-year dollars. For the CPI the base year is 1967, so the expenditure series has been deflated to 1967 dollars. In real terms—that is, after inflation is taken into account—the family's expenditures have increased each year. The numbers in the last column were obtained by dividing the expenditures by the corresponding CPI and then multiplying the result by 100.

TABLE 16.3

Expenditures for one household, CPI, and constant-dollar expenditures, 1981–1983

Year	Expenditures	CPI	Constant-dollar expenditures
1981	$14,563.30	272.4	$5,346.29
1982	17,890.28	289.1	6,188.27
1983	22,459.87	298.4	7,526.77

THE INDEX OF LEADING ECONOMIC INDICATORS

The index of leading economic indicators is one of the most widely watched federal government indexes. It was originally developed in 1938 by Wesley Mitchell and Arthur Burns, economists at the National Bureau of Economic Research, but it has since been taken over by the Bureau of Economic Analysis in the U.S. Department of Commerce. The index is made up of 12 time series that are supposed to move up or down before the rest of the economy starts to move in the same direction. The 12 components are shown in Table 16.4. Figure 16.2 shows the index from January 1948 through February 1987.

The index of 12 leading economic indicators has many critics. There is no theory behind the index which explains the lags between changes in the component series and changes in GNP. Further, the choice of which 12 series to put in the index is highly arbitrary. Series are added or dropped from time to time as government data-collection prac-

TABLE 16.4
The index of leading indicators: component series

Bureau of Economic Analysis series number	Description
1	Average work hours of production or nonsupervisory workers, manufacturing
5	Average weekly initial claims for unemployment, state programs
8	Manufacturers' new orders in 1982 dollars, consumer goods and materials industries
12	Index of net business formation
19	Index of stock prices, 500 common stocks (Standard & Poor)
20	Contracts and orders for plant and equipment in 1982 dollars
29	Index of private new housing units authorized by local building permits
32	Vendor performance, percent of companies receiving slower deliveries
36	Change in manufacturing and trade inventories on hand and on order in 1982 dollars
99	Change in sensitive materials prices
106	Money supply (M2) in 1982 dollars
111	Change in business and consumer credit outstanding

tices change, and there is no statistical rationale behind the selection of the 12 series selected in the sense that they have been shown to provide the best fit to the pattern of GNP.

The typical use of the index of leading economic indicators is to forecast changes of direction in the economy. A widely used rule of thumb is that three successive declines predict a recession. Figure 16.2 shows many three-month declines in the index. The rule that such a decline presages a recession is not foolproof. The rule predicted recessions in 1950, 1962, and 1966 which did not occur. The index also provides little guidance about when the recession will begin. Since 1945, it has peaked anywhere from 3 to 21 months before the economy began to enter a recession. The index has also been criticized for its emphasis on manufacturing, which has declined considerably in importance since the index was first formulated. Nevertheless, many economists consider the index useful as a rough guide to the future direction of unemployment and industrial production.

FIGURE 16.2
Monthly index of 12 leading economic indicators, January 1948– February 1987

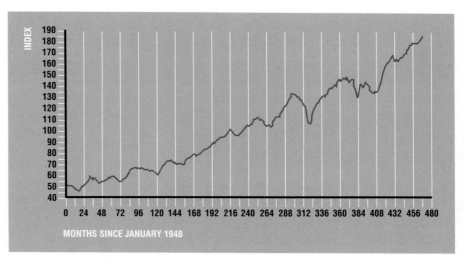

Source: *Business Conditions Digest,* various issues.

EXERCISES FOR SECTION 16.1

16.1 Calculate a Laspeyres, a Paasche, and a Fisher ideal index for the following data:

Prices, quantities, and expenditures for three items, 1987 and 1988

Item	1987			1988		
	P_{0i}	Q_{0i}	E_{0i}	P_{ti}	Q_{ti}	E_{ti}
Records	$7.00	20	$140.00	$6.00	20	$120.00
Movies	4.00	16	64.00	5.00	10	50.00
Lunches	3.00	300	900.00	2.90	310	899.00

16.2 Calculate a Laspeyres, a Paasche, and a Fisher ideal index for the following data:

Prices, quantities, and expenditures for four items, 1970 and 1980

Item	1970			1980		
	P_{0i}	Q_{0i}	E_{0i}	P_{ti}	Q_{ti}	E_{ti}
Tires	$36.00	4	$144.00	$56.00	4	$224.00
Engine tune-ups	32.50	2	65.00	44.00	1	44.00
Motor oil	1.25	10	12.50	2.11	6	12.66
Gasoline	0.40	452	180.80	1.10	213	234.30

16.3 Splice the following old and new index number series:

Year	Old index (1976 = 100)	New index (1982 = 100)	Year	Old index (1976 = 100)	New index (1982 = 100)
1975	99.6	—	1982	108.0	100.0
1976	100.0	—	1983	—	102.9
1977	101.7	—	1984	—	103.3
1978	104.3	—	1985	—	100.7
1979	107.6	—	1986	—	109.5
1980	105.2	—	1987	—	110.3
1981	104.2	—			

16.4 **(a)** Deflate the following sales series on the Ford Motor Company's annual sales in millions of dollars, given the values of the CPI and the new-car component of the CPI.

Year	Sales	CPI	New-Car CPI
1977	$37,841.50	181.5	142.9
1978	42,784.10	195.4	153.8
1979	43,513.70	217.4	166.0
1980	37,085.50	246.8	179.3
1981	38,247.10	272.4	190.2
1982	37,067.20	289.1	197.5
1983	44,454.60	298.4	202.4

(b) Which deflator for Ford's sales makes more sense, the overall CPI or the new-car component of the CPI? Explain.

16.5 A student has kept a record of the prices of the books she has purchased over the past eight semesters of college. She wants to use these data to construct a price index for college texts at her school and publish the results in the school paper. What problems might this index have?

SUMMARY

Index numbers are widely used and valuable in making business decisions. The main types of indexes discussed were the Laspeyres, the Paasche, and the Fisher ideal index. The Laspeyres index uses the base-period consumption bundle as the quantity weights while the Paasche uses the current-period consumption bundle as the quantity weights. The Laspeyres has the advantage of being cheaper to construct, but it can be misleading if the base period is very distant. Fisher's ideal index treats the base-period and current-period consumption bundles symmetrically. It has the disadvantage of having the highest cost of the three types. Few Fisher ideal indexes are published.

The chapter sketched out some interpretive problems with index numbers, such as quality changes, changes across geographical regions, random sampling errors, list prices instead of actual prices, and samples biased by including only large firms. Finally the chapter discussed the techniques of splicing old and new or revised series and deflating an expenditure series with a price index. The focus of the chapter was on price indexes, but the formulas for quantity indexes differ only slightly; most of the calculations, interpretations, and problems of price indexes carry over to quantity indexes.

SUGGESTED READING

U.S. Department of Labor. *The Consumer Price Index: Concepts and Content over the Years.* Bureau of Labor Statistics Bulletin no. 2134-2. Washington, D.C.: U.S. Government Printing Office, April 1984.

NEW TERMS

composite index An index that consists of more than one item.

expenditure relatives The ratio of expenditures on one commodity in the current period and the base period.

Fisher's ideal index An index that uses weights based on quantities of both the current period and the base period.

index number A measure of the value of a variable in relation to its value at some other location or in some other time period.

Laspeyres' index An index that uses weights based on some base period's quantities. A Laspeyres index can be used to determine what it would cost to buy the base period's market basket at the prices prevailing in the current period.

Paasche's index An index that uses weights based on the most recent period's quantities. A Paasche index can be used to determine what it would cost to buy the current period's market basket at the prices that prevailed in some prior period.

price relatives The ratio of prices of one commodity in the current period and the base period.

quantity relatives The ratio of quantities of one commodity in the current period and the base period.

simple unweighted index An index in which the sum of the prices in the current period is divided by the sum of the prices in the base period.

weighted aggregate index The ratio of current-period over base-period spending.

 NEW FORMULAS

Price relative at time t

$$PR_t = \frac{P_t}{P_0} \qquad (16.1)$$

Quantity relative at time t

$$QR_t = \frac{Q_t}{Q_0} \qquad (16.2)$$

Expenditure relative at time t

$$ER_t = \frac{P_t Q_t}{P_0 Q_0} \qquad (16.3)$$

Laspeyres' index

$$L_t = \frac{\sum_{i=1}^{n} Q_{0i} P_{ti}}{\sum_{i=1}^{n} Q_{0i} P_{0i}} \times 100 \qquad (16.4)$$

Paasche's index

$$\text{Paasche}_t = \frac{\sum\limits_{i=1}^{n} Q_{ti}P_{ti}}{\sum\limits_{i=1}^{n} Q_{ti}P_{0i}} \times 100 \qquad (16.5)$$

Fisher's ideal index

$$F_t = \sqrt{L_t \times \text{Paasche}_t} \qquad (16.6)$$

 ADDITIONAL EXERCISES

Asterisks designate exercises of greater than ordinary difficulty.

16.6 Construct a Laspeyres index for the following data:

Item	1976		1977	
	P_{0i}	Q_{0i}	P_{ti}	Q_{ti}
Beef tongue	$0.44	1	$0.64	0
Beef round	2.25	8	2.65	11
Beef tenderloin	4.78	2	6.43	4
Beef rib steak	3.99	5	4.06	6
Hamburger	0.58	36	0.64	32

If the above data represented one family's beef purchases for these years, what would you conclude about the family's income (assuming that the family size was constant)?

16.7 Construct a Paasche index for the data in exercise 16.6. How many items would go into the index?

16.8 Construct a Fisher ideal index for the data in exercise 16.6. Does the absence of any purchases of beef tongue in 1977 make a Fisher ideal index inappropriate? Try constructing the index with and without the tongue item.

16.9 Splice the following index so that the spliced series is a continuation of the old index.

Year	Old index (1932 = 100)	New index (1936 = 100)	Year	Old index (1932 = 100)	New index (1936 = 100)
1932	100.0	—	1939	—	92.2
1933	98.4	—	1940	—	88.4
1934	97.5	—	1941	—	87.1
1935	95.3	—	1942	—	80.3
1936	90.1	100.0	1943	—	78.1
1937	—	97.2	1944	—	55.4
1938	—	95.3			

16.10 Change the figures for the CPI so that the base year is 1983.

Year	CPI
1977	181.5
1978	195.4
1979	217.4
1980	246.8
1981	272.4
1982	289.1
1983	298.4

16.11 Create a Laspeyres index for the following data:

Prices and quantities per ton-mile, 1985–1987

	1985		1986		1987	
Item	P_{0i}	Q_{0i}	P_{1i}	Q_{1i}	P_{2i}	Q_{2i}
Railroad shipment	0.113	4,657.0	0.113	4,054.0	0.115	5,634.0
Barge shipment	0.069	2,985.0	0.074	872.0	0.075	533.0
Truck shipment	0.151	10,065.0	0.136	16,098.0	0.121	16,298.0
Air freight shipment	32.600	0.5	38.040	3.5	30.000	144.0

16.12 Construct a Paasche index for the data in exercise 16.11.

16.13 Construct a Fisher ideal index for the data in exercise 16.11. Which of the three types of index would make sense for these data?

16.14 Suppose you deflated your income over the past five years using the CPI and found that in constant dollars your current income is 108% of the level of income you had five years ago. What would you have to assume before you could state that you were 8% richer than you were five years ago?

16.15 A hospital administrator found that in one year the hospital's budget was $24 million and the next year the budget was $26 million. Suppose further that the number of patient days was the same over the two years. Explain why the percentage increase in the budget might be an inaccurate estimate of inflation in the goods and services that the hospital buys.

16.16 There have been substantial changes in new homes over the last 30 years. Lots have generally become smaller, the square footage of the median house has increased, and there have been shifts in the allocation of space within homes—more space for bathrooms and family rooms, for example, and less space for dining rooms and living rooms. In light of these changes, describe how you would construct an index for new-home construction.

16.17 Housing costs are an important component of the CPI. However, there are many problems in determining the cost of owner-occupied housing. Before 1978 the current cost of financing a new mortgage was an element of the CPI. If interest rates went up in one month, the CPI was adjusted. Implicitly, the adjustment assumed that all homeowners had to refinance their mortgages at the new higher rates month by month. Except for new variable-rate mortgages, this assumption is unrealistic. It has the effect of overstating the impact of any change in interest rates on the cost of living.

Similarly, before 1978 the prices of new homes were an element of the CPI. Here the implicit assumption was that everyone was buying a new home every month. Since 1978 the method used by the Bureau of Labor Statistics has been to estimate the rental value of owner-occupied housing—in effect, what you could get for your house if you tried to rent it. What are the problems with this approach to the costs of owner-occupied housing?

16.18 The broadest measure of inflation in the economy is the implicit GNP deflator. It represents all of the goods and services sold in the United States. It is used to deflate nominal GNP figures into a constant-dollar series. If you were evaluating the past real rates of return (i.e., net of inflation) on some alternative types of investments in order to decide which type to buy, would you be better off using the CPI or the implicit GNP deflator to deflate nominal rates of return to real rates of return?

16.19 Over the years 1984, 1985, and 1986 the rate of tuition increases at private colleges ran twice as high as the consumer price index. President Reagan's secretary of education, William Bennett, accused these colleges of being "greedy" and "charging what the traffic will bear." Bennett claimed that the extra money went into frills and administrators' salaries rather than into improvements in quality of education. None of the colleges countered the charges by stating that as private businesses they had every right to charge what the traffic would bear. The most common responses were that the colleges had been playing catch-up for years in which tuition went up more slowly than the rate of inflation and that professors' salaries were still substantially below what they had been in real terms in 1970. This brouhaha is an excellent example of the political controversies that surround the CPI. The assumption that the differences between the CPI and the costs of running a college are due to variations from year to year in the greediness of college administrators does not lead to any fruitful inquiry about what has in fact happened. Why might the CPI and the costs of running a college diverge? How would you construct an index for the costs of running a college?

16.20 The United Nations publishes statistics on per capita income in different countries. Evaluated at current exchange rates, the per capita incomes are usually stated in U.S. dollars. The per capita incomes in some of the poorest countries in the world are so low that they are hard to understand. In recent years Haiti and Bangladesh, for example, have had per capita annual incomes in the $100–200 range. In the United States it would be difficult to live a week, let alone a year, on $100. Obviously there are some differences in the cost of living in the United States versus Haiti and Bangladesh. What problems would you run into if you tried to construct a consumer price index for each of these countries so that you could make meaningful comparisons of the standards of living?

16.21 Every year the Hertz car rental company publishes figures on the per mile cost of operating a car. There are separate series for subcompact, compact, and full-size cars. The costs include repairs, insurance, gasoline, and maintenance. Are these figures an index or simply a sum of annual costs per mile per type of car?

16.22 You own a lumber products company that sells the products indicated in the following table. Under a LIFO (last in, first out) accounting system to value your inventories for tax purposes, you must use the beginning-of-year prices and the end-of-year quantities. What is the value of your inventory at the end of 1987? Construct an index of the price of the goods held in your inventory for the years 1986 and 1987. What type of price index would be consistent with the LIFO accounting rules?

Beginning-of-year prices and end-of-year inventories, 1985–1987

Item	1985 P	1985 Q	1986 P	1986 Q	1987 P	1987 Q
Plywood	$5.89	11,399	$6.16	14,988	$6.71	12,098
2-by-4's	0.61	25,434	0.68	42,543	0.71	45,210
4-by-4's	0.95	1,982	1.01	2,987	1.22	1,094

16.23* Consider the problem of creating an index for the level of violent crimes in some area. The crimes in the index might include murder, rape, armed robbery, and assault. The simplest way to construct an index would be to add up the total number of crimes in each category each year and divide this total by the total in some base year. As there are no "prices" attached to the quantities of crimes, there is no obvious way to weight the different violent crimes. A plausible way to weight the crimes would be the average sentence for each type of crime. Describe any problems that might arise with an index weighted in this way.

16.24* Insurance companies sometimes write policies on commercial buildings to cover the cost of replacing the building if it burns down. Suppose an old office building burned down in 1987. The building cost $550,000 to construct in 1934. Suppose an index of construction costs stood at 56 in 1934 and at 652 in 1987. If the insurance company relied on this construction cost index, what would it have to pay to replace the building? What might be wrong with this estimate of replacement costs?

16.25* A company that makes potato chips uses three types of vegetable oil to fry the chips: cottonseed oil, corn oil, and peanut oil. The proportions of the three oils can be varied to take advantage of changes in the prices of the oil. On the basis of the following data, what type of index would best measure the cost of vegetable oil used by the company? Assume that the three types of oil have the same properties for frying.

Item	1985 P	1985 Q	1986 P	1986 Q	1987 P	1987 Q
Cottonseed oil	$11.89	542	$11.72	651	$11.75	0
Corn oil	12.02	0	11.76	120	11.67	65
Peanut oil	13.87	0	13.24	0	11.51	751

16.26** The Dow Jones industrial average is a widely watched stock market index made up of the prices of stocks of 30 large industrial companies. Originally the index was simply the sum of those 30 stock prices. However, stock splits make the computation of the index more complicated. Here is a simple example of the computations needed after a stock split. The index consisted of stocks A, B, and C, and each stock started with a price of $100. The starting value of the index would be $300. Stock A splits two-for-one and the shares of stock A now sell for $50 each. To keep the index at $300 after the split, the sum of one share each of A, B, and C ($250) would have to be divided by 0.833.

Try this problem to test your understanding. Keep A at $50 per share and B at $100 per share. If C split four-for-one and the new shares of C sold for $40, what would be the value of the index?

17

Non-parametric Tests

etween 1972 and 1978 the Kroger grocery chain ran an advertising campaign, the Kroger Price Patrol, in 104 cities across the United States. In each city the theme of the campaign was that the Kroger stores had the lowest overall prices. Each week Kroger paid some homemakers to compare the prices of 150 items in Kroger stores with the prices charged for the same items by competing stores. The specific items in the list changed from week to week. The homemakers' findings were reported in television and newspaper ads. Kroger executives considered the campaign a great success because it raised Kroger's market share and it changed consumers' perception of Kroger's relative prices.

When the Price Patrol came to central Indiana, a student-run public interest group, INPIRG (for Indiana Public Interest Research Group) was coincidently conducting its own grocery-pricing survey as a public service. The students were puzzled when they found that Kroger always emerged as the store with the greatest number of lowest-priced items in the Kroger survey but was rarely the leader in the students' surveys. The students wanted to know if these results could be due to chance or if Kroger's weekly lists somehow failed to reflect overall prices. The students did not have the resources to survey every item in every store week after week—there are about 15,000 items in a supermarket—so instead they focused their efforts on the 150 items that Kroger selected for the weekly comparisons. The hypothesis that INPIRG wanted to test was that Kroger's management had "loaded" each weekly list in Kroger's favor by selecting items going on sale or "special" at Kroger.

For 27 weeks the students kept track of the proportion of the 150 items on Kroger's current Price Patrol list that had been reduced in price since the previous week. Over the same period they kept track of the proportion of price reductions on another list of about 500 items, chosen to represent all of the items in the store. The students' statistical sleuthing stopped at the 27th week, when Kroger ended its Price Patrol campaign in central Indiana. In each of the 27 weeks the proportion of items with price cuts since the previous week was about three times as high for the 150-item Kroger Price Patrol list than for the 500-item storewide list (roughly 15% versus 5%). A test for independence between the characteristics "having a price cut" and "being on Kroger's weekly list" would lead to the rejection of the null hypothesis of independence at any plausible level of significance. The chi-square test, discussed in Chapter 11, was the test used.

The students sent their findings to the Federal Trade Commission. When some FTC attorneys confronted the local Kroger manager with this evidence, he admitted that it was his practice to put 20 to 40 sale items on the Kroger Price Patrol list. The FTC later issued an order restraining Kroger from using surveys to claim lower prices unless some safeguards were used to ensure that the surveys were based on reasonable statistical methods and not subject to manipulation.

The chi-square test for independence is a nonparametric test. Nonparametric tests do not require the assumption that a population distribution is normal. Some nonparametric tests are also called "distribution-free" tests because they have no requirements in regard to the shape of the population distribution. The only requirement for the chi-square test of independence is that each expected frequency be at least 5. Figure 17.1 shows a population distribution of the two characteristics. The chi-square test is distribution-free because it places no restrictions on the heights of the bars in the figure.

This chapter introduces some additional nonparametric tests, primarily for measures of the center of continuous distributions. To see how these new tests differ from a test on proportions, consider a modification of the Kroger example.

Suppose Kroger's claim was not about the proportion of items that had lower prices, but rather about the amount of potential savings per week. Suppose Kroger claimed that on average you could save at least $15 if you bought a market basket of 150 items at Kroger. With this type of claim, the issue would be the mean savings per weekly shopping trip. An F test for equal means across a number of stores (Chapter 12) or a t test for the difference in means across two stores (Chapter 10) or a t test for one population mean (Chapter 9) requires the assumption of normality for the population distribution(s). However, the population distribution of prices in a grocery store is

FIGURE 17.1
A population distribution of nominal data from a 2-by-2 contingency table

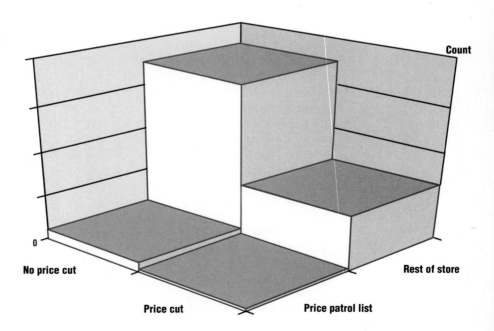

highly skewed, with most items under $1 and a few items over $20. The normality assumption does not hold. Another approach is to have enough observations so that the central limit theorem guarantees that the distribution of sample means will be approximately normal. In this example the number of weeks that the Kroger Price Patrol was run was not under the students' control. In other situations it could be too expensive to obtain enough data.

A variety of nonparametric tests have been devised for continuous data that involve small samples and nonnormal populations. There is no test for a single population mean, but the sign test can be used for a single population median. If two stores were being compared, the Wilcoxon signed rank test for the difference in population means could be used if the two populations were symmetrical. An alternative two-population test, the Wilcoxon-Mann-Whitney test for the difference in population means or medians, could be used if the two populations had the same shape. If the observations in the two populations were correlated, the Wilcoxon ranked sum test provides a nonparametric counterpart to the paired t test discussed at the end of Chapter 10. For multipopulation tests, the nonparametric counterpart to one-way analysis of variance is the Kruskal–Wallis test. The nonparametric counterpart to a t test for correlation between two random variables is the Spearman rank correlation test. In some circumstances, which we will describe in this chapter, some goodness-of-fit tests are more powerful than the chi-square test: the Kolmogorov-Smirnov test for any distribution and the Lilliefors test for normal distributions.

17.1 NONPARAMETRIC TESTS FOR THE POPULATION MEDIAN

Many of the nonparametric tests are relatively new—from 20 to 30 years old. These tests have gained wide acceptance among statisticians. If the underlying assumptions required by a parametric test cannot reasonably be made, the parametric test could be worthless. Moreover, nonparametric tests can be applied to types of data that parametric tests cannot handle. Most of the parametric tests require either interval or ratio data. The parametric tests that use rankings can work with ordinal data. A third advantage of the nonparametric tests is that they are less sensitive to extreme observations. An outlying observation poses difficulties when you are using a parametric test. Sometimes you cannot determine whether any outliers are due to recording errors or whether they really belong in the data set. You may be tempted to throw them out of the data set before doing the test, but you can't be sure if you are worse off by doing so. The nonparametric tests remove the temptation because they are insensitive to outliers.

738

■ **17.1A THE SIGN TEST FOR THE POPULATION MEDIAN**

Suppose you are stuck with a small sample (under 30) taken from a nonnormal population and you want to test the null hypothesis that the mean of the population from which the sample was drawn is a certain value. For example, you want to test the hypothesis that the mean time to failure of a particular type of marine engine is 2,000 hours of continuous operation. The distribution of times to failure is likely to be exponential rather than normal because the probability of surviving beyond any time x should decline continuously as x increases. In other words, the longer a machine has run, the less likely it is to continue to function. Suppose further that you have the resources to test only ten engines. There is no test for the population mean in this situation; the test that can be used, the **sign test,** is *a nonparametric test for the population median which requires the assumption that the population distribution is continuous.* The test against a particular value for the population median will be equivalent to a test for the mean if the population is symmetrical, as the mean and the median will be equal. In this example the population distribution is highly skewed. When the population is highly skewed, the median is a better measure of the center of a distribution than the mean, so a test against the median might be preferred to a test against the mean. If the population is not highly skewed but not normal either, you might prefer a test for the population mean. However, you have to use what is available. With small samples and nonnormal populations, you are limited to the sign test for the population median.

The sign test exploits the fact that, by definition, half of the observations in a continuous distribution must be on either side of the median. Thus if the null hypothesis of a median of 2,000 hours is true, the probability that a randomly selected observation will be above 2,000 is 0.5 and the probability that a randomly selected observation will be below 2,000 is also 0.5, no matter what the shape of the underlying distribution. If we assume that the observations are independent, the number of plusses (observations above 2,000 hours) and minuses (observations below 2,000 hours) should follow a binomial distribution with a Π of 0.5 and an n of 10. This reference to plusses and minuses shows you where the name "sign test" comes from. Because the numbers of plusses and minuses are discrete, you cannot precisely set the significance level of the test at one of the conventional values, such as 1%, 5%, or 10%. For example, in a two-tailed test of the null hypothesis that the population median is 2,000, the significance level can be set at 0.002, 0.022, 0.11. These significance levels correspond to the following sets of critical values: 10 plusses or 10 minuses; 9 or more plusses or 9 or more minuses; and 8 or more plusses or 8 or more minuses. These significance levels come directly from the binomial distribution. The probability of 0 successes in 10 trials with a Π of 0.5 is 0.001 and the probability of 10 successes in 10 trials with a Π of 0.5 is also 0.001. Hence the probability of getting 10 plusses or 10 minuses if the null hypothesis of a population median of 2,000 hours is true is 0.002. The significance levels of 0.022 and 0.11 would be found in a similar manner. Suppose 9 of the motors lasted more than 2,000 hours; in other words, you had 9 plusses and 1 minus. You could reject the null hypothesis of a population median of 2,000 hours at a significance level of 0.022.

Let's try a second example that uses the sign test. Suppose a gold mining company is willing to begin mining a particular site if the gold content per metric ton of ore is more than 12 ounces. The company has 20 samples of one metric ton of ore dug up from the site and analyzed for gold content. Fourteen of the samples have more than 12 ounces; that is, 14 are plusses. Note that for a continuous distribution the probability that an observation will have exactly 12 ounces is zero. At the 5% significance level, can the null hypothesis of 12 or fewer ounces be rejected? If it is not reasonable to assume that the distribution of ounces of gold per metric ton of ore is normal, the test will have to be against a median of more than 12 rather than against a mean of more than 12. In this example, the mean is a better measure of the center of the data than the median, even if the distribution is highly skewed, because the mining company wants to know the total yield of the mine. This yield is the mean number of ounces per metric ton times the number of tons of ore available. Unfortunately, a test for the mean is not available.

This mining example differs from the marine engine example in two other ways. Here you have a one-tailed test and a significance level specified in advance of the test. If the null hypothesis of a median of 12 ounces were true, the number of plusses would follow a binomial distribution with a Π of 0.5 and an n of 20. From the binomial table we find that the probability of exactly 14 or more successes is 0.0577. Fourteen plusses are not enough to allow you to reject the null at a 5% significance level. The 0.0577 is the p value of the test, the smallest significance level at which the null can be rejected. Figure 17.2

FIGURE 17.2
Binomial distribution with Π of 0.5 and n of 20

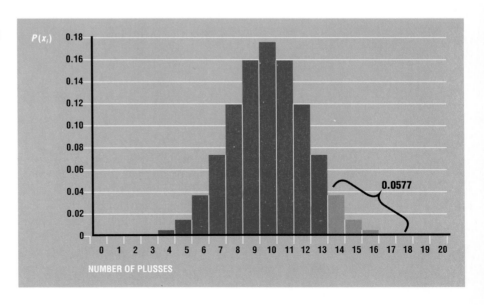

shows the binomial distribution with a Π of 0.5 and an n of 20 and marks off the area greater than or equal to 14. Rejection of the null hypothesis of a median of 12 or fewer ounces at the 5% significance level would require 15 or more observations above 12 ounces.

In both the marine engine and the gold mining examples the original data values were not listed. The only information used in the test was whether the observations were above or below the hypothesized median. Clearly, the sign test does not use all of the available information in the sample. A parametric test would use all of the information about the individual observations to calculate the sample mean and the sample standard deviation. For this reason a parametric test is more powerful than the corresponding nonparametric test. Recall from Chapter 9 that if one test is more powerful than another, it is less likely to accept a false null hypothesis when both tests are held to the same significance level. The problem is that if the assumption of normality required for the parametric test does not hold, the stated significance level of the parametric test will be wrong and the parametric test may be less powerful than the corresponding nonparametric test.

Let's try one last example of the sign test for a single median and this time show the corresponding parametric test. An actuary determines that a pension fund for a particular company will be able to cover its accrued liabilities if the mean age of covered workers at retirement is greater than 63 years. So far 25 covered workers have retired and the mean age among them is 65 with a sample standard deviation of five years. Sixteen of the retired workers were older than 63. The actuary would like to test at the 5% significance level the hypothesis that the mean age at retirement of all covered workers will be over 63. If the distribution of retirement ages is normal, the appropriate test of the null hypothesis of $\mu \leq 63$ is a t test. There are $n - 1$ or 24 degrees of freedom. The t score for 5% of the area in the right-hand tail with 24 degrees of freedom is 1.711. Substituting this t score into formula 9.4 yields

$$\text{C.V.} = \mu_0 + t_{\alpha/2}s_{\bar{x}} \tag{9.4}$$

$$\text{C.V.} = 63 + 1.711\frac{5}{\sqrt{25}}$$

$$\text{C.V.} = 64.711$$

Since the sample mean of 65 exceeds the critical value, the null hypothesis of a population mean of 63 or less is rejected at the 5% significance level.

If the assumption of a normal distribution for the ages of retirement is not tenable, the t test will not be accurate. Unless you can assume symmetry in the population distribution, the test will have to be against the hypothesized population median of 63. The critical value then is the number of successes that mark the upper 5% of a binomial distribution with a Π of 0.5 and an n of 25. In this example the normal approximation to the binomial can be used because

$n\Pi$ and $n(1 - \Pi)$ are both above 5. The expected number of successes (plusses) is $n\Pi$ or 12.5 and the standard deviation is

$$\sigma(X) = \sqrt{n\Pi(1 - \Pi)}$$

$$\sigma(X) = \sqrt{12.5(0.5)}$$

$$\sigma(X) = 2.5$$

The boundary line for the upper 5% of a normal distribution with a mean of 12.5 and a standard deviation of 2.5 is

$$x = \mu + z_\alpha\sigma$$

$$x = 12.5 + 1.64(2.5)$$

$$x = 16.6$$

Recall that only 16 of the retirees were plusses, or over 63 years of age. The null hypothesis of a median of less than 63 is accepted at the 5% significance level.

As you can see, the conclusions you reach from a set of data depend on whether you use a parametric or a nonparametric test. You can resolve the issue of which procedure to use by testing the data for normality, either with the chi-square goodness-of-fit test, discussed in Chapter 11, or the Lilliefors test, to be introduced later in this chapter. If the null hypothesis of a normal distribution is rejected, then the nonparametric test should be used.

The sign test can easily be extended to paired data. This test is the nonparametric counterpart to the paired difference test at the end of Chapter 10. Suppose your company had its salespeople take a course in selling techniques and you recorded their monthly sales levels before and after the course. Table 17.1 shows a data set that might result from such an experiment. The null hypothesis is that the median of the distribution of paired differences in monthly sales is

TABLE 17.1	Name	After training	Before training	Sign of difference in sales
Value of monthly sales per salesperson before and after training	Atlas	$16,101.00	$12,456.00	+
	Cohen	29,970.00	24,566.00	+
	Gold	18,371.00	19,543.00	−
	Jencks	15,637.00	10,941.00	+
	Kappel	24,802.00	23,388.00	+
	Lewis	25,010.00	23,176.00	+
	Moscow	19,700.00	17,902.00	+
	Schwartz	22,890.00	19,870.00	+
	Tubbs	32,000.00	30,124.00	+

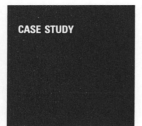

BLACKS IN TELEVISION COMMERCIALS

Black actors were rarely found in television commercials before the civil rights movement of the 1960s. Many companies thought they would lose more customers among whites than they would gain among blacks if they used black actors. Companies that marketed their products to black consumers were few and often too poor to buy air time. One of the effects of the civil rights movement was to break this color barrier. If you are under 30 years of age, all of this is pretty much ancient history. The presence of blacks in television commercials and shows causes little comment today. In the late 1960s this subject was under much closer scrutiny. Besides the obvious issue of employment opportunity, the increase in the use of black actors provides a measure of changing public attitudes. However, many advertisers have managed to increase the use of black actors without appearing to identify their products with blacks.

A simple percentage of the number of commercials with blacks would be a highly misleading measure of the gains that black actors have made in this period. If a black actor stood in the back of a crowd of ten actors, it would be recorded as a commercial that had some blacks. This example is not farfetched. In a study of the subject ("Three Seasons of Blacks on Television," *Journal of Advertising Research* 10 [April 1970]: 21–27), Joseph R. Dominick and Bradley S. Greenberg found that the median numbers of actors in a sample of commercials in which no blacks appeared during 1967, 1968, and 1969 were 1.6, 1.7, and 1.9. The corresponding medians for commercials in which black actors did appear were 10.0, 7.0, and 6.0. Using a sign test, the researchers found that the medians for commercials with and without blacks differed significantly at the 5% level for each of those years. The reason that they compared the median number of actors rather than the mean is that the distribution of the number of actors per commercial is highly skewed. Most commercials have one or two actors, but some have thousands.

Blacks tended not only to be lost in the crowds but also to appear in different types of commercials than whites. The percentages of prime-time product commercials in which black actors appeared in the sample from 1967 to 1969 were 3%, 6%, and 8%. For the same years the percentages of public-service commercials that showed blacks were 24%, 18%, and 21%. At the present time, some mass marketers clearly see no disadvantage in having blacks identified with their products: Michael Jackson and Lionel Richie are selling Pepsi, Bill Cosby is selling Coke and Jell-o Pudding Pops, Ella Fitzgerald is selling Memorex tapes, and Michael Jordan is selling Nike shoes.

zero. Suppose you wanted to test this hypothesis at the 5% significance level. If this null hypothesis were true, the number of plusses would follow a binomial distribution with a Π of 0.5 and an n of 9. The sample shows eight plusses. The probability of observing eight or more plusses is 0.02. Thus the null can be rejected at the 5% significance level. Note that even though the null hypothesis is rejected, you cannot be sure that the cause of the increase in sales is the training program. Perhaps a similar sample of nine salespeople who did not attend the training course would also show eight plusses when sales dollars were compared for the same two periods as those represented in Table 17.1. You can infer causality in an experimental setting only when every other factor is controlled. In this example, the increase in sales may be caused by improved business conditions rather than the training program.

17.1B CONFIDENCE INTERVALS FOR THE POPULATION MEDIAN

The fact that half of the observations in a continuous distribution must be on either side of the median can also be used to construct confidence intervals for the median. Here is an example that shows the logic behind the confidence interval. Suppose samples of size 10 were drawn from a continuous population and the points for each sample were plotted below the distribution. Figure 17.3 shows a skewed distribution with the observations plotted below as dots, each box of dots representing the observations in one sample, from the lowest to the highest. If the lowest and the highest points in each sample were used as the limits of a confidence interval, what proportion of all such confidence intervals would contain the population median? The confidence intervals would miss the population median in the samples in which all 10 observations were either above or below the population median. The probability that all 10 observations will be below the population median can be found in the binomial table; it is the probability of 0 successes in 10 trials for a Π of 0.5. The probability that all 10 observations in the sample will be below the population median is 0.001. The probability that all of them will be above the population median is also 0.001. The proportion of all possible confidence intervals that will miss the population median is $0.001 + 0.001$, or 0.002. The level of confidence for an interval is the proportion of all possible confidence intervals that could be constructed with the same procedure which would be expected to contain the parameter. Thus the level of confidence for an interval defined as the lowest to the highest observation in a sample of size 10 is 0.998.

It is not necessary to take the highest and lowest observations as the limits of the confidence interval for the population median. Suppose you took the second observation from the low end and the second observation from the high end as the limits of the interval. What would be the resulting level of confidence? The reasoning here is an extension of the reasoning in the paragraph above. A confidence interval would miss the population median if 9 or more observations in the sample of size 10 were above or below the population median. The probability of 9 or more successes in a binomial distribution with 10 trials and a Π of 0.5 is 0.0108. The level of confidence for intervals that run from the second lowest to the second highest observation in samples of size 10 is $1 - (0.0108 + 0.0108)$, or 0.9784. Similarly, the level of confidence for an

FIGURE 17.3
**Population distribution and
many samples of size 10**

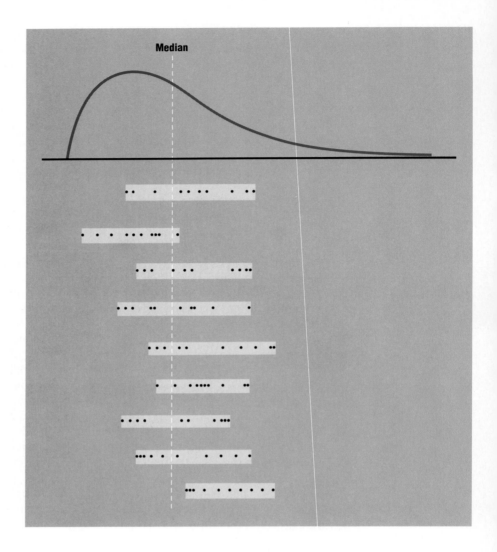

interval that ran from the third lowest to the third highest observation would be $1 - (0.0547 + 0.0547)$, or 0.8906. Figure 17.4 shows the same population and samples as Figure 17.3, but with the confidence intervals running from the third lowest to the third highest observation. Note that the end points of these confidence intervals do not have to be equidistant from the sample mean or the sample median. Note also that the confidence intervals do not have equal widths. In contrast, confidence intervals on the population mean are centered on the sample mean, and if the population standard deviation is known, they all have the same width for a given confidence level and sample size.

FIGURE 17.4
Population distribution and
many confidence intervals at
the 89% level of confidence

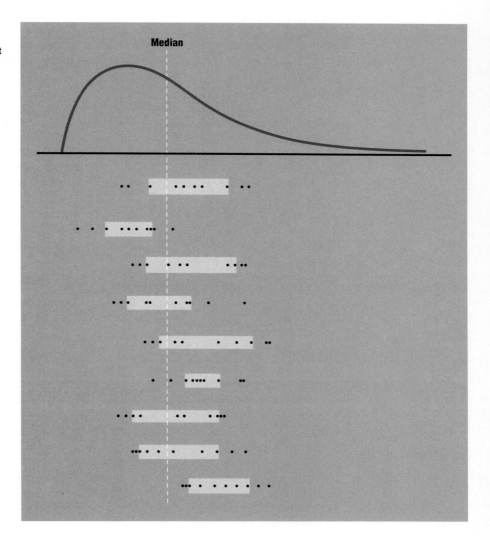

Here is the formula for finding the confidence level that runs from the xth observation from each end of an ordered list of observations:

(17.1)

$$1 - \alpha = 1 - 2\left(\sum_{i=n-x+1}^{n} \frac{n!}{(n-i)!i!}(0.5)^i(0.5)^{n-i} \right)$$

The "$1 - \alpha$" is the level of confidence. This formula applies only to confidence intervals on the population median for continuous populations. Another requirement is that the sample of n observations be random. The term in parentheses in formula 17.1 is just the binomial formula set to a Π of 0.5. Using the second, third, fourth, or higher observations from either end of the sample gives you a narrower confidence interval compared to using the first observation. Reductions in the width of the confidence interval come at the price of a lower level of confidence. This trade-off is the same as in the parametric confidence intervals discussed in Chapter 8.

Here's an example of a confidence interval on the population median. A union president is deciding on the union's bargaining position for a new compensation package for its members. The compensation package is made up of fringe benefits and wages. The president's personal preferences are to have a high ratio of fringe benefits to wages. However, the president cannot get too far away from the rank and file's preferences or he will not be reelected. Holding the total compensation fixed, the president wants to know the highest ratio of fringes to wages that would command a majority of the votes of union members. Young workers without families tend to prefer a large proportion of their compensation in wages while older employees tend to prefer a large proportion in pensions. Suppose the president could determine the median of the distribution of preferences among the members for ratios of wages to fringes. The median ratio could be used to determine the highest ratio of fringes to wages that would command a majority of the votes of union members. Note that 50% of the members must be on either side of the population median, so a ratio of fringes to wages at the median of the distribution of preferences must command a majority. This idea is called the "median voter rule." The theory of the median voter rule is that democratic bodies will reflect the preferences of the median voter.

The president could poll every member of the union to find the population median; but a sample is quicker and cheaper. Suppose the president took a random sample of 50 members and listed the results in ascending order. If the president constructed a confidence interval that ran between, say, the 18th observations from both ends, what level of confidence would this interval have? The 18th observation was chosen because it yields a confidence level close to 95%.

$$1 - \alpha = 1 - 2\left(\sum_{i=n-x+1}^{n} \frac{n!}{(n-i)!i!}(0.5)^i(0.5)^{n-i} \right) \tag{17.1}$$

$$1 - \alpha = 1 - 2\left(\sum_{i=50-18+1}^{50} \frac{n!}{(50-i)!i!}(0.5)^i(0.5)^{n-i} \right)$$

The expression in parentheses can be easily solved with the normal approximation to the binomial. The mean of the distribution is $n\Pi$, or 25. The standard deviation is

$$\sigma(X) = \sqrt{n\Pi(1 - \Pi)}$$ (5.11)

$$\sigma(X) = \sqrt{50(0.5)(0.5)}$$

$$\sigma(X) = 3.54$$

Substituting into the formula, we find the probability of 33 or more successes is

$$z = \frac{x - \mu}{\sigma}$$ (6.5)

$$z = \frac{32.5 - 25}{3.54}$$

$$z = 2.12$$

$$P(z \geq 2.12) = 0.0170$$

The continuity correction was the reason that 0.5 was subtracted from the limit of 33 successes in the z score formula. Substituting this probability back into formula 17.1 yields

$$1 - \alpha = 1 - 2(0.0170)$$

$$1 - \alpha = 0.9660$$

Note that the underlying distribution is supposed to be continuous. Each union member can prefer some proportion of the compensation package in fringes in the range of 0 to 1. While it is possible for the data to take any value in the range 0 to 1, the actual distribution may be discrete and very lumpy. Many union members might pick easy-to-remember proportions, such as 0.1 or 0.2. If the population distribution bunches up on such values, the level of confidence for the interval on the population median may not be accurate. With lumpy data it may no longer be true that 50% of the population data points are above the median. For example, with the distribution {0.1, 0.2, 0.3, 0.3, 0.3, 0.3, 0.3, 0.3, 0.4, 0.5} the proportion of the ten data points above the median of 0.3 is only 20%.

EXERCISES FOR SECTION 17.1

17.1 (a) Can the null hypothesis that the median of a continuous distribution equals 14 be rejected at a 10% significance level on the basis of the following random sample taken from that distribution: {12.4, 12.8, 14.6, 20.1, 24.5, 25.6, 34.5, 36.7, 39.0, 44.5}?

(b) Can the null hypothesis that the median of a continuous distribution is greater than or equal to 60 be rejected at a 1% significance level on the basis of the following random sample taken from that distribution: {60.1, 60.2, 60.3, 60.4, 60.6, 60.7, 60.8, 60.9, 61.0, 61.1}?

17.2 (a) Can the null hypothesis that the median of a continuous distribution is greater than or equal to 100 be rejected at a 5% significance level on the basis of the following random sample taken from that distribution: {80.1, 80.2, 80.3, 80.4, 80.6, 80.7, 80.8, 80.9, 90.0, 90.1, 90.1, 90.2, 90.3, 90.4, 90.6, 90.7, 90.8, 90.9, 100.1, 100.2}?

(b) Can the null hypothesis that the median of a continuous distribution is equal to 10 be rejected at a 10% significance level on the basis of the following random sample taken from that distribution: {1.1, 3.2, 4.3, 5.4, 10.6, 10.7, 10.8, 11.9, 12.0, 12.1, 13.1, 14.2, 16.3, 16.4, 16.6, 16.7, 17.8, 17.9, 18.1, 20.2}?

17.3 (a) What would be the confidence level for a confidence interval for the population median if the interval ran between the third observations from both ends of a sample of size 15?

(b) What would be the confidence level for a confidence interval for the population median if the interval ran between the fifth observations from both ends of a sample of size 20?

17.4 (a) What would be the confidence level for a confidence interval for the population median if the interval ran between the tenth observations from both ends of a sample of size 40?

(b) What would be the confidence level for a confidence interval for the population median if the interval ran between the 15th observations from both ends of a sample of size 50?

17.5 A national park wants to test at the 10% significance level the hypothesis that the median number of days spent in the park by backpackers is three. Use the following data to test this hypothesis:

0.5 0.5 0.5 1 1 1 2 2 2 2 3 3 3 3 3 4 4 4 4 5 5 5 6
6 7 8 8 9 10 12 14 16 20

17.6 An economist thinks that the stocks of companies that emphasize high dividends tend to be owned by relatively old stockholders and that the stocks of companies that emphasize reinvestment of earnings tend to be owned by relatively young stockholders. Data on the ages of a company's stockholders are difficult to obtain, and the economist has to rely on a small survey that has examined the portfolios of 12 individual stockholders at two points in their lives—age 40 and age 65. Each portfolio was assigned a composite rating for the emphasis placed by the companies on high dividends. In this rating a high number means that a portfolio favors high dividends and a low number means the portfolio favors reinvestment of earnings. Use these ratings to test at the 5% significance level the hypothesis that as individual stockholders aged, they moved to stock portfolios that favored current dividends.

Stockholder	Rating at age 40	Rating at age 65
1	2	9
2	1	1
3	8	7
4	1	6
5	2	4
6	3	2
7	9	10
8	4	5
9	5	5
10	6	8
11	7	8
12	10	10

17.7 A company that is planning to sell a new motorcycle aimed at the highway cruising market shows a prototype of the machine to 20 people whom it considers to be potential buyers. Each person is asked the price at which he or she would consider the motorcycle a good value. If the company were to construct a confidence interval at the 95% level of confidence for the population median of these prices, between what two observations in an ascending order of prices should the confidence interval run?

17.2 NONPARAMETRIC TESTS THAT USE RANKINGS

17.2A THE WILCOXON SIGNED RANK TEST FOR THE POPULATION MEAN OR MEDIAN

The Wilcoxon signed rank test is the third small–sample hypothesis test for the center of a population distribution with unknown standard deviation to be described in the text. It is also the last. The other two tests were the sign test and the *t* test. The reason for discussing three alternative hypothesis tests for small samples with unknown population standard deviations is that they require different assumptions about the population distribution and they use different amounts of information provided by the sample. The sign test has the least restrictive assumptions—simply that the distribution is continuous. The only information about the data set it uses is whether an observation is above or below the hypothesized median. The Wilcoxon signed rank test is more powerful than the sign test because it makes use of information about the distance between observations. The **Wilcoxon signed rank test** is *a nonpara-*

metric test for either the mean or the median of a single continuous distribution or for the difference between the means or medians of paired samples. In either case the test requires the assumption of a symmetrical distribution. If the population is symmetrical, the mean and median must be equal. The parametric test for the population mean with small samples and unknown population standard deviation, the *t* test, requires the most restrictive assumption of the three alternatives—that the population be normally distributed. Normal distributions are a subset of all symmetrical distributions. The point I am making by comparing the three tests for the center of data is that there is no statistical free lunch: more powerful tests require either a larger sample size or a population that fits more restrictive assumptions.

Here are the steps for the Wilcoxon signed rank test that the population mean/median equals a hypothesized value at a specified significance level. (1) A random sample is drawn from a symmetrical continuous population. (2) Any observations that fall on the hypothesized median/mean are dropped from the sample. (3) The hypothesized mean/median is subtracted from every remaining observation. (4) The differences are assigned ranks according to their *absolute values*. (5) The ranks are summed for all positive differences, T_+, and for all negative differences, T_-. The smallest of these two sums becomes the test statistic, T. (6) T is compared with the tabled critical values to decide if there is enough evidence to reject the null hypothesis. Note that the signs of the differences are ignored when assigning ranks but are considered when the sums (T_+ and T_-) are computed.

The Wilcoxon signed rank test requires the calculation of the **ranks of the differences between observations and the hypothesized mean/median.** *If the value of the xth difference is unique,* its rank is *the number of observed differences from the lowest up to and including the xth difference.* For example, suppose the absolute values of the differences between the hypothesized mean/median were {20, 36, 37, **38**, 40}. The rank of the difference "38" is 4. *If two or more differences have the same value,* we need to average their ranks. If the 4th and 5th differences are equal, their ranks would each be (4 + 5)/2 = 4.5. For example, if the absolute values of the differences were {20, 36, 37, **38, 38,** 40}, the rank of either difference "38" is 4.5. With the values of {20, 36, 37, **38, 38, 38,** 40}, the rank of each difference "38" is 5.

The reasoning behind the Wilcoxon signed rank test relies on both the property that 50% of the observations must be above and 50% must be below the median of a continuous distribution and the assumption of symmetry. A difference between an observation and the hypothesized mean/median of any given absolute size, $x_i - \mu_0$, is equally likely to be positive or negative because of the symmetry assumption. Thus, if the null hypothesis about the value of the mean/median is true, the sign attached to the rank of any absolute difference between the mean and an observation is equally likely to be positive or negative. The test uses the sum of the ranks assigned to the absolute values of the differences between observations *above* the hypothesized mean/median and the sum of the ranks assigned to the absolute values of the differences between observations *below* the hypothesized mean/median. If the null hypothesis is true, the values of these two sums are expected to be equal. If one of the sums is much smaller than the other, it is unlikely that the null hypothesis is true. Since for a given sample size, the sum of the ranks of all differences is fixed, knowing the sum of the ranks assigned to differences above the hypothesized mean/median implies that you know the sum of the ranks assigned to differences below the hypothesized mean/median. The Wilcoxon two-tailed test could be set up to work with either the larger or the smaller of the two sums of ranks—the choice is arbitrary. We'll use the smaller of the two sums of ranks as the test statistic. The probability that the smaller of the two sums of ranks will take any value can be determined from all of the combinations that result in a given sum. For example, consider a sample of size 5 after any ties are dropped. The total sum of ranks is $1 + 2 + 3 + 4 + 5$, or 15. If the lower of the positively or negatively signed sum of ranks is 4 (implying that the sum of ranks on the other side is 11), this sum might result from the fact that the observations ranked 1 and 3 are on one side of the hypothesized mean/median or from the fact that the observation ranked 4 is on one side of the hypothesized mean/median. There are 2^5 or 32 possible outcomes for the signed ranks for values below the hypothesized mean/median. Table 17.2 lists these 32 outcomes. Two of those 32 have the sum of 4, so the probability of a sum of 4 is 2/32, or 0.0625. This example is intended solely to show you how the probabilities for any sum of ranks are determined. In practice you will use Appendix Table D.16, to find the critical values for the test.

TABLE 17.2

Possible outcomes of ranks for a sample of size 5 without ties

	Rank below hypothesized mean/median	Sum of ranks		Rank below hypothesized mean/median	Sum of ranks
1	None	0	17	1, 2, 3	6
2	1	1	18	1, 2, 4	7
3	2	2	19	1, 2, 5	8
4	3	3	20	1, 3, 4	8
5	4	4	21	1, 3, 5	9
6	5	5	22	1, 4, 5	10
7	1, 2	3	23	2, 3, 4	9
8	1, 3	4	24	2, 3, 5	10
9	1, 4	5	25	2, 4, 5	11
10	1, 5	6	26	3, 4, 5	12
11	2, 3	5	27	1, 2, 3, 4	10
12	2, 4	6	28	2, 3, 4, 5	14
13	2, 5	7	29	1, 3, 4, 5	13
14	3, 4	7	30	1, 2, 4, 5	12
15	3, 5	8	31	1, 2, 3, 5	11
16	4, 5	9	32	1, 2, 3, 4, 5	15

TABLE 17.3
Yields of 15 acres and worksheet for Wilcoxon T statistic

Observation	Value	$\lvert x_i - \mu_0 \rvert$	Rank	Rank of observation above mean/median	Rank of observation below mean/median
1	215	35	14.0	—	14.0
2	219	31	13.0	—	13.0
3	220	30	12.0	—	12.0
4	233	17	11.0	—	11.0
5	234	16	9.5	—	9.5
6	236	14	8.0	—	8.0
7	237	13	7.0	—	7.0
8	240	10	6.0	—	6.0
9	250	.... Drop observation			
10	251	1	1.0	1.0	—
11	252	2	2.0	2.0	—
12	253	3	3.0	3.0	—
13	254	4	4.0	4.0	—
14	255	5	5.0	5.0	—
15	266	16	9.5	9.5	—
				24.5	**80.5**

Here's an example that illustrates these steps. A company that markets a herbicide claims that the median yield per acre of wheat treated with its product is 250 bushels for a given type of soil and amount of water. The distribution of yields per acre is assumed to be symmetrical but more tail-heavy than a normal distribution. Thus a parametric test cannot be used. An agricultural research station plants 15 acres of wheat in soil of the specified type and irrigates the field with the specified amounts of water. The observed yields after herbicide treatment are listed in ascending order in Table 17.3. Observation 9 is dropped from consideration because its value equals the hypothesized median. The value of T_+ is 24.5 and the value of T_- is 80.5. A simple check on the calculations is that the sum of all of the ranks must equal:

$$T_- + T_+ = \frac{n(n+1)}{2}$$

$$T_- + T_+ = \frac{14(15)}{2} = 105$$

(17.2)

Note that dropping the observation that equals the hypothesized mean/median reduces the sample size to 14.

TABLE 17.4

Decision rules for Wilcoxon signed rank test

Hypotheses		Decision rule
$H_0: \mu = \mu_0$	$H_1: \mu \neq \mu_0$	Reject H_0 if minimum of $(T_-, T_+) \leq$ C.V. (two-tailed)
$H_0: \mu \geq \mu_0$	$H_1: \mu < \mu_0$	Reject H_0 if $T_+ \leq$ C.V. (one-tailed)
$H_0: \mu \leq \mu_0$	$H_1: \mu > \mu_0$	Reject H_0 if $T_- \leq$ C.V. (one-tailed)

For a two-tailed test the lower of the two signed sums of ranks, 24.5, is used as the test statistic. Suppose the agricultural research station wanted to test the null hypothesis of a mean of 250 bushels per acre at a 5% significance level. The critical value for 14 observations (recall that one observation was dropped) and a 5% significance level from Table D.16 is 21. The decision rule is to reject the null hypothesis whenever the test statistic is equal to or lower than the critical value in the table. Here the null hypothesis of a mean yield of 250 bushels per acre is not rejected.

If the test were one-tailed instead of two-tailed, the procedure would be slightly different. Instead of taking the lower of the T_- and T_+ as the test statistic, the T_+ would be used for a left-tailed test and the T_- would be used for a right-tailed test. For example, if the null hypothesis was $\mu \geq 250$ (a left-tailed test because the alternative would be $\mu < 250$), the test statistic would be 24.5 and the null could be rejected at the 5% significance level because 24.5 is less than the critical value of 26. If the null hypothesis were $\mu \leq 250$ (a right-tailed test because the alternative would be $\mu > 250$), the test statistic would be 80.5 and the null would not be rejected at the 5% significance level because 80.5 is greater than the critical value of 26. Note that the critical values for the test at a 5% significance level taken from Table D.16 are different for the two-tailed test (21) and the one-tailed test (26). Table 17.4 summarizes the decision rules for hypothesis tests of both types.

■ 17.2B THE WILCOXON-MANN-WHITNEY TEST FOR TWO IDENTICAL DISTRIBUTIONS

The **Wilcoxon–Mann–Whitney test** is *a nonparametric test for the equality of two population means or medians which requires the assumption that the two populations have identically shaped distributions*. While the assumption of identically shaped distributions is restrictive, it is not necessary to assume that they have any particular shape; they can, for example, be asymmetrical. The Wilcoxon-Mann-Whitney test is the nonparametric counterpart to the two-independent-sample *t* test, which requires the assumption that both population distributions are normal. The two-sample *t* test is more powerful than the Wilcoxon-Mann-Whitney if the normality assumption is correct; but in many situations normality cannot be assumed. Figure 17.5 shows two panels with identically shaped distributions. In the left panel the two distributions have the same mean and in the right panel one of the distributions has been shifted to the right. The left panel shows the population distributions when the null hypothesis of equal

FIGURE 17.5
Population distributions when
null hypothesis of equal
means or medians is true
and when it is false

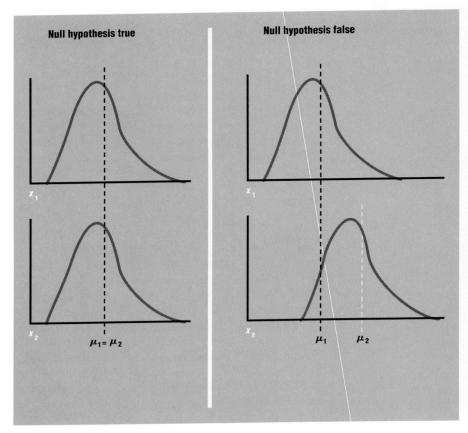

means or medians is true; the right panel shows the population distributions when the null hypothesis is false. The Wilcoxon-Mann-Whitney test is designed to distinguish between the two situations.

The Wilcoxon-Mann-Whitney test is very widely used, perhaps because two-population problems are so common in statistical analyses, perhaps because it is fairly robust for departures from the assumption of identical distributions. Also, relative to the parametric counterpart, it is one of the most powerful nonparametric tests. The probability of falsely accepting the null hypothesis is only a little bit higher with the Wilcoxon-Mann-Whitney test than with a two-sample t test when the samples have the same size and the population distributions are normal. The procedure for the Wilcoxon-Mann-Whitney test is to take independent samples from the two populations that are suspected to have equal means or medians, combine these samples, and then list all of the observations in ascending order. The ranks for the pooled obser-

TABLE 17.5

Portion of table of critical values for Wilcoxon-Mann-Whitney T statistic for a 2.5% significance level one-tailed test or a 5% significance level two-tailed test

		n_1 (SMALLER OF THE TWO SAMPLES)							
		3	4	5	6	7	8	9	10
n_2		$T_L–T_U$	$T_L–T_U$	$T_L–T_U$	$T_L–T_U$	$T_L–T_U$	$T_L–T_U$	$T_L–T_U$	$T_L–T_U$
3		5–16	6–18	6–21	7–23	7–26	8–28	8–31	9–33
4		6–18	11–25	12–28	12–32	13–35	14–38	15–41	16–44
5		6–21	12–28	**18–37**	19–41	20–45	21–49	22–53	24–56
6		7–23	12–32	19–41	26–52	28–56	29–61	31–65	32–70
7		7–26	13–35	20–45	28–56	37–68	39–73	41–78	43–83
8		8–28	14–38	21–49	29–61	39–73	49–87	51–93	54–98
9		8–31	15–41	22–53	31–65	41–78	51–93	63–108	66–114
10		9–33	16–44	24–56	32–70	43–83	54–98	66–114	79–131

vations are assigned the same way as in the Wilcoxon signed rank test except that observations that equal the hypothesized mean or median are not dropped. The ranks of observations in the smaller of the two samples are then summed. The sum of the ranks for the smaller sample is the test statistic, T. If the two samples are the same size, the choice of which sample to use for the test statistic is arbitrary for a two-tailed test. If the two samples are the same size and the test is one-tailed, the larger rank sum should be used for a right-tailed test and the smaller rank sum for a left-tailed test. This test statistic is compared with the upper and lower critical values in Table D.10. A portion of this table is reproduced in Table 17.5. In a two-tailed test the null will be rejected if the test statistic is outside the range $T_L–T_U$. In a left-tailed test the null will be rejected if the test statistic is smaller than T_L. In a right-tailed test the null will be rejected if the test statistic is larger than T_U.

The logic behind the Wilcoxon-Mann-Whitney test is similar to the reasoning behind the Wilcoxon signed rank test. Suppose the two samples were the same size. If the null of equal population means or medians is true and the two populations have the same shape, the lowest observation is as likely to have come from one population as from the other. The second lowest observation, too, could have come from either population. In fact, all ranks are equally likely to come from either population. Thus all possible sequences that contain $n_1 + n_2$ observations are equally likely. Determining the sums of ranks for every sequence is just a matter of listing the possible sequences and counting the ranks. The probability of a given rank sum for T_L or T_U is the number of sequences that have the given rank sum divided by the possible number of sequences. For example, let $n_1 = n_2 = 10$. What is the probability that T_L will be the rank sum of 55? This rank sum would occur if the ten lowest observations in the pooled sample all came from the same population, that is, $T_L = 1 + 2 + 3 + 4 + 5 + 6 + 7 + 8 + 9 + 10$, or 55. There are $_{20}C_{10} = 184,756$ possible sequences of the ten observations drawn from the first population. Thus the probability of the rank sum $T_L = 55$ is 1 divided by 184,756. The

TABLE 17.6
Number of pages produced by five high- and low-priced ribbons

High-priced ribbon	Low-priced ribbon
891	788
901	990
1,020	847
987	804
1,077	875

table of critical values was made by calculating the probability of every rank sum and then finding the ranks that mark the lower 5% of the distribution.

Here is an example. An office manager wants to test at the 2.5% significance level the claim that a high-priced printer ribbon lasts longer, on average, than a low-priced brand. The test consists of printing pages with each type of ribbon until the copy is judged too light. The test is done five times with ribbons of each type and the number of pages printed is recorded. While this distribution is discrete, the number of pages printed before a ribbon is exhausted is large enough so that the distribution can be treated as approximately continuous. Table 17.6 lists the results of the experiment. The ten observations then have to be listed in ascending order and assigned ranks. Table 17.7 shows the calculations.

Since this is a right-tailed test with equal sample sizes, the larger of the two rank sums is used as the test statistic. The critical value from Table D.10 is 37. Since the test statistic is also 37, the null hypothesis that the mean life of the high-priced ribbon is the same or less than that of the low-priced ribbon is rejected at the 2.5% significance level. Note that even with the small sample sizes of $n_1 = 5$ and $n_2 = 5$, it is possible to reject the null at the significance level of 2.5%. The assumption of identically shaped distributions is reasonable when you are comparing similar physical processes, such as printing, and it is a much weaker assumption than the normality of the two distributions.

TABLE 17.7
Worksheet for Wilcoxon-Mann-Whitney test

Number of pages	Sample	Rank	Rank of high-priced ribbon	Rank of low-priced ribbon
788	Low price	1.0	—	1.0
804	Low price	2.0	—	2.0
847	Low price	3.0	—	3.0
875	Low price	4.0	—	4.0
891	High price	5.0	5.0	—
901	High price	6.0	6.0	—
987	High price	7.0	7.0	—
990	Low price	8.0	—	8.0
1,020	High price	9.0	9.0	—
1,077	High price	10.0	10.0	—
			37.0	**18.0**

If the sample sizes are larger than the values in Table D.10, you can use the normal approximation to the sampling distribution of T. Note that the use of this normal approximation does not require the assumption that either population distribution in the Wilcoxon-Mann-Whitney test is normal. The expected value of the sampling distribution of T is

$$\mu_T = \frac{n_1(n_1 + n_2 + 1)}{2} \tag{17.3}$$

The sample with the subscript 1 is the smaller of the two samples. The standard deviation of the sampling distribution of T is

$$\sigma_T = \sqrt{\frac{n_1 n_2(n_1 + n_2 + 1)}{12}} \tag{17.4}$$

Here is an example of the Wilcoxon-Mann-Whitney test that uses the normal approximation to the sampling distribution of T. An automobile insurance company suspects that minivans suffer greater damage in collisions than full-sized station wagons. The company's records show 8 claims for minivans and 39 for full-sized station wagons. The data on claims are listed in Table 17.8.

Recall that whenever the sample sizes are unequal, the test statistic T is the sum of the ranks of the smaller of the two samples. It is easy to confuse "the smaller sum of ranks" and "the sum of the ranks of the smaller sample." The larger sample could have the smaller sum of ranks. The sum of the ranks of the smaller sample is the required test statistic. In this case the test statistic is 243. The histograms for the station wagons and the minivans (Figure 17.6) suggest that both population distributions are skewed to the right.

With skewed distributions, the appropriate test is the Wilcoxon-Mann-Whitney. The first step in using the normal approximation for this test is to find the expected value and standard deviation of T. The expected value of T is

$$\mu_T = \frac{n_1(n_1 + n_2 + 1)}{2} \tag{17.3}$$

$$\mu_T = \frac{8(8 + 39 + 1)}{2}$$

$$\mu_T = 192$$

	Count	Amount	Rank	Type of car	Rank of minivans
TABLE 17.8	1	$255.12	1.0	Station wagon	—
Value and rank of	2	260.45	2.0	Station wagon	—
claims in ascending	3	278.90	3.0	Station wagon	—
order by type of car	4	291.93	4.0	Station wagon	—
	5	296.00	5.0	Station wagon	—
	6	299.11	6.0	Station wagon	—
	7	300.18	7.0	Station wagon	—
	8	314.95	8.0	Station wagon	—
	9	315.88	9.0	Station wagon	—
	10	316.43	10.0	Station wagon	—
	11	319.52	11.0	Station wagon	—
	12	325.08	12.0	Station wagon	—
	13	326.01	13.0	Station wagon	—
	14	333.87	14.0	Station wagon	—
	15	350.00	15.0	Station wagon	—
	16	352.00	16.5	Minivan	16.5
	17	352.00	16.5	Station wagon	—
	18	365.77	18.0	Station wagon	—
	19	366.10	19.0	Station wagon	—
	20	369.01	20.0	Station wagon	—
	21	380.09	21.0	Station wagon	—
	22	381.99	22.0	Minivan	22.0
	23	382.20	23.0	Station wagon	—
	24	400.00	24.0	Station wagon	—
	25	412.11	25.0	Station wagon	—
	26	420.91	26.0	Minivan	26.0
	27	421.88	27.0	Station wagon	—
	28	435.00	28.0	Station wagon	—
	29	436.00	29.0	Minivan	29.0
	30	437.09	30.0	Station wagon	—
	31	439.23	31.0	Station wagon	—
	32	440.00	32.5	Station wagon	—
	33	440.00	32.5	Minivan	32.5
	34	444.78	34.0	Minivan	34.0
	35	446.01	35.0	Station wagon	—
	36	502.32	36.0	Station wagon	—
	37	512.97	37.0	Station wagon	—
	38	538.00	38.0	Station wagon	—
	39	548.12	39.0	Station wagon	—
	40	555.00	40.0	Minivan	40.0
	41	570.45	41.0	Station wagon	—
	42	589.12	42.0	Station wagon	—
	43	590.00	43.0	Minivan	43.0
	44	612.00	44.0	Station wagon	—
	45	612.02	45.0	Station wagon	—
	46	615.88	46.0	Station wagon	—
	47	625.00	47.0	Station wagon	—
					243.0

FIGURE 17.6

Histograms of dollar amounts
per collision claim for full-
sized station wagons and
minivans

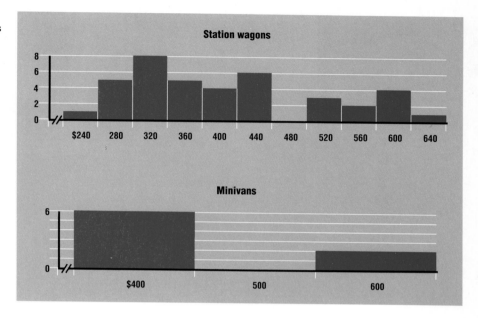

The standard deviation of T is

$$\sigma_T = \sqrt{\frac{n_1 n_2 (n_1 + n_2 + 1)}{12}} \tag{17.4}$$

$$\sigma_T = \sqrt{\frac{8(39)(8 + 39 + 1)}{12}}$$

$$\sigma_T = 35.33$$

The test is right-tailed because the company is interested in testing whether the average claim for minivans exceeds the average claim for full-sized station wagons. The formula for the critical value is

$$\text{C.V.} = \mu_{T_0} + z_\alpha \sigma_T \tag{17.5}$$

This formula for the critical value is the standard formula modified so that it begins with the hypothesized mean of T and the standard deviation term is the standard deviation of T. We'll use a 5% significance level. Substituting into formula 17.5 yields

$$\text{C.V.} = 192 + 1.64(35.33)$$

$$\text{C.V.} = 249.94$$

FIGURE 17.7
**Sampling distribution of *T*
and critical value of
hypothesis test that mean
claim per minivan is less
than or equal to mean claim
per station wagon at 5%
significance level**

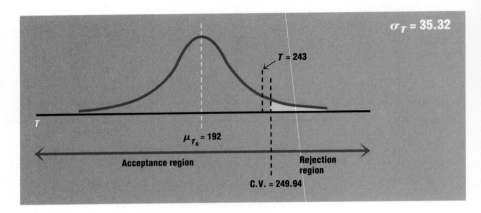

Since the value of the test statistic T is smaller than the critical value—that is, $243 < 249.94$—there is not enough evidence to reject the null hypothesis that the mean claim per minivan is less than or equal to the mean claim per full-sized station wagon. Remember that this test is valid only if the distributions of damage claims for minivans and full-sized station wagons have the same shape. Figure 17.7 shows the template for this hypothesis test.

17.2C THE WILCOXON TEST WITH PAIRED SAMPLES

The paired difference test discussed at the end of Chapter 10 requires the assumption that the two population distributions are normal if either sample size is small. The Wilcoxon signed rank test can be applied to paired data of any shape as long as the two population distributions have the same shape. This same-shape assumption is less restrictive than the normality assumption and it is quite reasonable in a paired-sample context. However, it uses more information about the differences between the paired observations than whether or not they are positive. Recall that this is the only information used by the sign test for paired differences.

Here is an example of the Wilcoxon signed rank test for paired data. A claims adjuster for an automobile insurance company suspects that the average repair costs of the two local body shops that specialize in repairing Mercedes-Benz cars are not equal. To test this hypothesis at the 5% significance level, she has the next 12 Mercedes-Benz cars that need body work and are covered by her company towed to both shops and gets estimates from both. These estimates are listed in Table 17.9. The data are paired because a car with heavy damage would be expected to have high repair costs at both shops. The differences with negative signs are shown in bold. Observations with zero differences are dropped from the analysis. The test is two-tailed because the adjuster

TABLE 17.9
Estimated repair costs, differences in estimates for same car, absolute values of differences, and ranks of absolute values

Car number	Estimate shop A	Estimate shop B	Difference in estimates	Absolute values of differences	Ranks of absolute differences
1	$ 455.12	$ 480.24	$−25.12	$ 25.12	3.0
2	874.32	851.09	23.23	23.23	2.0
3	235.16	230.08	5.08	5.08	1.0
4	1,085.00	961.99	123.01	123.01	8.0
5	925.00	925.00 ··················		Dropped from analysis ··············	
6	349.80	410.75	**−60.95**	**60.95**	**6.0**
7	869.00	540.21	328.79	328.79	11.0
8	1,236.84	1,183.40	53.44	53.44	5.0
9	652.99	610.20	42.79	42.79	4.0
10	894.29	790.37	103.92	103.92	7.0
11	286.00	125.00	161.00	161.00	9.0
12	545.28	320.61	224.67	224.67	10.0

has not specified which shop is suspected to have higher average prices. The test statistic, T, is the sum of the ranks of either the positive or negative differences, depending on which rank sum is the smallest. Here negative differences have the smallest rank sum, so the test statistic is 9, the sum of the ranks of the negative differences. The critical value from Table D.16 for the Wilcoxon signed rank test with an n of 11 and an α of 5% is 14. Thus the null hypothesis of equal mean repair costs is rejected at the 5% significance level.

The rules for deciding when to reject the null hypothesis in this paired difference test are the same as in the Wilcoxon signed rank test. In a right-tailed test the null hypothesis is rejected if the sum of the negatively signed ranks is less than or equal to the critical value. In a left-tailed test the null hypothesis is rejected if the sum of the positively signed ranks is less than or equal to the critical value. In a two-tailed test the null of equal population means or medians is rejected if the rank sum for the sign with the smaller rank sum is less than or equal to the critical value.

 17.2D THE KRUSKAL-WALLIS TEST FOR THE EQUALITY OF MORE THAN TWO POPULATION DISTRIBUTIONS

Recall that a one-way analysis of variance test for equal population means requires the assumption that all of the populations are normal and that they have the same standard deviation. The assumptions needed for the Kruskal-Wallis test are less restrictive. The **Kruskal-Wallis test** is *a nonparametric test for the equality of more than two population means or medians which requires the assumption that the distributions are identically shaped*. It can be considered an extension of the Wilcoxon-Mann-Whitney test. Both tests use the ranks of the pooled observations. The method of assigning ranks is also the same. The idea behind the Kruskal-Wallis test is that if the null hypothesis of equal population means/medians is true, the average of the ranks for each sample will tend to be the same. If there are n observations in the pooled sample, the average rank of the pooled sample will be $(n + 1)/2$. If the null hypothesis of equal means/medians is true, the average rank within each sample should be close to the average rank of the pooled sample. The expected value of the difference between the average rank in a sample and the overall average rank is zero. In math notation, this statement could be expressed as follows: If T_j is the sum of the ranks in the jth sample and n_j is the number of observations in that sample, then

$$E\left(\frac{T_j}{n_j} - \frac{n + 1}{2}\right) = 0$$

The expression in parentheses is the key part of the Kruskal-Wallis test statistic, H. The term in parentheses should be symmetrically distributed around the expected value of zero because the average rank for a sample is equally likely to be above or below the average rank for the pooled samples. The distribution of the term in parentheses should taper sharply as it diverges from zero because large deviations from the expected difference of zero are less likely than small deviations. Finally, if the sample size is large enough, the term in parentheses will be approximately normally distributed. Does this description sound familiar? Precisely the same points were made in Chapter 11 about the expression $(o_i - e_i)$, the key part of the test statistic in goodness-of-fit problems. As in Chapter 11, the expression in parentheses (average rank of the sample minus average rank of the pooled samples) can be scaled so that it follows a chi-square distribution.

The Kruskal-Wallis test statistic, H, is

(17.6)

$$H = \frac{12}{n(n + 1)}\left[\sum_{j=1}^{k} n_j \left(\frac{T_j}{n_j} - \frac{n + 1}{2}\right)^2\right]$$

Everything outside the parentheses in formula 17.6 has the purpose of scaling the terms within the parentheses so that the H test statistic will approximately follow a chi-square distribution with $k - 1$ degrees of freedom. The letter k stands for the number of samples. The approximation is considered good if each sample size, n_j, is at least 5.

Here is an example that illustrates the calculations needed for a Kruskal-Wallis test. A sewage treatment plant uses chemicals to kill microorganisms in

TABLE 17.10

Hours until all bacteria are killed by identical doses of three brands of sewage treatment chemicals

A	B	C
12.04	12.61	12.56
11.98	13.06	13.12
10.54	11.94	12.10
11.06	13.03	12.50
11.13	11.73	12.01
12.95	12.25	14.26
11.54	13.43	13.64
11.08	12.55	13.86
12.79	12.08	
12.47	13.22	

raw sewage. It tests identical amounts of three brands of chemicals on 50-gallon drums of raw sewage drawn from the same source. The tests record the time in hours required to kill all of the microorganisms in each drum. The treatment plant's experience with these chemicals suggests that the population distribution of times needed to kill all of the bacteria are not normal. Specifically, their experience is that the distributions are highly skewed to the right because treatment of occasional batches of sewage takes much longer to work than average. However, the assumption that the shapes of the distributions are the same for the three brands of chemicals seems tenable. The manager of the plant would like to test at the usual 5% significance level the hypothesis that the three brands, labeled A, B, and C, on average take the same time to kill all of the bacteria. Ten drums were used to test each brand, but the employees lost the results for the last two tests on brand C. Thus 28 tests were recorded across the three brands. Table 17.10 shows the results.

So that ranks may be assigned to the observations in the three samples, the observations have to be pooled and listed in ascending order. If there are any ties, the ranks will be assigned by the same method used in the Wilcoxon-Mann-Whitney test. Table 17.11 lists the 28 observations in ascending order and shows the brand and rank associated with each observation.

Now you have the information needed to substitute into formula 17.6:

$$H = \frac{12}{n(n+1)} \left[\sum_{j=1}^{k} n_j \left(\frac{T_j}{n_j} - \frac{n+1}{2} \right)^2 \right]$$

(17.6)

$$H = \frac{12}{28(28+1)} \left[10\left(\frac{86}{10} - \frac{28+1}{2} \right)^2 + 10\left(\frac{163}{10} - \frac{28+1}{2} \right)^2 + 8\left(\frac{157}{8} - \frac{28+1}{2} \right)^2 \right]$$

$$H = 8.73$$

TABLE 17.11
Worksheet for calculation of sum of ranks for each brand of chemicals

Brand	Pooled observations	Rank	Rank for A	Rank for B	Rank for C
A	10.54	1	1	—	—
A	11.06	2	2	—	—
A	11.08	3	3	—	—
A	11.13	4	4	—	—
A	11.54	5	5	—	—
B	11.73	6	—	6	—
B	11.94	7	—	7	—
A	11.98	8	8	—	—
C	12.01	9	—	—	9
A	12.04	10	10	—	—
B	12.08	11	—	11	—
C	12.10	12	—	—	12
B	12.25	13	—	13	—
A	12.47	14	14	—	—
C	12.50	15	—	—	15
B	12.55	16	—	16	—
C	12.56	17	—	—	17
B	12.61	18	—	18	—
A	12.79	19	19	—	—
A	12.95	20	20	—	—
B	13.03	21	—	21	—
B	13.06	22	—	22	—
C	13.12	23	—	—	23
B	13.22	24	—	24	—
B	13.43	25	—	25	—
C	13.64	26	—	—	26
C	13.86	27	—	—	27
C	14.26	28	—	—	28
			86	163	157

The test statistic is 8.73. At a 5% significance level the critical value for the test is the bound that marks the upper 5% of a chi-square distribution with $k - 1$ or 2 degrees of freedom, or 5.99. The test statistic exceeds this critical value, so the null of equal population means or medians is rejected.

After determining that all of the means are not equal, the manager of the sewage treatment plant may want to identify the brands that work significantly faster than others. The same issue arises here as the one that showed up after a one-way analysis of variance test. If all possible pairs of means are tested at a 5% significance level with a two-sample nonparametric test such as the Wilcoxon–Mann–Whitney, the overall significance level for all of the tests will be well above 5%. Recall that in the ANOVA context, Tukey's honest significant difference test was recommended to test the equality of all possible pairs of population means. With only three populations in this sewage treatment example, there is really no problem. If you had many more populations, there would indeed be a problem. The suggested readings at the end of the chapter discuss tests that can identify all pairs of means that are found to differ significantly by a Kruskal–Wallis test.

Another problem with the Kruskal–Wallis test is that the test statistic is not accurate if there are many ties in the data. Since the data are supposed to be from a continuous process, there should be few tied values. With the rounding that occurs when the data are recorded, however, it is possible to get many ties. The suggested readings also discuss an adjustment to the test statistic which compensates for any ties.

The Kruskal–Wallis test is less powerful than an ANOVA test if the assumptions of normality and equal standard deviations are valid. Nevertheless, the power of the Kruskal–Wallis test is very close (approximately 95%) to the power of the ANOVA test when these assumptions hold. In other words, you don't lose much by using a Kruskal–Wallis test. The Kruskal–Wallis is also robust to departures from the assumption of identical population shapes. It will still do a good job of identifying populations that have different means or medians when the population distributions do not have precisely the same shape. The Kruskal–Wallis test's relatively high power and robustness make it the preferred test for many problems.

◼ 17.2E THE SPEARMAN RANK CORRELATION

Chapter 13 discussed the coefficient of correlation, a measure of linear association between two random variables. The name of this measure is the Pearson product moment coefficient of correlation. The formula for the **Spearman coefficient of rank correlation,** *a measure of linear association between the ranks of two continuous random variables,* is the same as for the coefficient of correlation except that the ranks of the observations within the sample are substituted for the values of the observations. Hypothesis tests on the Pearson coefficient of correlation requires the assumption that both random variables are normally distributed, an assumption not required by the Spearman coefficient of correlation. The formula for the sample Spearman coefficient of correlation is

$$r_s = \frac{\sum\limits_{i=1}^{n} (u_i - \bar{u})(v_i - \bar{v})}{\sqrt{\sum\limits_{i=1}^{n} (u_i - \bar{u})^2 \sum\limits_{i=1}^{n} (v_i - \bar{v})^2}}$$

(17.7)

where r_s is the sample Spearman coefficient of correlation, u_i is the rank of the ith observation in the first sample, v_i is the rank of the ith observation in the second sample, and $\bar{u}$ and $\bar{v}$ are the averages of the ranks of the observations in the two samples. The terms $\bar{u}$ and $\bar{v}$ have to be equal because the two samples have the same size. If the values of the observations replaced the ranks in formula 17.7, it would be algebraically equivalent to the formula for the Pearson coefficient of correlation.

Here is an example that illustrates the calculations of the Spearman coefficient of correlation. A compensation consultant is interested in testing the hypothesis that the total compensation of corporate chief executive officers in firms with annual sales of $100 million to $200 million is positively correlated with the company's after-tax rate of return on investment. The direction of causation is uncertain in this example. A company may pay a high salary to a CEO who is thought to be largely responsible for the high rate of return or it may have a high rate of return because its high salary policy encouraged the CEO to improve the company's performance. When the direction of causation is uncertain, correlation is preferred to simple regression. And because the distribution of total compensation for CEOs is highly skewed, the Spearman coefficient of correlation is appropriate.

The compensation consultant gets the data in Table 17.12 from 15 firms with sales in the range of $100–200 million. These data have to be rearranged so that the observations appear in ascending order, and a rank has to be assigned to each observation. Table 17.13 shows these steps. The average rank of the 15 companies is 8, that is, $(15 + 1)/2$. With the information in Table 17.13 and the value of the average rank, you can substitute into formula 17.7:

$$r_s = \frac{\sum\limits_{i=1}^{n} (u_i - \bar{u})(v_i - \bar{v})}{\sqrt{\sum\limits_{i=1}^{n} (u_i - \bar{u})^2 \sum\limits_{i=1}^{n} (v_i - \bar{v})^2}}$$

(17.7)

$$r_s = \frac{(9 - 8)(7 - 8) + (5 - 8)(9 - 8) + \cdots + (13 - 8)(13 - 8)}{\sqrt{[(9 - 8)^2 + (5 - 8)^2 + \cdots + (13 - 8)^2][(7 - 8)^2 + (9 - 8)^2 + \cdots + (13 - 8)^2]}}$$

$$r_s = -0.036$$

TABLE 17.12	Company	Total compensation	Rate of return on investment
Total compensation of chief executive officers and rates of return on investment of 15 firms with annual sales of $100–200 million	A	$571,350	9.09%
	B	230,850	9.96
	C	105,423	6.07
	D	852,386	6.38
	E	196,533	7.59
	F	261,368	13.83
	G	53,574	10.52
	H	603,531	6.15
	I	685,776	11.55
	J	61,569	8.24
	K	954,010	6.62
	L	262,177	12.02
	M	441,635	13.55
	N	616,211	9.81
	O	840,515	13.43

TABLE 17.13
Worksheet for calculating ranks of compensations and rates of return

TOTAL COMPENSATION			RETURN ON INVESTMENT		
Listed in ascending order	Rank	Company	Listed in ascending order	Rank	Company
$ 53,574	1	G	6.07%	1	C
61,569	2	J	6.15	2	H
105,423	3	C	6.38	3	D
196,533	4	E	6.62	4	K
230,850	5	B	7.59	5	E
261,368	6	F	8.24	6	J
262,177	7	L	9.09	7	A
441,635	8	M	9.81	8	N
571,350	9	A	9.96	9	B
603,531	10	H	10.52	10	G
616,211	11	N	11.55	11	I
685,776	12	I	12.02	12	L
840,515	13	O	13.43	13	O
852,386	14	D	13.55	14	M
954,010	15	K	13.83	15	F

THE CORRELATION BETWEEN THE READABILITY OF MAGAZINE ADS AND THE EDUCATION LEVEL OF A MAGAZINE'S SUBSCRIBERS

The writers of magazine ads presumably are aware of their intended audience. An ad that was intended to appear in *Scientific American* would bear little resemblance to one slated for *True Confessions,* or so one would surmise. It seems plausible that the education level of a magazine's readers would be correlated with the readability of its ads. Readability is often measured with the Gunning Fog index, which purports to measure the number of years of education necessary to comprehend some material. The method of calculating the index is straightforward.

1 Determine the mean number of words per sentence in some passage.

2 In the same passage, determine the mean number of words with three or more syllables per 100 words, excepting proper names, combinations of short words such as *butterfly,* and three-syllable words created by the addition of *-es, -ed,* or *-ing.*

3 Add the mean number of words per sentence and the mean number of words of three or more syllables per 100 words and multiply the sum by 0.4.

Consider the second paragraph following formula 17.7, beginning "While the conclusion of no association . . ." The Fog index for this paragraph is 15.13, or a college junior reading level. F. Kelly Shuptrine and Daniel D. McVicker ("Readability Level of Magazine Ads," *Journal of Advertising Research* [October 1981]: 45–51) computed the Fog index for six randomly selected ads in each of nine magazines. The magazines chosen—*Scientific American, Fortune, The New Yorker, Sports Illustrated, Newsweek, People, National Enquirer, Grit,* and *True Confessions*—represent the full range of subscribers, from highly educated to hardly educated. The authors also created an index of the education levels of subscribers.

When you make the substitutions into formula 17.7, you have to be careful to match the ranks for the same observations. For example, the rank of company A's total compensation is 9 and the rank of company A's rate of return is 7. Thus the first term in the numerator, corresponding to company A, is $(9 - 8)(7 - 8)$. The resulting sample Spearman coefficient of correlation is -0.036. The possible values of Spearman's coefficient of correlation, like those of Pearson's, range from -1 to 1. Either extreme implies a perfect correlation in the sample data. The closer the sample Spearman coefficient of correlation is to 0, the less likely is an association between the two variables. In this example the Spearman coefficient of correlation is close to 0. Apparently among these

To measure the correlation between the indexes of readability and subscriber education level, the authors calculated the Spearman coefficient of rank correlation. The reason that the Spearman coefficient was used instead of the Pearson coefficient was that the indexes of readability and education are not normally distributed. The calculated Spearman coefficient was 0.183. The null hypothesis of no correlation could not be rejected at a 5% significance level. This finding may be due to the small size of the sample or to problems with either the index of readability or the education level of subscribers. Or perhaps the ads had not been prepared with specific magazines in mind. The authors found that

perhaps the most surprising aspect of reviewing the average Fog Index of the ads for all nine magazines is the fact that *none averaged up to or beyond a Junior-grade level in high school!* Advertisers may be designing their ads to meet the lowest common denominator (educational level of readership) regardless of what a magazine's general readership educational level is.

Referring to the nonsignificant Spearman coefficient of rank correlation, they commented:

These results imply that advertisers are not adapting their ads to the different educational levels of various magazines. It is recognized, however, that the advertisers may be deliberately designing their ads in this fashion in order to reach those with the least education in their reading audience. Perhaps if the ads were constructed and presented to reach the educational level of the magazines' audiences, the ads would be more effective.

A lot of thought usually goes into the copy (words) in magazine ads. The writers and account executives often agonize over every word. It is hard to believe that the entire industry is making a fundamental mistake. A sample much larger than six ads in each of nine magazines would enhance the credibility of these conclusions. Nevertheless, the case study shows an interesting and provocative use of the Spearman rank correlation coefficient.

firms the rate of return on investment is not related to the level of total compensation paid the CEO.

While the conclusion of no linear association is fairly obvious in this example, it may not be obvious in another data set. Here is a procedure to test the null hypothesis that the population Spearman coefficient of correlation, ρ_s, is 0. In words, the null hypothesis is that the ranks of the first variable are unrelated to the ranks of the second variable. The alternative hypothesis is that ρ_s is not equal to 0. The critical values for distributions of the Spearman correlation coefficient are listed in Table D.15. The table lists the upper critical value for a two-tailed test. The lower critical value is the same number with a negative sign. For example, at the 5% significance level with 15 observations the null hypothesis of no correlation will be rejected if the sample Spearman coefficient of correlation is either above 0.525 or below −0.525. Since the sample Spearman coefficient of correlation is −0.036, the null hypothesis cannot be rejected. Table D.15 goes to 30 pairs of observations. If you have a problem with more than 30 observations, you can approximate the critical value by dividing the z score by $\sqrt{n-1}$. For example, if you had 50 observations and you wanted to use a 5% significance level, the critical values would be $+1.96/\sqrt{50-1}$ and $-1.96/\sqrt{50-1}$, or 0.28 and −0.28.

EXERCISES FOR SECTION 17.2

17.8 (a) Use the Wilcoxon signed rank test to test the hypothesis that a population median equals 20 at the 5% significance level given the following random sample: {12, 14, 19, 20, 21, 24, 26, 27, 29, 30}.

(b) Use the Wilcoxon signed rank test to test the hypothesis that a population median exceeds 55 at the 5% significance level given the following random sample:

{11.3   20.3   55.0   36.4   81.7   99.9   134.1   55.0   67.6   98.7}

(c) Use the Wilcoxon signed rank test to test the hypothesis that a population median is less than 32 at the 5% significance level given the following random sample:

{10.12   15.69   22.89   21.50   21.50   32.00   38.02   49.43   51.63   79.37}

(d) What assumptions are required for these tests?

17.9 (a) Use the Wilcoxon-Mann-Whitney test to test at the 5% significance level the hypothesis that two populations have the same median given the following samples:

$A = \{12.3, 13.3, 13.3, 14.1, 14.9, 15.2\}$
$B = \{10.4, 13.3, 16.9, 18.0, 19.2, 21.7, 22.3, 25.8\}$

(b) Use the Wilcoxon-Mann-Whitney test to test at the 5% significance level the hypothesis that the mean of population A exceeds the mean of population B given the following samples:

$A = \{8, 8, 9, 10, 13, 14\}$
$B = \{7, 7, 8, 8, 8, 9, 10, 11\}$

(c) Use the Wilcoxon-Mann-Whitney test to test at the 5% significance level the hypothesis that the median of population A is less than the median of population B given the following samples:

$A = \{36, 38, 39, 40, 43, 44\}$
$B = \{37, 37, 38, 39, 39, 45, 47, 49\}$

(d) What assumptions are required for the above tests?

17.10 (a) Test at the 5% significance level the hypothesis that populations A and B have the same median. Assume that dependent samples were drawn from the populations and that the populations are not normally distributed. The sample data are:

Sample from A	Sample from B
80.7	91.6
221.4	219.2
82.3	65.0
346.1	255.6
125.7	123.8

(b) Use the same data and assumptions and test at the 5% significance level the hypothesis that the median of population B exceeds the median of population A.

17.11 Use the Wilcoxon-Mann-Whitney test to test the hypothesis that the mean of population A exceeds the mean of population B.

Observations from A	Observations from B	Observations from A	Observations from B
49	92	33	49
90	43	87	72
93	58	15	1
38	30	23	15
48	34	34	51
83	85	61	19
18	6	51	87
44	99	30	21
84	97	65	99
78	85	3	39
100	29	29	29
5	32	99	40
30	29	19	75
91	71	95	32
43	46	77	31

17.12 A state legislature passed a law setting the minimum and maximum sentences for rape at 5 years and 25 years, respectively. The law describes the circumstances that should prompt judges to impose longer or shorter sentences and it recommends that the average sentence be 15 years. Since the law was passed, 20 rape sentences have been handed down. If the 20 sentences are considered a random sample of all future possible sentences to be imposed by the state's judges, can the hypothesis that the judges are following the law's sentencing guidelines be rejected at the 5% significance level?

Sentence
lengths = {5, 5, 5, 5, 5, 8, 9, 12, 12, 13, 14, 15, 17, 18, 18, 19, 20, 21, 21, 25}

17.13 A movie theater has experimented with different ticket prices for its Friday- and Saturday-night screenings. For some movies it has charged $4 a ticket and for others it has charged $5 a ticket. The theater manager wants to test at the 5% significance level the hypothesis that the average sales at the concession stand are higher when the ticket price is $4 than when it is $5. The concession sales on $4 ticket nights were {1,243.90, 1,372.20, 1,686.40, 1,709.05, 1,713.25}. The concession sales on $5 ticket nights were {1,159.20, 1,242.20, 1,243.40, 1,297.95}.

17.14 You've entered a TV talent show called *Star Hunt*. After five minutes of singing, dancing, acting, playing a musical instrument, or telling jokes, your performance is scored by a panel of distinguished judges and by the live studio audience. The scoring by both the judges and the studio audience is on a 10-point scale with 10 as the highest possible score. If you score high enough, you will be eligible for various prizes and can go on to the next round of the contest. Each contestant is given a choice of taking the judges' score or the audience's score before he or she knows what those scores are. Ten contestants have performed before you are called onstage. The scores given these 10 contestants by the judges and the audience are listed below.

Contestant number	Judges' score	Audience's score
1	8.8	8.9
2	6.1	6.0
3	5.5	5.9
4	9.1	9.1
5	9.5	9.2
6	7.4	7.2
7	8.2	8.1
8	8.8	7.0
9	9.5	9.0
10	9.1	8.9

At the 5% significance level, can you reject the null that the median score given by the judges is less than or equal to the median score given by the audience? Which score should you take?

17.15 Test at the 5% significance level the hypothesis that the following three samples came from the same population. Assume that the population is not normally distributed.

A	B	C	A	B	C
25.99	27.65	26.99	26.56	25.15	26.92
26.52	30.68	26.62	29.20	30.65	25.74
27.50	31.80	26.02	27.84	31.37	
25.27	31.80	26.00	27.49	25.72	
25.27	29.59	25.40	29.02		

17.16 Test at the 5% significance level the hypothesis that the following four samples came from populations with identical medians. Assume that the shapes of the population distributions are the same.

A	B	C	D
5.37	5.90	6.27	6.51
5.34	5.81	6.03	6.28
5.69	5.97	6.04	6.15
5.16	5.97	6.22	6.65
5.68	5.66	6.40	6.03
5.93	5.02	6.08	6.25
5.81	5.81	5.95	6.16
5.68	5.91	6.15	6.39
5.67	5.10	5.92	
5.17	5.18		
5.78			
5.36			

17.17 Calculate the Spearman rank correlation coefficient for the following sample data and test at the 5% significance level the null hypothesis that the population coefficient of rank correlation equals zero.

X	Y	X	Y
96.45	24.23	84.10	46.76
95.06	26.46	77.01	51.37
81.07	35.32	74.72	49.79
88.44	30.47	83.69	50.12
90.57	37.90	74.60	57.49

17.18 Calculate the Spearman rank correlation coefficient for the following sample data and test at the 5% significance level the null hypothesis that the population coefficient of rank correlation equals zero.

X	Y	X	Y
16.72	383.65	18.92	386.45
17.40	382.36	19.77	386.48
17.58	382.69	19.52	381.41
17.95	387.90	20.48	390.44
18.65	381.95	20.85	385.53

17.19 Calculate the Spearman rank correlation coefficient for the following sample data and test at the 5% significance level the null hypothesis that the population coefficient of rank correlation equals zero.

X	Y	X	Y	X	Y
37.38	36.77	53.21	28.83	71.90	15.78
37.13	32.32	54.09	22.89	67.80	17.06
42.02	38.22	62.86	23.21	67.89	12.29
43.00	33.18	51.55	22.11	68.12	14.11
42.20	32.09	52.00	27.08	66.28	14.23
47.84	27.77	63.62	20.05	75.64	16.11
48.58	28.17	65.73	22.56	69.11	11.15
45.11	35.39	55.37	17.40	72.61	15.03
53.10	28.71	55.24	17.06	70.55	13.12
54.78	27.20	62.31	22.47	78.67	12.54
43.86	30.29	60.68	20.41	74.26	5.06
56.74	29.49	61.83	15.21	76.86	8.18
53.12	28.45	66.91	13.09	80.98	11.82
51.31	22.48				

17.20 The following prices are the spot market prices for 100-pound bags of potatoes over a 20-day period at three widely separated auction markets. Because shipping costs vary, the prices can vary substantially across the markets. Test at the 5% significance level the hypothesis that the average price over the period is the same in the three markets. Do not assume that the distribution of spot market prices is normal.

Idaho	Maine	Michigan	Idaho	Maine	Michigan
$4.29	$3.93	$3.97	$4.73	$4.61	$4.02
4.49	4.75	4.49	4.95	4.54	4.45
4.07	4.20	4.56	4.58	4.87	3.92
4.55	4.65	4.62	4.22	4.18	4.60
4.36	4.62	4.57	4.27	4.21	3.74
4.58	4.51	4.28	4.90	4.85	4.73
4.55	4.58	4.31	4.82	4.75	4.72
4.15	4.33	4.44	4.60	4.53	4.19
4.60	4.86	4.16	4.25	4.34	4.64
4.42	4.08	4.36	4.64	4.86	4.44

17.21 A professor who hires a graduate student to help her grade 100 term papers wants to compare the rankings she would give to the papers with those given by the graduate student. She reads ten of the papers and lists them from best to worst by the last name of the student. The graduate student reads the same ten papers and also lists them from best to worst according to his understanding of the professor's preferences. The following table shows these listings. Test at the 5% significance level the hypothesis that the rankings the professor and the graduate student will give to all future term papers are uncorrelated.

Professor	Graduate student	Professor	Graduate student
Smith	Smith	Juillerat	Perry
Kawasaki	Boone	Adams	Adams
Perry	Kawasaki	Morton	Juillerat
Boone	Morton	Shore	Begel
Rogers	Rogers	Begel	Shore

17.22 American Telephone & Telegraph is designing a computerized voice-recognition system for its long-distance information service. According to the plan, callers would tell the computer the city and the name of the person they were trying to reach and the computer would recognize the request and respond with the correct number. If the plan worked, it would reduce AT&T's costs and speed up the service. Regional accents pose a problem. A computer voice-recognition system designed to work with midwestern accents could fail to recognize words spoken in a southern or a New England accent. Suppose AT&T tried the following test to check the effectiveness of its voice-recognition algorithm across regional accents: 80 people, 20 with each of four regional accents, read the same city and name until it is recognized by the computer. The times in seconds until recognition are recorded below. Test at the 5% significance level the hypothesis that the mean times until recognition are the same for the four accents. Do not assume that the recognition times are normally distributed.

Midwestern	Southern	New England	New York City
0.18	0.50	0.50	0.37
0.72	0.05	0.73	0.99
0.01	0.37	0.19	0.28
0.37	0.32	0.53	0.79
0.25	0.14	0.40	0.36
0.01	0.66	0.13	0.68
0.43	0.25	0.43	0.65
0.10	0.32	0.98	0.06
0.98	0.48	0.88	0.79
0.71	0.94	0.53	0.45
0.91	0.05	0.17	0.68
0.77	0.85	0.98	0.56
0.32	0.19	0.52	0.91
0.04	0.04	0.14	0.06
0.26	0.29	0.34	0.15
0.07	0.33	0.85	0.98
0.57	0.40	0.00	0.56
0.14	0.43	0.27	0.97
0.93	0.93	0.59	0.21
0.42	0.31	0.77	0.05

17.3 GOODNESS-OF-FIT TESTS

 17.3A THE KOLMOGOROV-SMIRNOV TEST FOR GOODNESS OF FIT

The **Kolmogorov-Smirnov (K–S) test** is *a goodness-of-fit test that compares the sample cumulative density function with the expected density function to determine whether the sample data came from a specified continuous distribution.* The K–S test has several advantages over the chi-square goodness-of-fit test, discussed in Chapter 11. One advantage is that it works with small samples. The test can be done with as few as two observations, although with such small samples the hypothesized distribution would have to differ greatly from the sample data for the null to be rejected. A second advantage is that no information is lost by grouping the data into classes, as we must do for the chi-square test. A third advantage is that it is more powerful than the chi-square test; that is, at a given significance level the K–S test is less likely to accept a false null hypothesis. In view of these advantages, you may be wondering why you bothered with the chi-square test in Chapter 11. The K–S test applies only to continuous distributions with given parameters. The chi-square goodness-of-fit test can be used with discrete or continuous data and the parameters can be either given in the null hypothesis or estimated from the sample data. While the K–S test is often used for discrete distributions and with estimated parameters, in these cases the results are only approximate, and they can be less accurate than those obtained by the chi-square test. Also, a K–S test on a discrete distribution will have a lower significance level than the tabled value.

In a K–S test, as we have noted, you compare the sample cumulative density function with the expected cumulative density function. The concept of a density function was introduced in Chapter 6. Recall that the density of a continuous distribution is the height of the graph of the distribution. A **cumulative density function (CDF)** is *the area under the probability density function up to any point on the x axis.* Recall that the area under a probability density function must equal 1 and the area in between any two x values represents the probability that a randomly selected observation will fall in the range bounded by the two x values. Thus the cumulative density function for a given x value is the probability that a randomly selected observation will be less than or equal to that x.

The continuous uniform distribution is the easiest continuous distribution to illustrate these concepts. Take a continuous uniform distribution with a range of 0 to 10. The density or height of this distribution is 0.1. Figure 17.8 shows this density function. The cumulative density function of a continuous random variable x has the notation $F(x)$. In Figure 17.8 the cumulative density of $x = 2$ is shaded. $F(2)$ is the probability that a randomly selected observation from this continuous uniform distribution falls in the range 0 to 2. $F(2)$ is 0.2. Figure 17.9 shows the cumulative density function of the continuous uniform

FIGURE 17.8
Continuous uniform distribution with a range of 0 to 10

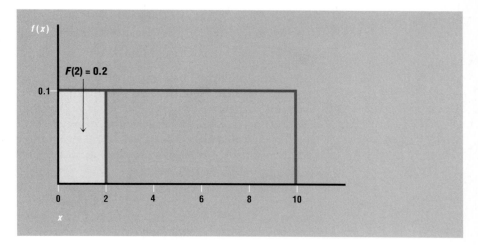

FIGURE 17.9
Cumulative density function of a continuous uniform distribution with a range of 0 to 10

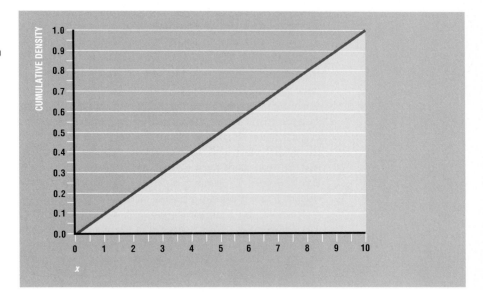

distribution that has the range 0 to 10. The CDF is a straight line. All continuous uniform distributions have straight-line CDFs.

A CDF can be defined for a sample as well as a population. The sample CDF is the proportion of all observations in a sample at or below some x value. It is similar to the cumulative relative frequency polygon, discussed in Chapter 2. The sample CDF is customarily drawn as a stair-step function. The K–S test statistic, D, is the maximum absolute deviation between the sample and the expected CDF. D depends only on the sample size, not on the shape of the hypothesized distribution. A large test statistic D would lead you to reject the null hypothesis that the sample came from the hypothesized distribution.

For the first example of the K-S test we'll stick to the continuous uniform distribution even though it rarely applies to real problems, because it is the easiest to describe. Here is an admittedly contrived example. A company that programs and markets video games buys some programming code that can be used to time events in a game. For example, it can set the time a rocket ship takes to explode on the screen anywhere from 0 to 100 seconds. Suppose the code was intended to make any time in that interval equally likely, hence a continuous uniform distribution. To see if the code works, the company takes a random sample of 40 events. The observations on the numbers of seconds are listed in ascending order in Table 17.14. The third column in the worksheet, the expected CDF, is the time for each observation divided by 100. Since the hypothesized distribution is uniform over the range 0 to 100, the probability of observing any time at or below one of the sample values is the sample value divided by 100. This expected CDF is compared with the sample CDF on the basis of the hypothesized population distribution. The sample CDF for the continuous uniform distribution is a straight line. With a sample of size 40, the sample CDF would rise 0.025 with each observation. The value 0.025 is 1 divided by 40. The test statistic D is the maximum absolute difference between the sample and expected CDFs. In this data set, the maximum absolute difference is at the 35th observation, D equals 0.0841. This test statistic is compared with the tabled K-S critical value in Table D.12, and the null hypothesis that the sample came from the hypothesized distribution is rejected if D exceeds the critical value. If the significance level is set at 5%, the critical value is 0.21. Thus the null hypothesis that the sample data came from a continuous uniform distribution with a range of 0 to 100 cannot be rejected at the 5% significance level. Figure 17.10 shows a plot of the expected and sample CDFs and the observation with the greatest absolute difference.

Because the computer example was a bit contrived, we'll cover two more examples of K-S tests, one against a normal distribution and one against a Poisson distribution, both with hypothesized parameters. Here is the normal distribution example. A university has a contract with a company to provide meals in the university's dining halls. The contract calls for an average of 10.5 ounces of meat per student per day. Some students have complained about the food, particularly about what they regard as inadequate servings. The students find a manual prepared by the food contractor which states that the meat portions per student per day should be normally distributed with a mean of 10.5 ounces and a standard deviation of 2 ounces. Several students agree to collect their meals over a few days, and a sample of 12 sets of meals is taken to a chemistry lab to be weighed. At the 5% significance level, is there enough evidence to reject the null hypothesis that the food contractor has followed his internal guidelines on meat portions?

TABLE 17.14
Data and worksheet for Kolmogorov-Smirnov test

Observation rank	Number of seconds	Expected cumulative density	Sample cumulative density	Absolute difference between observed and expected CDF
1	0.37	0.0037	0.0250	0.0213
2	1.95	0.0195	0.0500	0.0305
3	4.42	0.0442	0.0750	0.0308
4	13.45	0.1345	0.1000	0.0345
5	14.54	0.1454	0.1250	0.0204
6	14.85	0.1485	0.1500	0.0015
7	19.04	0.1904	0.1750	0.0154
8	19.29	0.1929	0.2000	0.0071
9	23.66	0.2366	0.2250	0.0116
10	28.76	0.2876	0.2500	0.0376
11	29.65	0.2965	0.2750	0.0215
12	31.85	0.3185	0.3000	0.0185
13	33.66	0.3366	0.3250	0.0116
14	36.44	0.3644	0.3500	0.0144
15	37.21	0.3721	0.3750	0.0029
16	41.02	0.4102	0.4000	0.0102
17	41.71	0.4171	0.4250	0.0079
18	47.19	0.4719	0.4500	0.0219
19	55.15	0.5515	0.4750	0.0765
20	55.47	0.5547	0.5000	0.0547
21	58.12	0.5812	0.5250	0.0562
22	58.65	0.5865	0.5500	0.0365
23	59.04	0.5904	0.5750	0.0154
24	59.73	0.5973	0.6000	0.0027
25	61.08	0.6108	0.6250	0.0142
26	63.56	0.6356	0.6500	0.0144
27	66.29	0.6629	0.6750	0.0121
28	71.00	0.7100	0.7000	0.0100
29	71.67	0.7167	0.7250	0.0083
30	72.06	0.7206	0.7500	0.0294
31	76.97	0.7697	0.7750	0.0053
32	78.49	0.7849	0.8000	0.0151
33	78.85	0.7885	0.8250	0.0365
34	86.10	0.8610	0.8500	0.0110
35	95.91	0.9591	0.8750	**0.0841**
36	96.79	0.9679	0.9000	0.0679
37	98.08	0.9808	0.9250	0.0558
38	99.21	0.9921	0.9500	0.0421
39	99.56	0.9956	0.9750	0.0206
40	100.00	1.0000	1.0000	0.0000

FIGURE 17.10

Expected and sample
cumulative density functions
for a continuous uniform
distribution and a sample of
40 observations

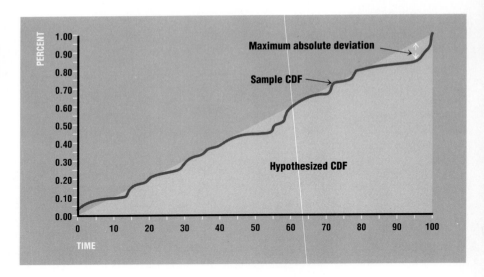

Table 17.15 shows the number of ounces of meat in the 12 sets of meals and the calculations for the K–S test statistic. The fourth column, "z score if null is true," is the z score for each observed number of ounces when a mean of 10.5 ounces and a standard deviation of 2 ounces are assumed. For example, the first z score of -1.61 equals $(7.29 - 10.5)/2$. The values in the column headed "Expected cumulative density" were found by using a normal table to get the

TABLE 17.15

Data and worksheet for Kolmogorov-Smirnov test against a normal distribution

Observation rank	Number of ounces	Sample cumulative density	z score if null is true	Expected cumulative density	Absolute difference between observed and expected CDF
1	7.29	0.08	−1.61	0.05	0.03
2	8.96	0.17	−0.77	0.22	0.05
3	9.25	0.25	−0.62	0.27	0.02
4	9.60	0.33	−0.45	0.33	0.00
5	9.69	0.42	−0.41	0.34	0.08
6	9.82	0.50	−0.34	0.37	0.13
7	10.07	0.58	−0.21	0.42	0.16
8	10.11	0.67	−0.20	0.42	0.25
9	10.22	0.75	−0.14	0.44	0.31
10	10.30	0.83	−0.10	0.46	**0.37**
11	10.90	0.92	0.20	0.58	0.34
12	11.64	1.00	0.57	0.72	0.28

areas to the left of each z score. For example, the area to the left of a z score of -1.61 is approximately 0.05. The last column shows the absolute deviations between the sample and expected cumulative densities. The test statistic D, the largest absolute deviation, is 0.37. At a 5% significance level, the critical value from the K-S table for 12 observations is 0.375. Since the test statistic is below the critical value, there is not enough evidence to reject the null that the sample came from the hypothesized distribution.

Figure 17.11 shows the expected and observed CDFs for this problem. The shaded area is the sample cumulative density function and the line is the expected cumulative density function if the null hypothesis of normal distribution with a mean of 10.5 and a standard deviation of 2 ounces were true.

Here is the Poisson distribution example. You are considering buying a business that repairs small aircraft engines. The owners of the business claim that they average eight requests for engine repairs per day. Rather than rely entirely on the company's books, you observe the business for 15 days and record the number of requests for engine repairs. If you assume that each request is independent, the distribution of occurrences should be Poisson with a mean of 8. Suppose you wanted to test whether the sample data came from a Poisson distribution with a mean of 8 at the 5% significance level. Table 17.16 lists the frequencies for 0 through 23 requests per day and shows the calculations for the K-S test statistic. The worksheet has been carried to 23 occurrences because a five-place cumulative Poisson distribution rounds to 1 by the 23d occurrence.

One difference between Table 17.16 and Tables 17.14 and 17.15 is that the word *density* has been replaced by *frequency*. The reason for this change is that

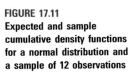

FIGURE 17.11
Expected and sample cumulative density functions for a normal distribution and a sample of 12 observations

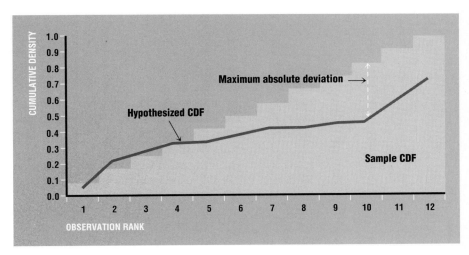

TABLE 17.16

Data and worksheet for Kolmogorov-Smirnov test against a Poisson distribution

Number of occurrences	Observed frequency	Running sum of observed frequencies	Relative cumulative observed frequency	Relative cumulative expected frequency	Absolute differences
0	0	0	0.00000	0.00033	0.00033
1	0	0	0.00000	0.00302	0.00302
2	2	2	0.13333	0.01375	0.11958
3	0	2	0.13333	0.04238	0.09095
4	0	2	0.13333	0.09963	0.03370
5	0	2	0.13333	0.19124	0.05791
6	1	3	0.20000	0.31337	0.11337
7	0	3	0.20000	0.45296	**0.25296**
8	3	6	0.40000	0.59255	0.19255
9	1	7	0.46667	0.71662	0.24995
10	2	9	0.60000	0.81589	0.21589
11	2	11	0.73333	0.88808	0.15475
12	0	11	0.73333	0.9362	0.20287
13	1	12	0.80000	0.96582	0.16582
14	1	13	0.86667	0.98274	0.11607
15	2	15	1.00	0.99177	0.00823
16	0	15	1.00	0.99628	0.00372
17	0	15	1.00	0.99841	0.00159
18	0	15	1.00	0.99935	0.00065
19	0	15	1.00	0.99975	0.00025
20	0	15	1.00	0.99991	0.00009
21	0	15	1.00	0.99997	0.00003
22	0	15	1.00	0.99999	0.00001
23	0	15	1.00	1	0.00000

the Poisson is a discrete distribution while the continuous uniform and the normal are continuous distributions. Otherwise the first three columns are similar to those in the earlier worksheets. The relative cumulative expected frequencies come from a Poisson distribution with a mean of 8. These values are just a running total of the values in the Poisson table for a mean of 8. The K-S test statistic is the maximum absolute deviation, $D = 0.25296$. The critical value for 5% and 15 observations is 0.338. There is not enough evidence to reject the null hypothesis that the population distribution is Poisson with a mean of 8. Since this distribution is discrete, the K-S is an approximation and the true significance level is slightly below the tabled value of 5%. Figure 17.12 shows the observed and expected cumulative relative frequencies for this Poisson example.

FIGURE 17.12

Expected and observed cumulative relative frequencies for a Poisson distribution with a mean of 8

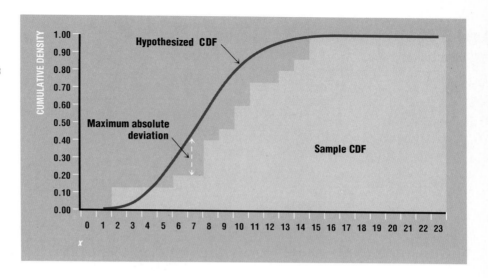

17.3B THE LILLIEFORS TEST FOR NORMAL DISTRIBUTIONS

The **Lilliefors test** is a *goodness-of-fit test for the normal distribution which compares the sample and expected cumulative densities when the population mean and standard deviation are not known.* In these circumstances the Lilliefors is a more powerful test than the chi square and the Kolmogorov-Smirnov. The procedure for the Lilliefors test is identical to the procedure for the K-S test. The only difference is in the critical values. The critical values for the Lilliefors test are given in Appendix Table D.14.

Here is an example. The research division of a bank holding company would like to estimate the mean time that elapses between the arrangement of financing for a new office building and the day the building is ready for occupancy. The bank can use this information in its calculations of the amount of money it can safely lend for office construction. The bank holding company has financed nine office buildings. Before it can use the data it has accumulated to construct a confidence interval with the *t* score, it must make two assumptions. One is that the nine projects are equivalent to a random sample of all future projects; the other is that the population distribution of times until occupancy is normally distributed. The first assumption cannot be tested, but the Lilliefors test can be used to check the second assumption. Since the population parameters are not known, the Lilliefors test is used instead of the K-S test.

Table 17.17 shows the data on the nine office buildings and has the worksheet for the test statistic. The sample mean and sample standard deviation were calculated from the observations in the second column. (Figure 17.13 shows the expected and observed cumulative relative frequencies for this normal distribution.) As there are nine observations, the sample cumulative density shown in the third column increases in steps of one-ninth with each obser-

TABLE 17.17
Data and worksheet for a Lilliefors test

Building	x_i	Sample CDF	z score	Expected CDF	Absolute difference
1	35.14	0.11	−1.81	0.04	0.07
2	35.51	0.22	−1.02	0.15	0.07
3	35.89	0.33	−0.20	0.42	0.09
4	35.90	0.44	−0.18	0.43	0.01
5	36.02	0.56	0.08	0.53	0.03
6	36.10	0.67	0.25	0.60	0.07
7	36.19	0.78	0.44	0.67	**0.11**
8	36.36	0.89	0.81	0.79	0.10
9	36.75	1.00	1.64	0.95	0.05

$\bar{x} = 35.98$

$s = 0.47$

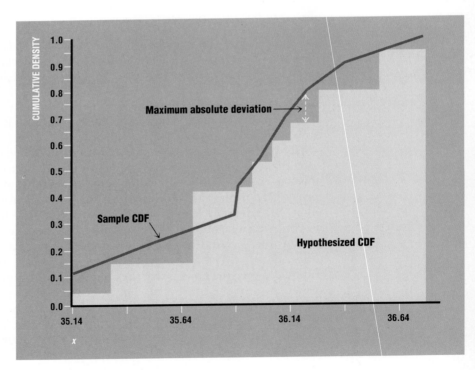

FIGURE 17.13
Expected and observed cumulative relative frequencies for a normal distribution with a mean of 35.98 and a standard deviation of 0.47

vation. The z scores in the fourth column are calculated from the sample mean and the sample standard deviation. The expected cumulative density is the area to the left of each z score under the standard normal distribution. The test statistic, the largest deviation between the sample and the expected cumulative densities, is 0.11. The critical value for a 5% significance level and 9 observations is 0.274. Thus there is not sufficient evidence to reject the null hypothesis that the population is normally distributed. The research division of the bank can proceed with its calculation of a confidence interval.

EXERCISES FOR SECTION 17.3

17.23 Test at the 5% significance level the hypothesis that the following sample came from a continuous uniform distribution with the range of 0 to 1:

0.455 0.027 0.539 0.844 0.939 0.980 0.397 0.845 0.955 0.347 0.782
0.373 0.930 0.921 0.649

17.24 Test at the 5% significance level the hypothesis that the following sample came from a normal distribution with a mean of 10 and a standard deviation of 2:

7.72 7.92 8.56 8.93 9.26 9.59 9.79 9.88 9.9 10.10 10.10 10.59
10.69 10.74 10.75 10.92 10.92 11.04 11.545 11.60

17.25 Test at the 5% significance level the hypothesis that the following sample came from a Poisson distribution with a mean of 2: {1, 4, 6, 5, 2, 3, 2, 5, 0, 4}.

17.26 Test at the 5% significance level the hypothesis that the following sample came from a normal distribution:

24.93 42.11 21.02 11.53 −11.17 24.54 32.31 38.72 17.14 26.30
34.48 44.80 13.52 19.76 34.15 41.74

17.27 An automobile manufacturer needs the gap between the trunk lid and the rear deck of the car to be within a certain range so that the trunk lid neither jams nor allows rain to enter. Robots are used to assemble the trunk lid. In a trial run of the assembly process, the following gaps in millimeters were observed:

16.20 28.48 86.96 81.82 46.99 66.91 74.46 56.13 81.81 85.29 68.15
91.10 30.66

Do these data provide sufficient evidence to reject at the 5% significance level the hypothesis that the population distribution is normal with a mean of 50 millimeters and a standard deviation of 15 millimeters?

17.28 A company places trainees in groups of ten for instruction on company procedures and policies. The expected proportion of successes is 0.8. If each trainee were an independent trial, would the following data on the number of successes in the last 20 groups contradict this expectation:

7 10 8 9 7 9 8 9 9 7 8 8 9 7 6 9 9 7 8 8

Set up the appropriate null and alternative hypotheses and conduct a test using a 10% significance level.

SUMMARY

Most of the parametric tests introduced in earlier chapters have a nonparametric counterpart in this chapter. The parametric tests usually require that the populations be normally distributed. The nonparametric tests have either no requirements in regard to the population distribution (in which case they are sometimes called distribution-free nonparametric tests) or less stringent requirements. They may, for example, require symmetry or identical shapes in the population distributions. The wider applicability of the nonparametric tests comes at the cost of a reduction in power. If the normality assumption holds, the parametric test will be less likely to accept a false null hypothesis than the corresponding nonparametric test. There is no free lunch in Statistics. However, some of the nonparametric tests, such as the Wilcoxon-Mann-Whitney and the Kruskal-Wallis, are very nearly as powerful as their corresponding parametric tests.

The chapter began with the sign test for the population median. This test relies on the fact that half of the observations in any continuous distribution are on either side of the median. You can use the sign test either to test for a specific value of the population median or to construct a confidence interval for the population median. The next section of the chapter presented five tests that use ranks. If you are willing to assume that the population distribution is symmetrical, the Wilcoxon signed rank test provides a more powerful test for the population mean/median than the sign test. The Wilcoxon signed rank test can also be extended to paired or dependent samples. The Wilcoxon-Mann-Whitney test for differences between two population means or medians requires the assumption of identical population shapes. The Kruskal-Wallis test for the equality of more than two population means or medians requires the assumption that all of the populations have identical shapes. The Spearman rank correlation test requires only that the two population distributions be assumed to be continuous.

The tests for independence, homogeneity, and goodness of fit discussed in Chapter 11 are nonparametric tests. They have been given a separate chapter because the tests for independence and homogeneity are so important. Goodness-of-fit tests can be done with either the chi-square distribution (as discussed in Chapter 11), the Kolmogorov-Smirnov test, or the Lilliefors test for normality. The chi-square version of the goodness-of-fit test is more powerful than the Kolmogorov-Smirnov test or the Lilliefors test for normality if the population parameters are given in the null hypothesis. However, the chi-square version of the goodness-of-fit test often requires more work, especially when you start with raw data and have to set up as many comparisons as possible that have an expected frequency of 5. If the parameters are not specified in the null hypothesis, the Kolmogorov-Smirnov and Lilliefors tests for normality are more powerful than the chi-square test.

SUGGESTED READINGS

Conover, W. J. *Practical Nonparametric Statistics.* 2d ed. New York: Wiley, 1980.

Lehmann, E. L. *Nonparametrics: Statistical Methods Based on Ranks.* San Francisco: Holden-Day, 1975.

NEW TERMS

cumulative density function (CDF) The area under the probability density function up to any point on the x axis.

Kolmogorov-Smirnov (K-S) test A goodness-of-fit test that compares the sample cumulative density function with the expected cumulative density function to determine whether the sample came from a specified continuous distribution.

Kruskal-Wallis test A nonparametric test for the equality of more than two population means or medians which requires the assumption that the distributions are identically shaped.

Lilliefors test A goodness-of-fit test for the normal distribution which compares the sample and expected cumulative density functions when the population mean and standard deviation are not known.

ranks of the differences between observations and the hypothesized mean/median If the value of the xth difference is unique, the number of observed differences from the lowest up to and including the xth difference; if two or more differences have the same value, $(k + 1)/2$ plus the number of differences with lower values, where k is the number of observations with the same value.

sign test A nonparametric test for the population median which requires the assumption that the population distribution is continuous.

Spearman coefficient of rank correlation A measure of linear association between the ranks of two continuous random variables.

Wilcoxon-Mann-Whitney test A nonparametric test for the equality of two population means or medians which requires the assumption that the two populations have identically shaped distributions.

Wilcoxon signed rank test A nonparametric test for either the mean or the median of a single continuous distribution or for the difference between the means or medians of paired samples. In either case the test requires the assumption of a symmetrical distribution.

 NEW FORMULAS

Confidence level for the population median

$$1 - \alpha = 1 - 2\left(\sum_{i=n-x+1}^{n} \frac{n!}{(n-i)!i!}(0.5)^i(0.5)^{n-i} \right) \tag{17.1}$$

Sum of ranks in one sample

$$T_- + T_+ = \frac{n(n+1)}{2} \tag{17.2}$$

Expected value of Wilcoxon-Mann-Whitney T statistic

$$\mu_T = \frac{n_1(n_1 + n_2 + 1)}{2} \tag{17.3}$$

Standard deviation of Wilcoxon-Mann-Whitney T statistic

$$\sigma_T = \sqrt{\frac{n_1 n_2 (n_1 + n_2 + 1)}{12}} \tag{17.4}$$

Critical value of Wilcoxon-Mann-Whitney T statistic

$$\text{C.V.} = \mu_{T_0} + z_\alpha \sigma_T \tag{17.5}$$

Kruskal-Wallis test statistic

$$H = \frac{12}{n(n + 1)} \left[\sum_{j=1}^{k} n_j \left(\frac{T_j}{n_j} - \frac{n + 1}{2} \right)^2 \right] \tag{17.6}$$

Spearman coefficient of rank correlation

$$r_s = \frac{\sum\limits_{i=1}^{n} (u_i - \bar{u})(v_i - \bar{v})}{\sqrt{\sum\limits_{i=1}^{n} (u_i - \bar{u})^2 \sum\limits_{i=1}^{n} (v_i - \bar{v})^2}} \tag{17.7}$$

ADDITIONAL EXERCISES

An asterisk designates an exercise of greater than ordinary difficulty.

17.29 A small bank has several thousand Visa accounts. The distribution of outstanding balances on these accounts is highly skewed. Most accounts have a zero outstanding balance and a few are at their credit limits. The bank takes a small monthly sample of the accounts and uses it to construct a confidence interval for the median of the distribution. Use the following data to construct a confidence interval for the median as close as possible to the 90% level of confidence:

0.00   1,171.94   796.22   306.74   437.92   459.05   982.69   1,052.32   415.46   755.09
0.00   0.00   902.01   954.66   0.00   1,143.48   0.00   681.71   1,175.94   1,083.96

17.30 Use the Kolmogorov–Smirnov test to test at the 5% significance level the null hypothesis that the following data were generated by a Poisson process with a mean of 2:

2   1   6   4   10   3   1   3   3   1   3   5   5   5   1   1   4   3   1   1   4   2

The data represent the number of canoes rented per hour by a canoe livery. What problems are caused by using the K-S for these data? Why might the number of canoes rented per hour not fit the assumptions required for a Poisson distribution?

17.31 Make a table that shows the levels of confidence for intervals on the population median from the third lowest to the third highest observation for sample sizes in the range of 10 to 20.

17.32 An executive relocation firm has been hired to assist in the relocation of the executives of a *Fortune* 1,000 company that is moving its headquarters from New York City to Miami, Florida. Most of the company's executives currently live in large four- or five-bedroom homes on Long Island. The relocation firm wants to test at the 5% significance level the hypothesis that the mean price of such homes is the same in both locations. Six of the executives' homes in Long Island and ten similar homes in Miami are appraised. Assume that the shapes of the distributions of home prices are the same in both locations. Can the null of equal population means be rejected on the basis of the following data? Long Island prices: $154,000, $175,000, $187,000, $199,000, $225,000, $245,000; Miami prices: $132,000, $145,000, $150,000, $151,000, $154,000, $157,000, $167,000, $187,000, $190,000, $199,000.

17.33 The owners of a unisex hair salon want to test at the 5% significance level the hypothesis that the mean time for women's haircuts exceeds the mean time for men's haircuts. Assume that the distributions of minutes required to cut the hair of persons of both sexes have the same shape. Use the following data to test the hypothesis: time in minutes for men, 12.1, 14.8, 16.7, 19.0, 22.5; time in minutes for women, 22.4, 22.7, 24.6, 26.9, 34.8.

17.34 The performances of college baseball teams vary from year to year because of the graduation of players, coaching changes, injuries, and random variation. (The last factor is usually called luck.) Use the following data on team rankings at the end of two successive seasons to test at the 5% significance level the hypothesis that there is no correlation between the rankings of the two years.

Year 1	Year 2
Michigan	Michigan
Iowa	Illinois
Indiana	Minnesota
Ohio State	Ohio State
Wisconsin	Iowa
Northwestern	Purdue
Michigan State	Indiana
Illinois	Wisconsin
Minnesota	Northwestern
Purdue	Michigan State

17.35 Healthcare Now! provides a health assessment course for employees. The course, paid for by the employer, is supposed to be cost-effective because it changes employees' attitudes and behavior in regard to health risks. These changes reduce absenteeism and the costs of training, recruiting, and medical insurance. To test the impact of this course on attitudes, an employer administered a ten-question survey to a group of 20 employees before and after they took the Healthcare Now! course. Each question was about attitudes toward health care; for example, "Your own behavior is the main determinant of how healthy you are." The possible responses were strongly disagree, disagree, uncertain, agree, strongly agree. The responses were coded as integers 1 through 5, with 5 being the most favorable attitude in regard to health. Use the data below on the mean response of the 20 employees to test the null hypothesis that attitudes toward health care were unchanged by the course.

Question	Mean response before course	Mean response after course
1	2.2	2.5
2	3.6	3.8
3	3.9	3.9
4	1.7	3.4
5	2.5	3.0
6	2.9	3.0
7	2.9	2.8
8	3.4	3.8
9	3.6	3.9
10	3.6	3.9

17.36 A library that owns more than a million books would like to construct a confidence interval at the 99% level of confidence for the median number of pages per book. What is the minimum number of books it should sample?

17.37 Years of experience waiting for a particular bus have convinced you that the distribution of minutes the bus is late is symmetrical but too heavy-tailed to be normal. You think the mean time late is five minutes. Use the following data to test this hypothesis at the 5% significance level:

4.33   5.56   6.59   5.42   5.85   5.32   5.42   4.76   4.10   4.66   4.27   3.76   4.31   3.86   4.34

17.38 Gambling casinos give gamblers what are known as "comps," or complimentary items, to encourage them to come to the casino and lose money. The comps include free rooms, tickets to floor shows, and free meals. One casino wanted to test the effect of these various comp packages on the amount wagered. The casino offered some gamblers a free room and others a free room with the tickets and the meals. The casino was uncertain whether the additional comps would encourage more gambling or if the gamblers would "waste" time eating and watching shows and thus spend less on gambling. The gamblers offered these packages were randomly selected from a pool of former guests at the casino. Use the following data on amounts wagered to test at the 5% significance level the hypothesis that the mean amount wagered is the same for both packages.

Free room only	Room, tickets, and meals	Free room only	Room, tickets, and meals
$264.94	$829.54	$162.16	$493.33
186.45	813.86	640.43	734.99
527.77	380.29	958.31	503.98
852.98	117.75	545.04	657.07
111.59	474.33	695.48	186.14
323.01	65.80	706.16	678.79
958.51	767.43	456.91	310.62
303.48	719.94		872.60
86.15	1.57		237.17
526.64	389.89		7.29
145.40	5.23		193.66
77.88	952.38		462.51
739.93	998.15		

17.39 The following data are the number of minutes taken by airline flight attendants to respond to passengers' requests for assistance. Test at the 5% significance level the hypothesis that this sample could be generated by an exponential process with a mean of 0.25 minute:

0.00   0.00   0.00   0.08   0.14   0.14   0.14   0.17   0.20   0.20   0.25   0.25   0.25
0.25   0.25   0.25   0.25   0.33   0.33   0.33   0.33   0.50   0.50   1.00   1.00

17.40 Herbal Living is a company that sells natural weight-loss products made from teas, spices, and herbs. To demonstrate the effectiveness of its weight-loss regimen, the company finds ten pairs of identical twins who are at least 30% over their ideal weight. One of each pair is put on the Herbal Living plan and the other is left to his or her own devices. At the end of a six-month test period, the weight gain or loss of each person is recorded. Use the Wilcoxon signed rank test to test at the 5% significance level the hypothesis that the mean weight loss on the Herbal Living plan exceeded the mean weight loss of twins who were not on the plan.

Weight change, Herbal Living twins	Weight change, other twins
−32.9	−14.5
0.0	+8.1
0.0	0.0
−17.7	−17.0
+8.7	+3.4
−23.0	−3.6
−18.7	−9.2
−17.0	−16.8
−39.1	−14.2
+19.0	+8.1

17.41 Jobs for the Handicapped is a for-profit firm that hires handicapped workers to package and sell light bulbs by phone. The firm relies on "cold calls," sales pitches made with no introduction or indication that the person being called is interested in the product. The amount sold depends largely on the scripts used in the cold calls. Jobs for the Handicapped has developed two new scripts, one that emphasizes the long-lasting qualities of the light bulbs and another that emphasizes the employment opportunities the company provides for handicapped workers. Use the following data to test at the 5% significance level the hypothesis that the median amount sold per call is the same for the two scripts. Assume that the distributions of amounts sold per call have the same shape for both scripts.

Script emphasizing light bulb life	Script emphasizing jobs for handicapped	Script emphasizing light bulb life	Script emphasizing jobs for handicapped
$0.00	$9.00	$0.47	$3.81
8.02	0.00	3.82	0.00
2.32	0.00	3.78	2.46
1.22	2.28	7.90	8.00
1.83	5.75	2.99	0.00
0.75	4.08	4.31	1.38
9.22	5.46	0.00	5.58
6.73	6.58	0.00	7.66
2.24	0.92	8.58	1.89
7.07	0.63	2.99	5.35
8.75	3.23		3.82
0.00	9.53		8.29
0.00	6.23		1.57
0.00	4.92		9.86
3.97	0.88		0.00

17.42 A company's human resources office uses tests and interviews to rank job candidates. The director wants to test at the 5% significance level the hypothesis that rankings based solely on test scores are uncorrelated with rankings based solely on interviews. Test this hypothesis with the following data on candidates for the same job:

Ranking by test score	Ranking by interview
Badger	Badger
Yamamoto	McGregor
Lee	Handfield
Jones	Harrison
Harrison	Jones
McGregor	Lee
Handfield	Yamamoto

17.43 The number of ounces of cornflakes that a packing machine puts in boxes is thought to be normally distributed. Use the following data to test this hypothesis at the 5% significance level.

11.95	11.95	11.96	12.03	11.88	12.11	11.93	12.13	12.02	11.93	11.91
12.05	12.02	11.99	11.97	12.06	12.01	11.80	11.94	11.75	11.95	11.92
11.98	11.96	12.12	11.99	11.86	11.99	12.04	11.95			

If the distribution turns out to appear to be normal, how could the cereal company use this information?

17.44 The distribution of distances needed to stop a train with a given weight and traveling at a given speed is highly skewed. Occasionally the brakes fail altogether and it takes several miles for the friction of the wheels on the rails to stop the train. Rather than throw out these extreme observations, you can estimate the median braking distance instead of the mean. You can, for example, construct a confidence interval on the population median at the 90% level of confidence. Do so for the following data in miles:

0.307	0.318	0.346	0.380	0.384	0.453	0.499
0.519	0.542	0.754	0.778	0.800	0.801	0.846
0.896	1.111	1.133				

17.45 A county prosecutor has received many complaints about a local transmission repair shop. To see if the shop charges for unnecessary repairs, the prosecutor has the transmissions of 12 cars adjusted so that they slip. The prosecutor then sends men with six of the cars to ask for a quote on the repair costs and women with the other six cars. The prosecutor suspects that the repair shop will view the women as less informed about cars and quote them higher prices. Use a 5% significance level to test the hypothesis that the median price quoted to the men is less than or equal to the median price quoted to the women.

Price quoted to men	Price quoted to women
$ 32.89	$1,554.21
56.90	786.00
310.48	1,209.13
143.99	667.98
445.01	398.00
276.00	441.87

17.46 Test at the 5% significance level the hypothesis that the mean number of shaves per razor blade is the same for the following four brands. The data below show the number of shaves per blade for a sample of men who regularly use the brand they tested. Do not assume that the distributions of number of shaves per brand are normal. Would a rejection of the null hypothesis indicate differences in the ability of the different blades to maintain a sharp edge?

Schick	Acta	Bic	Generic
12	19	4	9
0	8	5	1
12	2	2	6
7	11	19	3
2	13	13	17
15	5	6	1
1	4	19	18
9	16	7	14
12	3	10	5
14	15	0	18
8	20	8	17
7	1	9	16
19	7	18	18
18	8	17	2
5	7	10	14
17	17	14	7
2	1	4	17

17.47 The following data represent the time in minutes taken by a robot to change the oil in a car:

12.78   13.00   12.93   12.88   13.33   13.00   12.80   12.97   13.16   13.12   12.86
13.01   12.64   13.17

Test at the 5% significance level the hypothesis that these data were generated by a normal distribution with a mean of 13 and a standard deviation of 0.2. Which test would be the most powerful—a chi-square test, a Kolmogorov-Smirnov test, or a Lilliefors test?

17.48 After doing a Lilliefors test for the null hypothesis of normality at a 5% significance level for the following data, construct a histogram of the data set and use the appropriate test to decide if the center of the population distribution equals 48:

37.59   31.11   51.86   52.02   25.09   42.93   51.88   43.99   21.70   58.65   52.72
43.99   39.10   47.12   50.84   50.67   45.86   52.35

17.49* If you had done a sign test on some paired data and the test led to the rejection of the null hypothesis of equal medians, would there be any reason to do also a Wilcoxon signed rank test or a t test on the same data at the same significance level?

17.50 Test at the 5% significance level the hypothesis that the following data, which represent a sample of the spot market prices for titanium, follow a normal distribution:

102.44   94.74   144.11   139.05   111.92   137.78   128.32   125.74   106.32   110.34
94.79   116.23   110.25   139.34   147.42   112.45   78.50   111.36   97.49   138.73
128.39   159.45   122.47   129.07   139.29   116.79   116.18   114.16   119.59   104.92
98.11   130.31

18

Decision Theory

hen Procter & Gamble was developing its superabsorbent paper diaper, Ultra Pampers, the company was uncertain how consumers would react to the product. P&G did not know how much consumers would be willing to pay for a diaper that could prevent diaper rash and reduce the time spent in changing diapers. P&G also did not know if the health risks of toxic shock syndrome associated with its superabsorbent Rely tampons would be posed by the superabsorbent diapers as well. Even if no health problems were found, P&G was concerned that mothers would avoid the new diapers out of a misplaced fear that their babies would be at risk. P&G was the first company to market disposable diapers and the product was a huge success; annual sales for all disposable diapers topped $3 billion in 1987. Before 1987, however, P&G had been losing market share to Kimberly-Clark's Huggies brand diapers. Huggies' innovation was an elastic band to prevent leaks. P&G needed another innovation to regain its dominant position in the market.

Clearly the decision to market or not to market Ultra Pampers entailed tremendous payoffs or tremendous losses. The uncertainties surrounding this decision were reduced by laboratory product testing, expert advice, and consumer polls. P&G hired a panel of outside scientists to design and review laboratory tests of the new diapers. Some 10,000 babies tried the new diapers and their mothers were polled about their reactions. Each bit of information gleaned in this process was used to revise the estimated probability that the diaper would cause toxic shock syndrome and the estimated probabilities of various levels of sales. At this point P&G's choice to market Ultra Pampers seems to be a great success, and superabsorbent diapers may soon replace regular disposable diapers altogether. The generic disposables and Kimberly-Clark diapers, which were making inroads on P&G's market share, may take a beating.

The framework for incorporating additional information about probabilities when complex choices must be made is called statistical decision making. The Ultra Pampers example illustrates the issues addressed in statistical decision making. The examples introduced in this chapter will show how additional information can be incorporated in the decision-making process and how the optimal choice can be determined; but they will take into account fewer pieces of information than Procter & Gamble had when it decided its course of action in regard to Ultra Pampers. Nevertheless, by the end of the chapter you may agree that only an infantile mind could fail to see that statistical decision making, which can help businesses avoid rash choices, merits superabsorbent study.

18.1 INTRODUCTION TO DECISION MAKING

18.1A DECISION-MAKING CONCEPTS

Statistical decision making, or simply decision making, is *a formal framework for making choices when the possible results of the choices can be given numerical values and a probability can be assigned to the attainment of each result.* If you are an undergraduate, you may soon face the decision of working full-time after you receive your bachelor's degree or going to graduate school for an MBA. The possible results are the kinds of jobs to which both decisions will lead. The values assigned to those jobs could take the form of dollar values (such as the salary you would earn in each possible job) or be expressed in nonmonetary terms (such as the utility or pleasure afforded by each job). While assigning probabilities to all of the kinds of jobs you could have with an MBA degree or with a BA degree is difficult, you have some idea of the employment options open to holders of both degrees. Undoubtedly you have come up with some personal or subjective probabilities about your chances for various kinds of jobs given either degree.

Decision making is used to decide such business questions as how much inventory to hold, what size plant to build, what type of merchandise to carry, and how long a production run should be. In each of these situations some decision maker, such as a company's CEO, faces more than one choice, and an uncertain payoff is associated with each possible choice. There are several advantages to having a formal framework for problems of this sort. One advantage is that you can logically and objectively incorporate all of the information that you have in the decision-making process. A second advantage is that you can evaluate the potential benefit of buying more information about the probabilities of various payoffs and spend the optimal amount of resources in the decision-making process. A third advantage is that such a framework helps you make the best decision even when a problem is complex and the result of each decision is uncertain.

Before we get into the mechanics of decision making, the term "best" has to be clarified. The best choice is not necessarily the one with the highest dollar payoff. An MBA degree might provide you with the highest lifetime income, but it would not be your best choice if you hated the long work hours, frequent relocations, and constant pressure faced by some of the most highly paid MBAs. Economists describe people as utility maximizers rather than income maximizers. **Utility** is *the subjective satisfaction associated with any outcome.* Presumably business decision making is governed by the same utility maximization principle as all other decisions. Utility maximization may seem like a nebulous guide to decision making. Utility is subjective and difficult to measure. Advanced discussions of statistical decision making cited in the suggested readings at the end of this chapter describe how to incorporate the decision maker's utility in the analysis. For most business problems with uncertain outcomes, a simple assumption about the decision maker's utility function makes the choice with the highest expected money payoff the choice with the highest expected utility. This assumption is risk neutrality.

Recall that risk-neutral behavior was defined in Chapter 4 as a willingness to pay up to the expected value of a bet. The assumption of risk neutrality implies that the decision maker will regard the utility of $1 gained as equal to the disutility of $1 lost. A decision maker who is risk-neutral will try to maximize the expected profits of his or her firm. The assumption of risk neutrality simplifies the decision-making problem—all you have to consider is the money payoff associated with each decision rather than the value expressed in terms of utility. A decision maker who is risk-averse (who will pay only less than the expected value of a bet) would regard $1 gained as less valuable than $1 lost. A risk-averse decision maker would lean toward choices that were safer, even if they did not maximize the expected value of the firm's profits. A risk-seeking decision maker would lean toward decisions that might have big payoffs but would not maximize the firm's expected profits.

The three utility profiles in Figure 18.1 represent risk-neutral, risk-averse, and risk-seeking decision makers. Payoffs are measured in dollars; utility is measured in utiles, or units of pleasure. The numerical values of the utiles may be arbitrarily assigned on an interval or a ratio scale. For example, utiles could be numbered 1, 2, 3, 4, . . . or 50, 75, 100, 125. . . . As the numbering system for utiles is arbitrary, there is no way to compare one person's utility with another's. If I claim that I get 100 utiles from a profit of $1,000 a month and you claim that you get 5,000 utiles from the same profit, there is no basis for deciding if you are getting more utility than I. Each person provides his or her own standard of utility measurement.

Figure 18.1 shows that a given money gain and loss will be viewed differently by individuals who are risk-neutral, risk-averse, and risk-seeking. A risk-neutral person has a straight-line utility profile. Identical increases or decreases in profits will cause identical increases or decreases in utility. A risk-averse person will lose more utility for the drop in profits of a given dollar amount than he or she will gain in utility with the same dollar increase. The utility profile of a risk-averse individual is curved, with a slope that flattens as profits increase. Conversely, a risk-seeking individual will lose less utility for a drop in profits of a given amount than he or she will gain in utility with an increase of the same magnitude. The utility profile of a risk-seeking individual is curved with a slope that gets steeper as profits increase.

For most business decision makers and for most decisions, the assumption of risk neutrality is reasonably accurate. To see why, consider yourself making a small bet, say winning $1 if a coin comes up heads. You might be willing to pay up to 50 cents for the right to make this bet, perhaps more if you were a risk seeker. If you were offered a chance to flip a fair coin and win $100,000, you would be much less likely to offer the proportional amount, $50,000, for the right to make this bet. The reason is that you probably cannot afford to lose $50,000; you would be in debt for a long time. If a business decision is a

FIGURE 18.1
Utility profiles of risk-neutral, risk-averse, and risk-seeking decision makers

Suppose three individuals with the utility profiles shown here were offered the same bet. The bet has the expected value indicated. The bet would be equally likely to yield the amounts labeled "Lose" and "Win." The risk-neutral individual would be willing to pay up to the expected value for the right to make the bet. The risk-averse individual would not be willing to pay the expected value, and the risk-seeking individual would be willing to pay more than the expected value.

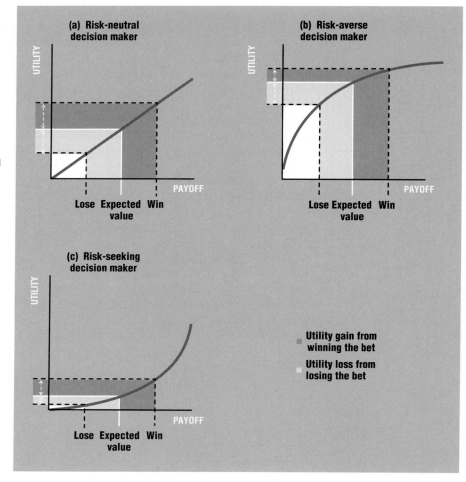

bet-the-family-farm decision—that is, the firm can go bankrupt if the results are unfavorable—a business decision maker may be risk-averse. Most business decisions involve potential losses or gains that are small in relation to the firm's current assets. These decisions are similar to your $1 coin-toss bet. Occasionally business people appear to be risk seekers even when the potential loss involved in a decision represents a large share of their assets. When the Hunt brothers of Texas tried to corner the world silver market in 1985, they gave the impression that the pleasure they would gain by winning $1 billion was much greater than the pain they would suffer if they lost $1 billion. In our examples we will use the assumption of risk neutrality and try to maximize our expected profits.

18.1B PAYOFF TABLES

Here is an example that illustrates the maximization of expected profits. An oil company is considering leasing an offshore drilling site. Whether the site contains oil is uncertain. The presence and absence of commercial quantities of oil are called "states of nature." This term, introduced in the section on Bayes' theorem in Chapter 4, refers to any unknown condition rather than a physical phenomenon in "nature." Rain tomorrow can be a state of nature in both physical and statistical senses; a company's announcement of a stock split in the next month can be a state of nature as well.

On the basis of geological studies commissioned by the oil company in our example, the probability that oil is present is 0.22. The asking price for the lease is $10 million. If the company pays this price and there is no oil at the site, the company will be out the lease price plus an additional $5 million for drilling. The company estimates that if it pays this price and strikes oil, the present value of the future stream of profits generated at the lease site will be $70 million. Table 18.1 lays out the choices facing the firm, the states of nature, the probabilities, and the possible payoffs. The expected value of the choice "Do not purchase lease" is obviously 0. The expected value of the choice "Purchase lease" can be found by a modification of the formula for expected value, introduced in Chapter 5:

$$E(X) = \sum_{i=1}^{n} x_i P(x_i) \qquad (5.1)$$

$$EP_i = \sum_{j=1}^{k} (x_i|s_j) P(s_j) \qquad (18.1)$$

EP_i is the expected payoff of the ith alternative. The k in formula 18.1 refers to the number of possible states of nature and the $x_i|s_j$ refers to the payoff x for choice i in the state of nature s_j. Note that the payoff of any choice given the

TABLE 18.1
Payoff table for oil-lease problem (millions of dollars)

Choice	STATE OF NATURE	
	Oil present (*probability* *0.22*)	Oil not present (*probability* *0.78*)
Purchase lease	55	(−15)
Do not purchase lease	0	0

state of nature is assumed to be fixed. To maximize the present value of future profits, the oil company must pick the choice with the highest expected payoff.

When we substitute the values in Table 18.1 in formula 18.1, we find the expected payoff of the choice "Purchase lease":

$$EP_1 = (-15)(0.78) + (55)(0.22),$$

$$= 0.4$$

and of the choice "Do not purchase lease":

$$EP_2 = (0)(0.78) + (0)(0.22)$$

$$= 0$$

In this case the expected value of the choice "Purchase lease" was greater than the expected value of the choice "Do not purchase lease"—$0.4 million versus $0, so the best choice is to purchase the lease. Under the assumption of risk neutrality, the best choice in this problem is the one that has an expected value above $0. Even if the expected value of the choice "Purchase lease" was only $1, it would still be the best choice, despite the risk of losing $15 million.

Our example could be made more realistic in a number of ways. There could be more than one potential lease site. There could be more than two states of nature; that is, the possible amount of oil at each site could have more than two discrete values or be a continuous random variable. The mathematics of statistical decision making is a good deal more complicated when the distribution of the states of nature is continuous than when it is discrete. Although many decision problems require the use of continuous distributions of the states of nature, we will stick to examples with discrete distributions. Another layer of uncertainty could be introduced by having sealed bids for the lease sites, so that the oil company would not know if its bid on any site would win. On the other hand, the company could reduce the uncertainty about the presence of oil at the site by purchasing more information, such as more extensive geological surveys. The decision-making process could be made sequential, with an initial decision for a small geological survey, a later decision for an extensive geological survey, and finally a decision for a bid on each site. Another complication is the possibility that the oil company is not risk-neutral.

Despite its simplicity, the oil-lease example involves three important concepts: the expected payoff given perfect information, the expected value of perfect information, and the minimization of opportunity cost. The **expected payoff given perfect information** is *the payoff expected if the states of nature are known in advance of any decision.* Perfect information in this case would be the knowledge that the site in our original problem did or did not contain oil. If the company knew there was no oil, it could avoid paying the lease price and avoid the drilling costs. If the company knew that oil was there, it could make a profit of $55 million. The expected payoff given perfect information *(EPGPI)* would equal

$$EPGPI = \sum_{j=1}^{k} \underset{i}{\text{Max}}(x_i|s_j)P(s_j)$$

(18.2)

EPGPI is the expected payoff given perfect information. The letter *s* refers to the state of nature, and there are up to *k* states of nature. The notation $\text{Max}(x_i|s_j)$ refers to the best payoff among the *n* possible payoffs given state s_j. If oil is present, the best payoff occurs if the oil company purchases the lease (55); if no oil is present, the best payoff results if the oil company does not purchase the lease (0). Substituting the numbers from the oil-lease example into formula 18.2 yields

$$EPGPI = (55)(0.22) + (0)(0.78)$$

$$EPGPI = 12.1$$

The $12.1 million is the amount that the oil company could expect to make if it could avoid any mistakes, that is, always bid if there is oil and never bid if there is no oil.

Suppose someone called the oil company's president and said that he had a secret report that revealed whether or not there was oil at the site; would this report be worth as much as $12.1 million to the oil company? No, because without the report the oil company has an expected profit of $0.4 million and after buying the report (but before reading it) the company has an expected profit of $12.1 million. The difference between these expected profits, $11.7 million, is the most the company would pay for the report. The $11.7 million is the **expected value of perfect information.** It is simply *the expected payoff given perfect information less the expected profits before any information is acquired.* Secret reports that tell the state of nature with certainty are rare. The point of these calculations is that $11.7 million provides an upper bound on what the oil company should pay for *any* information. In general, the expected value of perfect information is the most that a risk-neutral decision maker would pay for any information about the state of nature. We will discuss how much a decision maker would pay for information that is less than perfect later in the chapter.

The oil company wants to minimize its expected opportunity cost. **Oppor-tunity cost** is *the difference between the payoff of the best choice possible given a state of nature and the payoff of the choice actually made.* If the state of nature is a lack of

TABLE 18.2

Expected opportunity cost for oil-lease example

	STATE OF NATURE	
Choice	Oil present	Oil not present
Purchase lease	0	15
Do not purchase lease	55	0

oil at the site, the best decision would be not to purchase the lease. The actual choice was to go ahead and purchase the lease and drill for oil, so the opportunity cost is $15 million. If the actual state of nature is a deposit of oil at the lease site, there would be no opportunity cost because the firm's choice to purchase the lease and drill was the best given that state of nature. Table 18.2 shows the expected opportunity cost for each choice.

To minimize expected opportunity cost, the firm would take the choice with the lowest opportunity cost across all possible states of nature. Formula 18.3 is for the expected opportunity cost of a given choice, i:

$$EOL_i = \sum_{j=1}^{k} (L_i|s_j)P(s_j)$$

(18.3)

L stands for the opportunity cost and the notation $(L_i|s_j)$ refers to the opportunity cost for the ith choice given the jth state of nature. Substituting into formula 18.3 for the choice "Purchase lease" yields

$$EOL_1 = (0)(0.22) + (15)(0.78)$$

$$EOL_1 = 11.7$$

If the oil company had decided not to purchase the lease and drill, the expected opportunity cost would have been

$$EOL_2 = (55)(0.22) + (0)(0.78)$$

$$EOL_2 = 12.1$$

A comparison of the expected opportunity costs of the two possible choices indicates that going ahead with the purchase and drilling is the best choice. This choice has the lowest expected opportunity cost. You may have guessed that minimizing expected opportunity cost is equivalent to maximizing the expected payoff. The differences between the expected opportunity costs associated with the two alternatives is also equal to the difference between the

TABLE 18.3

Payoffs in dollars by number of trees ordered by student and number of trees demanded by customers

Number of trees ordered	NUMBER OF TREES DEMANDED							Expected payoff
	14	15	16	17	18	19	20	
14	10	10	10	10	10	10	10	$10.00
15	−1	25	25	25	25	25	25	21.29
16	−12	14	40	40	40	40	40	28.86
17	−23	3	29	55	55	55	55	32.71
18	−34	−8	18	44	70	70	70	32.86
19	−45	−19	7	33	59	85	85	29.29
20	−56	−30	−4	22	48	74	100	22.00

expected payoffs. It is just another way to look at the problem. The main advantage of looking at a problem in terms of opportunity cost is ease of interpretation. Opportunity cost is always positive or 0, whereas payoffs can be positive or negative. An opportunity cost table emphasizes the costs in forgone profits of any mistakes.

Here is another example that shows the advantage of viewing a decision problem in terms of opportunity cost. A student who wants to make some money over her Christmas break plans to sell live Christmas trees. She pays a gas station owner $200 to let her set up her trees on a corner of his lot and she sells the trees for $25 each. Each tree costs her $10, and if any trees are not sold by the day after Christmas, she has to pay $1 per tree to have them hauled to a dump. Her best estimate is that she is sure to sell at least 14 trees and that she cannot sell more than 20. In the range 14 to 20 each outcome is considered as likely as any other. We are assuming a small number of trees and equal probabilities to keep the problem simple. Realistic figures would measure the trees in truckloads. The numbers of trees that can be sold are the possible states of nature in this problem.

The payoffs for the various combinations of trees ordered and trees demanded were determined as follows. The payoff is the difference between the total revenue and the total cost. If the student ordered 14 trees and her customers demanded 14 trees, the total sales would be 14 × $25, or $350. The total cost would be 14 × $10 + $200 (the fixed cost), or $340. The payoff for the combination of 14 trees ordered and 14 trees demanded is $350 − $340, or $10.

The expected payoffs shown in Table 18.3 were obtained by substituting into formula 18.1:

$$EP_i = \sum_{j=1}^{k} (x_i|s_j)P(s_j) \qquad (18.1)$$

For the choice of ordering 14 trees,

$$EP_1 = (10)(1/7) + (10)(1/7) + \cdots + (10)(1/7)$$

$$EP_1 = 10.00$$

Because all of the seven states of nature are equally likely, the probability of each one is 1/7. If the student were risk-neutral, the best choice would be to order 18 trees, as this number has the highest expected payoff, $32.86.

We find the expected payoff given perfect information by substituting into formula 18.2:

$$EPGPI = \sum_{j=1}^{k} \text{Max}_i (x_i | s_j) P(s_j) \tag{18.2}$$

$$EPGPI = (10)(1/7) + (25)(1/7) + \cdots + (100)(1/7)$$

$$EPGPI = 55$$

The most the student should pay for any information about the demand for Christmas trees is the difference between the EPGPI and the expected payoff with only the prior information. The difference is $55.00 − $32.86, or $22.14. Except perhaps for Santa Claus, there may be no source of demand information in this situation. The maximum amount the student should pay for additional information was provided solely to review the concept.

Table 18.4 shows the opportunity cost for every combination of trees ordered by the student and demanded by customers. The choice with the opportunity cost of 0 is the best choice for each state of nature. The opportunity cost table emphasizes the cost of each mistake, either of ordering too few trees or of ordering too many. Table 18.4 shows that each tree that the student could sell but fails to order costs her a forgone profit of $15. Similarly, each tree that she orders but does not sell costs her $11—$1 for hauling and $10 to purchase the tree.

The expected opportunity cost for each state of nature can be found by substituting into formula 18.3. Again, the probability of each state of nature is 1/7. The choice with the lowest expected opportunity cost is 18 trees. The best choice is the same whether we focus on minimization of opportunity cost or maximization of expected payoff.

TABLE 18.4

Opportunity cost in dollars by number of trees ordered by student and number of trees demanded by customers

Number of trees ordered	NUMBER OF TREES DEMANDED							Expected opportunity cost
	14	15	16	17	18	19	20	
14	0	15	30	45	60	75	90	$45.00
15	11	0	15	30	45	60	75	33.71
16	22	11	0	15	30	45	60	26.14
17	33	22	11	0	15	30	45	22.29
18	44	33	22	11	0	15	30	22.14
19	55	44	33	22	11	0	15	25.71
20	66	55	44	33	22	11	0	33.00

EXERCISES FOR SECTION 18.1

18.1 For each situation below describe the possible states of nature and the payoffs, and indicate whether the distribution of payoffs across states of nature is discrete or continuous.

(a) The members of a local alumni club plan to travel by bus to attend the Rose Bowl game if its college wins the PAC-10. The buses have to be reserved in advance and the club is considering reserving one to five buses. Cancellation of a reservation for a bus costs $200. If a bus is filled, the club just breaks even on the money but earns considerable goodwill among its members.

(b) An automobile parts factory has an order for 120,000 seat-tilting mechanisms. After the machinery for a production run of these mechanisms is set up, the marginal cost goes down sharply as the length of the production run is increased. The firm is considering making more than 120,000 of the seat-tilting mechanisms in the hope that additional orders will come in. Beyond 120,000 units the marginal cost is $5.50 and the selling price (if there are additional orders) is $7.50.

(c) A manufacturer of washing machines is considering changing its warranty terms on new machines to make consumers aware of the improved reliability of the company's machines. The current warranty is in effect for one year. The time to failure for the machines follows an exponential distribution with a mean in the range of two to three years. An increase in the length of time a machine is covered by the new warranty will cause more consumers to buy the machine, but it will also increase the costs of honoring the warranty.

18.2 For each situation below describe the possible states of nature and the payoffs, and indicate whether the distribution of payoffs across states of nature is discrete or continuous.

(a) A local theater company is mounting a production of *Nicholas Nickleby*. The company rents a hall on a per performance basis and rents its scenery for a fixed fee for the run of the play. The actors are paid for their rehearsal time as well as for their performance. Before the show begins, the company has to decide the number of nights that *Nickleby* will run.

(b) A drug company wants to test the effect of variants of a new antibacterial drug on cultures of a drug-resistant strain of bacteria. Each variant is estimated to have a 0.0001 chance of killing the bacteria and 200 variants of the drug are known. Each test costs $500 and a successful drug would be worth $5 million. The drug company has to decide how many variants to test and it has to commit itself to a run of tests rather than decide sequentially.

(c) The operators of a nuclear power plant have to decide on the number of independent backup power supplies that should be installed in the plant's control room. If the control room's power is lost, the utility could lose $10 billion. Each power supply system costs $1 million and has a mean time to failure of 50 years. The operating life of the nuclear plant is 30 years.

18.3 Construct a payoff table for two choices, labeled *A* and *B*, and two states of nature, labeled *c* and *d*. The payoffs for *A* given *c*, *A* given *d*, *B* given *c*, and

B given *d* are 50, 10, −20, and 120, respectively. The probability of state *c* is 0.7.

(a) Calculate *EP, EPGPI,* and *EOL.*

(b) What is the best choice for a risk-neutral decision maker?

18.4 Construct an opportunity cost table for three choices, labeled *A, B,* and *C,* and two states of nature, labeled *d* and *e*. The payoffs for *A* given *d, A* given *e, B* given *d, B* given *e, C* given *d,* and *C* given *e* are −22, 30, −20, 44, −16, and 35, respectively. The probability of state *d* is 0.4.

(a) Calculate *EP, EPGPI,* and *EOL.*

(b) What is the best choice for a risk-neutral decision maker?

18.5 A company that runs a ferry boat for cars is considering replacing its present boat, which has a capacity of 10 cars. The company could retain its present boat or purchase a larger boat with a capacity of either 15 or 20 cars. The price that the company can charge is regulated and the ferry estimates that the future demand for its service at this price can be described by a discrete random variable that has equal probabilities over the range 8 to 22 cars per hour. If the profit per car carried was $1 and the loss due to overcapacity was 50 cents per car of overcapacity, what is the best choice among the available boats?

18.6 A company that markets frozen dinners is planning to fund a research and development project for a new low-cost package that can go from the freezer to the microwave oven to the table. The company estimates that if it spends $5 million on the project, it has a 0.5 probability of finding a usable package, and that if it spends $10 million, it has a 0.75 probability of finding a usable package. The company also estimates that the probability that another company will find a usable package before it does is 0.4. The outcomes of each company's research efforts are considered independent. If the frozen food company was risk-neutral and the payoff for being first with the new package was $40 million, should the company spend $5 million, $10 million, or nothing on research?

18.7 Suppose you were offered a choice of three bets, each at the price of $1: win $2 with a probability of 0.5; win $10 with a probability of 0.08; win $1.10 with a probability of 0.99.

(a) Which bet would you take if you were strongly risk-seeking?

(b) Which bet would you take if you were strongly risk-averse?

(c) Explain why you would not choose the bet with a $2 payoff even if you were risk-neutral.

18.8 Suppose you were offered your choice of three bets: (1) flip a fair coin and collect $1 if the coin comes up heads, lose $1 if the coin comes up tails; (2) flip a fair coin and collect $2 if the coin comes up heads, lose 50 cents if the coin comes up tails; (3) flip a fair coin and collect $10 if the coin comes up heads, lose $5 if the coin comes up tails. Explain why one of these bets would not be chosen whether you were risk-neutral, risk-averse, or risk-seeking. Do not use the argument that you have a moral objection to gambling.

18.2 DECISION MAKING WITH BAYES' THEOREM

18.2A DECISION TREES

The payoff tables in section 18.1 are inadequate for complicated decision problems. Decisions about purchasing information or decisions with sequential choices require a more powerful analytical tool than a payoff table. This tool is a decision tree. Probability trees were used in Chapter 5 to describe the probabilities of the possible outcomes of tossing a fair coin three times. A decision tree is similar to a probability tree, but some of the nodes or branching points in a decision tree are sets of choices while others are sets of states of nature. In these state-of-nature nodes the decision maker has no choice—the state of nature that exists is unknown and can only be assigned probabilities. In a probability tree all of the nodes are stochastic or random; they are similar to the state-of-nature nodes in a decision tree.

Say the Cartier jewelry company has a four-carat uncut diamond that is worth $100,000 if it is faceted as one large stone or $250,000 if it is successfully cut up into smaller stones. Cartier's chief diamond cutter estimates that there is a 1-in-5 chance that the diamond will shatter when it is cut. If it shatters, the pieces will be worth only $10,000. Figure 18.2 shows the decision tree for this problem.

In a decision tree the square nodes represent decisions and the circular nodes represent states of nature. If Cartier decided not to cut the diamond, it would get $100,000 with certainty. If Cartier decided to cut the diamond, the expected payoff would be $202,000. Substituting into formula 18.1 yields

FIGURE 18.2
Decision tree for Cartier diamond problem

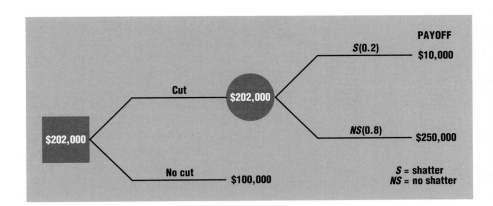

$EP = 250,000(0.8) + 10,000(0.2)$

$EP = 202,000$

The figure \$202,000 appears in the circle representing the state of nature to show the expected value at that point. If Cartier was risk-neutral, it would go ahead and cut the diamond. The same figure appears in the square representing the choice of cutting or not cutting the diamond because \$202,000 is the expected value of Cartier's best choice.

■ 18.2B PRIOR, POSTERIOR, AND PREPOSTERIOR ANALYSIS

So far, everything in this diamond example could have been done with a simple payoff table. Now we add a step that does not fit in a payoff table. For a \$5,000 fee Cartier can send the diamond to a laboratory that uses a scanning electron microscope to study the structure of the diamond and predict whether it will shatter when it is cut. According to the laboratory, the conditional probability of getting a prediction that the diamond will not shatter when it is cut given that it will indeed not shatter is 0.99. Also, the conditional probability of getting a prediction that the diamond will shatter when it is cut given that it will shatter is 0.95. Should Cartier pay the \$5,000 for the laboratory's prediction? If the answer is no, should Cartier go ahead and cut the diamond? If the answer to the question about paying \$5,000 is yes, what should Cartier do after a prediction that the diamond will shatter or after a prediction that the diamond will not shatter? Figure 18.3 shows the decision tree that includes these additional choices. The first node is a choice node for Cartier—whether to pay for the prediction. The second node in the lower main branch is also a choice node—whether to cut the diamond. As you can see, the lower main branch of this decision tree is the same as in Figure 18.2. The expected payoff for not cutting the diamond is \$100,000 and the expected payoff for cutting the diamond is \$202,000. Again, the choice with the highest expected value in the lower main branch is to go ahead with cutting the diamond.

Cartier must first determine if it is worthwhile to pay \$5,000 for the laboratory's prediction. We'll assume that the prior probabilities of Cartier's head cutter are also the probabilities that the laboratory will predict that the diamond will shatter or not shatter; that is, there is a 1-in-5 chance that the laboratory will predict that the diamond will shatter. The second node of the upper main branch of the decision tree is a chance or state-of-nature node. Cartier has no control over the prediction of the laboratory. The two branches labeled "Prediction = S [Shatter]" and "Prediction = NS [Not shatter]" have probabilities of 0.2 and 0.8. After this chance node, Cartier has a decision node. After either prediction it can choose to cut or not to cut the diamond. The terminal values in the upper main branch of the tree—\$5,000, \$245,000, and \$95,000—are the payoffs for a shattered, successfully cut and an uncut diamond net of the lab's fee. The revised probability that the diamond would

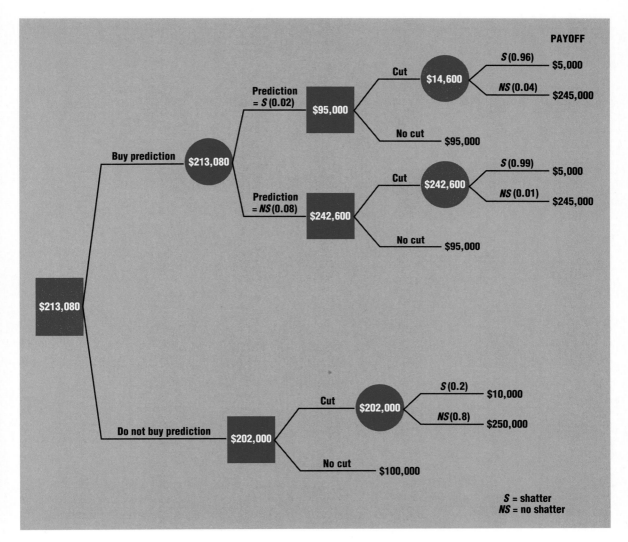

PAYOFF

FIGURE 18.3
Decision tree for Cartier diamond problem with choice of buying additional information

shatter given the information that the laboratory predicted that it would not shatter can be found with Bayes' theorem:

$$P(H_1|E) = \frac{P(H_1)P(E|H_1)}{\sum\limits_{i=1}^{n} P(H_i)P(E|H_i)}$$ (4.5)

Revised probabilities are also called posterior probabilities. To refresh your memory about these terms, you might review section 4.5, on Bayes' theorem, in Chapter 4. We'll use S for shattering and s for the prediction that the diamond will shatter. Similarly, NS and ns stand for not shattering and the prediction that the diamond will not shatter. Substituting into formula 4.5 yields

$$P(S|s) = \frac{P(S)P(s|S)}{P(S)P(s|S) + P(NS)P(s|NS)}$$

$$P(S|s) = \frac{(0.2)(0.95)}{(0.2)(0.95) + (0.8)(0.01)}$$

$$P(S|s) = 0.96$$

$$P(S|ns) = \frac{P(S)P(ns|S)}{P(S)P(ns|S) + P(NS)P(ns|NS)}$$

$$P(S|ns) = \frac{(0.2)(0.05)}{(0.2)(0.05) + (0.8)(0.99)}$$

$$P(S|ns) = 0.01$$

$$P(NS|s) = \frac{P(NS)P(s|NS)}{P(NS)P(s|NS) + P(S)P(s|S)}$$

$$P(NS|s) = \frac{(0.8)(0.01)}{(0.8)(0.01) + (0.2)(0.95)}$$

$$P(NS|s) = 0.04$$

$$P(NS|ns) = \frac{P(NS)P(ns|NS)}{P(NS)P(ns|NS) + P(S)P(ns|S)}$$

$$P(NS|ns) = \frac{(0.8)(0.99)}{(0.8)(0.99) + (0.2)(0.05)}$$

$$P(NS|ns) = 0.99$$

The revised probability that the diamond will shatter given the laboratory's prediction that the diamond will shatter is 0.96. The revised probability that the diamond will not shatter given the laboratory's prediction that it will shatter is 0.04. The revised probability that the diamond will shatter given the laboratory's prediction that it will not shatter is 0.01. And the revised probability that the diamond will not shatter given the laboratory's prediction that it will not shatter is 0.99. These revised probabilities can be used to calculate the expected value after either chance node. For example, the expected value of cutting the diamond after the laboratory's prediction that it will shatter is

$$EP_i = \sum_{j=1}^{k} (x_i|s_j)P(s_j) \tag{18.1}$$

$$EP_1 = (5{,}000)(0.96) + (245{,}000)(0.04)$$

$$EP_1 = 14{,}600$$

The expected payoff of this choice is $14,600. Since $95,000 is available with certainty if the diamond is not cut, Cartier will choose not to cut the diamond after the laboratory's prediction that it will shatter (recall that Cartier will have to pay $5,000 for the prediction). The choice node after the prediction that the diamond will shatter is labeled with a payoff of $95,000. If the laboratory predicts that the diamond will not shatter, the expected payoff is $242,600:

$$EP_2 = (5,000)(0.01) + (245,000)(0.99)$$

$$EP_2 = 242,600$$

Thus, if the laboratory predicts that the diamond will not shatter, Cartier will have the diamond cut. The choice node after the prediction that the diamond will not shatter is labeled with an expected payoff of $242,600. The next step is to calculate the expected payoff of the decision to purchase the laboratory's prediction:

$$EP_3 = (95,000)(0.2) + (242,600)(0.8)$$

$$EP_3 = 213,080$$

This expected payoff is $213,080. Since the expected payoff of the choice to purchase the laboratory's prediction exceeds the expected payoff of the choice of not purchasing the laboratory's prediction, Cartier will purchase the prediction.

This entire analysis with Bayes' theorem—*the calculation of the expected payoff of each choice before any information about the probability of each state of nature is received*—is called **preposterior analysis.** Because of all of the steps required, you may be thinking that a better name is preposterous analysis. All the effort can be justified by the money saved. The expected payoffs are based on *preposterior* probabilities because they are being analyzed before the probabilities that the diamond will and will not shatter have been predicted. Once the prediction is made, the expected payoff is either $95,000 or $242,600. After the laboratory makes a prediction, the expected payoff is called the posterior expected payoff and the analysis is called **posterior analysis,** which is defined as *the calculation of the expected payoff of each choice after the available information about the probability of each state of nature has been received.* On a preposterior basis the payoff is $213,080.

Here is another example that illustrates the use of decision trees and preposterior-posterior analysis. A farmer has to decide which crop to plant on a 500-acre field. His choices are to plant corn, to plant soybeans, or to let the field lie fallow. Letting the field lie fallow will give the farmer a profit of $30 per acre because of a government program (called PIK) that pays farmers to reduce planted acreage. The profits per acre of the other choices depend on the yields, which in turn depend on rainfall. We'll assume the price of each crop is fixed at the government support level. While rainfall is a continuous random

BIDDING FOR THE S.S. *KUNIANG*

The S.S. *Kuniang*, a British-built bulk cargo ship, ran aground off the coast of Florida in 1981. For insurance purposes the ship's owners declared the ship a total loss. The insurance company planned to hold an auction for the salvage rights to the ship. At the same time New England Electric Systems (NEES), a public utility company, was looking for a ship to carry coal to its generating plants. NEES was considering bidding for the salvage rights to the S.S. *Kuniang*, repairing the ship, and then using it primarily to carry coal. The problem was complex because of some legal restrictions on foreign-built ships. The Jones Act gives priority in loading to U.S.-built ships traveling between U.S. ports. A foreign-built ship may have to wait so long to load coal that it is uneconomical to operate. However, there is a second law that applies to salvaged vessels. A salvaged ship will be treated as American-built if its original owners have declared it a total loss (as the owners of the S.S. *Kuniang* did) and if the cost of repairs is at least three times the salvage value. Unfortunately, the salvage value, as determined by the Coast Guard, would not be known before the bid was due. If the Coast Guard set the salvage value of the S.S. *Kuniang* high enough, the repair cost would not be enough to permit the ship to be treated as American-built. There was little precedent for the Coast Guard setting the salvage value of a ship. NEES hoped that the salvage value would be set low, at the scrap value of the ship. Another possibility was that the Coast Guard would take the winning bid as the ship's salvage value.

Professor David Bell ("Bidding for the S.S. *Kuniang*," *Interfaces* 14 [March–April 1984]: 17–23) has described how a decision tree was used to analyze the following choices: (1) bidding for the *Kuniang*, (2) purchasing a new ship, (3) purchasing a tug and barge. Another complication was that if NEES bid and won, it could convert the S.S. *Kuniang* to a self-unloading ship. Such a conversion would speed up unloading at some cost in cargo capacity. If NEES needed to raise the repair costs of the ship to meet the requirement that repair costs must be at least three times the ship's salvage value, it could include the costs of the self-loader conversion as part of the repair costs. Figure 18.4 shows a portion of the decision tree for this problem. The number at each end point represents net present value. Net present value is the future profit assigned to each end point discounted to the date of the initial decision. These net present values are based on assumed values of future interest rates; the costs of operating each ship; the charges for shipping coal on ships not owned by NEES; the service lives of the *Kuniang*, the new ship, and the tug and barge; and the speed of each ship. The key factor determining the net present value of the end points that follow the chance branch "Win bid" is the amount bid for the ship. Raising the bid increases the chance of winning but lowers the present value of a successful bid. The numbers in the decision tree are based on a bid of $7 million.

Several points can be made about the decision tree. If after it won the auction the Coast Guard decided that the salvage value of the *Kuniang* was the bid price instead of the scrap value, NEES's best decision would be to spend an extra $6 million on the repairs to meet the Jones Act requirements. The only purpose of this added investment would be to meet the legal requirement. Another point is that the choice of a new ship dominates the tug-and-barge choice.

Professor Bell treated NEES as risk-neutral and tried to determine the bid that would give the company the highest expected payoff. The values in the decision tree were recalculated for a range of possible bids. He found that the optimal bid was in the range of $6 to $7 million. NEES went ahead with a bid of $6.7 million. The probabilities used in the calculation of the expected values were subjective. NEES had to guess the Coast Guard's behavior and the values of possible competing bids. The end result was that another bid of $10 million won the auction and that the Coast Guard used the scrap value of the *Kuniang* as the declared salvage value. NEES had assigned a value of only 0.3 to the probability that the Coast Guard would use the scrap value. If it had known this outcome ahead of time, its optimal bid would have been above the $10 million winning bid.

Professor Bell ended his article as follows: "Decision analysis is a remarkably effective tool. It provides a structure for complex problems, lending insight into the most relevant information to collect and into the rationale that explains the eventual solution. To be successful in using decision analysis one should have a firm grasp of elementary business mathematics, the ability to use a calculator, and a strong conviction that decision analysis can help. This article was written to convince the many people who have learned how to do decision analysis that it really does help."

FIGURE 18.4
Decision tree for New England Electric's bid on the S.S. *Kuniang*

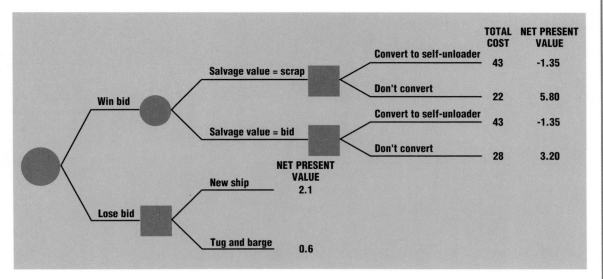

FIGURE 18.5
Decision tree for planting corn, planting soybeans, or collecting PIK payments

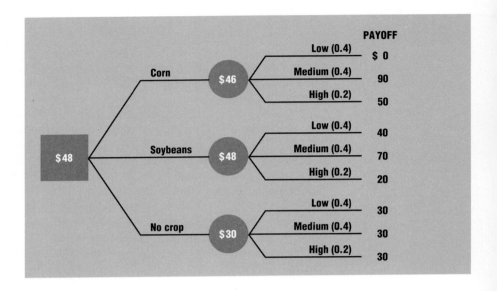

variable, we can simplify the problem by characterizing the rainfall as low, medium, and high. The estimated probabilities of these three states of nature are 0.4, 0.4, and 0.2. These probabilities are based on historical records of rainfall during the growing season. These probabilities are called prior probabilities because they are formed in advance of the current growing season and they are the starting point for estimating the probabilities of various levels of rainfall. The prior probabilities may be revised as more information about possible rainfall is gained.

The profits per acre of corn under the three conditions of rainfall are $0, $90, and $50. The profits per acre of soybeans under the three conditions of rainfall are $40, $70, and $20. Figure 18.5 shows the decision tree based on these numbers. The farmer's choices of corn, soybeans, and no crop branch out of a decision node. Each of the three state-of-nature nodes has branches labeled "Low," "Medium," and "High" to indicate the possible conditions of rainfall. Next to each label is the probability of each state of nature in parentheses. The expected payoff for each choice is shown inside the state-of-nature node at the end of the choice branch. These expected payoffs were found by formula 18.1:

$$EP_i = \sum_{j=1}^{k} (x_i|s_j)P(s_j) \qquad (18.1)$$

For the choice of planting corn:

$$EP_1 = (0)(0.4) + (90)(0.4) + (50)(0.2)$$
$$EP_1 = 46$$

For the choice of planting soybeans:

$$EP_2 = (40)(0.4) + (70)(0.4) + (20)(0.2)$$

$$EP_2 = 48$$

For the choice of no crop:

$$EP_3 = (30)(0.4) + (30)(0.4) + (30)(0.2)$$

$$EP_3 = 30$$

The choice of soybeans has the highest expected payoff, so it would be the best choice for a risk–neutral farmer. A strongly risk–averse farmer might take the choice "No crop" because each of the other choices has a chance of yielding a lower profit per acre.

The expected value given perfect information about rainfall is a profit of $62 per acre. We find this number by substituting into formula 18.2:

$$EPGPI = \sum_{j=1}^{k} \operatorname*{Max}_{i}(x_i|s_j)P(s_j) \tag{18.2}$$

$$EPGPI = (40)(0.4) + (90)(0.4) + (50)(0.2)$$

$$EPGPI = 62$$

On a per acre basis, the most that the farmer should pay for any information about rainfall is $62 − $48, or $14. This works out to $7,000 for the 500 acres the farmer has available.

Now we will add a second choice that brings a Bayesian aspect to the problem. The farmer is considering subscribing to a long–range weather service that charges $1,000 for its predictions. The service will predict whether the rainfall will be low, medium, or high. The service's track record is such that given either low or medium rainfall, it has a 0.9 probability of making the correct prediction and that it never predicted high rainfall when the actual rainfall was low or medium. Table 18.5 shows the joint probability distribution of the actual level of rainfall and the level predicted by the service.

TABLE 18.5
Joint probability distribution of actual and predicted rainfall

		PREDICTION			
		Low	Medium	High	
ACTUAL RAINFALL	Low	0.36	0.04	0	0.4
	Medium	0.04	0.36	0	0.4
	High	0	0	0.2	0.2
		0.4	0.4	0.2	

Now the farmer faces two choices: to subscribe or not to subscribe to the long-range forecast and to plant corn, soybeans, or nothing. Clearly, if the farmer decides to subscribe to the service, the choice of crop will depend on the prediction made by the service. Figure 18.6 shows the decision tree for this

FIGURE 18.6
Decision tree for subscribing to long-range weather service and for choice of crop

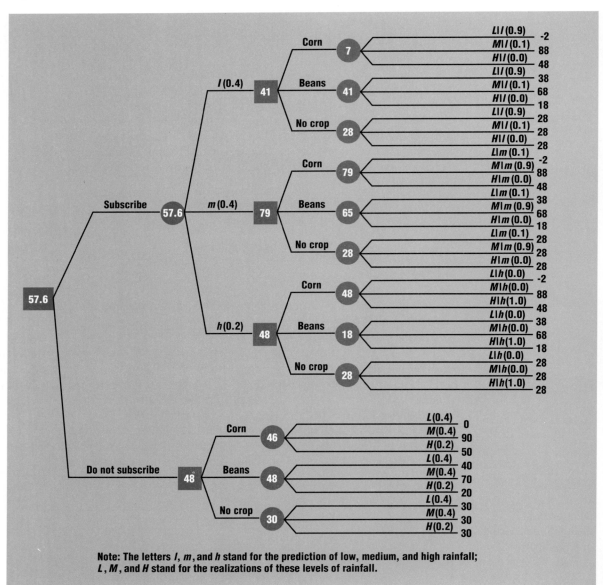

Note: The letters *l*, *m*, and *h* stand for the prediction of low, medium, and high rainfall; *L*, *M*, and *H* stand for the realizations of these levels of rainfall.

two-stage problem. The first choice in the decision tree is whether or not to subscribe to the service. The lower main branch of the decision tree is exactly the same as the decision tree for the original problem (Figure 18.5). If the farmer chooses not to subscribe to the service, the problem is the same as before. The upper main branch of the decision tree in Figure 18.6 shows the sequence once the decision to subscribe to the service has been made. This branch terminates in a state–of–nature node because the farmer has no control over which level of rainfall the service will predict. The three possible states of nature for this node are predictions of low, medium, and high rainfall. You can see in Table 18.5 that the probabilities of the three predictions are 0.4, 0.4, and 0.2, respectively. It is not necessary that the probabilities of these three predictions match the prior probabilities of the three levels of rainfall, but a good prediction service would not be too far off from the priors. Otherwise you would suspect that it systematically underestimates at least one of the levels.

After the state–of–nature node for the service's prediction there is a decision node for the farmer. Following each prediction the farmer has the choice of planting corn, soybeans, or nothing. Following these decision nodes are the last state–of–nature nodes for low, medium, and high levels of actual rainfall. With three possible predictions, three possible choices for planting, and three possible levels of actual rainfall, there are $3 \times 3 \times 3$ or 27 possible payoffs for the upper main branch of the decision tree. The $1,000 cost of the long-range weather service works out to $2 per acre over the farmer's 500 acres. Thus the payoffs in the upper main branch are $2 less than the corresponding payoffs in the lower main branch. For example, the payoff per acre of corn without a subscription to the weather service in conditions of low rainfall is $0. The payoff per acre of corn with a subscription to the weather service in conditions of low rainfall is $-2.

Once the service has made a prediction, the probabilities of the three levels of rainfall are no longer 0.4, 0.4, and 0.2. The farmer now has two pieces of information: the prior probability of each level of rainfall based on historical records and the conditional probability of each prediction. The revised estimate of the probability of each level of rainfall may be found with Bayes' theorem:

$$P(H_1|E) = \frac{P(H_1)P(E|H_1)}{\sum_{i=1}^{n} P(H_i)P(E|H_i)} \tag{4.5}$$

To simplify the notation before you substitute into formula 4.5, let the actual level of rainfall be L, M, and H for low, medium, and high and let the predicted levels of rainfall according to the weather service be l, m, and h. With three possible levels of rainfall and three possible predictions, there are nine possible revised probabilities: $P(L|l)$, $P(M|l)$, $P(H|l)$, $P(L|m)$, $P(M|m)$, $P(H|m)$,

$P(L|h)$, $P(M|h)$, $P(H|h)$. Substituting into formula 4.5 (only the first three substitutions are shown) yields

$$P(L|l) = \frac{P(L)P(l|L)}{P(L)P(l|L) + P(M)P(l|M) + P(H)P(l|H)}$$

$$P(L|l) = \frac{(0.4)(0.9)}{(0.4)(0.9) + (0.4)(0.1) + (0.2)(0.0)}$$

$$P(L|l) = 0.9$$

$$P(L|m) = \frac{P(L)P(m|L)}{P(L)P(m|L) + P(M)P(m|M) + P(H)P(m|H)}$$

$$P(L|m) = \frac{(0.4)(0.1)}{(0.4)(0.1) + (0.4)(0.9) + (0.2)(0.0)}$$

$$P(L|m) = 0.1$$

$$P(L|h) = \frac{P(L)P(h|L)}{P(L)P(h|L) + P(M)P(h|M) + P(H)P(h|H)}$$

$$P(L|h) = \frac{(0.4)(0.0)}{(0.4)(0.0) + (0.4)(0.0) + (0.2)(1.0)}$$

$$P(L|h) = 0.0$$

The revised probabilities are $P(L|l) = 0.9$, $P(M|l) = 0.1$, $P(H|l) = 0.0$, $P(L|m) = 0.1$, $P(M|m) = 0.9$, $P(H|m) = 0.0$, $P(L|h) = 0.0$, $P(M|h) = 0.0$, $P(H|h) = 1.0$. These revised probabilities are listed in the decision tree in parentheses next to the levels of actual rainfall in the 27 final branches of the upper half of the decision tree. These revised probabilities are used to calculate the expected payoff of each choice in regard to planting given the service's predictions. The expected payoffs are listed in the state-of-nature nodes for the levels of rainfall in Figure 18.6. Only the choice of corn after a prediction of low rainfall is shown. Substituting into formula 18.1 for the expected payoff yields

$$EP_i = \sum_{j=1}^{k} (x_i|s_j)P(s_j) \tag{18.1}$$

$$EP_4 = (-2)(0.9) + (88)(0.1) + (48)(0.0)$$

$$EP_4 = 7$$

Given the expected payoffs of $7, $41, and $28 for the choices of corn, soybeans, and no crop after a prediction of low rainfall, the farmer would plant soybeans, the choice with the highest expected payoff. Thus the choice node after the prediction of low rainfall has an expected payoff of $41. Similarly, given the expected payoffs of $79, $65, and $28 for the choices of corn, soybeans, and no crop after a prediction of medium rainfall, the farmer would plant corn. Thus the choice node after the prediction of medium rainfall has an expected payoff of $79. Given the expected payoffs of $48, $18, and $28 for the choices of corn, soybeans, and no crop after a prediction of high rainfall, the farmer would plant corn. Thus the choice node after the prediction of high rainfall has an expected payoff per acre of $48.

These expected payoffs given the possible rainfall predictions can be evaluated at the probabilities of each prediction, that is, 0.4 for low, 0.4 for medium, and 0.2 for high. Thus the expected payoff for the main branch of subscribing to the prediction service is

$$EP_i = \sum_{j=1}^{k} (x_i|s_j)P(s_j) \qquad \text{(18.1)}$$

$$EP_5 = (41)(0.4) + (79)(0.4) + (48)(0.2)$$

$$EP_5 = 57.6$$

The expected payoff per acre of subscribing to the weather service is $57.60, while the expected payoff without the subscription is $48.00. The net benefit of the subscription across the 500 acres is ($57.60 − $48)500, or $4,800. The $4,800 is the net benefit because the $1,000 cost of the weather service is included in the expected payoff. The farmer's best choice is to subscribe to the service.

EXERCISES FOR SECTION 18.2

18.9 You are faced with a choice of either A or B. If event x occurs, choice A has a payoff of $100 and choice B has a payoff of $20. If event y occurs, choice A has a payoff of $10 and choice B has a payoff of $200. If event z occurs, choice A has a payoff of $−40 and choice B has a payoff of $70. Events x, y, and z have the probabilities of 0.8, 0.1, and 0.1, respectively. Construct a decision tree for this problem and decide whether A or B is the best choice. What is the expected payoff of this best choice?

18.10 You are faced with an initial choice of either A or B. After you make this choice, outcome x or y will occur, with probabilities of 0.6 and 0.4, respectively. The payoff of choice B is $4,200 if event x occurs or $−1,229 if event y occurs. If you choose A, you have a second choice of either C or D after the chance event of either x or y has occurred. C has a payoff of $3,200 after the event x or a payoff of $246 after the event y. D has a payoff of $1,240 after the event x or $−150 after the event y. Construct a decision tree for this problem and decide whether A or B is the best initial choice. What is the expected payoff of this best choice?

18.11 Use the information in the following decision tree to determine the expected payoff at each choice node and the best choice.

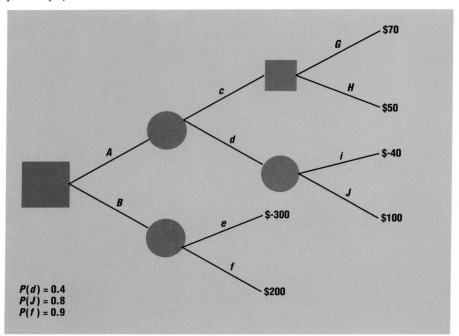

$P(d) = 0.4$
$P(J) = 0.8$
$P(f) = 0.9$

18.12 Use the information in the following decision tree to determine the expected payoff at each choice node and the best choice.

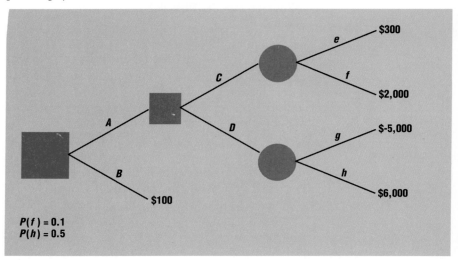

$P(f) = 0.1$
$P(h) = 0.5$

18.13 You are considering betting $100 on a horse named Bayes. If the horse wins, you will get your $100 back and an additional $900. If it loses, you will be out the $100. Your subjective probability of the horse's winning is 0.2. You could buy a tip from Mr. Oracle before making the bet. Given that a horse will lose, the probability that Mr. Oracle will predict that it will win is 0.05. Given that a horse will win, the probability that Mr. Oracle will predict a win is 0.5. The probabilities of Mr. Oracle's prediction that Bayes will win or lose are 0.1 and 0.9, respectively. Mr. Oracle's prediction costs $100. Construct a tree diagram for this problem and decide on the best course of action. What is the expected payoff of this best choice?

 SUMMARY

Statistical decision making allows you to handle complex problems in a systematic manner. You can use statistical decision making to decide if it is worthwhile to acquire more information or to determine what information would be the most worthwhile to acquire. It is a highly practical application of many of the tools covered in earlier chapters. For example, statistical decision making uses the concept of expected value, introduced in Chapter 5; Bayesian probability, introduced in Chapter 4; and sampling theory, introduced in Chapter 7. This chapter covered the most elementary techniques and concepts of decision making. The main concepts are expected payoff, expected opportunity cost, and the expected value of perfect information. The main tool used in the chapter is the decision tree. More advanced treatments discuss how to measure the decision maker's risk aversion and incorporate this factor into the analysis. More advanced treatments also discuss decision making with multiple objectives and continuous states of nature. Even the barebones treatment in this chapter, however, should give you a good idea of the power of decision analysis. Some typical business applications of decision analysis are choosing the optimal bid, setting the optimal production run, picking the optimal quantity to order when a supplier gives discounts for large orders, picking optimal inventory when demand for a product is uncertain, deciding on the optimal capacity for a new plant, and determining the optimal strategy in contract negotiations. Some of these applications will be illustrated in the exercises at the end of the chapter.

 SUGGESTED READINGS

Howard, Ronald A., James E. Matheson, and Katherine L. Miller, eds. *Readings in Decision Analysis*. 2d ed. Menlo Park, Calif.: Decision Analysis Research Group, Stanford Research Institute, 1976.

Schlaifer, R. *Analysis of Decisions under Uncertainty*. New York: McGraw-Hill, 1969.

Ulvila, Jacob W., and Rex V. Brown. "Decision Analysis Comes of Age." *Harvard Business Review* 60 (September–October 1982): 130–141.

NEW TERMS

expected payoff given perfect information The payoff expected if the states of nature are known in advance of any decision.

expected value of perfect information The expected payoff given perfect information less the expected profits before any information is acquired.

opportunity cost The difference between the payoff of the best choice possible given a state of nature and the payoff of the choice actually made.

posterior analysis The calculation of the expected payoff of each choice after the available information about the probability of each state of nature has been received.

preposterior analysis The calculation of the expected payoff of each choice before any information about the probability of each state of nature is received.

statistical decision making A formal framework for making choices when the possible results of the choices can be given numerical values and a probability can be assigned to the attainment of each result.

utility The subjective satisfaction associated with any outcome.

NEW FORMULAS

Expected payoff of the *i*th choice

$$EP_i = \sum_{j=1}^{k} (x_i|s_j)P(s_j) \tag{18.1}$$

Expected payoff given perfect information

$$EPGPI = \sum_{j=1}^{k} \operatorname*{Max}_{i}(x_i|s_j)P(s_j) \tag{18.2}$$

Expected opportunity cost of the *i*th choice

$$EOL_i = \sum_{j=1}^{k} (L_i|s_j)P(s_j) \tag{18.3}$$

ADDITIONAL EXERCISES

An asterisk designates an exercise of greater than ordinary difficulty.

18.14 The Environmental Science Service Administration (ESSA) has experimented with seeding hurricanes with silver iodide to increase rainfall and slow the speed of wind. The results are encouraging: most of the time the wind speeds of the hurricanes are substantially reduced. A few hurricanes, however, had increased wind speeds after the seeding. The probabilities of substantially reduced, unchanging, and increased wind speeds are 0.5, 0.4, and 0.1, respectively. A substantial reduction in wind reduces by $50 million the property damage caused by the average hurricane that makes a landfall. An increase in wind speed would add an expected $100 million to the damage. The cost of seeding hurricanes is $1 million. Construct a decision tree for this problem and determine the optimal choice.

18.15 Construct an expected opportunity cost table for exercise 18.14.

18.16 What is the expected value of perfect information in exercise 18.14?

18.17 Your company has ten sales representatives. You are considering a company motor pool for their use. The number of representatives who would need a car is a discrete random variable with equal probabilities in the range of 0 through 10. Each car in your motor pool costs you $10 a day whether or not it is used. Renting a car from an outside agency costs $30 a day. What is the optimal number of cars for your motor pool and what is the expected payoff for this number of cars relative to having no motor pool?

18.18 The Ford Motor Company is considering manufacturing the tires used on its new cars. It currently buys these tires at an average price of $25. If it ran its own tire factory at full capacity, the cost per tire could be reduced to $20. If the sales of Ford cars are not sufficient to keep the tire factory running full-time, the cost per Ford-made tire could go well above the current price of $25. Ford's probabilities for various number of tires per year and cost per tire are listed below.

Millions of tires	Probability	Cost per tire
4	0.2	$40
5	0.2	35
6	0.2	30
7	0.1	25
8	0.1	20
9	0.1	20
10	0.1	20

Should Ford make its own tires or continue to buy them from outside suppliers?

18.19 Your company has been sued for $5 million for patent infringement. The plaintiffs have offered to settle the case for $3 million. If you go to court, you will have to pay legal fees of $250,000. The value of the time your executives must spend to prepare the case is also estimated to be $250,000. The damage to the reputation of your company from a prolonged court case is real but difficult to value. Your subjective probability is that damage awards of $0, $1, $2, $3, $4, and $5 million are equally likely if the case goes to trial. Should you settle the case or take your chances in court?

18.20 You are a retailer of above-ground swimming pools. The manufacturer of a line of pools that you carry offers a quantity discount to its retailers. The first pool you buy per year costs $2,000 and the price decreases by $25 for every additional pool down to a price of $1,000 for the 41st pool. All pools beyond the 41st cost $1,000.

You make a profit of $250 per pool sold. If you have to keep an unsold pool in stock over the winter, the carrying cost is $100. If the distribution of pools sold is Poisson with a mean of 25, what is the optimal number of pools to order? (Hint: The normal approximation to the Poisson would be helpful.)

18.21 Pillsbury is considering switching from a box to a bag to package one of its cookie lines. The bag will save Pillsbury $5 million on a net present value basis compared to the cost of the box. However, consumers prefer the box because the cookies are less likely to be broken. Pillsbury estimates that if the lost sales are moderate, the net present value of lost profits from the lower sales would be $2 million. If the lost sales were severe, the net present value of lost profits from the lower sales would be $12 million. Pillsbury's subjective probability on these outcomes is 0.7 and 0.3.

(a) Should Pillsbury switch to the bag?

(b) Pillsbury could test-market the bags in a few cities to get more information about consumers' reactions. The test marketing would cost $3 million. The conditional probability of having a moderate loss of sales in the test marketing given a moderate loss in the entire market is 0.72. The conditional probability of having a severe loss of sales in the test market given a severe loss in the entire market is 0.66. Should Pillsbury go ahead with the bag or should it test-market first?

18.22 The National Forest Service uses controlled fires in national forests to remove underbrush and debris. The advantage of a controlled burn is that it prevents large buildups of debris that would cause major damage if they caught fire and it encourages tree growth. One of the disadvantages of controlled burns is that they are not always controlled. If the fire spreads beyond the intended boundary, valuable property can be destroyed. Another disadvantage is the cost of having firefighters on standby in case the fire gets out of hand. In actual controlled-burn problems, the Forest Service considers probability distributions for wind speeds, temperatures, and rain. We will ignore these factors and consider only the choices of not burning, burning, and ditching and burning. "Ditching" is digging trenches at the intended boundaries of the fire to reduce the probability that it will spread. Suppose a successful controlled fire in a particular forest had a payoff of $25,000. The probability that the fire would be successful—that is, contained within its present boundaries—was 0.85 for a fire without ditches and 0.96 for a fire with ditches. An unsuccessful fire would have a payoff of $−95,000. The cost of ditching is $12,000. What is the Forest Service's best choice?

18.23 You head the negotiating team for your company in contract talks with the International Brotherhood of Dissemblers. The contract with this union is about to expire and you want to use decision analysis to help formulate your strategy. Increases in the wage that you offer will reduce the probability that the union will go on strike. Strikes cost your company money but higher wage offers also cost your company money. You have narrowed the choices of wage offers to three: no increase, a 5% increase, and a 10% increase. The costs of these packages over the life of the contract are $80 million, $84 million, and $88 million, respectively. The probabilities you subjectively assign to a strike that will last long enough to force the union to accept these offers are 0.95, 0.80, 0.20, respectively. If there is a strike, the expected durations are 34 weeks, 8 weeks, and 2 weeks for the respective packages. Each week of the strike costs your company $250,000. Which offer would minimize your company's expected cost?

18.24 (This exercise continues exercise 18.23.) You could gain more information about the probability that the Dissemblers will strike by hiring a detective agency to send bogus workers to your factory and have them report back on the union members' feelings. The detective agency claims that among all of the cases in which there were strikes, they correctly predicted a strike 80% of the time. Also, among all of the cases in which there were no strikes, they correctly predicted that there would be no strike 90% of the time. The detective agency costs $500,000.

  (a) Is the agency worth hiring?

  (b) If you hired the agency and it predicted a strike for the 0% offer and no strikes for the 5% and 10% offer, which offer would be your optimal strategy?

  (c) If you hired the agency and it predicted a strike if you offered no increase or a 5% raise and no strike if you offered 10%, which offer would be your optimal strategy?

18.25 You run a software publishing company and you have been offered the marketing rights to a new spreadsheet program, code-named Lotus Eater, for the IBM-PS 80 computer. The fee for the marketing rights is $2 million. The value of the marketing rights depends on the number of copies of the program you will sell, which in turn depends on the number of IBM-PS 80 computers that will be sold and this program's performance in comparison with competing spreadsheet programs. The possible sales of the IBM-PS 80 can be characterized as low, medium, and high. Your subjective probabilities for these outcomes are 0.8, 0.1, and 0.1. If Lotus Eater is the best spreadsheet program available for this computer, you expect to make profits of $25 million if the machine's sales are high, $10 million if the machine's sales are medium, and $1 million if the machine's sales are low. If Lotus Eater is not the best among the available spreadsheet programs, you expect to make profits of $5 million if the machine's sales are high, nothing if the machine's sales are medium, and $-6 million if the machine's sales are low. Your subjective probability that Lotus Eater will be the best is 0.7. Before deciding to pay the $2 million fee, you want more information about potential competitors. You hire a panel of experts to test Lotus Eater and all of the available prerelease versions of known potential competitors. From past experience you estimate that the probability of a plus rating from the panel given that Lotus Eater is the best program is 0.9. The probability of a plus rating from the panel given that Lotus Eater is not the best program is 0.2. Construct a decision tree for this problem. The panel test costs $500,000. Is your best choice not to buy the marketing rights to the program, to buy the rights without forming the panel of experts, or to form the panel and wait for its opinion?

18.26 (This exercise continues exercise 18.25.) What is the most that you would be willing to pay for a panel of experts who would make perfect predictions about the probability that Lotus Eater is the best spreadsheet program?

18.27 You are planning to build a motel at a location where the expected number of customers per day is five. An empty hotel room costs you $20 a day and a filled hotel room brings in a net profit of $40 a day. If the distribution of the number of rooms filled is Poisson, what is the optimal size for the motel?

18.28 Construct an opportunity cost table for a motel of size 0 to 20 rooms and 0 to 20 customers, using the cost information in exercise 18.27.

18.29 What is the expected value of perfect information in the problem described in exercise 18.27?

18.30* You are a wholesaler of seafood products. You are offered 100 tins of beluga caviar at $50 per tin for the entire lot. Caviar is a highly perishable product, and if it is not refrigerated and handled properly, it can easily spoil. If the caviar is fresh, you can make a profit of $100 per tin. A tin of spoiled caviar is worth nothing. You know that if the caviar came from one source, the 100 tins are all fresh, but if it came from another source, half of the tins are spoiled. You ask the seller to let you test some of the tins. He replies that you have to buy any tin that you want to open and that the price for an individual tin is $250. If you start opening tins and all of them turn out to be fresh, how many should you open before deciding to buy the remaining tins at the $50 price?

Appendix A

Computer Applications

LEARNING OBJECTIVES

There is a computer application here for every chapter except Chapters 1 and 18. Each application is numbered to correspond with the chapter to which it applies; that is, Appendix A2 applies to Chapter 2, Appendix A3 applies to Chapter 3, and so on. The applications include exercises with real data. All of the data sets are available on disk.

The main objective of these appendixes is to help you understand how statistical computer packages can "open up" data analysis. The power of a good statistical package resides in the ability to help you quickly and easily view a data set from many perspectives. With a computer package you can ask and answer questions that would require hours of tedious work with a calculator or would be impossible with pencil and paper. Statistical packages open up data analysis because they make it easy to address many important questions that are too often ignored. Some of these questions are: Do subsets of the data appear to behave the same as the rest of the data set? Do the data appear to fit the assumptions required for certain statistical tests? Are there patterns or clusters in the data? Are there implausible observations or outliers in the data? What potential impact do missing observations have on any inferences made about the population from which the data were drawn?

A second objective of Appendix A is to give an overview of the mechanics of these packages. As each section of the appendix shows output or commands from four packages, it is not practical to cover all of their detailed commands. This appendix is not meant to replace the manuals that come with the various packages. Some of the most basic commands needed for each type of problem will be given, but you will have to refer to the manual for more details. After reading one of these computer applications, you should know at least which headings in the manual you need to look under for the additional details.

Because the notation and terminology vary from computer package to package, you can easily be tripped up. A third objective of Appendix A is to connect the terms and symbols in this text to the corresponding terms in a computer program. This "crosswalk" from the text to the packages will make using a computer package much easier. Note that Appendix E contains the variable definitions for the 30 data sets used in the computer exercises.

COMPUTER PACKAGES COVERED

Computer technology is improving rapidly and the statistical packages are changing as the hardware changes. One reason for giving examples from four packages is that colleges vary widely in their computer resources and no one package is available on all campuses. The four chosen for the illustrations are DATA DESK, MINITAB, SAS, and SPSSx. At least one of these packages should fit the computer resources on your campus. The older computer technology is found in the mainframe computer. Three of the packages used in the illustrations—MINITAB, SAS, and SPSSx—were originally written for mainframe computers. Microcomputer versions of these programs are now available, but they are the same as the original mainframe versions. The other package, DATA DESK, works only on the Macintosh.

Microcomputers offer many advantages to the student. One advantage is convenience—you don't have to go to a computer center to pick up output; you don't have to wait until you are assigned an account number and a password before you can use the machine. Students who own a microcomputer or have access to one can do computer-based problems at times that best suit them. Another advantage is that the newer programs tend to be more "user friendly." This phrase is computer jargon for "easy to learn and easy to use." On the other hand, the mainframe packages tend to have more statistical routines than the packages written for micros, and mainframe computers have the processing speed needed for large data sets and complex problems. Both mainframes and micros have their advocates. The various packages have camps of supporters as well. Appendix A focuses on what computer packages in general can do rather than on the merits of specific hardware or software.

For reasons of legal liability, computer software usually carries a disclaimer to the effect that there is no guarantee that the software is fit for any use whatsoever. The disclaimer that accompanies this computer appendix is modest in comparison. Computer packages change over time; some change every year. Any statement in this appendix to the effect that "package A cannot do such-and-such" is likely to become wrong in a few years. The instructions in this appendix apply to the packages available in the fall of 1987. Please regard the manual that accompanies a package as the final authority on what the package can and cannot do.

A2
Computer-Generated Graphs for Descriptive Statistics

A2.1 OVERVIEW

Graphing is a weak area for most statistical packages. SAS and SPSSx have related programs that can draw high-quality graphs on pen plotters, but the main programs are limited to what are called letter-quality graphs, which consist of letters, such as x's and O's, rather than lines. MINITAB is limited to letter-quality graphs. Since DATA DESK was written to take advantage of the graphics orientation of the Macintosh, it can make high-quality graphs. The best way to understand the distinction between letter-quality and high-quality graphs is to look at the outputs of the various programs. One high-quality histogram is in section A2.3A, on DATA DESK commands, and one letter-quality histogram is in section A2.3B, on MINITAB commands. Since the letter-quality histograms of SAS and SPSSx look about the same as the MINITAB version, they are not shown.

There are several reasons why graphs are a minor feature of most statistical programs. One reason is that graphs created by these programs are intended primarily as analytical tools rather than as illustrations that can be used in a presentation or a book. A second reason is that it is difficult to create complex graphs with letter graphics. SAS has pie and bar charts made with letters, but these graphs push against the limits of the practical. None of the programs has vertical line charts, pictographs, or frequency polygons. Finally, it should be noted that some other graphs are available in the four programs: all four programs can construct scatter diagrams (discussed in Chapter 13) and box plots (discussed in Chapter 3).

A2.2 AN APPLICATION

Figure A2.1 reproduces a histogram accompanying an article in an economics journal. The graph shows the real rates of return on 640 oil paintings sold at a London auction house between the years 1652 and 1961. The graph indicates that these paintings were poor investments. "Not only were rates of return on painting as an investment remarkably low," writes William J. Baumol, "they were also remarkably dispersed, meaning that this form of investment was quite risky." When Professor Baumol presented his findings at a conference on economic analysis of the arts, art dealers objected that art investment was indeed profitable. (I am indebted to Professor Leslie Singer of Indiana University–Northwest for a description of the conference.) The main objection was that the highly negative rates of return were for oil paintings originally attributed to a master but subsequently attributed to the master's students or to outright forgers. Clearly, lopping off the left tail of the histogram would make investment in oil paintings appear profitable. Is this statistical surgery justified? The

FIGURE A2.1

Histogram of rates of return on 640 oil paintings sold between 1652 and 1961

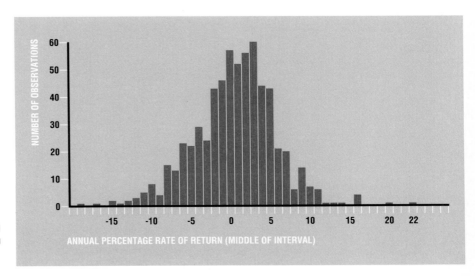

Source: William J. Baumol, "Unnatural Value; or Art Investment as a Floating Crap Game," *American Economic Review,* May 1986, p. 13.

answer is no. The buyers must have been confident that they were getting genuine old masters. Discovering that a previously undoubted painting is a forgery is part of the risk of investing in paintings. There can be no guarantee that a new technology for examining paintings or newly discovered historical information will not reveal a forgery or a student painting. Old masters' oil paintings are widely considered the least risky category of art investment. One reason is that a dead artist is unlikely to add to the supply of his paintings. A second reason is that public tastes are thought to be more stable regarding old masters. Presumably a histogram for rates of return on paintings by living artists would show even greater dispersion.

A2.3 COMMANDS

A2.3A DATA DESK

If you put the data for the oil paintings into DATA DESK and used the command HISTOGRAM, you would get Figure A2.2. With DATA DESK you have no control over the labeling of the x and y axes, and your control over the number of bars in the histogram is limited. You can increase or reduce the number of bars by holding down the option key as the window that contains the graph is dragged to the right or the left. There is no way to force the graph to have one bar for each integer rate of return. While Figure A2.2 does give you a picture of the overall pattern in the data set, it is difficult to determine the number of observations for a given rate of return. Also, all intervals have to have the same width. Recall that if the intervals all have the same width, it does not matter if the vertical axis is labeled with the frequency or the density.

FIGURE A2.2

Histogram created by DATA DESK

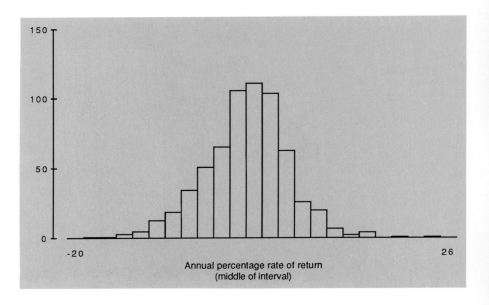

FIGURE A2.3
MINITAB histogram of rates of return on paintings

```
Histogram of C1    N = 640
Each * represents 10 obs.

Midpoint    Count
      -20       1   *
      -15       6   *
      -10      35   ****
       -5     111   ************
        0     257   *************************
        5     187   *******************
       10      33   ****
       15       7   *
       20       1   *
       25       2   *
```

A2.3B MINITAB

MINITAB created the letter-quality graphs in Figure A2.3. The MINITAB command for a histogram is HISTOGRAM followed by the variable name you have assigned to the data. MINITAB allows you to specify the common width for all of the intervals with the command INCREMENT = X. MINITAB also allows you to specify the midpoint of the first interval with the command START FIRST MIDPOINT AT X. For example, if the variable name for the rate-of-return data was RATE and you wanted a histogram with the first midpoint at −20 and interval widths of 2, the commands would be

```
MTB > HISTOGRAM RATE;
SUBC > INCREMENT = 2;
SUBC > START FIRST MIDPOINT AT −20.
```

The MTB > is MINITAB's prompt. The semicolons at the ends of the first and second lines tell MINITAB to expect a subcommand on the next line. The SUBC > is MINITAB's prompt for a subcommand. A period is required after the last subcommand.

A2.3C SAS

SAS can create letter-quality graphs similar to Figure A2.3. The SAS procedure you would invoke is CHART followed by the line HBAR (for "horizontal bar chart") or VBAR (for "vertical bar chart") plus the variable name. Even though they are called bar charts, the data can be in levels rather than categories. With SAS you can specify the limits of each interval, so the intervals can have different widths. If the rate-of-return data were stored under the variable name RATE, the commands to create a histogram with a first midpoint at −20 and interval widths of 2 would be

```
PROC CHART;
    HBAR RATE / MIDPOINTS=−20 TO 24 BY 2;
```

All SAS command lines must end with a semicolon. The slash in the second line indicates that an option follows the HBAR command. In this case it is the MIDPOINTS option.

 A2.3D SPSSx

SPSSx can create letter-quality graphs similar to Figure A2.3. With SPSSx you can ask for a histogram as part of the FREQUENCIES subprogram. SPSSx lets you specify the range of the histogram—that is, the highest and lowest values—and the width of the intervals. All intervals must have the same width. If the rate-of-return data were stored under the variable name RATE, the commands to create a histogram with the first midpoint at −20 and interval widths of 2 would be

```
FREQUENCIES     VARIABLES=RATE/
                HISTOGRAM=MIN(-20) INCREMENT(2)
```

EXERCISES FOR APPENDIX A2

A2.1 The third column of data set 1 contains the per capita income in thousands of dollars for the 43 largest U.S. standard metropolitan statistical areas (SMSAs) in 1985. An SMSA consists of a city and its suburbs. The source of the data is the *Statistical Abstract of the United States*. Create a histogram of the data. Does the distribution appear to be skewed or bimodal? In 1986 some congressmen introduced the term "bicoastal economy" to indicate their belief that the east and west coasts of the United States were doing well but the rest of the country was doing poorly in economic terms. The names of the 43 cities are in the first column of the data set. The east and west coast SMSAs are identified by a dummy variable. Create two histograms, one for the bicoastal economy and the other for the remaining cities, and decide whether the congressmen's argument is supported by the income data.

A2.2 The single column of data set 2 contains the horsepower of 392 car models sold in the United States over the years 1971 to 1983. The original source of the data is *Consumer Reports* magazine. Construct four histograms for the data set with different numbers of intervals, say 5, 7, 9, and 11. What additional information do you gain as the number of intervals is increased?

A2.3 The single column of data set 3 contains the tar content in milligrams for 203 brands of cigarettes sold in the United States in 1983. The source of the data is the Federal Trade Commission. Construct a histogram of the data. Does the distribution appear to be skewed or bimodal? What additional information about the brands of cigarettes would give you a better understanding of the distribution of tar content?

A3
Numerical Descriptors and Exploratory Data Analysis

A3.1 INTRODUCTION

Every statistical package has a command that will give you such basic descriptive statistics as the mean, median, and standard deviation. The same commands often give you many other descriptive statistics. To get a sense of the diversity of available descriptive statistics, consider that there are about 25 descriptive statistics that appear in only one of our four packages. Chapter 3 was necessarily limited to a small fraction of all descriptive statistics. The statistics discussed in Chapter 3 are the best known and have the widest application to business problems. The application in section A3.2 illustrates some of the more interesting or interpretive uses of computer packages. Instead of mechanically calculating summary measures, we will use some ratios and graphs to see what information can be gleaned from a data set. In section A3.3 we will concentrate on the more mundane aspects of using computers, the commands and terminology. If you compare the output of the various packages, you may be puzzled to find that the values of the kurtosis statistic differ. The different values are not due to an error. There are several commonly used formulas for kurtosis. Section A3.4 shows the commands for the optional material on box plots and stem-and-leaf displays.

Throughout this application we will use data set 4, on the sales and assets of the 25 largest foreign companies as compiled by *Fortune* magazine in 1985. Table A3.1 shows the first three observations in data set 4.

TABLE A3.1
Sales and assets in millions of dollars: First three observations in data set 4

Name	Sales	Assets
Royal Dutch/Shell Group	84,912	68,918
Mitsui and Co. Ltd*	68,464	25,093
Mitsubishi Corp.*	66,919	25,173

*Japanese companies.

A3.2 AN APPLICATION

The real power of computer packages comes into play when you try to manipulate the data and look at them in various ways. All of the packages allow transformations, such as adding or dividing two variables, or arithmetic operations on one variable, such as taking the logs of the observations. This application illustrates transformations that let you answer some interesting questions about the data set. After making the transformations, we will create two box plots and place them side by side.

The ratio of sales to assets is considered to be an indicator of a company's efficiency. Among firms of a given type, a high ratio of sales to assets suggests that a firm is making relatively efficient use of its capital. Retailers, for example, typically have a high ratio of sales to assets and manufacturing firms have a low ratio. So a manufacturing firm whose ratio of sales to assets is higher than that of another manufacturing firm is generally thought to be more efficient. The question that you can answer with this information is whether or not the large Japanese firms have higher ratios of sales to assets than other large foreign firms. Recall that Table A3.1 showed which firms were Japanese. You would have to split both the sales and the assets variables into Japanese and non-Japanese firms. Each package has its own command to split a data set. Next, you would compute the ratios of sales to assets for the Japanese and non-Japanese firms. Again the packages have different commands for this transformation.

An easy way to compare the distributions of these sets of ratios is to look at their box plots. (Box plots are described in the optional section of Chapter 3.) The boxes represent approximately the central 50% of the distribution and the line in the middle of the box is the median. Figure A3.1 contains the box plots created by DATA DESK. The first plot is for the non-Japanese firms and the second is for the Japanese firms. Almost all of the non-Japanese firms are within a narrow range of ratios of sales to assets. This narrow range is well below most of the distribution of the ratios of

FIGURE A3.1

Box plots of the ratios of sales to assets for the Japanese and non-Japanese firms in the *Fortune* data set

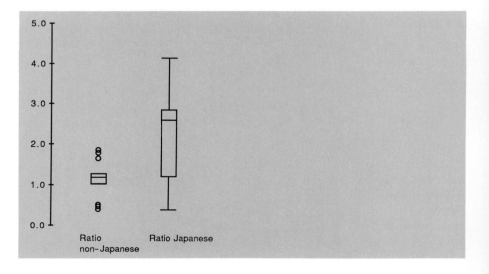

sales to assets of Japanese firms. The high outliers (the circles in the figure) for the non-Japanese firms are well below the median for the Japanese firms. Overall, the Japanese firms generate far more sales per dollar of assets. While this ratio is a rough guide to efficiency, the box plots clearly indicate that the large Japanese firms tend to be much more efficient than the large non-Japanese firms.

A3.3 COMMANDS FOR NUMERICAL DESCRIPTORS

A3.3A DATA DESK

If you click once on the icon of a variable to select that variable and then select the command Summary Reports in the Compute menu, DATA DESK will report the mean, the sample standard deviation, and the number of observations that have numerical values. DATA DESK always uses the $n - 1$ formula to compute a standard deviation. Nonnumerical observations can be names or categories, such as male and female. Many other descriptive statistics are available. If you pull down the Compute menu before clicking on the icon of a variable, the menu will have the command Select Summary Statistics. Many of the available statistics were not covered in the text: midrange, biweight mean, kurtosis, interquartile range, and most of the order statistics. Figure A3.2 shows the default output from DATA DESK.

To split the *Fortune* data set into Japanese and non-Japanese firms in DATA DESK, you would use the variable that equals 0 for every non-Japanese firm and 1 for every Japanese firm. Call this variable splitter. The name is arbitrary. Click on the icon of the sales variable once and, while holding the shift key down, click on the icon for the splitter variable once. Then click on the command Split into variables by group under the Manipulate menu. These actions will cause DATA DESK to create two new variables, one for the sales of all Japanese firms and one for the sales of non-Japanese firms. Similar actions will create two new variables for the assets of Japanese and non-Japanese firms. Next click on the command New Derived Variable under the FILE menu. DATA DESK will open a dialog box asking for the name of the new derived variable. Name it ratio non–Japanese for the ratio of sales to assets of the non-Japanese firms. A second dialog box will appear on the screen, asking for an algebraic expression that would define the new variable. If *sn* was the name assigned to the sales for non-Japanese firms and *an* was the name assigned to the assets of non-Japanese firms, the expression sn/an would create the desired ratio. Similar steps will create a new derived variable for the ratio of sales to assets for the Japanese firms.

FIGURE A3.2
Output from the DATA DESK
Summary Reports
command

```
Summary statistics for sales
  NumNumeric = 25
  Mean = 32185
  Median = 20815
  Standard Deviation = 19949
```

FIGURE A3.3
Output of the MINITAB
DESCRIBE **command for**
the sales data

```
MTB > DESCRIBE C1

              N     MEAN   MEDIAN   TRMEAN   STDEV   SEMEAN
SALES        25    32185    20815    30563   19949     3990

             MIN      MAX       Q1       Q3
SALES      16768    84912    18392    50271
```

A3.3B MINITAB

The command needed to create some descriptive statistics in MINITAB is
DESCRIBE. MINITAB refers to variables by the column that they occupy. Suppose
the sales data from data set 4 was in column 1 of a MINITAB worksheet. A NAME
command will cause MINITAB to print the name of the variable in any output. The
command NAME FOR C1 IS 'SALES', for example, will cause the output to refer
to the variable as SALES. You must have a quotation mark on each side of the word
SALES. Figure A3.3 shows the output of the DESCRIBE command as it appears on a
computer screen. The term TRMEAN stands for "trimmed mean" (a term not covered
in Chapter 3). The term SEMEAN stands for "standard error of the mean" (which will
be discussed in Chapter 7). Other than these two terms, there is little that requires
explanation in the MINITAB output. The letter Q stands for "quartile." Q1 is the
observation that marks the lower 25% of the data set and Q3 is the observation that
marks the lower 75% of the data set.

To split the *Fortune* data set into Japanese and non-Japanese firms in MINITAB, you
use a variable that equals 1 for Japanese firms and 0 otherwise. Suppose this variable
was in column 4. The CHOSE command selects a subset of data and stores it in another
column of the worksheet. For example, the commands

```
CHOSE 0 IN C4, CORRES ROWS OF C2, C3, INTO C5-C7
```

will put the sales of the non-Japanese firms into column 6 and the assets into column
7. The commands

```
CHOSE 1 IN C4, CORRES ROWS OF C2, C3, INTO C8-C10
```

will put the sales of the Japanese firms into column 9 and the assets into column 10.
The LET C11=C6/C7 command creates in column 11 a new variable that is the ratio
of sales to assets for the Japanese firms. Similarly, a LET C12=C9/C10 command
creates in column 12 a new variable that is the ratio of sales to assets for the Japanese
firms.

A3.3C SAS

After you have accessed data set 4 and named the SALES variable, the commands in
SAS needed to generate descriptive statistics for the SALES variable are

```
PROC UNIVARIATE;
VARIABLE  SALES;
```

Note that SAS command lines must end with a semicolon. The univariate procedure
is one of several procedures within SAS which create descriptive statistics. The output
presented in Figure A3.4 was created without any options specified. The user can

FIGURE A3.4
Default output from the SAS univariate procedure on the sales data

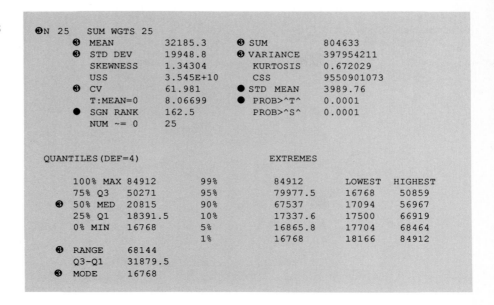

```
❸N 25    SUM WGTS 25
         ❸ MEAN       32185.3     ❸ SUM        804633
         ❸ STD DEV    19948.8     ❸ VARIANCE   397954211
           SKEWNESS   1.34304       KURTOSIS   0.672029
           USS        3.545E+10     CSS        9550901073
         ❸ CV         61.981      ● STD MEAN   3989.76
           T:MEAN=0   8.06699     ● PROB>^T^   0.0001
         ● SGN RANK   162.5         PROB>^S^   0.0001
           NUM ~= 0   25


    QUANTILES(DEF=4)                          EXTREMES

        100% MAX 84912       99%       84912      LOWEST   HIGHEST
        75% Q3   50271       95%       79977.5    16768    50859
      ❸ 50% MED  20815       90%       67537      17094    56967
        25% Q1   18391.5     10%       17337.6    17500    66919
        0% MIN   16768        5%       16865.8    17704    68464
                              1%       16768      18166    84912

      ❸ RANGE    68144
        Q3-Q1    31879.5
      ❸ MODE     16768
```

specify a weight for each observation so that the mean becomes a weighted mean. The default is that each observation has a weight of 1. This default value is the reason that the table reports a SUM WGTS (or sum of weights) of 25 for the 25 observations. The default formula for the standard deviation has $n - 1$ in the denominator. The command VARDEF=N tells SAS to use the number of observations in the data set in the denominator. In Figure A3.4 the statistics discussed in Chapter 3 are marked with a ❸ and any statistics that are discussed in later chapters are marked with a ●.

SAS gives you the 100th, 99th, 95th, 90th, 75th, 25th, 10th, 5th, 1st, and 0th percentiles. Figure A3.4 also shows the lowest five observations and the highest five observations. You have a choice of five formulas for calculating the percentiles. QUANTILES (DEF=4) tells the user that the fourth of the five definitions is being used in the calculations. Note also that the coefficient of variation (CV) in Figure A3.4 is in percent.

To split the *Fortune* data into Japanese and non-Japanese firms, you first have to use the dummy variable, number 7, that takes the value 0, for Japanese firms and 1 for all non-Japanese firms. See the SAS manual for details on reading data. Give this 0-1 variable the name SPLITTER. You can create new data sets by deleting observations from the *Fortune* data set. The commands

```
DATA JAPAN;
INFILE FORTUNE;
IF SPLITTER=0 THEN DELETE;
RATIO=SALES/ASSETS;
PROC UNIVARIATE;
VARIABLE SALES RATIO;
```

will give you descriptive statistics on the Japanese firms and the ratio of sales to assets for these firms. A similar set of commands will give you descriptive statistics on the non-Japanese firms.

FIGURE A3.5
Output from the SPSSx
FREQUENCIES **command**

```
        SALES


     ❸  MEAN        32185.320    ●  STD ERR  3989.758    ❸  MEDIAN    20815.000
     ❸  MODE        16768.000    ❸  STD DEV 19948.790    ❸  VARIANCE 397954211
        KURTOSIS    .672            S E KURT .902            SKEWNESS 1.343
        S E SKEW    .464         ❸  RANGE    68144.000   ❸  MINIMUM  16768.000
     ❸  MAXIMUM     84912.000    ❸  SUM      804633.000


        VALID CASES            25           MISSING CASES  0
```

 A3.3D SPSSx

After you have accessed data set 4 and named the SALES variable, SPSSx commands
for descriptive statistics for the SALES variable are:

```
FREQUENCIES  VARIABLES=SALES/
    STATISTICS=ALL
```

Note that subcommands in SPSSx such as STATISTICS=ALL must be indented.
Figure A3.5 shows the output when all statistics are requested. Statistics that were
discussed in Chapter 3 are marked with a ❸ and statistics that will be discussed in later
chapters are marked with a ●. SPSSx always uses the $n - 1$ formula for calculating the
standard deviation.

To split the *Fortune* data into Japanese and non-Japanese firms, you have to use the
variable that takes one value, 1, for all Japanese firms and another value, 0, for all
non-Japanese firms. Give this 0-1 variable the name SPLITTER. The commands

```
COMPUTE RATIO=SALES/ASSETS
SORT CASES BY SPLITTER
SPLIT FILE BY SPLITTER
FREQUENCIES VARIABLES=SALES RATIO
```

will give you descriptive statistics on the sales variable and on the ratio of sales to
assets, first for all Japanese firms and then for all non-Japanese firms.

A3.4 COMMANDS FOR EXPLORATORY DATA ANALYSIS

 A3.4A DATA DESK

The box plots in Figure A3.1 were created with DATA DESK. Other than their
vertical orientation, these box plots have the same layout as the box plots in
Chapter 3. The procedure to create a box plot begins when you click once on the icon
of the variable to select it. Next, click on the Box plots command in the Display
menu. If you click on several variables while holding down the shift key, all of them
will be selected. Then the Box plots command gives you juxtaposed box plots.

One last comment before we leave DATA DESK's exploratory data analysis routines. DATA DESK does not create stem-and-leaf displays, perhaps because these displays were originally conceived as paper-and-pencil techniques for analyzing data. The other packages have an advantage in this regard; they all have stem-and-leaf displays. Stem-and-leaf displays belong in computer packages because they can quickly highlight certain anomalies in data.

A3.4B MINITAB

MINITAB has a STEM-AND-LEAF command. If you have your data already entered in MINITAB, typing STEM-AND-LEAF and the column of the variable after a MINITAB prompt will create the display seen in Figure A3.6. This display is the same as the ones in Chapter 3 except that there is no vertical line to separate the stems from the leaves and there are no markers to distinguish the lower and upper stems of the various digits (e.g., 2·, 2★). Recall that the column of numbers to the left of the stem is the depth indicator. Also, on the row that contains the median there are parentheses around the number of observations on the row rather than a value for the depth. If there are enough data, MINITAB will automatically break each stem digit into more than two ranges. One limitation of MINITAB's stem-and-leaf routine is that the user cannot specify the column for the stems. If all observations have the same number of digits, the leftmost digit in the data is automatically made the stem.

The command BOXPLOTS followed by a variable name will create a box plot in MINITAB. If you want to juxtapose the box plots of more than one variable, you can command BOXPLOTS SALES BY ASSETS. You can use more than two variables and substitute any variable name for SALES and ASSETS. Figure A3.7 shows the box plot for the sales data. The plus marks the median of the data set. Within the limitations of letter-quality graphics, the MINITAB box plot is fairly clear and it conforms closely to the box plots in Chapter 3.

FIGURE A3.6
Stem-and-leaf display created by MINITAB

```
MTB > STEM-AND-LEAF 'SALES'

Stem-and-leaf of SALES       N  = 25
Leaf Unit = 1000

    10      1   6777888999
    (5)     2   00023
    10      2   55
     8      3
     8      3   5
     7      4
     7      4   9
     6      5   00
     4      5   6
     3      6
     3      6   68
     1      7
     1      7
     1      8   4
```

FIGURE A3.7
Box plot created by MINITAB

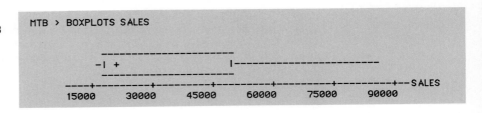

FIGURE A3.8
Stem-and-leaf display and box plot created by SAS

```
STEM LEAF                          #              BOXPLOT
   8 5                             1                 ^
   7                                                 .
   6 78                            2                 .
   5 0117                          4              +-----+
   4                                              .     .
   3 5                             1              . + .
   2 000112356                     9              *-----*
   1 77788899                      8              +-----+
     ----+----+----+----+
     MULTIPLY STEM.LEAF BY 10**+04
```

■ A3.4C SAS

The addition of one word to the command file that created the descriptive statistics is all that is needed to create both a stem-and-leaf display and a box plot in SAS. The word is PLOT and it is added immediately after the words PROC UNIVARIATE. The new line would read

     PROC UNIVARIATE PLOT;

SAS places the stem-and-leaf display and the box plot side by side. It is not possible to juxtapose several box plots. Figure A3.8 shows the stem-and-leaf display and the box plot of the sales data.

SAS's stem-and-leaf display has no depth indicators; the numbers to the right of the display indicate the number of observations per line. The user has no control over which column is the stem, and SAS will not split a stem digit into upper and lower ranges. The box plot is also crude. The line with an asterisk on either end marks the median and the lines with plusses on the ends mark the hinges. The hat symbols, ^, are the whiskers. The plus symbol in the middle of the box is not explained in the SAS manual; its purpose is not readily apparent.

■ A3.4D SPSSx

The MANOVA procedure in SPSSx has a PLOT subcommand with options for box plots and stem-and-leaf displays. The required commands for creating box plots and stem-and-leaf displays for the sales variable are

     MANOVA SALES
          /PLOT STEMLEAF BOXPLOTS

SPSSx will juxtapose several box plots if more than one variable is named after the MANOVA command. There is a slight difference between the box plots defined in

FIGURE A3.9

Stem-and-leaf display and box plot created by SPSSx

```
     Stem-and-leaf display for variable .. SALES

        0 . 77888899
        2 . 0001123556
        4 . 0117
        6 . 78
        8 . 5


   Box-Plot For variable .. SALES

            85        I    O
                      I
                      I
           KEY        I    X
        ----------    I    I
        * Median      I    I
        - 25%, 75%    I    I
        X High/Low    I    I
        O Outlier     I    I
        E Extreme     I    I
                      I   +-+-+
                      I   I   I
                      I   I   I
                      I   I   I
                      I   I   I
                      I   I * I
                      I   I   I
                      I   I   I
             7        I   +-+-+
```

Chapter 3 and an SPSSx box plot, seen in Figure A3.9. In the SPSSx box plots the whiskers extend to the smallest and largest observations in a group which are within one multiple of interhinge distance. Observations beyond the whiskers but within one and a half multiples of the interhinge are marked with an 0, for outliers. Observations beyond one and a half multiples of the interhinge are marked with an E, for extreme observations. The stem-and-leaf display in SPSSx is minimal. There are no depth indicators or markers for splitting a stem, and there is no provision for selecting the column that will be the stem.

EXERCISES FOR APPENDIX A3

A3.1 Many of the non-Japanese firms listed in data set 4 are oil companies (Royal Dutch/Shell Group, British Petroleum Co. Plc., ENI–Ente Nazionale Idrocarburi, Pemex–Petroleos Mexicanos, Elf Aquitane Group, TOTAL Group–Francaise des

Petroles, and Petrobas–Petroleo Brasileiro SA). The low ratio of sales to assets in the non-Japanese firms may be due to accounting rules on the value of petroleum reserves rather than any difference in efficiency. To address this question, create three box plots of the ratios of sales to assets, one for Japanese firms, one for foreign firms outside of the oil industry, and one for foreign oil firms. What conclusions can you draw?

A3.2 Record the amount of price change from the close of the previous trading day for the ten most actively traded stocks on the New York Stock Exchange for five days. Construct a box plot of these data. Do these most actively traded stocks tend to have positive or negative price changes?

A3.3 Data set 5 contains the ending dates, monthly volumes, and end-of-month closing prices of H & R Block's stock for 156 months. H & R Block is traded on the New York Stock Exchange. The first three lines of data set 5 are shown below.

(a) Determine the mean, mode, median, standard deviation, and range of the monthly volumes and the month-end prices.

(b) Construct a box plot of the monthly volumes.

(c) Construct a stem-and-leaf display of the month-end prices. What can you infer from this stem-and-leaf display about how stock prices are quoted?

Date	Monthly volume	Month-end price
10/31/74	502,200	5.75
11/29/74	282,200	5.4375
12/31/74	854,600	6.9375

A4
Probability Calculations and Joint Probability Tables

A4.1 INTRODUCTION

In general, statistics packages are not designed to solve probability problems. The packages focus on descriptive statistics and inferential statistics. Nevertheless, they can be used for some probability problems. All of the packages reviewed here can create joint probability tables from raw data sets. These tables are also called contingency tables and cross-tabulations. If you have a large data set, creating a table with the count or percentage for every intersection can be time-consuming. A computer-created table can be made in seconds, and it is much more likely to be accurate than a handmade table.

A4.2 AN APPLICATION

Consumer Reports magazine mails an annual survey to its subscribers. The survey covers cars, major appliances, and from time to time such services as homeowners' insurance. The 1986 survey received 277,000 responses. Overall, the subscribers to *Consumer Reports* are highly conscientious about filling out a long, detailed survey. The survey asks about the failures and repair costs of every major component of the two newest cars in the subscriber's household. As a service to its subscribers, *Consumer Reports* constructs a "trouble index" in which it classifies cars by model and year as either (1) much better than average, (2) better than average, (3) average, (4) worse than average, or (5) much worse than average. Data set 6 is a subset of the table printed in the April 1987 *Consumer Reports*. All trucks and vans have been eliminated as well as car models from Korea, Great Britain, and France. The remaining cars are identified by the home country of the company that sells the cars. Some Honda Accord models are built in the United States, for example, but they are identified as Japanese because Honda is a Japanese company. Table A4.1 shows the first three observations of data set 6.

Figure A4.1 was first created with DATA DESK and then modified to show the row and column totals. The grand total is 100.01% rather than 100% because of rounding. The table works the same as the joint probability tables discussed in Chap-

TABLE A4.1

First three observations of data set 6: Cars identified by model, year, home country, and *Consumer Reports* trouble index rating

Model	Year	Country	Rating*
AMC Eagle 6	1982	U.S.	5
AMC Eagle 6	1983	U.S.	5
Audi 4000	1981	Germany	3

*1 = much better than average, 2 = better than average, 3 = average, 4 = worse than average, 5 = much worse than average.

FIGURE A4.1

Percentages of car models by country and by level of trouble rating

```
Columns are levels of  country
Rows are levels of  rating

           U.S.    Germany    Japan     Sweden          total
    1      1.56     2.56      13.40      0.00            17.52
    2      4.13     2.13       4.55      1.14            11.95
    3     22.74     3.13       3.13      2.13            31.13
    4     18.04     0.57       0.00      0.28            18.90
    5     20.24     0.28       0.00      0.00            20.52
 total    66.71     8.67      21.08      3.55           100.01

table contents:  percent of table
```

ter 4. For example, the probability that a car model selected at random from the population of all car models in the list both is a U.S. model *and* has a much better than average trouble rating is the intersection term in the upper left corner, or 1.56%. The news reports of quality problems among American cars are supported by the table. Among all American models, 57.38% had ratings of worse than average or much worse than average. This value was found by dividing the sum of the percentages of American models with these ratings, 38.28%, by the percentage of American models in the survey, 66.71%. In contrast, none of the Japanese models was rated worse than average or much worse than average. Note that this statement does not mean that no individual Japanese cars had serious quality problems. The rating of a particular model and year represents the average experience of hundreds or even thousands of owners of that model. In one regard, Figure A4.1 understates differences in quality of American-made versus Japanese-made cars. All of the American car companies sell what are called captive imports—cars made abroad but sold with a domestic nameplate. Sometimes the domestic company emphasizes the foreign origin. The Plymouth Sapporo, for example, is unmistakably Japanese. One of the exercises for Appendix A4 is to redo the table with the captive imports classified by country of origin.

Another point that can be made about Figure A4.1 is that since not all data points consisting of a car model and year have the same sales volume, the percentages do not represent the probability that a randomly selected survey respondent will have a particular combination of country and trouble rating. For example, while the percentage of all passenger car models both sold by U.S. companies and having a much worse than average trouble rating is 20.24%, the proportion of respondents who own these cars should be much lower. Presumably *Consumer Reports* subscribers are reading about which models to avoid.

A4.3 JOINT PROBABILITY TABLES

A4.4A DATA DESK

Figure A4.1 was created with the DATA DESK Contingency tables command. You begin by clicking on the icon of the Country variable to select it and, while holding down the shift key, clicking on the icon of the Rating variable. Next click on the Contingency tables command under the Compute menu. After you have selected this command, a dialog box appears in which you can choose what will appear in the table. The only item selected in the dialog box was Percent of total table.

A4.4B MINITAB

MINITAB has a TABLE command that can take raw data and create the percentage for each cell. The command works with integer data. You could not record the countries by their names but would have to assign each country an integer code, such as U.S. = 1, Japan = 2, and so on. These codes are in data set 6. The MINITAB commands are

```
MTB > TABLE C1 VS C2;
SUBC > TOTPERCENT.
```

This example assumes that the rating variable was entered into column 1 and the country variable into column 2. The TOTPERCENT is a subcommand. In MINITAB the last subcommand must be followed by a period. A command line or any subcommand line that is followed by a subcommand must end with a semicolon.

 A4.4C SAS

SAS has a TABULATE procedure that can be used to create a table of counts for all intersections. You would have to divide the counts by the total number of observations to get the percentages. The observations can be recorded as names, such as U.S. and Japan, rather than coded with numerical values. The commands are

```
PROC TABULATE;
  CLASS RATING COUNTRY;
  TABLE RATING, COUNTRY;
```

 A4.4D SPSSx

SPSSx has a CROSSTABS procedure. The observations can be recorded as names rather than as numbers. Once the data are entered and the variable names assigned, the commands are

```
CROSSTABS VARIABLES = RATING BY COUNTRY
OPTIONS 5
```

The OPTIONS 5 gets you the percentages of the entire table.

EXERCISES FOR APPENDIX A4

(Exercises A4.1, A4.2, and A4.3 cannot be done with DATA DESK.)

A4.1 Many of the car models listed as U.S.-made in data set 6 are captive imports. The Plymouth Sapporo, Plymouth Champ, Merkur XR4Ti Turbo, Dodge Colt, Chevrolet Sprint, and Chevrolet Spectrum are made abroad. The Chevy Nova is difficult to classify as either U.S. or Japanese, as it is assembled in the United States, primarily from Japanese parts, in a plant jointly operated by General Motors and Toyota. To get a more accurate picture of joint probability distribution of country versus rating, redo Figure A4.1 after classifying the Plymouth Sapporo, Plymouth Champ, Dodge Colt, Chevrolet Sprint, and Chevrolet Spectrum as Japanese-made, classifying the Merkur XR4Ti Turbo as German-made, and dropping the Chevy Nova. What is the new probability that a randomly selected model year from the list will be U.S.-made *and* will have a much better than average trouble index rating?

A4.2 General Motors, Ford, and Chrysler have boasted in their advertisements about their commitment to high quality. Chrysler's slogan, for example, is "We build them best so we back them best"—a reference to its seven-year warranty. Ford's slogan is "The best-built cars in America." In light of the *Consumer Reports* data, Ford's slogan turns out to be a rather modest claim. Redo Figure A4.1 with the deletion and switches indicated in exercise A4.1, but with a separate table for each model year. Do these yearly tables show U.S. manufacturers gaining in quality on the Japanese manufacturers from 1981 to 1986? Note that it would take forever to do this problem by editing the data set line by line. Find a way to sort the observations by year or to eliminate observations for all but one year before creating each table.

A4.3 The 1936 *Literary Digest* poll is the most famous example of an inaccurate survey in the history of statistics. The poll was a mail survey in which 10 million people where asked whom they would vote for in the 1936 presidential election. The 10 million were identified by car registration, telephone ownership, or subscription to the *Literary Digest*. Approximately 2.5 million of the survey forms were returned. A substantial majority of the 2.5 million respondents said they would vote for the Republican candidate, Alf Landon. The landslide winner of the election was the Democratic candidate, Franklin Roosevelt. Exercise 7.1 in Chapter 7 discusses the *Literary Digest* poll and asks some questions about what may have gone wrong. One possible explanation is that Republicans were more likely than Democrats to own cars and phones in 1936.

In this computer exercise we will consider a Gallup survey from 1937. The Gallup survey asked a random sample of 2,277 Americans who had voted in the 1936 presidential election how they had voted and whether they owned a phone and/or car. Use the data in data set 30 to construct a joint probability table for the ownership categories (1) car and phone, (2) car and no phone, (3) phone and no car, and (4) neither car nor phone versus the voting categories (1) Roosevelt, (2) Landon.

> **(a)** If the sample accurately reflects the entire population, what is the conditional probability of a randomly selected voter having voted for Roosevelt, given that the voter owned both a car and a phone? What is the conditional probability of voting for Landon given that a randomly selected voter owned both a car and a phone?
>
> **(b)** If the sample accurately reflects the entire population, what is the conditional probability of a randomly selected voter having voted for Roosevelt given that the voter owned neither a car nor a phone? What is the conditional probability of voting for Landon given that a randomly selected voter owned neither a car nor a phone?
>
> **(c)** Can *Literary Digest*'s inaccurate prediction that Landon would win the election be blamed entirely on higher car and phone ownership by Landon voters? Explain.

A5
Determining Probabilities and Drawing Random Samples: Binomial, Poisson, and Hypergeometric Distributions

A5.1 INTRODUCTION

The computer packages we are reviewing can be applied to discrete distribution problems in two ways: to determine probabilities and to draw random samples. For determining probabilities, the main advantages of the computer packages are greater speed and precision over hand-held calculators. Though it takes no less time to use a package than to look up probabilities in a table, the packages are more flexible. They are

TABLE A5.1

Computer packages that can determine probabilities and draw random samples from binomial, Poisson, and hypergeometric distributions

Package	Binomial*	Poisson*	Hypergeometric*
DATA DESK	S	S	—
MINITAB	P, S	P, S	—
SAS	P, S	P, S	P
SPSSx	—	—	—

*P = determines probability; S = draws random sample.

not limited to the sample sizes and parameter values listed in a table. The primary reason for using a computer package to draw random samples from a binomial, Poisson, or hypergeometric distribution is that these samples give you a sense of how much the sample values can differ from the actual distribution. In other words, they give you a sense of what differences between a sample and a theoretical distribution may plausibly be due to chance.

The four packages vary widely in what they can do with discrete distributions. Table A5.1 shows which packages can determine probabilities and which can draw random samples from the binomial, Poisson, and hypergeometric distributions. After the applications, we will discuss determining probabilities with MINITAB and SAS. Next we will discuss drawing samples with DATA DESK, MINITAB, and SAS. SPSSx can neither determine probabilities nor draw samples from these distributions.

A5.2 SOME APPLICATIONS

Here is a binomial application. Often students incorrectly assume that a proportionality rule can be applied to probability problems. For example, when they are asked to find the probability of getting 110 or fewer successes in 200 trials with a Π of 0.5, they divide both the n and the x by 10 and use the binomial table to find the probability of 11 or fewer successes in 20 trials with a Π of 0.5. The probability of 110 or fewer successes in 200 trials is 0.98, but the probability of 11 or fewer in 20 trials is just 0.75. Figure A5.1 shows the MINITAB binomial routine output for 20 and 100 trials. In each case the Π is 0.5 and the number of successes is 10% above the mean of the distribution. The value labeled P(X LESS OR = K) is the cumulative probability of x or fewer successes. Clearly, the cumulative probabilities go up as the number of trials increases.

Here is a Poisson application. Many consumer product companies provide customers with an 800 number to call to register complaints or comments. TARP (Technical Assistance for Research and Policy) is a Washington, D.C.–based company that specializes in setting up such complaint lines. TARP has set up complaint lines for the Chevrolet division of General Motors, Goodyear Tire & Rubber, and Coors Beer, among others. Suppose a particular Chevrolet model, the 1987 Celebrity, has averaged 40 complaints per week. If each complaint about a particular car model is independent, the number of complaints per week should follow a Poisson process. In one week 70 complaints about the Celebrity are logged. Could this higher number be due

FIGURE A5.1
MINITAB output for the
binomial routine with Π of
0.5 and various *n* and *x*
values

```
MTB > set c1
DATA> 11
DATA> end
MTB > cdf c1;
SUBC> bino 20 .5 .
        K  P( X LESS OR = K)
    11.00                0.7483
MTB > set c1
DATA> 55
DATA> end
MTB > cdf c1;
SUBC> bino 100 .5 .
        K  P( X LESS OR = K)
    55.00                0.8644
```

```
MTB > set c1
DATA> 69
DATA> end
MTB > cdf c1;
SUBC> poisson mu = 40.
        K  P( X LESS OR = K)
    69.00                1.0000
```

to chance or should the company investigate a possible decline in quality at its Celebrity assembly plants? MINITAB and SAS can be used to find the probability of 69 or fewer complaints when the expected rate is 40. The complement of this probability is the chance of getting 70 or more complaints. Figure A5.2 shows the MINITAB output. The probability of 70 or more calls when the mean is 40 is approximately 0. With these results the appropriate action would be to check the quality of work at the assembly plants.

Here is a hypergeometric application. In big-money litigation cases, such as *Penzoil* v. *Texaco,* the attorneys representing both sides may investigate the backgrounds of all potential jurors. This information can help the attorneys determine the best strategy for presenting their case, the advisability of trying to settle out of court, and which potential jurors to keep off the jury if they can. Suppose a pool of potential jurors for a complex antitrust case has 100 members. The defense team thinks that if the jury can understand a highly technical economic argument, they will win their case. The defense attorneys thus consider a well-educated jury to be to their advantage. The background investigations of the pool of potential jurors have revealed that there are 40 college graduates among 100 potential jurors. What is the probability that a majority of the 12 selected for the jury will have a college education? If we assume that the final 12 are selected at random, this is a hypergeometric problem. Figure A5.3 shows the SAS commands and the one line of output. The probability of getting more than six jurors with a college education is only about 15%. The defense attorneys may want to negotiate a settlement and/or think of a simpler way to present their case.

Here is an application that draws a random sample from the binomial distribution. The University of Michigan has a scholarship program called MAG, for Michigan

FIGURE A5.3
SAS output for the hypergeometric routine

```
DATA;
  P=PROBHYPR(100,40,12,6);
  PUT P;
RUN;
0.8568906
```

Alumni Giving. MAG is a merit-based scholarship designed to encourage students with high grade point averages and SAT scores to attend the university. Local alumni clubs around the United States interview students who have expressed an interest in the University of Michigan and have been identified by the admissions office as meeting the minimum standards for the MAG scholarship. The university offers scholarships of $1,500 for the freshman year to students the alumni recommend. Although the dollar amount is modest, its benefit is enhanced by the fact that it is not counted as part of the university's contribution to a student's financial aid package. It is deducted from the parents' share of college expenses. The university has enough money for 225 scholarships. Over the past ten years the proportion of students who have accepted the scholarships has been 0.4. The admissions office limits the number of offers to 560, expecting about 225 to accept. Most years the number of acceptances has been close to 225. In 1987 only 197 students accepted the scholarships. The admissions office has to decide whether the drop has been due to chance or to some change that has made the university less competitive. If there has been some change, the admissions office can increase the number of offers in the hope of generating 225 acceptances. If the drop is due to chance, however, an increase in the number of offers may generate too many acceptances in subsequent years.

This problem fits the assumptions of the binomial distribution. The assumption that each student's decision is independent is reasonable because the students are scattered across the country. You could use the commands that will be described in section A5.3 to determine the probability of getting 197 or fewer successes in 560 trials with a Π of 0.4. Another approach to this problem is to simulate taking samples from a binomial distribution with a Π of 0.4 and an n of 560. DATA DESK, MINITAB, and SAS let you draw samples from binomial and Poisson distributions. To illustrate this simulation, DATA DESK was used to draw one sample with 100 observations from the binomial distribution with a Π of 0.4 and an n of 560. Figure A5.4 is the histogram for this sample. The histogram shows that the value 197 is unlikely to be due to chance. The admissions office may have to cast a wider net.

FIGURE A5.4

Histogram of a random sample of 100 observations from the binomial distribution with Π of 0.4 and n of 560

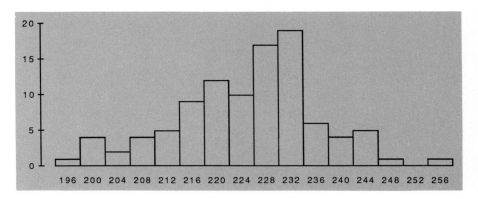

A5.3 DETERMINING PROBABILITIES FOR BINOMIAL, POISSON, AND HYPERGEOMETRIC DISTRIBUTIONS

A5.3A MINITAB

MINITAB finds probabilities for binomial and Poisson distributions with the PDF command. The commands

```
MTB > PDF;
SUBC > BINOMIAL WITH N = 100 AND P = 0.5.
```

would cause MINITAB to print the probabilities for 0 through 100 successes. A semicolon must follow the command PDF and a period must follow the subcommand BINOMIAL. N is the sample size and P is the probability of a success in a single trial. The commands for the Poisson are

```
MTB > PDF;
SUBC > POISSON PROBABILITIES MU = 40.
```

Again the semicolon and period are required. MINITAB does not allow you to specify a range of x values. If you want cumulative binomial or Poisson probabilities instead of the probability of each x value, replace the command PDF with CDF. The following commands would store the cumulative Poisson probabilities in column 2 of the MINITAB worksheet:

```
MTB > CDF;
SUBC > POISSON PROBABILITIES MU = 40, PUT INTO C2.
```

A5.3B SAS

SAS has a PROBBNML (probability binomial) function that will give you a cumulative binomial probability after you specify Π, n, and x. For example, the commands

```
DATA;
        P=PROBBNML(0.5,100,55);
        PUT P;
RUN;
```

result in the value 0.8643735.

The SAS POISSON function can find Poisson probabilities. Use the following commands to find the probability of 69 or fewer occurrences with a μ of 40:

```
DATA;
        P=POISSON(40,69);
        PUT P;
RUN;
```

Again, SAS gives you only the cumulative probability. Finally, the PROBHYPR function will find the cumulative probability for a hypergeometric distribution. The commands to find the probability of $x \leq 6$ given $\nu = 100$, $\gamma = 40$, and $n = 12$ are

```
DATA;
        P=PROBHYPR(100,40,12,6);
        PUT P;
RUN;
```

The value that results from these commands is 0.8568906.

A5.4 DRAWING RANDOM SAMPLES FROM BINOMIAL AND POISSON DISTRIBUTIONS

 ### A5.4A DATA DESK

To draw a random sample from any distribution, you begin by pulling down the menu under the `Manipulate` heading and selecting the command `Generate random numbers`. A dialog box appears that lets you select the number of samples, the size of each sample, and the distribution. If you select the binomial, you must specify Π and n. If you select the Poisson, you must specify μ.

 ### A5.4B MINITAB

MINITAB can draw random samples from binomial and Poisson distributions. The commands for drawing 100 observations from a binomial with a P of 0.4 and an n of 560 are

```
MTB > RANDOM 100 C1;
SUBC > BINOMIAL WITH N = 100, P = 0.4.
```

The `C1` specifies the column in which the results will be stored. You may use the command `PRINT C1` to see the data stored in column 1. To draw a sample with 100 observations from a Poisson with a μ of 10, the commands are

```
MTB > RANDOM 100 C1;
SUBC > POISSON MU = 10.
```

 ### A5.4C SAS

SAS generates a random sample from a binomial with the `RANBIN` function and a random sample from a Poisson with the `RANPOI` function. The commands for drawing 100 observations from a binomial with a Π of 0.4 and an n of 560 are

```
DATA A;
DO I=1 TO 100;
      X1=RANBIN(0,560,0.4)
      OUTPUT;
      END;
   PROC PRINT;
```

The `DO I=1 TO 100` tells SAS to draw 100 observations. The 0 that appears before n and Π specifies the seed or starting point for the random-number generator. The 0 tells SAS to use the computer's clock to obtain a seed value. For a sample of 100 observations from a Poisson distribution with a mean of 10, you would replace `RANBIN(0,560,0.4)` with `RANPOI(0,10)`.

EXERCISES FOR APPENDIX A5

A5.1 Jim and Tammy Rolex run a TV ministry called PTL—for Pass the Loot. The ministry generates an average of 45 letters bearing $25 checks per day. The Rolexes promise to read each letter and to pray personally for anyone who sends them $25. Some of their competitors have accused the Rolexes of being thespians, prodigals, and rodomontades. The day after these accusations surfaced, they received only 25 letters. The Rolexes claim that the reduction was due to chance and that their followers have flatly rejected the accusations. If each letter is independent, what is the probability of getting 25 or fewer letters when the mean rate is 45? You need either MINITAB or SAS to find the probabilities. If you are using DATA DESK, you can draw a sample of 100 observations from the Poisson distribution with a mean of 25 and use the proportion of the sample observations at or below 25 as an estimate of the population proportion.

(Exercise A5.2 cannot be done with SPSSx.)

A5.2 Microsoft Software in Bellevue, Washington, maintains an 800 help line for users of its software. Suppose Microsoft expected 20 calls per hour on this line and that it had a capacity of 30 calls an hour. Use a random sample of 100 observations from a Poisson distribution to estimate the proportion of hours in which there will be more than 30 calls. Plot the histogram of this sample.

(Exercise A5.3 cannot be done with DATA DESK or SPSSx.)

A5.3 Blood banks routinely test all donated blood for the antibody to the AIDS virus. The antibody test is highly reliable; the probability of the test's missing a blood sample that contains these antibodies is near zero. However, there can be up to a six-month delay between the start of an infection by the AIDS virus and the appearance of these antibodies. Blood drawn during this period can transmit AIDS but not be so identified by the antibody test. To screen out potential AIDS carriers from the donated blood pool, most blood banks ask donors if they are homosexual and if they use intravenous drugs—highly sensitive questions.

The Central Indiana Regional Blood Bank, the fifth largest supplier of blood in the United States, instituted a method of screening donors which it hopes will identify donors who have the AIDS virus but not the antibodies. Each donor is left alone with a card bearing two labels and the following instructions: "Peel off one of the barcodes and place on your Donor Registration Form in the designated area. Destroy unused barcode." One barcode is labeled "**Yes,** means use my unit for transfusion" and the other is labeled "**No,** means do not use my unit for transfusion. Use my unit for laboratory testing, and then discard." The barcodes cannot be distinguished without a barcode reader. This procedure gives donors an anonymous way of telling the blood bank that they suspect that their blood may carry the AIDS virus.

Suppose that during the first eight days of use of the barcodes the blood bank drew 1,820 units of blood and 7 of them had "No" barcodes. Of these 7, 6 tested negative for AIDS antibodies. On the basis of the proportion of all donors who test positive for the antibodies and the time between infection and the appearance of antibodies, the blood bank estimates that the population proportion of all donors who have the AIDS virus and no antibodies cannot be above 0.0005.

(a) What is the probability of getting 6 or more donors with "No" cards and negative antibody tests among the first 1,820 donors? Use the binomial distribution.

(b) Find the probability of getting 6 or more donors with "No" cards and negative antibody tests among the first 1,820 donors with the Poisson approximation to the binomial distribution.

(c) What is a simple explanation for the high number of "No" cards combined with negative antibody tests?

(Note: Since the blood bank began using this system it has changed the barcode responses to "**Use** my blood" and "**Do Not Use** my blood.")

A6
Determining Probabilities and Drawing Random Samples: Uniform, Normal, and Exponential Distributions

A6.1 INTRODUCTION

The computer applications for the continuous distributions discussed in Chapter 6 (the uniform, the normal, and the exponential) are basically the same as those for the discrete distributions discussed in Chapter 5. One advantage of using a computer to calculate probabilities for the normal and exponential distributions is that you can determine probabilities for z scores and for values of λx which are not listed in the normal and exponential tables and you can calculate probabilities to more places than you can with the tables. The main reason for drawing samples from the uniform, normal, and exponential distributions is that these samples sharpen your understanding of the distributions and give you a sense of how much random samples can differ from the theoretical distributions.

Table A6.1 shows which packages will calculate probabilities and which will draw samples from uniform, normal, and exponential distributions. As the table indicates, only MINITAB will calculate probabilities for a uniform distribution. Such calculations are so simple that it is unlikely that anyone would want them done by a package. The probability that a continuous uniform random variable will be less than or equal to some value x is $(x - a)/(b - a)$, where a and b are respectively the lower and upper limits of the distribution. For example, if X was a continuous uniform random variable with a range of 0 to 10, the probability of getting a randomly selected observation less than or equal to 1 is 0.1.

TABLE A6.1
Computer packages that can determine probabilities or draw random samples from uniform, normal, and exponential distributions

Package	Uniform*	Normal*	Exponential*
DATA DESK	S	S	—
MINITAB	P, S	P, S	P, S
SAS	S	P, S	S
SPSSx	S	S	—

*P = determines probability; S = draws random sample.

A6.2 SOME APPLICATIONS

Here is a normal distribution application. *The Wall Street Journal* conducts a semiannual survey, asking economists to predict interest rates, GNP, unemployment rates, and inflation over the next six months. The survey published on July 6, 1987, contained the predictions for the second half of 1987 and a review of the predictions for the first half of 1987, made the previous December. Figure A6.1 shows a histogram of the inflation predictions of 33 economists, as measured by the CPI, for the first half of 1987. The actual data are in data set 7. The histogram is reasonably close to a normal curve. A formal statistical test (discussed in Chapter 17) was performed on the 33 observations. The hypothesis that the population distribution of economists' predictions was normal could not be rejected. The mean of the 33 observations is 3.5 and the standard deviation is 0.7.

The annual rate of inflation for the first half of 1987, estimated on June 30, was 5.4%. As you can see from Figure A6.1, none of the economists' predictions were close to the actual rate. Inflation is notoriously difficult to forecast. The CPI can be influenced by changes in weather, shifts in OPEC policy, wars, international trade or currency agreements, and changes in Federal Reserve Board policy. The economists' predictions represent their best forecasts given the information available at the date they were made. The normal distribution with a mean of 3.5 and a standard deviation of 0.7 can be treated as the distribution of possible inflation rates for the first half of 1987, given all of the information available in December 1986.

Given the information available in December 1986, what is the probability of getting a rate of inflation greater than the observed rate of 5.4? MINITAB and SAS can determine probabilities for normal and exponential distributions. Figure A6.2 shows the MINITAB output. The density is the height of the normal curve when x is 5.4. Note that the listed value 0.0143223 is *not* the probability of having an observation exactly at 5.4. For all continuous distributions the probability of observing a point value of x is zero. Nonzero probabilities can be assigned only to a range of x values. The value 0.9967 is the probability of observing a rate of inflation at or below 5.4. The probability of getting a value greater than 5.4 is the complement, 0.003321. Given the information available in December 1986, it was extremely unlikely that the rate of inflation in the first half of 1987 would be above 5.4%.

Here is an exponential distribution application. Manufacturers of hard disk drives often include the mean time between failures (MTBF) in their ads. The ad for the Cutting Edge™ 30 megabyte hard disk drive in the August *MacWorld* magazine, for example, claimed an MTBF of 11,000 hours. The drive lists for $599 and has a one-year warranty. The ad for the 40 megabyte Priam™ hard disk drive in the August *MacUser* magazine claims an MTBF of 40,000 hours. This drive lists for $1,895 and also has a one-year warranty. If you ran these drives full time, what is the probability that they would fail during the warranty period?

The probabilities of failure during the warranty period can be determined with MINITAB or SAS. The mean times between failures for the two machines are 11,000 and 40,000 hours. A unit interval in this example is an hour. The λ, or expected rate per unit interval, would be 1/11,000 for the Cutting Edge™ hard disk and 1/40,000 for the Priam™ hard disk. There are $24 \times 365 = 8,760$ hours in a year. You are looking for the probability that either disk will fail in the first 8,760 hours of continuous use. Figure A6.3 shows the output for exponential probabilities from MINITAB. The

FIGURE A6.1
Histogram of 33 predictions for the annual rate of change in the CPI for the first half of 1987

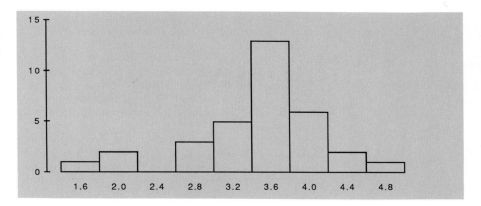

FIGURE A6.2
MINITAB output for normal probabilities

```
MTB > set c1
DATA> 5.4
DATA> end
MTB > cdf c1;
SUBC> normal mu = 3.5 sigma = .7 .
     5.4000    0.9967
```

probability that the Cutting Edge™ will fail during the first year is 0.55 and the probability that the Priam™ will fail during the first year is 0.20. While the Priam™ drive is much less likely to fail during the warranty period, it may not be cost-effective. You would need to know the cost of repairs and the cost of down time before you could make the best choice.

Here is an application that draws a random sample from a normal distribution. Architects who design sports stadiums and concert halls seem to despise women. During halftimes at sporting events and intermissions at concert halls, there are usually long lines outside of the women's restrooms and no lines outside of the men's restrooms. The common architectural practice is to allocate equal space to men's and women's restrooms, although the plumbing in the two restrooms differs. Because of

FIGURE A6.3
MINITAB output for exponential probabilities

```
MTB > set c1
DATA> 8760
DATA> end
MTB > cdf c1;
SUBC> exponential b = 11000 .
   8.76E+03     0.5490
MTB > cdf c1;
SUBC> exponential b = 40000 .
   8.76E+03     0.1967
```

the urinals in men's restrooms, men spend less time in restrooms than women do. To design a facility with enough stalls to eliminate lines in front of women's restrooms, architects need to know the distribution of women's time inside a stall and the number of women who will use the restrooms during a halftime or an intermission. Assuming independence, the number of women who will use the restrooms during a break would follow a binomial distribution in which n is the number of women in the audience. The distribution of women's time inside of a stall is likely to be approximately normal. It cannot be exactly normal because the distribution has a lower bound of zero.

Suppose a stopwatch was used to time discreetly a sample of 100 women during the halftime at a football game. The sample mean was 48 seconds and the sample standard deviation was 8 seconds. A random sample from a normal population with these parameters would give you an idea of the pattern of times you would be likely to find in a large sample taken from the actual population. All of the packages will draw random samples from normal distributions.

Figure A6.4 contains the histogram of a sample of 500 observations drawn with DATA DESK from the normal population with a mean of 48 and a standard deviation of 8. The sample data have a mean of 47.400 and a standard deviation of 7.6869. Even though the sample size is large and the sample mean and standard deviation are close to the specified parameters, the histogram does not look like a bell-shaped curve. The histogram is slightly skewed to the right and the bars do not taper smoothly in each direction. In a large sample you can be fairly certain that the sample mean will be close to the population mean. If an architect were designing a large concert hall or stadium, say of 5,000 seats or more, the key parameter would be the mean time. Given the expected number of users, an architect could eliminate lines by designing the facility with enough stalls to accommodate the average time in a stall. Designs for smaller halls have to make an allowance for random variation. In a hall with 500 seats, for example, 50 women visit the restroom during a 15-minute intermission. Three stalls would be enough as long as the sample mean time was under 54 seconds. A few chance observations at the high end of the distribution, however, could raise the sample mean above 54 seconds, create a line, and hold some women past the end of the intermission.

FIGURE A6.4
Histogram of a random sample of size 500 from a normal distribution with mean of 48 and standard deviation of 8

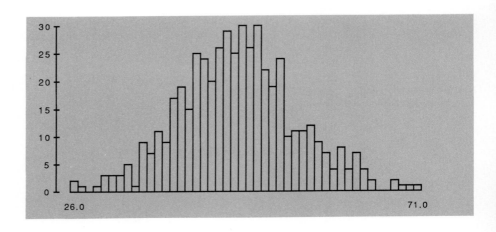

A6.3 DETERMINING PROBABILITIES FOR NORMAL AND EXPONENTIAL DISTRIBUTIONS

 A6.3A MINITAB

For a set of x values stored in column 1 of the worksheet, the MINITAB commands to calculate the density for each x and the area to the left of each x for a normal distribution with a mean of 48 and a standard deviation of 8 are

```
MTB > PDF C1, PUT INTO C2;
SUBJ > NORMAL WITH MU=48, SIGMA=8.
MTB > CDF C1, PUT INTO C3;
SUBJ > NORMAL WITH MU=48, SIGMA=8.
```

For a set of x values stored in column 1 of the worksheet, the MINITAB commands to calculate the density for each x and the area to the left of each x for a uniform distribution with a range of 1 to 10 are

```
MTB > PDF C1, PUT INTO C2;
SUBJ > UNIFORM 1 TO 10.
MTB > CDF C1;
SUBJ > UNIFORM 1 to 10.
```

The MINITAB commands to find the area to the left of an x value of 8,760 for an exponential distribution with a mean of 11,000 are

```
MTB > SET C1
DATA > 8760
DATA > END
MTB > CDF C1, PUT INTO C2;
SUBJ > EXPONENTIAL MU = 11000.
```

The x value and the parameter of 11,000 hours were taken from the application on the mean time between failures for hard disks.

 A6.3B SAS

The SAS commands for finding the probability that a normally distributed random variable with a mean of 3.5 and a standard deviation of 0.7 exceeds 5.4 are

```
DATA;
     X=PROBNORM((5.4-3.5)/0.7);
     PUT X;
RUN;
```

The term $(5.4-3.5)/0.7$ calculates the z score of 2.714286. SAS returns the probability of 0.9966791. These parameters were used in the application about economists' predictions of inflation rates.

A6.4 DRAWING RANDOM SAMPLES FROM UNIFORM, NORMAL, AND EXPONENTIAL DISTRIBUTIONS

 ### A6.4A DATA DESK

The procedure used to draw a random sample from a normal distribution is to select the command Generate random numbers from the Manipulate menu and then specify the number of samples, the size of each sample, the normal distribution, the mean, and the standard deviation.

The procedure used to draw a random sample from a uniform distribution is to select the command Generate random numbers from the Manipulate menu and then specify the number of samples, the size of each sample, and the uniform distribution. DATA DESK always draws samples from a uniform distribution with a range of 0 to 1.

 ### A6.4B MINITAB

The MINITAB commands to draw a sample of 500 observations from a normal distribution with a mean of 48 and a standard deviation of 8 are

```
MTB > RANDOM 500 OBSERVATIONS PUT INTO C1;
SUBJ > NORMAL MU = 48 SIGMA = 8.
```

The UNIFORM command will draw a sample from a uniform distribution over any range. If no range is specified, the default range is 0 to 1. The commands for drawing 500 observations from the range 0 to 10 and storing them in column 2 are

```
MTB > RANDOM 500 OBSERVATIONS PUT INTO C2;
SUBJ > UNIFORM 0 TO 10.
```

A6.4C SAS

The SAS commands to draw a sample of 500 observations from a normal distribution with a mean of 48 and a standard deviation of 8 are

```
DATA;
DO I=1 TO 500;
        X1=48+8*RANNOR(0);
        OUTPUT;
        END;
    PROC PRINT;
```

Of course you can substitute other values if you want means and standard deviations other than 48 and 8. The 0 after the RANNOR function is to tell SAS to use the computer's clock to get a seed value.

The following SAS commands will draw one observation from the uniform distribution over the 0 to 1 range:

```
DATA;
        X=RANUNI(0);
        PUT X;
    RUN;
```

The following SAS commands will draw one observation from an exponential distribution with a mean of 11,000:

```
DATA;
      X=RANEXP(0)/11000;
      PUT X;
RUN;
```

 A6.4D SPSSx

The following SPSSx commands will draw 500 observations from a normal distribution with a mean of 48 and a standard deviation of 8:

```
INPUT PROGRAM
LOOP #I=1 TO 500
COMPUTE X=48+NORMAL(8)
END CASE
END LOOP
END FILE
END INPUT PROGRAM
PRINT / X
EXECUTE
```

The argument of the NORMAL function is the standard deviation of 8. SPSSx automatically uses the computer clock to determine a seed value. To draw a sample with 500 observations from a uniform distribution over the range 0 to 1, substitute the following for the line beginning with COMPUTE in the above commands:

```
COMPUTE X=UNIFORM(1)
```

The argument of the UNIFORM function is the upper limit of the uniform distribution. The lower limit is always 0.

EXERCISES FOR APPENDIX A6

(Exercises A6.1 and A6.2 cannot be done with DATA DESK or SPSSx.)

A6.1 Use a computer package to find the probability of 50 or more successes in 100 trials with a Π of 0.532 directly from the binomial distribution. Next compute the same probability using the normal approximation to the binomial distribution. By how much does the approximation differ from the first figure? The purpose of this exercise is to check the accuracy of the normal approximation to the binomial. Do not forget the continuity correction.

A6.2 Use a computer package to find the probability of 80 or more occurrences with an expected rate of 72.3 directly from the Poisson distribution. Next compute the same probability using the normal approximation to the Poisson distribution. By how much does the approximation differ from the first figure? The purpose of this exercise is to check the accuracy of the normal approximation to the binomial. Do not forget the continuity correction.

A6.3 Draw a random sample of 250 observations from the standard normal distribution and plot the histogram of the sample. How many observations would you expect to be above +3 or below −3? How many do you get in your sample? How many observations would you expect to be above +2 or below −2? How many do you get in your sample?

(Exercise A6.4 cannot be done with DATA DESK or SPSSx.)

A6.4 You have purchased a video camcorder for $1,200. It comes with a 90-day warranty. The time between failures for this model is exponentially distributed with a mean of 600 days. The dealer offers you an extended warranty that will cover any repairs from the 91st day to the end of the third year. This extended warranty costs $250. What is the probability that the first failure will occur during the extended warranty period? You think that given the cost of repairs, the extended warranty is worthwhile only if there are two failures before the extended warranty expires. What is the probability of having two or more failures before the extended warranty expires? Assume that the repair time is zero and that a repaired machine is equivalent to a new machine.

(Exercise A6.5 cannot be done with DATA DESK.)

A6.5 Use a computer package to draw a random sample from a uniform distribution with a range of 0 to 100. Explain why multiplying the observations drawn from a uniform distribution with a range of 0 to 1 by the constant 100 will give you the sample you want.

A6.6 Draw samples of size 20, 30, 40, 50, 60, 70, 80, 90, and 100 from a standard normal distribution. Make a histogram of each sample. Does each sample look like a normal distribution? How do the histograms change as the sample size increases? If you are using the student version of DATA DESK, you will have to throw away each sample and empty the trash to keep enough room in the data file.

A6.7 Draw samples of size 20, 30, 40, 50, 60, 70, 80, 90, and 100 from a uniform distribution with a range of 0 to 1. Make a histogram of each sample. Does each sample look like a uniform distribution? How do the histograms change as the sample size increases? If you are using the student version of DATA DESK, you will have to throw away each sample and empty the trash to keep enough room in the data file.

A7
Simulating the Sampling Distribution of the Mean and Drawing Random Samples from a List

A7.1 INTRODUCTION

Reflecting the division of material in Chapter 7, Appendix A7 is in part theoretical and in part highly practical. The theoretical part is the simulation of sampling distributions. The practical part consists of instructions on how to draw simple random samples from lists. In the samples drawn in Appendixes A5 and A6, you considered only the distribution of individual observations. By taking multiple samples and plotting the histogram of the means of those samples, you can demonstrate the central limit theorem; that is, you can demonstrate that the distribution of sample means will be approximately normal. The central limit theorem is the most important theorem in statistical inference. This simulation will help you understand the theorem. You can also demonstrate that the normality approximation improves as the sample size increases. Another point made in Chapter 7 which you can demonstrate is that skewed population distributions require a larger sample size than symmetrical populations for the distribution of sample means to be approximately normal. Finally, you can demonstrate the version of the central limit theorem for the sampling distribution of the proportion; that is, the sample proportion will be approximately normal when $n\Pi$ and $n(1 - \Pi)$ are greater than 5.

Appendixes A5 and A6 discussed the use of statistical packages to draw random samples from binomial, Poisson, hypergeometric, normal, uniform, and exponential distributions. Computer packages are also used to draw random samples from lists of observations or sampling frames. Because of the cost savings, most sampling from large sampling frames is done with computers rather than tables of random numbers. Section A7.3 will discuss the mechanics of drawing a random sample with each of our packages from data set 5, the H & R Block stock data. Some of the packages have simple one-line commands that will draw the sample. Unfortunately, the procedures required by other packages are convoluted.

A7.2 AN APPLICATION

To illustrate the effect of larger sample sizes on the distribution of sample means, eight sets of 100 samples each were drawn from a continuous uniform distribution with a range of 0 to 1. The first set of 100 samples had 10 observations in each sample, the second set of 100 samples had 20 observations in each sample, and so on, until the eighth set had 80 observations per sample. With some of the computer packages, drawing eight sets of 100 samples can be very time-consuming. The exercises at the end of Appendix A7 are considerably smaller.

Figure A7.1 shows the box plots for the means within the eight sets of samples. Each box plot represents 100 sample means. The figure shows that the distributions of sample means tend to get tighter as the sample size increases. Both the whiskers and the hinges narrow as sample sizes increase from 10 to 80. The box plots also show a few outlying values for the sample mean. For example, in this simulation a sample mean near 0.35 was obtained from a sample of size 60. The expected value of the sample mean for a uniform distribution with a range of 0 to 1 is 0.5. The main points of the simulation are that as the sample size increases, the sample means tend to be closer to the expected value of 0.5, but even with a sample of as many as 60 observations, you are not *guaranteed* a sample mean close to the expected value.

Another point that can be made by these simulations is that the distribution of sample means will look increasingly like a normal curve as the sample size increases.

Box plots of eight sets of 100 sample means for samples drawn from a continuous uniform distribution with a range of 0 to 1 (the size of each sample varies from 10 in the first set to 80 in the eighth set)

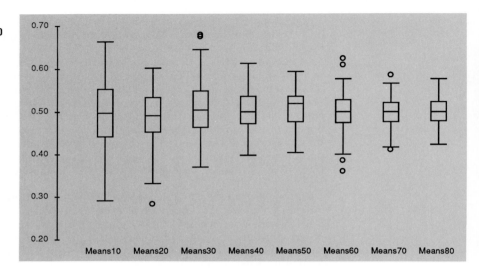

Histogram of the means of the 100 samples of size 10

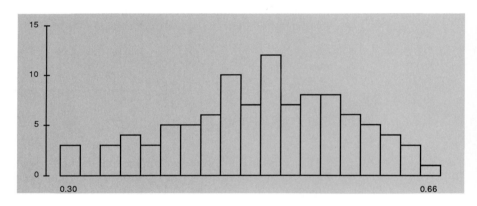

Figure A7.2 shows the histogram of the means of the 100 samples of size 10. It tapers slowly. Figure A7.3, in contrast, shows the more peaked histogram of the sample means when the sample size is 80.

The kind of sampling done with the uniform distribution can also be done with the other theoretical distributions discussed so far in the text: the binomial, the Poisson, the hypergeometric, the normal, and the exponential. Appendix A has already discussed the basic commands for drawing samples and for creating box plots and histograms. These simulations reinforce the idea that the sample mean, on average, does a good job of estimating the population mean when you have a large sample. The simulations also show that the distribution of sample means tends to tighten as the sample size increases. Lastly, the simulations show that the distribution of sample means will be approximately normal for large sample sizes.

FIGURE A7.3
Histogram of the means of the 100 samples of size 80

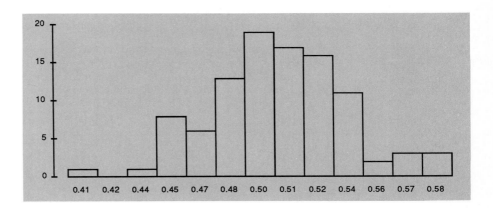

A7.3 DRAWING RANDOM SAMPLES FROM A LIST

 A7.3A DATA DESK

The procedure for drawing a random sample from a list of observations with DATA DESK is convoluted. Here is how you would draw a random sample of size 25 from the monthly volume variable in data set 5. Monthly volume is the second column of the H & R Block stock data set. Begin by drawing a random sample with the same number of observations as your sampling frame from the continuous uniform distribution. Data set 5 has 156 elements, so the sample from the uniform distribution should also have 156 observations. Click on the icon of this variable to select it and, while holding down the shift key, click on the icon of the monthly volume variable. Click on the command Sort on Y, Carry X's under the Manipulate menu. The observations in monthly volume will be sorted according to the order of the observations in the sample from the uniform distribution. To get a sample from monthly volume of size 25, delete all but the first 25 observations. You can avoid counting by hand by creating a variable with the Generate patterned data command under the Manipulate menu which runs from 1 to 25.

 A7.3B MINITAB

MINITAB's commands for getting a random sample from a list of numbers are simple. Here is how you would draw a random sample of size 25 from the monthly volume variable in data set 5. Monthly volume is the second column of the H & R Block stock data set. The MINITAB command for drawing the random sample is

```
MTB > SAMPLE 25 OBSERVATIONS FROM C2 AND PUT SAMPLE IN C4
```

If column 2 has the original list, the sample of 25 observations will appear in column 4.

A7.3C SAS

The SAS commands needed to generate a random sample of size 25 from 156 observations in data set 5 are somewhat more complex than the MINITAB commands but simpler than DATA DESK's. Monthly volume is the second column of the H & R Block stock data set. The 156 observations have been read into SAS and placed in a data set named HR. The SAS commands are

```
DATA SAMPLE;
     RETAIN N 156 K 25 ;
     SET HR;
     N = N - 1;
     IF UNIFORM(0) < K/N;
     K = K - 1;
PROC PRINT;
```

The RETAIN N 156 K 25 starts the value of the variable N at 156 (the population size) and value of the variable K at 25 (the sample size). The SET HR tells SAS to use the observations in the HR data set. The N = N - 1 steps down the variable N by units of one. The next line compares the ratio of K to N to a randomly drawn observation from the uniform distribution with a range of 0 to 1. If this ratio is less than the observation drawn from the uniform distribution, the variable K steps down by one unit and the observation becomes part of the data set SAMPLE. If the ratio is not less than the observation drawn from the uniform distribution, the variable K does not step down by one unit and the observation does not become part of the data set SAMPLE. SAS runs until it finds 25 observations to print in the data set SAMPLE.

 A7.3D SPSSx

The SPSSx commands for drawing a random sample from a list are simple. The data set with the list must be the active file, that is, the last file accessed. The SPSSx commands are

```
DATA LIST FILE = HR / OBS
SAMPLE 25 FROM 156
PRINT / OBS
EXECUTE
```

EXERCISES FOR APPENDIX A7

A7.1 To demonstrate the effect of a skewed versus a symmetrical population distribution on the distribution of sample means, take 25 samples of size 10 from the binomial distribution with a Π of 0.5 and an n of 50 and 25 samples of size 10 from the Poisson distribution with a μ of 1. The binomial with a Π of 0.5 is symmetrical and the Poisson with a μ of 1 is skewed to the right. Plot the histograms of the means of all samples. Are both histograms symmetrical?

A7.2 The box plots in Figure A7.1 show the medians of eight sets of 100 sample means. If the grand means of all sets of 100 sample means were shown in the figure, would they tend to be closer to, the same distance from, or farther from 0.5 than the medians? Explain.

A7.3 Data set 10 has 100 observations consisting of 25 0's and 75 1's. Draw 25 samples of size 10 from this population. For each of the 25 samples calculate the proportion of 1's. Plot the histogram of the sample proportion. Is the distribution of sample proportions symmetrical? Does it appear to be normally distributed? In this example, should the distribution of sample proportions be normal?

A8
Calculating Confidence Intervals

A8.1 INTRODUCTION

If you want to calculate confidence intervals, computer packages offer several advantages over the hand-held calculator. If the data set is large, there is a gain in time and accuracy. The gain in accuracy is not due solely to the elimination of simple keying errors on the calculator. Accuracy is also increased because the packages carry the calculations to many more digits than most calculators and provide t scores that are not listed in the t table. The most important advantage of using a computer package to calculate confidence intervals is that the ease of use encourages you to look more closely at the data. For example, you can construct confidence intervals for subsets of the data. You can make subsets by dropping outliers or by focusing on observations that have some common characteristic. You can also try a variety of confidence levels to see how the width of the interval is affected.

Our four computer packages offer different levels of support for calculating confidence intervals. The lowest level of support is provided by the inclusion of the sample mean and its standard error in the package's descriptive statistics routine. With these statistics and a t table, you can calculate the limits of a confidence interval. The highest level of support is provided when the package calculates limits of a confidence interval on the basis of a confidence level you specify, using either the t score or the z score—the choice depends on whether the population standard deviation is known. SAS and SPSSx are at the low end of support for calculating confidence intervals. MINITAB and DATA DESK are at the high end. The differences reflect the origins of the programs. MINITAB and DATA DESK were originally written to supplement instruction in statistics. Now both programs are widely used in professional work. SAS and SPSSx were originally mainframe packages aimed at university researchers and large businesses.

A8.2 AN APPLICATION

Data set 8 contains the records of 341 students who took a course in Principles of Microeconomics at Indiana University–Purdue University at Indianapolis (IUPUI) in 1986. The data set includes their course grade, high school ranking, ethnic status, math and verbal SAT scores, age, and instructor. The 341 students can be treated as a sample of all students to take this course at IUPUI in the future. One issue facing the instructors is the choice of textbook. Some textbooks are known to be harder than others. The instructors have some information on the textbooks that are used at other schools and the average SAT scores of freshmen entering those schools. While the entering freshmen are not strictly comparable to the students who took Principles of Microeconomics in 1986, they represent the only data available. The sample mean of the 341 students' combined math and verbal SAT scores would estimate the population mean for all future IUPUI students to take this course. A confidence interval around this point estimate would give the instructors a sense of how far off the point estimate might be. The data set has the SAT scores for 263 students (the other students did not take the SAT). The limits of a confidence interval at the 95% level of confidence are

$$L = 851.38 \qquad U = 897.57$$

If the instructors wanted to push the students a little, they might look for books that were used at schools that had average SAT scores at the high end of this range. IUPUI has a special admissions policy that lets students with low high school grades and/or SAT scores enter the university on a probationary basis. Those who do not achieve a certain GPA by the end of their first year have to leave. These probationary students make up the left tail of the distribution of SAT scores. Suppose the instructors of the course believed that the textbook decision should not be influenced by the SAT scores of the probationary students. As these students have been given a sink-or-swim chance to do college-level work, it would be inconsistent to use their SAT scores to define "college level." We'll take scores of less than 750 to represent probationary students. If every observation with a combined SAT score of 750 or less were stripped from the data set, how would the limits of a 95% confidence interval change? On the basis of the remaining 199 students, the limits of a confidence interval for the mean SAT score for all students with scores of 750 or more is

$$L = 931.04 \qquad U = 972.22$$

As you would expect, the limits of the confidence interval are raised when the lower tail of the sample is dropped. The width of the confidence interval should also be reduced. The reduction in the width, however, is modest—46.19 to 41.18. One point that can be noted about the data is the enormous variability in combined SAT scores. The range is from 420 to 1,420.

Another conclusion suggested by the last confidence interval is that the instructors should pick one of the easier books. Cutting off the lower tail of the sample *and* taking the upper limit of the confidence interval still leaves you with an estimated mean of only 972. Some IUPUI instructors have used Paul Samuelson and William Nordhaus's *Economics*, and others have used Richard Lipsey and Peter Steiner's *Eco-*

nomics, but these books are more typically used at schools whose students' SAT scores average between 1,100 to 1,200. Before these SAT scores were taken into consideration, the difficulties many students experienced were blamed on their laziness or their family responsibilities or their lack of study time because they held full-time jobs. While these factors may share the blame, inadequate high school preparation alone could account for their problems.

A8.3 COMMANDS

 ### A8.3A DATA DESK

DATA DESK can handle missing values and nonnumerical data. You can read the data into DATA DESK as they appear in the file. With DATA DESK you add the two variables SATMATH and SATVERB by writing an algebraic statement that defines a new derived variable. The New derived variable command is under the File menu. The algebraic statement would be

    SATMATH + SATVERB

The easiest way to eliminate the combined SAT scores below 750 is to sort the combined data with the Sort on Y, carry X's command under the Manipulate menu and then delete observations below 750. To create the confidence interval, use the Estimate command under the Compute menu. The program prompts you for the confidence level (95% is the default). Another prompt is whether to use the *t* score or the *z* score. If you select the *z* score, the program asks you for the population standard deviation. The program asks you if you want a two-sided or a one-sided confidence interval.

 ### A8.3B MINITAB

MINITAB will read data with missing values. Instead of the -99999 code in the data, an asterisk (*) is the MINITAB symbol for missing data. To combine the observations across two columns in MINITAB, use the following command:

    MTB > LET C3 = C1 + C2

The command for a confidence interval for the data in column 3 is

    MTB > TINTERVAL WITH 95 PERCENT CONFIDENCE, DATA IN C3

This command produces a 95% interval for the population mean when the *t* score is used. The WITH 95 PERCENT CONFIDENCE can be dropped because 95% is the default. You can specify another confidence level. For a 95% confidence interval with the *z* score the commands are

    MTB > ZINTERVAL SIGMA = 50, DATA IN C3

This example uses a population standard deviation of 50 for the sake of illustration. The OMIT command lets you delete a subset of the data. For example, the following commands will put the observations that have values 750 or above in column 4:

    OMIT ROWS WITH VALUES 0 UP THRU 749 IN C3 PUT IN C4

A8.3C SAS

Here are the SAS commands to read each student's SAT math and verbal scores, add them together, delete all missing observations and observations with combined scores of less than 750, calculate the sample mean and the standard error of the mean, and print the results. You have to calculate the limits of the confidence interval.

```
DATA STUDENTS;
INPUT SATVERB SATMATH;
SAT = SATVERB + SATMATH;
IF SATVERB EQ -99999 THEN DELETE;
IF SATMATH EQ -99999 THEN DELETE;
IF SAT LT 750 THEN DELETE;
PROC MEANS;
        VAR SAT;
```

◼ A8.3D SPSSx

SPSSx can read alphabetical data and handle missing data. Here are the SPSSx commands to read each student's SAT math and verbal scores, add them together, delete all observations with combined scores of less than 750, calculate the sample mean and the standard error of the mean, and print the results. You have to calculate the limits of the confidence interval. Note that missing observations are coded -99999.

```
DATA LIST FILE = STUDENTS/ SATVERB SATMATH
COMPUTE SAT = SATVERB + SATMATH
SELECT IF SAT GE 750
FREQUENCIES VARIABLES = SAT /
       STATISTICS = MEAN SEMEAN
FINISH
```

Another way to do this problem in SPSSx is to use the ONEWAY procedure for analysis of variance. This procedure will calculate confidence intervals at the 95% level.

EXERCISES FOR APPENDIX A8

A8.1 Data set 2 lists the horsepower of 392 car models sold in the United States over the years 1971 to 1983. Draw a sample of size 10 from this data set and construct confidence intervals at 70%, 80%, 90%, and 99% levels of confidence. Identify the intervals that include the population mean.

A8.2 Data set 3 lists the tar content in milligrams for 203 brands of cigarettes sold in the United States in 1983. The source of the data is the Federal Trade Commission. Draw 100 samples of size 10 from this data set. For each sample construct a confidence interval at the 95% level of confidence. How many of the 100 intervals include the population mean?

(Exercise A8.3 cannot be done with DATA DESK.)

A8.3 Draw a sample of 20 U.S. cars from data set 6. Construct a confidence interval at the 95% level of confidence for the population proportion of U.S. cars that are rated either worse than average or much worse than average. Does the interval include the population proportion?

A9
Hypothesis Tests with One Population

A9.1 INTRODUCTION

The main benefits of using a computer for the one-population hypothesis test discussed in Chapter 9 are speed and accuracy. These benefits are particularly important when you work with large data sets. DATA DESK and MINITAB have commands that directly do the hypothesis tests. SAS and SPSSx give you only the sample mean and the standard error of the mean. We have discussed these commands already. After you obtain the sample mean and the standard error of the mean, you need to calculate the critical value or values for the test.

A9.2 AN APPLICATION

The head of every SAE grade 8 bolt bears a symbol to show that the bolt meets the standards set by the Society of Automotive Engineers. These high-tension bolts are used in buildings, bridges, military vehicles, nuclear reactors, and other high-stress/high-temperature applications. About 1 billion grade 8 bolts are used by U.S. industry each year. Bogus grade 8 bolts have become a widespread problem. They are imported, primarily from Japan, by unscrupulous U.S. dealers and then mixed with shipments of genuine bolts. The bogus bolts have the same markings as the genuine grade 8 bolts, but because of differences in their chemical composition and tempering, the bogus bolts will fail at lower temperatures than the genuine bolts.

The Defense Industrial Supply Center in Philadelphia has tested random samples of bolts from its inventory and found that 20% of the bolts sampled were bogus. It is impractical to test every bolt before it will be used; besides the expense, the testing is destructive. And the bogus bolts have so deeply penetrated U.S. inventories that it will be impractical in the near future to set a standard of 100% genuine bolts for all but the most severe applications. The Department of Defense and many industrial firms are now setting up routine tests of bolts based on random sampling. A maximum tolerable proportion of bogus bolts has to be specified for each application and a formal hypothesis test must be conducted to determine whether a given shipment is below that maximum. Given the enormous number of bolts and fasteners involved, the record keeping and calculations will have to be done on computers. If we did not have this capability, the United States would really be screwed up.

A9.3 COMMANDS

A9.3A DATA DESK

The `Test` command under the `Compute` menu does single-population hypothesis tests. You can use a t score, or, if the population standard deviation is known, you can use the z score. You begin by clicking on the icon of the variable that contains the sample data and then clicking on the `Test` command. The program prompts you for the significance level and the hypothesized value of the population mean. It also prompts you to specify whether the test is two-tailed, left-tailed, or right-tailed. The output includes either the statement `Fail to reject Ho at alpha = 0.05` or the statement `Reject Ho at alpha = 0.05`.

If you have a large data set in which successes are coded 1 and failures 0, you can use the `Test` command for a hypothesis test on the population proportion. The critical value is approximate but quite close if there are more than 100 observations. An exact method for a hypothesis test on the population proportion is to use the `Summary reports` command under the `Compute` menu to calculate the sample mean. The sample mean is also the sample proportion of successes when the data are coded 0 or 1. With this sample proportion you can use formula 9.6 to calculate the critical value or values.

A9.3B MINITAB

The MINITAB commands for a right-tailed single-population hypothesis test with the z score are

```
MTB > ZTEST OF POPULATION MU = XX,SIGMA = XX,DATA IN C1;
SUBJ > ALTERNATIVE = 1.
```

The only thing that is not self-explanatory is the subcommand `ALTERNATIVE = 1`. It tells MINITAB to do a right-tailed test. A two-tailed test is the default. For a two-tailed test, the subcommand `ALTERNATIVE` is dropped and the semicolon is omitted from the first line. The commands for a left-tailed test with the t score are

```
MTB > TTEST OF POPULATION MU = XX,DATA IN CX;
SUBJ > ALTERNATIVE = -1.
```

The output for the TTEST or ZTEST commands is the p value of the test. Recall that if p is less than α for the test, you can reject the null hypothesis.

If you have a large data set in which successes are coded 1 and failures 0, you can use the TTEST command for a hypothesis test on the population proportion. The critical value is approximate but quite close if there are more than 100 observations. An exact method for a hypothesis test on the population proportion is to use the DESCRIBE command to calculate the sample mean. The sample mean is also the sample proportion of successes when the data are coded 0 or 1. With this sample proportion you can use formula 9.6 to calculate the critical value or values.

 A9.3C SAS

There are no simple commands in SAS for single-population hypothesis tests. The MEANS procedure reports the p value for the null hypothesis that a variable equals 0 (i.e., $H_0: \mu = 0$). If your null was that $\mu = 10$, you could use the MEANS procedure by creating a new variable with the statement

    Xl = X - 10;

The reported p value for the variable X1 would be the same as the p value for the null hypothesis of $\mu = 0$ for the variable X1. Recall that if the p value is less than the α for the test, you can reject the null hypothesis. SAS cannot easily do single-population hypothesis tests with the z score.

 A9.3D SPSSx

There are no simple commands in SPSSx to do one-population hypothesis tests. You can generate the sample mean and the standard error with the FREQUENCIES procedure and calculate the critical values for the test by hand.

EXERCISES FOR APPENDIX A9

(Exercise A9.1 cannot be done with DATA DESK.)

A9.1 Data set 10 consists of 75 1's and 25 0's. Let the 1's represent genuine bolts and the 0's bogus bolts. Treat this data set as a random sample from a large population of bolts. With this sample use a statistical package to test the null hypothesis that the population proportion of bogus bolts is greater than or equal to 0.3 at a 5% significance level. How far off is the reported p value from the correct p value?

A9.2 Data set 11 contains a one-in-twenty systematic sample from *Petersen's College Guide*. The first column is the enrollment per college. Use this sample to test the null hypothesis that the mean enrollment per college is less than or equal to 3,000 at a 5% significance level.

A9.3 The second column of data set 11 is the number of full-time faculty per college. Test the null hypothesis that the ratio of students to faculty at American colleges is less than or equal to 20 at the 5% significance level.

A10
Two-Population Confidence Intervals and
Hypothesis Tests

A10.1 INTRODUCTION

Some characteristics of the four packages' two-population tests and confidence intervals are common to all of them. First, none of the packages lets you specify σ_1 and σ_2. While this omission is not too serious—you rarely know both population standard deviations—it seems inconsistent to let the user specify the population standard deviation for a one-population test but not for a two-population test. Second, none of the packages constructs confidence intervals for the difference in population proportions. The reason may be that the calculations can easily be done by hand if the proportion of successes in each sample is known. A quick-and-dirty expedient for constructing difference-in-population-proportion confidence intervals is to use a package's routine for difference in means. This can be done if the successes and failures are coded 1 and 0, respectively. The difference-in-means confidence interval provides a rough approximation when the sample sizes are large (more than 100). Third, none of the packages does difference-in-population-proportion hypothesis tests. Again, the reason may be that the calculations can easily be done by hand given the proportion of successes in each sample.

A10.2 AN APPLICATION

During the 1940s and 1950s Pepsi-Cola ran advertisements with the theme that Pepsi's prices were lower than Coca-Cola's prices. Pepsi's slogan was "12 ounces for the price of 8." Coca-Cola was the established cola; Pepsi relied on the price differential to gain a foothold in the market. An unintended effect of these advertisements was the impression that Pepsi was a cheap substitute for Coca-Cola. Pepsi found that status-conscious consumers would disguise servings of Pepsi-Cola. A hostess would pour Pepsi into glasses in her kitchen, for example, and then serve her guests in the living room from a tray that held bottles of Coca-Cola plus the glasses filled with Pepsi-Cola. Pepsi's prospects for growth were stymied by the perception that it was a cheap substitute for Coca-Cola. In the late 1950s Pepsi began a new series of advertisements with the theme that Pepsi-Cola was *de rigueur* in high society. The slogan was "Be sociable." A typical ad featured models in dinner jackets and evening gowns drinking Pepsi-Cola on the deck of a yacht.

All of this must seem like ancient history, yet the ads that emphasized Pepsi's low price may have a lingering effect. Do families that usually buy Coca-Cola today have higher average incomes than families that usually buy Pepsi-Cola? In 1983 (before the New Coke debacle) Market Opinion Research conducted a telephone survey of 400 Indianapolis households on their soft-drink buying habits. The survey was sponsored by the local Pepsi-Cola bottling company. Among the 400 families, 54 were classified as Pepsi loyalists, 56 as Pepsi semiloyalists, 24 as Coke loyalists, and 19 as Coke semiloyalists. "Loyalists" were families that said they stuck with their preferred brand even if the other brand was $1 less per eight-pack. "Semiloyalists" were families that said they stuck with their preferred brand even if the other brand was 50 cents less per eight-pack. On the basis of this sample, Pepsi clearly dominates Coke in the Indianapolis area. The mean family income among the Pepsi loyalists and semiloyalists was approximately $19,592. The mean family income among the Coke loyalists and semiloyalists was approximately $25,549. The incomes are approximate because the data are classified into 11 income ranges with an open-ended top range of $50,000 or more. Four Coke and no Pepsi families were in this last range. The mean income in the last range was assumed to be $52,500. The null hypothesis of equal population mean incomes for the two groups could be rejected at a 5% significance level. The appropriate test is an unpooled difference-in-means hypothesis test. Since the sample sizes are large for both groups, the distributions of sample means will be approximately normal.

Differences in cola preference by income and by geographic area may account for a new puzzle in the cola wars. In 1987 both Pepsi-Cola and Coca-Cola were running advertisements claiming that their cola was the most preferred by consumers in blind taste tests. Both cannot be the most preferred. Either company can slant the test results toward its cola by picking the cities and the locations within the cities. For example, Pepsi-Cola would do better if its tests were conducted in such Pepsi strongholds as Indianapolis and in locations within those strongholds that were likely to be frequented by low-income families, such as K Marts. Coca-Cola would undoubtedly do well in some affluent suburbs of Atlanta, Coke's home base.

A10.3 COMMANDS

A10.3A DATA DESK

The `Test` command found under the `Compute` menu will do difference-in-means and paired difference-in-means hypothesis tests. The dialog box that appears after you have selected the `Test` command includes three choices under the heading `t test for` $\mu_1 - \mu_2$: a pooled variance test, an unpooled test, and a paired test. The program prompts you for the hypothesized difference between population means, the significance level, and whether the test is two-tailed, right-tailed, or left-tailed. The output states that the null hypothesis is rejected or that it is not rejected at the specified significance level.

DATA DESK does difference-in-means confidence intervals with the `Estimate` command found under the `Compute` menu. The options are similar to the ones for the difference-in-means hypothesis tests. You can specify the level of confidence, whether the interval is one- or two-sided, and whether the variances are pooled or not pooled.

 ### A10.3B MINITAB

The MINITAB commands for a right-tailed pooled variance test on the difference in population means are

```
MTB > TWOSAMPLE TEST FIRST SAMPLE IN C1, SECOND SAMPLE IN C2;
SUBJ > ALTERNATIVE = 1;
SUBJ > POOLED.
```

A minus 1 in place of the positive 1 would change the specification to a left-tailed test. For a two-tailed test the subcommand `ALTERNATIVE = 1` is dropped. For an unpooled test the subcommand `POOLED` is dropped. Note that if there are any subcommands, you must have a semicolon after the main command and that the last subcommand must be followed by a period. The output in all cases includes the p value of the test. MINITAB always does difference-in-means tests against the hypothesized difference of zero. If you wanted to test the null hypothesis that the difference between two population means was some other number, say 10, you would have to subtract 10 from the values in the first column. The commands would be

```
MTB > READ C1 C2
MTB > LET C3 = C1 - 10
MTB > TWOSAMPLE TEST C3, C1
```

Note that the two samples would be stored separately in column 1 and column 2. The `TWOSAMPLE TEST` command automatically reports the limits of a 95% confidence interval for the difference in population means. The commands to specify a 90% level of confidence are

```
MTB > TWOSAMPLE TEST PERCENT CONFIDENCE 90% C1, C2
```

Any number above 5 can be substituted for the 90. You can get a confidence interval based on a pooled estimate of the population variance by using the POOL subcommand.

To do a paired difference-in-means test in MINITAB, you first calculate the differences between pairs of observations and then do an ordinary hypothesis test for one population mean on the differences. The commands for a two-tailed paired difference test on a hypothesized difference of 10 are

```
MTB > LET C3 = C2 - C1
MTB > TTEST OF MU = 10, DATA IN C3
```

A10.3C SAS

The SAS procedure for a difference-in-means test is `TTEST`. The data have to be arranged so that all of the observations are in one column and an indicator variable, which can be numerical or alphabetical, is in another column. The indicator variable must take two values, such as 0 and 1, or "male" and "female." To test the hypothesis that the mean math SAT scores for men and women are the same, for example, the data could be arranged as follows:

Math SAT score	Sex
340	male
550	female
610	male
600	male

If the data are stored in the file SCORES, the commands are

```
DATA SCORES;
INPUT SCORES;
INPUT SATMATH SEX $;
PROC TTEST;
      CLASS SEX;
      VAR SATMATH;
```

The purpose of the $ in the input statement is to tell SAS that SEX is an alphabetical variable. The indicator variable is named in the CLASS statement and the variable containing the numerical values is named in the VAR statement.

The TTEST output includes the p values for both the pooled and unpooled tests of the null hypothesis that the population means are equal. The one-tailed p value is simply one-half of the reported two-tailed p value. The output also includes the results of an F test for the null hypothesis that the population variances are equal. This test is described in Appendix B. Do not automatically use the pooled version of the difference-in-means test just because the null hypothesis of equal population variances cannot be rejected. The pooled version of the test should be used only when there is a reason for treating the two population variances as equal. The most common reason is that you suspect the two samples came from the same population.

The TTEST procedure always does difference-in-means tests against the hypothesized difference of zero. If you wanted to test the null hypothesis that the difference between the means of the two populations' SAT math scores was some other number, say 10 points, you would have to subtract 10 from the observations for males. The following SAS command would be added after the INPUT command:

```
IF SEX EQ 'MALE' THEN SATMATH = SATMATH - 10;
```

The single quotes enclosing MALE tell SAS to look for a string of letters rather than a numerical value.

SAS will do a paired difference-in-means hypothesis test with the MEANS procedure. If the males and females in the SAT data were sets of brothers and sisters, the paired test would be appropriate. If the male and female math SAT scores were in different columns, you could create a new variable for the differences, for example, with the commands

```
DATA SCORES;
INPUT SATMALE SATFEM;
DIFF = SATMALE - SATFEM;
CARDS;
PROC MEANS;
      VAR DIFF;
```

SAS does not calculate the limits of a confidence interval for the difference in population means. To obtain a confidence interval you will have to calculate the limits from the sample means and sample standard deviations.

■ A10.3D SPSSx

SPSSx has a T-TEST procedure that does either independent sample or paired sample difference-in-means hypothesis tests. The data have to be arranged so that all of the observations are in one column and a numerical indicator variable is in another column. If the indicator variable takes only two values, such as 0 and 1, only the name of the indicator variable is required in the SPSSx GROUPS command. If the indicator variable takes more than two values, such as 1 and 2, the two values that will be used in the test have to be specified. For example, to test the hypothesis that the mean math SAT scores for men and women are the same, the data could be arranged as:

Math SAT score	Sex
340	2
550	1
610	2
600	2

The SPSSx commands are

```
T-TEST GROUPS = SEX / VARIABLES = SATMATH
```

The SEX variable can take only two values. Note that if the SEX variable was coded 0 and 1, the GROUPS = SEX command would not be sufficient. In this case the command has to be changed to GROUPS = SEX(0,1) to specify the two values of the SEX variable. Suppose CITY was an indicator variable coded 0 for Los Angeles, 1 for New York, 2 for Chicago, and so on. To test the null hypothesis that the mean Los Angeles SAT math scores equal the mean Chicago math scores, the commands are

```
T-TEST GROUPS = CITY (0,2) / VARIABLES = SATMATH
```

The T-TEST output includes p values for both the pooled and unpooled tests of the null hypothesis that the population means are equal. The one-tailed p value is simply one-half of the reported two-tailed p value.

The T-TEST procedure always does difference-in-means tests against the hypothesized difference of zero. If you wanted to test the null hypothesis that the difference between men's and women's SAT math scores means was some other number, say 10 points, you would have to subtract 10 from the observations for males. Use the following SPSSx command:

```
IF SEX EQ 1 SATMATH = SATMATH - 10
```

The T-TEST procedure has a command for paired tests. If the males and females in the SAT data were sets of brothers and sisters, the paired test would be appropriate. If the male and female math SAT scores are in different columns and assigned the names SATMALE and SATFEM, the commands for a paired test are

```
T-TEST PAIRS = SATMALE SATFEM
```

Again, the output is the p value for the null hypothesis of a zero mean difference.

SPSSx does not construct confidence intervals for the difference in population means. However, the T–TEST procedure does report the sample means, the degrees of freedom with the pooled and unpooled variances, and the estimated standard deviation of the difference in sample means for independent samples. With these terms you can easily construct confidence intervals.

EXERCISES FOR APPENDIX A10

A10.1 Test the hypothesis at the 5% significance level that the mean education levels of people in Indianapolis who prefer Coca-Cola and who prefer Pepsi-Cola are the same. Use data set 9. The first three lines of the data set are shown below:

Years of school completed by respondents and their Coke–Pepsi preference. Respondents who do not drink colas or are indifferent between Coca-Cola and Pepsi-Cola are dropped from the data set. 1 = Prefer Pepsi. 0 = Prefer Coke.

Education	Cola preference
8	1
8	1
8	1

A10.2 Data set 11 is a 1-in-20 systematic sample from a list of all public and private colleges in the United States in 1985–1986. The data set has the number of students in column 1, the number of full-time faculty in column 2, and a dummy variable that takes the value 1 for private schools in column 3. Test the hypothesis at the 5% significance level that the student–faculty ratio is the same across private and public schools.

A10.3 Use data set 11 to test the hypothesis at the 5% significance level that the student–faculty ratio is the same across large schools and small schools. Define a school as small if it has fewer than 5,000 students.

(Exercise A10.4 cannot be done with DATA DESK.)

A10.4 Data set 8 includes the SAT scores and grades for a set of students in an introductory economics class. Many of the students did not take the SAT test. Missing SAT scores are coded as −99999. These students may not have taken the test because they initially did not plan to attend college. Presumably the non-test-takers are, on average, less well prepared for college work than the test-takers. While no evidence is available as to what their SAT scores might have been, it is possible to test the null hypothesis that the mean grades of students who have SAT scores are less than or equal to the mean grades of students who do not have SAT scores. Use a 5% significance level.

A11
Using the Chi-Square Distribution for Tests of Homogeneity, Independence, and Goodness of Fit

A11.1 INTRODUCTION

All four packages offer good support for tests of homogeneity and independence. However, the support for chi–square goodness–of–fit tests is much weaker. For example, only SPSSx has a simple command to test whether a set of sample data could have been drawn from a binomial distribution. One solution to this lack of support for goodness–of–fit tests is to avoid the statistical packages altogether and instead use a spreadsheet program, such as Microsoft Excel™ or Lotus 1-2-3™. For a test against the binomial distribution, the sample proportion of successes, the expected frequencies, and the chi-square test statistic can easily be calculated with a spreadsheet. If you already know the expected frequencies (for example, when the population proportion is specified in the null hypothesis), you can use the chi-square test routines in SPSSx. These routines accept sets of expected and observed frequencies and report the chi-square test statistic. Another approach to goodness-of-fit problems is to program the chi-square formula into the package. This last approach is all that you can do with DATA DESK, MINITAB, and SAS.

There are some minor differences among the packages in the allowable data formats for tests of independence and homogeneity. DATA DESK works only with the raw data rather than the cell counts. MINITAB, SAS, and SPSSx can compute chi-square test statistics from either the raw data or the cell counts. MINITAB requires numerical data while DATA DESK, SAS, and SPSSx can handle alphabetical or numerical data.

A11.2 AN APPLICATION

The limitations of human memory guarantee that most of the advertisements consumers see and hear will be quickly forgotten. To sell a product, an advertisement has to be noticed and then create a lasting, favorable impression. Advertisers constantly experiment to find ways to make their ads memorable. Chi-square tests are widely used in such advertising research. For example, such attributes of print ads as color, size, and location within a newspaper or a magazine can easily be varied in an experimental setting. Readers can then be asked if they recall the test ad or if they agree with specific statements about the advertised product. The null hypotheses in tests of this type is that the distributions of recalling or agreeing are homogeneous across the tested attributes. If color was the test attribute, the null hypothesis could be that the proportion of readers who recall a print ad is the same across all of the colors tested.

Most of the research results on advertising are proprietary and do not appear in academic publications. An exception is an article by David A. Sheluga and Jacob Jacoby, "Do Comparative Claims Encourage Comparison Shopping? The Impact of Comparative Claims on Consumers' Acquisition of Product Information," which appeared in *Current Issues and Research in Advertising* in 1978 (pp. 23–37). This study relies on a small convenience sample, and its results may not hold up with larger probability samples. Nevertheless, the study is a good textbook application. The experimental design is simple, the main issue remains topical, and it illustrates a typical use of the chi-square distribution.

In the study, 60 students in a class on advertising were shown descriptions of eight competing pocket cameras. Seven of the brands were real cameras but one brand, Halina, was fictitious. Genuine slogans used in the advertising of the seven real cameras were placed below their respective names. Along with these displays, 30 students were shown the slogan "Introducing the Halina Readiload—Our care and craftsmanship bring you the *automatic* instamatic" beneath the name Halina. The other 30 students saw the slogan "Introducing the Halina Readiload—the pocket instamatic that *outfeatures* and *outperforms* Kodak." Each student was asked if he or she could recall the slogan for each of the eight cameras. The authors tested the null hypothesis that the number of students who could recall each slogan would be the same for every camera in the two sets of 30. The null hypothesis could be rejected at a 1% significance level. The comparative claim—"*outfeatures* and *outperforms* Kodak"—stuck in the students' minds. Fourteen were able to recall it. The next highest number for any slogan was 10 and the average number of recalls per slogan was 5.19. The investigators reported, "The findings support the contention that a comparative claim draws more attention to a brand. Significantly higher levels of claim recall were found for the comparative claim, both when compared with the conventional claims of the other brands, and with the conventional claim of the same brand. This finding suggests that the comparative claim was attended to, rehearsed, and stored in the shopper's long-term memory." Whenever you see a comparative claim in an advertisement—one that names competing brands—you can reasonably surmise that the ad has been tested for recall and favorable impact. Comparative ads often have a serious failing: they have good recall percentages but consumers wind up remembering the wrong brand. This phenomenon makes life more interesting for advertising agencies.

A11.3 COMMANDS

 A11.3A DATA DESK

DATA DESK has no routines for chi-square goodness-of-fit problems. If you know the expected and observed frequencies, you can use the New Derived Variable command to compute the chi-square test statistic

$$\sum_{i=1}^{n} \frac{(o_i - e_i)^2}{e_i}$$

If the letter O is assigned to the variable with the observed frequencies and E to the expected frequencies, you will type the following formula for the new derived variable:

```
SUM(((O - E)**2)/E)
```

After typing this formula, click in the close box in the upper left-hand corner of the formula window. Next, while holding down the command key, click on the icon of the derived variable and hit the equals key. A new icon for a variable containing the chi-square test statistic will appear.

Calculating expected frequencies in DATA DESK is complicated when the parameters have to be estimated from the sample data. Suppose you wanted to test the hypothesis that the number of successes in 50 sets of 10 trials followed a binomial distribution. If you entered the number of successes per set of 10 trials, you could use the Summary Reports command to calculate the sample mean. Dividing the mean by the number of trials per set (10 in this example) would give you the sample proportion of successes. For illustration, we will use 0.516 as the sample proportion. To calculate the expected frequencies, you first create a variable x that runs from 0 to 10 in integers and a variable C that contains the number of combinations of x that can be drawn from 10. C has the values {1, 10, 45, 120, 210, 252, 210, 120, 45, 10, 1}. Next create a new variable to contain the expected frequencies, E, with the New Derived Variable command. The algebraic statement for E is

```
50*C*((0.516)**X)*((1 - 0.516)**(10 - X))
```

This algebraic statement is simply the binomial formula for the probability of x successes in 10 trials multiplied by 50. The factor of 50 gives you the expected frequencies of x successes with 50 sets of 10 trials.

Next, click on the icon of E and select the Compute Derived Variable from the compute menu. The 11 expected frequencies may have to be collapsed so that each e_i is at least 5. Next, the observed frequencies have to be calculated from the 50 observations of successes per set of 10 trials. This can be done by sorting the observations and counting. These observed frequencies also have to be collapsed into cells to match the expected frequencies. Name the variable containing the observed frequencies O. At this point you can calculate the chi-square test statistic with a second derived variable containing the expression SUM(((O - E)**2)/E).

While all of these manipulations may seem like a great bother, if you had to work with larger numbers than sets of 10 trials, say more than 20 trials per set, DATA DESK could be a tremendous time saver. Note also that doing the same problem with a spreadsheet program would be much easier.

DATA DESK has the Contingency tables command under the Compute menu. In Appendix A4 the Contingency tables commands was used to create joint probability tables. After you have selected this command, a dialog box appears in which you can ask for the chi-square test statistic. The Contingency tables command works with either alphabetical or numerical codes for the categories. It does not work with cell counts, though.

■ **A11.3B MINITAB**

If you have a set of expected frequencies in column 1 and the corresponding observed frequencies in column 2, the following commands will compute a chi-square test statistic:

```
MTB > LET C3 = (C2 − C1)**2/C1
MTB > SUM C3
```

You should be able to recognize the formula for the chi-square test statistic embedded in these commands. Note that each expected frequency should be at least 5. If it is not, some of the cells should be collapsed.

If you are starting with a set of observed frequencies and have to estimate the parameter or parameters from the sample data, the commands are more involved. Suppose the observed frequencies are in column 1 and you want to test the null hypothesis that the number of successes in 50 sets of 10 trials follow a binomial distribution. The first step is to calculate the sample proportion of successes, or the total number of successes divided by the total of 500 trials. For illustration, we will use 0.516 as the sample proportion. The commands for finding the expected frequencies are

```
MTB > BINOMIAL PROB FOR N = 10, AND P = 0.516, PUT INTO C2
MTB > LET C3 = 50*C2
```

The 11 expected frequencies may have to be collapsed so that each e_i is at least 5. The observed frequencies also have to be collapsed into cells to match the expected frequencies. At this point you can calculate the chi-square test statistic with the commands

```
MTB > LET C4 = (C1 − C3)**2/C3
MTB > SUM C4
```

While all of these steps may seem like a great bother, if you had to work with larger numbers than sets of 10 trials, say more than 20 trials per set, MINITAB could be a tremendous time saver. Note also that doing the same problem with a spreadsheet program would be much easier.

Tests for homogeneity and independence can also be done with MINITAB. If you have the cell counts in columns 1 through 3 of a MINITAB worksheet, you can do a test for independence or homogeneity with the command `CHISQUARE ANALYSIS ON TABLE IN C1, C2, C3`. The MINITAB `TABLES` command will give you a chi-square test statistic if you have raw data in numerical form. The data format for the command `TABLES` was discussed in Appendix A4. As a reminder, the data must take integer values. Each row represents one observation, with the integer code for one attribute recorded in one column and the integer code for a second attribute recorded in another column. The subcommand `CHISQUARE` will give you a chi-square test statistic. For a simple illustration, consider a variable SEX with the values 0 for MALE and 1 for FEMALE and a variable EYECOLOR with the values 1 for BLACK, 2 for BROWN, 3 for BLUE, 4 for HAZEL, and 5 for GREEN. A test for independence between SEX and EYECOLOR would have the following commands. If the integer codes for the two attributes were stored in columns 1 and 2, the commands would be

```
MTB > NAME C1 = 'SEX'
MTB > NAME C2 = 'EYECOLOR'
```

```
MTB > TABLE 'SEX' BY 'EYECOLOR';
MTB > CHISQUARE.
```

A11.3C SAS

SAS does not have a procedure for chi-square goodness-of-fit tests. If you had a set of observed and expected frequencies under the variable letters O and E, the following commands would compute the chi-square test statistic:

```
CHISQ=((O-E)*(O-E))/E;
PROC MEANS SUM;
        VAR CHISQ;
```

Note that each expected frequency should be at least 5. If it is not, some of the cells should be collapsed.

If you are starting with a set of observed frequencies and have to estimate the parameter or parameters from the sample data, the SAS commands are more involved. Suppose the observed frequencies are in the variable O and you want to test the null hypothesis that the number of successes in 50 sets of 10 trials follow a binomial distribution. The first step is to calculate the sample proportion of successes, or the total number of successes divided by the total of 500 trials. For illustration, we will use 0.516 as the sample proportion. The commands for the expected frequencies are

```
DATA;
        DO FOR X = 0 TO 10;
                E=50*PROBBNML(0.516,10,X);
                PUT E;
        END;
```

The 11 expected frequencies may have to be collapsed so that each e_i is at least 5. The observed frequencies also have to be collapsed into cells to match the expected frequencies. Once you have the observed and the expected frequencies, you can proceed to calculate the chi-square test statistic with the commands in the first SAS example. While all of these steps may seem like a great bother, if you had to work with larger numbers than sets of 10 trials, say more than 20 trials per set, SAS could be a tremendous time saver. Note also that doing the same problem with a spreadsheet program would be much easier.

The SAS FREQ procedure does tests for homogeneity and independence. The attributes associated with each observation can be recorded in numerical codes or alphabetically. For a simple illustration, consider a variable SEX with the values MALE and FEMALE and a variable EYECOLOR with the values BLACK, BROWN, BLUE, HAZEL, and GREEN. A test for independence between SEX and EYECOLOR would have the following commands:

```
PROC FREQ;
        TABLES SEX*EYECOLOR/CHISQ;
```

A11.3D SPSSx

The SPSSx procedure does a variety of nonparametric tests, including chi-square tests for goodness of fit. If you have a set of observed frequencies stored in the variable O, the commands to compute a chi-square test statistic are

```
FILE HANDLE GFT/NAME='CHAP11.DAT'
DATA LIST FILE=GFT/CATEGORY 0
WEIGHTED BY 0
NPAR TEST CHISQUARE=CATEGORY/
      EXPECTED=E1,E2,...,EK
```

The E terms would be replaced by the set of expected frequencies. If no expected frequencies are specified, the default is equal expected frequencies.

SPSSx has a convenient command,

```
NPAR TESTS BINOMIAL(Π)=TESTDAT(1,2)
```

for a goodness-of-fit test against a binomial distribution in which the data are recorded numerically as successes or failures for individual trials. The Π is the hypothesized population proportion. TESTDAT is the name of a variable containing the sample data, which are coded 1 for successes and 2 for failures.

The lack of a function that will calculate binomial probabilities makes goodness-of-fit tests against sets of trials cumbersome. You would be much better off using a spreadsheet program. The following description sketches out how such a test can be done with SPSSx. If you wanted to test the hypothesis that the number of successes in 50 sets of 10 trials followed a binomial distribution and the variable O contained the number of successes per set of 10 trials, you could use the Frequencies procedure to calculate the sample mean. Dividing the mean by the number of trials per set (10 in this example) would give you the sample proportion of successes. For illustration, we will use 0.516 as the sample proportion. To calculate the expected frequencies, you first create a variable X that runs from 0 to 10 in integers and a variable C that contains the number of combinations of X that can be drawn from 10. C has the values {1, 10, 45, 120, 210, 252, 210, 120, 45, 10, 1}. Next create a new variable to contain the expected frequencies, E, with the following algebraic statement:

```
COMPUTE E = 50*C*((0.516)**X)*((1 - 0.516)**(10 - X))
```

This algebraic statement is simply the binomial formula for the probability of x successes in 10 trials multiplied by 50. The factor of 50 gives you the expected frequencies of x successes with 50 sets of 10 trials. The 11 expected and observed frequencies may have to be collapsed so that each e_i is at least 5. Name the variable containing the observed frequencies O. At this point you can calculate the chi-square test statistic with a second derived variable containing the expression

```
COMPUTE CHISQ = SUM(((O - E)**2)/E).
```

The SPSSx procedure for tests of homogeneity and independence is CROSSTABS, which was introduced in Appendix A4. For a simple illustration, consider a variable SEX with the values MALE and FEMALE and a variable EYECOLOR with the values BLACK, BROWN, BLUE, HAZEL, and GREEN. A test for independence between SEX and EYECOLOR would have the following commands:

```
CROSSTABS VARIABLES = SEX BY EYECOLOR
STATISTICS 1
```

EXERCISES FOR APPENDIX A11

(The following question was suggested by Dr. Kenton Juillen of Walker Research.)

A11.1 A manufacturer plans to reformulate his granola bar. Because of shifting conditions in agricultural markets which affect the availability and costs of certain ingredients, such products must be reformulated continuously to maintain a desirable product at a competitive price. Before proceeding with the reformulation, the manufacturer conducted a taste test of the original and the prototype granola bar. Consumers who regularly eat granola bars tasted one bar in a plain wrapper and rated it as follows:

Description	Code
Definitely will buy	5
Probably will buy	4
Might or might not buy	3
Probably won't buy	2
Definitely won't buy	1

In the test 200 consumers tasted the original bar and another 200 consumers tasted the prototype bar. In data set 12 the rating coded as above and the type of bar are coded 0 for original and 1 for prototype. Test the null hypothesis at a 5% significance level that the proportions of consumers who give each rating are the same for both types of bars. Does this test put the burden of proof on the new formulation? Should the test put the burden of proof on the new formulation? Explain.

A11.2 Byssinosis is a lung disease that can be caused by exposure to cotton dust. In 1973 a large cotton textile company tested workers who were exposed to cotton dust for the presence of this disease. In the first column of data set 13 1 indicates the presence of byssinosis among the tested workers and 0 indicates its absence. Data set 13 includes only white male smokers who worked in either the most dusty environment or the least dusty environment. A second variable is coded 1, 2, or 3 to show which workers had, respectively, under 10 years of employment, 10 through 19 years of employment, and 20 or more years of employment. A third variable is coded 0 for the high-dust environment and 1 for the low-dust environment.

(a) Test the hypothesis at a 5% significance level that incidence of byssinosis among white male smokers is independent of the three categories for years of employment.

(b) Test the hypothesis at a 5% significance level that incidence of byssinosis among white male smokers is independent of the three categories for years of employment for employees in the low-dust environment.

(c) Test the hypothesis at a 5% significance level that incidence of byssinosis among white male smokers is independent of the three categories for years of employment for employees in the high-dust environment.

(d) Describe your conclusions.

A11.3 Getting kicked in the head by a horse was one of the occupational hazards for officers in the Prussian Army in the nineteenth century. According to data published in 1898 by von Bortkewitsch there were, on average, 9.8 deaths per year caused by horse kicks in the Prussian Army over the years 1875–1894. Bortkewitsch's data (see data set 14) are the basis of a standard illustration of the Poisson distribution in statistics classes. Test the null hypothesis at a 5% significance level that the data are a sample from a Poisson distribution. Make sure that each expected frequency is at least 5.

A11.4 After the death of a reigning Roman Catholic pope, the College of Cardinals meets in Rome to elect a new pope. Beginning with St. Linus in 67 A.D. and ending with John Paul II in 1978, there were 262 elected popes. The average time between elections is 7.3218 years. Unless divine intercession occurs, the time from election to death of a pope can be considered a random variable with lower probabilities for longer reigns. Test the hypothesis at a 5% significance level that the time between papal elections follows an exponential distribution. Data set 15 has the observed and the expected frequencies for time between papal elections given the null hypothesis that the distribution is exponential with the observed mean of 7.3218. The first three observations in data set 15 are shown below. Which time periods cause the greatest additions to the chi-square test statistic? What physical or institutional constraint may account for the discrepancy from the exponential distribution?

Time between elections in years	Observed frequency	Expected frequency
Under 1	16	33.319810
1 and under 2	33	29.066133
2 and under 3	24	25.355489

A12
Analysis of Variance

A12.1 INTRODUCTION

Chapter 12 presented only the intuitive formulas for the F test statistic. The goal of the chapter was to provide an intuitive understanding of analysis of variance—a method of testing for the equality of multiple population means by comparing different estimates of the population variance. The exclusive use of intuitive formulas was intended to further this goal. If you had a large data set and had to do hand calculations, the computations based on these intuitive formulas would be exceedingly cumbersome. Facilitating hand calculations was not one of the goals of the chapter. Most statistics texts present analysis of variance in a less intuitive format meant to aid calculations. This format is an analysis-of-variance table. Analysis-of-variance tables help you with the calculations done by hand or on a hand-held calculator by showing the intermediate sums and by providing a step-by-step procedure to follow. All of the computer packages follow the traditional textbooks and present the results of analysis

TABLE A12.1
Data set of flavors listed by treatment

Sweet	Bitter	Salt	Sour	Sweet	Bitter	Salt	Sour
14	13	61	37	17	36	39	65
12	62	33	79	14	88	6	61
99	3	70	55	21	15	14	42
32	35	78	10	40	18	55	29
26	1	5	70	68	86	51	41
94	52	18	90	77	87	1	86
83	37	66	73	46	29	68	94
30	30	96	79	45	49	3	85
11	80	38	49	38	15	26	31
28	49	87	58	63	97	73	51
49	36	48	29	63	27	2	70
31	21	32	12	62	15	4	77
73	62	83	93				

of variance in this tabular form. The makers of computer programs currently follow the example set by early automobile manufacturers: they made cars that looked like horse-drawn carriages because they thought those designs would be more readily accepted. Here I am forced to present these contraptions. Nevertheless, the focus will be on the reported F-test statistics and the p values.

All of the computer packages can do one-way analysis of variance with unequal sample sizes. For two-way analysis of variance, DATA DESK and MINITAB must have equal cell counts. SAS and SPSSx can handle two-way analysis of variance with unequal cell counts, also called unbalanced designs.

A made-up data set will be used to illustrate the one- and two-way analysis of variance for all four packages. The data set has 100 observations. Each observation is associated with one of four treatments called sweet, bitter, salt, and sour. These names are used to make it easier to keep track of the experiment. There are 25 observations per treatment. Table A12.1 shows the data set.

To illustrate the output for a two-way test, the 100 observations and four flavors are used again along with a second factor called color. Again, the name is intended simply to help you keep track of the experiment. Color will be the treatment factor and flavor the blocking factor. There are now four levels of the blocking factor flavor—sweet, bitter, salt, and sour—and five levels of the treatment factor color—red, green, blue, yellow, purple. Each treatment-block cell has five observations. The total number of observations remains 100, that is, $5 \times 5 \times 4$. Table A12.2 shows the first ten observations in the data set.

TABLE A12.2
First ten observations in data set listed by treatment and block

Observation	Flavor	Color	Observation	Flavor	Color
14	sweet	red	62	bitter	red
13	bitter	green	33	salt	green
61	salt	blue	79	sour	blue
37	sour	yellow	99	sweet	yellow
12	sweet	purple	3	bitter	purple

In Greek mythology an Athenian brigand named Procustes was known for forcing unwary guests to fit into his special bed. Procustes either stretched short guests or cut the long ones down to the size of his bed. Most of the statistical programs offer a Procustean bed when it comes to accommodating data sets for one- or two-way analysis of variance. You have to make the data fit the required format. In some cases the rearrangement of the data can be both tedious and complex. MINITAB is a partial exception. It can do one-way analysis with the data either organized into columns by treatment or organized into a column for all observations and a second column for an indicator variable showing the treatments. DATA DESK must have a variable containing all of the observations and a second variable indicating the treatments for one-way analysis of variance. However, there is an easy method to convert a data set that has the observations listed by treatment into the required format. The other programs are less accommodating.

A12.2 AN APPLICATION

In 1985 and 1986 the Illinois Department of Employment Security conducted a formal experiment to test the effect of two incentive plans on the duration of insured unemployment. Employers in Illinois, as in all other states, pay a payroll tax to finance unemployment insurance benefits for laid-off workers. One objective of the experiment was to test whether an incentive plan could be cost-effective, that is, reduce the payroll taxes more than the cost of the incentives. One of the plans offered a $500 bonus to every laid-off worker who found a job within 11 weeks of initially filing an unemployment claim, provided that the job was for at least 30 hours a week and that the worker stayed on the new job at least four months. The other incentive plan had the same requirements but the $500 payment went to the employer who hired the laid-off worker. The 11-week requirement was chosen because it was less than the median duration of insured unemployment in Illinois. A pool of 12,101 new claimants were randomly assigned to either the $500 employee bonus plan, the $500 employer bonus plan, or the usual treatment of no bonus.

The large samples and the random assignments made sure that the three treatment groups would be closely matched on such observable characteristics as age, race, education, and pre-unemployment wage rate. Because of the expense involved—$750,000 in this study—such large formal social experiments are rare. However, formal experiments guarantee the validity of assumptions needed for statistical tests, such as equal variances and independent observations.

The appropriate technique to test the null hypothesis that the mean duration of insured unemployment over the year after the initial filing of an unemployment claim was the same for the three groups is one-way analysis of variance. The experimental results led to the rejection of this null hypothesis. The mean duration for the employee bonus plan was found to be significantly lower than the mean duration for the no-bonus group. There was no statistically significant effect for the employer bonus plan. The appropriate technique to test the null hypothesis that the mean wage rate over the year after the initial filing of an unemployment claim was the same for the three groups is also one-way analysis of variance. Here the suspicion is that the employee bonus might encourage some workers to accept a job with a lower wage in order to get the bonus. The null hypothesis of equal mean wages could not be rejected.

The bottom line on the project was that the employee bonus was cost-effective. The State of Illinois reduced its unemployment insurance payments by $2.32 for every $1.00 paid out in bonuses. It is not clear that this benefit-cost ratio could be maintained if the incentives became routine. Only 54% of the eligible employees in the experiment collected their money. If everyone in the employee payment plan who was eligible had collected the $500, the benefits would have dropped to $1.26 per $1.00 of cost. If the incentive payments became routine, it is likely that unemployed workers would become more familiar with its provisions and more likely to collect the $500. On the other hand, as the program became more familiar, unemployed workers might try harder to find jobs to become eligible for the payments and thus reduce the amount of unemployment insurance that the state would have to pay. The experiment provides strong evidence that a bonus plan for finding and keeping jobs would be cost-effective.

A more detailed description of this experiment can be found in Stephen A. Woodbury and Robert G. Spiegelman, "Bonuses to Workers and Employers to Reduce Unemployment: Randomized Trials in Illinois," *American Economic Review* 77 (September 1987), pages 513–530. The article describes many other statistical tests comparing wage rates and unemployment duration for various race-sex groups. The article is unusual in that all of the tests used are commonly discussed in introductory statistics courses. The reason that these relatively simple tests are appropriate is that the experimental design was elegantly simple. A last point about the study is the importance of statistical programs on computers. Large samples were used so that even a tiny difference in the behavior of unemployed workers across the treatments could be detected. With 12,101 unemployed workers in the sample, it would have been impossible to get accurate results without a computer.

A12.3 COMMANDS

A12.3A DATA DESK

The required data organization for one-way analysis of variance in DATA DESK is a column of observations plus a column of an indicator variable showing the treatment level for each observation. The indicator variables can be either numerical or alphabetical. Table A12.3 shows the first ten observations of our data set in the required format. If the observations are organized in columns for the five treatments, the Combine and make categories command from the Manipulate menu will reorganize the data into the required format.

TABLE A12.3
Flavor data
organized for one-way analysis of variance in DATA DESK

Observation	Indicator variable	Observation	Indicator variable
14	sweet	33	salt
13	bitter	79	sour
61	salt	99	sweet
37	sour	3	bitter
12	sweet	70	salt
62	bitter		

```
Analysis of Variance for    x


   Source        df    Sum of Squares    Mean Square    F-ratio       Prob

   x              3        4701.96          1567.32        2.0        0.1123
   Error         96       73472.0           765.333
   Total         99       78174.0
```

FIGURE A12.1
DATA DESK output for a one-way analysis of variance

To conduct a one-way analysis of variance with DATA DESK, select the icon for the variable containing the observations and then, while holding down the shift key, select the icon for the indicator variable. The last step is to select ANOVA from the Compute menu. The resulting output is shown in Figure A12.1. Here the F test statistic is labeled F-ratio. The numerator degrees of freedom are labeled with the name of the variable, x. The denominator degrees of freedom are labeled ERROR degrees of freedom. The p value of the test is labeled Prob. With the p value of 0.1123, you cannot reject the null hypothesis of equal means for any α lower than 11.23%. For example, there is not enough evidence to reject a null hypothesis of equal population means if the α level is set to 5%.

For a two-way analysis of variance in DATA DESK there have to be two indicator variables, one for the treatment levels and the other for the block levels. The indicator variables can be either numerical or alphabetical. Table A12.4 has the required data organization for the first ten observations of our data set.

The procedure for a two-way analysis is to select the icon for the variable containing the observations and then, while holding down the shift key, select the icons for the treatment and block indicator variables. The last step is to select ANOVA from the Compute menu. The resulting output is shown in Figure A12.2. Again the F-ratio terms are the F test statistics. The p values for the null hypotheses of no color effect, no flavor effect, and no interaction effect are labeled Prob. With the p values shown in Figure A12.2, the null hypotheses of no treatment effect, no block effect, and no interaction effect cannot be rejected at a 5% significance level.

TABLE A12.4

Data organization for two-way analysis of variance in DATA DESK

Observation	Treatment	Block
14	red	sweet
13	green	bitter
61	blue	salt
37	yellow	sour
12	purple	sweet
62	red	bitter
33	green	salt
79	blue	sour
99	yellow	sweet
3	purple	bitter

FIGURE A12.2
DATA DESK output for a two-way analysis of variance

```
Analysis of Variance for   x


Source       df    Sum of Squares    Mean Square    F-ratio      Prob

color        4         3228.26         807.065        1.1       0.3618
flavor       3         4701.96        1567.32         2.1       0.1018
Interaction 12        11616.9          968.078        1.3       0.2234
Error       80        58626.8          732.835
Total       99        78174.0


Analysis of Variance for   x
```

A12.3B MINITAB

MINITAB has two commands for one-way analysis of variance. The output for the two routines is identical; the difference is in the data formats. If the data are organized into columns by treatment, the command is AOVONEWAY. If the data are organized into one column for observations and a second column for a numerical indicator of treatment level, the command is ONEWAY. Figure A12.3 shows the commands and the output when AOVONEWAY is used.

The F test statistic is labeled F. The numerator degrees of freedom are labeled the FACTOR degrees of freedom. The denominator degrees of freedom are labeled ERROR degrees of freedom. No p value is given in the output. To decide if you would reject or accept the null hypothesis of equal treatment means, you would have to look up the critical value at 3 and 96 for a specified level in the F table. The critical value for 5% is approximately 2.70. There is not enough evidence to reject the null hypothesis of equal treatment means.

MINITAB automatically gives a set of 95% confidence intervals for the treatment means based on the pooled estimate of the population standard deviation. The formula for these confidence intervals is discussed in Chapter 12. Note that the level of confidence for all four intervals simultaneously including the respective treatment means is not 95%; it is 95% taken to the fourth power.

FIGURE A12.3
Commands and output for MINITAB's one-way analysis of variance with data organized into columns by treatment

```
MTB > AOVONEWAY C2,C3,C4,C5

ANALYSIS OF VARIANCE
SOURCE      DF         SS        MS          F
FACTOR       3       4702      1567       2.05
ERROR       96      73472       765
TOTAL       99      78174
                                   INDIVIDUAL 95 PCT CI'S FOR MEAN
                                   BASED ON POOLED STDEV
LEVEL        N       MEAN     STDEV    -----+---------+---------+---------+-
C2          25      45.44     26.53        (--------*--------)
C3          25      41.72     28.48    (--------*--------)
C4          25      42.28     30.48    (--------*--------)
C5          25      58.64     24.85                   (--------*--------)
                                       -----+---------+---------+---------+-
POOLED STDEV =      27.66             36        48        60        72
```

TABLE A12.5
Data organization for
two-way analysis of
variance in
MINITAB

C2 Values of x	C3 Level of flavor variable	C4 Level of color variable
14	1	1
13	2	2
61	3	3
37	4	4
12	5	1
62	1	2
33	2	3
79	3	4
99	4	1
3	5	2

For a two-way analysis of variance, MINITAB has to have the observations in one column and two numerical indicator variables, one for the levels of the treatments and one for the levels of the blocks. Table A12.5 shows the organization of the first ten observations. The indicator variable in column 3 takes the values 1, 2, 3, 4, and 5 for the colors red, green, blue, yellow, and purple, respectively. The indicator variable in column 4 takes the values 1, 2, 3, and 4 for the flavors sweet, bitter, salt, and sour. Note that while the order of the x observations is arbitrary, you must have the correct values of the indicator variables aligned with the observations. For example, the first x observation in the table, 14, could be placed farther down in the data set as long as it retained the indicator variables with the value 1 for the color red and the indicator variable with the value 1 for the flavor sweet.

The commands and the output of MINITAB's two-way analysis of variance are shown in Figure A12.4. The MINITAB two-way output is sparse. No F test statistics are provided. You have to calculate them by dividing the MS or mean square terms for colors and flavors and interaction by the mean square term for the denominator— labeled ERROR in Figure A12.4. For example, the F test statistics for the treatment variable colors, C3, is 807/733, or 1.10. After you calculate the F test statistics, you have to find the critical values in an F table to see if the null hypotheses of no treatment effects, no block effects, and no interaction effects are rejected.

FIGURE A12.4
MINITAB commands and
output for two-way analysis
of variance

```
MTB > TWOWAY ANALYSIS DATA IN C2, CLASSIFICATIONS IN C3 AND C4


ANALYSIS OF VARIANCE   C2

SOURCE          DF        SS        MS
C3               4      3228       807
C4               3      4702      1567
INTERACTION     12     11617       968
ERROR           80     58627       733
TOTAL           99     78174
MTB >
```

A12.3C SAS

For one-way analysis of variance the appropriate SAS procedure is ANOVA. Two-way analysis of variance can be done with the ANOVA procedure provided that the design is balanced. The ANOVA data format is a column for the dependent variable with one indicator variable for a one-way analysis of variance or two indicator variables for two-way analysis of variance. The indicator variables can be either numerical or alphabetical. Tables A12.3 and A12.4 in the DATA DESK section and A12.5 in the MINITAB section show allowed data formats. Here is a command file for a one-way analysis of variance:

```
PROC ANOVA;
        CLASS FLAVOR;
        MODEL X=FLAVOR;
        MEANS FLAVOR / TUKEY;
```

FLAVOR is the treatment variable, which is specified in the CLASS statement. The observations are stored under the variable X. The last line requests Tukey's HSD test. Figure A12.5 shows the output for this command file. The F test statistic and the p value of the test are shown in boldface. Generally the Tukey HSD test is not done unless the analysis of variance leads to the rejection of the null hypothesis of equal population means. It was done here to show the commands and the output. The formula used by SAS when there are unequal sample sizes differs from the formula shown in Chapter 12. The text formula is based on a recent paper cited at the end of the chapter; the formula in SAS is from the mid-1950s. For most problems this difference is a minor issue. The results of Tukey's HSD test are that none of the means are significantly different at the 5% level.

If you have a balanced two-way design, you can list a second indicator variable, such as color, in the CLASS and MODEL statements of the ANOVA procedure. For an unbalanced two-way design, the appropriate SAS procedure is GLM, or Generalized Linear Models. The data format is a column for the observations plus indicator variables for both the treatment and block levels. The indicator variables can be either numerical or alphabetical. Tables A12.4 and A12.5 show acceptable formats. The commands for a two-way analysis with the color and flavor classification variables are

```
PROC GLM;
        CLASS COLOR FLAVOR;
        MODEL X=COLOR FLAVOR COLOR*FLAVOR;
```

The term COLOR*FLAVOR specifies that the procedure should test for interactions between the treatment and block effects. The output for GLM's analysis of variance is nearly identical to the ANOVA procedure output except that it lists F test statistics and p values for COLOR, FLAVOR, and COLOR*FLAVOR. We will not show this output. One complication is that the output may list four sets of F test statistics and p values. What is going on is that when you are working with an unbalanced design, there are several ways of computing the F test statistics. No one method is unambiguously preferred to the others. A discussion of each method and the reasons for the ambiguity is beyond the scope of this text. The good news is that for most problems the four methods will result in the same decision regarding the acceptance of the null hypotheses of no treatment, block, or interaction effects.

FIGURE A12.5

Output for a one-way analysis of variance from the SAS ANOVA procedure

```
ANALYSIS OF VARIANCE PROCEDURE
CLASS LEVEL INFORMATION
CLASS      LEVELS     VALUES
FLAVOR        4        1 2 3 4
NUMBER OF OBSERVATIONS IN DATA SET = 100
DEPENDENT VARIABLE: X

SOURCE    DF   SUM OF SQUARES  MEAN SQUARE   F VALUE   PR > F    R-SQUARE    C.V.

MODEL      3    4701.96000000  1567.32000000   2.05    0.1123    0.060147  58.8359

ERROR     96   73472.00000000   765.33333333          ROOT MSE     X MEAN

CORRECTED
TOTAL     99   78173.96000000                        27.66465856  47.02000000


SOURCE    DF   ANOVA SS       F VALUE      PR > F

FLAVOR     3   4701.96000000   2.05        0.1123


TUKEY'S STUDENTIZED RANGE (HSD) TEST FOR VARIABLE: X
NOTE: THIS TEST CONTROLS THE TYPE I EXPERIMENTWISE ERROR RATE,
BUT GENERALLY HAS A HIGHER TYPE II ERROR RATE THAN REGWQ

ALPHA=0.05  DF=96  MSE=765.333
CRITICAL VALUE OF STUDENTIZED RANGE=3.698
MINIMUM SIGNIFICANT DIFFERENCE=20.459

MEANS WITH THE SAME LETTER ARE NOT SIGNIFICANTLY DIFFERENT.

TUKEY     GROUPING          MEAN     N  FLAVOR
             A             58.640    25  4
             A
             A             45.440    25  1
             A
             A             42.280    25  3
             A
             A             41.720    25  2
```

A12.3D SPSSx

ONEWAY is the best procedure for one-way analysis of variance in SPSSx because it provides tests for the difference between pairs of population means and single-population confidence intervals not available in other analysis-of-variance procedures. The required data format is one column containing the observations and a second column containing an indicator variable for the level of the treatments. The indicator variable must be numerical and take only integer values. The first two columns of Table A12.5 in section A12.3B illustrate the required data format for a one-way analysis of variance. The third column would be used in a two-way analysis of variance. The following command file reads in the data set with this organization, names the variables X, COLOR, and FLAVOR, and calls for the ONEWAY procedure with X as the observations and FLAVOR as the treatment. The column with the indicator variable for the color variable is not used in one-way analysis. The numbers (1,4) specify the lowest and highest values of flavor that will be included in the test. In this case all four flavors are used. The RANGES=TUKEY command requests Tukey's honest significant difference test for all possible pairs of population means. The STATISTICS 1 reports a series of 95% confidence intervals and descriptive statistics.

```
FILE HANDLE IN/ NAME='TWOWAY.DAT.1'
DATA LIST FILE=IN LIST/X COLOR FLAVOR
ONEWAY X BY FLAVOR (1,4)/RANGES=TUKEY
STATISTICS 1
```

A partial listing of the output generated by the command file above is shown in Figure A12.6. The F test statistic is called F ratio and the p value is called F PROB. With a p value of 0.1123 there is not enough evidence to reject the hypothesis of equal population means at any significance at or below 11.23%. The numerator and denominator degrees of freedom are labeled the BETWEEN GROUPS and WITHIN GROUPS degrees of freedom. These terms are used in Chapter 12. After the analysis-of-variance table the output gives some descriptive statistics. The confidence intervals shown in the output are *not* based on the pooled estimate of the population standard deviation. If the assumption of equal population standard deviations and variances underlying the analysis-of-variance test were correct, the reported confidence intervals would be incorrect. They would be too wide. Finally, the output shows the result of Tukey's HSD test. Ordinarily, Tukey's test is done only after the null hypothesis that all of the population means are equal has been rejected. It was done here to show

FIGURE A12.6
Output for the SPSSx
ONEWAY **procedure**

```
                         SUM OF       MEAN       F        F
 SOURCE           D.F.   SQUARES      SQUARES    RATIO    PROB.
 BETWEEN GROUPS    3     4701.9600    1567.3200  2.0479   .1123
 WITHIN GROUPS    96    73472.0000     765.3333
 TOTAL            99    78173.9600


                   STANDARD   STANDARD
GROUP  COUNT  MEAN  DEVIATION  ERROR    MINIMUM  MAXIMUM  95 PCT CONF INT FOR MEAN
Grp 1   25  45.4400  26.5268  5.3054   11.0000  99.0000   34.4903 TO 56.3897
Grp 2   25  41.7200  28.4847  5.6969    1.0000  97.0000   29.9621 TO 53.4779
Grp 3   25  42.2800  30.4775  6.0955    1.0000  96.0000   29.6995 TO 54.8605
Grp 4   25  58.6400  24.8477  4.9695   10.0000  94.0000   48.3834 TO 68.8966


MULTIPLE RANGE TEST
TUKEY-HSD PROCEDURE
RANGES FOR THE   .050 LEVEL -

        3.70   3.70   3.70

THE RANGES ABOVE ARE TABLE RANGES.
THE VALUE ACTUALLY COMPARED WITH MEAN(J)-MEAN(I) IS..
        19.5619 * RANGE * DSQRT(1/N(I) + 1/N(J))
NO TWO GROUPS ARE SIGNIFICANTLY DIFFERENT AT THE   .050 LEVEL
```

FIGURE A12.7
Output for SPSSx two-way analysis of variance with the ANOVA procedure

```
* * *   A N A L Y S I S   O F   V A R I A N C E   * * *

              X
        BY    COLOR
              FLAVOR
```

SOURCE OF VARIATION	SUM OF SQUARES	DF	MEAN SQUARE	F	SIGNIF OF F
MAIN EFFECTS	7930.220	7	1132.889	1.546	0.164
COLOR	3228.260	4	807.065	**1.101**	**0.362**
FLAVOR	4701.960	3	1567.320	**2.139**	**0.102**
2-WAY INTERACTIONS	11616.940	12	968.078	1.321	0.223
COLOR FLAVOR	11616.940	12	968.078	**1.321**	**0.223**
EXPLAINED	19547.160	19	1028.798	1.404	0.149
RESIDUAL	58626.800	80	732.835		
TOTAL	78173.960	99	789.636		

the commands and the output. The formula used by SPSSx when the sample sizes are unequal differs from the formula shown in Chapter 12. The text formula is based on a recent paper cited at the end of the chapter, while the formula in SPSSx is from the mid-1950s. For most problems this difference is a minor issue.

The command file and the analysis-of-variance table for a two-way analysis are almost the same as those shown above and in Figure A12.6. The new command file is

```
FILE HANDLE IN/ NAME='TWOWAY.DAT.1'
DATA LIST FILE=IN LIST/X COLOR FLAVOR
ANOVA X BY COLOR (1,5) FLAVOR (1,4)
```

The ANOVA procedure replaces the ONEWAY procedure. The list of indicator variables after the procedure now includes COLOR and FLAVOR. The numbers in parentheses specify the lowest and highest integer values of the indicator variables included in the analysis. Again the indicator variables must take integer values. The SPSSx ANOVA procedure can do an unbalanced design. Figure A12.7 shows the output for the command file above. It contains the F test statistics for some tests that are not discussed in Chapter 12. The F test statistics and the p values that are discussed there are printed in boldface type. Here the p value is called the SIGNIF OF F instead of the F. PROB in the ONEWAY routine. At a 5% significance level, the null hypotheses of no treatment effects, no block effects, and no interaction effects cannot be rejected.

EXERCISES FOR APPENDIX A12

A12.1 Exercise 3.18 asked you to compare the box plots of the rates of return on no-load and load mutual funds. The data source for that exercise is the July 1985 issue of *Consumer Reports*. Recall that a load is an up-front salesman's commission. Let us return to the same issue, given more recent data and information about the percentage load charged. Data set 16 lists the rates of return and loads for 289 mutual funds provided in the June 1987 issue of *Consumer Reports*. The rates of return are for the period April 1, 1982, through March 31, 1987. These rates would be much lower if the study had continued through the October 1987 stock market crash. The loads are recorded in integer values. Loads between 1% and 3% are coded 3 because there are few observations in this range. The figure below shows box plots for these data. Category 12b-1 refers to mutual funds that charge no explicit load but do charge a fee for advertising and marketing the mutual fund. The returns appear to be relatively high for the first three categories and then drift down with increasingly higher loads.

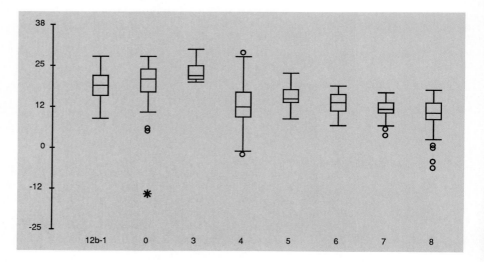

(a) Does the box plot support the assumption of homogeneity of variance?

(b) Can the null hypothesis of equal population mean returns across all load categories be rejected at a 5% significance level?

(c) Identify all pairs of load categories that have significantly different yields. Set α at 5%.

(d) Consumer Reports suggests that investors seeking high rates of return should avoid all funds with loads or 12b-1 marketing fees. Do the data support this advice?

A12.2 The first variable in data set 17 contains the coupon rates for the 304 bonds acquired by the pension fund TIAA–CREF in 1986. A second indicator variable is coded 0, 1, 2, 3, or 4 for bonds that have maturities under 10 years, from 10 to under 20 years, from 20 to under 30 years, from 30 to under 40 years, and 40 or more years, respectively.

(a) Treat the data as a random sample of all bonds sold in 1986 and test the hypothesis that the mean coupon rate is the same for the four categories of maturity.

(b) To check if the assumption of equal population variances is accurate, plot either the box plots for coupon rates for each category or a scatter diagram of coupon rates versus categories.

(c) As a further check on the assumption of equal variances, for each category of maturity plot the differences between the observations and the mean coupon rate within the category.

(d) Is the assumption of equal variances tenable? Which plot is best suited for analysis of this question? If the population variances appear to be unequal, what can you do?

(e) Identify all pairs of maturities that have significantly different coupon rates. Set α to 5%.

(f) Neither the coupon rate nor the maturity are random variables. Both terms can be set by the issuer of the bond. Does it make sense to test the hypothesis that the mean coupon rate is the same for the four categories of maturity? Explain.

(The following example was suggested by Professor Robert Kleyle of the IUPUI statistics department.)

A12.3 A large manufacturer of hardwood flooring tested a variety of compounds used to polish urethane coating on the flooring. The test consisted of varying the chemical binders (the treatments) that suspend the polishing agents as well as the types of hardwood (literally and figuratively, the blocks). Five chemical binders and four types of wood were tested with five repetitions of each treatment and block combination. The observations record the time, to the nearest second, needed to polish the wood in a high-speed polisher. The polisher automatically stops when the wood is polished.

(a) Test the null hypothesis at a 5% significance level that there are no treatment, block, or interaction effects. The data are in data set 18; the first column contains the observations, the second column the levels of the treatment factor, and the third column the levels of the blocking factor.

(b) Which combination of treatment and block has the lowest average polishing time?

(c) If the company had to use one type of rubbing compound for all types of hardwood, should it use the type identified in part b? Explain.

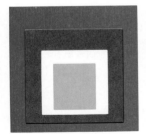

A13
Simple Regression and Correlation

A13.1 INTRODUCTION

When it comes to regression analysis, computer packages excel. Because of the importance of regression analysis, the four packages report extensive descriptive statistics, plot scatter diagrams of the observations, plot scatter diagrams of the residuals against the x values, and transform variables to fit nonlinear relationships. Some of the packages can even construct 95% confidence intervals for the mean and predicted values of y at every observed value of x. The statement made at the beginning of Appendix A, that computer packages open up data analysis, is exemplified by their treatment of regression analysis. You can go beyond simply estimating the effect of a unit change in x on μy. You can easily check if the relationship between x and μy is linear, if the overall fit is close, if any outliers strongly affect the estimated regression line, and if the assumptions of homoscedasticity and uncorrelated error terms are plausible.

A13.2 AN APPLICATION

The yield curve on bonds describes the relationship between yields with identical risk of default and time to maturity. The term "maturity" refers to the date at which a bond can be redeemed for cash. The shape of the yield curve varies with business conditions. When investors expect interest rates to rise over time, the yield curve will be an increasing function of maturity. When investors expect interest rates to fall over time, the yield curve may be a decreasing function of maturity. The qualification is that risk aversion among investors is a second factor widely thought to tilt the yield curve upward. Even if interest rates are expected to fall, the risk-aversion factor may be enough to cause the yield curve to have a positive slope. Much of the research into the determinants of yield curves has sought to identify the factors that influence expectations regarding future interest rates and to measure the extent of risk aversion among investors.

Long-term bonds are riskier than short-term bonds in the sense that they offer a greater opportunity both for loss and for gain. The prices of long-term bonds are more sensitive than the prices of short-term bonds to unanticipated changes in interest rates. To see why long-term bonds are riskier than short-term bonds, you have to understand the basics of bond pricing. Suppose you bought a new bond with a coupon rate of 10% and a one-year maturity. The cash value of the bond at redemption is $100. The selling price of the bond is its present value, the value today of a future payment. The formula for the present value of a payment of $100 x years in the future with an interest rate r is

$$PV = \frac{100}{(1 + r)^x}$$

The letters PV stand for present value. If the interest rate at the beginning of the year matched the coupon rate of 10%, the bond would sell for $100/1.1, or $90.91. Suppose you bought this bond for $90.91 in the morning and then found out that the current interest rate had changed to 5% in the afternoon. All new one-year bonds selling in the afternoon would cost $100/1.05, or $95.24 per $100 of redemption value. If we ignore the broker's commissions, you could turn around and sell the bond that you had bought for $90.91 in the morning for the new price of $95.24. The fall in the interest rate leads to a capital gain for the bondholder. This example can be turned around: a rise in interest leads to a capital loss for the bondholder.

FIGURE A13.1

Scatter diagram of yield versus time to maturity of U.S. Treasury bonds

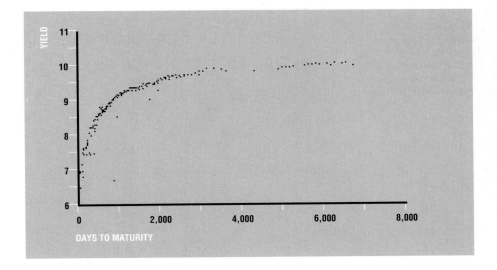

Now suppose you buy a new bond that will pay $100 at the end of ten years. In the morning the interest rate is 10%. The selling price or present value of this bond is

$$PV = \frac{100}{(1.1)^{10}}$$

$$PV = 38.55$$

Again, imagine that the interest rate falls to 5% in the afternoon. The new price of the bond or present value would be

$$PV = \frac{100}{(1.05)^{10}}$$

$$PV = 61.39$$

You could now sell the bond you had bought for $38.55 for $61.39. The capital gain for the same change in interest rates is much greater for the ten-year bond than for the one-year bond. The difference here is $22.84 versus $4.33. Similarly, the potential capital loss following an increase in interest rates is much greater with a long-term bond. Hence long-term bonds are riskier than short-term bonds.

To illustrate a yield curve, we will use the yields and maturities of U.S. Treasury bonds on October 8, 1987. Treasury bonds are considered to have zero risk of default. Hence a yield curve restricted to these bonds should show only the effect of changing maturity on yields. Figure A13.1 shows the scatter diagram of these bonds. The x axis in Figure A13.1 is the days to maturity. One reason that some of the bonds are outliers is that they have different tax treatments; citizens of countries other than the United States are not subject to withholding tax on the earnings of these bonds. Over all of the points, it is clear that the yield curve is nonlinear.

Scatter diagram of the natural logarithm of yield versus the natural logarithm of time to maturity of U.S. Treasury bonds and the estimated regression line

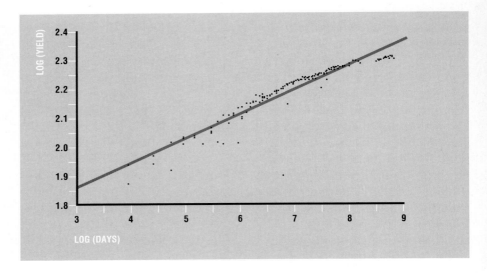

A least-squares line can be fitted by taking the natural logs of the original data. (Logarithmic transformations are discussed in section 13.4 of Chapter 13.) The logarithmic model is

$$\ln(y) = a + b(\ln(x))$$

where ln( ) stands for the natural log of the term in parentheses. Figure A13.2 shows the scatter diagram of the natural log of yields versus the natural log of days to maturity and an estimated regression line.

Figure A13.3 shows the DATA DESK output for the estimated regression line in Figure A13.2. According to the r^2 term, the log of days to maturity accounts for about 89% of the variation in the log of yields. The transformation to logs does a good job of changing the yield curve to a straight line. You can also tell from the output that the data cover 155 bonds, because the degrees of freedom are shown as $155 - 2$, or 153. The next part of the output shows an analysis-of-variance table. (The

FIGURE A13.3
DATA DESK output for regression of coupon rates on years to maturity

```
Dependent variable is:        log(yield)
R² = 88.9%      R²(adjusted) = 88.8%
s = 0.0335  with  155 - 2 = 153  degrees of freedom

Source   Sum of Squares    df      Mean Square    F-ratio
Regression   1.37644         1        1.3764       1227
Residual     0.171667       153       0.001122

Variable        Coefficient     s.e. of Coeff      t-ratio
Constant         1.60954          0.0168             95.7
log(days)        0.084022         0.0024             35.0
```

FIGURE A13.4

Scatter plot of the U.S. Treasury bond data and the antilog of the estimated regression line plotted against the days to maturity

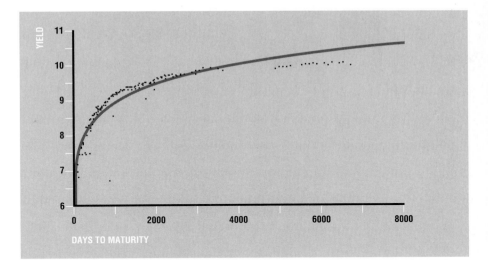

use of analysis of variance in a regression context is discussed in Chapter 14.) The last part of the output has the estimated regression coefficients and their standard errors. So far, this application has been basically a descriptive use of regression analysis. Suppose these data, which cover one day's sale of bonds, were considered a sample of bond yields under given expectations in regard to future interest rates. With this sample, the null hypothesis that the population slope coefficient for the log of days to maturity equals zero could be rejected at any conventional significance. The t ratio for the slope coefficient, 35, is a shortcut for testing this hypothesis. The t ratio is the slope coefficient divided by its standard error. If this ratio is greater than the $t_{\alpha/2}$, the null hypothesis of a zero population slope coefficient can be rejected at an α significance level. For example, since 35 exceeds 1.96, the null hypothesis could be rejected at a 5% significance level. The large sample size is the reason that you can use the z score of 1.96 in place of the t score.

To show how closely the estimated regression tracks the original data, the antilog of the estimated regression equation,

$$\ln(y) = 1.60954 + 0.084022 \ln(x)$$

is plotted against the days to maturity in Figure A13.4. Now that you have a yield curve, you may be wondering what you can do with it. Yield curves are used in theoretical studies of the determinants of interest rates. A more practical application is to compare yield curves for different dates to see if expectations in respect to interest rates are changing. Presumably the degree of risk aversion among investors is stable over time, so changes in the yield curve would reflect changing expectations. A yield curve that becomes steeper often indicates that investors anticipate higher inflation and consequently higher interest rates. The margin between near- and long-term interest rates is sometimes called the inflation premium, because higher inflation is associated with higher nominal interest rates.

A13.3 COMMANDS

A13.3A DATA DESK

DATA DESK has a `Regression` command under the `Compute` menu. The same command is used for simple regression (discussed in Chapter 13) and multiple regression (discussed in Chapter 14). To use the `Regression` command, first select the icon of the dependent variable and, while holding down the shift key, select the icon of the independent variable. Then click on the `Regression` command. The output consists of the usual descriptive statistics and an analysis-of-variance table. Additional statistics can be selected with the `Set regression options` command. This command shows up under the `Compute` menu if no variables are selected. The additional statistics available in this menu are not discussed in Chapter 13. DATA DESK automatically creates a variable to store the residuals, that is, the differences between the actual and predicted values of the y variable. You can plot the residuals against the x values with the `Scatterplot` command. The same command will create a scatter diagram of the x and y values of the original data. The variable that is selected first appears on the y axis. If the relationship between the x and y values appears to be nonlinear, you can transform the x or y values with the `New derived variable` command under the `File` menu. Some of the transformations are the square root of x, the natural log of x, and e to the x power. After the transformation you can immediately look at the scatter diagram of the transformed variable or variables and decide whether the transformed variable does a better job of meeting the linearity assumption required for least-squares regression. Figure A13.3 shows some DATA DESK output for the `Regression` command.

A13.3B MINITAB

The `REGRESS` command in MINITAB does single and multiple regression. You can store standardized residuals in a column. The residuals are the differences between the observed and the predicted values of y. Standardized residuals are the residuals divided by the standard error of the estimate. You can plot the standardized residuals against the x values to check for linearity with the `PLOT` command. Standardized residuals are convenient because under the normality assumption you can quickly identify outliers by looking for standardized residuals above 3. Recall that 0.3% of a normal distribution is beyond three standard deviations from the mean. Another purpose of the plot of standardized residuals is to let you check for nonlinearity and heteroscedasticity. You can also plot original x and y values with the `PLOT` command. The `LET =` command creates transformed variables. Functions that can be used with the `LET =` command include the square root of x (`SQRT`), the natural log of x (`LOGE`), and e to the x power (`ANTILOG`).

Figure A13.5 shows the output for a small hypothetical data set. Most of the terms and abbreviations should be self-explanatory. We will discuss the adjusted r^2—R–SQ (`adj`) in the output—in Chapter 14. The lowercase s on the same line is the standard error of the estimate. Analysis of variance in a regression context is also discussed in Chapter 14. The term `FIT` refers to the predicted values of y. The `NOBRIEF` command tells MINITAB to provide the predicted values of y (`FIT`), the residuals, the standard

FIGURE A13.5
MINITAB output for the
REGRESS **command**

```
MTB > NOBRIEF
MTB > REGRESS THE Y-VALUES IN C2 ON 1 PREDICTOR IN C1

The regression equation is
C2 = - 4.70 + 6.90 C1

Predictor        Coef        Stdev      t-ratio
Constant       -4.700        3.311       -1.42
C1             6.9000       0.9983        6.91

s = 3.157      R-sq = 94.1%      R-sq(adj) = 92.1%

Analysis of Variance

SOURCE          DF          SS           MS
Regression       1       476.10       476.10
Error            3        29.90         9.97
Total            4       506.00

Obs.       C1          C2        Fit  Stdev.Fit   Residual    St.Resid
  1       1.00        3.00       2.20      2.45       0.80        0.40
  2       2.00        6.00       9.10      1.73      -3.10       -1.17
  3       3.00       20.00      16.00      1.41       4.00        1.42
  4       4.00       21.00      22.90      1.73      -1.90       -0.72
  5       5.00       30.00      29.80      2.45       0.20        0.10
```

deviation of the mean of Y given x (Stdev.Fit), and the standard deviation of the residuals (St.Resid). The standard deviation of the mean of Y given x can be used to construct confidence intervals or to do hypothesis tests on the mean of Y given x.

To run the logarithmic regression as shown in section A13.2, use the following commands:

```
MTB > NAME C1 = 'X', C2 = 'Y'
MTB > LET C3 = LOGE(C1)
MTB > LET C4 = LOGE(C2)
MTB > REGRESS C4 ON 1 PREDICTOR IN C3
```

■ A13.3C SAS

The REG procedure in SAS does single and multiple regression. The REG procedure can create a list of residuals and 95% confidence intervals for the mean and predicted values of y. The REG can also do hypothesis tests for slope coefficients or the intercept. The PLOT procedure creates plots of residuals versus x values or y versus x values, which can be used as a check for linearity or homoscedasticity. Here is a listing of the commands needed to estimate a regression and plot the residuals:

```
DATA INDAT;
INPUT X Y;
PROC REG;
      MODEL Y=X;
      OUTPUT OUT=OUTDAT P=PRED R=RESID;
PROC PRINT;
PROC PLOT DATA = OUTDAT;
      PLOT RESID*X;
```

FIGURE A13.6
Output from SAS REG
procedure

```
DEP VARIABLE: Y
                                           ANALYSIS OF VARIANCE

                                 SUM OF              MEAN
                   SOURCE    DF   SQUARES            SQUARE        F VALUE   PROB>F

                   MODEL     1    476.10000          476.10000     47.76924  0.0000
                   ERROR     3     29.90000            9.97000
                   C TOTAL   4    506.00000

                   ROOT MSE     3.157002      R-SQUARE       0.9409
                   DEP MEAN           16      ADJ R-SQ       0.9210
                   C.V.        47.04723

                                    PARAMETER ESTIMATES

                          PARAMETER        STANDARD        T FOR H0:         PROB > |T|
         VARIABLE   DF    ESTIMATE         ERROR           PARAMETER=0

         INTERCEP   1     -4.7000000       3.31109200      -1.419            0.2508
         X          1      6.9000000       0.99833180       6.912            0.0062
```

The commands PROC REG and MODEL Y=X create the descriptive statistics and an analysis-of-variance table for the regression. The line OUTPUT OUT=OUTDAT P=PRED R=RESID creates a file to store the predicted values of y and the residuals. The last two lines create a plot of the residuals versus the x values.

Figure A13.6 shows the minimum output from the SAS REG procedure for a small hypothetical data set. Most of the terms and abbreviations should be self-explanatory. We will discuss the adjusted r^2—ADJ R-SQ in the output—in Chapter 14. Analysis of variance in a regression context is also discussed in Chapter 14. The term T FOR H0: PARAMETER=0 is the parameter estimate divided by its standard error. The PROB > |T| is the p value for a test of the null hypothesis that the parameter equals 0. The standard error of the estimate is ROOT MSE in the output.

To run the logarithmic regression shown in section A13.2, use the following commands:

```
DATA INDAT;
INPUT X Y;
X=LOG(X);
Y=LOG(Y);
PROC REG;
      MODEL Y=X;
```

The X = LOG(X) and Y = LOG(Y) replace the x and y variables with their logs. The rest of the commands are the same as in the previous example.

 A13.3D SPSSx

The SPSSx REGRESSION procedure is used for simple and multiple regression. For simple regression, the REGRESSION command must be followed by the VARIABLES,

```
Equation Number 1     Dependent Variable..    Y
Beginning Block Number  1.  Method: Enter     X
Variable(s) Entered on Step Number  1..     X


Multiple R              .97000     Analysis of Variance

R Square                .94091                   DF   Sum of Squares  Mean Square
Adjusted R Square       .92121     Regression     1      476.10000      476.10000
Standard Error         3.15700     Residual       3       29.90000        9.96667

                                   F =      47.76923      Signif F = .0062


------------------ Variables in the Equation ------------------

Variable              B          SE B        Beta        T   Sig T

X              6.900000      .998332     .970005     6.912   .0062
(Constant)    -4.700000     3.311092                -1.419   .2508
```

FIGURE A13.7
Output for SPSSx
REGRESSION procedure

DEPENDENT, and ENTER subcommands. If the independent variable is named x and the dependent variable named y, the commands for regression analysis are

```
REGRESSION      VARIABLES=X,Y/
         DEPENDENT=Y/
         ENTER X/
         SCATTERPLOT=(*RESID,X)
```

The last line creates a scatter diagram of the residuals against the x values. RESID is a temporary variable automatically created by SPSSx to hold the residuals. The asterisk before the RESID tells SPSSx to use this temporary variable in the scatter plot.

Figure A13.7 shows the minimum output for the SPSSx REGRESSION procedure for a small hypothetical data set. Most of the terms and abbreviations should be self-explanatory. We will discuss the adjusted r^2—ADJUSTED R SQUARE in the output—in Chapter 14. Analysis of variance in a regression context is also discussed in Chapter 14. The letter T is the parameter estimate divided by its standard error. This ratio is usually called the t ratio. The SIG T is the p value for a test of the null hypothesis that the parameter equals 0. The letter B refers to the estimated slope and the estimated intercept. SPSSx's use of B is anomalous.

To run the logarithmic regression as shown in section A13.2, use the following commands:

```
COMPUTE LOGX=LN(X)
COMPUTE LOGY=LN(Y)
REGRESSION      VARIABLES=LOGX,LOGY/
         DEPENDENT=LOGY/
         ENTER LOGX/
```

The command LOGX=LN(X) creates a new variable, LOGX, which equals the natural log of x. The rest of the commands are the same as in the previous example.

EXERCISES FOR APPENDIX A13

A13.1 Data set 1 was put together to predict the number of master's degrees that would be awarded by a proposed master's program in economics at IUPUI over a three-year period. The data set includes the following variables for the 44 largest cities in the United States: city name, population, per capita income, number of MA degrees in economics awarded during a three-year period, number of MBA degrees awarded during a three-year period, and a dummy variable that equals 1 if there is an institution that grants Ph.D. degrees in economics within 75 miles of the city. Run a simple regression of the number of MA degrees in economics on the population variable and answer the following questions.

   (a) What percentage of the total variation in the number of MAs awarded is accounted for by the population variable?

   (b) What is the predicted number of MA degrees for Indianapolis on the basis of this regression? The Indianapolis SMSA population is 1.19 million.

   (c) Construct confidence intervals for the mean and predicted values of the number of MA degrees in Indianapolis at a 95% level. Which type of confidence interval is appropriate in this context?

   (d) New York City is an extreme outlier in this data set, in terms of both the number of economics MAs and population. What effect does dropping New York from the data set have on the predicted number of MA degrees for Indianapolis?

   (e) Create a scatter diagram of the residuals versus the number of MA degrees. What assumption of the regression model do the data clearly violate?

A13.2 Compute the coefficient of correlation between the sales volume and the month-end prices for the H & R Block stock data in data set 5.

   (a) Test the null hypothesis at a 5% significance level that the data are a sample from a population with a zero correlation coefficient.

   (b) Is correlation analysis appropriate in this context? Explain.

A13.3 Data set 7 contains the predictions of a sample of prominent economists on near-term (six months) and longer-term (twelve months) GNP in columns 2 and 3, inflation in columns 4 and 5, and unemployment in columns 6 and 7. Calculate the coefficients of correlation between the near- and longer-term predictions for GNP, inflation, and unemployment. Compare correlations in economists' predictions for near and longer term for each variable and test the three hypotheses that each population coefficient of correlation is zero. Use a 5% significance level for each test.

A14
Multiple Regression Analysis

A14.1 INTRODUCTION

The commands for simple and multiple regression are the same in most programs. The DATA DESK, MINITAB, SAS, and SPSSx commands do not change when you go from simple to multiple regression. The emphasis here will be on the interpretation of multiple regression results and the application of such model-building techniques as logarithmic transformations and dummy variables.

A14.2 AN APPLICATION

At least half a dozen articles on the salaries of baseball players have appeared in economics journals. Economists primarily use regression analysis for such analyses. Data on baseball players are unusual because there are many measures of their productivity. There are statistics on hitters' lifetime batting averages, number of home runs, number of times at bat last season, fielding averages, bases stolen, total bases, and so on. The purpose of the economics articles was either to test a theory, such as the effect of monopsony power on salaries, or to suggest changes in public policy on professional sports. The purpose of our use of these rich and well-understood data is to illustrate the applications of multiple regression analysis.

Gerald W. Scully ("Pay and Performance in Major League Baseball," *American Economics Review* 64 [December 1974]: 915–930), experimented with all of the available performance measures to estimate the determinants of players' salaries. For hitters he settled on seven variables, of which four measure individual player performance and three measure characteristics of the city and home team. One player performance variable, x_1, was the lifetime slugging average, or the number of bases divided by the number of times at bat. The second variable, x_2, was a dummy variable equal to 1 if a player had a batting average above the average for all current players but a slugging average below the average for all current players. This dummy variable was intended to account for the productivity of players who are reliable hitters of singles but rarely get extra base hits. The third variable, x_3, was a player's total lifetime at-bats divided by the product of his number of years in the major leagues times 5,500 (the average number of at-bats in a season). This variable measures a hitter's percentage contribution to his team's performance. The fourth player variable, x_4, was the number of years he had played in the majors. The first team variable, x_5, was a dummy variable that equaled 1 if the team was in the National League. The National League was reputed to pay higher salaries. One city variable, x_6, was the population of the standard metropolitan statistical area (SMSA) of the player's team in 1970. An SMSA consists of a city and its suburbs. The second city variable, x_7, reflected a city's interest in baseball. It was the estimated increase in team revenue generated by an increase of one game in the number of the team's wins per season.

The dependent variable in the estimated regression equation was the base 10 log of players' salaries. The independent variables were in base 10 logs as well, except for the dummy variables. Professor Scully adopted this specification to account for the non-linearity in players' salaries. He argued that a star player may be only 20% better than an average player on some numerical measure of performance, such as slugging average, but that the star's salary may be 100% higher, because of the fans' interest in the stars.

Table A14.1 shows the output from Professor Scully's regression. The estimates are based on 93 observations of hitters' salaries over the 1968 and 1969 seasons. The implication of the r^2 of 0.81 is that 81% of the variation in the base 10 log of hitters' salaries is accounted for by the independent variables. The null hypothesis that the population slope coefficients equaled zero can be rejected for all of the player performance measures at a 5% significance level, except for the dummy variable for hitters with many singles. A shortcut for hypothesis tests against a null value of zero at a 5%

TABLE A14.1
Regression estimates on the base 10 logarithm of hitters' salaries measured in thousands of dollars

Variable definition	Variable	Coefficient	Standard error	t ratio
Intercept	a	0.6699	0.817	0.82
Log of lifetime slugging average	x_1	1.0716	0.225	4.76
Dummy variable equal to 1 if a player had a batting average above the average for all current players but a slugging average below the average for all current players	x_2	0.0579	0.036	1.61
Log of player's total lifetime at-bats divided by his number of years in the majors times 5,500	x_3	0.2746	0.089	3.10
Log of number of years in the majors	x_4	0.522	0.069	7.53
Dummy variable that equals 1 if the team is in the National League	x_5	0.194	0.031	0.62
Log of SMSA population in 1970	x_6	−0.0621	0.080	0.78
Log of estimated increase in team revenue generated by an increase of one game in the number of the team's wins per season	x_7	0.2645	0.094	2.81

$r^2 = 0.81$, $df = 85$

significance level is to check if the ratio of the slope coefficient divided by its standard error exceeds 1.96. If this condition is met, the null hypothesis can be rejected. The z score can be used here because there are 85 degrees of freedom. The coefficient divided by its standard error is called the t ratio. One of the city variables, x_7, is statistically significantly different from zero at the 5% level. In these data, the effects of the city's size and membership in the National League are not statistically significant. Overall, the salaries are closely tied to individual performance and loosely tied to city and team characteristics.

To interpret the particular coefficients, you have to know the mean value of salaries in the sample and the mean of the particular independent variable. Consider the impact of adding one year of experience to a player who has the sample average of eight years' experience in the majors. The average salary in the sample was \$48,100. The base 10 log of \$48,100 is 4.682. The estimated effect of an increase of one year in time in the majors is to raise the estimated log of wages by 0.522 ($\log(9) - \log(8)$), or to 4.733. The antilog of 4.733 is \$54,113. Other things being equal, an additional year of experience in the majors is worth \$54,113 minus \$48,100, or \$6,013. Recall that this figure is for 1968–1969. The salaries are considerably higher now because of inflation and free agency in the baseball market.

Although the dummy variables do not differ significantly from zero, we can calculate the effect of a change in a dummy variable on salaries for illustration. For a player with an average salary, for example, what is the estimated effect of going from the American to the National League? Changing leagues would raise the log of salaries from the mean value of 4.682 to 4.702 (4.682 plus the coefficient of the dummy variable for the National League of 0.0194). The antilog of 4.702 is \$50,297. The estimated increase in salary is \$2,197, or \$50,297 minus the average of \$48,100. Note that the dummy variable was defined as 1 for the National League and 0 for the American League. This figure is the estimated gain from changing leagues while everything else is held constant. Note also that while the best point estimate of the effect of changing leagues is \$2,197, there is not enough evidence in the data to reject the null hypothesis that the true effect is zero.

Now let's consider an article by Donald J. Cymrot ("Migration Trends and Earnings of Free Agents in Major League Baseball, 1976–1979," *Economic Inquiry* 21 [October 1983]: 545–556). Professor Cymrot used 13 independent variables to estimate the salaries of free agents. Three of these variables are team characteristics, six are personal characteristics (but not necessarily measures of players' productivity), and the remaining four are city characteristics. Table A14.2 shows the variables' names, the estimated regression coefficient, the standard errors, and the t ratios. The specification is semilogarithmic. The dependent variable is in natural logs but the independent variables are in the original levels. The semilog specification is common in salary studies because labor economists find that this specification does a good job of fitting measures of productivity to earnings data. The semilog specification leads to a handy interpretation of the coefficients: a coefficient is the estimated percentage change in the dependent variable for a unit change in the independent variable. The interpretation works because equal changes in logs represent equal percentage changes in levels. Take the series 10, 100, 1,000, 10,000, ... The corresponding logs in base 10 (the base chosen does not matter) are 1, 2, 3, 4, ... Equal changes in the logs (steps of 1) represent equal percentage changes in the levels (factors of 10).

To apply this interpretation to the estimated regression equation, consider the *PITCH* variable, a dummy variable that equals 1 if the player is a pitcher. The interpretation of the coefficient 0.38 is that, other things held equal, a free agent who is a

TABLE A14.2

Coefficients, standard errors, and t ratios for regression on the natural logarithm of free agents' salaries

Variable	Coefficient	Standard error	t ratio
INTERCEPT	11.58	0.444	26.06
MRP	0.17	0.031	5.47
PITCH	0.38	0.131	2.89
IMPROVE	0.34	0.181	1.88
ALLSTAR	0.04	0.038	1.05
RACE	0.02	0.143	0.14
EXPERIENCE	−0.02	0.026	−0.78
SAME	0.06	0.231	0.26
YR76	−0.67	0.171	−3.92
YR77	−0.28	0.184	−1.52
YR78	−0.42	0.157	−2.67
QUALITY	0.00	0.050	0.08
POP3	0.09	0.035	2.58
GROWTH2	0.03		2.49

$r^2 = 0.65$
$df = 64$
$s_e = 14.49$

pitcher makes 38% more than a nonpitcher. Consider the variable *YR78*. It is a dummy variable that equals 1 if the player was a free agent in 1978. The omitted year or base year is 1979. Other things being equal, a player who signed as a free agent got 42% less in 1978 than in 1979. To see how a nondummy variable works, consider *POP3*, a refinement of the x_6 (SMSA) variable used by Professor Scully. *POP3* is the population in millions of the SMSA or, for the largest urban areas, the SCSA (standard consolidated statistical area). When a city has both a National and an American League team, however, *POP3* is one-half of the SMSA or SCSA population. With this refinement, Professor Cymrot found that the null hypothesis of no population effect on salaries could be rejected at a 5% significance level. The interpretation of the coefficient is that, other things being equal, an increase of 1 million people in a city's population would raise a free agent's salary by 9%. Some of the other independent variables are complicated. Since our purpose is to interpret regression output rather than to study the baseball market, there is no reason to go into all 13 variables.

A14.3 PROBIT REGRESSION COMMANDS

A14.3A INTRODUCTION

The probit regression, as described in Chapter 14, is commonly used by economists. Computer packages for econometrics, such as SHAZAM, RATS, and TSP, have probit routines. Probit routines are less commonly found in general statistical packages. No such routine is currently available on DATA DESK or MINITAB. Given the omnibus nature of SAS and SPSSx, it is not surprising to find probit routines in those packages. Their routines are not quite the same as the probit method described in Chapter 14, however. The SAS and SPSSx procedures, both named PROBIT, are meant for what are called biological dosage-response models. In these models a group of subjects is given some specified level of a treatment and the number of subjects who respond to the treatment is recorded. Sets of 100 cockroaches may be given doses of a poison, for example, and the number that die in each set is recorded. The data for such an experiment would be recorded in two columns, one for the level of the poison dosage and the second for the number of cockroaches that die. If the sample size for each dose was not fixed, a third column could be used to record the sample size. The probit example in Chapter 14, in contrast, has individual data—the income and public television membership status of each person in the sample. If you have enough observations, you can aggregate the individual observations by dosage. Such aggregation is difficult with small data sets or with more than one independent variable from nonexperimental data.

SPSSx allows the use of individual data in its PROBIT procedure. The reported coefficients and standard errors are correct. When SPSSx encounters individual data, however, the reported χ^2 statistic that is used to test the overall fit of the probit regression is incorrect. SAS has a separate routine for probit problems with individual data, MPROBIT. If you have individual data, you should use MPROBIT instead of PROBIT. MPROBIT reports the correct χ^2 statistic. Note that MPROBIT is not described in the latest SAS manual, version 5 (1985). It was added to SAS after the manual was written. Your computer center may have a separate document on the MPROBIT procedure.

A14.3B SAS

The basic SAS commands for the public television probit example in Chapter 14 are shown below. The MEMBER variable is a 0–1 dummy that takes the value 1 if the person is a member of the television channel. The commands are basically the same as in the regression procedure. If there were more than one independent variable, they would be listed after the equals sign in the MODEL statement.

```
DATA PROBDAT;
INPUT ED MEMBER;
PROC MPROBIT;
        MODEL MEMBER=ED;
```

A14.3C SPSSx

The basic SPSSx commands for the public television probit example in Chapter 14 are shown below. The MEMBER variable is a 0–1 dummy that takes the value 1 if the person is a member of the television channel. The purpose of the COMPUTE NUMBER = 1 is to tell SPSSx that every group has one observation. Recall that SPSSx's PROBIT procedure is set up to work with groups of observations. If you had more than one independent variable, they would be listed after the word WITH in the line with the PROBIT command.

```
DATA LIST FREE / ED MEMBER
COMPUTE NUMBER = 1
PROBIT RESPONSE OF NUMBER WITH ED
BEGIN DATA
6 0
 .  .

 .  .
22 1
END DATA
```

EXERCISES FOR APPENDIX A14

A14.1 In exercise A13.1 you predicted the number of MA degrees in economics per three-year period in Indianapolis in a simple regression framework. Now you can try the same problem with multiple regression. The dependent variable is still the economics MA variable. The independent variables are population, income, MBAs, and the dummy variable for a nearby school that grants Ph.D.s in economics.

(a) Identify the independent variables that are statistically significantly different from 0 at a 5% level.

(b) What is the predicted number of MA degrees for Indianapolis? Use 0 for the value of the dummy variable and the respective sample means as the predicted Indianapolis values for the income and MBA variables. The Indianapolis SMSA population is 1.19 million.

(c) A few of the cities have zero MA degrees in economics. You could argue that these cities have a nonzero demand for MAs in economics which is met by importing people trained in other cities. Hence, including these cities in the regression would cause an underestimate of the number of degrees awarded in a city that has an MA program in economics. See what happens to the expected number of MA degrees in Indianapolis when these observations are dropped.

A14.2 Data set 19 contains the responses to five questions from a telephone survey of 627 randomly selected Indianapolis adults above the age of 65. All of the phone calls were made to private residences rather than hospitals or nursing homes. The survey was commissioned by Blue Cross of Indiana to study marketing issues for privately purchased health care insurance policies that supplement Medicare. The five responses included in the data set are: (1) the number of hospital days during the past year, (2) the number of regular doctor visits during the last year, (3) the number of visits to specialists or referral doctors during the last year, (4) the respondent's average monthly spending for prescription drugs, and (5) the respondent's age.

(a) Run a regression in which the dependent variable is the number of days in hospitals during the last year and the independent variables are the remaining four variables. What percentage of the variation in hospital days is accounted for by these four variables? Can the null hypothesis that all population slope coefficients equal zero be rejected at a 5% significance level?

(b) Create a scatter plot with the age variable on the x axis and the hospital days on the y axis. Which assumption of the regression model do the data clearly violate?

(c) What is wrong with the Blue Cross survey?

(Exercises A14.3 and A14.4 can be done with SAS and SPSSx.)

A14.3 Recall that data set 9 records the level of education and the preferences for Coca-Cola and Pepsi-Cola among families that have a cola preference. Use a probit regression to estimate the effect of changes in the level of education of the adult respondent on the probability that the family will prefer Coca-Cola.

(a) What is the probability that a family in which the respondent has a high-school education will prefer Coca-Cola?

(b) What is the probability that a family in which the respondent has a bachelor's degree will prefer Coca-Cola?

(c) What are the estimated mean and standard deviation of the latent variable representing Coca-Cola preference?

A14.4 Data set 20 contains the results of a statewide survey in Indiana on child car-seat usage. The survey used state troopers to stop 182 mothers while they were driving with small children so that the mothers could be questioned. The definitions of the variables in order of appearance in the data set are:

1. Age of the driver in years.

2. Hours per year available for housework by the driver. (Defined as 2,000 hours minus the annual hours of market work.)

3. Number of children four years of age and under in the car.

4. Dummy variable equal to 1 if the driver is married and equal to 0 otherwise.

5. Nonwage income of the driver (the mother in all cases). This was approximated by subtracting the projected annual earnings of the driver from the reported annual family income.

6. Years of education of the driver's spouse.

7. Hourly wage of the driver.

8. Driver's years of education.

9. Dummy variable equal to 1 if a federally approved car seat is rated as correctly used and equal to 0 otherwise.

(a) Use a probit regression to estimate the effect of the remaining variables on variable 9.

(b) Which explanatory variables have a statistically significant effect on child car-seat usage?

(c) Are mothers with higher wages more or less likely to use a child car seat?

A15
Forecasting Time Series

A15.1 INTRODUCTION

General statistical packages tend to have limited capabilities for time-series analysis. The reason is that time-series analysis is a peripheral, if not esoteric, topic for most users. To most business users and economists, however, time-series analysis is a central concern. If you plan to do extensive time-series analysis, you may need a specialized time-series package, such as TSP (Time Series Processor), SHAZAM, or RATS (Regression and Time Series). SAS markets a separate package, ETS (Econometrics and Time Series), which may be available on your university mainframe computer.

The forecasting techniques discussed in Chapter 15 can be classified by the type of analysis required: multiple regression, smoothing, or nonlinear estimation. All four of the packages discussed here can do forecasts that require multiple regression. The classical model is multiple regression with a time-trend variable and seasonal dummies as independent variables. The classical model is the only kind of analysis that can be done easily in DATA DESK. All four packages can also do autoregression. Autoregression is multiple regression with the lagged values of the forecast variable as independent variables. MINITAB and SPSSx have a lag operator that simplifies the creation of variables that are lagged values of the forecast variable. SAS has a special autoregression routine that automatically chooses the number of lags. Forecasts based on smoothing have much less support in the packages. SAS ETS has a routine for making forecasts based on exponential smoothing. MINITAB has a smoothing routine that relies on the medians of nearby values rather than on the average.

ARIMA forecasts require nonlinear estimation. As Chapter 15 indicates, in a few years ARIMA has become a highly popular forecasting technique. It is included in general statistical packages that otherwise do not have a strong time-series component. MINITAB, SAS, and SPSSx have ARIMA routines. Table A15.1 shows the forecasting techniques that are available in each package.

TABLE A15.1
Forecasting techniques available in four statistical packages

	DATA DESK	MINITAB	SAS	SPSSx
Classical model	Yes	Yes	Yes	Yes
Smoothing with moving averages	No	No	No	No
Exponential smoothing	No	No	Yes	No
Autoregressive model	Yes	Yes	Yes	Yes
ARIMA	No	Yes	Yes	Yes

A15.2 AN APPLICATION

Data set 27 contains the monthly average prices in dollars received by farmers per 100 pounds of live cattle. The source of the data is the Crop Reporting Board of the U.S. Department of Agriculture. The prices are shown in the first column, a time-trend variable appears in the second column, and dummy variables for the months of January through November occupy columns 3 through 13. The dummy variable for the month of December is omitted because the regression will not run if a dummy variable is included for every month. In this application, we will first make a 12-month forecast of cattle prices using the classical model, the autoregressive model, and an ARIMA model. This extended application will highlight the differences among the three forecasting models.

A plot of the price data against time, shown in Figure A15.1, suggests that these data are a good candidate for the classical forecast model. The plot shows a clear, downward trend and a stable seasonal pattern. Some background information on the cattle market will help you understand the plot. Cattle prices have shown a downward trend recently in response to a continuing decline in beef consumption. Consumers have been cutting back on red meat for health reasons. Second, cattle production and prices are strongly seasonal. Cattle births peak in the spring. It usually takes between 15 and 18 months to bring cattle to market weight. This pattern causes a seasonal peak in supply (and a low in prices) in the fall. Variations in demand are much smaller. There is a summer peak in demand for beef for outdoor barbecues.

Figure A15.2 shows the DATA DESK output for the regression of monthly prices on a time-trend variable and the monthly dummies. The time-trend variable is significantly different from zero at a 5% significance level. The average monthly change in prices is a drop of 11 cents. The prices in February, March, April, and May are significantly different (at a 5% significance level) from those of the omitted month of December.

FIGURE A15.1
Average prices received by farmers per 100 pounds of cattle, 1980–1987, by month

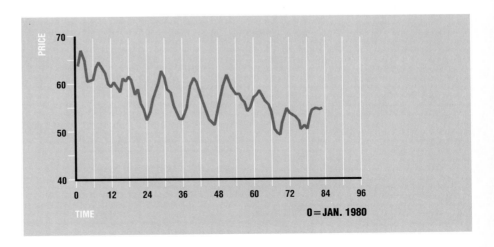

FIGURE A15.2
Regression output for the cattle price data based on the classical model

```
Dependent variable is:   price
R² = 60.4%      R² (adjusted) = 53.6%
s =  2.715  with   83 - 13 = 70   degrees of freedom

Source       Sum of Squares        df        Mean Square  F-ratio
Regression   785.478               12        65.5         8.88
Residual     515.852               70        7.36932

Variable    Coefficient   s.e. of Coeff     t-ratio
Constant    59.5774            1.227         48.6
time        -0.112558          0.0125        -8.98
Jan          1.70150           1.512          1.13
Feb          3.41406           1.511          2.26
Mar          3.71233           1.511          2.46
Apr          3.39631           1.510          2.25
May          3.40887           1.510          2.26
June         2.75000           1.510          1.82
July         2.26256           1.510          1.50
Aug          1.77512           1.510          1.18
Sept         1.07339           1.511          0.710
Oct          0.443088          1.511          0.293
Nov          0.355645          1.512          0.235
```

FIGURE A15.3
Cattle prices and a 12-month forecast based on the classical model

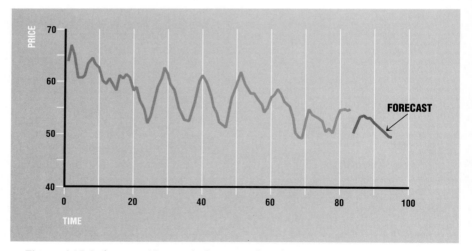

Figure A15.3 shows a 12-month forecast of cattle prices based on the coefficients estimated by the classical model. The peak in the data from February through May is repeated in the forecast. A continued downward trend is also clear in the forecast.

Let us consider the cattle data again, this time with an autoregressive model. Since there are 83 observations, the suggested procedure is to run the regressions with lags 1 through 19 (10 plus the nearest square root of 83). Table A15.2 shows the coefficients and the standard errors for the 19 regressions. The coefficient of any lag that is significantly different from zero at a 5% level is shown in boldface. The 12th lag is the highest one that is consistently significant.

TABLE A15.2
Regression coefficients and standard errors for lags 1 through 19

Constant	L1	L2	L3	L4	L5	L6	L7	L8	L9	L10	L11	L12	L13	L14	L15	L16	L17	L18	L19
6.548 (2.758)	**0.883** (0.048)																		
10.005 (2.465)	**1.264** (0.096)	**−0.443** (0.095)																	
9.674 (2.711)	**1.273** (0.113)	−0.380 (0.172)	−0.064 (0.107)																
7.869 (2.924)	**1.262** (0.114)	−0.270 (0.184)	−0.267 (0.177)	0.136 (0.107)															
6.633 (2.931)	**1.258** (0.111)	−0.205 (0.177)	−0.293 (0.177)	−0.104 (0.171)	**0.225** (0.103)														
7.235 (3.058)	**1.202** (0.118)	−0.117 (0.185)	−0.296 (0.178)	−0.182 (0.180)	0.230 (0.171)	0.034 (0.109)													
7.005 (3.128)	**1.164** (0.117)	−0.165 (0.182)	−0.189 (0.181)	−0.164 (0.177)	0.187 (0.178)	−0.151 (0.173)	0.191 (0.107)												
6.080 (3.025)	**1.054** (0.112)	−0.067 (0.170)	−0.290 (0.169)	−0.072 (0.169)	0.219 (0.165)	−0.098 (0.171)	−0.224 (0.163)	**0.366** (0.102)											
6.018 (3.183)	**1.003** (0.124)	−0.038 (0.174)	−0.269 (0.172)	−0.111 (0.174)	0.231 (0.171)	−0.078 (0.174)	−0.203 (0.175)	0.242 (0.168)	0.112 (0.112)										
5.694 (3.364)	**1.010** (0.127)	−0.021 (0.180)	−0.279 (0.176)	−0.117 (0.178)	0.249 (0.178)	−0.098 (0.181)	−0.205 (0.179)	0.254 (0.181)	0.144 (0.174)	−0.043 (0.115)									
4.053 (3.459)	**1.002** (0.126)	−0.019 (0.180)	−0.272 (0.178)	−0.102 (0.178)	0.258 (0.177)	−0.106 (0.184)	−0.261 (0.182)	0.283 (0.181)	0.252 (0.183)	−0.167 (0.173)	0.055 (0.114)								
4.624 (3.474)	**1.030** (0.127)	−0.058 (0.178)	−0.229 (0.176)	−0.031 (0.178)	0.202 (0.176)	−0.154 (0.182)	−0.169 (0.184)	0.266 (0.183)	0.204 (0.181)	−0.240 (0.182)	0.329 (0.171)	**−0.236** (0.112)							
5.257 (3.555)	**1.063** (0.131)	−0.070 (0.185)	−0.235 (0.177)	−0.054 (0.178)	0.160 (0.178)	−0.133 (0.182)	−0.145 (0.185)	0.213 (0.186)	0.265 (0.186)	−0.247 (0.182)	0.293 (0.184)	−0.337 (0.176)	0.130 (0.116)						
5.989 (3.739)	**1.071** (0.136)	−0.111 (0.199)	−0.183 (0.189)	−0.073 (0.182)	0.170 (0.181)	−0.123 (0.184)	−0.147 (0.187)	0.182 (0.192)	0.294 (0.192)	−0.233 (0.191)	0.278 (0.187)	**−0.397** (0.193)	0.254 (0.192)	−0.091 (0.125)					
5.676 (3.981)	**1.073** (0.139)	−0.095 (0.204)	−0.213 (0.202)	−0.071 (0.194)	0.182 (0.186)	−0.116 (0.188)	−0.135 (0.191)	0.175 (0.196)	0.290 (0.198)	−0.217 (0.200)	0.245 (0.197)	**−0.390** (0.197)	0.274 (0.208)	−0.059 (0.198)	−0.047 (0.132)				
7.585 (4.059)	**1.097** (0.137)	−0.139 (0.201)	−0.236 (0.200)	−0.027 (0.202)	0.239 (0.190)	−0.177 (0.186)	−0.141 (0.187)	0.154 (0.193)	0.295 (0.194)	−0.241 (0.197)	0.244 (0.198)	−0.298 (0.198)	0.251 (0.205)	−0.185 (0.207)	0.001 (0.199)	0.024 (0.130)			
7.169 (4.405)	**1.110** (0.144)	−0.158 (0.212)	−0.226 (0.206)	−0.019 (0.208)	0.225 (0.206)	−0.183 (0.199)	−0.131 (0.193)	0.161 (0.198)	0.297 (0.199)	−0.248 (0.203)	0.249 (0.204)	−0.295 (0.207)	0.236 (0.213)	−0.186 (0.212)	0.023 (0.220)	0.030 (0.203)	−0.018 (0.132)		
7.028 (4.771)	**1.115** (0.147)	−0.172 (0.221)	−0.209 (0.216)	−0.021 (0.212)	0.200 (0.212)	−0.127 (0.213)	−0.163 (0.206)	0.154 (0.205)	0.287 (0.203)	−0.268 (0.207)	0.267 (0.209)	−0.282 (0.213)	0.206 (0.220)	−0.170 (0.218)	0.033 (0.224)	0.046 (0.223)	−0.134 (0.206)	0.109 (0.135)	
7.013 (5.147)	**1.098** (0.149)	−0.151 (0.224)	−0.225 (0.225)	−0.006 (0.222)	0.200 (0.215)	−0.155 (0.218)	−0.094 (0.220)	0.103 (0.218)	0.287 (0.209)	−0.273 (0.210)	0.237 (0.213)	−0.260 (0.217)	0.225 (0.226)	−0.204 (0.225)	0.041 (0.229)	0.059 (0.227)	−0.108 (0.226)	−0.038 (0.209)	0.135 (0.138)

FIGURE A15.4

Cattle prices and 12-month forecasts based on the autoregressive model and the classical model

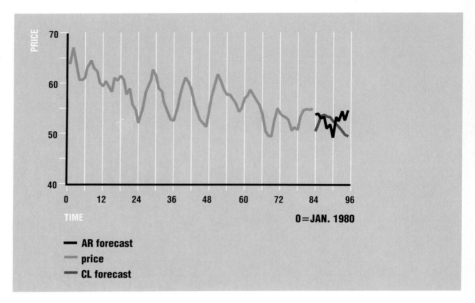

The regression with 12 lags will be used to make a 12-month forecast. Note that the number of months in the forecast is not tied to the number of lags in the estimating equation. The 12-month autoregressive forecast is shown in Figure A15.4. For comparison, the classical model forecast is shown in the same figure. An inspection of the two forecasts shows that they are quite different. After the spring peak, the values in the classical model forecast continue the downward trend observed in the data. The values in the autoregressive forecast drift upward when those forecast by the classical model are moving downward. A little reflection on the two models will help you see the basis for these differences. The classical model, with its embedded trend term, emphasizes the trend in the data. The autoregressive model repeats the pattern in the last 12 observations. If the autoregressive forecast were carried further, the estimated prices would stay in the range of the low to mid-50s. If the classical model forecast were carried further, the estimated prices would continue to fall, eventually reaching negative values.

Which forecast makes more sense? At this point you have to rely on your judgment and on research on the cattle market. Is it likely that consumers will cut their beef consumption further? Are ranchers unable to reduce their herds enough to compensate for the decrease in demand? If you think the answers to both questions are yes, then the trend toward lower prices is likely to continue and you should use the classical forecast. If you think the answer to either question is no, the autoregressive forecast may be better.

Lastly, consider a 12-month forecast based on an ARIMA model. There are seasonal ARIMA models that would be most appropriate for the cattle data. As seasonal ARIMA models are not discussed in Chapter 15, however, we will limit our analysis to a regular ARIMA model. The first step in ARIMA is to transform the data to make them stationary. Figure A15.5 shows a time plot of the first differences in cattle prices. Almost all of the downward trend is removed by first-differencing. Also, the differenced series appears to be homoscedastic.

First differences of the cattle prices plotted against time

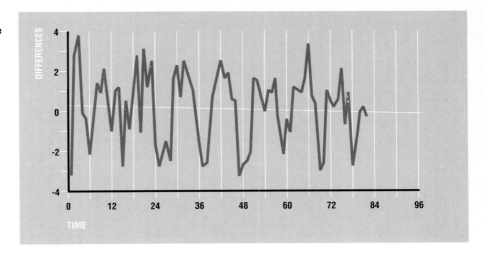

Autocorrelations of first differences of cattle prices

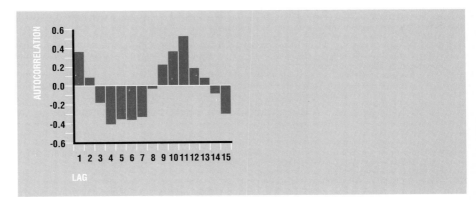

Partial autocorrelations of first differences of cattle prices

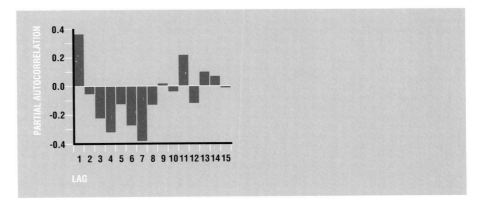

The second step is to examine the plots of the autocorrelation forecast (ACF) and the partial autocorrelation forecast (PACF) to determine the appropriate number of moving average and autoregressive terms. Figures A15.6 and A15.7 have the ACF

TABLE A15.3
Estimated
coefficients of the
ARIMA model

Term	Coefficient	Standard Error
AR1	1.198	0.101
AR2	−0.504	0.101
MA1	0.867	0.068

and PACF plots of the first-differenced series. The ACF exhibits the sinusoidal shape typical of a positive first and a negative second autocorrelation. The PACF has a long tail, which suggests that a moving average process is also present. The estimated model had two autoregressive terms and one moving average term. The estimated coefficients and standard errors are shown in Table A15.3.

Figure A15.8 compares the 12-month forecasts of the classical, autoregressive, and ARIMA models. So that the part of the plot containing the forecasts will be clearer, the original price series is shown beginning with the 48th month. In reflecting the downward trend in the original series, the ARIMA forecast falls between the classical and the autoregressive forecasts. Also, the ARIMA forecast predicts less month-to-month variation than either of the other forecasts. This difference is due in part to the use of a nonseasonal ARIMA model with patently seasonal data.

The relative accuracy of different forecast methods is currently the subject of strenuous debate (see S. Makridakis et al., *The Forecasting Accuracy of Major Time Series Methods* [New York: Wiley, 1984]). There are about a half-dozen contenders that are not described in Chapter 15. This debate is far beyond the scope of an introductory text; I mention it to point out that the comments on the three techniques discussed in this application by no means constitute an exhaustive treatment of the various forecast methods. One obvious point is that a simple ARIMA model will not work with seasonal data; the references at the end of Chapter 15 discuss seasonal ARIMA models. Another obvious point is that the emphasis on the trend in the data is greatest in

FIGURE A15.8
12-month forecasts of cattle
prices from the classical,
autoregressive, and ARIMA
models

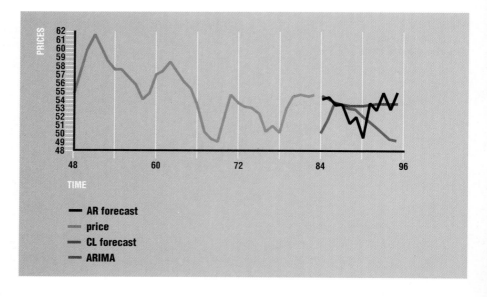

the classical model. As I mentioned before, whether or not this emphasis is appropriate must be a matter of judgment. Finally, simplicity aside, the autoregressive models offer no advantage over ARIMA models. AR models are subsets of ARIMA models. However, simplicity is often a persuasive advantage. You may want to use an autoregressive forecast for data without pronounced trends or for data that have trends that you think are unlikely to continue.

A15.3 COMMANDS

A15.3A DATA DESK

The classical model requires a data set that includes a time series, a time trend, and dummy variables for all but one of the seasons. Recall that "seasons" in a times series can be months, quarters of the year, or even days of the week. To run the regression, click on the icon of the variable containing the time series and, while holding down the shift key, click on the icons of the time trend and the dummy variables. Then click on the Regression command under the Compute menu.

With some difficulty, autoregressive models can be estimated with DATA DESK. The current student version of DATA DESK is limited to 15 variables, so you cannot estimate coefficients on more than 14 lags of the forecast variable. DATA DESK has no lag operator to create the lagged variables automatically. Lagged variables can be created manually in the following manner. Type command-n to create a new variable and give it the name L1 for lag 1. Open the variable containing the original time series and place the cursor at the left of the first observation. Scroll to the last observation in the series. While holding down the shift key, place the cursor to the right of the last observation and click on the mouse button. This operation will select (blacken) all of the observations in the original time series. Next type command-c to copy all of the original observations to the clipboard. Next click on the open variable, L1, so that the cursor appears inside L1. Type command-v to paste all of the original observations in L1. Add a blank line at the beginning of L1 by typing a return before the first observation and then delete the last observation in L1. To create the second lag, L2, you create a new variable from the original time series and then put two blank rows at the beginning and delete the last two observations. This procedure is extended to create the third and higher lags.

To run a least-squares autoregression, click on the icon of the time-series variable and, while holding down the shift key, click on the icons L1, L2, L3, and so on to the number of lags to be included. Then click on the Regression command under the Compute menu.

A15.3B MINITAB

If your data set contained the time series in column 1, a trend variable in column 2, and monthly dummy variables in columns 3 through 13, the following command would estimate a classical forecast:

```
MTB > REGRESS Y IN C1 ON  12 PREDICTORS IN C2-C13
```

The LAG BY 1 command creates the first lag of a variable. The commands to create two lags of a forecast variable stored in column 1 and then to run an autoregression on those two lags are:

```
MTB > LAG BY 1 OF C1, PUT INTO C2
MTB > LAG BY 2 OF C1, PUT INTO C3
MTB > REGRESS Y IN C1 ON 2 PREDICTORS IN C2, C3
```

These commands can be extended to higher lags.

MINITAB has some fairly easy commands for fitting ARIMA models. TSPLOT will plot a time series. A DIFFERENCES 1 command creates the first differences of a time series. The ACF and PACF commands plot the autocorrelation and partial autocorrelations. The ARIMA command estimates the ARIMA model. When you start with a time series in column 1, the following commands replicate the application on cattle prices. The three numbers after the ARIMA command refer respectively to the number of autocorrelations, the number of times the data are differenced, and the number of moving average terms. Note that the ARIMA command is run on the original data in column 1. If one difference is specified in the ARIMA command and the first-differenced data in column 2 are used in the model, the resulting estimate will be based on twice-differenced data. The FORECAST command is self-explanatory. The next line creates a new column of data, C4, which has the original observations followed by the forecasts. The last line creates a plot of the data in C4.

```
MTB > DIFFERENCE 1 FOR DATA IN C1, PUT INTO C2
MTB > ACF FOR SERIES IN C2
MTB > PACF FOR SERIES IN C2
MTB > ARIMA 2 1 1, DATA IN C1
MTB > FORECAST UP TO 12 LEADS AHEAD, STORE FORECASTS IN C3
MTB > STACK C3 ON TOP OF C1, PUT IN C4
MTB > TSPLOT PERIOD = 12 DATA IN C4
```

A15.3C SAS

If your data set has the time series in the first column, the trend variable in the second, and the monthly dummies in columns 3 through 13, the following SAS commands will estimate a classical model:

```
DATA INDAT;
INFILE INDAT;
INPUT SERIES TREND JAN FEB MAR APR MAY JUNE JULY
AUG SEPT OCT NOV;
PROC REG;
      MODEL SERIES=TREND JAN FEB MAR APR MAY JUNE
      JULY AUG SEPT OCT NOV;
```

The SAS autoregression routine in PROC FORECAST differs sharply from the autoregression procedure described in Chapter 15. The SAS procedure automatically detrends the time series and then searches for a set of statistically significant lags. After estimating the coefficients for these lags, the routine puts back the trend term and then makes a forecast. The following commands print a 12-month forecast of cattle data. Twelve months is the default forecast. A longer or a shorter time period can be specified. The data have to be recorded in two fields, the first containing the prices and the second the date, recorded as three letters for the month and two numbers for

FIGURE A15.9
Output from SAS
FORECAST procedure
showing a 12-month forecast

OBS	DATE	_TYPE_	_LEAD_	PRICE
1	DEC86	FORECAST	1	53.9009
2	JAN87	FORECAST	2	52.4831
3	FEB87	FORECAST	3	51.2975
4	MAR87	FORECAST	4	50.4070
5	APR87	FORECAST	5	50.3468
6	MAY87	FORECAST	6	50.9079
7	JUN87	FORECAST	7	51.6033
8	JUL87	FORECAST	8	52.1070
9	AUG87	FORECAST	9	52.4071
10	SEP87	FORECAST	10	52.3818
11	OCT87	FORECAST	11	51.9722
12	NOV87	FORECAST	12	51.3021

the year. For example, the first observation in the series is recorded as JAN80. Figure A15.9 shows the output created by these commands:

```
DATA CATTLE;
INFILE CATTLE;
INPUT PRICE DATE :MONYY.;
FORMAT DATE MONYY.;
PROC FORECAST OUT=A INTERVAL=MONTH METHOD=STEPAR;
ID DATE;
VAR PRICE;
PROC PRINT DATA=A;
```

PROC FORECAST can also create forecasts based on exponential smoothing. Instead of METHOD=STEPAR for stepwise autoregression, you specify METHOD=EXPO for exponential smoothing. Again, the implementation in SAS is more complex than the process described in Chapter 15. For example, you can specify double and triple exponential smoothing to forecast linear and quadratic trends. The SAS ETS manual provides the details.

An ARIMA forecast made by the ARIMA procedure in SAS requires three steps: IDENTIFY, ESTIMATE, and FORECAST. The IDENTIFY step prints the autocorrelations, the partial autocorrelations, and some other terms that are not covered in Chapter 15. You can first run an SAS job with only the IDENTIFY step and then decide on the appropriate model for estimation. You cannot run a job with the ESTIMATE step without having an IDENTIFY step precede it. In turn, you cannot run a job with the FORECAST step without having an ESTIMATE step precede it. Degrees of differencing are specified in the IDENTIFY step by numbers in parentheses. For example, PRICE(1) differences the price data once and PRICE(1,1) differences the price data twice. The following commands will duplicate the ARIMA forecast made in the application.

```
DATA CATTLE;
INFILE CATTLE;
INPUT PRICE DATE :MONYY.;
FORMAT DATE MONYY.;
PROC ARIMA;
```

```
IDENTIFY VAR=PRICE(1);
ESTIMATE P=2 Q=1;
FORECAST LEAD=12;
QUIT;
```

The P=2 and Q=1 in the ESTIMATE step specify, respectively, two autoregressive terms and one moving average term.

■ A15.3D SPSSx

If your data set has the time series in the first column, the trend variable in the second, and the monthly dummies in columns 3 through 13, the following SPSSx commands will estimate a classical model:

```
FILE HANDLE SERIES/ NAME='CATTLE.DAT'
DATA LIST FILE=SERIES/PRICE TREND JAN FEB MAR APR
MAY JUNE JULY AUG SEPT OCT NOV
REGRESSION
     /VARIABLES=PRICE TREND JAN FEB MAR
     APR MAY JUNE JULY AUG SEPT OCT NOV
     /DEPENDENT=PRICE
     /ENTER
```

An autoregressive model can be estimated with the lag function. The following example estimates the coefficients on the first three lags:

```
FILE HANDLE SERIES/ FILE='CATTLE.DAT'
DATA LIST FILE=SERIES/PRICE
COMPUTE LAG1=LAG(PRICE,1)
COMPUTE LAG2=LAG(PRICE,2)
COMPUTE LAG3=LAG(PRICE,3)
REGRESSION
     /VARIABLES=PRICE LAG1 LAG2 LAG3
     /DEPENDENT=PRICE
     /ENTER
```

The SPSSx procedure for ARIMA is BOX–JENKINS. A forecast based on this procedure requires three steps: IDENTIFY, ESTIMATE, and FORECAST. The IDENTIFY step prints the autocorrelations and the partial autocorrelations. You can begin with only the IDENTIFY subcommand and then proceed to the other steps. The commands to duplicate the application on cattle prices are as follows. The DIFFERENCE=1 subcommand specifies a first difference. The P=2, Q=1 subcommand specifies two autoregression terms and one moving average term. The NCONSTANT subcommand specifies that the model be estimated without a constant term. This is the customary procedure for differenced data. The LEAD=12 command specifies that the forecast be made for 12 periods.

```
FILE HANDLE SERIES/ FILE='CATTLE.DAT'
DATA LIST FILE=SERIES/PRICE
BOX–JENKINS VARIABLE=PRICE/DIFFERENCE=1/IDENTIFY/
     P=2/Q=1/NCONSTANT/ESTIMATE/LEAD=12/FORECAST/
```

EXERCISES FOR APPENDIX A15

A15.1 Data set 25 contains a monthly time series of the wholesale prices of frozen eviscerated turkeys sold in New York City from January 1973 through September 1986.

(a) Estimate a classical time series model with monthly dummy variables and forecast turkey prices for 12 periods. Omit January.

(b) Which months have the lowest prices? Are the prices during these months significantly lower than January prices? Set α at 5%.

(c) Is the trend in turkey prices significantly different than 0? Set α at 5%.

(Exercises A15.2 and A15.3 cannot be done with DATA DESK.)

A15.2 Column 4 of data set 26 contains a monthly time series on the value of the end–of–month raw materials inventories held by all durable goods manufacturers in the United States from January 1967 through January 1987. The data are seasonally adjusted and in constant dollars.

(a) Column 8 of the data set contains a price deflator for the raw materials inventories. Examine this price deflator series and determine the base period for the constant dollar valuation of the raw materials inventories.

(b) Use the classical model to make a 12-month forecast for raw materials inventories. Would quarterly or monthly seasonal dummies be appropriate in this model?

(c) Use a least squares autoregressive model to make a 12-month forecast for raw materials inventories. Check the first 20 lags for statistical significance.

(d) Use an ARIMA model to make a 12-month forecast for raw materials inventories. Decide how many differences are needed to remove any trend from the series and plot the ACF and PACF of the differenced series.

A15.3 Column 2 of data set 27 contains a monthly time series on end-of-month finished goods inventories for all nondurable manufactured goods in the United States from January 1967 through January 1987. The data are seasonally adjusted and given as constant dollars.

(a) Use the classical model to make a 12-month forecast for nondurable finished goods inventories. Would quarterly or monthly seasonal dummies be appropriate in this model?

(b) Use a least squares autoregressive model to make a 12-month forecast for nondurable finished goods inventories. Check the first 20 lags for statistical significance.

(c) Use an ARIMA model to make a 12-month forecast for nondurable finished goods inventories. Decide how many differences are needed to remove any trend from the series and plot the ACF and PACF of the differenced series.

A16
Index Numbers

A16.1 INTRODUCTION

Generating index numbers from price and quantity data requires only simple arithmetic on columns of data or variables. Appendix A16 introduces no new commands because you can create a Paasche or a Laspeyre or a Fisher index with the arithmetic commands built into the computer packages. These arithmetic commands were first discussed in Appendix A3. To simplify the illustration of the commands, we will use a small set of hypothetical data on the cost of attending a high school prom. Table A16.1 shows the prom cost data. The prices in each category are dollar average costs for the persons who purchase this item. The quantity weights are the proportions of couples who purchase the various items. There are eight categories of expenditures and 11 years.

The numbers in Table A16.1 were made up to produce a substantial difference between the Laspeyres and Paasche indexes. The sharp increase in the use of limousines is the main factor in this difference. Table A16.2 shows the Laspeyres and Paasche indexes for the data in Table A16.1. The interpretation of the Laspeyres index number of 151.98 for 1988 is that a 1978-style prom (with few limos or beauticians) would cost 51.98% more in 1988 than in 1978. The interpretation of the Paasche index number of 140.01 for 1988 is that a 1988-style prom (with most couples using limos and beauticians) would cost 40.01% more in 1988 than in 1978.

A16.2 AN APPLICATION

The Help-Wanted Index is a Laspeyres-quantity index that measures the level of classified job ads in relation to the base year of 1967. (See Noreen L. Preston, *The Help-Wanted Index: Technical Description and Behavioral Trends,* Conference Board Report no. 716 [New York, 1977].) Like the Index of Leading Economic Indicators, the Help-Wanted Index is used to forecast trends in the economy. The peak in the Help-Wanted Index generally leads peaks in the business cycle, as measured by GNP. The

TABLE A16.1
Prices and quantity weights for a high school prom, 1978–1988

Year	DRESS Price	DRESS Quantity	CORSAGE Price	CORSAGE Quantity	TUXEDO Price	TUXEDO Quantity	LIMOUSINE Price	LIMOUSINE Quantity
1978	$185	1.00	$ 8	1.00	$22	1.00	$90	0.02
1979	191	1.00	9	1.00	26	1.00	75	0.03
1980	202	1.00	9	1.00	28	1.00	75	0.01
1981	230	0.98	10	1.00	29	1.00	65	0.05
1982	254	0.97	12	1.00	32	1.00	65	0.16
1983	261	0.95	12	1.00	35	1.00	65	0.07
1984	286	0.97	12	1.00	34	1.00	55	0.08
1985	274	0.95	14	1.00	36	1.00	50	0.14
1986	264	0.92	15	0.98	39	1.00	50	0.24
1987	289	0.91	15	0.97	39	1.00	50	0.36
1988	275	0.93	15	0.96	40	1.00	45	0.58

Year	DINNER Price	DINNER Quantity	PROM TICKETS Price	PROM TICKETS Quantity	BEAUTICIAN Price	BEAUTICIAN Quantity	WOMEN'S SHOES Price	WOMEN'S SHOES Quantity
1978	$ 55	0.75	$25	1.00	$25	0.34	$45	1.00
1979	62	0.89	25	1.00	28	0.36	50	1.00
1980	75	0.97	25	1.00	34	0.45	56	0.99
1981	70	0.92	25	1.00	35	0.51	57	0.98
1982	82	1.00	25	1.00	36	0.40	52	0.97
1983	85	0.97	25	1.00	36	0.53	50	0.97
1984	90	0.96	25	1.00	38	0.55	51	0.92
1985	77	0.94	25	1.00	39	0.58	60	0.92
1986	90	0.98	25	1.00	40	0.62	62	0.89
1987	100	0.95	30	1.00	40	0.49	66	0.94
1988	100	0.97	30	1.00	40	0.58	62	0.91

TABLE A16.2
Laspeyres and Paasche indexes for the prom data in Table A16.1

Year	Laspeyres	Paasche	Year	Laspeyres	Paasche
1978	100.00	100.00	1984	145.45	145.03
1979	106.53	106.61	1985	142.92	140.33
1980	115.68	116.62	1986	144.73	140.91
1981	123.82	123.69	1987	157.06	149.37
1982	133.72	132.20	1988	151.98	140.01
1983	136.77	136.63			

lead varies from three to seven months. What this means is that if you observe a consistent fall in the Help-Wanted Index, you can be fairly confident that GNP will start to fall in the next three to seven months. You cannot be very confident about how long GNP will continue to fall or how much it will fall. The trough or bottom of the Help-Wanted Index is roughly coincident with the trough of GNP. This coincident pattern makes the Help-Wanted Index less useful as a tool to forecast expansions.

The Help-Wanted Index is maintained by the Conference Board, an independent, nonprofit business research organization. The monthly index is based on a complete count of want ads per day in the newspaper with the largest circulation in each of 51 SMSAs. These ad counts are modified in three ways. One way is to assign different weights to the totals for Sundays and for other days. Sunday papers have many more ads than weekday papers. Months can have either 28, 29, 30, or 31 days and either four or five Sundays. The weighting avoids a spurious increase in the index because a month had an extra Sunday or two or three extra weekdays. A second modification is to adjust the data seasonally. Spring and fall months usually have the highest volume of ads. The third adjustment is to weight each SMSA by its share of national nonagricultural employment. New York has the highest weight among the 51 SMSAs (12.2% in 1974) and Knoxville the lowest (0.5%). The formula for the Help-Wanted Index, HWI_t, is

$$HWI_t = \frac{\displaystyle\sum_{i=1}^{51} w_{it}Q_{it}}{\displaystyle\sum_{i=1}^{51} w_{i67}Q_{i67}}$$

where Q_{it} is the number of ads counted in the paper of the ith city in the tth month and w_{it} is the share of national nonagricultural employment of the ith city in the tth month. The subscript 67 refers to the average level in 1967. As you can see from the formula, a Laspeyres quantity index is similar to a Laspeyres price index.

The Conference Board report on the technical aspects of creating the index does not address one puzzle. The hundreds of thousands of ads that have to be counted every month must make for a very dull job and the tabulators have every conceivable opportunity to find out about alternative jobs. How do they keep anyone at this work?

A16.3 COMMANDS

A16.3A DATA DESK

The student version of DATA DESK has a limit of 15 variables, not counting derived variables. With eight variables for the prices of the various categories of expenditures and eight quantity weights, the prom data won't fit. If you are using the student version, you will have to work with fewer variables. We will go through the commands required to create the index. The variable names for the prices in the respective categories are dress, corsage, tuxedo, limousine, dinner, prom tickets, beautician, and women's shoes. The variable names for the quantity weights are, respectively, one, two, three, four, five, six, seven, and eight. DATA DESK's naming routine can be confused by either similar names or short names. For example, the names Q1, Q2, Q3, and so on for the quantity weights would not work. This is why distinct names were chosen. Another example of confusion in the naming routine is the word "hairdresser." Since the variable name dress is part of "hairdresser," DATA DESK will not accept hairdresser in the formula for a derived variable. So beautician was used instead.

The first step in creating a Laspeyres or a Paasche index is to create a variable representing the weighted sum of spending in each period.

$$\sum_{i=1}^{n} Q_{ti}P_{ti}$$

The above expression is given the name cost and created with a derived variable (type command d). The algebraic expression for cost is

```
(dress*one + corsage*two + tuxedo*three + limo rental*four
+ dinner*five + prom tickets*six + beautician*seven +
womens shoes*eight)
```

To get the values of the cost variable, select the icon of cost and then type command =. The value of cost for the first year is $336.55. This number is the denominator in the formula for the Laspeyres index. Recall formula 16.4 for the Laspeyres index:

$$L_t = \frac{\sum_{i=1}^{n} Q_{0i}P_{ti}}{\sum_{i=1}^{n} Q_{0i}P_{0i}}$$

A Laspeyres index is created with a new derived variable (type command d). The algebraic expression for the Laspeyres index is:

```
100*(1.00*dress + 1.00*corsage + 1.00*tuxedo + 0.02*
limousine + 0.75*dinner + 1.00*prom tickets + 0.34*
beautician + 1.00*womens shoes)/336.55
```

Note that the quantity weights, such as 0.02 for limousine, are the quantities for the first period. The variables, such as dress, consist of the prices at each period.

A Paasche index can also be created with a derived variable. Recall formula 16.5 for a Paasche index:

$$\text{Paasche}_t = \frac{\sum_{i=1}^{n} Q_{ti}P_{ti}}{\sum_{i=1}^{n} Q_{ti}P_{0i}}$$

To create the Paasche index, type command d and input the following algebraic expression:

```
100*(cost/(one*185 + two*8 + three*22 + four*90 + five*55 +
six*25 + seven*25 + eight*45))
```

Note that the cost variable must be calculated before the Paasche index can be calculated. The current version of DATA DESK does not allow one derived variable to be an argument of another derived variable. You calculate the cost variable by selecting its icon and typing command =.

A Fisher index is created with another new derived variable. The algebraic expression is

```
sqrt(Laspeyres*Paasche)
```

where Laspeyres and Paasche are the variable names assigned to the Laspeyres and Paasche indexes. Again, these variables have to be calculated before they can be used in another derived variable.

A16.3B MINITAB

To create indexes with MINITAB, the prom data are organized as in Table A16.1. The first column has the year and the other columns, 2 through 17, have the prices and quantity weights. The first step in creating a Laspeyres or a Paasche index is to create a variable representing the weighted sum of spending in each period.

$$\sum_{i=1}^{n} Q_{ti}P_{ti}$$

After the data are brought into MINITAB with a READ command, the following command will put the sum of spending in each period in column 18:

```
MTB > LET C18 = C2*C3 + C4*C5 + C6*C7 + C8*C9 + C10*C11 +
C12*C13 + C14*C15 + C16*C17
```

If you type the command PRINT C18, you will find that the value of cost for the first year is \$336.55. This number is the denominator in the formula for the Laspeyres index. Recall formula 16.4 for the Laspeyres index:

$$L_t = \frac{\sum_{i=1}^{n} Q_{0i}P_{ti}}{\sum_{i=1}^{n} Q_{0i}P_{0i}}$$

A Laspeyres index is created with the following command:

```
MTB >  LET C19 =100*(1.00*C2 + 1.00*C4 + 1.00*C6 + 0.02*C8
+ 0.75*C10 + 1.00*C12 + 0.34*C14 + 1.00*C16)/336.55
```

Note that the quantity weights, such as 0.02 for limousine, are the quantities for the first period. The variables, such as C2, consist of the prices at each period.

Recall formula 16.5 for a Paasche index:

$$\text{Paasche}_t = \frac{\displaystyle\sum_{i=1}^{n} Q_{ti}P_{ti}}{\displaystyle\sum_{i=1}^{n} Q_{ti}P_{0i}}$$

A Paasche index is created by the following commands:

```
MTB > LET C20 =100*C18/(C3*185 + C5*8 + C7*22 + C9*90 +
C11*55 + C13*25 + C15*25 + C17*45)
```

Recall that column 18 contains the sum of spending in each period.

A Fisher index is created with another transformed variable. The algebraic expression is

```
MTB > LET C21 = SQRT(C19*C20)
```

where C19 and C20 are the columns assigned to the Laspeyres and Paasche indexes.

A16.3C SAS

To create indexes with SAS, the prom data are organized as in Table A16.1. The first column has the year and the other columns, 2 through 17, have the prices and quantity weights. The first step in creating a Laspeyres or a Paasche index is to create a variable representing the weighted sum of spending in each period.

$$\sum_{i=1}^{n} Q_{ti}P_{ti}$$

The following commands will bring the prom data in SAS, calculate the sum of spending in each period under a variable named COST, and print the values of all variables.

```
DATA PROM;
INFILE PROM;
INPUT YEAR P1 Q1 P2 Q2 P3 Q3 P4 Q4 P5 Q5 P6 Q6 P7 Q7 P8 Q8;
COST = P1*Q1+P2*Q2+P3*Q3+P4*Q4+P5*Q5+P6*Q6+P7*Q7+P8*Q8;
PROC PRINT;
```

The value of COST for the first year is $336.55. This number is the denominator in the formula for the Laspeyres index. Recall formula 16.4 for the Laspeyres index:

$$L_t = \frac{\displaystyle\sum_{i=1}^{n} Q_{0i}P_{ti}}{\displaystyle\sum_{i=1}^{n} Q_{0i}P_{0i}}$$

A Laspeyres index is created by adding the following line before the PROC PRINT command:

```
LASPEYRE = 100*(1.00*P1+1.00*P2+1.00*P3+0.02*P4+0.75*P5+
1.00*P6+0.34*P7+1.00*P8)/336.55;
```

Note that the quantity weights, such as 0.02 for limousine, are the quantities for the first period. The variables, such as P1, consist of the prices at each period.

Recall formula 16.5 for a Paasche index:

$$\text{Paasche}_t = \frac{\sum\limits_{i=1}^{n} Q_{ti}P_{ti}}{\sum\limits_{i=1}^{n} Q_{ti}P_{0i}}$$

A Paasche index is created by adding the following line before the PROC PRINT command:

```
PAASCHE = 100*COST/(Q1*185+Q2*8+Q3*22+Q4*90+Q5*55+Q6*25+
Q7*25+Q8*45);
```

A Fisher index is created with another line added before the PROC PRINT command:

```
FISHER = SQRT(LASPEYRE*PAASCHE);
```

A16.3D SPSSx

To create indexes with SPSSx, the prom data are organized as in Table A16.1. The first column has the year and the other columns, 2 through 17, have the prices and quantity weights. The first step in creating a Laspeyres or a Paasche index is to create a variable representing the sum of spending in each period.

$$\sum_{i=1}^{n} Q_{ti}P_{ti}$$

The following program will create a listing of the sums of spending in the 11 periods.

```
FILE HANDLE INDEX / NAME='PROM COST. DAT'
DATA LIST FILE=INDEX LIST/YEAR P1 Q1 P2 Q2 P3 Q3 P4 Q4 P5
Q5 P6 Q6 P7 Q7 P8 Q8
COMPUTE COST=P1*Q1+P2*Q2+P3*Q3+P4*Q4+P5*Q5+P6*Q6+P7*Q7+
P8*Q8
LIST VARIABLES=YEAR COST
FINISH
```

The weighted average cost for first year is $336.55. This number is the denominator in the formula for the Laspeyres index. Recall formula 16.4 for the Laspeyres index:

$$L_t = \frac{\displaystyle\sum_{i=1}^{n} Q_{0i}P_{ti}}{\displaystyle\sum_{i=1}^{n} Q_{0i}P_{0i}}$$

A Laspeyres index is created by adding the following line after the COMPUTE statement in the SPSSx command file:

```
COMPUTE LASPEYRE=100*(P1+P2+P3+0.02*P4+0.75*P5+P6+0.34*P7+
P8) / 336.55
LIST VARIABLE=LASPEYRE
```

Note that the quantity weights, such as 0.02 for limousine, are the quantities for the first period. The variables, such as P1, consist of the prices at each period.

Recall formula 16.5 for a Paasche index:

$$\text{Paasche}_t = \frac{\displaystyle\sum_{i=1}^{n} Q_{ti}P_{ti}}{\displaystyle\sum_{i=1}^{n} Q_{ti}P_{0i}}$$

A Paasche index is created by adding the following line after the second COMPUTE statement in the SPSSx command file:

```
COMPUTE PAASCHE=100*COST/(Q1*185+Q2*8+Q3*22+Q4*90+Q5*55+
Q6*25+Q7*25+Q8*45)
LIST VARIABLE=PAASCHE
```

Recall that the variable COST contains the sum of spending in each period.

A Fisher index is created by adding another compute command to the command file:

```
COMPUTE FISHER= SQRT(LASPEYRE*PAASCHE)
```

EXERCISES FOR APPENDIX A16

A16.1 Data set 24 contains the number of life insurance policies for four types sold in the United States, in millions, and the total dollar benefits represented by all policies, in millions of dollars. The four types are ordinary, group, industrial, and credit. Ordinary life policies are sold to individuals. Group life insurance is usually sold to companies and offered as a fringe benefit to employees. Industrial life insurance policies cover the death benefits required by state workers' compensation systems. Credit life insurance policies are sold along with car and home mortgage contracts. The data set covers the years 1945 through 1984. The source of these data is the American Council of Life Insurance.

(a) Create four simple series of expenditure relatives using 1945 as the base year. In terms of the insurance industry's revenues, which types of life insurance have grown in relative importance and which have declined?

(b) Create a Laspeyres and a Paasche index for the average benefit per policy, again using 1945 as the base year. Note that the total dollar benefit of all policies equals the number of policies times the average benefit per policy. Describe what these indexes tell you about changes in the average benefit per policy.

A16.2 Data set 28 contains monthly time series on the wholesale prices and quantities of turkeys, hogs, and calves sold from January 1981 through September 1986.

    (a) Use these data to construct Paasche, Laspeyres, and Fisher price indexes for the three types of meat.

    (b) Which of the three types of meat has increased the most as a share of consumer's expenditures on meat?

    (c) Has consumer demand shifted toward the type of meat with relatively lower prices? Explain.

A16.3 Data set 29 contains annual U.S. electric utility sales in kilowatt-hours and prices in dollars for the following categories: residential, commercial, and industrial. The data are for the years 1966 through 1986.

    (a) Use these data to construct Paasche, Laspeyres, and Fisher price indexes for the three categories of customers.

    (b) Which category of customers has grown the most in terms of utility revenues since 1966?

    (c) Would utilities have more revenue if they could sell the 1986 quantities at 1966 prices or the 1966 quantities at 1988 prices?

A17
Nonparametric Tests

A17.1 INTRODUCTION

Eight nonparametric tests or confidence intervals are discussed in Chapter 17. Because there are so many tests, this appendix will treat each one as briefly as possible. Only the commands will be shown. The four computer packages vary widely in their support of nonparametric tests. About the only thing they have in common is that none of them does a Lilliefors test—perhaps because it is a fairly new test. Most of the packages focus exclusively on hypothesis testing. The exception is MINITAB, which has several nonparametric confidence intervals. Table A17.1 shows which tests and confidence intervals the various packages will do.

TABLE A17.1
Nonparametric tests available in four computer packages

	DATA DESK	MINITAB	SAS	SPSSx
Sign test for median	No	Yes	No	Yes
Confidence intervals for population median	No	Yes	No	No
Wilcoxon signed rank test for population median	*	Yes	*	Yes
Whitney-Mann-Wilcoxon test for two identical distributions	*	Yes	Yes	Yes
Wilcoxon test with paired samples	*	Yes	*	Yes
Kruskall-Wallis test for equality of more than two population means	*	Yes	Yes	Yes
Spearman rank correlation	*	Yes	Yes	Yes
Kolmogorov-Smirnov test for goodness of fit	No	No	Yes	Yes
Lilliefors test for normal distributions	No	No	No	No

*The test can be constructed with some difficulty.

The computer packages are open-ended and can often be set up to do tests for which they have no direct commands. In these cases the user is faced with the choice of trading his or her setup or programming effort against the time needed to perform the same test by hand. Where you stand on this trade-off depends on your computational and programming skills. The current version of DATA DESK is a complete bust as far as nonparametric tests are concerned; everything has to be set up by the user. More nonparametric tests have been promised in the next version. I have included DATA DESK instructions for most of the nonparametric tests, but they require considerable legerdemain. SAS is almost as bad. An asterisk (*) in Table A17.1 indicates that a test has to be constructed by the user.

A17.2 AN APPLICATION

The first and perhaps the most famous application of a nonparametric test is a sign test applied by John Arbuthnot to 82 years of London birth data ("An Argument for Divine Providence, taken from the constant Regularity observ'd in the Births of both sexes," *Philosophical Transactions of the Royal Society of London* 27 [1710]:186–190). Arbuthnot reasoned that if sex was determined by chance, approximately half of all babies would be males. The determination of the sex of a baby could be modeled as a throw of a fair coin. However, approximately equal numbers of male and female babies would result in fewer males than females at the age of marriage. Arbuthnot pointed out that the external Accidents to which males are subject (who must seek their Food with danger) do make a great havock of them, and that this loss exceeds far that of the other Sex, occasioned by Diseases incident to it, as Experience convinces us. To repair that Loss, provident Nature, by the Disposal of its wise Creator, brings forth more Males than Females; and that in almost constant proportion.

Arbuthnot argued that the surplus of male babies was just enough to make up for the "havock" and result in equal numbers of the two sexes at the age of marriage. Arbuthnot's anthropological arguments are debatable, but his mathematical reasoning is what concerns us here. Arbuthnot did not consider that chance variation could also be represented by an unbalanced coin. To Arbuthnot, "chance" in an experiment with two possible outcomes always results in a probability of 0.5. Under his null hypothesis of chance causation, the probability that male births in a year would exceed female births equals approximately one-half. It is approximate because there is a slight probability that male and female births in a year would be equal. An observed sex ratio far from the theoretical one-half would be evidence against the chance hypothesis. The London birth records Arbuthnot examined revealed that male births exceeded female births in 82 successive years. Call a year with more male than female births a plus. The probability of 82 successive plusses is $(0.5)^{82}$, or 2.07E-25, a number so small that it should convince the most obdurate skeptic of the workings of divine providence.
   Arbuthnot's sex-ratio application shows the long history of statistical tests for detecting divine intervention in human affairs. Other examples are tests for whether fire insurance is cheaper for churches than for commercial buildings and whether prayer promotes the recovery of the chronically ill. Exercise A11.4, on the duration of papal reigns, is a minor contribution to this tradition. These tests appear to have little effect on believers or atheists, but they do a nice job of illustrating the concepts behind statistical tests.

A17.3 COMMANDS

 A17.3A DATA DESK

WILCOXON SIGNED RANK TEST FOR THE POPULATION MEDIAN Begin with the data for one variable in an icon labeled AAAA. Suppose you are testing the null hypothesis that the mean/median of the population from which AAAA is drawn is 10. Eliminate all observations that equal 10. The next step is to create a new derived variable by typing command–d. Name this derived variable Difference. The algebraic statement for the Difference variable is AAAA − 10. If your null hypothesis value were a number other than 10, it would be used in place of the 10. To do this, click on the icon of the Difference variable, hold down the command key, and hit the equals key. These steps create a variable that contains the values of the Difference variable. You have to rank the differences according to their absolute values and then sum the ranks for the positive and negative differences. To keep track of which ranks come from positive and negative differences, create a second derived variable with the name Source. The algebraic statement for the Source variable is sign (Difference). Source will contain a positive 1 for every positive difference and a negative 1 for every negative difference. Select the icon of Source and type command = to compute the values of the derived variable. Next create a new derived variable named Absolutes containing the algebraic statement Abs (Difference). Next select the Absolutes variable and the Rank command under the Manipulate menu. The icon Rank: Absolutes contains the ranks of the absolute values of the differences. Now click on the icon of Rank: Absolutes and, while holding down the shift key, click on the icon of Source. Click on the command Split into variables by group under the Manipulate menu. Two new icons will appear, −1 and 1. The −1 will contain the ranks of all the negative differences and the 1 will contain the ranks of all the positive differences.

The test statistic is the smaller of these two sums of ranks. You can use DATA DESK to get these sums with the new derived variables command and the Sum (·) function. Before creating the new derived variables, change the names −1 and 1 because DATA DESK will treat them as the numbers −1 and 1. To change the names, select the icons −1 and 1 and type command–R. The new names can be negatives and positives. The new derived variables would have the algebraic statements Sum (negatives) and Sum (positives). Compare the two sums to find the smallest. The last step is to see if the smaller sum of the ranks falls within the acceptance region in Table E16.

WHITNEY-MANN-WILCOXON TEST FOR TWO IDENTICAL DISTRIBUTIONS Begin with data for two variables in two columns or icons. Say the first variable is AAAA and the second BBBB. Let AAAA have fewer observations than BBBB. Select both icons AAAA and BBBB and use the Append and make group variables command to create a single variable with all of the data. The default name for the combined data is Appended Data. The default name for the indicator variable that lists the variables from which all observations came is Group Names. Next, select the icon for Appended data and click on the Rank command under the Manipulate menu. This will create a new variable with the name Rank: Appended Data which contains the rankings for all of the observations. Next you have to split the rankings into two groups, one for each of the original variables. Select the icon for Rank: Appended Data and, while holding down the shift key, click on the icon for Group Names. Click on the Split into variables by group command under the Manipulate menu. Two new icons will appear, with the names AAAA and BBBB. These icons will hold the ranks rather than the original observations.

What you need for the Whitney-Mann-Wilcoxon test is the sum of the ranks of the original variable that has fewer observations. You can use DATA DESK to get this sum, but you first have to change the name of the rank variable so that it will not be confused with the original data. Use the Rename command under the Special menu to change the name of the icon for the ranks of the variable with fewer observations. Use the name RAAAA. Then click on the New derived variable command, give the derived variable the name Test statistic, and type in the expression sum (RAAAA). Then click in the close box, click on the icon Test statistic to select it, and then click on the command Compute derived variable under the Manipulate menu. A new icon will appear with the name Test statistic. It will contain the sum of the ranks of AAAA.

At this point you can compare the test statistic with the critical values in Table D10. If the two variables contain too many observations to use Table D10, you can use the normal approximation for the Whitney-Mann-Wilcoxon test. The calculations for the mean and the standard deviations with formulas 17.3 and 17.4 can be done on a hand calculator or with DATA DESK's New derived variable command.

WILCOXON TEST WITH PAIRED SAMPLES Start with two variables with the same number of observations, say AAAA and BBBB. Select both variables by clicking on one and then, while holding down the shift key, clicking on the other. The next step is to create a new derived variable by typing command–d. Name this derived variable Difference. The algebraic statement for the Difference variable is AAAA − BBBB. Drop any observations with a difference of zero. The rest of the test is the same as for the Wilcoxon signed rank test described above.

KRUSKALL-WALLIS TEST FOR EQUALITY OF MORE THAN TWO POPULATION MEANS For a test of the equality of three population means/medians, start with the sample data in three variables: AAAA, BBBB, CCCC. For the sake of illustration we will assume that the three samples have 50, 20, and 10 observations, respectively. Next select the icons for AAAA, BBBB, CCCC and click on the Append and make group variables command under the Manipulate menu. Use the default names of Appended data and Group names. Next select the icon of Appended data and click on the Rank command under the Manipulate menu. This will create a variable, Rank: Appended data, which will contain the ranks of all the observations. You need the sum of the ranks for each sample to substitute into formula 17.6. To split up the Rank: Appended data variable into ranks for the three samples, click on the icon of Rank: Appended data and, while holding down the shift key, click on the icon of the Group names variable. Next click on the command split into variables by group. This will create three new variables named AAAA, BBBB, CCCC, which now contain the ranks for the three samples. Rename these icons so that they will not be confused with the original AAAA, BBBB, CCCC. Type command–R and give them the names AAAArank, BBBBrank, CCCCrank. The next step is to get the sum of ranks for the three variables. Use three derived variables (command–d) and the algebraic expressions sum (AAAArank), sum (BBBBrank), sum (CCCCrank). Give the derived variables the names TAAAA, TBBBB, and TCCCC. Select the three icons TAAAA, TBBBB, TCCCC and type command =.

   At this point you can substitute the three sums of ranks in formula 17.6 and solve for the test statistic H by hand or use the new derived variable command to have DATA DESK solve for H.

$$H = \frac{12}{n(n+1)} \left[\sum_{j=1}^{k} n_j \left(\frac{T_j}{n_j} - \frac{n+1}{2} \right)^2 \right]$$

17.6

To use DATA DESK, create yet another new derived variable and type in the following algebraic statement for formula 17.6:

```
(12/(80*81))*((50*((TAAAA/50)-(81/2))**2) +
(20*((TBBBB/20)-(81/2))**2) + (10*((TCCCC/10)-(81/2))**2))
```

Recall that the three samples were assumed to have 50, 20, and 10 observations. The last step is to compare the test statistic with the critical value from a chi-square distribution with $k - 1$ or, in this example, 2 degrees of freedom.

SPEARMAN RANK CORRELATION Start with two variables having the same number of observations, say AAAA and BBBB. Select both variables by clicking on one and then, while holding down the shift key, clicking on the other. Next click on the Rank command under the Manipulate menu. Select both the Rank: AAAA and Rank: BBBB variables and click on the Correlation command under the Compute menu.

A17.3B MINITAB

SIGN TEST FOR THE MEDIAN The commands for a two-tailed sign test against a null value for the population median of 10 are

```
MTB > STEST MEDIAN = 10, DATA IN C1;
SUBC > ALTERNATIVE = 0.
```

The subcommands `ALTERNATIVE = -1` and `ALTERNATIVE = 1` specify respectively a left- and a right-tailed test. The output reports the p value of the test.

CONFIDENCE INTERVALS FOR THE POPULATION MEDIAN The commands for a sign confidence interval at the 90% level of confidence are

```
MTB > SINT WITH 90 PERCENT CONFIDENCE, DATA IN C1
```

WILCOXON SIGNED RANK TEST FOR THE POPULATION MEDIAN The commands for a two-tailed one-sample rank test against a population median of 10 are

```
MTB > WTEST MEDIAN = 10, DATA IN C1;
SUBC > ALTERNATIVE = 0.
```

The subcommands `ALTERNATIVE =-1` and `ALTERNATIVE = 1` specify respectively a left- and a right-tailed test. The output includes the p value of the test. There is also a command for a Wilcoxon confidence interval, `WINT`. The Wilcoxon test statistic and this confidence interval are not discussed in Chapter 17.

WHITNEY-MANN-WILCOXON TEST FOR TWO IDENTICAL DISTRIBUTIONS The commands for a two-sided Whitney-Mann-Wilcoxon test of equal population means/medians are

```
MTB > MANN-WHITNEY ALTE = 0, FIRST IN C1, SECOND IN C2
```

The word `ALTERNATIVE` has been abbreviated to fit the commands on one line. MINITAB recognizes a command by its first four letters. For a right- or a left-tailed test, substitute a 1 or a -1 in place of the 0. The `MANN-WHITNEY` command also does a confidence interval with a default level of confidence of approximately 95%. You can specify another confidence level by including the commands `XX PERCENT CONFIDENCE` on the same line. For example, the commands for a two-sided Whitney-Mann-Wilcoxon test plus a 90% confidence interval are

```
MTB > MANN-WHITNEY ALTE = 0, 90 PERCENT CONFIDENCE, C1, C2
```

Note that on most computers a single command line can be longer than the number of characters that will fit on one line of the screen. The text will automatically wrap around to a new line if you type enough characters, but the single command line does not end until you type a carriage return. The `MANN-WHITNEY` test reports a different test statistic, W, than the test statistic discussed in Chapter 17. The two test statistics are mathematically equivalent and result in the same p values.

WILCOXON TEST WITH PAIRED SAMPLES If the paired samples are in columns 1 and 2, you begin by subtracting column 2 from column 1 and storing the differences in column 3. The commands for a two-tailed paired sample test are

```
MTB > LET C3 = C1 - C2
MTB > WTEST MEDIAN = 0, DATA IN C3;
SUBJ > ALTERNATIVE = 0
```

The subcommands ALTERNATIVE =−1 and ALTERNATIVE = 1 specify respectively a left- and a right-tailed test. The output reports the p value of the test. The commands for a Wilcoxon 90% confidence interval are

```
MTB > WINT WITH 90 PERCENT CONFIDENCE, DATA IN C3
```

KRUSKALL-WALLIS TEST FOR EQUALITY OF MORE THAN TWO POPULATION MEANS
MINITAB requires that the data for a Kruskall-Wallis test of the equality of more than two population means be in two columns. The first column contains all of the observations and the second column contains a numerical indicator variable that shows which sample the observation came from. The commands are

```
MTB > KRUSKAL−WALLIS, DATA IN C1, SUBSCRIPTS IN C2
```

The output does not report the p value of the test. It includes the Kruskall-Wallis H test statistic.

SPEARMAN RANK CORRELATION There is no single MINITAB command for the Spearman rank correlation. Start with an equal number of observations in column 1 and column 2. The RANK command can be used to rank the observations in each column and then the CORRELATION command will calculate the Pearson coefficient of correlation on the ranks. Recall that the Spearman rank correlation is simply the Pearson coefficient of correlation applied to the ranks. The commands are

```
MTB > RANK C1, PUT RANKS INTO C3
MTB > RANK C2, PUT RANKS INTO C4
MTB > CORRELATION BETWEEN C3 AND C4
```

The resulting output is the Spearman rank correlation.

A17.3C SAS

WILCOXON SIGNED RANK TEST FOR THE POPULATION MEDIAN The SAS RANK procedure can be used to construct a Wilcoxon signed rank test. The data format is one column for the variable. In the example below, the variable containing the observations has the name X and the test is against a null hypothesis of a population median of 10. The line X = X − 10; redefines X as the difference between X and the hypothesized median. The SIGN function returns a value of either −1, 0, or 1, depending on whether the difference is negative, zero, or positive. The signs or Y's are sorted in order to get the sum of the ranks of the negative and the positive differences at a later step. The ABS function returns the absolute value.

```
DATA A; INFILE A;
     INPUT X;
     X = X − 10;
     Y = SIGN(X);

PROC SORT DATA=A OUT=B;
     BY Y;

DATA B; SET B;
     Z = ABS(X);
```

```
PROC RANK DATA=B OUT=C;
     VAR X;

DATA C; SET C; IF (Y = -1);

PROC PRINT DATA=C;
     VAR Z;
     SUM Z;


PROC PRINT DATA=A;
PROC PRINT DATA=B;
PROC PRINT DATA=C;
```

The term Z is the sum of the ranks of the negative differences between the observations and the hypothesized median of 10. Formula 17.2 can be used to obtain the sum of the ranks of the positive differences. The last step is to take the smaller sum of ranks and compare with the critical values in Table E10.

WHITNEY-MANN-WILCOXON TEST FOR TWO IDENTICAL DISTRIBUTIONS The SAS NPAR1WAY procedure does a Whitney-Mann-Wilcoxon test. The data format is one column for all of the observations and a second column for an indicator variable showing which sample the observation came from. In the example below, X is the variable for the data and SAMPLE is the indicator variable.

```
DATA A; INFILE A;
     INPUT X SAMPLE;

PROC NPAR1WAY DATA=A WILCOXON;
     CLASS SAMPLE;
     VAR X;

PROC PRINT DATA=A;
```

WILCOXON TEST WITH PAIRED SAMPLES The SAS RANK procedure can be used to construct a Wilcoxon test with paired samples. The data format is one column for the first variable and a second column for the second variable. In the example below, the variables containing the observations have the names R and S. The line X = R − S; defines X as the difference between R and S. The SIGN function returns a value of either −1, 0, or 1, depending on whether the difference is negative, zero, or positive. The signs or Y's are sorted in order to get the sum of the ranks of the negative and the positive differences at a later step. The ABS function returns the absolute value.

```
DATA A; INFILE A;
     INPUT R S;
     X = R - S;
     Y = SIGN(X);

PROC SORT DATA=A OUT=B;
     BY Y;

DATA B; SET B;
     Z = ABS(X);
```

```
PROC RANKS DATA=B OUT=C;
     VAR X;

DATA C; SET C; IF (Y= −1);
PROC PRINT DATA=C;
     VAR Z;
     SUM Z;

PROC PRINT DATA=A;
PROC PRINT DATA=B;
PROC PRINT DATA=C;
```

The term Z is the sum of the ranks of the negative differences. Formula 17.2 can be used to obtain the sum of the ranks of the positive differences. The last step is to take the smaller sum of ranks and compare with the critical values in Table E10.

KRUSKALL-WALLIS TEST FOR EQUALITY OF MORE THAN TWO POPULATION MEANS The SAS NPAR1WAY procedure does a Kruskall-Wallis test. The data format is one column for all of the observations and a second column for an indicator variable showing which sample the observation came from. In the example below, X is the variable for the data and SAMPLE is the indicator variable. Note that the WILCOXON option after the NPAR1WAY is not a mistake. If there are more than two levels for the indicator variable, the WILCOXON option automatically produces a Kruskall-Wallis test.

```
DATA A; INFILE A;
     INPUT X SAMPLE;

PROC NPAR1WAY DATA=A WILCOXON;
     CLASS SAMPLE;
     VAR X;

PROC PRINT DATA=A;
```

SPEARMAN RANK CORRELATION The SAS CORR procedure calculates the Spearman coefficient of rank correlation. For data stored in variables X and Y the commands are

```
DATA A; INFILE A;
     INPUT X Y;

PROC CORR SPEARMAN;

PROC PRINT DATA=A;
```

KOLMOGOROV-SMIRNOV TEST FOR GOODNESS OF FIT The NORMAL option in SAS's UNIVARIATE procedure will print the Kolmogorov-Smirnov D statistic and its p value against the null hypothesis of a normal distribution for variables with 50 or more observations: PROC UNIVARIATE is discussed in Appendix A3. Here are the required commands:

```
DATA A; INFILE A;
     INPUT X;
```

```
PROC UNIVARIATE NORMAL;
      VAR X;

PROC PRINT DATA=A;
```

SAS does not do a Kolmogorov-Smirnov test for distributions other than the normal.

 A17.3D SPSSx

SIGN TEST FOR THE MEDIAN The SIGN subcommand in the NPAR TESTS procedure does a sign test. In the following example the data are stored in a variable called X. For the null hypothesis that the population median equals 10, the commands are

```
COMPUTE Y = 10
NPAR TESTS SIGN = X WITH Y
```

The output reports the p value of the two-tailed sign test.

WILCOXON SIGNED RANK TEST FOR THE POPULATION MEDIAN The WILCOXON subcommand of the NPAR TESTS procedure performs a Wilcoxon test with paired samples. SPSSx can be tricked into doing a test on a single column of data against a hypothesized value of the population median. The data are stored in one column with the variable name X. In the following example, the hypothesized population median is 10. The commands are

```
COMPUTE Y = 10
NPAR TESTS WILCOXON = X WITH Y
```

The output includes the mean ranks of the smaller and larger groups and the p value for a two-tailed test.

WHITNEY-MANN-WILCOXON TEST FOR TWO IDENTICAL DISTRIBUTIONS The M–W subcommand in the NPAR TESTS procedure does a Whitney-Mann-Wilcoxon test. In the following example, the data are stored in variable X and an indicator showing which sample the observation came from is stored in Y. Y takes the values 0 and 1. The commands are

```
NPAR TESTS M–W = X BY Y (0,1)
```

The output includes the mean rank of the smaller and larger groups and the p value for a two-tailed test.

WILCOXON TEST WITH PAIRED SAMPLES The WILCOXON subcommand of the NPAR TESTS procedure performs a Wilcoxon test with paired samples. The data are stored in two columns with the variable names X and Y. The commands are

```
NPAR TESTS WILCOXON = X WITH Y
```

The output reports the p value for a two-tailed test.

KRUSKALL-WALLIS TEST FOR EQUALITY OF MORE THAN TWO POPULATION MEANS The subcommand K–W of the NPAR TESTS procedure performs a Kruskall-Wallis test. The data are stored in two columns, one column for the observations and the second for a numeric indicator variable showing which sample each observation came from.

In the following example, the observations are stored in variable X and the indicators in variable Y.

```
NPAR TESTS K-W = X BY Y (1,3)
```

The numbers in parentheses refer to the lowest and highest values of the indicator variable which are to be included in the analysis. For example, if the indicator variable took the values 1, 2, 3, and 4, the specification (1,3) would include the observations from samples 1, 2, and 3. The output includes the Kruskall-Wallis H test statistic (labeled CHI-SQUARE) and the p value for the test.

SPEARMAN RANK CORRELATION The NONPAR CORR procedure computes Spearman's coefficient of rank correlation. In the following example the data are stored in two columns with the variable names X and Y. The commands are

```
NONPAR CORR X Y
```

The output includes the Spearman coefficient of correlation, the number of pairs of cases, and the p value for the test of the null hypothesis of zero rank correlation.

KOLMOGOROV-SMIRNOV TEST FOR GOODNESS OF FIT The K-S subcommand in the NPAR TESTS procedure does Kolmogorov-Smirnov goodness-of-fit tests against the uniform, normal, and Poisson distributions. In each case the parameters can be estimated from the sample data or specified. The example below shows tests against all three distributions with specified parameters. The data are stored in the variables X, Y, and Z.

```
NPAR TESTS K-S (UNIFORM, 8, 12) = X
NPAR TESTS K-S (NORMAL, 8, 12) = Y
NPAR TESTS K-S (POISSON, 8) = Z
```

For the test against the uniform distribution, the 8 and 12 are the specified minimum and maximum values of the distribution. If no values are specified, the test is against the minimum and maximum values in the data. For the test against the normal distribution, the 8 and 12 are the specified mean and standard deviation of the distribution. If no mean and standard deviation are specified, the test is against the sample mean and sample standard deviation. For the test against the Poisson, the 8 is the mean of the distribution. If no mean is specified, the test is against the sample mean and maximum values in the data. The output includes the test statistic D (labeled ABSOLUTE in the output) and the p value for a two-tailed test.

EXERCISES FOR APPENDIX A17

A17.1 Data set 22 contains the number of professors and the level of GNP in the 30 countries that have the largest numbers of professors. The source is *The Book of World Rankings,* by George Thomas Kurian (1979), published by Facts on File.

(a) Construct a scatter plot with the number of professors on the x axis. Explain how the scatter plot indicates that for these data a Spearman rank correlation coefficient would be a better measure of association than Pearson's coefficient of correlation.

(b) What is the Spearman rank correlation coefficient for these data?

(c) Does a high value for the Spearman rank correlation coefficient prove that an increase in the number of professors in a country increases GNP? Explain.

A17.2 Data set 21 contains the prices of 13 textbooks sold in the college bookstores at IUPUI (col. 1) and IU–Bloomington (col. 2) in the fall of 1987. The books were selected from a list of business classes that require the same textbooks on both campuses. Since each row has the prices for the same book, the data are paired. Test the null hypothesis at a 5% significance level that the mean difference in book prices is zero. Use the sign test.

A17.3 Redo exercise A10.1 on the differences in education of people who prefer Coke and Pepsi with a Whitney-Mann-Wilcoxon test. Which test is appropriate for these data—the Whitney-Mann-Wilcoxon or the parametric test? Does the choice of tests make a difference in the results?

A17.4 Refer again to data set 21 and use a Wilcoxon test at a 5% significance level to test the null hypothesis that the mean difference in book prices is zero.

(Problem A17.5 cannot be done with DATA DESK or MINITAB.)

A17.5 Data set 23 contains Arbuthnot's data on male and female births (cols. 2 and 3, respectively) in London for the years 1629 to 1710. The years are in column 1. Construct a variable for the ratio of males to females in this data set. Use the Kolmogorov-Smirnov test at a 5% significance level for the null hypothesis that the ratios of male to female births are normally distributed.

A17.6 Data set 21 has a third column containing the prices of books sold by a private bookstore in Bloomington, Indiana. Use the Kruskall-Wallis test at a 5% significance level to test the null hypothesis that all three bookstores have the same mean price.

A17.7 Use the appropriate nonparametric test at a 5% significance level for the null hypothesis that the mean difference in male and female births in the Arbuthnot data (data set 23) is zero.

Appendix B

Confidence Intervals and Hypothesis Tests for the Population Variance

B.1
ONE-SAMPLE TESTS WITH A NORMAL POPULATION

B.2
ONE-SAMPLE TESTS WITH A NONNORMAL POPULATION: THE JACKKNIFE TECHNIQUE

B.3
COMPARING TWO POPULATION VARIANCES WITH NORMAL POPULATIONS

B.1 ONE-SAMPLE TESTS WITH A NORMAL POPULATION

In many business problems the variability of a distribution is the key issue. Recall the example from Chapter 3 (p. 51) of choosing the bottling machine with the least variability. The mean amount per bottle can be set by a machine's operator, but the variability from bottle to bottle is inherent in a machine's design. Another example is choosing, from among investments with equal expected rates of return, the one that has least variability. A third example in which variability is a key issue is choosing among weapons systems such as missiles. The mean or target point is set by the operator. It is easy to compensate for a weapon that consistently undershoots or overshoots the target. A weapon with high variability may literally and figuratively shoot you in the foot. Please do not think that statistical inference on variances is in an appendix because the subject is unimportant. The placement is for pedagogical reasons. These tests fit into the material on one- and two-population inference in Chapters 8 through 10, but they require the chi-square and F distributions that are discussed in Chapters 11 and 12. In this section of the appendix we discuss confidence intervals and a hypothesis test for the population variance, σ^2, given a normally distributed population. The next section discusses tests when the population is not normal.

Before discussing statistical inference on variances or standard deviations, one point can be made. The sample standard deviation is a biased estimate of the population standard deviation. For normally distributed populations the amount of bias is a known function of the sample size. If you simply need a point estimate of the population standard deviation and you know that the population is normal, you can use the factors in Appendix D, Table D.13, to scale up the sample standard deviation.

If a population is normally distributed, you can construct either a confidence interval or a hypothesis test by relying on the theorem that given a random sample of size n,

$(n - 1)S^2/\sigma^2$ is approximately chi-square distributed with $n - 1$ degrees of freedom. The S^2 stands for the random variable of all possible values of the sample variance taken from a normally distributed population. Here is a rough explanation of why the theorem holds. Recall from Chapter 10 that a chi-square distribution with $n - 1$ degrees of freedom is the sum of $n - 1$ independent squared standard normal distributions. For a given population the σ^2 is a constant. For normal populations the random variables, x_i and $\bar{x}$, are normally distributed. Thus, $(x_i - \bar{x})$ is also normally distributed because it is the difference between two normal random variables. The term $(x_i$ and $\bar{x})^2$ is a squared normal random variable. Since n is a constant the sample variance based on the term

$$\frac{(x_i - \bar{x})^2}{n - 1}$$

is also a squared normal random variable. Dividing by a constant does not change the normality of the distribution. Since the expected value of S^2 is σ^2, the ratio of the sample variance to the population variance has an expected value of 1. The ratio of the sample variance to the population variance has a lower bound of 0 and no upper bound. The ratio follows a chi-square distribution with $n - 1$ degrees of freedom. The reason for $n - 1$ degrees of freedom is that all but one of the observations in the sample are independent random variables. Each identical independent random variable, x_i, has a variance S^2.

A confidence interval is constructed with the following formula.

$$L, \ U = \frac{(n - 1)s^2}{\chi^2_{\alpha/2}(\text{right tail})}, \ \frac{(n - 1)s^2}{\chi^2_{\alpha/2}(\text{left tail})}$$

(B.1)

The degrees of freedom for the chi-square distribution are $n - 1$.

The s^2 term is variance for a particular sample. The terms in the denominators refer to the points in a chi-square distribution with $n - 1$ degrees of freedom marking $\alpha/2$ area in the left tail and $\alpha/2$ area in the right tail. Unlike the normal distribution and the Student's t, the chi-square distribution is not symmetrically distributed around a mean of 0. Thus, the chi-square term for the left tail does not equal a minus one times the chi-square term for the right tail.

Here is an example. Twenty circuit breakers which are designed to trip on average as the voltage reaches 144 volts are tested for their actual tripping voltage. Although the sample mean is close to 144, the sample variance is 60. What are the lower and upper limits of a 90% confidence interval for the population variance? Substituting into formula (B.1),

$$L, \ U = \frac{(n - 1)s^2}{\chi^2_{\alpha/2}(\text{right tail})}, \ \frac{(n - 1)s^2}{\chi^2_{\alpha/2}(\text{left tail})}$$

(B.1)

$$L, \ U = \frac{(20 - 1)60}{30.14}, \ \frac{(20 - 1)60}{10.12}$$

$$L = 37.82, \qquad U = 112.65$$

FIGURE B.1
Upper and Lower 5% Areas on a Chi-Square Distribution with 19 Degrees of Freedom.

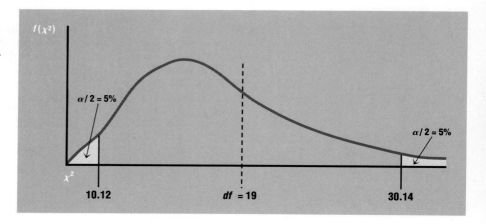

Note the point that marks the upper 5% of a chi-square distribution with 19 degrees of freedom is 30.14 and the point that marks the lower 5% is 10.12. Figure B.1 illustrates these two areas.

For some problems the confidence limits have to be found for the population standard deviation rather than the population variance. The exact formula for confidence intervals on the population standard deviation with small samples is complicated. An acceptable approximation is to take the square roots of the limits for the population variance. In this example the limits for a 90% confidence interval on the population standard deviation are approximately 6.15 to 10.61. If you had some expensive electrical machinery that would be destroyed by voltages above 160, the tested circuit breakers would not be adequate. Whereas they have a mean tripping point of 144 volts, the standard deviation is likely to be in the range of 6.15 to 10.61. Any value in this range would put the value 160 volts within 3 standard deviations of 144. Hence, there is better than a one-in-a-thousand chance that the circuit breaker will not trip when the voltage reaches 160. The one-in-a-thousand is approximately the area beyond 3 standard deviations to the right of mean on a normal curve.

Now we turn to hypothesis tests for the population variance with a normally distributed population. Given a hypothesized value for the population variance, the chi-square test statistic is given by formula (B.2)

$$\chi^2 = \frac{(n-1)s^2}{\sigma_0^2}$$

(B.2)

where σ_0^2 is the hypothesized population variance.

If the null hypothesis about the value of the population variance is true, the foregoing test statistic will be chi-square distributed with $n-1$ degrees of freedom. For a two-tailed test the critical values at an α significance level are the points in a chi-square distribution with $n-1$ degrees of freedom which marks $\alpha/2\%$ in the right tail and $\alpha/2\%$ in the left tail. For a right-tailed test the critical values at an α significance level are the points in a chi-square distribution with $n-1$ degrees of freedom which marks $\alpha\%$ in the right tail. Conversely, for a left-tailed test the critical values are the points in

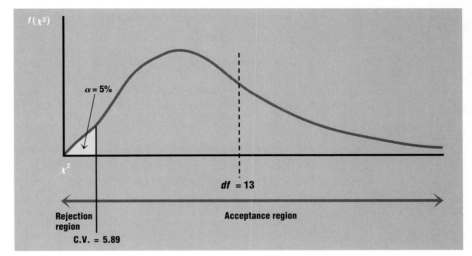

FIGURE B.2
Acceptance and Rejection Regions for a Left-Tailed Test on the Population Variance.

a chi-square distribution with $n - 1$ degrees of freedom which marks $\alpha\%$ in the left tail.

Here is an example of a left-tailed hypothesis test. A bank is considering marketing a new type of adjustable rate home mortgage in which the adjustments will be tied to changes in the six-month Treasury bill rate. The bank currently markets an adjustable home mortgage tied to changes in the prime rate of interest. The bank's finance officers think that the average spread between the two interest rate series is constant over a long period. Thus, tying changes in the adjustable rate mortgage to either of the series will not result in systematically higher average mortgage rates. However, the finance officers think that variability of the Treasury bill series will be lower than the variability of the prime rate series. Suppose the variance of the quarterly prime interest rate series expressed in integer percentage points was known to be 2. The finance officers' null hypothesis is that the Treasury bill series has a variance greater than or equal to 2. In notation, the null and alternative hypotheses are:

H_0: $\sigma^2 \geq 2$

H_1: $\sigma^2 < 2$

A sample of 14 quarters of observations on the six-month Treasury bill rates has a variance of 1.4. Is this lower variance enough evidence to reject the null hypothesis that Treasury bill variance is greater than or equal to 2? Substituting into formula (B.2),

$$\chi^2 = \frac{(n-1)s^2}{\sigma_0^2} \tag{B.2}$$

$$\chi^2 = \frac{(14-1)1.4}{2}$$

$$\chi^2 = 9.1$$

Assuming a 5% significance level, the critical value for the test is 5.89. There is not enough evidence to reject the null hypothesis that the population variance for the Treasury bill series is greater than or equal to 2. Figure B.2 shows the template for this test.

Right-tailed and two-tailed tests are easy modifications of the preceding example. For a right-tailed test the critical values would be to the right of the mean. For a two-tailed test the critical values would be on both sides of the mean. The mean of the chi-squared distribution is always the degrees of freedom.

B.2 ONE-SAMPLE TESTS WITH A NONNORMAL POPULATION: THE JACKKNIFE TECHNIQUE

A discussion of hypotheses tests and confidence intervals for the population standard deviation when the population is not normally distributed belongs in Chapter 17 on nonparametric tests. However, for convenience all of the tests on variances have been placed in this appendix. The jackknife technique is a clever way of applying statistical tests that rely on normality in the population distribution to nonnormal data. It is the statistical equivalent of making bricks without straw. This technique has many applications but we will discuss tests on the standard deviation only. Consider a sample of 10 independent observations from a nonnormal population. The sample has the values {2646, 1016, 2170, 1927, 1194, 1572, 1009, 197, 2636, 2646} and a sample standard deviation of 844.27. The problem is to construct a confidence interval at the 95% level for the population standard deviation.

The jackknife procedure begins by omitting one observation from the sample and calculating a standard deviation on the remaining nine observations. If you omitted each of the observations once, you could calculate 10 subsample standard deviations. These steps may appear to be silly. The best point estimate of the population standard deviation is obviously based on all of the data. The purpose of calculating 10 subsample standard deviations is not to get a point estimate of σ. It is to estimate a standard error term for the distribution of the sample standard deviation.

The 10 subsample standard deviations are not independent. On the contrary, most of their observations overlap. For most of the subsamples there would be little distance between the overall standard deviation and the subsample standard deviation. If the distances between the overall sample standard deviation and the 10 subsample standard deviations were scaled by multiplying them by some factor, these distances would increase. Here comes the hard part of the explanation. If the scaled distances are far enough apart, they can be treated *as if* they were independent random observations. Each scaled distance is treated as an independent estimate of the population standard deviation. These scaled distances are called *pseudovalues* and all 10 of them are called the *pseudosample*. The amazing result is that the pseudovalues will be approximately normally distributed no matter what the shape of the population distribution. Once you have your pseudosample you can use the procedure in Chapter 8 for constructing a confidence interval on the population mean with small samples and a normal population.

Here is a more formal description of the procedure. Let s be the overall standard deviation and s_{-i} be the subsample standard deviations based on omitting one observation at a time. The pseudovalues H_i equal

$$H_i = ns - (n-1)s_{-i} \tag{B.3}$$

Next compute the mean and the standard deviation of the pseudovalues, denoted $\overline{H}$ and s_H. The mean and standard deviation are computed in the usual way.

$$\overline{H} = \frac{\displaystyle\sum_{i=1}^{n} H_i}{n}$$

$$s_H = \sqrt{\frac{\displaystyle\sum_{i=1}^{n} (H_i - \overline{H})^2}{n - 1}}$$

Next compute the standard error of $\overline{H}$.

$$s_{\overline{H}} = \frac{s_H}{\sqrt{n}}$$

The last step is to construct the confidence interval for the population standard deviation, given the assumption of normality in the distribution of pseudovalues. The formula for the confidence interval is similar to formula (8.2) on page 278.

$$L, \ U = \overline{H} \pm t_{\alpha/2} s_{\overline{H}}$$

The degrees of freedom are $n - 1$.

Here are the calculations for the 10 observations {2646, 1016, 2170, 1927, 1194, 1572, 1009, 197, 2636, 2646}. The overall sample standard deviation s is 844.27. The 10 subsample standard deviations are {823.37, 858.29, 878.28, 891.52, 875.30, 894.18, 857.51, 698.28, 824.95, 823.37}. Substituting each of these subsample standard deviations into formula (B.3) creates the 10 pseudovalues of the pseudosample. The substitution for the first pseudovalue is

$$H_i = ns - (n - 1)s_{-i}$$ (B.3)

$$H_1 = 10(844.27) - (10 - 1)(823.37)$$

$$H_1 = 1032.37$$

The 10 pseudovalues are {1032.37, 718.08, 538.16, 418.98, 565.01, 395.02, 725.11, 2158.17, 1018.12, 1032.37}. Note that these pseudovalues are further apart than the differences between the overall standard deviation and the subsample standard deviations. The greatest difference between the overall standard deviation and a subsample standard deviation is 145.99. The greatest difference among the pseudovalues is 1763.15. Each pseudovalue is an independent estimate of the population standard deviation. The 10 psuedovalues are treated as data points in order to construct a confidence interval for the population standard deviation. Substituting into formulas (B.4), (B.5), (B.6), and (B.7) and using a 95% level of confidence,

$$\overline{H} = \frac{\sum\limits_{i=1}^{n} H_i}{n} \tag{B.4}$$

$$\overline{H} = \frac{1032.37 + 718.08 + \cdots + 1032.37}{10}$$

$$\overline{H} = 860.14$$

$$s_H = \sqrt{\frac{\sum\limits_{i=1}^{n} (H_i - \overline{H})^2}{n-1}} \tag{B.5}$$

$$s_H = \sqrt{\frac{(1032.37 - 860.14)^2 + (718.08 - 860.14)^2 + \cdots + (1032.37 - 860.14)^2}{10-1}}$$

$$s_H = 517.40$$

$$s_{\overline{H}} = \frac{s_H}{\sqrt{n}} \tag{B.6}$$

$$s_{\overline{H}} = \frac{517.40}{\sqrt{10}}$$

$$s_{\overline{H}} = 163.62$$

$$L, \ U = \overline{H} \pm t_{\alpha/2} s_{\overline{H}} \tag{B.7}$$

$$L, \ U = 860.14 \pm 2.262(163.62)$$

$$L = 490.04, \ U = 1230.24$$

The limits of a 95% confidence interval for the population standard deviation are from 490.04 to 1230.24. To do a hypothesis test with the jackknife procedure, formula (B.7) is modified to read

$$\text{C.V.}_L, \ \text{C.V.}_U = \sigma_0 \pm t_{\alpha/2} s_{\overline{H}} \tag{B.8}$$

where σ_0 is the hypothesized population standard deviation and the degrees of freedom are $n - 1$.

Formula (B.8) is for a two-tailed test. For a right-tailed test the $\pm$ changes to a $+$ and the $t_{\alpha/2}$ to a t_α. For a left-tailed test the $\pm$ changes to a $-$ and the $t_{\alpha/2}$ to a t_α.

You may be wondering about the origin of the term "jackknife procedure." The idea is that the procedure is a handy, multipurpose tool for robust statistical inference— as handy as a jackknife.

B.3 COMPARING TWO POPULATION VARIANCES WITH NORMAL POPULATIONS

You may encounter problems that involve standard deviations or variances of two independent populations. We will discuss confidence intervals on the ratio of two population variances and then a test for the equality of two population variances. Both the hypothesis tests and the confidence intervals assume that the two populations are normally distributed.

The confidence intervals are based on the theorem

$$\frac{S_1^2/\sigma_1^2}{S_2^2/\sigma_2^2} = F$$

where $df_{numerator} = n_1 - 1$ and $df_{denominator} = n_2 - 1$.

In words, the theorem is that the ratio of two independent sample variances each divided by its respective population variance follows an F distribution. Recall that an F distribution is defined as the ratio of two independent chi-square distributions each divided by its respective degree of freedom. The F distribution was introduced in Chapter 12. The discussion in section B.1 explained why $(n - 1)S^2/\sigma^2$ was chi-square distributed with $n - 1$ degrees of freedom. Given this result, the theorem is easy to derive as follows:

$$F = \frac{\chi^2/df_{numerator}}{\chi^2/df_{denominator}}$$

$$F = \frac{\dfrac{(n_1 - 1)(S_1^2/\sigma_1^2)}{n_1 - 1}}{\dfrac{(n_2 - 1)(S_2^2/\sigma_2^2)}{n_2 - 1}}$$

$$\frac{S_1^2/\sigma_1^2}{S_2^2/\sigma_2^2} = F$$

The terms in the theorem can be rearranged to find lower and upper limits for a $1 - \alpha$ confidence interval on the ratio of the population variances given two sample variances. The formula for the limits is

$$L, \ U = (F_{\alpha/2}^{n_2-1, m_1-1})\left(\frac{s_2^2}{s_1^2}\right), \ (F_{1-\alpha/2}^{n_2-1, m_1-1})\left(\frac{s_2^2}{s_1^2}\right)$$

(B.9)

Here is an example that uses formula (B.9). A commodities exchange uses two methods to quickly relay transactions to keypunchers who update the exchange's computerized quotation board. One method is through hand signals made by exchange employees who stand next to the trading pits. The other method is to use runners who scribble the details of the transactions on paper and carry the records to the keypunchers. The pits for the different commodities use either the hand signal or the runner method. At the end of each trading day all computerized records of transactions are cross-checked for accuracy against the written records of the commodity brokers. Occasionally the computerized records have errors: either the hand signal or the scribbled note carried by the runner does not match what the two brokers who were party to the transaction have recorded. The expected price error is zero—prices are recorded too high as frequently as they are recorded too low. The distribution of price errors for each method is also likely to be symmetric. Normality may be too strong an assumption because many errors are caused by transpositions of numbers. For example, recording a price of 19 as 91 would be a transposition. The confidence intervals created by formula (B.9) are fairly robust to small departures from normality. These unimodal, symmetric distributions should cause little problem.

The exchange wants an interval estimate, at the 90% confidence level, for the ratio of the variances of the amount of the price error per transaction generated by the two methods. Suppose the errors for 31 transactions recorded with the hand signal method have a variance of 0.2 dollar squared and the errors for 16 transactions recorded with the runner method have a variance of 0.1 dollar squared. The numerator and denominator degrees of freedom in this problem are 30 and 15 respectively. The boundary of the upper 5% of an F distribution with 30 numerator, and 15 denominator degrees of freedom is 2.25. This value is obtained from Appendix D, Table D.9. This table does not list boundaries that mark the lower 5%. However, there is a simple conversion. The boundary for the lower 5% equals 1 divided by the 95th percentile of an F distribution with 15 numerator degrees of freedom and 30 denominator degrees of freedom. Note the reversal of the numerator and denominator degrees of freedom. In notation the conversion is

$$F_{\alpha/2}^{n_2-1, m_1-1} = \frac{1}{F_{1-\alpha/2}^{n_1-1, n_2-1}}$$

(B.10)

The 95th percentile of an F distribution with 15 numerator and 30 denominator degrees of freedom is 2.01. The 5th percentile of an F distribution with 15 numerator and 30 denominator degrees of freedom is 1/2.25, or 0.444.

Now you have all of the terms needed to substitute into the formula for the confidence interval:

$$L, \ U = (F_{\alpha/2}^{n_2-1, n_1-1})\left(\frac{s_2^2}{s_1^2}\right), \ (F_{1-\alpha/2}^{n_2-1, n_1-1})\left(\frac{s_2^2}{s_1^2}\right) \tag{B.9}$$

$$L, \ U = (F_{0.05}^{15,30})\left(\frac{0.2}{0.1}\right), \ (F_{0.95}^{15,30})\left(\frac{0.2}{0.1}\right)$$

$$L, \ U = \left(\frac{1}{2.25}\right)\left(\frac{0.2}{0.1}\right), \ 2.01\left(\frac{0.2}{0.1}\right)$$

$$L = 0.89, \qquad U = 4.02$$

The limits of the confidence interval for the ratio of the standard deviations are approximately equal to the square roots of the limits of the confidence interval for the variances. Thus, a 90% interval on the ratio of the standard deviations would run from 0.94 to 2.00.

One conclusion that you can draw from these data is that since the interval for the variances included the value 1, the two methods of recording the transactions may have the same variability. The question could be settled easily with a larger data set. The exchange would certainly require more data before deciding to rely exclusively on one of the methods.

As a final comment on this example, consider the implication of using the variances of the distribution of errors as the criterion for selecting a recording method. Because of the squared term in the formula for the sample variance, this criterion places the greatest weight on the largest errors. If the exchange thought that all errors, large or small, were equally important, an interval estimate on the difference in population proportions of transactions recorded with errors would be the appropriate statistical technique.

The last topic in Appendix B is a test of the null hypothesis that two populations have the same variance or, equivalently, that they have the same standard deviations. The F test statistic is the ratio of the sample variances:

$$F = \frac{s_1^2}{s_2^2} \tag{B.11}$$

where $df_{\text{numerator}} = n_1 - 1$ and $df_{\text{denominator}} = n_2 - 1$.

Here is an example. The National Highway Traffic Safety Administration (NHTSA) is considering a rule about damage resistance for front bumpers in new automobiles. At the behest of the insurance industry, the rule is to be stated in terms of the dollar amount of damage rather than in the usual physical terms. A physical rule may state a limit for the amount of deflection in the bumper. Two methods for measuring dollar damage are proposed. One is to run a car into a concrete wall at 10 miles per hour and then have the damage appraised. The other method is to swing a 1,000-pound concrete pendulum into the bumper from a distance of five feet and then have the damage appraised. Since the limiting dollar amount of the damage can be set by the NHTSA, the key issue in choosing between the two methods is not whether they cause the same average amount of damage, but rather which method provides the most consistent measure. To test the two methods the agency has ten Ford Tauruses bashed by running into a wall and six bashed by the pendulum. The sample variance for the wall method is 80 squared dollars and the sample variance for the pendulum method is 16 squared dollars. The ratio of wall variance to pendulum variance is 5. Is this ratio high enough to reject the null hypothesis that the two methods have equal variances at a 10% significance level? The critical values for this two-tailed test are the boundary that marks the lower and upper 5% of an F distribution with $10 - 1$ numerator and $6 - 1$ denominator degrees of freedom. The critical value, found in Appendix D, Table D.9, is 4.77. The lower critical value is 1/95th percentile of an F distribution with 5 numerator degrees of freedom and 9 denominator degrees of freedom. The lower critical value is 1/3.48, or 0.29. The test statistic of 5 falls outside these critical values. The null hypothesis that the two methods have equal variances is rejected. Apparently it is easier to control the pendulum than to hit the wall at precisely 10 miles per hour. The pendulum method is how the NHTSA actually conducts its bumper tests.

EXERCISES FOR APPENDIX B

B.1 Construct a confidence interval for the population variance at the 95% level of confidence given the assumption that the population is normally distributed. The data are {12.54, 11.54, 11.38, 12.14, 11.34, 10.12, 16.10, 12.18, 12.49, 9.07, 15.17, 8.49, 13.11, 12.43, 12.31, 11.15, 14.18, 12.42, 9.18, 11.47}.

B.2 Construct a confidence interval for the population variance at the 95% level of confidence without the assumption that the population is normally distributed. The data are {2, 1, 3, 4, 2, 2, 5, 3}.

B.3 A sample of 20 mutual funds has a standard deviation for five-year rates of return of eight percentage points. Construct a confidence interval for the population variance at a 99% confidence level given the assumption that the distribution of the rates of return is normal.

B.4 Test the null hypothesis at a 10% significance level that two populations have equal variances. Assume that both populations are normally distributed. The data are $x_1 = \{10.25, 14.49, 13.52, 16.78, 16.33, 16.71, 10.61, 9.43, 14.82, 15.41, 17.32, 17.00\}$, $x_2 = \{13.23, 12.35, 16.36, 11.35, 23.11, 17.73, 21.19, 13.28, 11.74, 17.45\}$.

B.5 Construct a confidence interval at the 90% confidence level for the ratio of two populations variances. Assume that both populations are normally distributed. The data are $x_1 = \{1.01, -0.46, -1.20, -0.14, -0.96, 0.60, 1.58, -0.25, -2.62, 0.78\}$, $x_2 = \{0.14, 1.01, -2.27, 0.44, -0.62, 0.92, 1.62, 1.04, -0.68, -0.35\}$.

B.6 An international sanctioning body for gymnastic competitions is concerned about the consistency of scoring by its judges. The judges award points on a scale of 1 to 10 in increments of one-tenth of a point. From years of data the known variance of the scores of a group of 14 judges is 2.5 points squared. To reduce this variance, the sanctioning body invites the 14 judges to a training session on the mechanics of judging. After the training session all of the judges independently rate a gymnastic performance, giving it the following scores: {8.3, 10.0, 3.9, 7.2, 7.4, 6.7, 8.0, 5.9, 8.1, 7.4, 6.9, 6.9, 5.9, 5.8}. Test the hypothesis at a 5% significance level that the training session has reduced the variance of the judge's scores. Do not assume that the data are normally distributed.

B.7 Mr. Caveat Emptor, an investment advisor who publishes an investment newsletter, claims that among all advisors who have consistently beaten the rate of return on the Standard & Poor's Index of 500 stocks, his performance has been the most consistent. He offers evidence on his performance versus the other investment advisors who also publish such newsletters and who have beaten the Standard & Poor's Index. Over ten years the variance of returns on Mr. Emptor's recommended investments has been 13 percentage points squared. The same term for all other advisors is 15 percentage points squared. If Mr. Emptor's ten-year performance was considered a random sample from the distribution of his performance for all future years, could the null hypothesis that his variance is greater than or equal to the other investment advisors be rejected at a 5% significance level? Assume that the distribution of returns on Mr. Emptor's recommended investments is normal.

B.8 Test the hypothesis at a 10% significance level that the variance of the times between listing and sale for homes is the same for homes sold through a real estate agent and sold by the owner. Assume that both populations are normally distributed. The ten most recently real estate agent–sold homes in a city had a variance of 48 days squared. The ten most recently sold homes in the same city had a variance of 56 days squared. Is the assumption of normality reasonable in this context? Explain why the data may be biased.

Appendix C
Proofs

■ PROOF 1. (REFERENCE P. 40)

$$\sum_{i=1}^{n} (x_i - \bar{x}) = 0$$

$$\sum_{i=1}^{n} x_i - \sum_{i=1}^{n} \bar{x} = 0$$

From the rule for the summation of a constant, $\Sigma \bar{x} = n\bar{x}$:

$$\sum_{i=1}^{n} x_i - n\bar{x} = 0$$

$$\frac{n}{n} \sum_{i=1}^{n} x_i - n\bar{x} = 0$$

$$n \sum_{i=1}^{n} \frac{x_i}{n} - n\bar{x} = 0$$

$\bar{x}$ is defined to be $\Sigma x_i/n$, substituting $\bar{x}$ for $\Sigma x_i/n$:

$$n\bar{x} - n\bar{x} = 0$$

$$0 = 0 \quad ■$$

■ PROOF 2. (REFERENCE P. 53)

The text (page 53) states that the average value of sample variance equals the population variance. Although the term "average" is intuitive, the correct mathematical term is "expected value." The expected value of the sample variance is the average of all possible values of the sample variance weighted by frequency of each value. Expected values are discussed in Chapter 4. The notation $E(s^2)$ stands for expected

value of the sample variance. The proof below shows that the expected value of the sample variance equals the population variance.

$$E(s^2) = \sigma^2$$

Substituting the definition (with $n - 1$ in denominator) of the sample variance:

$$E\left[\sum_{i=1}^{n} \frac{(x_i - \bar{x})^2}{n - 1} \right] = \sigma^2$$

$$\frac{1}{n - 1} \left\{ E\left[\sum_{i=1}^{n} (x_i - \bar{x})^2 \right] \right\} = \sigma^2$$

$$\frac{1}{n - 1} \{[E(x_1 - \bar{x})^2 + E(x_2 - \bar{x})^2 + \cdots + E(x_n - \bar{x})^2]\} = \sigma^2 \quad (1)$$

The addition and offsetting subtraction of a constant from each term in the expressions $(x_i - \bar{x})^2$ leaves the square unchanged. Taking the first expectation as a start,

$$E(x_1 - \bar{x})^2 = E\{[(x_1 - \mu) - (\bar{x} - \mu)]^2\}$$

$$E(x_1 - \bar{x})^2 = E[(x_1 - \mu)^2 - 2(x_1 - \mu)(\bar{x} - \mu) + (\bar{x} - \mu)^2]$$

$$E(x_1 - \bar{x})^2 = E(x_1 - \mu)^2 - 2E(x_1 - \mu)(\bar{x} - \mu) + E(\bar{x} - \mu)^2 \quad (2)$$

Quadratic expression similar to the one above could be written for the observation 2 through n. The proof proceeds by substituting terms related to the population variance into the quadratic expression (2) and then substituting back into (1). By definition, the population variance is the average squared deviation between the observations and the population mean, or in expected value notation:

$$E(x_i - \mu)^2 = \sigma^2 \qquad \text{for } i = 1, 2, \ldots, n \qquad (3)$$

Proof number 9 will show that

$$E(\bar{x} - \mu)^2 = \frac{\sigma^2}{n} \qquad (4)$$

The remaining expectation term in (2) is $E(x_1 - \mu)(\bar{x} - \mu)$.

$$E[(x_1 - \mu)(\bar{x} - \mu)]$$
$$= E\left[(x_1 - \mu)\left(\frac{x_1 + x_2 + \cdots + x_n}{n} - \mu\right)\right]$$

$$E[(x_1 - \mu)(\bar{x} - \mu)]$$
$$= \frac{1}{n}E\{(x_1 - \mu)[(x_1 - \mu) +$$
$$(x_2 - \mu) + \cdots + (x_n - \mu)]\}$$

$$E[(x_1 - \mu)(\bar{x} - \mu)]$$
$$= \frac{1}{n}\{E(x_1 - \mu)^2 + E[(x_1 - \mu)(x_2 - \mu)]$$

$$+ E[(x_1 - \mu)(x_3 - \mu)] + \cdots$$
$$+ E[(x_1 - \mu)(x_n - \mu)]\}$$

The $n - 1$ nonsquared terms to the right of the equal sign all have a value of 0. The reason is that each observation is a random variable that, on average, equals the population mean. Think of x_1 as the random variable defined as the first observation drawn in a sample, rather than a specific number. The squared term is a variance of this random variable. This variance is the same as the population variance:

$$E[(x_1 - \mu)(\bar{x} - \mu)] = \frac{\sigma^2}{n} \qquad (5)$$

Substituting (3), (4), and (5) into (2),

$$E(x_1 - \bar{x})^2 = \sigma^2 - 2\frac{\sigma^2}{n} + \frac{\sigma^2}{n} \qquad (6)$$

$$E(x_1 - \bar{x})^2 = (n - 1)\frac{\sigma^2}{n}$$

Equation (6) is for the first observation in the sample. Identical terms could be written for observations 2 through n. Substituting n times formula (6) into (1) yields

$$\frac{1}{n - 1}\left[(n)(n - 1)\left(\frac{\sigma^2}{n}\right)\right] = \sigma^2$$

$$\sigma^2 = \sigma^2 \quad \blacksquare$$

The proof has shown that the sample variance defined with $n - 1$ in the denominator has an expectation or average value equal to the population variance.

■ PROOF 3. (REFERENCE P. 58)

$$P_{\text{tails}} \leq \frac{1}{k^2}$$

Figure C.1 shows a distribution divided into three areas, A, B, and C. Area A is bounded from above by line $k\sigma$ below the mean. Area C is bounded from below by line $k\sigma$ above the mean. Area B is between these two bounds. k can take any value above 1. Let ν_A, ν_B, and ν_C be the number of observations in areas A, B, and C. $\nu = \nu_A + \nu_B + \nu_C$.

Tchebychev's inequality can be rewritten as

$$\frac{\nu_A + \nu_C}{\nu} \leq \frac{1}{k^2}$$

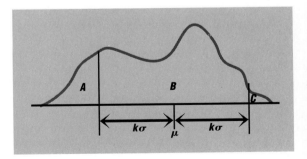

FIGURE C.1
Three Areas on an Arbitrary Distribution Separated by Boundaries of $\mu \pm k\sigma$.

The population variance can be calculated by separately summing the squared deviations for the three groups:

$$\frac{\Sigma_A(x_i - \mu)^2 + \Sigma_B(x_i - \mu)^2 + \Sigma_C(x_i - \mu)^2}{\nu} = \sigma^2 \quad (1)$$

In A and C, the absolute value of the smallest deviation from the mean is $k\sigma$. Thus

$$\nu_A k^2 \sigma^2 \le \sum_A (x_i - \mu)^2$$

$$\nu_C k^2 \sigma^2 \le \sum_C (x_i - \mu)^2$$

Substituting into (1) and changing the strict equality to a less than or equal sign,

$$\frac{\nu_A k^2 \sigma^2 + \Sigma_B(x_i - \mu)^2 + \nu_C k^2 \sigma^2}{\nu} \le \sigma^2$$

Since it is the sum of squared differences, dropping out the middle term in the numerator only makes the left side of the equation smaller:

$$\frac{\nu_A k^2 \sigma^2 + \nu_C k^2 \sigma^2}{\nu} \le \sigma^2$$

$$\frac{(\nu_A + \nu_C)}{\nu} k^2 \sigma^2 \le \sigma^2$$

$$\frac{(\nu_A + \nu_C)}{\nu} \le \frac{1}{k^2}$$

$$P_{\text{tails}} \le \frac{1}{k^2} \quad \blacksquare$$

■ PROOF 4. (REFERENCE P. 143)

Given that X and Y are independent random variables, we must prove

$$\sigma^2(X + Y) = \sigma^2(X) + \sigma^2(Y)$$

$$\sigma^2(X - Y) = \sigma^2(X) + \sigma^2(Y)$$

By definition of the variance

$$\sigma^2(X + Y) = \frac{\sum_{i=1}^{\nu} [(x_i + y_i) - (\mu_X + \mu_Y)]^2}{\nu}$$

Define **x** as $(X - \mu_X)$ and **y** as $(Y - \mu_Y)$. Then

$$\sigma^2(X + Y) = \frac{\sum_{i=1}^{\nu} (\mathbf{x}_i + \mathbf{y}_i)^2}{\nu}$$

$$\sigma^2(X + Y) = \frac{\sum_{i=1}^{\nu} (\mathbf{x}_i^2 + \mathbf{y}_i^2 + 2\mathbf{x}_i\mathbf{y}_i)}{\nu}$$

$$\sigma^2(X + Y) = \sigma^2(X) + \sigma^2(Y) + \frac{2\sum_{i=1}^{\nu} (\mathbf{x}_i\mathbf{y}_i)}{\nu}$$

Because of the assumption of independence between X and Y, the last term in the preceding equation equals zero. This can be shown as follows: group all $\mathbf{x}_i = a$, all $\mathbf{x}_i = b$, and so on, and call these classes A, B, and so on. Let the number of items in each class be ν_A, ν_B, ... The last term can be written

$$\frac{2\sum_{i=1}^{\nu} (\mathbf{x}_i\mathbf{y}_i)}{\nu} = \frac{2}{\nu}\left[a \sum_A \mathbf{y}_i + b \sum_B \mathbf{y}_i + \cdots \right]$$

$$\frac{2\sum_{i=1}^{\nu} (\mathbf{x}_i\mathbf{y}_i)}{\nu} = \frac{2}{\nu}[a\nu_A\mu_{\mathbf{y}|A} + b\nu_B\mu_{\mathbf{y}|B} + \cdots]$$

where $\Sigma_A\mathbf{y}_i$ is the total of the **y** observations in class A, $\mu_{\mathbf{y}|A}$ is the mean of the **y** observations in class A, and ν_A is the number of observations in class A. Since **x** and **y** are independent, the mean of the **y** values is the same for any specified value of **x**. Thus, $\mu_{\mathbf{y}|A} = \mu_{\mathbf{y}|B} = \mu_{\mathbf{y}} = \Sigma\mathbf{y}_i/\nu = 0$. This completes the proof. The proof for a difference instead of a sum of independent random variables is the same except that the term that drops out, the cross-products of the deviations from the means or $\Sigma\mathbf{x}_i\mathbf{y}_i$, has a minus sign instead of a plus sign.

PROOF 5. (REFERENCE P. 153)

$$\sigma(X) = \sqrt{n\Pi(1 - \Pi)}$$

Let Y_1 be a random variable that takes the value 0 with probability $1 - \Pi$ if the first trial is a failure and 1 with probability Π if the first trial is a success. Y_2 is defined similarly for the second trial, and so on out to Y_n. From the definitions of the expected value and variance of a random variable, the expected value and variance of Y_1 is

$$E(Y_1) = \sum_{i=1}^{2} y_i P(y_i)$$

$$= 1(\Pi) + 0(1 - \Pi)$$

$$= \Pi$$

$$\sigma^2(Y_1) = \sum_{i=1}^{2} (y_i - E(Y_1))^2 P(y_i)$$

$$= (1 - \Pi)^2 \Pi + (0 - \Pi)^2 (1 - \Pi)$$

$$= \Pi(1 - \Pi)^2 + \Pi^2(1 - \Pi)$$

$$= \Pi[(1 - \Pi)^2 + \Pi(1 - \Pi)]$$

$$= \Pi(1 - \Pi)(1 - \Pi + \Pi)$$

$$= \Pi(1 - \Pi)$$

The variances of the random variables Y_2 through Y_n are also equal to $\Pi(1 - \Pi)$. From the definition of the binomial distribution, each trial is independent. Next we use the additive property of the variances of independent random variables:

$$\sigma^2(X) = \sigma^2(Y_1 + Y_2 + \cdots + Y_n)$$

$$= \sigma^2(Y_1) + \sigma^2(Y_2) + \cdots + \sigma^2(Y_n)$$

$$= \Pi(1 - \Pi) + \Pi(1 - \Pi) + \cdots + \Pi(1 - \Pi)$$

$$= n\Pi(1 - \Pi)$$

$$\sigma(X) = \sqrt{n\Pi(1 - \Pi)} \quad \blacksquare$$

PROOF 6. (REFERENCE P. 157)

$$P(x) = \frac{\mu^x e^{-\mu}}{x!}$$

The proof uses the formula for binomial probability distribution as a starting point. Let Π be the probability of success in a binomial distribution and n be the number of trials. Let $n\Pi = \mu$, the mean of a Poisson process. The idea behind the proof is that a Poisson distribution can be described, for sufficiently small time intervals, as a binomial process with either 1 or 0 possible occurrences:

$$P(x) = \frac{n!}{(n - x)!x!} \Pi^x (1 - \Pi)^{n-x}$$

Substituting μ/n for Π,

$$P(x) = \frac{n!}{(n - x)!x!} \left(\frac{\mu}{n}\right)^x \left(1 - \frac{\mu}{n}\right)^{n-x} \tag{1}$$

$$P(x) = \frac{n!}{(n - x)!x!} \left(\frac{\mu}{n}\right)^x \left(1 - \frac{\mu}{n}\right)^{n} \left(1 - \frac{\mu}{n}\right)^{-x}$$

$$P(x) = \frac{\mu^x}{x!} \left(1 - \frac{\mu}{n}\right)^{n} \left(1 - \frac{\mu}{n}\right)^{-x}$$

$$\frac{n(n - 1) \cdots (n - x + 1)}{n^x}$$

$$P(x) = \frac{\mu^x}{x!} \left(1 - \frac{\mu}{n}\right)^{n} \left(1 - \frac{\mu}{n}\right)^{-x} \cdot 1 \cdot$$

$$\left(1 - \frac{1}{n}\right)\left(1 - \frac{2}{n}\right) \cdots \left(1 - \frac{x - 1}{n}\right) \tag{2}$$

Breaking the right side of (2) into four parts and taking the limit as n approaches infinity of each part,

$$\lim_{n \to \infty} \frac{\mu^x}{x!} = \frac{\mu^x}{x!} \tag{3}$$

$$\lim_{n \to \infty} \left(1 - \frac{\mu}{n}\right)^{n} = e^{-\mu} \tag{4}$$

$$\lim_{n \to \infty} \left(1 - \frac{\mu}{n}\right)^{-x} = 1 \tag{5}$$

$$\lim_{n \to \infty} 1\left(1 - \frac{1}{n}\right)\left(1 - \frac{2}{n}\right) \cdots \left(1 - \frac{x - 1}{n}\right) = 1 \tag{6}$$

Substituting (3), (4), (5), and (6) into (1),

$$\lim_{n \to \infty} \frac{n!}{(n - x)!x!} \left(\frac{\mu}{n}\right)^x \left(1 - \frac{\mu}{n}\right)^{n-x} = \frac{\mu^x e^{-\mu}}{x!}$$

$$P(x) = \frac{\mu^x e^{-\mu}}{x!} \quad \blacksquare$$

■ PROOF 7. (REFERENCE P. 183)

For a continuous uniform distribution with a range of a to b, the variance equals

$$\sigma^2(X) = \frac{(b-a)^2}{12}$$

The proof requires calculus. The expected value of a continuous random variable X is defined as an integral from $-\infty$ to ∞ of $xf(x)\ dx$. The integral serves the same function as the summation in the formula for discrete random variables. The variance of a continuous random variable is defined as $E[X - \mu]^2$. The density function of a continuous uniform distribution, $f(x)$, is $1/(b-a)$ for $a \le x \le b$ and 0 otherwise.

$$E(X) = \int_{-\infty}^{\infty} xf(x)\ dx$$

$$= \int_{-\infty}^{a} xf(x)\ dx + \int_{a}^{b} xf(x)\ dx + \int_{b}^{\infty} xf(x)\ dx$$

$$E(X) = \int_{-\infty}^{a} x(0)\ dx + \int_{a}^{b} x\left(\frac{1}{b-a}\right) dx + \int_{b}^{\infty} x(0)\ dx$$

$$E(X) = 0 + \left(\frac{1}{b-a}\right)\frac{x^2}{2}\bigg|_{a}^{b} + 0$$

$$E(X) = \left(\frac{1}{b-a}\right)\left[\frac{b^2}{2} - \frac{a^2}{2}\right]$$

$$E(X) = \frac{1}{2}\left(\frac{1}{b-a}\right)(b^2 - a^2)$$

$$E(X) = \frac{1}{2}\left(\frac{1}{b-a}\right)(b-a)(b+a)$$

$$E(X) = \frac{b+a}{2} \qquad (1)$$

$$\sigma^2(X) = E(X - E(X))^2 = \int_{-\infty}^{\infty} (x-\mu)^2 f(x)\ dx$$

$$\sigma^2(X) = \int_{-\infty}^{\infty} (x)^2 f(x)\ dx - \mu^2$$

$$\sigma^2(X) = \int_{-\infty}^{a} (x)^2 f(x)\ dx + \int_{a}^{b} (x)^2 f(x)\ dx$$

$$+ \int_{b}^{\infty} (x)^2 f(x)\ dx - \mu^2$$

$$\sigma^2(X) = \int_{-\infty}^{a} (x)^2(0)\ dx + \int_{a}^{b} (x)^2\left(\frac{1}{b-a}\right) dx$$

$$+ \int_{b}^{\infty} (x)^2(0)\ dx - \mu^2$$

$$\sigma^2(X) = 0 + \frac{1}{b-a}\frac{x^3}{3}\bigg|_{a}^{b} + 0 - \mu^2$$

$$\sigma^2(X) = \frac{1}{b-a}\left[\frac{b^3}{3} - \frac{a^3}{3}\right] - \mu^2$$

$$\sigma^2(X) = \frac{1}{3}\left(\frac{1}{b-a}\right)(b^3 - a^3) - \mu^2$$

$$\sigma^2(X) = \frac{1}{3}\left(\frac{1}{b-a}\right)(b-a)(b^2 + ab + a^2) - \mu^2$$

$$\sigma^2(X) = \frac{1}{3}(b^2 + ab + a^2) - \mu^2$$

Substituting for μ from (1),

$$\sigma^2(X) = \frac{1}{3}(b^2 + ab + a^2) - \left[\frac{b+a}{2}\right]^2$$

$$\sigma^2(X) = \frac{1}{3}(b^2 + ab + a^2) - \frac{1}{4}(b^2 + 2ab + a^2)$$

$$\sigma^2(X) = \frac{4(b^2 + ab + a^2) - 3(b^2 + 2ab + a^2)}{12}$$

$$\sigma^2(X) = \frac{(4b^2 + 4ab + 4a^2) - (3b^2 + 6ab + 3a^2)}{12}$$

$$\sigma^2(X) = \frac{b^2 - 2ab + a^2}{12} \qquad (2)$$

$$\sigma^2(X) = \frac{(b-a)^2}{12} \qquad ■$$

■ PROOF 8. (REFERENCE P. 235)

$$E(\bar{x}) = \mu$$

Note that $\bar{x}$ is a random variable of the possible values taken by the mean of a sample of size n rather than the realized value from one sample. By definition,

$$E(\bar{x}) = E\left(\frac{\sum_{i=1}^{n} x_i}{n}\right)$$

$$E(\bar{x}) = E\left(\frac{x_1}{n} + \frac{x_2}{n} + \cdots + \frac{x_n}{n}\right)$$

$$E(\overline{x}) = E\left(\frac{x_1}{n}\right) + E\left(\frac{x_2}{n}\right) + \cdots + E\left(\frac{x_n}{n}\right)$$

$$E(\overline{x}) = \frac{1}{n}[E(x_1) + E(x_2) + \cdots + E(x_n)] \qquad (1)$$

for all $iE(x_i) = \mu$. Substituting μ into (1),

$$E(\overline{x}) = \frac{1}{n}(\mu + \mu + \cdots + \mu)$$

$$E(\overline{x}) = \frac{1}{n}(n\mu)$$

$$E(\overline{x}) = \mu \quad \blacksquare$$

■ PROOF 9. (REFERENCE P. 236)

$$\sigma_{\overline{x}} = \frac{\sigma}{\sqrt{n}}$$

We will square both sides in order to use the additivity property of variances:

$$\sigma_{\overline{x}}^2 = \frac{\sigma^2}{n}$$

By definition $\overline{x} = (1/n)(x_1 + x_2 + \cdots + x_n)$. Consider each observation a random variable:

$$\sigma_{\overline{x}}^2 = \text{variance of } \left[\frac{1}{n}(x_1 + x_2 + \cdots + x_n)\right]$$

Recall the property $\sigma^2(kX) = k^2\sigma^2(X)$ where k is any constant and X is any random variable:

$$\sigma_{\overline{x}}^2 = \frac{1}{n^2}[\text{variance of } (x_1 + x_2 + \cdots + x_n)]$$

Since the x_i's are drawn in a random sample, each x_i is independent. Hence the variances of x_i's can be added. Furthermore, each x_i is drawn from the same population, so each one has a variance of σ^2:

$$\sigma_{\overline{x}}^2 = \frac{1}{n^2}(n\sigma^2) = \frac{\sigma^2}{n}$$

$$\sigma_{\overline{x}} = \frac{\sigma}{\sqrt{n}} \quad \blacksquare$$

■ PROOF 10. (REFERENCE P. 253)

$$\sigma^2(\text{sample median}) = 1.57\sigma_{\overline{x}}^2$$

This ratio is correct for normally distributed populations. However, the relative efficiency of the sample mean compared to sample median varies across population distributions. For some lighter tailed distributions, the sample mean is more than 1.57 times as efficient as the sample median. Conversely, for some tailheavy distributions, such as the Laplace and Cauchy, the sample median is more efficient than the sample mean. One continuous distribution that was discussed in the text, the continuous uniform, can be used to illustrate a ratio much higher than 1.57. A large sample formula for the variance of the sample median for any continuous distribution is

$$\sigma^2(\text{sample median}) = \frac{1}{4n[f(\text{median})]^2} \qquad (1)$$

where $f(\text{median})$ is the density or height of the distribution at the median. Consider a continuous uniform distribution with a width of 10. The density over the range of the distribution is 0.1 and thus the density at the median is also 0.1. The variance of the sample median for this distribution is

$$\sigma^2(\text{sample median}) = \frac{1}{4n(0.1)^2}$$

$$\sigma^2(\text{sample median}) = \frac{1}{0.04n}$$

Recalling proof 6, the variance of the sample mean for the same continuous distribution is

$$\sigma_{\overline{x}}^2 = \frac{\sigma^2}{n} = \frac{(b-a)^2/12}{n}$$

$$\sigma_{\overline{x}}^2 = \frac{(10)^2}{12n}$$

Comparing variance of the sample median to that of the sample mean,

$$\frac{\sigma^2(\text{sample median})}{\sigma_{\overline{x}}^2} = \frac{1/0.04n}{100/12n}$$

$$\sigma^2(\text{sample median}) = 3\sigma_{\overline{x}}^2$$

Now we turn to the standard normal distribution. The density function of the standard normal is

$$f(x) = \frac{1}{\sqrt{2\pi}} e^{-x^2/2}$$

The median of the standard normal distribution is 0. The density function takes the value of $1/\sqrt{2\pi}$ at 0. Substituting into (1),

$$\sigma^2(\text{sample median}) = \frac{1}{4n(1/\sqrt{2\pi})^2}$$

$$\sigma^2(\text{sample median}) = \frac{2\pi}{4n}$$

Recall that the standard normal distribution has a variance of 1. Comparing the variance of the sample median to the variance of the sample mean for the standard normal population,

$$\frac{\sigma^2(\text{sample median})}{\sigma_{\bar{x}}^2} = \frac{2\pi/4n}{1/n}$$

$$\sigma^2(\text{sample median}) = 1.57\sigma_{\bar{x}}^2 \quad \blacksquare$$

PROOF 11. (REFERENCE P. 258)

The proof is that the mean square error equals the variance of an estimator plus the square of its bias. The symbol θ stands for the population parameter and $\hat{\theta}$ for the estimator.

Mean square error $= E(\hat{\theta} - \theta)^2$

Mean square error $= E[\hat{\theta} - E(\hat{\theta}) + E(\hat{\theta}) - \theta]^2$

Mean square error $= E\{[\hat{\theta} - E(\hat{\theta})] + [E(\hat{\theta}) - \theta]\}^2$

Mean square error $= E[\hat{\theta} - E(\hat{\theta})]^2 + E[E(\hat{\theta}) - \theta]^2$
$\qquad\qquad\qquad + 2E[\hat{\theta} - E(\hat{\theta})][E(\hat{\theta}) - \theta] \quad (1)$

Consider the last term

$2E[\hat{\theta} - E(\hat{\theta})][E(\hat{\theta}) - \theta]$
$\quad = 2\{[E(\hat{\theta})]^2 - [E(\hat{\theta})]^2 - \theta E(\hat{\theta}) + \theta E(\hat{\theta})\} = 0$

Note that for a given population θ is a constant and the expected value of a constant equals the constant. Substituting 0 for the last term in (1),

Mean square error $= E[\hat{\theta} - E(\hat{\theta})]^2 + [E(\hat{\theta}) - \theta]^2$

The term $E[\hat{\theta} - E(\hat{\theta})]^2$ is the variance of $\hat{\theta}$ while the term $[E(\hat{\theta}) - \theta]^2$ is the squared bias. $\quad \blacksquare$

PROOF 12. (REFERENCE P. 508)

The proof is that α and β as defined in the following equations minimize the sum of squared deviations from the regression line.

$$\beta = \frac{\sum\limits_{i=1}^{\nu} x_i y_i - \nu \mu_x \mu_y}{\sum\limits_{i=1}^{n} x_i^2 - \nu \mu_x^2}$$

$$\alpha = \mu_y - \beta \mu_x$$

We need to show that α and β minimize $\sum_{i=1}^{\nu}(y_i - \hat{y}_i)^2$.

Substituting the formula for $\hat{y}$,

$$\sum_{i=1}^{\nu} (y_i - \hat{y}_i)^2 = \sum_{i=1}^{\nu} (y_i - \alpha - \beta x_i)^2 \qquad (1)$$

The α and β that minimize (1) will also minimize (1) divided by ν. This expression is the mean squared deviation.

$$\frac{\sum\limits_{i=1}^{\nu} (y_i - \hat{y}_i)^2}{\nu} \quad \frac{\sum\limits_{i=1}^{\nu} (y_i - \alpha - \beta x_i)^2}{\nu} \qquad (2)$$

Equation (2) can be expressed as

$$\frac{\sum\limits_{i=1}^{\nu} (y_i - \hat{y}_i)^2}{\nu} =$$

$$\frac{\sum\limits_{i=1}^{\nu} \{y_i + (\mu_y - \mu_y) - \alpha - \beta[x_i + (\mu_x - \mu_x)]\}^2}{\nu}$$

$$\frac{\displaystyle\sum_{i=1}^{\nu}(y_i - \hat{y}_i)^2}{\nu} =$$

$$\frac{\displaystyle\sum_{i=1}^{\nu}[(y_i - \mu_y) - \beta(x_i - \mu_x) + (\mu_y - \alpha - \beta\mu_x)]^2}{\nu}$$

$$\frac{\displaystyle\sum_{i=1}^{\nu}(y_i - \hat{y}_i)^2}{\nu} = \sigma_y^2 + \beta^2\sigma_x^2 - 2\beta\mathrm{cov}(x, y)$$

$$+ \frac{(\mu_y - \alpha - \beta\mu_x)^2}{\nu} \quad \text{(3)}$$

where σ^2 is the variance and $\mathrm{cov}(x, y)$, the covariance of x and y, is defined as

$$\frac{\displaystyle\sum_{i=1}^{\nu}(y_i - \mu_y)(x_i - \mu_x)}{\nu}$$

Equation (3) can be expressed as

$$\sum_{i=1}^{\nu}(y_i - \hat{y}_i)^2 = \sigma_y^2(1 - \rho^2) + (\beta\sigma_x - \rho\sigma_y)^2$$

$$+ (\mu_y - \alpha - \beta\mu_x)^2 \quad \text{(4)}$$

where ρ is the population correlation coefficient. Equation (4) is the key step. The right side of (4) is minimized if the second and third terms are zero. This occurs when $\beta = \rho\sigma_y/\sigma_x$ and $\alpha = \mu_y - \beta\mu_x$. When α and β take these values, the remaining minimum variance of the y values around the regression line equals $\sigma_y^2(1 - \rho^2)$. Note that $\alpha = \mu_y - \beta\mu_x$ is the standard formula for α and $\rho\sigma_y/\sigma_x$ can be shown to be equivalent to the formula for the slope coefficient found in the text.

$$\beta = \rho\frac{\sigma_y}{\sigma_x}$$

$$\beta = \left(\frac{\mathrm{cov}(x, y)}{\sigma_x\sigma_y}\right)\left(\frac{\sigma_y}{\sigma_x}\right)$$

$$\beta = \left(\frac{\mathrm{cov}(x, y)}{\sigma_x^2}\right)$$

$$\beta = \frac{\displaystyle\sum_{i=1}^{\nu}(x_i - \mu_x)(x_i - \mu_x)/\nu}{\displaystyle\sum_{i=1}^{\nu}(x_i - \mu_x)^2/\nu}$$

$$\beta = \frac{\displaystyle\sum_{i=1}^{\nu}x_iy_i - \nu\mu_x\mu_y}{\displaystyle\sum_{i=1}^{n}x_i^2 - \nu\mu_x^2} \quad \blacksquare$$

■ PROOF 13. (REFERENCE P. 519)

$$\rho = \frac{\displaystyle\sum_{i=1}^{\nu}z_{x_i}z_{y_i}}{\nu} \quad \text{(1)}$$

The proof is that the numerator equals ν or $-\nu$ when correlation between X and Y perfect:

$$\nu = \sum_{i=1}^{\nu}z_{x_i}z_{y_i}$$

If there is perfect positive correlation between X and Y, the z scores for the pairs of observations would be identical. Replacing the z_{y_i} with z_{x_i} and writing out the z score formula,

$$\nu = \sum_{i=1}^{\nu}\left[\left(\frac{x_i - \mu}{\sigma}\right)\left(\frac{x_i - \mu}{\sigma}\right)\right]$$

$$\nu = \sum_{i=1}^{\nu}\left(\frac{x_i - \mu}{\sigma}\right)^2$$

$$\nu = \frac{\displaystyle\sum_{i=1}^{\nu}(x_i - \mu)^2}{\sigma^2}$$

$$\nu = \frac{\displaystyle\sum_{i=1}^{\nu}(x_i - \mu)^2}{\displaystyle\sum_{i=1}^{\nu}(x_i - \mu)^2/\nu}$$

$$\nu = \nu$$

It there was perfect negative correlation, the numerator of the right-hand expression in (1) would equal $-\nu$. The proof would have a minus sign in front of one of the terms $(x_i - \mu)/\sigma$ in (2). The rest of the proof would be the same. ∎

■ PROOF 14. (REFERENCE P. 583)

The proof is that the following "normal" equations minimize the sum of square deviations between the observed values of y and the estimated regression plane. This proof requires calculus.

$$\sum_{i=1}^{\nu} y_i = \nu\alpha + \beta_1 \sum_{i=1}^{\nu} x_i + \beta_2 \sum_{i=1}^{\nu} z_i \tag{1}$$

$$\sum_{i=1}^{\nu} x_i y_i = \alpha \sum_{i=1}^{\nu} x_i + \beta_1 \sum_{i=1}^{\nu} x_i^2 + \beta_2 \sum_{i=1}^{\nu} x_i z_i \tag{2}$$

$$\sum_{i=1}^{\nu} z_i y_i = \alpha \sum_{i=1}^{\nu} z_i + \beta_1 \sum_{i=1}^{\nu} x_i z_i + \beta_2 \sum_{i=1}^{\nu} z_i^2 \tag{3}$$

Equation (1) is found by minimizing the squared deviations, $\Sigma(y_i - \hat{y}_i)^2$, with respect to α. Taking the partial derivative with respect to α and setting it equal to 0,

$$\frac{\partial(y_i - \hat{y}_i)^2}{\partial\alpha} = \frac{\partial \sum_{i=1}^{\nu} (y_i - \alpha - \beta_1 x_i - \beta_2 z_i)^2}{\partial\alpha} = 0$$

$$-2 \sum_{i=1}^{\nu} (y_i - \alpha - \beta_1 x_i - \beta_2 z_i) = 0$$

$$\sum_{i=1}^{\nu} y_i = \nu\alpha + \beta_1 \sum_{i=1}^{\nu} x_i + \beta_2 \sum_{i=1}^{\nu} z_i$$

Equation (2) is found by minimizing the squared deviations, $\Sigma(y_i - \hat{y}_i)^2$, with respect to β_1. Taking the partial derivative with respect to β_1 and setting it equal to 0,

$$\frac{\partial(y_i - \hat{y}_i)^2}{\partial\beta_1} = \frac{\partial \sum_{i=1}^{\nu} (y_i - \alpha - \beta_1 x_i - \beta_2 z_i)^2}{\partial\beta_2} = 0$$

$$-2 \sum_{i=1}^{\nu} x_i(y_i - \alpha - \beta_1 x_i - \beta_2 z_i) = 0$$

$$\sum_{i=1}^{\nu} x_i y_i = \alpha \sum_{i=1}^{\nu} x_i + \beta_1 \sum_{i=1}^{\nu} x_i^2 + \beta_2 \sum_{i=1}^{\nu} x_i z_i$$

Equation (3) is found by minimizing the squared deviations, $\Sigma(y_i - \hat{y}_i)^2$, with respect to β_2. Taking the partial derivative with respect to β_2 and setting it equal to 0,

$$\frac{\partial(y_i - \hat{y}_i)^2}{\partial\beta_2} = \frac{\partial \sum_{i=1}^{\nu} (y_i - \alpha - \beta_1 x_i - \beta_2 z_i)^2}{\partial\beta_2} = 0$$

$$-2 \sum_{i=1}^{\nu} z_i(y_i - \alpha - \beta_1 x_i - \beta_2 z_i) = 0$$

$$\sum_{i=1}^{\nu} z_i y_i = \alpha \sum_{i=1}^{\nu} z_i + \beta_1 \sum_{i=1}^{\nu} z_i x_i + \beta_2 \sum_{i=1}^{\nu} z_i^2 \quad ∎$$

■ PROOF 15. (REFERENCE PP. 664–65]

$$E(DW) = 2$$

The expected value of the Durbin–Watson statistic is approximately 2 rather than exactly 2. The explanation of why the value is approximate will also shed some light on why the test has inconclusive regions. Recall the formula for the Durbin–Watson test statistic,

$$DW = \frac{\sum_{t=1}^{n-1} (e_{t+1} - e_t)^2}{\sum_{t=1}^{n} e_t^2} \tag{2}$$

where e_t are the residuals of the least squares regression. Suppose (incorrectly) that the e's are standard normal random variables. The expected value of a squared standard normal variable is 1. Expanding the numerator of (1),

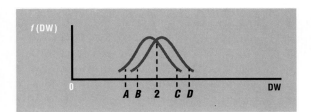

FIGURE C.2
Limiting Distributions of the
DW Statistic under the Null
Hypothesis.

$$DW = \frac{\sum\limits_{t=1}^{n-1} (e_{t+1}^2 - 2e_{t+1}e_t + e_t^2)}{\sum\limits_{t=1}^{n} e_t^2}$$

$$= \left(\frac{\sum\limits_{t=1}^{n-1} e_{t+1}^2}{\sum\limits_{t=1}^{n} e_t^2}\right) + \left(\frac{\sum\limits_{t=1}^{n-1} e_t^2}{\sum\limits_{t=1}^{n} e_t^2}\right) - 2\left(\frac{\sum\limits_{t=1}^{n-1} e_{t+1}e_t}{\sum\limits_{t=1}^{n} e_t^2}\right)$$

The first and second terms in the preceding equation are approximately equal to 1. The ratio of the sums in the last term is an estimate of the correlation between e_t and e_{t+1}. Thus, if there is no autocorrelation,

$$DW \cong 1 + 1 - 2(0), \qquad \text{or } 2$$

If there is perfect positive autocorrelation,

$$DW \cong 1 + 1 - 2(1), \qquad \text{or } 0$$

If there is perfect negative autocorrelation,

$$DW \cong 1 + 1 - 2(-1), \qquad \text{or } 4.$$

The problem is that the residuals are *not* normal random variables. The distribution of the residuals depends on the actual sample data. Durbin and Watson show that actual distribution of DW must lie between two limiting distributions, shown in Figure C.2. Five percent tail areas for the two limiting distributions are indicated by the letters A, B, C, and D. No matter what the actual distribution of DW is for a given set of sample data, the null hypothesis of no autocorrelation will be rejected whenever DW lies to the right of D or to the left of A. The null hypothesis cannot be rejected when DW falls in the range of B to C. However, the test is inconclusive whenever DW falls inside of the ranges of A to B or C to D.

Appendix D

Statistical Tables

Entry is probability $P(X = x) = \dfrac{n!}{(n - x)!x!}\Pi^x(1 - \Pi)^{n-x}$

TABLE D.1
Binomial probabilities

n	x	.01	.02	.03	.04	.05	.06	.07	.08	.09
2	0	0.9801	0.9604	0.9409	0.9216	0.9025	0.8836	0.8649	0.8464	0.8281
	1	0.0198	0.0392	0.0582	0.0768	0.0950	0.1128	0.1302	0.1472	0.1638
	2	0.0001	0.0004	0.0009	0.0016	0.0025	0.0036	0.0049	0.0064	0.0081
3	0	0.9703	0.9412	0.9127	0.8847	0.8574	0.8306	0.8044	0.7787	0.7536
	1	0.0294	0.0576	0.0847	0.1106	0.1354	0.1590	0.1816	0.2031	0.2236
	2	0.0003	0.0012	0.0026	0.0046	0.0071	0.0102	0.0137	0.0177	0.0221
	3	0.0000	0.0000	0.0000	0.0001	0.0001	0.0002	0.0003	0.0005	0.0007
4	0	0.9606	0.9224	0.8853	0.8493	0.8145	0.7807	0.7481	0.7164	0.6857
	1	0.0388	0.0753	0.1095	0.1416	0.1715	0.1993	0.2252	0.2492	0.2713
	2	0.0006	0.0023	0.0051	0.0088	0.0135	0.0191	0.0254	0.0325	0.0402
	3	0.0000	0.0000	0.0001	0.0002	0.0005	0.0008	0.0013	0.0019	0.0027
	4	0.0000	0.0000	0.0000	0.0000	0.0000	0.0000	0.0000	0.0000	0.0001
5	0	0.9510	0.9039	0.8587	0.8154	0.7738	0.7339	0.6957	0.6591	0.6240
	1	0.0480	0.0922	0.1328	0.1699	0.2036	0.2342	0.2618	0.2866	0.3086
	2	0.0010	0.0038	0.0082	0.0142	0.0214	0.0299	0.0394	0.0498	0.0610
	3	0.0000	0.0001	0.0003	0.0006	0.0011	0.0019	0.0030	0.0043	0.0060
	4	0.0000	0.0000	0.0000	0.0000	0.0000	0.0001	0.0001	0.0002	0.0003
	5	0.0000	0.0000	0.0000	0.0000	0.0000	0.0000	0.0000	0.0000	0.0000
6	0	0.9415	0.8858	0.8330	0.7828	0.7351	0.6899	0.6470	0.6064	0.5679
	1	0.0571	0.1085	0.1546	0.1957	0.2321	0.2642	0.2922	0.3164	0.3370
	2	0.0014	0.0055	0.0120	0.0204	0.0305	0.0422	0.0550	0.0688	0.0833
	3	0.0000	0.0002	0.0005	0.0011	0.0021	0.0036	0.0055	0.0080	0.0110
	4	0.0000	0.0000	0.0000	0.0000	0.0001	0.0002	0.0003	0.0005	0.0008
	5	0.0000	0.0000	0.0000	0.0000	0.0000	0.0000	0.0000	0.0000	0.0000
	6	0.0000	0.0000	0.0000	0.0000	0.0000	0.0000	0.0000	0.0000	0.0000
7	0	0.9321	0.8681	0.8080	0.7514	0.6983	0.6485	0.6017	0.5578	0.5168
	1	0.0659	0.1240	0.1749	0.2192	0.2573	0.2897	0.3170	0.3396	0.3578
	2	0.0020	0.0076	0.0162	0.0274	0.0406	0.0555	0.0716	0.0886	0.1061
	3	0.0000	0.0003	0.0008	0.0019	0.0036	0.0059	0.0090	0.0128	0.0175
	4	0.0000	0.0000	0.0000	0.0001	0.0002	0.0004	0.0007	0.0011	0.0017
	5	0.0000	0.0000	0.0000	0.0000	0.0000	0.0000	0.0000	0.0001	0.0001
	6	0.0000	0.0000	0.0000	0.0000	0.0000	0.0000	0.0000	0.0000	0.0000
	7	0.0000	0.0000	0.0000	0.0000	0.0000	0.0000	0.0000	0.0000	0.0000
8	0	0.9227	0.8508	0.7837	0.7214	0.6634	0.6096	0.5596	0.5132	0.4703
	1	0.0746	0.1389	0.1939	0.2405	0.2793	0.3113	0.3370	0.3570	0.3721
	2	0.0026	0.0099	0.0210	0.0351	0.0515	0.0695	0.0888	0.1087	0.1288
	3	0.0001	0.0004	0.0013	0.0029	0.0054	0.0089	0.0134	0.0189	0.0255
	4	0.0000	0.0000	0.0001	0.0002	0.0004	0.0007	0.0013	0.0021	0.0031
	5	0.0000	0.0000	0.0000	0.0000	0.0000	0.0000	0.0001	0.0001	0.0002
	6	0.0000	0.0000	0.0000	0.0000	0.0000	0.0000	0.0000	0.0000	0.0000
	7	0.0000	0.0000	0.0000	0.0000	0.0000	0.0000	0.0000	0.0000	0.0000
	8	0.0000	0.0000	0.0000	0.0000	0.0000	0.0000	0.0000	0.0000	0.0000

(continued)

TABLE D.1

Binomial probabilities (continued)

						Π				
n	x	.01	.02	.03	.04	.05	.06	.07	.08	.09
9	0	0.9135	0.8337	0.7602	0.6925	0.6302	0.5730	0.5204	0.4722	0.4279
	1	0.0830	0.1531	0.2116	0.2597	0.2985	0.3292	0.3525	0.3695	0.3809
	2	0.0034	0.0125	0.0262	0.0433	0.0629	0.0840	0.1061	0.1285	0.1507
	3	0.0001	0.0006	0.0019	0.0042	0.0077	0.0125	0.0186	0.0261	0.0348
	4	0.0000	0.0000	0.0001	0.0003	0.0006	0.0012	0.0021	0.0034	0.0052
	5	0.0000	0.0000	0.0000	0.0000	0.0000	0.0001	0.0002	0.0003	0.0005
	6	0.0000	0.0000	0.0000	0.0000	0.0000	0.0000	0.0000	0.0000	0.0000
	7	0.0000	0.0000	0.0000	0.0000	0.0000	0.0000	0.0000	0.0000	0.0000
	8	0.0000	0.0000	0.0000	0.0000	0.0000	0.0000	0.0000	0.0000	0.0000
	9	0.0000	0.0000	0.0000	0.0000	0.0000	0.0000	0.0000	0.0000	0.0000
10	0	0.9044	0.8171	0.7374	0.6648	0.5987	0.5386	0.4840	0.4344	0.3894
	1	0.0914	0.1667	0.2281	0.2770	0.3151	0.3438	0.3643	0.3777	0.3851
	2	0.0042	0.0153	0.0317	0.0519	0.0746	0.0988	0.1234	0.1478	0.1714
	3	0.0001	0.0008	0.0026	0.0058	0.0105	0.0168	0.0248	0.0343	0.0452
	4	0.0000	0.0000	0.0001	0.0004	0.0010	0.0019	0.0033	0.0052	0.0078
	5	0.0000	0.0000	0.0000	0.0000	0.0001	0.0001	0.0003	0.0005	0.0009
	6	0.0000	0.0000	0.0000	0.0000	0.0000	0.0000	0.0000	0.0000	0.0001
	7	0.0000	0.0000	0.0000	0.0000	0.0000	0.0000	0.0000	0.0000	0.0000
	8	0.0000	0.0000	0.0000	0.0000	0.0000	0.0000	0.0000	0.0000	0.0000
	9	0.0000	0.0000	0.0000	0.0000	0.0000	0.0000	0.0000	0.0000	0.0000
	10	0.0000	0.0000	0.0000	0.0000	0.0000	0.0000	0.0000	0.0000	0.0000
12	0	0.8864	0.7847	0.6938	0.6127	0.5404	0.4759	0.4186	0.3677	0.3225
	1	0.1074	0.1922	0.2575	0.3064	0.3413	0.3645	0.3781	0.3837	0.3827
	2	0.0060	0.0216	0.0438	0.0702	0.0988	0.1280	0.1565	0.1835	0.2082
	3	0.0002	0.0015	0.0045	0.0098	0.0173	0.0272	0.0393	0.0532	0.0686
	4	0.0000	0.0001	0.0003	0.0009	0.0021	0.0039	0.0067	0.0104	0.0153
	5	0.0000	0.0000	0.0000	0.0001	0.0002	0.0004	0.0008	0.0014	0.0024
	6	0.0000	0.0000	0.0000	0.0000	0.0000	0.0000	0.0001	0.0001	0.0003
	7	0.0000	0.0000	0.0000	0.0000	0.0000	0.0000	0.0000	0.0000	0.0000
	8	0.0000	0.0000	0.0000	0.0000	0.0000	0.0000	0.0000	0.0000	0.0000
	9	0.0000	0.0000	0.0000	0.0000	0.0000	0.0000	0.0000	0.0000	0.0000
	10	0.0000	0.0000	0.0000	0.0000	0.0000	0.0000	0.0000	0.0000	0.0000
	11	0.0000	0.0000	0.0000	0.0000	0.0000	0.0000	0.0000	0.0000	0.0000
	12	0.0000	0.0000	0.0000	0.0000	0.0000	0.0000	0.0000	0.0000	0.0000
15	0	0.8601	0.7386	0.6333	0.5421	0.4633	0.3953	0.3367	0.2863	0.2430
	1	0.1303	0.2261	0.2938	0.3388	0.3658	0.3785	0.3801	0.3734	0.3605
	2	0.0092	0.0323	0.0636	0.0988	0.1348	0.1691	0.2003	0.2273	0.2496
	3	0.0004	0.0029	0.0085	0.0178	0.0307	0.0468	0.0653	0.0857	0.1070
	4	0.0000	0.0002	0.0008	0.0022	0.0049	0.0090	0.0148	0.0223	0.0317
	5	0.0000	0.0000	0.0001	0.0002	0.0006	0.0013	0.0024	0.0043	0.0069
	6	0.0000	0.0000	0.0000	0.0000	0.0000	0.0001	0.0003	0.0006	0.0011
	7	0.0000	0.0000	0.0000	0.0000	0.0000	0.0000	0.0000	0.0001	0.0001
	8	0.0000	0.0000	0.0000	0.0000	0.0000	0.0000	0.0000	0.0000	0.0000
	9	0.0000	0.0000	0.0000	0.0000	0.0000	0.0000	0.0000	0.0000	0.0000
	10	0.0000	0.0000	0.0000	0.0000	0.0000	0.0000	0.0000	0.0000	0.0000
	11	0.0000	0.0000	0.0000	0.0000	0.0000	0.0000	0.0000	0.0000	0.0000
	12	0.0000	0.0000	0.0000	0.0000	0.0000	0.0000	0.0000	0.0000	0.0000
	13	0.0000	0.0000	0.0000	0.0000	0.0000	0.0000	0.0000	0.0000	0.0000
	14	0.0000	0.0000	0.0000	0.0000	0.0000	0.0000	0.0000	0.0000	0.0000
	15	0.0000	0.0000	0.0000	0.0000	0.0000	0.0000	0.0000	0.0000	0.0000

TABLE D.1
Binomial
probabilities
(continued)

						Π				
n	*x*	.01	.02	.03	.04	.05	.06	.07	.08	.09
20	0	0.8179	0.6676	0.5438	0.4420	0.3585	0.2901	0.2342	0.1887	0.1516
	1	0.1652	0.2725	0.3364	0.3683	0.3774	0.3703	0.3526	0.3282	0.3000
	2	0.0159	0.0528	0.0988	0.1458	0.1887	0.2246	0.2521	0.2711	0.2818
	3	0.0010	0.0065	0.0183	0.0364	0.0596	0.0860	0.1139	0.1414	0.1672
	4	0.0000	0.0006	0.0024	0.0065	0.0133	0.0233	0.0364	0.0523	0.0703
	5	0.0000	0.0000	0.0002	0.0009	0.0022	0.0048	0.0088	0.0145	0.0222
	6	0.0000	0.0000	0.0000	0.0001	0.0003	0.0008	0.0017	0.0032	0.0055
	7	0.0000	0.0000	0.0000	0.0000	0.0000	0.0001	0.0002	0.0005	0.0011
	8	0.0000	0.0000	0.0000	0.0000	0.0000	0.0000	0.0000	0.0001	0.0002
	9	0.0000	0.0000	0.0000	0.0000	0.0000	0.0000	0.0000	0.0000	0.0000
	10	0.0000	0.0000	0.0000	0.0000	0.0000	0.0000	0.0000	0.0000	0.0000
	11	0.0000	0.0000	0.0000	0.0000	0.0000	0.0000	0.0000	0.0000	0.0000
	12	0.0000	0.0000	0.0000	0.0000	0.0000	0.0000	0.0000	0.0000	0.0000
	13	0.0000	0.0000	0.0000	0.0000	0.0000	0.0000	0.0000	0.0000	0.0000
	14	0.0000	0.0000	0.0000	0.0000	0.0000	0.0000	0.0000	0.0000	0.0000
	15	0.0000	0.0000	0.0000	0.0000	0.0000	0.0000	0.0000	0.0000	0.0000
	16	0.0000	0.0000	0.0000	0.0000	0.0000	0.0000	0.0000	0.0000	0.0000
	17	0.0000	0.0000	0.0000	0.0000	0.0000	0.0000	0.0000	0.0000	0.0000
	18	0.0000	0.0000	0.0000	0.0000	0.0000	0.0000	0.0000	0.0000	0.0000
	19	0.0000	0.0000	0.0000	0.0000	0.0000	0.0000	0.0000	0.0000	0.0000
	20	0.0000	0.0000	0.0000	0.0000	0.0000	0.0000	0.0000	0.0000	0.0000

						Π				
n	*x*	.10	.15	.20	.25	.30	.35	.40	.45	.50
2	0	0.8100	0.7225	0.6400	0.5625	0.4900	0.4225	0.3600	0.3025	0.2500
	1	0.1800	0.2550	0.3200	0.3750	0.4200	0.4550	0.4800	0.4950	0.5000
	2	0.0100	0.0225	0.0400	0.0625	0.0900	0.1225	0.1600	0.2025	0.2500
3	0	0.7290	0.6141	0.5120	0.4219	0.3430	0.2746	0.2160	0.1664	0.1250
	1	0.2430	0.3251	0.3840	0.4219	0.4410	0.4436	0.4320	0.4084	0.3750
	2	0.0270	0.0574	0.0960	0.1406	0.1890	0.2389	0.2880	0.3341	0.3750
	3	0.0010	0.0034	0.0080	0.0156	0.0270	0.0429	0.0640	0.0911	0.1250
4	0	0.6561	0.5220	0.4096	0.3164	0.2401	0.1785	0.1296	0.0915	0.0625
	1	0.2916	0.3685	0.4096	0.4219	0.4116	0.3845	0.3456	0.2995	0.2500
	2	0.0486	0.0975	0.1536	0.2109	0.2646	0.3105	0.3456	0.3675	0.3750
	3	0.0036	0.0115	0.0256	0.0469	0.0756	0.1115	0.1536	0.2005	0.2500
	4	0.0001	0.0005	0.0016	0.0039	0.0081	0.0150	0.0256	0.0410	0.0625
5	0	0.5905	0.4437	0.3277	0.2373	0.1681	0.1160	0.0778	0.0503	0.0312
	1	0.3280	0.3915	0.4096	0.3955	0.3601	0.3124	0.2592	0.2059	0.1562
	2	0.0729	0.1382	0.2048	0.2637	0.3087	0.3364	0.3456	0.3369	0.3125
	3	0.0081	0.0244	0.0512	0.0879	0.1323	0.1811	0.2304	0.2757	0.3125
	4	0.0004	0.0022	0.0064	0.0146	0.0283	0.0488	0.0768	0.1128	0.1562
	5	0.0000	0.0001	0.0003	0.0010	0.0024	0.0053	0.0102	0.0185	0.0312
6	0	0.5314	0.3771	0.2621	0.1780	0.1176	0.0754	0.0467	0.0277	0.0156
	1	0.3543	0.3993	0.3932	0.3560	0.3025	0.2437	0.1866	0.1359	0.0938
	2	0.0984	0.1762	0.2458	0.2966	0.3241	0.3280	0.3110	0.2780	0.2344
	3	0.0146	0.0415	0.0819	0.1318	0.1852	0.2355	0.2765	0.3032	0.3125
	4	0.0012	0.0055	0.0154	0.0330	0.0595	0.0951	0.1382	0.1861	0.2344
	5	0.0001	0.0004	0.0015	0.0044	0.0102	0.0205	0.0369	0.0609	0.0938
	6	0.0000	0.0000	0.0001	0.0002	0.0007	0.0018	0.0041	0.0083	0.0156

(continued)

**TABLE D.1
Binomial
probabilities
(continued)**

		Π								
n	x	.10	.15	.20	.25	.30	.35	.40	.45	.50
7	0	0.4783	0.3206	0.2097	0.1335	0.0824	0.0490	0.0280	0.0152	0.0078
	1	0.3720	0.3960	0.3670	0.3115	0.2471	0.1848	0.1306	0.0872	0.0547
	2	0.1240	0.2097	0.2753	0.3115	0.3177	0.2985	0.2613	0.2140	0.1641
	3	0.0230	0.0617	0.1147	0.1730	0.2269	0.2679	0.2903	0.2918	0.2734
	4	0.0026	0.0109	0.0287	0.0577	0.0972	0.1442	0.1935	0.2388	0.2734
	5	0.0002	0.0012	0.0043	0.0115	0.0250	0.0466	0.0774	0.1172	0.1641
	6	0.0000	0.0001	0.0004	0.0013	0.0036	0.0084	0.0172	0.0320	0.0547
	7	0.0000	0.0000	0.0000	0.0001	0.0002	0.0006	0.0016	0.0037	0.0078
8	0	0.4305	0.2725	0.1678	0.1001	0.0576	0.0319	0.0168	0.0084	0.0039
	1	0.3826	0.3847	0.3355	0.2670	0.1977	0.1373	0.0896	0.0548	0.0312
	2	0.1488	0.2376	0.2936	0.3115	0.2965	0.2587	0.2090	0.1569	0.1094
	3	0.0331	0.0839	0.1468	0.2076	0.2541	0.2786	0.2787	0.2568	0.2188
	4	0.0046	0.0185	0.0459	0.0865	0.1361	0.1875	0.2322	0.2627	0.2734
	5	0.0004	0.0026	0.0092	0.0231	0.0467	0.0808	0.1239	0.1719	0.2188
	6	0.0000	0.0002	0.0011	0.0038	0.0100	0.0217	0.0413	0.0703	0.1094
	7	0.0000	0.0000	0.0001	0.0004	0.0012	0.0033	0.0079	0.0164	0.0312
	8	0.0000	0.0000	0.0000	0.0000	0.0001	0.0002	0.0007	0.0017	0.0039
9	0	0.3874	0.2316	0.1342	0.0751	0.0404	0.0207	0.0101	0.0046	0.0020
	1	0.3874	0.3679	0.3020	0.2253	0.1556	0.1004	0.0605	0.0339	0.0176
	2	0.1722	0.2597	0.3020	0.3003	0.2668	0.2162	0.1612	0.1110	0.0703
	3	0.0446	0.1069	0.1762	0.2336	0.2668	0.2716	0.2508	0.2119	0.1641
	4	0.0074	0.0283	0.0661	0.1168	0.1715	0.2194	0.2508	0.2600	0.2461
	5	0.0008	0.0050	0.0165	0.0389	0.0735	0.1181	0.1672	0.2128	0.2461
	6	0.0001	0.0006	0.0028	0.0087	0.0210	0.0424	0.0743	0.1160	0.1641
	7	0.0000	0.0000	0.0003	0.0012	0.0039	0.0098	0.0212	0.0407	0.0703
	8	0.0000	0.0000	0.0000	0.0001	0.0004	0.0013	0.0035	0.0083	0.0176
	9	0.0000	0.0000	0.0000	0.0000	0.0000	0.0001	0.0003	0.0008	0.0020
10	0	0.3487	0.1969	0.1074	0.0563	0.0282	0.0135	0.0060	0.0025	0.0010
	1	0.3874	0.3474	0.2684	0.1877	0.1211	0.0725	0.0403	0.0207	0.0098
	2	0.1937	0.2759	0.3020	0.2816	0.2335	0.1757	0.1209	0.0763	0.0439
	3	0.0574	0.1298	0.2013	0.2503	0.2668	0.2522	0.2150	0.1665	0.1172
	4	0.0112	0.0401	0.0881	0.1460	0.2001	0.2377	0.2508	0.2384	0.2051
	5	0.0015	0.0085	0.0264	0.0584	0.1029	0.1536	0.2007	0.2340	0.2461
	6	0.0001	0.0012	0.0055	0.0162	0.0368	0.0689	0.1115	0.1596	0.2051
	7	0.0000	0.0001	0.0008	0.0031	0.0090	0.0212	0.0425	0.0746	0.1172
	8	0.0000	0.0000	0.0001	0.0004	0.0014	0.0043	0.0106	0.0229	0.0439
	9	0.0000	0.0000	0.0000	0.0000	0.0001	0.0005	0.0016	0.0042	0.0098
	10	0.0000	0.0000	0.0000	0.0000	0.0000	0.0000	0.0001	0.0003	0.0010
12	0	0.2824	0.1422	0.0687	0.0317	0.0138	0.0057	0.0022	0.0008	0.0002
	1	0.3766	0.3012	0.2062	0.1267	0.0712	0.0368	0.0174	0.0075	0.0029
	2	0.2301	0.2924	0.2835	0.2323	0.1678	0.1088	0.0639	0.0339	0.0161
	3	0.0852	0.1720	0.2362	0.2581	0.2397	0.1954	0.1419	0.0923	0.0537
	4	0.0213	0.0683	0.1329	0.1936	0.2311	0.2367	0.2128	0.1700	0.1208
	5	0.0038	0.0193	0.0532	0.1032	0.1585	0.2039	0.2270	0.2225	0.1934
	6	0.0005	0.0040	0.0155	0.0401	0.0792	0.1281	0.1766	0.2124	0.2256
	7	0.0000	0.0006	0.0033	0.0115	0.0291	0.0591	0.1009	0.1489	0.1934
	8	0.0000	0.0001	0.0005	0.0024	0.0078	0.0199	0.0420	0.0762	0.1208
	9	0.0000	0.0000	0.0001	0.0004	0.0015	0.0048	0.0125	0.0277	0.0537
	10	0.0000	0.0000	0.0000	0.0000	0.0002	0.0008	0.0025	0.0068	0.0161
	11	0.0000	0.0000	0.0000	0.0000	0.0000	0.0001	0.0003	0.0010	0.0029
	12	0.0000	0.0000	0.0000	0.0000	0.0000	0.0000	0.0000	0.0001	0.0002

n	x	.10	.15	.20	.25	.30	.35	.40	.45	.50
15	0	0.2059	0.0874	0.0352	0.0134	0.0047	0.0016	0.0005	0.0001	0.0000
	1	0.3432	0.2312	0.1319	0.0668	0.0305	0.0126	0.0047	0.0016	0.0005
	2	0.2669	0.2856	0.2309	0.1559	0.0916	0.0476	0.0219	0.0090	0.0032
	3	0.1285	0.2184	0.2501	0.2252	0.1700	0.1110	0.0634	0.0318	0.0139
	4	0.0428	0.1156	0.1876	0.2252	0.2186	0.1792	0.1268	0.0780	0.0417
	5	0.0105	0.0449	0.1032	0.1651	0.2061	0.2123	0.1859	0.1404	0.0916
	6	0.0019	0.0132	0.0430	0.0917	0.1472	0.1906	0.2066	0.1914	0.1527
	7	0.0003	0.0030	0.0138	0.0393	0.0811	0.1319	0.1771	0.2013	0.1964
	8	0.0000	0.0005	0.0035	0.0131	0.0348	0.0710	0.1181	0.1647	0.1964
	9	0.0000	0.0001	0.0007	0.0034	0.0116	0.0298	0.0612	0.1048	0.1527
	10	0.0000	0.0000	0.0001	0.0007	0.0030	0.0096	0.0245	0.0515	0.0916
	11	0.0000	0.0000	0.0000	0.0001	0.0006	0.0024	0.0074	0.0191	0.0417
	12	0.0000	0.0000	0.0000	0.0000	0.0001	0.0004	0.0016	0.0052	0.0139
	13	0.0000	0.0000	0.0000	0.0000	0.0000	0.0001	0.0003	0.0010	0.0032
	14	0.0000	0.0000	0.0000	0.0000	0.0000	0.0000	0.0000	0.0001	0.0005
	15	0.0000	0.0000	0.0000	0.0000	0.0000	0.0000	0.0000	0.0000	0.0000
20	0	0.1216	0.0388	0.0115	0.0032	0.0008	0.0002	0.0000	0.0000	0.0000
	1	0.2702	0.1368	0.0576	0.0211	0.0068	0.0020	0.0005	0.0001	0.0000
	2	0.2852	0.2293	0.1369	0.0669	0.0278	0.0100	0.0031	0.0008	0.0002
	3	0.1901	0.2428	0.2054	0.1339	0.0716	0.0323	0.0123	0.0040	0.0011
	4	0.0898	0.1821	0.2182	0.1897	0.1304	0.0738	0.0350	0.0139	0.0046
	5	0.0319	0.1028	0.1746	0.2023	0.1789	0.1272	0.0746	0.0365	0.0148
	6	0.0089	0.0454	0.1091	0.1686	0.1916	0.1712	0.1244	0.0746	0.0370
	7	0.0020	0.0160	0.0545	0.1124	0.1643	0.1844	0.1659	0.1221	0.0739
	8	0.0004	0.0046	0.0222	0.0609	0.1144	0.1614	0.1797	0.1623	0.1201
	9	0.0001	0.0011	0.0074	0.0271	0.0654	0.1158	0.1597	0.1771	0.1602
	10	0.0000	0.0002	0.0020	0.0099	0.0308	0.0686	0.1171	0.1593	0.1762
	11	0.0000	0.0000	0.0005	0.0030	0.0120	0.0336	0.0710	0.1185	0.1602
	12	0.0000	0.0000	0.0001	0.0008	0.0039	0.0136	0.0355	0.0727	0.1201
	13	0.0000	0.0000	0.0000	0.0002	0.0010	0.0045	0.0146	0.0366	0.0739
	14	0.0000	0.0000	0.0000	0.0000	0.0002	0.0012	0.0049	0.0150	0.0370
	15	0.0000	0.0000	0.0000	0.0000	0.0000	0.0003	0.0013	0.0049	0.0148
	16	0.0000	0.0000	0.0000	0.0000	0.0000	0.0000	0.0003	0.0013	0.0046
	17	0.0000	0.0000	0.0000	0.0000	0.0000	0.0000	0.0000	0.0002	0.0011
	18	0.0000	0.0000	0.0000	0.0000	0.0000	0.0000	0.0000	0.0000	0.0002
	19	0.0000	0.0000	0.0000	0.0000	0.0000	0.0000	0.0000	0.0000	0.0000
	20	0.0000	0.0000	0.0000	0.0000	0.0000	0.0000	0.0000	0.0000	0.0000

TABLE D.2
Poisson probabilities

Entry is probability $P(X = x) = \dfrac{\mu^x e^{-\mu}}{x!}$

					μ				
x	0.1	0.2	0.3	0.4	0.5	0.6	0.7	0.8	0.9
0.0	0.9048	0.8187	0.7408	0.6703	0.6065	0.5488	0.4966	0.4493	0.4066
1.0	0.0905	0.1637	0.2222	0.2681	0.3033	0.3293	0.3476	0.3595	0.3659
2.0	0.0045	0.0164	0.0333	0.0536	0.0758	0.0988	0.1217	0.1438	0.1647
3.0	0.0002	0.0011	0.0033	0.0072	0.0126	0.0198	0.0284	0.0383	0.0494
4.0	0.0000	0.0001	0.0003	0.0007	0.0016	0.0030	0.0050	0.0077	0.0111
5.0	0.0000	0.0000	0.0000	0.0001	0.0002	0.0004	0.0007	0.0012	0.0020
6.0	0.0000	0.0000	0.0000	0.0000	0.0000	0.0000	0.0001	0.0002	0.0003

					μ				
x	1.0	1.5	2.0	2.5	3.0	3.5	4.0	4.5	5.0
0.0	0.3679	0.2231	0.1353	0.0821	0.0498	0.0302	0.0183	0.0111	0.0067
1.0	0.3679	0.3347	0.2707	0.2052	0.1494	0.1057	0.0733	0.0500	0.0337
2.0	0.1839	0.2510	0.2707	0.2565	0.2240	0.1850	0.1465	0.1125	0.0842
3.0	0.0613	0.1255	0.1804	0.2138	0.2240	0.2158	0.1954	0.1687	0.1404
4.0	0.0153	0.0471	0.0902	0.1336	0.1680	0.1888	0.1954	0.1898	0.1755
5.0	0.0031	0.0141	0.0361	0.0668	0.1008	0.1322	0.1563	0.1708	0.1755
6.0	0.0005	0.0035	0.0120	0.0278	0.0504	0.0771	0.1042	0.1281	0.1462
7.0	0.0001	0.0008	0.0034	0.0099	0.0216	0.0385	0.0595	0.0824	0.1044
8.0	0.0000	0.0001	0.0009	0.0031	0.0081	0.0169	0.0298	0.0463	0.0653
9.0	0.0000	0.0000	0.0002	0.0009	0.0027	0.0066	0.0132	0.0232	0.0363
10.0	0.0000	0.0000	0.0000	0.0002	0.0008	0.0023	0.0053	0.0104	0.0181
11.0	0.0000	0.0000	0.0000	0.0000	0.0002	0.0007	0.0019	0.0043	0.0082
12.0	0.0000	0.0000	0.0000	0.0000	0.0001	0.0002	0.0006	0.0016	0.0034
13.0	0.0000	0.0000	0.0000	0.0000	0.0000	0.0001	0.0002	0.0006	0.0013
14.0	0.0000	0.0000	0.0000	0.0000	0.0000	0.0000	0.0001	0.0002	0.0005
15.0	0.0000	0.0000	0.0000	0.0000	0.0000	0.0000	0.0000	0.0001	0.0002

					μ				
x	5.5	6.0	6.5	7.0	7.5	8.0	9.0	10.0	11.0
0.0	0.0041	0.0025	0.0015	0.0009	0.0006	0.0003	0.0001	0.0000	0.0000
1.0	0.0225	0.0149	0.0098	0.0064	0.0041	0.0027	0.0011	0.0005	0.0002
2.0	0.0618	0.0446	0.0318	0.0223	0.0156	0.0107	0.0050	0.0023	0.0010
3.0	0.1133	0.0892	0.0688	0.0521	0.0389	0.0286	0.0150	0.0076	0.0037
4.0	0.1558	0.1339	0.1118	0.0912	0.0729	0.0573	0.0337	0.0189	0.0102
5.0	0.1714	0.1606	0.1454	0.1277	0.1094	0.0916	0.0607	0.0378	0.0224
6.0	0.1571	0.1606	0.1575	0.1490	0.1367	0.1221	0.0911	0.0631	0.0411
7.0	0.1234	0.1377	0.1462	0.1490	0.1465	0.1396	0.1171	0.0901	0.0646
8.0	0.0849	0.1033	0.1188	0.1304	0.1373	0.1396	0.1318	0.1126	0.0888
9.0	0.0519	0.0688	0.0858	0.1014	0.1144	0.1241	0.1318	0.1251	0.1085
10.0	0.0285	0.0413	0.0558	0.0710	0.0858	0.0993	0.1186	0.1251	0.1194
11.0	0.0143	0.0225	0.0330	0.0452	0.0585	0.0722	0.0970	0.1137	0.1194
12.0	0.0065	0.0113	0.0179	0.0263	0.0366	0.0481	0.0728	0.0948	0.1094
13.0	0.0028	0.0052	0.0089	0.0142	0.0211	0.0296	0.0504	0.0729	0.0926
14.0	0.0011	0.0022	0.0041	0.0071	0.0113	0.0169	0.0324	0.0521	0.0728
15.0	0.0004	0.0009	0.0018	0.0033	0.0057	0.0090	0.0194	0.0347	0.0534
16.0	0.0001	0.0003	0.0007	0.0014	0.0026	0.0045	0.0109	0.0217	0.0367
17.0	0.0000	0.0001	0.0003	0.0006	0.0012	0.0021	0.0058	0.0128	0.0237

TABLE D.2
Poisson probabilities (continued)

					μ				
x	5.5	6.0	6.5	7.0	7.5	8.0	9.0	10.0	11.0
18.0	0.0000	0.0000	0.0001	0.0002	0.0005	0.0009	0.0029	0.0071	0.0145
19.0	0.0000	0.0000	0.0000	0.0001	0.0002	0.0004	0.0014	0.0037	0.0084
20.0	0.0000	0.0000	0.0000	0.0000	0.0001	0.0002	0.0006	0.0019	0.0046
21.0	0.0000	0.0000	0.0000	0.0000	0.0000	0.0001	0.0003	0.0009	0.0024
22.0	0.0000	0.0000	0.0000	0.0000	0.0000	0.0000	0.0001	0.0004	0.0012
23.0	0.0000	0.0000	0.0000	0.0000	0.0000	0.0000	0.0000	0.0002	0.0006
24.0	0.0000	0.0000	0.0000	0.0000	0.0000	0.0000	0.0000	0.0001	0.0003
25.0	0.0000	0.0000	0.0000	0.0000	0.0000	0.0000	0.0000	0.0000	0.0001

					μ				
x	12	13	14	15	16	17	18	19	20
0.0	0.0000	0.0000	0.0000	0.0000	0.0000	0.0000	0.0000	0.0000	0.0000
1.0	0.0001	0.0000	0.0000	0.0000	0.0000	0.0000	0.0000	0.0000	0.0000
2.0	0.0004	0.0002	0.0001	0.0000	0.0000	0.0000	0.0000	0.0000	0.0000
3.0	0.0018	0.0008	0.0004	0.0002	0.0001	0.0000	0.0000	0.0000	0.0000
4.0	0.0053	0.0027	0.0013	0.0006	0.0003	0.0001	0.0001	0.0000	0.0000
5.0	0.0127	0.0070	0.0037	0.0019	0.0010	0.0005	0.0002	0.0001	0.0001
6.0	0.0255	0.0152	0.0087	0.0048	0.0026	0.0014	0.0007	0.0004	0.0002
7.0	0.0437	0.0281	0.0174	0.0104	0.0060	0.0034	0.0019	0.0010	0.0005
8.0	0.0655	0.0457	0.0304	0.0194	0.0120	0.0072	0.0042	0.0024	0.0013
9.0	0.0874	0.0661	0.0473	0.0324	0.0213	0.0135	0.0083	0.0050	0.0029
10.0	0.1048	0.0859	0.0663	0.0486	0.0341	0.0230	0.0150	0.0095	0.0058
11.0	0.1144	0.1015	0.0844	0.0663	0.0496	0.0355	0.0245	0.0164	0.0106
12.0	0.1144	0.1099	0.0984	0.0829	0.0661	0.0504	0.0368	0.0259	0.0176
13.0	0.1056	0.1099	0.1060	0.0956	0.0814	0.0658	0.0509	0.0378	0.0271
14.0	0.0905	0.1021	0.1060	0.1024	0.0930	0.0800	0.0655	0.0514	0.0387
15.0	0.0724	0.0885	0.0989	0.1024	0.0992	0.0906	0.0786	0.0650	0.0516
16.0	0.0543	0.0719	0.0866	0.0960	0.0992	0.0963	0.0884	0.0772	0.0646
17.0	0.0383	0.0550	0.0713	0.0847	0.0934	0.0963	0.0936	0.0863	0.0760
18.0	0.0255	0.0397	0.0554	0.0706	0.0830	0.0909	0.0936	0.0911	0.0844
19.0	0.0161	0.0272	0.0409	0.0557	0.0699	0.0814	0.0887	0.0911	0.0888
20.0	0.0097	0.0177	0.0286	0.0418	0.0559	0.0692	0.0798	0.0866	0.0888
21.0	0.0055	0.0109	0.0191	0.0299	0.0426	0.0560	0.0684	0.0783	0.0846
22.0	0.0030	0.0065	0.0121	0.0204	0.0310	0.0433	0.0560	0.0676	0.0769
23.0	0.0016	0.0037	0.0074	0.0133	0.0216	0.0320	0.0438	0.0559	0.0669
24.0	0.0008	0.0020	0.0043	0.0083	0.0144	0.0226	0.0328	0.0442	0.0557
25.0	0.0004	0.0010	0.0024	0.0050	0.0092	0.0154	0.0237	0.0336	0.0446
26.0	0.0002	0.0005	0.0013	0.0029	0.0057	0.0101	0.0164	0.0246	0.0343
27.0	0.0001	0.0002	0.0007	0.0016	0.0034	0.0063	0.0109	0.0173	0.0254
28.0	0.0000	0.0001	0.0003	0.0009	0.0019	0.0038	0.0070	0.0117	0.0181
29.0	0.0000	0.0001	0.0002	0.0004	0.0011	0.0023	0.0044	0.0077	0.0125
30.0	0.0000	0.0000	0.0001	0.0002	0.0006	0.0013	0.0026	0.0049	0.0083
31.0	0.0000	0.0000	0.0000	0.0001	0.0003	0.0007	0.0015	0.0030	0.0054
32.0	0.0000	0.0000	0.0000	0.0001	0.0001	0.0004	0.0009	0.0018	0.0034
33.0	0.0000	0.0000	0.0000	0.0000	0.0001	0.0002	0.0005	0.0010	0.0020
34.0	0.0000	0.0000	0.0000	0.0000	0.0000	0.0001	0.0002	0.0006	0.0012
35.0	0.0000	0.0000	0.0000	0.0000	0.0000	0.0000	0.0001	0.0003	0.0007
36.0	0.0000	0.0000	0.0000	0.0000	0.0000	0.0000	0.0001	0.0002	0.0004
37.0	0.0000	0.0000	0.0000	0.0000	0.0000	0.0000	0.0000	0.0001	0.0002
38.0	0.0000	0.0000	0.0000	0.0000	0.0000	0.0000	0.0000	0.0000	0.0001
39.0	0.0000	0.0000	0.0000	0.0000	0.0000	0.0000	0.0000	0.0000	0.0001

TABLE D.3 The hypergeometric probability distribution

The table shows combinations of ν, n, γ, and x in which $0 \le x \le \gamma \le n \le 0.5\nu$. Other values of ν, n, γ, and x can be found by use of the following identities:

$P(\nu, n, \gamma, x) = P(\nu, \gamma, n, x)$

$P(\nu, n, \gamma, x) = P(\nu, n, \nu - \gamma, n - x)$

$P(\nu, n, \gamma, x) = P(\nu, \nu - n, \gamma, \gamma - x)$

$P(\nu, n, \gamma, x) = P(\nu, \nu - n, \nu - \gamma, \nu - n - \gamma + x)$

For example, to find $P(16, 12, 10, 6)$ use the second identity to convert to $P(16, 4, 6, 0)$ and the third identity to convert to $P(16, 6, 4, 0)$.

ν	n	γ	x	$P(x)$	ν	n	γ	x	$P(x)$	ν	n	γ	x	$P(x)$	ν	n	γ	x	$P(x)$
2	1	1	0	0.5	7	2	2	2	0.0476	8	4	4	4	0.0143	10	3	2	0	0.4667
2	1	1	1	0.5	7	3	1	0	0.5714	9	1	1	0	0.8889	10	3	2	1	0.4667
3	1	1	0	0.6667	7	3	1	1	0.4286	9	1	1	1	0.1111	10	3	2	2	0.0667
3	1	1	1	0.3333	7	3	2	0	0.2857	9	2	1	0	0.7778	10	3	3	0	0.2917
4	1	1	0	0.75	7	3	2	1	0.5714	9	2	1	1	0.2222	10	3	3	1	0.525
4	1	1	1	0.25	7	3	2	2	0.1429	9	2	2	0	0.5833	10	3	3	2	0.175
4	2	1	0	0.5	7	3	3	0	0.1143	9	2	2	1	0.3889	10	3	3	3	0.0083
4	2	1	1	0.5	7	3	3	1	0.5143	9	2	2	2	0.0278	10	4	1	0	0.6
4	2	2	0	0.1667	7	3	3	2	0.3429	9	3	1	0	0.6667	10	4	1	1	0.4
4	2	2	1	0.6667	7	3	3	3	0.0286	9	3	1	1	0.3333	10	4	2	0	0.3333
4	2	2	2	0.1667	8	1	1	0	0.875	9	3	2	0	0.4167	10	4	2	1	0.5333
5	1	1	0	0.8	8	1	1	1	0.125	9	3	2	1	0.5	10	4	2	2	0.1333
5	1	1	1	0.2	8	2	1	0	0.75	9	3	2	2	0.0833	10	4	3	0	0.1667
5	2	1	0	0.6	8	2	1	1	0.25	9	3	3	0	0.2381	10	4	3	1	0.5
5	2	1	1	0.4	8	2	2	0	0.5357	9	3	3	1	0.5357	10	4	3	2	0.3
5	2	2	0	0.3	8	2	2	1	0.4286	9	3	3	2	0.2143	10	4	3	3	0.0333
5	2	2	1	0.6	8	2	2	2	0.0357	9	3	3	3	0.0119	10	4	4	0	0.0714
5	2	2	2	0.1	8	3	1	0	0.625	9	4	1	0	0.5556	10	4	4	1	0.381
6	1	1	0	0.8333	8	3	1	1	0.375	9	4	1	1	0.4444	10	4	4	2	0.4286
6	1	1	1	0.1667	8	3	2	0	0.3571	9	4	2	0	0.2778	10	4	4	3	0.1143
6	2	1	0	0.6667	8	3	2	1	0.5357	9	4	2	1	0.5556	10	4	4	4	0.0048
6	2	1	1	0.3333	8	3	2	2	0.1071	9	4	2	2	0.1667	10	5	1	0	0.5
6	2	2	0	0.4	8	3	3	0	0.1786	9	4	3	0	0.119	10	5	1	1	0.5
6	2	2	1	0.5333	8	3	3	1	0.5357	9	4	3	1	0.4762	10	5	2	0	0.2222
6	2	2	2	0.0667	8	3	3	2	0.2679	9	4	3	2	0.3571	10	5	2	1	0.5556
6	3	1	0	0.5	8	3	3	3	0.0179	9	4	3	3	0.0476	10	5	2	2	0.2222
6	3	1	1	0.5	8	4	1	0	0.5	9	4	4	0	0.0397	10	5	3	0	0.0833
6	3	2	0	0.2	8	4	1	1	0.5	9	4	4	1	0.3175	10	5	3	1	0.4167
6	3	2	1	0.6	8	4	2	0	0.2143	9	4	4	2	0.4762	10	5	3	2	0.4167
6	3	2	2	0.2	8	4	2	1	0.5714	9	4	4	3	0.1587	10	5	3	3	0.0833
6	3	3	0	0.05	8	4	2	2	0.2143	9	4	4	4	0.0079	10	5	4	0	0.0238
6	3	3	1	0.45	8	4	3	0	0.0714	10	1	1	0	0.9	10	5	4	1	0.2381
6	3	3	2	0.45	8	4	3	1	0.4286	10	1	1	1	0.1	10	5	4	2	0.4762
6	3	3	3	0.05	8	4	3	2	0.4286	10	2	1	0	0.8	10	5	4	3	0.2381
7	1	1	0	0.8571	8	4	3	3	0.0714	10	2	1	1	0.2	10	5	4	4	0.0238
7	1	1	1	0.1429	8	4	4	0	0.0143	10	2	2	0	0.6222	10	5	5	0	0.004
7	2	1	0	0.7143	8	4	4	1	0.2286	10	2	2	1	0.3556	10	5	5	1	0.0992
7	2	1	1	0.2857	8	4	4	2	0.5143	10	2	2	2	0.0222	10	5	5	2	0.3968
7	2	2	0	0.4762	8	4	4	3	0.2286	10	3	1	0	0.7	10	5	5	3	0.3968
7	2	2	1	0.4762						10	3	1	1	0.3	10	5	5	4	0.0992

TABLE D.3 The hypergeometric probability distribution (continued)

ν	n	γ	x	$P(x)$
10	5	5	5	0.004
11	1	1	0	0.9091
11	1	1	1	0.0909
11	2	1	0	0.8182
11	2	1	1	0.1818
11	2	2	0	0.6545
11	2	2	1	0.3273
11	2	2	2	0.0182
11	3	1	0	0.7273
11	3	1	1	0.2727
11	3	2	0	0.5091
11	3	2	1	0.4364
11	3	2	2	0.0545
11	3	3	0	0.3394
11	3	3	1	0.5091
11	3	3	2	0.1455
11	3	3	3	0.0061
11	4	1	0	0.6364
11	4	1	1	0.3636
11	4	2	0	0.3818
11	4	2	1	0.5091
11	4	2	2	0.1091
11	4	3	0	0.2121
11	4	3	1	0.5091
11	4	3	2	0.2545
11	4	3	3	0.0242
11	4	4	0	0.1061
11	4	4	1	0.4242
11	4	4	2	0.3818
11	4	4	3	0.0848
11	4	4	4	0.003
11	5	1	0	0.5455
11	5	1	1	0.4545
11	5	2	0	0.2727
11	5	2	1	0.5455
11	5	2	2	0.1818
11	5	3	0	0.1212
11	5	3	1	0.4545
11	5	3	2	0.3636
11	5	3	3	0.0606
11	5	4	0	0.0455
11	5	4	1	0.303
11	5	4	2	0.4545
11	5	4	3	0.1818
11	5	4	4	0.0152
11	5	5	0	0.013
11	5	5	1	0.1623
11	5	5	2	0.4329
11	5	5	3	0.3247
11	5	5	4	0.0649
11	5	5	5	0.0022
12	1	1	0	0.9167
12	1	1	1	0.0833
12	2	1	0	0.8333
12	2	1	1	0.1667
12	2	2	0	0.6818
12	2	2	1	0.303
12	2	2	2	0.0152
12	3	1	0	0.75
12	3	1	1	0.25
12	3	2	0	0.5455
12	3	2	1	0.4091
12	3	2	2	0.0455
12	3	3	0	0.3818
12	3	3	1	0.4909
12	3	3	2	0.1227
12	3	3	3	0.0045
12	4	1	0	0.6667
12	4	1	1	0.3333
12	4	2	0	0.4242
12	4	2	1	0.4848
12	4	2	2	0.0909
12	4	3	0	0.2545
12	4	3	1	0.5091
12	4	3	2	0.2182
12	4	3	3	0.0182
12	4	4	0	0.1414
12	4	4	1	0.4525
12	4	4	2	0.3394
12	4	4	3	0.0646
12	4	4	4	0.002
12	5	1	0	0.5833
12	5	1	1	0.4167
12	5	2	0	0.3182
12	5	2	1	0.5303
12	5	2	2	0.1515
12	5	3	0	0.1591
12	5	3	1	0.4773
12	5	3	2	0.3182
12	5	3	3	0.0455
12	5	4	0	0.0707
12	5	4	1	0.3535
12	5	4	2	0.4242
12	5	4	3	0.1414
12	5	4	4	0.0101
12	5	5	0	0.0265
12	5	5	1	0.221
12	5	5	2	0.4419
12	5	5	3	0.2652
12	5	5	4	0.0442
12	5	5	5	0.0013
12	6	1	0	0.5
12	6	1	1	0.5
12	6	2	0	0.2273
12	6	2	1	0.5455
12	6	2	2	0.2273
12	6	3	0	0.0909
12	6	3	1	0.4091
12	6	3	2	0.4091
12	6	3	3	0.0909
12	6	4	0	0.0303
12	6	4	1	0.2424
12	6	4	2	0.4545
12	6	4	3	0.2424
12	6	4	4	0.0303
12	6	5	0	0.0076
12	6	5	1	0.1136
12	6	5	2	0.3788
12	6	5	3	0.3788
12	6	5	4	0.1136
12	6	5	5	0.0076
12	6	6	0	0.0011
12	6	6	1	0.039
12	6	6	2	0.2435
12	6	6	3	0.4329
12	6	6	4	0.2435
12	6	6	5	0.039
12	6	6	6	0.0011
13	1	1	0	0.9231
13	1	1	1	0.0769
13	2	1	0	0.8462
13	2	1	1	0.1538
13	2	2	0	0.7051
13	2	2	1	0.2821
13	2	2	2	0.0128
13	3	1	0	0.7692
13	3	1	1	0.2308
13	3	2	0	0.5769
13	3	2	1	0.3846
13	3	2	2	0.0385
13	3	3	0	0.4196
13	3	3	1	0.472
13	3	3	2	0.1049
13	3	3	3	0.0035
13	4	1	0	0.6923
13	4	1	1	0.3077
13	4	2	0	0.4615
13	4	2	1	0.4615
13	4	2	2	0.0769
13	4	3	0	0.2937
13	4	3	1	0.5035
13	4	3	2	0.1888
13	4	3	3	0.014
13	4	4	0	0.1762
13	4	4	1	0.4699
13	4	4	2	0.3021
13	4	4	3	0.0503
13	4	4	4	0.0014
13	5	1	0	0.6154
13	5	1	1	0.3846
13	5	2	0	0.359
13	5	2	1	0.5128
13	5	2	2	0.1282
13	5	3	0	0.1958
13	5	3	1	0.4895
13	5	3	2	0.2797
13	5	3	3	0.035
13	5	4	0	0.0979
13	5	4	1	0.3916
13	5	4	2	0.3916
13	5	4	3	0.1119
13	5	4	4	0.007
13	5	5	0	0.0435
13	5	5	1	0.272
13	5	5	2	0.4351
13	5	5	3	0.2176
13	5	5	4	0.0311
13	5	5	5	0.0008
13	6	1	0	0.5385
13	6	1	1	0.4615
13	6	2	0	0.2692
13	6	2	1	0.5385
13	6	2	2	0.1923
13	6	3	0	0.1224
13	6	3	1	0.4406
13	6	3	2	0.3671
13	6	3	3	0.0699
13	6	4	0	0.049
13	6	4	1	0.2937
13	6	4	2	0.4406
13	6	4	3	0.1958
13	6	4	4	0.021
13	6	5	0	0.0163
13	6	5	1	0.1632
13	6	5	2	0.4079
13	6	5	3	0.3263
13	6	5	4	0.0816
13	6	5	5	0.0047
13	6	6	0	0.0041
13	6	6	1	0.0734
13	6	6	2	0.3059
13	6	6	3	0.4079
13	6	6	4	0.1836
13	6	6	5	0.0245
13	6	6	6	0.0006
14	1	1	0	0.9286
14	1	1	1	0.0714

(continued)

ν	n	γ	x	P(x)
14	2	1	0	0.8571
14	2	1	1	0.1429
14	2	2	0	0.7253
14	2	2	1	0.2637
14	2	2	2	0.011
14	3	1	0	0.7857
14	3	1	1	0.2143
14	3	2	0	0.6044
14	3	2	1	0.3626
14	3	2	2	0.033
14	3	3	0	0.4533
14	3	3	1	0.4533
14	3	3	2	0.0907
14	3	3	3	0.0027
14	4	1	0	0.7143
14	4	1	1	0.2857
14	4	2	0	0.4945
14	4	2	1	0.4396
14	4	2	2	0.0659
14	4	3	0	0.3297
14	4	3	1	0.4945
14	4	3	2	0.1648
14	4	3	3	0.011
14	4	4	0	0.2098
14	4	4	1	0.4795
14	4	4	2	0.2697
14	4	4	3	0.04
14	4	4	4	0.001
14	5	1	0	0.6429
14	5	1	1	0.3571
14	5	2	0	0.3956
14	5	2	1	0.4945
14	5	2	2	0.1099
14	5	3	0	0.2308
14	5	3	1	0.4945
14	5	3	2	0.2473
14	5	3	3	0.0275
14	5	4	0	0.1259
14	5	4	1	0.4196
14	5	4	2	0.3596
14	5	4	3	0.0899
14	5	4	4	0.005
14	5	5	0	0.0629
14	5	5	1	0.3147
14	5	5	2	0.4196
14	5	5	3	0.1798
14	5	5	4	0.0225
14	5	5	5	0.0005
14	6	1	0	0.5714
14	6	1	1	0.4286
14	6	2	0	0.3077
14	6	2	1	0.5275
14	6	2	2	0.1648
14	6	3	0	0.1538
14	6	3	1	0.4615
14	6	3	2	0.3297
14	6	3	3	0.0549
14	6	4	0	0.0699
14	6	4	1	0.3357
14	6	4	2	0.4196
14	6	4	3	0.1598
14	6	4	4	0.015
14	6	5	0	0.028
14	6	5	1	0.2098
14	6	5	2	0.4196
14	6	5	3	0.2797
14	6	5	4	0.0599
14	6	5	5	0.003
14	6	6	0	0.0093
14	6	6	1	0.1119
14	6	6	2	0.3497
14	6	6	3	0.373
14	6	6	4	0.1399
14	6	6	5	0.016
14	6	6	6	0.0003
14	7	1	0	0.5
14	7	1	1	0.5
14	7	2	0	0.2308
14	7	2	1	0.5385
14	7	2	2	0.2308
14	7	3	0	0.0962
14	7	3	1	0.4038
14	7	3	2	0.4038
14	7	3	3	0.0962
14	7	4	0	0.035
14	7	4	1	0.2448
14	7	4	2	0.4406
14	7	4	3	0.2448
14	7	4	4	0.035
14	7	5	0	0.0105
14	7	5	1	0.1224
14	7	5	2	0.3671
14	7	5	3	0.3671
14	7	5	4	0.1224
14	7	5	5	0.0105
14	7	6	0	0.0023
14	7	6	1	0.049
14	7	6	2	0.2448
14	7	6	3	0.4079
14	7	6	4	0.2448
14	7	6	5	0.049
14	7	6	6	0.0023
14	7	7	0	0.0003
14	7	7	1	0.0143
14	7	7	2	0.1285
14	7	7	3	0.3569
14	7	7	4	0.3569
14	7	7	5	0.1285
14	7	7	6	0.0143
14	7	7	7	0.0003
15	1	1	0	0.9333
15	1	1	1	0.0667
15	2	1	0	0.8667
15	2	1	1	0.1333
15	2	2	0	0.7429
15	2	2	1	0.2476
15	2	2	2	0.0095
15	3	1	0	0.8
15	3	1	1	0.2
15	3	2	0	0.6286
15	3	2	1	0.3429
15	3	2	2	0.0286
15	3	3	0	0.4835
15	3	3	1	0.4352
15	3	3	2	0.0791
15	3	3	3	0.0022
15	4	1	0	0.7333
15	4	1	1	0.2667
15	4	2	0	0.5238
15	4	2	1	0.419
15	4	2	2	0.0571
15	4	3	0	0.3626
15	4	3	1	0.4835
15	4	3	2	0.1451
15	4	3	3	0.0088
15	4	4	0	0.2418
15	4	4	1	0.4835
15	4	4	2	0.2418
15	4	4	3	0.0322
15	4	4	4	0.0007
15	5	1	0	0.6667
15	5	1	1	0.3333
15	5	2	0	0.4286
15	5	2	1	0.4762
15	5	2	2	0.0952
15	5	3	0	0.2637
15	5	3	1	0.4945
15	5	3	2	0.2198
15	5	3	3	0.022
15	5	4	0	0.1538
15	5	4	1	0.4396
15	5	4	2	0.3297
15	5	4	3	0.0733
15	5	4	4	0.0037
15	5	5	0	0.0839
15	5	5	1	0.3497
15	5	5	2	0.3996
15	5	5	3	0.1499
15	5	5	4	0.0167
15	5	5	5	0.0003
15	6	1	0	0.6
15	6	1	1	0.4
15	6	2	0	0.3429
15	6	2	1	0.5143
15	6	2	2	0.1429
15	6	3	0	0.1846
15	6	3	1	0.4747
15	6	3	2	0.2967
15	6	3	3	0.044
15	6	4	0	0.0923
15	6	4	1	0.3692
15	6	4	2	0.3956
15	6	4	3	0.1319
15	6	4	4	0.011
15	6	5	0	0.042
15	6	5	1	0.2517
15	6	5	2	0.4196
15	6	5	3	0.2398
15	6	5	4	0.045
15	6	5	5	0.002
15	6	6	0	0.0168
15	6	6	1	0.151
15	6	6	2	0.3776
15	6	6	3	0.3357
15	6	6	4	0.1079
15	6	6	5	0.0108
15	6	6	6	0.0002
15	7	1	0	0.5333
15	7	1	1	0.4667
15	7	2	0	0.2667
15	7	2	1	0.5333
15	7	2	2	0.2
15	7	3	0	0.1231
15	7	3	1	0.4308
15	7	3	2	0.3692
15	7	3	3	0.0769
15	7	4	0	0.0513
15	7	4	1	0.2872
15	7	4	2	0.4308
15	7	4	3	0.2051
15	7	4	4	0.0256
15	7	5	0	0.0186
15	7	5	1	0.1632
15	7	5	2	0.3916
15	7	5	3	0.3263
15	7	5	4	0.0932
15	7	5	5	0.007
15	7	6	0	0.0056
15	7	6	1	0.0783
15	7	6	2	0.2937
15	7	6	3	0.3916

TABLE D.3 The hypergeometric probability distribution (continued)

ν	n	γ	x	P(x)	ν	n	γ	x	P(x)	ν	n	γ	x	P(x)	ν	n	γ	x	P(x)
15	7	6	4	0.1958	16	5	4	2	0.3022	16	7	5	2	0.4038	16	8	7	6	0.0196
15	7	6	5	0.0336	16	5	4	3	0.0604	16	7	5	3	0.2885	16	8	7	7	0.0007
15	7	6	6	0.0014	16	5	4	4	0.0027	16	7	5	4	0.0721	16	8	8	0	0.0001
15	7	7	0	0.0012	16	5	5	0	0.1058	16	7	5	5	0.0048	16	8	8	1	0.005
15	7	7	1	0.0305	16	5	5	1	0.3777	16	7	6	0	0.0105	16	8	8	2	0.0609
15	7	7	2	0.1828	16	5	5	2	0.3777	16	7	6	1	0.1101	16	8	8	3	0.2437
15	7	7	3	0.3807	16	5	5	3	0.1259	16	7	6	2	0.3304	16	8	8	4	0.3807
15	7	7	4	0.3046	16	5	5	4	0.0126	16	7	6	3	0.3671	16	8	8	5	0.2437
15	7	7	5	0.0914	16	5	5	5	0.0002	16	7	6	4	0.1573	16	8	8	6	0.0609
15	7	7	6	0.0087	16	6	1	0	0.625	16	7	6	5	0.0236	16	8	8	7	0.005
15	7	7	7	0.0002	16	6	1	1	0.375	16	7	6	6	0.0009	16	8	8	8	0.0001
16	1	1	0	0.9375	16	6	2	0	0.375	16	7	7	0	0.0031	17	1	1	0	0.9412
16	1	1	1	0.0625	16	6	2	1	0.5	16	7	7	1	0.0514	17	1	1	1	0.0588
16	2	1	0	0.875	16	6	2	2	0.125	16	7	7	2	0.2313	17	2	1	0	0.8824
16	2	1	1	0.125	16	6	3	0	0.2143	16	7	7	3	0.3855	17	2	1	1	0.1176
16	2	2	0	0.7583	16	6	3	1	0.4821	16	7	7	4	0.257	17	2	2	0	0.7721
16	2	2	1	0.2333	16	6	3	2	0.2679	16	7	7	5	0.0661	17	2	2	1	0.2206
16	2	2	2	0.0083	16	6	3	3	0.0357	16	7	7	6	0.0055	17	2	2	2	0.0074
16	3	1	0	0.8125	16	6	4	0	0.1154	16	7	7	7	0.0001	17	3	1	0	0.8235
16	3	1	1	0.1875	16	6	4	1	0.3956	16	8	1	0	0.5	17	3	1	1	0.1765
16	3	2	0	0.65	16	6	4	2	0.3709	16	8	1	1	0.5	17	3	2	0	0.6691
16	3	2	1	0.325	16	6	4	3	0.1099	16	8	2	0	0.2333	17	3	2	1	0.3088
16	3	2	2	0.025	16	6	4	4	0.0082	16	8	2	1	0.5333	17	3	2	2	0.0221
16	3	3	0	0.5107	16	6	5	0	0.0577	16	8	2	2	0.2333	17	3	3	0	0.5353
16	3	3	1	0.4179	16	6	5	1	0.2885	16	8	3	0	0.1	17	3	3	1	0.4015
16	3	3	2	0.0696	16	6	5	2	0.4121	16	8	3	1	0.4	17	3	3	2	0.0618
16	3	3	3	0.0018	16	6	5	3	0.206	16	8	3	2	0.4	17	3	3	3	0.0015
16	4	1	0	0.75	16	6	5	4	0.0343	16	8	3	3	0.1	17	4	1	0	0.7647
16	4	1	1	0.25	16	6	5	5	0.0014	16	8	4	0	0.0385	17	4	1	1	0.2353
16	4	2	0	0.55	16	6	6	0	0.0262	16	8	4	1	0.2462	17	4	2	0	0.5735
16	4	2	1	0.4	16	6	6	1	0.1888	16	8	4	2	0.4308	17	4	2	1	0.3824
16	4	2	2	0.05	16	6	6	2	0.3934	16	8	4	3	0.2462	17	4	2	2	0.0441
16	4	3	0	0.3929	16	6	6	3	0.2997	16	8	4	4	0.0385	17	4	3	0	0.4206
16	4	3	1	0.4714	16	6	6	4	0.0843	16	8	5	0	0.0128	17	4	3	1	0.4588
16	4	3	2	0.1286	16	6	6	5	0.0075	16	8	5	1	0.1282	17	4	3	2	0.1147
16	4	3	3	0.0071	16	6	6	6	0.0001	16	8	5	2	0.359	17	4	3	3	0.0059
16	4	4	0	0.272	16	7	1	0	0.5625	16	8	5	3	0.359	17	4	4	0	0.3004
16	4	4	1	0.4835	16	7	1	1	0.4375	16	8	5	4	0.1282	17	4	4	1	0.4807
16	4	4	2	0.2176	16	7	2	0	0.3	16	8	5	5	0.0128	17	4	4	2	0.1966
16	4	4	3	0.0264	16	7	2	1	0.525	16	8	6	0	0.0035	17	4	4	3	0.0218
16	4	4	4	0.0005	16	7	2	2	0.175	16	8	6	1	0.0559	17	4	4	4	0.0004
16	5	1	0	0.6875	16	7	3	0	0.15	16	8	6	2	0.2448	17	5	1	0	0.7059
16	5	1	1	0.3125	16	7	3	1	0.45	16	8	6	3	0.3916	17	5	1	1	0.2941
16	5	2	0	0.4583	16	7	3	2	0.3375	16	8	6	4	0.2448	17	5	2	0	0.4853
16	5	2	1	0.4583	16	7	3	3	0.0625	16	8	6	5	0.0559	17	5	2	1	0.4412
16	5	2	2	0.0833	16	7	4	0	0.0692	16	8	6	6	0.0035	17	5	2	2	0.0735
16	5	3	0	0.2946	16	7	4	1	0.3231	16	8	7	0	0.0007	17	5	3	0	0.3235
16	5	3	1	0.4911	16	7	4	2	0.4154	16	8	7	1	0.0196	17	5	3	1	0.4853
16	5	3	2	0.1964	16	7	4	3	0.1731	16	8	7	2	0.1371	17	5	3	2	0.1765
16	5	3	3	0.0179	16	7	4	4	0.0192	16	8	7	3	0.3427	17	5	3	3	0.0147
16	5	4	0	0.1813	16	7	5	0	0.0288	16	8	7	4	0.3427	17	5	4	0	0.208
16	5	4	1	0.4533	16	7	5	1	0.2019	16	8	7	5	0.1371					

(continued)

TABLE D.3 The hypergeometric probability distribution (continued)

ν	n	γ	x	P(x)	ν	n	γ	x	P(x)	ν	n	γ	x	P(x)	ν	n	γ	x	P(x)
17	5	4	1	0.4622	17	7	5	1	0.2376	17	8	7	5	0.1037	18	5	4	0	0.2337
17	5	4	2	0.2773	17	7	5	2	0.4072						18	5	4	1	0.4673
17	5	4	3	0.0504	17	7	5	3	0.2545	17	8	7	6	0.013	18	5	4	2	0.2549
17	5	4	4	0.0021						17	8	7	7	0.0004	18	5	4	3	0.0425
17	5	5	0	0.128	17	7	5	4	0.0566	17	8	8	0	0.0004	18	5	4	4	0.0016
					17	7	5	5	0.0034	17	8	8	1	0.0118					
17	5	5	1	0.4	17	7	6	0	0.017	17	8	8	2	0.0968	18	5	5	0	0.1502
17	5	5	2	0.3555	17	7	6	1	0.1425						18	5	5	1	0.4173
17	5	5	3	0.1067	17	7	6	2	0.3563	17	8	8	3	0.2903	18	5	5	2	0.3338
17	5	5	4	0.0097						17	8	8	4	0.3628	18	5	5	3	0.091
17	5	5	5	0.0002	17	7	6	3	0.3394	17	8	8	5	0.1935	18	5	5	4	0.0076
					17	7	6	4	0.1273	17	8	8	6	0.0415					
17	6	1	0	0.6471	17	7	6	5	0.017	17	8	8	7	0.003	18	5	5	5	0.0001
17	6	1	1	0.3529	17	7	6	6	0.0006						18	6	1	0	0.6667
17	6	2	0	0.4044	17	7	7	0	0.0062	17	8	8	8	0	18	6	1	1	0.3333
17	6	2	1	0.4853						18	1	1	0	0.9444	18	6	2	0	0.4314
17	6	2	2	0.1103	17	7	7	1	0.0756	18	1	1	1	0.0556	18	6	2	1	0.4706
					17	7	7	2	0.2721	18	2	1	0	0.8889					
17	6	3	0	0.2426	17	7	7	3	0.3779	18	2	1	1	0.1111	18	6	2	2	0.098
17	6	3	1	0.4853	17	7	7	4	0.216						18	6	3	0	0.2696
17	6	3	2	0.2426	17	7	7	5	0.0486	18	2	2	0	0.7843	18	6	3	1	0.4853
17	6	3	3	0.0294						18	2	2	1	0.2092	18	6	3	2	0.2206
17	6	4	0	0.1387	17	7	7	6	0.0036	18	2	2	2	0.0065	18	6	3	3	0.0245
					17	7	7	7	0.0001	18	3	1	0	0.8333					
17	6	4	1	0.416	17	8	1	0	0.5294	18	3	1	1	0.1667	18	6	4	0	0.1618
17	6	4	2	0.3466	17	8	1	1	0.4706						18	6	4	1	0.4314
17	6	4	3	0.0924	17	8	2	0	0.2647	18	3	2	0	0.6863	18	6	4	2	0.3235
17	6	4	4	0.0063						18	3	2	1	0.2941	18	6	4	3	0.0784
17	6	5	0	0.0747	17	8	2	1	0.5294	18	3	2	2	0.0196	18	6	4	4	0.0049
					17	8	2	2	0.2059	18	3	3	0	0.5576					
17	6	5	1	0.32	17	8	3	0	0.1235	18	3	3	1	0.386	18	6	5	0	0.0924
17	6	5	2	0.4	17	8	3	1	0.4235						18	6	5	1	0.3466
17	6	5	3	0.1778	17	8	3	2	0.3706	18	3	3	2	0.0551	18	6	5	2	0.3852
17	6	5	4	0.0267						18	3	3	3	0.0012	18	6	5	3	0.1541
17	6	5	5	0.001	17	8	3	3	0.0824	18	4	1	0	0.7778	18	6	5	4	0.021
					17	8	4	0	0.0529	18	4	1	1	0.2222					
17	6	6	0	0.0373	17	8	4	1	0.2824	18	4	2	0	0.5948	18	6	5	5	0.0007
17	6	6	1	0.224	17	8	4	2	0.4235						18	6	6	0	0.0498
17	6	6	2	0.4	17	8	4	3	0.2118	18	4	2	1	0.366	18	6	6	1	0.256
17	6	6	3	0.2666						18	4	2	2	0.0392	18	6	6	2	0.4
17	6	6	4	0.0667	17	8	4	4	0.0294	18	4	3	0	0.4461	18	6	6	3	0.237
					17	8	5	0	0.0204	18	4	3	1	0.4461					
17	6	6	5	0.0053	17	8	5	1	0.1629	18	4	3	2	0.1029	18	6	6	4	0.0533
17	6	6	6	0.0001	17	8	5	2	0.3801						18	6	6	5	0.0039
17	7	1	0	0.5882	17	8	5	3	0.3258	18	4	3	3	0.0049	18	6	6	6	0.0001
17	7	1	1	0.4118						18	4	4	0	0.3271	18	7	1	0	0.6111
17	7	2	0	0.3309	17	8	5	4	0.1018	18	4	4	1	0.4758	18	7	1	1	0.3889
					17	8	5	5	0.009	18	4	4	2	0.1784					
17	7	2	1	0.5147	17	8	6	0	0.0068	18	4	4	3	0.0183	18	7	2	0	0.3595
17	7	2	2	0.1544	17	8	6	1	0.0814						18	7	2	1	0.5033
17	7	3	0	0.1765	17	8	6	2	0.2851	18	4	4	4	0.0003	18	7	2	2	0.1373
17	7	3	1	0.4632						18	5	1	0	0.7222	18	7	3	0	0.2022
17	7	3	2	0.3088	17	8	6	3	0.3801	18	5	1	1	0.2778	18	7	3	1	0.4718
					17	8	6	4	0.2036	18	5	2	0	0.5098					
17	7	3	3	0.0515	17	8	6	5	0.0407	18	5	2	1	0.4248	18	7	3	2	0.2831
17	7	4	0	0.0882	17	8	6	6	0.0023						18	7	3	3	0.0429
17	7	4	1	0.3529	17	8	7	0	0.0019	18	5	2	2	0.0654	18	7	4	0	0.1078
17	7	4	2	0.3971						18	5	3	0	0.3505	18	7	4	1	0.3775
17	7	4	3	0.1471	17	8	7	1	0.0346	18	5	3	1	0.4779	18	7	4	2	0.3775
					17	8	7	2	0.1814	18	5	3	2	0.1593					
17	7	4	4	0.0147	17	8	7	3	0.3628	18	5	3	3	0.0123	18	7	4	3	0.1258
17	7	5	0	0.0407	17	8	7	4	0.3023						18	7	4	4	0.0114

TABLE D.3 The hypergeometric probability distribution (continued)

ν	n	γ	x	$P(x)$
18	7	5	0	0.0539
18	7	5	1	0.2696
18	7	5	2	0.4044
18	7	5	3	0.2247
18	7	5	4	0.0449
18	7	5	5	0.0025
18	7	6	0	0.0249
18	7	6	1	0.1742
18	7	6	2	0.3733
18	7	6	3	0.3111
18	7	6	4	0.1037
18	7	6	5	0.0124
18	7	6	6	0.0004
18	7	7	0	0.0104
18	7	7	1	0.1016
18	7	7	2	0.3049
18	7	7	3	0.3629
18	7	7	4	0.1815
18	7	7	5	0.0363
18	7	7	6	0.0024
18	7	7	7	0
18	8	1	0	0.5556
18	8	1	1	0.4444
18	8	2	0	0.2941
18	8	2	1	0.5229
18	8	2	2	0.183
18	8	3	0	0.1471
18	8	3	1	0.4412
18	8	3	2	0.3431
18	8	3	3	0.0686
18	8	4	0	0.0686
18	8	4	1	0.3137
18	8	4	2	0.4118
18	8	4	3	0.183
18	8	4	4	0.0229
18	8	5	0	0.0294
18	8	5	1	0.1961
18	8	5	2	0.3922
18	8	5	3	0.2941
18	8	5	4	0.0817
18	8	5	5	0.0065
18	8	6	0	0.0113
18	8	6	1	0.1086
18	8	6	2	0.3167
18	8	6	3	0.362
18	8	6	4	0.1697
18	8	6	5	0.0302
18	8	6	6	0.0015
18	8	7	0	0.0038
18	8	7	1	0.0528
18	8	7	2	0.2217
18	8	7	3	0.3695
18	8	7	4	0.264
18	8	7	5	0.0792
18	8	7	6	0.0088
18	8	7	7	0.0003
18	8	8	0	0.001
18	8	8	1	0.0219
18	8	8	2	0.1344
18	8	8	3	0.3225
18	8	8	4	0.3359
18	8	8	5	0.1536
18	8	8	6	0.0288
18	8	8	7	0.0018
18	8	8	8	0
18	9	1	0	0.5
18	9	1	1	0.5
18	9	2	0	0.2353
18	9	2	1	0.5294
18	9	2	2	0.2353
18	9	3	0	0.1029
18	9	3	1	0.3971
18	9	3	2	0.3971
18	9	3	3	0.1029
18	9	4	0	0.0412
18	9	4	1	0.2471
18	9	4	2	0.4235
18	9	4	3	0.2471
18	9	4	4	0.0412
18	9	5	0	0.0147
18	9	5	1	0.1324
18	9	5	2	0.3529
18	9	5	3	0.3529
18	9	5	4	0.1324
18	9	5	5	0.0147
18	9	6	0	0.0045
18	9	6	1	0.0611
18	9	6	2	0.2443
18	9	6	3	0.3801
18	9	6	4	0.2443
18	9	6	5	0.0611
18	9	6	6	0.0045
18	9	7	0	0.0011
18	9	7	1	0.0238
18	9	7	2	0.1425
18	9	7	3	0.3326
18	9	7	4	0.3326
18	9	7	5	0.1425
18	9	7	6	0.0238
18	9	7	7	0.0011
18	9	8	0	0.0002
18	9	8	1	0.0074
18	9	8	2	0.0691
18	9	8	3	0.2419
18	9	8	4	0.3628
18	9	8	5	0.2419
18	9	8	6	0.0691
18	9	8	7	0.0074
18	9	8	8	0.0002
18	9	9	0	0
18	9	9	1	0.0017
18	9	9	2	0.0267
18	9	9	3	0.1451
18	9	9	4	0.3265
18	9	9	5	0.3265
18	9	9	6	0.1451
18	9	9	7	0.0267
18	9	9	8	0.0017
18	9	9	9	0
19	1	1	0	0.9474
19	1	1	1	0.0526
19	2	1	0	0.8947
19	2	1	1	0.1053
19	2	2	0	0.7953
19	2	2	1	0.1988
19	2	2	2	0.0058
19	3	1	0	0.8421
19	3	1	1	0.1579
19	3	2	0	0.7018
19	3	2	1	0.2807
19	3	2	2	0.0175
19	3	3	0	0.5779
19	3	3	1	0.3715
19	3	3	2	0.0495
19	3	3	3	0.001
19	4	1	0	0.7895
19	4	1	1	0.2105
19	4	2	0	0.614
19	4	2	1	0.3509
19	4	2	2	0.0351
19	4	3	0	0.4696
19	4	3	1	0.4334
19	4	3	2	0.0929
19	4	3	3	0.0041
19	4	4	0	0.3522
19	4	4	1	0.4696
19	4	4	2	0.1625
19	4	4	3	0.0155
19	4	4	4	0.0003
19	5	1	0	0.7368
19	5	1	1	0.2632
19	5	2	0	0.5322
19	5	2	1	0.4094
19	5	2	2	0.0585
19	5	3	0	0.3756
19	5	3	1	0.4696
19	5	3	2	0.1445
19	5	3	3	0.0103
19	5	4	0	0.2583
19	5	4	1	0.4696
19	5	4	2	0.2348
19	5	4	3	0.0361
19	5	4	4	0.0013
19	5	5	0	0.1722
19	5	5	1	0.4304
19	5	5	2	0.313
19	5	5	3	0.0783
19	5	5	4	0.006
19	5	5	5	0.0001
19	6	1	0	0.6842
19	6	1	1	0.3158
19	6	2	0	0.4561
19	6	2	1	0.4561
19	6	2	2	0.0877
19	6	3	0	0.2951
19	6	3	1	0.483
19	6	3	2	0.2012
19	6	3	3	0.0206
19	6	4	0	0.1845
19	6	4	1	0.4427
19	6	4	2	0.3019
19	6	4	3	0.0671
19	6	4	4	0.0039
19	6	5	0	0.1107
19	6	5	1	0.3689
19	6	5	2	0.3689
19	6	5	3	0.1342
19	6	5	4	0.0168
19	6	5	5	0.0005
19	6	6	0	0.0632
19	6	6	1	0.2846
19	6	6	2	0.3953
19	6	6	3	0.2108
19	6	6	4	0.0431
19	6	6	5	0.0029
19	6	6	6	0
19	7	1	0	0.6316
19	7	1	1	0.3684
19	7	2	0	0.386
19	7	2	1	0.4912
19	7	2	2	0.1228
19	7	3	0	0.227
19	7	3	1	0.4768
19	7	3	2	0.2601
19	7	3	3	0.0361
19	7	4	0	0.1277
19	7	4	1	0.3973

(continued)

ν	n	γ	x	P(x)	ν	n	γ	x	P(x)	ν	n	γ	x	P(x)	ν	n	γ	x	P(x)
19	7	4	2	0.3576	19	8	7	1	0.0734	19	9	8	2	0.1	20	5	3	0	0.3991
19	7	4	3	0.1084	19	8	7	2	0.2567						20	5	3	1	0.4605
19	7	4	4	0.009	19	8	7	3	0.3668	19	9	8	3	0.2801	20	5	3	2	0.1316
19	7	5	0	0.0681	19	8	7	4	0.2292	19	9	8	4	0.3501	20	5	3	3	0.0088
19	7	5	1	0.298						19	9	8	5	0.2					
19	7	5	2	0.3973	19	8	7	5	0.0611	19	9	8	6	0.05	20	5	4	0	0.2817
					19	8	7	6	0.0061	19	9	8	7	0.0048	20	5	4	1	0.4696
19	7	5	3	0.1987	19	8	7	7	0.0002						20	5	4	2	0.2167
19	7	5	4	0.0361	19	8	8	0	0.0022	19	9	8	8	0.0001	20	5	4	3	0.031
19	7	5	5	0.0018	19	8	8	1	0.0349	19	9	9	0	0.0001	20	5	4	4	0.001
19	7	6	0	0.0341						19	9	9	1	0.0044					
19	7	6	1	0.2043	19	8	8	2	0.1712	19	9	9	2	0.0468	20	5	5	0	0.1937
					19	8	8	3	0.3423	19	9	9	3	0.191	20	5	5	1	0.4402
19	7	6	2	0.3831	19	8	8	4	0.3056						20	5	5	2	0.2935
19	7	6	3	0.2838	19	8	8	5	0.1223	19	9	9	4	0.3437	20	5	5	3	0.0677
19	7	6	4	0.0851	19	8	8	6	0.0204	19	9	9	5	0.2864	20	5	5	4	0.0048
19	7	6	5	0.0093						19	9	9	6	0.1091					
19	7	6	6	0.0003	19	8	8	7	0.0012	19	9	9	7	0.0175	20	5	5	5	0.0001
					19	8	8	8	0	19	9	9	8	0.001	20	6	1	0	0.7
19	7	7	0	0.0157	19	9	1	0	0.5263	19	9	9	9	0	20	6	1	1	0.3
19	7	7	1	0.1284	19	9	1	1	0.4737						20	6	2	0	0.4789
19	7	7	2	0.3301	19	9	2	0	0.2632	19	9	9	9	0	20	6	2	1	0.4421
19	7	7	3	0.3438						20	1	1	0	0.95					
19	7	7	4	0.1528	19	9	2	1	0.5263	20	1	1	1	0.05	20	6	2	2	0.0789
					19	9	2	2	0.2105	20	2	1	0	0.9	20	6	3	0	0.3193
19	7	7	5	0.0275	19	9	3	0	0.1238	20	2	1	1	0.1	20	6	3	1	0.4789
19	7	7	6	0.0017	19	9	3	1	0.418						20	6	3	2	0.1842
19	7	7	7	0	19	9	3	2	0.3715	20	2	2	0	0.8053	20	6	3	3	0.0175
19	8	1	0	0.5789						20	2	2	1	0.1895					
19	8	1	1	0.4211	19	9	3	3	0.0867	20	2	2	2	0.0053	20	6	4	0	0.2066
					19	9	4	0	0.0542	20	3	1	0	0.85	20	6	4	1	0.4508
19	8	2	0	0.3216	19	9	4	1	0.2786	20	3	1	1	0.15	20	6	4	2	0.2817
19	8	2	1	0.5146	19	9	4	2	0.418						20	6	4	3	0.0578
19	8	2	2	0.1637	19	9	4	3	0.2167	20	3	2	0	0.7158	20	6	4	4	0.0031
19	8	3	0	0.1703						20	3	2	1	0.2684					
19	8	3	1	0.4541	19	9	4	4	0.0325	20	3	2	2	0.0158	20	6	5	0	0.1291
					19	9	5	0	0.0217	20	3	3	0	0.5965	20	6	5	1	0.3874
19	8	3	2	0.3179	19	9	5	1	0.1625	20	3	3	1	0.3579	20	6	5	2	0.3522
19	8	3	3	0.0578	19	9	5	2	0.3715						20	6	5	3	0.1174
19	8	4	0	0.0851	19	9	5	3	0.3251	20	3	3	2	0.0447	20	6	5	4	0.0135
19	8	4	1	0.3406						20	3	3	3	0.0009					
19	8	4	2	0.3973	19	9	5	4	0.1084	20	4	1	0	0.8	20	6	5	5	0.0004
					19	9	5	5	0.0108	20	4	1	1	0.2	20	6	6	0	0.0775
19	8	4	3	0.1589	19	9	6	0	0.0077	20	4	2	0	0.6316	20	6	6	1	0.3099
19	8	4	4	0.0181	19	9	6	1	0.0836						20	6	6	2	0.3874
19	8	5	0	0.0397	19	9	6	2	0.2786	20	4	2	1	0.3368	20	6	6	3	0.1878
19	8	5	1	0.227						20	4	2	2	0.0316					
19	8	5	2	0.3973	19	9	6	3	0.3715	20	4	3	0	0.4912	20	6	6	4	0.0352
					19	9	6	4	0.209	20	4	3	1	0.4211	20	6	6	5	0.0022
19	8	5	3	0.2649	19	9	6	5	0.0464	20	4	3	2	0.0842	20	6	6	6	0
19	8	5	4	0.0662	19	9	6	6	0.0031						20	7	1	0	0.65
19	8	5	5	0.0048	19	9	7	0	0.0024	20	4	3	3	0.0035	20	7	1	1	0.35
19	8	6	0	0.017						20	4	4	0	0.3756					
19	8	6	1	0.1362	19	9	7	1	0.0375	20	4	4	1	0.4623	20	7	2	0	0.4105
					19	9	7	2	0.18	20	4	4	2	0.1486	20	7	2	1	0.4789
19	8	6	2	0.3406	19	9	7	3	0.3501	20	4	4	3	0.0132	20	7	2	2	0.1105
19	8	6	3	0.3406	19	9	7	4	0.3001						20	7	3	0	0.2509
19	8	6	4	0.1419	19	9	7	5	0.1125	20	4	4	4	0.0002	20	7	3	1	0.4789
19	8	6	5	0.0227						20	5	1	0	0.75					
19	8	6	6	0.001	19	9	7	6	0.0167	20	5	1	1	0.25	20	7	3	2	0.2395
					19	9	7	7	0.0007	20	5	2	0	0.5526	20	7	3	3	0.0307
19	8	7	0	0.0065	19	9	8	0	0.0006	20	5	2	1	0.3947	20	7	4	0	0.1476
					19	9	8	1	0.0143	20	5	2	2	0.0526	20	7	4	1	0.4132

TABLE D.3 The hypergeometric probability distribution (continued)

ν	n	γ	x	P(x)
20	7	4	2	0.3381
20	7	4	3	0.0939
20	7	4	4	0.0072
20	7	5	0	0.083
20	7	5	1	0.3228
20	7	5	2	0.3874
20	7	5	3	0.1761
20	7	5	4	0.0293
20	7	5	5	0.0014
20	7	6	0	0.0443
20	7	6	1	0.2324
20	7	6	2	0.3874
20	7	6	3	0.2583
20	7	6	4	0.0704
20	7	6	5	0.007
20	7	6	6	0.0002
20	7	7	0	0.0221
20	7	7	1	0.155
20	7	7	2	0.3486
20	7	7	3	0.3228
20	7	7	4	0.1291
20	7	7	5	0.0211
20	7	7	6	0.0012
20	7	7	7	0
20	8	1	0	0.6
20	8	1	1	0.4
20	8	2	0	0.3474
20	8	2	1	0.5053
20	8	2	2	0.1474
20	8	3	0	0.193
20	8	3	1	0.4632
20	8	3	2	0.2947
20	8	3	3	0.0491
20	8	4	0	0.1022
20	8	4	1	0.3633
20	8	4	2	0.3814
20	8	4	3	0.1387
20	8	4	4	0.0144
20	8	5	0	0.0511
20	8	5	1	0.2554
20	8	5	2	0.3973
20	8	5	3	0.2384
20	8	5	4	0.0542
20	8	5	5	0.0036
20	8	6	0	0.0238
20	8	6	1	0.1635
20	8	6	2	0.3576
20	8	6	3	0.3179
20	8	6	4	0.1192
20	8	6	5	0.0173
20	8	6	6	0.0007

ν	n	γ	x	P(x)
20	8	7	0	0.0102
20	8	7	1	0.0954
20	8	7	2	0.2861
20	8	7	3	0.3576
20	8	7	4	0.1987
20	8	7	5	0.0477
20	8	7	6	0.0043
20	8	7	7	0.0001
20	8	8	0	0.0039
20	8	8	1	0.0503
20	8	8	2	0.2054
20	8	8	3	0.3521
20	8	8	4	0.2751
20	8	8	5	0.0978
20	8	8	6	0.0147
20	8	8	7	0.0008
20	8	8	8	0
20	9	1	0	0.55
20	9	1	1	0.45
20	9	2	0	0.2895
20	9	2	1	0.5211
20	9	2	2	0.1895
20	9	3	0	0.1447
20	9	3	1	0.4342
20	9	3	2	0.3474
20	9	3	3	0.0737
20	9	4	0	0.0681
20	9	4	1	0.3065
20	9	4	2	0.4087
20	9	4	3	0.1907
20	9	4	4	0.026
20	9	5	0	0.0298
20	9	5	1	0.1916
20	9	5	2	0.3831
20	9	5	3	0.298
20	9	5	4	0.0894
20	9	5	5	0.0081
20	9	6	0	0.0119
20	9	6	1	0.1073
20	9	6	2	0.3065
20	9	6	3	0.3576
20	9	6	4	0.1788
20	9	6	5	0.0358
20	9	6	6	0.0022
20	9	7	0	0.0043
20	9	7	1	0.0536
20	9	7	2	0.2146
20	9	7	3	0.3576
20	9	7	4	0.2682
20	9	7	5	0.0894

ν	n	γ	x	P(x)
20	9	7	6	0.0119
20	9	7	7	0.0005
20	9	8	0	0.0013
20	9	8	1	0.0236
20	9	8	2	0.132
20	9	8	3	0.3081
20	9	8	4	0.3301
20	9	8	5	0.165
20	9	8	6	0.0367
20	9	8	7	0.0031
20	9	8	8	0.0001
20	9	9	0	0.0003
20	9	9	1	0.0088
20	9	9	2	0.0707
20	9	9	3	0.2311
20	9	9	4	0.3466
20	9	9	5	0.2476
20	9	9	6	0.0825
20	9	9	7	0.0118
20	9	9	8	0.0006
20	9	9	9	0
20	10	1	0	0.5
20	10	1	1	0.5
20	10	2	0	0.2368
20	10	2	1	0.5263
20	10	2	2	0.2368
20	10	3	0	0.1053
20	10	3	1	0.3947
20	10	3	2	0.3947
20	10	3	3	0.1053
20	10	4	0	0.0433
20	10	4	1	0.2477
20	10	4	2	0.418
20	10	4	3	0.2477
20	10	4	4	0.0433
20	10	5	0	0.0163
20	10	5	1	0.1354
20	10	5	2	0.3483
20	10	5	3	0.3483
20	10	5	4	0.1354
20	10	5	5	0.0163
20	10	6	0	0.0054
20	10	6	1	0.065
20	10	6	2	0.2438
20	10	6	3	0.3715
20	10	6	4	0.2438
20	10	6	5	0.065
20	10	6	6	0.0054
20	10	7	0	0.0015
20	10	7	1	0.0271

ν	n	γ	x	P(x)
20	10	7	2	0.1463
20	10	7	3	0.3251
20	10	7	4	0.3251
20	10	7	5	0.1463
20	10	7	6	0.0271
20	10	7	7	0.0015
20	10	8	0	0.0004
20	10	8	1	0.0095
20	10	8	2	0.075
20	10	8	3	0.2401
20	10	8	4	0.3501
20	10	8	5	0.2401
20	10	8	6	0.075
20	10	8	7	0.0095
20	10	8	8	0.0004
20	10	9	0	0.0001
20	10	9	1	0.0027
20	10	9	2	0.0322
20	10	9	3	0.15
20	10	9	4	0.3151
20	10	9	5	0.3151
20	10	9	6	0.15
20	10	9	7	0.0322
20	10	9	8	0.0027
20	10	9	9	0.0001
20	10	10	0	0
20	10	10	1	0.0005
20	10	10	2	0.011
20	10	10	3	0.0779
20	10	10	4	0.2387
20	10	10	5	0.3437
20	10	10	6	0.2387
20	10	10	7	0.0779
20	10	10	8	0.011
20	10	10	9	0.0005
20	10	10	10	0

TABLE D.4

Areas under the standard normal probability distribution

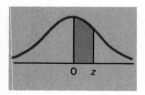

Example: For z = 2.05, shaded area is .4798 out of the total area of 1.

z	.00	.01	.02	.03	.04	.05	.06	.07	.08	.09
0.0	.0000	.0040	.0080	.0120	.0160	.0199	.0239	.0279	.0319	.0359
0.1	.0398	.0438	.0478	.0517	.0557	.0596	.0636	.0675	.0714	.0753
0.2	.0793	.0832	.0871	.0910	.0948	.0987	.1026	.1064	.1103	.1141
0.3	.1179	.1217	.1255	.1293	.1331	.1368	.1406	.1443	.1480	.1517
0.4	.1554	.1591	.1628	.1664	.1700	.1736	.1772	.1808	.1844	.1879
0.5	.1915	.1950	.1985	.2019	.2054	.2088	.2123	.2157	.2190	.2224
0.6	.2257	.2291	.2324	.2357	.2389	.2422	.2454	.2486	.2518	.2549
0.7	.2580	.2611	.2642	.2673	.2703	.2734	.2764	.2794	.2823	.2852
0.8	.2881	.2910	.2939	.2967	.2995	.3023	.3051	.3078	.3106	.3133
0.9	.3159	.3186	.3212	.3238	.3264	.3289	.3315	.3340	.3365	.3389
1.0	.3413	.3438	.3461	.3485	.3508	.3531	.3554	.3577	.3599	.3621
1.1	.3643	.3665	.3686	.3708	.3729	.3749	.3770	.3790	.3810	.3830
1.2	.3849	.3869	.3888	.3907	.3925	.3944	.3962	.3980	.3997	.4015
1.3	.4032	.4049	.4066	.4082	.4099	.4115	.4131	.4147	.4162	.4177
1.4	.4192	.4207	.4222	.4236	.4251	.4265	.4279	.4292	.4306	.4319
1.5	.4332	.4345	.4357	.4370	.4382	.4394	.4406	.4418	.4429	.4441
1.6	.4452	.4463	.4474	.4484	.4495	.4505	.4515	.4525	.4535	.4545
1.7	.4554	.4564	.4573	.4582	.4591	.4599	.4608	.4616	.4625	.4633
1.8	.4641	.4649	.4656	.4664	.4671	.4678	.4686	.4693	.4699	.4706
1.9	.4713	.4719	.4726	.4732	.4738	.4744	.4750	.4756	.4761	.4767
2.0	.4772	.4778	.4783	.4788	.4793	.4798	.4803	.4808	.4812	.4817
2.1	.4821	.4826	.4830	.4834	.4838	.4842	.4846	.4850	.4854	.4857
2.2	.4861	.4864	.4868	.4871	.4875	.4878	.4881	.4884	.4887	.4890
2.3	.4893	.4896	.4898	.4901	.4904	.4906	.4909	.4911	.4913	.4916
2.4	.4918	.4920	.4922	.4925	.4927	.4929	.4931	.4932	.4934	.4936
2.5	.4938	.4940	.4941	.4943	.4945	.4946	.4948	.4949	.4951	.4952
2.6	.4953	.4955	.4956	.4957	.4959	.4960	.4961	.4962	.4963	.4964
2.7	.4965	.4966	.4967	.4968	.4969	.4970	.4971	.4972	.4973	.4974
2.8	.4974	.4975	.4976	.4977	.4977	.4978	.4979	.4979	.4980	.4981
2.9	.4981	.4982	.4982	.4983	.4984	.4984	.4985	.4985	.4986	.4986
3.0	.4987	.4987	.4987	.4988	.4988	.4989	.4989	.4989	.4990	.4990
3.1	.4990	.4991	.4991	.4991	.4992	.4992	.4992	.4992	.4993	.4993
3.2	.4993	.4993	.4994	.4994	.4994	.4994	.4994	.4995	.4995	.4995
3.3	.4995	.4995	.4995	.4996	.4996	.4996	.4996	.4996	.4996	.4997
3.4	.4997	.4997	.4997	.4997	.4997	.4997	.4997	.4997	.4998	.4998
3.5	.4998	.4998	.4998	.4998	.4998	.4998	.4998	.4998	.4998	.4998
3.6	.4998	.4998	.4999	.4999	.4999	.4999	.4999	.4999	.4999	.4999
3.7	.4999	.4999	.4999	.4999	.4999	.4999	.4999	.4999	.4999	.4999

TABLE D.5
Areas under the exponential distribution

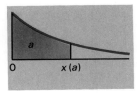

Entry is area *a* under the exponential curve from 0 to *x*(*a*).

μ	0.00	0.01	0.02	0.03	0.04	0.05	0.06	0.07	0.08	0.09
0.0	0.0000	0.0100	0.0198	0.0296	0.0392	0.0488	0.0582	0.0676	0.0769	0.0861
0.1	0.0952	0.1042	0.1131	0.1219	0.1306	0.1393	0.1479	0.1563	0.1647	0.1730
0.2	0.1813	0.1894	0.1975	0.2055	0.2134	0.2212	0.2289	0.2366	0.2442	0.2517
0.3	0.2592	0.2666	0.2739	0.2811	0.2882	0.2953	0.3023	0.3093	0.3161	0.3229
0.4	0.3297	0.3363	0.3430	0.3495	0.3560	0.3624	0.3687	0.3750	0.3812	0.3874
0.5	0.3935	0.3995	0.4055	0.4114	0.4173	0.4231	0.4288	0.4345	0.4401	0.4457
0.6	0.4512	0.4566	0.4621	0.4674	0.4727	0.4780	0.4831	0.4883	0.4934	0.4984
0.7	0.5034	0.5084	0.5132	0.5181	0.5229	0.5276	0.5323	0.5370	0.5416	0.5462
0.8	0.5507	0.5551	0.5596	0.5640	0.5683	0.5726	0.5768	0.5810	0.5852	0.5893
0.9	0.5934	0.5975	0.6015	0.6054	0.6094	0.6133	0.6171	0.6209	0.6247	0.6284
1.0	0.6321	0.6358	0.6394	0.6430	0.6465	0.6501	0.6535	0.6570	0.6604	0.6638
1.1	0.6671	0.6704	0.6737	0.6770	0.6802	0.6834	0.6865	0.6896	0.6927	0.6958
1.2	0.6988	0.7018	0.7048	0.7077	0.7106	0.7135	0.7163	0.7192	0.7220	0.7247
1.3	0.7275	0.7302	0.7329	0.7355	0.7382	0.7408	0.7433	0.7459	0.7484	0.7509
1.4	0.7534	0.7559	0.7583	0.7607	0.7631	0.7654	0.7678	0.7701	0.7724	0.7746
1.5	0.7769	0.7791	0.7813	0.7835	0.7856	0.7878	0.7899	0.7920	0.7940	0.7961
1.6	0.7981	0.8001	0.8021	0.8041	0.8060	0.8080	0.8099	0.8118	0.8136	0.8155
1.7	0.8173	0.8191	0.8209	0.8227	0.8245	0.8262	0.8280	0.8297	0.8314	0.8330
1.8	0.8347	0.8363	0.8380	0.8396	0.8412	0.8428	0.8443	0.8459	0.8474	0.8489
1.9	0.8504	0.8519	0.8534	0.8549	0.8563	0.8577	0.8591	0.8605	0.8619	0.8633
2.0	0.8647	0.8660	0.8673	0.8687	0.8700	0.8713	0.8725	0.8738	0.8751	0.8763
2.1	0.8775	0.8788	0.8800	0.8812	0.8823	0.8835	0.8847	0.8858	0.8870	0.8881
2.2	0.8892	0.8903	0.8914	0.8925	0.8935	0.8946	0.8956	0.8967	0.8977	0.8987
2.3	0.8997	0.9007	0.9017	0.9027	0.9037	0.9046	0.9056	0.9065	0.9074	0.9084
2.4	0.9093	0.9102	0.9111	0.9120	0.9128	0.9137	0.9146	0.9154	0.9163	0.9171
2.5	0.9179	0.9187	0.9195	0.9203	0.9211	0.9219	0.9227	0.9235	0.9242	0.9250
2.6	0.9257	0.9265	0.9272	0.9279	0.9286	0.9293	0.9301	0.9307	0.9314	0.9321
2.7	0.9328	0.9335	0.9341	0.9348	0.9354	0.9361	0.9367	0.9373	0.9380	0.9386
2.8	0.9392	0.9398	0.9404	0.9410	0.9416	0.9422	0.9427	0.9433	0.9439	0.9444
2.9	0.9450	0.9455	0.9461	0.9466	0.9471	0.9477	0.9482	0.9487	0.9492	0.9497
3.0	0.9502	0.9507	0.9512	0.9517	0.9522	0.9526	0.9531	0.9536	0.9540	0.9545
3.1	0.9550	0.9554	0.9558	0.9563	0.9567	0.9571	0.9576	0.9580	0.9584	0.9588
3.2	0.9592	0.9596	0.9600	0.9604	0.9608	0.9612	0.9616	0.9620	0.9624	0.9627
3.3	0.9631	0.9635	0.9638	0.9642	0.9646	0.9649	0.9653	0.9656	0.9660	0.9663
3.4	0.9666	0.9670	0.9673	0.9676	0.9679	0.9683	0.9686	0.9689	0.9692	0.9695
3.5	0.9698	0.9701	0.9704	0.9707	0.9710	0.9713	0.9716	0.9718	0.9721	0.9724
3.6	0.9727	0.9729	0.9732	0.9735	0.9737	0.9740	0.9743	0.9745	0.9748	0.9750
3.7	0.9753	0.9755	0.9758	0.9760	0.9762	0.9765	0.9767	0.9769	0.9772	0.9774
3.8	0.9776	0.9779	0.9781	0.9783	0.9785	0.9787	0.9789	0.9791	0.9793	0.9796
3.9	0.9798	0.9800	0.9802	0.9804	0.9806	0.9807	0.9809	0.9811	0.9813	0.9815

μ	0.0	0.1	0.2	0.3	0.4	0.5	0.6	0.7	0.8	0.9
4.0	0.9817	0.9834	0.9850	0.9864	0.9877	0.9889	0.9899	0.9909	0.9918	0.9926
5.0	0.9933	0.9939	0.9945	0.9950	0.9955	0.9959	0.9963	0.9967	0.9970	0.9973
6.0	0.9975	0.9978	0.9980	0.9982	0.9983	0.9985	0.9986	0.9988	0.9989	0.9990
7.0	0.9991	0.9992	0.9993	0.9993	0.9994	0.9994	0.9995	0.9995	0.9996	0.9996
8.0	0.9997	0.9997	0.9997	0.9998	0.9998	0.9998	0.9998	0.9998	0.9998	0.9999
9.0	0.9999	0.9999	0.9999	0.9999	0.9999	0.9999	0.9999	0.9999	0.9999	0.9999

**TABLE D.6
Table of
random digits**

229339	384399	756256	489181	812133
931289	649291	753524	552642	289215
257049	946090	067504	628875	541351
476211	547088	124160	905258	845718
817459	210357	543792	706802	295013
066726	388854	506423	585235	114364
582183	330200	800689	192565	523620
943908	661972	873214	167724	194595
385314	598190	988586	414672	450500
015838	717117	606979	591384	480804
396224	010971	602539	087982	902755
378768	700149	618942	724594	264938
082412	829544	186447	813858	527770
198944	380294	852981	117752	111587
729522	474334	323013	065795	958511
282394	767425	303482	719940	086151
289642	001571	526641	389892	145401
103530	005233	077880	952377	739929
057815	998153	162155	493331	640426
852966	734985	958312	503982	545043
929199	657073	695480	186141	706161
103195	678787	456909	310623	899429
430908	872604	981674	237167	248931
524032	007293	724105	193664	960922
637512	462509	602005	937973	523117
793792	116638	366165	162612	168304
368530	698516	116607	044784	826416
094726	746948	148147	095550	923721
110931	033340	525085	199844	801162
464904	177902	238861	595077	565261
669525	506042	098876	831695	484451
747406	368392	763244	872299	883804
772781	336563	854614	612213	595901
234115	529495	462219	673019	622833
007522	091644	391403	354934	637695
540817	892410	809616	353881	871154
621414	669540	990570	651855	891479
275222	267120	618881	683334	951446
748306	075057	538299	329513	321792
798965	376388	975585	806732	829940
421234	823944	091217	112289	346633
912231	017654	870956	321273	888656
109695	849731	567993	385910	156661
882278	196762	084869	524597	027633
651260	621231	167739	274902	438980
776931	010406	082321	707885	676284
038940	378829	101104	420516	829467
479843	100860	338790	269134	479156
929641	242248	507965	599914	828323
622360	797866	748886	768844	041870

TABLE D.7
Percentiles of Student's *t*-distribution

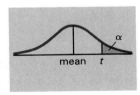

The *t* value for 12 degrees of freedom that bounds a right-tail area of 0.025 is 2.179.

Degrees of Freedom	RIGHT-TAIL AREA, α				
	0.100	0.050	0.025	0.010	0.005
1	3.078	6.314	12.706	31.821	63.657
2	1.886	2.920	4.303	6.965	9.925
3	1.638	2.353	3.182	4.541	5.841
4	1.533	2.132	2.776	3.747	4.604
5	1.476	2.015	2.571	3.365	4.032
6	1.440	1.943	2.447	3.143	3.707
7	1.415	1.895	2.365	2.998	3.499
8	1.397	1.860	2.306	2.896	3.355
9	1.383	1.833	2.262	2.821	3.250
10	1.372	1.812	2.228	2.764	3.169
11	1.363	1.796	2.201	2.718	3.106
12	1.356	1.782	2.179	2.681	3.055
13	1.350	1.771	2.160	2.650	3.012
14	1.345	1.761	2.145	2.624	2.977
15	1.341	1.753	2.131	2.602	2.947
16	1.337	1.746	2.120	2.583	2.921
17	1.333	1.740	2.110	2.567	2.898
18	1.330	1.734	2.101	2.552	2.878
19	1.328	1.729	2.093	2.539	2.861
20	1.325	1.725	2.086	2.528	2.845
21	1.323	1.721	2.080	2.518	2.831
22	1.321	1.717	2.074	2.508	2.819
23	1.319	1.714	2.069	2.500	2.807
24	1.318	1.711	2.064	2.492	2.797
25	1.316	1.708	2.060	2.485	2.787
26	1.315	1.706	2.056	2.479	2.779
27	1.314	1.703	2.052	2.473	2.771
28	1.313	1.701	2.048	2.467	2.763
29	1.311	1.699	2.045	2.462	2.756
30	1.310	1.697	2.042	2.457	2.750
40	1.303	1.684	2.021	2.423	2.704
60	1.296	1.671	2.000	2.390	2.660
∞	1.282	1.645	1.960	2.326	2.576

Source: E. S. Pearson and H. O. Hartley, *Biometrika Tables for Statisticians*, vol. I, Table 12, published for the Biometrika Trustees of the University Press, Cambridge. Reprinted by permission of the Biometrika Trustees.

TABLE D.8
Chi-square distribution

Degrees of Freedom	\multicolumn{5}{c}{RIGHT-TAIL AREA}
	0.100	0.050	0.025	0.010	0.005
1	2.70554	3.84146	5.02389	6.63490	7.87944
2	4.60517	5.99147	7.37776	9.21034	10.5966
3	6.25139	7.81473	9.34840	11.3449	12.8381
4	7.77944	9.48773	11.1433	13.2767	14.8602
5	9.23635	11.0705	12.8325	15.0863	16.7496
6	10.6446	12.5916	14.4494	16.8119	18.5476
7	12.0170	14.0671	16.0128	18.4753	20.2777
8	13.3616	15.5073	17.5346	20.0902	21.9550
9	14.6837	16.9190	19.0228	21.6660	23.5893
10	15.9871	18.3070	20.4831	23.2093	25.1882
11	17.2750	19.6751	21.9200	24.7250	26.7569
12	18.5494	21.0261	23.3367	26.2170	28.2995
13	19.8119	22.3621	24.7356	27.6883	29.8194
14	21.0642	23.6848	26.1190	29.1413	31.3193
15	22.3072	24.9958	27.4884	30.5779	32.8013
16	23.5418	26.2962	28.8454	31.9999	34.2672
17	24.7690	27.5871	30.1910	33.4087	35.7185
18	25.9894	28.8693	31.5264	34.8053	37.1564
19	27.2036	30.1435	32.8523	36.1908	38.5822
20	28.4120	31.4104	34.1696	37.5662	39.9968
21	29.6151	32.6705	35.4789	38.9321	41.4010
22	30.8133	33.9244	36.7807	40.2894	42.7956
23	32.0069	35.1725	38.0757	41.6384	44.1813
24	33.1963	36.4151	39.3641	42.9798	45.5585
25	34.3816	37.6525	40.6465	44.3141	46.9278
26	35.5631	38.8852	41.9232	45.6417	48.2899
27	36.7412	40.1133	43.1944	46.9630	49.6449
28	37.9159	41.3372	44.4607	48.2782	50.9933
29	39.0875	42.5569	45.7222	49.5879	52.3356
30	40.2560	43.7729	46.9792	50.8922	53.6720
40	51.8050	55.7585	59.3417	63.6907	66.7659
50	63.1671	67.5048	71.4202	76.1539	79.4900
60	74.3970	79.0819	83.2976	88.3794	91.9517
70	85.5271	90.5312	95.0231	100.425	104.215
80	96.5782	101.879	106.629	112.329	116.321
90	107.565	113.145	118.136	124.116	128.299
100	118.498	124.342	129.561	135.807	140.169

Source: C. M. Thompson, "Tables of the Percentage Points of the χ^2-Distribution," *Biometrika,* 1941, vol. 32, pp. 188–189. Reproduced by permission of the Biometrika Trustees.

TABLE D.8
Chi-square distribution (continued)

DEGREES OF FREEDOM	LEFT-TAIL AREA				
	0.005	0.010	0.025	0.050	0.10
1	.0000393	.0001571	.0009821	.0039321	.0157908
2	.0100251	.0201007	.0506356	.102587	.210720
3	.0717212	.114832	.215795	.351846	.584375
4	.206990	.297110	.484419	.710721	1.063623
5	.411740	.554300	.831211	1.145476	1.61031
6	.675727	.872085	1.237347	1.63539	2.20413
7	.989265	1.239043	1.68987	2.16735	2.83311
8	1.344419	1.646482	2.17973	2.73264	3.48954
9	1.734926	2.087912	2.70039	3.32511	4.16816
10	2.15585	2.55821	3.24697	3.94030	4.86518
11	2.60321	3.05347	3.81575	4.57481	5.57779
12	3.07382	3.57056	4.40379	5.22603	6.30380
13	3.56503	4.10691	5.00874	5.89186	7.04150
14	4.07468	4.66043	5.62872	6.57063	7.78953
15	4.60094	5.22935	6.26214	7.26094	8.54675
16	5.14224	5.81221	6.90766	7.96164	9.31223
17	5.69724	6.40776	7.56418	8.67176	10.0852
18	6.26481	7.01491	8.23075	9.39046	10.8649
19	6.84398	7.63273	8.90655	10.1170	11.6509
20	7.43386	8.26040	9.59083	10.8508	12.4426
21	8.03366	8.89720	10.28293	11.5913	13.2396
22	8.64272	9.54249	10.9823	12.3380	14.0415
23	9.26042	10.19567	11.6885	13.0905	14.8479
24	9.88623	10.8564	12.4011	13.8484	15.6587
25	10.5197	11.5240	13.1197	14.6114	16.4734
26	11.1603	12.1981	13.8439	15.3791	17.2919
27	11.8076	12.8786	14.5733	16.1513	18.1138
28	12.4613	13.5648	15.3079	16.9279	18.9392
29	13.1211	14.2565	16.0471	17.7083	19.7677
30	13.7867	14.9535	16.7908	18.4926	20.5992
40	20.7065	22.1643	24.4331	26.5093	29.0505
50	27.9907	29.7067	32.3574	34.7642	37.6886
60	35.5346	37.4848	40.4817	43.1879	46.4589
70	43.2752	45.4418	48.7576	51.7393	55.3290
80	51.1720	53.5400	57.1532	60.3915	64.2778
90	59.1963	61.7541	65.6466	69.1260	73.2912
100	67.3276	70.0648	74.2219	77.9295	82.3581

TABLE D.9 Percentiles of the *F* distribution

a. 5% in the right tail

d_{fn}

d_{fd}	1	2	3	4	5	6	7	8	9	10	12	15	20	24	30	40	60	120	∞
1	161.4	199.5	215.7	224.6	230.2	234.0	236.8	238.9	240.5	241.9	243.9	245.9	248.0	249.1	250.1	251.1	252.2	253.3	254.3
2	18.51	19.00	19.16	19.25	19.30	19.33	19.35	19.37	19.38	19.40	19.41	19.43	19.45	19.45	19.46	19.47	19.48	19.49	19.50
3	10.13	9.55	9.28	9.12	9.01	8.94	8.89	8.85	8.81	8.79	8.74	8.70	8.66	8.64	8.62	8.59	8.57	8.55	8.53
4	7.71	6.94	6.59	6.39	6.26	6.16	6.09	6.04	6.00	5.96	5.91	5.86	5.80	5.77	5.75	5.72	5.69	5.66	5.63
5	6.61	5.79	5.41	5.19	5.05	4.95	4.88	4.82	4.77	4.74	4.68	4.62	4.56	4.53	4.50	4.46	4.43	4.40	4.36
6	5.99	5.14	4.76	4.53	4.39	4.28	4.21	4.15	4.10	4.06	4.00	3.94	3.87	3.84	3.81	3.77	3.74	3.70	3.67
7	5.59	4.74	4.35	4.12	3.97	3.87	3.79	3.73	3.68	3.64	3.57	3.51	3.44	3.41	3.38	3.34	3.30	3.27	3.23
8	5.32	4.46	4.07	3.84	3.69	3.58	3.50	3.44	3.39	3.35	3.28	3.22	3.15	3.12	3.08	3.04	3.01	2.97	2.93
9	5.12	4.26	3.86	3.63	3.48	3.37	3.29	3.23	3.18	3.14	3.07	3.01	2.94	2.90	2.86	2.83	2.79	2.75	2.71
10	4.96	4.10	3.71	3.48	3.33	3.22	3.14	3.07	3.02	2.98	2.91	2.85	2.77	2.74	2.70	2.66	2.62	2.58	2.54
11	4.84	3.98	3.59	3.36	3.20	3.09	3.01	2.95	2.90	2.85	2.79	2.72	2.65	2.61	2.57	2.53	2.49	2.45	2.40
12	4.75	3.89	3.49	3.26	3.11	3.00	2.91	2.85	2.80	2.75	2.69	2.62	2.54	2.51	2.47	2.43	2.38	2.34	2.30
13	4.67	3.81	3.41	3.18	3.03	2.92	2.83	2.77	2.71	2.67	2.60	2.53	2.46	2.42	2.38	2.34	2.30	2.25	2.21
14	4.60	3.74	3.34	3.11	2.96	2.85	2.76	2.70	2.65	2.60	2.53	2.46	2.39	2.35	2.31	2.27	2.22	2.18	2.13
15	4.54	3.68	3.29	3.06	2.90	2.79	2.71	2.64	2.59	2.54	2.48	2.40	2.33	2.29	2.25	2.20	2.16	2.11	2.07
16	4.49	3.63	3.24	3.01	2.85	2.74	2.66	2.59	2.54	2.49	2.42	2.35	2.28	2.24	2.19	2.15	2.11	2.06	2.01
17	4.45	3.59	3.20	2.96	2.81	2.70	2.61	2.55	2.49	2.45	2.38	2.31	2.23	2.19	2.15	2.10	2.06	2.01	1.96
18	4.41	3.55	3.16	2.93	2.77	2.66	2.58	2.51	2.46	2.41	2.34	2.27	2.19	2.15	2.11	2.06	2.02	1.97	1.92
19	4.38	3.52	3.13	2.90	2.74	2.63	2.54	2.48	2.42	2.38	2.31	2.23	2.16	2.11	2.07	2.03	1.98	1.93	1.88
20	4.35	3.49	3.10	2.87	2.71	2.60	2.51	2.45	2.39	2.35	2.28	2.20	2.12	2.08	2.04	1.99	1.95	1.90	1.84
21	4.32	3.47	3.07	2.84	2.68	2.57	2.49	2.42	2.37	2.32	2.25	2.18	2.10	2.05	2.01	1.96	1.92	1.87	1.81
22	4.30	3.44	3.05	2.82	2.66	2.55	2.46	2.40	2.34	2.30	2.23	2.15	2.07	2.03	1.98	1.94	1.89	1.84	1.78
23	4.28	3.42	3.03	2.80	2.64	2.53	2.44	2.37	2.32	2.27	2.20	2.13	2.05	2.01	1.96	1.91	1.86	1.81	1.76
24	4.26	3.40	3.01	2.78	2.62	2.51	2.42	2.36	2.30	2.25	2.18	2.11	2.03	1.98	1.94	1.89	1.84	1.79	1.73
25	4.24	3.39	2.99	2.76	2.60	2.49	2.40	2.34	2.28	2.24	2.16	2.09	2.01	1.96	1.92	1.87	1.82	1.77	1.71
26	4.23	3.37	2.98	2.74	2.59	2.47	2.39	2.32	2.27	2.22	2.15	2.07	1.99	1.95	1.90	1.85	1.80	1.75	1.69
27	4.21	3.35	2.96	2.73	2.57	2.46	2.37	2.31	2.25	2.20	2.13	2.06	1.97	1.93	1.88	1.84	1.79	1.73	1.67
28	4.20	3.34	2.95	2.71	2.56	2.45	2.36	2.29	2.24	2.19	2.12	2.04	1.96	1.91	1.87	1.82	1.77	1.71	1.65
29	4.18	3.33	2.93	2.70	2.55	2.43	2.35	2.28	2.22	2.18	2.10	2.03	1.94	1.90	1.85	1.81	1.75	1.70	1.64
30	4.17	3.32	2.92	2.69	2.53	2.42	2.33	2.27	2.21	2.16	2.09	2.01	1.93	1.89	1.84	1.79	1.74	1.68	1.62
40	4.08	3.23	2.84	2.61	2.45	2.34	2.25	2.18	2.12	2.08	2.00	1.92	1.84	1.79	1.74	1.69	1.64	1.58	1.51
60	4.00	3.15	2.76	2.53	2.37	2.25	2.17	2.10	2.04	1.99	1.92	1.84	1.75	1.70	1.65	1.59	1.53	1.47	1.39
120	3.92	3.07	2.68	2.45	2.29	2.17	2.09	2.02	1.96	1.91	1.83	1.75	1.66	1.61	1.55	1.50	1.43	1.35	1.25
∞	3.84	3.00	2.60	2.37	2.21	2.10	2.01	1.94	1.88	1.83	1.75	1.67	1.57	1.52	1.46	1.39	1.32	1.22	1.00

Source: Table 18, E. S. Pearson and H. O. Hartley, *Biometrika Tables for Statisticians*, vol. I, published for the Trustees at the University Press, Cambridge. Reprinted by permission of the Biometrika Trustees.

TABLE D.9   Percentiles of the F distribution (continued)

b. 2.5% in the right tail

d_{fn}

d_{fd}	1	2	3	4	5	6	7	8	9	10	12	15	20	24	30	40	60	120	∞
1	647.8	799.5	864.2	899.6	921.8	937.1	948.2	956.7	963.3	968.6	976.7	984.9	993.1	997.2	1001	1006	1010	1014	1018
2	38.51	39.00	39.17	39.25	39.30	39.33	39.36	39.37	39.39	39.40	39.41	39.43	39.45	39.46	39.46	39.47	39.48	39.49	39.50
3	17.44	16.04	15.44	15.10	14.88	14.73	14.62	14.54	14.47	14.42	14.34	14.25	14.17	14.12	14.08	14.04	13.99	13.95	13.90
4	12.22	10.65	9.98	9.60	9.36	9.20	9.07	8.98	8.90	8.84	8.75	8.66	8.56	8.51	8.46	8.41	8.36	8.31	8.26
5	10.01	8.43	7.76	7.39	7.15	6.98	6.85	6.76	6.68	6.62	6.52	6.43	6.33	6.28	6.23	6.18	6.12	6.07	6.02
6	8.81	7.26	6.60	6.23	5.99	5.82	5.70	5.60	5.52	5.46	5.37	5.27	5.17	5.12	5.07	5.01	4.96	4.90	4.85
7	8.07	6.54	5.89	5.52	5.29	5.12	4.99	4.90	4.82	4.76	4.67	4.57	4.47	4.42	4.36	4.31	4.25	4.20	4.14
8	7.57	6.06	5.42	5.05	4.82	4.65	4.53	4.43	4.36	4.30	4.20	4.10	4.00	3.95	3.89	3.84	3.78	3.73	3.67
9	7.21	5.71	5.08	4.72	4.48	4.32	4.20	4.10	4.03	3.96	3.87	3.77	3.67	3.61	3.56	3.51	3.45	3.39	3.33
10	6.94	5.46	4.83	4.47	4.24	4.07	3.95	3.85	3.78	3.72	3.62	3.52	3.42	3.37	3.31	3.26	3.20	3.14	3.08
11	6.72	5.26	4.63	4.28	4.04	3.88	3.76	3.66	3.59	3.53	3.43	3.33	3.23	3.17	3.12	3.06	3.00	2.94	2.88
12	6.55	5.10	4.47	4.12	3.89	3.73	3.61	3.51	3.44	3.37	3.28	3.18	3.07	3.02	2.96	2.91	2.85	2.79	2.72
13	6.41	4.97	4.35	4.00	3.77	3.60	3.48	3.39	3.31	3.25	3.15	3.05	2.95	2.89	2.84	2.78	2.72	2.66	2.60
14	6.30	4.86	4.24	3.89	3.66	3.50	3.38	3.29	3.21	3.15	3.05	2.95	2.84	2.79	2.73	2.67	2.61	2.55	2.49
15	6.20	4.77	4.15	3.80	3.58	3.41	3.29	3.20	3.12	3.06	2.96	2.86	2.76	2.70	2.64	2.59	2.52	2.46	2.40
16	6.12	4.69	4.08	3.73	3.50	3.34	3.22	3.12	3.05	2.99	2.89	2.79	2.68	2.63	2.57	2.51	2.45	2.38	2.32
17	6.04	4.62	4.01	3.66	3.44	3.28	3.16	3.06	2.98	2.92	2.82	2.72	2.62	2.56	2.50	2.44	2.38	2.32	2.25
18	5.98	4.56	3.95	3.61	3.38	3.22	3.10	3.01	2.93	2.87	2.77	2.67	2.56	2.50	2.44	2.38	2.32	2.26	2.19
19	5.92	4.51	3.90	3.56	3.33	3.17	3.05	2.96	2.88	2.82	2.72	2.62	2.51	2.45	2.39	2.33	2.27	2.20	2.13
20	5.87	4.46	3.86	3.51	3.29	3.13	3.01	2.91	2.84	2.77	2.68	2.57	2.46	2.41	2.35	2.29	2.22	2.16	2.09
21	5.83	4.42	3.82	3.48	3.25	3.09	2.97	2.87	2.80	2.73	2.64	2.53	2.42	2.37	2.31	2.25	2.18	2.11	2.04
22	5.79	4.38	3.78	3.44	3.22	3.05	2.93	2.84	2.76	2.70	2.60	2.50	2.39	2.33	2.27	2.21	2.14	2.08	2.00
23	5.75	4.35	3.75	3.41	3.18	3.02	2.90	2.81	2.73	2.67	2.57	2.47	2.36	2.30	2.24	2.18	2.11	2.04	1.97
24	5.72	4.32	3.72	3.38	3.15	2.99	2.87	2.78	2.70	2.64	2.54	2.44	2.33	2.27	2.21	2.15	2.08	2.01	1.94
25	5.69	4.29	3.69	3.35	3.13	2.97	2.85	2.75	2.68	2.61	2.51	2.41	2.30	2.24	2.18	2.12	2.05	1.98	1.91
26	5.66	4.27	3.67	3.33	3.10	2.94	2.82	2.73	2.65	2.59	2.49	2.39	2.28	2.22	2.16	2.09	2.03	1.95	1.88
27	5.63	4.24	3.65	3.31	3.08	2.92	2.80	2.71	2.63	2.57	2.47	2.36	2.25	2.19	2.13	2.07	2.00	1.93	1.85
28	5.61	4.22	3.63	3.29	3.06	2.90	2.78	2.69	2.61	2.55	2.45	2.34	2.23	2.17	2.11	2.05	1.98	1.91	1.83
29	5.59	4.20	3.61	3.27	3.04	2.88	2.76	2.67	2.59	2.53	2.43	2.32	2.21	2.15	2.09	2.03	1.96	1.89	1.81
30	5.57	4.18	3.59	3.25	3.03	2.87	2.75	2.65	2.57	2.51	2.41	2.31	2.20	2.14	2.07	2.01	1.94	1.87	1.79
40	5.42	4.05	3.46	3.13	2.90	2.74	2.62	2.53	2.45	2.39	2.29	2.18	2.07	2.01	1.94	1.88	1.80	1.72	1.64
60	5.29	3.93	3.34	3.01	2.79	2.63	2.51	2.41	2.33	2.27	2.17	2.06	1.94	1.88	1.82	1.74	1.67	1.58	1.48
120	5.15	3.80	3.23	2.89	2.67	2.52	2.39	2.30	2.22	2.16	2.05	1.94	1.82	1.76	1.69	1.61	1.53	1.43	1.31
∞	5.02	3.69	3.12	2.79	2.57	2.41	2.29	2.19	2.11	2.05	1.94	1.83	1.71	1.64	1.57	1.48	1.39	1.27	1.00

Source: Table 18, E. S. Pearson and H. O. Hartley, *Biometrika Tables for Statisticians*, vol. I, published for the Trustees at the University Press, Cambridge. Reprinted by permission of the Biometrika Trustees.

TABLE D.9 Percentiles of the F distribution (continued)

c. 1% in the right tail

d_{fm}

d_{fd}	1	2	3	4	5	6	7	8	9	10	12	15	20	24	30	40	60	120	∞
1	4052	4999.5	5403	5625	5764	5859	5928	5981	6022	6056	6106	6157	6209	6235	6261	6287	6313	6339	6366
2	98.50	99.00	99.17	99.25	99.30	99.33	99.36	99.37	99.39	99.40	99.42	99.43	99.45	99.46	99.47	99.47	99.48	99.49	99.50
3	34.12	30.82	29.46	28.71	28.24	27.91	27.67	27.49	27.35	27.23	27.05	26.87	26.69	26.60	26.50	26.41	26.32	26.22	26.13
4	21.20	18.00	16.69	15.98	15.52	15.21	14.98	14.80	14.66	14.55	14.37	14.20	14.02	13.93	13.84	13.75	13.65	13.56	13.46
5	16.26	13.27	12.06	11.39	10.97	10.67	10.46	10.29	10.16	10.05	9.89	9.72	9.55	9.47	9.38	9.29	9.20	9.11	9.02
6	13.75	10.92	9.78	9.15	8.75	8.47	8.26	8.10	7.98	7.87	7.72	7.56	7.40	7.31	7.23	7.14	7.06	6.97	6.88
7	12.25	9.55	8.45	7.85	7.46	7.19	6.99	6.84	6.72	6.62	6.47	6.31	6.16	6.07	5.99	5.91	5.82	5.74	5.65
8	11.26	8.65	7.59	7.01	6.63	6.37	6.18	6.03	5.91	5.81	5.67	5.52	5.36	5.28	5.20	5.12	5.03	4.95	4.86
9	10.56	8.02	6.99	6.42	6.06	5.80	5.61	5.47	5.35	5.26	5.11	4.96	4.81	4.73	4.65	4.57	4.48	4.40	4.31
10	10.04	7.56	6.55	5.99	5.64	5.39	5.20	5.06	4.94	4.85	4.71	4.56	4.41	4.33	4.25	4.17	4.08	4.00	3.91
11	9.65	7.21	6.22	5.67	5.32	5.07	4.89	4.74	4.63	4.54	4.40	4.25	4.10	4.02	3.94	3.86	3.78	3.69	3.60
12	9.33	6.93	5.95	5.41	5.06	4.82	4.64	4.50	4.39	4.30	4.16	4.01	3.86	3.78	3.70	3.62	3.54	3.45	3.36
13	9.07	6.70	5.74	5.21	4.86	4.62	4.44	4.30	4.19	4.10	3.96	3.82	3.66	3.59	3.51	3.43	3.34	3.25	3.17
14	8.86	6.51	5.56	5.04	4.69	4.46	4.28	4.14	4.03	3.94	3.80	3.66	3.51	3.43	3.35	3.27	3.18	3.09	3.00
15	8.68	6.36	5.42	4.89	4.56	4.32	4.14	4.00	3.89	3.80	3.67	3.52	3.37	3.29	3.21	3.13	3.05	2.96	2.87
16	8.53	6.23	5.29	4.77	4.44	4.20	4.03	3.89	3.78	3.69	3.55	3.41	3.26	3.18	3.10	3.02	2.93	2.84	2.75
17	8.40	6.11	5.18	4.67	4.34	4.10	3.93	3.79	3.68	3.59	3.46	3.31	3.16	3.08	3.00	2.92	2.83	2.75	2.65
18	8.29	6.01	5.09	4.58	4.25	4.01	3.84	3.71	3.60	3.51	3.37	3.23	3.08	3.00	2.92	2.84	2.75	2.66	2.57
19	8.18	5.93	5.01	4.50	4.17	3.94	3.77	3.63	3.52	3.43	3.30	3.15	3.00	2.92	2.84	2.76	2.67	2.58	2.49
20	8.10	5.85	4.94	4.43	4.10	3.87	3.70	3.56	3.46	3.37	3.23	3.09	2.94	2.86	2.78	2.69	2.61	2.52	2.42
21	8.02	5.78	4.87	4.37	4.04	3.81	3.64	3.51	3.40	3.31	3.17	3.03	2.88	2.80	2.72	2.64	2.55	2.46	2.36
22	7.95	5.72	4.82	4.31	3.99	3.76	3.59	3.45	3.35	3.26	3.12	2.98	2.83	2.75	2.67	2.58	2.50	2.40	2.31
23	7.88	5.66	4.76	4.26	3.94	3.71	3.54	3.41	3.30	3.21	3.07	2.93	2.78	2.70	2.62	2.54	2.45	2.35	2.26
24	7.82	5.61	4.72	4.22	3.90	3.67	3.50	3.36	3.26	3.17	3.03	2.89	2.74	2.66	2.58	2.49	2.40	2.31	2.21
25	7.77	5.57	4.68	4.18	3.85	3.63	3.46	3.32	3.22	3.13	2.99	2.85	2.70	2.62	2.54	2.45	2.36	2.27	2.17
26	7.72	5.53	4.64	4.14	3.82	3.59	3.42	3.29	3.18	3.09	2.96	2.81	2.66	2.58	2.50	2.42	2.33	2.23	2.13
27	7.68	5.49	4.60	4.11	3.78	3.56	3.39	3.26	3.15	3.06	2.93	2.78	2.63	2.55	2.47	2.38	2.29	2.20	2.10
28	7.64	5.45	4.57	4.07	3.75	3.53	3.36	3.23	3.12	3.03	2.90	2.75	2.60	2.52	2.44	2.35	2.26	2.17	2.06
29	7.60	5.42	4.54	4.04	3.73	3.50	3.33	3.20	3.09	3.00	2.87	2.73	2.57	2.49	2.41	2.33	2.23	2.14	2.03
30	7.56	5.39	4.51	4.02	3.70	3.47	3.30	3.17	3.07	2.98	2.84	2.70	2.55	2.47	2.39	2.30	2.21	2.11	2.01
40	7.31	5.18	4.31	3.83	3.51	3.29	3.12	2.99	2.89	2.80	2.66	2.52	2.37	2.29	2.20	2.11	2.02	1.92	1.80
60	7.08	4.98	4.13	3.65	3.34	3.12	2.95	2.82	2.72	2.63	2.50	2.35	2.20	2.12	2.03	1.94	1.84	1.73	1.60
120	6.85	4.79	3.95	3.48	3.17	2.96	2.79	2.66	2.56	2.47	2.34	2.19	2.03	1.95	1.86	1.76	1.66	1.53	1.38
∞	6.63	4.61	3.78	3.32	3.02	2.80	2.64	2.51	2.41	2.32	2.18	2.04	1.88	1.79	1.70	1.59	1.47	1.32	1.00

Source: Table 18, E. S. Pearson and H. O. Hartley, *Biometrika Tables for Statisticians*, vol. I, published for the Trustees at the University Press, Cambridge. Reprinted by permission of the Biometrika Trustees.

TABLE D.10
Critical values of the Wilcoxon–Mann–Whitney T distribution

Test statistic is the rank sum associated with the smaller sample (if equal sample sizes, either rank sum can be used).

a. $\alpha = .025$ one-tailed; $\alpha = .05$ two-tailed

		n_1														
n_2	3		4		5		6		7		8		9		10	
	T_L	T_U	T_L	T_U	T_L	T_U	T_L	T_U	T_L	T_U	T_L	T_U	T_L	T_U	T_L	T_U
3	5	16	6	18	6	21	7	23	7	26	8	28	8	31	9	33
4	6	18	11	25	12	28	12	32	13	35	14	38	15	41	16	44
5	6	21	12	28	18	37	19	41	20	45	21	49	22	53	24	56
6	7	23	12	32	19	41	26	52	28	56	29	61	31	65	32	70
7	7	26	13	35	20	45	28	56	37	68	39	73	41	78	43	83
8	8	28	14	38	21	49	29	61	39	73	49	87	51	93	54	98
9	8	31	15	41	22	53	31	65	41	78	51	93	63	108	66	114
10	9	33	16	44	24	56	32	70	43	83	54	98	66	114	79	131

b. $\alpha = .05$ one-tailed; $\alpha = .10$ two-tailed

		n_1														
n_2	3		4		5		6		7		8		9		10	
	T_L	T_U	T_L	T_U	T_L	T_U	T_L	T_U	T_L	T_U	T_L	T_U	T_L	T_U	T_L	T_U
3	6	15	7	17	7	20	8	22	9	24	9	27	10	29	11	31
4	7	17	12	24	13	27	14	30	15	33	16	36	17	39	18	42
5	7	20	13	27	19	36	20	40	22	43	24	46	25	50	26	54
6	8	22	14	30	20	40	28	50	30	54	32	58	33	63	35	67
7	9	24	15	33	22	43	30	54	39	66	41	71	43	76	46	80
8	9	27	16	36	24	46	32	58	41	71	52	84	54	90	57	95
9	10	29	17	39	25	50	33	63	43	76	54	90	66	105	69	111
10	11	31	18	42	26	54	35	67	46	80	57	95	69	111	83	127

Source: F. Wilcoxon and R. A. Wilcox, "Some Rapid Approximate Statistical Procedures," 1964, pp. 20–23. Reproduced with the permission of American Cyanamid Company.

TABLE D.11 Critical values of the Durbin-Watson test statistic

$$\alpha = .05$$

Number of Independent Variables k

	1		2		3		4		5		6		7	
n	DW_L	DW_U	DW_L	DW_U	DW_L	DW_U	DW_L	DW_U	DW_L	DW_U	DW_L	DW_U	DW_L	DW_U
15	1.08	1.36	0.95	1.54	0.82	1.75	0.69	1.97	0.56	2.21	0.56	2.22	0.45	2.47
16	1.10	1.37	0.98	1.54	0.86	1.73	0.74	1.93	0.62	2.15	0.62	2.16	0.50	2.39
17	1.13	1.38	1.02	1.54	0.90	1.71	0.78	1.90	0.67	2.10	0.66	2.10	0.55	2.32
18	1.16	1.39	1.05	1.53	0.93	1.69	0.82	1.87	0.71	2.06	0.71	2.06	0.60	2.26
19	1.18	1.40	1.08	1.53	0.97	1.68	0.86	1.85	0.75	2.02	0.75	2.02	0.65	2.21
20	1.20	1.41	1.10	1.54	1.00	1.68	0.90	1.83	0.79	1.99	0.79	1.99	0.69	2.16
21	1.22	1.42	1.13	1.54	1.03	1.67	0.93	1.81	0.83	1.96	0.83	1.96	0.73	2.12
22	1.24	1.43	1.15	1.54	1.05	1.66	0.96	1.80	0.86	1.94	0.86	1.94	0.77	2.09
23	1.26	1.44	1.17	1.54	1.08	1.66	0.99	1.79	0.90	1.92	0.90	1.92	0.80	2.06
24	1.27	1.45	1.19	1.55	1.10	1.66	1.01	1.78	0.93	1.90	0.93	1.90	0.84	2.04
25	1.29	1.45	1.21	1.55	1.12	1.66	1.04	1.77	0.95	1.89	0.95	1.89	0.87	2.01
26	1.30	1.46	1.22	1.55	1.14	1.65	1.06	1.76	0.98	1.88	0.98	1.87	0.89	1.99
27	1.32	1.47	1.24	1.56	1.16	1.65	1.08	1.76	1.01	1.86	1.00	1.86	0.93	1.97
28	1.33	1.48	1.26	1.56	1.18	1.65	1.10	1.75	1.03	1.85	1.03	1.85	0.95	1.96
29	1.34	1.48	1.27	1.56	1.20	1.65	1.12	1.74	1.05	1.84	1.05	1.84	0.98	1.94
30	1.35	1.49	1.28	1.57	1.21	1.65	1.14	1.74	1.07	1.83	1.07	1.83	1.00	1.93
31	1.36	1.50	1.30	1.57	1.23	1.65	1.16	1.74	1.09	1.83	1.09	1.83	1.02	1.92
32	1.37	1.50	1.31	1.57	1.24	1.65	1.18	1.73	1.11	1.82	1.11	1.82	1.04	1.91
33	1.38	1.51	1.32	1.58	1.26	1.65	1.19	1.73	1.13	1.81	1.13	1.81	1.06	1.90
34	1.39	1.51	1.33	1.58	1.27	1.65	1.21	1.73	1.15	1.81	1.14	1.81	1.08	1.89
35	1.40	1.52	1.34	1.58	1.28	1.65	1.22	1.73	1.16	1.80	1.16	1.80	1.10	1.88
36	1.41	1.52	1.35	1.59	1.29	1.65	1.24	1.73	1.18	1.80	1.18	1.80	1.11	1.88
37	1.42	1.53	1.36	1.59	1.31	1.66	1.25	1.72	1.19	1.80	1.19	1.80	1.13	1.87
38	1.43	1.54	1.37	1.59	1.32	1.66	1.26	1.72	1.21	1.79	1.20	1.79	1.15	1.86
39	1.43	1.54	1.38	1.60	1.33	1.66	1.27	1.72	1.22	1.79	1.22	1.79	1.16	1.86
40	1.44	1.54	1.39	1.60	1.34	1.66	1.29	1.72	1.23	1.79	1.23	1.79	1.18	1.85
45	1.48	1.57	1.43	1.62	1.38	1.67	1.34	1.72	1.29	1.78	1.29	1.78	1.24	1.84
50	1.50	1.59	1.46	1.63	1.42	1.67	1.38	1.72	1.34	1.77	1.34	1.77	1.29	1.82
55	1.53	1.60	1.49	1.64	1.45	1.68	1.41	1.72	1.38	1.77	1.37	1.77	1.33	1.81
60	1.55	1.62	1.51	1.65	1.48	1.69	1.44	1.73	1.41	1.77	1.41	1.77	1.37	1.81
65	1.57	1.63	1.54	1.66	1.50	1.70	1.47	1.73	1.44	1.77	1.44	1.77	1.40	1.81
70	1.58	1.64	1.55	1.67	1.52	1.70	1.49	1.74	1.46	1.77	1.46	1.77	1.43	1.80
75	1.60	1.65	1.57	1.68	1.54	1.71	1.51	1.74	1.49	1.77	1.49	1.77	1.46	1.80
80	1.61	1.66	1.59	1.69	1.56	1.72	1.53	1.74	1.51	1.77	1.51	1.77	1.48	1.80
85	1.62	1.67	1.60	1.70	1.57	1.72	1.55	1.75	1.52	1.77	1.53	1.77	1.50	1.80
90	1.63	1.68	1.61	1.70	1.59	1.73	1.57	1.75	1.54	1.78	1.54	1.78	1.52	1.80
95	1.64	1.69	1.62	1.71	1.60	1.73	1.58	1.75	1.56	1.78	1.56	1.78	1.54	1.80
100	1.65	1.69	1.63	1.72	1.61	1.74	1.59	1.76	1.57	1.78	1.57	1/78	1.55	1.80

Source: J. Durbin and G. S. Watson, "Testing for Serial Correlation in Least Squares Regression. II," *Biometrika*, 1951, vol. 38, pp. 173, 175. Reproduced by permission of the Biometrika Trustees. Values for k = 6 and k = 7 taken from N. E. Savin and K. J. White, "The Durbin-Watson Test for Serial Correlation with Extreme Sample Sizes or Many Regressors." *Econometrica*, 45 (1977): 1989–1996. With permission from The Econometric Society.

TABLE D.11
Critical values of the Durbin-Watson test statistic (continued)

$\alpha = .01$

CT	Number of Independent Variables k									
	1		2		3		4		5	
n	DW_L	DW_U	DW_L	DW_U	DW_L	DW_U	DW_L	DW_U	DW_L	DW_U
15	0.81	1.07	0.70	1.25	0.59	1.46	0.49	1.70	0.39	1.96
16	0.84	1.09	0.74	1.25	0.63	1.44	0.53	1.66	0.44	1.90
17	0.87	1.10	0.77	1.25	0.67	1.43	0.57	1.63	0.48	1.85
18	0.90	1.12	0.80	1.26	0.71	1.42	0.61	1.60	0.52	1.80
19	0.93	1.13	0.83	1.26	0.74	1.41	0.65	1.58	0.56	1.77
20	0.95	1.15	0.86	1.27	0.77	1.41	0.68	1.57	0.60	1.74
21	0.97	1.16	0.89	1.27	0.80	1.41	0.72	1.55	0.63	1.71
22	1.00	1.17	0.91	1.28	0.83	1.40	0.75	1.54	0.66	1.69
23	1.02	1.19	0.94	1.29	0.86	1.40	0.77	1.53	0.70	1.67
24	1.04	1.20	0.96	1.30	0.88	1.41	0.80	1.53	0.72	1.66
25	1.05	1.21	0.98	1.30	0.90	1.41	0.83	1.52	0.75	1.65
26	1.07	1.22	1.00	1.31	0.93	1.41	0.85	1.52	0.78	1.64
27	1.09	1.23	1.02	1.32	0.95	1.41	0.88	1.51	0.81	1.63
28	1.10	1.24	1.04	1.32	0.97	1.41	0.90	1.51	0.83	1.62
29	1.12	1.25	1.05	1.33	0.99	1.42	0.92	1.51	0.85	1.61
30	1.13	1.26	1.07	1.34	1.01	1.42	0.94	1.51	0.88	1.61
31	1.15	1.27	1.08	1.34	1.02	1.42	0.96	1.51	0.90	1.60
32	1.16	1.28	1.10	1.35	1.04	1.43	0.98	1.51	0.92	1.60
33	1.17	1.29	1.11	1.36	1.05	1.43	1.00	1.51	0.94	1.59
34	1.18	1.30	1.13	1.36	1.07	1.43	1.01	1.51	0.95	1.59
35	1.19	1.31	1.14	1.37	1.08	1.44	1.03	1.51	0.97	1.59
36	1.21	1.32	1.15	1.38	1.10	1.44	1.04	1.51	0.99	1.59
37	1.22	1.32	1.16	1.38	1.11	1.45	1.06	1.51	1.00	1.59
38	1.23	1.33	1.18	1.39	1.12	1.45	1.07	1.52	1.02	1.58
39	1.24	1.34	1.19	1.39	1.14	1.45	1.09	1.52	1.03	1.58
40	1.25	1.34	1.20	1.40	1.15	1.46	1.10	1.52	1.05	1.58
45	1.29	1.38	1.24	1.42	1.20	1.48	1.16	1.53	1.11	1.58
50	1.32	1.40	1.28	1.45	1.24	1.49	1.20	1.54	1.16	1.59
55	1.36	1.43	1.32	1.47	1.28	1.51	1.25	1.55	1.21	1.59
60	1.38	1.45	1.35	1.48	1.32	1.52	1.28	1.56	1.25	1.60
65	1.41	1.47	1.38	1.50	1.35	1.53	1.31	1.57	1.28	1.61
70	1.43	1.49	1.40	1.52	1.37	1.55	1.34	1.58	1.31	1.61
75	1.45	1.50	1.42	1.53	1.39	1.56	1.37	1.59	1.34	1.62
80	1.47	1.52	1.44	1.54	1.42	1.57	1.39	1.60	1.36	1.62
85	1.48	1.53	1.46	1.55	1.43	1.58	1.41	1.60	1.39	1.63
90	1.50	1.54	1.47	1.56	1.45	1.59	1.43	1.61	1.41	1.64
95	1.51	1.55	1.49	1.57	1.47	1.60	1.45	1.62	1.42	1.64
100	1.52	1.56	1.50	1.58	1.48	1.60	1.46	1.63	1.44	1.65

TABLE D.12

The Kolmogorov–Smirnov statistic (parameters specified)

		RIGHT-TAIL AREAS		
n	.10	.05	.025	.005
1	.900	.950	.975	.995
2	.684	.776	.842	.929
3	.565	.642	.708	.829
4	.494	.564	.624	.734
5	.446	.510	.563	.669
6	.410	.470	.521	.618
7	.381	.438	.486	.577
8	.358	.411	.457	.543
9	.339	.388	.432	.514
10	.322	.368	.409	.486
11	.307	.352	.391	.468
12	.295	.338	.375	.450
13	.284	.325	.361	.433
14	.274	.314	.349	.418
15	.266	.304	.338	.404
16	.258	.295	.328	.391
17	.250	.286	.318	.380
18	.244	.278	.309	.370
19	.237	.272	.301	.361
20	.231	.264	.294	.352
25	.21	.24	.264	.32
30	.19	.22	.242	.29
35	.18	.21	.23	.27
40			.21	.25
50			.19	.23
60			.17	.21
70			.16	.19
80			.15	.18
90			.14	
100			.14	

Source: L. H. Miller, "Table of Percentage Points of Kolmogorov Statistics." *Journal of the American Statistical Association,* 1956, vol. 51, pp. 111–121, Table 1. Adapted with permission.

TABLE D.13

Bias correction factors for estimating σ when population is normal

For $n > 10$, a satisfactory approximation, accurate to three decimal places, may be obtained from the formula

$$c \approx 1 + \frac{1}{4(n-1)}$$

A technical explanation of this factor is given by A. Hald, *Statistical Theory with Applications* (New York: John Wiley & Sons, 1952), pp. 299–300.

SAMPLE SIZE, n	CORRECTION FACTOR, c
2	1.253
3	1.128
4	1.085
5	1.064
6	1.051
7	1.042
8	1.036
9	1.032
10	1.028

Source: Park J. Ewart, James S. Ford, and Chi-Yuan Lin, *Applied Managerial Statistics* (Englewood Cliffs, N.J.: Prentice-Hall, 1982), p. 588.

**TABLE D.14
Lilliefor's test
statistic for
normality**

SAMPLE SIZE	RIGHT-TAIL AREAS					
	0.200	0.150	0.100	0.050	0.010	0.001
4	.303	.321	.346	.376	.413	.433
5	.289	.303	.319	.343	.397	.439
6	.269	.281	.297	.323	.371	.424
7	.252	.264	.280	.304	.351	.402
8	.239	.250	.265	.288	.333	.384
9	.227	.238	.252	.274	.317	.365
10	.217	.228	.241	.262	.304	.352
11	.208	.218	.231	.251	.291	.338
12	.200	.210	.222	.242	.281	.325
13	.193	.202	.215	.234	.271	.314
14	.187	.196	.208	.226	.262	.305
15	.181	.190	.201	.219	.254	.296
16	.176	.184	.195	.213	.247	.287
17	.171	.179	.190	.207	.240	.279
18	.167	.175	.185	.202	.234	.273
19	.163	.170	.181	.197	.228	.266
20	.159	.166	.176	.192	.223	.260
25	.143	.150	.159	.173	.201	.236
30	.131	.138	.146	.159	.185	.217
40	.115	.120	.128	.139	.162	.189
100	.074	.077	.082	.089	.104	.122
400	.037	.039	.041	.045	.052	.061
900	.025	.026	.028	.030	.035	.042

Source: G. E. Dallas and Leland Wilkinson, "An Analytic Approximation to the Distribution of Lilliefors's Test Statistic for Normality," *The American Statistician,* November 1986, vol. 40, no. 4, p. 295.

**TABLE D.15
Critical values of
Spearman's rank
correlation
coefficient**

The α values correspond to a one-tailed test of H_0: $\rho_s = 0$. The value should be doubled for two-tailed tests.

n	$\alpha = .05$	$\alpha = .025$	$\alpha = .01$	$\alpha = .005$	n	$\alpha = .05$	$\alpha = .025$	$\alpha = .01$	$\alpha = .005$
5	.900				18	.399	.476	.564	.625
6	.829	.886	.943		19	.388	.462	.549	.608
7	.714	.786	.893		20	.377	.450	.534	.591
8	.643	.738	.833	.881	21	.368	.438	.521	.576
9	.600	.683	.783	.833	22	.359	.428	.508	.562
10	.564	.648	.745	.794	23	.351	.418	.496	.549
11	.523	.623	.736	.818	24	.343	.409	.485	.537
12	.497	.591	.703	.780	25	.336	.400	.475	.526
13	.475	.566	.673	.745	26	.329	.392	.465	.515
14	.457	.545	.646	.716	27	.323	.385	.456	.505
15	.441	.525	.623	.689	28	.317	.377	.448	.496
16	.425	.507	.601	.666	29	.311	.370	.440	.487
17	.412	.490	.582	.645	30	.305	.364	.432	.478

Source: E. G. Olds, "Distribution of Sums of Squares of Rank Differences for Small Samples," *Annals of Mathematical Statistics,* 1938, 9. Reproduced with the permission of the Editor, *Annals of Mathematical Statistics.*

TABLE D.16

Critical values of T in the Wilcoxon signed-rank test

One-Tailed	Two-Tailed	$n=5$	$n=6$	$n=7$	$n=8$	$n=9$	$n=10$
$\alpha=.05$	$\alpha=.10$	1	2	4	6	8	11
$\alpha=.025$	$\alpha=.05$		1	2	4	6	8
$\alpha=.01$	$\alpha=.02$			0	2	3	5

One-Tailed	Two-Tailed	$n=11$	$n=12$	$n=13$	$n=14$	$n=15$	$n=16$
$\alpha=.05$	$\alpha=.10$	14	11	21	26	30	36
$\alpha=.025$	$\alpha=.05$	11	14	11	21	25	30
$\alpha=.01$	$\alpha=.02$	7	10	13	16	20	24
$\alpha=.005$	$\alpha=.01$	5	7	10	13	16	19

One-Tailed	Two-Tailed	$n=17$	$n=18$	$n=19$	$n=20$	$n=21$	$n=22$
$\alpha=.05$	$\alpha=.10$	41	47	54	60	68	75
$\alpha=.025$	$\alpha=.05$	35	40	46	52	59	66
$\alpha=.01$	$\alpha=.02$	28	33	38	43	49	56
$\alpha=.005$	$\alpha=.01$	23	28	32	37	43	49

One-Tailed	Two-Tailed	$n=23$	$n=24$	$n=25$	$n=26$	$n=27$	$n=28$
$\alpha=.05$	$\alpha=.10$	83	92	101	110	120	130
$\alpha=.025$	$\alpha=.05$	73	81	90	98	107	117
$\alpha=.01$	$\alpha=.02$	62	69	77	85	93	102
$\alpha=.005$	$\alpha=.01$	55	61	68	76	84	92

One-Tailed	Two-Tailed	$n=29$	$n=30$	$n=31$	$n=32$	$n=33$	$n=34$
$\alpha=.05$	$\alpha=.10$	141	152	163	175	188	201
$\alpha=.025$	$\alpha=.05$	127	137	148	159	171	183
$\alpha=.01$	$\alpha=.02$	111	120	130	141	151	162
$\alpha=.005$	$\alpha=.01$	100	109	118	128	138	149

One-Tailed	Two-Tailed	$n=35$	$n=36$	$n=37$	$n=38$	$n=39$
$\alpha=.05$	$\alpha=.10$	214	228	242	256	271
$\alpha=.025$	$\alpha=.05$	195	208	222	235	250
$\alpha=.01$	$\alpha=.02$	174	186	198	211	224
$\alpha=.005$	$\alpha=.01$	160	171	183	195	208

One-Tailed	Two-Tailed	$n=40$	$n=41$	$n=42$	$n=43$	$n=44$	$n=45$
$\alpha=.05$	$\alpha=.10$	287	303	319	336	353	371
$\alpha=.025$	$\alpha=.05$	264	279	295	311	327	344
$\alpha=.01$	$\alpha=.02$	238	252	267	281	297	313
$\alpha=.005$	$\alpha=.01$	221	234	248	262	277	292

One-Tailed	Two-Tailed	$n=46$	$n=47$	$n=48$	$n=49$	$n=50$
$\alpha=.05$	$\alpha=.10$	389	408	427	446	466
$\alpha=.025$	$\alpha=.05$	361	379	397	415	434
$\alpha=.01$	$\alpha=.02$	329	345	362	380	398
$\alpha=.005$	$\alpha=.01$	307	323	339	356	373

Source: F. Wilcoxon and R. A. Wilcox, "Some Rapid Approximate Statistical Procedures," 1964, p. 28. Reproduced with the permission of American Cyanamid Company.

TABLE D.17
Percentage points of the Studentized range

Upper 5% points

						k				
$\Sigma n_J - k$	2	3	4	5	6	7	8	9	10
1	17.97	26.98	32.82	37.08	40.41	43.12	45.40	47.36	49.07
2	6.08	8.33	9.80	10.88	11.74	12.44	13.03	13.54	13.99
3	4.50	5.91	6.82	7.50	8.04	8.48	8.85	9.18	9.46
4	3.93	5.04	5.76	6.29	6.71	7.05	7.35	7.60	7.83
5	3.64	4.60	5.22	5.67	6.03	6.33	6.58	6.80	6.99
6	3.46	4.34	4.90	5.30	5.63	5.90	6.12	6.32	6.49
7	3.34	4.16	4.68	5.06	5.36	5.61	5.82	6.00	6.16
8	3.26	4.04	4.53	4.89	5.17	5.40	5.60	5.77	5.92
9	3.20	3.95	4.41	4.76	5.02	5.24	5.43	5.59	5.74
10	3.15	3.88	4.33	4.65	4.91	5.12	5.30	5.46	5.60
11	3.11	3.82	4.26	4.57	4.82	5.03	5.20	5.35	5.49
12	3.08	3.77	4.20	4.51	4.75	4.95	5.12	5.27	5.39
13	3.06	3.73	4.15	4.45	4.69	4.88	5.05	5.19	5.32
14	3.03	3.70	4.11	4.41	4.64	4.83	4.99	5.13	5.25
15	3.01	3.67	4.08	4.37	4.59	4.78	4.94	5.08	5.20
16	3.00	3.65	4.05	4.33	4.56	4.74	4.90	5.03	5.15
17	2.98	3.63	4.02	4.30	4.52	4.70	4.86	4.99	5.11
18	2.97	3.61	4.00	4.28	4.49	4.67	4.82	4.96	5.07
19	2.96	3.59	3.98	4.25	4.47	4.65	4.79	4.92	5.04
20	2.95	3.58	3.96	4.23	4.45	4.62	4.77	4.90	5.01
24	2.92	3.53	3.90	4.17	4.37	4.54	4.68	4.81	4.92
30	2.89	3.49	3.85	4.10	4.30	4.46	4.60	4.72	4.82
40	2.86	3.44	3.79	4.04	4.23	4.39	4.52	4.63	4.73
60	2.83	3.40	3.74	3.98	4.16	4.31	4.44	4.55	4.65
120	2.80	3.36	3.68	3.92	4.10	4.24	4.36	4.47	4.56
∞	2.77	3.31	3.63	3.86	4.03	4.17	4.29	4.39	4.47

						k				
$\Sigma n_J - k$	11	12	13	14	15	16	17	18	19	20
1	50.59	51.96	53.20	54.33	55.36	56.32	57.22	58.04	58.83	59.56
2	14.39	14.75	15.08	15.38	15.65	15.91	16.14	16.37	16.57	16.77
3	9.72	9.95	10.15	10.35	10.52	10.69	10.84	10.98	11.11	11.24
4	8.03	8.21	8.37	8.52	8.66	8.79	8.91	9.03	9.13	9.23
5	7.17	7.32	7.47	7.60	7.72	7.83	7.93	8.03	8.12	8.21
6	6.65	6.79	6.92	7.03	7.14	7.24	7.34	7.43	7.51	7.59
7	6.30	6.43	6.55	6.66	6.76	6.85	6.94	7.02	7.10	7.17
8	6.05	6.18	6.29	6.39	6.48	6.57	6.65	6.73	6.80	6.87
9	5.87	5.98	6.09	6.19	6.28	6.36	6.44	6.51	6.58	6.64
10	5.72	5.83	5.93	6.03	6.11	6.19	6.27	6.34	6.40	6.47
11	5.61	5.71	5.81	5.90	5.98	6.06	6.13	6.20	6.27	6.33
12	5.51	5.61	5.71	5.80	5.88	5.95	6.02	6.09	6.15	6.21
13	5.43	5.53	5.63	5.71	5.79	5.86	5.93	5.99	6.05	6.11
14	5.36	5.46	5.55	5.64	5.71	5.79	5.85	5.91	5.97	6.03
15	5.31	5.40	5.49	5.57	5.65	5.72	5.78	5.85	5.90	5.96
16	5.26	5.35	5.44	5.52	5.59	5.66	5.73	5.79	5.84	5.90
17	5.21	5.31	5.39	5.47	5.54	5.61	5.67	5.73	5.79	5.84
18	5.17	5.27	5.35	5.43	5.50	5.57	5.63	5.69	5.74	5.79
19	5.14	5.23	5.31	5.39	5.46	5.53	5.59	5.65	5.70	5.75
20	5.11	5.20	5.28	5.36	5.43	5.49	5.55	5.61	5.66	5.71
24	5.01	5.10	5.18	5.25	5.32	5.38	5.44	5.49	5.55	5.59
30	4.92	5.00	5.08	5.15	5.21	5.27	5.33	5.38	5.43	5.47
40	4.82	4.90	4.98	5.04	5.11	5.16	5.22	5.27	5.31	5.36
60	4.73	4.81	4.88	4.94	5.00	5.06	5.11	5.15	5.20	5.24
120	4.64	4.71	4.78	4.84	4.90	4.95	5.00	5.04	5.09	5.13
∞	4.55	4.62	4.68	4.74	4.80	4.85	4.89	4.93	4.97	5.01

Upper 1% points

k

$\Sigma n_J - k$	2	3	4	5	6	7	8	9	10
1	90.03	135.0	164.3	185.6	202.2	215.8	227.2	237.0	245.6
2	14.04	19.02	22.29	24.72	26.63	28.20	29.53	30.68	31.69
3	8.26	10.62	12.17	13.33	14.24	15.00	15.64	16.20	16.69
4	6.51	8.12	9.17	9.96	10.58	11.10	11.55	11.93	12.27
5	5.70	6.98	7.80	8.42	8.91	9.32	9.67	9.97	10.24
6	5.24	6.33	7.03	7.56	7.97	8.32	8.61	8.87	9.10
7	4.95	5.92	6.54	7.01	7.37	7.68	7.94	8.17	8.37
8	4.75	5.64	6.20	6.62	6.96	7.24	7.47	7.68	7.86
9	4.60	5.43	5.96	6.35	6.66	6.91	7.13	7.33	7.49
10	4.48	5.27	5.77	6.14	6.43	6.67	6.87	7.05	7.21
11	4.39	5.15	5.62	5.97	6.25	6.48	6.67	6.84	6.99
12	4.32	5.05	5.50	5.84	6.10	6.32	6.51	6.67	6.81
13	4.26	4.96	5.40	5.73	5.98	6.19	6.37	6.53	6.67
14	4.21	4.89	5.32	5.63	5.88	6.08	6.26	6.41	6.54
15	4.17	4.84	5.25	5.56	5.80	5.99	6.16	6.31	6.44
16	4.13	4.79	5.19	5.49	5.72	5.92	6.08	6.22	6.35
17	4.10	4.74	5.14	5.43	5.66	5.85	6.01	6.15	6.27
18	4.07	4.70	5.09	5.38	5.60	5.79	5.94	6.08	6.20
19	4.05	4.67	5.05	5.33	5.55	5.73	5.89	6.02	6.14
20	4.02	4.64	5.02	5.29	5.51	5.69	5.84	5.97	6.09
24	3.96	4.55	4.91	5.17	5.37	5.54	5.69	5.81	5.92
30	3.89	4.45	4.80	5.05	5.24	5.40	5.54	5.65	5.76
40	3.82	4.37	4.70	4.93	5.11	5.26	5.39	5.50	5.60
60	3.76	4.28	4.59	4.82	4.99	5.13	5.25	5.36	5.45
120	3.70	4.20	4.50	4.71	4.87	5.01	5.12	5.21	5.30
∞	3.64	4.12	4.40	4.60	4.76	4.88	4.99	5.08	5.16

k

$\Sigma n_J - k$	11	12	13	14	15	16	17	18	19	20
1	253.2	260.0	266.2	271.8	277.0	281.8	286.3	290.4	294.3	298.0
2	32.59	33.40	34.13	34.81	35.43	36.00	36.53	37.03	37.50	37.95
3	17.13	17.53	17.89	18.22	18.52	18.81	19.07	19.32	19.55	19.77
4	12.57	12.84	13.09	13.32	13.53	13.73	13.91	14.08	14.24	14.40
5	10.48	10.70	10.89	11.08	11.24	11.40	11.55	11.68	11.81	11.93
6	9.30	9.48	9.65	9.81	9.95	10.08	10.21	10.32	10.43	10.54
7	8.55	8.71	8.86	9.00	9.12	9.24	9.35	9.46	9.55	9.65
8	8.03	8.18	8.31	8.44	8.55	8.66	8.76	8.85	8.94	9.03
9	7.65	7.78	7.91	8.03	8.13	8.23	8.33	8.41	8.49	8.57
10	7.36	7.49	7.60	7.71	7.81	7.91	7.99	8.08	8.15	8.23
11	7.13	7.25	7.36	7.46	7.56	7.65	7.73	7.81	7.88	7.95
12	6.94	7.06	7.17	7.26	7.36	7.44	7.52	7.59	7.66	7.73
13	6.79	6.90	7.01	7.10	7.19	7.27	7.35	7.42	7.48	7.55
14	6.66	6.77	6.87	6.96	7.05	7.13	7.20	7.27	7.33	7.39
15	6.55	6.66	6.76	6.84	6.93	7.00	7.07	7.14	7.20	7.26
16	6.46	6.56	6.66	6.74	6.82	6.90	6.97	7.03	7.09	7.15
17	6.38	6.48	6.57	6.66	6.73	6.81	6.87	6.94	7.00	7.05
18	6.31	6.41	6.50	6.58	6.65	6.73	6.79	6.85	6.91	6.97
19	6.25	6.34	6.43	6.51	6.58	6.65	6.72	6.78	6.84	6.89
20	6.19	6.28	6.37	6.45	6.52	6.59	6.65	6.71	6.77	6.82
24	6.02	6.11	6.19	6.26	6.33	6.39	6.45	6.51	6.56	6.61
30	5.85	5.93	6.01	6.08	6.14	6.20	6.26	6.31	6.36	6.41
40	5.69	5.76	5.83	5.90	5.96	6.02	6.07	6.12	6.16	6.21
60	5.53	5.60	5.67	5.73	5.78	5.84	5.89	5.93	5.97	6.01
120	5.37	5.44	5.50	5.56	5.61	5.66	5.71	5.75	5.79	5.83
∞	5.23	5.29	5.35	5.40	5.45	5.49	5.54	5.57	5.61	5.65

Appendix E

Data Sets—Definitions and Descriptions

DATA SET	VARIABLE DEFINITIONS	N*	EXERCISES
1. Economic and education characteristics for U.S. cities (Source: *Statistical Abstract of the United States*)	1. City name up to 10 letters 2. Population in millions in 1979 3. Per capita income in 1979 in thousands of dollars 4. Number of economics masters degrees awarded between June 30, 1982, and July 1, 1985 5. Number of business and management masters degrees awarded between June 30, 1982 and July 1, 1985 6. Dummy variable that equals 1 if there is an institution that grants Ph.D. degrees in economics within 75 miles 7. Dummy variable that equals 1 if the city is a coastal city. Any city within 50 miles of either coast is defined as a coastal city	44	A2.1 A13.1 A14.1
2. Automobile horsepower (Source: Federal Trade Commission)	1. The horsepower of 392 car models sold in the United States during the years 1971 to 1983	392	A2.2 A8.1
3. Tar content of cigarettes	1. Tar content by brand of cigarette in milligrams	203	A2.3 A8.2
4. Sales and assets for large foreign firms (Source: *Fortune* Magazine)	1. Name of company 2. Sales in millions of dollars in 1985 3. Assets in millions of dollars at the end of 1985 4. Dummy variable that equals 1 if the company is Japanese 5. Dummy variable that equals 1 if the company is in the oil business	25	A3.1
5. H. & R. Block stock price and volume data (Source: Compuserve Information Service)	1. Dates recorded as month/day/year 2. Monthly sales volumes in shares traded 3. End-of-month per share prices	156	A3.3 A13.2
6. Reliability index for cars by national origin (Source: *Consumer Reports*)	1. Code for car model: A = AMC, AU = Audi, BM = BMW, B = Buick, C = Cadillac, CH = Chevrolet, CHR = Chrysler, DA = Datsun, D = Dodge, F = Ford, H = Honda, HY = Hyundai, I = Isuzu, J = Jaguar, L = Lincoln, MA = Mazda, ME = Mercedes, M = Mercury, MEK = Merkur, MI = Mitsubishi, N = Nissan, O = Oldsmobile, P = Plymouth, PO = Pontiac, R = Renault, S = Saab, SU = Subaru, T = Toyota, V = Volkswagon, VO = Volvo, 6 = 6 cylinders, 8 = 8 cylinders 2. Year of car model 3. Code for home country: 1 = U.S., 2 = Japan, 3 = Germany, 4 = Sweden 4. Model reliability index 5. Dummy variable that equals 1 if the model is from 1981 6. Dummy variable that equals 1 if the model is from 1982 7. Dummy variable that equals 1 if the model is from 1983 8. Dummy variable that equals 1 if the model is from 1984 9. Dummy variable that equals 1 if the model is from 1985 10. Dummy variable that equals 1 if the model is from 1986	703	A4.1† A4.2† A8.3†
7. Economists' forecasts (Source: *Wall Street Journal*)	1. Names of each economist, up to 10 letters 2. Rate of growth forecast for GNP for 1st half of 1987 made in December 1986 3. Rate of growth forecast for GNP for 2nd half of 1987 made in June 1987 4. Percentage change in CPI forecast for 1st half of 1987 made in December 1986	35	A13.3

*Number of observations.
†Cannot be performed using Data Desk.
‡Cannot be performed using MINITAB.

DATA SET	VARIABLE DEFINITIONS	N*	EXERCISES
	5. Percentage change in CPI forecast for 2nd half of 1987 made in June 1987 6. Unemployment rate forecast for May 1987 made in December 1987 7. Unemployment rate forecast for November 1987 made in June 1987 Missing observations are coded as -99999		
8. SAT scores, high school class rankings, and grades (Source: Indiana University—Purdue University at Indianapolis registrar)	1. Letter grades coded A+ = 98, A = 95, A− = 92, B+ = 88, B = 85, B− = 82, C+ = 78, C = 75, C− = 72, D+ = 68, D = 65, D− = 62, F = 55, fail on pass/fail option = -99999, pass on pass/fail option = -99999 2. High school graduating class rank in percentiles. Missing data coded as -99999 3. Ethnic status. 0 = white, 1 = nonwhite 4. SAT score for math portion. Missing data coded as -99999 5. SAT score for verbal portion. Missing data coded as -99999 6. Age in years 7. Dummy variable for two instructors 8. Dummy variable coded 1 for the presence of an SAT score	340	A10.4†
9. Coke versus Pepsi preference and educational levels (Source: Pepsi-Cola Bottling Co., Inc., of Indianapolis)	1. Education levels of respondents in years 2. Dummy variable coded as 1 for Pepsi loyalist and 0 for Coke loyalist	295	A10.1 A14.3†‡ A17.3
10. 75 1's and 25 0's for sampling problem	1. 75 1's and 25 0's for sampling problem	100	A7.3 A9.1†
11. A 1-in-20 sample of colleges enrollment and faculty (Source: *Peterson's Guides*)	1. Student enrollment for the 1985–1986 academic year 2. Full-time faculty. Missing data coded as -99999 3. Dummy variable coded 1 for private institution and 0 for public	70	A9.2 A10.2 A10.3
12. Preferences over granola bar formulations (Fabricated data)	1. Dummy variable coded 0 for original formulation of the granola bar and 1 for the prototype formulation 2. Ratings coded as 5 = definitely will buy, 4 = probably will buy, 3 = might or might not buy, 4 = probably won't buy, and 5 = definitely won't buy	400	A11.1
13. Incidence of byssinosis by duration of employment (Source: *International Statistics Review,* vol. 45, pp. 51–62)	1. Dummy variable coded 1 for the presence of byssinosis and 0 for its absence among white male smokers who were employed in factory with exposure to cotton dust 2. Years of employment coded 1 for under 10 years, 2 for 10 and under 20 years, and 3 for more than 20 3. Dummy variable coded 0 for a high dust environment, and 1 for a low dust environment	1132†	A11.2
14. Deaths per year from horse kicks in Prussian Army (Source: D. F. Andrews and A. M. Herzberg, *Data,* New York: Springer-Verlag, 1985)	1. Year (1875–1894) 2. Number of deaths observed in each year 3. Categories of 0 through 18 or more deaths per year 4. Observed frequency of each category 5. Expected frequency based on the null hypothesis that the number of deaths follows a Poisson distribution with the sample mean of 9.8	20	A11.3
15. Observed and expected frequency of Papal elections (Source: *The World Almanac and Book of Facts,* 1987)	1. Years between Papal elections from 0 through 14 or more 2. Observed number of Papal elections for each number of years 3. Expected number of Papal elections per year based on the null hypothesis that the time between elections follows an exponential distribution with a mean of 7.328 years	15	A11.4

*Number of observations.
†Cannot be performed using Data Desk.
‡Cannot be performed using MINITAB.

DATA SET	VARIABLE DEFINITIONS	N^*	EXERCISES
16. Mutual funds rates of return and loads (Source: *Consumer Reports*)	1. Annual rates of returns on mutual funds for the period April 1, 1982 through March 31, 1987 recorded as integers, i.e., 24% as 24 2. Loads for each mutual fund expressed as integers, i.e., 5% as 5. The Category 12b-1 mutual funds that charge no load but do charge a fee for advertising and marketing are coded as 1000 to accommodate programs that require integer category identifiers	288	A12.1
17. TIAA-CREF bond purchases, coupon rate versus maturity (Source: TIAA-CREF 1988 Annual Report)	1. Coupon rates for 304 bonds acquired by TIAA-CREF in 1986 2. Maturities coded as 0 for bonds that mature in under 10 years, 1 for bonds that mature from 10 to under 20 years, 2 for bonds that mature from 20 to under 30 years, 3 for bonds that mature from 30 to under 40 years, and 4 for bonds that mature in more than 40 years	304	A12.2
18. Rubbing compounds for hardwood flooring (Fabricated data)	1. Time in seconds for a high-speed polisher to polish a block of hardwood 2. Codes of 1, 2, 3, 4, and 5 for five chemical binders used as treatments in the test 3. Codes of 1, 2, 3, and 4 for the four types of hardwood used as a blocking factor in the test	100	A12.3
19. Blue Cross survey on health care for elderly (Source: Blue Cross of Indiana)	1. Number of hospital days during the past year 2. Number of regular doctor visits during the past year 3. Number of visits to specialists or referral doctors during the past year 4. Respondent's average monthly spending for prescription drugs 5. Respondent's age	621	A14.2
20. Child car seat usage survey (Source: Center for Auto Safety, Riley's Children's Hospital)	1. Age of the driver in years 2. Hours per year available for housework by the driver. (Defined as 2000 hours minus the annual hours of market work) 3. Number of children four years of age and under in the car 4. Dummy variable equal to 1 if the driver is married and equal to 0 otherwise 5. Nonwage income of the driver (the mother in all cases), approximated by subtracting the projected annual earnings of the driver from the reported annual family income 6. Years of education of the driver's spouse 7. Hourly wage of the driver 8. Driver's years of education 9. Dummy variable equal to 1 if a federally approved car seat is rated as correctly used and equal to 0 otherwise	187	A14.4
21. Bookstore prices (Source: IUPUI faculty survey)	1. Prices of business books at the IUPUI bookstore 2. Prices of the same books at the IU-Bloomington bookstore 3. Prices of the same books at a private bookstore	13	A17.2 A17.4
22. Number of professors versus GNP (Source: Thomas Kurean, *The World Book of Rankings,* 1979)	1. Names of the 30 countries that had the largest numbers of professors in 1978, up to 10 letters 2. Number of professors per country 3. GNP per country in 1978	30	A17.1
23. Arbuthnot's data on sex ratios (Source: *Philosophical Transactions of the Royal Society of London,* vol. 27, 1710)	1. Year from 1629 to 1710 2. Number of male births 3. Number of female births	82	A17.5 A17.7

* Number of observations.
† Cannot be performed using Data Desk.
‡ Cannot be performed using MINITAB.

DATA SET	VARIABLE DEFINITIONS	N*	EXERCISES
24. Life insurance in force in the U.S. (Source: American Council of Life Insurance)	1. Number of ordinary life insurance policies sold in millions of policies 2. Year beginning with 1945 and ending with 1984 3. Dollar benefits of ordinary life insurance policies sold in millions of dollars 4. Number of group life insurance policies sold in millions of policies 5. Dollar benefits of group life insurance policies sold in millions of dollars 6. Number of industrial life insurance policies sold in millions of policies 7. Dollar benefits of industrial life insurance policies sold in millions of dollars 8. Number of credit life insurance policies sold in millions of policies 9. Total dollar benefits of credit life insurance policies sold in millions of dollars	40	A16.1
25. Wholesale price of turkeys sold in New York City from January 1973 through September 1986 (Source: Economic Research Service, U.S.D.A.)	1. Trend term running from 1 to 165 in integers 2. Wholesale prices of ready-to-cook turkeys (hens, 8–16 lb) sold in New York City. Prior to 1980 the prices are for frozen eviscerated tom turkeys 3. Dummy variable that equals 1 for tom turkeys and 0 for hens 4. Dummy variable set to 1 for the months April, May, and June 5. Dummy variable set to 1 for the months July, August, and September 6. Dummy variable set to 1 for the months October, November, and December	165	A15.1
26. Inventories held by all durable goods manufacturers in the United States from January 1967 through December 1986 (Source: Bureau of Economic Analysis, U.S. Department of Commerce)	1. Dates recorded as year-month. For example, January 1967 is coded 6701. The beginning date is January 1967 and the ending date is January 1987 2. Value of finished goods inventories in dollars (end of month) 3. Value of work-in-process inventories in dollars (end of month) 4. Value of raw material inventories in dollars (end of month) 5. Sales per month in dollars 6. New orders 7. Price deflator 8. Raw material price index 9. Commercial paper rate All data are in constant dollars and seasonally adjusted	240	A15.2
27. Inventories for all non-durable manufactured goods in the United States from January 1967 through December 1986 (Source: Bureau of Economic Analysis, U.S. Department of Commerce)	1. Dates recorded as year-month. For example, January 1967 is coded 6701. The beginning date is January 1967 and the ending date is January 1987 2. Value of finished goods inventories in dollars (end of month) 3. Value of work-in-process inventories in dollars (end of month) 4. Value of raw material inventories in dollars (end of month) 5. Sales per month in dollars 6. New orders 7. Price deflator 8. Raw material price index 9. Commercial paper rate All data are in constant dollars and seasonally adjusted	240	A15.3
28. Wholesale prices and quantities of turkeys, hogs, and calves from January 1981 through September 1986 (Source: Economic Research Service, U.S.D.A.)	1. Date recorded as month/year 2. Wholesale price of turkeys in cents per pound 3. Certified federally inspected turkey slaughter in the United States in millions of pounds 4. Average prices received by farmers for hogs in the United States in cents per pound 5. Federally inspected hogs slaughtered in the United States in thousands of head 6. Average price received by farmers for calves in dollars per 100 pounds 7. Federally inspected slaughter of calves and vealers in the U.S. in thousands of head	69	A16.2

* Number of observations.
† Cannot be performed using Data Desk.
‡ Cannot be performed using MINITAB.

DATA SET	VARIABLE DEFINITIONS	N*	EXERCISES
29. U.S. electric utility sales in kilowatt-hours and prices in dollars for the categories: residential, commercial, and industrial (Source: *Statistical Yearbook of the Electric Utility Industry,* 1986)	1. Year from 1966 to 1986 2. Number of kilowatt-hours sold by all U.S. utilities to residential customers 3. Average price of a kilowatt-hour sold by all U.S. utilities to residential customers 4. Number of kilowatt-hours sold by all U.S. utilities to commercial customers 5. Average price of a kilowatt-hour sold by all U.S. utilities to commercial customers 6. Number of kilowatt-hours sold by all U.S. utilities to industrial customers 7. Average price of a kilowatt-hour sold by all U.S. utilities to industrial customers	21	A16.2
30. 1937 Gallup survey on the 1936 presidential election and phone or automobile ownership (Source: American Institute of Public Opinion)	1. Vote in 1936 presidential election coded as 1 for Roosevelt, 2 for Landon 2. Ownership status coded as 1 for car and phone, 2 for car and no phone, 3 for phone and no car, 4 for neither car nor phone	2277†	A4.3†

*Number of observations.
†Cannot be performed using Data Desk.
‡Cannot be performed using MINITAB.

Appendix F
Answers to Odd-Numbered Exercises

■ CHAPTER 2

2.1(a) Birth dates are interval type data. The zero is not the beginning of time, but, in the Christian calendar, the birth of Christ. The starting point in the Moslem and the Jewish calendars are at different times. Birth dates are discrete; the minimum recorded increment is usually a day.

(b) Dress sizes are interval type data. A zero size dress exists. It is not invisible. Dress sizes are discrete data recorded in integer or half units.

(c) Grade point averages are ratio type data. A zero grade point average is the absence of any "honor" points. Grade point averages are continuous data. Some schools compute G.P.A.s to two decimal places on a four-point scale, others to three decimal places on a four-point scale. While there are intervals between possible recorded values, the divisions are fine enough to treat the data as continuous.

(d) I.Q. scores are interval type data. The zero point is an arbitrary level of intelligence. I.Q. scores are discrete data. They are recorded in integer values.

2.3(a) Treating each candidate as an observation, the number of votes is ratio data. If each voter was an observation, the data would be nominal.

(b) The order of finish is ordinal data.

(c) The categories of loser and winner are nominal data.

2.5(a) Ratio data on the time taken by each antacid to provide relief.

(b) Treating each doctor as an observation, the number of doctors recommending each type of pain pill is nominal data. Treating each type of pain pill as an observation, the number of doctors recommending it is discrete-ratio data.

(c) Puffery.

(d) This borders on puffery since "fine automobiles" is not defined. If fine automobile was defined as a Mercedes, the claim would be self-evident. If a larger set of automobiles, say Lincoln, Cadillac, Mercedes, and Rolls-Royce, were defined as "fine automobiles" and their owners were asked to name their preferred model, the resulting data would be nominal. Each fine automobile owner would be an observation.

2.7 The answer to this exercise will depend on the specific frequency classes chosen by the student. One possibility is shown here.

Class	Frequency	Relative frequency	Cumulative frequency
30 and under 40	2	0.03	2
40 and under 50	4	0.06	6
50 and under 60	24	0.35	30
60 and under 70	13	0.19	43
70 and under 80	10	0.14	53
80 and under 90	11	0.16	64
90 and under 100	5	0.07	69

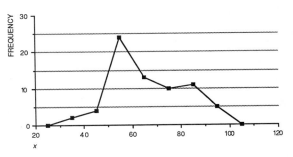

Frequency Polygon

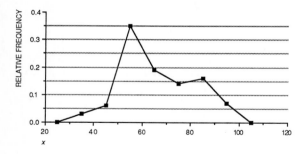

Relative Frequency Polygon

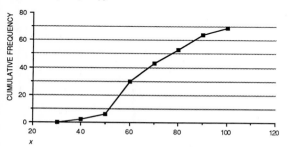

Cumulative Frequency Polygon

2.9

Class	Frequency	Width	Density
0 through 9	20	9	2.22
10 through 19	50	9	5.56
20 through 29	80	9	8.89
30 through 35	50	5	10.00

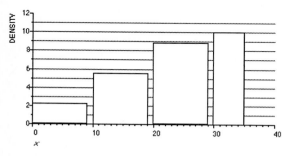

2.11(a) A cumulative frequency polygon can be used to show the level of annual sales needed to be in the top ten. A bar chart that names each representative on the x axis and shows the sales level on the y axis could

also be used. If the bars are ordered by sales level, it will be easier to see the level needed to be in the top ten.

(b) You have to use a vertical line chart to preserve all of the information in the raw data. The x axis would be hours in integer units and the y axis the frequency.

(c) Since the question calls for the percentage gain in sales for each representative, you will have to use a bar chart that names each representative on the x axis and shows the percent gain in sales on the y axis. If any of the percentage changes are negative, the chart may be easier to read with the axes reversed.

(d) The best chart for showing the proportion in each range of sales is a relative frequency polygon.

2.13 A vertical line chart would preserve all of the original information. Fourteen levels for the x axis is high for a vertical line chart. A histogram would be easier to read and provide an adequate representation at the price of losing some of the original information.

2.15(a) If you treat each head of cattle as an observation, these are nominal type data.

(b) Treating each month as an observation, these are discrete–ratio type data.

(c) The revenue from each head of cattle is continuous ratio data. The data are likely to be recorded to the penny. This is fine enough to be considered continuous.

(d) These are nominal data.

2.17 Two juxtaposed bar charts will facilitate such a comparison.

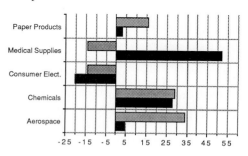

■ 1985
▦ 1986

2.19(a) The financial pages of most newspapers carry partial listings of bond prices. Standard & Poor's publishes a detailed monthly listing. Discrete ratio data.

(b) The financial pages of most newspapers give the complete listing of NYSE closing prices. Discrete ratio data.

(c) The *Statistical Abstract of the United States* has a historical listing. Continuous ratio data.

(d) The *Federal Reserve Bulletin*. Continuous ratio data.

2.21

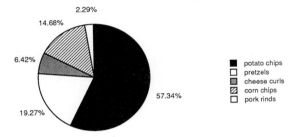

Proportion of Total Sales by Product Category for the Better Chip Company

2.23 Worksheet

Class	Frequency	Cumulative frequency
0 and under 50,000	45	45
50,000 and under 75,000	15	60
75,000 and under 90,000	11	71
90,000 and under 105,000	16	87
105,000 and under 120,000	8	95
120,000 and under 135,000	3	98
135,000 and under 150,000	1	99
150,000 or more	1	100

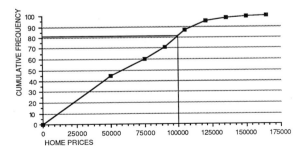

There were approximately 19 homes selling for more than $100,000. Note the accompanying graph is not accurate for the 100th home. The price of the 100th home is not known, but it was assumed to be below $165,000 to close off the graph. This assumption does not affect the estimate of the number of homes selling above $100,000.

2.25

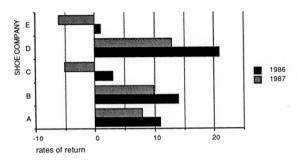

2.27 The answer to this exercise will depend on the specific frequency classes chosen by the student. One possibility is shown here. The following worksheet is used for problems 2.27 and 2.28.

Class	Frequency	Cumulative frequency
above 1.0 to 1.5	8	8
above 1.5 to 2.0	6	14
above 2.0 to 2.5	5	19
above 2.5 to 3.0	9	28
above 3.0 to 3.5	4	32
above 3.5 to 4.0	7	39
above 4.0 to 4.5	5	44
above 4.5 to 5.0	6	50

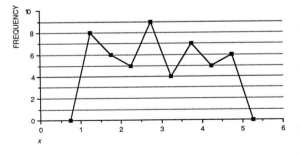

2.29 Worksheet

Salary range	Frequency	Width	Density
0 and under 15,000	2	15,000	0.00013
15,000 and under 20,000	5	5,000	0.00100
20,000 and under 25,000	24	5,000	0.00480
25,000 and under 30,000	30	5,000	0.00600
30,000 and under 35,000	12	5,000	0.00240
35,000 or more	8	19,000	0.00042

Note: The density for each range is the frequency divided by the width of the range. For example, the density for the first range is 2 divided by 15,000. The last range has a width of 19,000 because the highest observation is $54,000.

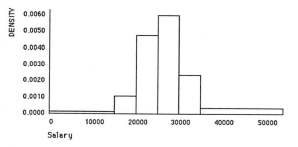

Histogram of Professors' Salaries

2.31(a)

Range of minutes	Frequency	Width	Density
0 and under 30	8	30	0.27
30 and under 35	59	5	11.80
35 and under 40	98	5	19.60
40 and under 45	67	5	13.40
45 and under 50	43	5	8.60
50 and under 55	30	5	6.00
55 and under 60	2	5	0.40
60 and under 65	8	5	1.60
65 and under 70	3	5	0.60

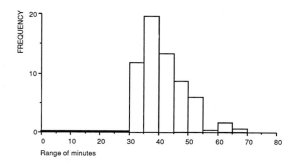

Histogram of Compact Disc Playing Times

(b) There are 318 compact discs in the data set. Of these, 165, or a proportion of 0.52, have less than 40 minutes of music.

2.33(a) A bar chart would work well for 10 years of data. If there are many years, a line chart for a time series (shown in Chapter 15) would be better.

(b) Nine slices can be shown with a pie chart. A few more and it would become unreadable. You would have to go to a bar chart.

(c) A cumulative frequency polygon would show the number of drivers convicted for driving at or below any speed.

2.35

Salary range	Frequency	Width	Density
under $10,000	58	10,000	0.0058
$10,000 and under $12,500	44	2,500	0.0176
$12,500 and under $15,000	34	2,500	0.0136
$15,000 and under $17,500	46	2,500	0.0184
$17,500 and under $20,000	68	2,500	0.0272
$20,000 and under $22,500	78	2,500	0.0312
$22,500 and under $25,000	109	2,500	0.0436
$25,000 and under $27,500	125	2,500	0.0500
$27,500 and under $30,000	56	2,500	0.0224
$30,000 and over	78	14,000	0.0056

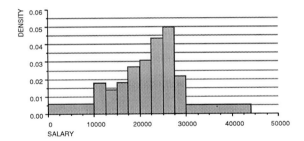

Note that the lower bound of the distribution is $0. The density of the first interval, from $0 to $10,000, is 58 divided by 10,000, or 0.0058. The last interval has an upper bound of $44,000. The density of the last interval is the frequency of 78 divided by the range of $14,000 ($30,000 to $44,000), or 0.0056.

■ CHAPTER 3

3.1

$$\bar{x} = \frac{21 + 90 + 11 + \cdots + 55}{40} = \frac{1,524}{40} = 38.1$$

There are 40 observations, so the median is halfway between the 20th and the 21st observation. The data have to be sorted to determine the 20th (34) and the 21st (37) observations. The median is 35.5. The mode is the most frequently occurring value. The mode is 28 with a frequency of 5.

3.3

$$\bar{x} \cong \frac{50(4.5) + 150(14.5) + 100(24.5) + 50(34.5)}{350}$$

$$\cong \frac{6,575}{350} \cong 18.79$$

The median class is from 10 through 19 because this class contains the 175th and 176th observations.

$$m \cong \left[\frac{\left(\frac{350 + 1}{2}\right) - (50 + 1)}{150 - 1} \right] 9 + 10$$

$$\cong 17.52$$

The mode is approximated by the midpoint of the modal class. The mode is approximately 14.5.

3.5(a) The long history of the company suggests that the distribution of the number of years of employment will be skewed to the right, i.e., have a few employees with many years of employment. The best measure of the center of the data for a skewed distribution is the median.

(b) The mode would help the company decide on a location.

(c) If distribution of pension checks was not highly skewed, the mean would be the best measure of the center of the distribution. The mean has the advantage of making it easy to compute total pension payments.

(d) The chapter mentions that the sample mode and sample median are poor estimators of a center of a population. The sample mean is the preferred estimator for the center of a distribution for statistical inference.

3.7 The mean is the center of balance of a data set. The replica would balance on the median only if the data set was symmetrical.

3.9 First, we assume the data represent a population.

$$\mu = \frac{39 + 62 + \cdots + 147}{5}$$

$$= 90.2$$

$$\sigma = \sqrt{\frac{(39 - 90.2)^2 + (62 - 90.2)^2 + \cdots + (147 - 90.2)^2}{5}}$$

Next we consider the data to be a sample.

$$\bar{x} = \frac{39 + 62 + \cdots + 147}{5}$$

$$= 90.2$$

$$s = \sqrt{\frac{(39 - 90.2)^2 + (62 - 90.2)^2 + \cdots + (147 - 90.2)^2}{4}}$$

$$= \sqrt{1,727.7} = 41.57$$

3.11 The relative variability can be compared using the coefficient of variation. For data set A

$$\bar{x} = \frac{1 + 3 + 6 + 8 + 13}{5}$$

$$= \frac{31}{5} = 6.2$$

$$s = \sqrt{\frac{(1 - 6.2)^2 + (3 - 6.2)^2 + \cdots + (13 - 6.2)^2}{4}}$$

$$= \sqrt{\frac{86.8}{4}} = \sqrt{21.7}$$

$$= 4.658$$

$$w_A = \frac{4.658}{6.2}$$

$$= 0.751$$

For data set B

$$\bar{x} = \frac{1,030 + 2,976 + 5,931 + 7,463 + 12,652}{5}$$

$$= \frac{30,052}{5} = 6,010.4$$

$$s = \sqrt{\frac{(1,030 - 6,010.4)^2 + \cdots + (12,652 - 6,010.4)^2}{4}}$$

$$= 4,478.816$$

$$w_B = \frac{4,478.816}{6,010.4}$$

$$= 0.745$$

Data set B has greater relative variability than A, but the calculations have to be taken to three places to the right of the decimal point to show a difference.

3.13 The hot dog stand desires the probability of running out of hot dogs to be less than 0.01. This is the same as requiring $P_{tails} \leq 0.01$. Using equation 3.14 we can find the corresponding value of k:

$$0.01 = \frac{1}{k^2}$$

$$k = 10$$

We can now use the value of k to determine the number of hot dogs to start the day with:

$$B = k\sigma + \mu$$

$$B = 10(10) + 300$$

$$= 400$$

3.15 Using the stem and leaf display from exercise 3.14, we determine the median to be 237, the lower hinge is 223, and the upper hinge is 254. The distance between hinges is 254 minus 223, or 31. Three halves of 31 is 46.5. There are no outliers as 223 minus 46.5 equals 176.5, which is less than all of the observations. For the right side 254 plus 46.5 equals 300.5, which is greater than all of the observations. Thus, the fences mark the minimum (201) and the maximum (281) observed values in the box plot below.

x

3.17 The median is 5,200, the lower hinge is 3,800, and the upper hinge is 5,800. The distance between the hinges is 5,800 minus 3,800, or 2,000. Three halves of the interhinge distance is 3,000. There are no observations less than 800 (3,800 − 3,000) or greater than 8,800 (5,800 + 3,000), so the fences are set at the minimum observation (2,880) and the maximum observation (6,600). This results in the following box plot:

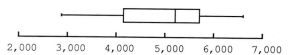

The box plot indicates the median of the sample data is above $5,000 and that at least 75% of the observations are above $4,000. While the left whisker of the box plot does extend below $3,000, this car may not be in as good condition as your Camaro was before being wrecked. The data suggest that even after some bargaining to lower the asking price, it would be difficult to replace your Camaro for $3,000.

3.19 The median is halfway between the 2000th and 2001st observations. This point falls between the boundaries of the classes 5,000–24,999 and 25,000–49,999. You do not need to use formula 3.5. The median unit sales per title in the catalogue is approximately 24,999.5. The result would be exact only if the highest observation in the 5,000–24,999 class was 24,999 and the lowest observation in the 25,000–49,999 class was 25,000.

3.21

$$\bar{\bar{x}} = \frac{5,000(10) + 4,000(11) + 1,000(12)}{5,000 + 4,000 + 1,000}$$

$$= 10.6$$

The known population sizes for the three types of sheep are the best weights for estimating the mean yield of the 10,000 sheep, which as shown is 10.6 pounds.

3.23 The coefficient of variation can be used to compare relative variability. For country A

$$\bar{\bar{x}} = \frac{10 + 15 + \cdots + 9}{10}$$

$$= \frac{108}{10} = 10.8$$

$$s = \sqrt{\frac{(10 - 10.8)^2 + (15 - 10.8)^2 + \cdots + (15 - 10.8)^2}{9}}$$

$$= \sqrt{\frac{167.6}{9}} = 4.32$$

$$w = \frac{4.32}{10.80}$$

$$= 0.40$$

For country B

$$\bar{\bar{x}} = \frac{200 + 220 + \cdots + 205}{10}$$

$$= \frac{2,077}{10} = 207.7$$

$$s = \sqrt{\frac{(200 - 207.7)^2 + \cdots + (205 - 207.7)^2}{9}}$$

$$= \sqrt{\frac{2,592.1}{9}} = 16.97$$

$$w = \frac{16.97}{207.70}$$

$$= 0.08$$

Country A has the greater relative variability. Country A's absolute changes would be easier to predict because it has had the smallest absolute changes from year to year. Methods of forecasting are discussed in Chapter 15. The percentage changes would be easier to predict for country B.

3.25 The midpoint of the last class can be found by substituting into formula 3.4.

$$13,000 = \frac{3,000(2,500) + \cdots + 500x_6}{33,000}$$

$$x_6 = \frac{33,000(13,000) - \cdots - 1,500(22,500)}{500}$$

$$= 65,500$$

This midpoint is the average of the lower class limit of 25,000 and the maximum salary. Thus,

$$x_6 = \frac{25,500 + \text{maximum}}{2}$$

$$\text{maximum} = 2(65,500) - 25,000$$

$$= 106,000$$

3.27

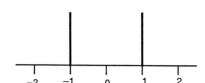

3.29 The appropriate values used to draw the box plots are shown in the table below.

	Treasury bonds	State bonds	Municipal bonds
Median	10.0	9.85	8.5
Upper hinge	11.0	10.4	12.5
Lower hinge	10.0	8.25	5.5
Inner-hinge distance	1.0	2.15	7.0
Three halves of the inner-hinge distance	1.5	3.225	10.5
Upper outlier limit	12.5	13.625	23.0
Lower outlier limit	8.5	5.025	−5.0
Upper fence	12.0	13.0	15.0
Lower fence	9.0	7.0	5.0

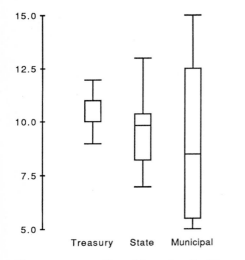

There are no outliers. Note that for Treasury bonds the lower hinge and the median take the same value.

3.31 The modal observation is $1.099. The distribution drops very quickly above this modal value. The gas stations seem to favor ending with a leaf of 9. The high prices all end with a leaf of 0. If the purpose of the ending in the left 9 is to appear cheaper (the tenth of a cent is displayed in a small size compared to the dollars and cents), the highest price gas stations may not care about appearing cheaper. Again, this is sheer speculation. In the stem and leaf display the stem is two places to the right of the decimal.

depth		
1	107*	2
6	107•	68899
10	108*	2344
24	108•	55556667779999
(14)	109*	01222233333444
25	109•	5566666788999999999999
4	110*	00
	110•	
2	111*	0
	111•	
1	112*	0

3.33 The stem is in the ones column.

depth		
2	640	00
6	641	6666
	642	
8	643	22
17	644	888888888
	645	
	646	
	647	
(4)	648	0000
19	649	666
	650	
16	651	222
13	652	8
	653	
12	654	44
	655	
10	656	000
7	657	6
	658	
6	659	22222
	660	
1	661	6

The data are highly granular with gaps between some of the stems and all leaves being even numbers. One explanation is that the original data were recorded in integer miles and a 1.6 miles per kilometer conversion was subsequently applied to data.

3.35

$$\text{Percentile}_{60} \cong \left(\frac{0.6n - f_b}{f_{60}}\right) w_{60} + L_{60}$$

In the above formula the subscript 60 stands for the 60th percentile. w_{60} is the width of the class containing the 60th percentile. L_{60} is the lower limit of the class containing the 60th percentile. f_{60} is the frequency of this class. f_b is the frequency of all classes below the class containing the 60th percentile. n is the number of observations in the data set. This formula is not as complex as formula 3.5, but it suffices. Applying this formula to the data in exercise 3.25, we find that with 33,000 observations, the 60th percentile is the 19,800th observation. This observation falls in the third class. Thus f_b equals 3,000 plus 8,000, or 11,000. Finally, we compute our approximate 60th percentile value:

$$\text{Percentile}_{60} \cong \left[\frac{0.6(33,000) - 11,000}{11,000} \right] 5,000 + 10,000$$

$$\cong \left(\frac{8,800}{11,000} \right) 5,000 + 10,000$$

$$\cong 4,000 + 10,000$$

$$\cong 14,000$$

3.37 Using the weighting scheme advocated by Alphonse, the mean scores for Alphonse and Bonard are respectively 3.85 and 3.54. Using the weighting scheme advocated by Bonard, the mean scores for Alphonse and Bonard are respectively 3.19 and 3.23. (Supporting calculations follow.) Not surprisingly, their positions are reversed under the two weighting schemes, each doing better under the scheme he advocates. With ordinal data there is no best way to weight the rankings. The Bonard weighting is commonly used in academia, but it is not obvious why a "strongly agree" should count precisely 25% more than an "agree." Using Alphonse's weighting scheme

$$\bar{\bar{x}}_{\text{Alphonse}} = \frac{10(10) + 5(5) + \cdots + 3(0)}{10 + 5 + \cdots + 3}$$

$$= 3.85$$

$$\bar{\bar{x}}_{\text{Bonard}} = \frac{2(10) + 23(5) + \cdots + 3(0)}{2 + 23 + \cdots + 3}$$

$$= 3.54$$

Using Bonard's weighting scheme

$$\bar{\bar{x}}_{\text{Alphonse}} = \frac{10(4) + 5(4) + \cdots + 3(1)}{10 + 5 + \cdots + 3}$$

$$= 3.19$$

$$\bar{\bar{x}}_{\text{Bonard}} = \frac{2(5) + 23(4) + \cdots + 3(1)}{2 + 23 + \cdots + 3}$$

$$= 3.23$$

3.39 The coefficient of variation for their offensive performance (points scored by the team) is 20/80 or 0.25. The coefficient of variation for their defensive performance (points scored by their opponents) is 18/65 or 0.28. Their relative variability over the 20 games is slightly higher for defense than offense.

3.41 The stem is in the tens column.

Computer Stuff

depth		
2	1	89
10	2	14444478
14	3	0245
15	4	0
15	5	8
14	6	9999
10	7	4444
	8	
	9	
6	10	8
	11	
	12	
5	13	0279
1	14	0
	15	
	16	

Peripherals Galore

depth		
	1	
5	2	89999
14	3	001234556
15	4	0
15	5	1
	6	
14	7	0899
10	8	444
7	9	4
	10	
6	11	8
	12	
5	13	077
2	14	9
	15	
1	16	0

Byte Wholesalers

depth		
3	1	999
14	2	05555588889
(3)	3	089
13	4	02
11	5	9
10	6	4444
6	7	89
	8	
4	9	27
2	10	9
	11	
1	12	1
	13	
	14	
	15	
	16	

3.43(a) The box plots, labels, and values are shown below. The box plot was drawn with DATA DESK, which uses an asterisk to mark extreme outliers (more than three times the interhinge distance beyond either hinge) and a circle to mark regular outliers (between one and a half and three interhinge distances beyond either hinge).

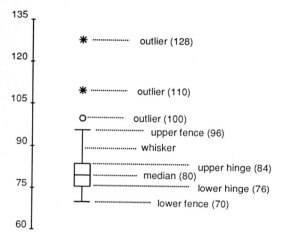

(b) The median fare has the value $80. The proportion of fares at or above the value $80 is 32/54, or 0.59.

(c) The most unusual feature is all of the leaves are even numbers. An explanation is that the one-way fares are in integer dollars and the round-trip fares are two times a one-way fare.

3.45(a) The stem is in the tens column.

```
depth
    6   1*   889999
   32   2•   00000011111111222233334444
 (35)   2*   55566777788888999999999999999999999999
   41   3•   001344
   35   3*   56899999
   27   4•   0123
   23   4*   7899
   19   5•   013
   16   5*   789
   13   6•   0014
    9   6*   5
    8   7•   01
    6   7*   7
    5   8•   01
    3   8*   59
    1   9•   0
```

(b) For the bank data, the median is 29, the upper hinge is 39.5, and the lower hinge is 23. The inter-hinge distance is 16.5. Three halves of the interhinge distance is 24.75. This results in an upper fence of 64.25 and a lower fence of 18. All values above 64.25 are outliers. There are nine of these values (65, 70, 71, 77, 80, 81, 85, 89, 90). The completed box plot is shown below.

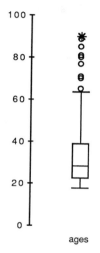

ages

The box plot was drawn with DATA DESK, which uses an asterisk to mark extreme outliers (more than three times the interhinge distance beyond either hinge) and a circle to mark regular outliers (between one and a half and three interhinge distances beyond either hinge).

(c) The data are highly skewed to the right with many outliers. It seems implausible that the bank would have so many older job applicants. Perhaps the bank is located in a retirement community. Another explanation is recording errors. The data set is rich in leaves of 9. The modal value of the applicants' ages is 29. Job applicants usually don't lie about their ages because any false statements on an application can be grounds for being fired. Nevertheless, it is implausible that the bank would attract so many applicants who were 29 years old.

CHAPTER 4

4.1(a) {Fred-Betty, Fred-Buck, Fred-Harold, Betty-Buck, Betty-Harold, Buck-Harold}.

(b) One sixth.

(c) Zero.

4.3(a) 0.1.

(b) 0.1.

4.5 There is one explanation that would be consistent with rational behavior. If the person thought that the probability of the Republicans winning the election was exactly 0.3, then he would be indifferent.

4.7 If both A and B have nonzero probabilities and they are mutually exclusive, they cannot be statistically independent. The reason is that with statistical independence the probability of A intersect B must equal $P(A)P(B)$. This result is from the multiplication rule. But the product $P(A)P(B)$ cannot be zero if both terms are nonzero and mutually exclusive events are defined as events that have a zero intersection. The converse is also true: if C and D are statistically independent, they cannot be mutually exclusive.

4.9 The most common mistake students make on this exercise is to add the probabilities of dying on each jump, $0.001 + 0.001 + 0.001$ equals 0.003. This answer is incorrect. If the class required 1,000 jumps, would the probability of dying be 1.00? If the class required 2,000 jumps, would the probability of dying be 2.00? The probabilities of dying on successive jumps cannot be added because they are conditional probabilities: the probability of dying on the second jump is 0.001 given that the first jump was successful. The easiest way to do the problem is to calculate the probability of surviving all three jumps and then subtract the probability of surviving from 1 to find the probability of dying. The probability of surviving the first jump is 0.999; given that the first jump is survived, the probability of surviving the second jump is 0.999; and given that the second jump is successful, the probability of surviving the third is 0.999. From the multiplication rule for dependent events the probability of surviving three jumps would be 0.999^3, or 0.997003, and the probability of not surviving all three jumps would be 0.002997.

4.11(a) Values in boldface indicate facts given in the problem.

	B	$\overline{B}$	
A	**0.2**	0.3	**0.5**
$\overline{A}$	0.1	0.4	0.5
	0.3	0.7	1.00

(b)

$$P(A \mid B) = \frac{0.2}{0.3} = 0.67$$

$$P(A \mid \overline{B}) = \frac{0.3}{0.7} = 0.43$$

$$P(B \mid A) = \frac{0.2}{0.5} = 0.40$$

$$P(B \mid \overline{A}) = \frac{0.1}{0.5} = 0.20$$

$$P(\overline{A} \mid B) = \frac{0.1}{0.3} = 0.33$$

$$P(\overline{A} \mid \overline{B}) = \frac{0.4}{0.7} = 0.57$$

$$P(\overline{B} \mid A) = \frac{0.3}{0.5} = 0.60$$

$$P(\overline{B} \mid \overline{A}) = \frac{0.4}{0.5} = 0.80$$

(c)

$$P(A \mid B) + P(\overline{A} \mid B) = 1.00$$
$$P(B \mid A) + P(\overline{B} \mid A) = 1.00$$
$$P(A \mid \overline{B}) + P(\overline{A} \mid \overline{B}) = 1.00$$
$$P(B \mid \overline{A}) + P(\overline{B} \mid \overline{A}) = 1.00$$

4.13 Values in boldface indicate facts given in the problem.

	instant	disc	slr	other		
0 and under 10,000	0.04	0.08	0.06	0.02	**0.20**	$\frac{100}{500}$
10,000 and under 20,000	0.04	0.08	0.06	0.02	**0.20**	$\frac{100}{500}$
20,000 and under 30,000	0.04	0.08	0.06	0.02	**0.20**	$\frac{100}{500}$
30,000 or more	0.08	0.16	0.12	0.04	**0.40**	$\frac{200}{500}$
	0.20 $\frac{100}{500}$	**0.40** $\frac{200}{500}$	**0.30** $\frac{150}{500}$	**0.10** $\frac{50}{500}$	1.00	

4.15(a) Values in boldface indicate facts given in the problem.

	color	b & w	
under 10 years	0.2	0.0	0.2
10 years or more	**0.4**	0.4	**0.8**
	0.6	0.4	**1.0**

(b) There are several ways to check for statistical independence. One is to note that for statistical independence to hold, the probability of the replaced set being color must equal the conditional probability of being color given that the set was 10 years or older. This is the definition of statistical independence. The probability of being color is 0.6. The probability of being color given 10 years or older is 0.5. There is statistical dependence.

4.17 The change from the exact order of the first three finishers in exercise 4.16 to the set of the first three finishers in this exercise makes this problem a combination rather than a permutation.

$$_{33}C_3 = \frac{33!}{(33-3)!3!}$$
$$= 5{,}456$$

The probability of sharing in the prize is 1/5,456, or 0.00018.

4.19(a)

$$_8P_6 = \frac{8!}{(8-6)!}$$
$$= 20{,}160$$

At $2 per ticket the total cost of the tickets would be $40,320.

(b) The probability of having to share the winnings is 5,000/20,160, or 0.248.

4.21

$$P(H_1 \mid E) = P(H_1 \mid E) = \frac{(0.9)(0.5)}{(0.9)(0.5)+(0.1)(0.5)}$$
$$= 0.9$$

$$P(H_1 \mid E) = P(H_1 \mid \overline{E}) = \frac{(0.9)(0.5)}{(0.9)(0.5)+(0.1)(0.5)}$$
$$= 0.9$$

$P(H_1 \mid E) = 0.9$ and $P(H_1 \mid \overline{E}) = 0.9$. The experiment provides no information.

4.23 Let T be a takeover bid and R be a rumor.

$$P(T \mid R) = \frac{(0.3)(0.7)}{(0.3)(0.7)+(0.7)(0.2)}$$
$$= 0.6$$

4.25 Let G be a genuine painting and N be the curator deciding that it is a forgery.

$$P(\overline{G} \mid N) = \frac{(0.8)(0.85)}{(0.8)(0.85) + (0.2)(0.1)}$$

$$= 0.97$$

4.27(a)

$$_{36}P_3 = \frac{36!}{(36-3)!}$$

$$= 42,840$$

(b) The number of sequences of three numbers is $36 \times 36 \times 36$, or 46,656.

4.29 Let F be a flood, T a tornado, and E an earthquake. In any one year the probabilities of each disaster are approximately $P(F) = 0.05$, $P(T) = 0.1$, and $P(E) = 0.125$. The exact probabilities can be found with the exponential distribution discussed in Chapter 6. The Red Cross office will run out of resources if any two or if all three emergencies occur. Note that the equations below rely on formula 4.2b for the intersection of independent events.

$$P[(F \cap T \cap E) \cup (F \cap T \cap \overline{E}) \cup (F \cap \overline{T} \cap E) \cup$$
$$(\overline{F} \cap T \cap E)]$$
$$= P(F)P(T)P(E) + P(F)P(T)P(\overline{E}) + P(F)P(\overline{T})P(E) +$$
$$P(\overline{F})P(T)P(E)$$

$$P[\ldots] = (0.05)(0.1)(0.125) + \cdots + (0.95)(0.1)(0.125)$$

$$= 0.0225$$

4.31 Let W be your firm winning the contract. Let E be the presence of a bid by your greatest competitor.

$$P(W \mid E) = \frac{(0.25)(0.1)}{(0.25)(0.1) + (0.75)(0.4)}$$

$$= 0.077$$

4.33 If you knew as few as six of the intersection terms and all of the marginal probabilities, you could determine the probabilities of the remaining six inter-

section terms by adding across the columns or down the rows. In general, you need a minimum of $(r - 1)(c - 1)$ intersection terms, where r is the number of rows and c is the number of columns.

4.35(a) The relative frequency definition of probability.

(b) The subjective definition of probability.

(c) The relative frequency definition of probability.

4.37 Let S be the event the missing plans are stolen as opposed to misfiled. Let A be the absence from work of the person who last checked out the plans.

$$P(S \mid A) = \frac{(0.01)(1.00)}{(0.01)(1.00) + (0.99)(0.02)}$$

$$= 0.34$$

4.39(a)

	seen at show	not seen at show	
visit by representatives	**0.50**	0.10	**0.60**
no visit by representatives	0.20	0.20	0.40
	0.70	0.30	**1.00**

The values shown in boldface in the joint probability table were given in the exercise. The remaining values can be found with simple addition or subtraction. The percentage of Pure Sound's retailers that neither were visited by a representative nor saw the speakers at the consumer electronics show is 20%.

(b) The probability that a retailer who has been on the list of those visited by a manufacturer's representative will be among the retailers who see the speakers at the next annual show is 0.5/0.6, or 0.833.

4.41 There are combinations of 2 taken from 20 for the A list and combinations of 3 taken from 25 from the B list.

$$_{20}C_2 = \frac{20!}{(20-2)!2!}$$

$$= 190$$

$$_{25}C_3 = \frac{25!}{(25-3)!3!}$$

$$= 2{,}300$$

The number of combinations can be found by substituting into formula 4.4. The total number of sets of 5 choices is a product of these combination terms, or $190 \times 2{,}300 = 437{,}000$.

4.43 You can use Bayes' theorem to combine the two pieces of information—the proportion of all orders coming from each restaurant and the conditional probability of observing a Monday order from each restaurant.

$$P(A \mid M) = \frac{(0.3)(0.05)}{(0.3)(0.05) + \cdots + (0.05)(0.7)}$$

$$= 0.06$$

Using the same procedure, you can find the revised probabilities for restaurants B through F. These are 0.12, 0.32, 0.24, 0.12, 0.14. The best order for calling the restaurants is C, D, F, B or E, and A.

4.45(a) Purchasers of service contracts have a higher probability of a major repair than all car purchasers.

(b) The conditional probability of a major repair given no service contract purchase is 250/600, or 0.417. The conditional probability of a major repair given a service contract purchase is 250/400, or 0.625.

4.47 Let SP be a month-long rise in the Standard & Poor's index. Let $X+$ be a month-long rise in the price of steel stocks and $X-$ be a month-long fall in the price of steel stocks. Let $A+$ be a month-long rise in the price of auto stocks and $A-$ be a month-long fall in the price of auto stocks. Then $P(X+ \mid SP) = 12/15$, $P(A+ \mid X-) = 5/30$, $P(A- \mid X-) = 25/30$, $P(A+ \mid X+) = 40/45$, $P(A- \mid X+) = 5/45$. We can use Bayes' theorem to combine the information that Standard & Poor's index has risen for the past month and the price of auto stocks has fallen. The $P(X \mid SP) = 12/15$ is the prior probability of $X+$.

$$P(X+ \mid A-) = \frac{(0.8)(0.111)}{(0.8)(0.111) + (0.2)(0.833)}$$

$$= 0.35$$

4.49 The probability that at least one student has the same birthday as you is 1 minus the probability that none of the students have your birthday. Assuming independence among birthdays (for example, no twins in the class) and 365 days in a year (ignoring leap years), the probability that none of the other 49 students shares your birthday is $(364/365)^{49}$, or 0.87. This result comes from formula 3.2b for the probability of the intersection of a series of independent events and the assumption that every possible date is equally likely. The probability that at least one student has the same birthday as you is 0.13.

4.51 You have 25 choices for your brother and once that choice is made you have 24 choices for the friend. The total number of arrangements is 25×24, or 600. This is a permutation problem because switching the dates of the two men would be a different arrangement.

$$_{25}P_2 = \frac{25!}{(25-2)!}$$

$$= (25)(24)$$

$$= 600$$

4.53(a) The sequence of an order by the first of the five customers, an order by the second customer, and no orders from the remaining three customers has a probability of $(0.2)(0.2)(0.8)(0.8)(0.8)$, or 0.02048. This result comes from formula 3.2b for the probability of a sequence of independent events. There are 10 mutually exclusive and equally probable sequences that have exactly two orders. This is the number of combinations of two items that can be taken from a list of five items. Using the addition rule for mutually exclusive events, the probability of getting exactly two orders from the five households is 10(0.02048), or 0.2048.

(b) You must assume independence between the customers.

4.55 Naming one of the two losers does not change the prior probability of 1/3. The city manager does not pick the name of the losing firm at random. If your firm wins the bid, the city manager can arbitrarily choose either Acme or Excelsior as the firm to name. If your firm loses the bid, the city manager is forced to name the other losing firm. This result can be analyzed in a Bayesian framework. Let Y be your firm winning the bid, E be the naming of Excelsior as the losing firm. The same calculations would apply to the naming of Acme as the losing firm.

$$P(Y \mid E) = \frac{(\frac{1}{3})(\frac{1}{2})}{(\frac{1}{3})(\frac{1}{2}) + (\frac{2}{3})(\frac{1}{2})}$$

$$= \frac{1}{3}$$

4.57 The reciprocals of the payoffs are no longer equal to the subjective probabilities of each horse winning the race because the posted payoffs are net of the 25% of the betting pool or handle skimmed off by the track. If the payoffs are scaled up to reflect all of the money bet, the reciprocals of these scaled payoffs would equal the subjective probabilities. If x is 75% of y, then $1.333x$ equals y. The subjective probabilities are Alhambra, 0.4; Black Beauty, 0.09; Flicka, 0.01; Secretariat, 0.05; Silver, 0.1; Trigger, 0.2; Victory, 0.05; Whimsy, 0.1.

4.59 You have to use Bayes' theorem to find the revised probability of the parent with the negative test result carrying the recessive gene and the revised probability of the parent with the positive test result carrying the recessive gene. The probability of both parents carrying the recessive gene is the intersection term, or the product of the posterior probabilities. Finally, the probability that the child will have Tay-Sachs disease is one fourth the probability that both parents are carriers. Let $T1$ and $T2$ be the first and second parent carrying the recessive gene and + be a positive test and − a negative test.

$$P(T1 \mid +) = \frac{(0.2)(0.98)}{(0.2)(0.98) + (0.8)(0.01)}$$
$$= 0.9607843137$$

$$P(T2 \mid -) = \frac{(0.2)(0.02)}{(0.2)(0.02) + (0.8)(0.99)}$$
$$= 0.0050251256$$

$$P(T1 \cap T2) = P(T1)P(T2)$$
$$= (0.9607843137)(0.0050251256)$$
$$= 0.0048280619$$

$$P(T\text{child}) = \frac{(0.0048280619)}{4}$$
$$= 0.0012070155$$

$$P(T\text{child}) \cong 0.001$$

CHAPTER 5

5.1(a) Discrete integer.

(b) The underlying process is continuous.

(c) Discrete, integer.

(d) Discrete, fractional.

5.3 Probability distribution of X, the sum of two fair dice.

Values of X	Probability
2	1/36
3	2/36
4	3/36
5	4/36
6	5/36
7	6/36
8	5/36
9	4/36
10	3/36
11	2/36
12	1/36

5.5 You will run out of equipment (the literal meaning of run out is to have none left) if three or more orders for moving pianos are received in one day. This probability is 0.1. Some students may interpret run out to mean four or more orders, which has a probability of 0.05. You will have idle equipment if two or fewer orders are received in one day. This probability is 0.9.

5.7 Assuming that the insurance rate was actuarially fair, before the war the probability of losing a tanker was

$E(X) = \$250{,}000(1 - P_{loss}) - \$24{,}775{,}000P_{loss}$

$0 = \$250{,}000(1 - P_{loss}) - \$24{,}775{,}000P_{loss}$

$0 = \$250{,}000 - \$25{,}000{,}000P_{loss}$

$\$25{,}000{,}000P_{loss} = \$250{,}000$

$P_{loss} = 0.01$

During the war the probability of losing a tanker was

$E(X) = \$1{,}000{,}000(1 - P_{loss}) - \$24{,}000{,}000P_{loss}$

$0 = \$1{,}000{,}000(1 - P_{loss}) - \$24{,}000{,}000P_{loss}$

$0 = \$1{,}000{,}000 - \$25{,}000{,}000P_{loss}$

$\$25{,}000{,}000P_{loss} = \$1{,}000{,}000$

$P_{loss} = 0.04$

The probability of losing a supertanker increased from 0.01 to 0.04.

5.9 From the symmetry of the two probability distributions it is clear that both distributions have expected values equal to a zero price change. A risk-neutral or risk-averse investor would prefer stock B because it has less variance than stock A. A risk-seeking investor might prefer stock A because it has a higher probability of increasing in price than stock B. The missing information is the degree of risk aversion of the client.

$E(A) = -\$2.00(0.2) + (-\$1.00)(0.2) + \cdots + \$2.00(0.2)$

$\quad = 0.00$

$\sigma^2(A) = (-2 - 0)^2(0.2) + \cdots + (2 - 0)^2(0.2)$

$\quad = 2.00$

$E(B) = -2.00(0.1) + (-1.00)(0.1) + \cdots + 2.00(0.1)$

$\quad = 0.00$

$\sigma^2(B) = (-2 - 0)^2(0.1) + \cdots + (2 - 0)^2(0.1)$

$\quad = 1.00$

5.11 The only differences between the calculations of the expected value of the mail-in offer shown on page 138 of the textbook and the calculations here are the inclusion of the garnet with a value of $0 and the exclusion of the 44 cents postage.

$E(X) = \$500(0.0001) + \cdots + 0(0.995996)$

$\quad = 0.2852$

Ignoring the garnet and the soap leaves an expected value of 28.52 cents. While hardly enough to cause anyone to make a special trip to the grocery store, this expected value might sway someone who preferred a competing brand to try one of the Procter & Gamble products.

5.13(a) Let A be an audit by the IRS and D be the yacht deduction being disallowed. $P(A) = 0.25$ and $P(D \mid A) = 0.9$. By the multiplication rule, $P(D \cap A) = P(A)P(D \mid A)$, or $0.25(0.9) = 0.225$. If you take the deduction and it is allowed, you gain 33% of the $10,000 you would have owed. If the deduction is disallowed, you lose the fine of $10,000. Substituting into the expected value formula,

$E(X) = \$3{,}333.33(1 - 0.225) + (-10{,}000)(0.225)$

$\quad = 333.33$

If you were a risk-neutral and morally obtuse individual, you would take the yacht deduction because it has a positive expected value.

(b) If the IRS set the fine high enough to make the expected value of taking the deduction 0, you would be indifferent between taking and not taking the deduction.

$E(X) = \$3{,}333.33(1 - 0.225) + (-\text{fine})(0.225)$

$\$0 = 2{,}583.33 - \text{fine}(0.225)$

$\text{fine}(0.225) = \$2{,}583.33$

$\text{fine} = \dfrac{\$2{,}583.33}{0.225}$

$\text{fine} = \$11{,}481.47$

5.15 The following steps rely on formula 5.3 for the addition of expected values and formulas 5.5 and 5.6 for multiplicative and additive constants.

$E[5X + (Y - 100)] = E(5X) + [E(Y - 100)]$

$\quad = 5E(X) + [E(Y) - 100]$

$\quad = 5E(X) + E(Y) - 100$

$\quad = 5(19.5) + 30.5 - 100$

$\quad = 28$

5.17 Let X be the labor time to assemble one tailwing.

$E(X) = 20(0.1) + 25(0.2) + 30(0.3) + 35(0.2) + 40(0.2)$

$\quad = 31$

$\sigma^2(X) = (20 - 31)^2(0.1) + \cdots + (40 - 31)^2(0.2)$

$\qquad = 39.0$

Let Y be the cost in dollars for the labor to assemble one tailwing. Using formula 5.5 for a multiplicative constant,

$E(Y) = E(25X)$

$\qquad = 25E(X)$

$\qquad = 25(31)$

$\qquad = 775$

Note that the variance of Y is measured in dollars-squared. Using formula 5.8a for the variance of a random variable with a multiplicative constant,

$\sigma^2(Y) = \sigma^2(25X)$

$\qquad = 25^2\sigma^2(X)$

$\qquad = 625(39.0)$

$\qquad = 24{,}375$

Let Z be the total cost of making one tailwing. Z is five times the labor cost plus an overhead of $1,000.

$E(Z) = E(5Y + 1{,}000)$

$\qquad = E(5Y) + E(1{,}000)$

$\qquad = 5E(Y) + 1{,}000$

$\qquad = 5(775) + 1{,}000$

$\qquad = 4{,}875$

To determine the variance of Z, use formulas 5.8 and 5.8a for variances with additive and multiplicative constants.

$\sigma^2(Z) = \sigma^2[5Y + 1{,}000]$

$\qquad = \sigma^2(5Y) + \sigma^2(1{,}000)$

$\qquad = 5^2\sigma^2(Y) + 0$

$\qquad = 25(24{,}375)$

$\qquad = 609{,}375$

Let Z_1 through Z_{100} be the costs of the 100 tailwings. A common mistake is to consider the number 100 to be a multiplicative constant. Each tailwing is a separate random variable. The expected costs for 100 tailwings are

$E(Z_1 + Z_2 + \cdots + Z_{100}) = E(Z_1) + E(Z_2) + \cdots + E(Z_{100})$

$\qquad = 4{,}875 + 4{,}875 + \cdots + 4{,}875$

$\qquad = 487{,}500$

To find the variance of the cost of the 100 tailwings, we have to assume that the cost for each tailwing is an independent random variable. This assumption may not be accurate in this context because the workers learn as the tailwings are built. This learning curve should reduce the expected costs and the variability in costs as more tailwings are assembled.

$\sigma^2(Z_1 + Z_2 + \cdots + Z_{100}) = \sigma^2(Z_1) + \cdots + \sigma^2(Z_{100})$

$\qquad = 609{,}375 + \cdots + 609{,}375$

$\qquad = 60{,}937{,}500$

$\sigma(Z_1 + Z_2 + \cdots + Z_{100}) = \sqrt{60{,}937{,}500}$

$\qquad = 7{,}806.25$

5.19(a)

$P(8) = \dfrac{20!}{(20 - 8)!(8)!}\, 0.25^8(1 - 0.25)^{(20-8)}$

$\qquad = 0.0609$

(b)

$P(x \geq 8) = 1 - P(x \leq 7)$

$P(x \geq 8) = 1 - 0.8981$

$\qquad = 0.1019$

(c)

$P(x \leq 12) = 1 - P(x \geq 13)$

$P(x \leq 12) = 1 - 0.0002$

$\qquad = 0.9998$

(d) The problem asks for 12 or fewer failures rather than successes. Having 12 or fewer failures in 20 trials is equivalent to having 8 or more successes in 20 trials. The answer to part d is the same as part b, or 0.1019.

5.21 You will be short of space if five or more riders need to use the sag wagon. Being short of space implies more people want to ride in the sag wagon than there are spaces. I use the term "run out" of space in some other exercises to mean having none left. The parameters are $\Pi = 0.1$, $n = 20$, and $x \geq 5$. $P(x \geq 5) = 1 - P(x \leq 4)$.

$P(x \geq 5) = 1 - 0.9569$, or 0.0431.

5.23 From the binomial table with $\Pi = 0.5$ and $n = 10$,

$P(x \geq 7) = 0.1719$

5.25 $P(x \geq 7) = 1 - P(x \leq 6)$. From the Poisson table with $\mu = 15$,

$P(x \geq 7) = 1 - 0.0075 = 0.9925$

5.27 If the average number of blemishes is one per square yard, two-square-yard pieces will average two blemishes. The μ for this problem is 2. A two-square-yard piece of cloth will be accepted if it has three or fewer blemishes. Using the Poisson table with $\mu = 2$,

$P(x \leq 3) = 0.8571$

5.29 X is a binomial random variable. $\Pi = 0.01$, $n = 100$, and $x < 3$. Since $n \geq 20$ and $\Pi \leq 0.05$, we can use the Poisson approximation to the binomial. $\mu = n\Pi$ or $100(0.01) = 1$. From the Poisson table for $\mu = 1$,

$P(x < 3) = 0.9197$

5.31 $\Pi = 0.001$, $n = 10,000$, $x \geq 15$. Since $n \geq 20$ and $\Pi \leq 0.05$, we can use the Poisson approximation to the binomial distribution. $\mu = n\Pi$, or $10,000(0.001) = 10$. From the Poisson table with $\mu = 10$,

$P(x \geq 15) = 0.0835$

5.33 $\Pi = 0.001$, $n = 1,000$, $x \geq 3$. Since $n \geq 20$ and $\Pi \leq 0.05$, we can use the Poisson approximation to the binomial distribution. $\mu = n\Pi$, or $1,000(0.001) = 1$. $P(x \geq 3) = 1 - P(x \leq 2)$. From the Poisson table with $\mu = 1$,

$P(x \geq 3) = 1 - 0.9197 = 0.0803$

5.35 $\nu = 80$, $\gamma = 10$, $x \geq 4$, $n = 6$. The solution can be found with the hypergeometric formula. The table stops with $\nu = 20$.

$$P(4) = \frac{{}_{10}C_4 \ {}_{70}C_2}{{}_{80}C_6}$$

$$= \frac{\left[\dfrac{10!}{(10-4)!(4!)}\right]\left[\dfrac{70!}{(70-2)!(2!)}\right]}{\dfrac{80!}{(80-6)!(6!)}}$$

$$= 0.0016988$$

$$P(5) = \frac{{}_{10}C_5 \ {}_{70}C_1}{{}_{80}C_6}$$

$$= \frac{\left[\dfrac{10!}{(10-5)!(5!)}\right]\left[\dfrac{70!}{(70-1)!(1!)}\right]}{\dfrac{80!}{(80-6)!(6!)}}$$

$$= 0.000059$$

$$P(6) = \frac{{}_{10}C_6 \ {}_{70}C_0}{{}_{80}C_6}$$

$$= \frac{\left[\dfrac{10!}{(10-6)!(6!)}\right]\left[\dfrac{70!}{(70-0)!(0!)}\right]}{\dfrac{80!}{(80-6)!(6!)}}$$

$$= 0.000001$$

$P(x \geq 4) = P(4) + P(5) + P(6)$

$= 0.001688 + 0.000059 + 0.000001$

$= 0.001748$

5.37 $\nu = 20$, $\gamma = 5$, $x \geq 1$, $n = 6$. The solution can be found with the hypergeometric formula or the hypergeometric table. Note that $P(x \geq 1) = 1 - P(x = 0)$.

$$P(0) = \frac{{}_{5}C_0 \ {}_{15}C_6}{{}_{20}C_6}$$

$$= \frac{\left[\dfrac{5!}{(5-0)!(0!)}\right]\left[\dfrac{15!}{(15-6)!(6!)}\right]}{\dfrac{20!}{(20-6)!(6!)}}$$

$$= 0.1291$$

$P(x \geq 1) = 1 - 0.1291 = 0.8709$

5.39 Assuming that you continued to average $1,000 in medical expenses, under plan A your annual costs would equal $400 plus 20% of $600, or $520. Under plan B your annual costs would equal $600. However, an increase in your medical expenses could make the costs under plan B lower than plan A. The point at which the two plans have the same costs can be found algebraically.

$$400 + 0.2(x - 400) = 600$$
$$0.2x - 80 = 200$$
$$0.2x = 280$$
$$x = 1,400$$

If your annual medical expenses go any higher than $1,400, plan *B* becomes cheaper than plan *A*. The problem provides no information about the standard deviation of your past health costs. A higher standard deviation would increase the attractiveness of plan *B*. Any anticipated medical expenses would also increase the attractiveness of plan *B*. If you were at all risk averse, plan *B* would be preferred because it shelters you from possibly substantial costs for the expected price of $80 per year.

5.41 The possible outcomes are that Michigan wins by more than 10 points, exactly 10 points, or less than 10 points. From your point of view as a bookie an outright loss by Michigan is the same as a win by less than 10 points. Your cash position is $1,000 if Michigan wins by more than 10 points, $3,000 if Michigan wins by exactly 10 points, and −$1,000 otherwise. Applying the subjective probabilities to these cash positions,

$$E(X) = 1,000(0.4) + 3,000(0.1) + (-1,000)(0.5)$$
$$= 200$$

5.43 $\Pi = 0.7$, $x \geq 10$, $P(x \geq 10) = 0.72$, n is unknown. This is a binomial problem because each subject in the tests either develops a sore throat or does not. The probability is the same for all subjects, 0.7. The assumption of independence is reasonable as long as the subjects are kept apart. The unknown is the minimum sample size needed to guarantee at least 10 sore throats. The solution can be found by trying different values of n until $P(x \geq 10) = 0.72$. To be able to use the binomial table the problem can be expressed in terms of the probability of not getting a sore throat: 0.3 is the probability of *not* getting a sore throat. If x is the number of people in the test not getting a sore throat, then $P(n - 10) = 0.72$, the number of healthy people minus 10 must have a probability of 0.72. The probabilities through $n = 15$ were taken from the binomial table. The rest were calculated with the formula. The problem was set up so that it could be done solely with the binomial table. The sums at the bottom of the table are the probabilities of having enough sore throats. For example, in a sample of size 12 if 2 or fewer subjects were healthy, you would have the required 10 or more sore throats. The probability of having 2 or fewer healthy subjects in a sample of size 12 is 0.2528. Clearly, a sample size

of 12 is too small to guarantee 10 sore throats with a probability of 0.72. The minimum sample is 15. With a sample of 14 the probability of 10 or more sore throats is still below 0.72.

x	12	13	14	15	16	17	18	19
0	0138	0097	0068	0047	0033	0023	0016	0011
1	0712	0540	0407	0305	0228	0169	0126	0093
2	1678	1388	1134	0916	0732	0581	0458	0358
3		2181	1943	1700	1465	1245	1046	0869
4			2290	2186	2040	1868	1681	1491
5				2061	2099	2081	2017	1916
6					1649	1784	1873	1916
7						1201	1376	1525
8							0811	0981
9								0514
sum	0.2528	0.4206	0.5842	0.7215	0.8246	0.8952	0.9404	0.9674

5.45 Find $P(x \geq 10)$ for $\Pi = 0.05$, $n = 20$. These parameters are in the binomial table. The probabilities for 10 through 20 successes (audits) with a $\Pi = 0.05$ are all zero. Therefore, $P(x \geq 10) \cong 0$.

5.47 $\Pi = 0.75$, $n = 6,500$, $x > 5,000$. Find $P(x > 5,000)$. To use Tchebychev's inequality start with the formulas for the mean and the standard deviation of a binomial distribution.

$$E(X) = (6,500)(0.75)$$
$$= 4,875$$
$$\sigma(X) = \sqrt{6,500(0.75)(1 - 0.75)}$$
$$= 34.91$$
$$k = \frac{5,000 - 4,875}{34.91}$$
$$= 3.581$$
$$P_{tails} \leq \frac{1}{(3.581)^2}$$
$$\leq 0.078$$

The probability of getting more than 5,000 is actually smaller than 0.078. The normal approximation to the binomial (discussed in Chapter 6) provides a more accurate estimate of the correct probability.

5.49 Suppose the shop opens with no customers present. At least one chair will be empty for 30 minutes if no more than three customers arrive during the first 30 minutes. The independent arrivals with a constant expected rate implies that flow of customers to the barbershop follows a Poisson process. $\mu = 4$ per half-hour and $x < 4$. The answer can be found with the aid of the Poisson table.

$$P(x < 4) = 0.4335$$

5.51 $P(x = 1) = 0.1$, $\lambda = 8$ per 24 hours or 0.333 per hour. The problem is to find a t that yields the required probability. Look in the Poisson table (holding x to 1) for the μ that has results in the closest probability to 0.1. A μ of 3.5 has $P(1) = 0.1057$. With $\mu = 3.5$ as the starting point, you can substitute other values for μ into the Poisson formula to hunt for a μ with a probability closer to 0.1. A μ of 3.5772 yields a $P(1) = 0.1000$. Since $\mu = \lambda t$, $3.5773 = 0.333t$. Thus t equals 10.732 hours. This solution is not unique. A μ of 0.1 has $P(1) = 0.0905$. The precise μ at which $P(1) = 0.1$ is 0.111833. The corresponding t is 0.3355.

5.53 $\lambda = 0.001$ particle per foot, $t = 200$ feet, $x \geq 4$, $P(x \geq 4) = 1 - P(x \leq 3)$

$$\mu = \lambda t$$

$$= 0.001 \frac{\text{particle}}{\text{foot}} \, 200 \text{ feet}$$

$$= 0.2 \frac{\text{particle}}{200 \text{ feet}}$$

From the Poisson table the entries under a μ of 0.2 for 0 through 3 occurrences are 0.8187, 0.1637, 0.0164, and 0.0011. $P(x \geq 4) = 1 - (0.8187 + 0.1637 + 0.0164 + 0.0011)$, or 0.0001. Thus, 1 in 10,000 tapes would have to be destroyed.

5.55 $\Pi = 0.0003$, $n = 10{,}000$, $x = 0$. The Poisson approximation to the binomial distribution can be used since $\Pi \leq 0.05$ and $n \geq 20$. $\mu = n\Pi$, or $10{,}000(0.0003) = 0.3$. From the Poisson table under a μ of 0.3 the probability of 0 occurrences equals 0.7408.

5.57 This is a hypergeometric problem because the sample of size 10 is drawn from the fixed population of 20. $\nu = 20$, $n = 10$, $\gamma = 2$, $x \geq 1$. $P(x \geq 1) = 1 - P(x = 0)$. Substituting into the hypergeometric formula,

$$P(0) = \frac{{}_2C_{0(20-2)}C_{(10-0)}}{{}_{20}C_{10}}$$

$$= \frac{\left[\frac{2!}{(2-0)!0!}\right]\left[\frac{18!}{(18-10)!10!}\right]}{\frac{20!}{(20-10)!10!}}$$

$$= 0.2368$$

$$P(x \geq 1) = 1 - 0.2368$$

$$= 0.7632$$

5.59 Assuming that the eight applicants are equally qualified lathe operators and the selection process was random, the probability that only men would be selected can be found with the hypergeometric distribution. $\nu = 11$, $n = 3$, $\gamma = 8$, $x = 3$. From the hypergeometric table the answer is 0.3394. Note that the table gives probabilities were $n \geq \gamma$. However, $\{\nu = 11, n = 3, \gamma = 8, x = 3\}$ is equivalent to $\{\nu = 11, n = 3, \gamma = 3, x = 0\}$, which is listed in the table.

5.61(a) $\lambda = 1$ per 20 minutes, $\mu = 3$ per hour. From the Poisson table $P(0) = 0.0498$.

(b) $P(0) = 0.001$. The problem is to find a μ such that the probability of 0 occurrences equals 0.001.

$$P(0) = 0.001$$

$$0.001 = \frac{\mu^0 e^{-\mu}}{0!}$$

$$0.001 = e^{-\mu}$$

$$\ln(0.001) = \ln(e^{-\mu})$$

$$-6.908 = -\mu$$

$$\mu = 6.908$$

$$\lambda = 3/\text{hour}$$

$$\mu = \lambda t$$

$$6.908 = (3/\text{hour})t$$

$$t = \frac{6.908}{3}$$

$$= 2.303 \text{ hours}$$

5.63 The optimal number of sole can be found by calculating the expected value for ordering 10, 11, 12, 13, 14, or 15 sole. The calculations ignore any loss of goodwill from running short of sole. The payoffs for every number are \$8 times the number sold minus \$2 times the number ordered. These payoffs ignore any other costs of running the restaurant. Since these other costs do not vary with the number of dinners sold, they are irrelevant to the optimization problem. The profit maximizing quantity of sole is 14.

5.65(a) The expected value of the mail-in lottery net of the costs of two 22-cent stamps is

$$E(X) = \$0.00(0.9763) + 500\left(\frac{1}{85,340}\right) - 0.44$$

$$= \$0.0447 - 0.44$$

$$= -\$0.3953$$

(b) Since the expected value of the payoff is substantially less than the costs of the stamps, the rules about hand lettering and a limit of one per day must serve a function other than making the lottery have a negative overall value. There are several possibilities. One is that Ralston wanted to further discourage people from participating in the mail-in survey. By making it enough of a bother, Ralston can discourage even risk-seeking people who enjoy lotteries. A second possible reason for the restrictions is to convey the false impression that the lottery is valuable. People who cannot calculate expected values may treat the restrictions as a signal that the mail-in or the cereal package lottery is a good deal.

(c) The expected value of the in-package currency cannot be calculated exactly. According to the rules, 4.2 million packages of the cereal have foreign currency with an average exchange rate of less than 1¢. We can find an upper bound for the expected value of the in-package currency by treating the foreign currency as having exactly a 1¢ value. The 100,550 specially marked packages that contain the United States currency have the listed probabilities. Note that the probability of getting 1¢ is 4.2 million divided by the sum of 4.2 million and 100,550.

$$E(X) = \$0.01(0.9763) + 500\left(\frac{1}{85,340}\right)$$

$$= \$0.0545$$

CHAPTER 6

6.1 $a = 30$, $b = 40$.

$$f(x) = \frac{1}{40 - 30}$$

$$= 0.1$$

$$E(x) = \frac{40 + 30}{2}$$

$$= 35$$

$$\sigma^2(X) = \frac{(40 - 30)^2}{12}$$

$$= 8.33$$

6.3 $a = 6$, $b = 12$, from formula 1, $f(x) = 1/6$. $P(7 < x < 12)$ equals the area in the range 7 to 12, or $5(1/6)$, or $5/6$ (or 0.833).

6.5 Substituting the value of 20 into the formula for the mean of a continuous uniform random variable,

$$20 = \frac{(b + a)}{2}$$

$$40 = b + a \tag{1}$$

Substituting the value of 12 into the formula for the variance of a continuous uniform random variable,

$$12 = \frac{(b - a)^2}{12}$$

$$(b - a)^2 = 144$$

$$b - a = 12 \tag{2}$$

Subtracting (2) from (1),

$$(b + a) - (b - a) = 40 - 12$$

$$2a = 28$$

$$a = 14$$

Substituting $a = 14$ into the expected value formula,

$$20 = \frac{(b + 14)}{2}$$

$$40 = b + 14$$

$$b = 26$$

$$f(x) = \frac{1}{b - a}$$

$$= \frac{1}{26 - 14}$$

$$= \frac{1}{12}$$

$$P(x > 22) = (26 - 22)\left(\frac{1}{12}\right)$$

$$= \frac{4}{12}$$

$$= \frac{1}{3} \text{ or } 0.333$$

6.7(a)

$$z = \frac{125 - 120}{3}$$

$$= 1.67$$

$$P(z > 1.67) = 0.5 - 0.4525$$

$$= 0.0475$$

(b)

$$z = \frac{121 - 120}{3}$$

$$= 0.33$$

$$z = \frac{124 - 120}{3}$$

$$= 1.33$$

$$P(0.33 < z < 1.33) = 0.4082 - 0.1293$$

$$= 0.2789$$

(c)

$$z = \frac{118 - 120}{3}$$

$$= -0.67$$

$$z = \frac{124 - 120}{3}$$

$$= 1.33$$

$$P(-0.67 < z < 1.33) = 0.2486 + 0.4082$$

$$= 0.6568$$

(d)

$$z = \frac{115 - 120}{3}$$

$$= -1.67$$

$$z = \frac{116 - 120}{3}$$

$$= -1.33$$

$$z = \frac{122 - 120}{3}$$

$$= 0.67$$

$$z = \frac{124 - 120}{3}$$

$$= 1.33$$

$$P(-1.67 < z < -1.33) + P(0.67 < z < 1.33)$$
$$= (0.4525 - 0.4082) + (0.4082 - 0.2486)$$
$$= 0.0443 + 0.1596$$
$$= 0.2039$$

6.9 $\mu = 30$, $\sigma = 10$, X is approximately normally distributed. Find x such that $P(X > x) = 0.01$. $P(X > x) = 0.01$ implies that $P(\mu < X < x) = 0.49$. The closest probability to 0.49 in the standard normal table is $z(2.33) = 0.4901$.

$$x = \mu + z\sigma$$
$$= 30 + 2.33(10)$$
$$= 53.3$$

Thus, 53 diapers will yield slightly more than a 1% chance of running out and 54 diapers will yield slightly less than 1%.

6.11 Given the parameters in exercise 6.10 and the assumption of independence, find

$$P[(x_1 < 30) \cup (x_2 < 30) \cup \cdots \cup (x_{90} < 30)]$$
$$= 1 - P[(x_1 > 30) \cap (x_2 > 30) \cap \cdots \cap (x_{90} > 30)]$$
$$= 1 - P(x_1 > 30)P(x_2 > 30) \cdots P(x_{90} > 30)$$
$$= 1 - 0.9938^{90}$$
$$= 1 - 0.5714$$
$$= 0.4286$$

6.13

$$z = \frac{100 - 78}{15}$$

$$z = 1.47$$

$$P(z > 1.47) = 0.5 - 0.4292$$

$$= 0.0708$$

For any day on which Northern agrees to buy power, the probability of needing the power is less than 0.2. The probability is too low to convince the regulatory commission that the contract is worthwhile.

6.15(a)

$$z = \frac{400 - 800}{150}$$

$$= -2.67$$

$$P(z < -2.67) = 0.50 - 0.4962$$

$$= 0.0038$$

(b) $P(X < x) = 0.2$ implies that $P(\mu > X > x) = 0.3$. The closest probability to 0.3 in the standard normal table is $z(0.2995) = 0.84$. Note that the z score is -0.84 because the boundary is in the left tail.

$$x = \mu + z\sigma$$

$$= 800 + (-0.84)(150)$$

$$= 674$$

6.17(a) $\Pi = 0.25$, $n = 100$, X is a binomial random variable. The words "between 22 and 30" can be interpreted two ways, either inclusive or exclusive of the endpoints. I consider the latter to be the more accurate interpretation of the wording. However, the solutions for both interpretations are shown below. Excluding the endpoints, we must find $P(22 < x < 30)$ with the normal approximation to the binomial. $n\Pi = 0.25(100)$, or 25, and $n(1 - \Pi) = 0.75(100)$, or 75. The requirement that both $n\Pi$ and $n(1 - \Pi)$ be greater than or equal to 5 is met. 23 is the lower bound and 29 is the upper bound. The values substituted for x in the z score formula are 22.5 and 29.5. Including the endpoints, we must find $P(22 \leq x \leq 30)$ with the normal approximation to the binomial. 22 is the lower bound and 30 is the upper bound. The values substituted for x in the z score formula are 21.5 and 30.5.

$$\mu = 100(0.25)$$

$$= 25$$

$$\sigma = \sqrt{100(0.25)(0.75)}$$

$$= 4.33$$

Including the endpoints,

$$P(-0.81 < z < 1.27) = 0.2910 + 0.3980$$

$$= 0.6890$$

Excluding the endpoints,

$$P(-0.58 < z < 1.04) = 0.2190 + 0.3508$$

$$= 0.5690$$

(b)

$$z = \frac{25.5 - 25}{4.33}$$

$$= 0.12$$

$$P(z < 0.12) = 0.5 + 0.0478$$

$$= 0.5478$$

(c)

$$z = \frac{28.5 - 25}{4.33}$$

$$= 0.81$$

$$P(z > 0.81) = 0.5 - 0.2910$$

$$= 0.2090$$

(d)

$$z = \frac{28.5 - 25}{4.33}$$

$$= 0.81$$

$$P(z < 0.81) = 0.50 + 0.2910, \text{ or } 0.7910$$

6.19

$$\mu = 10,000(0.01)$$

$$= 100$$

$$\sigma = \sqrt{10,000(0.01)(0.99)}$$

$$= 9.95$$

$$z = \frac{89.5 - 100}{9.95}$$

$$= -1.06$$

$$z = \frac{120.5 - 100}{9.95}$$

$$= 2.06$$

$$P(-1.06 < z < 2.06) = 0.3554 + 0.4803$$

$$= 0.8357$$

6.21

$\mu = 3,482(0.05)$

$\quad = 174.1$

$\sigma = \sqrt{3,482(0.05)(1 - 0.05)}$

$\quad = 12.86$

$z = \dfrac{245.5 - 174.1}{12.86}$

$\quad = 5.55$

$P(z > 5.55) \cong 0$

6.23(a) Π is the probability that a student will pass the exam to enter college. $\Pi = 0.3$. n is the number of students per teacher, or 60. X, the number of students who pass per teacher, is a binomial random variable. We can use the normal approximation for the distribution of X because $n\Pi$ and $n(1 - \Pi)$ are both greater than or equal to 5.

$x = \mu + z\sigma$

$\quad = n\Pi + z\sqrt{n\Pi(1 - \Pi)}$

$\quad = 60(0.3) + 1.28\sqrt{60(0.3)(0.7)}$

$\quad = 18 + 1.28(3.5496)$

$\quad = 22.54$

If the test results followed a binomial distribution with a Π of 0.3, a teacher would need to have 23 out of 60 students get passing scores in order to get the bonus.

(b) The binomial distribution requires independence between trials. If there is a "teacher effect" on student scores (which is what the bonus is intended to reward), the assumption of independence would be incorrect. Presumably the students of the best teachers have a much better than 30% chance of passing the test and 23 out of 60 passes may be too low for a teacher bonus.

6.25(a) $\mu = 10$, X is a Poisson random variable. $\sigma = \sqrt{10}$, or 3.162. Find $P(x \leq 5)$ with the Poisson table and the normal approximation to the Poisson. 5 is the upper bound so the value substituted for x in the z score formula is 5.5. The normal approximation to the Poisson can be used because $\mu > 5$. From the Poisson table with $\mu = 10$,

$P(x \leq 5) = 0.0671$

$z = \dfrac{5.5 - 10}{3.162}$

$\quad = -1.42$

$P(z < -1.42) = 0.5 - 0.4222$

$\quad\quad\quad\quad\quad = 0.0778$

(b) Find $P(x \geq 12)$ with both the Poisson table and the normal approximation to the Poisson. Note that $P(x \geq 12) = 1 - P(x \leq 11)$. From the Poisson table with $\mu = 10$,

$P(x \geq 12) = 1 - 0.6968$, or 0.3032

12 is a lower bound, so $x = 11.5$.

$z = \dfrac{11.5 - 10}{3.162}$

$\quad = 0.47$

$P(z > 0.47) = 0.5 - 0.1808$

$\quad\quad\quad\quad\quad = 0.3192$

(c) Find $P(8 < x < 11)$ with both the Poisson table and the normal approximation to the Poisson. From the Poisson table with $\mu = 10$,

$P(8 < x < 11) = 0.2502$

9 is the lower bound and 10 the upper bound. $x = 8.5$ and 10.5.

$z = \dfrac{8.5 - 10}{3.16}$

$\quad = -0.47$

$z = \dfrac{10.5 - 10}{3.16}$

$\quad = 0.16$

$P(-0.47 < z < 0.16) = 0.1808 + 0.0636$

$\quad\quad\quad\quad\quad\quad\quad\quad\quad = 0.2444$

(d) Find $P(x \leq 20)$ with both the Poisson table and the normal approximation to the Poisson. From the Poisson table with $\mu = 10$,

$P(x \leq 20) = 0.9985$

20 is an upper bound, so $x = 20.5$.

$$z = \frac{20.5 - 10}{3.162}$$

$$= 3.32$$

$$P(z < 3.32) = 0.5 + 0.4995$$

$$= 0.9995$$

6.27 $\lambda = 25$ per ton, $\mu = \lambda t$, $\mu = 25(100)$, or 2,500, $\sigma = \sqrt{2,500}$, or 50, X is Poisson random variable. Find $P(x \le 2,300)$. Since μ is greater than or equal to 5, we can use the normal approximation to the Poisson. 2,300 is the upper bound, so $x = 2,300.5$.

$$z = \frac{2,300.5 - 2,500}{50}$$

$$= -3.99$$

$$P(z < -3.99) \cong 0$$

6.29(a) $\lambda = 0.5$ per year. X is a Poisson random variable. $\mu = \lambda t$, $\mu = 0.5(5)$, or 2.5. Find $P(x > 2.5)$. From the exponential table, $P(x > 2.5) = 1 - 0.9179$, or 0.0821.

(b) This is the same problem as part a. With the exponential distribution the probability that the next occurrence will be more than t is the same as the probability that the time between the next two occurrences will be more than t.

(c) $\mu = \lambda t$, $\mu = 0.5(10)$, or 5. Find $P(x < 5)$. From the exponential table, $P(x < 5) = 0.9933$.

(d) $\mu = \lambda t$, $\mu = 0.5(4)$, or 2. $\mu = 0.5(6)$, or 3. Find $P(4 < x < 6)$. From the exponential table, $P(2 < x < 3) = 0.9502 - 0.8647$, or 0.0855.

6.31 $\lambda = 1$ per week. X is a Poisson random variable. $\mu = \lambda t$, $\mu = 1(2)$, or 2. Find $P(x < 2)$. From the exponential table, $P(x < 2) = 0.8647$.

6.33 $\lambda = 0.00001$ per mile. X is a Poisson random variable. $\mu = \lambda t$, $\mu = 0.00001(50,000)$, or 0.5. Find $P(x < 0.5)$. From the exponential table, $P(x < 0.5) = 0.3935$.

6.35 $\lambda = 1$ per 20 years or 0.05 per year. X is a Poisson random variable. $\mu = \lambda t$, $\mu = 0.05(12)$, or 0.67. Find $P(x < 0.6)$. From the exponential table, $P(x < 0.6) = 0.4512$.

6.37 X is a normal random variable, $\sigma = 0.5$. Find μ such that $P(x < 16) = 0.01$. $P(x < 16) = 0.01$ implies

$P(16 < X < \mu) = 0.49$. The closest probability to 0.49 in the standard normal table is 0.4901. $z(0.4991) = 2.33$. Since the boundary is in the left tail the z score is -2.33. Substituting this z score into formula 6.5,

$$-2.33 = \frac{16 - \mu}{0.50}$$

$$-2.33(0.50) = 16 - \mu$$

$$\mu = 16 + 1.165$$

$$= 17.165$$

6.39 $\mu = 1,080$, $\sigma = 48$, X is a normal random variable. Find x such that $P(X > x) = 0.1$. $P(X > x) = 0.1$ implies $P(\mu < X < 16) = 0.4$. The closest probability to 0.4 in the standard normal table is 0.3997. $z(0.3997) = 1.28$. Substituting this z score into formula 6.5,

$$1.28 = \frac{x - 1,080}{48}$$

$$x = 1,080 + 1.28(48)$$

$$= 1,141.44, \text{ or } 1,141$$

The lowest SAT score that the college will accept is 1,141. SAT scores are expressed in integers, so the result was rounded down to 1,141. With your SAT score of 1,300, your score is high enough.

6.41 To earn more than \$8,200 at \$4 per yes answer, you need more than 8,200/4 or 2,050 yes answers. $n = 10,000$, $\Pi = 0.2$, X is a binomial random variable. We can use the normal approximation to the binomial because $n\Pi$ and $n(1 - \Pi)$ are both greater than or equal to 5. 2,051 is a lower bound. $x = 2,050.5$.

$$\mu = 10,000(0.20)$$

$$= 2,000$$

$$\sigma = \sqrt{10,000(0.20)(0.80)}$$

$$= 40$$

$$z = \frac{2,050.5 - 2,000}{40}$$

$$= 1.26$$

$$P(z > 1.26) = 0.5 - 0.3962$$

$$= 0.1038$$

6.43 $\Pi = 0.5$, $n = 50$, X is a binomial random variable where X represents the number of matched couples that do not marry. Find $P(x \geq 30)$. We can use the normal approximation to the binomial because $n\Pi$ and $n(1 - \Pi)$ are both greater than or equal to 5. 30 is a lower bound. $x = 29.5$. $\mu = n\Pi$, $0.5(50) = 25$. $\sigma = \sqrt{n\Pi(1 - \Pi)}$, $\sigma = \sqrt{50(0.25)(0.25)}$, $\sigma = 3.54$.

$$z = \frac{29.5 - 25}{3.54}$$

$$= 1.27$$

$$P(z > 1.27) = 0.5 - 0.3980$$

$$= 0.1020$$

6.45 $\Pi = 0.04$, X is a binomial random variable. Find an n such that $P(x > 2,500) = 0.05$. The normal approximation to the binomial can be used as $n\Pi \geq$ and $n(1 - \Pi) \geq 5$. The z score for 5% in the right tail is 1.64. 2,500 is a lower bound. $x = 2,499.5$.

$$1.64 = \frac{2,499.5 - n(0.96)}{\sqrt{n(0.04)(0.96)}}$$

$$0 = 0.3427n^2 - 1784.3335n + 2,322,836.2024$$

$$n = \frac{-b \pm \sqrt{b^2 - 4ac}}{2a}$$

$$= \frac{-(-1784.3335) \pm \sqrt{(-1784.3335)^2 - 4(0.3427)(2,322,836.2024)}}{2(0.3427)}$$

$$= 2,586.6201, \text{ or } 2,620.7836$$

The roots were calculated without rounding. The calculations are shown to four places so that the equations would fit on the page. If the actual calculations were done to only four places, the value under the radical would be negative. Taking the calculations to six places results in roots that are 2,586.5162 and 2620.8903. The lower root shown above, 2,586.6201, is the most tickets you can sell while keeping the probability of running out to 5%. Since you cannot sell a fractional ticket, the solution is 2,586. The result that you can sell only 86 tickets over the capacity of a 2,500-seat hall is surprising. Note also that the assumption of independence is unlikely to be accurate because people come in groups to concerts.

6.47 $\lambda = 14$ per day. $\mu = \lambda t$, $\mu = 14(365)$, or 5,110. $\sigma = \sqrt{5,110}$, or 71.48. Find $P(x > 5,000)$. We can use

the normal approximation to the Poisson because $\mu \geq 5$. 5,001 is a lower bound. $x = 5,000.5$.

$$z = \frac{5000.5 - 5110}{71.48}$$

$$= -1.53$$

$$P(z > -1.53) = 0.5 + 0.4370$$

$$= 0.9370$$

The second part of exercise 6.47 asks for the probability of your funeral home having more than 450 funerals per year given the assumption that it gets one tenth of the city's funerals. If there were 4,510 or more funerals in the city, your funeral home would have more than 450. Find $P(x \geq 4,510)$.

$$z = \frac{4509.5 - 5110}{71.48}$$

$$= -8.40$$

$$P(z > -8.40) \cong 0$$

Given the assumptions in the problem, your funeral home is certain to get more than 450 funerals per year. Note that if you work this problem scaled down to the distribution of funerals for the one funeral home, [$\lambda = 1.4$ per day, $\mu = \lambda t$, $\mu = 1.4(365)$, or 511, $\sigma = \sqrt{511}$, or 22.605], you will get a different answer, primarily because scaling changes the problem. Since the standard deviation equals the square root of μ, there is proportionally less variability around a mean of 5,110 than a mean of 510. The correct answer is found when the problem is done in the original scale.

6.49 $\lambda = 1$ per 10 miles, or 0.1 per mile. $\mu = \lambda t$, $\mu = 0.1(60)$, or 6. X is a Poisson random variable. Find $P(x > 6)$. From the exponential table with $\mu = 6$, $P(x > 6) = 1 - 0.9975$, or 0.0025.

6.51 $\Pi = 0.75$. $n = 6,500$. X is a binomial random variable. Find $P(x > 5,000)$. Since $n\Pi$ and $n(1 - \Pi)$ are greater than or equal to 5, you can use the normal approximation to the binomial. $\mu = n\Pi$, $\mu = 6,500(0.75)$, or 4,875. $\sigma = \sqrt{n\Pi(1 - \Pi)}$, $\sigma = \sqrt{6,500(0.75)(0.25)}$, or 34.91. Since 5,001 is the lower bound, $x = 5,000.5$.

$$z = \frac{5,000.5 - 4,875}{34.91}$$

$$= 3.59$$

$$P(z > 3.59) = 0.5 - 0.4998, \text{ or } 0.0002$$

With Tchebychev's inequality in exercise 5.47, the probability of having more than 5,000 students accept was less than or equal to 0.078. The actual probability is approximately 0.0002.

6.53 $\mu_1 = 3$, $\sigma_1 = 0.01$, $\mu_2 = 2.95$, $\sigma_2 = 0.009$. $n_1 = 1,000$, $n_2 = 1,000$. A door and frame match if the frame is 0.05 foot wider than the door. Clearly, if the two standard deviations were equal, all of the doors could be matched to a frame. The door-frame example would be the same as the example that began the chapter on worms and rollers. Since the standard deviations are not equal, there are some doors that cannot be matched to a frame. A diagram will help you visualize where the mismatches are and the solution. Suppose 3 feet was subtracted from the width of each frame and 2.95 feet from each door. This subtraction does not change the problem. Both distributions would now be centered on zero. The accompanying figure shows the two normal distributions centered on zero. The more peaked distribution drawn with a black line represents the doors and the less peaked distribution drawn with a gray line represents the frames. Four areas are labeled a, b, c, and d. All of these areas are on the right side of the mean. By symmetry, there are identical areas on the left side. Area a represents doors that do not have a matching frame. Area b represents door and frames that match. Area c also represents door and frames that match. Area d represents frames that do not have a match. Clearly, $a = d$. There must be as many unmatched doors as frames. If we can find a and multiply it by 2 (for the left side), we would have the proportion of the 1,000 doors that are unmatchable. The same figure would be the proportion of frames that are unmatchable.

The solution involves finding the value of x where the two normal curves cross on the right side. We can find this x by substituting into formula 6.4 the density function of the normal distribution. The value substituted for μ is 0 and the values for the standard deviations are 0.01 and 0.009. We equate the density functions and solve for x. The penultimate step is to find the z scores and areas between the mean and these z scores. The last step is to take the difference between these areas, multiply by 2, and then multiply by 1,000.

$$f(x_{frames}) = \frac{1}{\sqrt{2\pi}(0.01)}e^{-1/2[(x-0)/0.01]^2}$$

$$f(x_{doors}) = \frac{1}{\sqrt{2\pi}(0.009)}e^{-1/2[(x-0)/0.009]^2}$$

$$x = 0.00947806$$

$$z_{doors} = \frac{0.00947806}{0.009}$$

$$= 1.05$$

$$P(0 < z < 1.05) = 0.3531$$

$$z_{frames} = \frac{0.00947806}{0.01}$$

$$= 0.95$$

$$P(0 < z < 0.95) = 0.3289$$

$$a = 0.3531 - 0.3289$$

$$= 0.0242$$

Given $a = 0.024$, the area representing unmatched frames for both sides is 0.048. Applying this proportion to the 1,000 frames results in approximately 48 unmatched frames. There are also 48 unmatched doors.

6.55 $\lambda = 1$ per 100 patient-years or 0.01 per patient-year, $\mu = \lambda t$, $\mu = 0.01(10,000)$, or 100. X is a Poisson random variable. $\sigma = \sqrt{100}$, or 10. We can use the normal approximation to the Poisson distribution because $\mu \geq 5$. The central 50% of the distribution is approximately bounded by the z scores of -0.67 and 0.67. The area between the mean and a z score of 0.67 is 0.2486. Substituting into the z score formula,

$$z = \frac{x - 100}{10}$$

$$0.67 = \frac{x - 100}{10}$$

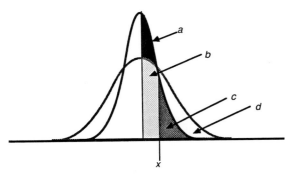

$x = 106.7$

$x \cong 107$

The central 50% of the distribution of the number of Jarvik III mechanical heart failures among 10,000 patient-years is from 93 to 107.

6.57 Four quarters yield a minimum of 4 times 6, or 24 minutes of drying time. Thus, four quarters guarantees you the 19 minutes. Three quarters can result in a minimum of 18 minutes or a maximum of 36. While the time for each quarter is a continuous uniform random variable, the time for three quarters is not. Thus, the probability of needing four quarters is not 1/18 (the 1 minute below 19 minutes divided by the range of minutes for three quarters). We will use Tchebychev's inequality. $E(X) = (b + a)/2$, $E(X) = (12 + 6)/2$, or 9. $E(X_1 + X_2 + X_3) = 9 + 9 + 9$, or 27.

$$\sigma^2(X) = \frac{(b - a)^2}{12}$$

$$= \frac{(12 - 6)^2}{12}$$

$$= 3$$

$$\sigma^2(X_1 + X_2 + X_3) = \sigma^2(X_1) + \sigma^2(X_2) + \sigma^2(X_3)$$

$$= 3 + 3 + 3$$

$$= 9$$

$$\sigma(X_1 + X_2 + X_3) = \sqrt{9}$$

$$= 3$$

$$k = \frac{B - \mu}{\sigma}$$

$$= \frac{19 - 27}{3}$$

$$= -2.67$$

$$P_{\text{tails}} \leq \frac{1}{k^2}$$

$$\leq \frac{1}{(-2.67)^2}$$

$$\leq 0.1406$$

The maximum probability of needing more than three quarters is 0.1406.

CHAPTER 7

7.1 There are two possible problems. The pool of potential respondents (car owners, phone owners, and magazine subscribers) may have not been representative of all voters (particularly in 1936). This pool may have been richer and more Republican than the set of all voters. Second, the low response rate suggests the presence of self-selection bias. People who favored Landon may have been more willing to return the survey form than people who favored Roosevelt. Exercise A4.3 in the computer appendix shows that both of these problems were present.

7.3 I picked pet grooming shops because there were more than 50 of them in the *Indianapolis Yellow Pages*. There ought to be enough pet grooming shops to do this exercise in all but the smallest towns.

7.5 You have 95% of 5,000, or 4,750 respondents. Of these 4,750, Venetian blinds are favored by 82%, or 3,895. If all of the 250 nonrespondents favored Venetian blinds, there would be 3,895 + 250, or 4,145 people in the pool of 5,000 who favored Venetian blinds. The highest possible proportion favoring Venetian blinds is 4,145/5,000, or 82.9%. At the opposite extreme, all 250 of the nonrespondents favor drapes. This results in 3,895/5,000, or 77.9% of the possible sample favoring Venetian blinds. The 77.9% is further from 82% than the 82.9%. The maximum difference between the observed proportion of 0.82 and the possible proportion favoring Venetian blinds is 0.82 − 0.799 = 0.021, or 2.1 percentage points.

7.7 Most schools would have a small proportion of all students bothering to answer the survey. At *real* party schools the students would be too busy going to parties to answer the survey. Cal Tech is a small school and a few extra favorable responses to the survey would boost the observed proportion of all students answering the question yes. The survey has no mechanism to prevent padding or ballot box stuffing. One prankster who mailed in a bunch of yes responses might have raised Cal Tech to the top spot. Perhaps Isaac Schlimazel will learn to be more skeptical as he pursues his studies at Cal Tech.

7.9 The exercise is solved two ways. The first solution is approximate because odd-numbered days are treated as half of all days in a year. The second solution uses the exact proportion of odd-numbered days.

First Solution
Half of all respondents, or 100 whites and 100 blacks, would be expected to have the last digit of their social security number be 5 or more. Among these, half would be expected to have a birth date with an even number. Thus 50 white and 50 black respondents would be expected to answer "yes" whether or not they believe racial prejudice is a serious problem in the workplace. Among the 200 black respondents there were 120 yes answers, or 70 more than expected solely from the social security and birthday questions. The estimated proportion of blacks who believe racial prejudice is a serious problem in the workplace is these 70 excess yes answers divided by the 100 who were expected to answer the question on racial prejudice. Thus the estimated percentage is 70%. Among the 200 white respondents there were 101 yes answers, or 51 more than expected solely from the social security and birthday questions. The estimated proportion of whites who believe racial prejudice is a serious problem in the workplace is these 51 excess yes answers divided by the 100 who were expected to answer the question on racial prejudice. Thus the estimated percentage for whites is 51%.

Second Solution
While at first glance one would expect a 50/50 breakdown between even and odd birthdates, that is not

Month	Total days	Odd days	Even days
Jan	31	16	15
Feb*	28.25	14.25	14
Mar	31	16	15
Apr	30	15	15
May	31	16	15
June	30	15	15
July	31	16	15
Aug	31	16	15
Sept	30	15	15
Oct	31	16	15
Nov	30	15	15
Dec	31	16	15
		186.25	179

*Adjusted for leap year.

quite correct, as the table shows below (assuming a uniform distribution of births).

$P(\text{odd}) = 186.25/365.25 = 0.509925$, or approximately 0.51
$P(\text{even}) = 179/365.25 = 0.490075$, or approximately 0.49

Thus 49 white and 49 black respondents would be expected to answer "yes" on the basis of having an even birthday. From those 100 with a true yes/no choice, 71 blacks and 52 whites are the estimated number who believe prejudice is a serious problem (71% and 52%).

7.11(a) X is normally distributed, $\mu = 89$, $\sigma = 12$, $n = 19$. Find $P(85 < \bar{x} < 93)$. Since the population is normally distributed, $\bar{x}$ is normally distributed. The finite population correction factor cannot be applied because no population size is mentioned.

$$\sigma_{\bar{x}} = \frac{12}{\sqrt{19}}$$
$$= 2.753$$

$$z = \frac{85 - 89}{2.753}$$
$$= -1.45$$

$$z = \frac{93 - 89}{2.753}$$
$$= 1.45$$

$$P(-1.45 < z < 1.45) = 0.4265 + 0.4265$$
$$= 0.8530$$

(b)

$$z = \frac{80 - 89}{2.753}$$
$$= -3.27$$

$$z = \frac{98 - 89}{2.753}$$
$$= 3.27$$

$$P(-3.27 < z < 3.27) = 0.4995 + 0.4995$$
$$= 0.9990$$

(c)

$$\sigma_{\bar{x}} = \frac{12}{\sqrt{40}}$$

$$= 1.897$$

$$z = \frac{91 - 89}{1.897}$$

$$= 1.05$$

$$P(z > 1.05) = 0.5000 - 0.3531$$

$$= 0.1469$$

(d)

$$\sigma_{\bar{x}} = \frac{12}{\sqrt{5}}$$

$$= 5.367$$

$$z = \frac{91 - 89}{5.367}$$

$$= 0.37$$

$$P(z < 0.37) = 0.5000 + 0.1443$$

$$= 0.6443$$

7.13 If the average of the five orders is less than 240, the print shop will not exhaust its stock of paper. X is approximately normally distributed. $\mu = 200$, $\sigma = 40$, $n = 5$. Find $P(\bar{x} < 240)$. Since the population is normally distributed, $\bar{x}$ is normally distributed. The finite population correction factor is not required because no population size is mentioned. The population can be considered all future orders.

$$\sigma_{\bar{x}} = \frac{40}{\sqrt{5}}$$

$$= 17.889$$

$$z = \frac{240 - 200}{17.889}$$

$$= 2.24$$

$$P(z < 2.24) = 0.5000 + 0.4875$$

$$= 0.9875$$

7.15

$$\sigma_{\bar{x}} = \frac{2}{\sqrt{100}} \sqrt{\frac{800 - 100}{800 - 1}}$$

$$= (0.2)(0.9360)$$

$$= 0.1872$$

$$z = \frac{5.11 - 4.55}{0.1872}$$

$$= 2.99$$

$$P(z > 2.99) = 0.5000 - 0.4986$$

$$= 0.0014$$

7.17

$$\sigma_{\bar{x}} = \frac{1}{\sqrt{50}} \sqrt{\frac{500 - 50}{500 - 1}}$$

$$= 0.1343$$

$$z = \frac{12 - 13}{0.1343}$$

$$= -7.45$$

$$P(z < -7.45) \cong 0$$

7.19 $n = 25$, $\mu = 5$, $\sigma = 1$. $\sigma_{\bar{x}} = 1/\sqrt{25}$, or 0.2. X, the weights of the salmon, is normally distributed. Thus $\bar{x}$ is normally distributed. The fisherman's revenue is $(25)\bar{x}(\$1.50)$, or $\$37.5\bar{x}$. To cover the out-of-pocket costs $\$37.5\bar{x} \geq \$125$. Hence, $\bar{x} \geq 3.33$. Find $P(\bar{x} < 3.33)$.

$$z = \frac{3.33 - 5}{0.2}$$

$$= -8.35$$

$$P(z < -8.35) \cong 0$$

7.21(a)

$$\sigma_{\bar{p}} = \sqrt{\frac{0.30(1 - 0.30)}{100}}$$

$$= 0.0458$$

$$z = \frac{0.35 - 0.30}{0.0458}$$

$$= 1.09$$

$P(z > 1.09) = 0.5000 - 0.3621$

$\qquad = 0.1379$

(b)

$z = \dfrac{0.40 - 0.30}{0.0458}$

$\quad = 2.18$

$z = \dfrac{0.60 - 0.30}{0.0458}$

$\quad = 6.55$

$P(2.18 < z < 6.55) \cong P(2.18 < z)$

$\qquad\qquad\qquad \cong 0.5000 - 0.4854$

$\qquad\qquad\qquad \cong 0.0146$

(c)

$z = \dfrac{0.20 - 0.30}{0.04858}$

$\quad = -2.18$

$P(z < -2.18) = 0.5000 - 0.4854$

$\qquad\qquad = 0.0146$

(d)

$1.28 = \dfrac{x - 0.30}{0.0458}$

$x = 1.28(0.0458) + 0.30$

$\quad = 0.3586$

7.23

$\sigma_{\bar{p}} = \sqrt{\dfrac{0.025(1 - 0.025)}{1,000}}$

$\quad = 0.0049$

$z = \dfrac{0.048 - 0.025}{0.0049}$

$\quad = 4.69$

$P(z > 4.69) \cong 0$

7.25

$\sigma_{\bar{p}} = \sqrt{\dfrac{0.1(1 - 0.1)}{500}}$

$\quad = 0.0134$

$z = \dfrac{0.14 - 0.1}{0.0134}$

$\quad = 2.99$

$P(z > 2.99) = 0.5000 - 0.4986$

$\qquad\qquad = 0.0014$

7.27 $\Pi = 0.000001$, $n = 50,000,000$. Find $P(\bar{p} \geq 30/50,000,000 = 0.0000006)$. This problem could be done using the sample proportion, but there is a much quicker way to do it. The number of seeds that will bear a blue tomato is a binomial random variable. With such a small Π and large n, the Poisson closely approximates the binomial. 50 successful seeds would be expected in 50 million trials. The standard deviation of the Poisson distribution is the square root of the mean, or 7.07. Applying the normal approximation to the Poisson distribution,

$z = \dfrac{30 - 50}{7.07}$

$\quad = -2.83$

$P(z > -2.83) = 0.5000 + 0.4977$

$\qquad\qquad = 0.9977$

7.29 Sampling with replacement results in $5 \times 5 \times 5$, or 125 possible samples of size 3. Similarly, there are $5 \times 5 \times 5 \times 5$, or 625 possible samples of size 4 with replacement. The exercise asks for a plot of the sampling distribution of the sum of the highest and lowest observation in a sample divided by 2. Shown on p. 1042 are summary tables for the frequency distribution of each estimator and its variance. Also shown, on p. 1043, are the vertical line charts for $n = 3$ and $n = 4$, for both $\bar{x}$ and the estimator (min + max)/2.

For $(min + max)/2$ estimator

	for $n = 3$			for $n = 4$	
$(min + max)/2$	frequency	$f_i(x - \mu)^2$	$(min + max)/2$	frequency	$f_i(x - \mu)^2$
1.0	1	4.0	1.0	1	4.0
1.5	6	13.5	1.5	14	31.5
2.0	13	13.0	2.0	51	51.0
2.5	24	6.0	2.5	124	31.0
3.0	37	0.0	3.0	245	0.0
3.5	24	6.0	3.5	124	31.0
4.0	13	13.0	4.0	51	51.0
4.5	6	13.5	4.5	14	31.5
5.0	1	4.0	5.0	1	4.0
sum	125	73.0		625	235.0

$$\sigma^2_{(min+max)/2} = 73/125, \text{ or } 0.584 \qquad \sigma^2_{(min+max)/2} = 235/625, \text{ or } 0.376$$

For the sample mean estimator

	for $n = 3$			for $n = 4$	
$\bar{x}$	frequency	$f_i(x - \mu)^2$	$\bar{x}$	frequency	$f_i(x - \mu)^2$
1.000	1	4.000	1.00	1	4.0
1.333	3	8.333	1.25	4	12.25
1.667	6	10.667	1.50	10	22.5
2.000	10	10.000	1.75	20	31.25
2.333	15	6.667	2.00	35	35.00
2.667	18	2.000	2.25	52	29.25
3.000	19	0.000	2.50	68	17.00
3.333	18	2.000	2.75	80	5.00
3.667	15	6.667	3.00	85	0.00
4.000	10	10.000	3.25	80	5.00
4.333	6	10.667	3.50	68	17.00
4.667	3	8.333	3.75	52	29.25
5.000	1	4.000	4.00	35	35.00
sum	125	83.333	4.25	20	31.25
			4.50	10	22.5
			4.75	4	12.25
			5.00	1	4.00
				625	312.50

$$\sigma^2_{\bar{x}} = 83.333/125, \text{ or } 0.667$$

$$\sigma^2_{\bar{x}} = 312.50/625, \text{ or } 0.5$$

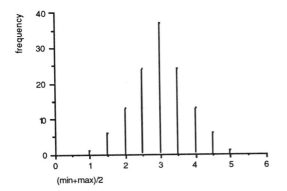

Vertical Line Chart Showing the Frequency of Each Value of the Estimator (min + max)/2 for the Population {1, 2, 3, 4, 5} with $n = 3$

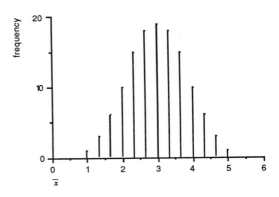

Vertical Line Chart Showing the Frequency of Each Value of the Estimator $\bar{x}$ for the Population {1, 2, 3, 4, 5} with $n = 3$

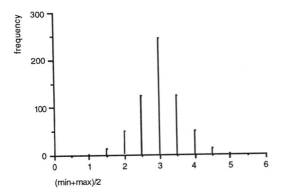

Vertical Line Chart Showing the Frequency of Each Value of the Estimator (min + max)/2 for the Population {1, 2, 3, 4, 5} with $n = 4$

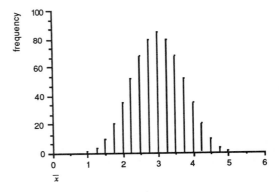

Vertical Line Chart Showing the Frequency of Each Value of the Estimator $\bar{x}$ for the Population {1, 2, 3, 4, 5} with $n = 4$

Given the population {1, 2, 3, 4, 5}, both estimators are unbiased. The (min + max)/2 is more efficient. It has a lower variance than $\bar{x}$ at both $n = 3$ and $n = 4$. Both estimators are consistent—their variances decrease as the sample size increases. You may be wondering why not use (min + max)/2 all of the time in place of $\bar{x}$ if it is more efficient, and also unbiased and consistent. Note that these results hold only for the population {1, 2, 3, 4, 5}. Different estimators can do better or worse on the criteria of lack of bias, effi-

ciency, and consistency for different populations. The better performance of (min + max)/2 if an artifact of the population chosen. If the population was not uniformly distributed, these results would not hold up. For example, given the population {1, 2, 3, 4, 100}, the $E(\bar{x}) = \mu$, or 22. The $E[(\text{min} + \text{max})/2]$ is 31.12 for $n = 4$. With this population (min + max)/2 also has a higher variance than $\bar{x}$ [at $n = 4$, the variance of (min + max)/2 is 572.3216 and the sample mean is 380.5].

7.31(a) Estimate, it is the result of some sample.

(b) Estimator, the sample mean is a random variable used to estimate the population mean.

(c) Estimator, this statement is about the properties of an estimator.

(d) Estimate, it is the result of some sample.

7.33 Let the subscript 1 refer to the mail survey and the subscript 2 refer to the phone survey. n_1 is a random variable that depends on the number of people who return mailed questionnaires, which in turn depends on the population proportion of customers who favor computerized itineraries. n_2 equals 50. Let Π be the unknown population proportion of customers who favor computerized itineraries. According to Chapter 7, the mean square error can be decomposed into the variance of the estimator and the square of the bias. Since the telephone survey is unbiased, its mean square error is solely the variance of the sample proportion. From formula 7.4, the variance of $\bar{p}$ for a sample of size 50 equals

$$\text{Mean square error}_2 = \sigma_{\bar{p}}^2 = \frac{\Pi(1-\Pi)}{50}$$

The mean square error of the mail survey consists of the variance of the sample proportion plus the square of the bias. The variance of the sample proportion equals $\Pi(1-\Pi)/n_1$, but n_1 is a random variable in the mail survey. If Π equaled 0, 40% of the 250 potential respondents would mail back the survey. At the other extreme, if Π equaled 1, 80% of the 250 potential respondents would mail back the survey. The range of n_1 is 0.4(250) to 0.8(250), or 100 to 200. n_1 is a linear function of Π. $n_1 = 100 + 100\Pi$. Thus, the variance of the sample proportion is $\Pi(1-\Pi)/(100 + 100\Pi)$. The squared bias equals the square of the difference between the expected value of the sample proportion and Π. The expected value of the sample proportion for the mail survey is the expected number of yes answers divided by the expected number of returned surveys. The expected number of yes answers is $250(0.8)\Pi$, or 200Π. The expected sample size is $100 + 100\Pi$. The mean square error of the mail survey is

$$\text{Mean square error}_1 = \frac{\Pi(1-\Pi)}{100 + 100\Pi} + \left(\frac{200\Pi}{100 + 100\Pi} - \Pi\right)^2$$

$$= \frac{\Pi(1-\Pi)}{100 + 100\Pi} + \left(\frac{2\Pi}{1+\Pi} - \Pi\right)^2$$

Both mean square errors equal 0 when Π equals either 0 or 1. If everyone in the population either says yes or says no, you cannot make an error in estimating the population proportion. If for all Π between 0 and 1, one of the estimators had a mean square error that was smaller than that of the other estimator, it would obviously be preferred. The accompanying graph plots the mean square error of the mail survey minus the mean square error of the phone survey. The mean square error of the mail survey is almost always greater than the mean square error of the phone survey. The exceptions are for Π greater than 0.94 and Π below 0.02. Thus, the phone survey is preferred over the mail survey of equal cost.

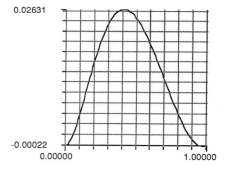

Another approach to the mail versus phone survey problem is to not use $\bar{p}$ as an estimator of Π. If you are willing to use the subjective probabilities of responding to the survey in estimating Π as well as deciding on which method of sampling, it is possible to construct an unbiased estimate of Π from the mail survey. The value of $\bar{p}$ is a function of Π.

$$\bar{p} = \frac{0.8n\Pi}{0.8n\Pi + 0.4n(1-\Pi)}$$

$$= \frac{2\Pi}{\Pi + 1}$$

Solving the last expression for Π,

$$\Pi = \frac{\bar{p}}{2 - \bar{p}}$$

The term $\bar{p}/(2 - \bar{p})$ is an unbiased estimator of Π provided that your subjective probabilities of the respondents returning the survey are correct. With this estimator the mail survey is preferred to the phone survey. The mail survey has no bias and a larger sample size.

7.35

$$\sigma_{\bar{x}} = \frac{4}{\sqrt{40}}$$

$$= 0.6325$$

$$z = \frac{28 - 30}{0.6325}$$

$$= -3.16$$

$$P(z < -3.16) = 0.5000 - 0.4992$$

$$= 0.0008$$

7.37 $\mu = 12$, $\sigma = 3$, $n = 10$. Find $P(\bar{x} > 55)$. The exercise does not provide the assumption of normality. There are two ways to proceed. One way is to use Tchebychev's inequality. The other is assume normality. Tchebychev's inequality is preferred because it requires no assumption. The chance of getting an $\bar{x}$ of 55 or more for $n = 10$ is remote with either method.

$$\sigma_{\bar{x}} = \frac{3}{\sqrt{10}}$$

$$= 0.9487$$

$$k = \frac{55 - 12}{0.9487}$$

$$= 45.31$$

$$P_{\text{tails}} \le \frac{1}{k^2}$$

$$\le \frac{1}{45.33^2}$$

$$\le 0.000486$$

If the assumption of normality was made, the z score would equal 45.33. The formula for z is the same as for k. The probability of observing values to the right of a z score of 45.33 is approximately 0.

7.39

$$\sigma_{\bar{x}} = \frac{3}{\sqrt{5}}$$

$$= 1.342$$

$$z = \frac{(1 + \mu) - \mu}{\sigma_{\bar{x}}}$$

$$= \frac{1}{1.342}$$

$$= 0.75$$

$$P(z < 0.75) = 0.2734$$

By symmetry, the probability of being within plus or minus one bandage is twice 0.2734, or 0.5468.

7.41 If the survey includes some categorical or yes/no questions you could try those questions out in a small phone survey and see if the proportions in the phone survey are substantially different from the proportions in the large mail survey. Presumably, if there appears to be little or no difference on these questions, nonresponse bias would have little impact on the more complex questions in the mail survey.

7.43 For systematic sampling to be equivalent to a simple random sample there must be no pattern in the data. In this case corn acreage must not be related to the position of the farm owner's name in the alphabetical list. In one county this assumption could be easily violated. Suppose the first family to settle in the county was the Sandys. The various Sandy descendants have the largest farms and the highest acreage in corn. If your skip pattern of taking every 30th farm by chance jumped over all of the Sandy farms, you would miss most of the larger farms in the county. Blindly applying a technique such as systematic sampling can be dangerous. A stratified sample that allocated observations to small, medium, and large farms would be more accurate.

Another plausible explanation is that the distribution of farms is highly skewed with a few very large farms. Even if there were no pattern to the names of the farmers with the very large farms, the survey could easily miss all of them. Again, the solution is a stratified sample.

7.45 Since the sample mean is the center of gravity for a given sample, $\Sigma(x_i - a)^2$ is minimized with $a = \bar{x}$. Hence, $\Sigma(x_i - \bar{x})^2$ is less than $\Sigma(x_i - \mu)^2$ for all $\bar{x}$ not equal to μ. The implication of this result is that $\Sigma(x_i - \bar{x})^2$ is, on average, too low. The solution is to use $n - 1$ in place of n in the denominator of the formula for the sample variance. This change makes the sample variance an unbiased estimator of the population variance. This result is proved in appendix C.

7.47 Note that even if each book borrower acted independently, each book would not be an independent trial. Borrowers usually take several books. Treating each book as an independent trial,

$$\sigma_{\bar{p}} = \sqrt{\frac{0.1(1 - 0.1)}{300}}$$

$$= 0.0173$$

$$z = \frac{0.08 - 0.1}{0.0173}$$

$$= -1.16$$

$$P(z < -1.16) = 0.5000 - 0.3770$$

$$= 0.1230$$

7.49 X is the observed speed. The observed speed is equal to μ (the average speed) plus the effect of the radar error, the speedometer error, and random variation in driving speed. Since each source of variation is normally distributed, X is normally distributed. Since each source of variation has a zero mean, the errors do not shift the distribution of X. $\mu = 65$, $\bar{x} = 67.1$, $n = 1,000$. $\sigma_1 = 4$, $\sigma_2 = 8$, $\sigma_3 = 12$. Find $P(\bar{x} > 67.1)$. The assumption of independence among the sources of error is reasonable because the radar should not affect the car's speedometer and the speedometer error should not affect the random variation in the drivers' speeds. Given the assumption of independence, the standard deviation of X is

$$\sigma(X) = \sqrt{\sigma_1^2 + \sigma_2^2 + \sigma_3^2}$$

$$= \sqrt{16 + 64 + 144}$$

$$= 14.967$$

With the standard deviation of X we can proceed to find $P(\bar{x} > 67.1)$:

$$\sigma_{\bar{x}} = \frac{14.967}{\sqrt{1,000}}$$

$$= 0.4733$$

$$z = \frac{67.1 - 65}{0.4733}$$

$$= 4.44$$

$$P(z > 4.44) \cong 0$$

Even with the three sources of variation, there is an approximately zero probability of observing a sample mean of 67.1 if the population mean was 65, for a sample of 1,000.

CHAPTER 8

Note that in the solutions I assume that students know the z scores for the most commonly used tail areas: 10%, 5%, 2.5%, 1%, and 0.5%. These z scores are 1.28, 1.64, 1.96, 2.33, and 2.57.

8.1(a)

$$L, U = 26 \pm 1.96\left(\frac{5}{\sqrt{10}}\right)$$

$$L = 22.90, \ U = 29.10$$

(b)

$$L, U = 26 \pm 1.96\left(\frac{5}{\sqrt{30}}\right)$$

$$L = 24.21, \ U = 27.79$$

(c)

$$L, U = 26 \pm 1.64\left(\frac{5}{\sqrt{10}}\right)$$

$$L = 23.41, \ U = 28.59$$

(d)

$$L, U = 26 \pm 2.57\left(\frac{5}{\sqrt{30}}\right)$$

$$L = 23.65, \ U = 28.35$$

8.3(a)

$$L, U = 23 \pm 1.96\left(\frac{8}{\sqrt{300}}\right)$$

$$L = 22.09, \ U = 23.91$$

(b)

$$L, U = 23 \pm 2.57 \left(\frac{8}{\sqrt{300}} \right)$$

$$L = 21.81, \ U = 24.19$$

(c)

$$L, U = 23 \pm 1.64 \left(\frac{8}{\sqrt{2,000}} \sqrt{\frac{10,000 - 2,000}{10,000 - 1}} \right)$$

$$L = 22.74, \ U = 23.26$$

8.5 $n = 100$, X is normally distributed, $\sigma = 68$, $1 - \alpha = 95\%$, $L = \$56.15$, $U = \$93.85$. A 95% confidence interval should have a range of $\pm 1.96(68/\sqrt{100})$, or 26.66. The range of the confidence interval reported by the survey firm is $93.85 - 56.15$, or 37.70. Either the firm made a computational error or the sample was smaller than 100. An n of 50 is consistent with the reported interval.

8.7(a)

$$L, U = 4.3 \pm 2.57 \left(\frac{0.25}{\sqrt{1,000}} \right)$$

$$L = 4.28, \ U = 4.32$$

(b) The average duration of a call is sensitive to the method of billing. Consumers will avoid long calls under a measured service system. The confidence interval constructed from data collected when measured billing was not in effect should, other things being equal, overestimate revenue under the new system.

8.9(a)

$$L, U = 24 \pm 2.201 \frac{5}{\sqrt{12}}$$

$$L = 20.82, \ U = 27.18$$

(b)

$$L, U = 24 \pm 2.015 \frac{5}{\sqrt{6}}$$

$$L = 19.89, \ U = 28.11$$

(c)

$$L, U = 24 \pm 1.96 \frac{5}{\sqrt{36}}$$

$$L = 22.37, \ U = 25.63$$

(d)

$$L, U = 24 \pm 1.28 \frac{5}{\sqrt{36}}$$

$$L = 22.93, \ U = 25.07$$

8.11 The interesting issue in this exercise is whether the 200 titles currently displayed in the video store should be considered the population from which the 20 in the sample were drawn. If the answer were yes, the finite population correction would be required. The correct answer is no. The relevant population is all of the future titles you could stock rather than the 200 that happen to be in the store last month. This population is large enough to not require the finite population correction factor.

$$L, U = 8 \pm 2.093 \frac{3}{\sqrt{20}}$$

$$L = 6.60, \ U = 9.40$$

8.13

$$L, U = 2 \pm 1.96 \frac{3}{\sqrt{100}} \sqrt{\frac{1,000 - 100}{1,000 - 1}}$$

$$L = 1.442, \ U = 2.558$$

The limits of 95% confidence interval for the total number of coins in the 10,000 square yards is 1,000 times the limits for mean number of coins in one square yard, or 1,442 to 2,558.

8.15(a)

$$L, U = 14 \pm 2.807 \left(\frac{4}{\sqrt{24}} \right)$$

$$L = 11.71, \ U = 16.29$$

(b) Nonresponse bias may be present. Less than half of the potential pool of respondents kept the diaries. While the direction of the bias is uncertain, the presumption is that the more serious camera users kept the diaries and that the sample mean overestimates the population mean number of rolls of film used in the first year. The assumption of normality is suspect because the distribution of rolls of film used in the first year has a lower bound of zero but no upper bound. Someone who got the camera as a gift may not use it. A professional photographer may use thousands of rolls in the first year. The distribution is likely to be highly skewed. The easiest solution for this problem is to work with a larger sample. The solution to the nonresponse bias may be to pay the respondents for their diaries and verify the number of pictures taken by checking the prints.

8.17(a)

$$L, U = 0.4 \pm 1.96\sqrt{\frac{0.40(1 - 0.40)}{100}}$$

$$L = 0.304, \ U = 0.496$$

(b)

$$L, U = 0.4 \pm 1.64\sqrt{\frac{0.40(1 - 0.40)}{100}}$$

$$L = 0.320, \ U = 0.480$$

(c)

$$L, U = 0.6 \pm 1.64\sqrt{\frac{0.60(1 - 0.60)}{100}}$$

$$L = 0.520, \ U = 0.680$$

(d)

$$L, U = 0.4 \pm 1.96\sqrt{\frac{0.40(1 - 0.40)}{100}}\sqrt{\frac{1,000 - 100}{1,000 - 1}}$$

$$L = 0.309, \ U = 0.491$$

8.19

$$L, U = 0.8167 \pm 1.96\sqrt{\frac{0.8167(1 - 0.8167)}{300}}$$

$$L = 0.773, \ U = 0.860$$

8.21

$$L, U = 0.03 \pm 1.64\sqrt{\frac{0.03(1 - 0.03)}{1,000}}$$

$$L = 0.021, \ U = 0.039$$

8.23(a)

$$L, U = 0.1 \pm 1.96\sqrt{\frac{0.1(1 - 0.1)}{1,000}}\sqrt{\frac{10,000 - 1,000}{10,000 - 1}}$$

$$L = 0.082, \ U = 0.118$$

At a 95% level of confidence the range of the number of defective shirts among the 9,000 that have not been released to the stores is from 738 to 1,062. These figures were obtained by multiplying the above proportions by 9,000.

(b) The required assumption is that each shirt is an independent trial with the same probability of being defective. If there were runs of bad shirts (runs of runs?), the confidence interval for the population proportion could be way off.

8.25(a)

$$n = \frac{1.96^2 40}{10^2}$$

$$= 1.5366$$

The sample size rounds up to 2. The sampling distribution of $\bar{x}$ would be normal for this small a sample if the population was normally distributed. Without the assumption of normality for the population, a sample size of at least 30 is required to guarantee approximate normality in the distribution of $\bar{x}$.

(b)

$$n = \frac{1.64^2 40}{10^2}$$

$$= 1.0758$$

(c)

$$n = \frac{1.96^2 40}{1^2}$$

$$= 153.664$$

The sample size rounds up to 154. Because $n \geq 30$, it is not necessary to assume that the population is normally distributed.

(d)

$$n = \frac{2.57^2 40}{1^2}$$

$$= 264.196$$

The sample size rounds up to 265.

8.27

$$n = \left[\frac{10(1.96)}{1}\right]^2$$

$$= 384.16$$

$$= 385$$

8.29

$$n = \frac{0.5(0.5)(1.96)^2}{(0.01)^2}$$

$$= 9,604$$

The benefit of using 0.5 as a planning value is that you are guaranteed that the sample size is large enough to satisfy the specified tolerance and confidence levels. As the actual proportion is likely to be well below 0.5, the disadvantage is that the sample size is larger than is necessary. The recent proportion of women of child-bearing age who have had babies is a reasonable planning value.

8.31

$$n = \frac{0.005(1 - 0.005)(1.96)^2}{(0.0001)^2}$$

$$= 1,911,196$$

8.33 The sample of current residents of mobile homes misses any former resident who died from cancer caused by the formaldehyde. If there are any such deaths, the estimated proportion would be biased downward.

$$L, U = 0.01333 \pm 1.96\sqrt{\frac{0.01333(1 - 0.01333)}{3,000}}$$

$$L = 0.009, \ U = 0.017$$

8.35

$$n = \left[\frac{1.96(50)}{1}\right]^2$$

$$= 9,604$$

8.37

$$L, U = 900 \pm 2.262\left(\frac{45}{\sqrt{10}}\right)$$

$$L = 867.81, \ U = 932.19$$

The limits of a confidence interval on the per mile consumption are 1/300 the limits of consumption per 300 mile trip, or from 2.893 to 3.107 gallons per mile.

8.39

$$L, U = \$254 \pm 1.96\left(\frac{14}{\sqrt{50}}\right)$$

$$L = \$250.12, \ U = \$257.88$$

8.41

$$L, U = 24 \pm 1.96\left(\frac{6}{\sqrt{50}}\right)$$

$$L = 22.337, \ U = 25.663$$

8.43

$$L, U = \$950 \pm 2.093\left(\frac{246}{\sqrt{20}}\right)$$

$$L = \$834.86, \ U = 1,065.14$$

8.45 The implication of the standard deviation being greater than the mean along with a lower bound of zero is that the distribution is highly skewed to the right. However, with a sample of size 350, the distribution of $\bar{x}$ would be very close to normal no matter how skewed the population distribution is.

$$L, U = 12 \pm 1.96\left(\frac{14}{\sqrt{350}}\right)$$

$$L = 10.533, \ U = 13.467$$

8.47 The limits are approximately 0.06 to 0.45. These limits are as close as you can get with the values in the binomial table without using interpolation.

8.49 The figures below are the histograms of the original data and the logs of the values.

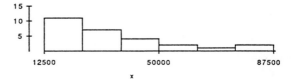

Histogram of the Original Data

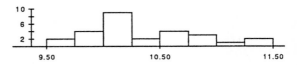

Histogram of the Natural Logarithms of the Data

We assume that the distribution of the log values is approximately normal. The limits of a 95% confidence interval for the log values are

$$L, U = 10.349 \pm 2.056\left(\frac{0.47728}{\sqrt{27}}\right)\left(\sqrt{\frac{345-27}{345-1}}\right)$$

$$L = 10.167, \quad U = 10.531$$

The last step is to take the antilogs of the above limits so that the interval is expressed in dollars. The antilogs are $26,029.87 to $37,458.91.

8.51(a)

$$L = \bar{x} - t_\alpha s_{\bar{x}}$$

$$L = 0.92 - 1.383\frac{0.09}{\sqrt{10}}$$

$$= 0.88$$

(b) Obviously, the assumption of normality may not be accurate. A more subtle problem is that the observations may not be independent. Some conditions affecting the robot's performance may persist across the trials. Finally, the robot may be programmed to learn its task. If this were the case, it would become more proficient as it did more motorcycles. For example, the robot could be programmed to remember the locations of the parts that it picks up. Any of these problems would invalidate the confidence interval.

CHAPTER 9

9.1(a)

$$\mathrm{C.V.}_L, \mathrm{C.V.}_U = 25 \pm 1.96\left(\frac{3}{\sqrt{15}}\right)$$

$$\mathrm{C.V.}_L = 23.482, \quad \mathrm{C.V.}_U = 26.518$$

The sample mean of 26 is inside the acceptance region bounded by 23.482 and 26.518. There is not enough evidence to reject the null hypothesis that the population mean equals 25 at a 5% significance level.

(b)

$$\mathrm{C.V.}_L, \mathrm{C.V.}_U = 25 \pm 1.64\left(\frac{3}{\sqrt{40}}\right)$$

$$\mathrm{C.V.}_L = 24.222, \quad \mathrm{C.V.}_U = 25.778$$

The sample mean of 23 is outside the acceptance region bounded by 24.222 and 25.778. The null hypothesis that the population mean equals 25 is rejected at a 10% significance level.

(c)

$$\mathrm{C.V.}_L, \mathrm{C.V.}_U = 25 \pm 2.57\left(\frac{3}{\sqrt{60}}\right)$$

$$\mathrm{C.V.}_L = 24.005, \quad \mathrm{C.V.}_U = 25.995$$

The sample mean of 24.6 is inside the acceptance region bounded by 24.005 and 25.995. There is not enough evidence to reject the null hypothesis that the population mean equals 25 at a 1% significance level.

(d)

$$\mathrm{C.V.}_L, \mathrm{C.V.}_U = 25 \pm 1.28\left(\frac{3}{\sqrt{10}}\right)$$

$$\mathrm{C.V.}_L = 23.786, \quad \mathrm{C.V.}_U = 26.214$$

The sample mean of 24 is inside the acceptance region bounded by 23.786 and 26.214. There is not enough evidence to reject the null hypothesis that the population mean equals 25 at a 20% significance level.

9.3 (a) $H_0: \mu = 10$, $H_1: \mu \neq 10$, a two-tailed test.
(b) $H_0: \mu = 10$, $H_1: \mu \neq 10$, a two-tailed test.
(c) $H_0: \mu \leq 10$, $H_1: \mu > 10$, a right-tailed test.
(d) $H_0: \mu \geq 10$, $H_1: \mu < 10$, a left-tailed test.
(e) $H_0: \mu \geq 10$, $H_1: \mu < 10$, a left-tailed test.
(f) $H_0: \mu \leq 10$, $H_1: \mu > 10$, a right-tailed test.

9.5 The acceptance of the null hypothesis in a left-tailed test implies that the sample mean was to the right of the critical value. Switching from a left-tailed to a two-tailed test while holding the significance level constant would move the lower critical value further to the left. The reason is that spreading the fixed α across both tails requires a higher z score. Clearly, if the sample mean was to the right of the left-tailed critical value, it will also be to the right of the lower critical value for the two-tailed test. However, the two-tailed test imposes an upper critical value which the left-tailed test did not. If the sample mean lies to the right of this upper critical value, the null hypothesis would be rejected. The accompanying figure shows both possibilities.

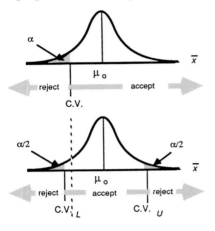

If $\bar{x}$ is to the right of C.V.$_U$, the null hypothesis in a two-tailed test will be rejected. If $\bar{x}$ is to the left of C.V.$_U$, the null hypothesis in a two-tailed test cannot be rejected.

9.7(a) The null hypothesis is that the sales of pickles in the more expensive plastic jar are the same or less than the sales in the glass jar. The alternative hypothesis is that the sales of pickles in the more expensive plastic jar are greater than the sales of pickles in the glass jar. The pickle processor would need strong evidence that the sales are higher for the unbreakable jars before accepting the higher expense.

(b) The null hypothesis is that the cheaper material has the same or better effect in melting ice than the salt. The alternative hypothesis is that the more expensive material has a better effect in melting ice. In this example there is a conflict between two rationales for setting up hypothesis tests. Ordinarily, any established treatment is given the benefit of the doubt over a new treatment. The other rationale is that the cheapest treatment be given the benefit of the doubt. By making the null hypothesis be that the cheaper material has the same or better effect in melting ice than the salt, the city manager is more willing to accept the risk of mistakenly using a less effective treatment than to accept the risk of mistakenly spending more

money than necessary. There is no clear-cut rule when two rationales conflict. For example, a doctor considering a new and cheaper cancer treatment would be likely to put the burden of proof on the new treatment in spite of it being cheaper.

(c) The null hypothesis is that schools that are visited by the former convicts and drug addicts have the same or worse levels of drug abuse than the schools that are not visited. The alternative hypothesis is that these schools have lower levels of drug abuse. The treatment has some cost; at a minimum, it takes some of the time of the students and staff. The burden of proof should be on the new program. For this test to be valid, the schools would have to be randomly selected for visits or at least paired so that schools with similar incidence rates were treated and not treated.

(d) The null hypothesis is that the veteran quarterback is a better player than the two rookies. The alternative hypothesis is that one of the rookies is the best quarterback. The coach has more information about the veteran. Of course, if one of the rookies was a high school all-American and the veteran did poorly the previous season, the coach may look at the problem differently.

9.9(a)

$$\text{C.V.} = 30 + 1.64\left(\frac{5}{\sqrt{40}}\right)$$

$$= 31.297$$

$$z = \frac{31.297 - 32}{0.7906}$$

$$= -0.89$$

$$P(z < -0.89) = 0.1867$$

(b)

$$z = \frac{31.297 - 31}{0.7906}$$

$$= 0.38$$

$$P(z < 0.38) = 0.6480$$

(c)

$$z = \frac{31.297 - 30.5}{0.7906}$$

$$= 1.01$$

$$P(z < 1.01) = 0.8438$$

(d)

$$z = \frac{31.297 - 30.25}{0.7906}$$

$$= 1.32$$

$$P(z < 1.32) = 0.9066$$

9.11(a)

$$C.V. = 30 + 1.64\left(\frac{5}{\sqrt{60}}\right)$$

$$= 31.059$$

$$z = \frac{31.059 - 32}{0.6455}$$

$$= -1.46$$

$$P(z < -1.46) = 0.0721$$

(b)

$$z = \frac{31.059 - 31}{0.6455}$$

$$= 0.09$$

$$P(z < 0.09) = 0.5359$$

(c)

$$z = \frac{31.059 - 30.50}{0.6455}$$

$$= 0.87$$

$$P(z < 0.87) = 0.8078$$

(d)

$$z = \frac{31.059 - 30.25}{0.6455}$$

$$= 1.25$$

$$P(z < 1.25) = 0.8944$$

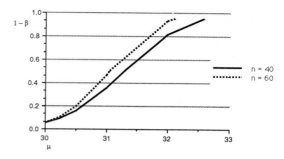

Power Curves with n Set to 40 and 60

β is 0.5 when μ_1 equals the critical value, 31.059. β approaches 0.95 as μ_1 approaches the null hypothesis value of 30. β is 0.05 when $z = -1.64$. Substituting into the z score formula: $-1.64 = (31.059 - \mu_1)/0.6455$, or $\mu_1 = 32.118$. The power equals $1 - \beta$, so the power is 0.5 at $\mu_1 = 31.059$ and it approaches 0.05 as μ_1 approaches 30.

9.13 Because of the symmetry property of the normal distribution, the β's for means of 28, 29, 29.5, and 29.75 are equal respectively to the β's for 32, 31, 30.5, and 30.25. Note that as the mean approaches 30 from either direction, β approaches 0.90 and the power approaches 0.10.

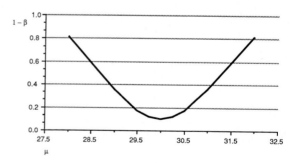

Power Curve with n Set to 40

9.15

$$n = \frac{(1.64 + 0.84)^2(5)^2}{(30 - 31)^2}$$

$$= 153.76, \text{ or } 154 \text{ rounded up}$$

9.17(a)

$$\text{C.V.}_L, \text{C.V.}_U = 60 \pm 2.262\left(\frac{5}{\sqrt{10}}\right)$$

$$\text{C.V.}_L = 56.424, \text{C.V.}_U = 63.576$$

The sample mean of 58 falls within the acceptance region. There is not enough evidence to reject the null hypothesis that the population mean equals 60.

(b)

$$\text{C.V.}_L, \text{C.V.}_U = 60 \pm 2.145\left(\frac{5}{\sqrt{15}}\right)$$

$$\text{C.V.}_L = 57.231, \text{C.V.}_U = 62.769$$

The sample mean of 58 falls within the acceptance region. There is not enough evidence to reject the null hypothesis that the population mean equals 60.

(c)

$$\text{C.V.}_L, \text{C.V.}_U = 60 \pm 2.262\left(\frac{10}{\sqrt{10}}\right)$$

$$\text{C.V.}_L = 52.848, \text{C.V.}_U = 67.152$$

The sample mean of 58 falls within the acceptance region. There is not enough evidence to reject the null hypothesis that the population mean equals 60.

(d)

$$\text{C.V.}_L, \text{C.V.}_U = 60 \pm 2.262\left(\frac{20}{\sqrt{10}}\right)$$

$$\text{C.V.}_L = 45.693, \text{C.V.}_U = 74.307$$

The sample mean of 58 falls within the acceptance region. There is not enough evidence to reject the null hypothesis that the population mean equals 60.

9.19 $H_0: \mu \geq 100$, $H_1: \mu < 100$, $n = 10$, $\bar{x} = 115$, $s = 8$, $\alpha = 5\%$. For any $\alpha \leq 50\%$, the critical value of the test will be below 100. The sample mean of 115 is above 100, so it is clear that the null hypothesis cannot be rejected.

9.21

$$\text{C.V.} = \$0 + 1.711\left(\frac{175}{\sqrt{25}}\right)$$

$$= \$59.885$$

The null hypothesis that the population mean is less than or equal to 0 is rejected at a 5% significance level. The customers have more of an incentive to report errors that go against them than errors in their favor. If the bills in the sample reflected corrections made by the customers, you would expect that the remaining errors would be more in the customers' favor. An average error of zero dollars does not guarantee zero kickbacks. The clerk who is under suspicion could make extra errors in favor of the fastener company involving companies not participating in the kickback scheme to offset the errors in favor of those companies participating in the kickback scheme.

9.23(a)

$$\text{C.V.} = 0.09 + 2.33\sqrt{\frac{0.09(1 - 0.09)}{200}}$$

$$= 0.137$$

The sample proportion of 0.1 falls within the acceptance region. There is not enough evidence to reject the null hypothesis that the population proportion is less than or equal to 0.09.

(b) Note that the standard error of the proportion is based on the null hypothesis rather than the sample proportion. Thus, the critical values are the same as in part a. The sample proportion of 0.13 falls within the acceptance region. There is not enough evidence to reject the null hypothesis that the population proportion is less than or equal to 0.09.

(c)

$$\text{C.V.} = 0.09 + 2.33\sqrt{\frac{0.09(1 - 0.09)}{1,000}}$$

$$= 0.111$$

The sample proportion of 0.095 falls within the acceptance region. There is not enough evidence to reject the null hypothesis that the population proportion is less than or equal to 0.09.

(d)

$$C.V. = 0.09 + 2.33\sqrt{\frac{0.09(1 - 0.09)}{2,000}}$$

$$= 0.1049$$

The sample proportion of 0.092 falls within the acceptance region. There is not enough evidence to reject the null hypothesis that the population proportion is less than or equal to 0.09.

9.25

$$C.V. = 0.15 - 1.64\sqrt{\frac{0.15(1 - 0.15)}{300}}$$

$$= 0.1162$$

The sample proportion of 0.1167 falls within the acceptance region. There is not enough evidence to reject the null hypothesis that the population proportion is greater than or equal to 0.15. While it is below the expected 15%, the sample proportion is not low enough to reject the possibility that the reduction in the proportion of infertile eggs is due to chance.

9.27 Setting up a left-tailed test puts the burden of proof on your conjecture that the competing radio station has overstated its audience share. Some students key in on the word "claim" in the first sentence of the exercise and argue that since WXIX claims to have an audience share of 40%, the test should be right tailed. In other words, believe that WXIX is lying unless you have strong proof to the contrary. Ordinarily, the burden of proof would be on the advertising manager of the competing station.

$$C.V. = 0.4 - 1.64\sqrt{\frac{0.4(1 - 0.4)}{1,000}}$$

$$= 0.375$$

The sample proportion of 0.38 falls within the acceptance region. There is not enough evidence to reject the null hypothesis that the population proportion is greater than or equal to 0.4.

9.29

$$C.V. = 0.98 - 1.64\sqrt{\frac{0.98(1 - 0.98)}{1,000}}\sqrt{\frac{10,000 - 1,000}{10,000 - 1}}$$

$$= 0.973$$

The sample proportion of 0.965 is below the critical value of 0.973. The null hypothesis that the proportion of good bulbs is greater than or equal to 0.98 is rejected at a 5% significance level.

9.31(a)

$$z = \frac{95 - 100}{20/\sqrt{100}}$$

$$= -2.5$$

$$P(z < -2.5) = 0.0062$$

Since this is a two-tailed test, the P value equals the area in the hypothesized distribution of $\bar{x}$ to the left of 95 plus the corresponding area to the right of 105. The P value is twice 0.0062, or 0.0124.

(b) By the symmetry of the normal distribution, the P value for $\bar{x} = 105$ for a two-tailed test with the same sample size and the same standard deviation would equal the P value in part a, or 0.0124.

(c)

$$z = \frac{99 - 100}{10/\sqrt{50}}$$

$$= -0.71$$

$$P(z < -0.71) = 0.2389$$

Since this is a two-tailed test, the P value equals the area in the hypothesized distribution of $\bar{x}$ to the left of 99 plus the corresponding area to the right of 101. The P value is twice 0.2389, or 0.4778.

(d) By the symmetry of the normal distribution, the P value for $\bar{x} = 101$ for a two-tailed test with the same sample size and the same standard deviation would equal the P value in part c, or 0.4778.

9.33

$$z = \frac{0.5714 - 0.8}{\sqrt{0.8(1 - 0.8)/70}}$$

$$= -4.78$$

$$P(z < -4.78) \cong 0$$

Since this is a two-tailed test, the P value equals the area under the hypothesized distribution of $\bar{p}$ to the left of 0.5714 plus the corresponding area in the right tail. 0.5714 is 0.2286 below the hypothesized population proportion of 0.8, and 0.8 plus 0.2286 is 1.0286. The highest value the proportion can take is 1.000. The area to the right of 1.0000 is 0. The P value for the test is approximately zero.

9.35

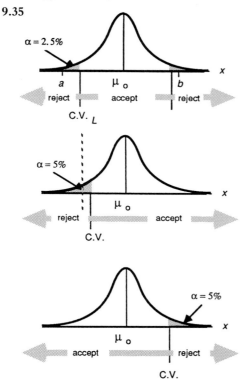

If the null hypothesis is rejected because $\bar{x}$ is to the left of C.V.$_L$ in the first panel (such as point a), it will also be rejected in the second panel. However, if the null hypothesis is rejected because $\bar{x}$ is to the left of C.V.$_L$ in the first panel, it will be accepted in the third panel. The other possibility is that the two-tailed null hypothesis is rejected because $\bar{x}$ is in the right tail (as in point b in the upper panel). The null hypothesis will be rejected in the bottom panel.

9.37 Note that $n\Pi < 5$. The normal approximation to the sampling distribution of the proportion cannot be applied. The distribution of the number of defective picture tubes per set of 5,000 is hypergeometric. However, since $n < 0.05\nu$, the distribution is approximately binomial. This problem can be solved directly with the binomial distribution or with the Poisson approximation to the binomial. Using the

Poisson approximation to the binomial the solution can be found with the aid of the Poisson table. $\mu = n\Pi$, $\mu = 100(0.005)$, or 0.5. The $P(x \geq 2) = 1 - [P(0) + P(1)]$. $P(x \geq 2) = 1 - (0.6065 + 0.3033)$, or 0.0902.

9.39 Find the P value of the test. Either the normal approximation to the sampling distribution of the proportion or the Poisson approximation to the binomial can be used for this problem. $n\Pi$ and $n(1 - \Pi) \geq$ 5. Also, $\Pi \leq 0.05$ and $n \geq 20$. The Poisson approximation is easier for this exercise and is more accurate than the normal approximation for small values of Π. $\mu = n\Pi$, $\mu = 1,000,000(0.00001)$, or 10. The P value of the test is the probability of 13 or more leaks. This probability can be found with the Poisson table. $P(x \geq 13) = 1 - P(x \leq 12)$, $P(x \geq 13) = 1 - 0.7916$, or 0.2084. You can get close to the same answer using the normal approximation to the sampling distribution of the proportion provided that you use a continuity correction, i.e., use 0.0000125 in place of 0.000013 in the z score formula. With the continuity correction the z score is 0.79 and the P value is 0.2148. The continuity correction is usually not used on proportion problems but it makes a substantial difference here because of the small number of successes.

9.41 Given the assumption of independence among events, the distribution of number of jams in a fixed period would be Poisson. $\mu = \lambda t$, $\mu = (1 \text{ per hour}) \times (2 \text{ hours})$, or 2. The P value of the test is probability of observing 10 or more jams. The solution can be found with the Poisson table. $P(x \geq 10) = 1 - P(x \leq 9)$. The P value of the test is $1 - 0.9999$, or 0.0001.

9.43 If the tasters could not distinguish between Löwenbräu and Miller, approximately one-third of the tasters would pick the Löwenbräu from among the three glasses. If they could distinguish the Löwenbräu, the proportion picking the Löwenbräu glass should be higher than one-third. If the test statistic is the sample proportion picking Löwenbräu, the test should be right tailed. This puts the burden of proof on Löwenbräu being distinguishable from Miller.

$$\text{C.V.} = 0.33 + 1.64\sqrt{\frac{0.33(1 - 0.33)}{25}}$$

$$= 0.4842$$

9.45 The problem can be done directly with the binomial distribution or with the Poisson approximation to the binomial. The P value is the probability of finding 3 or more defective smoke alarms among the 60 tested, given the probability that any one is defective is 0.01. Note that even though the population is fixed at 1,000, it is not known if there are exactly 10 defectives among the 1,000. The $\Pi = 0.01$ comes from the production process for smoke alarms. This is why the distribution is binomial rather than hypergeometric. The Poisson approximation is the easiest way to solve the problem. $\mu = n\Pi$, $\mu = 60(0.01)$, or 0.6. $P(x \geq 3) = 1 - P(x \leq 2)$. From the Poisson table with $\mu = 0.6$, $P(x \geq 3) = 1 - 0.9769$, or 0.0231.

9.47

$$\text{C.V.}_L, \text{C.V.}_U = 0.5 \pm 1.96\sqrt{\frac{0.5(1 - 0.5)}{25}}$$

$$\text{C.V.}_L = 0.304, \text{C.V.}_U = 0.696$$

The sample proportion of 0.68 is within the acceptance region. There is not enough evidence to reject the null hypothesis at a 5% significance level that the coin is fair.

9.49

$$\text{C.V.}_L = \text{C.V.}_U = 2 \pm 2.57\left(\frac{0.3}{\sqrt{10}}\right)$$

$$\text{C.V.}_L = 1.756, \text{C.V.}_U = 2.244$$

The sample mean of 3.12 is above the upper critical value of 2.244. The null hypothesis that the mean of the distribution of dough weights equals 2 pounds is rejected at a 1% significance level. The administrative decision is to recalibrate the scaling machine.

9.51

$$z = \frac{448,952.03 - 478,000}{120,000/\sqrt{30}}$$

$$= -1.33$$

$$P(z > -1.33) = 0.9082$$

9.53

$$\text{C.V.} = 2,160 - 1.64\sqrt{2,060}$$

$$= 2,083.780$$

The observed rate of 2,060 occurrences per 36-hour period is below the critical value of 2,083.780. The null hypothesis that the mean rate has not gone down is rejected at a 5% significance level.

■ CHAPTER 10

10.1(a)

$$L, U = (34 - 25) \pm 1.96\sqrt{\frac{20}{10} + \frac{30}{20}}$$

$$L = 5.33, \ U = 12.67$$

(b)

$$L, U = (44 - 65) \pm 1.96\sqrt{\frac{20}{20} + \frac{30}{20}}$$

$$L = -24.10, \ U = -17.90$$

(c)

$$L, U = (44 - 65) \pm 1.96\sqrt{\frac{20}{50} + \frac{30}{70}}$$

$$L = -22.78, \ U = -19.22$$

(d)

$$L, U = (44 - 65) \pm 1.96\sqrt{\frac{20}{200} + \frac{30}{200}}$$

$$L = -21.98, \ U = -20.02$$

10.3

$$L, U = (\$2,000 - 1,500) \pm 1.96\sqrt{\frac{300^2}{200} + \frac{300^2}{300}}$$

$$L = \$446.32, \ U = \$553.68$$

10.5

$$L, U = (\$125,000 - 90,000) \pm 1.96\sqrt{\frac{20,000^2}{50} + \frac{20,000^2}{50}}$$

$$L = \$27,160, \ U = \$42,840$$

The limits of a 95% confidence interval for the total compensation for 2,000 executives would be 2,000 times the limits of the 95% confidence interval for the mean compensation per executive, or $L = \$54,320,000$ and $U = \$85,680,000$.

10.7(a)

$$\text{C.V.} = 0 + 1.64 \sqrt{\frac{20^2}{20} + \frac{30^2}{15}}$$

$$= 14.669$$

There is not enough evidence to reject the null hypothesis that the mean of X is less than or equal to the mean of Y because the observed difference in sample means of 1 is on the acceptance side of the critical value of 14.669.

(b)

$$\text{C.V.} = 0 + 1.28 \sqrt{\frac{20^2}{20} + \frac{30^2}{15}}$$

$$= 11.449$$

There is not enough evidence to reject the null hypothesis that the mean of X is less than or equal to the mean of Y because the observed difference in sample means of 3 is on the acceptance side of the critical value of 11.449.

(c)

$$\text{C.V.} = 0 + 1.64 \sqrt{\frac{20^2}{30} + \frac{30^2}{50}}$$

$$= 9.180$$

There is not enough evidence to reject the null hypothesis that the mean of X is less than or equal to the mean of Y because the observed difference in sample means of -1 is on the acceptance side of the critical value of 9.18. Since this is a right-tailed test and the observed difference in means is negative, there is no reason to compute the critical value. You know in advance that the observed difference in means is on the acceptance side of the critical value.

(d)

$$\text{C.V.} = 0 + 1.28 \sqrt{\frac{20^2}{30} + \frac{30^2}{50}}$$

$$= 7.165$$

There is not enough evidence to reject the null hypothesis that the mean of X is less than or equal to the mean of Y because the observed difference in sample means of 2 is on the acceptance side of the critical value of 7.165.

10.9

$$\text{C.V.} = 0 + 1.64 \sqrt{\frac{15^2}{20} + \frac{15^2}{30}}$$

$$= 7.101$$

The difference in sample means is 6. There is not enough evidence to reject the null hypothesis at a 5% significance level that the mean number of kernels of corn in the new variety is less than or equal to the mean number of kernels in the old variety.

10.11

$$\text{C.V.} = 0 + 1.64 \sqrt{\frac{40^2}{50} + \frac{35^2}{50}}$$

$$\text{C.V.} = 12.327$$

The difference in sample means of 10 is not high enough to reject the null hypothesis that the new design will sell the same or fewer bottles per store at a 5% significance level. Even if the difference in sample means exceeds the critical value, the conclusion that the new design will outsell the old design is suspect. The novelty of the new design may hold consumers' attention for only a short time.

10.13 The null hypothesis is that the new switches will, on average, outlast the old switches by 100 hours or less. A difference in sample means of at least 100 hours in favor of the new switches would be needed to provide evidence to reject this null hypothesis. The observed difference is 50 hours. The test is easy because there is no reason to calculate a critical value. (If you are curious, it equals 125.931.) You know that the null hypothesis is accepted at a 5% significance level because the critical value for the test has to be to the right of 100.

10.15(a)

$$df = \frac{[(9^2/12) + (3^2/14)]^2}{\frac{(9^2/12)^2}{12 - 1} + \frac{(3^2/14)^2}{14 - 1}}$$

$$= 13.09, \text{ or } 13 \text{ rounded down}$$

$$L, \ U = (24 - 25) \pm 3.012 \sqrt{\frac{9^2}{12} + \frac{3^2}{14}}$$

$$L = -9.190, \ U = 7.190$$

(b)

$$s_{pool}^2 = \frac{(14-1)6^2 + (11-1)1^2}{14 + 11 - 2}$$

$$= 20.78$$

$$df = 14 + 11 - 2$$
$$= 23$$

$$L, \; U = (44 - 20) \pm 2.807\sqrt{\frac{20.78}{14} + \frac{20.78}{11}}$$

$$L = 18.844, \; U = 29.156$$

(c)

$$L, \; U = (20 - 14) \pm 2.57\sqrt{\frac{15^2}{50} + \frac{18^2}{250}}$$

$$L = -0.187, \; U = 12.187$$

(d)

$$df = \frac{[(4^2/3) + (7^2/8)]^2}{\frac{(4^2/3)^2}{3-1} + \frac{(7^2/8)^2}{8-1}}$$

$$= 6.70, \text{ or } 6 \text{ rounded down}$$

$$L, \; U = (15 - 32) \pm 3.707\sqrt{\frac{4^2}{3} + \frac{7^2}{8}}$$

$$L = -29.548, \; U = -4.452$$

10.17

$$L, \; U = (125 - 155) \pm 1.96\sqrt{\frac{8^2}{120} + \frac{10^2}{100}}$$

$$L = -32.427, \; U = -27.573$$

As the sample size increases the confidence interval becomes narrower. Also, with large samples the z can be used in place of the t score.

10.19(a)

$$df = \frac{[(4^2/10) + (8^2/8)]^2}{\frac{(4^2/10)^2}{10-1} + \frac{(8^2/8)^2}{8-1}}$$

$$= 9.78, \text{ or } 9 \text{ rounded down}$$

$$C.V._L, \; C.V._U = 0 \pm 2.262\sqrt{\frac{4^2}{10} + \frac{8^2}{8}}$$

$$C.V._L = -7.009, \; C.V._U = 7.009$$

The difference in sample means is -2. There is not enough evidence to reject the null hypothesis that the mean of X equals the mean of Y at a 5% significance level.

(b)

$$s_{pool}^2 = \frac{(10-1)4^2 + (8-1)8^2}{10 + 8 - 2}$$

$$= 37.00$$

$$df = 10 + 8 - 2$$
$$= 16$$

$$C.V._L, \; C.V._U = 0 \pm 2.120\sqrt{\frac{37}{10} + \frac{37}{8}}$$

$$C.V._L = -6.117, \; C.V._U = 6.117$$

The difference in sample means is -2. There is not enough evidence to reject the null hypothesis that the mean of X equals the mean of Y at a 5% significance level.

(c)

$$df = \frac{[(7^2/6) + (8^2/8)]^2}{\frac{(7^2/6)^2}{6-1} + \frac{(8^2/8)^2}{8-1}}$$

$$= 11.63, \text{ or } 11 \text{ rounded down}$$

$$C.V._L, \; C.V._U = 0 \pm 2.201\sqrt{\frac{7^2}{6} + \frac{8^2}{8}}$$

$$C.V._L = -8.850, \; C.V._U = 8.850$$

The difference in sample means is -5. There is not enough evidence to reject the null hypothesis that the mean of X equals the mean of Y at a 5% significance level.

(d)

$$s_{pool}^2 = \frac{(6-1)7^2 + (8-1)8^2}{6 + 8 - 2}$$

$$= 57.75$$

$$df = 6 + 8 - 2$$
$$= 12$$

$$C.V._L, \; C.V._U = 0 \pm 2.179\sqrt{\frac{57.75}{6} + \frac{57.75}{8}}$$

$$C.V._L = -8.943, \; C.V._U = 8.943$$

The difference in sample means is -5. There is not enough evidence to reject the null hypothesis that the mean of X equals the mean of Y at a 5% significance level.

10.21

$$df = \frac{[(0.2^2/10) + (0.4^2/15)]^2}{\dfrac{(0.2^2/10)^2}{10 - 1} + \dfrac{(0.4^2/15)^2}{15 - 1}}$$

$ = 21.72$, or 21 rounded down

$$\text{C.V.} = 0 - 1.721\sqrt{\frac{0.2^2}{10} + \frac{0.4^2}{15}}$$

$\phantom{\text{C.V.}} = -0.208$

There is not enough evidence to reject the null hypothesis that the mean time at McDonald's is equal to or greater than the mean time at Burger King. The observed difference of 0.1 is positive. It is not necessary to calculate the critical value because the test is left tailed.

10.23

$$\text{C.V.}_L,\ \text{C.V.}_U = 0 \pm 1.96\sqrt{\frac{3^2}{100} + \frac{8^2}{100}}$$

$$\text{C.V.}_L = -1.675,\ \text{C.V.}_U = 1.675$$

The difference in sample means is -2. The null hypothesis that the mean time is the same in both directions is rejected at a 5% significance level.

10.25

$$s^2_{\text{pool}} = \frac{(10 - 1)\$0.50^2 + (15 - 1)\$0.75^2}{10 + 10 - 2}$$

$\phantom{s^2_{pool}} = 0.4063$

$df = 10 + 10 - 2$

$ = 18$

$$\text{C.V.} = -\$0.10 - 1.734\sqrt{\frac{0.4063}{10} + \frac{0.4063}{10}}$$

$\phantom{\text{C.V.}} = -0.594$

The observed difference is $-\$0.25$. There is not enough evidence to reject the null hypothesis that the chocolates add 10 cents or less to the maids' tips. The executive may have to clean some more rooms be-

cause with the small sample sizes and the high variability in tips, the chocolateless rooms need to average 59 cents less than the chocolate rooms before the null hypothesis can be rejected. Increasing the maids' tips may not be the prime objective of the chocolate program. The hotel is more concerned with building up good will and repeat customers. A more plausible test is to see if the proportion of repeat customers is higher in the chocolate rooms.

10.27(a)

$$\sigma_{\bar{p}_X - \bar{p}_Y} = \sqrt{\frac{0.5(1 - 0.5)}{50} + \frac{0.49(1 - 0.49)}{100}}$$

$ = 0.08660$

$$z = \frac{0 - (0.5 - 0.49)}{0.08660}$$

$ = -0.12$

$P(z > -0.12) = 0.5478$

(b)

$$\sigma_{\bar{p}_X - \bar{p}_Y} = \sqrt{\frac{0.5(1 - 0.5)}{20} + \frac{0.49(1 - 0.49)}{30}}$$

$ = 0.14433$

$$z = \frac{0 - (0.5 - 0.49)}{0.14433}$$

$ = -0.07$

$P(z < -0.07) = 0.4721$

(c)

$$\sigma_{\bar{p}_X - \bar{p}_Y} = \sqrt{\frac{0.5(1 - 0.5)}{50} + \frac{0.49(1 - 0.49)}{100}}$$

$ = 0.08660$

$$z = \frac{0.01 - (0.5 - 0.49)}{0.08660}$$

$ = 0.00$

$$z = \frac{-0.01 - (0.5 - 0.49)}{0.08660}$$

$ = -0.23$

$P(-0.23 < z < 0.00) = 0.091$

(d)

$$\sigma_{\bar{p}_X - \bar{p}_Y} = 0.13092$$

$$z = \frac{0.05 - (0.5 - 0.49)}{0.13092}$$

$$= 0.31$$

$$z = \frac{-0.05 - (0.5 - 0.49)}{0.13092}$$

$$= -0.46$$

$$P[(z < -0.46) \cup (z > 0.31)] = 0.7011$$

10.29

$$L, U = (0.0244 - 0.0398) \pm$$
$$1.96 \sqrt{\frac{0.0244(1 - 0.0244)}{5,580} + \frac{0.0398(1 - 0.0398)}{5,630}}$$

$$L = -0.022, \ U = -0.009$$

10.31

$$L = (0.025 - 0.05) -$$
$$1.64 \sqrt{\frac{0.025(1 - 0.025)}{1,000} + \frac{0.05(1 - 0.05)}{1,100}}$$

$$L = -0.038$$

One way to interpret the confidence interval is that at a 95% level of confidence the population proportion of old homes that have fires may be 0.038 higher than the proportion of new homes.

10.33

$$\bar{p}_{pool} = \frac{100(0.3) + 200(0.6)}{100 + 200}$$

$$= 0.5$$

$$C.V._L, \ C.V._U = 0 \pm 1.96 \sqrt{\frac{0.5(1 - 0.5)}{100} + \frac{0.5(1 - 0.5)}{200}}$$

$$C.V._L = -0.120, \ C.V._U = 0.120$$

The difference in sample proportions is -0.3. The null hypothesis that the difference in proportions between men and women who use grocery store coupons equals zero is rejected at a 5% significance level.

10.35(a)

$$\bar{d} = \frac{-30 + (-20) + (-24) + (-9) + (-30)}{5}$$

$$= -22.6$$

$$s_d = \sqrt{\frac{[-30 - (-22.6)]^2 + \cdots + [-30 - (-22.6)]^2}{5 - 1}}$$

$$= 8.7063$$

$$s_{\bar{d}} = \frac{8.7063}{\sqrt{5}}$$

$$= 3.8936$$

$$C.V. = -20 - 2.132(3.8936)$$

$$= -28.301$$

There is not enough evidence to reject the null hypothesis that population difference is greater than or equal to -20.

(b)

$$C.V._L, \ C.V._U = -20 \pm 2.776(3.8936)$$

$$C.V._L = -30.809, \ C.V._U = -9.191$$

There is not enough evidence to reject the null hypothesis that population difference equals -20.

(c)

$$C.V. = -20 + 2.132(3.8936)$$

$$= -11.699$$

There is not enough evidence to reject the null hypothesis that the mean difference is less than or equal to -20. Note that this is a right-tailed test and the observed mean difference is less than -20. It was not necessary to calculate the critical value of the test. **(d)** $H_0: \delta = -20$, $H_1: \delta \neq -20$. $\alpha = 5\%$. This is the same test as part b stated in terms of the alternative hypothesis (the inequality) rather than the null hypothesis (the equality).

10.37

$$\bar{d} = \frac{0.2 + 0.1 + \cdots + 0.4}{10}$$

$$= 0.25$$

$$s_d = \sqrt{\frac{(0.2 - 0.25)^2 + (0.1 - 0.25)^2 + \cdots + (0.4 - 0.25)^2}{10 - 1}}$$

$$= 0.2718$$

$$s_{\bar{d}} = \frac{0.2718}{\sqrt{10}}$$

$$= 0.0860$$

$$\text{C.V.} = 0 + 1.833(0.0860)$$

$$= 0.158$$

The null hypothesis that the difference in tread wear of the old minus new designs is less than or equal to zero is rejected at a 5% significance level.

10.39

$$\bar{d} = \frac{(-0.7) + (-0.7) + \cdots + (-0.3)}{5}$$

$$= -0.44$$

$$s_d = \sqrt{\frac{[(-0.7) - (-0.44)]^2 + \cdots + [(-0.3) - (-0.44)]^2}{5 - 1}}$$

$$= 0.2608$$

$$s_{\bar{d}} = \frac{0.2608}{\sqrt{5}}$$

$$= 0.1166$$

$$\text{C.V.} = 0 - 2.132(0.1166)$$

$$= -0.249$$

The null hypothesis that the time to complete the 100 miles is the same or higher for the Powercam™ is rejected at a 5% significance level. Note that the test does not specify which type of crank is used first. If the riders are tired from the first 100 miles, the second will take longer. If the riders have sufficient time to rest, they may do better on the second trip because they are more familiar with the route. The order should be varied across riders. Also, the odd number of observations makes it difficult to control for the effects of the order. With an odd number of riders, more riders must use one of the cranks first. An even sample size would avoid this problem.

10.41

$$\bar{d} = \frac{(-1) + (-2) + \cdots + (-3)}{6}$$

$$= -1.4$$

$$s_d = \sqrt{\frac{[(-1) - (-1.4)]^2 + \cdots + [(-3) - (-1.4)]^2}{5 - 1}}$$

$$= 1.1402$$

$$s_{\bar{d}} = \frac{1.1402}{\sqrt{5}}$$

$$= 0.5099$$

$$\text{C.V.} = 0 - 2.132(0.5099)$$

$$= -1.087$$

The null hypothesis that Wordbuster is the same speed or slower than Ineptwriter is rejected at a 5% significance level. The possible flaw is that the secretaries will type the letter faster the second time because they are more familiar with the letter. Recall that the same letter is used for each program. This learning effect is difficult to eliminate with a small and odd-numbered sample. If the order of the programs was alternated between Wordbuster and Ineptwriter, three of the five secretaries would be using one of the programs second. Using an even number of secretaries, with half following one order and half the alternative order, would avoid this problem.

10.43

$$L, U = (\$4 - 1) \pm 1.96 \sqrt{\frac{(0.55)^2}{40} + \frac{(0.24)^2}{40}}$$

$$L = \$2.81, \ U = \$3.19$$

10.45

$$s_{\text{pool}}^2 = \frac{(10 - 1)(0.4)^2 + (15 - 1)(0.2)^2}{10 + 15 - 2}$$

$$= 0.0870$$

$$U = (9.8 - 9.6) + 1.3198 \sqrt{\frac{0.0870}{10} + \frac{0.0870}{15}}$$

$$= 0.359$$

At a 90% level of confidence, the upper bound for the difference between regular shoes and the Cheeta Hot

Sole is 3.59 seconds. This represents a 3.7% reduction on the base time of 9.8 seconds. Advertising the highest percentage reduction consistent with a 90% level of confidence is disingenuous. The average percentage reduction observed in the test was only 2%. This 2% figure is much more supportable. A conservative advertising claim would be the 0.4% reduction represented by the lower limit of a 90% interval.

10.47 The exercise states that you know that you will sell more mattresses at the lower price. However, the sample mean was 26 mattresses at the stores selling at the high price and 21 mattresses at the stores selling at the low price. With these results you have no reason to consider the lower price. The confidence interval is unnecessary.

10.49 If both types of ketchup are assumed to have the same standard deviation, the most efficient estimate of the difference in the population mean pouring times would use two equal samples. The tolerance of 1 second equals the half-width of the confidence interval.

$$1 = 1.96 \sqrt{\frac{30^2}{n} + \frac{30^2}{n}}$$

$n = 6,914.88$, or 6,915 rounded up

10.51 The effects of the two medications on the opposite sides of a teenager's face should be correlated. Teenagers with many pimples would be expected to show a large reduction in the number of pimples for both medications and teenagers with few pimples would be expected to show a small reduction for both medications. The correct confidence interval would be based on the paired differences.

10.53(a)

$$C.V._L, \ C.V._U = 0 \pm 1.96 \sqrt{\frac{(15)^2}{10} + \frac{(8)^2}{12}}$$

$$C.V._L = -10.34, \ C.V._U = 10.34$$

The difference in sample means is -5. There is not enough evidence to reject the null hypothesis that the two rivers have the same PCB levels at a 5% significance level.

(b)

$$df = \frac{[(15^2/10) + (8^2/12)]^2}{\frac{(15^2/10)^2}{10-1} + \frac{(8^2/12)^2}{12-1}}$$

$$= 13.17, \text{ or } 13 \text{ rounded down}$$

$$C.V._L, \ C.V._U = 0 \pm 2.160 \sqrt{\frac{(15)^2}{10} + \frac{(8)^2}{12}}$$

$$C.V._L = -11.396, \ C.V._U = 11.396$$

There is not enough evidence to reject the null hypothesis that the two rivers have the same PCB levels at a 5% significance level.

(c)

$$s^2_{pool} = \frac{(10-1)15^2 + (12-1)8^2}{10+12-2}$$

$$= 136.45$$

$$df = 10 + 12 - 2$$

$$= 20$$

$$C.V._L, \ C.V._U = 0 \pm 2.086 \sqrt{\frac{136.45}{10} + \frac{136.45}{12}}$$

$$C.V._L = -10.433, \ C.V._U = 10.433$$

There is not enough evidence to reject the null hypothesis that the two rivers have the same PCB levels at a 5% significance level.

10.55 Evaluated at the sample means an hour of labor produces 1.4 tons of steel in the United States and costs $28 in labor. Thus $28 divided by 1.4, or $20, is the per ton U.S. average labor cost. Similarly, $21 divided by 1.7 tons per hour, or $12.35, is the per ton Japanese average labor cost. Ignoring all other costs, the Japanese could sell a ton of steel for $7.65 less than the comparable U.S. price, without being guilty of dumping. To get a sense of how far off this estimated percentage may be from the true labor cost differential, 95% confidence intervals can be constructed for differences in hourly labor costs and differences in output per hour in both countries. A price ratio based on the upper limit of the hourly labor cost difference and the lower limit of the productivity difference would show the lowest relative price the Japanese could sell steel at without dumping. Taking these two limits would give the Japanese steel manufacturers the benefit of any doubt due to sampling error in estimating either parameter. A 95% confidence interval for the differences in hourly labor costs is

$$L, U = (\$28 - 21) \pm 1.96 \sqrt{\frac{3^2}{500} + \frac{9^2}{400}}$$

$$L = \$6.08, \quad U = \$7.92$$

A 95% confidence interval for the differences in output per manhour is

$$L, U = (1.4 - 1.7) \pm 1.96 \sqrt{\frac{0.2^2}{500} + \frac{0.5^2}{400}}$$

$$L = -0.35204, \quad U = -0.24796$$

The lower limit (which represents the largest productivity difference) of the confidence interval for the difference in productivity is -0.35204 ton. The difference in sample means was 0.3 ton. Suppose we allocate the additional difference of $0.35204 - 0.3 = 0.05204$ equally between the countries. This would give an approximate productivity of $1.4 - 0.02602 = 1.37398$ tons per labor hour for the U.S. and $1.7 + 0.02602 = 1.72602$ tons per labor hour for Japan. The upper limit (which represents the largest hourly labor cost difference) of the difference in hourly labor costs is \$7.92. The difference in sample means was \$7. Suppose we again split the difference of $\$7.92 - \$7.00 = \$0.92$ equally between the two countries. This results in an adjusted hourly labor cost of $\$28.00 + \$0.46 = \$28.46$ for the U.S. and $\$21.00 - \$0.46 = \$20.54$ for Japan. Labor cost per ton in the U.S. would be $\$28.46/1.37398 = \$20.71$. Labor cost per ton in Japan would be $\$20.54/1.72602 = \$11.90$. If all other costs were equal, the Japanese could sell a ton of steel for $\$20.71 - \$11.90 = \$8.81$ less than the U.S. price and still not be guilty of dumping. Of course, this differential was calculated under assumptions most favorable to the Japanese steel producers regarding sampling error.

10.57 A planning value of 0.5 for each population proportion guarantees that the sample size will be large enough to meet the desired tolerance at the specified level of confidence. The tolerance for the difference in population proportions is 0.01 and the level of confidence is 95%. The n shown below is the sample size for each group. Setting the tolerance equal to the half-width of the confidence interval,

$$0.01 = 1.96 \sqrt{\frac{0.5(1 - 0.5)}{n} + \frac{0.5(1 - 0.5)}{n}}$$

$$n = 19{,}208$$

10.59 The problem is to find the reserves needed to cover any potential losses from 100 independent contracts. We begin by using the sample values to estimate a one-sided confidence limit for the population mean difference.

$$L = (\$4.10 - 4.00) - 1.64 \sqrt{\frac{(0.15)^2}{40} + \frac{(0.15)^2}{40}}$$

$$= -\$0.048$$

Suppose we use this "worst case" estimate for the mean difference and use the sample standard deviations as estimates of the population standard deviations. The standard error for 100 contracts is

$$s_{\bar{x}-\bar{y}} = \sqrt{\frac{(0.15)^2}{100} + \frac{(0.55)^2}{100}}$$

$$= 0.057009$$

Relying on the sampling distribution of the difference in means of two independent samples, 95% of all contracts are expected to have a difference between the contract and spot price that exceeds L.

$$L = -0.048 - 1.64(0.057009)$$

$$= -0.141495$$

There is a 5% chance that the average loss per contract will be equal to or less than \$0.141495 per bushel or \$141,495 for the million-bushel contract. Since there are 100 of these contracts, the required cash reserve would be 100(\$141,495), or \$14,149,500.

10.61 You would gain the most information on the difference between the population means by allocating the sample so that the standard error of the difference in means is minimized. This is the same as minimizing the variance of the distribution of the difference in means. The solution can be found with the use of calculus or by a search. Start with an equal allocation and vary the sample in each direction.

$n_X + n_Y = 6,000$

$$\sigma^2_{\bar{x}-\bar{y}} = \frac{\sigma^2_x}{n_X} + \frac{\sigma^2_y}{6,000 - n_X}$$

$$= \frac{(1.5)^2}{n_X} + \frac{(1.1)^2}{6000 - n_X}$$

The following table shows the results of setting n_1 equal to 2,999 through 3,006.

n_1	Variance of the difference in means
2,999	0.001153449
3,000	0.001153333
3,001	0.001153218
3,004	0.001152873
3,005	0.001152759
3,006	0.001152645

The variance of the difference in sample means as a portion of the sample allocated to the group with the highest population variance (younger women) is increased. This reduction in the variance of the differ-

ence in sample means does not continue until n_X reaches 6,000. The accompanying graph shows the variance of the difference in sample means as a function of n_X. A further search shows that the minimum variance occurs at n_X equal to 3,462.

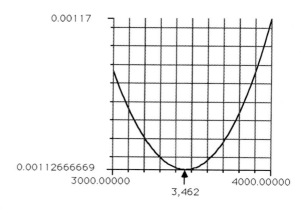

CHAPTER 11

11.1(a) 18.3070   **(b)** 15.9871   **(c)** 31.4104   **(d)** 50.8922

11.3

Range	o_i	Lower z	Upper z	Probability	e_i	Collapsed o_i	Collapsed e_i	$(o_i - e_i)^2/e_i$
under 5	10		-2.5	0.0062	0.9796			
5 and under 7	20	-2.5	-1.5	0.0606	9.5748	30	10.5544	35.8269
7 and under 11	70	-1.5	0.5	0.6247	98.7026	70	98.7026	8.3467
11 and under 14	35	0.5	2.0	0.2857	45.1406	58	48.7430	1.7580
14 and under 15	15	2.0	2.5	0.0166	2.6228			45.9316
15 and over	8	2.5		0.0062	0.9796			
	158							

$df = k - h - 1$, or $3 - 0 - 1 = 2$. C.V. $= 5.99147$. The null hypothesis that the population is normally distributed with a mean of 10 and a standard deviation of 2 is rejected at a 5% significance level.

11.5

Number of occurrences	o_i	Probability	e_i	Collapsed o_i	Collapsed e_i	$(o_i - e_i)^2/e_i$
0	0	0.0067	0.3216			
1	1	0.0337	1.6176			
2	5	0.0842	4.0416	6	5.9808	0.00006164
3	7	0.1404	6.7392	7	6.7392	0.01009269
4	9	0.1755	8.4240	9	8.4240	0.03938462
5	10	0.1755	8.4240	10	8.4240	0.29484520
6	6	0.1462	7.0176	6	7.0176	0.14755896
7	4	0.1044	5.0112	4	5.0112	0.20404802
8	3	0.0653	3.1344	6	6.4032	0.02538891
9	2	0.0363	1.7424			0.72138004
10	1	0.0181	0.8688			
11	0	0.0137	0.6576			
	48					

$df = k - h - 1$, or $7 - 0 - 1 = 6$. C.V. $= 12.5916$. There is not enough evidence to reject the null hypothesis that X is Poisson distributed with a mean of 5.

11.7 The exercise does not specify the distribution for the time to sale of a home other than a mean of 60 days and that every home is independent. However, the time to an occurrence with a constant expected rate is usually modeled with an exponential distribution. This exercise is similar to the example of time to failure for the bank machines case study in the chapter. H_0:X is exponential with a mean of 60, H_1:X is not exponential with a mean of 60, $\alpha = 5\%$. Here is a worksheet showing the calculations:

Range	o_i	Lower λt	Upper λt	Probability	o_i	Collapsed e_i	Collapsed o_i	χ^2
0 and under 20	20	0.00	0.33	0.2811	14.055	20	14.055	2.5146
20 and under 40	12	0.33	0.67	0.2072	10.360	12	10.360	0.2596
40 and under 60	6	0.67	1.00	0.1438	7.190	6	7.190	0.1970
60 and under 80	4	1.00	1.33	0.1034	5.170	8	8.985	0.1080
80 and under 100	4	1.33	1.67	0.0763	3.815			
100 and over	4	1.67		0.1882	9.410	4	9.410	3.1103
	50				50.000			6.1895

$df = k - h - 1$, or $5 - 0 - 1 = 4$. C.V. $= 9.48773$. There is not enough evidence to reject the null hypothesis that the mean time to sale is 60 days. The real estate agent should not advise clients that the average time to sale has been reduced. Note that the observations in the range 80 and under 100 could be collapsed into the range above or the one below. Either way the null hypothesis would not be rejected. The usual practice is to put the extra observations in the class with the smallest expected frequency.

11.9

		1	2	3	4	5	Total
successes	o_i	50	110	86	125	125	496
	e_i	50	99	99	124	124	
failures	o_i	50	90	114	125	125	504
	e_i	50	101	101	126	126	
total		100	200	200	250	250	1,000

		1	2	3	4	5	Total
successes	$(o_i - e_i)^2/e_i$	0.003	1.176	1.756	0.008	0.00	2.935
failures	$(o_i - e_i)^2/e_i$	0.003	1.157	1.729	0.008	0.008	2.889
							5.824

$df = (r - 1)(c - 1)$, or $(2 - 1)(5 - 1) = 4$. C.V. = 9.48773. There is not enough evidence to reject the null hypothesis that the population proportions of successes are the same for the five populations.

11.11

		1	2	3	4	Total
malpractice suit	o_i	10	24	5	10	49
	e_i	11.3087	22.615	7.538	7.538	
no malpractice suit	o_i	290	576	195	190	1,251
	e_i	288.692	577.385	192.462	192.462	
total		300	600	200	200	1,300

		1	2	3	4	Total
malpractice suit	$(o_i - e_i)^2/e_i$	0.1511	0.0849	0.8545	0.8041	1.8947
no malpractice suit	$(o_i - e_i)^2/e_i$	0.0059	0.0033	0.0335	0.0315	0.0742
						1.9689

$df = (r - 1)(c - 1)$, or $(2 - 1)(4 - 1) = 3$. C.V. = 7.81473. There is not enough evidence to reject the null hypothesis that the population proportions of malpractice lawsuits are the same for the four types of operations.

11.13

		U.S.	Greece	France	G.B.	Total
hate goat cheese	o_i	800	200	300	600	1,900
	e_i	441.860	397.674	397.674	662.791	
like goat cheese	o_i	100	500	300	500	1,400
	e_i	325.581	293.023	293.023	293.022	488.372
other	o_i	100	200	300	400	1,000
	e_i	232.558	209.302	209.302	348.837	
total		1,000	900	900	1,500	4,300

		U.S.	Greece	France	G.B.	Total
hate goat cheese	$(o_i - e_i)^2/e_i$	290.283	98.259	23.990	5.949	418.481
like goat cheese	$(o_i - e_i)^2/e_i$	156.295	146.198	0.166	0.277	302.936
other	$(o_i - e_i)^2/e_i$	75.558	0.413	39.303	7.504	122.778
						844.195

$df = (r - 1)(c - 1)$, or $(3 - 1)(4 - 1) = 6$. C.V. $= 12.5916$. The null hypothesis that goat cheese preference is homogeneous across the four populations is rejected at a 5% significance level.

11.15

		A	B	C	D	F	Total
sophisticated	o_i	110	130	205	135	20	600
	e_i	90	120	210	120	60	
general	o_i	20	35	80	44	21	200
	e_i	30	40	70	40	20	
simple	o_i	20	35	65	21	59	200
	e_i	30	40	70	40	20	
total		150	200	350	200	100	1,000

		A	B	C	D	F	Total
sophisticated	$(o_i - e_i)^2/e_i$	4.444	0.833	0.119	1.875	26.667	33.938
general	$(o_i - e_i)^2/e_i$	3.333	0.625	1.429	0.400	0.050	5.837
simple	$(o_i - e_i)^2/e_i$	3.333	0.625	0.357	9.025	76.050	89.390
							129.165

$df = (r - 1)(c - 1)$, or $(3 - 1)(5 - 1) = 8$. C.V. $= 15.5073$. The null hypothesis that type of calculator is independent of course grade is rejected at a 5% significance level.

11.17

		Det.	Dallas	S.F.	Total
prefer Classic Coke	o_i	300	325	350	975
	e_i	325	325	325	
do not	o_i	700	675	650	2,025
	e_i	675	675	675	
total		1,000	1,000	1,000	3,000

		Det.	Dallas	S.F.	Total
prefer Classic Coke	$(o_i - e_i)^2/e_i$	1.923	0.000	1.923	3.846
do not	$(o_i - e_i)^2/e_i$	0.926	0.000	0.926	1.852
total					5.698

$df = (r - 1)(c - 1)$, or $(2 - 1)(3 - 1) = 2$. C.V. $= 5.99147$. There is not enough evidence to reject the null hypothesis that the population proportions of people who prefer Classic Coke are the same for the four cities.

11.19

o_i	e_i	$(o_i - e_i)^2/e_i$
812	800	0.180
165	170	0.147
23	30	1.633
1,000	1,000	1.960

$df = k - h - 1$, or $3 - 0 - 1 = 2$. For $\alpha = 5\%$, C.V. = 5.99147. There is not enough evidence to reject the null hypothesis that the population proportions are 0.8, 0.17, and 0.03. This conclusion also holds for $\alpha = 10\%$, 2.5%, 1%, or 0.5%.

11.21(a)

		1	2	3	4	5
acceptable	o_i	3,960	4,895	5,400	6,300	7,200
	e_i	4,086.44	4,994.54	5,448.59	6,356.69	7,264.79
not acceptable	o_i	540	605	600	700	800
	e_i	413.56	505.46	551.41	643.31	735.21
total		4,500	5,500	6,000	7,000	8,000

		6	7	8	9	10	Total
acceptable	o_i	7,280	7,280	7,360	7,360	7,440	64,475
	e_i	7,264.79	7,264.79	7,264.79	7,264.79	7,264.79	
not acceptable	o_i	720	720	640	640	560	6,525
	e_i	735.21	735.21	735.21	735.21	735.21	
total		8,000	8,000	8,000	8,000	8,000	71,000

		1	2	3	4	5
acceptable	$(o_i - e_i)^2/e_i$	3.912	1.984	0.433	0.506	0.578
not acceptable	$(o_i - e_i)^2/e_i$	38.657	19.602	4.282	4.996	5.710

		6	7	8	9	10	Total
acceptable	$(o_i - e_i)^2/e_i$	0.032	0.032	1.248	1.248	4.226	14.199
not acceptable	$(o_i - e_i)^2/e_i$	0.315	0.315	12.330	12.330	41.755	140.292
							154.491

$df = (r - 1)(c - 1)$, or $(2 - 1)(10 - 1) = 9$. C.V. = 16.919. The null hypothesis that the population proportions of acceptable briefs are the same for the 10 inspectors is rejected at a 5% significance level.

(b) The quality of the briefs may have varied across the 10 samples. This source of variation would be eliminated if the inspectors checked the same briefs.

11.23(a)

		Calendar	Pen set	Nothing	Total
increasing	o_i	45	50	55	150
	e_i	50.000	50.000	50.000	
constant	o_i	15	12	20	47
	e_i	15.667	15.667	15.667	
decreasing	o_i	40	38	25	103
	e_i	34.333	34.333	34.333	
total		100	100	100	300

		Calendar	Pen set	Nothing	Total
increasing	$(o_i - e_i)^2/e_i$	0.500	0.000	0.500	1.000
constant	$(o_i - e_i)^2/e_i$	0.028	0.858	1.198	2.084
decreasing	$(o_i - e_i)^2/e_i$	0.935	0.392	2.537	3.864
					6.948

$df = (r - 1)(c - 1)$, or $(3 - 1)(3 - 1) = 4$. C.V. $= 9.48773$. There is not enough evidence to reject the null hypothesis that order category is homogeneous across the three groups receiving the different gifts.

(b) The test is for homogeneity because the office supply firm has selected the column totals.

(c) The company should give up making Christmas presents to influence sales. There is no apparent association between presents and sales. Note that this conclusion may not hold for more valuable presents or may not reflect buyers' reactions if they have come to expect a gift and are now cut off.

11.25

Failures per box	o_i	Probability	e_i	Collapsed o_i	Collapsed e_i	$(o_i - e_i)^2/e_i$
0	950	0.9416	941.6	950	941.6	0.075
1	40	0.0568	56.8	50	58.4	1.208
2	10	0.0015	1.5			1.283
3 plus	0	0.0001	0.1			
	1,000					

$df = k - h - 1$, or $2 - 1 - 1 = 0$. C.V. $= ?$ There are no degrees of freedom left after estimating the population proportion with the sample data and collapsing to get each expected frequency equal to at least 5. There are too few observations to draw the conclusion that the null hypothesis is either rejected or accepted.

11.27(a)

		Blue	Red	Brown	Yellow	Total
excellent	o_i	12	12	12	14	50
	e_i	14.356	12.376	12.376	10.891	
good	o_i	10	10	6	14	40
	e_i	11.485	9.901	9.901	8.713	
fair	o_i	14	14	16	16	60
	e_i	17.228	14.851	14.851	13.069	
poor	o_i	22	14	16	0	52
	e_i	14.931	12.871	12.871	11.327	
total		58	50	50	44	202

		Blue	Red	Brown	Yellow	Total
excellent	$(o_i - e_i)^2/e_i$	0.387	0.011	0.011	0.888	1.297
good	$(o_i - e_i)^2/e_i$	0.192	0.001	1.537	3.208	4.938
fair	$(o_i - e_i)^2/e_i$	0.605	0.049	0.089	0.657	1.400
poor	$(o_i - e_i)^2/e_i$	3.347	0.099	0.761	11.327	15.534
						23.169

$df = (r - 1)(c - 1)$, or $(4 - 1)(4 - 1) = 9$. C.V. = 16.919. The null hypothesis that rating is independent of color is rejected at a 5% significance level.

(b) The initial choice of a chocolate bar in a colored wrapper or a Hershey bar may confuse the results. You are not sure if you are sampling from the population of shoppers who dislike Hershey bars or shoppers who particularly like the color they were first offered. Also, the design introduces an unnecessary source of variation—the four samples may differ in their tastes regarding chocolate rather than their perceptions of color. An alternative design is to offer every subject all four colored bars and have them rate each bar. The order of tasting would have to be rotated to remove any order effects on ratings.

11.29

		S.P.	Regular	Total
A	o_i	3,346	3,027	6,373
	e_i	3,102.513	3,270.487	
B	o_i	1,148	1,361	2,509
	e_i	1,221.435	1,288.565	
C	o_i	456	830	1,286
	e_i	626.052	660.948	
total		4,950	5,218	10,168

		S.P.	Regular	Total
A	$(o_i - e_i)^2/e_i$	19.109	18.128	37.237
B	$(o_i - e_i)^2/e_i$	4.415	4.188	8.603
C	$(o_i - e_i)^2/e_i$	46.190	43.818	90.009
				135.849

$df = (r - 1)(c - 1)$, or $(3 - 1)(2 - 1) = 2$. C.V. = 5.99147. The null hypothesis that rating is homogeneous across both types of audiences is rejected at a 5% significance level.

11.31 The test statistic is 4.424. The degrees of freedom are $k - h - 1 = 10 - 1 - 1$, or 8. C.V. = 15.5073. There is not enough evidence to reject the null hypothesis that the data are drawn from a normal distribution with a mean of 0.

Lower x	Upper x	Lower z	Upper z	o_i	e_i	Collapsed o_i	Collapsed e_i	χ^2
$-\infty$	under -8	$-\infty$	-2.77	0	0.28			
-8	under -7	-2.77	-2.42	3	0.50			
-7	under -6	-2.42	-2.08	0	1.10			
-6	under -5	-2.08	-1.73	0	2.30			
-5	under -4	-1.73	-1.38	4	4.20	7	8.38	0.227
-4	under -3	-1.38	-1.04	8	6.54	8	6.54	0.326
-3	under -2	-1.04	-0.69	9	9.59	9	9.59	0.036
-2	under -1	-0.69	-0.35	8	11.81	8	11.81	1.229
-1	under 0	-0.35	0.00	11	13.68	11	13.68	0.525
0	under 1	0.00	0.35	16	13.68	16	13.68	0.393
1	under 2	0.35	0.69	14	11.81	14	11.81	0.406
2	under 3	0.69	1.04	12	9.59	12	9.59	0.606
3	under 4	1.04	1.38	5	6.54	5	6.54	0.363
4	under 5	1.38	1.73	7	4.20	10	8.38	0.313
5	$+\infty$	1.73	$+\infty$	3	4.18			4.424
				100				

11.33(a) There are combinations of 2 taken from 50, or 1,225 possible pairwise tests for the 50 states. If the null hypothesis that the population proportions are the same across states with and without can deposit laws, the probability of falsely rejecting the null hypothesis in 1,225 pairwise tests each with a 1% significance level is $1 - (0.99)^{1,225}$, or 0.9999955. Thus, the probability of making a Type I error is virtually certain.

(b) A chi-square test of the null hypothesis that all 50 states have the same proportion of consumers who buy soft drinks in cans is one possibility. Another design is a simple difference-in-proportions test on the individual data (each of the 100 respondents per state) in which the null hypothesis is that the proportions who buy soft drinks in cans are the same for states with and without deposit laws. The proportion of population in each state who buy soft drinks in cans may be sensitive to other variables, for example, state income levels or mean temperatures. A one-variable approach may not be adequate to control for these other factors. Probit analysis (discussed in Chapter 13) is a more general framework for analyzing categorical variables that are thought to be influenced by several other variables, either categorical or in levels.

CHAPTER 12

12.1(a) 5.41 **(b)** 9.01 **(c)** 7.71 **(d)** 3.84 **(e)** 1.00

12.3(a) In these and subsequent examples, the calculation of the within-sample means and within-sample standard deviations is not shown.

APPENDIX F ANSWERS TO ODD-NUMBERED EXERCISES

$$\bar{\bar{x}} = \frac{9(4.2222) + 8(4.8750) + 7(5.2857)}{9 + 8 + 7}$$

$$= 4.7500$$

$$F = \frac{[9(4.2222 - 4.7500)^2 + 8(4.8750 - 4.7500)^2 + 7(5.2857 - 4.7500)^2]/(3 - 1)}{(0.94444 + 0.69643 + 0.57143)/3}$$

$$= 3.15$$

The degrees of freedom are $k - 1$, and $\Sigma n_j - k$, or 2 and 21. C.V. = 3.47. There is not enough evidence to reject the null hypothesis that the population means are equal.

(b) The assumptions are that the populations are normally distributed and that the three population variances are equal.

12.5(a)

$$\bar{\bar{x}} = \frac{5(43.4) + 5(51.8) + 5(41) + 5(50.8)}{5 + 5 + 5 + 5}$$

$$= 46.75$$

$$F = \frac{[5(43.40 - 46.75)^2 + 5(51.80 - 46.75)^2 + 5(41 - 46.75)^2 + 5(50/8 - 46.75)^2]/(4 - 1)}{(24.3 + 61.7 + 10 + 10.7)/4}$$

$$= 5.39$$

The degrees of freedom are $k - 1$, and $\Sigma n_j - k$, or 3 and 16. C.V. = 3.24. The null hypothesis that the population mean levels of cholesterol are equal across the four breeds is rejected at a 5% significance level.

12.7(a) The degrees of freedom are $\Sigma n_j - k$, or 21. The limits of a 95% confidence interval for the first population are

$$L, U = 7.8889 \pm 2.080\sqrt{\frac{25.47333}{9}}$$

$$L = 4.390, \ U = 11.387$$

The limits of a 95% confidence interval for the second population are

$$L, U = 7 \pm 2.080\sqrt{\frac{25.47333}{8}}$$

$$L = 3.289, \ U = 10.711$$

The limits of a 95% confidence interval for the third population are

$$L, U = 7.5714 \pm 2.080\sqrt{\frac{25.47333}{7}}$$

$$L = 3.603, \ U = 11.540$$

(b) The overall level of confidence for the three intervals holding simultaneously is 0.95^3, or 0.857.

12.9 The null hypothesis of equal population means was not rejected in exercise 12.4. If the null hypothesis is not rejected, Tukey's honest significant difference test should not be used to identify pairs of populations that are significantly different. The test may give spurious results (say there is a significant difference when there is none) in these circumstances.

12.11 For the null hypothesis of no treatment effects,

$$F = \frac{3(3)[(62.222 - 60.75)^2 + (64.444 - 60.75)^2 + (57 - 60.75)^2 + (59.333 - 60.75)^2]/(4 - 1)}{[(34 - 56.667)^2 + (54 - 40)^2 + \cdots + (93 - 90)^2]/3(4)(2)}$$

$$= 0.633$$

There are $h - 1$ and $kh(m - 1)$ degrees of freedom, or 3 and 24. At a 5% significance level the critical value for the test is 3.01. There is not enough evidence to reject the null hypothesis of no treatment effect. For the null hypothesis of no block effect,

$$F = \frac{4(3)[(56.083 - 60.75)^2 + (36.167 - 60.75)^2 + (90 - 60.75)^2]/(3 - 1)}{151.194}$$

$$= 58.799$$

There are $k - 1$ and $kh(m - 1)$ degrees of freedom, or 2 and 24. At a 5% significance level the critical value for the test is 3.40. The null hypothesis of no block effects is rejected at a 5% significance level. For the null hypothesis of no interaction effect,

$$F = \frac{3[(56.667 - 62.222 - 56.083 + 60.75)^2 + \cdots + (90 - 59.333 - 90 + 60.75)^2]/(3 - 1)(4 - 1)}{151.194}$$

$$= 0.696$$

There are $(h - 1)(k - 1)$ and $kh(m - 1)$ degrees of freedom, or 6 and 24. At a 5% significance level the critical value for the test is 2.51. There is not enough evidence to reject the null hypothesis of no interaction effect.

12.13 For the null hypothesis of no treatment effects,

$$F = \frac{4[(20.75 - 21.438)^2 + (21 - 21.438)^2 + (20.5 - 21.438)^2 + (23.5 - 21.438)^2]/(4 - 1)}{[(5 - 20.75 - 5.75 + 21.438)^2 + \cdots + (50 - 23.5 - 47.250 + 21.438)^2]/(4 - 1)(4 - 1)]}$$

$$= 4.33$$

The degrees of freedom are $(h - 1)$ and $(h - 1)(k - 1)$, or 3 and 9. The critical value with a 1% significance level is 6.99. The null hypothesis of no treatment effect is rejected because the F test statistic of 4.33 exceeds the critical value of 6.99. For the null hypothesis of no block effects,

$$F = \frac{4[(5.75 - 21.438)^2 + (24.25 - 21.438)^2 + (8.5 - 21.438)^2 + (47.25 - 21.438)^2]/(4 - 1)}{1.785}$$

$$= 812.453$$

The degrees of freedom are $(k - 1)$ and $(h - 1)(k - 1)$, or 3 and 9. The critical value with a 1% significance level is 6.99. The null hypothesis of no treatment effect is rejected because the F test statistic of 812.453 exceeds the critical value of 6.99.

12.15 For the null hypothesis that the dyes have the same mean lasting power,

$$F = \frac{4[(28.5 - 30.312)^2 + (31.5 - 30.312)^2 + (30.5 - 30.312)^2 + (30.75 - 30.312)^2]/(4-1)}{[(18 - 28.5 - 24.250 + 30.312)^2 + \cdots + (50 - 30.75 - 46 + 30.312)^2]/(4-1)(4-1)}$$

$$= 0.420$$

The degrees of freedom are $(h-1)$ and $(h-1)(k-1)$, or 3 and 9. The critical value with a 5% significance level is 3.86. There is not enough evidence to reject the null hypothesis the four dyes have the same mean lasting power because the F test statistic of 0.42 is less than the critical value of 3.86. Note that the exercise does not ask for a test of a block effect. The four dyes are used on each woman's hair simply to control for the variation in lasting power due to differences in hair types.

12.17 For the null hypothesis of no treatment effect,

$$F = \frac{4(5)[(3,659.4 - 4,508.6)^2 + (5,691.2 - 4,508.6)^2 + (4,175.1 - 4,508.6)^2]/(3-1)}{[(3,208 - 3,370.2)^2 + (4,526 - 3,370.2)^2 + \cdots + (8,765 - 4,321.4)^2]/4(3)(4)}$$

$$= 1.779$$

There are $h-1$ and $kh(m-1)$ degrees of freedom, or 2 and 48. At a 5% significance level the critical value for the test is between 3.15 and 3.23. There is not enough evidence to reject the null hypothesis of no treatment effects because the F test statistic of 1.78 is to the left of both of these critical values. For the null hypothesis of no block effect,

$$F = \frac{3(5)[(4,279.2 - 4,508.6)^2 + (7,898.5 - 4,508.6)^2 + \cdots + (4502.7 - 4,508.6)^2]/(4-1)}{(3,208 - 3,370.2)^2 + (4,526 - 3,370.2)^2 + \cdots + (8,765 - 4,321.4)^2/4(3)(4)}$$

$$= 6.16$$

There are $k-1$ and $kh(m-1)$ degrees of freedom, or 3 and 48. At a 5% significance level the critical value for the test is between 2.76 and 2.84. The null hypothesis of no block effects is rejected at a 5% significance level because the F test statistic of 6.16 is above these values. For the null hypothesis of no interaction effect,

$$F = \frac{5[(3,370.2 - 3,659.4 - 4,279.2 + 4,508.6)^2 + \cdots + (4,321.4 - 4,175.1 - 4,502.7 + 4,508.6)^2]/(3)(2)}{[(3,208 - 3,370.2)^2 + (4,526 - 3,370.2)^2 + \cdots + (8,765 - 4,321.4)^2]/4(3)(4)}$$

$$= 0.641$$

There are $(h-1)(k-1)$ and $kh(m-1)$ degrees of freedom, or 6 and 48. At a 5% significance level the critical value for the test is between 2.25 and 2.34. There is not enough evidence to reject the null hypothesis of no interaction effect.

12.19 Only the calculation for the treatment is shown below:

$$F = \frac{3[(84,376 - 79,230.33)^2 + (76,772.67 - 79,230.33)^2 + (76,542.33 - 79,230.33)^2]/(3-1)}{(30,987 - 84,376 - 27,284.67 + 79,230.33)^2 + \cdots + (126,540 - 76,542.33 - 127,597.33 + 79,230.33)^2/(2)(2)}$$

$$= 8.014$$

The degrees of freedom are $(h-1)$ and $(h-1)(k-1)$, or 2 and 4. The critical value with a 5% significance level is 6.94. The null hypothesis that the three payment regimes result in the same mean consumption is rejected at a 5% significance level.

12.21 The first model had the best accuracy in the artillery test. The point estimate of the population mean for the first model is 248 feet, based on eight observations. The variance, as estimated from the average within-sample variance, is 11,010. There were 27 observations. The degrees of freedom are 24. Substituting into formula 12.6,

$$L, \ U = 248 \pm 2.064\sqrt{\frac{11,010}{8}}$$

$$L = 171.43, \ U = 324.57$$

12.23 Since the null hypothesis of equal population means was rejected in exercise 12.18, we can proceed with Tukey's HSD test. The absolute values of differences between these pairs are

1–2	$|122.4 - 117.6|$	$= 4.8$
1–3	$|122.4 - 135|$	$= 12.6$
2–3	$|117.6 - 135|$	$= 17.4$

$$HSD = 3.31\sqrt{\frac{205.153}{100}}$$

$$= 4.74$$

All three pairs have significant differences.

12.25 The point estimate of the population mean for the second plan is \$16,744, based on 10 observations. The variance, as estimated from the average within-sample variance, is 2,697,473 dollars squared. There were 30 observations. The degrees of freedom are 27.

$$L, \ U = \$16,744 \pm 2.052\sqrt{\frac{2,697,473}{10}}$$

$$L = \$15,678.25, \ U = \$17,809.75$$

12.27 Since the null hypothesis of equal population means was rejected in exercise 12.24, we can proceed with Tukey's HSD test. The absolute values of differences between these pairs are

1–2	$|11,400 - 16,744|$	$= 5,344$
1–3	$|11,400 - 11,432|$	$= 32$
2–3	$|16,744 - 11,432|$	$= 5,312$

The Studentized range has k and $\Sigma n_j - k$ degrees of freedom, or 3 and 27. The value for q falls between 3.49 and 3.53 (for 30 and 24 degrees of freedom). As 27 is halfway between 24 and 30, we will split the difference and use 3.51.

$$HSD = 3.51\sqrt{\frac{2,697,473}{10}}$$

$$= \$1,823.00$$

Plans 1 and 2 and plans 2 and 3 have significant differences.

12.29 The suspect assumption is equal variances across the blocks. Inexperienced mechanics often make mistakes and have to redo their work. By this argument, the variances should be higher for the inexperienced mechanics. Analysis of variance is fairly robust to departures from the equal variance assumption.

12.31 The assignment to the experiment was not random. The parents who were most concerned about their children's education may have opted for the year-round schools. The improvements in reading scores by the end of the third grade may be due to factors outside the experiment, such as extra parental attention. If this is the case, apparent treatment effects would not carry over to the rest of the system's children.

CHAPTER 13

13.1

$$b = \frac{[(30) + 3(45) + \cdots + 10(55)] - 5(5.8)(45.6)}{(1^2 + 3^2 + \cdots + 10^2) - 5(5.8)^2}$$

$$= 2.2721$$

$$a = 45.6 - 2.2721(5.8)$$

$$= 32.4218$$

$$r^2 = \frac{(34.6939 - 45.6)^2 + \cdots + (55.1428 - 45.6)^2}{(30 - 45.6)^2 + \cdots + (55 - 45.6)^2}$$

$$= 0.8358$$

$$r = \frac{(-1.2519)(-1.6371) + \cdots + (1.0954)(0.9865)}{5 - 1}$$

$$= 0.9142$$

13.3(a) Causality must go from age to time needed to run a mile. For example, it is impossible that running faster will make an adult younger. Age can be considered a fixed variable rather than random variable and the time to run a mile is a random variable that should be inversely correlated with adult age. Thus, regression analysis is more appropriate than correlation analysis.

(b) Regression analysis is more appropriate. Unless the company is very large, there is little feedback from the growth of the company's sales to the growth of GNP.

(c) Rabbits do not eat deer and vice versa. There is no apparent direct effect between the populations. Both populations are influenced by common factors such as weather and the feed left in the preserve. Correlation analysis is most appropriate.

(d) More hunters should reduce the number of deer. However, an increase in the deer population could attract more hunters. If the number of hunters was entirely controlled by the number of hunting licenses issued by the state, regression analysis would be appropriate. If the number of hunters was a random variable, correlation analysis would be appropriate.

13.5 The acquisition of any company that had earnings inversely correlated with Happy Time's earnings would tend to reduce the variability of the combined earnings. From the viewpoint of stabilizing earnings, the Six Banners Theme Park would appear to be the worst acquisition, the Grim Auction and Salvage best, and the highly stable Acme Bread Bakers somewhere in between. This presumption can be checked by comparing the coefficients of correlation. For Happy Time and Acme,

$$r = \frac{(-0.2778)(0) + (0.1588)(0) + \cdots + (1.9051)(0)}{11 - 1}$$

$$= 0.1105$$

For Happy Time and Grim Auction and Salvage,

$$r = \frac{(-0.2778)(-0.0868) + \cdots + (1.9051)(-1.2010)}{11 - 1}$$

$$= -0.8965$$

For Happy Time and Six Banners Theme Park,

$$r = \frac{(-0.2778)(0.0267) + \cdots + (1.9051)(1.4953)}{11 - 1}$$

$$= 0.9391$$

The highest *negative* correlation in earnings is between Happy Time and Grim Auction and Salvage. The best acquisition for stabilizing earnings is Grim Auction and Salvage.

13.7 $a = -0.0060$, $b = 0.1193$.

$$s_e = \sqrt{\frac{(0.4 - 0.3518)^2 + (0.9 - 0.8290)^2 + \cdots + (0.0 - 0.4711)^2}{6 - 2}}$$

$$= 0.2759$$

13.9

C.V. $= 0 + 3.747(0.0428)$

$\qquad = 0.1604$

The sample slope is 0.1193. There is not enough evidence to reject the null hypothesis that the population slope is less than or equal to 0.0 at a 1% significance level.

13.11(a) The lake will begin to flood over the retaining wall when the number of inches between the lake and the top of the wall is zero. The estimated regression line is $\hat{y} = 3{,}676.58 - 1.8452x$, where $\hat{y}$ is in inches and x is the year. Setting $\hat{y}$ to zero and solving for x yields the value 1,992.51. The observations were made in July, so the predicted date for flooding is January 1993. This prediction ignores any seasonal variations in the level of the lake or flooding caused by high winds and tides.

(b)

$L, U = -1.8452 \pm 2.447(0.1390)$

$\qquad L = -2.185, \ U = -1.505$

13.13(a) The estimated effect of adding one employee per shift on the average output per shift is simply the slope coefficient of the regression of output on the number of employees.

$$b = \frac{[5(5{,}435) + \cdots + 9(8{,}971)] - 19(7.2105)(7{,}059.3158)}{(5^2 + 5^2 + \cdots + 9^2) - 19(7.2105)^2}$$

$\qquad = 754.308$

(b)

$L, U = 754.308 \pm 2.110(90.40)$

$\qquad L = 563.564, \ U = 945.052$

(c) This economic law is variously called the law of diminishing returns, the law of diminishing marginal product, or the law of variable proportions.

13.15(a) The estimated regression line is $\hat{y} = 16.8262 + 0.7839x$. The standard error of the estimate is 3.577.

$L, U = [16.8262 + 0.7839(20)] \pm$

$2.776(3.577) \times$

$$\sqrt{1 + \frac{1}{6} + \frac{(20 - 22.3333)^2}{(15^2 + 17^2 + \cdots + 32^2) - 6(22.3333)^2}}$$

$\qquad L = 21.671, \ U = 43.337$

(b)

$L, U = [16.8262 + 0.7839(20)] \pm$

$4.604(3.577)\sqrt{\dfrac{1}{6} + \dfrac{(20 - 22.3333)^2}{(15^2 + 17^2 + \cdots + 32^2) - 6(22.3333)^2}}$

$\qquad L = 25.322, \ U = 26.674$

(c)

C.V. $= 30 -$

$2.132(3.577)\sqrt{\dfrac{1}{6} + \dfrac{(20 - 22.3333)^2}{(15^2 + 17^2 + \cdots + 32^2) - 6(22.3333)^2}}$

C.V. $= 26.674$

The predicted value of y given an x of 20 is 32.5042. Since this value is on the high side of 30, it was not necessary to calculate the critical value for a left-tailed test. There is not enough evidence to reject the null hypothesis at a 5% significance level.

13.17 The estimated regression line is $\hat{y} = -17.0856 + 0.4959x$. The standard error of the estimate is 1.838. The predicted value for the mean number of defects at an assembly-line speed of 60 transmissions per hour is 12.6684. The limits of a 95% confidence interval for the mean of Y at $x = 60$ are

$L, U = 12.6684 \pm$

$2.306(1.838)\sqrt{\dfrac{1}{10} + \dfrac{(60 - 64.5)^2}{(55^2 + 55^2 + \cdots + 75^2) - 10(64.5)^2}}$

$\qquad L = 10.038, \ U = 14.299$

13.19 The estimated regression line is $\hat{y} = \$156.4936 + 54.7916x$. The standard error of the estimate is 445.5710. The predicted value for the mean level of sales at 40 hours per week is $\$2{,}348.1573$. The limits of a 90% confidence interval for the mean of Y at $x = 60$ are

$L, U = \$2{,}348.1576 \pm$

$1.740(445.571) \times$

$$\sqrt{\frac{1}{19} + \frac{(40 - 34.9474)^2}{(20^2 + 20^2 + \cdots + 56^2) - 19(34.9474)^2}}$$

$\quad L = \$2{,}153.10, \ U = \$2{,}543.22$

The limits of a 90% confidence interval for the one week's sales (a confidence interval on the predicted value of Y) with x set at 40 are

$L, U = \$2{,}348.1576 \pm$

$1.740(445.571) \times$

$$\sqrt{1 + \frac{1}{19} + \frac{(40 - 34.9474)^2}{(20^2 + 20^2 + \cdots + 56^2) - 19(34.9474)^2}}$$

$\quad L = \$1{,}548.70, \ U = \$3{,}147.61$

13.21(a) $r = 0.9623$.

$$\text{C.V.}_L, \ \text{C.V.}_U = 0 \pm 2.447\sqrt{\frac{1 - (0.9623)^2}{8 - 2}}$$

$$\text{C.V.}_L = -0.272, \ \text{C.V.}_U = 0.272$$

The null hypothesis that the population correlation coefficient equals zero is rejected at a 5% significance level.

(b) Rejecting the null hypothesis implies that there is some association between the Dow Jones Industrial Average and the length of skirts. This association may be due to chance, to other factors that cause both variables to move together, or to high stock values causing women to buy shorter skirts. Only under the circumstance of the causation running from skirts to the Dow Jones Industrial Average would the promotion of micromini skirts affect the average.

13.23 The estimated regression line in terms of the transformed variables is $y' = 0.945 + 23.3771x'$, where $y' = y/x$ and $x' = 1/x$. The estimated regression line expressed in terms of the original variables is $y = 23.3771 + 0.945x$.

13.25 Working with the transformed variables, $y' = 0.945 + 23.377x'$, $y' = 0.945 + 23.377(1/60)$, or 1.3346. The standard error of the estimate is 0.3481. The limits of a 95% confidence interval for the population mean given an x of 60 are

$L, U = 1.3346 \pm$

$$2.306(0.3481)\sqrt{\frac{1}{10} + \frac{(1/60 - 0.02163)^2}{0.00575 - 10(0.02163)^2}}$$

$\quad L = 1.0531, \ U = 1.6161$

Expressed in terms of the untransformed y, the limits of a 95% confidence interval for the population mean, with x equal to 60, are 60 times the above limits. $L = 63.186, \ U = 96.966$. The limits of a 95% confidence interval for the predicted value of Y'' given an x of 60 are

$L, U = 1.3346 \pm$

$$2.306(0.3481)\sqrt{1 + \frac{1}{10} + \frac{(1/60 - 0.02163)^2}{0.00575 - 10(0.02163)^2}}$$

$\quad L = 0.4839, \ U = 2.1853$

Expressed in terms of the untransformed y, the limits of a 95% confidence interval for the predicted value of Y, with x equal to 60, are 60 times the above limits. $L = 29.034, \ U = 131.118$.

13.27(a) In the first plot the relationship between Y and x is clearly not linear. The second plot shows the log of Y plotted against x. This relationship appears to be much closer to linear.

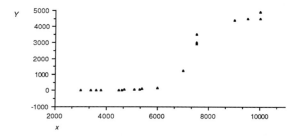

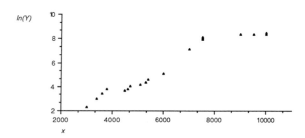

(b) Using the natural log of Y as the dependent variable, the regression line is est. $\ln(Y) = -0.2894 + 0.000958x$. The standard error of the estimate is 0.637.

13.29

$\text{C.V.}_L, \text{C.V.}_U = [6.8024] \pm$

$$2.101(0.6370)\sqrt{\frac{1}{20} + \frac{(5,000 - 6,020)^2}{828,720,000 - 20(6,020)^2}}$$

$\text{C.V.}_L = 6.4745, \text{ C.V.}_U = 7.1303$

These critical values can be compared to the predicted log of vacation spending for a family that has a per capita income of $5,000, or 4.5006. The null hypothesis that the mean spending for families with per capita incomes of $5,000 equals $900 is rejected at a 5% significance level.

13.31 The coefficient of determination is the percentage of the total variation in presidents' salaries accounted for by the number of employees. $\text{SST} = 17,724,245,967$, $\text{SSR} = 13,828,426,663$, $r^2 = 78.02\%$.

$r^2 =$

$$\frac{(\$79,734.89 - 124,861.67)^2 + \cdots + (185,090.06 - 124,861.67)^2}{(\$88,000 - 124,861.67)^2 + \cdots + (180,000 - 124,861.67)^2}$$

$= 0.7802$

13.33 This exercise requires a confidence interval for the predicted value of Y given x.

$L, \ U = \$117,239.62 \pm$

$$2.228(19,738)\sqrt{1 + \frac{1}{12} + \frac{(1,642 - 1,856)^2}{52,237,608 - 12(1,856)^2}}$$

$L = \$71,379.01, \ U = \$163,100.23$

13.35 The plot of Y versus x appears to be reasonably linear, but the second plot of the residuals of the regression shows high residuals when x is either low or high and low for intermediate values of x. This pattern suggests that the relationship between Y and x violates the assumption of linearity.

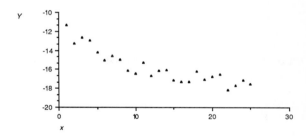

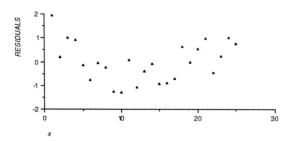

13.37 Since the mean mileage for the 584 already-sold trucks and the mileage of each of the 100 additional trucks is one million miles, the term $(x^* - \bar{x})^2$ equals 0. The predicted selling price for a truck with one million miles is $45,000 - 0.015(1,000,000)$, or $30,000. The critical value for the hypothesis test is

$$\text{C.V.} = \$30,500 - 1.64(1.450)\sqrt{\frac{1}{584}}$$

$\text{C.V.} = \$30,401.60$

At a 5% significance level the null hypothesis is rejected. The predicted selling price of $30,000 is well below the critical value of $30,401.60. You should not buy the 100 trucks.

13.39 The data appear to be heteroscedastic, with little variance for low x values and a high variance for large x values.

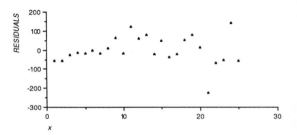

13.41 Economics textbooks usually have the axes reversed from the convention of having the independent variable on the y axis. The quantity demanded is a function of the price, which ought to be on the x axis. Regression analysis would be appropriate if you had observations of the quantity demanded at various prices. To trace out a demand curve there must be variations in supply with a stable demand. Observed market prices and quantities may be due to variations in demand or supply or both.

13.43 This exercise calls for a test of the null hypothesis that the population slope coefficient is 0. The critical values are

$$C.V._L, C.V._U = 0 \pm 2.179(0.1871)$$
$$C.V._L = -0.4077, C.V._U = 0.4077$$

The estimated slope coefficient is -1.5855. The null hypothesis that β equals zero is rejected at a 5% significance level.

13.45 This exercise calls for a 95% confidence interval on the predicted value of Y given an x of 55 cents.

$$L, U = 125.390 \pm$$
$$2.179(4.196)\sqrt{1 + \frac{1}{14} + \frac{(55 - 53.7143)^2}{40,896 - 14(53.7143)^2}}$$
$$L = 115.91, \ U = 134.87$$

13.47

$$C.V._L, C.V._U = 0 \pm 1.782\sqrt{\frac{1 - 0.8568}{14 - 2}}$$
$$C.V._L = -0.195, C.V._U = 0.195$$

The estimated coefficient of correlation is -0.9256. The null hypothesis that ρ equals zero is rejected at a 10% significance level.

13.49 This exercise calls for a test of the null hypothesis that the population slope coefficient is 0. The critical values are

$$C.V._L, C.V._U = 0 \pm 2.179(0.0031)$$
$$C.V._L = -0.0068, C.V._U = 0.0068$$

The estimated slope coefficient is 0.0104. The null hypothesis that β equals zero is rejected at a 5% significance level.

13.51 This exercise calls for a 95% confidence interval on the predicted value of Y given an x of 10 days.

$$L, U = 0.1265 \pm$$
$$2.179(0.03179)\sqrt{1 + \frac{1}{14} + \frac{(10 - 8.0714)^2}{1,021 - 14(8.0714)^2}}$$
$$L = 0.054, \ U = 0.199$$

13.53 This exercise calls for a test of the null hypothesis that the population correlation coefficient is 0. The critical values are

$$C.V._L, C.V._U = 0 \pm 1.782\sqrt{\frac{1 - 0.4925}{14 - 2}}$$
$$C.V._L = -0.366, C.V._U = 0.366$$

The estimated coefficient of correlation is 0.7018. The null hypothesis ρ = zero is rejected at a 10% significance level.

13.55 This exercise calls for a test of the null hypothesis that the population slope coefficient is 0. The critical values are

$$C.V._L, C.V._U = 0 \pm 2.179(0.0615)$$
$$C.V._L = -0.134, C.V._U = 0.134$$

The estimated slope coefficient is 0.2582. The null hypothesis that average daily temperature has no effect on electricity consumption is rejected at a 5% significance level.

13.57 The expected consumption on a day with a 95° average temperature is 13.8407 millions of kilowatts. This exercise asks for the probability that the consumption will exceed 17 million kilowatts. Substituting into the z score formula,

$$z = \frac{17 - 13.8407}{(1.704)\sqrt{1 + \frac{1}{14} + \frac{(95 - 87.4286)^2}{107,780 - 14(87.4286)^2}}}$$
$$z = 1.73$$

$$P(z > 1.73) = 0.0418$$

Since the sample standard deviation is used instead of the population standard deviation, the ratio would follow a Student's distribution rather than a normal distribution. The tail area to the right of $t = 1.73$ is approximately 5%.

13.59 The sample coefficient of correlation is 0.7712. The critical values for a test of $\rho = 0$ are

$$\text{C.V.}_L, \text{C.V.}_U = 0 \pm 1.782\sqrt{\frac{1 - 0.5947}{14 - 2}}$$

$$\text{C.V.}_L = -0.327, \text{C.V.}_U = 0.327$$

The null hypothesis that $\rho = 0$ is rejected at a 10% significance level.

13.61 Using weighted least squares, y' given an x of \$215,000 is 1.1801. The expected selling price given an appraisal of \$215,000 is 215,000(1.1801), or \$253,721.50. The following calculation of the z score is based on the transformed variables.

$$z = \frac{400{,}000/215{,}000 - 1.1801}{(0.2842)\sqrt{1 + \frac{1}{40} + \frac{(1/215{,}000 - 0.00001258)^2}{0.00000001 - 40(0.00001258)^2}}}$$

$$z = 2.35$$

$$P(z > 2.35) = 0.0094$$

Since the sample standard deviation is used instead of the population, the ratio would follow a student's distribution rather than a normal distribution. The tail area to the right of is approximately 1%.

CHAPTER 14

14.1

$$b_1 = \frac{59.445(671.995) - 91.945(431.636)}{143(59.445) - [91.945]^2}$$

$$= 5.56$$

$$b_2 = \frac{143(431.636) - 91.945(67.1995)}{143(59.445) - [91.945]^2}$$

$$= -1.34$$

$$a = 41.102 - 5.56(6.500) - (-1.34)(13.264)$$

$$= 22.74$$

$$R^2 =$$
$$\frac{(14.900 - 41.102)^2 + (19.790 - 41.102)^2 + \cdots + (66.546 - 41.102)^2}{(15.178 - 41.102)^2 + (21.480 - 41.102)^2 + \cdots + (66.386 - 41.102)^2}$$

$$= 0.993$$

14.3

$$b_1 = \frac{40(65) - (-29)(-68)}{28(40) - [-29]^2}$$

$$= 2.25$$

$$b_2 = \frac{28(-68) - (-29)(65)}{28(40) - [-29]^2}$$

$$= -0.068$$

$$a = 16 - 2.25(7) - (-0.068)(7)$$

$$= 0.73$$

$$R^2 =$$
$$\frac{(8.914 - 16)^2 + (13.686 - 16)^2 + \cdots + (25.208 - 16)^2}{(4 - 16)^2 + (6 - 16)^2 + \cdots + (11 - 16)^2}$$

$$= 0.992$$

14.5(a) There is no apparent nonlinearity or heteroscedasticity in either plot. The effects would have to be quite pronounced to show up in such a sample.

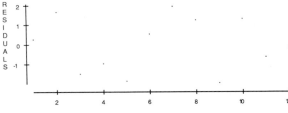

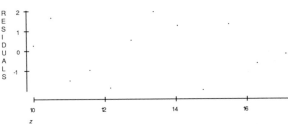

(b) $r = 0.997$. There appears to be multicollinearity.
(c) Again, with a small sample size it is difficult to draw conclusions. However, the second plot suggests the presence of heteroscedasticity.
(d) $r = 0.905$. There appears to be multicollinearity.

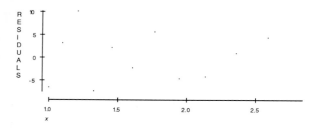

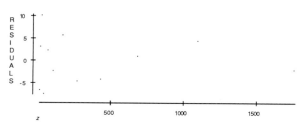

14.7 The estimated multiple regression has the slope coefficients 2.75 and 0.106 for x and z respectively. The simple regressions have the slope coefficients 88.6 and 0.108. The coefficient on x is highly sensitive to the presence of the z variable. This effect is common in the presence of multicollinearity. Ordinarily, you would keep both variables in the regression because controlling for the effect of x should improve the estimate of the effect of z. However, with high multicollinearity and a small sample, x should be dropped.

14.9

$$F = \frac{38,209.9/2}{317.612/9}$$

$$= 541.37$$

The critical value for the test with 2 and 9 degrees of freedom is 4.26. The null hypothesis that independent variables have no effect on the dependent variable is rejected at a 5% significance level.

14.11

$$s_{b_z} = \sqrt{\frac{(0.5947)^2(28)}{28(40) - [-29]^2}}$$

$$= 0.1884$$

$$\text{C.V.}_L, \text{ C.V.}_U = 0 \pm 3.182(0.1884)$$

$$\text{C.V.}_L = -0.599, \text{ C.V.}_U = 0.599$$

The estimated slope coefficient for z is -0.068. There is not enough evidence to reject the null hypothesis that the population slope coefficient for z is zero at a 5% significance level.

14.13

$$b_1 = \frac{5(220.15) - 3.50(-17.48)}{119.75(5) - [3.50]^2}$$

$$= 1.981$$

$$b_2 = \frac{119.75(-17.48) - 3.50(220.15)}{119.75(5) - [3.50]^2}$$

$$= -4.883$$

$$a = 8.843 - 1.981(5.250) - (-4.883)(0.500)$$

$$= 0.884$$

$$R^2 = \frac{(6.827 - 8.843)^2 + \cdots + (6.827 - 8.843)^2}{(6.37 - 8.843)^2 + \cdots + (6.61 - 8.843)^2}$$

$$= 0.990$$

$$\bar{R}^2 = \left[\frac{521.476/(20 - 2 - 1)}{526.668/(20 - 1)}\right] - \left(\frac{2}{20 - 2 - 1}\right)$$

$$= 0.989$$

$$s_e = \sqrt{\frac{5.1918}{n - 2}}$$

$$= 0.553$$

$$s_{b_x} = \sqrt{\frac{(0.553)^2(5)}{119.75(5) - [3.5]^2}}$$

$$= 0.051$$

$$s_{b_z} = \sqrt{\frac{(0.553)^2(119.75)}{119.75(5) - [3.5]^2}}$$

$$= 0.250$$

14.15(a) $\hat{y} = 114.29 + 20.05(10) + 5.08(15) - 16.66(10)(15)$, $\hat{y} = -2,108.010$.
(b) $\hat{y} = 114.29 + 20.05(20) + 5.08(25) - 16.66(20)(25)$, $\hat{y} = -7,687.710$.
(c)

$$L, U = 20.05 \pm 1.701(5.11)$$

$$L = 11.358, \ U = 28.742$$

(d)

$L, U = 5.08 \pm 1.701(2.09)$

$\qquad L = 1.52, \ U = 8.64$

(e)

C.V.$_L$, C.V.$_U = 0 \pm 2.048(9.95)$

$\qquad$ C.V.$_L = -20.378$, C.V.$_U = 20.378$

The estimated slope coefficient for the interaction term is -16.66. There is not enough evidence to reject the null hypothesis at a 5% significance level that the population interaction term has a slope coefficient of zero.

14.17

$\ln(\hat{y}) = 0.2026 + 1.231 \times \ln(x) + 2.209 \times \ln(z)$.

14.19(a) SALES $= 300 - 200(1) + 0.00045(1,000)$, or 104.5.

(b) With 100 observations, the z score for a two-tailed test at a 5% significance level is approximately 1.96. With a large sample a shortcut for testing against an hypothesized slope coefficient of zero is to divide the slope coefficient by its standard error. If the absolute value of this ratio is less than 1.96, the null hypothesis cannot be rejected. The dummy variables for Tuesday and Wednesday are not significantly different from 0. In other words, the expected sales on a Tuesday or a Wednesday are not significantly different at a 5% level from sales on a Monday.

14.21(a)

$\ln$(barrels) $= 9.48 + 0.392(12) + 0.000237(1,540)$
$\ln$ (barrels) $= 9.48 + 0.392(12) + 0.00237(1,540)$
$\qquad\qquad = 17.8338$
$\exp(17.8338) = 55,405,907$

(b) The natural log of production expressed as rate per day increases by 0.392 for every \$1 increase in the price of a barrel of petroleum. Thus the log of production is a linear function of price. However, production in barrels is not a linear function of price. The change in production for a \$1 increase in price depends on the initial price.

14.23 $z = \{-2, -1.33, -0.67, 0, 2\}$, probabilities of observed responses $= \{0.0228, 0.0918, 0.2514, 0.5, 0.0228\}$, likelihood $= 0.0000059986$, or 0.000006.

14.25(a) $z = (75 - 80)/20$, or -0.25. $P(\text{yes}) = 0.4013$.

(b) $z = (45 - 80)/20$, or -1.75. $P(\text{no}) = 0.9599$.

14.27 $\hat{y} = 11.5608 + 0.1292\text{TRUCKS} + 4.4398\text{FREEZES}$.

14.29

$$F = \frac{4,390.81/2}{2,163.59/7}$$

$$= 7.10$$

The critical value for 2 and 7 degrees of freedom and a 5% significance level is 4.74. The null hypothesis that all of the independent variables account for none of the variation of the dependent variable is rejected at a 5% significance level.

14.31 $\hat{y} = -50.6239 + 0.06677\text{SPACE} + 46.7347\text{LOT}$.

14.33

$$F = \frac{4,940.59/2}{895.51/7}$$

$$= 19.31$$

The critical value for 2 and 7 degrees of freedom and a 5% significance level is 4.74. The null hypothesis that all of the independent variables account for none of the variation of the dependent variable is rejected at a 5% significance level.

14.35 $\hat{y} = 122.748 + 0.9191\text{INCOME} + 0.0710\text{SERVICES}$.

14.37

$$F = \frac{6,868.67/2}{8,311.73/12}$$

$$= 4.96$$

The critical value for 2 and 12 degrees of freedom and a 5% significance level is 3.89. The null hypothesis that all of the independent variables account for none of the variation of the dependent variable is rejected at a 5% significance level.

14.39

$$\overline{R}^2 = \frac{0.8/(20 - 5 - 1)}{1/(20 - 1)} - \frac{5}{20 - 5 - 1}$$

$$= 0.729$$

14.41(a) $\hat{y} = 211 + 0.13(100) + 18(16) +$ 0.00295(100)(16), or 516.72.

(b)

C.V.$_L$, C.V.$_U = 0 \pm 1.96(0.00048)$

$\quad$ C.V.$_L = -0.00094$, C.V.$_U = 0.00094$

The estimated slope coefficient of 0.00295 is above the upper critical value of 0.00094. The null hypothesis that the population slope coefficient on the interaction term is zero can be rejected at a 5% significance level.

(c) $\hat{y} = 211 + 0.13(110) + 18(16) +$ 0.00295(110)(16), or 518.492. The increase in yield is $518.492 - 516.72 = 1.772$.

14.43(a) This exercise calls for a test of the null hypothesis that the slope coefficient for the dummy variable MAN is less than or equal to 0.

C.V. = \$0 + 1.64(356.09)

C.V. = \$583.99

The estimated slope coefficient is above the critical value of 583.99. The null hypothesis that slope coefficient on the MAN dummy variable is less than or equal to 0 is rejected at a 5% significance level.

(b) Proof in a statistical sense is difficult in this context. You cannot know if unobserved factors correlated with gender (for example, total years of labor market experience) influenced salaries or if chance variation led to the higher salaries for men.

14.45 The estimated probit regression is $\hat{y} = -2.2209 + 0.13781x$. The corresponding μ is 16.1157 and the corresponding σ is 7.2564. The number of hours of baking time that results in a 0.99 probability of success is 33.023.

14.47 The estimated probit regression is $\hat{y} = -5.2471 + 0.012831x$. The corresponding μ is 408.9393 and the corresponding σ is 77.9362. The probability of a company with 300 employees having a plane is 0.808. The probability of a company with 400 employees having a plane is 0.4562. The increase in employment raises the probability of having a plane by 0.3754.

■ CHAPTER 15

15.1 $\hat{y} = 34.8417 + 3.1053t$, $\quad \hat{y} = 34.8417 +$ 3.1053(20), or 96.948.

15.3

$$DW = \frac{613.7989}{369.4736}$$

$$= 1.661$$

With 19 observations and an α of 5%, $DW_L = 1.18$ and $DW_U = 1.40$. The test statistic, $DW = 1.661$, is above DW_U. There is not enough evidence to reject the null hypothesis of no positive serial correlation.

15.5(a) A trend in the number of patients per day implies that the population slope coefficient on the trend variable does not equal zero. The t score of 1.96 is based on a 5% significance level and the assumption of a large sample size. Other significance levels could be used. The sample size is given as 140 in part c.

C.V.$_L$, C.V.$_U = 0 \pm 1.96(1.61)$

$\quad$ C.V.$_L = -3.1556$, C.V.$_U = 3.1556$

The estimated slope coefficient of 0.4 is within the acceptance region. There is not enough evidence to reject the null hypothesis that the population slope coefficient on the trend term is zero.

(b) A shortcut for testing whether a slope coefficient is significantly different from 0 at 5% significance level is to compare the absolute value of the ratio of the estimated slope coefficient to 1.96. Again, this t score assumes a large sample. The null hypothesis of a zero population slope coefficient is rejected whenever the absolute value of the ratio exceeds 1.96. In the exercise Monday, Friday, and Saturday have significantly different expected numbers of patients than the day with an omitted dummy variable, Sunday.

(c) The table stops at 100 observations. Taking the last value as an approximation for 140 observations, DW_L and DW_U equal 1.55 and 1.80 respectively with an α of 5% and a **k** of 7. The test statistic, $DW = 2.91$, is above $4 - DW_L$, or 2.45. The null hypothesis of no serial correlation is rejected at a 5% significance level. Note that as n increases $4 - DW_L$ approaches 2. Thus if you reject the null hypothesis at table value taken at $n = 100$, you must also reject the null hypothesis for higher values of n.

15.7

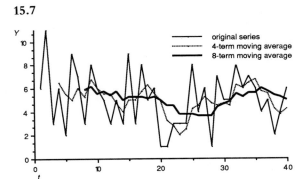

The forecast values of the series for either a one-step forecast or for all future steps are 4.25 based on the four-term moving average and 5 based on the eight-term moving average. The first eight and the last three data points are shown below.

t	Y	4-Term moving average	8-Term moving average
1	6		
2	11		
3	3		
4	6	6.5	
5	2	5.5	
6	9	5.0	
7	7	6.0	
8	3	5.25	5.875
.	.	.	.
.	.	.	.
.	.	.	.
38	2	4.25	5.375
39	4	3.75	5.25
40	6	4.25	5.0

15.9

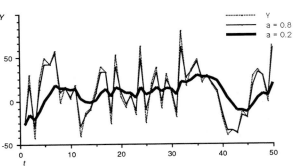

The smoothed series with a set to 0.2 sticks closely to the original time series. The smoothed series with a set to 0.8 drifts upward because its values are largely determined by the upward trend at the beginning of the series. The forecast value with $a = 0.8$ is 111.80. The forecast value with $a = 0.2$ is 4.98. The first three and the last three data points for each series are shown below.

t	Y	Exponentially smoothed series with $a = 0.8$	Exponentially smoothed series with $a = 0.2$
1	−27	−27.00	−27.00
2	30	18.60	−15.60
3	−43	−30.68	−12.06
.	.	.	.
.	.	.	.
.	.	.	.
48	28	28.90	8.28
49	−1	4.98	6.42
50	75	61.00	20.14

15.11

Period	Forecast
1	−47.60
2	47.32
3	−19.12
4	27.38
5	−5.17
6	17.62
7	1.67
8	12.83
9	5.02
10	10.49

15.13 The best equation for forecasting contains the constant and the slope coefficients on the first 12 lags. The twelfth lag is always statistically significant and none of the higher lags are consistently statistically significant.

15.15 The panel in the upper left corner appears to be an exploding time series. If this were the only problem, the series could be forecast with an autoregressive model in which some of the slope coefficients were greater than 1. The real problem for an autoregressive model is not that the series has an increasing amplitude (the swings on the vertical axis get wider as *t* increases) but that the periodicity increases (the peak to peak distance on the *t* axis increases as *t* increases). The increasing periodicity implies a nonconstant lag structure.

The upper right panel exhibits a quadratic trend. It could be estimated with an autoregressive model in which the slope coefficient on the first lag was positive (reflecting the upward trend) and the slope coefficients on higher order lags were negative (reflecting the concavity of the curve). An alternative estimation method for these data would be multiple regression in which the independent variables are the *t* and t^2. Differencing the data and estimating a linear trend would be a third alternative. Differencing is discussed in the next section.

The middle left panel is similar to the upper right panel. The only difference is that some of the higher order lags would have positive slope coefficients. The comments about two alternative models made in the previous paragraph apply to this panel as well.

The middle right panel shows a damped rather than an exploding time series. This panel shows a typical response of an autoregressive system to an initial shock. The estimated constant term would not be significantly different from zero and the coefficients on the first two lags would have opposite signs.

The lower left panel shows a clear trend and what appears to be a seasonal pattern. With these data a forecast could be with either the classical regression model or with an autoregressive model.

The lower right panel could be estimated with an autoregressive model. Only the constant term would be significant. If all of the observations had the same *y* value, none of the lags would have a significant coefficient. In fact, the slope coefficients on the lags could not be estimated.

15.17 The three figures show the residuals of regressions of the time series, the first differences, and the second differences. Almost all of the trend is removed by second differencing.

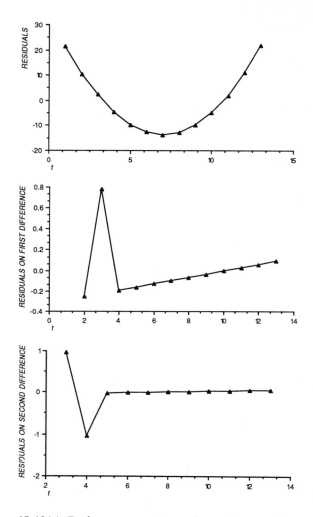

15.19(a) Both autoregressive and moving average. Magazine advertising over long periods is a function of the number of readers and their demographics. Rich readers count more heavily with advertisers than poor ones. Gains in readership due to shifting reader interests (for example, increased interest in fitness) or new editorial style should follow an autoregressive process. Short-term increases in ads should follow a moving-average process because advertisers tend to buy blocks of ads for up to six months to get the discount.

(b) An autoregressive process. Deer in a period are, in economic terminology, an input to the production of deer in the next period. Thus a shock such as severe weather will affect the deer population for many subsequent years.

(c) An autoregressive process. Each wave of buyers brings in a smaller group of friends in the subsequent week.

15.21(a) Setting the first observation to $t = 1$, $\hat{y} = 514.867 - 8.1394(11)$, or 425.334.

(b)

$$L, U = 425.334 \pm 2.306(2.629)\sqrt{1 + \frac{1}{10} + \frac{(11 - 5.5)^2}{385 - 10(5.5)^2}}$$

$$L = 417.992, \quad U = 432.676$$

15.23 For an n of 100 and an α of 5%, $DW_L = 1.65$. For an n of 100 and an α of 1%, $DW_L = 1.52$. At either significance level the null hypothesis of no positive serial correlation is rejected. The forecasts would be unbiased but the standard error of the slope coefficient would be biased downward. You would be more likely to make a Type II error, rejecting the null hypothesis of a zero slope coefficient on the trend term, if in fact the true trend was zero.

15.25 The exponentially smoothed series is 5.00, 10.50, 12.75, 12.38, 15.19, 12.10, 7.55, 9.28, 11.14, 11.57, 9.29, 6.15, 10.58, 11.29, 7.15, 12.08, 7.04, 6.02, 10.01, 5.51. The one-step-ahead forecast is 5.51. The exponentially smoothed series is more sensitive to the last observation because 50% of the value of the exponentially smoothed series depends on the last observation whereas 20% of the five-item moving average depends on the last observation.

15.27 The forecast values are $24,539.53$, $-14,303.91$, $-13,462.15$, $-42,354.24$, $-42,441.39$, $-52,169.59$.

15.29 The steady decay in the autocorrelations in the first figure suggest a first-order autoregressive process; the two spikes in the second figure suggest positive first and second autoregressive process. The estimation should be one difference term, no moving average terms, and both one and two autoregressive terms.

15.31 $\hat{y} = 1,034 + 121(27) - 456$, or $3,845$.

15.33 The three-item moving average appears to provide the best representation of the pattern in the data.

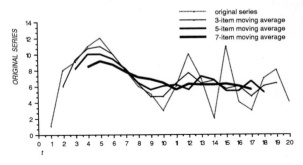

15.35 The estimated slope coefficient is 0.4353. The standard error of the slope coefficient is 0.2657. There are 18 degrees of freedom.

$$C.V._L, C.V._U = 0 \pm 2.101(0.2657)$$
$$C.V._L = -0.5582, \quad C.V._U = 0.5582$$

There is not enough evidence to reject the null hypothesis that the population slope coefficient is zero (i.e., that there is no trend).

15.37 With 36 observations $k = 5$, and a 5% significance level, $4 - DW_L$ equals 2.76 and $4 - DW_U$ equals 2.26. At a 1% significance level, $4 - DW_L$ equals 3.01 and $4 - DW_U$ equals 2.41. At a 5% significance level there is not enough evidence to reject the null hypothesis of either no serial correlation or positive correlation. At a 1% significance level the test is inconclusive because $DW = 4 - DW_U = 2.26$.

15.39(a) $\hat{y} = 0.5344 + 0.8642(4.10)$, or 4.0776.

(b) The standard error of the slope coefficient is 0.0907. There are 17 degrees of freedom.

$$C.V._L, C.V._U = 0.2 \pm 2.110(0.0907)$$
$$C.V._L = 0.0086, \quad C.V._U = 0.3914$$

The estimated slope coefficient on the first lag is 0.8642. The null hypothesis that the population slope coefficient is 0.2 is rejected at a 5% significance level.

(c)

$$DW = \frac{7.8925}{3.0152489}$$

$$= 2.618$$

At 19 observations and a 15% significance level, $4 - DW_L = 2.82$, and $4 - DW_U = 2.6$. The DW test statistic is between these critical values. The test is inconclusive, the hypothesis of no serial correlation can be neither accepted nor rejected.

(d) Recall that when lagged values of the time series are used as independent variables, the DW test statistic is biased toward the value 2 (and acceptance of the null hypothesis of no serial correlation). Hence, the "inconclusive" conclusion in part c is suspect. If the null hypothesis of no serial correlation was false, a first-order autoregression would not capture all of the systematic motion in the series. A higher order autoregression would be appropriate, or a more advanced estimation technique such as Cochrance-Orcutt (discussed in the econometrics textbooks cited at the end of Chapter 14).

15.41 The two spikes in the correlogram suggest a positive first- and positive second-order moving average process. The steady decay in the partial correlogram also suggests a moving average process. The estimated model should be first and second moving average, one difference term, and no autoregressive terms.

15.43 Ignoring leap years, the first day of the eleventh year of the series is the 3,651st day from the beginning of the series. Since every fourth year is a leap year, there can be either two or three leap years in the 10-year series. Hence, the first day of the eleventh year of the series is either the 3,653rd or the 3,654th day from the beginning of the series. The estimated number of people attending the museum is either $\hat{y} = 345 + 2(3,653)$, or $\hat{y} = 345 + 2(3,654)$. These estimates are 7,6531 and 7,653. Given the available information, the best estimate is 7,652.

15.45 Listed below are the original series, the estimated trend line, the smoothed series using a four-item last-term moving average, the exponentially smoothed series with $a = 0.3$, and the forecast values for each period based on a first-order autoregression for the 20 observations. The bottom row has the forecast for each method for the next month.

t	Original series	Trend	Four-item moving average	Exponentially smoothed $a = 0.3$	Auto-regressive forecasts
1	59	55.2715		59.00	
2	51	55.2745		56.60	54.3663
3	52	55.2775		55.22	55.9263
4	54	55.2805	54.00	54.85	55.7313
5	59	55.2835	54.00	56.10	55.3413
6	58	55.2865	55.75	56.67	54.3663
7	55	55.2895	56.50	56.17	54.5613
8	55	55.2925	56.75	55.82	55.1463
9	52	55.2955	55.00	54.67	55.1463
10	60	55.2985	55.50	56.27	55.7313
11	54	55.3015	55.25	55.59	54.1713
12	52	55.3045	54.50	54.51	55.3413
13	57	55.3075	55.75	55.26	55.7313
14	56	55.3105	54.75	55.48	54.7563
15	60	55.3135	56.25	56.84	54.9513
16	53	55.3165	56.50	55.69	54.1713
17	53	55.3195	55.50	54.88	55.5363
18	57	55.3225	55.75	55.52	55.5363
19	52	55.3255	53.75	54.46	54.7563
20	57	55.3285	54.75	55.22	55.7313
forecasts 21		55.3315	54.75	55.22	54.7563

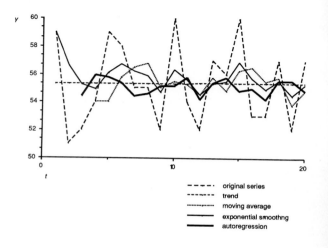

- - - - - original series
- - - - - - trend
· · · · · · · · moving average
———— exponential smoothng
━━━━ autoregression

The plot suggests that the trend line is the least useful for forecasting this series. The trend line is essentially flat (the estimated slope coefficient is 0.003). The first-order autoregression does a poor job of tracking the original series. The ratio of the estimated slope coefficient to the standard error (commonly called the t ratio) for the first lag is -0.851. A ratio above 2.11 would be required for the first lag to be statistically significant at a 5% level. It is difficult to pick between the moving average and the exponentially smoothed forecast methods. Note that none of the four forecast methods picks up the pattern that the values usually drop sharply after each peak. A moving average may be better suited to these data, although 20 observations are too few for ARIMA. Note also that all four methods come up with nearly the same forecast for the 21st period. The forecasts would diverge for longer patterns.

15.49 The correlogram shows a dampened sine wave pattern which suggests a positive first-order autoregressive process and a negative second-order autoregressive process. The two spikes in the partial correlogram also suggest both a first-order and a second-order autoregressive process. The estimated model would have one difference term, two autoregressive terms, and no moving average terms.

■ CHAPTER 16

16.1

$$L_{1988} = \frac{20(\$6) + 16(\$5) + 300(\$2.90)}{20(\$7) + 16(\$4) + 300(\$3)} \times 100$$

$$= 96.92$$

$$\text{Paasche}_{1988} = \frac{20(\$6) + 10(\$5) + 310(\$2.90)}{20(\$7) + 10(\$4) + 310(\$3)} \times 100$$

$$= 96.31$$

$$F_{1988} = \sqrt{96.92(96.31)}$$

$$= 96.61$$

16.3

Year	Old index (1976 = 100)	New index (1982 = 100)	Spliced index (1982 = 100)
1975	99.6		92.2
1976	100.0		92.6
1977	101.7		94.2
1978	104.3		96.6
1979	107.6		99.6
1980	105.2		97.4
1981	104.2		96.5
1982	108.0	100.0	100.0
1983		102.9	102.9
1984		103.3	103.3
1985		100.7	100.7
1986		109.5	109.5
1987		110.3	110.3

16.5 The student's book purchases would tend to be within her major. If books in her field were revised rapidly and old books were often supplanted by new editions, her personal price index would show more inflation than all college textbooks. The opposite effect could also occur. The question does not state whether she bought any used books. Used book purchases may distort the index, especially if she started with mostly used books in her freshman classes and had to buy mostly new books for her senior classes. Another distortion could be caused by upper level textbooks being longer and having smaller production runs and more technical typesetting than lower level textbooks. This progression may cause her personal price index to overstate actual inflation. The number of books may vary from semester to semester, which also distorts the index.

16.7 Only four items would go into the index because the quantity of beef tongue in the ending year is zero. A Paasche index is based on the ending period's quantity weights.

$$\text{Paasche}_{1977} =$$

$$\frac{0(\$0.64) + 11(\$2.65) + 4(\$6.43) + 6(\$4.06) + 32(\$0.64)}{0(\$0.44) + 11(\$2.25) + 4(\$4.78) + 6(\$3.99) + 32(\$0.58)} \times 100$$

$$= 115.45$$

16.9

Year	Old index (1932 = 100)	New index (1936 = 100)	Spliced index (1932 = 100)
1932	100.0		100.0
1933	98.4		98.4
1934	97.5		97.5
1935	95.3		95.3
1936	90.1	100.0	90.1
1937		97.2	87.6
1938		95.3	85.9
1939		92.2	83.1
1940		88.4	79.6
1941		87.1	78.5
1942		80.3	72.4
1943		78.1	70.4
1944		55.4	49.9

16.11 The values of a Laspeyres index for the years 1985, 1986, and 1987 are 100.0, 94.1, and 87.8.

16.13 The values of a Fisher ideal index for the years 1985, 1986, and 1987 are 100.0, 93.5, and 88.5. The greatest percentage change in either prices or quantities over the years 1986 to 1988 is the jump in air freight from 0.5 ton-mile to 144 ton-miles. This jump in air freight may be due to a high elasticity of demand for air freight. Note that the price of air freight falls slightly from $32.60 to $30.00 per ton-mile over the years 1985 to 1987. Another explanation is changes in the mix of goods being shipped toward more perishable items. In other words, the changing pattern in method of shipment may be generated by changing relative prices or by exogenous shifts in the demand for the different methods of shipping. If the latter is the case, a Paasche index would be appropriate since you would continue to use relatively high amounts of air freight in the future even if its price rose. If the former were true, a Laspeyres index would be appropriate. You would be interested in how much it would cost at current prices to reproduce the original ton-mile mix. If you have no idea why the ton-mile mix is changing, the Fisher ideal index is a reasonable compromise.

16.15 The increase in the hospital's budget may reflect changes in the quality of the goods and services that it purchases or technical changes embodied in new goods and services. For example, the hospital may have added a CAT scanner or a similar piece of expensive medical technology.

16.17 Suppose the rental market changes, with rents going up. All homeowners now face a higher opportunity cost for living in their own homes. That is, they are giving up more money by not renting out their homes. However, most homeowners will continue to live in their own homes and have exactly the same mortgage payments as before the increase in rental prices. An accurate measure of housing cost would reflect actual rents paid, increases in mortgage payments due to higher interest rates on variable-rate mortgages and new mortgages, and increases in the prices of newly purchased homes. This mix of data is more difficult to collect than any other of the alternatives mentioned in the exercise.

16.19 The costs of running a college and the CPI might diverge if colleges purchased expensive new technology for research or teaching. Also, if the method of instruction shifted to more professors and less graduate assistants, costs would go up. Costs are also sensitive to class size. Finally, colleges could maintain the same level of inputs and mix of inputs but find that the costs of particular items they buy have gone up faster than the CPI. For example, the cost of books purchased for the library may go up faster in price than the CPI.

One solution for constructing an index is to base it on a fixed set of inputs. Most of the inputs are fairly consistent over time such as teaching staff, custodial staff, food, books, and computers. The big-ticket technological items such as cyclotrons often come from grant money.

16.21 The figures should be an index. There are thousands of types of repairs possible and maintenance tasks vary from model to model even within the categories subcompact, compact, and full-size. Hertz could sum its own repair, insurance, gasoline, and maintenance costs for its rental fleet for each type of car and divide by the number of cars in each category. This procedure would provide a poor measure of the costs of operating each type of car. For example, the resulting estimates would be sensitive to variations in maintenance due to changes in corporate policy. The results would also be sensitive to random variation in the damage sustained by Hertz rental cars (admittedly small in this large a pool). There may also be systematic changes in the accident rates due to changing traffic laws.

16.23 An immediate problem is that the weight that should be assigned to capital punishment is less than obvious. Another difficulty is that average sentences served per current crime are unclear because of the possibility of parole. Also, if sentences were increased across all crimes, the crime index would increase with no actual increase in the number of crimes or even a decrease in the number of crimes. Shifts in sentences among crimes as well as shifts in criminal activity across jurisdictions could also change the index. Some of these "problems" may not be problems if the length of a sentence reflected the "price" or social cost of a crime.

16.25 Under the assumption that the three oils have the same properties for making potato chips, the company should use only the cheapest oil. In 1985 only the cheapest oil, cottonseed, was used. In 1986 most of the oil used was the cheapest, cottonseed at 651 units (probably 40 gallon barrels), while a small amount of the next cheapest oil was used, corn oil at 120 units. The use of two types of oil in the same year may be due to price variations within a year, or to a firm having to honor long-term contracts to purchase oil. 1987 is similar to 1986 in that most of the consumption is the cheapest oil, peanut at 751 units, with a small amount of the next cheapest oil, corn at 65 units. The index that would best measure the potato chip company's cost for oil is simply the average price per unit of oil purchased. If three oils are considered equivalent by the potato chip company, there is no separate demand for each oil. The only costs that matter are per unit purchased. The average costs for each year are $11.89, $11.73, and $11.52. With 1985 as the base the indexes are 100.0, 98.7, and 96.9.

CHAPTER 17

17.1(a) There are 8 plus values in 10 observations. The probability of 8 or more successes in 10 trials with a Π of 0.5 is 0.0547. The null hypothesis that the population median equals 14 is rejected at a 10% significance level.

(b) None of the observations take values below 60. There is no evidence against the null hypothesis.

17.3(a)

$$1 - \alpha = 1 - 2(0.0032 + 0.0005 + 0.0000)$$
$$= 0.9926$$

(b)

$$1 - \alpha = 1 - 2(0.0046 + 0.0011 + 0.0002 + 0.0000 + 0.0000)$$
$$= 0.9882$$

17.5 The results of a sign test are suspect because, as recorded, the data clearly violate the assumption of a continuous distribution. We will proceed with the sign test to show the calculations. Five of the 33 observations are for 3 days. There are 10 minuses and 18 pluses. The expected number of either pluses or minuses is 0.5(33), or 16.5. Since 10 is farther from 16.5 than 18, we will find the probability of 10 or fewer minuses in 33 trials. Applying the normal approximation to the binomial,

$$\sigma(X) = \sqrt{33(0.5)(0.5)}$$
$$= 2.8723$$

$$z = \frac{10.5 - 16.5}{2.8723}$$
$$= -2.09$$

$$P(z < -2.09) = 0.0183$$

17.7 Since the binomial table runs to 20 trials, it is not necessary to use the normal approximation to the binomial. An interval running from the sixth to the fifteenth observation has a confidence level of 0.9586. An interval running from the seventh to the fourteenth observation has a confidence level of 0.8846. The interval of the 6th to the 15th is the closest to a 95% level of confidence.

17.9(a) The sum of ranks for the smaller sample, A, is 31. The critical values from the Wilcoxon-Mann-Whitney table with samples of sizes 6 and 8 are 29–61. There is not enough evidence to reject the null hypothesis at a 5% significance level that the two populations have the same median.

(b) The sum of ranks for the smaller sample, A, is 56. The critical values from the Wilcoxon-Mann-Whitney table with samples of sizes 6 and 8 are 29–61. There is not enough evidence to reject the null hy-

pothesis at a 5% significance level that the two populations have the same median.
(c) The sum of ranks for the smaller sample, A, is 42.5. The critical values from the Wilcoxon-Mann-Whitney table with samples of sizes 6 and 8 are 29–61. There is not enough evidence to reject the null hypothesis at a 5% significance level that the two populations have the same median.

(d) The assumptions are that the two populations have identical distributions except for their mean/medians and that the distributions are continuous.

17.11 Since the two samples are the same size and this is a right-tailed test, the test statistic is the larger rank sum, 949. With 30 observations per sample, the normal approximation to the Wilcoxon-Mann-Whitney T is required. No significance level is specified. 5% is used in the calculations below. Students may choose another significance level.

$$\mu_T = \frac{30(30 + 30 + 1)}{2}$$
$$= 915$$

$$s_T = \sqrt{\frac{30(30)(30 + 30 + 1)}{12}}$$
$$= 67.6387$$

$$\text{C.V.} = 915 + 1.64(67.6387)$$
$$= 1,025.927$$

With a test statistic of 949 and a critical value of 1,025.927, there is not enough evidence to reject the null hypothesis that the mean of A is less than or equal to the mean of B at a 5% significance level.

17.13 Since the sample sizes are unequal, the test statistic is the sum of ranks of the smaller sample—the observations of concession sales with $5 ticket prices. This sum of ranks is 11. The critical value with samples of sizes 4 and 5, $\alpha = 5\%$, and a left-tailed test is 12. To reject the null hypothesis the test statistic would have to be less than or equal to 12. There is enough evidence to reject the null hypothesis that concession sales are the same or less on a $4 ticket night as on a $5 ticket night.

17.15 If the three samples came from the same population, they would have identical means/medians and identical distributions. Applying the Kruskal-Wallis test,

$$H = \frac{12}{26(26 + 1)}\left[10\left(\frac{121}{10} - \frac{26 + 1}{2}\right)^2 + 9\left(\frac{164}{9} - \frac{26 + 1}{2}\right)^2 + 7\left(\frac{66}{7} - \frac{26 + 1}{2}\right)^2\right]$$
$$= 5.7492$$

Under the null hypothesis of equal means/medians, H is chi-square distributed with $k - 1$ or 2 degrees of freedom. The critical value is 5.99147. There is not enough evidence to reject the null hypothesis at a 5% significance level.

17.17

$$r_s = \frac{(10 - 5.5)(1 - 5.5) + (9 - 5.5)(2 - 5.5) + \cdots + (1 - 5.5)(10 - 5.5)}{\sqrt{[(10 - 5.5)^2 + (9 - 5.5)^2 + \cdots + (1 - 5.5)^2][(1 - 5.5)^2 + (2 - 5.5)^2 + \cdots + (10 - 5.5)^2]}}$$
$$= -0.855$$

The critical values for a two–tailed test at a 5% significance level of the null hypothesis of no linear correlation for the ranks of X and Y are ± 0.648. The null hypothesis is rejected.

17.19

$$r_s = \frac{(2 - 20.5)(39 - 20.5) + (1 - 20.5)(36 - 20.5) + \cdots + (40 - 20.5)(4 - 20.5)}{\sqrt{[(2 - 20.5)^2 + (1 - 20.5)^2 + \cdots + (40 - 20.5)^2][(39 - 20.5)^2 + (36 - 20.5)^2 + \cdots + (4 - 20.5)^2]}}$$

$$= -0.898$$

The table stops at an n of 30. There are 40 observations in the sample. However, the last value in the table can be used for the test. The critical value for a two–tailed test at a 5% significance at an n of 30 is 0.364. The critical value for n of 40 would have an absolute value below 0.364. With a test statistic of -0.898, the null hypothesis of no linear association in the ranks is rejected.

17.21 Ranks of 1 through 10 can be assigned to either the professor's order or the graduate student's order. In the following example the ranks are assigned according to the professor's order. For example, Kawasaki has a rank of 2 in the professor's column and a rank of 3 in the graduate student's column.

$$r_s = \frac{(1 - 5.5)(1 - 5.5) + (2 - 5.5)(3 - 5.5) + \cdots + (10 - 5.5)(9 - 5.5)}{\sqrt{[(1 - 5.5)^2 + (2 - 5.5)^2 + \cdots + (10 - 5.5)^2][(1 - 5.5)^2 + (3 - 5.5)^2 + \cdots + (9 - 5.5)^2]}}$$

$$= 0.782$$

The critical value for a two–tailed test at a 5% significance level of the null hypothesis of no linear correlation for the ranks of X and Y is 0.648. The null hypothesis is rejected.

17.23

Observation rank	x	Expected cumulative density	Sample cumulative density	Absolute difference between observed and expected CDF
1	0.027	0.027	0.067	0.040
2	0.347	0.347	0.133	0.214
3	0.373	0.373	0.200	0.173
4	0.397	0.397	0.267	0.130
5	0.455	0.455	0.333	0.122
6	0.539	0.539	0.400	0.139
7	0.649	0.649	0.467	0.182
8	0.782	0.782	0.533	**0.249**
9	0.844	0.844	0.600	0.244
10	0.845	0.845	0.667	0.178
11	0.921	0.921	0.733	0.188
12	0.930	0.930	0.800	0.130
13	0.939	0.939	0.867	0.072
14	0.955	0.955	0.933	0.022
15	0.980	0.980	1.000	0.020

The critical value at a 5% significance level is 0.304. This critical value exceeds the test statistic of 0.249. There is not enough evidence to reject the null hypothesis that the population is continuous uniform with a range of 0 to 1.

17.25

Number of occurrences	Observed frequency	Running sum of observed frequencies	Relative cumulative observed frequency	Relative cumulative expected frequency	Absolute differences
0	1	1	0.100	0.135	0.035
1	1	2	0.200	0.406	0.206
2	2	4	0.400	0.677	0.277
3	1	5	0.500	0.857	**0.357**
4	2	7	0.700	0.947	0.247
5	2	9	0.900	0.983	0.083
6	1	10	1.000	0.995	0.005
7	0	10	1.000	0.999	0.001
8	0	10	1.000	1.000	0.000

The test statistic is 0.357. At 10 observations and a 5% significance level, the critical value for the KS distribution is 0.368. There is not enough evidence to reject the null hypothesis that the population distribution is Poisson with a mean of 2.

17.27 Although a chi-squared goodness of fit test can be more powerful when the parameters are given, the small sample size in this exercise precludes its use. A KS test is used instead.

Car	x_i	Sample CDF	z score	Expected CDF	Absolute differences
1	16.20	0.0769	−2.25	0.0122	0.0647
2	28.48	0.1538	−1.43	0.0764	0.0774
3	30.66	0.2308	−1.29	0.0985	0.1323
4	46.99	0.3077	−0.20	0.4207	0.1130
5	56.13	0.3846	0.41	0.6591	0.2745
6	66.91	0.4615	1.13	0.8708	**0.4093**
7	68.15	0.5385	1.21	0.8869	0.3484
8	74.46	0.6154	1.63	0.9484	0.3330
9	81.81	0.6923	2.12	0.9830	0.2907
10	81.82	0.7692	2.12	0.9830	0.2138
11	85.29	0.8462	2.35	0.9906	0.1444
12	86.96	0.9231	2.46	0.9931	0.0700
13	91.10	1.0000	2.74	0.9969	0.0031

The test statistic is 0.4093. The critical value from the KS table at 5% and 13 observations is 0.325. The null hypothesis that the population is normally distributed with a mean of 50 and a standard deviation of 15 is rejected.

17.29 There are 20 observations. Intervals constructed from the sixth, seventh, and eighth observations from the ends have confidence levels of 0.9586, 0.8846, and 0.7368 respectively. The closest interval to a 90% level is the seventh observation from each end of the data listed in ascending order. This interval runs from 415.46 to 954.66.

17.31

n	Level of confidence
10	0.8906
11	0.9344
12	0.9616
13	0.9776
14	0.9868
15	0.9926
16	0.9960
17	0.9978
18	0.9986
19	0.9994
20	0.9996

17.33 The assumption that the two distributions have the same shape allows a Wilcoxon-Mann-Whitney test. This is a left-tailed test with identical sample sizes. The test statistic can be either of the two sum of ranks. In this case the smaller sum of ranks is 16 for men. The critical value at a 5% significance level and five observations for both samples is T_L, or 19. The null hypothesis that the haircut time for men is greater than or equal to the time for women is rejected.

17.35 The underlying data are ordinal, so a nonparametric test is appropriate. The data are also paired, so either a sign test or a Wilcoxon signed rank test could be used. Since the assumption of identical shapes is reasonable in a paired context, the Wilcoxon signed rank test will be used. The question asks for a test of the null hypothesis that the course leaves health attitudes "unchanged." This wording implies a two-tailed test. The test statistic is the smaller of the two rank sums. In this case it is a rank sum of 2 for the one positive difference. The question does not state a significance level. Five percent is used here. Students may use different significance levels. The critical value at 5% and 9 observations (one tie was deleted) is 6. The null hypothesis that average ratings on the health questions are unchanged by the program is rejected at a 5% significance level.

17.37 The assumption of a symmetrical distribution allows the Wilcoxon signed rank test. The test statistic for a two-tailed test is the smaller of the sum of ranks, in this case 43 for the positively signed ranks. The critical value for 15 observations and a 5% significance level is 25. There is not enough evidence to reject the null hypothesis that the mean time the bus is late is 5 minutes.

17.39

Observation	x_i	Sample CDF	$\mu = \lambda x$	Expected CDF	Absolute differences
1	0.00	0.04	0.00	0.0000	0.040
2	0.00	0.08	0.00	0.0000	0.080
3	0.00	0.12	0.00	0.0000	0.120
4	0.08	0.16	0.32	0.2739	0.114
5	0.14	0.20	0.56	0.4288	**0.229**
6	0.14	0.24	0.56	0.4288	0.189
7	0.14	0.28	0.56	0.4288	0.149
8	0.17	0.32	0.68	0.4934	0.173
9	0.20	0.36	0.80	0.5507	0.191
10	0.20	0.40	0.80	0.5507	0.151
11	0.25	0.44	1.00	0.6321	0.192
12	0.25	0.48	1.00	0.6321	0.152
13	0.25	0.52	1.00	0.6321	0.112
14	0.25	0.56	1.00	0.6321	0.072
15	0.25	0.60	1.00	0.6321	0.032
16	0.25	0.64	1.00	0.6321	0.008
17	0.25	0.68	1.00	0.6321	0.048
18	0.33	0.72	1.32	0.7329	0.013
19	0.33	0.76	1.32	0.7329	0.027
20	0.33	0.80	1.32	0.7329	0.067
21	0.33	0.84	1.32	0.7329	0.107
22	0.50	0.88	2.00	0.8647	0.015
23	0.50	0.92	2.00	0.8647	0.055
24	1.00	0.96	4.00	0.9817	0.022
25	1.00	1.00	4.00	0.9817	0.018

The test statistic is 0.229. The critical value of the KS distribution at a 5% significance level and 25 observations is 0.24. There is not enough evidence to reject the null hypothesis that the distribution of time until an airline flight attendant responds to a request for assistance is exponential with a mean of 0.25 minute.

17.41 The assumption of identical shapes for the distributions of sales for the two scripts allows a Wilcoxon-Mann-Whitney test. The test is two-tailed at a 5% significance level. With samples of size 25 and 30, the normal approximation to the sampling distribution of T is required. The test statistic is the sum of ranks of the smaller sample, or 661.5 for the first sample.

$$\mu_T = \frac{25(25 + 30 + 1)}{2}$$

$$= 700$$

$$\sigma_T = \sqrt{\frac{25(30)(25 + 30 + 1)}{12}}$$

$$= 59.1608$$

C.V.$_L$, C.V.$_U$ = 700 ± 1.96(59.1608)

C.V.$_L$ = 584.045, C.V.$_U$ = 815.955

The test statistic of 661.5 is within the acceptance region. There is not enough evidence to reject the null hypothesis at a 5% significance level that the two scripts have the same population medians.

17.43

Box	Ounces	Sample CDF	z score	Expected CDF	Absolute differences
1	11.75	0.0333	−2.62	0.0044	0.0289
2	11.80	0.0667	−2.03	0.0212	0.0455
3	11.86	0.1000	−1.32	0.0934	0.0066
4	11.88	0.1333	−1.08	0.1401	0.0068
5	11.91	0.1667	−0.73	0.2327	0.0660
6	11.92	0.2000	−0.61	0.2709	0.0709
7	11.93	0.2333	−0.49	0.3121	0.0788
8	11.93	0.2667	−0.49	0.3121	0.0454
9	11.94	0.3000	−0.37	0.3557	0.0557
10	11.95	0.3333	−0.26	0.3974	0.0641
11	11.95	0.3667	−0.26	0.3974	0.0307
12	11.95	0.4000	−0.26	0.3974	0.0026
13	11.95	0.4333	−0.26	0.3974	0.0359
14	11.96	0.4667	−0.14	0.4443	0.0224
15	11.96	0.5000	−0.14	0.4443	0.0557
16	11.97	0.5333	−0.02	0.4920	0.0413
17	11.98	0.5667	0.10	0.5398	0.0269
18	11.99	0.6000	0.22	0.5871	0.0129
19	11.99	0.6333	0.22	0.5871	0.0462
20	11.99	0.6667	0.22	0.5871	**0.0796**
21	12.01	0.7000	0.45	0.6736	0.0264
22	12.02	0.7333	0.57	0.7157	0.0176
23	12.02	0.7667	0.57	0.7157	0.0510
24	12.03	0.8000	0.69	0.7549	0.0451
25	12.04	0.8333	0.81	0.7910	0.0423
26	12.05	0.8667	0.92	0.8212	0.0455
27	12.06	0.9000	1.04	0.8508	0.0492
28	12.11	0.9333	1.63	0.9484	0.0151
29	12.12	0.9667	1.75	0.9599	0.0068
30	12.13	1.0000	1.87	0.9693	0.0307

$\bar{x}$ = 11.9717

s = 0.0847

The test statistic is 0.0796. The critical value from the Lilliefors table at a 5% significance level and 30 observations is 0.159. There is not enough evidence to reject the null hypothesis that the population distribution of ounces of cornflakes per box is normal.

17.45 This problem requires the assumption of identically-shaped distribution of prices for both men and women in order to do a Wilcoxon–Mann–Whitney test. Since the data are not paired, a sign test would be inappropriate. The null hypothesis is that the median price quoted to men is less than or equal to the median price quoted to women. Thus the test is right-tailed. With equal sample sizes the test statistic is the larger rank sum, or 55 for women. The critical value with 6 observations in each sample and a 5% significance level is 50. The null hypothesis, that the distributions of prices quoted to men and women for transmission repairs have the same median, is rejected.

17.47

Car	Time	Sample CDF	z score	Expected CDF	Absolute differences
1	12.64	0.0714	−1.80	0.0359	0.0355
2	12.78	0.1429	−1.10	0.1357	0.0072
3	12.80	0.2143	−1.00	0.1587	0.0556
4	12.86	0.2857	−0.70	0.2420	0.0437
5	12.88	0.3571	−0.60	0.2743	0.0828
6	12.93	0.4286	−0.35	0.3632	0.0654
7	12.97	0.5000	−0.15	0.4404	0.0596
8	13.00	0.5714	0.00	0.5000	0.0714
9	13.00	0.6429	0.00	0.5000	0.1429
10	13.01	0.7143	0.05	0.5199	**0.1944**
11	13.12	0.7857	0.60	0.7257	0.0600
12	13.16	0.8571	0.80	0.7881	0.0690
13	13.17	0.9286	0.85	0.8023	0.1263
14	13.33	1.0000	1.65	0.9505	0.0495

The test statistic is 0.1944. The critical value taken from the KS table at a 5% significance level and 14 observations is 0.314. There is not enough evidence to reject the null hypothesis that the distribution of the time taken by the robot to change oil is normal with a mean of 13 minutes and a standard deviation of 0.2 minute. The KS test is the most powerful because the sample size is too small for more than two cells with a chi-square test.

17.49 If the sign test rejects the null hypothesis of equal medians for the paired data, a Wilcoxon signed rank test and a paired t test will also reject the null hypothesis. The assumption for both the Wilcoxon signed rank test and a paired t test implies that the population median of the paired differences and the population mean of the paired differences are equal, i.e., both assume symmetry. Thus, testing for difference in means rather than medians would not be a valid reason for using these tests. The question asks for "any reason." A possible reason is practice doing these tests, but this is a lame answer.

CHAPTER 18

18.1(a) The states of nature are first the college winning or not winning the PAC-10 and second the number of buses that alumni would fill given that the college was going to the Rose Bowl. The payoffs are minus $200 for every bus reserved if the state of nature is that the team does not win the PAC-10. In the state of nature the team wins the PAC-10 the payoffs depend on both the number of alumni taking the bus trip and on the value assigned to the goodwill. If none of the alumni are interested in taking the trip, the worst possible payoff is minus $200 for every bus reserved. The best possible payoff is $0 if five buses are filled plus the value of the goodwill generated by five buses full of alumni. This last term may depend on another state of nature, which team wins the Rose Bowl. The distribution of payoffs is discrete; the alumni club cannot order a fractional bus.

(b) The states of nature are the number of seat-tilting mechanisms that will be sold by the company beyond the original order of 120,000. The payoffs are $2 for every mechanism past 120,000 that is built and sold or minus $5.50 for every mechanism built and not sold. The latter payoff assumes a zero scrap value for unsold mechanisms. The distribution is discrete because there are no fractional seat mechanisms. However, the numbers are large enough to model the probabilities for the states of nature as continuous.

(c) One state of nature is the mean time to failure of the washing machines. It is described in the exercise as being in the range of two to three years. The company can affect this mean by increasing the inspections of the machines it builds or using more reliable parts, but the mean time to failure is not entirely in the company's control. For example, intensity of use by consumers would affect the mean time to failure. Another state of nature is the number of additional machines it would sell for a given increase in the warranty. On a per machine basis, the increased costs in honoring the warranty for a given increase in the warranty's duration are determined by the mean time to failure. The payoffs for a given increase in the warranty's duration are the added sales less the added repair costs. The distribution of payoffs is continuous because one of the states of nature, the mean time to failure, is continuous.

18.3(a)

	c	d
A	50	10
B	−20	120

$$EP_A = 50(0.7) + 10(0.3)$$
$$= 38$$

$$EP_B = -20(0.7) + 120(0.3)$$
$$= 22$$

$$EPGPI = 50(0.7) + 120(0.3)$$
$$= 71$$

$$EOL_A = 0(0.7) + 110(0.3)$$
$$= 33$$

$$EOL_B = 70(0.7) + 0(0.3)$$
$$= 49$$

(b) The best choice for a risk-neutral decision maker is A. A has the highest expected payoff and, equivalently, the lowest expected opportunity loss.

18.5 The expected hourly payoffs for the ferry boats with capacities of 10, 15, and 20 cars are $9.70, $12.20, $12.20. Given the information in the exercise, either the 15- or the 20-car boat would be chosen. The calculations do not allow for the cost of the boat, but presumably the larger boat costs more. On the other hand, the ferry boat company may prefer to bear the cost of the larger boat if it thinks demand will increase some time in the future.

18.7(a) Assuming that the dollar you pay to make each bet is not returned if you win, the respective expected values of the three bets are $0.00, −$0.20, and $0.089. A strongly risk-seeking individual would take the second bet. The high potential payoff would compensate for negative expected value.

(b) A strongly risk-averse individual would take the third bet. The high probability of winning would compensate for the low payoff.

(c) A risk-neutral individual would choose, among a set of bets with equal cost, the best with the highest expected payoff. The $1.10 payoff bet dominates the $2 payoff bet.

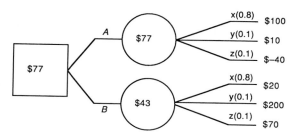

18.9 The best choice is *A*. It has an expected payoff of $77.

18.11 The choice node *GH* has an expected payoff of $70. Once the *GH* node is reached, the $70 is available with certainty. The chance node after choosing *A* has an expected payoff of $70.80. The chance node after choosing *B* has an expected payoff of $150.00. The best choice at the *AB* choice node is *B*.

18.13 Let *w* and *l* be the prediction of winning and losing by Mr. Oracle and *W* and *L* be a win or a loss by the horse Bayes. $P(W|w) = 0.7143$, $P(L|w) = 0.2857$, $P(W|l) = 0.1163$, $P(L|l) = 0.8837$. The best course of action is to make the bet but not pay Mr. Oracle for a tip. At $100 Mr. Oracle's advice is considerably overpriced.

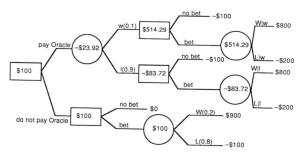

18.15 The figures in the opportunity cost table are in millions of dollars.

	seeding reduces wind speed	seeding does not change wind speed	seeding increases wind speed
seed	0	1	101
do not seed	49	0	0

18.17 If you had no motor pool, you would average five rental cars a day and have an expected expense of $150. The optimal number of cars in the motor pool is seven. With seven cars your expected daily expenses for rental cars plus the fixed cost of the motor pool cars is $86.36. The expected payoff for having a seven–car motor pool relative to having no motor pool is $150 – 86.36, or $63.64.

18.19 Settlement has a known cost of $3 million. The expected cost of litigation is also $3 million. The uncertain costs to the company's reputation from a prolonged trial should tip the decision toward settlement, unless you are strongly risk seeking.

18.21(a) The expected gain from switching to bags is $0. Presumably, Pillsbury would not make a move that had a zero gain on an expected net present value basis.

(b) If the two possible states of the world after changing from a box to a bag are the moderate and the severe loss of sales, the best decision is to stay with the box. The best possible return after testing is when the sales' loss is moderate. This return equals the gain of $5 million minus the $2 million in lost sales and the $3 million for the cost of the test marketing, or a net of $0. Although it is not necessary to calculate the expected payoffs given the decision to test, the decision tree is shown to illustrate the calculations. The values in the decision tree are in millions of dollars. Let *M* and *S* be the states of nature moderate and severe loss of sales, and let *m* and *s* be the predictions from the test marketing. $P(S|m) = 0.1683$, $P(S|s) = 0.5025$, $P(M|m) = 0.8317$, $P(M|s) = 0.4975$.

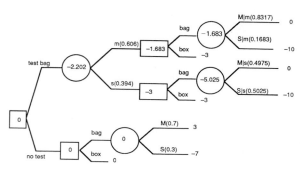

18.23 Let S be a strike and NS be no strike. The figures in the decision tree are in millions of dollars. The best strategy is to offer a 5% increase in wages.

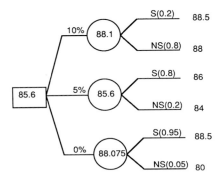

18.25 Let H, M, and L be high, medium, and low sales respectively for the IBM-PS 80. Let B and NB be respectively the Lotus Eater program among the best and not among the best spreadsheet programs for the IBM-PS 80. Let *plus* and *minus* refer to the rating by the panel of experts. $P(B|plus) = 0.9130$, $P(B|minus) =$

0.0870, $P(NB|plus) = 0.2258$, $P(NB|minus) = 0.7742$. The best choice is to use the test panel and let their rating determine whether to market the product.

18.27 The optimal number of rooms is six. The expected daily profit for six rooms is \$150.38.

18.29 Perfect information in this example is not knowing the true mean of the Poisson distribution. The mean of 5 was given in the exercise. Perfect information is knowing how many people will show up each day. If the hotel could be built anew each day at the correct number of customers, the expected daily profits would be \$200. The value of perfect information is thus \$200 − 150.38 = \$49.62.

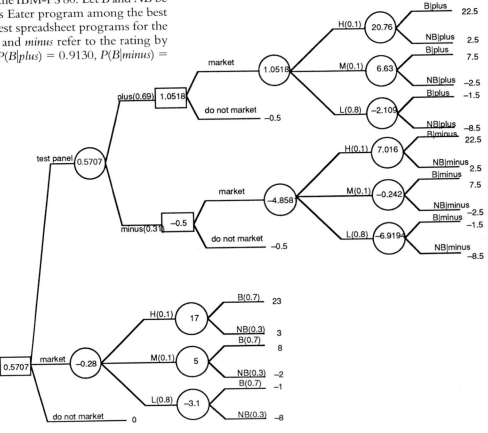

Index

Page references in *italic* refer to formulas.

ABOUT THE AUTHOR

Robert Sandy is Associate Professor of Economics at Indiana University–Purdue University at Indianapolis, where he has taught a course in Statistics for Business and Economics since 1974. He received his B.A. in Economics from the University of Michigan and his Ph.D. in Economics from Michigan State University. His primary research area is the economics of occupational safety and health. He has written a number of articles in professional journals, such as *Applied Economics* and the *American Business Law Journal*. Robert Sandy has also served as a consultant to businesses, not-for-profit agencies, and government agencies at the local, state, and national levels.

Population Distribution	Mean	Variance	$P(x)$ or $f(x)$	Estimator of the Mean
Difference Between Observations Drawn from Two Populations with Identical σ^2's For problems in which independent samples are drawn from two populations. σ^2 is the common variance of the two populations.	$\mu_1 - \mu_2$	$2\sigma^2$	If both populations are normally distributed, the distribution of differences between observations drawn from the two populations will be normally distributed.	$\overline{x}_1 - \overline{x}_2$
Difference Between Observations Drawn from Two Populations with Different σ^2's For problems in which independent samples are drawn from two populations that do not have the same variance.	$\mu_1 - \mu_2$	$\sigma_1^2 + \sigma_2^2$	If both populations are normally distributed, the distribution of differences between observations drawn from the two populations will be normally distributed.	$\overline{x}_1 - \overline{x}_2$
Difference Between Observations Drawn from Dependent Samples Applies to tests with paired or matched samples. d_i are paired differences and n is the number of pairs.	δ	σ_d^2	The distribution of paired differences will be normal if both of the populations are normal.	$\overline{d} = \dfrac{\sum\limits_{i=1}^{n} d_i}{n}$
Difference Between Two Proportions with Independent Samples Applies to the difference in the proportion of successes in two independent binomial processes.	$\Pi_1 - \Pi_2$	$\Pi_1(1 - \Pi_1) + \Pi_2(1 - \Pi_2)$		$\overline{p}_1 - \overline{p}_2$
Chi-square Distribution Applies to goodness of fit problems, tests of independence, and tests of homogeneity. The degrees of freedom, df, is the sole parameter. o refers to the observed frequency and e to the expected frequency.	df	$2df$	The sum of df independent squared standard normal distributions.	$\chi^2 = \sum\limits_{i=1}^{k} \dfrac{(o_i - e_i)^2}{e_i}$
F Distribution Used for tests of the null hypothesis that more than two populations have the same mean. Also used for tests of the null hypothesis that all slope coefficients in a multiple regression are zero. df_n is the numerator degrees of freedom and df_d is the denominator degrees of freedom.	for $df_d > 2$, $\dfrac{df_d}{df_d - 2}$	for $df_d > 4$, $\dfrac{2df_d^2(df_n + df_d - 2)}{df_n(df_d - 2)^2(df_d - 4)}$	The ratio of two independent chi-square distributions divided by their degrees of freedom.	